Energy and Resource Efficiency
Without the tears

Volume I - *Framework*

Niall Enright

IWIK Publishing, Manchester, England

Publication Information:

IWIK Publishing, an imprint of SustainSuccess Ltd

Address: 19 Harboro Road, Sale, Manchester M33 5AF, UK
Tel: +44 7812 519965
web: www.sustainsuccess.co.uk/iwik
email: iwik@sustainsuccess.co.uk

Editor: Louise Bolotin, SfEP
Maths Editor: Jane Thompson

ISBN: 978-912012-00-8 (Paperback)
978-912012-71-8 (Hardback)

Updated September 2018 (text 1.3)

Images © fotolia.com unless otherwise stated. **Front Cover**. Top L-R: *antiksu; pruscha; unknown; naj*. Middle L-R: *KittyKat; michaeljung; luna; gavran333*. Bottom L-R: *Niall Enright; HamsterMan; Peel Medial Ltd.*
Back Cover. Top *Niall Enright*. Bottom L-R: unknown; *Niall Enright; Niall Enright; Niall Enright; Luchsen.*

IWIK I Wish I Knew It's What I Know

Contents

Dedications

This book is dedicated to the late Prof David JC MacKay, whose outstanding book *Sustainable Energy — without the hot air*[491] inspired me to start writing. I have followed David's example and made this book freely available in PDF form. David was also generous with his advice and encouragement on some of my early work on this book, but sadly he died in 2016, before I was able to send him a completed draft.

© JJAVA, Fotolia.com

I hope that in the text I have properly acknowledged the many great people and organizations that have done so much to lay the foundations for successful resource efficiency. I feel truly humble and privileged to be able to share their work with my readers. Giants in the field include Amory Lovins and the Rocky Mountain Institute, which have tirelessly led the charge on energy efficiency; Ellen MacArthur, the around the world sailor who has established a foundation which passionately promotes the circular economy; Christopher Russell, one of the most practical exponents of industrial energy efficiency I know, and his previous outfit, the Alliance to Save Energy; Walt Patterson who eloquently reminds us not to repeat the mistakes of the past; John "Skip" Laitner; both the American and European Councils for an Energy Efficient Economy; Advanced Energy Economy. Thanks, too, to the Association of Energy Engineers, the Energy Institute and Energy Managers Association, which have made a massive contribution to professionalize our craft. Dr Steven Fawkes in the UK and Donald Gilligan at NAESCO in the US who have done such outstanding work promoting the business, financial and commercial case for efficiency. Don was a great mentor and guide in my first professional forays into the US energy efficiency scene.

At the heart of this book is the notion that resource efficiency is about change and I must salute the inspiring work that Bob Doppelt has done in this area, along with other sustainability advocates such as Bob Willard and Mark Epstein as well as more general thinkers on change such as Paul Gibbons. We must also applaud a cadre of academics in the field of human behaviour and psychology who have brought this important research out of the lab and into practical programmes: Doug McKenzie-Mohr, Daniel Kahneman and Robert Cialdini are just some of the leading figures here.

The experience I have to offer is a product of the great clients and organizations I have worked for. I particularly want to single out David Glover at Peel Land and Property Group, one of the most inspiring senior executives I have been privileged to work for. Thanks to the many other folks in Peel with whom it has been a joy to work with delivering some great efficiency improvements: Derek Elliott, Chris Foran, Nick Poole, Chris Dunham, Kellie Naylor, Paul Chappels, Dale Mullane, Andrew Dutton at Liverpool John Lennon Airport, Alex Pepper and Phil Hall at Peel Ports, Phil Harris and the great team at Engie, and Jenny Lawless who kept me organized throughout!

How this book is made

© Guy Shapira - Foltolia.com

In an ideal world, no one will buy this book! I have made a PDF version available free of charge.

I have done this because my intention when writing this book was never to make money, but to give something back and help, in some small way, our common efforts on sustainability.

If on the other hand, you are old-fashioned, like me, and you want to use a print copy of the book, you can order a copy through the website:

www.sustainsuccess.co.uk/iwik

When your order is placed (whether on my site or other retailer's site such as Amazon), the book will be printed from a print on demand service, Lightning Source. This will reduce waste in the manufacturing process by matching the number of copies produced to the demand.

Moreover, because Lightning Source operates in the US, UK and Australia, the printing will take place closer to the final destination and transport resources will be reduced.

Another advantage of this method of production is that I can rapidly incorporate changes in the text into both the electronic and print versions. If you do see any errors or have any suggestions for improvement, then do please let me know (my contact details are on Page 2).

My career started at March Consulting Group, which became Enviros, then part of SKM and finally Jacobs. I also had senior roles with RPS and ERM, which are global consultancies addressing some of the most pressing environmental challenges of our time. I would like to express thanks to all the colleagues I worked alongside in these organizations who taught me so much about my craft. Many of you will be acknowledged in the real-world stories in this book, but I would like to also mention Ray Gluckman, Keith Webster, Bob Bailey, Chris Stubbs, Alan Couch, Peter Cohen, Peter Young, Gary Armstrong, and Julie Gartside from my time at Enviros; Richard Wise, Charles Allison, Peter Rawlings, Scott Foster, Braulio Pikman, Massimo Bettanin, Ben Boardman, Peter Temesvary, Peter Fink, James Spurgeon, Walter Heinz at ERM, as well as Zomo Fisher, now with the great Accenture sustainability team led by Peter Lacy.

A special mention, too, for Caroline Robertson-Brown, an associate in my own small consultancy, and her partner Nick Robertson-Brown, for their encouragement as fellow authors. Thanks, too, to Phil Kilburn for his help accessing many of the scientific papers quoted here. Arne Springorum and the team at HEC Consulting in the Czech Republic are awe-inspiring in their commitment and skill delivering value from resource efficiency for their customers. A dedication, too, to Paul Stepan at Verco, a terrific sustainability professional and role model for others.

I pay tribute, too, to the many institutions driving resource efficiency around the world. The European Union should be applauded for their work in this area, the previous US government for their use of stimulus funding to support efficiency, the UK government who has shown leadership on many policy fronts such as legally binding carbon budgets, the World Bank's programmes such as ESMAP, the Intergovernmental Panel on Climate Change who negotiated the Paris Accord. Organizations such my alma mater Cambridge University's Programme for Sustainability Leadership, the World Business Council for Sustainable Development, The Carbon Trust, UNEP Finance Initiative, Environmental Defence Fund, CSRIO in Australia, the Green Buildings Councils around the world, and the Pew Centre, are just some of the organizations doing great work driving change into the business sector.

Writing a book forces one to consider the origins and rationale for a particular fact or concept which is taken as given. This has given me the delight of looking into the academic work underpinning many aspects of my profession. Important figures (not mentioned previously) are Donella Meadows, Edward Daly, Steve Sorrell, James Hansen, Michael Grubb an old university colleague, Charles Keeling of the famous CO_2 records, the team at the University of East Anglia Climatic Research Institute led by Phil Jones and the Tyndal Centre based in my home town of Manchester.

Finally, I dedicate this book to my wife Jane, daughter Amy and son Connor, who have shown great patience as this project has taken over my life. Thanks so much for encouraging me and supporting me in countless ways.

The companion materials

Real World: *Practical examples*

These items highlight key points in the text with examples drawn from the real world, warts and all.

Exploration: *More depth*

Here a topic is covered in more depth or we go off on a tangent to a related point.

In Numbers: *Useful maths*

These side items will show you how to calculate key values related to the main text (using Excel if appropriate).

Standards: *Relevant standards*

Items in this style link the text to the requirements of related standards like ISO 50001.

Summary: *Exercises and Reading*

Green boxes have reference material such as a chapter summary, further reading and questions for students.

Key **concepts** are shown like this.

Sidebars summarize *important points.*

1.1 **An example of a caption**
Explanatory text is in regular font.
Source: contains information about copyright.

This book has an extensive set of companion files. This zipped file contains many resources that will help efficiency practitioners to drive improvements in their organization, and provide students and teachers with powerful learning materials. Resources provided include:

- Over 70 spreadsheet models that build on the data and financial analysis techniques set out in this book. These Excel spreadsheets extend the learning points in this book, and enable practitioners to employ the analysis and presentation techniques set out in this book.
- 5 Business case models, as used by the author in investment appraisals.
- Over 200 illustrations and cartoons for use in educational materials and presentations; and 9 posters, summarizing key concepts.
- A lighting hours tool to create daily, weekly or monthly lighting demand profiles based on a place name or coordinates.
- A complete set of certified ISO 50001 process documentation (provided courtesy of Peel Land and Property Group).
- MACCBuilder Pro, a tool to draw Marginal Abatement Cost Curves (version 3, previously sold for £85).
- MATOD a simple M&T spreadsheet system for non-commercial use.
- A software specification document that can form the basis for the selection and procurement of a resource-efficiency tool.

The companion file download can be found at:

www.sustainsuccess.co.uk/iwik

Simply add the companion files to your basket and then, at checkout, use the COUPON CODE EREWTTbuyer17 which will allow the companion files to be downloaded free of charge. **Do not share this code or the files with others - the licence to use these files is strictly for you, the owner of the book.**

Please note the disclaimer on your use of these materials set out on the web page and page 2 of this book. Documents are largely in Microsoft® Excel®, Word®, PowerPoint®, PNG or Abobe® PDF formats.

Introduction

© Delphotostock - Foltolia.com

"It is not the strongest of the species who survive, nor the most intelligent, but the ones that are **most responsive to change"**

- Charles Darwin.

Today there are just over 7 billion people alive on the earth. Our species has been staggeringly successful because of our ability to shape the environments which we occupy. We have been able to master the energy locked into fossil fuels from centuries-old sunlight. Our agriculture has changed almost half of the surface of the earth. We produce 200 kg of steel and 400 kg of cement for each person each year. These abundant material resources and our planet's capacity to deal with the wastes that we create are the foundations of our civilization.

Global population will grow by a further 31% to 9.3 billion by 2050. Population growth is not our greatest challenge, however, since our economy will increase threefold over the same period due to rising wealth. Most of this growth will not be from the trivial consumption that we associate with developed lifestyles. It will be for building and running the schools, hospitals, housing, roads, water supplies, power stations and places of work that every human being deserves. Delivering the energy and material needed to lift billions out of poverty is good, just and fair.

Only there is a problem. There is convincing scientific evidence that today's energy and resource consumption has dangerous environmental consequences. The biggest worry is the rising concentration of greenhouse gases in the atmosphere, in particular, CO_2, which contributes to climate change. Today natural processes remove just half of the CO_2 we are putting into the atmosphere each year, so the gas is accumulating and causing our planet to warm. The science around the warming effect of CO_2 is evident. The consequences of this warming are less certain, but just about every predicted effect is very undesirable. These include a large loss of biodiversity; reduced agricultural production and increased famine; mass migration and social upheaval; to more frequent and severe extreme weather events and rising sea levels.

Climate change is not the only environmental risk posed by our patterns of resource use. A team from the Stockholm Institute of Resilience published a paper in Nature in 2009,[625] which concluded that there are nine fundamental planetary boundaries beyond which we face "*unacceptable environmental change*". Of these boundaries, we have already crossed three: climate change, the rate of biodiversity loss and the nitrogen cycle.

Despite glossy case studies to the contrary, the resource efficiency landscape is littered with disappointments and failures.

We also face some serious economic challenges. Scarcer and more difficult to extract resources become more expensive. Despite the respite given by the recent financial downturn, the long-term trend is for *cheap* resources to run out. Many organizations are making investment decisions around assets with a lifetime of 25 or even 50 years: in building, factories, infrastructure, power generation, transport and so on. It is inevitable that resource availability will have an impact on the value of these assets in future. Neither should those organizations whose value is mostly intangible – like Apple, Nike and Coca-Cola – have cause for complacency. Their future values depend on consumer goodwill and entire categories of products, such as bottled water, can go out of fashion because of concern about resource availability.

No choice exists that can preserve the *status quo*, least of all *doing nothing*. We are depleting finite resources and natural services. We are already living beyond our planet's capacity to support us. Our organizations have to choose between being passive bystanders in the change that is heading our way or charting a course that gives us some measure of control over events.

But what path to follow? Despite glossy case studies to the contrary, the resource efficiency landscape is littered with disappointments, false starts, blind alleys and premature declarations of victory. Far too many organizations, being charitable, put a *spin* on their achievements, preferring *greenwash* to the effort and dedication that real change involves. Agencies promoting efficiency often fail to recognize the real-world constraints that organization face and so pitch unrealistic or oversimplified approaches. Formal education in resource efficiency lags far behind the scale of change needed, and so people responsible for improvement are often inadequately prepared. Manufacturers push technology as the *silver bullet* solution when in reality management and organization are critical to success. In short, energy and resource efficiency is not easy.

Fortunately a different, more focused, attitude is now emerging, driven by the growing realization that the environmental challenges we face are the biggest business opportunity in this century. An unprecedented wave of innovation is starting to shake up old business models and provide extraordinary opportunities to those who can deliver superior performance with fewer resources.

This book draws on my experience over almost three decades in hundreds of energy and resource-efficiency programmes for large multinationals, public institutions and small organizations around the world. I will tell you why some programmes fail, to help you avoid the most common mistakes. I will also reveal what makes me hugely optimistic that organizations can and will rise to the challenge. This book sets out proven ideas and methods which will enable your organization to achieve Energy and Resource Efficiency without the tears.

Niall Enright

1 Getting Started

*"How do you eat an elephant? Why, **one bite at a time**, of course!"*

-Traditional proverb

Most publications dealing with contemporary environmental, social and sustainability issues present our current situation in alarming terms. They are right.

If we objectively consider the challenges that face us it is easy to become overwhelmed by a sense of powerlessness. The scale of the problems is such that, as individuals or as organizations, it seems that we can do little. So there is a tendency to ignore the "*elephant in the room*". We, us, our organizations, are in what psychologists call *denial* - a subconscious defence mechanism against things that worry us or make us anxious.[158]

This paralysis is wrong. Yes, we are facing some significant challenges, but from these come great opportunities. Our generation has the potential to create a world where resources like energy are plentiful, affordable, do not harm the environment and are available to all. We can transform our economies to recover the huge percentage of materials that we use once only and discard. We can reverse the damage to habitats. We can do this while greatly improving the quality of life for the majority of people on our planet and restoring a balance with nature.

The solution to our paralysis is recognition that we *can* make a difference. There is not one amazing single silver bullet that will deliver the necessary change, but of lots of different actions which individually seem insignificant but add up to a radical change.

The following pages set out the nature of the challenge we face. There are lots of graphs pointing in the wrong direction. The intention is to demonstrate that *change is inevitable whether we like it or not*. However, in this chapter and throughout this book I also hope to convey my optimism in our ability to tackle the challenges we face. In my experience, when people set themselves the task reducing the resources they use, with dedication, effort and focus they invariably far exceed their own expectations.

In these pages, we will see that the practical steps organizations need to take to transform their impact on the environment are already well understood. I will demonstrate that those organizations that take these steps will create an enormous amount of value for themselves and their stakeholders. Far from being a threat, those organizations willing to take bold, transformational action have a very great deal to gain.

1.1 What is resource efficiency?

Resource efficiency is a continual improvement process to reduce an organization's absolute impact on material resources and natural services while delivering value to the organization and its stakeholders.

Exploration: *Conservation vs efficiency*

In our definition of resource efficiency, we should use common sense about of the meaning of the word *efficiency*. This is because the priority should always be *conserving* resources rather than using them more efficiently.

For example, I am conserving electricity when I switch a light off, while I am using electricity more efficiently when I swap from an incandescent lamp to a LED lamp. Clearly the former is better than the latter.

There is also a finer point about the words *use* or *consumption*. If we consider energy, it is never actually consumed; rather it is *converted* from one form to another, usually less useful, form.

Similarly, materials are rarely consumed but rather changed and dissipated in ways which make them difficult to reuse, or which cause harm to the environment.

For simplicity, this book will consider the term *efficiency* as meaning both *conservation* and *efficiency* and the words *use* or *consumption* will be taken to mean both *conversion* and *consumption* of resources.

From time to time I will examine the words we use, not in order to be pedantic or dogmatic, but because exploring meanings can give us insight and greater understanding.

A good way to explore what is meant by resource efficiency is to consider how it is described by those involved in promoting it.

Angela Cropper of the United Nations Environment Programme (UNEP) defines resource efficiency as "*reducing the environmental impact of the consumption and production of goods and services over their full life cycle*".[176] It is interesting to note that this definition describes a process, rather than an outcome. In other words, we can never *be* resource efficient — except to the extent that we may at some point use no resources at all. All we can do is apply the resource-efficiency process to optimize what we do in environmental terms.

Taking into consideration the scale of transformation needed (for example to reduce CO_2 emission by 80% by 2050) leads us to the conclusion that resource efficiency is a long-term activity, a continual improvement process rather than a one-off exercise.

Janez Potočnik, the EU Commissioner for the Environment, echoes this definition but also expands on the word resource to mean "*material resources such as metals, minerals and food, and natural resources which provide services, including clean air, land and water*".[601] So here is the concept of environmental or ecosystem services as resources which should be used in an efficient or sustainable manner. Building on the definitions above we can state that for our purposes a resource is a physical material or an environmental service on which our organization depends.

Angela Cropper used the term "*life cycle*" in her definition of resource efficiency. By that, she is stating the obvious fact that it is no good optimizing the efficiency of one stage of a process, only to create greater inefficiencies elsewhere. Centralising manufacturing may well lead to the more efficient use of energy, but the benefits of this could be lost in the increased transportation costs to and from the factory. Thus resource efficiency is essentially about "*doing more with less*" over the whole life cycle of a product or service.

In the real world, however, few organizations set themselves a whole life objective for resource efficiency. Often it is not desirable, or even possible, for an organization to fully quantify or manage the life cycle impacts of their resource use. In most circumstances, organizations reasonably assume that they are applying resource-efficiency thinking if they seek to reduce inputs or waste at any given stage in product manufacture or service delivery, regardless of the wider ramifications. So for practical reasons, I will leave out the term *"life cycle"* from our definition, although we will not ignore this in the pages to come.

Another consideration in defining resource efficiency lies around the misconception that the only aim of resource efficiency is reducing environmental harm. This viewpoint is reinforced by the fact that many of the strongest advocates of resource efficiency come from the environmental field, like Angela and Janez. Presenting resource efficiency as a *purely* environmental issue can paradoxically reduce the impetus to act – as it obscures the fact that resource efficiency provides substantial value to the organization.

Exploration: ***Continual vs continuous***

Continuous and continual have the same linguistic root in the word continue, but they have subtly different meanings.

Continuous means *"without cessation or interruption"*. It occurs five times more frequently in written English that the word continual, which means *"something that recurs frequently"*. Thus if I have a continuous cough for a year, then I have coughed non-stop over that period while if I have a continual cough for a year it means that I have been coughing, off and on, over the year.

I have chosen to use the word continual because it better reflects the way resource efficiency works in organizations. The aim is to get ever more efficiency, but organizations will sometimes stand still or even retreat. By implementing a continual process, we commit to coming back time and again to drive activities that make our resource use better. We will not give up!

The international standard for energy management systems, ISO 50001, takes a similar view and refers to continual improvement as *"a recurring process which results in enhancement of energy performance and the energy management system"*.

In truth, both words are used fairly interchangeably, and we should not get too hung up on which is correct.

Because a chemical process might be fed from a sustainable feedstock which does not deplete natural resources, it does not mean to say that we should be blind to the economic or other benefits of further reducing our consumption of that feedstock. Sustainability, in its widest sense, encompassing profit motives and social benefits as well as environmental outcomes, is the best justification for seeking resource efficiency.

To be completely honest, my observation is that few organizations commit to resource efficiency because of the environmental benefits it provides. They do so because resource efficiency delivers lower operating costs, greater profits, good brand image, a more motivated workforce and it reduces risks. In short, resource efficiency creates value for shareholders, service users, stakeholders, employees and customers.

To encourage organizations to adopt resource efficiency, we should do everything that we can to highlight the benefits that it offers. That means including value in our definition. Here value is used to mean not just financial value, but also the ability of an organization to achieve its core purpose. For a university, value is the amount of research it does or the number of students it educates. For a hospital, it is the number of patients treated. For a theatre, it is the number and quality of shows it can put on. For a private company, it happens to be profit.

So, let's look at our definition:

> Resource efficiency is a continual improvement process to reduce an organization's absolute impact on material resources and natural services while delivering value to the organization and stakeholders.

The observant reader may have noticed the word "*absolute*" has crept into the definition. If I had stated that the improvement needs to be "*per unit of output*" this could mean that, if the organization grows, all it does is to *slow the rate of increase of harm*. Later, in Chapter 11 on Goals, we will explore relative improvement objectives in some depth.

In the definition above, the notion of "*delivering value to the organization*" may rankle some folks. It is not intended to mean that resource efficiency should only be contemplated if there is a financial return. Far from it. What this

Exploration: *Efficiency vs productivity*

In one of his regular blogs,[275] Dr Steven Fawkes, a great proponent of energy efficiency, suggests that the term *energy productivity* might better describe the goal we are seeking rather than *energy efficiency*.

The word efficiency, he argues, has connotations of frugality or sacrifice. It implies a threat to suppliers of energy and it is plain boring! Productivity, on the other hand, implies progressiveness, the creation of value and wealth. It links to the core objectives of firms or governments. It is much sexier.

Like most of Steven's blogs, this was quite thought-provoking, and I considered if I should not use productivity rather than efficiency in this book. In the end, I chose to stick with the word efficiency. Although Steven and I agree on the need to emphasize the positive value that efficiency brings, I felt the word productivity had its drawbacks.

A productive activity is one that creates greater output for a given input (i.e. efficiently) but also in *absolute* terms. Car Plant A is said to produce more than Car Plant B if it makes more cars than B. Because of this connection with making more, I have steered away from using the expression energy productivity despite its attractiveness at many levels. Resource efficiency should question our use of goods and the paradigm of growth.

Also, on a practical level, I want folks who are interested in this topic to find this book on the shelves or the internet!

Real World: *UK emissions goals*

One of the most ambitious examples of a very real and urgent resource-efficiency target exists because we need to stabilize and then reduce CO_2 emissions to our atmosphere. In the UK, this objective has led to a goal, set out in an Act of Parliament, to reduce in absolute terms, emissions in 2050 by 80%, compared with 1990 levels. This goal has been translated into a series of interim legally binding *"Carbon Budgets"* by the Climate Change Committee, shown in the graph below.

This target, by the way, is just to *stabilize* CO_2 in the atmosphere. At a global level, since only about half our emissions are disposed of by nature, we need to halve our output of greenhouse gases. The 80% target comes about because developed countries agree they need to go further than 50% to permit emerging economies to emit mode greenhouse gases as they develop. To begin to reduce CO_2, we will need to go even further beyond 2050.

The UK economy, as expressed by Gross Domestic Product (GDP), is likely to grow threefold in the period 1990-2050. Thus the 80% target emissions reduction *per unit of GDP* is actually 93%. At first, this appears to be a hugely ambitious – possibly unattainable - goal. But let's consider the time involved and the progress made so far. In the most recent Carbon Budget report,[161] the Climate Change Committee pointed out that emissions have already decreased by over a quarter. On a compound basis, the remaining target reduction is around 3.8% each year. Some of this improvement will come from reducing the carbon intensity of the energy supply (e.g. through renewable energy) and some from decreased demand for energy. This combination of many sources of improvement, spread over many sectors over many years, is what we call a *continual improvement process*, where the sum of many different actions over an extended time frame results in a large change.

Since 1990, for example, UK manufacturing has increased its energy intensity by 46%.[73] Thus the value of goods produced per unit of energy consumed has almost doubled. This improvement has been achieved by adopting a broad range of different approaches from better design and leaner operations through to technology improvements. The driver has been competitiveness, rather than a marked change in attitude towards the environment.

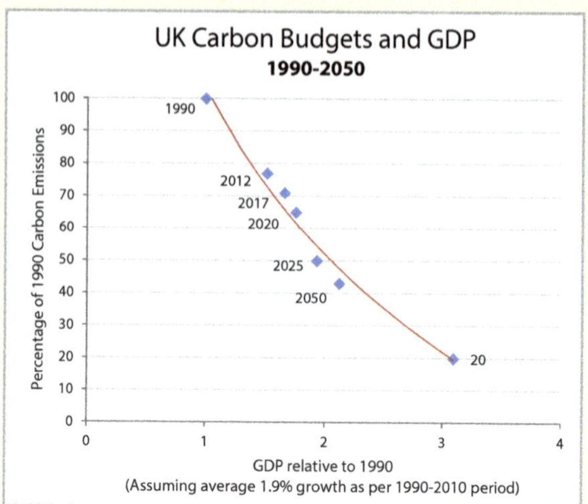

1.2 **The UK Carbon Budgets illustrate the reduction in CO_2 emissions sought in the UK, during a period when GDP is expected to rise threefold.**
Source: Niall Enright, data from CCC and World Bank. Spreadsheet in companion file pack

This decrease is evidence that it is possible to decouple energy use from growth. It shows us too that value-creation is an effective driver for resource efficiency. Organizations have delivered improvement because it makes business sense, so presumably can do so again in future.

1.3 **Business leaders are generally more engaged with the value basis for action than the environmental case**

Source: Niall Enright, drawn using Pixton. All cartoons are included in the companion file pack

states is that in every case where an organization is undertaking resource efficiency it is because it creates value. If not, they would not do it. So it is self-evident that the *process* of resource efficiency, where it is carried out, is value-increasing in some way.

I am not saying that resource efficiency must always create *financial* value (although it does so on most occasions). For example, an organization may be forced to adopt expensive resource-efficiency measures because of regulations. The resource-efficiency process nevertheless provides value because it allows the organization to stay in business. It provides a licence to operate. In this example, compliance may have diminished immediate financial returns, but has undoubtedly increased value compared to the alternative of shutting down.

Throughout this book, there is an emphasis on value. That is because value is a shorthand for the rationale or justification for action. Value is what motivates and enables organizations to act. If we can crack the value proposition for our organization, we will be a long way towards delivering a resource-efficiency programme that will succeed.

In this book, I am making a conscious distinction between a resource-efficiency *project* and a resource-efficiency *programme*. If we take energy efficiency, for example, many organizations approach this as an engineering activity. When energy prices rise, they ask the engineering department to find the cheapest ways of reducing energy use. This is the traditional technical approach to many resource efficiency challenges. However, this project approach is usually limited in scope and time. Once the immediate problem has been solved, the need to do more comes to an end. Furthermore, the improvement focus is often limited functionally. Projects often involve a *"quick fix"* or *"end-of-pipe"* improvement, especially in the area of waste.

In order to encourage organizations to adopt resource efficiency, we should do everything that we can to highlight the ***benefits*** *that it offers.*

Many organizations that say they are undertaking continual improvement around energy and waste are in reality implementing a series of projects one after another. This approach has successfully delivered most of the resource improvements for organizations to date, so we must not be overly critical.

Continual improvement requires much more than a series of projects. It requires the organization to look much more systematically at resource use and to focus on the root causes of waste. Operators need to be involved in reducing the waste at source. Product designers need to *"design out"* the embedded energy or waste. Procurement and finance teams should incorporate resource efficiency in purchasing and capital allocation. Marketing and communications should involve customers and stakeholders in the desired improvement. This book emphasizes the *programme* approach to resource efficiency, although there is plenty of material to support those who are working in a project-oriented way.

1.2 A privilege and a pleasure

Energy and resource efficiency tackles some very serious problems. In many cases it is also difficult and challenging to get right. That does not mean to say that it should not be fun. My experience is that enabling organizations to address their resource use is not only a great privilege but it is also hugely stimulating, exciting, interesting and personally rewarding.

1.4 **Working with folks at the *"coalface"* finding how to reduce resource use is hugely rewarding**
Source: Image © ndoeljindoel - Fotolia.com

Following on from our definition, we can see that an energy and resource efficiency programme aims to:

- achieve long-term, substantial decrease in resource use;
- deliver a tremendous amount of value;
- involve many aspects of the organization as well as external stakeholders;
- address pressing environmental issues.

As one can imagine, these are quite challenging objectives. There are pitfalls, barriers and difficulties that mean those working on a resource efficiency programme need to have dedication, effort and tenacity.

On the other hand, it is enormously exhilarating to be part of a successful programme. The variety of organizations and facilities means that no two programmes are identical, so there is much creativity involved. The wide range of skills required, from change management, design, engineering, finance to communications means that this is usually a great team effort and a fantastic learning opportunity. If the right measurement systems are in place one can see *the dial* move quite quickly. The breadth of the engagement needed means going from boardroom to shop floor, so there is little risk of boredom.

Many folks tasked with driving resource efficiency in their organizations come to see this as one of the best opportunities in their professional careers. The exposure, learning, camaraderie and satisfaction in overcoming obstacles and contributing to a better planet all make this a fantastically rewarding activity.

As we cover the challenges we face and the barriers that need to be overcome, bear in mind that working on resource efficiency should be fun and *will be* rewarding. With the benefit of almost three decades working in this field, I am more optimistic than ever of our ability to deliver the required change. Improvements in technology are revolutionizing our need for resources. The power of capital is awakening to the tremendous opportunity that resource efficiency offers. Politicians and policymakers are beginning to live up to their responsibilities. Moreover, organizations all over the world are taking action, because resource efficiency adds value and enables the organization to achieve its core objectives better.

1.3 Change is inevitable

Our capacity to use the resources provided by our planet, both material resources and environmental services, will inevitably have to undergo huge transformation in the coming decades. Change is unavoidable.

People can foresee the future only when it coincides with their own wishes, and the most grossly **obvious facts can be ignored when they are unwelcome.**

- George Orwell

Science is informing us that we are exhausting many of our planet's natural resources faster than they can be replenished. Our actions are changing the balance of nature in ways that are potentially harmful to our future. We already exceed our planet's capacity to absorb waste CO_2 by a factor of two. The accumulation of this gas in the atmosphere is not disputed. Nor are the mechanisms why it acts like a greenhouse to reflect heat back into our environment. Although the effects of global warming are less sure and expressed as a range of probabilities, these are highly undesirable, if not catastrophic.

CO_2 absorption is just one of the ecosystem's services on which life on earth depends. There are many others such as the water, nitrogen and phosphorous cycles. Our fishers rely on the fisheries from which they make their livelihoods. Drugs companies develop many new drugs from the rich store of molecules produced by biological processes.

The natural world is in constant flux. Many different components of the system influence each other to create the benign, relatively stable environment on which our present prosperity is based. But nature is not always so kind; there is evidence that changes in the environment have caused the collapsed of previous civilizations, such as the Maya in Central America.[519, 501] We may feel that we are immune from such weather effects because of our more advanced technology, but recent events such as hurricanes Katrina and Sandy in the US have reminded us of our vulnerability. The Third National Climate Assessment[756] in 2014, compiled by over 300 experts opens with:

> *"Climate change, once considered an issue for a distant future, has moved firmly into the present. Corn producers in Iowa, oyster growers in Washington State, and maple syrup producers in Vermont are all observing climate-related changes that are outside of recent experience. So, too, are coastal planners in Florida, water managers in the arid Southwest, city dwellers from Phoenix to New York, and Native Peoples on tribal lands from Louisiana to Alaska. This National Climate Assessment concludes that the evidence of human-induced climate change continues to strengthen and that impacts are increasing across the country."*

In many ways, our present-day societies, with their intricate just-in-time global supply chain interdependencies for food and resources, and the complex

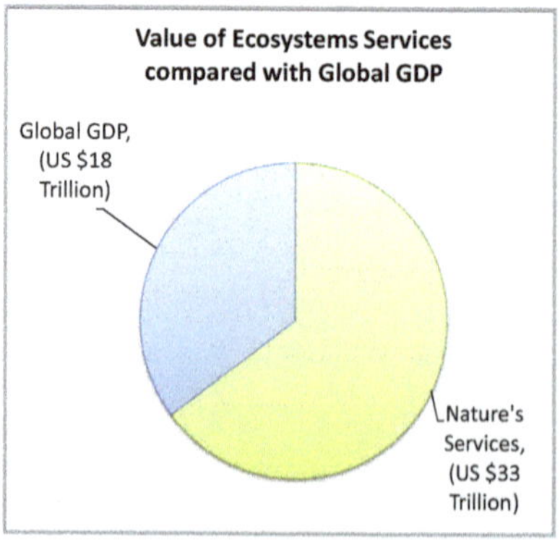

1.5 **Nature's services are worth almost twice as much as global GDP**
Source: Adapted from "The Value of the World's Ecosystem Services and natural capital", Constanza et al, Nature Vol 387, p256.[170]

financial markets which underpin our economies, make us more vulnerable to environmental shocks, not less. The Kobe earthquake in Japan put several manufacturers halfway around the world in California out of business just because the supply of critical components was interrupted.

The evidence shows that we have already exceeded the limits of many of the natural resources on which our businesses, economies and societies depend. One authoritative source is the 2005 Millennium Ecosystems Assessment[515] This report drew from a panel of over 1,300 experts worldwide, was co-chaired by the director of the United Nations University and the chief scientist of the World Bank and had at its heart a process of peer review from a wide range of academic, governmental and non-governmental organizations:

> *"Humans are fundamentally, and to a significant extent irreversibly, changing the diversity of life on Earth, and most of these changes represent a loss of biodiversity...... approximately 60% (15 out of 24) of the ecosystem services evaluated in this assessment ... are being degraded or used unsustainably."*

A team from the Stockholm Institute of Resilience published a paper in Nature in 2009[625] which concluded that there are nine fundamental planetary "boundaries" beyond which we face "unacceptable environmental change". Of these limits, we have already crossed three: climate change, the rate of biodiversity loss and the nitrogen cycle.

It is important that our organizations understand that the environment is not something distant and irrelevant. It is the place where we live and where our organizations derive all the material and services on which we fundamentally depend. These boundaries are not abstract lines over which some scientist may debate. The impact of crossing these barriers is not a single isolated event such as the loss of species such as polar bears, which while sad has no direct connection to our businesses. Crossing these boundaries could bring about the failure of the fundamental systems underpinning our wealth and our wellbeing.

Our economy depends on the services which nature provides. Nature is like a bank account from which we can draw minerals, food, air, water, energy, medicines as well as services, like CO_2 removal, water purification, waste disposal, recreation and tourism. The value of this account was conservatively estimated to be US$33 trillion in 1997,[170] almost twice the entire global GDP. While we can place a financial value on these natural services, we must not fool ourselves that we can buy our way out of the problems we are creating at some point in the future. For most of the ecosystems services that our planet provides there are no known substitutes at any price.

Our economy is a subset of our environment, not the other way round. Even as far ago the 18th century, Adam Smith, considered the father of modern economics and capitalism, in his great work, *An Inquiry into the Nature and Causes of the Wealth of Nations,* understood that the primary source of wealth was "*soil, climate and territory*". It is from this primary wealth that secondary wealth can be created by transforming these resources into food or steel or shelter and then from these, in turn, the basis for all tertiary wealth arises – the paper money, debts, stock, bonds, derivatives and other abstractions. When we talk of financial value we measure this in terms of tertiary wealth, but, in reality, all this tertiary wealth is simply a future claim on primary or secondary sources of wealth; it is not wealth in itself.

Money permits us to acquire a share of the resources that our planet provides, but the future purchasing power of money is not a fixed quantity. There is an intimate link between the availability of this natural wealth and its cost. If wheat or oil or gold are in short supply, their price rises and we have inflation. If they suddenly become more plentiful we have less money per unit of natural resource and the prices fall, so we have deflation. By and large, our relationship with nature is entering a protracted inflationary phase. As resources become scarcer, their prices rise. This inflation pressure will have profound impacts on the competitiveness and value of organizations unless there is a compensatory improvement in resource efficiency.

A name given to the combined resources and services which create all human wellbeing is natural capital. In his outstanding book, *Beyond Growth*[181] Herman E. Daly provides a remarkable insight into just how much humanity depends on this natural capital.

> *"Probably the best index of the scale of the human economy as part of the biosphere is the percentage of human appropriation of the total world production of photosynthesis. Net primary production (NPP) is the amount of solar energy captured in photosynthesis by primary producers, less the energy used in their own growth and reproduction. NPP is thus the basic food resource for everything on earth not capable of photosynthesis.*[763] *Vitousek calculates that 25% of potential global (terrestrial and aquatic) NPP is now appropriated by human beings. If only terrestrial NPP is considered the fraction rises to 40%. Taking the 25% figure for the entire world, it is apparent that two more doublings of the human scale will give 100%. Since this would mean zero energy left for all non-human and non-domesticated species, and since humans cannot survive without the services of ecosystems (which are made up of other species) it is clear that two more doublings of the human scale is an ecological impossibility, although arithmetically possible".*

Our economy is a subset of the environment, not the other way around. The environment is the **primary source** of all our wealth.

More recent work[111] has lowered the estimate for the terrestrial human appropriation of net primary production (HANPP) to around 25%. This revision in no way detracts from Daly's message that exponential growth is simply a physical impossibility unless we can dramatically increase the earth's photosynthetic output, which is improbable. The big new driver of HANPP is deceivingly called green energy or biomass, which uses photosynthesis-derived molecules as a power source ranging from fast-growing wood, through to palm oil, algae-derived diesel or sugar cane ethanol, so we can expect this number to go up. ⇒ page 21.

Exploration: *The Limits to Growth*

One of the most influential books on the subject of resource efficiency in the 20th century was *The Limits to Growth*[517] by Donella H. Meadows, Dennis L. Meadows, Jørgen Randers and William W. Behrens III. Commissioned in 1972 by the Club of Rome, a think tank the study used computer systems – an innovation at the time – to model the interaction of population, pollution, investment, agriculture and natural resources. Each of these drivers had positive and negative feedbacks with each other – e.g. as agriculture improved so did food per capita, which boosted population; at the same time as population increased so too did pollution and natural resource consumption, while a decline in resources decreased industrial output.

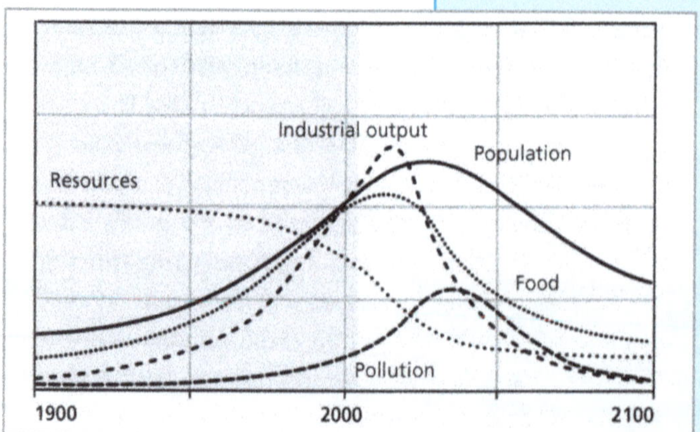

1.6 **The *"Standard Run"* of the limits to growth painted a picture of resource decline, reduced industrial output and less food per capita leading to declining population**
Source: Scenario 1 in "Limits To Growth - The 30 year update"[518] Meadows et al, page 169.

This was the first application of *"systems dynamics"* simulation at a global level. The model created, called *World 3*, was not intended to provide precise predictions (after all, the world is much more complex than a computer model) but was designed to explore whether there were limits to growth, as the title implied, and what these might be.

The results were quite dramatic - starting with an approximation of conditions in 1970 the base case scenario indicated that growth would decline markedly from the early 2010s onwards. In this base scenario, shown left, pollution was not the most significant constraint to growth; rather it was the need to divert more and more capital investment to maintain the production of both food and services.

What the team did next was to explore whether changes in the assumptions in the model would lead to different outcomes, for example, if the rate of population growth was decreased or the level industrial output was reduced. They discovered that changes in single parameters merely postponed the collapse – with more natural resources, the population and industrial production simply climb higher before they collapse due to the effect of rising pollution and the increasing cost of extraction of the resources. However, what the model did show is that it is possible to achieve a steady state, which does not imply a substantial loss of material standards, by controlling population and growth and by realizing big technological breakthrough in resource efficiency, switching to renewable resources, in pollution prevention and degrading less agricultural land. In other words, the book signalled that the kinds of change needed to avoid collapse must be economy-wide.

The book had an enormous impact and was translated into more than 30 languages with over 1 million copies sold. The success of the title led to a heated debate, in which many of the most vociferous critics were economists. There are some understandable reasons for the criticism. First, there are some who had an initial gut reaction against the *"Models of Doom"*, which seemed to challenge the prevalent view of ever-increasing prosperity. Others felt that these questions were better dealt by economists modelling the *"real economy"*, i.e. econometrics, rather than being left to outsiders using an unknown (and unproven) technique. Finally, others misunderstood how the World 3 model worked and failed to realize that it was never intended to provide a precise prediction.

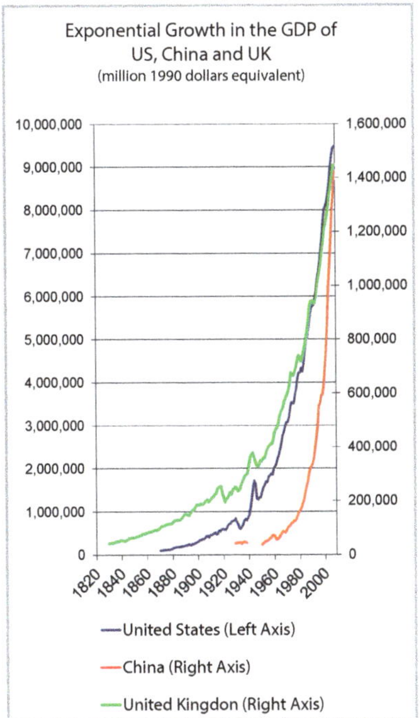

1.7 **Although the UK, US and China have all started their economic growth at different times with different starting conditions we can see that they follow an exponential curve, with historic rates of growth around 8-9% per year**
Source: Niall Enright based on data from Angus Maddison "Historical Statistics of World Economy 1-2008".[492]

Ugo Bardi's highly readable *The Limits to Growth Revisited*[56] is a fascinating tale of the passionate argument sparked by the book and explains what conclusions we can properly draw from the work.

Contrasting with the pessimistic perspective of Meadows and others is a cadre of economists who believe that declines in resources will be more than compensated by humanity ingenuity, thus enabling us to maintain our current levels of exploitation for the foreseeable future. Such optimists would include Howard Barnett and Chandler Morse who, in their 1963 classic *Scarcity and Growth*,[57] argued that resource scarcity did not threaten economic growth. A follow-up investigation in the late 1970s, *Scarcity and Growth Reconsidered*, reached mostly the same conclusions. More recently, *The Sceptical Environmentalist*[479] by Bjørn Lomborg, stated that fears about diminishing resources were exaggerated, although the basis for the findings in the books have since been questioned, notably in a series of articles in *Scientific American*, called *Misleading Math About the Earth.*[619] Of course, there is the adage that "*a pessimist is simply an optimist who is in full possession of the facts*", and the most recent follow-up to *Scarcity and Growth*[710] concludes that:

> *"The message of Scarcity and Growth that depletion of market resources was not a problem has given way to a concern that 'new scarcities' of environmental quality, global climate, and biological diversity are emerging."*

In other words, it is not the availability of resources per se that is the critical issue, but the capacity of natural services to deal with the pollution and other damage that our consumption is creating. It is the *"Sinks"* rather than the *"Sources"* that are the primary concern.

The core idea that *The Limits to Growth* revealed is the concept that exponential growth in a finite system will inevitably lead to diminishing returns. This idea seems completely obvious to biologists, who will be very familiar with the experiment that shows yeast or bacterial populations doubling every few hours/days in a petri dish only to eventually decline as resources are exhausted, or waste products accumulate to toxic levels. Economists, on the other hand, see only the data on the upward part of the curve of human progress to date. This data leads them to either ignore the resource limits altogether or to incorporate a technology factor (formally known as Solow's residual in the economic production function models) to explain how GDP rises faster than expected from growth in the capital investment and labour inputs alone. This residual is not insignificant and accounts for about 1.2% of the US's average growth rate during the period 1950-2000. Naturally, it is the subject of much debate by economists with some suggesting that it reflects resource efficiency – particularly concerning the way energy is converted to useful work. One of the key questions that this assumption gives rise to is whether innovation itself is subject to the law of diminishing returns – in other words are incremental efficiency improvements going to become increasingly difficult to achieve?

In exponential growth, you get growth on top of growth. Exponential growth is what our economies are entirely based on. The idea of perpetual growth has become the central assumption of our current economic paradigm. Those brief periods of time when growth falters or goes into reverse are referred to as recessions and are something to be avoided at all costs. Recessions are bad as the lack of growth means, among other things, a decline in prosperity, reduced sales, falling incomes, and an inability to service the accumulated debts that have powered the growth in the first place. Hence our economic systems – and private

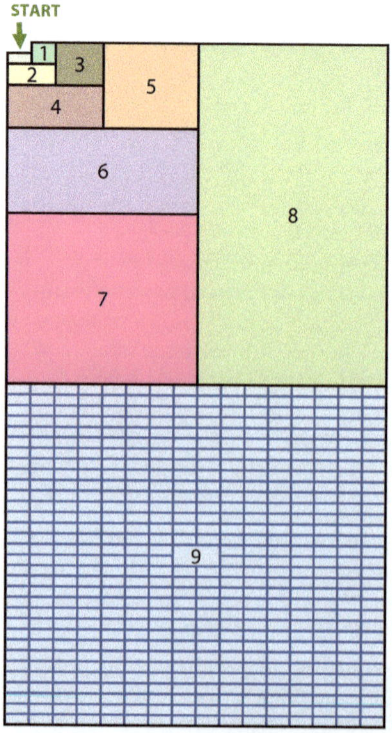

1.8 **Illustration of the concept of exponential growth in resource consumption**

Each rectangle is double the area of the previous one, and is larger than the size of all the previous rectangles put together. The illustration shows the effect of nine doublings leading to a 512-fold increase in resource use. Just three more doublings will fill this page.
Source: Niall Enright, available in companion file pack

business as a whole – are set up to deliver regular year-on-year growth, something which they have managed to do relatively consistently since the Industrial Revolution. The attribute of exponential growth is that it starts slowly, but after a few cycles the rate of increase can be enormous, as shown by the diagram left. Let's imagine that we begin with the box labelled "start" representing just one unit of resource, and then double this to get to box 1, and then we double at each step all the way through to box 9. We can see that after nine doublings we are consuming 512 times our original resource (each rectangle represents the initial resources we were using at the outset). In just three more steps our consumption will be over 4,000 times the original rate. Another important observation is that each step consumes more than *all the preceding steps added together*.

Of course global GDP has not doubled every year – in fact, its average since 1960 has been 3.5% per year.[805] That does not mean to say that global GDP is not growing exponentially; it just means that the doubling time is greater than one year – it is 20.3 years to be precise. So at the current rate of growth in 20 years' time, we will be using twice the resources that we are today, and *in that period we will have consumed more resources we have used in all history up to today*.

Some economies like China's have been growing at an average of 10% per year since 1990; that is a doubling every seven years, so that in 2015 they are eight times the size they were in 1994. In a mere seven more years, to maintain current rates of growth, China will need twice as many power stations, producing twice as many manufactured goods from twice as many factories, doubling the transportation of goods, consuming twice as many natural resources and emitting twice the pollution. See *Why China is key* (page 68) for more on this issue.

Perhaps the next global doubling will be achieved easily, and maybe the one after that or even the one after that. But clearly, unarguably, this process cannot continue forever if resources are finite. That is the key message in *The Limits to Growth*.

The original team have produced follow-up works to *The Limits*, the most recent in 2004,[518] where they reiterate that the book is not a prediction, but rather a declaration that exponential growth cannot continue forever and that we can either manage our transition away from growth or face collapse. Several independent studies have shown that the original predictions in *The Limits to Growth* correspond well with trends observed in the intervening 40 years, the most recent by Graham Turner at CSRIO in Australia.[327]

We should consider *The Limits of Growth* as a warning. Unless we can radically reduce resource consumption and industrial output, through resource efficiency, and significantly reduce damage to the environment, and reduce population growth and improve food yields, we run a risk of uncontrolled decline in human welfare. It is not all bad news - since 1980 we have increased the US$ global GDP per kg oil equivalent 1.64 times (from US$4.20 to US$6.84[805]) and this book will demonstrate that there exists a tremendous potential for further improvement. However, it is evident that our rate of improvement in resource efficiency needs to match the growth in GDP simply to stabilize our absolute rate of resource use – and in all probability, we need to reduce our rate of resource consumption much more dramatically. This statement does not originate in an ideological perspective – rather it is simply what the maths and science are telling us. Those organizations that can grasp this are the ones which are more likely to prosper as society (or nature) starts to impose these limits.

1.4 Ecological footprint

To sustain today's level of resource use we will need a planet one and half times the size of our own. Rising affluence means that, at the current rate of growth in resource use, this will increase to four planets by 2050.

One measure of the capacity of our planet to meet our needs is provided by the Global Footprint Network, which produces an ecological footprint value that calculates the land area required to sustainably provide all the natural services upon which human life depends. These services include land to grow our food crops or oceans for our fisheries, forests to absorb the CO_2 we emit, watersheds to gather, purify and collect the water we need, and so forth. Today, humanity uses the equivalent of 1.5 planets[287] to provide the resources we use and absorb our waste. Returning to our bank account analogy, then, in 2011 we would have started to go overdrawn on the 27th September, and each year we will go overdrawn earlier and earlier. Just as with a real bank account, the effect of this overconsumption is cumulative – our overdraft is getting bigger and bigger, the harm we are doing by exceeding these boundaries is accumulating and increasing the risk of irreversible, abrupt change which would threaten our wellbeing. The debt that we are incurring is growing – in fact, if we were a business we would be trading insolvently.

The Ehrlich Equation[243] states that three factors can determine the human impact on the environment. First is the total Population. Next is Affluence, a measure of how many material goods and services each person will consume. Finally there is Technology, a measure of how resource-intensive our delivery of the goods and services is – in effect a measure of resource efficiency.

Impact = Population x Affluence x Technology

If our planet's resources are like a bank account, then today we are growing overdrawn on the 27th September each year. By 2050 that it will be the 1st April.

If we look at forecasts for each of these parameters in the future, we can estimate how our impact on nature is likely to grow. For example, the earth's Population is expected to increase from 7.4 billion in 2016 to 9.3 billion in 2050, an increase of 31%. Affluence is anticipated to rise threefold from 2012 to 2050, using per capita GDP as an equivalent value. We can assume that Technology improvements will flow at the current rate, where global resource intensity per unit GDP has decreased about 1/3rd from 1980 to 2008. Plugging these values into our formula we get:

Impact = 1.3 x 3 x 0.67 = 2.6

So our impact in 2050 will be around two and a half times what it is today. Translating this into the ecological footprint, since we are already using up

In order to achieve balance, we need to achieve the same economic growth with one-tenth the resources. We need a **Factor Ten** improvement.

1.5 planets worth of resources each year, by 2050 we would need four planets worth. Our bank account will start to go overdrawn on the 1st April each year. This impact is clearly not a sustainable proposition.

These data show that Technology, i.e. resource efficiency, can offset Population growth. The real challenge (and opportunity) is the increased consumption that the rise in Affluence will bring, which far outstrips our past rate of improvement in resource intensity. China's growth is not because its population is doubling every seven years, it is because millions of people are becoming more affluent and demanding more consumer goods and services - see *Why China is key (page 68)*. Unless we in the developed world are willing to reduce our standard of life dramatically, it is unrealistic to deny this lifestyle to people in emerging economies. Lifestyle is as much about hospitals, schools and other services, as it is about personal material consumption. This insight leads us to the inevitable conclusion that if Affluence and Population are set to grow, then the only way to reduce our global impact on the environment is through improved Technology i.e. resource efficiency.

This rationale has led to the notion of material decoupling, which proposes that the only sustainable path is to achieve economic growth without consuming more natural resources. Early thinking on this emerged from Ernst Ulrich von Weizsäcker, Amory Lovins and Hunter Lovins as far ago as 1998[777] with their book, *Factor Four: Doubling Wealth, Halving Resource Use*. Today, even higher levels of dematerialization are proposed by Friedrich Schmidt-Bleek and others, who argue for a Factor Ten improvement. The rationale is simple, based on the numbers above. If the affluent 20% of the world's population are consuming 1.5 planets worth of resources today, then if everyone shares their same lifestyle we would consume 7.5 planet's worth of resources (1.5 *5=7.5). Add in the 30% growth in population and we will need almost 10 earth's resources (7.5 *1.3 = 9.75). Thus a factor 10 decrease is required by 2050.

As we have seen, the UK government's target is a five-fold decrease in absolute CO_2 emissions between 1990 and 2050 which, allowing for growth, equates to a 15-fold improvement in the carbon intensity per unit of GDP over the same time. These are the orders of magnitude improvement that we need to achieve equilibrium with the natural processes that manage carbon, called *stabilization* in the jargon of climate change.

Just as an individual cannot continue to build up debt forever, consequences will accumulate for organizations as a result of our collective over-exploitation of natural resources. This excessive consumption poses significant risks, as well as great opportunities. Organizations are central to the response as it is through organizations of all types that change can be achieved: institutions, public service providers, private companies all have a role to play. The bottom line is that all these organizations will gain value by addressing resource use, both directly and in the influence that they exert over others.

⇒ page 27.

Real World: *Mineral scarcity*

So, just how much of our planet's resources are we using up? Estimating the availability of mineral resources is notoriously difficult. However, we can get a good impression if we examine a range of forecasts and data for these resources.

The table on the next page illustrates some of the possible scarcity indicators for a range of economically valuable resources. The data is drawn from a variety of sources shown by the references in the column headings. Risk or scarcity is shown by the cell shading: red (high) and yellow (medium).

The minerals are ranked by the number of years' reserves available, calculated as the ratio of current production to the reserves stated. The US Geological Survey (USGS) refers to *"reserves"* as that part of the total *"resource"* of a mineral that is economic to extract, because it meets the physical, chemical and economic criteria to make extraction economic[757]

- The remaining years of reserves are shown in column 3. This is likely to be an over-estimate because demand for most minerals is growing rapidly. Indeed, another complicating factor is that the reserves themselves may not be immediately available as the USGS note that the *"term reserves need not signify that extraction facilities are in place and operative"* so we may see scarcity even earlier than the data indicate, particularly as some mining operations can take a decade or more to become operational.
- The next column in the table gives risk index for the minerals. This index is based on seven criteria: scarcity; production concentration; reserve distribution; recycling rate, substitutability and the governance of the top producing and top reserve-hosting nations in turn.
- The next column has a data on the end-of-life recycling rate. Here there is a lot of uncertainty so the highest and lowest recycling rates are shown. The data for gold and silver excludes jewellery recycling.
- Column 6 shows the reliance of the US on imports of these minerals.
- The next column, shows the major producing country. It is remarkable how many times China is the lead producer.
- Dependency refers to whether the mineral is mined in its own right or is a by-product of other mineral extractions. Clearly, if it is mined as a by-product, then its availability may be more challenging if the primary mineral becomes scarce or expensive to mine.
- Finally, the column indicates how easy it is to substitute the mineral with alternatives.

Uranium data is taken from the International Atomic Energy Agency,[399] although there is a lot of argument around the availability of sufficient uranium to enable significant quantities of future electricity generation in nuclear power stations.

Determining the risks associated with minerals shortages is made even more complicated because of the actions of states to protect supplies of resources. China restricted the exports of rare earth elements in 2010, which led to a surge in the prices for these minerals which are vital for a range of industrial goods. One of the rare earth minerals is neodymium, which is essential to the manufacture of powerful magnets and is used in a wide variety of applications from headphones

1.9 **(following page) Table illustrating the availability and risk of a number of important minerals**
Source: Niall Enright using data from the US Geological Survey,[757] British Geological Survey,[87] the United Nations Environment Program[729] and the International Atomic Energy Agency[399]
This table is available as an A3 poster in the companion file pack

Mineral	Group	Reserves (Years)[757]	BGS Risk Index[87]	EOL Recycling Rate[729]	US Import Reliance[757]	Top Producer[86]	Dependency[757]
Indium	Speciality	Low	7.6	<1	100%	China (55%)	High
Antimony	Speciality	11	9.0	5-85%	84%	China (82%)	Low
Lead	Non-Ferrous	16	6.2	52-95	31%	China (54%)	Low
Chromium	Ferrous	18	6.2	87-93	66%	Kazakhstan (20%)	Low
Zinc	Non-Ferrous	19	4.8	19-60	82%	China (36%)	Low
Gold	Precious	20	5.7	15-20*	0%	China (15%)	Low
Silver	Precious	20	6.2	30-50*	72%	Mexico (21%)	Med
Tin	Non-Ferrous	20	6.7	91	75%	China (40%)	Low
Cadmium	Speciality	23	6.7	15	0%	China (32%)	High
Titanium	Non-Ferrous	23	4.8	91	0%	Canada (23%)	Low
Strontium	Speciality	28	8.6	<1	100%	China (80%)	Low
Nickel	Ferrous	30	6.2	57-63	37%	Philippines (17%)	Low
Manganese	Ferrous	34	5.7	53	100%	China (31%)	Low
Copper	Non-Ferrous	39	4.3	43-53	36%	Chile (32%)	Low
Molybdenum	Ferrous	41	8.6	30	0%	China (41%)	Low
Barium	Speciality	41	8.1		70%	China (45%)	Low
Bismuth	Speciality	42	9.0	<1	95%	China (50%)	High
Tungsten	Speciality	49	9.5	10-66	49%	China (83%)	Low
Mercury	Speciality	52	8.6	1-62%	0%	China (71%)	Low
Iron	Ferrous	58	5.2	52-90	0%	China (44%)	Low
Cobalt	Non-Ferrous	60	7.6	68	75%	Congo, D R (68%)	High
Niobium	Ferrous	84	7.6	50-56	100%	Brazil (94%)	Low
Gallium	Speciality	100	7.6	<1	100%	China (n/a)	High
Bauxite/Aluminium	Non-Ferrous	108	4.8	42-70	100%	Australia (31%)	Low
Uranium[399]	Non-metal	111	5.7			Kazakhstan (36%)	Low
Platinum Group	Precious	164	7.6	60-701	90%	South Africa (59%)	Med
Vanadium	Ferrous	184	6.7	<1%	100%	China (52%)	High
Phosphorous	Non-metal	299			4%	China (44%)	Low
Lithium	Speciality	371	6.7	<1	60%	Chile (49%)	Low
Rare Earths	Speciality	1,273	9.5	<1	100%	China (90%)	Low
Magnesium	Non-Ferrous	Large	7.1	39	26%	China (65%)	Low
Nitrogen (Ammonia)	Non-metal	Large			29%		Low

Mineral	Substitutes[757]	Uses[757]
Indium	High	LCD Displays, Electrical Components
Antimony	High	Ammunition; Lead; Flame retardant; Batteries
Lead	Med	Lead-acid Batteries
Chromium	Low	Stainless Steel
Zinc	High	Galvanizing; Brass and Bronze; Other Alloys
Gold	Med	Electronics, Jewellery, Coins, Dental
Silver	High	Electrical and Electronics; Coins and Medals; Photography; Jewellery
Tin	High	Cans and Containers; Construction; Transportation; Electrical
Cadmium	Med	Alloys; Coatings; Batteries; Pigments; Plastics Stabilizers
Titanium	Low	Aerospace; Armour; Chemical Processing; Marine; Medical; Power Generation
Strontium	Med	Pyrotechnics and Signal Flares; Magnets; Alloys; Pigments; Glass
Nickel	Med	Transportation; Fabricated Metal Products; Electrical Equipment; Petroleum
Manganese	Low	Alloys in Construction, Machinery and Transportation; Steel
Copper	Med	Buildings, Electric Products, Transportation, Consumer Equip., Machinery
Molybdenum	Low	Iron, Steel and Alloys
Barium	High	Oil and Gas Extraction; Paints and Plastics; Automotive; Medical
Bismuth	High	Lead replacement; Metallurgy; Sprinkler Systems
Tungsten	High	Carbide Materials for Construction, metalworking, mining, oil and gas; Alloys
Mercury	High	Chlorine Production; Electronics; Fluorescent Lighting
Iron	Low	Construction, Transportation, Cans and Containers
Cobalt	Med	Alloys; Carbides; Metals; Chemicals
Niobium	Med	Steels; Alloys
Gallium	Med	LED, Solar Panels
Bauxite/Aluminium	Med	Transportation, Packaging, Buildings, Electrical, Machinery
Uranium	Low	Power Generation
Platinum Group	Low	Catalysts; Computer Equipment; Glass; Displays
Vanadium	Med	Steel Alloys
Phosphorous	Low	Agriculture as a fertiliser
Lithium	High	Ceramics and Glass; Batteries; Lubricants; Casting Powders; Air Treatments
Rare Earths	Med	Catalysts; Alloys; Magnets
Magnesium	High	Alloys in Transportation, Packaging; Cast products; Desulfurisation of Steel
Nitrogen (Ammonia)	Low	Agriculture as a fertiliser

to car parts. Prices increased 10-fold in just a year after the export restrictions, from US$50 per lb to US$500. Although most manufacturers were able to absorb these price increases because the rare earths are used in trace amounts, one exception was Toyota, whose Prius Hybrid uses a kilogram of neodymium and which was forced to raise prices as a result.[427]

Another mineral that has been in the news recently is tantalum, which is used for the manufacture of capacitors for the use in electronics, especially cell phones and high-temperature alloys in wind turbines. Tantalum, in the form of the mineral coltan, found in the Democratic Republic of Congo, has been blamed for bringing about conflict, corruption and environmental degradation. The problem with coltan arises because artisanal mining is possible – groups of men create a crater in a stream in which they wash muds, and the coltan ore settles out. A group of people can mine a kilo a day, which means that they can earn US$200 a month compared with $10 for the average Congolese worker.[155] UN reports[728] on the cause of the conflict in Congo cite access to coltan, and other minerals, as a major driving force for the conflict, and in particular for the involvement of foreign troops from Rwanda. Since the primary coltan deposits are in Kahuzi Biega National Park, home to mountain gorillas, there has also been a very negative impact on the gorilla population, which is said to have been nearly cut in half from 258 to 130.[155] This impact has led the US SEC to mandate that US companies declare their use of so-called conflict minerals that are produced in the Democratic Republic of the Congo and neighbouring countries,[497] with companies such as Apple reporting that all their coltan is ethically sourced.[496,466]

What is clear from this data, and the broader issues raised by the rare earth and coltan cases, is that organizations cannot treat many mineral resources as if they are infinite and without economic and reputational risk. Over half the mineral reserves listed in the table will be exhausted in under 50 years at current rates of production. Although human ingenuity is likely to overcome some limitations in the future, it is by no means certain that this will enable costs of these resources to be kept low.

The table above shows the expected lifetime of the minerals at the *current rate* of extraction. In fact the demand for many of these minerals is growing rapidly. The chart below shows world production of iron ore. In 2000 this was around 1,000 million tonnes. By 2005 this had increased by 50% to 1,500 million tonnes. By 2010 this has doubled to 3,000 million tonnes: a threefold increase in just a decade.

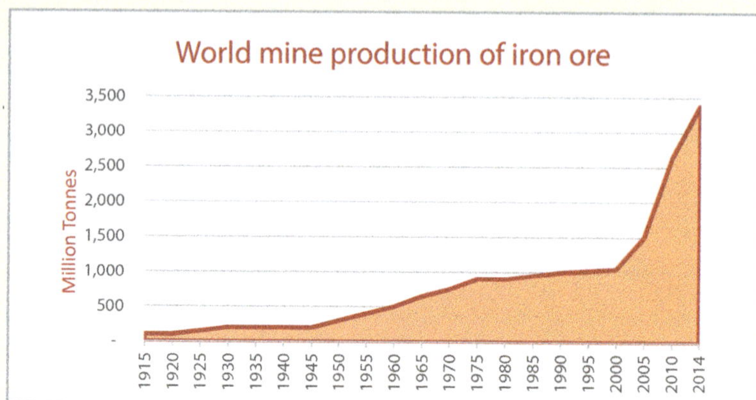

1.10 **World mine production of iron ore, 1915-2010**
Source: Niall Enright based on data from British Geological Survey[86]

1.5 Climate change

The evidence for climate change is overwhelming. Many organizations remain in denial about the implications for them. Here we consider the scale of change that is needed just to stabilize our emissions.

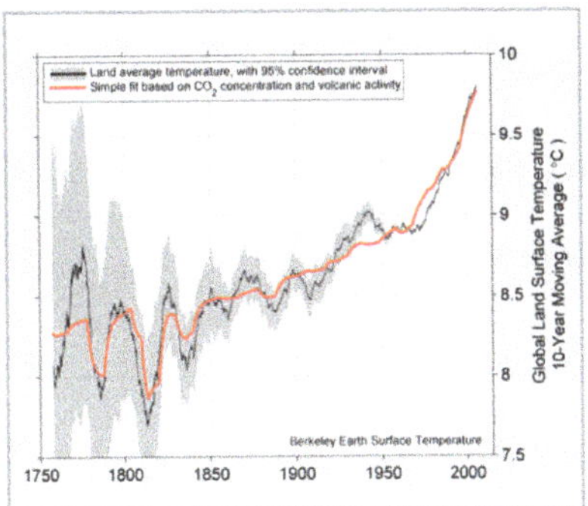

1.11 **The annual and decadal land surface temperature from the Berkeley Earth average, compared to a linear combination of volcanic sulfate emissions (responsible for the short dips) and the natural logarithm of CO2 (responsible for the gradual rise).** The grey area is the 95% confidence interval. *Source: Reproduced from "A New Estimate of the Average Earth Surface Land Temperature" Robert Rohde, Richard A. Muller et al., 2012.*[628].

Arguably the most dangerous boundary faced by the human race is climate change. There is now no debate over the basic fact that the earth is warming at an unprecedented rate. The most comprehensive study of weather station data from around the world so far, by the Berkeley Earth Surface Temperature project, published in 2012,[628] has concluded that "*the rise in average world land temperature globe is approximately 1.5° C in the past 250 years, and about 0.9° C in the past 50 years*".

The Berkeley team looked at over 1.6 billion temperature records and concluded that there are two drivers for the observed temperature fluctuations: volcanic activity, which had the effect of reducing temperature, and atmospheric CO_2 which increases the temperature. In a blow to theories proposed by many of those that believe that climate change is not a result of human activity and atmospheric CO_2, the study observed that "*solar variation does not seem to impact the temperature trend*". Rather than use complex climate models to demonstrate the linkage, the Berkeley team combined the effect of CO_2 and volcanoes into one single figure and saw if this would fit with the observed temperature rise, as illustrated left by the line in red on the chart left.

So, if CO_2 is the primary cause of global average temperature increases, then what causes the increase in atmospheric CO_2? Well, the rise coincides with the industrialization of our economies from the 18th century, which was characterized by the combustion of ever-increasing quantities of fossil fuels – first coal, but then gas and oil. As the carbon in these fuels is burnt it is converted to carbon dioxide, CO_2, which enters the atmosphere. Some of the CO_2 in the atmosphere is used by plants in photosynthesis, or absorbed by oceans, natural "*sinks*" which provide a natural service of "*system regulation*". A recent study in the journal *Nature*[53] has shown that these sinks have almost doubled in their capacity to absorb CO_2 in the last 50 years, increasing their carbon extraction from 2.4 ± 0.8 to 5.0 ± 0.9 billion tonnes of carbon per year between 1960 and 2010, primarily through increased uptake by the oceans. Unfortunately, this increase has not managed to keep pace with the sheer volume of CO_2 we are putting into the atmosphere so that today the planet is only able to absorb around half of the 34 billion tonnes of CO_2 we are emitting each year[573] Because we are exceeding nature's capacity to absorb CO_2, it has accumulated in the atmosphere, with concentrations rising from the pre-industrial level of 280 parts per million (ppm) to today's figure of just over 400 ppm.

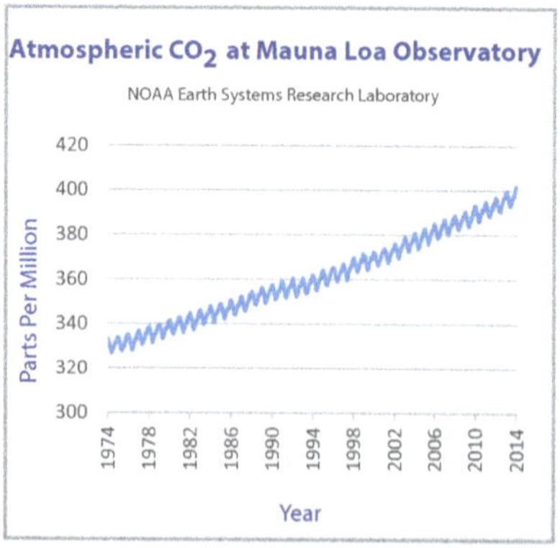

1.12 **CO_2 records from the Hawaiian Mauna Loa Observatory show the rise of CO_2 over the last 50 years, from 320 parts per million to over 400 parts per million**
Source: Dr. Pieter Tans NOAA Earth Systems Research Laboratory.[558]

Although the rise shown in the graph, above left, looks gradual, in environmental terms the change is very abrupt when considered at over a longer timescale. Ice-core data for the last 800,000 years shows that CO_2 levels have ranged from 170 to 300 ppm.[488] Plotted on a chart below left, we can see that the current level of CO_2 is 30% above the highest levels observed in the recent past (for comparison, our species Homo Sapiens has been around for around 200,000 years). Plotted on this graph are the lower and higher emissions scenarios of the International Institute for Applied System Analysis (IIASA) Greenhouse Gas Initiative Database,[393] which suggest, in the absence of specific policies on climate change, that emissions could rise to between 550ppm and 900ppm by the end of the century.

It is worth noting that the changes in surface temperature that we see today are lagging behind the CO_2 increases. This lag is because it takes more time for the vast oceans (500 times the mass of the atmosphere) to warm up and to release their heat, an effect that is called thermal inertia. According to a spate of papers in *Science Magazine* around 2005[348,520,781] the pent-up increase in temperature, even if we were to stabilize our emissions at the current level, is around 0.6° C. The time-lag due to thermal inertia is difficult to calculate because it involves issues such as the mixing rate of surface water due to winds etc., but the best estimate is that it is around 40 years. In some ways, this lag is to be welcomed because it means that the warming that we are seeing at the surface today is lower than it would otherwise have been – giving us more of a breathing space to adapt. But the key point to take away is that the temperatures we are experiencing today are as a result of the CO_2 concentrations in the 1970s.

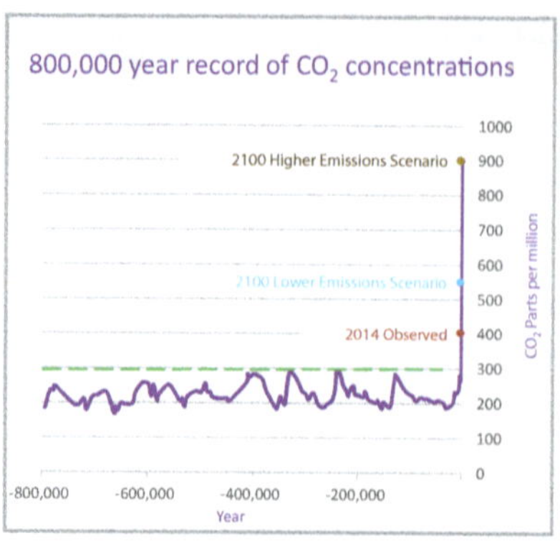

1.13 **Antarctic ice-core sample show the fluctuations in atmospheric CO_2 over the last 800,000 years.** During this time concentrations have varied from around 170 ppm to 300 ppm and these changes are well correlated with temperature data. The rise to 400 ppm in the last couple of centuries is unprecedented, as is the forecast concentrations of 550 ppm or 900 ppm by the end of the century.
Source: Niall Enright based on ice-core data from Lüthi, et al.[488] *and on IIASA emissions projections.*[165]

Because of concerns about climate change, and the huge amount of science that needed to be consolidated to understand what the likely effects are, the United Nations established the Intergovernmental Panel on Climate Change (IPCC) in 1988 to provide scientific advice on the issue. The IPCC publishes a comprehensive assessment report ever five to seven years or so. The IPCC itself does not carry out original research, but rather gathers together all the published work in science journals and through a process of subject-matter expert review and debates, produces conclusions which are then subject to line-by-line review by over 120 governments before the final reports are agreed. As a mechanism to reduce errors and present a balanced view of the science, the process works well, although there have been criticisms that it inevitably leads to a conservative assessment of risk and omits the latest science, for example around sea level rises. Nevertheless, the latest IPCC report, the 5th Assessment Report published in 2013,[683] paints a picture of a likely temperature rise from 1990 under most scenarios exceeding 2° C by the end of the century and possibly as much as 4-6° C.

The problem with these temperature rises is that they could lead to some very severe consequences. Our ecological systems struggle to adapt quickly to temperature changes. For example, the breeding cycle of birds is no longer synchronized to when the insects they feed on emerge; or habitats like rainforests retreat further up mountains at a rate faster than the amphibians who depend on them can keep up; or corals become stressed and lose the algae that they need for nutrition, which leads to bleaching. The IPCC concludes that: "*Approximately 20% to 30% of plant and animal species assessed so far are likely to be at increased risk of extinction if increases in global average temperature exceed 1.5° C to 2.5° C*". Table 4.1 in Chapter 4 of Working Group II Report on "*Impacts, Adaptation and Vulnerability*" in 2007[586] lists 78 scientific papers published on the theme of the effect of temperature changes on species makes for particularly sobering reading.

Foundations

1.14 **There is a large body of literature estimating the impact of temperature rises on our ecosystems or biodiversity**
It is remarkable how seemingly small temperature increases compared to pre-industrial levels can have very detrimental impacts on ecosystems.
Source: Adapted from table 4.1 in the IPCC Working Group II Report: Impacts, Adaptation and Vulnerability[586]

Temperate increase from pre-industrial levels °C	Effect on biodiversity or ecosystems
1.1-1.6	8% loss freshwater fish habitat, 15% loss in Rocky Mountains, 9% loss of salmon [N. America]
1-2.3	All coral reefs bleached [Great Barrier Reef, S.E. Asia, Caribbean]
1.6-2.4	7-14% of reptiles, 8-18% of frogs, 7-10% of birds and 10-15% of mammals committed to extinction as 47% of appropriate habitat in Queensland lost [Australia]
2.1-2.3	15-37% (mean 24%) of species committed to extinction [Globe]
2.2	3-16% of plants committed to extinction [Europe]
2.4	63 of 165 rivers studied lose >10% of their fish species [Globe]
2.5	Sink service of terrestrial biosphere saturates and begins turning into a net carbon source [Global]

It is the last entry in the table above that gives us a hint of one of the most worrying aspects of climate change. That is the issue of feedback loops. At the moment the earth is absorbing around half of the CO_2 we are emitting, thus reducing the potential temperature rise from our emissions. However, there is concern that these natural sinks may become saturated – for example, as the ocean absorbs CO_2 it becomes acidic[614] and its ability to absorb further CO_2 is reduced. If a decline in the sinks' capacity occurs then the rate of increase in the CO_2 concentration in the atmosphere will inevitably accelerate even if we maintain our current emissions levels. As if that is not enough, there is evidence that some positive feedback loops exist in the climate system that could lead to irreversible temperature rises, even if we bring our CO_2 emissions back into equilibrium with natural sinks. One example of a possible positive feedback loop relates to Arctic ice cover, the vast sheet of sea ice that covers the Arctic ocean from Greenland to Russia, and which grows in winter and shrinks in summer due to natural seasonal warming of the oceans. Only the summer shrinking is getting greater as surface temperatures have increased, so that on average the sea-ice cover has decreased by 12% each decade since the late 1970s.[687] Arctic sea ice helps regulate the climate by reflecting the sun's radiation back into space, whereas ocean not covered by ice absorbs heat.

The presence of debate does not invalidate the conclusion that human activity is leading to an accumulation of greenhouse gases, which in turn will have severe consequence for the environment and our economies.

Thus as ice melts the oceans get warmer, which reduces ice formation, which leads to further warming of the sea. This cycle is known as a positive feedback loop. Another example of a positive feedback loop is the release of methane from frozen lakes[771] and tundra, which increases as temperature rises. Since methane, in turn, is a very potent greenhouse gas its release contributes to further warming, which in turn can lead to further releases of methane.

The ecological impacts of climate change and concern about these *"runaway"* feedback cycles have led scientist and governments in the IPCC and beyond to conclude that future global warming should be restricted to no more than 2° C above pre-industrial levels.[585] Setting a limit of temperature rise implies a two-stage process, the first stabilization where we stop increasing the CO_2 concentration, and then reduction where we work to bring the concentration back down. The stabilization of CO_2 concentration is quoted in ppm and around the time of the publication of the third IPCC report a target of 550 ppm was thought appropriate to limit temperature rise to 2° C. This stabilization concentration was revised to 400-450 ppm in a subsequent conference in 2005, "*Avoiding Dangerous Climate Change: A Scientific Symposium on Stabilization of Greenhouse Gases*". It was the emerging evidence that 450 ppm represented an unacceptable risk that led the UK government to tighten its CO_2 targets from a 60% reduction in 2050 to 80%. Today, some scientists have argued that the safe target is really 350 ppm,[349] which is also the boundary set by the Stockholm Resilience Institute paper mentioned earlier, a threshold which we have already exceeded by reaching 400 ppm today.

While there is a healthy ongoing debate about whether the IPCC is excessively alarmist about the possible effects of climate change or whether it is over-conservative in its projections, it is evident that it would be prudent to take action on our emissions of CO_2 and other greenhouse gases. The nature of science is that there will always be debate – that is the process that lies at the very heart of scientific enquiry. But the presence of discussion does not invalidate the conclusion that human activity (primarily emitting fossil fuels and damaging land husbandry practices) is leading to an accumulation of greenhouse gases, which in turn will have severe consequences for the environment. Nor should we allow uncertainty about some details to cause a delay in our response – see the item on the precautionary principle (page 76).

Remember the earlier observation that the environment is here - it is the place where businesses and organizations operate and derive all their value. Climate change is not something remote, but rather it represents one of the most powerful forces, possibly the most powerful force, on our organizations' capacity to deliver future value. Climate change is not abstract because it is happening now. The findings of a report by the Global Humanitarian Forum in Geneva, in 2009[312] estimated that every year climate change leaves over 300,000 people dead, 325 million people seriously affected and economic losses of US$125 billion. Four billion people are vulnerable, and 500 million people are at extreme risk, mainly in poor countries.

Though we cannot attribute single unusual weather events to climate change (because there are have been and always will be weather extremes as a result of natural phenomena like the La Niña and El Niño ocean cycles), we can attribute a series of exceptional climate events to climate change according to a recent paper by James Hansen.[422] The deadly heatwave in Europe in 2003 that killed over 35,000, the Russian heatwaves of 2010,[580] the droughts in Texas and Oklahoma in 2011 which cost the economy US$1.5 billion,[293] and 2012's heatwave in the US, whose costs have yet to be quantified, demonstrate that these unusual weather events are global phenomena.

All over the world, weather-related disasters seem to be growing – no sooner did Australia come out of *"the big dry"* decade-long drought, but it had to endure the worst flooding in history. Hurricane Katrina caused damage estimated at US$125 billion and South East Asia experienced some of its worst floods in history. And these events are having a real, material impact on the ability of organizations to generate value. Just ask the many thousands of businesses in Thailand, where floods closed hundreds of factories and severely impacted exports, or the farmers of the US, Russia and Australia who have seen their crops decimated, or the coal-miners of Western Australia whose open-cast mines were flooded or the Health agencies in France which were overwhelmed by ill, elderly people in the 2003 heatwave, or the city of New Orleans which has to pay for the recovery from Katrina.

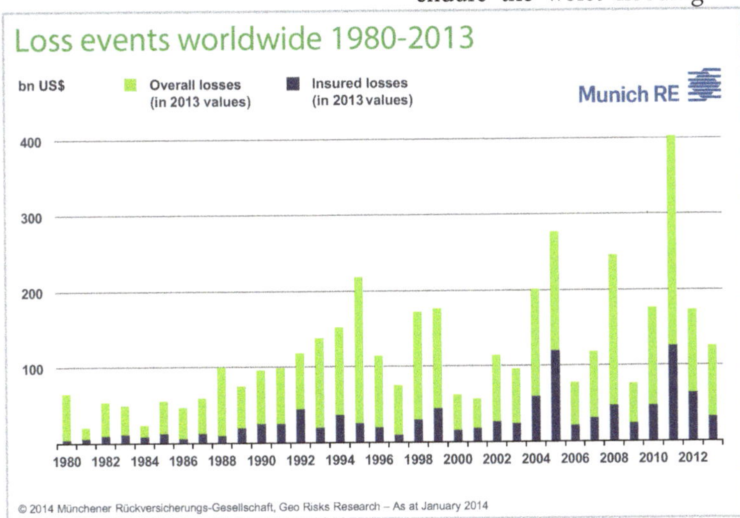

1.15 **Natural catastrophes are leading to large losses worldwide**
Source: © Münchener Rückversicherungs-Gesellschaft, Geo Risks Research – As at Jan 2014.

Of course, the rising cost of insurance is just one of the impacts of climate change on the operating expenses that organizations face. The payouts to those directly affected by severe weather translate to a cost increase for all organizations needing weather-related insurance, such as cover for flood, storms and business interruption. Thus, in our interconnected world, we don't have to be directly affected by the severe weather to feel its effect, as supply-chain disruption and impacts on the price of commodity foods can send shockwaves through industries, markets and countries. Rioting and social unrest has been linked to food price increases caused by abnormal weather[462] (and possibly the diversion of corn to produce ethanol – a second-order effect of our response to climate change).

So far we have discussed the difficulties that climate change is bringing. However, it is equally valid to think of climate change, and resource inefficiency, in a much more positive light. Our responses to these challenges provide organizations with an unprecedented opportunity to add value, create wealth and address the needs of stakeholders. For the private sector, climate change has been called *"the largest wealth creation opportunity of all time"*.

Just about every existing category of expenditure in our economy is likely to change in one way or another. This presents ***a huge opportunity.***

The demand created by increasing affluence will be increased by the need to replace or retrofit existing infrastructure on an almost unimaginable scale. Public sector institutions will thrive from the innovation and cost reduction that will emerge. Shareholders will, overall, see value rise and stakeholders will see service improve. Millions will be lifted from poverty, and we will secure the wellbeing of future generations.

Nicholas Stern, in his review of the economics of climate change, states that the cost of preventing dangerous climate change will be around 1% of global GDP.[680] A more recent analysis by the McKinsey Global Institute[63] concludes that a 10-fold increase in productivity per unit of carbon will be required (from US$740 / tCO_2 today to US$7,300 / tCO_2), and that the cost of this would be 0.6–1.4% of global GDP by 2030. They compared this cost – arguably a form of insurance - with global expenditure on insurance, which is running at 3.3% of GDP in 2005. McKinsey looked at the range of mitigation options and concluded that around 2 Gt of Carbon (7 Gt of CO_2e) could be abated at a negative cost to society (i.e. with savings greater than costs), 35-45% of the required abatement in wealthy countries. These are primarily the value-adding resource-efficiency activities that this book focuses on.

In a separate paper,[272] McKinsey concludes that cumulative new investments in the US to capture 0.8 Gt C per year (3 Gt CO_2e), would cost US$1.1 trillion, which while a large number, nevertheless represents just 1.5% of the anticipated capital spend during this period. Timing is critical. In an update to its report[512] McKinsey points out that the recent economic downturn has reduced the forecast emission from several countries, however:

> *Timing remains of the essence. Policymakers need to act now if they want to reduce emissions to the levels that the climate scientists believe are necessary to stabilize global warming below 2 degrees Celsius. Our model shows that a delay of 10 years, with action on abatement starting in 2020 instead of 2010, would cut the abatement potential by half from 38 $GtCO_2e$ to 19 $GtCO_2e$. Under such a "delayed" abatement path, the emissions trajectory would exceed a 550 ppm stabilization path – as laid out by IPCC authors – in some years. This would make it challenging to limit global warming to the 3 degree Celsius threshold associated with a 550 ppm stabilization.*

If the expenditure on climate change is very large, but modest in comparison with existing expenditure on activities such as insurance, why then the "hype" about it presenting a huge business opportunity? Well, the truth is that climate change is a game changer because it will drive innovation in so many different goods and services – it is not that huge amounts of *new* spend will be required, but that just about *every existing category of spend is likely to change* in one way or another. This is the source of the opportunity. Climate change is disruptive. For those organizations whose business models cannot adapt, it will be disaster, but for others, the changes will prove, in the long run, to add significant value.

In Numbers: *Carbon quantities*

CO_2 (or CO_2 equivalent CO_{2e}) is often expressed as a quantity of carbon. Since the molecular mass of carbon is 12 and oxygen is 16, the total molecular mass of CO_2 is 44 (12+(16*2)). So 44kg of CO_2 has 12kg of Carbon (and 32kg of Oxygen).

To convert 1kg CO_2 to 1kg C multiply by 0.273 (12/44). To go from 1kg C to 1kg CO_2 multiply by 3.67 (44/12).

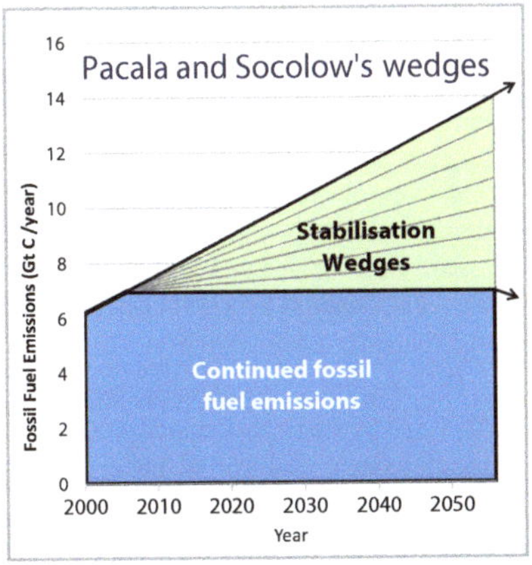

1.16 **Pacala and Socolow's "stabilization wedges" represent activities which together could stabilize CO_2 emissions at 7 Gt per year, by effectively eliminating any growth in emissions due to "Business As Usual".** These wedges start small and then build up over 50 years to achieve their full potential. *Source: "Stabilization wedges: solving the climate problem for the next 50 years with current technologies", Science 2004.*[582]

Real World: ***Reducing to climate change - a huge opportunity?***

The scale of change required just to stabilize CO_2 emissions (in other words to arrive at a point where nature can absorb our annual emissions), is truly quite breathtaking.

Stephen Pacala and Robert H. Socolow published a paper in Science[582] in 2004 with the idea of 15 *"stabilization wedges"* that each would reduce emissions over 50 years by 1 gigaton carbon. A gigaton, Gt, is 1 billion tonnes of carbon, equivalent to 3.67 billion tonnes of CO_2.

At the time, they argued that we would need to fully implement seven of these wedges to achieve stabilization, as shown on the graph to the left. The full list of *"wedges"* is shown in the table overleaf. We should bear in mind that not all the wedges are additive, so more than seven would be needed in practice.

It is important to note that Pacala and Socolow were not prescribing the precise mitigation path to stabilize emissions, but demonstrating that there were many actions, using existing technologies, that are capable of dramatically reducing emissions. There have been lots of useful critiques of these wedges – for example, the *"hydrogen economy"* approach to replacing petrol seems to be less effective than using electric vehicles.

Palaca and Socolow note that *"Improvements in efficiency and conservation probably offer the greatest potential to provide wedges"*, although they go on to observe that *"...efficiency and conservation options are less tangible than those from the other categories. Improvements in energy efficiency will come from literally hundreds of innovations..."* Clearly, efforts around resource efficiency are very important.

If we look at the solutions suggested by Pacala and Socolow, we cannot fail to be struck by the scale of effort needed. We may need to use 1/6th of the world's cropland for biofuels. We may need to install 4 *million* wind turbines to replace petrol used by cars. We may need to increase the efficiency of *all* our building by 25%. That is for just three of the gigaton reductions.

When colleagues ask me *"is resource efficiency combined with large-scale renewables best placed to address climate change or should we focus on more centralized supply-side nuclear power or carbon capture to solve the problem?"*, my answer is that we almost certainly need *all* of these technologies in combination, if we are likely to have the impact we require. It is not *"either/or"* but *"and"*.

Some commentators have likened the effort needed to that made in the Second World War[313] – an analogy that has been adopted by a group of entrepreneurs, including Richard Branson, who founded the *Carbon War Room*. The scale is staggering – upgrading or replacing huge amounts of the world's energy generation infrastructure, massively overhauling energy efficiency in homes, the public sector and businesses, dramatically changing the face of transportation and agriculture. No sector of society will remain untouched by the effort. Others state that, while the financial implications are not excessive, we will be limited by practicalities like skills shortages and capacity to implement the technologies at the scales required; yet others reflect that the absence of political will is the greatest barrier, something not likely to be resolved soon. For those, like the Carbon War Room, with faith in the market's ability to deliver the necessary investment, climate change is seen as *"The Wealth Creating Opportunity of Our Generation."*[117]

Foundations

1.17 **Pacala and Socolow's 15 Strategies to reduce emissions by 1 Gt C each over a 50-year period**

Source: "Stabilization wedges: solving the climate problem for the next 50 years with current technologies", Science 2004.[582]

Type	Option	Description
Efficiency and Conservation	Efficient Vehicles	Increase the fuel economy for 2 billion cars from 30 to 60 miles per US gallon (i.e. increasing it from 13 to 26 kilometres per litre). When they wrote their paper, there were 500 million cars on the planet. They expected that by 2054 this number would quadruple, but that efficiency would double. The current EU standard for petrol cars is 5L/100km i.e. 20 km per litre.
	Reduce Vehicle Use	Halve travel by 2 billion vehicles from 10,000 miles a year to 5,000 miles. Clearly this wedge and the previous wedge are not additive, although we could do some of both.
	Efficient Buildings	Reduce carbon in all buildings and appliances by 25%. This is for industrial, commercial and residential buildings.
	Efficient Coal	Double the efficiency of all coal plants from 30% to 60%.
Fuel Switch	Coal to Gas	Replace 1,400 coal plants with gas plants (gas produces half the CO_2 as coal per unit of energy). That is four times the worldwide total of gas plants in 2004.
Carbon Capture and Storage (CCS)	CCS at Baseload Power Plants	Capture CO_2 from power plant at either 800 GW of coal plants of 1,600 GW of gas plants (by comparison there was a total of 1,060 GW of coal power generation in 1999, so this is a large proportion of plants)
	CCS for Hydrogen	Create hydrogen gas using CCS from 250 Mt/y from coal or 500 Mt/y from gas, which can be burnt for power or used to displace liquid fuels.
	CCS for Coal to Synthetic Fuels	Capture CO_2 at plants which create 30 million barrels of oil equivalent a day (1.8 teralitres per year) synthetic fuels from coal a year. This is 40% of total oil production in 2007.
Non-fossil Fuel Power Generation	Nuclear Power	Replace 700 GW of coal fired plant with nuclear power, which is twice the capacity in 2004.
	Wind Power	Add 2 million 1 MWp wind turbines (to replace 700 GW coal plants). The reason you need more wind capacity than coal is that wind is intermittent. In 2011 we had just over 10% of this capacity installed worldwide (238,351 MW).
	Solar Photovoltaic (PV) Power	Replacing 700 GW of coal-fired power plants by 2000 GW of peak photovoltaic solar power. This would occupy 2 million acres.
	Wind for Hydrogen	Add 4 million 1MWp wind turbines to provide hydrogen to replace petrol in vehicles.
	Biomass replaces Fossil Fuels	Replace fossil fuels with biomass by using about 1/6th of the world's cropland.
Land Use	Reverse Deforestation	Decrease tropical deforestation to zero and establish 300 Mha of new plantation.
	Change Agricultural Practices	Change to agriculture to use a no-till approach, which reduces emissions of CO_2 from soil dramatically. This represents a 10-fold increase in this type of soil management.

1.6 Biodiversity

Foundations

Biodiversity is a measure of the complexity and variety of living organisms in an ecosystem. Maintaining biodiversity is important because this diversity and abundance helps all organisms to thrive and adapt to change.

The second of the boundaries that we have crossed, according to the Stockholm Institute of Resilience paper,[625] mentioned earlier, is the rate of biodiversity loss. Just as we rely on energy resources to power our economies, we also rely on biodiversity for our fundamental economic wellbeing. Take, for example, the value of bees in pollinating crops which has been estimated to be worth around US$14.6 billion in the US and €153 billion worldwide,[301] which represents 9.5% of the value of world agricultural production! Around 70% of all new drugs introduced in the US in the past 25 years have been derived from natural products.[551] Plant-derived medicines include aspirin from the willow tree, pain-killing morphine from poppies, cancer treatment Taxol from the Pacific yew – nature is magnificent at creating complex molecules which we would not easily produce in laboratories.

Perturbations in biodiversity can have remarkable and unforeseen impacts on humans – for example, the use of the drug diclofenac to treat livestock in India led to a dramatic decline in vulture populations, which allowed their competitors, feral dogs, to move into the niche they previously occupied. As the dog population boomed the incidence of dog bites and rabies increased – leading to 48,000 additional deaths (from a paper in Ecological Economics,[498] quoted in The Guardian).[163]

Understanding and responding to what nature does for our organization and what our organization does to nature is common sense.

Measuring biodiversity loss is tricky because we haven't come close to measuring the total number of species on the planet. The paper by the *Stockholm* Institute of Resilience used data from the *Millennium Ecosystem Assessment: Current States and Trends*, Chapter 4,[311] which concluded that "*Over the past few hundred years humans may have increased the species extinction rate by as much as three orders of magnitude*". i.e. 1,000-fold, and that between 12% and 52% of higher-order species are threatened with extinction. Because of the difficulty in measuring species, the World Wide Fund for Nature has established a *Living Planet Index* which instead tracks the population of 2,600 vertebrate species all over the world. The index for 2012[817] has declined by 30% since 1970. The picture was also quite mixed; there has been some considerable improvement in species populations in the temperate regions of 31%, masking a sharp decline of -61% in tropical regions, where most biodiversity loss is taking place today.

To explore the importance of biodiversity, let's examine one small part of the planet's contribution to our annual bank account: fish. In contrast to

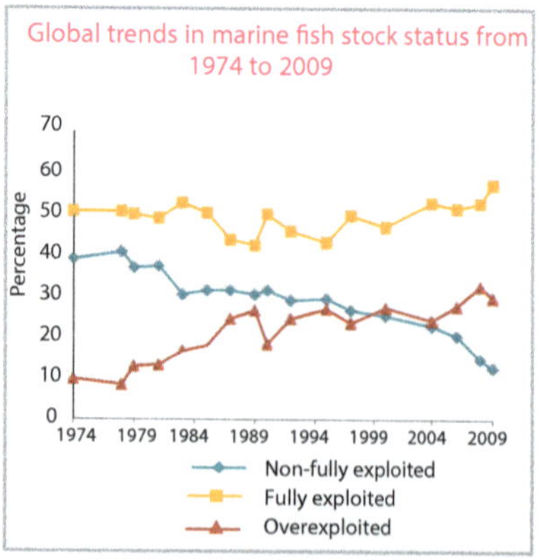

1.18 **Global trend in marine fish stocks 1974 to 2009**
Source: "Review of the state of world marine fishery resources", FAO, 20011.[727]

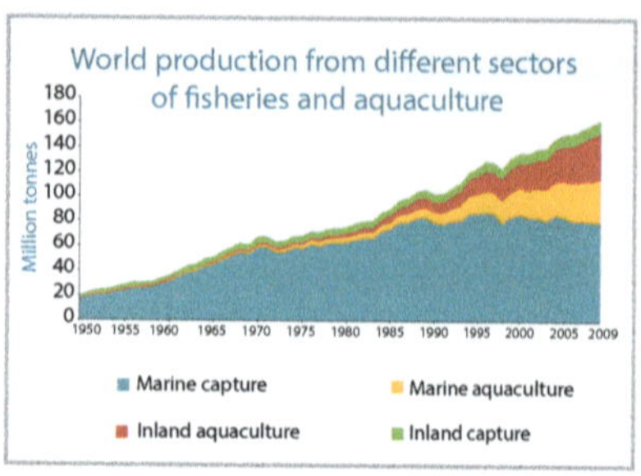

1.19 **Total fish production from various sources, 1974 to 2009**
Source: "Review of the state of world marine fishery resources", FAO, 20011.[727]

the depletion of non-renewable resources, such as minerals, which we have described earlier, there has been plenty of opportunities to *reverse* the decline in fisheries. While fossil fuels take millions of years to produce, fish stocks can be replaced within a few generations, if appropriately managed. They are truly *renewable* resources *so long as the underlying ecology is not damaged.*

Unfortunately, we have not been good at responding to warnings of ecological harm. One of the most dramatic examples of our failure to react to evidence was the collapse of the Atlantic northwest cod fisheries off Newfoundland in Canada. A thriving industry which had supported 35,000 fishers and related workers was wiped out because of over-fishing, which led to a sudden total collapse in the cod population. Despite a complete moratorium on fishing imposed 20 years ago, in 1992, the population of cod has still not returned and may never do so, as the ecosystem may have changed irrevocably.

A study in Science[814] stated that, in 2003, 29% of all global fisheries could be defined as *"collapsed"* – that is to say, that catches had dropped to 10% or less of the recorded maximum. Crucially, the study found that it is those populations where there are high levels of biodiversity that seem to be able to recover best from over-fishing – perhaps explaining why North Sea cod has fared better than the Atlantic northwest cod, despite the high level of over-fishing. These figures tally with the Food and Agriculture Organizations (FAO) annual survey of marine fish stocks in 2009,[727] which shows that around 30% of fish stocks are *"overexploited"*, *"depleted"* or *"recovering"*. As can be seen from the chart above left, there is a consistent general movement of fish stocks from *"not fully exploited"* through to *"exploited"* and *"overexploited"*.

With over 500 million people depending on fish and a billion people relying on it as the primary source of protein, the state of the world's fish stocks is crucial. Unfortunately, the over-exploitation of fish stocks is having a direct effect on marine fish catches, as illustrated by the blue series in the graph below left, showing that a peak was reached in 1996; we have since seen a decline of 10%.

The World Bank and FAO estimated in 2009[794] that the difference between the potential and actual economic benefits of fisheries amounts to a staggering US$50 billion a year, or US$2 trillion dollars over the preceding two decades. This gap is equivalent to more than half the value of the global seafood trade. In its green economy report, the United Nations Environment Programme (UNEP)[731] suggested an investment of US$8 billion per year to end over-fishing and allow fish stocks to recover – this would involve removing up to 13 million of the 22 million boats currently in service and retraining 22 million fishers for other work.[598] If you are in the fishing industry or a supplier to the fishing industry, resource scarcity is certainly going to impact future business value if proposals like this are implemented.

To deplete non-renewable stocks for present-day needs could be considered unfortunate but unavoidable, but to exhaust entirely renewable resources and deprive future generations of their benefits is positively criminal.

Exploration: *Sources, sinks and "peak oil"*

The economy which we operate today is largely based on a one-way journey of natural resources - from source to sink.

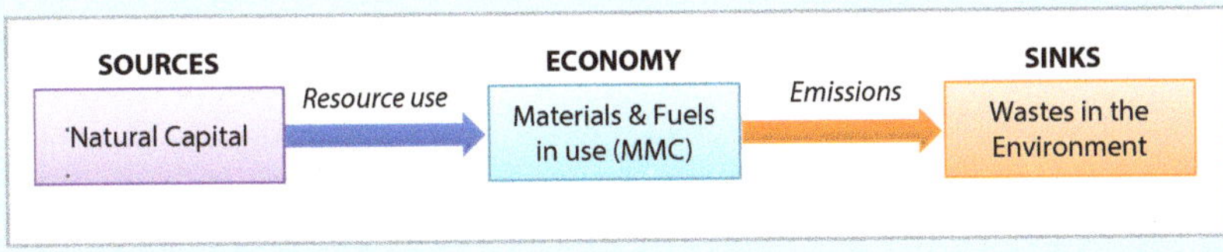

Among all the resources that we use oil is particularly important. It has been the availability of huge quantities of very cheap oil that has powered the unprecedented economic growth of the last century, just as coal was the energy source that underpinned the industrial revolution in the years before. Oil has wondrously been described as *"fossilized sunlight"* because it is derived from the sun that fell on plants over thousands of years.

Oil is an incredible material; it is very portable, its derivatives – gasoline/petrol, diesel, etc. - are volatile and so can drive combustion engines and, as chemical feedstock, oil underpins a vast range of industries. Oil also has around one and half times the energy density of coal and over four times that of wood; one litre of gasoline/petrol contains 32 MJ or 8.5 kWh, which is equivalent to about 8.5 days of human labour. It is because we have had a plentiful supply of cheap, transportable energy that we have been able to create surpluses of wealth to reinvest in more resource extraction and so put in motion the remarkable economic growth seen this century. Cheap oil (and other fossil fuels) are so important to the global economy, accounting for about 5% of global expenditure as a percentage of GDP,[538, Fig. 5] that governments have spent US$400 billion subsidizing these in 2010, through tax breaks and other mechanisms, according to the International Energy Agency[389] and Cowe.[172]

However, oil, like other fossil fuels, is a non-renewable resource, in other words, it is not replenished by nature – it can only be used once. The source of oil is finite. Furthermore, the use of oil leads to emissions of CO_2 whose natural sinks are already saturated. Given the importance of oil to the economy, it is not surprising that there has been a lot of debate about whether the source or the sink would ever limit our use of this resource, and when those limits would arrive.

The data on oil reserves has never been entirely transparent – oil company share prices and OPEC country voting rights are both based on stated reserves, so there is clearly an incentive to be optimistic about the amount of extractable oil. However, in 2010 the International Energy Agency finally acknowledged what many commentators had been saying for years – that *"peak oil"* production had arrived several years ago:

> *"Crude oil output reaches an undulating plateau of around 68-69 Mb/d by 2020, but never regains its all-time peak of 70 Mb/d reached in 2006."*[389]

The IEA did go on to say there would be another 25 or so years where synthetic fuels and unconventional oils, such as those obtained by *"fracking"*, may grow slightly to fill the gap, but these too will peak in 2035 and possibly as early as 2020.

It is important to note that *"peak oil"* doesn't mean that we are physically running out of the stuff. What we are running out of is cheap oil from existing oil fields, and so we rely increasingly on new fields like Macondo in the Gulf of Mexico, scene of the Deepwater Horizon tragedy and spill. Here, the oil is in a relatively small pocket of 50m barrels located 40 miles offshore in the Gulf of Mexico under 1,500 metres of water and a further 4,000 metres of sediment and rock. Another plentiful supply is tar sands and oil shales where much larger amounts of energy and water are required to extract the oil. Peak oil means that the oil which remains is becoming much harder to find, it will need much more capital to extract and so will inevitably cost much more to provide than current oil supplies. It is these cost factors which has limited production.

Similar issues relate to a supposedly abundant fossil fuel – coal. For example, in the US the premier "quality" coal, black anthracite, almost pure carbon, has to all extents, been depleted. This decline means that now the major production is of bituminous coal, providing slightly less energy per kg than anthracite, but this too has plateaued. The next coal types to be extracted are sub-bituminous and last of all lignite, known as brown coal, which is low in energy and high in moisture. As access to the better grades of coal became more difficult, we have had to resort to the poorer ones, as illustrated in figure 1.20, overleaf.

So we can see that the issue of sources is really about the availability of *cheap* resources. As we deplete the best quality fossil fuels, those that remain require more energy to extract and the energy return on energy invested (EROEI) declines, so the net energy falls. Not only does this increase the cost but for the same net amount of energy to enter the economy, we will need to extract ever-increasing quantities of the fossil fuel. As more and more energy is diverted into producing resources and meeting

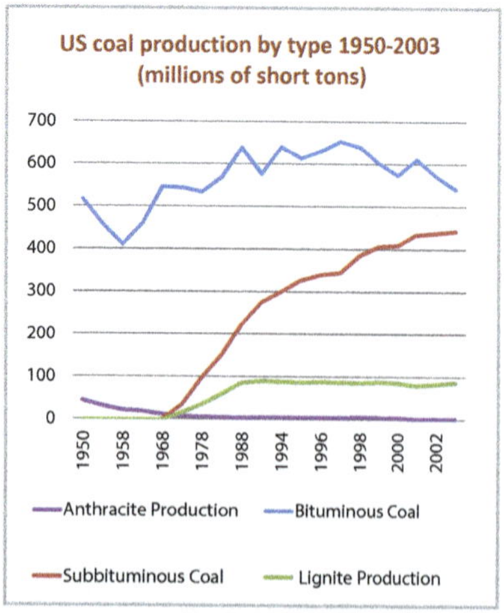

1.20 **Historic trend in US coal production 1950 to 2002 showing the decline in better quality types and the resulting increase in production of poorer grades**

Source: Niall Enright, data from "Coal Production in the United States – a Historic Overview", Energy Information Agency.[752]

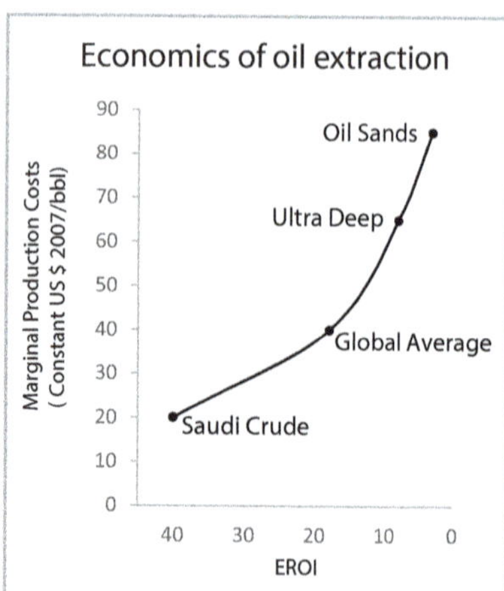

1.21 **Oil production costs in comparison with the energy return on investment for various sources**

Note that the production costs are not the same as the total cost of oil, which includes processing, transportation etc.

Source: Niall Enright, based on Murphy and Hall.[538]

the basic needs of life the surplus which is available to invest in more productive capital (factories etc.) or used in discretionary consumption (TV's, fridges, cars, etc.), becomes smaller.

There is another, potentially more significant, limit on our use of fossil fuels, the availability of sinks which remove the waste CO_2 that their combustion produces. In simple terms, there is a finite budget of carbon that we can emit to the atmosphere over the next 40 years if we are to achieve our stabilization objectives. The implication of this is that, unless we can capture and store the CO_2 emissions from fossil fuels, half of all known reserves will have to stay in the ground.[522] We simply cannot keep global warming anywhere near 2° C if all that carbon is released. Avoiding climate change has huge implications for the future valuation of all fossil-fuel industries, for it seems that the market has not yet factored in the possibility that half the assets of these businesses are potentially going to remain stranded under the earth. If carbon capture and storage is viable, and we hope that it is, it will add significantly to the energy used at the power plant (11-40% depending on the technology used[567]) as well as to the overall cost of the electricity produced.

A consequence of a fixed budget of carbon is that we need to think about the amount of CO_2 that the various fossil fuels emit per unit of electricity generated. This emissions intensity varies depending on the technology and the grade of fossil fuel, but in the UK, taking into account extraction and transportation of the fuel, the figures in 2016[200] are as follows:

Fuel	kg CO_2e emitted per kWh electricity generated	kWh electricity generation per kg CO_2e emitted
Coal	0.360	2.77
Fuel Oil	0.317	3.15
Petrol	0.288	3.47
Natural Gas	0.209	4.78
Wood Logs/Chips	0.013	76.92

So we can see that we can get almost twice the amount of electricity from natural gas per kg CO_2 emitted as we can from coal (or in other words natural gas produces almost half the CO_2 as coal per unit electricity generated). Consequently, we can make our limited carbon budget stretch further by using natural gas for power generation in preference to coal. It clearly makes sense for us to focus on gas as the bridging fossil fuel during the transition to low-carbon electricity generation based entirely on renewables. Ideally, new gas plant will only come on-stream where it can be shown to displace coal.

Policy is now broadly favouring natural gas over coal, although issues such as security of supply and market price may still support coal in certain regions. The US has benefited from a sharp increase in natural gas supply as a result of hydraulic fracturing (fracking), which has boosted energy security and also lowered costs significantly. There are concerns that methane leakage[20] from fracking, where older technology is used, may offset the benefits of shale gas compared to coal. This is because methane is a powerful greenhouse gas, whose warming effect over 20 years is 86 times greater than CO_2 (much more than the cited 21 times over 100 years).[683]

Foundations

Summary:

1. Resources include material resources and natural services (such as waste disposal) on which we depend.

2. Resource efficiency is a continual improvement process.

3. You can never *be* resource efficient (unless you use zero resources), all you can do is apply energy and resource efficiency techniques in order to improve.

4. Current sources of supply of resources and sinks for disposal of waste are not sustainable. Change is inevitable.

5. Despite the assumption of conventional economics, indefinite growth is an impossibility.

6. Our impact on the natural environment is the product of Population, Affluence and Technology. Affluence is the dominant factor in our increased demand for resources.

7. There is strong evidence that many minerals and fuels are becoming scarcer and more expensive. This provides another powerful reason to reduce resource use, alongside the environmental impact.

8. Climate change is real and is a serious threat to our future wellbeing.

9. There are sufficient technological solutions available today to reduce our emissions of CO_2 to a safe level. The scale of these is very large and we still lack the political and economic will to act.

10. Biodiversity loss threatens our economy in many ways. Experience from fisheries and other ecosystems shows that perturbations in populations can have unforeseen and dangerous consequences.

11. Addressing climate change will cost between 1%-1½% of GDP. This is entirely affordable, but early action is required if we are going to keep the cost down.

12. It is not the scale of expenditure that makes climate change a significant business opportunity, but the fact that it will drive innovation and change in almost all categories of expenditure.

13. Energy and resource efficiency is a very rewarding and enjoyable career choice.

Further Reading:

Meadows, Donella. Randers, Jorgen and Meadows, Dennis. 2004. *Limits to Growth - the 30- year update.* Earthscan. ISBN 978-1-844407-144-9. This title revisits the original models using more up-to-date data.

Daly, Herman, E. 1996. *Beyond Growth: the economics of sustainable development*. Beacon Books. ISBN 978-0-8070-4709-5. A profound, accessible and ground-breaking investigation of the limits of orthodox economic and development thinking.

Heinberg, Richard. 2011. *The End of Growth: Adapting to our new economic reality.* Clairview Books. ISBN 978-1-905570-33.

MacKay, David JC. 2009. *Sustainable Energy — without the hot air*. UIT. ISBN 978-0-9544529-3-3 (pbk). A comprehensive, highly accessible guide to meeting energy demand in a sustainable way. Thoroughly recommended.

Gilding, Paul. 2011. *The Great Disruption - how the climate crisis will transform the global economy.* Bloomsbury. ISBN 978-1-4088-2218-0. One perspective on the scale of change needed to address climate change. Both frightening and optimistic.

Questions:

1. What evidence is there that we are reaching the limits of our natural resources?
2. Describe the difference between a source and a sink. Give examples of environmental issues caused by human impact on each of these.
3. Describe some of the concerns about the future availability of mineral resources.
4. What is climate change and what are its potential effects?
5. Consider the UK government's Scenario for Climate Change Mitigation, illustrated below (available at http://2050-calculator-tool.decc.gov.uk). This model permits you to choose from a number of supply-side and demand-side strategies to reduce CO_2 emissions.

   a) Is it possible to reach the required 80% emissions reduction with just supply-side or just demand-side strategies or are both required?

   b) List your chosen strategies to reduce emissions by 80%. Which of these do you think will be easiest to achieve? Which do you think will be most difficult? Why?

   c) The model includes a financial analysis of the costs of each strategy. To what extent should cost form the basis for strategy selection? Would cost change your own choices?

1.22 **2050 Pathways is a tool which illustrates possible pathways to reduce the UK's emissions by 80% by 2050. It is a sophisticated model which brings home the extensive changes that will be needed to achieve the target emissions reductions. On the other hand , it confirms that these changes are possible. As well as enabling you to create your own pathway, sample pathways are provided from a diverse range of organisations, such as Friends of the Earth or the Sustainable Energy Association, illustrated below. Niall Enright's choices can be found at** https://bit.ly/2p5di1k

Source: © *Department of Energy and Climate Change.*

2 Contemporary Ideas

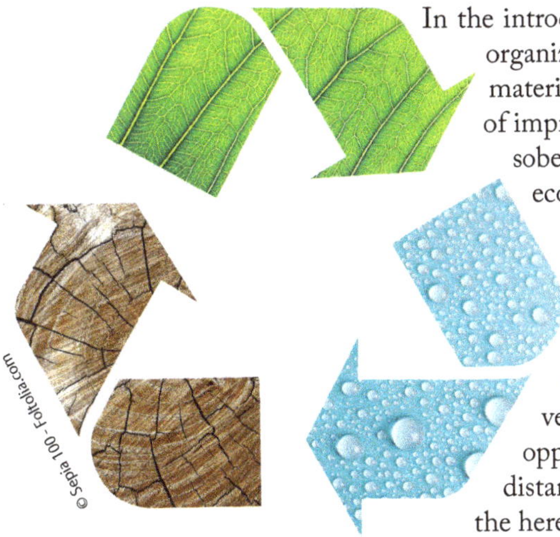

© Sepia 100 - Fotolia.com

In the introductory chapter, I have characterized resource efficiency in organizations as a change process which reduces the extraction of material resources and the use of natural services. The importance of improving our resource efficiency has been highlighted by some sobering evidence that we are at a critical threshold in our ecosystem's ability to support us.

One key conclusion is that change is *inevitable,* simply because our demand for resources cannot grow indefinitely.

In mobilizing ourselves and our organizations to confront this change, we can either evoke fear or opportunity. I very much favour positive messages around immediate opportunities - as we humans are very poor at responding to distant threats, however large. We are programmed to think in the here-and-now. Furthermore, our organizations, especially listed companies, are also often equally limited in their decision-making horizons.

So the emphasis in this book will be very much on the opportunity that exists today for all organizations to develop and prosper through greater resource efficiency.

Some commentators, such as Heck and Roger in *Resource Revolution,*[361] have called the transition to a resource-efficient economy the biggest business opportunity in a century. Groups like *The Carbon War Room*[117] see "*solving climate change as the greatest wealth-generating opportunity of our time*".

In the next chapter, we will be exploring the many sources of value that this change offers. However, before we turn our attention to the specific benefits that resource efficiency can bring organizations, it makes sense to explore some of the contemporary ideas about how we might tackle resource efficiency at a global, economy-wide or organization level.

One cannot manage change, one can only be **ahead of it**.

-Peter Drucker

These ideas will help us to understand better how we can respond to the breathtaking scale of change that has already begun and place our own efforts in a broader context. The consistent message is that there are many points in the flow of materials from nature to our economy where we can intervene - and each of these points represents an outstanding opportunity for organizations to innovate and deliver value.

2.1 A Sequence of efficiencies

There are multiple points where we can improve efficiency in the way we extract, transform, consume and return resources to the environment.

2.1 **Although these fruit tarts have needed labour (a cook) and capital (a kitchen) to produce they would not be possible without natural capital (flour, water, fruit etc.).** The flour, eggs, yeast, fruits and sugars all derive from the natural environment. These require soil and sunlight and rain. They need microbes to release nutrients and birds to keep pests at bay. In short, all human wealth is ultimately derived from natural resources in one form or another.
Source: Image © Philip Kinsey - Fotolia.com.

To start our exploration, I am going to turn to an economist who became famous for developing ways to bring nature and resources into an economic theory, which he did from within a pillar of modern financial institutions: the World Bank.

Herman E. Daly's *Beyond Growth,*[181] published in 1997, was a landmark in a new approach to economics, called ecological economics, and challenged conventional economic orthodoxy and the possibility that growth can continue infinitely. In his thinking, Daly dismisses the notion that the environment and natural capital are externalities which can just be ignored when considering how wealth and prosperity are created, and he gives us some useful insights to understand how the real economy works and how and why we should tackle resource efficiency.

First of all, let us consider value-added. In traditional economics thinking, value is created through the combination of labour and capital – thus when we bake a tart, as Daly describes in his book, we combine our labour with the capital of our kitchen to change flour and other ingredients into a product which has greater value – hence *"added value"*. Conventional economics is all about the flow of these goods (tarts) or services (tart-making) among firms and households and the added value that is created at each stage. Fiscal and monetary policies can encourage this value creation, for example by supporting debt, or diverting the value formed to meet wider needs of societies, for example through taxation. Politics for most of our recent history has been essentially about how labour, capital and governments divided the spoils of this added value.

Continuing with the tarts analogy, Daly digresses from orthodox economics by stating that we can't simply ignore the importance of natural capital. To make a tart we need sugar and flour, butter and fruit. Before that, we need sugar cane plants, and wheat, and cows and apple trees. Before that we need a gene pool of living organisms from which to develop our cow and crop varieties, and before that we need soil and water and sunlight to support all these plants and animals. Then there are the worms and microbes and nutrients in the ground which maintain its consistency and enable the grasses and crops to grow strong. We also need the birds that keep the pests at bay and the gases in the atmosphere which maintain the temperature at the surface of the earth and which enable water vapour to rise into clouds and fall over vast catchments which purify it and deliver it to our plants. Finally, we need the sunlight which drives the whole process and creates the precursor molecules which form our tart. In short, the other essential ingredient in the value adding process is natural capital. Without flour and fruit and butter there would be no tart, no matter how much labour and capital we have at our disposal. This point is critical as orthodox economics tends to state that capital and natural capital are interchangeable whereas, with very few exceptions, there are no man-made substitutes for natural services.

In an empty world – in other words, one which has few humans – this natural capital, although finite, will not be affected by the amount that we take for our use. Unfortunately, we do not live in an empty world. Earlier, in Section 1.4 of this book, we saw that our ecological footprint is already 1.5 planets, which indicates that we are now approaching what Daly would call a full world.

In these circumstances, the natural capital that we use to create value-added in our human economy will reduce the pool of natural capital that is available for the environment itself. In a full world, our capacity to create value-added starts to be limited by the available natural capital as much as by labour or human capital – as the fishers of the Great Banks fisheries in Newfoundland found to their costs when cod stocks collapsed. In a full world, when we create a palm-oil plantation in Indonesia to provide us with oils for foodstuffs, cosmetics or biofuels, we may have destroyed a tropical ecosystem which provided natural services (food, shelter, etc.) for animals such as orangutans. Thus in a full world, creating man-made capital (MMC) reduces natural capital.

We create MMC to provide services – such as cooking, in the case of our oven – or transportation in the case of a car. It is not the capital per se which creates value, but the services that the capital stock provides. Human development has been about increasing the stock of MMC, and so meeting the needs of a growing population and at the same time providing greater material wealth. However, that progress comes at a price in terms of natural capital.

Orthodox economics states that natural capital and capital are interchangeable whereas, with few exceptions, **there are no man-made substitutes for natural services.**

Daly used the equation below to describe the relationship between MMC and the decrease in natural capital, which consists of a series of ratios. As you can see, the first ratio on the left-hand side is essentially a description of resource efficiency – the quantity of MMC services that we can achieve per unit of natural capital services sacrificed.

$$\frac{\text{MMC Services Gained}}{\text{NC Services Sacrificed}} = \frac{\text{MMC Services Gained}}{\text{MMC Stock}} \times \frac{\text{MMC Stock}}{\text{throughput}} \times \frac{\text{throughput}}{\text{NC Stock}} \times \frac{\text{NC Stock}}{\text{NC Services Sacrificed}}$$

Daly's formula describes the flow of natural capital into MMC, such as cars or kitchens. However, there are some flows of natural capital which do not turn into MMC per se, but which nevertheless are used to provide services

Our first strategy in resource efficiency is to ***eliminate or reduce the demand*** *for a service which uses natural capital.*

for people. I am thinking here of things like fossil fuels which provide energy, or the fish in the sea which give us food. So I have taken Daly's formula and generalized it to reflect these additional human demands for services derived from natural capital.

$$\frac{\text{MMC Services Gained}}{\text{NC Services Sacrificed}} = \frac{\text{MMC Services Gained}}{\text{Human Service Provider}} \times \frac{\text{Human Service Provider}}{\text{Natural Resource}} \times \frac{\text{Natural Resource}}{\text{NC Stock}} \times \frac{\text{NC Stock}}{\text{NC Services Sacrificed}}$$

I have substituted MMC stock with *"human service provider"* i.e. an item (like a car or a quantity of energy) that can provide a human service. In Daly's original formula, the second and third terms on the right of the equal sign have a measure of throughput, i.e. the quantity of the MMC that needs to be created over time. In my revised version of the formula, I have dispensed with throughput because the numerator on the left is essentially the consumption of a service over a given period, and thus all the ratios in the equation have a time component – they represent a rate of conversion of one element to another over a constant time.

Looking at these flows, we can see that they are driven by the need for human services gained. Thus our first strategy in resource efficiency is to reduce the overall demand for the service, which will lead to a decrease in all the other factors, and thus ultimately reduce the natural capital services sacrificed. So if the service requirement is for heat to keep our homes warm, the first strategy is not to deliver the heat more efficiently but rather to dispense with our need for heat altogether (woolly jumper, anyone?).

The modified formula shows the flow of material or services from the source to the human economy, and if one were to substitute real values into the equation, these would need to be for the same time frame. Below are some simple examples of these resource flows:

$$\frac{\text{Person-miles}}{\text{NC Services Sacrificed}} = \frac{\text{Person-miles}}{\text{Car}} \times \frac{\text{Car}}{\text{Iron}} \times \frac{\text{Iron}}{\text{Iron Ore}} \times \frac{\text{Iron Ore}}{\text{NC Services Sacrificed}}$$

$$\frac{\text{Food Energy}}{\text{NC Services Sacrificed}} = \frac{\text{Food Energy}}{\text{Bread}} \times \frac{\text{Bread}}{\text{Wheat}} \times \frac{\text{Wheat}}{\text{Land \& Water}} \times \frac{\text{Land \& Water}}{\text{NC Services Sacrificed}}$$

$$\frac{\text{Lighting}}{\text{NC Services Sacrificed}} = \frac{\text{Lighting}}{\text{Electricity}} \times \frac{\text{Electricity}}{\text{Wind}} \times \frac{\text{Wind}}{\text{Land}} \times \frac{\text{Land}}{\text{NC Services Sacrificed}}$$

We can see that this formula consists of a series of ratios, which I have listed below. The first term, services per unit of the service provider is what I will call the service efficiency or (MMC services gained)/(human service provider). Remember that it is the service gained from the service provider which is critical to meeting human needs, not the service provider stock itself.

$$\frac{\text{Resource}}{\text{Efficiency}} = \frac{\text{Service}}{\text{Efficiency}} \times \frac{\text{Production}}{\text{Efficiency}} \times \frac{\text{Extraction}}{\text{Efficiency}} \times \frac{\text{Ecological}}{\text{Efficiency}}$$

There are essentially three strategies to increase service efficiency:

- Increase the conversion efficiency from provider to service
- Increase the utilization of the service provider by better matching the demand
- Increase the life of the service provider

An example of an efficiency improvement would be to use LED lights to get more light or lumens (the service) per kWh of electricity (the service provider).

Secondly, we could better match capacity to demand. Here we want to ensure that we do not have surplus service provider stock (e.g. ovens or cars) laying idle as this stock requires natural capital to produce. By *"stock"* I also mean the physical condition of utilities, such as the pressure in our steam system or the temperature of our hot water or the intensity and colour temperature of our lighting, which needs to match the demand from our processes.

There are many ways we can align capacity to demand. First, we could eliminate surplus capacity entirely so that the stock of service providers is smaller. This strategy generally applies to service providers, like cars, which are discrete items rather than service providers like energy or water, which can be allocated precisely. If we halve the number of cars we have in stock, we double the service efficiency as the remaining cars will need to make twice as many journeys or carry twice as many passengers to meet the same demand. As a result, we will halve the steel required for each unit of service and, everything else being equal, we will also halve the amount of iron ore needed. A technique for capacity improvement is shared utilization, such as occurs when people switch from poorly utilized personal transportation to much better utilized public transportation (see case study on the next page).

*We can increase **efficiency, match supply and demand** or **increase the lifetime** of the goods that provide a service.*

Not only does mismatched capacity and demand lead to equipment being idle, and thus physical resources being wasted, but it often also leads to conversion inefficiency in the equipment. A boiler or chiller or motor operating at part-load tend to be less efficient at converting the input energy into heat, coolth or motion respectively. Thus there are often multiple benefits from aligning demand to capacity (or avoiding oversizing of equipment).

Real World: *The incredibly low service efficiency of our cars*

An example taken from Allwood and Cullen's excellent book, *Sustainable Materials — without the hot air,*[22] illustrates the concept of service efficiency. In the UK we have 28 million cars, with an average of four seats each, so there is a staggering total of 28,000,000x4x24x365 or 981,120,000,000 seat-hours service available each year from our stock of cars. In practice, there are 60 million of us, and we each spend an average of 225 hours a year in cars, so we are using up 60,000,000x225 or 13,500,000,000 seat-hours, which is just 1.4% of the total available. Cars are the second largest expenditure we ever make and yet they spend the vast majority of their lives sitting in our driveways or in parking spaces. By anyone's reckoning, our use of our stock of cars in the UK is staggeringly inefficient. In the US, car capital of the world, the utilization isn't much higher at just 4%.

We could improve the efficiency in several ways. Our first strategies involve increasing use of the cars; for example, we could carpool and so increase the average number of passengers in each journey, which would have the effect of decreasing the number of cars needed. In some countries, such as the US, there are considerable incentives to carpool such as access to separate lanes on highways for multiple-occupancy vehicles. An alternative to sharing a journey is to share a car. Companies such as Zipcar are providing an option whereby individuals in the major cities can enjoy the freedom of personal transportation without the cost of outright ownership while resulting in much greater utilization of the cars in the scheme (and thus a lower cost per mile for the consumer as well as greater sustainability). Research by Frost and Sullivan in 2010 predicted that, by 2016, there will be 5.5 million members of car-sharing schemes in Europe with a total of 77,000 cars,[297] which will replace 1 million driver-owned cars.

Instead of car-sharing we can replace inefficiently used cars with more efficiently used public transport, thereby delivering the required service, the desired number of seat-hour journeys, with less capital stock. This transport mode shift is a form of substitution where a more efficient alternative displaces and existing service provider.

An alternative strategy where we have a current surplus of stock is simply to decrease the total capacity. In Paris recently they alternately banned odd and even-numbered licence-plate cars from the city's streets during peak summer smog days. This policy reduced the total car capacity by half, led to a greater utilization of the remaining cars and reduced pollution.

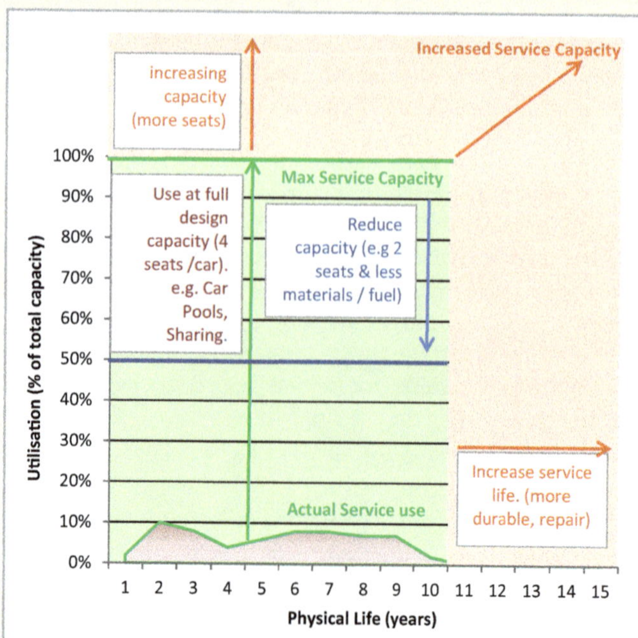

2.2 **Intensity/time chart for car**
Source: Allwood and Cullen[22] Adapted by Niall Enright.

Another option would be to reduce the replacement rate of cars by increasing their lifespan through better maintenance (perhaps the taxation on vehicles should decrease over time to encourage retention and reuse). Herman Daly makes the point in his book that people can overlook some important distributive considerations regarding matching capacity to demand. An example of this is that where poorer families have access to a car they tend to have just one car which is more highly utilized and whose life is prolonged, whereas richer families tend to have multiple vehicles which are replaced more frequently and which are more often idle. Clearly, among those with fewer assets, the car is much more valued.

On the design front, there are two directions we can go, both with the aim to align capacity with demand. Assuming we can fill them, we can get more from our cars by adding even more seats, so that each journey carries more people (if we extend this to its natural conclusion we will end up with a bus on a public transport network rather than a car). Or we could go in the other direction and reduce the number of seats to match the actual demand – thus a two-seater car will require less material than a four-seater.

The more steps, or conversions, in the transformation of natural capital to man-made capital, the more inefficient system becomes.

Foundations

Another way to match demand to capacity is to distribute service providers efficiently. This brings us to the observation that the distribution of service providers (cars, fuel, etc.) in society, and in organizations, is an important consideration in their service efficiency. That is to say, the increase in human services gained for each increment in service providers will vary depending on the location of the service provider, partly because of distribution losses and partly because some processes require more inputs that others.

Our third strategy to improve service efficiency is to extend the service provider life, which reflects the fact that replacing a piece of equipment in a take-make-waste economic model means extracting resources to make the replacement. Many consumer service providers, like cars or TVs, are replaced even before they come to a natural end of life. Other service providers are replaced as their service provision degrades, so factors like durability and maintainability need to be improved to extend their lifetimes. In fact, this preservation of *"stock"* is the central concept in the notion of the *"circular economy"*, which we will explore shortly. In Daly's original equation the second ratio (MMC stock / throughput), was the *maintenance efficiency* of the stock, but this has merged into service efficiency in the modified version.

In these examples, service efficiency is a single ratio in our flow. In practice, any step in the flow may be expanded into one or more intermediate steps. A simple example is compressed air in manufacturing plants, which provides the service tool power, is produced in a compressor from another service provider, electricity, as shown in the equation below:

$$\frac{\text{Tool Power}}{\text{NC Services Sacrificed}} = \frac{\text{Tool Power}}{\text{Compressed Air}} \times \frac{\text{Compressed Air}}{\text{Electricity}} \times \frac{\text{Electricity}}{\text{Wind}} \times \frac{\text{Wind}}{\text{Land}} \times \frac{\text{Land}}{\text{NC Services Sacrificed}}$$

The more steps, or conversions, the more inefficient systems become. One question for our manufacturer might be whether it is better to provide tool power directly from electricity rather than from compressed air.

The next ratio in our resource efficiency formula is what I call the production efficiency. This is the number or quantity of the service provider that can be created from our resource or (human service provider)/(natural resource). In the examples above, this reflects the amount of natural resource, e.g. iron (or steel) that each service provider, e.g. car requires; or the amount of electricity we can generate for a given amount of wind, or the amount of bread we can make from a quantity of wheat. The word production reflects that in this step, human activities are applied to a natural resource to change it into a form that enables it to become a service provider.

$$\text{Resource Efficiency} = \text{Service Efficiency} \times \text{Production Efficiency} \times \text{Extraction Efficiency} \times \text{Ecological Efficiency}$$

Real World: *How car-sharing unlocked value for Croydon, London*

It is easy to see why individuals find car-sharing clubs attractive. Considerable financial savings can result because the cost of depreciation, insurance, maintenance and road tax are spread over many club members. Another attraction is the reduced time and effort involved in many aspects of car ownership, such as renewing insurance, organizing services and purchasing and disposing of cars.

For organizations, too, car-sharing schemes can be very attractive. In a closed scheme cars are only shared within the organization, a model often referred to as a *"pool car"* facility. What is becoming more interesting is the concept of an open scheme where the vehicles are shared with people outside the organization. With the increasing popularity of public car-sharing schemes, organizations can now consider joining these public schemes to meet their transport needs, rather than establishing their own scheme.

A case in point is Croydon Council in London which, following a review in 2010, replaced its car fleet with 23 dedicated cars operated by Zipcar. Clearly, the council needed to be confident that its staff would have access to the necessary number of vehicles and so, between 8 am and 6 pm on working days, the cars were exclusively for the use of the council employees. However, outside these hours the local community could also access the vehicles. The headline results, as quoted in Zipcar's press release[824] were very positive:

- Employee car use dropped by 42%, which also reduced emissions;
- The council achieved savings of £544,000; and
- 23 additional vehicles were made available to Croydon Zipcar members at peak evening and weekend hours.

The key figure here is the decrease in actual mileage by employees. Overall reductions of car miles driven are common among car-club members, as members use cars less for some types of journeys or trips at some times of the day. It seems that when we *own* a car, we feel compelled to use it even when there may be more practical alternatives available.

One of the additional benefits of car-sharing by organizations is a reduction in car parking spaces, which can be very costly to provide. In fact, parking is a real plus for many open schemes in cities such as London, where parking is notoriously difficult to obtain or expensive. The UK's Department for Transport emphasizes:

> *"The importance of on-street spaces cannot be underestimated both for open and closed schemes; not least because they provide a very visible image of the presence of a car club, and demonstrate direct benefits for potential users.*
>
> *The provision of dedicated parking spaces is a major incentive for the uptake of community car clubs, particularly in urban areas."*

Those organizations that are located in built-up areas well-served by public transport and which operate their own fleet of vehicles could well find that joining an open car-club scheme offers very considerable benefits.

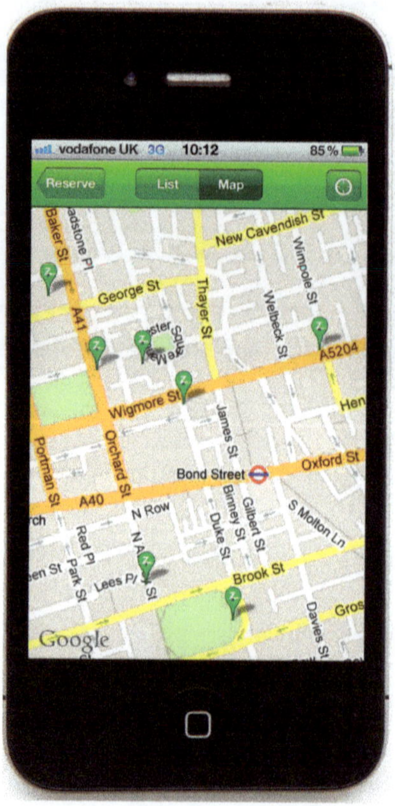

2.3 **Smartphone apps are making car-sharing schemes much easier and convenient for individuals**
In this example, a Zipcar member in London can quickly locate available vehicles near them, and can even filter the map by vehicle type. The process of reserving and unlocking the car can also be completed though the same app.
Source: Image courtesy of Zipcar press department.

2.4 **This is the primary product, by weight, of an aircraft factory**
It is called swarf and is the waste material milled from aluminium panels.
Source: Image © nicknick_ko - Fotolia.com

Foundations

For capital goods like cars, trains, aeroplanes, kettles, etc., the production efficiency reflects the overall efficiency of conversion from the base raw material to the capital good. In some cases the inefficiency of the production process is staggering; the yield loss (i.e. the ratio of the total resource used compared to that in a final product) by weight of liquid aluminium metal to the final panel of an aircraft wing is a staggering 90%.[22, p193] If we measure the main product of an aircraft manufacturer by weight it would be scrap, not planes! Just as in the case of service efficiency, there may be multiple conversion steps in the production process, each of which introduces yield loss.

Among the largest challenges facing us concerning production efficiency relates to yield loss in electricity production. Here we have a dominant model of generation, using large-scale centralized power stations which use fossil fuels, and which throw away their primary energy output, so-called *"waste"* heat. This waste results in a conversion efficiency of just 36% on average - for every 100 kWh of fossil fuel energy input we produce just 36 kWh of electricity. Coupled with the additional losses in transforming the power to high voltage for transmission over large distances and then transforming it back to the normal voltages used in homes and businesses, we end up with less than 30% of the energy value of the primary fuel being delivered in the service provider.

The root causes of yield loss are many and complex. Clearly, manufacturing defects are an obvious cause of yield loss, but this is actually a surprisingly small percentage compared to other losses. A more insidious root cause of yield loss is due to excessive standardization. For example, steel is initially produced in the form of rectangular billets, whose surfaces have imperfections which need to be cut off before the billets are then rolled into a standard rectangular sheet, whose ends and edges in turn may need trimming. These rectangular sheets are easy for the manufacturer to handle, but if the final product is a tin can, then cutting out the circular discs at either end of the tin will inevitably lead to further losses. For the aircraft manufacturer, the problem of standardization is greater because they not only need wedge-shaped pieces of aluminium which don't match the supplied sheets, but these pieces need to have variable thicknesses. At present, all the manufactures can do is to machine down sheets whose thickness matches the thickest part of the wing – wasting considerable metal in the process.

Of course, much of this waste material can be recycled if it is segregated and resmelted, although it may well find itself in a final material of lower quality than the original because of impurities that have been introduced at later stages (such as surface coatings or different elements which form alloys). In fact, more than one-quarter of all molten steel and nearly half of all molten aluminium never make it into a component and circulate around and around an internal loop, using more energy and creating more emissions.

Yield loss in these two metals is a greater driver of the embodied energy than the energy of downstream manufacturing. Embodied energy is all the energy that has gone into the finished material or product to that point in its manufacture. It is important that this last point is understood. The energy required to produce liquid metal is huge – say 100 GJ per tonne for aluminium, and the subsequent stages in the manufacture of metal products are, comparatively, much lower in their energy use, say 20 GJ per tonne to cast, roll and machine the aluminium into an aircraft wing. One would be forgiven for thinking that, from a resource efficiency perspective, one should focus on the liquid metal phase. Wrong! If we look at the embodied energy of the wing it rises from 100 GJ per tonne at the liquid metal stage to 1200 GJ/tonne at the end of the process. This is because, for each kilo of aluminium that remains in our wing, we have thrown away over 9 kg of aluminium and with it all the energy that was used to melt the liquid aluminium in the first place and to get it to the point at which it was discarded. It is because of multiplier effects like these, that, from a carbon and emissions perspective (and also from a value perspective), the optimization of yield should be a very high priority.

The family of methods to improve yield are collectively called waste minimization. Within waste minimization, we will encounter many techniques: for example, product design can select materials to use on the basis of their eventual yield – so we may use cast parts where there is usually much less waste rather than machined parts. Process optimization will ensure that we organize our cutting to get the most out of a standard sheet of material - something that the fabric industry does extremely well with sophisticated computer cutting machines to get the absolute maximum usable cloth out of a roll of fabric. Some process optimization involves the integration of process steps so that, for example, waste heat at one point in a process can be used to preheat a heat-requiring step elsewhere in a process.

One remarkable technology that is being developed now, and which holds the potential for the elimination of yield losses, is called *"additive manufacturing"*, or 3D printing, which creates very little waste compared to conventional "subtractive manufacturing" and which is becoming possible with metals as well as plastics.

Additive manufacturing, *also known as 3D printing, is a remarkable new technology which may dramatically decrease yield loss.*

Operational efficiency initiatives such as Monitoring and Targeting (M&T), Six Sigma, Lean and TPM will reduce wasted material and energy in the manufacturing process by ensuring that plant is run at optimum load and conditions. Quality initiatives ensure that there is less end-product discarded through defects and so the yield improves.

The next step in our resource flow equation is what I have called the extraction efficiency. This term reflects the efficiency with which we obtain the precursor resource to our production process from the natural capital from which it arises (natural resource)/(natural capital stock):

$$\text{Resource Efficiency} = \text{Service Efficiency} \times \text{Production Efficiency} \times \text{Extraction Efficiency} \times \text{Ecological Efficiency}$$

2.5 **Some parts of southern Spain, such as Almeria Province, are described as a "sea of plastic", referring to the numerous plastic-covered greenhouses that produce early season fruit and vegetables.**
Source: Image © bright - Fotolia.com

2.6 **This pair of satellite images shows the impact of massive and rapid agricultural development in Almeria Province along Spain's southern coast**
In the earlier 1974 image, left, the landscape reflects rather typical rural agricultural land use. In the 2000 image, right, much of the same region—an area covering roughly 20,000 hectares (49,421 acres)—has been converted to intensive greenhouse agriculture for the mass production of market produce. The reliance of this industry on underground water aquifers has led to it being described as a *"water mining"* rather than farming. The problem here is that the rate of abstraction of water from the aquifers can exceed the natural replenishment rate, so the ecological efficiency of this industry is very poor, compared with traditional agricultural methods.
Source: Image from "UNEP 2005 One Planet Many People, atlas of our changing environment", which has been kindly made available online at http://www.grid.unep.ch/activities/global_change/atlas/atlas_examples.php, (and also in the companion files for this book).

Extraction efficiency is something that can be difficult to control. For example, the extraction *inefficiency* of oil and coal has been rising over time as the best reserves have been depleted and those that remain are less accessible - see *Sources, sinks and "peak oil"* (page 37). An individual mine will demonstrate the same characteristic decline over time.

In the case of our equations, we are concerned with the extraction efficiency per unit of natural capital used. Thus we might be interested in the amount of wheat we can obtain per unit of land and water since this is the natural capital we are exploiting. In this case, one strategy to improve the extraction efficiency would be to use a higher-yielding variety of wheat, or perhaps one that required less water. Similarly, if our resource is timber, then we would be better using a fast-growing variety like pine rather than a slow-growing one like mahogany. Or if it is a wind resource (which also depends on the natural capital of land), then we might increase our turbine size or density to get more energy from a given acre of land. Daly called this term the *Growth Efficiency* to reflect the idea that, over time, natural processes create the resource that is being extracted (timber, wind, etc.).

Another strategy to improve the extraction efficiency is to recycle waste material into the flow to substitute for the supply that comes from natural capital. By recycling, we can meet the input resource requirements of the Production step using less natural-capital derived resource (ideally none).

Our final ratio is the degree to which the use or exploitation of the natural capital leads to a sacrifice in natural capital services, (NC stock)/(NC services sacrificed). For want of a better expression, I have called this ecological efficiency. From a sustainability perspective it is the denominator, natural capital services sacrificed that we wish to minimize.

$$\text{Resource Efficiency} = \text{Service Efficiency} \times \text{Production Efficiency} \times \text{Extraction Efficiency} \times \text{Ecological Efficiency}$$

It is in here we get into some complex considerations, as few uses of natural capital have one single ecological impact. For example, creating an open-cast mine can affect biodiversity, watershed and water purification processes, animal migration, and – through the use of energy – the planet's CO_2 absorbing services.

It is also important to understand whether the impact is a sustainable one or not. Extracting fresh water from snowmelt in Greenland may have an ecological impact close to zero and be sustainable in perpetuity (at modest volumes), while extracting (or *"mining"*) fresh water from an aquifer in southern Spain may reduce the supply to rivers, increase soil salinity, and only be sustained for a few years. More subtly, extracting a resource like timber from a forest may have a comparatively small impact until the point at which it becomes unsustainable and the harm rises significantly. Thus the rate of extraction influences ecological efficiency considerably.

Real World: *Competitive advantage*

Daly's notions of a cascade of efficiencies will only drive change if buyers know the impacts on natural capital of alternative sources of their materials.

There are several schemes, such as Environmental Product Declarations, which are intended to provide this information in an independently certified and consistent manner.

The pressure on materials suppliers to report their impact is growing, even for *"commodity"* materials.

For example, research that I was involved in recently for the WBCSD Cement Sustainability Initiative strongly suggests that high-level building rating schemes, like BREEAM and LEED, are moving towards a life cycle assessment of the materials used in construction. These rating systems are important to property developers, so they are likely to pressure their supplier for this information.

Some organizations are moving even more quickly as they perceive that they have a superior product and want to gain a competitive advantage today. One such example is CEMEX, which is the first cement manufacturer to label the CO_2 content of its Portland cement.

2.7 **CEMEX UK introduced a carbon label in 2010, based on the Carbon Trust Carbon Footprint, on 25kg bags of cement with a high proportion of fly-ash aimed at smaller builders and end-users.** *Source unknown.*

For an organization wishing to be more resource-efficient, assuming that the desired resource is indispensable, the way to improve this ratio is to source responsibly. There are myriad labelling schemes to help the buyers differentiate resources that have a small ecological impact from those that have a large impact. These schemes are particularly relevant when it comes to materials like timber, which can have a broad range of ecological impacts. Labelling schemes proliferate for consumer-facing goods, such as line-caught tuna, free range eggs, organic foods and so forth. However, for bulk materials the story is different.

Just five global mega resources – steel, cement, paper, plastic and aluminium – account for over half of all CO_2 emissions from manufacturing.[22] Yet there are no labelling schemes that permit buyers to source those supplies which have the lowest carbon emissions easily. We know that when aluminium is produced using electricity from a renewable resource such as hydro, it will have a lot fewer embodied emissions than that using fossil fuel sources. The same can be said of energy itself – for example, different sources of oil have different embodied emissions of CO_2 related to the extraction energy intensity. Oil extraction from tar sands, according to Shell's sustainability report,[659] requires 10 times more energy than conventional sources. Furthermore, Shell has a particularly poor record on flaring, especially in its oil fields in Nigeria, an activity in which surplus gas extracted alongside the oil cannot be marketed and so is simply burnt in flares, which gives Nigerian oil a greater embodied carbon than similar oils where the gas is captured and marketed.

A key challenge for sourcing of commodities from sustainable sources is the very fact that they are commodities. A tonne of aluminium at a given purity is the same as any other tonne of aluminium of the same composition. There is no *"brand"* effect here. Consumers of the commodity are interested in price alone. This price is often determined by a complex mix of factors in the free market, such as the location of the supply, economic effects (or distortions) such as subsidies, tariffs and trade barriers. Indeed, these commodities are traded in huge volumes globally, with secondary and futures markets set up to hedge the supply risk and enable downstream users of these resources to establish the future prices for their products with some certainty. These market intermediaries or speculators make the connection between the supplier and consumer of these resources even more tenuous. While these structures are economically efficient, they lead to a homogenization of the products and create a process where the purchasing decision is based exclusively on price.

Pressure to change this is coming from the supply chain. Manufacturers and end-consumers are demanding that their suppliers provide them with information to help them make informed choices. The particular emphasis is on information about carbon emissions, although the impact on water resources is also a growing area of concern. The common name used to describe the process of assessing the impact of extraction, production, use and disposal of materials is life cycle assessment (LCA) (see page 440).

Of course, there is potentially a much better way than labelling to drive improvement in our ecological efficiency, and that is cost. If we examine the ratios above we can see that our financial systems create disincentives for the first three inefficiencies: we are financially penalised for poor service efficiency, production efficiency and extraction efficiency. However, there is no equivalent direct financial penalty for poor ecological efficiency, because the sacrifice of natural capital services has no price (or rather it is a cost that is "externalised" and borne by society as a whole). Thus material waste has a cost – but reduction of biodiversity has none. It seems that cost only enters the equation once we have ownership of a resource and, since no one owns the CO_2 absorption capabilities of the oceans, no charge arises for reducing this service. Moves to create a price for carbon emissions would enable ecological efficiency to have a financial impact – although we need to be clear, any cost driver only really works where the cost is material. The aircraft wing manufacturer can afford to throw away so much waste because the value-added swamps the value wasted.

By looking at the flow of resources, we can see that inefficiency takes many different forms. Resource inefficiency can arise because we deliver the service we need from our resources inefficiently, or from the under-utilization of a resource such as a car, which means that we need more of that resource to meet our needs. Inefficiency can arise because we are disposing of equipment earlier than necessary, which leads to greater replacement rates. Inefficiency can also be due to errors in production, which mean we scrap the product, or because we have to work with standardized raw materials that have greater yield losses. Resource inefficiency can arise because of the tendency of designers to over-specify materials, or because our buying processes aren't set up to source the resources with the lowest environmental impact.

A key lesson that arises from these equations is that inefficiencies cascade through each step of the chain. In the example below we can see that compound losses mean that 100 units of energy available in the form of coal become fewer than 10 units of energy available to provide our service (pumping water). The total efficiency of the system is, starting at the origin with generation, (1-0.7) x (1-0.09) x (1-0.1) x (1-0.02) x (1-0.25) x (1-0.2) x (1-0.33) = 9.68%. The message here is that saving at the end of the system will save 10 times as much natural capital, as saving at the start of the system.

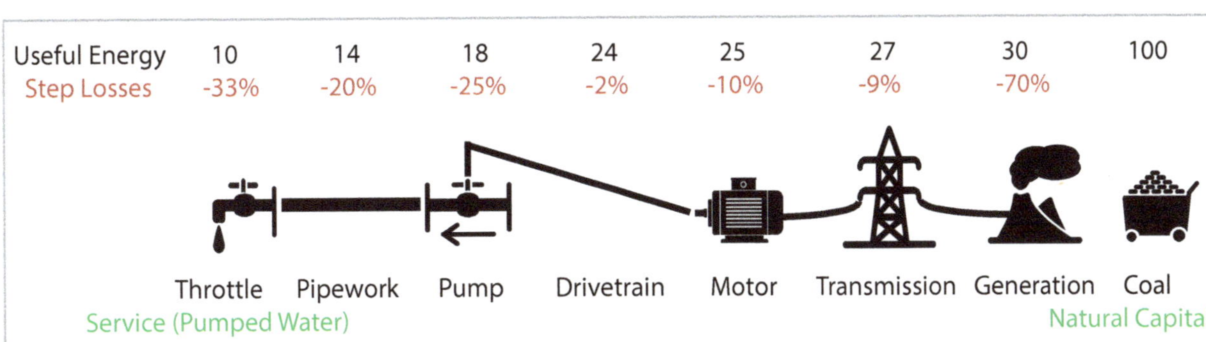

2.8 **End-use reduction of human services will lead to significantly greater savings in natural capital due to the compounding effect of losses (numbers rounded for simplicity)**
Source: Niall Enright adapted from Amory Lovins.[486]
Vector images © Strezhnev Pavel & Anthonycz - Fotolia.com

2.2 Ending "take, make, waste"

The one-way journey from source to sink is very wasteful. Building on Herman Daly's work is a recognition that if resources can be maintained longer in circulation then the impact on natural capital will be reduced.

Herman E. Daly's work emphasizes the one-way flow of natural resources in our current economic system and gives us real insight into the sequence of inefficiencies that can waste a significant proportion of our natural capital. However, this work only considered resources up until the point at which they provide human services. We have seen in our introductory chapter that an important concern is also the end-of-life disposal of the resource, the impact that the waste has on natural capital.

If we add this dimension to the flow we have described, then we get a *"take, make, waste"* flow, often described as the linear economy.

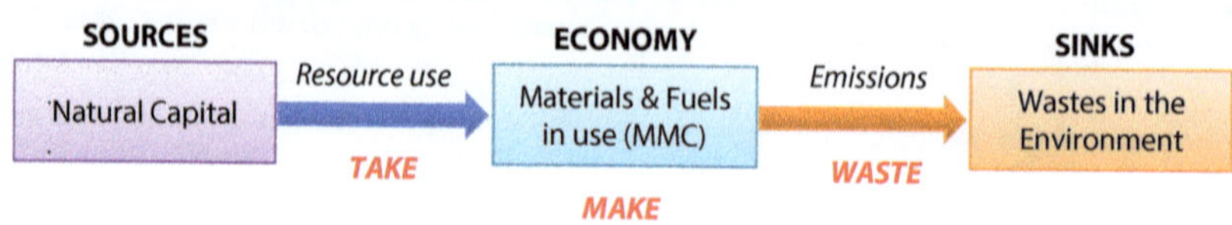

2.9 **In the "take, make, waste" linear economy, materials flow on a one-way journey through the economy to disposal**
Source: Niall Enright

So, just how big is this flow? In volume terms, around 65 billion tonnes of raw materials entered the global economic system in 2010, a figure expected to grow by 26% to about 82 billion tonnes in just 10 years.[249] Even in advanced economies, the majority of this material passes through the economy just once. In Europe, for example, rates of recovery of materials after the end of their (first) functional life are quite low – of the 2.7 billion tonnes of waste generated, only 40% were recycled, composted or reused according to the Eurostat waste statistics for 2011[249]

In an infinite world, the sources and sinks would be limitless, and this one-way journey will be without consequences. However, we don't live in an infinite world and sooner or later (about now, according to many commentators) we need to change the approach in a fundamental way.

One vision for the change needed has been set out by a designer and a chemist. William McDonough and Michael Braungart in their book *Cradle to Cradle.*[506.] Playing on the expression *"Cradle to Grave"*, which described the linear economy, McDonough and Braungart emphasize the importance of recirculating materials within the economy, in a *"borrow, use, return"* approach where, when they are eventually returned, materials have a positive effect on natural capital.

2.10 **In the "borrow, use, return" approach, materials are maintained for as long as possible within the economy or technosphere and when disposed of they will have a positive environmental impact**
Source: Niall Enright

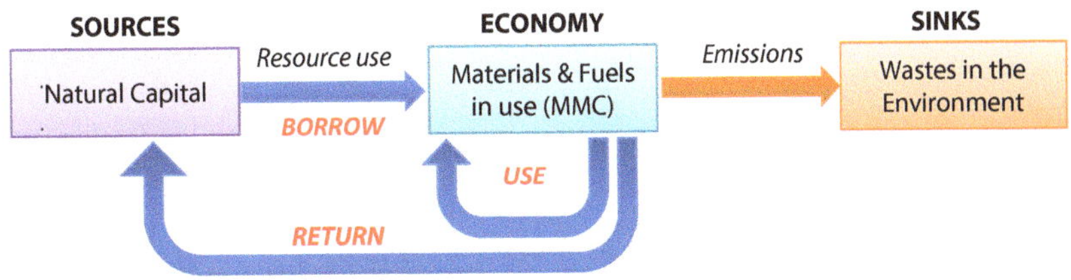

Real World: ***Sweden's tax break***

© Kadmy, Fotolia.com

The Swedish government has identified the embedded carbon in consumer goods as a growing problem.[575]

In order to encourage reuse, tax breaks on the repair of a wide range of goods, ranging from appliances to bicycles are being introduced from January 2017.

Two changes are proposed. First, the valued added tax (VAT, a sales tax) on repairs will be reduced from 25% to 12%. Second, people will be able to claim back from income tax half the cost of repairs to appliances.

The result is that a repair that would have cost SEK400 including VAT (SEK320 excluding VAT) will end up costing only SEK179 (SEK41 saved on VAT and half the balance saved in income tax).

Not only will this extend the service life of many goods but it should also stimulate a growing repair sector, which will offer employment to many workers who lack formal education.

McDonough and Braungart see design as the core discipline in a resource-efficient world. If we can design our products to be recirculated in the economy or to be returned safely to nature, then much of the damage that we do will be eliminated.

This mindset produces some unexpected outcomes – for example, their book is not produced on paper – ostensibly a renewable material, but one which needs lots of energy and chemicals to make – but instead on a synthetic material made of plastic resin. This material is very durable (so the book will last much longer), it is waterproof (so it is better than paper in that it can survive getting wet) and can be reused to create more synthetic paper. It is the characteristic of virtually infinite reuse which makes this material much better than paper, which generally can only be *"downcycled"* a few times into poorer quality materials before eventual disposal (for example, as a fuel in energy from waste or as toilet paper).

Drawing heavily from the language that we use to explain how nature manages flows of resources, McDonough and Braungart describe the world economy as a large complex of metabolic processes, collectively the technosphere, which are fed by *"technical nutrients"*. Ideally, these technical nutrients, like the plastic paper in their book, can cycle around infinitely in the economy. While it is preferable if these technical nutrients are not hazardous, it does not particularly matter if they are, as they are never intended to return to the environment, but to remain in the technosphere. Of course science, namely the law of entropy, tells us that this recycling of materials cannot continue forever without some losses or degradation of the materials.

There are materials which cannot be designed as perpetual feedstuffs for the technosphere and in this case the design needs to deliberately take into account the return phase, by ensuring that the material is biodegradable. In effect, all materials that return to the environment should be *"biological nutrients"* for a natural metabolic process in the *"biosphere"*, which will nourish and enhance the environment. In *Cradle to Cradle*, as in nature *"all waste is food"*. Resource efficiency is about ensuring that we align our materials choices with suitable metabolic processes in the biosphere and the technosphere, and also that we make our products in a way that makes it possible to separate out the various *"nutrients"* at the end of their lives.

Exploration: *Why the debt-fuelled linear economy is the current paradigm*

It is surprising how deeply imprinted the linear economy is in our current social, political, economic and financial structures. It all starts with value-added. The baker in Herman Daly's world takes flour and water and fruit and heat and creates a product (a pie) whose value is greater than the value of the inputs. This pie is the added value which drives economic activity in one direction (value-destroying activities are both illogical and should be self-limiting by the unsustainable cumulative losses). At a national level, we recognize this value-added in the form of the main measure of economic activity, Gross Domestic Product or GDP, which is the sum of all the financial transactions from farmer to baker to the consumer. For-profit organizations recognize value-creating activities as the difference in their costs and their sales, in other words, their profits. GDP and profit represent the wealth creation of organizations and nations - they are the fundamental "scorecard" for human endeavour.

For the baker to make more money, they need to bake and sell more pies – profit is largely a product of volume. The majority of businesses are built on the fundamental premise that the greater the flow through the system, the greater the accumulated value. Companies also believe in the ideas of economies of scale, that is to say that production and distribution become more efficient as the organization grows. So again, ever-greater flow through the organization is considered positive.

But, in a linear economy any increase in the flow leads to greater extraction of natural capital. Despite this obvious negative consequence of increased flow, our societies are based on ever-increasing growth. Much of this increase is to satisfy the aspirations of the emerging middle classes in countries such as India or China, whose needs and wants cannot (and should not) be easily denied. By 2030, the global middle class will almost double in size to 5 billion people largely from emerging economies[705] – growth that took over a century to accomplish in the developed world will now be compressed into just 20 years.

Anticipating this huge surge in need for human services, our financial system has created vast amounts of money to fund this development. Of course, these funds are just a human construct. It is, in fact, debt, that is to say, that the money is being made available on the basis that there will be a repayment. The assets created (homes, factories, schools, hospitals, etc.) will have their initial costs (the principal) and a return for the investors (the interest) repaid out of the future income (or taxes) that the assets will make possible.

Thus growth is the central paradigm of our economic system: to meet the needs of a growing (and increasingly affluent) population and to service the massive amount of debt that the global economy has accumulated and will need. This emphasis on growth leads policymakers to create incentives that encourage the wasteful linear economy. An example is the *"cash for clunkers"* policy which, in the recent economic downturn, was implemented in 22 countries to provide incentives to trade in old, perfectly serviceable, automobiles and replace them with brand new vehicles. From a resource efficiency perspective, this makes no sense at all. The policy is designed to shorten the life of our existing stock of vehicles and, by lowering the cost of new vehicles it disincentivises a switch away from private ownership to shared ownership or public transport modes described earlier in *Real World: The incredibly low service efficiency of our cars* (page 46).

Supporting this debt-based growth permeates our economic and financial systems. In private companies, interest on debt can be written off against profits (in other words, debt-holders have the first call on profits compared to society as a whole). Furthermore, retained profits (not used for economic expansion or growth) attract no additional taxation and so can serve to provide further debt (directly as loans by the organizations or indirectly through loans by banks, which leverage these balances). Everywhere there is a bias towards borrowing – but spending money today means repaying tomorrow with natural resources (after all, money is simply and fundamentally no more than a call on future natural capital).

Organizations, too, fall under the sway of this dominant paradigm. Managers and leaders of profit-making organizations are expected to deliver year-on-year increases in profit. However much they recognize the ultimate unsustainability of this model, it is tough to justify a radical departure from the linear approach. Private businesses see increasing flow or volume as their key mission. There are many measures of this basic goal, from market share, unit shipments, available reserves, subscribers or profits – volume flow is the fundamental objective. So businesses borrow in turn to create their own growth. And why not, given that the global customer base is growing at the fastest rate in history!

The challenge for global resource efficiency is not merely a matter of improving the current status quo but of being able to meet a doubling of demand *and* servicing the current and future debt that has accumulated in our economies. What is clear is that natural capital, already stretched to the limit, cannot meet these needs without a rethink of the "take, make, waste" model. In short, we need to decouple economic growth from resource extraction – we need a radical change in our resource efficiency.

2.3 The waste hierarchy

Eliminating our demand for a resource is clearly the most effective way to reduce our impact on nature. Here, an updated version of the waste hierarchy prioritizes the techniques that we can employ to minimise impacts on natural capital.

Notions that some ways of dealing with waste are preferable to others have been around for a long time. In the 1980s, the idea of the 3Rs: *reduce*, *reuse* and *recycle* became commonplace. This phrase sets out the order, or hierarchy, for dealing with waste in which reduction is preferable to reuse, which in turn is better than recycling.

Such ideas gained extra significance when they started to form the basis of regulation. In the US, the 3Rs were introduced into federal waste rules in the 1990s. The European Waste Directive in 2008 formalized a similar five-stage waste hierarchy: *prevention*, *prepare for reuse*, *recycle*, *recover* (e.g. extract energy from the waste) and *disposal*.

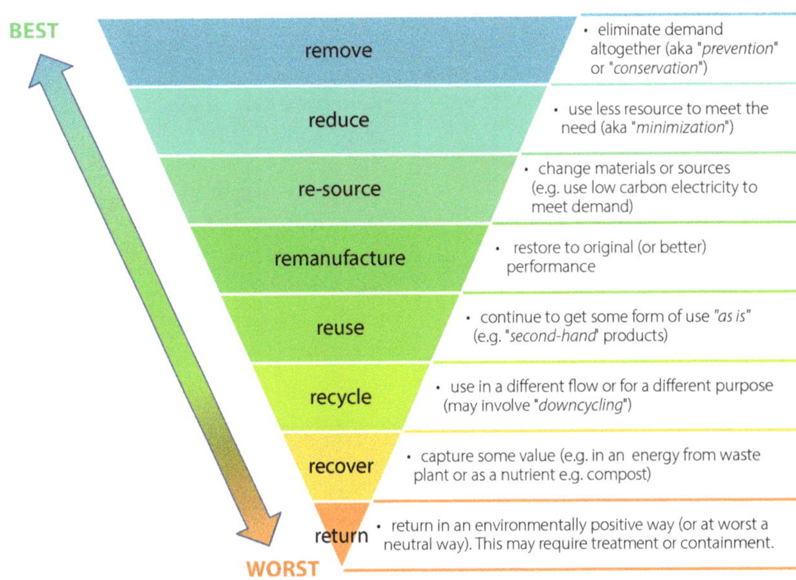

2.11 **The resource efficiency hierarchy shows, from top to bottom, eight resource efficiency methods with increasing environmental impact and decreasing cost-effectiveness.** Please note that these methods are not exclusive - one would start at the top to remove the need for the resource, then move on to sourcing and finally develop end-of-life strategies. *Source: Niall Enright. Image available in the companion file pack.*

These waste hierarchies convey the idea that waste is a valuable resource, rather than a burden. Reducing waste in the first place clearly avoids the costs of the raw materials that would have ended up in the waste. For that waste which cannot be avoided, one may as well get some income by recycling it, rather than paying costs associated with disposal.

This concept of waste as value has been remarkably successful. After early investment in capacity and market development, developed countries now have a huge and thriving area of economic activity based on extracting value from a wide range of waste streams.

Because the waste hierarchy is such an important part of environmental regulation today, it is worth revisiting in light of contemporary thinking on resource efficiency. Three recent developments come to mind.

1. We now consider the upstream effects of resource use as much as the downstream impact of wastes arising and end-of-life disposal. This more holistic approach is called *resource efficiency*, where extraction and issues of scarcity determine our selection of materials as much as disposal considerations.

2. There is a growing rejection of the linear economy or take, make, waste model for a borrow, use, return or cradle to cradle approach. The idea is that materials, once removed from the environment, should be managed such that they can recirculate permanently in the economy. Ideally, materials never return to the natural environment, but if they do, they should serve as a nutrient to improve the environment.

3. Broadly speaking, individuals and organizations are now more willing to do things because of environmental benefits rather than just because they generate financial value.

When we take these three trends into account, what emerges is an updated version of the hierarchy, or more simply, the 8Rs, shown on the previous page.

At first glance, little seems to have changed. At the top of our hierarchy we still have the *prevent* approach, although now in two methods: remove and reduce. By creating a remove method, we are more clearly signposting product design and formulation as a means of cutting out resource need altogether. *Remove* reminds us that our first approach should be: "*Why do I need this resource at all? What function does it fulfil? Can we design this requirement out? Can I deliver the service that this product provides in a different way?*"

Below *reduce* there is a new method, to re-source. In this step, once we have removed or minimized our need for a resource, we then consider alternative sources or materials with a lower ecological impact. For example, if our resource is electricity, we should first start by employing every practical option to reduce its use and only then find out if we can source the electricity from an alternate supply, such as a renewable source. Using harvested rainwater instead of treated town's water is another example of re-sourcing.

Some organizations behave as if the *re-source* choice is the only method they should use, so they will buy 100% renewable energy and declare "*problem solved*". Because they are not affecting demand and because substitute sustainable sources are often more expensive, this approach is often the least

Real World: *L'Oreal policies in the US*

It is sometimes helpful to look at organizational policies around waste, to align these with the resource efficiency hierarchy.

For example, I worked with French cosmetics giant L'Oreal on some resource efficiency audits in the US and discovered that it was the policy that any non-product materials that left the site were categorized as waste.

L'Oreal are to be applauded for using this strict definition of waste in their reporting to encourage the sites to eliminate all forms of waste.

However, this definition had the unintended effect of making it preferable to burn waste in a biomass boiler on site as the emissions to air are not counted in waste reports.

Before adopting this definition, the waste from the shampoo lines was sold in bulk to nearby car wash companies who would use it to clean cars. Not only did this displace the need for virgin detergents but it probably led to some very classy finishes on the local cars!

© Alexander Marushin, Fotolia.com

This form of recycling, *reuse*, is higher up the resource efficiency hierarchy than the incineration option, but L'Oreal's definition of waste unwittingly led to the lower *recover* choice being favoured. We shall see later that the categorization of waste can often impede more sustainable resource use.

Exploration: *Terminology*

Different terms are used to describe how materials and products fit into the waste hierarchy. The following pairs of words relate first to the source of the material and second to the sink, and are often confused.

- A recycled product will contain material that has been sourced from material that has been used previously;
- A product with recyclable material can be returned into the economy instead of being discarded into a natural sink;

and:

- A renewable product is made of resources that nature can replenish. However, for a product to be renewable it strictly should also be recyclable (or at least returned to the environment in a form that has a neutral impact);
- If a product is recoverable, it means that its material can be separated *and* recycled at the end of life.

Again, another pair of terms that leads to confusion:

- A product with biobased materials has been derived from biological sources; while
- A biodegradable product is one that can be returned to nature and will decompose into harmless (or nutritious) materials through the action of natural organisms or processes.

It is important to note that a biobased product is not necessarily biodegradable, nor is a biodegradable product necessarily biobased.

Similarly, a recycled product is not necessarily recyclable, or vice versa.

cost-effective option. Also, in a world where renewable supplies do not satisfy total demand, the extra resource that this organization gets from renewable sources reduces that available to other organizations. Regrettably, this approach persists because it is seen as being lower effort than the alternatives, especially where the resources involved are a small part of the organization's total costs.

After *re-source,* turn to end-of-life considerations. Remanufacture involves returning the resource or equipment to its original specification, or better. Reuse in the same process is usually better in value and environmental terms than recycling or *downcycling* into a different process. Here the circular economy proponents would argue that all products should be designed with *reuse* in mind. Products should be taken back by their manufacturers, be easy to disassemble and be reincorporated into the next batch of goods.

It may be possible to recover some part of the resource where recycling is not feasible. Examples include converting embedded energy into heat by burning waste or extracting nutrients from the waste by composting. This recovery reduces the need for virgin resources - gas or fertilisers in these cases.

Finally, we can return the waste to the environment. Here the use of language reflects updated thinking. The old term, *disposal*, has connotations of discharge, dumping or throwing away, whereas *return* addresses the "*circular economy*" thinking that we are borrowing a resource and then returning it carefully to nature. Ideally, the material returned will be a nutrient to enhance natural capital. If our waste is not a nutrient, it should at least be inert. If this is not possible then we must contain, treat or dilute it to remove its effect on the environment.

When making the judgements about which techniques to employ it is important that the actual availability of the infrastructure to deliver the promised environmental benefit is taken into consideration. It is common, for example, to see plastics marked as recyclable when in fact few recycling facilities exist.

In devising this new hierarchy, I contemplated an additional "R" at the bottom of the diagram, which was going to be remediate. This step reflects the result of harmful return of waste to the environment and the obligation to put that right. However, I decided against this because it could imply that dumping followed by remediation at a later date is a legitimate waste management method.

Traditionally, multiple methods in the resource efficiency hierarchy would be applied in sequence, so the use of the resource is first minimized, then some material is recycled and the remaining waste returned to the environment. What is clear, though, is that we need to think more about the overall system rather than just one individual waste stream. Thus the new – but infinitely reusable - material used for the book *Cradle to Cradle* by McDonough and Braungart may require more inputs to produce than conventional paper, but the ability to reuse the material time after time leads to much less waste overall. We should not think about the single journey of an isolated waste stream – but the overall flow of many materials over many cycles.

2.4 Remanufacture

Although many organizations are exploring the possibility of remanufacture for the first time, this actually has a long history.

Real World: *A hidden gem*

Remanufacturing is a very widespread activity. According to the US database of remanufacturers,[487] there are almost 7,000 firms in North America engaged in remanufacturing.

The extent of the industry is largely unappreciated because the activities go under many names.

Industry	Term
Aircraft	Overhaul
Automotive	Rebuild
Automotive	Recondition
Electric Motors	Rewind
Furniture	Restore
Industrial Valves	Repair
Medical Imaging	Refurbish
Military Equipment	Reset
Musical Instruments	Restore
Rail	Rebuild
Tires	Recap, Retread
Toner cartridges	Recharge

A study by the US International Trade Commission[754] states that the total value of the US remanufacturing market is US$43 billion, with 180,000 employees, a figure that they say may be an underestimate. An older study in 2004[590] estimated that the UK market for remanufactured goods was worth £5 billon and provided 50,000 jobs.

Remanufacturing involves taking used equipment or material and restoring it to at least original specification. It is one of the most promising areas for action to reduce resource use. It is important to note that remanufactured equipment is not *second-hand* or *repaired*, but has been returned to *at least* the Original Equipment Manufacturer (OEM) performance and supplied with a warranty that is at the *same as or better* than the warranty on a comparable new product.

Of course, remanufacturing is not new. In the US, for example, companies were providing motor rewinding services as far back as the turn of the century. These businesses have thrived because of the economic benefits that remanufacturing offers. Remanufactured equipment has the same performance as the original, but typically can be sold to customers for 30-40% less than the original cost to purchase and costs 40-65% less to produce.[315] The financial case is compelling: the (re)manufacturer has a higher profit and the customer a lower price.

In environmental and social terms, too, there are significant benefits. For example, because the input of remanufacturing is largely labour, these businesses provide lots of jobs. Resource requirements are very low since the vast majority of the materials used are already present. In the case of photocopiers, for example, 93% of a remanufactured machine by weight is composed of reused parts[503] and the remanufacturing of a product typically only consumes 15% of the energy needed to make it from scratch.[315]

Remanufacture can be undertaken by the OEM, contracted agents or by third parties. It has been estimated that the ratio of employment in OEM to contracted to third-party remanufacturers is 1:2:4. The vast majority of remanufacturers are small, third-party businesses specializing in the remanufacture of specific components: motor rewinding, tyre retreading and toner cartridge refilling are good examples. This small company size is one reason why it's hard to quantify and develop the remanufacturing sector as a whole.

Successful remanufacture requires three things.

1. Systems to collect back equipment or material at the end of life (sometimes called reverse supply chains). These can be quite challenging to put in place. For example, Caterpillar relies on its distributors to drive the return of cores.

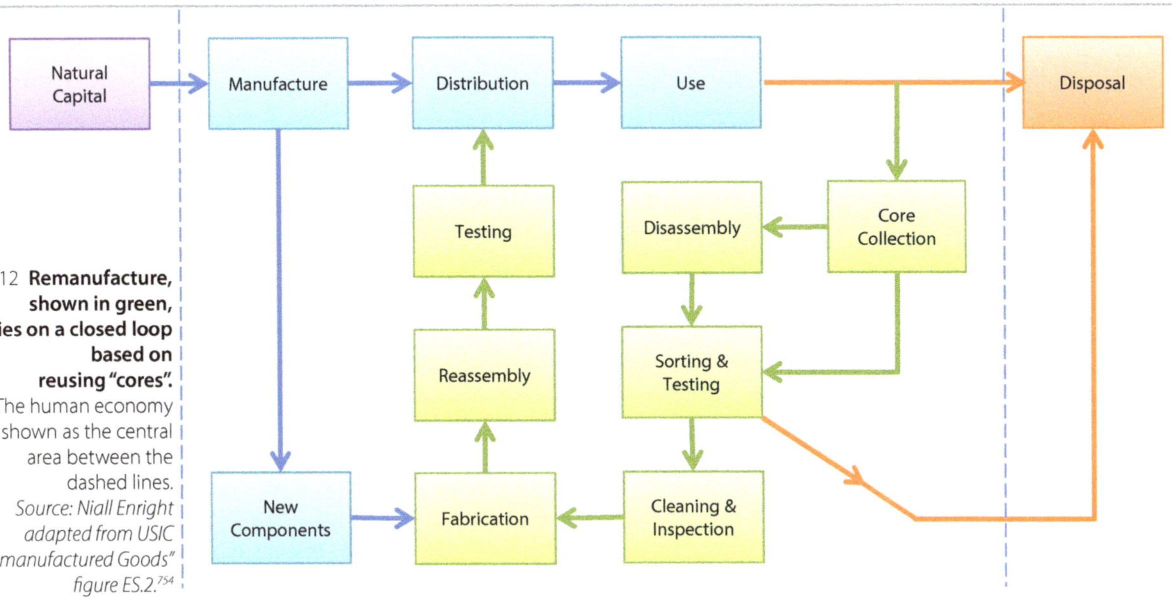

2.12 **Remanufacture, shown in green, relies on a closed loop based on reusing "cores".** The human economy is shown as the central area between the dashed lines. *Source: Niall Enright adapted from USIC "Remanufactured Goods" figure ES.2.*[754]

2. Processes to disassemble, inspect, restore or replace, test and assure the equipment or material. Here, design for remanufacture is paramount so that goods can be easily disassembled, tested, and refurbished.

3. Demand for remanufactured products. It is notable that some consumers perceive remanufactured products as inferior and this resistance needs to be overcome. In some markets, like toner cartridge refilling, some OEM will claim that remanufactured products are sub-standard as they wish to protect their markets for new products.

Standards: *BS 8887-220:2010 The process of remanufacture*

This is one of a series of standards in the design for manufacture, assembly, disassembly and end-of-life (MADE).

The standard sets out the steps required to convert a used product to an "as new" condition, which has at least the equivalent performance and warranty as a new product.

The standard is not applicable to consumables such as foods and fuels, commodity materials like chemicals, or to digital media.

The standard covers the remanufacture of a whole product or a component part.

BS 8887-1 is aimed at the design stage while BS 8887-2 has a list of terms and definitions.

Because of these challenges, there remains considerable further scope for the development of remanufacturing. The USITC study concluded that current remanufacturing intensity in the US as a whole, defined as the total value of shipments of remanufactured goods as a share of total sales of all products within that sector, is only 2%. For example, in Caterpillar's category of heavy-duty off-road equipment, the remanufacturing intensity is only 3.8%[754] This is one reason perhaps why the company continues to see considerable potential for growth in this market.

In 2000, policymakers in the EU introduced the End of Life Vehicles (ELV) Directive. It is based on the notion of extended producer responsibility where the manufacturer is obliged to deal with their products at the end of their life. These regulations require the manufacturers to meet specific targets:

	From 1 Jan 2006	From 1 Jan 2015
Reuse and Recycling	80%	85%
Reuse and Recovery	85%	95%

Real World: *CAT Reman*

Caterpillar, the iconic US manufacturer of earth-moving and other equipment, has developed a highly successful remanufacture business model, called CAT® Reman. Caterpillar's remanufacturing activities began in 1973 and now employ over 3,600 people.[248]

Their successful approach is based on keeping the components, or *"cores"* as they are called, within the Caterpillar network. Thus the basic offer for a Caterpillar customer is a one-for-one exchange at the end of the service life of the equipment.[88]

The first incentive for customers to buy CAT® Reman is that the equipment is offered at a price well below new but has the same guaranteed service life and warranties as new. At the same time, the customer is also asked to pay a deposit (approximately equal to the cost of the unit itself). When the customer returns their used components (cores), Caterpillar refunds their deposit.

In 2010, over 2 million cores were salvaged, over 60,000 tonnes (134 million lbs) of material remanufactured or recycled, and 85% of the energy *"value added"* was preserved. For equipment such as locomotive and engines, remanufacture can involve an improvement in the fuel efficiency, noise and particulate emissions of the equipment. It is "*better for old*".[124]

The difference between these targets is the amount, by weight, of the vehicle that must go to an incineration facility. Thus, before 2015, if over 85% of the vehicle is reused, then there is no requirement for further incineration. The ELV regulations have created a thriving vehicle disposal, reuse and recovery sector throughout the EU, although it has been criticized for considering recycling and reuse as equivalent, so does not explicitly drive reuse or remanufacture.

Remanufacturing is becoming a global business. For example, Caterpillar stated in its affidavit to the USITC study that it sells remanufactured products in 170 countries. However, they also identified many barriers to this trade, most notably the lack of standard definition for a remanufactured product, which means that customs officials in some countries treat remanufactured products as used products, which may be banned from importation.

The British Standards Institution published BS 8887-220:2010, which defines the process of remanufacturing. However, steps to develop this as an ISO standard have faltered.[131]

OEMs also put up legal challenges to prevent third parties from remanufacturing their goods, usually citing patent infringement. The legal debate centres around the question as to whether an owner has a legitimate right to repair their products vs the rights of the patent holder to prevent others from making parts that use their unique design[132] By and large, the courts have favoured the rights of remanufacturers, although the fact that these are cases that are usually treated on their individual merits means that those who wish to enter the field need to consider the patent implication carefully. Indeed, the inability for manufacturers to use Freedom of Information to access technical specifications adds yet another barrier as they have to reverse-engineer products, even when out of patent, to remanufacture them, adding to the time and cost involved, according to a report by the All-Party Parliamentary Sustainable Resource Group in the UK.[21]

These barriers are likely to reduce as policymakers, consumers and manufacturers appreciate the value of remanufacturing. From an organizational perspective remanufacturing offers an opportunity to lower costs of input materials and equipment. If the organization is a manufacturer then they should consider incorporating remanufacturing into their business model; not only does this offer significantly greater levels of profitability per item sold, but it also provides a great way to retain relationships with customers and to send a positive signal to stakeholders and society.

Organizations also have a vital role in influencing attitudes around remanufactured products. A case study[249] on the remanufacturing activities of Ricoh UK has shown that "*it is the public sector in particular that has shown considerable interest in remanufactured product, while commercial sector uptake has been slower*". Clearly, public sector organizations have a role to play in demonstrating that remanufactured goods are equivalent to new, which will hopefully encourage private firms to follow suit.

2.5 The circular economy

Contemporary ideas of resource efficiency are often described in terms of a circular economy, where materials and goods are designed to exist in perpetuity, thus radically decreasing the demand on natural capital.

The ideas of Herman E. Daly, Willam McDonough, Michael Braungart and many others (notably Walter Stahel, Ken Webster, Amory Lovins and E.F. Schumacher) are now coalescing into the concept of the circular economy. These ideas are being actively promoted by institutions such as the Ellen MacArthur Foundation, consultancies such as Lavery/Pennell and McKinsey & Co and universities like Harvard, and are seen as the blueprint for growth without environmental degradation.

These ideas are hugely relevant to our own organization's approach to resource efficiency because they describe the change that we will (voluntarily or not) need to make to address the scarcity of (cheap) resources and the saturation of sinks. In fact, these ideas, largely couched in terms of economy-wide changes, provide a great insight into the practical organizational changes that we can adopt. These are not theoretical or abstract concepts. Rather these ideas have formed the basis for success in many industries and underpin a significant, and growing, wave of innovation and policymaking worldwide.

The core concept is the objective of keeping materials in the human economy for as long as possible. Not only does this reduce the damage that extracting the resource from nature entails, as well as all the intermediate steps in production, but is also eliminates the impacts that eventual disposal can cause. The circular economy is one where materials cycle forever in the human economy.

This is sometimes described as an economy based on the "*maintenance of stock*".[675] The notion is not new. It is something that parts of our societies are already doing, for example in:

- the conservation of cultural heritage and art;
- the preservation of health and wellbeing;
- the management and preservation of human knowledge;
- the management and conservation of land (biodiversity, agriculture).

In simple terms paintings, people, books, forest and fields are not material things that we use a few times and discard – rather, we make tremendous efforts to preserve them while making them available to more and more people and for years to come. A circular economy applies the same value to all

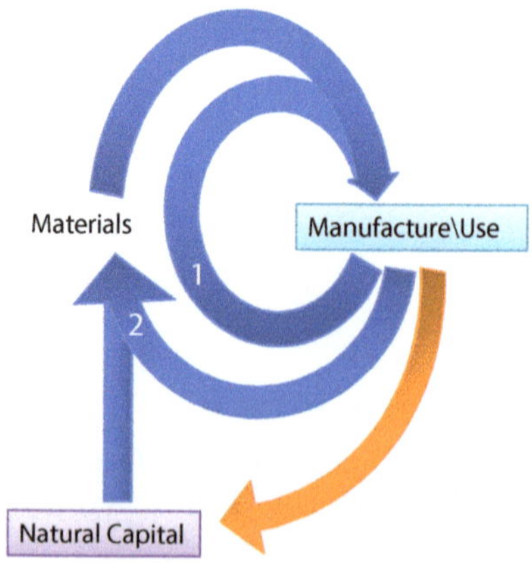

2.13 **Walter Stahel and Genevieve Reday described two essential flows for a circular economy**
Loop 1 involves life extension followed by the reuse, remanufacture, repair and reconditioning of goods so that they provide service for as long as possible. Loop 2 involves the recycling of molecules and materials once the goods come to end of their productive life. Ideally the recycled molecules and materials will be cheaper than those extracted from nature (blue). I have added a third loop, in orange, to show that molecules that cannot be recycled should be returned to nature as a nutrient, which increases natural capital (e.g. as compost) *Source: Niall Enright, adapted from "The Business Angle of a Circular Economy" by Walter Stahel.*[675]

the material resources and objects that we use in our society - recognizing that the value of a toaster and a Picasso are different!

Of course, the first strategy of resource efficiency is to eliminate the demand for the human services in the first place, which is rarely mentioned in circular economy publications. First, we try to use less, *then* we ensure that what we use has little or no impact on natural capital.

The circular economy concerns itself with those services which we do require, and is oriented to the notion of reducing flows. It is basic common sense.

- The life extension of equipment through better initial design and construction as well as better maintenance;
- Greater utilization of equipment by distributing and sizing the equipment effectively, as well as consolidating the demand for the service to fit the machine availability;
- If the equipment becomes redundant to its owner, then reuse by others becomes a vital part of life extension (another form of distribution);
- Where the equipment has failed and no longer provides its service, then repair and remanufacture represent efficient ways of reducing flow compared to outright replacement;

All of the above represent what Walter Stahel refers to as the product-specific Loop 1 cycle for goods. These goods could involve not just equipment but complex materials such as engine oil for which the notion of refurbishment (filtering, cleaning) apply just as readily as they do for equipment. In the product-specific loop, the challenges are around the design quality of the product (in terms of reliability, durability and repairability of the product) and the way in which its use is optimized. In many cases, this loop is governed by the relationship between the equipment user, the supplier or distributor and the manufacturer.

The second loop is the material-specific Loop 2 cycle, which involves the recovery of molecules and base materials from equipment whose lifetimes cannot be extended or which have become redundant. This is about enabling the processing systems that created the products and materials to receive back the molecules at the end of life and involves take-back systems (sometimes called *"reverse supply chains"*), manufacturers and raw materials producers. Here again, design is an important characteristic in terms of ease of disassembly, choice of materials that can be readily recycled or which can be returned (worst case) to nature in a positive or benign form. Loop 2 strategies include:

- Where repair is not possible then recycling should be used to recover the molecules in the service provider so that new service providers can be created from the old. Recycling here means returning the molecules into their original function.

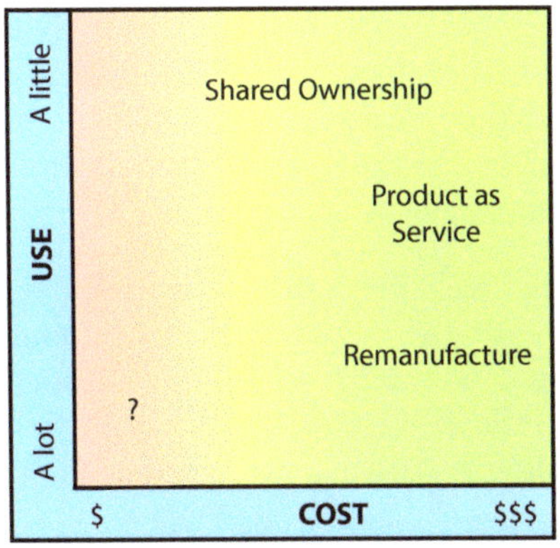

2.14 **New business models in the circular economy have been largely oriented to higher value goods** Some examples of shared ownership are appearing for lower valued and infrequently used goods. *Source: Niall Enright*

- Downcycling, on the other hand, involves reusing the material in other service providers, which may deliver a lower net natural resource requirement (for example, the use of ground green bottle glass in road core rather than in new bottle manufacture, which reduces the need to extract aggregate from natural sources).
- Recovery involves the partial recuperation of some of the embodied natural capital in the material – the typical example being incineration to extract energy or composting to obtain biological nutrients.

The principles above relate to items of equipment or material service providers. Where the service provider is a utility such as energy or water, there is not such a clear demarcation between Loop 1 and Loop 2 strategies, although many of the same principles apply.

- Life extension and utilization of utilities, as for equipment, involves better design and maintenance of the interface between the resource and the process (in other words, the resource-consuming equipment such as heat exchangers, motors, etc.), so that a greater proportion ends up providing the desired service (heating, cooling, motion, etc.).
- Higher utilization of utilities can also be achieved by distributing and sizing the utility effectively. Any energy and resource efficiency practitioner will know that significant losses occur in distribution systems (examples include compressed air leaks, water leaks, network and transformer losses). What is less commonly appreciated is that the *"sizing"* of the resource is also critical to resource efficiency (examples include chilled water delivered at an unnecessarily low temperature to a process, compressed air at an unnecessarily high pressure, an over-sized motor running at part-load). In the utility world, the *condition* of the utility is a significant influence on efficiency – its temperature, pressure, voltage, purity and so forth govern the efficiency of the desired service.
- Reuse elsewhere applies to utilities. Heat recovery systems and water recycling are all examples where a utility is reused for its original purpose. Here, the general notion is that the less transformation in the resource the greater the potential recovery of its service potential (thus recovering heat through a heat exchanger directly into another material flow is better than recovering heat through a steam turbine, which generates electricity but has a lower efficiency (analogous to downcycling rather than reuse). In reality, the process needs often dictate that transformation is the only means of recuperating some of the useful value in the utility.

For energy, we are reminded that the laws of entropy mean that perpetual cycling of energy is not possible as losses and transformation from a more useful form, such as electricity, to a less useful form, such as heat, are inevitable. Indeed the same applies to all materials – there will be inevitable losses which make their eventual replacement essential. Gold is a material for which we practice exceptional levels of *"stock maintenance"*. However, we continue to lose significant quantities each year largely through electronic waste. Thus underpinning the circular economy is also a recognition that the inevitable extraction of resources from nature and their disposal into sinks must be harmless. In the circular economy, renewable resources are essential – in other words, the outer loop with the natural environment must be sustainable forever.

The strategies above clearly relate to the techniques that an organization might use to reduce its resource consumption. Avoiding leaks, maintaining equipment, setting correct temperate setpoints, reusing wastes are all obvious, albeit slightly unglamorous, strategies. Indeed, we shall see that achieving resource efficiency is down to methodically applying the common-sense principles set out above.

However, as one more closely examines the implication of the circular economy we can see that it proposes a much more sweeping change to the business models that organizations will (voluntarily or not) adopt in the future. First of all, dramatic life extension and a large increase in the utilization of equipment such as cars will potentially lead to huge declines in the manufacture of new cars (as intended!). What does that mean in turn for automobile component manufacturers, for car dealers, for road-builders? Clearly, there will be potential winners and losers from a substantial shift to a circular economy.

The thinking today is that the biggest winners from a circular economy are likely to be workers – that is to say that all the repair, refurbishment, reuse activities will require a much greater use of labour. Earlier we mentioned Adam Smith, who identified land (or resources) as the primary source of wealth, but he had two other essential ingredients for prosperity – labour and capital. The first industrial revolution was about maximizing the efficiency of labour through the adoption of technology. In 1790 a farm worker could harvest one-quarter of an acre a day with a scythe; by 1890 a man could harvest 10 acres with a steam-powered threshing machine. The second industrial revolution was about further increasing the efficiency of labour and also maximizing the use of capital through the creation of sophisticated finance systems (which pooled resources via shareholding and created money via debt). Thus coal mines, mills and factories could be developed with debt repaid out of future profits. In fact, the financial markets have grown to become sources of prosperity in themselves, rather than enablers of material wealth creation such as factories. For example, in the foreign-exchange markets only 9% of transactions are for the *"real economy"* and 91% of transactions are as a result of speculation on the movement of price.[55]

Historically there has been little fundamental change in terms of the services we obtain per unit of resource we take from Nature.... at least not in comparison to the improvement in labour and capital efficiency.

The first of Adam Smith's ingredients for prosperity, land or resources, have not been revolutionized in the same way as labour and capital. The massive increases in efficiency, several magnitudes in size (or hundreds to thousands of times), in labour and capital have not been matched by equivalent increases in resource efficiency. Undoubtedly, the yield from agriculture has increased substantially, but when we look at resources as a whole, we have seen little fundamental change in our use of these in relation to the natural services we obtain. We do not see our managers looking at *"financial return per kg iron employed"* – in fact, our systems barely acknowledge resource efficiency as an input cost rather than a fundamental part of the wealth-creation engine.

Our lack of concern for husbanding resource stocks almost certainly reflects the historical decline in resource costs through the 20th century. It seems that technology, new discoveries and human ingenuity have been able to keep pace with demand so effectively that, with the exception of man-made events such as wars or the 1970s oil shocks, prices have almost halved over the century, despite GDP growing from US$1.9 trillion to US$36.6 trillion (an 18-fold increase, in 1990 dollar equivalent terms). See the green line in the chart below.

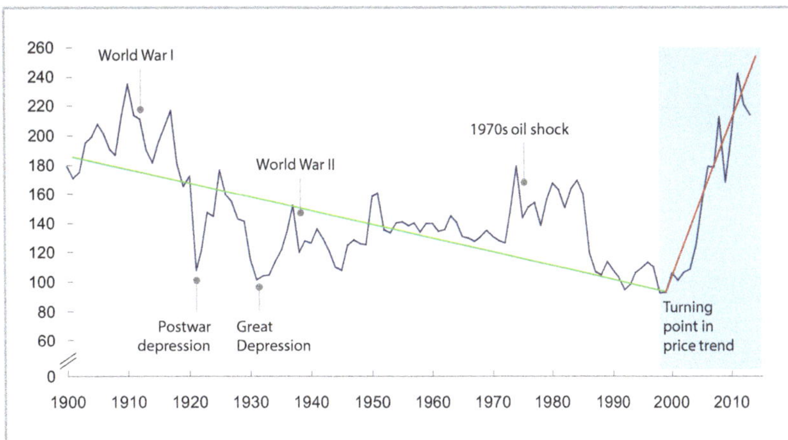

2.15 **Resource prices have increased significantly since the turn of the century. McKinsey Commodity price Index. Retail Price Index 1999-2000 =100.** *Source: McKinsey Global Institute.[215] Based on arithmetic average of four commodity sub-indexes: food, non-food agricultural raw materials, metals and energy. Data: Grilli and Yang; Pfaffenzeller; World Bank; International Monetary Fund; Organization for Economic Co-operation and Development statistics; Food and Agriculture Organization of the United Nations; UN Comtrade; McKinsey Global Institute analysis.*

Since the turn of the century, however, resource prices have soared - wiping out all the gains made in the previous century. The red line in the chart above shows this increase. There are two causes for this rise: supply and demand. On the supply side, new discoveries and technologies are no longer able to keep the cost of supply down, shale gas apart, hence the availability of cheap resource is dwindling. On the demand side, the huge increase in affluence of consumers, mainly in emerging economies, is leading to much greater consumption of natural resources.

Nearly a third of profit warnings issued by FTSE 350 companies in 2011 were attributed to rising resource prices[19]

Clearly, resources will be the critical factor in enabling or impeding the forecast growth in the global economy. Already we are seeing sovereign states buying or optioning huge tracts of land in Africa and South Asia and establishing long-term commodity contracts in a bid to secure future resources. However, the solution is not ownership of limited resources but a transition to an economy that recognizes that these resources are limited and are likely to rise in cost. What is needed is an economic model which can deliver the required improvement in material wellbeing and income, at an unprecedented scale, with much more limited inputs of natural resources and demands for natural sinks – in other words, a circular economy. ⇒ page 70.

2.16 **The aspirations of hundreds of millions of new middle-class consumers, largely in emerging economies such as China, are driving a huge increase in resource use.**
Source: © chinaview - Fotolia.com

Exploration: *Why China is key*

China is mentioned time and again as a key player in terms of future resource efficiency. This is because the scale of development there is so vast.

- In the three years between 2008 and 2010, China emplaced 4.9 Gt of cement, more than the US laid in the *entire 20th century* (4.6 Gt), and between 2009 and 2011 it used even more.[667, p91]
- China will build 1- 2.5 cities the population of Dallas (pop. 2.5m) *every year* for the foreseeable future.[797] Today the US has nine cities with over 1 million inhabitants[785] and Europe has 18.[784] By 2030, China will have 221 cities that size.[797]
- This construction is as a result of a huge mobilization of people taking place. In the 20 years to 2030, China's cities will have added 350 million people to its urban population, which is more than the entire population of the US and the largest migration of people in history.[306]
- The UK needed 154 years to 1850 in the first industrial revolution to double per capita GDP for 9 million people from US$1,300 to US$2,600. It took the US 53 years to 1870 to double the income of 10 million people in the second industrial revolution. China took just 12 years to 2000 to achieve this doubling to a staggering 1,020 million people. India doubled in the 16 years to 2002, starting with 822 million people.[776]
- The economies of the BRIC countries (Brazil, Russia, India and China), which account for 40% of the world's population, are growing at breakneck speed. According to the World Bank data[806] to 2012, China's GDP has doubled in just five years, Russia and Brazil in six years, and India's in seven years. In 2013, China's economy was worth US$9.3 trillion. In another five years, at the present rate, it would be worth US$20 trillion (which is greater than the US's US$16 trillion today) and then by 2025 US$40 trillion, more than double the US in just five years![570] (This compares with GDP-doubling taking 16 years in the US and 18 years in the EU).
- This GDP growth reflects the growing affluence of the Chinese as they quite reasonably aspire to the material quality of life enjoyed by people in the developed counties. One remarkable demonstration of this growing consumerism is *Singles Day* symbolized by the lonely 1s on 11/11, (11 November), which online retailer Alibaba has turned into a massive online sales event. On 11 November 2013, Alibaba's companies had 402 million unique visitors to their sites – more than a third of China's entire adult population - and shipped US$5.75 billion of products, two and half times the equivalent of the *Cyber Monday* sales in the US.[800]

There are powerful forces that are driving this development. There is a pent-up desire for the population to improve their quality of life. Population growth is a factor, but more particularly *mobilization*, which has released hundreds of millions of workers, which has underpinned the expansion of manufacturing. There have been massive direct inwards investments from the developed economies which have transferred large parts of their manufacturing base to China (and also technology, know-how and expertise). The availability of debt to fund this progress is also significant. It is unlikely that these forces will diminish.

The Chinese are gravely concerned about the scale of resource demand from their economy and are taking steps to improve their own resource use significantly. There are many reasons to be positive about the long-term trend in China.

- According to the United Nations Environment Programme (UNEP),[644] improving resource efficiency has been the main brake on materials use. Going back to the Ehrlich Equation (*Impact = Population x Affluence x Technology*), from 1995 to 2005 the factor for T for China (which is a measure of resource efficiency) was -43%, while the impact of Population growth was +13% and the impact of Affluence was +130%. Efficiency really is having a positive impact now and can (must) continue to do so in the future.
- Real improvement is being seen in other indicators. For example, domestic material consumption per $ value created, also quoted in the UNEP report, halved between 1978 and 2003.
- Although it is too early to draw firm conclusions, China appears to be falling out of love with coal. Not only is coal generation losing money but the public is increasingly resistant to new coal generation because of pollution concerns.[818] In 2012, a third of all coal-fired power stations were delayed[137] and for the first time investment in wind exceeded that in coal-fired thermal generation plants,[474] which have halved since 2005.[818]
- The circular economy promotion law in 2009 and substantial investments in circular economy pilots, many of which have exceeded their targets, demonstrates the commitment of the authorities to resource efficiency.[807]
- Between 2000 and 2005, the energy intensity of the Chinese economy improved by 20%[138] and has the potential to match the best in the developed world through the adoption of just 79 key technologies.[604, 138]

It is important that those based in the developed economies do not see the scale of the resource use in China and the other emerging economies as an excuse for inaction on the basis that our impact is trivial by comparison. There are many reasons why it is in our interests to work hard to drive our emissions.

- China (and the other emerging economies) have the potential to bypass our inefficient stages of industrialization and move immediately to the best available technologies. It is very much in our own interests that we continue to develop and showcase these technologies in the developed countries.
- China is not *"separate"* from our economies. In fact, 22% of Chinese emissions in 2004 were because of the demand of developed economies for export goods. We have a responsibility to reduce those emissions we can directly influence to offset those outside our control.
- China has signed up to the 2016 Paris Climate Agreement, ahead of many developed countries, and reports indicate that this is seen as an important way to increase economic activity.[596]

We shall see later that Chinese and other emerging market consumers are among the most environmentally aware in the world. This sensitivity around environmental issues provides further grounds for optimism. Clearly, those organizations that can meet the needs of these Chinese consumers in the most resource-efficient way will have a huge competitive advantage.

By transforming our own industries we can set a powerful example to Chinese organizations.

Real World: *Oil price volatility*

There has been a great deal of coverage about the fall in oil prices in late 2014. Does this mean that the long-term trend in prices is reversing?

While predicting oil prices is difficult, commentators like Citi indicate that the average cost of production in 2020 will be around US$60 per barrel,[150, p39] with many new projects costing over US$80 per barrel. This figure is well above the long-term average costs of oil and so it seems clear that we will have to become accustomed to a world where energy costs are high, and volatile.

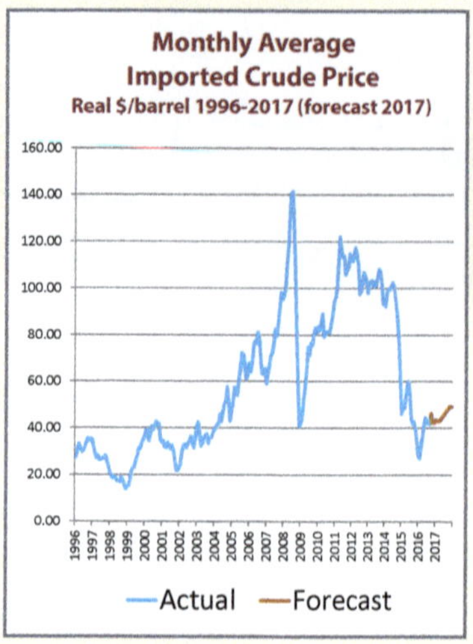

2.17 **Historical and forecast US crude oil import costs** Source: EIA[245]

Proponents of a circular economy sometimes refer to this as the third industrial revolution, which will be based on a massive increase in resource efficiency just as the first and second industrial revolutions were achieved through transformations in labour and then capital efficiency. It is important to note, like previous revolutions, the move to a circular economy is not a conscious act. Nations don't wake up one day and say "*let's have a circular economy*" and all aspects of society change to accommodate this. Circular economy activities have been with us for many decades and there are plenty of examples of successful business models based on conserving resources in operation today – these have emerged as a response to the limits on resources or the demands of customers for more sustainable alternatives. The transition to a predominantly circular economy is expected to arise spontaneously as a result of many different drivers:

- Rising resource costs;
- Increasing consumer concern about the environment;
- Investor recognition that existing business models pose increasing risk and decreasing returns;
- Policymakers' efforts to change market behaviours to reduce risk, maintain growth and deliver stability;
- Innovators and entrepreneurs who see huge opportunity in the disruptive nature of the change and the scale of opportunity;
- Continuing scientific and technological changes which will transform how human services are delivered.

Although the forces driving us towards better resource efficiency appear inexorable, the move to the circular economy will certainly be assisted by specific actions by policymakers. Among these stimuli could be changes to the financial system: for example a shift from taxing labour to taxing resources; the phased removal of subsidies for fossil fuels; the introduction of fiscal incentives for stock management such as adjustments to depreciation allowances that encourage the retention of equipment. Other areas of support could involve early-stage research and development and support for transitioning technologies from pilot to deployment; training and educating the required workforce; new fiscal measures of resource efficiency that provide further transparency to investors in private companies.

Supporters of a circular economy refer to this as the third industrial revolution.

There will be many organizations that will resist the notion of a circular economy because of the threat that this poses to their businesses and investments. Although the boom in primary resource extraction driven by increasing affluence will initially greatly favour extraction businesses, in the longer term the clear losers from a shift to the circular economy will be resource industries such as oil and gas, mining and metals, forestry products, energy and water utilities. But it is not just the primary resource businesses

Resource scarcity is becoming the contemporary evolutionary force for organizations... at the same time, it represents **the biggest business opportunity in a century.**

that will be affected but also a vast swathe of intermediate businesses – the distributors that ship the resources around the world, chemicals companies that convert these basic raw materials into more complex molecules which may be readily recuperated. In fact, it is likely that no sector of the economy is likely to be unaffected.

This process is likely to lead to considerable change in the business landscape as old business fail and new ones take their place. The lifespan of organizations is surprisingly short. Studies of the Fortune 500 largest corporations show that a third of all companies listed in 1970 had vanished by 1983 through acquisition, merger, or by being broken to pieces. The average lifespan of a multinational corporation is between 40 and 50 years.[202] Earlier work on European and Japanese companies indicates that the mean lifespan for all firms, regardless of size, is 12.5 years.[208]

These industry upheavals are not new. The textiles revolution which gave us mass-produced cotton clothing in the first industrial revolution killed off the furs industry, which at the time made John Jacob Astor one of the richest men of all time with a fortune equivalent to US$176 billion in today's terms.[361] The second industrial revolution put paid to the huge whale-oil industry (fortunately, before all whales were killed); the guano and saltpetre industry, which was a important economic sector in the 19th and early 20th century, particularly for South America, collapsed when the Haber-Bosch process enabled nitrogen fixation and hence artificial fertilisers. All these once-booming industries were subject to dramatic shocks as a result of the development of superior substitutes for the natural services that they provided.

Resource scarcity is becoming the contemporary evolutionary force for organizations (alongside other pressures such as globalization and unprecedented advances in technology). The main challenges for existing organizations will be the threat from new entrants to the market that can experiment with leaner business models which are exposed to fewer resource risks that existing businesses. These trends are affecting all sorts of industries – who could imagine a decade ago, for example, that the modern smartphone was going to displace a whole raft of electronics goods: cameras, videos cameras, calculators and portable music players, as well as other products such as dictionaries and maps. From a resource efficiency perspective, the consolidation of multiple functions into one device is clearly a positive development. From the user's perspective, the same service is being provided for in a more convenient package. Now all we need to do is to design smartphones to last longer!

Organizations will need to address this change by using fewer resources themselves *and* by assessing what these changes will mean to their business models and their supply chains. Indeed, little will remain unchanged. By anticipating this change, some organizations will harness the competitive advantage that the transition to the new economy will offer. Far from representing a threat to many organizations, the forthcoming resource revolution represents a source of great value, as the next chapter will show.

A service must be significantly **superior** to the product it displaces.

Real World: *Opportunities presented by products-as-services*

One of the most significant implications of the circular economy is a shift away from owning things to sharing things, as we have seen in the earlier example of car-sharing. If we examine what the car user wants, it is to get from A to B as cheaply, rapidly, safely and comfortably as possible. In order to achieve this objective, ownership is not necessary. In fact, ownership brings lots of hassle; the car needs taxing, insuring, servicing, cleaning, disposal at its end of life and so forth. Shared ownership holds out the promise that other people will deal with all these inconveniences and leave the user to access the service they need at a lower cost.

The classic example of a business model built on providing a service rather than a product is in the photocopiers and printers sector. When copiers were first developed, they were very expensive pieces of technology. Recognizing this, companies like Xerox developed service-based models where the end-user only had to pay for each copy they produced, and Xerox took care of the maintenance of the copiers, replacing them with new or reconditioned units when they came to the end of their lives. Although the initial cash flow for Xerox was smaller than it would have been if they simply sold their copiers, the market was considerably larger and, over time, the returns were greater than for a straightforward sale.

There are many other examples of products-as-service, spanning a broad range of industries.

- Michelin provides the US military with tyres on a *"pay per landing"* (aircraft tyres) and a *"pay per 100 miles"* for vehicle tyres.
- Lyonaisse des Eaux is one of many companies to offer industrial clients *"chauffage"* contracts for their utilities – taking ownership of boilers and distribution networks and simply charging the users for the amount of hot water they consume.
- DESSO leases its carpet to users, retaining ownership of the product. At the end of the useful life, the carpet is collected and 100% recycled into new carpet which is leased again.
- Rolls-Royce jet engines can be purchased on a *"per thrust hour"* basis with ownership, all servicing and eventual disposal/replacement being taken care of by Rolls-Royce.
- Phillips Lighting has experimented on a *"per lux"* service model where the user simply pays for the amount of light they require; this incentivized Philips to provide longer-lasting and high-efficiency lighting solutions.
- Mud Jeans is leasing fashionable jeans for €5.95 a month with any repairs in that time included free of charge. At the end of the year the customer can return the jeans or buy them for another four payments; or they can switch to a new pair of jeans for €10.
- Ecolab Hygiene in Serbia provides a chemicals leasing service for production line lubrication chemicals, which turns the traditional payment per litre of chemicals supplied into a payment per hour of operation of the packaging line. This lowered the cost for the customer, replaced toxic chemicals with environmentally friendly alternatives and increased recycling rates as Ecolab is now incentivized to supply fewer chemicals and reduce waste disposal costs.

Note that these product-as-service models offer the opportunity for the manufacturers to make more money than their traditional sales model. This is because, by retaining ownership of the product, the manufacturer is in a position to keep the value inherent in the equipment or materials when they come to be replaced at the end of their life, as we have seen in the example of Caterpillar.

These service models have to overcome an idea of ownership, which is something that is deeply ingrained in us. Over decades, advertisers and marketers have conditioned us to think that acquisition is linked with happiness. Changing the habits in consumers, in particular, is not easy.

One reason for this is that when one buys a service one is not buying a product, whose attributes are readily discernible, but one is buying a relationship with the service provider. Later, in *Why certainty drives the resource efficiency proposal* (page 185), we shall see why this creates a bias against services, because, generally speaking, products are search goods while services are credence goods which require a greater leap of faith on the part of the buyer before they can commit.

Because of the inherent bias towards ownership, one of the key principles of any organization seeking to develop a product-as-service proposition is that the service must be significantly superior to the product that it displaces. This concept, which also applies to much more straightforward materials substitution, means that the performance of the service must be superior to that of the product in some dimensions: it must be cheaper, more reliable, more attractive and more convenient, and so forth. Ideally, it will be markedly "cooler" as well. These improvements must be considerable to get us to change our choices: an offer based on a like-for-like substitution, or a small incremental 10% gain, will struggle to gain acceptance and overcome the inertia behind the incumbent solution.

Product-as-service providers need to convince customers and consumers that access is more important than ownership. We can see that some consumers, particularly younger people, are already embracing this notion, which underpins the success of services like Spotify, Netflix and Kindle. For users of these services, owning the physical media (CD, DVD or book) is not as important as the ability to access a huge volume of content through a wide range of devices. These services offer some real advantages over ownership: the content is virtually unlimited; content is available very quickly; search and match features enable new content to be discovered; it is convenient and portable and it is much cheaper than buying the physical media oneself.

One of my favourites[160] examples of collaborative consumption is *Pley* which rents expensive Lego® sets (and now many other toys) to 15,000 parents in the US for a modest monthly subscription.

In the music and movie industries, physical goods have been replaced by *"virtual"* goods. The Service itself, entertainment, has remained unchanged. But we have seen many physical goods moving successfully to a service model: from diggers to cars, chemicals to jeans, engines to carpets. These new approaches are springing up, not because of the undoubted resource efficiency benefits they offer, but because they make business sense to both the manufacturers and the customers. It is this additional value that leads to the conclusion that products-as-services are likely to grow significantly and in so doing, ensure much better utilization of the resources that flow through our economies.

2.18 ***Pley* shows there is virtually no market where shared ownership cannot be made to work**

For a monthly fee, members can access over 500 toys in a simple *"Play. Return. Repeat"* process. The proposition is that children will be less bored and parents won't accumulate cupboards full of unloved toys.
Source: Pley Media relations.

2.6 A wider ecology

The circular economy is based on collaboration between individuals and between organizations.

A circular economy pushes organizations to think more carefully about their relationships with customers, distributors and other manufacturers. In the previous section, we saw that shared ownership is leading to *collaborative consumption* by consumers willing to share ownership. The importance of core recycling for remanufacturers means there are new forms of *collaborative distribution* between manufacturers and distributors to improve the return of used products. Manufacturers also need to come together in new forms of *collaborative production* which involve several different manufacturers partnering to optimize the use of resources. Usually, by-products or wastes of one manufacturer are the input into another process. Partnerships don't just have to involve materials, but could reflect the use of shared services or facilities. These groupings are called ecoparks, and there are many these around the world.

2.19 **Kalundborg Symbiosis in Denmark** Over 50 years, 29 major material flows have linked the participants. *Source: Kalundborg Symbiosis,[434] reproduced with kind permission.*

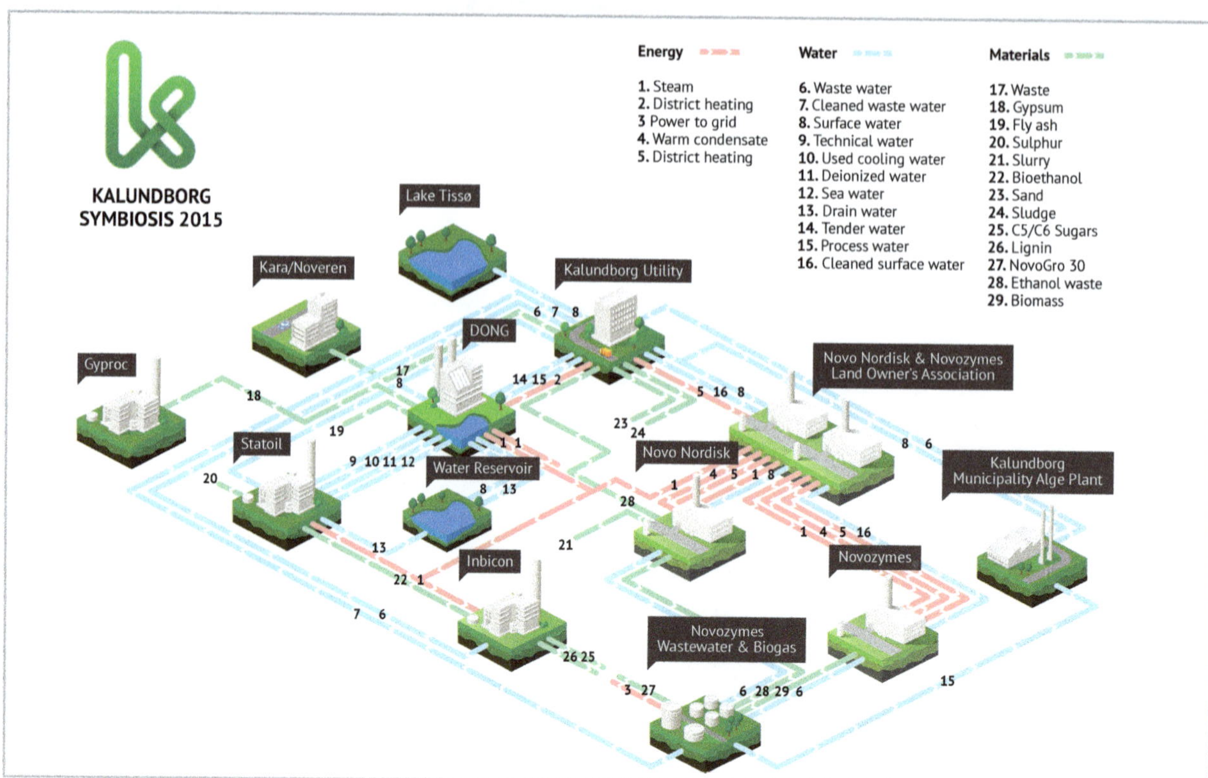

"We make a lot of money on this."

- Niels Christian Kjær, DONG Energy, one of the leading partners in the Kalundborg Symbiosis.

Foundations

One of the oldest ecoparks is the Kalundborg Symbiosis in Denmark, shown by the figure left. In this diagram, we can see how the waste from one facility becomes the input for another. For example, the waste heat from the Asnaes Power Station run by DONG Energy supplies a community district heating scheme (Flow 2 in the diagram), the Novo Nordisk pharmaceuticals plant (1) and the Statoil Refinery (1). In total, there are 29 material and resource flows connecting the partners at Kalundborg. Overall the environmental benefits of the collaboration are considerable:[433]

- CO_2 emission reduced by 240,000 tonnes annually;
- 3 million m^3 of water saved through recycling and reuse;
- Recycling of 150,000 tons of gypsum from desulphurization of flue gas (SO_2) replaces import of natural gypsum ($CaSO_4$).

It appears that the motivator for the development of the ecopark was not environmental performance, but simply good business sense:

> *By all accounts the Kalundborg industrial symbiosis was not designed by consultants or financed by Danish government officials, but rather was the result of several distinct bilateral deals between company employees seeking, on the one hand, to reduce waste treatment and disposal costs, and, on the other, to gain access to cheaper materials and energy while generating income from production residue.*[209]

It was only in the late 1980s that the environmental benefits of the ecopark began to gain attention[136] and the term industrial ecology was first coined.[296] Although Kalundborg has received much publicity it is possible the Kwinana Industrial Area in Western Australia, operating since the 1950s, may lay to claim to being older and larger. Here, in 2000 there were 106 collaborations (68 involving materials and 38 services or facilities) between 21 process industries.[758]

Extending the idea of industrial ecology to nature gives us the idea of *"waste = food"*. Thus the waste that we dispose of, an input to nature, should be a valuable feedstock for a biological process which increases natural capital. In a circular economy, organizations should choose materials for the positive effect they have on the natural environment at the end of their lives.

Collaboration across more complex networks is the inevitable consequence of a move to a circular economy. Clearly, much greater thought needs to go into the design of products, supply chains and processes. The payoff for this increased effort is more rewarding and long-lasting partnerships between organizations. These partnerships will result in lower resource costs, reduced supply risks, differentiation and competitive advantage, better environmental performance and significantly greater value for stakeholders. The evolution of Kalundborg over the years is proof that these forms of collaboration do not impede innovation; indeed, they encourage it.

2.7 The precautionary principle

When an activity raises threats of harm to human health or the environment, precautionary measures should be taken even if some cause and effect relationships are not fully established scientifically.

The English language is rich with aphorisms that warn us to take care before we proceed: "*look before you leap*", "*better safe than sorry*", "*first, do no harm*" and "*prevention is better than cure*". This notion of caution, originating in the German concept of Vorsorgeprinzip (loosely "*forecaring principle*"), lies at the heart of the debate about roles and responsibilities for resource efficiency.

The precautionary principle states:

> When an activity raises threats of harm to human health or the environment, precautionary measures should be taken even if some cause and effect relationships are not fully established scientifically.

The key here is whether there is an obligation to act in the absence of absolute scientific evidence pointing to possible harm. Clearly, if such an obligation exists, then organizations today that are contributing to climate change by, for example, emitting CO_2 to the atmosphere or reducing forests have a duty to take measures to lessen the threat to the environment brought about by their actions. And the same principle applies to other forms of resource depletion, around water or biodiversity and so forth.

The statement of the precautionary principle above comes from the *Wingspread Consensus Statement on the Precautionary Principle*, a conference in 1998 in the US. Thirty-two participants including treaty negotiators, activists, scholars and scientists from the US, Canada and Europe came together, under the auspices of the Science & Environmental Health Network, to discuss the principle.[655] Not only did they emphasize the notion that the principle "*incites us to take anticipatory action in the absence of scientific certainty*", but also that "*in this context, the proponent of an activity, rather than the public, should bear the burden of proof*". In other words, it is up to the proposer to demonstrate that no harm will arise from their proposal rather than for the public to prove otherwise.

No wonder then that this concept is hotly debated. In some jurisdictions like the European Union, the principle has been explicitly incorporated into the Lisbon Treaty as the basis for policymaking on the environment. The principle has also underpinned some international treaties, like the Montreal Protocol, which banned ozone-depleting substances, and the UN Conference on Development and the Environment in 1992. It is a concept that environmentalists support because it speaks to the idea that much harm to the environment cannot be easily undone so a cautious approach is rational.

Just as there are very committed proponents of the precautionary principle, there are also strong opponents. Some arguments against the principle point out that is it far too vague and so would be difficult to apply in law, while others resist the idea that organizations should bear the burden of risks that are not yet fully proven or believe that the slavish adoption of the principle would paralyse innovation and progress.

What is clear is that the precautionary principle undermines the strategy of climate-change deniers who deliberately cherry-pick uncertainties in the science to convey the idea that human-induced climate change is merely an opinion. If the precautionary principle is applied, then the presence of doubt or uncertainty cannot be used as the basis for inaction. No wonder then that Rex Tillerson, now US Secretary of State, ex-CEO of ExxonMobil, an organization which is known to have actively supported lobby groups which brief against climate change,[9] is so opposed to the principle. In a recent article on shale gas extraction by Fortune and CNN he is quoted as follows:[568]

> *"What's happened is the tables have been turned around now to where we have to prove it's not going to happen," he says. "Well, that is a very dangerous exchange to get into because where it leads you from a regulatory and policy standpoint is to govern by the precautionary principle. And the precautionary principle will absolutely undermine the economy."* He adds, *"If you want to live by the precautionary principle, then crawl up in a ball and live in a cave."*

This emphasis on the proposer having to prove the absence of harm, rather than the opponent having to prove harm, is what makes the precautionary principle so significant. Any organization operating within the precautionary principle should establish positively that its actions will not harm, rather than proceed as if this is the case in the absence of data to the contrary. This shift in the burden of proof should enable us to make better-informed decisions around issues of resource use by explicitly requiring us to examine the potential for harm in circumstances where there is doubt about the science.

That is not to say that the critics do not have some valid points about the precautionary principle. If the principle were applied slavishly to every decision, then it would certainly be an excessive burden on progress, so it needs to be used only where the risk of harm is believed to be great. In other words there has to be some scientific basis for the existence of significant danger, a threshold where it is triggered. Rex Tillerson's fear-mongering characterization of the precautionary principle as a destroyer of civilization does no favours to businesses as a whole, or for that matter to the levels of trust that stakeholders have in ExxonMobil.

The precautionary principle ***protects and enhances value for shareholders.***

In fact, the precautionary principle as a common-sense approach to risk, with the kind of legal nuances set out above, protects and enhances value for shareholders as much as it protects the environment. What would the value of the asbestos industry be today if the precautionary principle had been applied? Hundreds of companies make products much, much more inherently

Real World: *Courts and the precautionary principle*

The precautionary principle is being enshrined in law in a wide range of jurisdictions. The effectiveness of this will depend on the interpretation that is given through judgements and case law.

For example, a recent effort in the US courts to prevent the Large Hadron Collider being switched on because of fear that it would create an artificial black hole that would devour the world was dismissed because the plaintiff could not demonstrate a "*credible threat of harm*".[736]

These judgements are difficult. On the one hand, society has a need to be protected and on the other hand, there is also a need for development. In general, it seems that the courts can balance these requirements in a way that makes the precautionary principle a sound basis for decision-making. There is an excellent commentary on these various rulings by the Chief Judge of the Land and Environment court in Australia, Justice Brian Preston, in his presentation to the Law Society of New South Wales in 2006.[603] Particularly useful is the explanation, on page 15 forwards, of the Australian court's decision in the case of *Telstra Corporation Limited v Hornsby Shire Council*,[562] which is one of the most comprehensive judicial expositions of the practical application of the precautionary principle.

The case in itself was not particularly exceptional. A developer had requested permission to install a mobile telephony base station on the roof of a local sports club. An objection was submitted by the local council due to concern that the station would emit harmful electromagnetic energy, using the precautionary principle. In the end, the case was lost by the objectors on appeal because "*the claimed effects are unsubstantiated and without reasonable evidentiary foundation*".

It was this appeal that enabled the precautionary principle to be clarified further. The ruling effectively stated that there are two preconditions for the application of the precautionary principle: a threat of serious or irreversible environmental damage and scientific uncertainty as to the environmental damage. Clearly, if there was scientific certainty then preventative action would inevitably need to be taken, so the precautionary principle would be irrelevant. The ruling went on to describe that the evidence of harm needed to be based on data (not just a hypothesis or speculation), that unanimous scientific support for the scenario envisaging harm is not required, and then describes the level of uncertainty that is needed for this evidence. If the two preconditions, threat and uncertainty, are met, then two further things flow: there is a shift in the burden of proof as the decision-maker must now assume that the threat of serious damage is a reality – putting the onus on the proponent of the development to prove otherwise. The second is the principle of preventative anticipation whereby action can be taken to prevent the threat before the reality and seriousness of the threat are entirely known. In assessing what steps to take, the risks associated with various alternative options should be considered, and prudence would suggest that a margin of error should be applied in favour of the environment where there is uncertainty. Importantly, the action taken should be sensible and proportionate to the risk identified.

The interpretation of the precautionary principle within national jurisdictions is now well established. A challenge remains its application in international trade agreements, where there are diverging views from the US and EU, for example.

If the two preconditions, ***threat and uncertainty,*** *are met, then: 1) there is* ***a shift in the burden of proof from the objector to the proposer*** *and 2)* ***action must be taken to mitigate the risk.***

*What destroyed 98% of shareholder value in the **asbestos industry** was denial of the threat posed by the product, resulting in ineffective, late, complacent and reactive approach to scientific uncertainty.*

Foundations

dangerous than asbestos – explosives, poisons, radioactive materials – and their businesses prosper to this day. What destroyed 98% of shareholder value in the asbestos industry was the denial of the threat posed by the product, resulting in ineffective, late, complacent and reactive approach to scientific uncertainty.[657] The antidote to this denial is the precautionary principle, which forces executives to examine the implications of decisions which could lead to great, but uncertain, harm. Surely this examination can only be in the interest of shareholders. If it can be demonstrated that no harm arises, then the business can proceed with its plans – on the other hand, if damage does occur, it is in the interest of the shareholder that this potential risk is reduced.

One of the most significant regulatory examples of the precautionary principle can be seen in the REACH (Registration, Evaluation, Authorization and Restriction of Chemicals) regulations in Europe. These regulations require producers (and importers) of chemicals to register every chemical they produce and assess its toxicity. Effectively what this regulation - and the precautionary principle as a whole - does is to shift the burden of ensuring public safety from the government to businesses. Rather than regulators having to prove potential harm from substances, companies have to prove that their products are safe and if there are toxicity or other issues, that the benefits outweigh the risks.

In thinking about how the precautionary principle applies to our organizations, we could fall into the trap of thinking of it as a purely legal issue. In practice, it is much more. Our attitude to future risk will determine whether we emerge positively from the inevitable change that is coming. There clearly is a lot of uncertainty about this change, but the precautionary principle tells us that uncertainty is no excuse for inaction.

In the first chapter of this book, we described the very considerable risks posed by our use of resources. In this and the following chapters, the focus has changed to opportunities and value. As a decision support tool, the precautionary principle should also be about the positive choices that resource efficiency presents:

> When offered a value-adding choice that reduces threats of harm to our organization, a precautionary approach suggests that we should proceed even if the value or harm cannot be fully quantified at present.

In other words, our duty to our stakeholders requires us to act on the opportunity. That is not to say that every investment in resource efficiency, however slender the return, should be favoured. The general principle that resource use will be a determinant of future success does, however, hold true and so action to reduce reliance on resources and to mitigate the harmful effects that arise are very much in the interest of the organization and its stakeholders. In the next chapter on value, the scale of the benefits of resource efficiency will become clear.

Real World: *Arguments for action on climate change*

The precautionary principle provides us with a very compelling argument to mitigate the environmental risks we confront in the absence of absolute certainty of the threat. One illustration I have used on many occasions with audiences who are not yet convinced of the reality of climate change is reproduced below.

2.20 **A rational assessment of the risks around climate change concludes that urgent and effective action is needed, despite there being some uncertainty about the precise consequences.** In this table we must select between option A, shown on the third row, and option B on the fourth. *Source: Niall Enright. This image is available in the companion files pack.*

		Dangerous Climate Change is.....	
		Not Real	Real
Our Response to Climate change is.... {Our Choice}	A. Ineffective	**That was lucky** Although we did not respond effectively, there was no problem in the first place. Life continues as today.	**We have blown it** By not responding to the threat we have allowed a catastrophe. Life as we know it ends.
	B. Effective	**That was good nevertheless** Our response has created a better environment, considerable cost savings for business, and less dependency on fossil fuels. Life continues better than today.	**Thank goodness** By responding effectively, we have avoided a potential catastrophe, and we have the benefits of resource efficiency. Life continues better than today.

The important thing to convey in using this illustration is that we can only influence which row we choose (i.e. whether to make an effective response or not). The columns represent what we are uncertain about (whether climate change is real or not). Clearly, we want to avoid the red box – where life as we know it comes to an end. Thus the only rational choice is "B", to take action on climate change, despite uncertainties about the consequences of inaction.

The reader will note that there is no *downside* portrayed in the bottom left box, where dangerous climate change is not real, but we have nevertheless substantially transformed our organizations to reduce emissions and adapt to rising temperatures. While individual businesses, such as those based on the unabated combustion of fossil fuels, may well see a substantial reduction in their value unless they change the core business model, the majority of organizations will gain from resource efficiency to address climate change. That is because using less energy and creating less waste reduces costs. Delivering more efficient products will provide a competitive advantage. The new technology gold rush to mitigate carbon emissions will create countless business opportunities and thousands of jobs. Anticipating rather than reacting to regulation will create greater degrees of freedom for business operations extending, rather than diminishing, their licence to operate and innovate.

2.21 **What if it is a big hoax? Even if the effects of climate change are not as serious as thought, action to reduce fossil fuel use will bring many benefits.** This reinforces a precautionary approach to emissions since the preventative response is sensible and proportionate to the threat (and in fact brings positive benefits). *Source: Niall Enright, drawn using Pixton.*

Foundations

Summary:

1. Our impacts on the natural environment arise from a series of transformations that convert natural capital into goods that provide us with services. Herman Daly describes natural capital as the source of all services and hence all wealth.

2. Some of these transformations do no harm. Others deplete the natural resources other living organisms require. We need to examine the full life cycle of our resource use to assess our impact, including the return of materials to the environment at the end of life.

3. Natural capital and financial capital are not interchangeable as in most cases there is no substitute for the natural capital.

4. Some goods which provide us with services, such as cars, are very poorly utilized. New business models based on shared ownership offer a way to improve utilization dramatically, and so reduce the amount of goods needed.

5. For a service to displace a product it usually needs to be significantly superior in several respects.

6. Buying a products-as-service involves entering into a long-term relationship with the supplier. It is a *credence* sale, based on trust, and so the supplier needs to adjust their offer to build the confidence of the customer.

7. Generally speaking, reducing resource use at the final point where it provides a service leads to greater benefits as the cumulative inefficiencies and damage done by all the intervening transformations from the original natural capital are eliminated.

8. The 8 Rs in the waste hierarchy guide us on the relative merits of different improvement strategies: remove, reduce, re-source, remanufacture, reuse, recycle, recover and return.

9. In the circular economy, the aim is to extend the lifetime of service-providing goods and at the end of life to remanufacture, reuse, and recover parts and materials.

10. Reverse supply chains or core collection systems are key to remanufacture.

11. Remanufacture business models have the potential to generate significantly greater profits that linear manufacture.

12. Design is the key skill in the circular economy.

13. Resource costs have increased significantly in the last decade, wiping out all the gains of the 20th century. This price increase is largely as a result of the demand of the emerging economies, coupled with more expensive production costs. How organizations respond to this scarcity and cost will determine their fortunes in the 21st century.

14. Meeting the legitimate aspiration of China's people for a better quality of life has led to unprecedented levels of resource use. China's policymakers and citizens recognize the need for sustainable development, and it is in the interests of organizations in the developed economies to do all that we can to support this aspiration.

15. Resource scarcity can shape organizations and industries. A move towards a circular economy will happen, not because of government edict but because it makes business sense.

16. In a circular economy, there is a much greater level of collaboration between organizations in the supply chain. Collaborative production becomes the norm where the inputs to one organization's process is the waste from another's.

17. The precautionary principle states that if a proposed course of action could cause harm, then the burden of proof falls on the proposer to demonstrate the absence of harm and to take measures to mitigate the possible damage. The precautionary principle enhances shareholder value by reducing the risk of future catastrophic liability.

Further Reading:

Allwood, J. M. et al. *Sustainable Materials — without the hot air* (UIT Cambridge Ltd, 2015). An outstanding and highly accessible exploration of the potential and limits of material resource efficiency. 2nd Edition ISBN 978-1906860301.

On new business models: McDonough, William and Braungart, Michael, 202 *Cradle to Cradle - remaking the way we make things* North Point Press. ISBN 978-0-86547-587-0 or Gatnsky, Lisa, 2010 *The Mesh - why the future of business is sharing.* Portfolio Penguin. ISBN 978-1-59184-430-3 (pbk.)

On the circular economy: Heck, Stefan and Rogers, Matt, 2014, *Resource Revolution - how to capture the biggest business opportunity in a century.* Melcher Media. ISBN 9781477801192 or Ellen MacArthur Foundation, 2013. *A New Dynamic - effective business in a circular economy.* ISBN 978-0-9927784-1-5.

Randall, Alan, 2011. *Risk and Precaution.* Cambridge University Press. ISBN978-0-521-75919-9, covers the concept of the precautionary principle very well.

Questions:

1. What are the advantages of "product-as-service" business models? What kind of goods are suitable? Are there any disadvantages or barriers to their adoption?

2. Using the waste hierarchy, describe how the 8Rs apply for a specific resource such as form of energy, water or a raw material.

3. Describe the precautionary principle. How are legislators ensuring that it is applied in a common-sense way? Why can it enhance shareholder value?

4. Can resource efficiency be considered "the business opportunity of a century"?

5. Explain the differences between the following pairs of words: recycled and recyclable; biodegradable and biobased; renewable and recoverable.

6. What is industrial ecology? Give examples successful ecoparks and the barriers to their development in the US (see page 155).

7. Consider the *Greenhouse Gamble*™ game from the MIT Joint Programme on the Science and Policy of Climate Change (illustration and link left).

a) What do these wheels show? Why are the sections on the wheels different sizes? Does the game help people understand risk?

b) In the NO POLICY option, is there an outcome where warming is limited to 2° C? How have the outcomes changed over time (see http://globalchange.mit.edu/gamble/more/comparison)?

2.22 ***Greenhouse Gamble***™ **is a game designed to help the public appreciate the risks of climate change, with and without policies to prevent warming**

Source: Massachusetts Institute of Technology Joint Programme on the Science and Policy of Climate Change. Available at *http://globalchange.mit.edu/focus-areas/uncertainty/gamble.*

3 Value

From the preceding chapter, we can see that the social, economic and environmental benefits of resource efficiency are not in doubt. However, if organizations are to act on these, they need to internalize these wider benefits and so justify, to their various constituents, why effort and money should be devoted to improvements in resource use.

Central to this call to action is the notion of value. This includes what we all traditionally think of as value: money, or financial value, in the form of cost reduction or higher asset or share valuations. Value, however, also includes many other desirable features such as reduced risk, enhanced service delivery, competitiveness, improved stakeholder engagement, wider societal good and many other aspects which all contribute to the organization's ability to achieve its core mission.

Value, in our context, is best understood as the increase in an organization's capacity to achieve its primary goals, arising from improved resource efficiency.

Value is shorthand for the increased capacity for an organization to achieve its goals that is brought about by improved energy and resource efficiency.

This notion of value is entirely compatible with the aspirations of public sector organizations, whose mission is to serve a particular community. In a university, for example, value is anything that can improve the quantity and quality of teaching or research outcomes. Resource efficiency can deliver this value by releasing additional funds for education, or by improving the comfort of the spaces in which learning takes place, or by making the institution more attractive to students and funding agencies. If an organization can demonstrate that resource efficiency supports its core mission - whether to make profit or to serve - then it will be able to justify efforts in this area.

This chapter illustrates the many sources of value that are available and how these can be quantified. We shall see that organizations initiate energy and resource efficiency programmes for a variety of reasons and that the primary driver for action can change from individual to individual or unit to unit.

Where energy and resource efficiency are closely aligned with the organization's core mission, the commitment to improvement is more certain. Every successful resource efficiency programme starts by defining the value it brings and aligning this to the core objectives of the organization.

3.1 Types of value

We can describe value in the form of a pyramid which connects an organization to its customers, regulators and shareholders.

Why would shareholders contemplate a resource efficiency programme? What arguments would persuade them to act? What is the rational basis for this investment in a profit-driven, short-termist economy like ours? Why should we preferentially invest in this activity rather than the myriad other investment choices we have? Shareholders - or rather markets - are not, unfortunately, moral creatures. Arguments like "*it's for the good of the next generation*" don't wash here. Understanding the way that markets work and what drives organizations make certain decisions is, therefore, important if we are to build a case for action.

It is clear that our current focus on short-term, immediate value - as well as a political process in many democracies geared to constant re-election - is leading to destructive behaviours by some organizations and institutions. Capitalism and politics often fail to achieve their stated objectives – the maximization of value. Executives make decisions on behalf of shareholders who reject longer-term value creation in place of immediate profit maximization, in some cases seeking improvement over a period as short as a quarter. Yet this behaviour is absolutely against the interest of the shareholders as it will lead to stranded assets, devalued brands, lower competitiveness and greater risk. Eloquent voices are articulating a need to shift to a new form of capitalism, *sustainable capitalism*,[309] which takes these factors into account, using the maximization of value as the basis for change. Even insiders, such as Andrew Haldane, Executive Director for Financial Stability at the Bank of England, are setting out the case,[343] with substantial supporting evidence, that this short-termism, a recent phenomenon, is making financial markets act irrationally. We have seen in the introductory chapter that our pursuit of immediate profit to the detriment of longer-term value is totally irrational from a global perspective and equally foolish from an investor's perspective.

Resource efficiency is a sound investment choice. It can usually take place in a sufficiently fast time frame and with lower opportunity costs and greater returns than many alternative investments.

Notwithstanding the need to overhaul our markets and our politics, we can nevertheless present a compelling case to act on resource efficiency even in the current situation, by demonstrating that value creation can take place in a sufficiently fast time frame and with lower opportunity costs or greater returns than other competing investment opportunities. We shall demonstrate that resource efficiency is a sound investment choice; in fact, it is an essential choice. The core argument we shall develop for resource efficiency can be summarized as "*achieving improved returns by better meeting the needs of three constituencies: shareholders, legislators and stakeholders*". This insight has led organizations like

3M, GE, AkzoNoble, Interface and others to commit so wholeheartedly to aspects of resource efficiency. Not because this is a moral decision (although at the level of individuals and even organizations, morals do matter), but because it makes business sense.

The core notion underpinning capitalism is that funds will be preferentially attracted to those investments that provide the greatest return. In this chapter, we shall demonstrate that accurately quantified resource efficiency represents the one of the most attractive options open to investors and so is entirely consistent with the capital allocation priorities of most organizations.

We can visualize these compelling arguments for action in the form of a pyramid. In a private sector business, the shareholder interests would be placed at the top, emphasizing that the delivery of shareholder value (in the form of dividends or share price) is the core mission of the business. The base of the pyramid is formed by legislators and customers who can determine the success of the business' value creation based on the manner in which the organization responds to their needs. It is from this base that the organization can create value, so it is fitting that this is the foundation of the pyramid. Finally, at the centre, we have disclosure, which connects the three constituencies and drives the interactions between them.

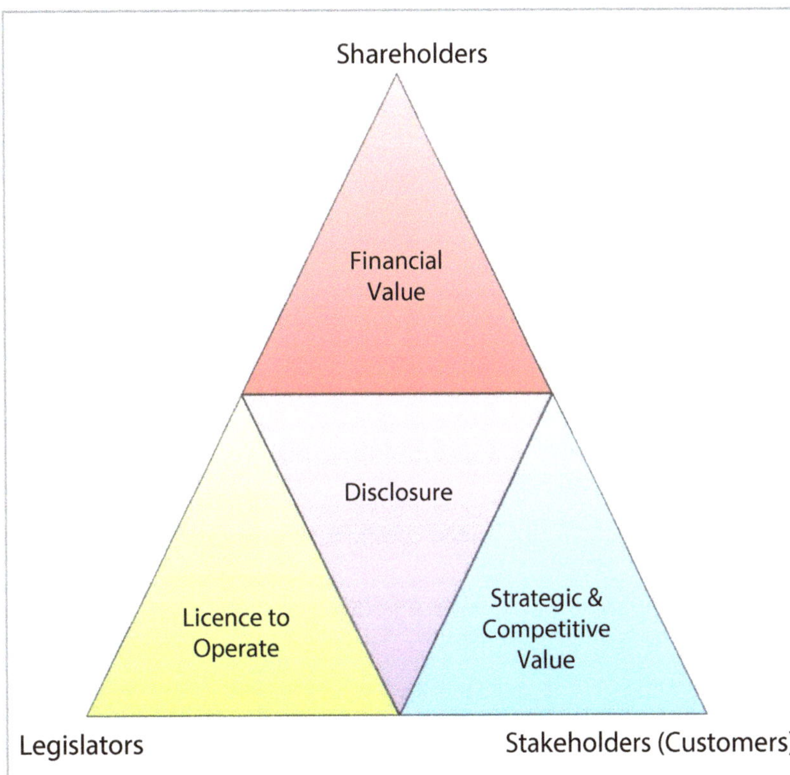

3.1 **Financial arguments for resource efficiency programmes in private sector companies tend to focus on enhancing shareholder value through increased profit generation, as well as engagement of legislators and stakeholders, accompanied by effective disclosure, to increase competitive advantage.**
Source: Niall Enright

In a public sector organization, the nomenclature is somewhat different – but we still have the same three axes: financial performance, licence to operate and the satisfaction of stakeholder needs. The rate of change and competition in the public space makes resource efficiency just as important as in the private sector – but the dynamics and drivers may be different.

If you imagine the pyramid for a publicly funded hospital, the pyramid would appear as shown below, where we have now set the needs of patients, aka stakeholders, at the top. However, as in the case of the private sector organization, the ability to deliver this core objective is determined by the base of the pyramid. Value is created by the efficiency with which the organization uses the resources of the *shareholder*, i.e. the organization or body that pays for the services (such as the taxpayer) while satisfying the needs of regulators.

As in the private sector, these three elements are bound together by disclosure. We shall see that increasing transparency, through increased stakeholder interest in sustainability, as well as more formal mandatory reporting requirements, are very powerful drivers of change. The legitimate interests of these three key constituencies in an organization's performance around resource efficiency, coupled with these much greater reporting requirements, is forcing organizations to see resource efficiency as a requirement of good governance.

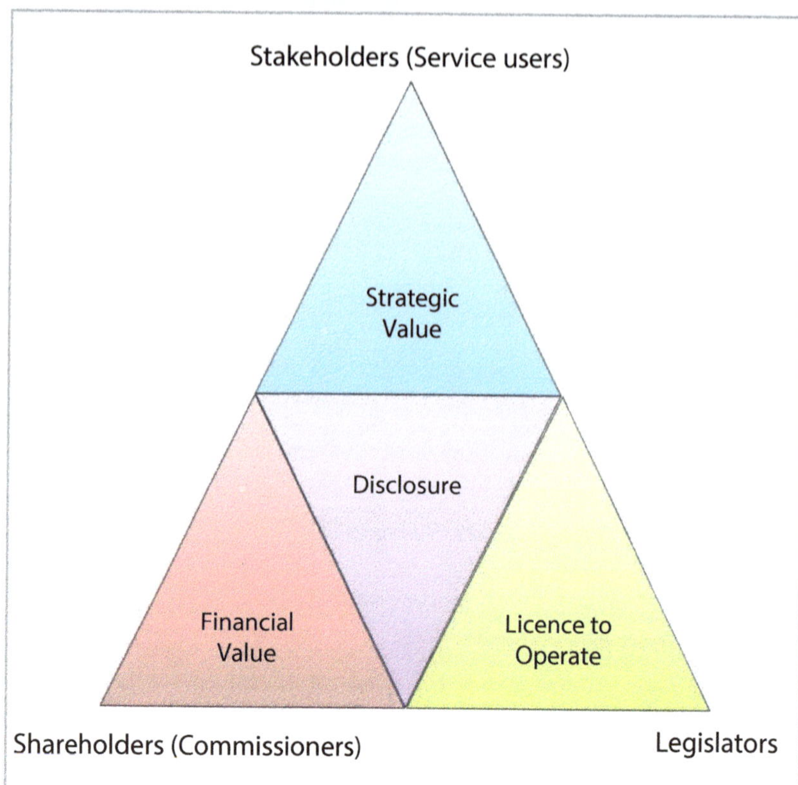

3.2 **Rational arguments for a resource efficiency programme in public sector organizations tend to speak to the way that resource efficiency can help the public body better deliver its core services. However, the needs of the commissioning agencies and taxpayers (aka the "shareholders") as well as legislators are important.**
Source: Niall Enright

Real World: *Executive drivers*

Many surveys on the subject of sustainability or energy efficiency highlight the importance of value creation as the basis for corporate action.

The American Council for an Energy Efficient Economy undertook a survey on energy efficiency in 48 large companies (ranging from US$8 billion to US$99 billion turnover).[605] This study showed that the main drivers of action were a corporate commitment to reduce carbon (45% of respondents), followed by cost reduction (32%), both of which have a strong value basis for action. Interestingly, the anticipation of legislation was rather low on the list of motivators – perhaps reflecting the disinclination of US legislators to tackle emissions at a federal level.

Another recent survey by Accenture[460] asked what motivated CEOs to take action on the broader topic of sustainability, and came to similar conclusions – the preservation and enhancement of value are far more important than regulatory drivers at this time. For these CEOs, the single most important factor was "Brand, trust and reputation" (72% of respondents), followed by "Potential for revenue growth/cost reduction" (44%). Interestingly "Personal motivation" scored third at 42%.

Surprisingly the Accenture survey identified that "Pressure from investors/shareholders" was not seen as a strong driver, despite the many efforts to introduce sustainability performance into the assessment of share value. Only 12% of respondents cited this as a factor driving action on sustainability issues. This disconnect between investors and performance is a real concern and is discussed in the next chapter on Barriers.

Surveys can provide some indication of the relative importance ascribed to each of the four drivers for resource efficiency (see left). From these we can see that there is a strong appreciation of the need to reduce carbon emissions (which in many markets have a direct financial as well as a reputational impact), to address brand and to reduce operating costs. At senior levels, the link between resource efficiency and the organization's core objectives of delivering value (profit or service) is understood.

This emphasis on value does not mean to say that the stakeholder and legislation influences are not appreciable. Many organizations understand that long-term value can only be created when both the needs of the shareholder and the needs of the stakeholders and legislators are met. By stakeholders I principally mean customers, but also include suppliers, the local community, employees, non-governmental organizations (NGOs) – any external entity that has an interest in and potential to influence the business. Among the champions for this *hard-nosed* mutual benefit has been Michael Porter at Harvard Business School with the concept of sustainable value.[600] This presumption of aligned benefit (organizations and society) provides the rational justification for action on resource efficiency and sustainability – in fact, it implies that *not acting* on the needs of stakeholders represents a failure of duty in respect of the shareholder as it will ultimately destroy value.

It seems self-evident that if a supplier sells a product that harms his customers, they are not likely to stay in business for long. Despite a few obvious exceptions to this premise, such as alcohol and tobacco - where there are deeper physiological and social aspects at work - the avoidance of harm to customers is generally recognized as a sound principle to adopt in order preserve value.

Of course, many businesses do not set out to create products that lead to harm – quite the opposite – but they nevertheless have the potential to damage as a result of the impacts of their own activities or their supply chains. In particular, many businesses have come to believe that their emissions of greenhouse gases, primarily CO_2, are seen by their customers as a legitimate concern. Where the company is facing the end-consumer these reputational aspects are becoming powerful motivators for action in their own right, but, as we shall see later, this influence is also extending to companies earlier in the supply chain.

From surveys of executives (such as those quoted left), it seems clear that the most significant rational argument for undertaking resource efficiency in the private sector is the financial value achieved by cost reduction combined with meeting the needs of stakeholders and shareholders. In the public sector, there is less data on the motivation for action and many institutions no doubt, given their service ethos, would have little disagreement about the mutuality of interest between the organization and its stakeholders. However, public sector bodies still operate in a market (they have to attract funds and clients and are exposed to reputational risk) and they run a profit and loss account which makes them as interested in cost as any private sector organization.

3.2 Direct cost savings

The most obvious, and often the most compelling, argument for energy and resource efficiency is that reducing waste saves money, which creates financial value.

Resource efficiency can deliver financial value in many ways: from direct cost reduction, greater reliability, lower capital equipment requirements, lower maintenance, lower shipping costs (less packaging and weight), improved sales, lower risk (e.g. compliance and supply-chain risk), enhanced asset values, brand or share values. These benefits, taken together, can provide a compelling argument for action on resource efficiency.

The most immediate benefit of energy and resource efficiency is the direct cost savings that arise when waste is eliminated. Every pound, yen or dollar saved goes straight to the bottom line and will enhance share value or allow for further service delivery. In the later section, on discovery, we will describe the processes of quantifying cost savings from resource efficiency at the site or facility level. However, it is helpful to consider what the scale of direct cost savings is for a variety of industries.

The US Department of Energy has developed a series of energy footprints for different sectors of the US industrial economy.[219] These studies considered the flows of energy to and within factories in the US and concluded that, across all manufacturing industries, only 40% of primary energy used actually went into useful work. In 2006, of 21,972 TBtu (6.44 TWh) of primary energy used for manufacturing only 8,701 TBtu (2.55 TWh) (39%) were usefully used in process energy – the rest were either lost or used in non-process activities.

The first major source of losses is in electricity generation, where a staggering 68% of the primary energy is thrown away as waste heat in coal-fired power stations. No other industry in the world would be able to operate on the basis of consuming a finite resource and then throwing away almost 70% of its value as waste, but that is exactly what generation based on fossil fuels is doing in the US (and in many other countries).

*The most immediate benefit of energy and resource efficiency is the direct **cost savings** that arise when waste is eliminated.*

I hear the protests that this is not the responsibility of manufacturing businesses, or other large organizations, as they simply have to tie into the available energy supply. Well, that is correct – up to a point. Many organizations, not just in the manufacturing sector, can install on-site generation which in the form of combined heat and power or co-generation can reuse the heat that is otherwise thrown away and so achieve efficiencies greater than 80%. Distributed generation, which brings generation at a smaller scale to locations which can use the heat, is seen as a key strategy to reduce carbon emissions – and the size of the systems available makes it suitable for a broad range of facilities – not just factories, but large offices, hotels, etc.

If we exclude the off-site generation losses, what proportion of the energy that arrives at a US manufacturing facility gate actually performs useful work? The answer is just 43%, averaged across all industries in 2014 (which is worse than the 2006 MECS data figure of 56%, although there may have been system changes - such as greater on-site generation - which may invalidate direct comparison).

3.3 **Proportion of energy entering US factory gates that performs useful process work, compared to non-process work**
Source: US Department of Energy, Advanced Manufacturing Office, Energy Footprints for a range of industries based on 2010 data adjusted to 2014.[219]
The spreadsheet model used to calculate this table is available in the companion file pack.

	Energy by end-use in US industry		
	Process Energy	Other Energy	Losses
Aluminium and Alumina	44%	4%	51%
Cement	56%	1%	43%
Chemicals	47%	5%	48%
Computers, Electronics, Appliances	27%	29%	44%
Fabricated Metal Products	40%	15%	45%
Food and Beverage	31%	9%	61%
Forest Products	21%	5%	74%
Foundries	44%	11%	44%
Glass	41%	5%	55%
Machinery	28%	27%	45%
Petroleum Refining	65%	1%	34%
Plastics	42%	13%	44%
Iron and Steel	45%	6%	50%
Textile	30%	11%	59%
Transportation Equipment	28%	24%	47%
All industries	43%	6%	51%

Some of the energy that arrives at the factory gate goes into *"other"* activities – for example, for automotive manufacturers the facility needs to have lighting and heating in the assembly halls; this energy is not going directly into the product, but it is nevertheless necessary for the production to take place. Undoubtedly there are offices, a canteen, security cabins and myriad other consumers of energy which contribute to the *"other"* figure. One approach to resource efficiency in industry would be to challenge this non-process use and to see if those peripheral consumers can be eliminated. Later on in the techniques section on data analysis (page 460) we can see how regression analysis, a statistical technique, can help quantify this non-production energy use and target its reduction. If your industry is one of those that is listed in the table above then a good starting point to assessing the potential for energy efficiency would be to look at the energy footprint for your sector.

Real World: *An illusion?*

A number of critics of government support for energy efficiency, largely economists or politicians outside the energy efficiency space, claim that the US has already achieved its potential for cost reduction and therefore incentives to stimulate greater levels of efficiency are misguided.

Amory Lovins, a long-time proponent of energy efficiency, systematically rebuts this assumption in his 2007 paper, *Energy Myth Nine – Energy Efficiency Improvements Have Already Reached their Potential*,[483] which provides concrete evidence that innovation, design, policy changes and renewable energy sources together provide tremendous potential for further savings.

> "Indeed a closer examination suggests that the potential for energy efficiency is actually growing because of the following four factors: breakthroughs in energy-saving equipment and in integrative design, better marketing and advances in transportation".

This message of increasing potential with time reinforces the view that resource efficiency should be a continual improvement process, where opportunities that might not be cost-effective one year become so as a result of innovation and market changes (such as rising resource cost) and so need to be regularly re-examined.

Amory Lovins famously developed the concept of the Negawatt, the idea that a "negative watt" (i.e. a reduction in energy use) is a tangible resource which is cheaper and more rapid to deploy than an actual watt of new energy supply.

In these footprints, the "*losses*" represent the energy that did not do useful work, which falls into two broad categories. The energy lost in a distribution process, such as compressed air and steam leaks; and the energy lost in a conversion process, such as the waste heat generated as electrical energy is converted to motive energy in a motor, or to light in a lamp. In fact, the losses stated in these footprint reports would be higher, but the authors were not able to quantify the amount waste heat that came out of the processes themselves.

Also, these footprints did not fully consider the energy lost due to unnecessary work. For example: leaving conveyors running with nothing going down the line; running two boilers on part-load when one boiler on full load will do; forgetting to switch the lights off; setting the temperature too high. These operational wastes are in additional to the waste that arises from the equipment efficiencies themselves and can represent a sizeable percentage the factory's energy use. Unnecessary losses also stem from design decisions. For example, if a building is poorly insulated additional cooling or heating might be needed throughout its whole life. The poor design of a part could mean that there is a large offcut of metal when it is stamped out, with wasted energy tied up in the discarded metal, even if it is recycled.

So we can see that there is a theoretical potential for energy efficiency in manufacturing of *at least* 50% if you could eliminate *all* those losses. Are these savings plausible? If we look at the case studies in *Real World: Energy efficiency savings claimed* (page 94), it certainly seems to be the case that, when they put their minds to it, organizations *can* achieve very large improvements. Numerous other reports, cited in Table 4.1 on page 153, confirm this potential.

So, assuming that significant energy savings are possible, the next question is just how *material* are these savings? Can they have a meaningful impact on the financial performance of the firm? Again the US provides a useful dataset: the Manufacturing Energy Consumption Survey[244] (MECS), which was last updated in 2010. The MECS includes information on energy consumption per US$ of value added, which is calculated by subtracting the cost of materials, supplies, containers, fuel, purchased electricity and contract work from the value of shipments of both products manufactured and services delivered. One can think of value added as the profit generated in the manufacturing process.

By combining this data with the average cost per unit of energy for each of the manufacturing sectors, together with the losses data already described in Figure 3.3, then we can arrive at Figure 3.4, opposite. From this, we can conclude that for many industries, energy efficiency represents a large potential source of value, with the biggest potential beneficiaries, not unexpectedly, the primary processing industries, such as cement, paper, metals and refining. If we consider that the percentage of value added in these tables *excludes* the cost of labour, then, the percentage savings as a percentage of gross profit are likely to be double the figure shown, giving an overall potential direct impact on profit of energy efficiency of more than 6%.

Industry Sector	US NAICS (National Industrial Classification Codes)	Footprint Data Process Energy	 Other Energy	 Losses (A)	MECS Data Energy Cons. per US$ Value Added (kBtu)	 Energy Cons. per US$ Value Shipped (kBtu)	 Average Energy costs for all purchased energy (US$/ Mbtu)	 Energy Cost Per US$ value added (% share) (B)	 Energy Cost Per US$ value Shipped (% share)	Value (% of Value Add - could be double as a % of Gross Profit) (A) x (B)
Cement	327310	56%	1%	43%	83.30	43.30	US$5.05	42.1%	21.9%	18.2%
Paper	322	21%	5%	74%	26.40	12.10	US$6.93	18.3%	8.4%	13.6%
Aluminium and Alumina	3313	44%	4%	51%	25.70	6.80	US$8.86	22.8%	6.0%	11.7%
Wood Products	321	21%	5%	74%	16.30	6.90	US$8.92	14.5%	6.2%	10.8%
Iron and Steel Mills	331111	45%	6%	50%	29.80	9.30	US$7.24	21.6%	6.7%	10.7%
Glass Containers	327213	41%	5%	55%	20.20	12.20	US$8.41	17.0%	10.3%	9.3%
Mineral Wool	327993	41%	5%	55%	15.20	8.50	US$10.61	16.1%	9.0%	8.8%
Petroleum Refining	324110	65%	1%	34%	39.50	5.40	US$6.03	23.8%	3.3%	8.1%
Textile Mills	313	30%	11%	59%	7.10	3.30	US$12.69	9.0%	4.2%	5.3%
Foundries	3315	44%	11%	44%	7.10	3.70	US$13.35	9.5%	4.9%	4.2%
Chemicals	325	47%	5%	48%	7.30	3.90	US$9.39	6.9%	3.7%	3.3%
Steel Products	3312	45%	6%	50%	6.10	2.20	US$8.71	5.3%	1.9%	2.6%
Food	311	31%	9%	61%	4.30	1.80	US$9.12	3.9%	1.6%	2.4%
Plastics and Rubber Products	326	42%	13%	44%	3.20	1.60	US$16.04	5.1%	2.6%	2.3%
Textile Product Mills	314	30%	11%	59%	2.40	1.20	US$13.33	3.2%	1.6%	1.9%
Beverage	3121	31%	9%	61%	1.70	0.90	US$13.18	2.2%	1.2%	1.4%
Fabricated Metal Products	332	40%	15%	45%	1.90	0.90	US$14.86	2.8%	1.3%	1.3%
Apparel	315	30%	11%	59%	0.90	0.50	US$19.64	1.8%	1.0%	1.0%
Leather and Allied Products	316	30%	11%	59%	1.00	0.40	US$17.67	1.8%	0.7%	1.0%
Electrical Equipment and Appliances	335	27%	29%	44%	1.70	0.80	US$13.24	2.3%	1.1%	1.0%
Transportation Equipment	336	28%	24%	47%	1.00	0.50	US$14.30	1.4%	0.7%	0.7%
Machinery	333	28%	27%	45%	0.90	0.50	US$16.10	1.4%	0.8%	0.7%
Computers	334	27%	29%	44%	0.70	0.40	US$18.27	1.3%	0.7%	0.6%
Pharmaceuticals	3254	43%	6%	51%	0.70	0.50	US$13.00	0.9%	0.7%	0.5%
All industries		**43%**	**6%**	**51%**	**6.40**	**2.80**	**US$9.12**	**5.6%**	**2.4%**	**2.8%**

3.4 **Estimation of value added potential from energy efficiency for a range of US manufacturing sectors. The red line is the average.**
Source: DOE. US Manufacturing Footprint, (2014)[219] and MECS (2010),[244] DOE and Niall Enright. (1kBTU = 0.293 kWh and 1 MBTU = 293 kWh). The spreadsheet model used to calculate this table is available in the companion file pack.

When considering the impact of energy on value added it is important to recognize that for high-value products, like medicines, computers or cars, the percentage contribution of energy to value may be relatively small in comparison to other inputs (such as labour or parts). However, the total energy cost may still be very high; for example, in 2008 in the US the automotive industry spent around US$3.6 billion on energy.[300] Just because the energy efficiency value available in automotive plants is around 0.7% of value added (see Transportation Equipment in Figure 3.4) does not mean that one should never contemplate energy efficiency! The example of Ford's US$130 million cumulative savings (page 95) illustrates this well. As the rate of inflation for energy outstrips the rate of inflation for other inputs, the impact from energy efficiency on value added will inevitably rise.

Of course, those organizations with high energy intensity will have different strategies for managing their energy use – they will often want to consider the efficiency of their core processes and technologies to avoid or recover large quantities of wasted energy. For those organizations with low energy intensity, the options are usually more around the energy that conditions the workplace such as cooling, heating, lighting and ventilation.

There is evidence from a recent study in the UK[373] that waste savings are worth around three to four times more than energy savings in total, while the savings potential from water efficiency is somewhat less. The total no and low-cost savings identified in the study were valued at £23 billion of which £18 billion are from waste reduction and £4 billion from energy.

3.5 **Relative value of resource efficiency savings opportunity across the UK economy**
Note, figures have been rounded.
Source: Reproduced from Table 1 "Further Benefits of Business Resource Efficiency", Defra, March 2011.[373]

Type	Resource	Estimated Savings Opportunity	
		£ bn.	Mt CO_2
No-cost / low-cost less than 1 year payback	Energy	4	13
	Waste	18	16
	Water	1	0
	Subtotal	23	29
Payback greater than 1 year	Energy	7	30
	Waste	22	29
	Water	4	1
	Subtotal	33	61
Total Savings		**55**	**90**

We should note that the figures from the UK study are not directly comparable to the US data in Figure 3.4, as the UK data is based on an assessment of feasible projects, while the US data is based on energy flows.

Focusing on the value of the no and low-cost resource savings more closely in Figure 3.6, opposite, we can see that, for the industrial sector, they are potentially worth a staggering 15.5% of gross profits (i.e. when employment costs are taken into account). For the service sector the impact on gross profits is 1.8%. Just to emphasize that this is for those opportunities we would consider to be "*no-brainers*" with a payback of less than a year, not the total value of all resource efficiency opportunities available.

Across all the sectors the longer-payback opportunities add a further 1.5 times the value of the no and low-cost savings. Thus the overall value of resource efficiency opportunities across the private sector is around 13% of gross profits (5.2% of no and low-cost savings, plus a further 7.8% of longer-payback opportunities assuming a 1.5 multiplier). These are numbers which should make any finance director sit up and pay attention.

So, just how consistent and realistic are these savings projections? On the following pages, we shall see that many organizations, when they put their minds to it, can achieve double-digit improvements in energy efficiency.

3.6 **Sector summary of no/low-cost resource efficiency savings opportunities in UK economy**
Please note that totals on this table do not correspond with the totals in Figure 3.5 because public sector savings are not shown.
Source: Reproduced from Table 60 "Further Benefits of Business Resource Efficiency", Defra, March 2011.[373]

Sector, £m	Waste	Water	Energy	Total	GVA (£m)	Employ-ment Costs (£m)	Gross Profit (£m)	**Savings as % of Gross Profit.**
Construction	2,601	2	-	2,603	67,991	37,312	30,679	**8.5**
Chemicals and non-metallic mineral products	4,396	11	105	4,512	18,445	14,399	4,046	**111.5**
Food, drink & tobacco	219	76	32	327	26,076	11,633	14,443	**2.3**
Metal manufacturing	3,675	5	40	3,720	15,171	12,340	2,831	**131.4**
Other industrial	1,847	42	207	2,096	73,545	40,218	33,327	**6.3**
Industrial total	**12,738**	**123**	**384**	**13,245**	**201,228**	**115,902**	**85,326**	**15.5**
Retail & wholesale	111	-	140	251	149,867	83,786	66,081	**0.4**
Hotels & catering	5	7	99	111	29,900	19,234	10,666	**1.0**
Other service	633	65	188	886	318,864	143,988	174,876	**0.5**
Transport & storage	912	-	2,842	3,754	56,163	35,541	20,622	**18.2**
Service (private) total	**1,661**	**71**	**3,269**	**5,001**	**554,794**	**282,549**	**272,245**	**1.8**
Agriculture	362	84	-	446	1,499	621	878	**50.8**
Total private sector savings	**14,761**	**278**	**3,653**	**18,691**	**757,521**	**399,072**	**358,449**	**5.2**

If we move away from thinking of our organization in isolation and consider how we can collaborate with others to reduce our resource costs, we will find very significant direct cost savings opportunities from adopting the circular economy approaches described in the last chapter. In its report *Towards a Circular Economy*,[249] the Ellen MacArthur Foundation calculated "*that, in the medium-lived complex products industries, the circular economy represents a net materials cost savings opportunity of US$340 to US$380 billion p.a. at an EU level for a 'transition scenario' and US$520 to US$630 billion p.a. for an 'advanced scenario,' net of the materials used in reverse-cycle activities in both cases*". A more recent report for the World Economic Forum[807] in 2014 states that the annual global value of resource efficiency is US$1 trillion. ⇒ (page 96)

Real World: *Energy efficiency savings claimed*

Many energy efficiency case-studies, in my experience, need to be taken with a pinch of salt in particular where organizations are reporting relative, per unit improvements, or emissions reductions are being claimed which do not meet a materiality test (see section 11.5 on page 374). Nevertheless, the selection of savings claimed below point to a considerable potential for reduced energy use across a wide range of sectors, particularly where organizations take a long-term approach.

DuPont launched its *Bold Energy Plan* in 2008, which built upon already impressive efforts to reduce energy use since 1990. The plan targets a 5% annual reduction (US$50 million) and in the first year achieved US$47 million savings and a 4% increase in productivity in relation to energy use. These savings were generated through 266 projects that required little or no capital expenditure and 74 funded projects which collectively provided a rate of return of 65% (i.e. better than a two-year payback). This improvement was on top of the savings made since the 1990s of 39% per unit production.[231] From an investment perspective this initiative achieved outstanding returns. In 2015 DuPont reported that they made US$2.6 billion revenues from products that help customers improve energy efficiency and/or reduce greenhouse gas emissions.

3.7 **Many organizations are claiming that substantial energy efficiency improvements are achievable**

*Denotes a Greenhouse Gas (GHG) reduction

Source: Publicly available case studies - see individual references.

Organization	Absolute improvement	Per unit improvement
3M		81% (1973-2010)[1] 46% (2000–2010)[1] 30% (2005-2016)[3]
Anheuser-Busch InBev		8.8% (2012-2015)[4]
Bank of America	26% (2010-2014)[54]	
Department of Energy and Climate Change (UK)[189]	26% (2010-2015)	29% (2010-2015)
DuPont[231]	18% (1990-2010)[231]	44% (1990-2010)[231] 11% (2010-2014)[233]
Ford		40% both globally and in US on a per vehicle basis (2000-2010)[289] 25% reduction on manufacturing energy per vehicle (2010-2015)[290]
Gap Inc.	8.5% (2011 to 2014)[303]	
General Motors Corporation[85]		14% (2010-2015)[308] per vehicle
IBM Corporation[383]	63% (1990-2015) 6.3% (2015)	
Lafarge[461]		26% (1990-2014)
Marks & Spencer		39% (2007-2016) per sq ft. DD adjusted, UK and Ireland[499]
Johnson & Johnson[426]	23% (1990-2010)[426] 5.2% (2010-2015)[425]	>50% vs revenues (1990-2006)[426] 17% vs revenues (2010-2015)[425]
Pfizer[85]	12% (2012 -2015)	
Toyota		36% per unit (2001-2015)[712]
Unilever		39%* per US$ tonne (2008-2015)[733]

3M is another leader in energy efficiency, which has achieved a staggering reduction of 80% of energy use per net sales value between 1973 and 2010. A case study in 2010[241] describes an employee motivation and engagement programme which has led to the implementation of 1,900 projects between 2005 and 2009 worth US$100 million in energy savings over the period. This programme used a simple scoring system around three metrics to rank facilities: Btu per pound of product emissions reduction, a rating of the energy programme effectiveness, and the projects delivered compared to plant spend. Plants were ranked as platinum, gold, silver or bronze – an excellent example of how suitably designed peer pressure can help motivate action (more on this later in Section 19.7 on page 666). These improvements do not seem to be coming to an end, with a reduction in energy use per unit production of 30% across 3M between 2005 and 2016.

Even in the cement industry, which according to Table 3.3 has one of the highest proportions of process energy compared to non-process energy at 56%, savings can be very considerable. Lafarge claims a reduction of 26% of its net CO_2 emissions per tonne of cement between 1990 and 2015.[461]

Bank of America has achieved an absolute emissions reduction of 24% in just four years, partly through a reduction in the total office space used, but also due to energy efficiency programmes. Cumulative savings from energy efficiency since 2004 are estimated at US$301 million.[54] IBM is another company that has tracked the value of energy conservation measures and calculated the total from 1990 to 2015 at US$579 million.[383]

Johnson & Johnson pledged to reduce its greenhouse gas emissions from all facilities worldwide to 7% below 1990 levels by 2010. By 2006, Johnson & Johnson cut its emissions to 16.8% below 1990 levels, despite more than tripling revenue over the same period (in other words the effective improvement per unit revenue was over 50%). The company's energy efficiency programme resulted in an estimated US$30 million annualized savings over the last 10 years, and the greenhouse gas reduction projects delivered an average 16% internal rate of return.[426] Between 2010 and 2015 a 5.2% absolute reduction was achieved showing that further improvements can be made on the back of historically good performance.

UK retailer Marks & Spencer achieved a 39% energy use per sq ft reduction in their UK and Republic of Ireland estate between 2007 and 2016, well on the way to a 50% reduction by 2020.[499]

Ford demonstrates how even manufacturers of high value-added products where energy is a relatively small element of total cost, can reap substantial benefits from increased efficiency. Between 2000 and 2010 they reduced energy use per vehicle manufactured by 40% both globally and in the US.[290] The cumulative value of the five years' savings to 2011 alone exceeds US$130 million at today's prices.[289] The pace is not lessening, with the most recent reports citing a further 25% improvement between 2010 and 2015.

Public sector organizations are also able to achieve significant reductions over short periods of time if they put their minds to it. The UK's Department of Energy and Climate Change (DECC) signed up to the 10:10 campaign for individuals and organizations to reduce emissions by 10% in 2010. In fact, they were able to achieve a 20% reduction in that year. Since then they have made a further 26% absolute improvement in the energy used.

Where organizations take a long-term approach, very significant energy savings can be achieved.

3.3 Indirect cost savings

If we examine the embedded cost of inefficiency in the materials that we purchase we soon realize that the savings from energy and resource efficiency are even greater than at first thought.

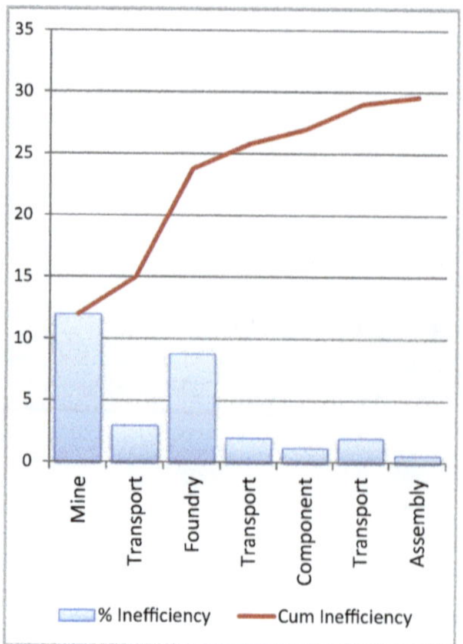

3.8 **Illustration of the accumulation of inefficiencies through the supply chain**
Source: Niall Enright.

Direct cost savings, although very considerable in their own right, are like the tip of the iceberg in terms of the true value that resource efficiency can bring organizations.

We can start by considering the fact that the cost of inefficiencies accumulates through supply chains. Inefficiencies and the associated costs are aggregated several-fold as materials move from one manufacturing stage to another, as the output from one factory serves as the input for another. Assuming that the manufacturers maintain their margins, the costs of inefficiencies become embedded in the unit cost of the materials at each stage and so are eventually borne in full by the final recipient.

Let us consider a simple example, illustrated left, moving from mineral ore to refined metal, to component manufacturer, to assembly in the automotive sector. The press hall stamping body panels in a car factory may appear to have a low energy intensity, but actually it is paying for any wasted energy inputs and inefficiencies that have arisen in the foundry and rolling mills which have refined and shaped the steel sheets that are used for the panels, and which are embodied, quite literally, in the cost it pays for the steel. Indeed they are also paying the cost associated with any inefficiencies in the transportation from the foundry to the press shop. In a similar fashion, the foundry is paying, hidden in the price of the ore it receives, the costs associated with any inefficiencies in the mining process and the transport inefficiencies from mine to foundry.

These hidden costs are one reason why supply-chain resource efficiency is particularly important to those manufacturers at the end of the chain. It could be argued from Table 3.4 that the assembly plant's potential value from energy efficiency is a direct 0.6% of the car manufacturing margin *plus* 1.2% of the steel component manufacturing margin *plus* 8.8% of the iron and steel refining margin and as well as the unquantified mining and multiple transportation inefficiencies at each step.

There is another factor which further increases the embedded cost of inefficiency in the inputs to the final stage of a supply chain. Sticking with the example of energy in the car plant we need to recognize that the waste of energy can additionally accumulate through material wastage at each stage of manufacture. The cumulative effect of the energy inefficiencies described above assumes that 100% of the inputs into one process find their way into the

🔑 *The costs of inefficiencies accumulate in the supply chain and are paid for by the end-user of the resource.*

outputs, but of course, there is also waste at each stage of the process, which has the effect of increasing the amount of energy at each stage which has not done useful work. These losses, in turn, add to the price of the material that is sold on.

For example, the offcuts of steel in the foundry have embedded energy and materials which will require further processing to be reused. So a programme focused on resource efficiency, which reduces both the energy used *and* the physical waste created, will deliver greater value to each stage of the supply chain than one which focuses on energy alone.

From this understanding, we can formulate a couple of general concepts when considering resource efficiency in supply chains. Bearing in mind that gross value added (GVA) represents the net profit from manufacture (value of the product less input costs) then as we move further along the supply chain:

- the direct cost of energy as a proportion of GVA generally *declines*; and
- the cumulative embedded cost of energy *rises*.

Producers of goods with high added value (cars, medicines, perfumes) have often told me that resource efficiency is not particularly interesting to them because the direct savings potential as part of their GVA is very low (see, for example, pharmaceutical at the bottom of Table 3.4). In fact, the true benefit of improved resource efficiency is very much greater because they are at the end of a long supply chain.

If we look, for example, at aircraft manufacturers where energy is a tiny proportion of GVA, it turns out, according to Allwood and Cullen's *Sustainable Materials — without the hot air*, that over 90% of the high-grade aluminium they receive is turned into waste swarf. In fact, it could be argued that it is swarf that is their primary product, not aircraft (see page Figure 2.4 on page 49)!

If we think in terms of *avoidable* costs, and thus potential benefits of resource efficiency, for this manufacturer we can see that these fall into two categories:

- The cumulative embedded cost of resource inefficiencies arising at all the preceding steps in the manufacture of the aluminium that *does* find its way into the final product; and
- The full cost of all the aluminium used that *does not* find its way into the final manufactured product and becomes waste in their own process.

Thus the way we quantify the effect of inefficiency depends on whether the resource we are assessing find its way into the final product or service, or whether it becomes waste itself. Clearly, the material that is used has lower potential avoidable cost than that which is wasted.

Missing this embedded cost is not unusual. In the area of waste reduction, it is common to base the business case for improvement on the reduced

Real World: ***The gold standard***

Contrary to common presumption, conforming to the most rigorous or *gold* standard can often save money.

For example, Hewlett-Packard anticipated in the 1990s that lead would be one day be banned from solders and by 2006 had successfully developed alternatives.[554] When the EU, as anticipated, banned lead HP was able to comply more rapidly and at lower costs that other electronics manufacturers.

By adopting the most rigorous and stringent environmental standards, multinational manufacturers can save a lot of money because their products will comply in all markets. The alternative, of complying with the least stringent standards for each market, incurs additional costs associated with separate design, supply chain, logistics, testing and compliance. This is just another example of indirect costs savings from resource efficiency.

direct costs associated with managing the waste such as transportation and disposal etc. In fact, the real value is that disposal cost *plus* the full cost of the resource in the first place. This is the *value with hidden savings*, which takes into account the direct costs of waste disposal as well as the hidden cost of the raw materials, energy, water, packaging, labour and overheads that went into the wasted material in the first place.

In some cases the hidden value can be as much as 20 times the direct costs – a simple example to illustrate this point is white office paper which costs around £65 per tonne to dispose of (the UK landfill tax) but actually costs £1,200 per tonne to buy in the first place. Figure 34 shows that the average ratio between direct and hidden costs is around three times the direct cost. As in the case of our earlier examples, the further along the supply chain the greater this hidden value usually is.

3.9 **Hidden savings as a multiple of the direct waste disposal costs for a selection of UK sectors**
Source: Adapted by the author from Table 7.4 in "Quantification of the business case for Resource Efficiency, Defra 2007".[468]

	Sector	Ratio of Hidden Savings to Direct Costs
Industrial	Construction	1.0
	Mining & quarrying	1.0
	Food & drink	9.1
	Energy supply	1.3
	Base metals / Mechanical engineering	1.6
	Machinery, electrical & transport equipment	7.5
	Chemicals, rubber & plastics	5.0
	Paper, printing & publishing	2.0
	Other industrial	3.3
Commercial	Retail et al.	4.1
	Travel agents et al.	3.4
	Hotels & catering	1.0
	Transport	1.0
	Education	1.0
	Misc. service industries	1.0
	Commercial services other	2.7

One of the most enjoyable aspects of working in the field of energy and resource efficiency is the satisfaction of finding hidden sources of value that enable a compelling business case for an investment to be made which otherwise would not have been viable based on direct cost savings alone. The panel opposite provides some real-world examples of these hidden benefits.

One of the most enjoyable aspects of developing the business case for resource efficiency is uncovering the ***hidden sources of value*** *that make an otherwise marginal proposition compelling.*

Real World: ***Valuing hidden savings is critical to success***

It is usual for waste treatment or reduction opportunities to be valued just on the basis of the avoided cost of disposal, particularly when the inputs into the waste are not appreciated.

An example of this would be effluent (wastewater) from a dairy. Typically the business case for reducing this waste is calculated in terms of the reduced sewerage charges that result from having to treat less effluent. In fact, there is usually a much larger financial benefit from reducing effluent, which comes about from reducing the amount of valuable milk that forms part of the effluent and preventing hot water entering the wastewater system and thus avoiding unnecessary heating and treatment of this water. It can be difficult to get approval for wastewater projects unless these types of second-order hidden benefits are quantified.

Another example of missed value is often found in lighting projects, where the business case is typically made on the basis of the energy consumption of the old technology versus that of the new technology. The value is usually calculated simply by comparing the wattage of the old and the new lamps and working out what the savings in electricity costs will be.

In fact, it is often the cost of labour that is associated with the new lighting system that is an even more important issue. A real-world example is intu Trafford Centre, one of the UK's largest retail and leisure destinations near Manchester, which I have helped on energy efficiency and sustainability for many years.

The malls in the Trafford Centre are large, elaborate, lavish, public spaces. They are lit with a wide variety of different feature lights, some of which are on during the day to create the right ambience and others which come on when the natural light level is too low. As you would expect with the different types of lamps and usage patterns, they need to be replaced at different times, depending on their rated life expectancy. Replacement often requires the use of a *cherry picker* or *elevated platform* so that the maintenance technician can safely access the light fitting. The process of re-lamping is thus expensive and time-consuming.

3.10 **A hidden cost saving from a lighting project**
Source: © Big Face - Fotolia.com

In considering the business case for changing the lighting, over half of the value was because the proposed LED lamps had a lifetime over twice as long as the lamps they were replacing, and so this considerable labour cost was more than halved. Quantifying this value was essential as it turned an unlikely four-year payback project to a less than two years project, which gained the full support of the tenants who would approve and fund the improvement.

3.4 Share or asset value

It is becoming increasingly easy to establish how energy and resource efficiency can preserve or increase the value of assets - whether these are physical assets like buildings or financial assets such as shares.

Much economic activity is driven by asset valuations. The primary basis for valuing most assets are:

- the future cash flow that it will generate; and
- the degree of risk to that cash flow.

If we consider physical assets first, we can think of things like buildings, power stations, trains and aeroplanes. These each have a future cash flow associated with them: rent, electricity sales and transport revenues respectively. We also can see how energy and resource efficiency can influence each of these income streams.

Total occupancy costs will be higher in a building with poor energy efficiency and so the rental potential will be lower.

- An inefficient power station will generate less electricity per unit of input and so will deliver a smaller free cash flow.
- Similarly, an inefficient vehicle will be less profitable than an efficient one because of the larger expenditure on fuel.

In fact, almost every physical asset we can think of will have its ability to generate cash influenced by energy and resource efficiency. The obvious exception being *goods* like art, fine wine, jewels, pork bellies, etc., which are held in anticipation of a future price rise due to *demand* factors.

Assets like buildings and power stations are often valued in multiples of this future cash flow. A commercial building may be valued at 18 times its annual income, while a power plant may be valued at 25 times its annual earnings. The difference in these valuations reflects the fact that the power station offers a more certain source of revenue than the office, which could have periods of vacancy, and so investors in power plants are willing to accept a lower annual return for the lower risk. Of course, there are many other factors to consider in these valuations, such as how *liquid* or easy to sell the asset is, what comparable assets are selling for, what alternative uses of the land are possible and suchlike.

Asset valuation is, in simple terms, a calculation of the ***risk-weighted income*** *that the asset will produce in the future.*

However, the principle that asset valuation is a reflection of a risk-weighted income stream is broadly accepted.

Each dollar saving from resource efficiency will increase the asset value by the **multiple of earnings** of the organization or asset class.

An important conclusion we can draw from this simple model is that the net annual value of resource efficiency savings will impact on the asset's valuation in line with the multiplier of income applied to that asset type. Thus a resource efficiency programme that increases the profitability of a portfolio of offices or a power station by US$1 million will potentially increase the value of the assets by US$18 million or US$25 million, respectively.

The same principle of a multiplier effect also applies to non-physical financial assets - notably stocks in companies. In simple terms, stocks have a price to earnings ratio, which varies from sector to sector and company to company based on risk and other factors, but which reflects the multiple of earning that investors are willing to pay for that stock. The calculations for turning efficiency savings into share value are explored in more detail in the piece *In Numbers: Translating operational savings to value (page 108)*.

If the immediate efficiency savings of US$1 million in an office have led to an increase in the asset value of US$18 million, then the asset owner will be ahead if the cost of implementing the savings is less than US$18 million. Thus the multiplier can give a crude indication of the investment to savings ratio that can be justified.

Here our explanation takes a twist because it is important to note that most investments in resource efficiency are funded from profits or debt, not from fresh capital provided by the asset owner. In the case of funding savings from profits, assuming that a programme has a greater than one-year payback, the cost in year one will outweigh the benefit, profit will decrease and the asset value will *fall*. The asset owner may still justify this investment, because of the longer-term rise in profit and asset value. However, if the initial investment were to be borrowed from an external lender and spread over several years, such that the repayment of interest and capital is less than the annual savings, the organization would immediately increase its profitability and the programme would have a positive benefit on asset value from day one.

The US Chemical Giant DuPont developed the "DuPont Formula" in the 1920s to describe how an investor can increase the return on their investment or *return on equity*. Firstly, they can reduce their operating costs or increase their sales - both of which lead to an increase in *net operating margin*. Another way that investors can increase their return is to increase the output from their existing factories, their *asset turnover*. Finally, if the investor can borrow funds for expansion (rather than use their own funds), so long as those investments deliver a greater profit then they will see their asset values rise. This is known as *financial leverage*. The illustration on the next page shows how resource efficiency has effects well beyond simple cost reduction which contribute to an increased return for investors. For example, the positive brand benefits of resource efficiency can support increased sales; lower risk can increase the market multiplier investors will give the company's stock; efficiency can lead to better throughput and asset turnover.

⇒ page 104.

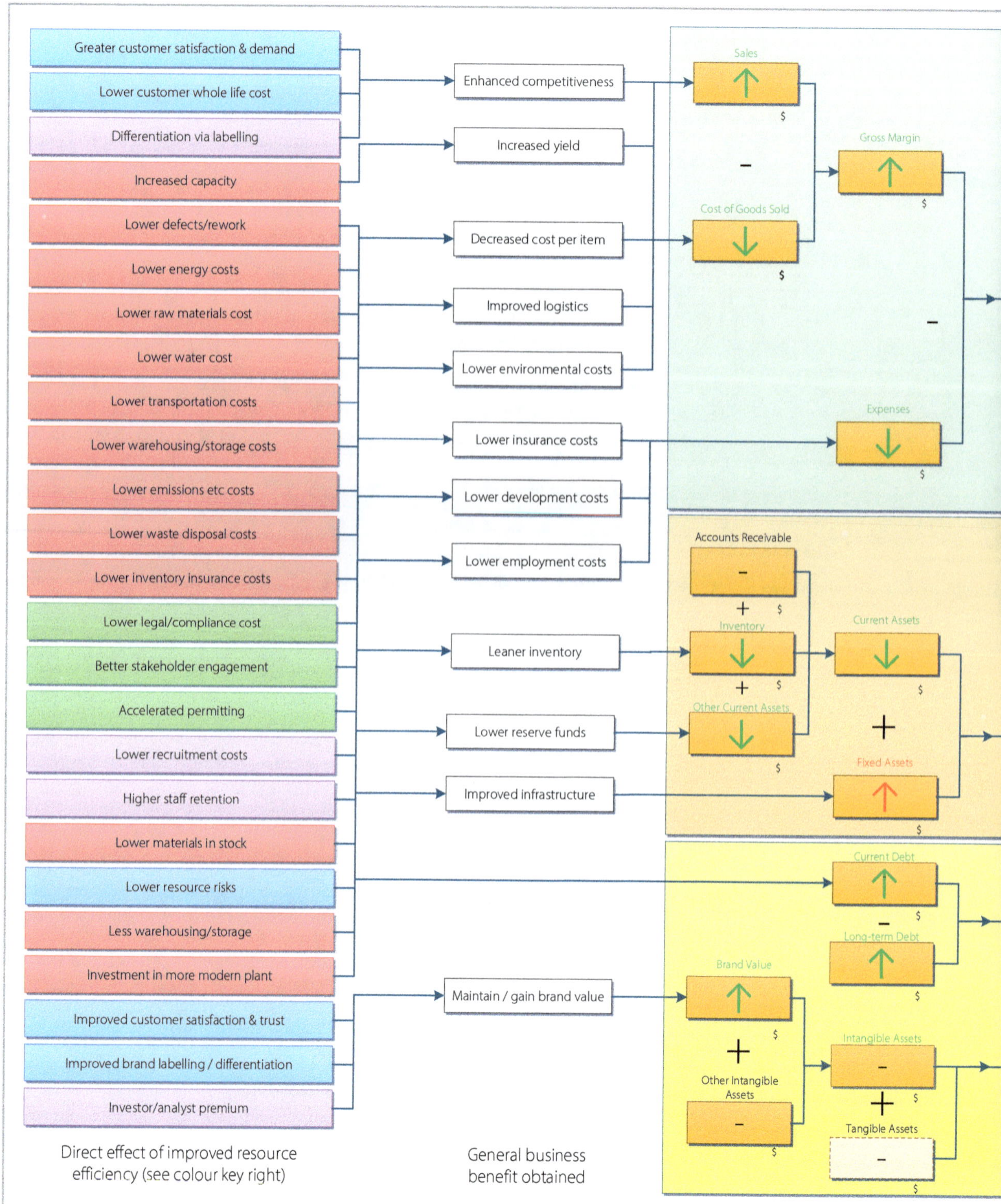

Greater customer satisfaction & demand
Lower customer whole life cost
Differentiation via labelling
Increased capacity
Lower defects/rework
Lower energy costs
Lower raw materials cost
Lower water cost
Lower transportation costs
Lower warehousing/storage costs
Lower emissions etc costs
Lower waste disposal costs
Lower inventory insurance costs
Lower legal/compliance cost
Better stakeholder engagement
Accelerated permitting
Lower recruitment costs
Higher staff retention
Lower materials in stock
Lower resource risks
Less warehousing/storage
Investment in more modern plant
Improved customer satisfaction & trust
Improved brand labelling / differentiation
Investor/analyst premium
Enhanced competitiveness
Increased yield
Decreased cost per item
Improved logistics
Lower environmental costs
Lower insurance costs
Lower development costs
Lower employment costs
Leaner inventory
Lower reserve funds
Improved infrastructure
Maintain / gain brand value
Sales
Cost of Goods Sold
Gross Margin
Expenses
Accounts Receivable
Inventory
Other Current Assets
Current Assets
Fixed Assets
Current Debt
Long-term Debt
Brand Value
Other Intangible Assets
Intangible Assets
Tangible Assets
Direct effect of improved resource efficiency (see colour key right)
General business benefit obtained

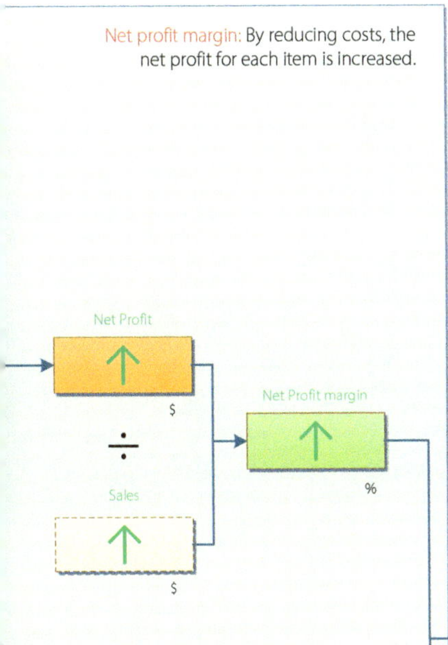

DuPont Formula:

Return on equity = net profit margin * asset turnover * financial leverage

Net profit margin: It is obvious that we can make money if we increase our profit margin on every item we sell.

Asset turnover: We can also make money by producing more items from our existing plant, i.e. if we are more productive.

Financial leverage: Shareholders can make a greater return if someone else (e.g. a bank) pays for the investment in equipment. However too high a leverage will lead to a worry about the company's ability to service the debt.

This is known as the DuPont Formula because DuPont was the first corporation to adopt this in the 1920s.

This illustration shows the many ways that resource efficiency can improve the return on equity.

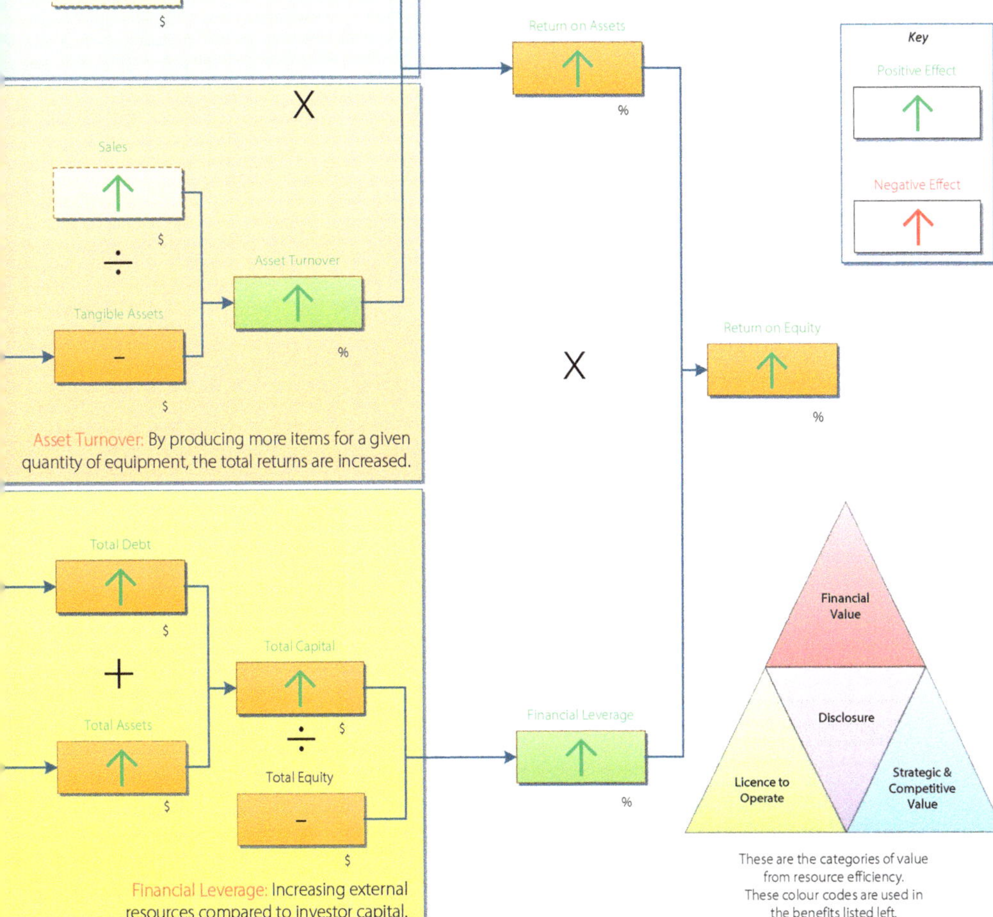

3.11 **The DuPont Formula provides a rationale for how resource efficiency can deliver improved returns for shareholders in private companies** *Source: Niall Enright. A poster version of this spread is available in the companion file pack.*

> **Real World:** *Building values*
>
> There is increasing evidence that energy efficiency directly translates into asset value in commercial property.
>
> One of the most cited studies on the link between energy efficiency and office value was carried out in the US in 2009.[246] The conclusion of Piet Eichholz and his colleagues could not be clearer:
>
> > *"The results suggest that an otherwise identical commercial building with an Energy Star certification will rent for about three per cent more per square foot; the difference in effective rent* [taking into account different vacancy rates] *is estimated to be about six per cent. The increment to the selling price may be as much as 16 per cent. These effects are large, and they are consistent."*
>
> Similar evidence is emerging in the European markets where, for example, Nils Kok and his colleagues used Energy Performance Certificates to show that efficient buildings (those rated A-C) had a 6.5% rental premium compared to those rated D or worse.[445] The willingness of tenants to pay a premium for energy efficient property seems widespread. The Australian study, *Building Better Returns*[550] cites a 9% increase in value for properties that have the highest energy rating.
>
> The critical business issue here is that the costs to achieve the high efficiency and certification are usually much less than the value generated, making investment in energy efficiency in commercial property a very sound proposition.
>
> Significantly, many studies show the sale premium is even greater than the increased income alone explains.

One particular class of assets where valuation is critical is commercial property. Here valuation tends to be based on three approaches:

- A cost approach which values the asset at the cost to create or purchase an equivalent (e.g. the land, building costs, depreciation etc);
- A sales comparison approach which looks at what price buyers have been willing to pay for comparable assets in the same location; and
- An income approach, such as the ones we have described, where the valuation is based on the future income the property will generate.

We have already seen how resource efficiency can lead to an increase in an asset value in an *income-based approach* to valuation simply because reduced resource costs can result in reduced landlord costs and lower total occupancy costs which will support larger rents and increase operating profit.

However, the most common form of valuation methodology is based on sales comparison which states, in essence, that if others have been willing to pay £x/sq ft for a similar property, then it is reasonable to assume that £x/sq ft is a good estimate of the real worth of the property. In the UK the most common valuation technique, called the *investment approach*, appears superficially to be an income method. However, it is driven by a market yield factor to calculate the value of the future income and because this is determined by recent sales data it makes this essentially a sales comparison approach.

In terms of resource efficiency, this poses a problem, however, and that is that the *"efficiency"* of the asset is rarely used as a characteristic of the property when establishing a sales comparison. This is changing with the introduction of voluntary (e.g. *Energy Star* in the US) or mandatory (e.g. *Energy Performance Certificates* in the EU) property rating schemes. These schemes, alongside more general sustainability ratings such as LEED or BREEAM, will enable the resource efficiency of the building to be categorized and so help understand the market premium, if any, that this offers to the asset owners.

Research is now emerging that is enabling the effect of resource efficiency and sustainability to be isolated from other property characteristics using these rating schemes (see box left). The real challenge remains, however, in getting conservative valuation professionals to take full account of these trends in their valuations. The influential Royal Institution of Chartered Surveyors (RICS) has issued *Valuation Information Paper # 13* (1st Ed). on *Sustainability and Property Valuation*[621] which states:

> *"The valuer has a responsibility to the client to ensure that a valuation reflects the material factors that may influence value. Markets appear to be moving towards a requirement for greater recognition of sustainability issues. Accordingly, as sustainability issues grow in relevance to the market place, it becomes increasingly important that the valuer is aware of them and can reflect them in the advice given."*

So just how exactly do energy efficiency and performance ratings lead to an increase in a building's value? Below is a model valuation of a multiple occupancy commercial property in which we can see an overall increase in value of 16%. In the example below, there are several effects on value:

1. The first effect is an increase in the rental of 4% from £20/month/m^2 to £20.80. The tenant will see a decrease in energy costs which will entirely offset this increase (if we use a typical figure of £30/year/m^2 energy costs which is reduced by 40% for best practice, this results in a net cost reduction of £1.00/month/m^2, which is greater than the £0.80 rent premium).
2. Vacancies decrease by 50% because of the attractiveness of the building.
3. Landlord administration costs reduce by 5% due to lower turnover of tenants and lower landlord energy costs of voids.
4. Together, these three effects lead to a total increase in the effective rental income of the office of 6%, which matches the US/EU research data opposite.
5. If we consider the selling price of the property, we can see that this is even greater than the income increase. This could reflect a higher assumed yield (because of the lower risk associated with *"quality"* tenants and lower voids) or possibly because of a *"sustainability premium"*. A premium increase of 3% would bring the value increase to 9% as per the Australian study, while a premium of 8% would be required to match the US study.

This example shows that for buildings we may need to factor in a hidden *"premium"* if we are to account for research data on sales. ⇒ page 111.

3.12 **Hypothetical Building Valuation based on observed impact of energy efficiency labelling from research studies (opposite)**
Source: Niall Enright

Valuation	Normal Building	Efficient Building	Assumption / Observation
Lettable Floor Area (m^2)	10,000	10,000	
Rental per m^2 (£/month)	20.00	20.80	4% increase due to favourable building rating
Total Rental Value (£/year)	**2,400,000**	**2,496,000**	
Less Vacancy Costs @ 5%	120,000	60,000	Assume vacancy reduces 50%
Less Admin Costs	31,506	22,054	Assume resulting 10% reduction in admin cost
Net Rent (income)	**2,248,494**	**2,413,946**	A 7% increase in rent, in line with research
Less Interest on Site Value	34,500	34,500	
Net Building Rent	**2,213,994**	**2,379,446**	
Multiplier assuming a 5.8% yield	17.19	17.24	Assume no impact on Covenant Strength / Yield
Value of Building	38,058,557	41,024,928	
Value of the Site	600,000	600,000	
Sustainability Increment		3,281,994	Assume lower risk / green premium of 8%
Total Market Value	**38,658,557**	**44,906,922**	**A 16% increase overall, in line with research**

Real World: *Good environmental performance correlates with share value*

Executives in private sector companies are employed to run the organizations in such as way that the shareholders benefit - usually by increasing the share value or increasing the income (or dividend) distribution. Unfortunately, there is an assumption among many executives that resource efficiency destroys stock value. This myth arises from the general, outdated presumption that any environmental investment, other than the minimum requirement to comply with environmental legislation, reduces corporate profitability. The question is what evidence is there that there is a connection between resource efficiency and financial performance, and is this relationship positive or negative? Fortunately, there is a growing body of evidence that shows that above average environmental performance correlates strongly with superior financial results.

In the first instance, we can see that bad environmental performance destroys value. A study of firms in the US S&P 500 index in 2001[447] assessed the impact of poor environmental performance (measured by the amount of toxic emissions produced) and concluded that the "*intangible liability*" of toxic emissions was around 9% of the replacement value of the firms' tangible assets. In other words, poor resource efficiency has a direct effect on a firm's value. The adverse effects of poor environmental performance can be seen in other markets, such as in India,[341] while the positive impact of good environmental news on share prices in Japan has also been observed.[819]

More recently, because of the large proportion of the Global 500 companies participating in the Carbon Disclosure Project (CDP), we have been able to establish if there is a connection between carbon reporting and financial performance. In the 2013 CDP Global 500 Report[608] (page 17), the chart reproduced below is used to demonstrate the relationship. This data suggests that indicators of superior carbon reporting and management are also indicators of superior financial performance in the longer term. Companies in the Carbon Disclosure Leadership Index (CDLI), on average, provided approximately two-thirds greater returns (including interest, capital gains, dividends and distributions) of other Global 500 businesses in the period 2005 to 2013. For companies rated top in carbon management the average improvement was smaller, but nevertheless 25% above the other Global 500 businesses. These improved returns are clearly of great interest to current and potential shareholders.

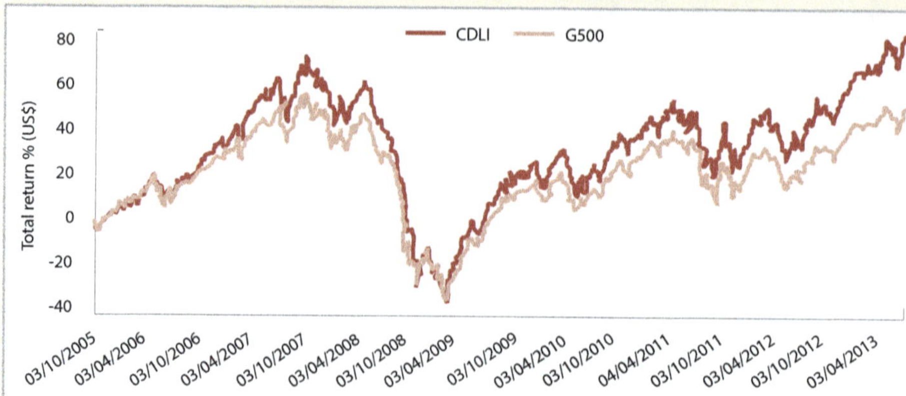

3.13 **Carbon Disclosure Leadership Index [2005 - 2013] returns against overall Global 500**[209]

CDLI companies are those which achieve a top 10% score for reporting of emissions. G500 companies are the 500 largest listed companies globally.

Source: Reproduced with kind permission of the Carbon Disclosure Project

Given the difficulties in obtaining a clear indicator of environmental performance, as well as the varied options for measuring financial performance, it is not surprising that the literature on the link between resource efficiency and value remains mixed. However, a review of 32 major studies by a team in Alicante, Spain,[528] concluded that the majority of papers supported the link between "*green management*" and financial performance.

Osmosis Investment Management has presented much more definitive evidence linking resource efficiency and investment returns. They compared the investment performance of the 10% most resource-efficient companies (in a group they call MoRE World, for Model of Resource Efficient) with that of the remaining, less resource-efficient rivals (MSCI World).[577] It seems that, on average, between 2007 and 2015 the resource-efficient companies yielded an average return of 12.2% compared to the yield from the less resource-efficient companies of 5.7%. In the latest years, 2014 and 2015, for the first time, the efficient companies under-performed slightly in markets with relatively little overall growth, although the fund managers report that this is due primarily to currency exchange rates, not the underlying stock performance (see table opposite).

ANNUAL RETURNS (%)									
	2007	2008	2009	2010	2011	2012	2013	2014	2015
MoRE World	15.54%	-40.49%	59.62%	22.39%	-3.71%	23.60%	33.03%	3.82%	-3.63%
MSCI World	9.04%	-40.71%	29.99%	11.76%	-5.54%	15.83%	26.68%	4.94%	-0.87%
Excess Return	6.51%	0.23%	29.63%	10.63%	1.83%	7.77%	6.35%	-1.12%	-2.75%

MoRE World is a globally diversified portfolio of large cap companies made up of the top decile of resource-efficient companies from each sector of the world economy, ex-financials. Data as at end Dec 15.

3.14 **Osmosis Investment Management suggest that more resource-efficient companies generate superior returns** *Source: Reproduced with kind permission from Osmosis Investment Management*

The connection between resource efficiency and financial performance is sometimes clouded by confusion over the performance of stocks in *green technologies*, such as renewable energy, biofuels and low-carbon services, as opposed to the performance of companies in more traditional sectors, which are *also* more resource-efficient than their peers. The green technology group will naturally show a high degree of volatility and variability in performance, as would be expected from investments in newly emerging fields with unproven technologies. In judging the impact of resource efficiency on stock performance, it is the type of analysis carried out by CDP and Osmosis which is more relevant, as it illustrates the impact of resource efficiency across all sectors.

Arising from this research is a key question: "*Is the improved stock market performance of these businesses directly as a consequence of the improved resource efficiency or is the efficiency merely a proxy for other characteristics of the businesses which will independently lead to better returns (e.g. the quality of management, expertise available or better finance).*" In other words, do good management and good results precede good environmental performance, or does good environmental performance precede good results. What is cause and what is effect? A paper by Clarkson and colleagues[151] supports the intuitive response that both factors are probably at play, but observes that the management effect is likely to be more significant. Thus the commonly held view among investors that good environmental performance is a proxy for superior management overall is justified. This view is echoed in the CDP results above where they state that "*the capability of the management team or the company's broader approach to identifying and capitalizing on opportunities or managing risks*"[608, p13] could be influencing the relationship between carbon disclosure and financial performance – it is not the carbon management alone that is causing the increased returns.

Significantly, resource efficiency does also appear to predict performance and so provide the basis for stock selection in a portfolio of investments. The attractiveness of resource efficiency as an indicator is that it is a true measure, at one level, of the economic effectiveness of the businesses and is relatively easy to quantify on a like-for-like basis. At another level, resource efficiency can be considered a measure of the "real" extent of sustainability within organizations – a measure that may be much more revealing than some of the other indicators of sustainability which are inconsistent and are often more qualitative check-box assessments.

Statistical analysis of the performance of their MoRE companies by Osmosis – called a Brinson performance attribution – shows that stock selection, i.e. resource efficiency, had the greatest impact on the return of the portfolio. As Osmosis put it:

> *"This represents verifiable, statistical evidence that the market, directly or indirectly, values companies that are more efficient at converting their resources to goods. And what's more, the market has priced this concept for many years. Resource efficiency, in fact, does lead to shareholder value."*[578]

Although we have not yet fully untangled the linkages between shareholder value associated and good environmental performance – and we may well find that the answer is varied and complex – the evidence is nevertheless mounting that action on the environment is linked with financial performance. It is time to slay the myth of value-destruction once and for all.

Senior executives need to understand that many investors are turning to environmental performance as a proxy measure for management competence – and so a visible resource efficiency programme should have the additional benefit of increasing management standing in the eyes of shareholders, with the additional advantage that resource efficiency also contributes strongly to the bottom line. Longer-term investors, possibly recognizing the inevitability of change, are also beginning to understand the emerging risk around resource efficiency and so are increasingly likely to place a premium on business models that are less exposed to rising costs, scarcity and risks around sources and sinks.

In Numbers: *Translating operational savings to value*

So just how do we calculate the effect of resource efficiency on the share price of a business? Let's use an example to illustrate this, with simple numbers. Suppose our company Widgets Inc. has sales of US$100 million, on which it makes a profit of US$10 million. The share price is trading at a sector-average multiplier of eight times earnings, so the company's total stock is worth US$80 million (eight times US$10 million profits), and if 10 million shares have been issued then each share is worth US$8.00 (US$80 million value/10 million shares).

If we know that the cost of manufacturing is US$50 million (the remaining US$40 million costs are marketing, distribution, overheads etc.) and that the potential improvement in the manufacturing value added from no/low-cost resource efficiency is 5%, then we can calculate the share price impact as follows:

Additional profit from resource efficiency = 5% * US$50 million = US$2.5 million
Total profit = US$10 million + US$2.5 million = US$12.5 million
Total enterprise value = 8 * US$12.5 million = US$100 million
Share price = US$100 million value / 10 million shares = US$10.00 per share
Increase in share price = US$10.00 - US$8.00 = US$2.00 / share
Percentage increase in share price = US$2.00 / US$8.00 = 25%

So, in this very simple example, a reduction in manufacturing costs of 5% (which is equivalent to a decrease in total costs of 2.8%) has led to a 25% share price increase for Widget Inc. If we examine the calculations above, we can see that the key driver of value is that manufacturing costs are five times the current profit. In other words, the impact of a small percentage cost reduction can be multiplied several times over in terms of impact on share price, according to the following formula:

$$\text{Share Price Increase (\%)} = \frac{\text{Manuf. Costs (US\$)}}{\text{Profit (US\$)}} \; x \; \text{Manuf. Cost Savings (\%)}$$

Or more generally

$$\text{Share Price Increase (\%)} = \frac{\text{Costs (US\$)}}{\text{Profit (US\$)}} \; x \; \text{Cost Savings (\%)}$$

The important fact to take away from this example is that for businesses where manufacturing costs are greater than the profits (as is the case for most manufacturing businesses), a given percentage decrease in production costs will have a greater percentage effect on profit and hence stock value. This insight should influence how we interpret Table 3.4 as the share value impact may be several multiples of the energy efficiency percentage value added.

Shares, as with other forms of assets, are priced on the investor's view of the future cash that the investment will generate and the level of risk associated with the investment. Clearly, investors appreciate that there are environmental, supply chain, market and other risks related to resource use and may therefore be inclined to apply a larger multiple on the share price if they can see that Widgets Inc. is managing its resources better than the average company in its sector. If the effect of this were to give Widgets Inc. a multiple of 8.5 times earnings as a result of visible management of resource risk, then the impact of the resource efficiency improvement of 5% of manufacturing costs would now be almost 33%.

Remember DuPont's US$47 million saving from energy efficiency in 2008/09 (*Real World: Energy efficiency savings claimed (page 94)*)? Well, that represented around 0.38% of the variable costs of US$12,500 million (the closest we have to manufacturing costs in published data).[233] This figure may seem insignificant, but let's see what that could mean in terms of share value, using the above formula plugging in the 2009 earnings before interest and tax (called EBIT) of US$2,578 million:

$$\text{Share Price Increase (\%)} = \frac{\text{Manuf. Costs (US\$)}}{\text{Profits (US\$)}} \; x \; \frac{\text{Manuf. Costs Savings (US\$)}}{\text{Manuf. Costs (US\$)}}$$

$$\text{Share Price Increase (\%)} = \frac{\text{US\$12,500 million}}{\text{US\$2,578 million}} \; x \; \frac{\text{US\$47 million}}{\text{US\$12,500 million}} = 4.81 \text{ x } 0.38\% = 1.8\%$$

While many factors influenced the share price of DuPont in 2009, from the analysis above we can see that the energy efficiency efforts in 2008/09 ought to have had a positive impact of around 1.8% or almost five times the 0.38% proportion of variable costs affected. As in the earlier example, this is because the variable costs are nearly five times the declared profits.

Rearranging the formula above, and eliminating the manufacturing costs term, we get an alternative means to determine percentage impact of the resource efficiency programme on share value, by using just the US$ savings and the US$ profits:

$$\text{Share Price Increase (\%)} = \frac{\text{Costs Savings (US\$)}}{\text{Profit (US\$)}}$$

If the organization is sales-focused, we may choose to describe the effect of the resource efficiency savings in terms of the equivalent sales required to achieve the same profit impact. In the case of DuPont, the 2009 earnings (EBIT) of US$2,578 million was earned on sales of US$26,109 million, so we can calculate the sales equivalent of the energy efficiency efforts as:

$$\text{Sales Equivalent (US\$)} = \frac{\text{Total Sales (US\$)}}{\text{Profit (US\$)}} \times \text{Cost Savings (US\$)}$$

$$\text{Sales Equivalent (US\$)} = \frac{\text{US\$26,109 million}}{\text{US\$2,578 million}} \times \text{US\$47 million} = \text{US\$476 million}$$

In other words, to make the US$47 million saved from energy efficiency, DuPont would have had to sell US$476 million more of product (with all the marketing and cash flow implications that would involve)! Saying that the energy efficiency programme is equivalent to almost "half a billion dollars of sales" will certainly have an impact.

Unfortunately, little research has been done to correlate share price movements with investments in resource efficiency. However, a very recent study by Wingender and Woodroof[795] suggests that there is a link between the announcement of an energy management project and abnormal share price increases over and above that expected from movements in the market as a whole in a period of 30 days after the announcement. For announcements made in the daily press, the abnormal increase in share price within 10 days of the announcement was 3.75%, with a high level of statistical significance (<0.01). Although the sample was small and used announcements from some time ago, it certainly seems that investors recognize that resource efficiency will inevitably translate into improved share price and so mark up the stock.

For organizations in the public or not-for-profit sector, this share value focus may appear irrelevant. However, there is an equivalent multiplier effect regarding service delivery, which relates to the marginal cost or the proportion of fixed costs to variable costs in their organization. Let us consider a hospital. There are many fixed costs – like the maintenance costs, heating and lighting, administration services and so forth, which are usually unaffected by patient numbers. Then there are variable costs like the drugs bill, patient catering costs, etc. which directly depend on the number of patients being treated. Finally, there are some costs which are called semi-variable, such as nursing costs which are unlikely to rise immediately if just a few additional patients are admitted but would do so once a certain level was exceeded (e.g. through the employment of additional agency staff).

Let's imagine a fictitious eye surgery unit in a hospital which carries out an average of 10,000 cataract operations a year (this is a good example as it is the most common operation in the UK totalling 260,000 treatments a year at the cost of £200 million – an average of £770 per treatment in 2006). If we consider the costs to run the hospital unit, shown in Table 3.15 on the next page, we can see that there are some fixed, some semi-variable and some variable costs which altogether add up to £7.7 million. To arrive at a marginal cost, we consider how these costs will rise if we increase our patient numbers by, say, 500 or 5%. We can see that, for the variable cost elements, the costs will rise by the full 5%. On the other hand, for some of the fixed-cost elements, there may be no cost increase at all (e.g. ward costs are largely determined by the property cost rather than the number of occupied beds), while the semi-variable costs will rise, but not by the full 5%. We can then add up these costs and see that the total cost is £140,000 to treat the additional 500 patients or just £280 per treatment. This marginal cost of £280 is considerably lower than the average costs per treatment of £770 because some cost elements do not rise with a modest increase in patient numbers. A substantial increase, such as doubling the treatment numbers, could mean that some costs we have considered fixed would also increase – e.g. new wards would need to be built.

	Current Cost		5% Increase in Treatments	
	£'000s	Cost type	% Additional cost	£ '000s Additional cost
Wards (rent, cleaning, maintenance etc.)	1,400	Fixed	0%	0
Utilities (electricity, gas, water)	440	Fixed	0%	0
Theatres (rent, cleaning, equipment etc.)	1,600	Semi-variable	2%	32
Diagnostics (Pathology/X-ray etc.)	700	Semi-variable	3%	21
Drugs and lenses	860	Variable	5%	43
Patient catering	200	Variable	5%	10
Medical staff	500	Semi-variable	2%	10
Nursing and other staff	800	Semi-variable	3%	24
Admin and overheads	1,200	Fixed	0%	0
Total Costs (£ '000s)	**7,700**			**140**
Total # operations	10,000			500
Average Cost per Operation £	**£770**		**Marginal Cost £**	**£280**

3.15 **Fictitious cost breakdown for a UK eye surgery unit treating 10,000 patients a year**
Alongside the current costs an estimate is made of the incremental cost to treat an additional 500 (5%) patients, from which a marginal cost can be established.
Source: Niall Enright

So what does this mean for our resource efficiency business case?

Well, let's imagine that we can reduce the utilities cost of this eye hospital by 20% which would be worth £88,000. Although this is a saving of just 1.14% of the total £7.7 million cost of the hospital, we could actually deliver £88,000/£280 or 314 additional treatments for this sum, an increase of 3.14%.

Thus the service delivery impact of resource efficiency is multiplied wherever the marginal cost of additional service provision is lower than the overall average cost (which is usually the case). The formula to use is:

$$\text{Service Increase (\%)} = \frac{\text{Average Service Cost (US\$)}}{\text{Marginal Service Cost (US\$)}} \times \text{Cost Savings (\%)}$$

There are many types of public services in addition to healthcare where there is a significant fixed cost element: for example in education (schools, colleges and universities), in social services, in policing, in public administration or voluntary sector organizations. If it is suspected that the marginal cost of service delivery is below the current average price, then it makes sense to calculate the additional service provision made possible by resource efficiency on a marginal cost basis, rather than an average cost basis.

We should note, of course, that the same marginal cost effects apply in private sector organizations and if we are building a resource efficiency business case around increased output or production, then we should do so at a marginal cost per unit output, rather than the average cost per unit of output. Naturally, this marginal cost depends on the size of the increase in production and factors such as the under-utilization or capacity of the existing plant.

3.5 Stranded assets and risk

Although there is no standard formula to calculate the effect of resource efficiency risk on asset value we can estimate this important source of value by either looking at the premium investors are willing to pay for certain assets over and above cash flow valuations, or by modelling the effect of resource risk on the asset.

Value

We finished the last item on the valuation of commercial property by indicating that there may be a hidden premium that asset owners are willing to pay for resource efficiency over and above the value that the income stream alone would justify.

If asset valuation relies on the two fundamentals of cash flow and risk, then it must be the risk component that leads the buyer to pay extra for an efficient asset. If one accepts the earlier argument on the inevitability of change in resource efficiency, then one can see why a long-term investor would conclude than an energy-efficient office would be less risky than an inefficient one. If energy prices rise at above the inflation rate then the efficient property will become increasingly attractive compared to the standard building. Indeed, future regulations may require inefficient building to be up-rated to higher efficiency standards (as is happening in the UK where it will not be possible to sell or rent buildings with the bottom EPC ratings of F and G from 2018). As new buildings are required to meet ever-improved standards of efficiency, what is today's high performer will be average and what is average will be poor. All of which represents a risk to our future cash flow and asset value.

We call any asset that suddenly becomes unusable, obsolete or non-performing a stranded asset. Energy and resource scarcity, climate change and the resultant change in consumer attitudes all pose quite significant risks to the future performance of assets. Without water, a golf course cannot maintain the fairways; if power stations are required to use carbon capture and storage, their costs will increase dramatically; if fish stocks collapse then the value of a fishing vessel does so too.

In discussing the business drivers for sustainability with a board director of one of the UK's biggest property portfolios, I was told that their entire approach was based on "*avoiding accelerated depreciation*". By this, the property company means minimizing the requirement to invest in building refurbishments at a greater level than originally anticipated because the property was underperforming in sustainability terms. By anticipating the future trends in building performance and bringing forward improvements in resource efficiency, the number and cost of future refurbishments could be minimized and the asset value maintained for longer. In essence, their intention is to put off the day when the building has to be knocked down and rebuilt - the ultimate form of asset revaluation.

Real World: ***Unburnable carbon***

Possibly the largest of all stranded assets are the coal, oil and gas reserves of energy businesses.

The total reserves of carbon in fossil fuels amount to 2,860 $GtCO_2$. All this carbon has a value on the balance sheets of major energy corporations as well as in the estimated wealth of many sovereign nations which are effectively fossil fuel businesses.

The Grantham Research Institute on Climate Change and the Environment at the London School of Economics and Political Science has calculated that only a small proportion of the carbon in these reserves can be released to the atmosphere if we are to limit temperature rises to 2° C. Our budget is 900 $GtCO_2$ for an 80% probability to stay below 2° C and 1,075 $GtCO_2$ for a 50% probability.[113] Carbon capture and storage would increase these numbers by only 12-14%.

Taking into account the valuation of carbon reserves of US$27 trillion[511] then, in the best case scenario, we would have to write off over half of this value. Indeed, it is not just the *equity* that is at risk but also the many trillions of US$ of bonds and debts that these companies have amassed.

For those who think that any correction in value will be borne by a small group of wealthy investors, think again. The largest part of this wealth sits in pension plans (where in the EU it accounts for about 5% of the total value of these funds).[778]

"Those corporations that continue to invest in new fossil fuel exploration, are really in flagrant breach of their fiduciary duty," according to UNFCC chief Christiana Figueres,[509] *"because the science is abundantly clear* [that these assets cannot be exploited]."

Property fund managers are not sentimentalists. They can see that an early response to the issues of resource efficiency will lead to a portfolio that is better performing in the long run than that of their competitors. If there is an opportunity to get rid of their "*scabby dogs*" (in other words, those properties in which the improvement costs outweigh the value increase) then they will do so.

The most common approach to assessing risk is called probabilistic risk assessment. This analysis is often used for climate change impact assessment - for example, a soft-drinks manufacturer might want to factor in the effect of water scarcity on their business. They would consider all the ways that changing climate could affect their business: the cost of corn syrup would rise, bad weather could affect distribution, hot weather might increase sales, water scarcity may cause production problems. Each of these individual effects is given a cost and a probability. Thus if there is a 1 in 100 probability that an event leading to a loss of £1 million could occur, then the impact of this risk could be said to be £10,000 (£1 million/100). If the *same* event leads to other losses, these can be added together to arrive at a total risk value.

Risk, such as supply risks have a direct effect on the cost of resources. Other risks manifest themselves in value terms in more indirect ways, such as customer perception which influences sales. One project I was involved in was for a global titanium dioxide (TiO_2) manufacturer whose main client used this energy-intensive white powder as an essential ingredient in consumer paints. The key business risk was the poor rating that the paint manufacturer was receiving in the Walmart Sustainability Index, as a result of the relatively high carbon intensity of their source of titanium dioxide. In fact, the manufacturer assessed the risk as being one-quarter of their entire sales. This cost was central to the business case to improve energy efficiency and decrease emissions.

Often the risk related to a resource depends on an estimate of the future costs of that resource. For example, in deciding to build a combined heat and power plant (which is usually more efficient than imported electricity because waste heat is used, not thrown away), one needs to consider the difference between the price of electricity and the cost of the input fuel (the *spark price*). In these circumstances, one would usually carry out an analysis where one would choose a number of alternative future price scenarios and work out the net savings or costs. This is called a sensitivity analysis, and it is often used making a business case for equipment or technologies which have a long lifetime (see page 593). Thus rather than looking at today's prices, it is much better to take a view on the future trend of costs and developed the business case using the long-term, risk-adjusted value the investment will bring.

Of course, there is one risk that every organization will bear (to a greater or lesser extent) but which they *alone* cannot eliminate. That is the danger of climate change, which requires individuals, organizations and institutions to work collectively to overcome. Although we cannot treat climate change mitigation as a straight cost-benefit proposition, we can nevertheless introduce a sense of its financial impact as a means of gaining support for greater efficiency.

3.6 Productivity

Asset Turnover is a key term in the DuPont Formula. It reflects the positive effect that increased productivity has on value. This effect is often overlooked when the sources of value from resource efficiency are quantified.

Another hidden benefit related to energy and resource efficiency, often overlooked in developing a call to action, is productivity. It is common for manufacturing organizations to have process *bottlenecks* which limit their operations. In many cases, these are directly resource-related. For example, I have lost count of the number of times I have seen plants which are constrained by their refrigeration capacity, many in the brewing industry. In these circumstances a project to increase the efficiency of the refrigeration system can, and should, be valued not in terms of the direct energy savings but in terms of the value of the additional product, beer, that can be produced as a result (using the marginal production costs to work out the profitability).

An alternative way of valuing the efficiency improvement through capacity improvement is to work out the capital cost for additional refrigeration equipment to deliver the equivalent increase in coolth to the plant. This avoided capital cost is the valuation process that underpins demand side management programmes - that is to say the programmes that electricity companies run to reduce electricity demand in a particular network and so *avoid* having to build new power-generation plant. As we shall see later (page 566), every business case relies on *comparing* a base case and an efficient case and often the base *business as usual* case may require significant capital expenditure that the efficient case does not.

A study by Ernst Worrell and colleagues, published in *Energy*,[815] found that the average payback of 52 manufacturing energy efficiency projects improved from four years, when only energy savings were included, to less than two years when both energy and productivity benefits were included. In many organizations today, this would make the difference between approval and rejection of the project.

Lest we fall into the trap of thinking of productivity benefits only apply in manufacturing, Fisk[282] estimates that the annual productivity gain from office workers through working in an energy efficiency environment is between 0.5% and 5%, worth between US$16 billion and US$160 billion in 1996 values.

The DuPont formula captures the notion of productivity in the term asset turnover. Productivity means efficient use of our available resources (inventory, equipment and people). It is quite easy to see just how resource efficiency can improve all these elements.

Real World: ***Diavik, diamonds and diesel***

Trevor MacInnis - Wikipedia

In a remote part of northern Canada is a diamond mine, owned by RioTinto, located on an island in a lake 220 km south of the Arctic Circle.

The equipment and materiel needed to operate the mine mainly comes by road from Yellowknife, 353 km away, during a narrow window of 10 weeks when the ground and lakes are sufficiently frozen to permit trucks to carry heavy loads to the mine.[623]

One of the critical inputs to the site is diesel, which fuels the power plant, mine vehicles and heating systems. Over 18 million litres of diesel are stored on site and need to be replenished in the short road transport window.

In developing a business case for the efficient use of the diesel, a key business driver will be the reduction in risk associated with a lower fuel requirement. Indeed, in 2006 a mild winter meant that the Diavik mine was not fully resupplied in time and the mine had to resort to costly air shipment to make up the difference.

Although this may seem to be an extreme case of resource risk, the reality is that many organizations could identify resources without which they would be severely affected. Reducing dependency or building resilience around these critical resources can form a sound value basis for an energy and resource efficiency programme.

3.7 Strategic and brand value

We have seen that the value of tangible assets, such as buildings, can be enhanced by resource efficiency. The value of intangible assets, such as brands, can also be increased through the adoption of resource efficiency best practices which provide strategic advantage.

The degree to which we will employ cost savings compared to competitiveness to justify resource efficiency depends on the nature of the organization we are seeking to persuade. For many organizations, the focus is not on making things more efficiently, but on bringing products to market more successfully. Recently, Apple became the world's most valuable company, surpassing oil giant ExxonMobil. Apple is a virtual business which outsources its manufacturing to third parties. Unlike ExxonMobil, Apple owns few physical assets like oil wells, tankers, refineries - what it does own are intangible assets like designs, patents and, above all, its brand, which according to Brand Finance is the world's most valuable, worth a staggering US$145 billion in 2016, making up almost 24% of Apple's US$586 billion market capitalization. That is to say the brand contributes to one-quarter of the Apple share price.

So, how do we begin to articulate the brand benefits of resource efficiency? Well, here we need to understand how companies value brands. Since a brand is something that a company owns, we need to turn our attention to the company balance sheet, which lists all the things the company owns. These can be its assets, as well as what it owes, its liabilities, which are calculated at a given moment in time, usually on the final day of company financial year. An asset might be a tangible asset such as a factory or cash in the bank, or an intangible asset like a brand, such as the Apple trademark and logo, a patent or a customer database.

One way that we can get to the value of intangible assets such as brands is to determine what investors are willing to pay for these whenever a business is sold. At that time the business will have tangible assets worth a certain amount (called the book value or net asset value of the company), but if the price paid for the business is higher than the sum of the tangible assets, then the difference is shown in the balance sheet as goodwill. This represents the value that people have placed on the intangible parts of the business, such as reputation, brands, customer databases and patents, etc., all of which can contribute to the future earnings of the business (hence the rather quaint name, goodwill, given to these intangible characteristics which influence future earnings).

*80% of the stock value on the US S&P 500 Index is in intangible assets such as **brand value**.*

Historically, goodwill would only be determined when a business was purchased or sold, but since 2001 US accounting rules (SFAS 142) require that US businesses should regularly assess the goodwill on their balance sheet

Standards: ***ISO 10668 brand valuation***

This new standard sets out the methods that an organization might use to determine brand value in a more consistent way.

This standard describes three broad approaches:

- The income approach – which considers the future earnings of the brand over its economic life compared to a non-branded competitor;
- The market approach – looks at the price paid for similar brands when businesses have been acquired;
- The cost approach – takes into account how much it has cost or would cost to develop the brand and assumes that a prudent investor would not pay more than this amount for a brand.

An important aspect of the ISO standard is that it requires the "*behavioural aspects of the brand*" to be considered when a valuation is being undertaken.

The valuer needs to understand and form an opinion on the behaviour of stakeholders in all the geographic and customer segments where the brand operates.

It is here that linkage between environmental performance and brand value can be incorporated into an assessment of the value that resource efficiency can bring through enhanced sales opportunities or diminished risk.

For an excellent introduction to brand valuation, see Gabriella Salina's *The International Brand Valuation Manual.*[638]

and determine if it is impaired – in other words, if its value has declined – in which case some of it has to be written off. This test applies at a business unit or brand level and is based either on the future cash flow the unit/brand will generate, or on a market value, e.g. if a similar brand has been sold. This is similar to earlier income or sales approach for valuing property (see page 104). When the accounting rules on impairment were introduced, there were some spectacular reductions of goodwill, with one-third of the companies with the largest goodwill writing off about 30% of it. The impact on some of the Balance sheets of these companies was huge; AOL Time Warner reported a write-off of US$54 billion.[381]

Value

This transparency on the value of brands on businesses' balance sheets is vital for investors as the market capitalization of many companies becomes increasingly dominated by intangible assets, particularly brands. Research by Ocean Tomo shows that the proportion of the intangible asset value of all the companies on the US S&P 500 Index in the US, has risen from 17% in 1975 to 80% in 2010.[572]

One of the criticisms around brand valuation is that there is a huge range of different methodologies to arrive at a brand value, many of which are proprietary, having been developed by brand consultancies. In the case of L'Oreal, whose case study is on the following pages, the Balance sheet value of intangible assets (goodwill and other intangible assets) is around €9.1 billion,[457] while the US brand valuation firm Brand Finance, valued L'Oreal's brands at over US$28 billion.[82] The difference lies in the Brand Finance approach, which is based on what an organization might pay to obtain the advantages that the brand confers.

Another source of variation is the different approach in valuations undertaken for financial purposes, which tend to consider attitudinal aspects less, compared to the valuation methods for marketing purposes, which sometimes do not adhere to established valuation best practice. The new ISO 10668 standard sets out the methods that an organization might use to determine brand value in a more consistent way (see left).

What the L'Oreal case study has highlighted is the importance of consumers in emerging markets to leading brands. Many of these consumers associate western brands with quality, luxury and trustworthiness. These are characteristics that the brand must preserve at all costs as they chase the rapidly growing demand for consumer products which are essential to the firm's success.

Brands which have invested heavily in a sustainable and ethical position, such as The Body Shop (see page 119), cannot afford to be complacent just because they have built up a solid set of values in the west over many years. They need to demonstrate their commitment to these new consumers and align with their dominant concerns, which may not be the same as the western customers'.

⇒page 118.

For some companies, like L'Oreal, cost saving does not drive efficiency.

There are much more **powerful** reasons to achieve ambitious improvements.

Real World: *L'Oreal enhancing brand value through resource efficiency*

There are many different drivers for energy and resource efficiency. I have, for example, recently worked on energy, waste and water efficiency with the global multinational L'Oreal, the world's largest cosmetics group. It is interesting to note that, for this company, the commitment to resource efficiency was not driven by cost considerations but by much more important strategic issues.

My involvement was with L'Oreal manufacturing sites in the US and Europe. The objective was to identify projects which could deliver a huge step-change in efficiency. For example, the goal for CO_2 emissions reductions was a 50% decrease by 2015 in absolute terms compared to 2005. Similar ambitious targets were in place for waste and water. The expert team which undertook the site audits were able to identify the projects that could deliver these saving - there was no single "silver bullet" solution for any of the resources, but rather a portfolio of projects which together exceeded the goals - and with a reasonably attractive payback. However, we were repeatedly advised during the site audits that cost was not the primary consideration - manufacturing, if I recall correctly, is just about 14% of L'Oreal's costs, with advertising and promotion being a much larger share. So if cost savings were not a driver, then what was? Simply put, today L'Oreal serves one billion customers worldwide. They would like to serve two.

> *In L'Oreal's case the objective of serving two billion consumers is inseparable from the ambition of making L'Oreal a model corporate citizen, based on sustainable, responsible and shared growth.*[459]

For L'Oreal is a business all about managing brands. And these brands represent significant value. Their flagship brand is L'Oreal, which is the second most valuable cosmetics brand in the world (after Olay) and was worth, in 2013, US$8,696 million or 34% of the value of the company.[82] All in all, the 13 major brands owned by L'Oreal listed in Figure 3.16 are valued at over US$28 billion by Brand Finance and represent 35% of the entire value of the company.

It is worth noting that, in addition to the visibility of the brand value arising from the impairment tests under SFAS 142, mentioned above, third parties such as Brand Finance have been producing independent valuation on many global brands for several years.[81] The Brand Finance methodology considers the royalty that a third party would be willing to pay to secure the marketing benefits associated with the brand – an approach that is supported by tax authorities as it is based on actual prices paid for royalties as well as publicly available financial information.

So how are the ambitious resource efficiency objectives of L'Oreal linked to the very considerable brand value of the company? In a presentation by L'Oreal's Director of Sustainability, Francis Quinn, made in June 2011,[609] we can get some insight into the motivation for resource efficiency. Quinn describes the basis for future sustainable growth as:

- *The strategic management of our raw materials*
- *The optimization of our intangible value drivers*

The first point reflects the fact that L'Oreal products depend on a vast range of raw materials, many of biological origins, which paradoxically puts future business growth at risk due to reduced availability, as well as health and environmental risks.

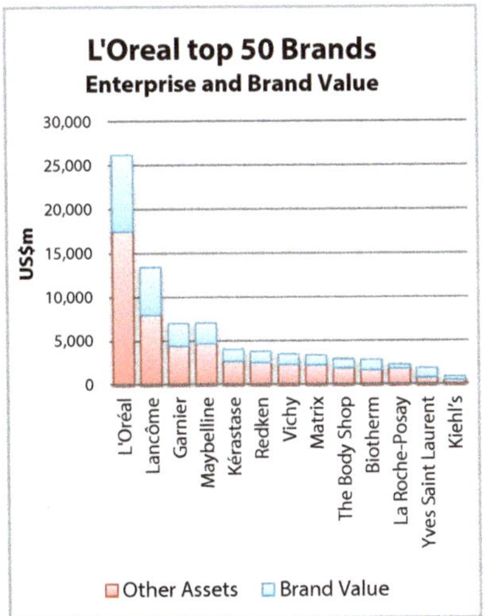

3.16 **L'Oreal owns 13 of the top 50 cosmetics brands in the world. The value of these brands is considerable.**
Source: Niall Enright based on data published by Brand Finance Global Top 50 Cosmetics Brands 2013[82]

This can be overcome, argues Quinn, by carefully selecting the raw materials which are just necessary and sufficient to the performance of the products, by facilitating the management of these *"premium choice"* resources internally, by simplifying manufacturing processes, particularly the supply chain and by maximizing product safety.

The second aspect of L'Oreal's sustainable growth is to *"optimize the intangible value drivers"* which contribute increasingly to the competitive advantage of L'Oreal. These include the human and intellectual capital which will help L'Oreal innovate products that are less resource-intensive and the *"relationship with stakeholders"*, which describes the fundamental trust that the consumer has with the L'Oreal brands.

So, we can see how managing resources and environmental impacts align to the core strategy of L'Oreal. No wonder then that L'Oreal is determined to deliver such ambitious – perhaps even market-leading – resource efficiency in its factories. No surprise then that the payback of these initiatives is not the key issue. We are, after all, talking about underpinning and growing US$28 billion of value in the business, the value associated with the brands, a value which is reviewed annually and for which any decline needs to be reported to investors. L'Oreal's insight is that its ability to continue to meet the needs of and retain the trust of customers depends in part on understanding and reducing the risk related to raw materials. L'Oreal sees resource efficiency as an integral part of its drive to generate the next one billion customers, and thus a source of competitive advantage.

L'Oreal's perspective is backed up by research that indicates that consumers are concerned about the environmental performance of brands. In a survey in 2007, a TANDBERG/Ipsos MORI poll of 16,800 people[698] showed:

> *"More than half of global consumers interviewed said they would prefer to purchase products and services from a company with a good environmental reputation, and almost 80% of global workers believe that working for an environmentally ethical organization is important. That amounts to one billion consumers and over 700 million workers worldwide."*

Interestingly, it was consumers from China who demonstrated the greatest sensitivity to corporate behaviour, with 67% of respondents agreeing with the statement: *"I would be more likely to purchase products and services from companies with a good reputation for environmental responsibility."* In the same poll, 45% of all respondents indicated that they had taken personal action on climate change with notably high proportions of respondents in Canada (56%), Australia (55%) and China (52%) stating they had taken personal steps to reduce their carbon footprint. If individuals consider a topic important enough to take personal action, then it is reasonable to conclude that these same consumers would modify their purchasing preferences if a brand were to be seen as having a positive or negative impact on climate change. The study concluded that China has the *"greenest population"*, with individuals in Australia, Canada, Sweden, the US and Great Britain also showing a keen interest in the environmental performance of companies, as illustrated in Figure 3.17 on the next page.

Clearly, L'Oreal has done its homework well. This study provides evidence that consumers in markets such as China are concerned about the business' environmental performance and supports the proposition that the ambitious resource efficiency initiatives, if effectively communicated, will indeed provide a source of competitive advantage.

3.17 **Rank of 15 countries with the "greenest populations", based on responses to workplace preference, purchasing preference and personal action on climate change**
Source: 2007 Tandberg/IpsosMORI.[698]

Rank	Country	Percentage preferring green workplaces	Percentage preferring green purchases	Percentage taking personal action	Total	Average percentage responding to green indicators
1	China	84	67	52	203	68
2	Australia	87	52	55	194	65
3	Canada	86	34	56	176	59
4	Sweden	86	46	42	174	58
5	US	81	42	41	164	55
6	Great Britain	74	27	46	147	49
7	Norway	79	30	37	146	49
8	Japan	57	40	45	142	47
9	Brazil	81	32	29	142	47
10	Netherlands	69	35	32	136	45
11	Italy	81	33	17	131	44
12	Spain	77	18	35	130	43
13	France	72	23	33	128	43
14	Russia	69	32	21	122	41
15	Germany	56	28	33	117	39

One reason that Chinese consumers may be particularly keen on the "green credentials" of products or companies, is the very high level of pollution that they encounter in many large cities - this would suggest that they would be supportive of brands which claim to have lower levels of waste, emissions and energy use.

Changing consumer sentiment opens up opportunities for goods and services to be delivered in innovative ways. As a result, otherwise mature markets are opened up to innovation and disruption. Tesla, a US manufacturer, took advantage of the opportunity that this offers to deliver an electric vehicle that is not only more resource-efficient than conventional cars but also surpasses them in several key respects. The fuel, for example, costs around one-fifth the price of gasoline; the torque of an electric motor offers much greater acceleration; the motor creates a near-silent ride and allows a top-end sound system to be installed; and the absence of many moving parts means that servicing is no longer required to maintain the car's warranty.

Consumers in **China** state the greatest intent to make green purchases.

Consumer anxiety about the environment creates the space for new business models such as products as services. As an appreciation of the challenges we face increases, dissatisfaction with existing models will grow. Consumers, on the whole, don't want to curtail their lifestyles, but they do want to enjoy their consumption in a way that does not lead to guilt or anxiety. This is why climate change offers the opportunity for radical change in many categories of spend.

For a consumer to have a **feel-good experience** when purchasing a product they need to know that the purchase will support their own **values**.

Real World: ***The Body Shop***

One of the key brands in the L'Oreal portfolio is The Body Shop, accounting for about 4% of turnover in 2011. Acquired by L'Oreal in 2006 this business, created back in 1976 by Anita Roddick, is famed for its ethical stand and emphasis on natural ingredients, over 60% of which it sources itself through its Community Fair Trade programme. Present in over 60 countries, The Body Shop is a case study of how being first to market with a strong sustainability proposition can differentiate a product and create competitive advantage.

The values which underpin The Body Shop's brand are clearly set out on their website under the banner *"living our values"*:

- Against Animal Testing
- Community Fair Trade
- Activate Self-esteem
- Defend Human Rights
- Protect The Planet

They go on to articulate these values as follows:

> *"We believe there is only one way to be beautiful, nature's way.*
>
> *We've believed this for years and still do. We constantly seek out wonderful natural ingredients from all four corners of the globe, and we bring you products bursting with effectiveness to enhance your natural beauty. While we're doing this we always strive to protect this beautiful planet and the people who depend on it. We don't do it this way because it's fashionable. We do it because, to us, it's the only way."*

The key to The Body Shop's historical and current success is that these values are not simply window-dressing; they are acted upon by the organization. The regular campaigns it mounts tackle some challenging issues such as human sex trafficking, HIV transmission or violence in the home.

On resource efficiency The Body Shop is reducing carbon emissions, using bottles made from recycled materials and has phased out plastic bags for paper ones. In the decade between 2010 and 2020, they have set a goal to reduce CO_2 emissions by 50%, reduce waste by 50% and water consumption by 25%.

This evidence that the organization is truly *living* its values is what underpins the feel-good consumer experience when shopping at The Body Shop. The customer is not buying nostalgia for past good deeds, but buying into the idea that their individual purchase will deliver some good for indigenous peoples, the environment or society now. The Body Shop's brand needs to constantly prove the fact that a purchase leads to additional action in support of its values.

Consumers associate beauty products with naturalness, cottage rather than industrial manufacturing, ethically sourced oils, etc. etc. The allure of the products would be diminished if they were linked with waste, pollution or oppression. We are not speaking here of ethical consumption, rather of brand association built around the essential trust between the brands and the, mainly female, customers for beauty products. Similar sentiments drive all of the L'Oreal brands.

Value

3.8 Consumer sentiment

It is not just brands that can be influenced by resource efficiency issues, but also whole categories of products.

3.18 **Bottled water is a product category that has come under fire as a result of resource efficiency issues. It is in danger of losing its *"cool"*.**
Source: © Luchshen - Fotolia.

It is not just individual companies but sometimes whole classes of products that are subject to resource efficiency criticisms. Bottled water is one of the fastest-growing, and most profitable, categories of drink sales but has come under considerable flak from environmentalist, consumer groups and even religious groups.[339] There is a very real risk that bottled water will fall out of fashion and become *"uncool"* as consumers recognize that shipping a bottle of Evian or Perrier or Pellegrino halfway around the world to restaurants in New Delhi, New York or New Zealand simply does not make environmental sense. In the US, cities from New York to Seattle to Fayetteville have banned single-serving bottled water from official functions[238, 467] on environmental and cost grounds. With 38 billion plastic bottles discarded each year in the US alone, manufacturers are reacting to the consumer backlash by reducing the quantity of plastic they put in each bottle, so-called *"lightweighting"*; for example, Pepsi reduced its Aquafina brand bottle weight by 40% to just 15 grams.[820] In a recent study for a major sports events promoter, I was asked to identify the most significant environmental effects of the events – and the large quantities of bottled water used at the venues turned out to be one of the biggest impacts. The promoter is now considering reducing bottled water in favour of water-fountains and refillable bottles.

Resource efficiency is proving central to consumer attitudes in this category, and this example demonstrates how the industry could have done a much better job in anticipating the reputational aspects linked to resources. For example, suppliers could have developed recycled packaging or sourced the water locally or encouraged reuse of their bottles or put in place recycling schemes. The problem is that it is much harder to repair a damaged reputation that it is to create a positive one. If these trends continue, no doubt investor sentiments will be impacted and share prices will begin to suffer.

It is no surprise then that across the majors in global bottled soft drinks, such as Pepsi, Coca-Cola and Nestle, there is an enormous amount of effort being put into defending their brands' impact on water resources. Water is their Achilles heel, and so water efficiency efforts are recognized to be of strategic importance to the future of their businesses.

The combination of increasing supply-chain disclosure and the water stress due to climate change mean that the pressure on bottled water producers is unlikely to diminish in the foreseeable future.

Some product manufacturers may feel immune from consumer or brand sentiment because they are far down the supply chain providing commodity ingredients. However as brand owners realize that initiatives like Walmart's Sustainability Index[770] may impact on customer sentiment, they are increasingly turning to life cycle assessment to understand if their products have high *embedded carbon* content. This disclosure means that high-emissions raw materials producers are under pressure to modify their processes or risk their customers' reducing their use of the raw material, or seeking alternative products with similar attributes but lower emissions.

Real World: ***Pepsi gets into deep water***

In 2003 Pepsi faced allegations that its operations in India were consuming too much water. This is a country where water holds special significance for many people – bathing in the Ganges is an act of purification – and where frequent scarcity has sensitized many Indians to the value of water.

The impact of these allegations increased when pesticides were found in many of Pepsi's products bottled in India. The fact that these were present in minute amounts and that other bottlers such as Coca-Cola also exhibited similar levels of contamination could not deflect critics who pointed out that water quality standards in the US banned any trace of pesticides.

So, Pepsi embarked on an effective, if not expensive, damage-limitation exercise over many years, addressing the water quality issues and investing in a wide range of community agriculture and other water conservation programmes which enabled them to claim:

> *"In our 2009 report we announced that PepsiCo's manufacturing facilities in India not only conserved 3 billion litres of water, but achieved a positive water balance – giving back more water than our facilities consumed. And they did that again in 2010."*[594]

This claim doesn't just appear on Pepsi's corporate web pages but it is also boldly repeated on Pepsi's Aquafina bottled water sold in India:[407]

> *"Giving Back MORE WATER Than We Take*
>
> *We call it 'Positive Water Balance'. To help save a precious resource that is fast depleting in India. Through rain-water harvesting, community water-sheds, and water conservation in agriculture, we at PepsiCo India saved 836 million litres* more water than we consumed in 2009.*
>
> *To know more, log on to www.tomorrowbetterthantoday.com*
>
> **As confirmed by an independent audit"*

Despite these undoubtedly positive efforts, PepsiCo continues to be challenged by activists in India on its claim of sound water stewardship. The critics point out that the calculations by the auditors, Deloitte Touche Tohmatsu India, omit the water needed to cultivate the large volumes of sugar added to its drinks and has only counted that directly used by the bottling plants.[407]

PepsiCo India, to coin a phrase, is "*not out of the woods yet*" and further work will need to be carried out to address the supply-chain impacts of their operations.

Sometimes resource efficiency is driven by an urgent strategic concern, such as PepsiCo's need to demonstrate good stewardship of water resources in India.

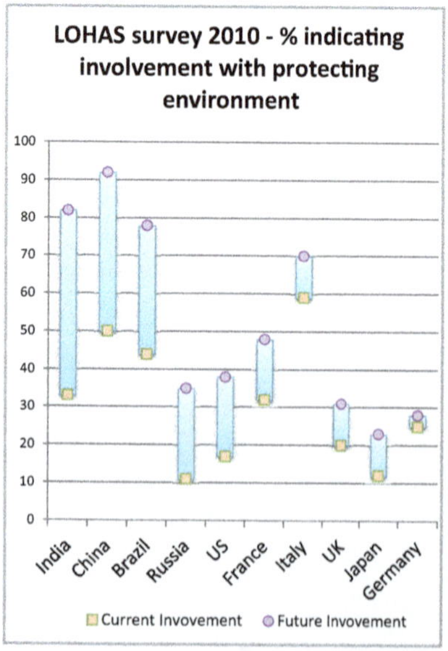

3.19 **The LOHAS research by NMI in 2010 demonstrates that consumers intend to increase their involvement with protecting the environment** This intent is particularly strong in three of the emerging BRIC economies of Brazil, Russian, India and China. *Source: Natural Marketing Institute (NMI) 2010 LOHAS Global.*[557]

The view that resource efficiency is a positive opportunity for businesses, rather than a defensive response to criticism, is supported by data on the attitudes of consumers by the Natural Marketing Institute (NMI), whose LOHAS (Lifestyles of Health and Sustainability) database has been tracking consumer attitudes to the environment for over 15 years. In this model, consumers are divided into five segments. First, there are the *"LOHAS consumers"* who have especially strong views about personal and planetary health (14% of US consumers). At the other extreme are the *"unconcerneds"* who have little interest in the environment (17%). Finally, there are three further groups, collectively the *"Sustainable Mainstream"*, who exhibit some degree of concern about sustainability (together 69%). Since the start of the LOHAS research, interest in sustainability has risen to the point that it has become mainstream, and this *"makes sustainability much more attractive to brands, and in fact, increases the business imperative since consumers are already engaged"*.[557]

The key message here is that green consumers are not a niche segment of the market when over 80% of US consumers are interested in green products and services. These green consumers are motivated by a range of factors: personal health, environmental stewardship, lower costs or the *"cool"* image or trendiness of green products. The NMI research again reinforces the view from the Tandberg/IpsosMORI poll mentioned on page 118 that green consumerism is not just a trend restricted to more prosperous economies, but is particularly strong in some fast-growing markets.

In her excellent exposition of environmental marketing[579] Jacquelyn Ottman describes one of four environmental consumers types as *"Resource Conservers"*:

> *"Resource Conservers hate waste. Spot them wearing classically styled clothing, toting cloth shopping bags and sipping from reusable water bottles. Avid recyclers of milk jugs and Tide bottles, they drop off old electronics at Best Buy. They read news online to save trees, and are quick to reuse their Reynolds wrap. Ever watchful of saving their 'drops' and 'watts', they install low-flow showerheads and compact fluorescent bulbs branded with EPA's Energy Star and WaterSense labels. Shunning over-packaged products, they only turn on the lights when they have to, and they plug their appliances into power strips for easy shut-off when they leave for work."*

As a Resource Conserver myself, I can relate to this description. We are the consumers who will be particularly attuned to the resource efficiency of the products we purchase. Having made personal efforts to live a lifestyle that conserves resources, we are highly unlikely to favour products associated with waste. Resource Conservers are the ideal marketing target for a whole range of products: from eco-friendly dishwasher tablets (which I use), to hybrid vehicles (like the Toyota Prius which I drive), through to compact fluorescent and LED lamps (which occupy almost every light fitting in my home) and domestic solar panels (3.3 kWp of which are fitted to my roof). Resource efficiency can bring tremendous value to those companies who can effectively tap into this consumer sentiment which strongly shapes product selection.

Real World: *Method*

One example of a highly successful breakthrough company using sustainability and resource efficiency as a key selling feature is Method Products Inc. This business has placed eco-innovation at the heart of its success, as demonstrated by a raft of industry firsts, from ultra-concentrated detergents, 100%-recycled PET bottles to the first compostable wipe.

Method has been successful because it got both the product right as well as the other aspects of marketing – the brand is funky, the packaging is modern and very attractive,[767] and cleaning is made a *"fun experience"*. All of which allowed the company to take away market share from long-established, more conventional brands by focusing on a specific group of customers: *"progressive domestics: a younger, professionally-employed, female-skewed customer. This customer tends to view home as a refuge, thus spending short periods of time cleaning different surfaces in the home on several days during the week as needed"*, according to a business school case study on Method's early success.[188]

By 2012, revenues were over US$100 million and the company merged with leading Belgian producer of environmentally friendly cleaning products, Ecover. Today Method products are available in more than 40,000 retail locations throughout North America, Europe, Australia and Asia.

Method makes a virtue of its resource-efficient production processes.

3.20 **Method illustrates its sustainability credentials under three headings: Products, Process (shown here) and Company**
The Method soap manufacturing plant in Chicago was the world's first industrial facility to achieve LEED Platinum certification and was designed by *Cradle to Cradle* author and architect, William McDonough.[507]
Source: Method Inc. press information.[524]

The Method story is about products built on resource efficiency, and then superbly marketed. The above case study describes the early sustainability objectives:

1. Zero waste › recycling industrial scrap;
2. Eliminate toxics › creating biodegradable products;
3. Renewable energy › purchasing renewable energy for office and facilities;
4. Close the loop › using biodegradable packaging;
5. Inform and educate › participating in a lecture circuit;
6. Redefine commerce › developing laundry detergent three times as strong as others.

The aim wasn't just to avoid environmental harm but to deliver change through the business, hence the emphasis on education in the fifth objective and on revolutionizing their industry in the sixth. The founders of Method see the purpose of their organization as creating good by using innovation to drive change.

3.9 Licence to operate

Every organization is constrained in what it is permitted to do. These constraints are often imposed by regulations or the needs of stakeholders. Increasingly, resource issues are influencing organizations' liberty to conduct their business.

In January 2014, the Chinese news agency Xinhua announced the closure of 8,300 heavily polluting companies in Hebei province. This is not the first such action on resource efficiency grounds; in 2010, the Chinese government announced the closure of over 2,000 highly energy-inefficient plants in 18 industries, including 762 cement factories, 279 paper mills, 175 steel mills, 192 coking plants and an unspecified number of aluminium mills. These examples provide a compelling demonstration of the link between resource efficiency and the ability to operate. China has announced its intention of reducing the emissions intensity per unit GDP by 40-45% by the end of 2020 and this will no doubt involve further closures of inefficient and polluting plant.

Compliance with environmental and social regulation is considered a prerequisite of our ability to function as organizations. We call these combined regulations, codes, permits, standards, etc., the licence to operate. This complex web of regulation significantly impacts on the core mission of many organizations: the capacity to make profits in private enterprises, or service delivery and stakeholder perception for public ones. For the Chinese factories, their licence to operate was summarily revoked.

Organizations can only exist if they have a ***licence to operate****. This requires compliance with of formal environmental and social regulations as well as standards imposed by customers.*

On closer examination, our licence to operate not only includes mandatory regulations but, for many organizations, we also often have an additional tier of self-imposed or voluntary standards, which shape what they can or cannot do. These voluntary standards may reflect industry efforts to self-regulate rather than to be regulated, or they may arise from stakeholder disquiet about issues such as CO_2 emissions.

Our capacity to operate is also affected by the licences to operate of our core suppliers, for changes to them can also affect us. While many of the Chinese firms mentioned above were not owned and operated by multinationals, their closure will undoubtedly have big implications for their customers, many of which may be multinationals or are contract manufacturing for multinational companies. In the complex web of supply chains, a perturbation caused by regulations on one part of the chain has the potential to influence many other parts.

Compliance with regulations represents the *"must have"* or minimum level of resource efficiency that we need to adopt as organizations. It is not optional, but it is true to say that it is sometimes begrudgingly applied in a *"tick box"* way as some organizations look for loopholes to avoid the additional burden or distraction they perceive this brings to their primary tasks (of making money or delivering services).

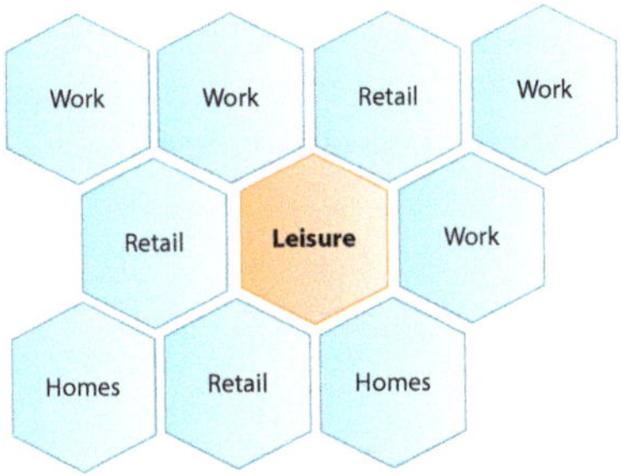

3.21 **Then: it used to be that the anchor of a development would be a leisure, work or retail destination**

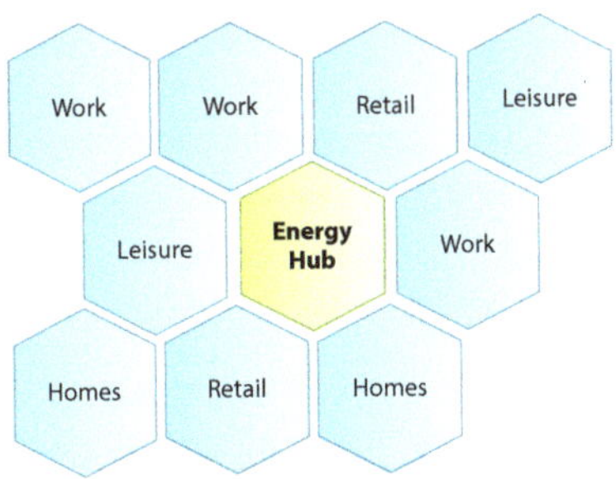

3.22 **Now: an available low-carbon energy source is becoming increasingly important in any large development**
Source: Niall Enright

So, how extensively is the licence to operate influenced by our resource efficiency? Very much, it seems. Regulations are increasingly determining the fundamental relationship between the resources we use and the value we create. This is true across all five principal areas of resource efficiency: energy and emissions, natural resources, water, waste and land use.

In simple terms, we can think of these regulations as being designed to impose:

- obligations to work within limits or standards in our resource use;
- requirements to measure or report resource use;
- obligations to pay a levy for resource use, including taxation and participation in markets such as the EU Emissions Trading Scheme.

Historically, regulation used to be all about the first approach – for example, enforcing limits on the amount of emissions organizations could put into the atmosphere, or creating minimum standards for the energy efficiency of new vehicles (a form of resource efficiency). Increasingly, though, we are seeing the second and third types of measures being introduced, in an effort to create additional incentives for organizations to take resource efficiency seriously or to use markets to deliver efficiency for the lowest cost. That does not mean to say that authorities aren't still heavily relying on the first category to drive change.

One area where regulation is the tool of choice to drive efficiency is in the property sector. This emphasis is no surprise, given that buildings are responsible for 40% of all CO_2 emissions and building is a process already heavily controlled through standards and codes. Back in 2009, the European Parliament voted to strengthen the Energy Performance in Buildings Directive so that all new buildings would have to meet tough *"nearly zero carbon"* standards from 2019 (public buildings from 2016). In the US, the groundwork for *"Net Zero Energy Buildings"* is being laid, as it is in many other jurisdictions. For organizations in the business of constructing, owning or operating buildings the resource efficiency writing is on the wall, to coin a phrase, and we have already seen how the arrival of a new generation of high-performing, high-efficiency buildings is driving property values.

For property developers, such as Peel Land & Property Group in the UK, where I have worked on sustainability for many years, these regulations are encouraging the incorporation of low-carbon generation as a key feature of new developments, as illustrated left. Peel has a somewhat unique advantage in this respect, in having both a renewable energy company, Peel Energy, and a supply business, Peel Utilities, within the wider group of companies. No doubt other property businesses will follow this approach in time.

Failure to comply with vehicle emissions regulations can be costly. Just look at VW, where the "dieselgate" scandal has led to provisions of €16.2 billion.

Property is not the only major sector where the imposition of standards is achieving radical transformation. Another example is the automotive industry, where emissions standards are also driving change. In 1998, the EU and the car industry (ACEA) entered into a voluntary agreement with legislators to achieve an average emissions level of 140g CO_2 per kilometre by 2008. This target was not achieved and so in 2008 the European Parliament passed legislation around a revised target of 130g CO_2 per km by 2015 and floated a longer-term target of 95g CO_2 per km by 2020. These emissions targets are not absolute, but are adjusted according to the mass of each vehicle (so that heavier vehicles are permitted to have higher than the average 130g CO_2 emissions per km target while lighter vehicles have a lower emissions allowance).

This mass adjustment was introduced to meet the needs of manufacturers and member states – French and Italian manufacturers tend to produce lighter vehicles with lower emissions, while the UK and German producer's vehicles are heavier with higher emissions. Once the portfolio is adjusted for mass, where a manufacturer's average emissions exceed the mandated level, they will pay a €95 per gram CO_2 per vehicle *"emissions premium"*. While the mass adjustment makes for a fairer allocation of emissions across manufacturers, it paradoxically works against materials resource efficiency by eliminating the incentive for manufacturers to reduce the weight (hence material) in each car as well as the absolute emissions per vehicle. Complementing the fleet emissions requirements are obligations for energy labelling of new cars, which informs consumers of the grams CO_2 per kilometre and which band (A to G) the vehicle lies in.

Real World: *Regulations with impact*

The EU regulation on the Registration, Evaluation, Authorization and Restriction of Chemicals (REACH) demonstrates just how far determined policymakers can go to regulate, restrict and even ban substances that have harmful impacts on citizens and the environment. This piece of regulation is probably the EU's most complex piece of legislation, covering some 850 pages:[790]

- A huge list of substances are covered (in their pure forms, in mixtures or in products);
- It applies equally to imported as well as domestically produced substances so has impacted organizations worldwide;
- The manufacturers have to take responsibility for assessing the risk of these substances and making the assessment public;
- All parts of the supply chain are affected; for example, retailers need to inform customers of the composition of a product within 45 days of receiving a request.

This regulation has caused a profound change in the global chemicals, manufacturing and retail sectors. No longer is it the case that regulators need to prove that a substance is harmful, but rather the manufacturers need to provide evidences (using strict scientific methods) that a material is not harmful. Where a material does have a negative impact, they must remove this from production, or where no immediate alternative exists they must demonstrate that they are investing in finding alternatives.

It is not entirely implausible to envisage policymakers using climate and resource impacts as criteria for future evaluation under REACH as these are material to human wellbeing.

Exploration: *Choice editing*

As described opposite, regulators are ensuring that new UK house-buyers have no choice but to purchase super energy-efficient homes from 2016. The option to choose a brand-new, poorly performing home will no longer be available to consumers.

This regulatory control over market availability of goods on the basis of their efficiency is quite widespread. Minimum energy performance (MEP) standards for appliances can be found in many countries. A case in point is that of incandescent lamps, which are being banned in many jurisdictions worldwide because of their poor efficiency compared to alternatives. As a result, manufacturers like GE, Toshiba and Phillips are getting out of the incandescent lamp business[396] and retailers like IKEA are not far behind.[29]

Standards do not just apply to business-to-consumer products. There are proposals to ban materials like oil derived from Canadian tar sands or unsustainable palm oil, or to outlaw emissions-intensive electricity generation plant, purely on the basis of the resources they consume.

Restricting availability of products is not just a question of regulation. Increasingly, retailers are removing products with poor environmental performance from their shelves. This is called choice editing, where the consumer's choice of a product has been reduced by the supplier. While it is not illegal to buy a refrigerator with a poor *"G"* rating, it is almost impossible to do so because none of the large retailers stock them. The ability for retailers to make these decisions is strongly influenced by mandatory rating and labelling schemes which enable products to be differentiated. The question of why the retailers refuse to stock the poor-performing products is interesting; is it because these products do not align with their own sustainability aspirations, or because the retailer feels they deliver poor environmental performance or value for money for their customers, or simply because they feel that once customers are able to differentiate (via labelling) the demand for those products will be low and so they do not deserve precious shelf-space? The answer is probably a combination of all of the above.

Choice editing is profoundly influencing a wide range of product categories from refrigerators and electrical goods, where the operational performance of the product is in question, through to responsibly sourced timber, sustainable fish, peat products and organic cotton, where the embedded resource and sustainability impacts are the basis for differentiation.

In putting together a resource efficiency business case, it would be prudent for most organizations to determine the degree of risk that they face as a result of choice editing, whether in the form of regulatory standards or as a result of retailer choice editing. See page 138 for a case study on the significant impact on the supply chain retailers can have regarding resource efficiency.

3.23 **Incandescent lamps, banned in many jurisdictions and no longer stocked by retailers**
Source: © photology1971, fotolia.com

Limits on resource extraction and use have a great potential to affect the ability of organization to achieve its primary goals.

In a similar fashion, limits on resource extraction imposed by authorities are forcing organizations to rethink and redesign their resource-consuming processes. Not long ago I was meeting with food manufacturers in the Spanish region of Cataluña, who were confronting sudden and dramatic decreases in the quantity of water they were allowed to abstract from aquifers using boreholes. Historically, they could take as much water as they needed – indeed, their requirements were prioritized because of their important status as employers and taxpayers. Now they were facing potentially ruinous costs to implement water recycling processes in their factories. Similarly, Australian industry was dramatically affected by a decades-long drought, and many other regions of the world are seeing the availability of free natural resource (particularly water, but also biomass) being severely constrained. With resource scarcity biting, a fundamental part of organizations' licence to operate is being rewritten through controls on what were perceived to be free, limitless resources.

Resource availability is not just influenced by limits on extraction, but also by export controls. With as much of 95% of the world's supply of rare earth elements being provided by China, there was a dramatic rise in prices in 2010 when China reduced export quotas by 72% to limit the environmental effect of rare earth extraction and to create a more sustainable resource.[59] This, in turn, has led to China being taken to the World Trade Organization by the US and Japan.[60] While, in time, other suppliers will enter the market because of the higher prices, the short-term effect may be to give Chinese manufacturers a competitive advantage as they are not affected by the export controls.

Resource supply constraints are thus not an issue just for the developed economies, like Cataluña or Australia, but touch all corners of the globe. We have already cited the example of China, which has become the manufacturing engine-room for many western companies and it is worth expanding on what is happening there regarding energy efficiency. In the 2006-2010 Five Year Plan, the Chinese government set a goal of reducing energy use per GDP production by a massive 20%. Why would the Chinese government act in this way, when the effect may be to create a barrier to manufacturing? According to the World Bank, Chinese industries use 20% to 100% more energy per unit of output than their US, Japanese and other counterparts, and they are acutely aware that this represents an enormous additional cost, environmental burden and long-term energy insecurity. From the government's perspective resource efficiency is in the national interest. Beyond 2010, China is committed to reducing its carbon emissions even further per unit of economic output, by 40% to 45% by 2020, compared with 2005 levels,[709] which will undoubtedly be translated into further regulation.

With the transfer of manufacturing to China and India etc., we are exacerbating the energy intensity of our current lifestyles. As people realize this, there is likely to lead to a greater scrutiny on resource efficiency in the emerging economies. This pressure on product efficiency is another reason to expect increasing policies around resource efficiency in the newly emerging manufacturing superpowers.

Thus in building a resource efficiency case it is important to understand what resources are regulated, directly or indirectly (via the supply chain), and also limited by their potential scarcity. The willingness of regulators to subject these resources to further constraints in the pursuit of policy objectives is a very real and convincing argument for early action on resource efficiency. These restrictions reinforce worries about *energy security* or *security of supply*, which are driving many organizations to diversify their resource supplies.

The final aspect of regulation is the imposition of a requirement to pay for resource use. Most organizations are familiar with abstraction licences, for example, to extract water from a river, which is often a set price per unit. Examples of taxation to do with resource use abound, such as the aggregates tax, which is a levy on sand, gravel and rock that is dug from the ground or dredged from the sea in UK waters. The tax addresses the environmental damage caused by these business activities in the form of noise, dust and loss of biodiversity. Note that the policy objective is intended to internalize an externality – that is to say to make aggregate users pay an environmental cost that they currently do not bear. Similar taxation can apply at *end-of-pipe* rather than at the extraction point – for example, the UK has imposed the landfill tax to encourage waste minimization, resource recycling and reuse.

One of the most closely observed examples of cost as a driver for resource efficiency has been the EU Emissions Trading Scheme (the EU-ETS). This scheme started in 2005 and put an obligation on over 11,000 large CO_2 point sources (that is to say process or generation plant with over 20MW thermal capacity) across Europe to surrender allowances for their emissions. The process of achieving emissions reductions in most of these emissions trading schemes is termed cap and trade, where each year the number of allowances made available is reduced, forcing organizations to either lower their emissions or to pay penalties for excess allowances. The trade component meant that the emissions allowances (which were initially allocated free of charge based on historical emissions, but are now being increasingly sold by auction) could be bought and sold between participants. This provides a revenue stream for those organizations which are able to most rapidly reduce their emissions, so directing investment in energy efficiency towards the organizations where it could be most cheaply delivered.

Market mechanisms to reduce resource use, such as CO_2 emissions trading, are adding costs to resource-inefficient organizations.

There are many critics of the EU-ETS Phases I and II. On the one hand, there is the fundamental concern that the scheme is merely providing *permits to pollute* (akin to medieval *indulgences*, which allowed sinners to continue to sin - as wittily observed by the English journalist, George Monbiot in the Guardian newspaper).[531] On the other hand, critics maintain that it was poorly designed and so has failed to provide a consistent price signal which is essential for long-term investment. What cannot be denied is that the experience gained has proved hugely valuable for policymakers globally and is underpinning the next generation of trading schemes which are now being implemented in markets like the US, China, and Korea.

Emissions trading schemes enable projects to happen which would not otherwise have been economic. For some organizations this has been hugely **value-creating.**

The idea of emissions trading in cap and trade has been used long before the current focus on carbon. The granddaddy of all the programmes successfully aimed to reduce sulphur dioxide emissions that cause acid rain in the US in 1990 through the Clean Air Act. Many cities also have a long history of local emissions trading for different substances. Examples include Santiago Chile from 1992 (focusing on PM10 particulates in the *"Emission Offset Programme of Supreme Decree No 4"*); Los Angeles from 1994 (NOx and SOx in the "Regional Clean Air Incentives Market – RECLAIM") and Chicago from 2000 (Reducing Ozone in the *"Emissions Reduction Market System"*).

2013 was a pivotal year for emissions trading schemes. The EU-ETS entered its third phase, albeit with a significant over-allocation of allowances, which has led to a collapse in the price of carbon allowances, which regulators are in the process of correcting. At the same time, nine new emissions trading schemes were launched worldwide: five in China and California in the US, Switzerland, Kazakhstan and Quebec in Canada. According to the report of the International Carbon Action Partnership,[385] two other schemes were launched in 2014, one more in 2015. In 2016, these schemes were putting a price to 4.5 billion tonnes of greenhouse gas emissions (9% of the world total). These figures are due to rise to 6.8 billion tonnes (16% of emissions) as new emissions trading schemes come online in China and Canada.[385]

The early forecasts of a common global approach to emissions trading have given way to a patchwork of legislation as policymakers find international consensus through forums such as the United Nations Framework Convention on Climate Change (UNFCC) difficult to achieve. With the solitary exception of Australia, which repealed its emissions trading legislation in 2014, organizations should interpret these schemes as signalling the strong intent of policymakers to rein in future carbon emissions.

Within environmental and policymaking circles there is a lot of debate about the use of cap and trade schemes to manage another resource: water. Already there are programmes in place in the Ohio River Basin and Chesapeake Bay in the US – just two of 14 *"Water Quality Trading"* schemes in place in the US. Worldwide, a recent survey by Ecosystems Marketplace[676] identified 113 Payment for Watershed Services (PWS) schemes worldwide. The largest number of schemes is in Latin America, but by far the greatest trading value is in China. The majority of schemes involve government payments and schemes with private sector participation are few and far between, but the report concludes:

> *"While relatively speaking this type of payment is still small, or so it seems, that this is perhaps one of the areas where we are likely to see tremendous growth in the years to come. After all, if the private sector does not start paying for watershed services, then we are missing an important potential solution to this problem."*

A variation on the cap and trade approach is the UK's Climate Change Levy, which taxes the use of electricity and gas (0.568 and 0.198 pence/kWh respectively in

Country / Region	Name	Sectoral Focus	Status / Date
UK	Climate Change Agreements (Climate Change Levy)	Large, non-domestic electricity and gas users	2001(closed 2009 as EU-ETS replaced it)
Australia (New South Wales)	NSW Greenhouse Gas Reduction Scheme	Electricity retailers and certain large electricity users	2003 (closed 2012)
EU	Emissions Trading Scheme	Industry and Power Generation, >20MW, 11,000 installations	2005
Canada (Alberta)	Greenhouse Gas Reduction Programme	Large emitters of CO_2 (>100,000 t pa)	2007
Canada (British Columbia)	Greenhouse Gas Reduction Targets	government operations	2008
New Zealand	Emissions Trading Scheme	Power, industry, transport, waste, forestry	2008
US (9 states)	Regional Greenhouse Gas Initiative	Electricity generators, >25 MW sources	2009
UK	Carbon Reduction Commitment	Large energy users: > 6.000 MWh electricity pa	2011
Japan (Tokyo)	Emissions Trading System	13,000 facilities over 1,500 kL crude oil equivalent p.a.	2011
Japan (Saitama)	Saitama Prefecture ETS	600 large facilities	2012
US & Canada (7 states and 4 provinces)	Western Climate Initiative (includes the California AB-32 initiative)	Electricity, industry, transportation, residential, and commercial fuel use	2012/2013
Kazakhstan	Kazakhstan ETS	Power sector and industry	2013
Switzerland	Swiss ETS	Power sector and industry. In force since 2008 but mandatory from 2013	2008/2013
Canada (Quebec)	Quebec ETS	Power sector, industry, with transport and buildings later	2013
China (Beijing, Guandong, Shanghai, Shenzen, Tianjin)	Emissions Trading Scheme	Power sector and industry. Initial pilot in advance of national scheme planned for 2015	2013
China (Chongqing)	Emissions Trading Scheme	Industry	2014
China (Hubei)	Emissions Trading Scheme	Power sector and industry	2015
South Korea	Emissions Trading Scheme	400+ large producers of CO_2 (>25,000 t pa): ~60% of emissions	2015
China (National)	Emissions Trading Scheme	Power generation, metallurgy and non-ferrous metals, building materials, chemicals, and aviation (> 26,000 tCO2/year)	Proposed for 2017
Canada (Ontario)	Ontario ETS	Power sector, industry, transport, buildings, agriculture and waste	Proposed for 2017
Ukraine	Emissions Trading Scheme	Similar to EU-ETS, details not yet available	Proposed for 2017

3.24 **Carbon Emissions Markets worldwide**
Source: Niall Enright and ICAP.[384]

Environmental Market	Market Value (US$M) 2008
Regulated Carbon	117,600
Water Quality	9,250
Biodiversity	2,900
Voluntary Carbon	705
Forest Carbon	37

3.25 **In 2008 the global market for environmental services was valued at around US$130 billion**
In 2014, US$141 billion was traded in the various schemes with the EU-ETS representing the lion's share of activity. The table above shows that carbon is not the only maker, although it is by far the dominant one. Whatever way we choose to look at it, resource use (and environmental services) are likely to have an increasing influence on the value our organizations can deliver.
Source: Ecosystems Marketplace[676]

2017, in the order of 5-8% of the energy costs for large industrial users). *Energy intensive* industries can reduce the levy charges by 90% for electricity and 65% for gas if they enter into a Climate Change Agreement, or CCA, with the government. The CCA describes an emissions reduction commitment which can incorporate a *product mix algorithm* and so permit absolute emissions to rise if production or some other measure of activity has increased. There is a total of about 5,000 CCAs in the UK, which are organized through 54 sector agreements (usually administered by trade associations). These agreements are an example of modifying a tax by the "carrot" of a rebate for emissions reduction.

As if two emissions pricing schemes were not enough (the EU-ETS and CCL/CCA) the UK has also introduced the Carbon Reduction Commitment Energy Efficiency Scheme (CRC). This scheme is designed to incentivize large energy users not in the EU-ETS or CCA schemes to reduce energy use through a straightforward levy on carbon emissions of £16 per tonne, which has the effect of raising energy costs by 8-12% depending on fuel mix and tariffs. Originally, the scheme had a novel recycling mechanism which would reward those organizations which made the greatest emissions reductions, but this was scrapped as part of the austerity drive and the £700 million pounds raised kept by the government. The burden of multiple overlapping schemes has led the government to consider merging these into one single system in the future.

Businesses are largely in favour of market-based emissions reduction schemes as they see the regulatory alternative as the imposition of fixed emissions limits across industry or the imposition of a flat carbon tax. The trade element of cap and trade enables individual businesses to decide whether to go to the market to buy emissions allowances or to invest in reducing emissions. In theory, the market will deliver the cheapest reductions first, thereby increasing the overall efficiency of the scheme and reducing the impact on profitability. In 2012, 150 global businesses called on governments to develop "*clear, transparent, and consistent price signals*" through "*creation of a long-term policy framework*" that would include all major producers of greenhouse gases – in other words to develop a global emissions trading system.

Policymakers like the cap element of cap and trade, which gives them assurance that the targeted sector will operate within the (tightening) limits imposed by the cap. At the same time, the links between these schemes and the Kyoto Mechanisms, which allow rich countries to invest in emissions reductions in developing countries (so-called Annex 1 states) rather than at home, is popular with the developing countries which receive inward investments as a result. Although there are many legitimate concerns about the effectiveness of these emissions trading schemes, it seems that the political and business weight behind them suggests that they will be a feature in resource efficiency for some time. While the price of the allowances is in the order of 10-20% of the cost of the energy itself, this figure can be expected to rise in time.

Exploration: *Kraft and Unilever*

Shortly before publication of this book, Kraft made an audacious bid to take over Unilever, which was soon withdrawn. This bid raises some interesting questions about the effect of sustainability on corporate value.

It seems to me that there is a clash of visions here, between those who see sustainability as adding value in the long-term and those who consider that these are non-essential luxuries that drag on the short-term returns of the business.

The private equity investors behind Kraft have a track record of aggressive consolidation of businesses and cost-cutting. The problem for them was that, if sustainability does add value, then stripping out this activity from the company may not have the desired effect on the stock price.

The bet the private equity investors were making is that the short-term profit increase by cost-cutting would exceed the premium that the current shareholders place on the business because of its social responsibility.

On a bid of this size, there are inevitably big political considerations on each side of the deal. Unilever rebuffed Kraft's bid on the basis of *"strategic"* as well as financial reasons, which I take to be code for the difference between short-term and long-term thinking about value. It seems that these messages resonated with policymakers and markets, causing Kraft to withdraw.

This experience may suggest that Unilever's record on sustainability has had a protective effect against the hostile bid by Kraft. Whether this is as a result of investor sentiment or wider political support is not yet clear. But, either way, this protective effect is, arguably, another source of value for sustainable companies.

Real World: *An observation on investor sentiment*

A discussion on share and asset value would not be complete without noting that, in the real world, there is evidence that some markets continue to react *unfavourably* to resource efficiency efforts such as voluntary emissions reduction initiatives[421, 223] or to participation in programmes like EPA Climate Leaders.[281]

It seems that not all investors are as wholly enthusiastic or convinced about the benefits of energy and resource efficiency as I am.

The question is whether this adverse reaction is due to investor bias against these initiatives or whether it is due to an underlying negative aspect of environmental leadership - such as the diversion of capacity from more productive work.

Clearly, there will be cases where overzealous efforts on environmental performance will reduce the competitiveness of an organization - particularly when these are not mandated on all market participants. However, this is likely to be the exception rather than the norm, and energy and resource efficiency are predominantly value adding rather than value destroying, if sensibly adopted.

What seems to be at work is sentiment. It appears that some investors continue to overestimate the costs or underestimate the benefits of resource efficiency, at least, from the evidence above, in respect of emissions by US-listed firms. This perception does not mean that resource efficiency is bad per se, just that we need to better ensure that the net financial benefits that flow from such voluntary actions are properly quantified and communicated to the market. It may be that proponents of resource efficiency are not articulating the benefits adequately.

This bias is given false legitimacy by the narrow interpretation of fiduciary duty in which the sole responsibility of institutional investors and executives is considered to be to maximize financial returns for the shareholders. Thus any investment going beyond the requirements of minimum compliance represents an abuse of position. See *Why fiduciary duty and shareholder rights need to be reinterpreted* (page 218) for a fuller exploration of this topic. Unfortunately, this discredited and outdated *"zero sum"* perspective, in which the choice is stated as either profit or the environment, persists widely today, especially in the finance community.

Another factor at play may be skills. Many financial decision-makers lack training in science or technology, so they do not grasp the issues presented to them by engineering, environmental or design experts. Finance functions, with some notable exceptions, tend to be conservative and to stick with what they know, and putting a price on resource sufficiency is not usually in their repertoire, especially if they do not understand the detail.

We shall see in the next chapter on the barriers to resource efficiency that there are some further psychological and behavioural reasons for this investor bias.

All of these factors mean that proponents of resource efficiency need to work doubly hard to quantify the value on offer (direct and hidden) and to establish the evidence base that can overcome this innate scepticism. We have already seen that there is evidence that environmental performance impacts positively on shareholder return, as demonstrated in the piece *Real World: Good environmental performance correlates with share value* (page 106). The studies on asset value, although at an early stage, are also suggesting a premium for energy efficiency and, possibly, sustainability. In due course, hopefully, this will be widely accepted.

3.10 Disclosure

Transparency and disclosure are seen by many organizations as threats to their image and reputation. However, for organizations which are demonstrating leadership on resource efficiency, consistent reporting will enable them to gain competitive advantage over those organizations which are late to respond.

At the centre of our resource efficiency value pyramid is the concept of disclosure which represents all the communications, voluntary or involuntary, that an organization makes - or are made about an organization - to customers, shareholders and stakeholders about its resource efficiency performance. Because of disclosure, shareholders can assess the returns on investment, legislators can regulate, commentators can pass judgement, the public can exercise choice and lobby groups can bring influence to bear. Disclosure allows organizations to benchmark themselves and it couples the needs of shareholders to the needs of stakeholders. Because of the potential for value creation or destruction, the nature and impacts of resource efficiency statements made by the organization should be considered carefully.

Disclosure drives resource efficiency most effectively where it provides a means of comparison between choices. Supplier A's milk has embedded carbon of 1kg CO_2 per litre while supplier B's is only 0.9 kg per litre. All other things being equal, vendor B now has a competitive advantage over supplier A and may, over time, be favoured by progressive retailers. Supplier A, because of disclosure, can see that B has this advantage so may now choose to invest in order to match, or even improve on, supplier B's performance. If the changes needed by a vendor to improve performance are relatively low-cost (or may even generate savings), then the case for action becomes quite compelling. Public availability of data supports the notion of informed decision-making, which is the key to sustained resource efficiency (see page 271).

The requirement to ***disclose performance*** *may prove to be the single most powerful driver for improvements in resource efficiency.*

Disclosure of resource efficiency performance, in many forms, is being increasingly mandated. In the US, for example, *Part 98* rules enacted by the Environmental Protection Agency (EPA) require suppliers of fossil fuels or industrial greenhouse gases (GHGs), manufacturers of vehicles and engines, and facilities that emit 25,000 tonnes or more per year of GHG emissions, to submit annual GHG emission reports to the EPA. In due course, these thresholds are expected to drop. Many states also have their own reporting requirements that go beyond the EPA's.[658]

Similar provisions exist in Europe under the EU-ETS for facilities with a thermal capacity of 20MW. In the UK, the recent Carbon Reduction Commitment mandates reporting for all organizations that consume over 6,000 MWh of electricity a year, which captures organizations not reporting under the EU-ETS and the Climate Change Agreement regulations. In

Mandatory disclosure in companies' annual reports and accounts is growing.

Australia, the 2007 National Greenhouse and Energy Reporting Act introduced a single National Greenhouse and Energy Reporting Scheme, NGERS, for the reporting and dissemination of information about greenhouse gas emissions, greenhouse gas reduction projects, and energy use and production of corporations. The limits here are, like in the US, expressed in tCO_2 equivalent emissions and have been set much lower, at 5,000 tonnes for facility thresholds and 125,000 tonnes for corporations, dropping to 50,000 tonnes in 2012. Many other countries including Japan have mandatory GHG reporting for some sectors of industry.

Another form of mandatory disclosure is the obligation for companies to provide GHG reporting in their annual reports and accounts. In the UK, the Climate Change Act 2008 mandates that companies listed on the London Stock Exchange report greenhouse gas emissions annually. Internationally, the IASB/FASB, which governs accountancy standards in many jurisdictions, is also working on standards for carbon emissions trading to form part of the International Financial Reporting Standards (IFRS). While it may appear arcane, the nuances of how rights to resources (e.g. allowances to emit carbon into the atmosphere) are treated financially may have a significant impact on the business case for resource conservation. Should an allowance be treated as income, especially when it is "free issued" by a government? What about the liability that is brought about by the act of emitting carbon? What happens when allowance values change in a market like the EU-ETS where these are freely traded? A paper on the subject by the Association of Chartered and Certified Accountants in 2010[482] showed that for large EU-ETS participants, carbon allowances could average some 2.4% of total liabilities – a material figure for businesses like power generation where margins are small.

Value

There is evidence that current carbon reporting does not meet the needs of investors or fiduciaries (boards of governors etc.). A recent survey[342] of investors summarized:

> *"The outcomes are unexpected. We identify an absence of a general market momentum towards environmental investing while at the same time strong demand for company reports on environmental matters. Although most questionnaire respondents and interviewees had collected company-issued reports on greenhouse gases emissions levels and environmental management programmes, all were dissatisfied with that information, and none had used it to guide portfolio allocation levels."*

Although carbon (and increasingly other resources such as water) are material to some businesses' financial performance, the quality of reporting and the poor engagement by the market mean that resource exposure is not being properly translated into the assessment of future share value. There is a *chicken and egg* problem – until reporting is reliable and consistent, investors will not make decisions based on it, but the effort will not be put into improving reporting until it is seen to be material to investor decisions.

Businesses are anticipating that this logjam will be resolved by mandatory reporting and are seeking to influence the nature of these requirements. One key initiative is from the Climate Disclosure Standards Board, "*a consortium of business and environmental organizations formed for the purpose of jointly advocating a generally-accepted international framework for companies to disclose information about climate change-related risks and opportunities, carbon footprints, carbon reduction strategies, and their implications for shareholder value*". Its sample climate change report, which can be found as an annex to the draft Climate Change Reporting Framework,[130] runs to 26 pages and covers all aspects of "*Typico plc's*" emissions. I would strongly urge any CEO or CFO or FD reading this to have a look at this glimpse of potential future reporting requirements.

Another reporting initiative is led by the International Integrated Reporting Committee (IIRC), which describes itself as "*a powerful, international cross section of leaders from the corporate, investment, accounting, securities, regulatory, academic and standard-setting sectors as well as civil society*". The IIRC's objective is to develop a new approach to corporate reporting that integrates financial and social reporting in one document. The thinking behind integrated reporting is that it will reveal the linkages between an organization's strategy, governance and financial performance and the social, environmental and economic context within which it operates. In its first discussion document[395] the IIRC set out some ambitious goals:

> *"Integrated reporting reflects and builds upon existing developments in reporting, including the following.*
>
> - *The ongoing international convergence of accounting standards through the collaborative efforts of the International Accounting Standards Board (IASB) and the US-based Financial Accounting Standards Board (FASB) to improve both IFRS and US GAAP, and to eliminate the differences between them*
> - *The work of The Prince's Accounting for Sustainability Project, the Global Reporting Initiative, the World Business Council for Sustainable Development, the World Resources Institute, the World Intellectual capital Initiative, the Carbon Disclosure Project, the Climate Disclosure Standards Board, the European Federation of Financial Analysts, the United Nations (UN) Conference on Trade and Development, the UN Global Compact, the International Corporate Governance Network, the Collaborative Venture on Valuing Non-Financial Performance, and many others to develop principles, methodologies, guidelines and standards for the accounting and reporting of non-financial information."*

The thinking behind **integrated reporting** is that it will reveal the linkages between the organization's strategy, governance, and financial performance and the social, environmental and economic context in which it operates.

It is interesting to note that the four corporate reports cited in the discussion paper as examples of integrated reporting are from companies which depend heavily on resources in their business – AkzoNobel, BHP Billiton, SASOL, and Anglo American – which have wide-ranging environmental, social and financial impacts in the communities in which they operate.

Indicator
B Corporation Ratings
Bloomberg Sustainability Reporting Initiative (1)
Business in the Community (BITC)
Carbon Disclosure Project Leadership Index (4)
Ceres Water Risk Benchmark
Corporate Knights Capital Global 100
Corporate Responsibility Magazine 100 Best Corporate Citizens
Dow Jones Sustainability Index (2)
ECPI Sustainability Index
EIRiS Company Sustainability Ratings
Ethibel
Footsie4Good Index (5)
Fortune's Most Admired Companies
GoodGuide
Inrate Sustainability Assessment
Maplecroft Sustainability Performance Benchmark (MSPB)
MSCI ESG Indices (3)
Newsweek's Green Rankings
Oekom Corporate Ratings
Sustainalytics Company Ratings (6)
Thomson Reuters Asset4 ESG Rating
Trucost Corporate Environmental Data and Profiles
Underwriters Laboratory UL880 for Manufacturers
Vigeo ASPI
Walmart Sustainability Index (see page 138)

3.26 **Disclosure includes the many corporate and product ratings systems that have sprung up in response to consumer and investor demand for data on sustainability and resource efficiency**
The top six most used schemes according to SustainAbility[692] are shown in brackets.
Source: Niall Enright.

A paper by Michael Blanding at the Harvard Business School[73] identified 16 countries with mandatory corporate social responsibility (CSR) reporting. A linked paper[234] suggested that while the numbers of companies implementing integrated reporting is currently small – with most of those businesses that produce CSR reports doing so in a separate document – there is a growing demand not just for reporting but for integrated reporting. From 1 March 2010, 450 companies on the Johannesburg Stock Exchange were required to file an integrated report or explain why they have not done so, and in France, the Grenelle II Act, passed in July 2010, makes integrated reporting mandatory for about 2,500 businesses and a few hundred state-owned companies. For the reader interested in exploring the subject of integrated reporting, the IIRC discussion document[395] and Harvard Business School's *The Landscape of Integrated Reporting*[234] are good starting points. What is clear is that disclosure of corporate performance on resource efficiency is becoming increasingly mandated, transparent and standardized.

The pressure for disclosure is growing. In the US, for example, powerful investor groups such as CalPERS (the Californian Public Employees Retirement System) and the Investor Network on Climate Risk, which together represent US$7 trillion in assets, have been lobbying for companies to make much more transparent assessments of their exposure to climate change – in particular in their SEC 10-K reports. While not strictly a resource efficiency reporting requirement, this does throw a spotlight corporately on related issues such as water scarcity, power prices and supply-chain pressure.

As the linkage between investment performance and social and environmental issues becomes clearer, and to support the demand from consumers to have independent assessments of company performance, third-party rating systems have proliferated (see left). Some of these systems depend on voluntary disclosure, such as the Carbon Disclosure Project, while many others will rate the organization regardless of whether or not they wish to be measured. Some, such as the GoodGuide rate individual products and so influence consumer perceptions of brands as well as companies. SustainAbility recently carried out a research programme on these rating schemes, *Rate the Raters*, and catalogued over 100 different schemes.

The rating systems themselves have come in for criticism, however, where the ratings do not align with public sentiment about organizations. For example inclusion of Halliburton in the Dow Jones Sustainability Index raised some eyebrows among commentators,[662] because of the relatively small weight put on social and environmental factors in the index. As SustainAbility pointed out in their study, transparency around the scoring criteria is essential if these rating schemes are to enjoy the trust of investors and the public alike.

The next wave of rating systems, arguably the most rigorous to date, are being driven by a different group of concerned stakeholders: retailers. They are forcefully driving product sustainability appraisal down into their supply chains because of the considerable financial and strategic value this offers them.

Real World: ***Walmart Sustainability Index sends shockwaves down the supply chain***

Since 2012 Walmart has been encouraging its suppliers to participate in a sustainability programme that rates the products Walmart buys from each supplier in up to 15 key metrics, many of which are related to energy and resource efficiency.

When Walmart encourages a vendor to take a particular course of action, they tend to do so. Walmart is the world's largest retailer, the second-largest public corporation, the world's biggest private employer. Their influence is enormous, not just in the US but the 27 countries they operate in worldwide.

Initially, the programme was measuring the overall sustainability of the supplier. However, in 2012, the first set of questionnaires were developed around highly specific categories of supplies (as diverse as computers and toys, down to particular foodstuffs such as pasta, apples or bread). These questionnaires have been developed by The Sustainability Consortium, which means that this approach is now used by many other global retailers such as Marks & Spencer and Ahold, as well as many consumer goods suppliers from Unilever through to McDonald's.

The important thing to note about these questionnaires is that they have the potential to inform purchasing decisions within Walmart (and the other participants):[749]

"Every one of our buyers, and their leadership, are required to set annual performance objectives related to sustainability topics in their categories. The scorecards are the primary tool our buyers use to prioritize improvement opportunities and set their annual performance objectives."[770]

In other words, sustainability performance, including climate impacts and resource use, will be a factor influencing the selection of suppliers to Walmart.

Walmart made it clear that *"responses to this questionnaire will be accepted in good faith, relying on the integrity of the supplier. Violation of that good faith will be considered very serious by Walmart. Merchandising teams will be trained in how to identify inaccuracies."*[747] No wonder then that when these questionnaires were first introduced the consulting firm I was working for (ERM) had many anxious calls from Walmart vendors concerned to respond accurately and to score well. Indeed there has been a ripple effect down the entire supply chain so that companies that are distant commodity producers have felt the effects of the questionnaires (such as the titanium dioxide producer mentioned on page 112).

It is important to note that Walmart sees resource efficiency as completely aligned with its core purpose *"to keep our promise to our customers of delivering great products that will help them save money and live better"*. The sustainability initiative is expected to drive down the costs – savings which no doubt benefit Walmart shareholders as well as customers. This commercial driver is good, for it means that sustainability is not just *"nice to have"* but *"must have"*.

3.27 **Suppliers wishing for their goods to feature on the shelves of Walmart will need to deliver good sustainability performance** The desire to sell more sustainable products features prominently in Walmart's communications. *Source: Courtesy of Walmart Press Office.*

3.11 Product certification

As well as increasing levels of corporate disclosure, transparency around the resource efficiency of products is growing, from both voluntary initiatives and mandatory labelling and declarations.

Value

We have seen how stakeholder and consumer sentiment represents a huge source of value for organizations that can demonstrate leadership in resource efficiency. This value can then cascade down into the brands that an organization owns, as in the case of L'Oreal. This positive brand effect adds further value to individual products which have positive environmental attributes.

Where a producer has a product that they know possesses positive features in comparison to their competitors they may choose to advertise the fact, usually through some form of product label. This type of label is called an ecolabel or endorsement label and today there is a plethora of such environmental endorsement labels in the marketplace. It is fair to say that many of these labels are lacking in rigour and credibility, in particular, those which are narrow in scope or have no third-party verification: around one-third of labels, according to the survey by the World Resources Institute and Big Room Inc.[816]

The sheer number of product labels making environmental claims is breathtaking, so apparently manufacturers feel that these must confer some value. On the Ecolabel Index website[237] there are currently over 465 labels in 199 countries and 25 industry sectors, at least 80 of which relate to energy, water, waste or resource use. In the WRI survey mentioned above, 62% of the 118 respondee ecolabels said that they could be utilized globally, which suggests that consumers have the potential to be exposed to many hundreds of labels spanning an extensive range of goods and even some services. Using the same data, researchers at Duke[323] have highlighted the speed of development of ecolabels; in 2004, there were virtually no products certified, in 2005 there were 510 products and today there are over 13,600.

Clearly, this situation will need to change. First of all, consistent assessment `methodologies are being developed which will give consumers greater confidence, such as for Life Cycle Assessment (ISO 14040) and for Environmental Product Declarations (ISO 14025), which covers statements of the environmental impacts of goods based on ISO 14040. There is more on these standards in Volume II of this book.

As well as greater standardization, the consultancy SustainAbility has identified a number of other trends that will influence the future development of ecolabels.[693] First, there is the increasing interest in the business-to-business sector as *green procurement policies* expand. There is the consolidation of overlapping labels. Finally, organizations are moving away from a passive label to a more active engagement in the issues behind the labels.

In France, the Grenelle II Environment Bill was passed by the French Assembly on 11 May 2010 and set the world's first programme for mandatory environmental/carbon declaration for consumer products (as opposed to appliances). Once again, we see disclosure as a tool to drive resource efficiency.

In the UK, independence, standardization and rigour have been provided by the labelling scheme run by the Carbon Trust. The largest adopter of this approach was the dominant food retailer, Tesco, which is applying the label to many of its own-brand products. The example (Figure 3.28, left) of Tesco's soap powder is particularly effective because it is simple to understand the emissions *per wash*. The label provides a comparison with alternative products (the non-biological detergent) and it provides the consumer with some practical advice on how to reduce their impact (wash at 30°C). However, Tesco announced in 2012 that it was suspending its labelling initiative, citing the lack of uptake by other supermarkets and the cost required to assess each product.[759]

3.28 **The UK retailer Tesco voluntarily labelled many own-brand products with clear and helpful information based on third-party verified data**
Source: Unknown

Where there are mandatory performance criteria there are almost always labels, which build on the evaluation processes required by the standards. As far back as 1966,[780] France set mandatory standards for refrigerators and freezers and accompanied this with a labelling scheme on these and other domestic appliances including televisions. After the oil crisis in the 1970s a range of other countries developed minimum energy performance standards and labelling schemes, a process which has been continuing through to today. A popular form of label uses a *star* approach which originated in the Far East and Australia, initially on appliances such as air conditioning units, and has been adopted by other countries such as India.[353] This form of mandatory disclosure has had an enormous impact on manufacturers as consumers either voluntarily select the better performing appliances, or authorities use the performance indicators to mandate minimum performance or retailers simply choice edit the poor performers out. The more widely adopted labels and standards become, the greater the incentive for other countries to take these up, as the fear of *dumping* of poorer performing equipment drives policymakers to protect domestic consumers.

Another example of mandatory labelling is the EU Energy Labelling Directive. This legislation requires most large household appliances (white goods such as fridges, freezers and washing machines), cars and light bulbs to display a prominent A (good) to G (poor) rating at the point of sale or on the product/packaging.

These labels have permitted some retailers to choice edit (see page 127) out some of the efficient products, so the consumer is simply not being offered the less efficient models. The A-G rating scheme is applied to a wide range of goods from appliances to cars and even buildings (page 147), and many countries outside the EU, such as Brazil, Iran, South Africa and China, have adopted labels with a similar appearance. There is, however, one problem with this particular form of label, which is essentially *grade inflation* – that is to say that because A is the best score, it's hard then to move to a higher score as efficiency improves or standards tighten. In the EU there is now an A+, A++ and A+++ score for some products (see page 658), for example. In the Australian labelling scheme, this problem is overcome by adding another 4 "*Superefficiency*" stars to appliances which meet the six basic star ratings.

The range of goods that are being rated is growing. Earlier we mentioned that automobiles are now rated on the amount of CO_2 that they emit. From 2012, in the UK, the energy performance of car tyres, on a scale of A-G, has also had to be displayed on the tyres at the point of sale.

It is not surprising to learn that the success with energy labelling has led to similar approaches for water. Policymakers in many jurisdictions are concerned about water scarcity and where simple measures such as efficient showerheads can have a significant impact on use, these are being promoted through labelling, or mandatory standards (such as the US toilet flush maximum of 3.4 US gallons per flush or 6.1 litres). The Australian and New Zealand Water Efficiency Labelling and Standards scheme (WELS) has been in place in Australia since 2006 and in New Zealand since 2011.

3.29 **Even though this refrigerator label from Macedonia uses the Cyrillic alphabet, it is evident that the appliance is a good performer with a clear "A" rating reinforced by the green bar indicating lower energy use.** Interestingly, the EU label also reports the noise level of the appliance in dB at the bottom of the label. The strong branding with the EU flag provides consumer confidence in the label. *Source: Bjankuloski06en, Wikipedia*

Another form of product certification, albeit at a much larger scale relates to buildings or even whole facilities. Here we can see a range of disclosure relating to mandatory audits and certification. According to the World Energy Council, 38 countries now have energy audit programmes for industry in place.[808] To maximize uptake, some of these programmes are compulsory, while others are voluntary, with full or partial subsidies. In some countries, such as the Netherlands, the regulations stipulate that projects which meet a certain minimum payback must be implemented. Often the audits have to be undertaken by certified external auditors, to ensure consistency and independence. In South Korea, the government has instructed that over 300 of the largest industrial CO_2 emitters set energy saving and emissions reduction targets or face fines.[74]

In Europe, the member states of the EU are required to undertake mandatory energy audits as a result of the EU Energy Efficiency Directive (2012/27/EU). This requires all organizations with either 1) over 250 employees or 2) a turnover more than €50 million *and* a balance sheet over €43 million, to undertake energy audits covering over 90% of the total energy used by the organization. Some notable features of the scheme are the requirement not only to identify energy savings opportunities, but to develop an economic appraisal (based on whole life costing) of these opportunities.

The notion behind these compulsory audits is that many organizations have significant opportunities to reduce energy use, but fail to develop these because energy is a relatively low-cost component. If senior management is presented with well-considered opportunities to reduce cost, it is argued, then they are much more likely to act on these.

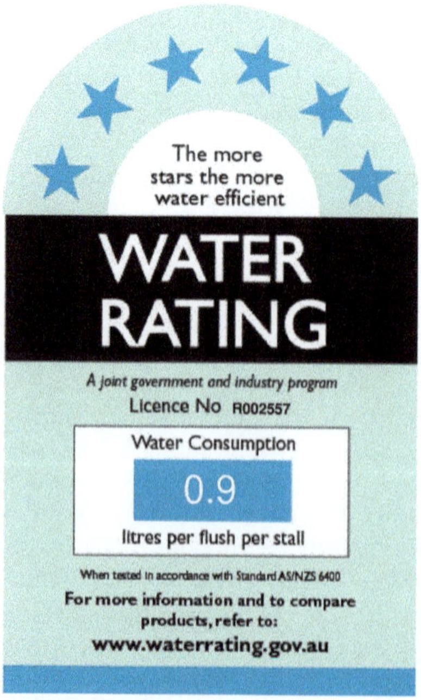

3.30 **Which water-consuming appliance would you buy?**
Source: http://enviro-tech.com.au/ & http://wels.brandfm.com/home/welcome

In Australia, the similar Energy Efficiency Opportunities Act 2006 (EEO) was one of the most comprehensive mandated programmes for energy efficiency anywhere in the world, until its repeal in 2014. The analysis of the first five years of this programme identified some positive results:

- 54% of the savings identified in the mandatory audits went on to be implemented, saving Australian businesses over AU$880 million.
- A conservative estimate is that the EEO programme was responsible for approximately 40% of energy efficiency improvements within the Australian industrial sector, so the programme clearly increased savings.

My direct experience of both schemes is that while energy users are treating these as a compliance issue, it is having some effect on the attitudes to energy use, at least at a senior level. The inclusion of the energy management system ISO 50001 as a route to compliance in the EU scheme and emphasis on human factors in the EEO have both helped to communicate the fact that energy efficiency is about much more than technical projects.

In countries where there is water scarcity, it is not surprising to hear that mandatory water audits are also, no pun intended, in the pipeline. The first candidates for these audits, not unnaturally, are water utilities, where accurate data can inform regulatory limits on losses. Companies, too, are in the spotlight; for example in India, mandatory water audits for certain industries are likely and incentives for efficiency may support these. Globally, large water-consuming industries as diverse as hotels, golf course operators, agriculture and power companies should be preparing for much greater scrutiny of their water use.

While the quality and effectiveness of these mandatory audits can be sometimes questionable, in my experience, and the penalties for failing to undertake the audits (when they are compulsory) are not always enforced, there is clearly a desire on the part of regulators to achieve some minimum level of resource awareness within organizations. This mandatory approach is particularly relevant to large resource users where the resource costs are a relatively small part of their overall operating costs, and so there may not historically have been many efforts to quantify and manage that resource.

Unfortunately, these prescriptive audits can occasionally hinder our progress on resource efficiency. They may fail to take into account the operational efficiencies, instead focusing on equipment and technology alone and so create a weak business case for action. It is possible for the facility involved to say "*we had an audit, but it demonstrated that there isn't much that we can cost-effectively do*".

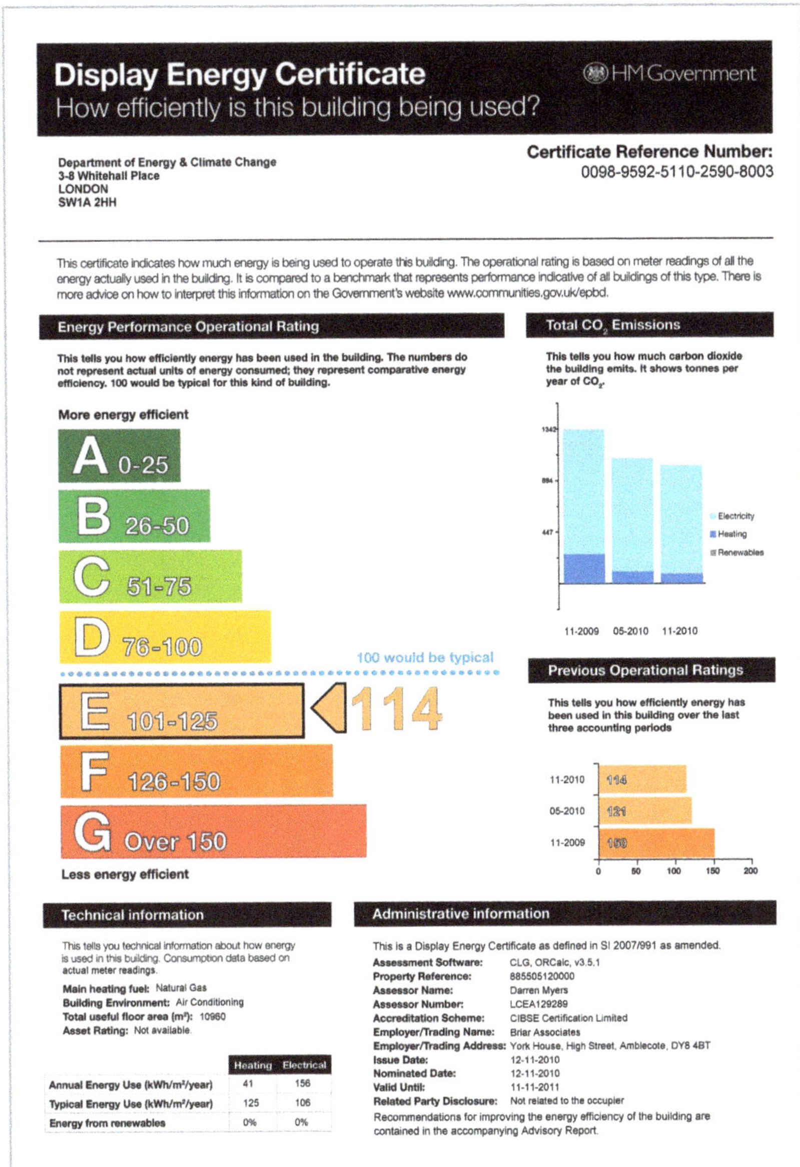

Display Energy Certificate

How efficiently is this building being used?

HM Government

Department of Energy & Climate Change
3-8 Whitehall Place
LONDON
SW1A 2HH

Certificate Reference Number:
0098-9592-5110-2590-8003

This certificate indicates how much energy is being used to operate this building. The operational rating is based on meter readings of all the energy actually used in the building. It is compared to a benchmark that represents performance indicative of all buildings of this type. There is more advice on how to interpret this information on the Government's website www.communities.gov.uk/epbd.

Energy Performance Operational Rating

This tells you how efficiently energy has been used in the building. The numbers do not represent actual units of energy consumed; they represent comparative energy efficiency. 100 would be typical for this kind of building.

Total CO_2 Emissions

This tells you how much carbon dioxide the building emits. It shows tonnes per year of CO_2.

Previous Operational Ratings

This tells you how efficiently energy has been used in this building over the last three accounting periods

Technical information

This tells you technical information about how energy is used in this building. Consumption data based on actual meter readings.

Main heating fuel: Natural Gas
Building Environment: Air Conditioning
Total useful floor area (m^2): 10960
Asset Rating: Not available.

	Heating	Electrical
Annual Energy Use (kWh/m²/year)	41	156
Typical Energy Use (kWh/m²/year)	125	106
Energy from renewables	0%	0%

Administrative information

This is a Display Energy Certificate as defined in SI 2007/991 as amended.

Assessment Software: CLG, ORCalc, v3.5.1
Property Reference: 885505120000
Assessor Name: Darren Myers
Assessor Number: LCEA129289
Accreditation Scheme: CIBSE Certification Limited
Employer/Trading Name: Briar Associates
Employer/Trading Address: York House, High Street, Amblecote, DY8 4BT
Issue Date: 12-11-2010
Nominated Date: 12-11-2010
Valid Until: 11-11-2011
Related Party Disclosure: Not related to the occupier

Recommendations for improving the energy efficiency of the building are contained in the accompanying Advisory Report.

3.31 **Display Energy Certificates are mandatory for almost all publicly accessible buildings in the EU**
Source: Niall Enright

It is much better that the organization itself decides to undertake the audits and so there will be greater ownership of the results, which are then more likely to be implemented. Another drawback with highly subsidized audits is that because they are *free* or because they have been *mandated*, they are treated as valueless. That is to say that because the organization has not had to approve the budget, they are sanctioned at a low level and get little management attention as a result. Furthermore, as a compliance issue, they are not seen as a business opportunity but as a piece of legislation to meet at the lowest possible cost and inconvenience. Compulsion devalues the benefits that good audits can deliver and risks organizations adopting the cheapest route to compliance.

Value

Another example of facility reporting is the introduction of Display Energy Certificates (DEC) and Energy Performance Certificates (EPCs) for buildings across the EU, as part of the Energy Performance of Buildings Directive. Both of these incorporate an A to G rating scheme, similar to the white goods labels described earlier, which will let occupants or purchasers know the performance of the building. The DEC is intended to be prominently displayed in all publicly accessible buildings (with some exceptions such as places of worship) and is based on the recent historical performance of the building, a so-called operational rating. An example of a DEC can be seen in Figure[3.31], left. It is a performance rating of the efficiency of building operation.

Because EPCs are most often used to rate new buildings or refurbishments where there is no operating history, the assessment is on the efficiency of the design. EPCs are mandatory asset ratings whenever a building is built, sold or a new lease is entered into.

Real World: *That much*

Disclosure can sometimes be useful simply by bringing to the attention of people in an organization just what resources cost.

On a few occasions, while working on the Carbon Reduction Commitment in the UK, I have been able to observe a moment of astonishment – the time when the senior executive team for the first time appreciate just how much they spend on energy.

It is quite remarkable just how many organizations do not know at a senior level just what the cost of energy and resources is to their businesses.

When I was working at a huge US refinery with BP on energy efficiency programmes just a few years ago, I could find no one who knew just what the cost of energy was: *"probably US$1 billion"* was as close as it got!

The UK government has announced that EPCs are going to form the basis for choice editing. From 2018, no commercial building may be let or sold if it has an EPC rating of F or G.

We have already seen in Section 3.4 the substantial evidence linking building energy ratings with value. The mandatory nature of the EPC and DEC regimes means that building owners will be compelled to make the performance data available on a new letting or the sale of the property, which are the critical points in establishing the value of the building. Another way, albeit pure speculation at the moment, in which the EPC and DEC could affect value would be if the building rates (i.e. the local property tax) are determined by the certification, with better-rated buildings being taxed at a lower rate. There is a precedent for this in the taxation of cars in the UK which was based on their CO_2 emissions.

The US has an equivalent, but voluntary, building labelling system, Energy Star. Australia has the mandatory NABERS commercial building labelling scheme (which is currently voluntary in New Zealand). Indeed, there are a large number of sustainability labelling schemes worldwide, such as BREEAM and LEED, which take into account energy, water and waste aspects of the construction or operation of the building. As planning authorities increasingly impose requirements to meet levels of performance and as the market begins to link certification to the valuation of the assets, the take-up of these rating schemes is growing.

For some organizations I have worked with, the most powerful form of reporting driving change are league tables involving public disclosure or ranking in comparison with peers. For businesses whose brand and reputation attract a premium price, the maintenance of a position of leadership in respect of the environment is seen as important. The reputational impact in this case only works where the league table is public. There are many other voluntary league tables, such as the Energetics One-to-Five ranking or industry-specific tables like the Solomon Energy Intensity Index for Refineries, which serve a benchmarking purpose, and can produce a *call to action*, but which publish competitor data on an anonymous basis.

In their book *Embedded Sustainability: the next big competitive advantage*, Chris Lazlo and Nadya Zhexembayeva[464] describe disclosure as one of the three forces that will transform the nature of business in the 21st century (the other two are scarcity of resources and increasing expectations by consumers, investors, employees and other constituents of business). What Laszlo and Zhexembayeva refer to as "*radical transparency*", and I call disclosure, is in many ways a product of their other two forces (or the three other drivers in our pyramid). Because of resource scarcity (such as our planet's limited ability to absorb CO_2), regulators are compelling organizations to disclose more and more performance data. Because of rising expectations of consumers and investors, there is greater demand for disclosure and willingness by leading organizations to disclose more in order to create competitive advantage. If we compare the state of disclosure today with that a decade ago, the word "*radical*" is an entirely appropriate adjective to use.

3.12 Legal challenges

Legal risks related to organizations' impact on the environment or use of natural resources are increasing. Citizens, pressure groups, jurisprudists and legislators are working hard to incorporate these matters into existing laws or to develop new laws.

A central economic paradigm of the early 21st century has been that the free flow of goods and services leads to the greatest efficiency, which in turn generates the greatest wealth and wellbeing. This neoliberal orthodoxy is coming under increasing challenge, however.

Notions of *"free trade at all cost"* are being questioned as legal experts and States consider whether the *process and production methods* (PPMs) under World Trade Organization (WTO) law can form the basis for states to incentivize goods produced with environmentally friendly or low-carbon PPMs. The test here has been whether products which are similar but have different PPMs can be considered to be *"like"* (in which case discrimination is not permitted), or whether the products do have some inherent difference from which a distinction can be made. Early findings under the General Agreement on Tariffs and Trade (GATT), the predecessor to the WTO, found against efforts to distinguish tuna imports on the basis of the *"dolphin-friendly"* methods of fishing. However, the situation appears to be changing. A report from the International Bar Association[400] indicates that the Appellate Body in the WTO is now prepared to consider the *regulatory purpose* of the measure to restrict imports when weighing up whether a measure is discriminatory. In the example cited, the US was seeking to bar the import of flavoured cigarettes and the objective of promoting public health justified the ban. The extension of this principle to encompass regulatory objectives such as mitigating climate change would seem relatively straightforward.

Because free-trade agreements are now seen as a significant barrier to states being able to regulate the environmental performance of corporations through measures such as *choice editing*, trade legislation is coming under increasing scrutiny from legislators, pressure groups and politicians. The cosy *back-room* deals of the past, weighted heavily towards investors, are highly unlikely in the future.

A recurring question that arises in the legal aspects of resource efficiency relates to whether plaintiffs (such as members of the public) can sue organizations for damages resulting from resource inefficiencies. This question is a matter of *standing*, and in many jurisdictions citizens have little prospect of suing private corporations under common law for resource-related harm. The recent significant ruling in the use case of *Massachusetts vs Environmental Protection Agency* heard by the Supreme Court in the US, hinged in part on whether the

Real World: ***A programme of work***

The International Law Association[469] has listed the legal and procedural issues that need to be overcome to develop climate change legislation:

(i) the actionable rights affected by climate change;
(ii) clarification of the role and definition of legal standing;
(iii) issues regarding causation, including appropriate standards for proving a legally recognisable causal link between greenhouse gas emissions and relief sought;
(iv) whether knowledge, including foreseeability of harm, is relevant to liability or judicial relief;
(v) development of methods for awarding remedies and relief as warranted by the circumstances, including uniform standards by which to apportion damages, and the provision of declaratory, interim and/or injunctive relief;
(vi) issues regarding standards of liability;
(vii) the interrelationship of competing claims from states, communities and individuals;
(viii) limitation periods for claims;
(ix) the availability of pre-trial and interim applications for disclosure and discovery;
(x) guidelines on costs awards in climate change cases; and (xi) guidelines for the jurisdictional reach of domestic and international courts to adjudicate climate change related claims.

> **Real World:** *Liability?*
>
> Research has suggested that some two-thirds of the carbon dioxide (and methane) emitted globally since the mid-19th century can be traced to just 90 fossil fuel producers and cement manufacturers.
>
> The findings, compiled by Richard Heede, a researcher at the Climate Accountability Institute in Colorado, were published in 2014 in the journal Climatic Change.[363] Of these 90 *majors*, 81 are companies, (50 owned by investors and 31 owned by states).
>
> Currently, international legislation around climate change is framed in terms of the obligations of the rich (so-called Annex I) states to reduce emissions and to compensate poor nations for the damage caused.
>
> The question that Heede's research poses is whether liability for these emissions should lie not with states, as emitters of the CO_2 but with these *majors*, as the producers of the fossil fuels and cement:
>
>> *"Shifting the perspective from nation-states to corporate entities – both investor-owned and state-owned companies – opens new opportunities for those entities to become part of the solution rather than passive (and profitable) bystanders to continued climate disruption."*
>
> In terms of liability, it may be pertinent that half of the carbon emissions in the products of these organizations have arisen since 1986 when the dangers of climate change were from man-made emissions were well established.
>
> Heede also notes that these organizations, by nature of their production capacity and reserves are most likely to contribute the lion's share of future emissions.

state has standing because the potential injuries from global warming were not concrete or particularized (individual and personal). In their 5:4 ruling, the judges agreed that the EPA should regulate emissions of carbon dioxide and other gases that contribute to global warming from new motor vehicles.

Here we see an instance of a state suing an agency to amend the regulation of corporations by the agency. Thus, while citizens do not have the power to regulate, they can pursue legal routes to influence those who do. The recent success of Urgenda in compelling the Dutch government to increase its climate change targets[83] demonstrates that citizens can force governments to regulate effectively on their behalf.

However, a bigger concern to business is the possibility for private claims that may be brought by citizens or pressure groups. This is described thus in the excellent book *Climate Change Liability*:[480]

> *Private law claims envisage one person, C, who alleges he/she has suffered damage from climate change, suing D, who is allegedly responsible in part for it, for compensation, or for an order to make D change his/her behaviour. C might be a person who suffered in a heatwave, or had his/her house flooded. D might be an oil company or power generator. The claim will be brought in 'tort' or 'delict'. In common law systems a specific tort has to be alleged, and those most commonly discussed in this context are 'nuisance' and 'negligence'. Establishment of this type of liability has been seen as a kind of holy grail by environmental campaigners and as an unacceptable disaster scenario by sectors of industry which might have to bear the cost. The numbers of potential claimants and defendants in this type of action, and the scale of potential compensation, are all huge, and indeed the very wide scope of such claims is one policy factor against their being permitted. No action of this type has yet succeeded.*

While successful legal challenges around resource use remain largely framed in terms of regulatory efficacy (and so tend to be directed to agencies) that still does not stop interested actors from using the threat of legal action to challenge corporate behaviour. For example, a recent letter[334] from some leading environmental pressure groups has raised the issue of personal liability for climate change emissions by the board directors of leading fossil fuel companies. Significantly, the insurers of the companies involved were advised that a claim may be pursued against the directors and officers (D&O) insurance policy held by the companies.

Given the enormous costs of climate change, it is clear that at some point, the question of who bears that cost (liability) will go to court. There are many issues that will need to be resolved along the way – see *Real World: A programme of work* (page 145) – but with 90 organizations believed to be responsible for over half of all emissions since the 1980s (see box left), it seems that some organizations may be more vulnerable to challenge than others. Decisions that these organizations are making today may prove costly if liability for harm can be demonstrated in the future.

3.13 The value case

There is no doubt that the ability of organizations to achieve their core objectives is being influenced increasingly by their environmental and sustainability performance. Understanding these influences is key to managing change.

Value

The preceding sections paint a compelling case for the influence that resource efficiency can have on the ability of an organization to achieve its core mission. In some cases, such as reducing costs, the impacts are positive, in others, such as outright bans on some products, they can be very damaging to the organizations affected. This brings us back to the notion that preserving the status quo is not an option - change is inevitable and organizations need to anticipate and respond to these influences if they wish to realize their goals.

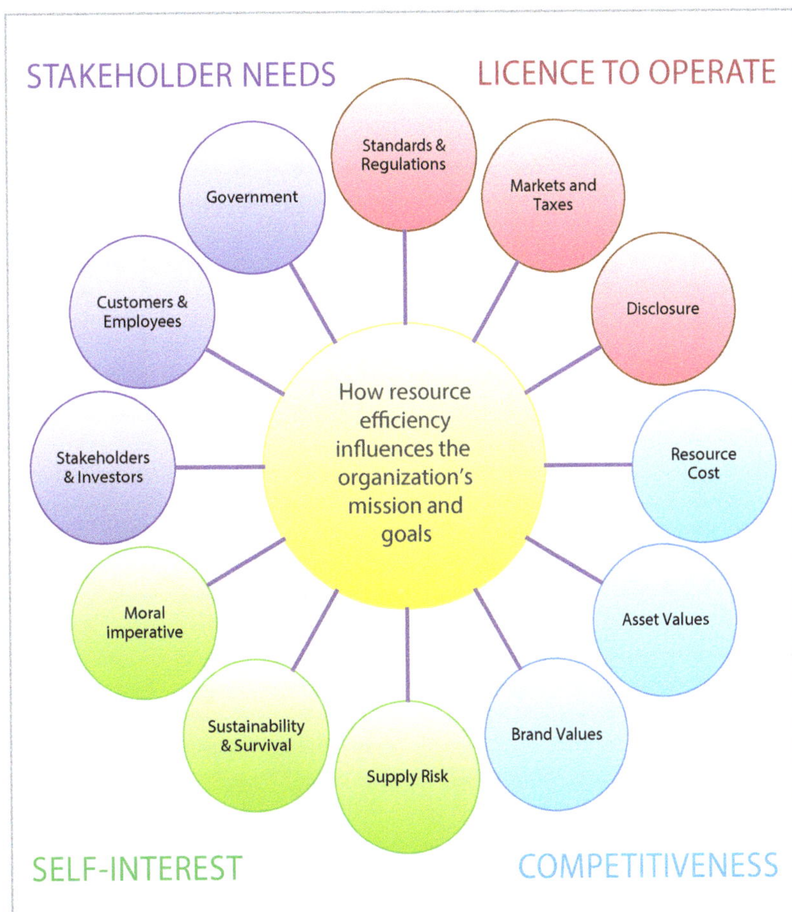

3.32 **Drivers for resource efficiency are both opportunities and threats**
Source: Niall Enright. This image is available in the companion file pack.

Exploration: *Value or values?*

This chapter is unashamedly entitled *Value*. It speaks of the myriad ways that financial value, brand value, or the capacity to deliver service may be influenced by the drivers for change in resource use. The emphasis has been on the tangible, *justifiable* basis for improved efficiency.

But what of the notion of *values* - that is to say the moral imperative to act, the need to remain faithful to fundamental principles of justice, fairness and decency. Should we not recognize that some issues we face, such as climate change, require us all to act together and *urgently*. Is it not true that everyone, especially those of us with some authority or influence, must show leadership and rise above our own interests, or indeed even make sacrifices, to address a very real danger.

There is undoubtedly an overwhelming moral basis for action. For some people this will be based on their religion, for others on social or political perspectives, or maybe on compassion or love, or simply the joy of the life we have today. These are powerful, generally positive, motivators which enable human beings to pursue difficult paths with tenacity and to accomplish great changes.

The focus of this book is on organizational transformation. In my experience, conversations with leaders in organizations about their values are, in reality, no more than a discussion about the *branding* of their institution, not about core principles or a moral compass to guide the organization's behaviour. *"Don't be evil"*, a guiding principle at Google, was initially a defiant dig at other large dominant software corporations, but would appear not to have been a commitment to base all decisions on the basis of morals, if recent controversies over tax, privacy, intellectual property and market abuse are to be believed.

The reality is that organizations are largely amoral (which may make them behave in evil ways in some cases). This book does not set out to change this. I have deliberately *not* advanced arguments that organizations should act because of their moral obligations.

However, values do have an enormous impact on the interests of the organization, and any decision-maker would be very foolish if they ignore the moral basis for action on resource efficiency. Values are hugely important in the sense that they drive the perception and behaviours of staff, customers and stakeholders.

But rather than attempt to crystallize these moral sentiments, which are not considered *actionable* by organizations, I have instead focused on the very real second-order effects: brand image, staff retention, investor sentiment, regulation and risk, which will enhance or impair the organization's ability to deliver. These second-order effects can form the basis for the *justification* for action. When engaging organizations in the need for change, I advise a conversation about value as this has the potential to drive change much more effectively than talking about values.

However, when talking to individuals as individuals, whether friends and neighbours, politicians or executives, we should not be afraid to share the values that drive us to action. Nor should we shy away from demonstrating these values through the purchasing decisions we make, the pressure groups we support, the demands we make on our legislators or the vociferousness with which we stand up to those organizations who are causing harm. Organizations do not sit in a vacuum, and they will not last long if they reject values of the societies in which they operate.

Any decision-maker would be foolish to ignore the moral basis for action on resource efficiency because it has a huge impact on the interests of the organization.

3.33 **Value infographic**
Note that the decline in emissions is smaller than the decline in energy use, because a good proportion of the EU's energy is already low or zero carbon. This kind of "vivid" communication of value is very powerful. Similar icon-based infographics can be easily created to communicate the direct and indirect benefits of energy and resource efficiency within an organization.
Source: Niall Enright. Adapted from a Coalition for Energy Savings Poster.[154] *The data comes from a European Commission impact assessment of the benefits of energy efficiency.*

Summary:

1. Value is shorthand for an improved capacity by the organization to achieve its core mission.

2. The value available to organizations from resource efficiency is enormous.

3. The value provided by resource efficiency extends far beyond the simple financial savings from using fewer resources. It ranges from the fundamental licence to operate, through better engagement with the needs of consumers and stakeholders, to radical competitive advantage.

4. The average direct financial value of resource efficiency opportunities in the private sector is around 13% of gross profits. This figure is much greater if we take into account the hidden costs associated with embedded inefficiency incorporated into the prices of goods for which the organization is paying.

5. The percentage increase in the share price resulting from resource efficiency is determined by the costs savings achieved divided by profit.

6. There is evidence from equity markets that more resource-efficient companies deliver superior returns to investors.

7. Resource efficiency reduces the risk that assets may become stranded or require expensive upgrades in the future. In the case of commercial buildings, there is a clear rationale for substantially increased asset values.

8. Intangible brand value accounts for around 80% of the S&P index. Brand value is an asset on many companies balance sheets which is strongly affected by consumer sentiment. Increasing transparency around brand valuation is shining a spotlight on the consumer behaviour towards the brand.

9. Consumers in many markets, especially China, express an intention to formulate purchasing decision on the basis of product green credentials.

10. "Choice editing" by retailers or regulations by policymakers are removing the most resource-intense products from the market.

11. Mandatory disclosure will give consumers and stakeholders precise information about the real performance of organizations in environmental terms. Product certification is bringing this disclosure to the point of sale.

12. Organizations are under tremendous and growing pressure from policymakers who want to drive change, from investors who wish to understand risk and opportunity, and from customers or service users who want to buy sustainably.

13. Although moral values are a difficult platform on which to create a justification for action on resource efficiency, they cannot be ignored as they shape the decisions, aspirations and actions of customers, service users and stakeholders.

Further Reading:

McDonough, William and Braungart, Michael, 2009. *Cradle to Cradle - remaking the way we make things.* North Point Press. ISBN 978-0-86547-587-0 (pbk)

Gatnsky, Lisa, 2010. *The Mesh - why the future of business is sharing.* Portfolio Penguin. ISBN 978-1-59184-430-3 (pbk.)

Heck, Stefan and Rogers, Matt, 2014. *Resource Revolution - how to capture the biggest business opportunity in a century.* Melcher Media. ISBN 9781477801192

Ellen MacArthur Foundation, 2013. *A New Dynamic - effective business in a circular economy.* ISBN 978-0-9927784-1-5. An introduction to some of the key concepts in the circular economy.

Questions:

1. Using your own organization, or another one of your choice, write down the primary purpose of the organization and then go on to describe how this purpose can be enhanced or impeded by resource efficiency.

2. For your named organization, order the sources of resource efficiency value in Figure 3.32, by their relative impact on the organization and justify this ranking with examples or evidence.

3. Expanding on the previous question, go on to discuss what factors enhance or impede these drivers and how these can be changed.

4. For your named organization, write a one-page call to action on resource efficiency aimed at convincing the chief executive to act.

5. What evidence is there that the share value of companies is influenced by resource efficiency? Consider effects on both the profit and loss of the business and the balance sheet. Can these effects be quantified?

6. Do morals have a role to play in organizations? Discuss.

7. My organization's operating profit is US$1 million a year on a turnover of US$20 million. If the price-equity multiplier in my sector is 10 and there are five million shares in circulation, what is the current share price? Manufacturing costs are 50% of my organization's costs, and resource efficiency can reduce this by 5%. Given this, what will the new share price be, and by what percentage will it increase?

8. Consider if you have made a purchasing decision based on resource efficiency. Describe the information that you had available at the time and why you made the choice that you did. For the particular item in question, consider if there is a "threshold" value below which the impetus to act ceases to influence your choice.

4 Resource Efficiency is not Easy

We have seen from the previous chapters that there are some very compelling reasons to implement energy and resource efficiency programmes.

- For most organizations, there is a significant amount of value to be unlocked. This value is often expressed in financial terms, but can also be described in terms of competitive advantage, a continuing licence to operate, enhanced brand value, or a greater ability to deliver service to stakeholders.
- Resource inefficiency represents a major threat to our survival. Society's demand for improvement is bringing about rapid change which we can either actively shape or passively observe. To be a winner we need to engage with this change and use it to competitive advantage.
- There is a moral imperative to do so. Our organizations are not islands; we have obligations to many stakeholders not just in the present but also in the future.

Day by day, it *seems* that more and more organizations are appreciating these facts and responding vigorously to the risks and opportunities of energy and resource efficiency. If one were to examine what many of the world's largest corporations are saying, one would be forgiven for thinking that the problem is solved.

The energy and resource efficiency landscape is littered with disappointments, premature declarations of victory and outright failures.

Unfortunately, this is not the case. Despite self-congratulatory case studies to the contrary, the reality is that energy and resource efficiency efforts are proving difficult to sustain in organizations. Many efforts remain superficial while what is published by some organizations is downright misleading. The landscape is littered with disappointments, premature declarations of victory and outright failures. In my professional experience, fewer than one-third of programmes achieve their objectives, and a much smaller proportion approach their true potential for improvement.

There are many reasons for this. Resource efficiency is complex. It requires many parts of the organization to be engaged for protracted periods of time. Sometimes it needs third-parties to act in concert with us. It seems never-ending; no sooner has some improvement been made but there is a demand for more – whether to satisfy regulators, stakeholders or just to remain competitive.

Real World: *A puzzling resistance*

It seems that even when firms are explicitly shown the savings potential, there is a failure to fully realize the value identified. An interesting study[262] by consultants SKM Enviros looked at the uptake of recommendations made following energy efficiency audits in manufacturing industry in the US, and concluded:

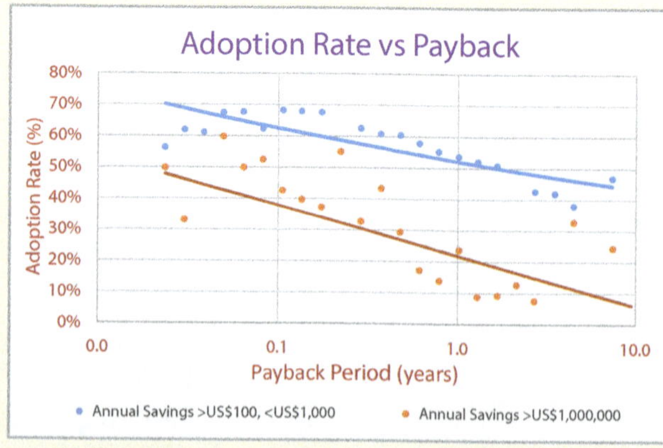

- The adoption rate for projects depended on the payback. As expected, longer payback projects were less likely to be implemented, as shown by the fact that both the lines in the chart slope downwards; and
- The greater the savings potential for a project, the less likely it was to be adopted. Here we can see that the blue series for project saving up to US$1,000 are above (have a higher adoption rate) the orange series for projects saving >US$1 million.

It is the latter point that is somewhat counterintuitive. The study indicated that if you have a project saving US$1,000 with a one-year payback, then there was a 50% probability of adoption, but for a US$1,000,000 project, the adoption probability drops to around 20%. Surely the very fact of the size of the savings available would act as an encouragement to adoption, not the opposite? The data tells us that the greater the value, the less likely organizations are to act – why is that?

If we tapped someone on the shoulder and said "*look there's a £10 note on the floor over there, all you need to do is to bend down and pick it up*", it would be reasonable to expect them to pick that note up. And yet the reality is that organizations don't stoop down to pick up the note, or at least not all of them. Why not? One explanation could be that there are actually lots of £10 notes and even £20 and £50 notes around and these others were the ones being picked up, or perhaps the organizations simply don't believe the note is worth £10. Clearly, decisions to invest in resource efficiency are not straightforward. We shall explore many of these barriers and solutions in this chapter.

4.1 **Declining adoption rate with increasing value**

Source: SKM Enviros, 2005[262] adapted by Niall Enright, using data in the US Department of Energy's Industrial Assessment Centre database. The data extracted from the original chart is available in a spreadsheet in the companion file pack.

People often start an energy and resource efficiency programme with the viewpoint that this is primarily a technological challenge. After all, we are constantly reminded that the solution to carbon emissions is better kit - whether that be solar panels, or more efficient boilers, or electric vehicles. It does not take long, however, for it to dawn on those tasked with delivering results that the main obstacles are not technological, but organizational. It is this failure to perceive of energy and resource efficiency as a change management process which often results in the disappointing outcomes.

In this part of the book, we will focus on understanding why energy and resource efficiency is difficult. We will learn that there are many practical barriers that prevent most organizations from achieving their potential. We will learn that the least significant issue is technology - most of the challenges are to do with motivation, capability, organization and finance - in short, they are management issues.

Real examples of difficulties faced by organizations will be explored, not to criticize, but so that we may learn from and avoid those pitfalls. This will help us to better understand a framework in the next chapter, designed to overcome many of the obstacles. First, we will start by seeing if the barriers are real.

4.1 The evidence for barriers

Many studies of energy efficiency indicate that there is a large potential for improvement. Because this potential has not been realized, we can reasonably conclude that there must be some barriers to adoption and extrapolate these to broader resource efficiency, which shares similar characteristics.

The most visible evidence of barriers to energy efficiency is the difference between the observed efficiency and potential efficiency. If this difference is small, then one could argue that there are not many barriers to achieving the full potential, whereas if this difference is large, one could conclude that the obstacles are considerable. First of all, let us look at a range of studies on the potential for energy efficiency:

4.1 **Summary of the energy efficiency potential identified in a number of studies**
Source: Niall Enright

Author	Year	Geography	Sector	Potential
Stern[680]	2007	Global	All	20%, 2050
WWF[493]	2007	Global	All	39%, 2050
McKinsey[351]	2008	European Union	All	20%, 2020
McKinsey[328]	2009	US	All	23%, 2020
NAS[541]	2010	US	All	17-19%, 2020
Lovins[485]	2011	US	All	40%, 2050
ACEE[605]	2012	US	All	42-59%, 2050
GEA[404]	2012	Global	Buildings	46%, 2050
WEC[811]	2013	Global	All	18%, 2020
CITI[423]	2013	Europe	All	20%, 2033
EPRI[264]	2014	US	All	11%, 2035
ICF[386]	2015	Europe	Industry	20%, 2050 (technical)

It is interesting to note that, for 2020, there is a consensus that energy efficiency potential is 17-23%. On the other hand, if we consider the longer term there is a considerable divergence. One reason for this is that some of the studies, such as the Stern review, are forecasts of the proportion of emissions reductions that will arise from efficiency, rather than the potential per se. Furthermore, growth clouds the picture. Thus the Lovins analysis predicts a 40% absolute reduction in emissions as a result of efficiency improvements, which means that the underlying change is considerably greater if one takes into account growth in output of the economy.

Although there are these differences between the various studies, it is difficult not to conclude that there is significant potential for energy efficiency in the global economy. Our earlier analysis of the savings potential indicated that, if anything, the potential for water and raw materials is even greater. Because of this unrealized potential, we can safely assert that barriers do exist.

4.2 Top down vs. bottom up

There are two approaches to evaluating the potential for efficiency measures - we can look at the feasibility of individual measures and add these up, or we can look at economic factors like energy intensity (energy/GDP) and compare economies. While these two approaches produce different values for the potential, they don't undermine the idea of barriers.

Country / Region	Total emissions reduction %	% share of total emissions reduction from to efficiency
Nigeria	53%	64%
Saudi Arabia	10%	58%
EU	12%	40%
Mexico	36%	32%
Brazil	11%	31%
US	29%	27%
Vietnam	25%	27%
South Africa	54%	23%
Argentina	32%	21%
Turkey	21%	21%
India	14%	17%
Ethiopia	64%	14%
China	20%	8%

4.2 **Analysis of the Intended Nationally Determined Contribution (INDC) of 13 Signatories to the Paris Climate Change Agreement in 2015**
This study shows that energy efficiency is an important component of the emissions reductions commitments offered as part of the Paris Climate Agreement. The INDCs set out medium-term actions to 2030. Collectively these commitment represent an improvement in energy productivity of 1.8% per annum.
Source: Energy Transitions Commission / Ecofys, Pathways from Paris, 2016.[258]

From an economics perspective, all decisions are rational. The savings mentioned in the previous sector, if true, should attract large numbers of investors seeking good returns. The fact that we don't see this needs explanation.

Orthodox economists looking from the *top down* argue that since the markets for energy-using technologies are broadly efficient, then the only possible reason that these efficiency opportunities are not being taken up is that they do not, as claimed in the studies, offer a net positive return on investment. They argue that far from there being barriers specific to resource efficiency, the failure of adoption is due to the more general case that the costs of efficiency are understated and so the investment is, in truth, unattractive. These *hidden and missing* costs could account for the apparent lack of investment, and a perception of greater risk explains the higher hurdle rates that efficiency investments are required to achieve.

From the *bottom up* perspective of people like myself, who regularly encounter many highly attractive opportunities within organizations, this explanation does not ring true. We see lots of *no and low-cost* savings, called *"no regrets"* savings since they don't take any investment away from alternative options. According to the orthodox economists, this category of savings simply should no longer exist - they cost nothing and take no money away from alternative investments, and so should have been implemented in full.

So, who is right? As in all debates, there is almost certainly some truth in both arguments. For example, common errors in the bottom up aggregation of opportunities are an underestimation of installed cost and the failure to take into account the interaction of the savings (a 20% saving on top of a 20% saving adds up to 36% savings, not 40% – see page 606). On the other hand, the assumption that decisions are entirely rational and that there are no barriers specific to resource efficiency flies in the face of much research evidence and the personal observations of many efficiency practitioners like myself. There are genuine challenges associated with getting organizations to invest time and money in efficiency and denying that these exist means denying the potential, which could rob us of the opportunity to deliver what is almost certainly the cheapest and most rapid method of addressing climate change. For those who want to understand these economics arguments in greater depth, Steve Sorrell's *The Economics of Energy Efficiency* is very informative (see Recommended Reading).

4.3 Theoretical vs feasible

Some opportunities may be technically feasible but impractical for a number of reasons, such as cost, timing and so forth. In analysing the barriers, we are primarily interested in those factors that prevent otherwise feasible projects from proceeding.

The figure below shows that currently achievable efficiency may only be a small proportion of the theoretical savings possible. This is due to factors such as the availability of technology, costs and practical operational issues - for example, some opportunities in a continuous process plant may have to wait for a *shut-down* before they can be implemented, which may be several years. In practice, organizations have three types of opportunities - those that are immediately available, those that may be available in the future and those that are not economic and will only be implemented if there is some form of market intervention or other business incentive (e.g. regulation, subsidy or customer demand).

When considering barriers to the adoption of efficiency programmes it is common to focus on the reasons why organizations do not implement the first of these types of savings - those that appear to have a good financial return and no practical or technical obstacles. However, we will also touch in this chapter on the barriers to achieving the other potential savings.

Barriers

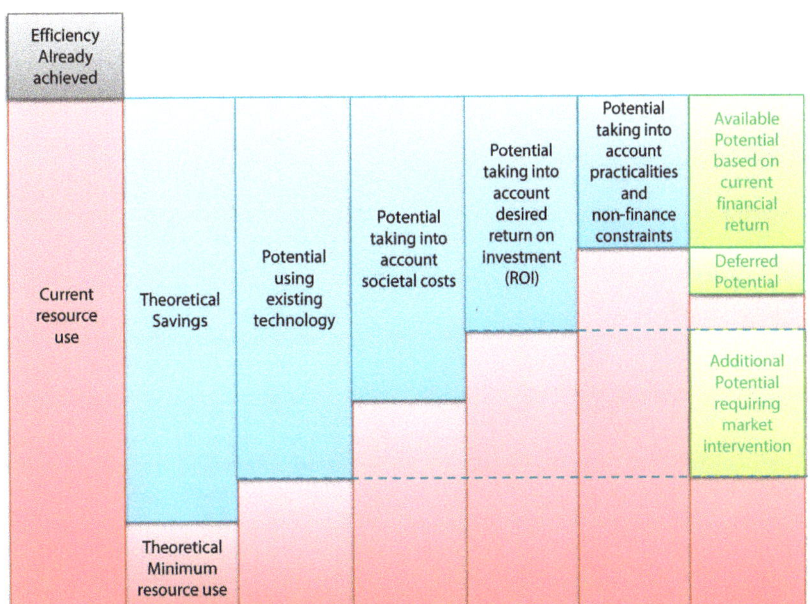

4.3 **Illustration of the various types of resource efficiency savings**
Source: Niall Enright, inspired by Steve Fawkes' outstanding book "Energy Efficiency".[274]

4.4 A classification of barriers

The considerable body of research into barriers to energy (and resource) efficiency has enable us to place these into a number of categories.

There have been a large number of studies and surveys which explain the reasons why individuals and organizations fail to exploit the opportunities for greater efficiency that are presented to them, even when they appear compelling.

The table below lists the most common barriers that have been identified. First of all, there are some broad characteristics, or general barriers, of energy and resource efficiency that make it inherently more difficult for folks to support. Then there are three types of barriers that are specific to the adoption of opportunities: structural barriers, behavioural barriers and availability barriers.

Resource efficiency opportunities face a series of barriers, any one of which could prevent an otherwise technically and economically sound proposition from proceeding.

GENERAL BARRIERS	
	Fragmented: There may be a large number of opportunities in many parts of the organization. **Concentrated:** Where resources are concentrated, peripheral savings seem inconsequential. **Lack of proof**: It is difficult to measure and to prove savings (i.e. what *would* have happened). **Low impact**: Efficiency may not be material to the core objective of many organizations. **Available effort or finance**: These may be required elsewhere or in short supply. **Low mind-share**: Other issues are considered more important in the organization.
OPPORTUNITY-SPECIFIC BARRIERS	
Structural	**Hidden transaction costs:** The cost of efficiency is higher than stated. **Split incentive:** The situation where costs sit in one entity and benefits in another. **Term issues**: Ownership is not guaranteed for long enough to realize the benefit. **Irreversibility:** Some technical measures cannot be "undone", which reduces their appeal. **Reversibility** of behaviour-based savings. **Pricing distortions**: Regulations, subsidies, taxes or other market distortions.
Behaviour	**Psychological**: Inaccurate estimation; loss aversion; inertia and status quo bias, anchoring. **Habit and standards:** Practices that inhibit efficiency such as "like for like replacement". **Information barriers**: Insufficient data or knowledge to make correctly informed decision. **Prestige and other incentives:** Perceived low impact of efficiency on advancement. **Credence good:** Lack of experience means that resource efficiency requires a "leap of faith".
Availability	**Lack of capital:** The most common reason quoted for rejection of opportunities. **Lack of human capital**: Inability to release people's time to deliver the opportunity. **Product availability**: A solution with the right return on investment is not available. **Adverse bundling**: The opportunity benefits are offset by undesirable effects. **Timing**: The improvement requires a shut-down or such like which is not feasible.

4.4 **A classification of barriers to resource efficiency**
Source: Niall Enright, format adapted from McKinsey & Co. Study on US opportunities.[328]

Considering these barriers, it is a wonder that anything ever gets done.

4.5 General barriers

General barriers to resource efficiency are related to the broad nature of these programmes compared to other types of investments, such as the fragmented nature of projects or difficulties in measuring outcomes. Not all programmes exhibit these characteristics, but it is useful to know how to deal with them if they do occur.

Real World: ***What do people say?***

In Willliam Prindle's 2009 survey of energy efficiency in US manufacturing for the ACEEE and Pew Centre,[604] around half of 48 companies said they had failed to meet their goals. The top five reasons reported were:

- limited capital for investment;
- limited leadership buy-in;
- improving efficiency was harder than expected;
- there were competing priorities and resources; and
- lagging momentum and employee interest.

The shortage of capital to pay for projects was reported to be the single greatest challenge, outnumbering any other single item by a four to one ratio.

A more recent survey by Noesis, concurs, stating that *"more than half the time, 'not budgeted' is the reason projects do not get internal approval. One-quarter of the projects are derailed by a 'lack of certainty,' of their estimated savings".*[559]

This survey reinforces the difficulties getting approval for energy efficiency projects: one in four in-house energy managers got less than 25% of projects approved while only half of consultants got 25% or more of projects funded.

The general barriers are characteristics of energy and resource efficiency programmes which makes them more challenging to implement, compared to other forms of investment. Not all programmes share these characteristics, but the ones covered here are common enough to warrant examination.

The first aspect relates to the often diverse and fragmented nature of energy and resource efficiency opportunities within organizations. Where resources are consumed across an organization, it goes without saying that efforts to reduce the consumption may need to be equally widespread. This fragmented nature is particularly common in programmes that are focused on no and low-cost opportunities, which usually depend on a large number of relatively small actions to be undertaken.

There are many techniques that we can use to address the drawbacks of such a diversified use of resources, such as:

- We can establish where the largest resource use occurs and only focus effort there (in other words, follow the *"80:20 rule"* or Pareto).
- We can design the programme as a rolling initiative that progressively tackles the opportunities location by location or function by function and so spreads the effort. This approach also clearly demonstrates the benefits to participants in the next stage of the programme and reduces the uncertainty that they may have about the outcome.
- We can standardize the tasks that have to be repeated, e.g. we can specify and select the most efficient motors centrally so that the process of procuring these is greatly simplified and the wheel is not constantly being reinvented. Standardization is a strategy of organizations with large but consistent operations, such as supermarkets, which will devote a lot of time to selecting one type of efficient display cabinet, or a lower weight packaging solution, and then deploy these multiple times within their organization.
- If our organization is not uniform, we can take the opposite approach and empower local operations to find their own solutions using *"management by objectives"* rather than adopting a prescriptive approach. This permits the local operations to define their own approach and optimize the effort that they put into the savings delivery.

Just to be clear, small and repeatable savings are usually a positive feature of resource efficiency programmes despite the challenges that they bring. These activities are attractive because, in the aggregate, they can lead to substantial improvements, often with a good financial return. Unfortunately, these types of savings are often overlooked in audits because they appear individually to offer little return on the effort needed to implement them. This is a common mistake which auditors should take care to avoid.

In contrast to organizations with a fragmented resource use, there are others where resource use is highly concentrated. An example would be aluminium plants, where a very large proportion of the on-site electricity use is for the electrolysis of bauxite into alumina (for US plants this is 120 out of 153 TBtu, or 78%).[216] As one would expect, these concentrated consumers of resources tend to be well managed.

The problem from a resource efficiency perspective is that the sophisticated control in this one area of the process leads to a *blind spot* about other forms of energy and resource use, which can be significant. My experience working in some steel *mini-mills* in the US found huge value in the ancillary areas such as the rolling mills and air handling equipment. Some of the techniques that I found helpful to overcome the inevitable *"so what?"* effect were mainly to do with how savings were presented:

- I would exclude electricity for electrolysis when calculating the percentage savings, arguing that this is an exceptional direct process input.
- I would relate the savings to jobs - at the time there was a lot of *downsizing*, which was perceived to be the major opportunity for cost efficiency. If our savings represented the average cost for 10 positions, then that would make an impression.

Also inducing a *"so what"* effect are two other similar characteristics. First of all, there is the fact that some categories of resource efficiency will have a low impact on the organization, simply because resources are a very small input cost. This, in turn, can lead to a little awareness of resource issues and thus a lack of commitment to resource efficiency. In these situations, promoting resource efficiency can very much feel like pushing a boulder up a hill. Again some general tips can help:

- Money is not the only benefit that resource efficiency can bring. We have seen from the previous chapter that there are many other sources of value such as positive stakeholder perception, brand enhancement, employee engagement and so forth. Identifying an otherwise difficult-to-achieve non-financial benefit can be helpful.
- Translating the modest financial benefit at the bottom line to the top-line equivalent of the business (e.g. sales) can multiply the perceived value many-fold.

Real World: *So what?*

An audit of a large aluminium smelting plant identified electricity use of over 1.8 TWh a year, 95% of which is utilized by the *"potlines"* to convert alumina and carbon to aluminium and carbon dioxide.

The study identified opportunities to save 3.5 GWh of electricity with a value of US$150,000 requiring an investment of USD$193,000 (an IRR over 60% in 5 years or a simple payback of 1.3 years). The savings represented 0.2% of the site's electricity usage.

Unfortunately, the investment was not made, despite its attractiveness from a purely financial perspective. The saving was just not material enough.

The lesson here is that a sound financial return is not the only requirement for many resource efficiency projects. Organizations are also conscious that they have a limited workforce and want to focus this on those activities that add the greatest value. Delivering the lighting projects would have had an "opportunity cost" elsewhere in the business and so the choice not to proceed is logical.

Real World: ***Utility theory***

There is a scientific explanation of the *"so what?"* reaction in the box opposite. It argues that it is not the size of the benefit that matters, but the relative change in wealth.

Utility theory dates all the way back to 1738 and work by Daniel Bernoulli, who wanted to explain why people make choices that are not based on the size of a payout alone.

The insight that Bernoulli brought was that the value that a decision-maker gives to alternative choices (or risks) depends on the extra benefit that the choices bring. In other words, if we are poor then we would value a US$1 gain much more than if we are rich. A dollar has a much greater utility to a poor person than to a millionaire. Utility increases with wealth at a diminishing rate, which is why the utility function is curved, as illustrated below.

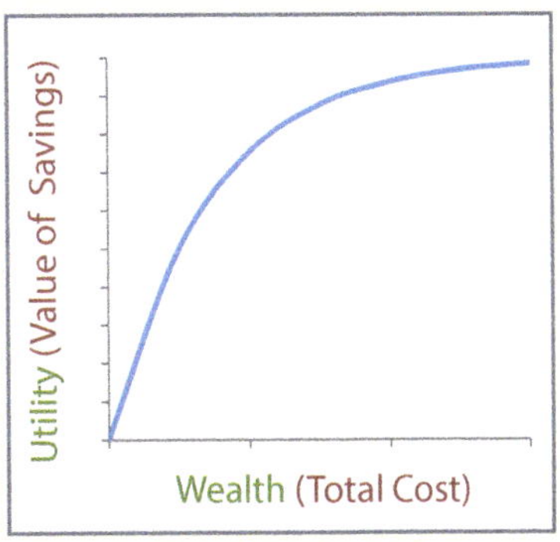

Looking at the example opposite, the savings of US$150,000, despite being substantial, have a very low utility as they represent just 0.2% of US$75 million spend on electricity. The plant can gain greater utility from using its capital and workforce in other ways.

In presenting the business case for efficiency we need to be aware that decision-makers are unconsciously making choices between alternatives so as to maximize the utility that a given effort can produce.

- Aligning the resource efficiency benefits to the core purpose of the organization can make a difference - e.g. a school may be motivated to act because resource efficiency is about contributing to a better future for the children they serve. This alignment with the core purpose is something that is recommended in all energy and resource efficiency programmes, even if there is also a compelling financial case.

- Comparing the organization to competitors can also compensate for a lack of financial incentive, as some organizations position themselves as being quality providers and would not want to be unfavourably benchmarked.

- It is entirely feasible to undertake a resource efficiency programme even if there is little real interest at a senior level and its impact on the organization is relatively small. The key here is to make the effort involved as little as possible. Here, one needs to be typically thinking about modifying existing systems rather than introducing new ones. One example would be introducing lower emissions standards into a vehicle fleet - rather than driver education which involves greater effort. In a similar vein, one would tend to concentrate on quick, one-off *"fire and forget"* technical fixes, such as changing the settings on building management systems or using voltage reduction equipment, which can be fitted in the switchroom and deliver improvement with little or no disruption to the organization.

- There is a significant proportion of the economy which is not driven by the financial benefits of resource efficiency, so governments often put in place other incentives to encourage improvement to overcome this failure of demand. One such incentive is direct taxation, such as the Landfill Tax in the UK, which penalizes organizations for disposing of waste to landfill as opposed to recycling. This tax changes the financial case for action by making the *business as usual* option more expensive than the recycling option. Another mechanism to encourage organizations to act even where there may not be an overwhelming financial case is mandatory reporting. This disclosure has had a big effect on consumer-facing organizations such as retailers that, for reputational reasons, do not wish to be seen to be the poorest performer in their sector. Thus in low impact or awareness situations, promoters of resource efficiency are advised to look at regulations to see if these offer a compelling basis for action. Compliance activities often do not have to meet a financial hurdle rate and are not regarded as discretionary.

Another challenge that can arise relates to a lack of available effort or finance usually due to competing initiatives. For example, an organization may be undergoing a restructuring or implementing a new management system, which means that another initiative will not be sanctioned by management, or the degree of uncertainty surrounding the change makes a programme unpredictable. In these circumstances, those desiring to achieve some resource efficiency improvements may need to *fly under the radar*, focusing on interventions that do not attract lots of attention but which can nevertheless move the organization forward.

One of the most common problems with resource efficiency is the difficulty in proving the benefit achieved. Often resource usage is determined by drivers like weather, production or activity, so the consumption of the resource could rise because of these external factors, despite underlying efficiency improvements. This difficulty can be compounded by the fact that investors are often looking at the financial return they are achieving, rather than the unit of consumption, which means that price movements can also mask any gains.

- A good way to overcome this problem is to carry out a controlled assessment with two identical resource consumers, where one is improved and the other not, and then analyse the results. It is quite easy to test just one light fitting and use temporary *clip-on* metering to measure the change in energy consumption. Similar comparisons are often easy to make in transport fleets where an operator may have a large number of identical vehicles.

- Sometimes a controlled test is not desirable. In this case, it is possible to use statistical techniques like regression analysis (see page 460) to compensate for the influence of variables like weather, temperature or activity, so revealing the underlying performance. The accuracy of this approach depends on there being sufficient *before* and *after* data. It is important to stress that it is rarely justifiable to delay a programme purely to obtaining the *before* data to verify improvement. I have seen quite a few examples of organizations taking a year or more to get metering *"just right"* before embarking on obvious housekeeping improvements and so denying themselves a considerable benefit (see *"Why resource efficiency is like the hotel business"* on page 333).

- For many technical improvement opportunities, there are formalized methods to evaluate the outcome. The most important of these is the International Performance Measurement and Verification Protocol® (IPMVP®)[406] which provides a very detailed approach to calculating the energy and water savings made in specific projects. The IPMVP is widely used in the US, and increasingly internationally, as the basis for verification of savings delivered in Energy Performance Contracts (EPCs) where a third party funds an improvement and is repaid from savings made by the resource users. Because this Measurement and Verification protocol can underpin substantial contract values, it is very thorough and detailed, with a high level of credibility attached (see page 519).

Given the importance of measurement in resource efficiency programmes, there is an entire chapter dedicated to this later in the book (page 413). Here, we will see that concern about an inability to prove savings is largely misplaced, and there is a wide range of techniques which can provide an accurate assessment of the return on investment from resource efficiency measures.

Real World: *Peel Land & Property Group proving delivered savings*

Peel Land & Property Group is a major developer of commercial, retail and industrial space in the UK, with over nine million square feet of property and 33,000 acres of land. I have been working closely with Mark Whittaker, Paul Chappels and his team of asset managers. Over several years, Peel has established itself as one of the leading property companies in the UK in terms of sustainability, through concerted efforts to improve the energy efficiency of the existing property portfolio. In fact, the independent Carbon Trust benchmarked Peel's Carbon Management as the best of 29 UK property developers and managers in 2012, and they were the first major UK property company to achieve ISO 50001 in 2015 (see page 733).

One of the biggest challenges for Chris Foran, the Energy Champion, was to communicate to tenants and the Peel Land & Property Board the fact that improvements were occurring even though the energy bills were rising.

The report shown right, for an office block, *Venus*, illustrates how this was achieved. Since the management team and tenants were focused on bills, the report did so as well.

- The latest bill data was always shown in blue.
- The actual bill for the same month a year ago is shown in orange.
- Last year's bill adjusted to reflect today's energy cost is shown in yellow.
- Savings (yellow - blue) are in green.

This report enabled the Board and tenants to understand the avoided cost which resulted from the energy management programme. Even though bills have increased, there has been a saving every month.

Subsequent pages of the report provide an analysis against weather for both heating and cooling, which further increases the credibility of the savings reported. These reports are available for all Peel's offices.

Venus Energy Saving Report — **31 August 2013**

Saving Summary

Latest Bill: August (↓ in chart)		
Latest actual electricity bill	£ 25,735	A
Bill last year @ last year's cost 9.46 p/kWh	£ 25,552	B
Bill last year @ today's cost 10.6 p/kWh	£ 28,578	C

Real savings in August		
Electricity cost savings (C-A)	£ 2,842	10%

Based on prev. year consumption at today's cost compared to latest bill

YTD Savings		Months
Electricity	£ 19,557	8

Real month by month savings for the 12 months To August

Estimated Annualized savings		
Electricity	£ 29,336	10%

YTD Savings extrapolated to whole year: i.e (12/8) of FYTD value above.

Procurement

Electricity Price Analysis		Change
Today's cost p/kWh	10.60	12%
Last year cost p/kWh	9.46	

We locked out the wholesale element of the electricity prices in April 2010 through to 31/10/2012. At that time, Summer 11 was locked out at £42.45/MWH and Summer 12 at £46.80/MWH. There is a differential of 10% between the prices. Summer 12 prices peaked at £62/MWH, some 30% above where we locked out so the decision to lock when we did was a very good one. The other elements of the price are regulated by Ofgem and beyond our control (CCL costs transmission and distribution costs, triad charges) and these increase annually every April. On average these were increased by 17% in April 2012

Unit cost include all bill elements.

Bill summary analysed by actual/paid bills

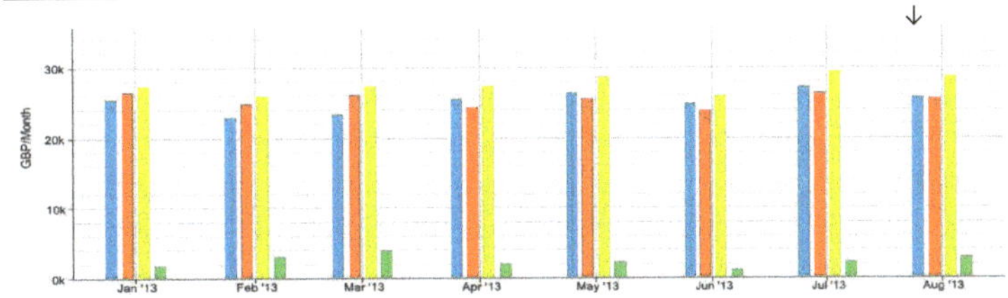

Latest actual electricity bill
Bill last year @ last year's cost 9.46 p/kWh
Bill last year @ today's cost 10.6 p/kWh
Cost savings

Property Summary

Building	90,912 Sq Ft		
Voids	0 sq ft	% Void	0%
Tenants	No tenant issues		

20/10/2013 — Page 1

4.5 **Report highlighting avoided electricity costs**

Source: reproduced with kind permission from Peel Land & Property Group.

4.6 Structural barriers

The first hurdles for many specific investment opportunities are structural. These can be distortions to the markets (like subsidies), which make otherwise viable projects unattractive, or perhaps because the people who pay for the improvement don't receive the benefit, such as the landlord and the tenant.

Structural barriers are characteristics of the market or the organization which impede investment in the efficiency proposal, despite its apparent technical feasibility and attractiveness.

The first of these barriers, which has the effect of making projects uneconomic, are subsidies. An example of a subsidy that acts against water conservation is in India, where there are around 15 million tube wells that allow small farmers to irrigate their crops. Initially, the electricity used by the pumps was metered, from which a power charge and a water extraction charge could be levied from the farmer. However, difficulties in maintaining the metering meant that the electricity companies moved to flat tariffs based on the horsepower of the pump. The government heavily subsidizes these tariffs and Worldwatch estimates that the farmers only pay 13% of the real cost of the electricity.[122] As a result of the flat-rate tariffs, farmers were not penalized for over-extracting water or over-irrigating. Neither was there an incentive to reduce electricity consumption.

Market-distorting subsidies involving resources are very common. The IEA's latest estimates indicate that fossil fuel consumption subsidies worldwide amounted to US$523 billion in 2011, up from US$412 billion in 2010, with subsidies to oil products representing over half of the total.[810] In fact, these subsidies only represent the direct discounting of the fossil fuels for domestic markets in relation to the global market or reference price. There are many other forms of support which act to suppress the world market price itself. One example is the cost of defending the Straits of Hormuz to allow oil exports to the US, which is not reflected in the price of oil. Another is the cost of externalities, i.e. the costs of damage to the environment which is caused by CO_2 emissions arising from fossil fuel use, which is borne by society as a whole. Favourable tax concessions, such as those recently proposed by the UK Chancellor in support of unconventional gas or *"fracking"*, also represent subsidies, as is the minimum price guarantee that has been given to the developers of the Hinkley 2 Nuclear Power Plant.

Any organization that treats resource costs as an overhead is essentially ***subsidizing*** *its end-users.*

Of course, subsidies are also used to encourage organizations to take up the most resource-efficient technologies. It is a common policy measure to encourage the adoption of energy-efficient equipment through direct subsidies or other fiscal mechanisms such as enhanced capital allowances. Many forms of renewable energy would not be economical if it were not for the subsidies they enjoy. However, globally subsidies for renewables, at just US$88 billion in 2011,[391] are small compared to those for fossil fuels.

Although market subsidies represent a significant distortion of the economics of resource efficiency and sometimes work against the adoption of more efficient technologies, it would be easy for organizations to assume that these are issues beyond their control. However, if we consider a subsidy to be any situation where an end-user of a resource does not pay the full price, then we can see that those organizations that treat resource costs as an overhead are essentially subsidizing the resource users. The problem with this is that the end-users are much less likely to want to reduce their resource use or invest in efficiency measures. Several principles flow from this insight.

- End-users should pay the full cost of the resources that they use. This usually takes the form of departmental cost allocation (see page 624).
- Wherever possible, this cost should be based on actual metered resource use, rather than on a per sq ft., or another arbitrary allocation method. In this way, the users will directly benefit from their own improvement actions, rather than rely on the facility as a whole improving.
- The allocation of cost should seek to include all the costs associated with the resource - for example, emissions charges, procurement costs, delivery costs and so forth. It can even add an element of cost for the maintenance and operation of the central plant that may provide or condition the resource on behalf of the end-users.

Sometimes the suggestion that there should be departmental cost allocation is strongly resisted. I have heard many arguments against: the end-users are much too busy worrying about their core duties (e.g. doctors and nurses), the end-users have little ability to influence resource use compared to the property team (e.g. university departments) and end-users are not going to get engaged (e.g. hourly-paid workers on the shop floor).

By and large, these are not arguments against the notion of making users pay, but more generally against the relative impact that the end-users may have on the resource consumption compared to the impact that a specialist function like the facilities management, maintenance or engineering team can have. Clearly, there are cost-benefit considerations to take into account when considering departmental cost allocation, not least the cost of metering (see page 426).

In response to the argument that specialist inputs are required to drive improvements, I find that the most effective approach is usually to keep the primary responsibility with the end-user. This enables us to create the incentives to drive the no and low-cost savings at the demand end of the spectrum. At the same time, specific goals should also be set for the engineers or production schedulers to reduce resource use. In the Framework, later, there is a strong emphasis on multidisciplinary working to address resource use.

Real World: *Decentralizing costs*

Often, resource and utility costs are considered overheads, paid for out of a central, usually site-level or business-level, budget. The problem is that this reduces the incentive for resource end-users to make savings.

Peel Ports Group is a diverse organisation, with ports in the north and south, east and west of the UK. The largest port location in the portfolio is the Port of Liverpool which houses hundreds of tenants, 3 km of quays and covers over 500 acres.

Given the variety, size and scale of the operations, the approach adopted by the Group Energy Manager for Peel Ports Group is to delegate the ownership of energy costs to the operations teams, who are ultimately responsible for securing improvements.

This enables the ports to harness the enormous expertise of the marine operations and engineering teams on the ground. Responsibility for the energy costs is empowering – enabling the teams to decide how existing practices or equipment can be changed to drive improvement.

It is important to note that this approach aligns with the broader culture of the organization. Delegation and empowerment lie at heart of the Peel Ports Group formula for success, and explains the sector-leading innovation, operational discipline and service flexibility in a safe working environment, which are a hallmark of the ports.

Although there may be a few occasions when it is not justifiable on a cost-benefits basis, aligning resource use to the existing management structure of an organization, through departmental cost allocation is beneficial. It means that end-users will have a conversation about resource use at least once a year (at budget time), they will be aware of the cost of the resources they consume and have an incentive to reduce usage. Individuals who are motivated to act will be in a position where the result of their actions can be seen, and they can receive recognition.

Another form of distortion also relates to the pricing of resources. In many markets, the more of a resource you consume, the cheaper it is. These declining tiered tariffs mean that large resource consumers often have a lower incentive to conserve than small users. Even at the level of a single facility, tiered tariffs mean that the first band of consumption saved is worth least. Thus the marginal cost of the resource saved may be substantially below the overall average price. It is interesting to note that in some markets, such as California, declining tiers are no longer used (See *Energy Efficiency*,[274] page 21). Here, rising block tariffs means that large consumers pay more for energy, and so have a higher incentive to save, as shown in the item below.

Another form of pricing barrier can occur when resource users have entered into a take-or-pay contract. These contracts are not uncommon in the case of third-party-funded central plant upgrades (e.g. water treatment plants, steam boilers, chillers, combined heat and power plant, air compressors, etc.).

Real World: ***Rising block tiers***

Throughout much of the history of electricity supply, the idea that bigger is better has prevailed: bigger power stations could deliver more electricity at a cheaper unit cost. Increased energy use, it was argued, reflects an increase in economic activity, output, jobs and taxation and so should be encouraged. As a result, the pricing models, which are generally determined by governments or regulators, have been designed to promote growth in consumption.

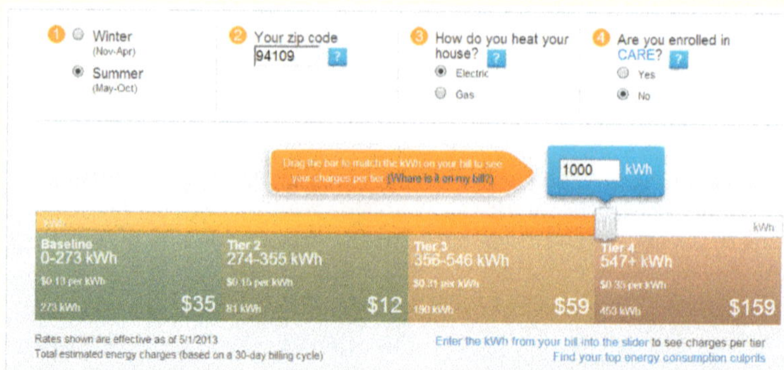

Following the energy crisis in 2001, California legislators have required the utility PG&E to introduce rising tiered blocks for residential electricity users to incentivize energy efficiency measures and solar PV. The image above is taken from PG&E's website[595] where a consumer can enter their monthly electricity usage (in this case 1000 kWh) and is then informed of their total bill. The first tier is set at US$0.13/kWh up until the baseline of 273 kWh. For the next 30% consumption above the baseline, the cost rises to US$0.15/kWh. These first two tiers have been fixed, so price increases fall more heavily on the next two tiers, from 30-70% above the baseline where the cost is US$0.31/kWh and from 100%+ of the baseline where the cost is US$0.35/kWh.

For the example consumer above, they can see that they are paying US$159 for the final 453 kWh of electricity they use (i.e. US$0.35/kWh). It is interesting to note that at around US$0.31/kWh solar photovoltaics (PV) becomes cost-competitive. Thus the consumer has a strong incentive to install PV (or undertake efficiency measures) to at least offset this last slice of expensive electricity.

*Most of the market barriers to resource efficiency have their equivalents **within** organizations - ranging from incomplete price signals through to split incentives and regulatory barriers.*

In these circumstances, the investor in the upgrade will often expect the end-user to commit to a minimum quantity of the resource (water, heat, coolth, electricity, compressed air and so forth). If the end-user falls below that consumption they have to pay regardless, thus eliminating any financial returns from efficiency measures beyond this minimum.

- Because decision-makers are often aware of the pricing complexities around many resources, it is essential to establish which cost should be used in calculating the return on efficiency investments. Lack of appreciation of these subtleties can undermine the credibility of a proposal.

- Those involved in negotiating resource supply contracts should ideally ensure that the cost rises with volume rather than declines. This rate structure provides a real incentive to reduce resource use. However, this approach is contrary to everything that most procurement managers have been taught: emphasizing volume in order to get a discount.

Other structural barriers relate to market bias. For example, the way the electricity market works globally acts as a major impediment to efficiency. The core paradigm is that electricity demand is uncontrollable and so generators need to be incentivized to provide sufficient capacity to meet the needs of the peak periods. This leads to large amounts of excess capacity which run at partial load and so are inefficient. It also adds significantly to the cost of electricity. Network operators are often rewarded from the capital investments they make to deliver new capacity, so they have absolutely no incentive to support energy efficiency. If we were to treat the network as a single system, it would become apparent that investment in reducing the peak demand through energy efficiency is much more cost-effective than building more capacity. In a perfect market, investors should be much more attracted to the returns from demand reduction that capacity increase.

Few markets have more widespread regulatory barriers to efficiency than the US electricity supply market. In their devastating critique of regulation, Casten and Ayres[123] point out that "*the most important barrier to competition... is the legal prohibition against private wires crossing public streets*". This seemingly reasonable regulation, designed to stop the wasteful duplication of cabling that would arise if a second utility company wanted to distribute electricity, has the perverse effect of putting the US power distribution networks under the control of the existing utilities. They subsequently use extortionate charging to access these wires as a very effective barrier to prevent smaller, distributed electricity generators from building local capacity. From a resource efficiency perspective this is dreadful news, because these local generators could make full use of their waste heat and so achieve efficiencies of 80% or more compared to 36% for central plant. They would also be able to charge less for their electricity as a result of selling the heat.

As with subsidies, it would be easy for organizations to assume that these issues are external market distortions beyond their control. However, there are analogies to all these market features operating within organizations.

Real World: *Inflexible regulation*

Sometimes well-intentioned regulations have a perverse effect.

In his 2003 article on *Regulatory Roadblocks to Turning Waste to Wealth,*[209] Pierre Desroches cites a range of issues in the US which act as barriers for companies wishing to turn waste into useful materials. Desroches suggests that an industrial Ecopark such as Kalundborg (see page 74) would not be permitted in the US. The flue gases that Statoil pipes to Gyproc for plasterboard production and the liquid sulfur that Statoil sells to Kemira for fertilizer production would not be approved as these materials would be considered *hazardous waste*. Also, these materials would violate *90-day storage rules*, which prevent the accumulation of hazardous materials.

Furthermore, the *"mixtures that derive from"* rules would mean that any materials or products that are derived from substances deemed to be hazardous wastes are also considered to be hazardous wastes themselves.

It seems that there are two perspectives on industrial waste. On the one hand, one can treat it as something undesirable with negative environmental and health consequences, and so regulate against its production and for its destruction. On the other hand, one can treat industrial waste as a valuable feedstock from which economically valuable products can be produced and which reduces the demand for virgin resources.

Unfortunately, many environmental statutes (notably in the EU) treat waste in such a way as to provide persistent biases against recycling, reclamation and technological innovation, rather than taking the Danish approach of considering each scheme on merit.

Just as external markets have regulatory barriers to efficiency, so do organizations. A common example is the standardization by organizations around a few fixed electric motor sizes, which makes sense from a maintenance and ease of replacement perspective, but which works against overall system efficiency. Another example is the regulatory framework imposed by the Food and Drugs Administration on pharmaceuticals manufacturers, which means that they are severely restricted in the changes that they can make to their approved manufacturing processes. Although this is an external requirement, my experience with pharmaceuticals companies is that they then reinterpret these regulations to prevent many changes to systems outside the scope of the regulated process and so perpetuate inefficiencies.

Another parallel between organizations and markets is the separation of the supply and the demand function. In many organizations, the response to increasing demand is simply to increase supply, adding an air compressor, or water treatment unit or boiler or chiller to the existing services, without considering the alternative, and possibly cheaper, option to reduce demand. Unfortunately, some engineering functions can become complicit in this as they see their primary function in terms of installing and operating equipment. The result is that they will dismiss or undermine opportunities for end-user demand reduction as these would potentially lead to reductions in their budgets. This is a structural barrier where budgets or power considerations impede efficiency opportunities which otherwise would be feasible.

It is advisable when developing a proposal for resource efficiency to understand the structural barriers within organizations which would lead to opposition to such a programme. These often revolve around misaligned incentives, budgets and power relationships. A technique called pairwise comparison (see opposite) can help to reveal these barriers.

A further, very well-researched, structural barrier is known variously as split incentives, the agency barrier or the principal-agent barrier. This barrier exists when the person able to make the investment in the efficiency measure does not gain the benefit. The typical example is that of a landlord who owns a building and a tenant who pays the utility bills. Usually, the cheapest improvement to a building in terms of energy efficiency is the addition of insulation. However, since this involves altering the fabric of the building it falls to the landlord to make the necessary changes, but they have no desire to incur the cost since it is the tenant who would gain the savings.

Split incentives are much more widespread than most people imagine. For example, just about every set-top box provided by satellite or cable TV services is a case of split incentives. The service provider has no incentive to purchase more efficient models since it is the customer who pays the electricity bill. According to the most definitive study on the scale of the principal-agency barrier,[719 page 145] 95% of US set-top boxes are affected, consuming, in total 68.4 PJ of electricity (19 TWh). To put that figure into context, it represents just under 1.5% of US residential energy use and 76% of the output of the

Real World: *The power of pairwise comparison*

Where there may be multiple people influencing decisions around a resource efficiency programme, it is useful to explore the individual objectives of each person using a technique called pairwise comparison. This is a quick and straightforward process which can provide some insight into why there might be some resistance to a programme or project, and is based on the idea that different decision-makers have different drivers, some of which may be complementary and others incompatible.

In the example below, I have illustrated a pairwise comparison for an energy efficiency programme at BP where some resistance was being observed (for the full story, see *Spanning the intent gap – enManage™ at BP* (page 239)).

In the comparison, each key role is shown on the top row and left-hand column, alongside a word describing the key priority for that position. Thus the second column refers to the corporate team's original objective of reducing carbon, and relates this to the other decision-makers' objectives:

- the site business unit leaders (BULs) saw carbon management as a distraction from the reliability priority (hence a -1 score);
- the finance director felt it would lead to unproductive investment, and so reduce profit (another -1 score);
- the procurement team felt that carbon reduction would result in more CAPEX being spent (another -1 score); and
- the only supportive person was the engineering manager, who quite liked the prospect of new CAPEX that the carbon reduction objective could bring (so they score +1).

All of these scores are shown in the column as a series of scores: +1 where objectives are complementary or -1 where they are in conflict (or perceived to be in conflict). The integer values can be changed to express the degree of effect if desired but in the example below, I kept the comparison simple. Having completed all the columns we can see that the reduced carbon objective, as stated, was poorly aligned to key decision-makers' goals.

The enManage™ team modified the objective by changing the goal to decreasing energy use (= lower cost or = greater profit). In addition, much emphasis was placed on the value of energy management as a form of condition monitoring of equipment (= greater reliability). We then assessed this change using the pairwise comparison, as shown in the bottom row of the table. The change has aligned the programme with the goals of the BUL and the finance director, while the procurement team have moved from being hostile to being broadly neutral, on the basis that the CAPEX will yield to productive cost reductions.

	Corporate Team ↓ Carbon (*Original Goal*)	Site BUL ↑ Reliability	Finance Director ↑ Profit	Procurement ↓ CAPEX	Engineering ↑ CAPEX
Corporate Team ↓ Carbon (*Original Goal*)					
Site BUL ↑ Reliability	-1				
Finance Director ↑ Profit	-1	+1			
Procurement ↓ CAPEX	-1	-1	+1		
Engineering ↑ CAPEX	+1	+1	-1	-1	
enManageTM ↓ Energy (=↑ Reliability & ↑ Profit)	+1	+1	+1	0	+1

enManage™ is a trademark of Jacobs

4.6 **Pairwise comparison indicating the compatibility between the objectives of different decision-makers and an efficiency programme**

Source: Niall Enright. A blank pairwise comparison template is available in the companion file pack.

Barriers

new Hinkley C nuclear reactor in the UK. The problem continues to this day with a 2011 study by the NRDC[561] estimating that set-top box energy consumption has risen to 27 TWh at a cost to consumers of US$3 billion, two-thirds of which is incurred when the boxes are not actively being used.

If we look at another technology, mobile phones, we see quite a different picture. Here, the service providers have gone to great lengths to improve the efficiency of their equipment even though it is the end-user who pays for the energy. The reason for this is that efficiency provides *utility*, that is to say that it leads to longer battery life, something users value and which will influence purchasing decisions. Section 4.8 discusses adverse bundling, which involves the selection of equipment on the basis of attributes other than efficiency, which can be considered a form of split incentive, as the resource consumer effectively has little or no choice about the efficiency of the equipment.

The split incentive situation can be simplified as three of four cases, where the principal is the end-user and the third party is the agent.

4.7 **Three types of principal-agent problems**
Source: Mind the gap, T'serclaes, P and Jollands, N, IEA, 2007.[719]

	End-user can select technology	End-user cannot choose the technology
End-user pays the bill	**Case 1:** (Principal and agent the same so no split incentive)	**Case 2: Efficiency Problem** (Agent decides on technology and principal pays the bill)
End-user does not pay bill	**Case 3: Use and Efficiency Problem** (Principal can select technology and agent pays the bill)	**Case 4: Use Problem** (Agent can select the technology and also pays the bill)

We tend to think of most landlord-tenant situations as being Case 2, where the disadvantaged party is the tenant, who pays higher than necessary energy bills because of the lack of incentive on the landlord to invest capital in the property. This is said to give rise to an *efficiency problem*, because the equipment will be more likely to be inefficient due to the lower capital expenditure. However, while this is the usual situation in the UK and Dutch office markets, in Japan it is very common for office tenants to pay rent inclusive of energy costs.[719] This is a Case 4 situation and gives rise to a *use problem*, because the tenant has little incentive to curtail usage, as the savings made will be spread over all the tenants in the building.

Although organizations should ideally behave as one single entity from an efficiency perspective, the reality is that each cost and profit centre often acts independently in terms of decision-making. This compartmentalization means that, as well as being a significant concern across organizations, the principal/agent barriers described above are also very common within organizations. These internal conflicts are one reason why, later in this book, I emphasize the value of the support of a sufficiently senior sponsor or executive in our programme, to resolve any misaligned objectives among the participants.

Conflicting objectives are possibly the single most important structural barrier reducing the uptake of efficiency opportunities.

Real World: *Split incentives are everywhere*

Split incentives is a term that is given to any situation where the risks and rewards associated with resource efficiency are not aligned between the participants. This conflict of objectives is possibly the single most important structural barrier reducing the implementation of viable resource efficiency programmes.

Just within the last 12 months, I have seen the following cases:

- The finance function is rewarded for minimizing CAPEX regardless of any operational benefits, so they starve the organization of investment (they see efficiency as *non-core* and so direct available capital to growth or service objectives even where the return is lower than the efficiency project).
- Construction managers are rewarded by bringing in projects to specification, on time and under budget. As soon as there is pressure on budgets, resource efficiency investments are treated as *optional extras* and *value-engineered* out (after all the construction managers will be long gone once the building has been completed).
- Shift supervisors who don't pay for energy, so are quite happy to run unnecessary equipment in standby *"just in case"* something fails.
- A third party has an *"all-inclusive operations and maintenance"* contract on a CHP plant and refuses to run it at high load (i.e. most efficiently) as this may increase their maintenance costs.
- An IT Director won't contemplate raising the temperature in the server room by just 1°C, despite extensive evidence that this will not affect reliability and comprehensive safeguards to rapidly increase cooling if needed.
- A hotel general manager who turned down a lighting improvement project on the grounds of *"ambience"*, even though the replacement lamps had the same output and a slightly warmer (i.e. improved) colour temperature.
- A main board director rejecting a resource efficiency programme because they interpret their *fiduciary duty* as the delivery of short-term profit regardless of the longer-term value being lost.

All these decision-makers were making perfectly rational choices. As they see it they are fulfilling their responsibility to their organizations diligently and competently.

Although energy and resource efficiency are rarely objectives to be achieved *"at all costs"*, progress is less likely if folks have an absolute veto and no incentive to support the programme.

Later, we will see that there are many techniques that can help to align people's performance and resource efficiency. For example, by using whole life costing, financial objectives can be reconciled better with lower resource use. We can make the connection between reliability and resource efficiency. We redefine fiduciary duty to take into account medium to long-term performance (see page 218).

In the examples cited above, in most cases, the only solution was to engage with the leadership to ensure that all parts of the organization were working together to a common goal. This shows why a strong mandate for energy and resource efficiency and teamwork is so important in any programme.

Real World: *Over my dead body*

One of my most disappointing experiences in an efficiency programme involved an oil producer. Detailed audits led by my colleague Richard Wise and colleagues at ERM had identified a large savings potential resulting from woefully under-investment in equipment, some of which was over 80 years old.

Based on these findings, we developed a proposal, with very compelling supporting evidence, that would deliver rapid improvements through many simple measures (such as lagging steam lines, isolating unused lines, operational improvements for boilers, training and so forth).

All in all, the programme across the organization would cost €4.7 milllion but save €10.1 million, a return on investment of under six months. These were conservative estimates; in fact, we expected to achieve much more.

What we did not count on was the extreme hostility to the proposal from the central engineering function in this company. They argued - with some justification - that improving an old boiler's efficiency from 30% to 50% through better operational control was counterproductive when a new boiler could achieve 85%+ efficiencies. The only solution they would accept was large-scale capital investment, and they pulled every trick they could out of the bag to successfully kill the proposal.

However, it was a pyrrhic victory. There was no capital available for improvement and much to my deep regret the organization continues to pollute and throw away shareholder value on an unimaginable scale. This is an example of a structural barrier within an organization whereby the interest of one internal function acts against resource efficiency.

There are some solutions to split incentives.

- The first is to just remove the split - that is to ensure that the same party pays for the resource and determines the equipment selection/efficiency. This approach was the solution in the case of beverage vending machines in Japan, where the government regulated contracts so that the beverage distributor is responsible for both the vending machine choice and the payment of the utility bill.
- The second is for both parties to enter into a gain-share agreement whereby the principal or energy end-user, for example, a tenant, agrees to reimburse the agent, or landlord, for an investment that the agent makes. These agreements are often called Green Leases (see box opposite).

In addition to incentive issues, there are also contractual barriers to resource efficiency. A typical example nowadays relates to the need for guaranteed fuel supplies to enable investment in biomass generation. This is an example of asset specificity where the viability of a technology is integrally linked to a particular commercial relationship. Once a biomass plant is built it can potentially be held hostage by its local suppliers, who are free to increase the cost of fuel knowing that the plant has few alternative sources. Another example involves not inputs but outputs; a facility may produce a waste stream for which there may be limited markets or processing options. In this situation, the downstream recipients of the waste could exploit their monopoly position.

The solution to these structural barriers often involves some form of vertical integration where contractual, partnership or ownership structures are created to safeguard all the parties, particularly the investors.

I have seen many similar contractual arrangements as a prerequisite to improved efficiency within organizations. For example, arrangements in BP petrochemicals facilities for the *off-gas* from one business unit to be used as a fuel for another required a long-term contract so that the arrangement endures ownership changes. Difficulties associated with setting up contracts can act as a powerful structural barrier to otherwise perfectly viable efficiency opportunities.

The next obstacles that we will consider are term issues, which concern the problems that can arise from the time frame associated with a resource efficiency opportunity. Coincidentally, in the same month that I was writing this chapter, I was involved in discussions about the future of a building and whether or not a *deep retrofit* should be undertaken, which would address some of its shortcomings concerning energy consumption. The barrier to this investment is the likelihood that the building will simply be demolished and a new one built in the next 10-15 years, which would not enable the investment on the retrofit to be recovered through savings.

A very common term issue relates to the installation of solar PV. The business case for third-party-funded PV in the UK has at times been quite attractive.

Real World: ***Green leases***

A partial solution to split incentives between landlords and tenants involves entering into an agreement to share costs.

These agreements have expanded to include non-financial collaboration on sustainability issues. For example, the landlord may agree to give the tenants information on energy usage for the building, or to provide space for the segregation of waste in service yards.

The name given to this type of agreements is a Green Lease, but in my experience working in property it is much better to refer to these as a Memorandum of Understanding, or if less "legalese" wording is required, I use the expression Green Partnership.

The London Better Buildings Partnership has done a lot of excellent work on Green Leases and has made available an outstanding Green Lease Toolkit[69] which I highly recommend.

The main stumbling block is the requirement that the property owner gives the investor a *lease* over the roof for 25 or even 30 years. It is rare that organizations can see that far ahead, however much they claim to be planning for the long term, so this often becomes the barrier to implementation.

I have seen many resource efficiency opportunities in industrial plants fail just because the company was not certain enough of the long-term operations of the facility. Many factories don't know what they are likely to be doing in five years time, let alone 10 or more. This mismatch between the necessary contract length, or tenor, and the lack of clarity over future resource demand, is a big reason why the ESCO third party funding models (see page 634), which were so successful in municipalities, universities, schools and hospitals (the so-called *"MUSH"* market) have struggled in industrial markets.

There is little that can be done about term barriers to resource efficiency in the case where a fundamental change in an organization is likely to take place - in other words, where the demand for the resource is likely to change significantly over the period in which savings are needed to fund investment. However, if the term issue is related to ownership of an asset such as a building whose demand is not expected to change significantly, there may be some solutions.

- In the UK, the Green Deal is a funding mechanism in which the cost of the project is linked to the building and repaid through energy bills. Because the debt is tied to the electricity meter, the investor is assured a return even if the ownership of the property changes.
- A similar scheme in the US, the Property Assessed Clean Energy financing scheme (PACE), ties the repayment to the property. It is a form of "senior debt", which means that it takes precedence over the mortgage on the building so that even if a building owner goes into administration, the investor in the efficiency measure can be assured the prospect of repayment.

Linked with timing issues are two other opposed structural barriers to resource efficiency. On the one hand is the reversibility of savings.

4.8 **This illustrates "split incentives"**
Source: Niall Enright, drawn using Pixton Image is in the companion file pack.

Real World: ***Credit card trust***

In some organizations, the purchasing and approvals processes can take a long time and require considerable effort. Sometimes those making the decisions may not be technically competent to assess the finer aspects of the request. These factors can act as a powerful disincentive to make small investments, which are often the ones that produce the highest return.

An excellent way of reducing these transaction costs and empowering folks to *"just get on and do it"* is the provision of corporate credit cards for minor expenditure.

I've seen this working well in some organizations, mainly in manufacturing in the US.

The idea is that an appropriately senior person at an operating level (e.g. an engineering manager or shift supervisor) is given a corporate credit card with which they can make payments for minor items of equipment to do with energy efficiency (like meters, ultrasonic leak detectors, motion detectors and suchlike). There is usually a limit for each item (I've seen US$2,000 and US$5,000 limits), as well a monthly limit. At the end of each month, the individual is, of course, expected to account for and justify their expenditure.

So empowering people to act as soon as they see an opportunity for improvement can short-circuit the delays and encourage folks at the operating level to take personal ownership for delivering projects.

For example, improvements based on behaviour changes may be judged by decision-makers not to be permanent, and so they will reduce the value that they perceive these types of projects can offer.

The best way to ensure that reversible changes are sustained is to use measurement to track performance and so permit management intervention if a return to the *bad old ways* is detected. Allied to this, the performance measurement should be embedded within the regular management processes of the organization so that it cannot easily be discarded in the future. Clearly, these design elements need to be communicated to the decision-maker if they are to overcome their concern.

The opposite situation occurs with changes that are irreversible, in which case the decision-makers may be concerned that they are *locking in* a particular choice in perpetuity. A typical example of this occurs in investment decisions for combined heat and power (or co-generation as it is known in some countries). The financial viability of these CHP plants depends on what is called the *"spark gap"*, that is to say, the difference between the value of the electricity generated and the cost of the fuel source (often natural gas but increasingly biofuels). These prices can be quite finely poised, and many decision-makers may well be aware that there have been periods of time when the gap made CHP uneconomic.

- If a decision is truly irreversible (and many are not) then the best way to assess the risk concerned is to look at the primary drivers for the business case and carry out a sensitivity analysis (see page 593) A range of scenarios should be employed, which the decision-maker would consider plausible. Many governments, for example, have useful forward energy cost projections (typically high, medium and low scenarios). If the investment can be shown to yield an acceptable return under a wide range of situations, the decision-maker may not be so concerned about irreversibility.
- It is important that the same sensitivity analysis, using the same scenarios, is carried out on the business as usual (BAU) option. In some cases the risks associated with inaction are greater than those with a decision to proceed, but unless the BAU analysis is undertaken only the risk related to the investment decision will be considered.

We started this chapter with the example from SKM Enviros, which demonstrated that adoption of opportunities declines with the apparent scale of benefit. The explanation for this brings us to the final structural barrier, the presence of hidden or missing transaction costs (in this case in the way the opportunities are recorded in the IAC database used in the study).

The transaction costs are all the costs associated with completing a project. Often the fixed cost - the purchase price of the equipment or technology - is just a small part of the true project costs. There are a lot of other potential costs to consider. First of all, there are the costs of identifying the opportunity, the cost of initial investigation and design and the consideration of alternatives. These are often referred to as search costs and tend to be higher when new (to the host) technologies are involved. Then there are the costs associated with appraisal of the business case and gaining acceptance for the project. Next, there are the costs related to procurement: the specification and search for potential suppliers and negotiations with shortlisted vendors. This category of costs is generally understood so is not usually "hidden", although for novel applications these costs can be underestimated.

The true **installed cost** of a piece of technology is often as much as **double** the cost of the equipment alone.

Then there are the commissioning costs, which may involve direct costs such as permitting and operator training, as well monitoring costs to determine if the improvement has achieved the desired goal. Allied to the commissioning costs are the costs associated with disruption of production or services during the installation of the project.

The largest category of missing and hidden costs can be described as additional engineering costs. These might be the costs of extra design work to integrate the solution within existing production processes, or additional cabling for a power supply or pipe work for materials, or maybe it is the notional value of the land required to accommodate the project. These additional engineering costs are over and above any quotation from the supplier and are frequently missing, or understated, in the formal the business case. Where the supplier provides an "*as installed*" price, it is easy to fall into the trap of assuming that all the installation costs have been considered.

The next category of cost is related to the potential for the project to underperform in terms of the expected benefits, described by SKM Enviros as the *"risk of delivery"* costs. Although this is expressed as a cost, it is a reduction of savings or other benefits. Clearly, these are likely to be hidden as the vendors are not going to highlight their potential shortcomings, and rarely, too, will the proponent of the project.

Finally, there are ongoing operational costs for the equipment which may not be fully quantified, nor may the expenses associated with initial teething problems which may occur over the first year or two of operation of the project. I would add to this category loss of utility costs - that is to say that the improvement may impact negatively on some aspect of the organization. Many people, for example, argue that compact fluorescent lights provide a poorer quality of light than incandescent lights and a cost could be attributed to this loss of lighting quality.

The table overleaf summarizes some of the transaction costs associated with an opportunity. This table is by no means complete, and the scale of the impact of these costs set out by SKM Enviros has a limited academic evidence base. Nevertheless, the estimates align well with my own personal experience, although I should confess here to being a former SKM Enviros employee, but not involved in producing this report. In short, the cost of a piece of technology is often as much as double the CAPEX cost of the equipment alone.

4.9 **Different types and impact of hidden and missing costs on commercial and industrial energy efficiency projects**
Source: Niall Enright, adapted from SKM Enviros study[262] page 40.

	Stage or Activity	Cost	Commercial	Industrial
	Generic costs usually included	*"Normal"* engineering and procurement overhead	10% to 30% of cost	10% to 30% of cost
Hidden and Missing Costs	Project Identification	Information and search costs	0.5 to 2 hours	3 to 10 hours
	Project Appraisal	Carrying out the investment appraisal	0.5 to 5 hours	0 to 100 hours
	Commissioning	Monitoring/managing the installation	0.5 to 5 hours	1% to 6% of cost
	Disruption Costs	E.g. lost production during installation	0% to 2.5% of cost	0% to 5% of cost
	Additional Engineering	Site costs for the host, e.g. power supply, land, drainage, integration.	5 to 30% of cost	5 to 20% of cost
	Risk of Delivery	Potential not to achieve expected savings	5% to 20% of savings	5% to 20% of savings
	Ongoing Management	Managing the efficiency measure once installed	0 to 12 hours	0 to 60 hours

By their very nature, these hidden or missing costs are not explicitly considered by the decision-makers when they appraise an investment opportunity, but they do influence project uptake. That is to say, that the decision-makers, suspecting that there are hidden costs, perhaps based on the evidence that similar projects in the past ran over budget or underachieved, will tend to discount the benefits of these types of opportunities. The most common way that decision-makers do this is by raising the required investment performance, or hurdle rate, for resource efficiency projects as a whole by demanding a shorter payback period or a higher rate of return.

The key to overcoming this barrier is thoroughness in the preparation of the business case for the opportunity.

To eliminate **hidden and missing costs**, all the cost associated with an opportunity must be incorporated into the business case and a post-implementation review completed.

- Most audits only provide a very high-level estimate of costs, typically based on the capital cost of the technology proposed. The real cost is usually at least twice the equipment-only cost. It is thus essential that all the costs associated with the opportunity are estimated, and that these are visibly incorporated into any business case so that they can be explicitly discussed with the decision-makers (for more on this, see Chapter 17 on page 555). Equally, care should be taken to justify the benefits in the business case and a conservative position adopted, particularly concerning vendor claims.
- Post-implementation reviews are also critical to establishing if the original business case was sufficiently accurate. If particular categories of costs appear to be regularly understated, then efforts should be made to improve these. The results of these reviews should be made available to the decision-makers to increase their confidence in the process.

As well as the direct transaction cost associated with the investment, there may well be opportunity costs. That is to say that our resource efficiency measure does not sit in isolation and there will at any time be many other organizational improvement opportunities available which compete with the efficiency proposal. Decision-makers tend to think in an *either/or* fashion, and so there will often be an assumption that our efficiency project will be incompatible with other initiatives affecting the same part of the organization.

- In presenting the case for a resource efficiency project, it is advisable to establish if other projects will compete for some of the core resources (e.g. time, money, land, etc.) which impact the same part of the organization. If these are found then the efficiency case needs to demonstrate how it will integrate, coexist with or even enhance the other projects. For more on procrastination, see *Why now?* (page 209).
- It is also worth thinking about competing initiatives not just as a threat to efficiency projects, but as potential opportunities. That is to say that changes which are made for other reasons can nevertheless enable us to *piggy-back* improvements in resource use or management. Thus the introduction of a new finance system could allow us to implement departmental cost allocation, or a reorganization may enable people's new job descriptions to include greater accountability for resource use. In fact, there are few changes where is it not possible to think of some aspect of improved efficiency that can also be incorporated.

This section has covered a lot of thinking from several conventional schools of economics, such as transaction cost economics. We shall see in the next section that there are also schools of behavioural economics that can shed light on why seemingly compelling opportunities for resource efficiency are declined. When presented with an opportunity to improve efficiency we always have a choice - on the one hand, we can do nothing and, on the other, we can choose to proceed. Thus every investment decision requires a comparison of the consequences of business as usual and the improvement opportunity. In short, we are always comparing alternatives (see page 566).

Real World: ***Operating budgets vs financial appraisal***

Because of their *operational* nature, most resource efficiency investment recommendations are made within a cost-centre budget. This might be a product category, departmental-level budget or a facility budget.

The problem that arises is that budgeting at this level often fails to take into account many important factors such as the cost of capital, the organization's required growth rate and average return on capital employed, or the impact of taxation. All of these can make a huge difference to the viability of the opportunity.[637] Sadly, the most common form of appraisal used at an operating level is the simple payback period, which has many drawbacks (see page 565).

Take taxation which is a *missing cost* in many business cases. In the first instance, there may be tax breaks associated with resource efficiency projects, like the UK's Enhanced Capital Allowances Scheme, which can allow a business to offset the entire cost of the equipment against taxable profits in year one. Thus, if £1,000 is spent on a high-efficiency motor, the organization may be able to reduce its tax bill by £200 (assuming corporation tax is 20%) in the first year, rather than spreading this saving over a five-year period. On the other hand, taxation can also decrease the free cash flow that arises from the savings achieved and so reduce the overall benefits. In either case, the effects of tax should be assessed where the investment is large.

4.7 Behavioural barriers

There are some psychological reasons why our investment choices can be biased against the change demanded by energy and resource efficiency projects. Habits, customs and a tendency to simplify through standardization can also lead to inefficiencies. The issue of certainty in the outcome is central.

This section focuses primarily on the decision-making process. We shall see that there are a number of psychological traits that bias decisions against resource efficiency. Sometimes the systems we create for decision-making act against change. We shall also see that economists have some insights on why we appear to act irrationally when presented with decisions.

According to one of the leading thinkers on decision-making, Nobel prize-winner for economics in 1978, Herbert Simon, decision-making involves three steps: intelligence, design and choice.

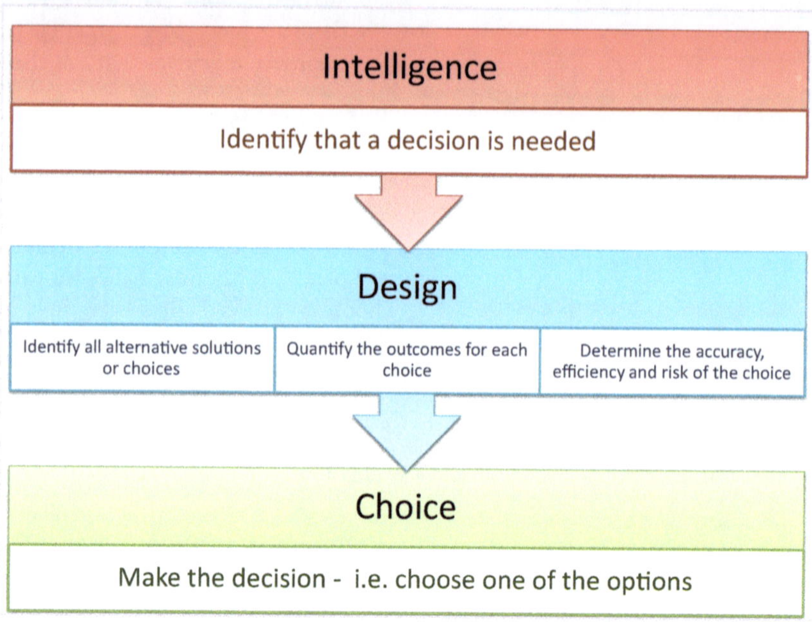

Simon's work in the 1940s and 1950s forms the basis for much management consulting today.[365] He pointed out that humans rarely make optimum decisions - the effort or mental capabilities involved in getting and analysing all the necessary data is simply too great - so instead we *"satisfice"*, i.e. we settle for a decision-making process that gets us to what we believe is close to the correct answer. This meets our need for a speedy conclusion with sufficient accuracy: hence *satisfice*, formed from joining *satisfy* and *suffice*.

Real World: ***Rational or irrational***

The return on investment from energy and resource efficiency projects required in many organizations appears irrational.

A good return on investment in most private organizations would be in the order of 15-20% - taking into account the cost of money and factoring in an element of risk and effort.

However, energy and resource efficiency investments in the same organizations are usually required to achieve significantly higher return - often a payback of two years or less, equivalent to a simple IRR of 50%.

Similarly, at the level of many economies, the cost of reducing electricity demand by 1 kW through energy efficiency is often very much cheaper than building 1 kW of new generation capacity. Clearly, then, if we behaved entirely rationally, we should exploit the efficiency options first, but often we do not.

If we assume that there is no difference in the cost of money for either type of investment, then the only rational explanation why the rates of return expected of energy and resource efficiency are much greater is because the decision-maker feels that there are significantly greater risks and effort involved. It doesn't matter that the reality is different, it is the *perception* that counts.

Decisions of all types are subject to shortcuts: even in big formal selection processes with significant implications, not all alternatives are truly considered. For smaller, "instant" decisions, the influence of instinct and gut feeling is even greater.

These shortcuts are a natural product of millions of years evolution of human decision-making, which helps us rapidly assess a situation and respond in the way that best ensures our survival. Our brains have developed tricks, called heuristics, that speed up decision-making. Some heuristics involve consciously applying a rule of thumb, following "gut instinct" based on previous experience, taking an educated guess or applying common sense. Other types of mental shortcuts involve processes that we use unconsciously. The term that is given to the abbreviated decisions using these heuristics is bounded rationality because our intention is still to arrive at a correct answer.

4.10 **Simon's three-step decision-making process (opposite)** *Source: Niall Enright*

Prior experience or knowledge influence many decisions, so the fact that many decision-makers asked to approve energy and resource efficiency opportunities are simply not educated in sciences and engineering can cause a problem. Even at the most basic level, these decision-makers don't understand the units of measure (who among us can say they *really* understand how big a kWh, MWh, or GWh are or a hectolitre, m^3 and so forth). So the rules of thumb or common sense that may guide these decision-makers are rooted in a very shallow pool of experience. The result is that there is a bias against resource efficiency due to inaccurate estimation.

- The solution to this is to describe the decision in terms that the decision-maker can better understand. Usually, this involves expressing the choices in terms of financial alternatives as most people understand money data. Proposal documents should avoid jargon, place overly technical details into appendices and provide some scale equivalents for units or other measures that may not be readily understood.

> In the absence of information about the risk and effectiveness of an opportunity, decision-makers operating in an unfamiliar domain will tend to **assume the worst.**

I often see proposals that are good at setting out the choices available, and the outcomes in terms of the improvement and financial benefit, but which are completely silent on the third component of the design stage: effectiveness and risk. In the absence of any commentary about the ease (or not) of implementation or the risk associated with the opportunity (e.g. the impact of future price changes or the reliability of the technology), decision-makers, operating in an unfamiliar domain, will tend to assume the worst. Thus, clear communication of risk and effectiveness is important. The Exploration section at the end of this chapter provides an insight into the impact of certainty on decision-makers.

The lack of technical and dimensional awareness in decision-makers makes them even more prone to a common heuristic called anchoring. This is where we have a tendency to give prominence to the first piece of information we receive.

The research in the real world piece on the next page shows the effect of anchoring - 505 people were asked to rank the relative energy use and saving

Real World: *Perception is everything*

It seems that people are not particularly good at estimating the quantity of energy used by equipment or the savings that can be achieved from some common energy conservation measures. Shahzeen Attari and her colleagues at Columbia University in the US invited 505 members of the public across America to fill out a questionnaire which asked them to compare the energy use and energy savings potential for some common appliances and conservation actions.

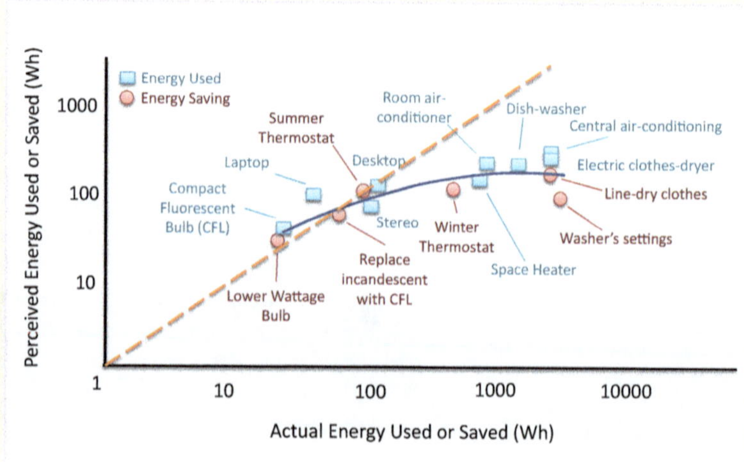

4.11 **Mean perception of energy used or saved as a function of actual energy used or saved for 15 devices and activities**
Source: Attari et al, "Public perception of energy consumption and saving", PNAS, 2010[44]

If we look at the chart above, we will see on the vertical axis what the volunteers thought the energy use or saving is. On the horizontal axis is the actual usage or saving. If the volunteers correctly assessed the energy use then the results would all be plotted along the dashed orange line.

In fact, we can see that the energy use is habitually underestimated, as shown by the points below the dashed line. This underestimation was out by a factor of as much as 10 for the points on the right of the chart. It seems that participants were correct in estimating the relative energy use of appliances, for example, by correctly stating that a desktop computer uses more energy than a laptop, but were poor at assessing the actual difference (on average participants said the desktop used 1.2 times more energy rather than the actual 2.9 times). This may have been the result of an effect called anchoring (see main text).

In addition to the participants' inability to quantify relative differences in energy use, the study also picked up a bias towards what were referred to as curtailment activities, such as switching a light off, compared to efficiency activities such as replacing the existing lamp with a more efficient type. This bias towards switching off energy-consuming devices rather than improving the efficiency of the devices, the authors speculate, may be "*because efficiency improvements almost always involve research, effort and out-of-pocket costs (e.g., buying a new energy-efficient appliance), whereas curtailment may be easier to imagine and incorporate into one's daily behaviours without any up-front costs.*" These are themes to which we will return to several times in this chapter.

from 15 items of equipment or improvement actions. It is very important to note the way in which they were asked the question: "*A 100-watt incandescent light bulb uses 100 units of energy in one hour. How many units of energy do you think each of the following devices typically uses in one hour?*"[45] This initial quantity of energy was what the participants focused on, and they then adjusted their estimates for the other equipment using this base unit. The study showed that for energy uses which are much larger than this initial *anchor* value, the participants adjusted insufficiently and so underestimated energy consumption by as much as a factor of 10. If a different anchor value were to be proposed, then there would almost certainly be a different spread in the estimates provided by the participants.[32]

Anchoring tells us that the *elevator pitch* or summary of our recommendation to a decision-maker is very important. It suggests that we should put the scale of the benefit right up front. We should not be afraid to start our request with a clear statement such as: "*This proposal describes how we can permanently reduce our annual operating costs by US$1.5 million, increase reliability and throughput by 5%, for an initial investment of just US$2.3 million.*" Clearly, the more closely aligned this statement is to the core purpose of the organization, the better. It is desirable that the decision-maker is correctly primed with the scale of the opportunity so that they will be anchoring around this value and relating this proposal to other investment opportunities that have similar costs and benefits.

Real World: *Convincing IBM*

In his book, *The Next Sustainability Wave,*[791] Bob Willard recounts how when he was at IBM in 1997 he tried to convince the then CEO, Lou Gerstner, about the importance of sustainability.

In his letter, which took six months to craft, he said IBM should "*use our leadership and ingenuity in leading the world towards a sustainable global economy*" and closed by asking Gerstner "*would you be willing to apply your leadership to help save our future?*" Passionate stuff.

The outcome was that the letter was treated as a charitable or philanthropic request and directed to the corporate community affairs director.

On reflection, what Bob Willard says he should have said is "*Dear Lou, I have some thoughts on how IBM could increase its profits by 38%. Interested? Yours truly.....*"

This is an excellent example of a short "elevator pitch" anchoring around the scale of the benefits.

Another form of bias originating from the experience of the decision-maker is called confirmation bias. This bias is the tendency for individuals to give greater weight to evidence that supports their existing beliefs. One mechanism for this is when, rather than considering the evidence for an investment in a neutral way, the decision-maker unconsciously selectively recalls the results of previous projects to reinforce their own beliefs. If they are sceptical of the benefits of resource efficiency, they will remember previous project outcomes in a more negative fashion. It is important to note that this is often an unconscious bias - the decision-maker will often feel that they are completely objective in their assessment. Confirmation bias not only works against the adoption of projects but can also lead to over-confidence and a tendency to approve marginally acceptable projects for those decision-makers whose beliefs favour resource efficiency.

There are a number of techniques that can be used to overcome negative confirmation bias (in other words, a situation where the decision-maker is clearly predisposed against a project).

- The proposal could include additional supporting information for an investment (e.g. case studies) which can reduce the effect of selective recall.
- The decision-making process can involve an intermediary "technical review" involving an expert panel who will not display the same bias.

Real World: ***Confirmation bias and consultants***

A form of confirmation bias can occur when organizations seek external advice from consultants.

Most consulting assignments start with a brief in which expectations are set out. This will define the scope of the advice, but it may also reveal the beliefs of the individual client who has commissioned the work.

The reality is that many (but not all) consultants are reluctant to give their clients feedback which would contradict these beliefs. After all, if they go around upsetting customers, then they are not likely to remain in business for long.

One extreme example of this confirmation bias was when I was instructed *twice* by a leading consulting firm, once in Europe and once in the US, that when meeting with employees in ExxonMobil the subject of climate change was never to be raised. I am not sure whether this was an ExxonMobil requirement or a decision of the consultancy.

Surely the reason for employing consultants is to bring in additional expertise and alternative, objective, perspectives?

The desire to avoid controversy means that consultants may simply reinforce the client's existing bias, giving advice that is, at best, a narrow analysis of a client's situation, or at worst, ignores real risk in their approach to resource use. This effect is greatest when strategic advice is sought, rather than technical advice which tends to be less controversial. The solution is for the client to make it clear that they expect open and honest advice. If they do not do so, the default behaviour of most consultants will be not to challenge the existing view.

- It may be possible for the decision-maker to be reassured by the inclusion of "breaks" into the project, where progress only occurs if the investment achieves certain predefined objectives at various stages.
- Finally, the best way to remove the bias is to change the decision-maker, although this is easier said than done!

Unfortunately, prejudice against resource efficiency is, in my experience, surprisingly common within senior decision-making cadres in large organizations. Many factors bring about this bias: it may have something to do with the innate conservatism (in the political sense) of senior post-holders in multinational corporations. Linked to this may be the baggage of climate change denial. There may be a tendency to categorize efficiency as an environmental issue with a presumption that it is value-destroying rather than value-creating. There could also be aspects of denial of the scale of change required in resource use, particularly when the organization itself is a significant contributor to the problem, such as a fossil fuel business. Finally, there may be *"macho"* cultural issues where the decision-makers see their sole duty as delivering short-term returns for shareholders and aggressively resist anything that could give the impression of a wider social benefit, even if that action makes sense financially (see the piece on Fiduciary Duty on page 218).

For those promoting resource efficiency within an organization, it is tough to overcome the resistance of an individual decision-maker.

- One possible solution is to take advantage of yet another form of bias called champion bias, which occurs when a decision-maker gives undue weight to evidence from a proponent who is trusted. By gaining an endorsement from a trusted confidant we may overcome the bias of the decision-maker.

Of course, champion bias can also distort decision-making in a negative way. It may be the case the Champion has their own agenda concerning the resource efficiency opportunity and reasons to undermine the proposal. This behaviour is often caused by misaligned incentives or timescales between the Champion and the organization. For example, in consumer goods companies, brand managers tend to rotate rapidly so may be disinclined to consider longer-term changes to their brands. Short tenures mean that changes such as product reformulation, which could support resource efficiency and higher margins, are rejected in favour of shorter-term marketing investments that are likely to have an impact, and generate a bonus, during their own tenure.

This bias against resource efficiency explains why regulators have, at times, had to resort to directly mandating organizations to undertake specific actions, such as carry out energy audits or report their CO_2 emissions in their annual accounts. Despite all the evidence of an overwhelming benefit associated with resource efficiency, the view is that only by compulsion can some of the most recalcitrant organizations be forced to act in the interests of their shareholders and society.

We can turn to another Nobel prize-winning economist, (2002), Daniel Kahneman, who is considered the father of behavioural economics, to demonstrate another aspect of decision-making which could influence choices around resource efficiency opportunities. This heuristic is called loss aversion, where how we describe a choice greatly influences the outcome. In simple terms if people are given a choice A) to definitely receive £100 or B) a 50% probability of receiving £200 or nothing, the vast majority will choose A. On the other hand, when offered the choice C) of definitely losing £100 compared to option D) a 50% probability of losing £200 or nothing, the overwhelming majority of respondents will take the risk and go for choice D).

The psychology of these choices appears to be related to a greater desire to avoid the pain associated with a loss than to experience the joy of a gain. In the first case, our choice A) is avoiding the 50% probability that we would lose the £100 which was certain. In the second case, choice D) gives us a 50% possibility of preventing any loss, which is better than C) where the loss is definite, albeit smaller. It seems we really don't want to face the recriminations of making a choice that led to a loss.

Tversky and Kahneman's 1981 paper[723] in *Science* gives numerous examples of just how strong this effect is, and how framing influences choice. I have adapted an example in that paper to resource efficiency. Let us imagine that there is a new waste disposal tax being introduced next year which will cost our site £60,000 a year. We have a choice of two projects that can help us to reduce the impact of this tax, both of them requiring the same investment. One of the projects, A, involves a proven, but old, technology which can reduce the waste being taxed by a third. Project B is an unproven technology with the potential to save all the tax, but also a high element of risk. The do-nothing approach would not be acceptable to our stakeholders, so we need to choose either A or B.

barriers

We could present the choices to our decision-makers in two different ways:

People place greater weight on what they may be asked to give up than on what they are gaining.

This creates an **inherent bias against change.**

Framing in terms of savings:

1. Technology A will definitely save £20,000
2. Technology B has 1/3 chance of saving all £60,000 but 2/3 chance of saving £0

Framing in terms of losses:

3. For Technology A our losses will definitely be £40,000
4. Technology B has 1/3 chance of £0 losses but 2/3 chance of our losses being £60,000

From an economist's perspective, both projects offer the same potential risk-adjusted benefit so neither should be favoured. Couching the options in terms of savings would favour the choice of Project A, whereas if the options were framed in terms of losses, then the decision would be biased towards Project B.

Real World: *Berkshire Brewery*

One of my first energy efficiency assignments in the early 1990s was to program a refrigeration expert system for a brewery ammonia refrigeration system. This plant had four enormous 1 MW compressors, which sounded just like a jet engine when they started up and shook the whole of the utilities building to the foundations. In fact, if two of the chillers were started up simultaneously they would trip the local grid and pitch parts of the nearby town of Reading into darkness.

You would have thought that given this risk, and the fact that the brewery production was entirely dependent on the cooling, that a sophisticated control system would have managed the chillers.

Indeed, there was just such a system, to which the new expert system was going to be linked, but in the course of our project, we discovered that for some shifts, particularly at night, the operators bypassed the automatic control system and ran the plant entirely in manual mode. They were setting the compressor loadings and choosing which of the compressors to run at any given time. This was a very inefficient way to operate the plant, but what was even more surprising is that these operators had been running the plant manually in this way for years.

When asked why they did this, the operators in question confessed to a lack of confidence in the control logic and a belief that they could run the plant more reliably and efficiently. They felt that they knew the system and would make better choices. I wonder how many supposedly automated systems are overridden in this way unbeknown to the system designers and chief engineers.

The example above is artificial, of course, and real-world decisions are never so clear-cut, but we do need to bear in mind that the bias against losses is real and does influence decisions. In essence, foregone gains are less painful than perceived losses, which explains why some folks focus disproportionately on the cost and risks of the efficiency opportunity rather than the benefits.

An opportunity lost (i.e. a missed opportunity to gain) is valued less than a real out-of-pocket cost, even though from an orthodox economics perspective it shouldn't be. A bird in the hand is really worth two in the bush. This bias is reflected in tort law, where judges make the distinction between *"loss by way of expenditure and failure to make gain"* and in contract law, where a party that breaches a contract is more likely to be held to the original terms if the action is taken to make an unforeseen gain, than if it is taken to avoid a loss.[156]

This desire not to incur losses contributes to inertia or status-quo bias.[640] Since most decisions to make a change involve some sort of gain and some loss, and individuals place greater weight on the losses, then there is a tendency to favour retaining the status quo. Status-quo bias is very prevalent in terms of consumer choices about energy supply; despite the deregulation of markets in many economies, consumers tend to remain with their existing supplier for extended periods, even when switching is made easy and competitors offer substantially better rates.

This resistance to novelty also underlies another of our barriers to resource efficiency: habit and custom. One example of this negative behaviour occurs in homes and offices where thermostat setpoints and timings are overridden to suit individuals; if people are accustomed to control, they often resist automation. The real-world example on the left shows that this behaviour also occurs in industrial settings with equipment that has a high resource use.

Another example of a poor custom is the widespread tendency for maintenance staff to prefer like-for-like replacement. There are sound reasons why this may be a sensible approach; not only does this involve equipment with which the maintenance team are familiar and which has known performance, but it also means that there is potentially a greater pool of spare parts and expertise to deal with problems when things go wrong. Indeed, the procurement costs of the equipment are lower since there is no requirement to select from multiple options. All these factors may well make like-for-like replacement a reasonable choice, but it also potentially misses a golden opportunity to gain efficiency improvements where capital is already committed.

Although we will see later that standardization is a tool to drive improvement, it also accounts for a significant amount of material waste. In their excellent exploration of material waste[22] Julian Allwood and Jonathan Cullen provide a revealing insight into the many ways that custom and convenience lead to large-scale material losses. One example they cite is the standard industrial I-beam illustrated opposite, left. In practice, the optimum load-bearing shape would be a tapering beam as shown on the right, which uses less steel.

4.12 **Standard versus optimized I-beam design**
Source: Reproduced with permission from: Sustainable Materials — without the hot air, Julian Allwood and Jonathan Cullen[22]

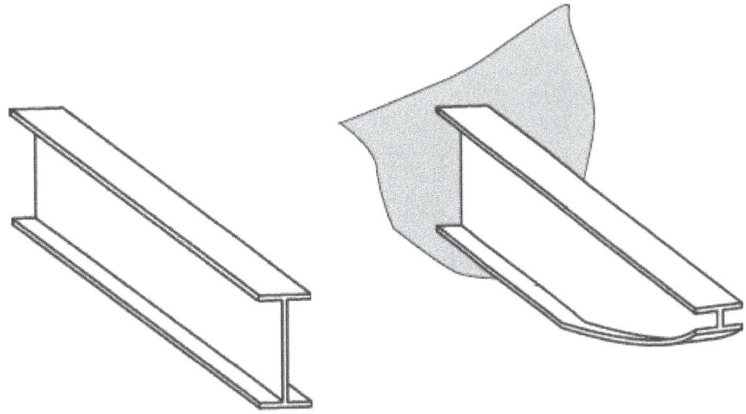

The resource inefficiency in the standard I-beam arises because more material is used than is necessary for the core purpose, supporting weight. In fact, over-specification of materials is rife in construction. The building codes in the UK describe the minimum requirements for reinforcement to carry certain loads. Given that this is a legal obligation, and contractors could face severe penalties for not meeting the standard (as well as penalties from their customer), there is a clear incentive to exceed the specification - using a *belts and braces* approach to reduce risk. However, since the building codes will already have a redundancy margin for weight-bearing, it would be much better if the requirements were described as a target, with the clear implication that exceeding this is undesirable. Language really does matter!

The penalty for component failure is very high compared to the additional cost of over-design. This asymmetric risk means that over-design is rife. Not only are the materials in most buildings over specified, but so too are many heating, ventilation and air conditioning (HVAC) components, such as chillers and boilers. Because these are over-sized and designed to perform in excess of the hottest or coldest day of the year, their standard mode of operation can be quite inefficient. The additional cost of equipment and energy that this entails is small in comparison to the cost of fixing an existing under-sized system. Anyway, it is not the mechanical and electrical (M&E) design contractors that bear the additional operating costs, it is their client, while the cost of under-specification in the design would fall on their shoulders. This is a form of the principal-agent barrier described in the previous section. No wonder then that the design contractors are so conservative - they have nothing to gain by sizing the system more precisely, and potentially a lot to lose.

We must not fall into the trap of assuming that all material use as such is bad. Increasing some material use significantly (e.g. insulation) can lead to a substantial decrease in another resource use (gas for heating). In the chapter on design, (page 755), we make the point that the optimum resource efficiency in buildings depends on an integrative design approach, where all materials are considered in balance.

Real World: ***Concrete conservatism***

The manufacture of cement contributes to 5% of global CO_2 emissions,[112] or a staggering 19% of industrial carbon emissions.[22]

No wonder then that finding an alternative to the emissions-intensive Portland Cement is a top priority.

There are some promising prospects too, among them.[774]

- **Novachem**: based on magnesium silicates, which absorbs CO_2 from the atmosphere and offers the possibility of *carbon-negative* cement.
- **Calera**: which uses waste CO_2 to create the cement, thus providing a valuable carbon capture and storage technique.
- **Calix**: which creates cement in a way that enables CO_2 emissions to be captured.
- **Geopolymer cements**: which use industrial waste like fly ash to create various types of cement.

The key to the adoption of these new types of cement will be their structural performance and costs. The final barrier, however, will be standards.

Steven Kosmatka of the Portland Cement Association is quoted as saying: *"It took PCA about 25 years to get the standards changed to allow 5% limestone* [in the Portland cement mix]. *So things move kind of slowly."*[72]

I have seen this myself. On a recent project, I found it impossible to get the client to consider a mix with more recycled materials. Neither the architects or construction company were willing to warrant long-term performance for anything other than the traditional mix.

Our next example of custom as a barrier to efficiency is the widespread habit of using standardized electric motors and transformers, which has a large impact on electricity consumption at a global scale. By only stocking a limited number of standard motors, many organizations then have to oversize the installed motors, which can lead to considerable inefficiency. Given that motors consume about 65% of the electricity used in industry, this habit is very damaging indeed.

The last behavioural barrier to resource efficiency lies in the relatively low prestige currently associated with this topic. In many organizations, it is the development of new products and services that bring kudos, not the optimization and refinement of existing activities. This bias disinclines decision-makers from investing effort and reputation in efficiency programmes, which can be addressed by:

- Communicating the importance of resource efficiency, its role in the design of products and services, and the value derived.
- Demonstrating to the decision-maker at the outset that their active participation will enhance their prestige and prospects. The most common way of doing this is by having senior management visibly supporting the programme.
- Incorporating features like league tables and rewards, which can help elevate the status of participation in a programme.

I have observed that this cultural aversion to championing resource efficiency tends to be strongest in larger organizations. One explanation for this might be something to do with career progression in these types of organizations in both the private and the public sector, which is based on a series of relatively short placements, from three to five years. As the individual moves up the *"greasy pole"*, they seem to become increasingly concerned with *"not messing up"*, so that they can successfully transition to their next assignment. This form of progression makes individuals excessively risk-averse. Because resource efficiency involves change, and change is uncomfortable and requires challenging the conventional way of doing things, those middle managers who are concentrating on *not rocking the boat* are further disinclined to engage with the subject. This can lead to the classic *middle management squeeze* where mid-level individuals in positions of authority impede change (see page 239 *Spanning the Intent Gap at BP*, for a real-world example of this).

Of course, there will always be exceptions to the rule. I have seen folks at all levels of organizations give their careers a significant boost by championing resource efficiency. It is often the case that a far-sighted facilities or departmental manager can catch the attention of their senior colleagues by delivering a really effective resource efficiency programme. Indeed, senior management are often the first to recognize that people able to deliver a successful resource efficiency demonstrate talents that are of great value to the organization.

⇒page 189.

It goes without saying that confidence in the outcome greatly influences a decision to proceed.

Different levels of certainty influence our strategies to get decision-makers to "yes".

Exploration: *Why certainty drives the resource efficiency proposal*

One of most significant challenges in obtaining commitment for energy and resource efficiency initiatives is due to the certainty attributed to the programme outcomes. Economists talk about anything we buy, whether it is a programme – such as our resource efficiency project – or a piece of equipment, as falling into one of three groups.

1. The first group are search goods – items where the buyer, through research, can establish the characteristics or benefits that it will bring before purchase. An example of a search good is a new car where trustworthy independent fuel-consumption data is available to inform the choice in advance.
2. The second category is experience goods, where the consumer will only really know if the goods have delivered the expected benefits after purchase. For example, a second-hand car is an experience good because, despite published data being available, the owner will only know how efficient and reliable this particular used car is after they have bought it and driven it for some time.
3. Finally, we have credence goods where there is no way of determining clearly before or after purchase if the goods are delivering the expected benefit, and so consumers have to rely on some other form of evaluation to assess their performance. Keeping with the cars analogy, an example of a credence good would be the repairs that a garage says are needed on your car, which we do not know are required in the first place and which may or may not deliver a benefit.

   In credence goods the buyer is almost entirely dependent on the seller's superior expertise and may consequently be cheated – in the early 1990s, 53% of car repairs in the US were found to be unnecessary.[798] Some credence goods such as medicines or legal services are subject to government regulations so that consumers are protected in the absence of clear (or easily understood) evidence of benefit.

These three categories represent a risk spectrum – from the lower risk search goods through to the much higher risk credence goods which are essentially bought *"on faith"*.

So where does our resource efficiency programme fit on this spectrum of risk? Well, individual items, such as buying a new boiler or a new water treatment plant, can be considered search goods. However, when we look at a programme as a whole, there is a much higher level of uncertainty. This uncertainty relates to the performance of all the individual technologies in combination, variability in the hours of operation, the behavioural response of staff, the high search costs to quantify the investments, future resource prices, and so forth. Taken together, these uncertainties mean that the programme as a whole can be considered, at best, as an experience good. That is to say that the decision-maker doesn't really know what the outcome will be until after they have embarked on the programme.

In practice, the true results in many resource efficiency programmes are difficult to assess even after implementation, unless the reader adopts some of the techniques, such as Monitoring and Targeting (M&T), or Measurement and Verification (M&V), set out later in this book. This because of the numerous external factors such as weather, production volumes and mix, organizational boundary changes, resource costs, etc., that influence resource use and so mask the impact of the changes. In these circumstances, resource efficiency is often perceived as a credence good. In other words, *the decision-maker may not be confident that they can determine the benefits that it will bring, even after the programme has been put in place* and so will have to make a decision on the basis of other factors, in particular, the reputation or expertise of the proponent for the programme.

In general, products are easier to evaluate than services. A new motor will have comprehensive manufacturer's data on operation at different loads and so is easy to evaluate, compared to, say, a behaviour change programme asking people to switch off the lights, whose impact is relatively unpredictable. The more our resource efficiency programme engages with people, the more likely it is to achieve continual improvement and to deliver the preferred low-cost demand improvements rather than the more expensive supply-side changes. However, because people aspects are harder to predict, incorporating these into our programme has the effect of making the project as a whole more of a credence good and thus more difficult to sell to decision-makers.

From an economic perspective,[367] resource supply and resource efficiency are substitutes for each other – each can be used to meet the organization's needs for goods and services (materials, water, heating, cooling, lighting, etc.). Organizations have a choice between simply purchasing the necessary level of resource (the *status quo supply* option), or buying a new service (the resource efficiency option) which may be cheaper but, as an experience or credence good, is essentially a *leap of faith*.

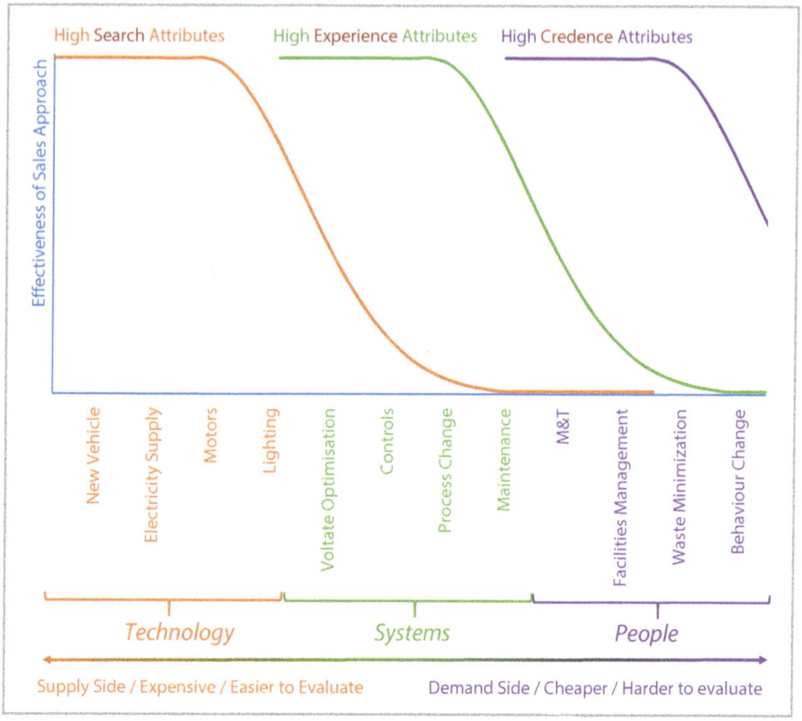

4.13 **Economists can explain the observed bias towards technical fixes for resource efficiency as a consequence of the ease of evaluation or certainty of these solutions** Easy to evaluate opportunities can be sold using a search goods approach, shown in orange. As the effectiveness of this approach decreases because there is less certainty, an experience goods approach is suggested, shown in green. Finally, a credence goods sales approach is recommended, shown in purple, where there is very little certainty of the benefits, even after the project has been completed. *Source: Niall Enright. This illustration and a poster version are available in the companion file pack, along with a spreadsheet model to "tweak" the contents.*

If we can place our energy and resource efficiency opportunity in the **correct economic good category** *and respond appropriately, then the probability of success is greatly* **increased.**

From the decision-makers' perspective, resource supply is a simple and easily understood product – with vendors who are long-established and trustworthy, product attributes which are well-known and with an established buying process. Thus the supply option consists of low-risk, search goods whose costs and benefits can be quantified through a standard procurement process. A resource efficiency programme, on the other hand, involves a number of relatively unknown suppliers with technologies that may not be understood or proven. The individual items of equipment that form part of the resource efficiency option may only be procured occasionally, so the research costs in quantifying their benefits may be high. Thus the overall certainty of the benefit is considerably lower in the efficiency option than in the supply option, and the transaction costs (overheads) are higher. Given these two alternatives, it is not surprising that decision-makers often choose *the devil they know* rather than what is perceived as the riskier option of resource efficiency.

If we can place our energy and resource efficiency programme in the correct economic goods category, and respond appropriately, then the probability of success is increased. Table 4.14 on the next page sets out each of the suggested approaches. In an ideal world, we will present the decision-maker with a proposal that fully evidences the outcome to be achieved (i.e. it is a search good) and so it competes on an equal footing with the status quo, the supply option. In practice, assembling the necessary evidence for such a business case can be difficult and time-consuming. As a result, the initial proposal may well be a high-level conceptual pitch, which is much more of a credence good. Such a proposal will require the endorsement of a trusted sponsor or expert so that the decision-maker feels justified in giving approval. For a credence approach, the request to the decision-maker will usually be not to commit to a full-blown a programme, but perhaps to approve some further investigation which will enable the proposal originator to return to the decision-maker a second time with a proposal that is less uncertain (much more along the lines of an experience good). This effort to actively increase the certainty of a proposal at each stage is common in seeking commitments for resource efficiency programmes.

If the decision-maker has some initial uncertainty about the outcomes - i.e. they are being asked to commit to an experience good - then we could ask for a pilot project to be approved to confirm the effectiveness of the process. Using the data gathered during the pilot, we would then return for support for the expansion of the programme, but this time with much higher levels of certainty. In effect, we are doing what any grocer would do when introducing a cheese that consumers do not know from previous experience - offering a "*try before you buy*" sample of the cheese to customers, so that they can decide if they like it or not.

Many of the techniques set out later in our method for resource efficiency exist to address the challenges of verifying programme outcomes. We shall see, for example, that the continual measurement of programme performance using M&T minimizes the effect of external factors; while the Opportunities Database provides for constant reassessment of costs and benefits from discrete projects, thus lowering ongoing transaction costs and validating the return on existing investments. Collectively, these techniques make our programme an experience good rather than a credence good. These techniques guarantee that the outcome of the programme will be measured effectively. Thus, when we describe the benefits of our resource efficiency programme to a decision-maker, we may need to communicate some of the techniques that we will use, like measurement, that make it an experience or search good rather than a credence good.

The same considerations of certainty that surround a programme approval decision, also apply to discrete investments or projects. Where the investment is in a proven technology running for a fixed number of hours, like a variable-speed drive, then this is very much a search good whose benefits can be easily quantified in advance of the decision to proceed. Where we are considering less tangible projects, such as behaviour change, it is important that we deploy the best possible quantification techniques, such as regression analysis, to make a case for the benefits so that we can increase the certainty in the approver's mind for these softer projects. These techniques are set out later in this book.

4.14 **The effect of certainty on programme and proposal design**
Source: Niall Enright

Economic category	Search Good	Experience Good	Credence Good
Usual type of proposal	Detailed business case	← in between →	High -level or conceptual
Uncertainty	Low	← in between →	High
Key Success Factor	**Clear Cost-Benefit Quantification**: A search good enables the buyer to be confident of the pros and cons of the alternatives in advance of making a decision. Provide detailed cost-benefit case based on systematic audit to fully quantify resource use and applicable technologies for improvement. Give clear evidence for improvement (and performance guarantees where possible) for key technologies to be used in the programme. Use statistical techniques to quantify scale of behavioural savings possible. Provide full costs of all programme elements including design, procurement, training, installation and commissioning costs for technologies and indirect costs such as cost of capital, management time and opportunity costs (such as production interruptions).	**Measurement and Control**: For an experience good, the buyer needs to be confident that they can evaluate their choice after making the purchase. In order to achieve this we should incorporate measurement at the centre of the programme, using proven techniques such as Monitoring and Targeting (M&T) or Measurement and Verification (M&V) to give the decision-maker confidence that they will be able to assess the programme's performance at any point and control the inputs in line with the benefits.	**Endorsement**: For a credence good, the buyer will be looking at attributes such as the reputation of the seller to provide assurance that the project will achieve the claimed benefits. Thus if the initiator of the resource efficiency proposal does not have sufficient credibility with the decision-maker, they need to engage with people the decision-maker trusts to gain their endorsement for the programme. This situation is common for a high-level or conceptual proposal, so it is important that this type of proposal receives the strongest possible support from multiple "experts" around the decision-maker.
Programme Design	**Regular Review**: As we ask for ongoing approval for the programme or projects, we need to maintain the credibility of our assessment by updating the business case in light of (usually rising) resource costs, (usually falling) technology costs, and newly emerged technologies. Our decision-maker needs to be confident that a robust quantification of benefits is being undertaken at every stage.	**Staged Implementation**: Not quite "try before you buy", but we should introduce every opportunity to confirm the programme's success - either through a pilot project or through a phased rollout, at which point the decision-maker can verify the savings (using M&T or M&V) prior to more confidently committing to further work.	**Trustee Led**: Given the level of uncertainty, a decision-maker may well put in place a governance structure that involves people they trust. Typically we would create a steering group where the decision-maker or his trusted experts can regularly oversee progress and take corrective action if they feel the programme is not performing.
Initial Request	**Full Programme Request** (usually with clear hurdle rate): Here it is customary, having credibly demonstrated the benefits in advance of a decision to proceed, to ask for approval for the full programme, albeit with an agreement on the total spend, IRR, maximum payback or other hurdles that the project should meet.	**Staged Request**: Ask for support just until the first milestone, e.g. propose a pilot programme or create a definite threshold for further commitment (e.g. "if we achieve 5% savings with a year's payback, then we can roll this out further").	**Deferred Decision**: For this kind of proposal we will usually seek agreement to actions that can decrease uncertainty, such as undertaking an audit process or establishing an expert panel to design the programme, prior to seeking approval for an initial pilot or milestone-based programme.

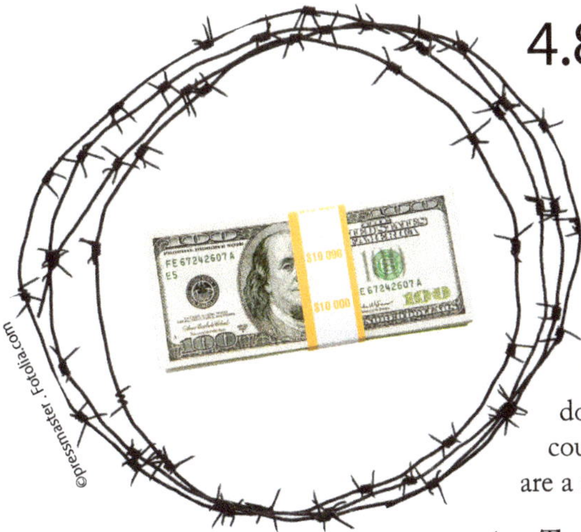

4.8 Availability barriers

Even when an energy and resource efficiency opportunity gets over the many general, structural or behavioural barriers, it may still be rejected because key resources of time, money or technology are not accessible. Most of these availability barriers can be overcome if they are anticipated.

"*We don't have the money*" is the most common reason given for turning down energy and resource efficiency proposals.[605] This statement, of course, needs to be deconstructed if we are to find a solution, and there are a number of possible interpretations.

1. The project meets the required financial return, but there isn't enough money available *now*. This is a budget barrier due to available finance being committed to other activities. In other words: "*We don't have the money now*".

2. The project does not currently meet the required financial return so money cannot be released to support it. This is a hurdle barrier in which the project, as currently described, will not receive funding because other projects, which perform better financially, will receive the available finance, i.e. "*The investment return is insufficient*".

3. The project meets the required financial returns, but the effort associated with accessing the necessary funding is too great. This is a complexity barrier, which could relate to the scale of benefits compared to the approval process. This is simply a case of "*It is too much hassle to get the money*".

4. The project meets the required financial return, but the decision-maker is not inclined to support the project. In other words, lack of finance is being used as an excuse for rejecting the project. This is a support barrier. There can be a variety of reasons why the decision-maker wants to hide the true reasons for rejecting a project. Perhaps the proposal is weak, but they don't want to hurt the feelings of the proponent. Maybe the rejection is based on perceived risk or uncertainty in the decision-maker's mind, but they don't want to spend time exploring the basis for this. Or, possibly, their decision is based on instinct or some other psychological factor.

There are two pots of money in most organizations: OPEX and CAPEX. In order to understand the availability of finance barrier, we need to consider which type of finance is unavailable.

Firstly, let us consider OPEX, which stands for *OPerating EXpenses*. Many programmes targeting no/low-cost opportunities can be funded entirely from within an OPEX budget – project costs can be met entirely by savings made in resources. In many cases, unit or department managers can spend their OPEX budgets with a reasonable degree of discretion, once these are allocated in an annual budgeting cycle, so decisions to *"spend to save"* may require a lower level of approval. For opportunities focused on OPEX funds, we may find that timing and cash flow issues dictate our ability to progress the projects – for example, it may be essential to ensure that savings fully cover during the budget year in which they occur. Thus, a project with a six-month payback would be unlikely to be implemented in the second half of the budget year. There are several techniques we can use to overcome an OPEX budget constraint.

- In looking at opportunities with a relatively quick return, say under three years simple payback, it is often possible to develop a business case around the notion that the budget already exists, but is hidden.[637] In other words, before implementing the project, the organization has already committed some of the OPEX budgets to paying for resources which will be *wasted*. What is needed is simply to divert this expenditure towards a more useful investment in greater efficiency.
- For opportunities focused on OPEX budgets, it is often important to think about the cash flow (see page 570) associated with the project. For example, it may be possible to structure a deal with suppliers of the more efficient equipment so that the timing of their invoices ensures a positive cash flow, where the cumulative savings outweigh costs at all times.
- In some organizations, such as retail and commercial properties, costs can be passed on by the landlord, via a service charge, to tenants. Since the expense of the measure is fully recoverable, there may not be a budget issue. However, there could still be some cash flow implications, particularly if the landlord chooses to defer part of the costs from one service charge budget year to the next, in order to make the measure cost-neutral for the tenants. Once again, a deal with suppliers for delayed payment may help align the costs and benefits.

Whereas operating expenses are the routine costs necessary for the day-to-day operation of the organization, *CApital EXpenses*, known as CAPEX or capital, are usually more strategic and have longer-lasting effects. Capital expenses tend to be larger; the benefits (usually in the form of savings or additional revenues) tend to be spread over a period of years; and, once made, the investment is often irreversible, which means that a mistake or change of mind can often lead to significant costs. Capital expenses usually involve the purchase of tangible goods, such as equipment or software which will be used for years, and so are treated as assets of the organization. Assets must be shown on the organization's balance sheet, which may also give rise to special tax considerations and additional investor scrutiny.

For all these reasons, organizations manage their capital budget in a different way to the control imposed over operational expenses. Typically, decisions about CAPEX are taken at a more senior level in the organization, with

Real World: ***Unobtainable savings***

I recently had a call from an energy consultant friend, who will remain anonymous for reasons which will soon become apparent.

In the call, he told me about a project that he had just finished at a gas liquefaction facility somewhere in the Middle East. This facility uses huge *"trains"* of compressors to convert natural gas to a liquid which can then be shipped around the world.

After some very detailed analysis, my friend came to the conclusion that the conservative savings potential for the plant was - wait for it - US$ 2,000,000 a year. Yes, US$2 million each year. And, even better, this is with a great payback of under two years.

One would have thought that such a huge potential for improvement would be greeted with delight. But there is a catch. The savings would be in the form of reduced natural gas consumption and because of the design of the plant there would be no increase in output by making it more efficient. So although the *"fuel"*, natural gas, has a value on the open market, making the efficiency improvement would not make a real saving for the operator ***today***. All that would happen is that in 20 or 30 years' time, as the gas fields reached the end of life, there would be a little more gas left in the ground to extract. Needless to say, nothing was done.

I have countless other examples, many from the oil, gas and chemicals sectors, where use of *"own fuels"* creates a barrier to investment as the fuel is treated as *"free"*, or even worse, the combustion process is considered a form of waste disposal for by-gases (aka flaring). The fact that these processes lead to huge and unnecessary CO_2 emissions and complacency around energy consumption is lamentable.

greater scrutiny and with the involvement of functions, such as finance, which sit outside the requesting department.

Another characteristic of CAPEX is that it is often limited by the organization's ability to borrow or to ask shareholders to invest further. Thus most organizations are short of capital, which leads to rationing. This restriction on capital is a common reason why resource efficiency projects which appear to be *"no brainers"* in terms of their financial return can fail to gain support because other projects, perhaps even with a lower financial return but addressing a higher organizational priority, have used up the available capital.

It is important to re-emphasize this point. Other investments with a *lower* rate of return may receive funding *ahead* of our resource efficiency opportunity. Although private sector businesses can be considered profit-driven, their allocation of resources involves more than just establishing a financial return. Thus an organization may feel that its priority is to increase market share and invest heavily in activities that have little or no immediate return but support their long-term strategic objectives.

- It is important to align the resource efficiency proposal with the core objectives of the organization and to demonstrate that the proposed investment supports the organization's priorities (e.g. market size, service delivery, stakeholder needs, brand image and so forth).

Barriers

Having said that CAPEX represents larger items of expenditure, some organizations will treat all payments of a particular type as CAPEX. Thus, even small expenditure on waste separation bins, tap flow regulators, simple controls or steam traps each costing a modest amount, may be treated as CAPEX and subject to capital approval rules. Thus, understanding what category of spend falls under the heading capital is important even if we believe we are only looking at a programme focused on immediate operational savings.

It is important to **align the resource efficiency proposal with the core objectives of the organization** and to demonstrate that the proposed investment supports the organization's priorities.

Some strategies can be adopted when CAPEX is not available for a resource efficiency project or programme.

- If the organization does not have access to its own CAPEX, then third-party funding may well be possible. This funding can range from an agreement with the equipment supplier to pay them out of the savings made, through project finance from a bank, to much more complex special purpose vehicles and partnerships established around large investments with a long time frame.
- One important consideration for companies contemplating external finance is whether or not the borrowing costs and debt appear on the balance sheet. As investors are looking for companies not to exceed a certain amount of debt per unit revenue, this can be a very real hurdle for organizations wishing to make investments in resource efficiency. Luckily, mechanisms do exist to enable external funding to appear

off balance sheet. This might involve equipment being leased to the client and remaining the property of the supplier until a given point in the future when it has been fully depreciated and is transferred to the client for a *peppercorn* sum (e.g. US$1). The Sarbanes-Oxley Act has had a big impact on some of these schemes, as it seeks to drive much greater transparency into corporate accounting and many organizations have an outright ban on any form of disguised debt.

- If low-cost CAPEX items are required, drawing from maintenance budgets, essentially a source of pre-approved CAPEX, may be an effective way to get these smaller items in place rapidly.

- One way of guaranteeing the availability of CAPEX for an efficiency programme is to create a dedicated fund. In this case, the organization will estimate the amount of capital that will be needed to achieve an overall improvement and set aside this CAPEX specifically for energy and resource efficiency.

Obtaining finance for resource efficiency projects is described in more depth in Chapter 18 on funding. The key point to take away here is that funding can usually be found if there is sufficient will. The cause is rarely the lack of money, but a lack of management commitment, a lack of sufficient priority given to resource use, or simply a lack of knowledge on the part of the proponent about how to present the case and the availability of finance.

There are also solutions to the other financial barriers.

- A hurdle barrier exists when the business case for the investment falls below the minimum required return. The simplest way to overcome this obstacle is to go away and reformulate the scope and design of the proposed opportunity to improve its attractiveness. Sometimes it is possible to bundle projects which sit above the desired hurdle rate with other projects which fall below, and so arrive at a basket of activities which can be approved.

- We should not let the hurdle barrier put us off altogether. The reality is that within most organizations there will be a significant amount of investment taking place which does not meet hurdle rate. For example, investments in compliance (e.g. health and safety or reporting) are often regarded as *unavoidable overheads* and not expected to make a return. I have seen many situations where middle managers have been able to get resource efficiency projects through by labelling them as a compliance action. I am not suggesting this as a universal solution, as it creates an impression in the mind of decision-makers that resource efficiency is a negative environmental cost rather than something which is more often highly value-adding to organizations.

- Some of the third-party funding or deferred-payment options mentioned earlier may also help address complexity barriers. If a third party can bring finance and technology to the table, then that may prove attractive - although one should be careful not simply to replace an investment challenge with a legal or contractual one.

- It may be the case that the small size of the project makes the effort associated with finding funding unattractive and so increasing the scope may be the only solution. Alternatively, where a project is small but has the potential to act as a pilot for a much bigger investment, the effort may be justifiable if the decision-maker can see the potential to replicate the project and the larger benefit that could be gained.

The next challenge for organizations seeking to become more efficient is the availability of human capital, that is the availability of people's time to progress the opportunities. Today, organizations in the public and private sectors are usually lean operations where employees are highly utilized. This shortage of labour means that projects that require considerable amounts of time and effort may be turned down simply because folks are already very busy.

The nature of energy and resource efficiency as change management processes means that the kind of people who are needed to deliver the improvement are those individuals who are considered key to the organization. They will be people who have demonstrated that they can get things done, who are willing to challenge convention, who can motivate and lead a team. They will be folks who are considered good and effective at whatever they turn their hands to. They might also be quite technically proficient, with engineering, process or data analysis skills. These kinds of people tend to be in short supply.

The most effort in a resource efficiency programme is at the beginning, especially if this involves setting up management, measurement and reporting systems.

4.15 **Most energy and resource efficiency programmes require an initial peak of effort, which can act as a barrier to adoption**
Source: Niall Enright, on the website.

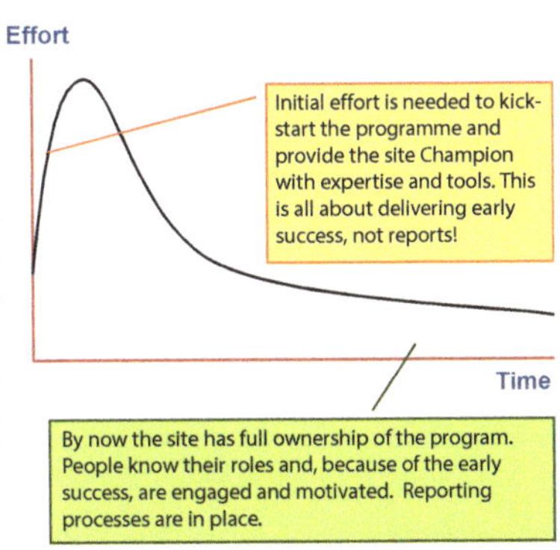

- The obvious solution to the initial effort peak is to bring in external resources to get the programme up and running. This resource usually consists of consultants who are hired for a defined period and may help with setting up systems, identifying opportunities or working with the Governance team to put in place the management processes. The benefit of using consultants is that they can bring the experience of previous programmes and so help the in-house personnel avoid some of the more common pitfalls around change management programmes in general and energy and resource efficiency in particular. There are some disadvantages in using consultants in that they are likely to be more expensive than in-house staff, their presence could reduce ownership and learning, and they may not be as motivated as the site staff.

- Another way of addressing the requirement for people's time is to involve a lot of people for a little time, rather than just a few people for a lot of time. Although I have seen some examples of successful *hit squad* approaches to resource efficiency, such as in ExxonMobil and Shell in the 1990s, most organizations will favour spreading and embedding effort

One of the most effective strategies to deal with lack of time is to **involve a lot of people for a little time, rather than just a few people for a lot of time.**

within their existing organization. The idea here is that people are working smarter, not harder. Later on, we will explore the best way to organize a programme and the role of a dedicated resource or Champion. However, even where there is one individual taking day-to-day responsibility, they should be acting as enablers helping to channel the efforts of many other people rather than doing the work themselves.

Sometimes resource efficiency proposals fail because the proposer has not quantified the amount of people's time is required and shown to decision-makers how this can be made available. This is particularly the case where the project focuses on behaviours or changes to systems and processes. The over-emphasis on the financial aspects of a proposal is another factor that can lead to the availability of people to be overlooked.

The real-world example, opposite, illustrates a conversation with a senior manager in which they were asked to contribute a specific amount of time to a programme. Being explicit about the effort required at all levels is important so that there is an agreement to make that time available. A depressingly common cause of programme failure occurs when key individuals working part-time on the project get sucked back into their "day job" for one crisis or another and the programme withers through lack of effort.

The next barrier is called product availability. Here, we are referring to the fact that a project may be financially or technically viable but the specific equipment needed is not readily available. One example would be the difficulty in obtaining triple-glazed windows in the UK compared to, say, Germany, where there are plenty of manufacturers. Sometimes product availability barriers are due to regulatory issues so that a particular technology has not been approved for use or is subject to import tariffs which affect its economic viability (such as the recent tariffs applied to the import of Chinese PV panels into the EU to protect European manufacturers).

Adverse bundling refers to a barrier to resource efficiency in which inefficiencies are introduced as a result of other technologies. One example that comes to mind is the requirement of manufacturers who have high concentrations of volatile organic compounds (VOCs) in their exhaust air to reduce this pollution. The most common technical solution involves incinerating the waste which leads to lost resource (the VOCs which are not recovered), additional resource input (the gas for the combustion) and additional pollution (CO_2 rather than VOCs). People often confuse the word *new* with *improved* - actually, the water power that we used in our mills a century ago was better from a resource perspective than the coal-based steam power that replaced it.

Another form of adverse bundling occurs when the buyer is unable to ascertain the effectiveness of a technology fully. This uncertainty may lead to a situation where the seller can take advantage of the buyer and present the technology as being more effective than it really is. Unfortunately, I have encountered many such *snake oil salesmen* in the area of energy efficiency, making dubious or

Real World: *The "15 minutes is all" pitch*

When asking for senior management commitment to a resource efficiency programme, it always pays to be clear about the scale of effort that you expect from them and their teams.

If folks aren't told how much effort is involved, they are bound to assume the worst.

Below you can see a conversation I would have with a site manager. It may appear a little artificial, but this is quite a good distillation of the very many real conversations I have had on this topic over the years.

Niall (to site manager): *I've got some good news for you!*

Manager: *Oh really! What's that?*

Niall: *Our programme audit has shown we can cut US$1 million a year off your costs – and we can do this at low risk, with an average of a two-year payback, a positive cash flow after just six months and a return on investment twice our hurdle rate.*

Manager: *Sounds too good to be true. What's the catch?*

Niall: *Here's the deal – I need you to commit to 15 minutes of your time once a month. If you give the programme this, I am confident of success.*

Manager: *Is that all? How does that work?*

Niall: *Well, you need to spend 15 minutes of your monthly management meeting discussing energy and resource efficiency with your department heads. We'll set up all the honest and fair progress indicators in advance. If the departments are doing well, you let them know and if they are struggling, you see what help they need.*

Manager: *So how are the US$1 million savings made?*

Niall: *The business case from the audit has already identified the specific projects and behaviour changes that can deliver the saving – we just need to crank the handle.*

Now that you've got this focus on resources at your meeting, the department heads are going to devote some time to this at their own meeting – maybe 15 minutes a month also.

Manager: *OK.*

Niall: *The team leaders will then start covering this at their Monday morning meetings, working out what they need to do that week.*

That way we get everyone involved in delivering the projects, operational improvements and maintenance changes. They'll get started on the stuff the audit picked up but it won't be long before we have a new heap of projects and ideas.

Manager: *But my folks are really busy already.*

Niall: *That's right! That's why we're going to involve a lot of people for a little time each and we're going to help them with a bunch of neat tools to make better-informed decisions around energy and resources. It's not a question of working harder but working smarter. The key is to get better-informed decision-making and let folks make the better decisions. And remember, there's US$1 million a year available for us!*

It's not a question of working harder, but of working ***smarter.***

overstated claims for magical pieces of technology such as *"magnetic fuel efficiency enhancers"*, which is utter claptrap, or *"voltage optimization"* which works, but rarely delivers the claimed benefits, in my experience. The problem with these products is that they undermine the confidence of buyers, and so contribute to the adverse responses towards energy and resource efficiency opportunities.

The final availability constraint relates to timing barriers, where the timing is not right to implement a resource efficiency opportunity. In some cases this might be because of the physical operation of a process. For example, many large continuous process chemicals plants operate 24 hours a day, 7 days a week and only ever shut down on scheduled turnarounds, which might be years apart. If an efficiency opportunity requires a change to a process, it might be that this has to wait for the next turnaround. Carrying out a major refurbishment of an office often has to wait until the end of a tenancy so that the landlord can gain access to the space. Maybe a retailer needs to wait until after the busy Christmas shopping period before they launch a staff awareness and motivation campaign around waste and recycling. The timing barrier is closely linked with the human capital barrier, in that an individual may be required to install an item of equipment, but they are occupied with other tasks.

We have seen just how many different barriers there are to resource efficiency programmes and projects. Many of these barriers can be overcome with some thought. Some obstacles require patience and a revisiting of the opportunity when circumstances change. Other barriers cannot be overcome easily, and so we need to guard against setting simplistic and unrealistic expectations for resource efficiency based purely on technical feasibility.

In the following chapter, we will explore a framework that anticipates many of these barriers to energy and resource efficiency. However, first, we need to consider an unintended consequence of success in efficiency programmes - the rebound effect or Jevons effect.

4.16 **What my PV installer told me.** When I had solar photovoltaic panels installed on my roof (see page 580), the contractor advised me to use as much electricity as I wanted during the day, because it was *"free"*. This thinking is an example of how resource efficiency can encourage greater use of a resource leading to a *"rebound effect"*, discussed next. The correct behaviour, from a resource efficiency perspective, is to just consume the electricity that I need and export the rest to the grid for others to use. The reason the electricity was considered *"free"* is that I do not have an electricity meter capable of recording the electricity exported, so it is *"deemed"* that 50% of the generation is exported regardless, and so the amount used does not affect the payment received. *Source: Niall Enright, drawn using Pixton. Image is included in the companion file pack.*

4.9 The rebound or Jevons effect

A desirable consequence of greater efficiency in a resource-consuming process is lower cost. This lower cost can, however, lead to greater utilization of the process - after all, price drives demand. Here, we will explore the significance of this effect.

4.17 **William Stanley Jevons**
Jevons was a Victorian economist who in 1865 published *The Coal Question*, in which he argued that greater efficiency in the use of coal, at a national level, would result in greater use of coal, not less.
Source: Portrait of W. Stanley Jevons at 42, by G. F. Stodart, Public domain via Wikipedia.

"Whatever, therefore, conduces to increase the efficiency of coal, and to diminish the cost of its use, directly tends to augment the value of the steam-engine, and to enlarge the field of its operations."

W Stanley Jevons, The Coal Question, 1865

In the previous chapter on Value, we saw that one of the great benefits of energy and resource efficiency is that it saves money. The rebound effect reflects the fact that this money may be used in some other ways which may increase resource use.

For example, a householder may use the energy cost savings they make on a city break, and so take a resource-consuming flight that they otherwise wouldn't have. These are the income effects of resource efficiency.

By becoming more efficient, a particular process may be favoured or gain a competitive advantage over its alternatives. For example, if my factory becomes more efficient, my organization may choose to divert production to my site. This is called an output or substitution effect.

Although not strictly a barrier to investment, the rebound effect can lead to lower than expected results from resource efficiency programmes, so we need to take this into account. A critical question, when we think about the rebound effect, is just how big an effect this is. If the net result is that we go on to consume more resources as a result, then this is called a backfire, where the ratio of energy added to energy saved is greater than one. This notion of a backfire is referred to as the Khazzoom-Brookes postulate, after the two researchers who proposed this.

However, if the rebound effect is less than unity, then efficiency measures do still lead to lower resource use, although we do need to take into account the size of the effect if we are to predict the outcome of our programme or set an achievable goal.

Measuring the rebound effect is, as one would expect, quite difficult. Just consider the consequences of saving money. Some organizations pass savings on to shareholders in the form of increased dividends, which they may or may not spend on increased consumption. Other organizations may spend the savings internally on increased production. Yet other organizations may reinvest the saving on further efficiency investments and so decrease their consumption of resources even further.

Furthermore, the debate has become obscured by politics. There are some economists, generally of the *neoliberal school*, who go so far as to say that the rebound effect is a reason why we should *not* support energy or resource

Study	Year	Effect %
Schipper/Grubb	2000	5-15
Hass/Biermyer	2000	20-30
Berkhout et al	2000	0-15
Greening et al	2000	v. small
Allan et al	2007	30-50

4.18 **The size of the rebound effect from selected studies**
Source: Collated from The Myth of Resource Efficiency - the Jevons Paradox, Polimeni, Mayumi, Giampetro and Alcott, Earthscan 2008.[599]

efficiency measures, through policies or subsidies and suchlike, as this will *inevitably* lead to a rise in use. This rationale fits in with a narrative that is against state interventions in people's lives, however well-meaning.

Because of the complexities involved, the evidence on the size of the rebound effect is quite mixed, as shown in the table on the left. There are, however, a few general observations that we can consider from the various studies.

First of all, the rebound effect in developed economies seems to be much smaller than in developing economies. We can understand why this might be by looking back to our Ehrlich Equation (page 21):

Impact = Population x Affluence x Technology

In developing countries, Affluence is comparatively low, so there is demand for savings to be directed to increasing the *quantity* of things: car ownership or car journeys made, or hospitals, or education, or meat consumption. In developed countries, where there is some degree of saturation of energy services (i.e. the basic needs of people are generally met) a greater proportion of the savings will go into increasing the *quality* of things. For example, getting a new television (which will be more efficient than the one replaced), or improving the quality of wine one consumes, or spending more on drugs to treat illnesses, all of which have a lesser rebound effect on resource use, as they are substitutional effects.

All things being equal, it is not unreasonable to assume, as Mark Diesendorf does,[212] that income effects (i.e. spare cash generated by savings) will be spent in the same proportion to energy expenditure in the economy, i.e. around 8-10%, in a developed economy, which is quite a small effect.

So in our planning for our resource efficiency programmes *within* our own organizations, we should always take a cautious view on the absolute savings that we can achieve and allow for a modest rebound effect in the order of 10%. External rebound effects are much harder to predict, but largely outside our control or of direct concern to us.

There is certainly no compelling argument for curtailing efficiency investments because of the rebound effect. I have stated right from the outset that resources are needed to improve the sum of human happiness, and even if there is a *"backfire"* because we are working to reduce the intensity of resources used, the availability of more resources to meet these needs is clearly desirable. To argue the opposite is like stating that the best way to deliver energy savings is to shut our factories down - logical in the sense we will use less resources, but missing the point of the resources altogether.

Understanding that rebound effects do exist means that policymakers can design support for energy efficiency that doesn't lead to income effects, e.g. through revenue-neutral forms of taxation. Organizations could also ring-fence savings into revolving funds so that savings are reinvested into achieving further savings to avoid them being spent on increased activity elsewhere.

Summary:

1. There are many barriers to energy and resource efficiency.

2. In setting expectations for a programme, a distinction needs to be made between what is technically feasible and what is practical given financial, operational and market realities.

3. We can categorize barriers into the following types:

   - General barriers are aspects of our efficiency programme, such as complexity, that make it less attractive than other activities.

   - Structural barriers are biases in the system that impede efficiency, such as subsidies for fossil fuels, or split incentives.

   - Behavioural barriers are psychological biases which work against change, such as loss aversion or status-quo bias.

   - Availability barriers describe features such as the lack of money or people to take a programme forward.

4. There are solutions for most of the barriers, particularly if the problems are anticipated in advance.

5. Certainty is a recurring theme for energy and resource efficiency projects. Great care needs to be taken in articulating the costs, effort and outcomes in every proposal.

6. Understanding whether the decision-maker sees the proposal as a search, experience or credence good can transform the type of proposal and likelihood of success.

7. While there is often a focus on the money aspects of a proposal, we must not overlook the fact that the effort required also plays a huge role in any decision. It is important to credibly quantify this effort and show how it can be delivered, whether through in-house or external resources.

8. It is usually better to ask many people to contribute a small amount of time to the programme, than to ask a few people to make a large time contribution.

9. Aligning the proposal with the core objectives of the organization is always a good thing. It can avoid a "so what" response and helps to position the opportunity against other initiatives which are competing with your programme.

10. Be aware that the Rebound Effect can reduce programme outcomes, but this is either a small effect, or largely external to our own organizations

11. Despite all these barriers there are two key drivers in favour of energy and resource efficiency: organizations are under huge pressure to respond to resource efficiency and the benefits that it can bring beats many other investments hands down! Time and opinion are firmly on the side of advocates for change.

Further Reading:

Sorrell, S. et al. 2004. *The economics of energy efficiency: barriers to cost-effective investment*. Edward Elgar Pub. ISBN-13: 978-1840648898.

This highly recommended book is rather expensive and most of the material is to be found online in an earlier piece of work for the EU: Sorrell, S. et al. 2000. *Reducing barriers to energy efficiency in public and private organizations*. Joule III Report 8: 2007. http://www.sussex.ac.uk/Units/spru/publications/reports/barriers/final.html

Questions:

1. Consider two recent proposals which you have put forward, one successful and one not [it does not have to be a resource efficiency proposal but could be any situation where you needed to persuade someone of the merits of an idea, e.g. getting a group of people to share a flat, deciding to go to a film]:
    - Set out how the proposal was described and who were the decision-makers.
    - What was their view of the proposal (was it a search, experience or credence good)? Discuss if the approach taken suits the decision-maker's perception.
    - Complete a pairwise comparison on the proposal and the decision-makers. Does this change your approach in any way?
2. Describe three psychological barriers to resource-efficiency and how these might be overcome.
3. List the most common structural barrier to resource efficiency and give some examples and how they may be overcome.
4. We have asserted that barriers exist because organizations are not achieving their full potential. Consider three studies (either those cited here or new studies) and comment on the reliability of the calculated potential. How may this be over or understated?
5. How does *anchoring* influence the presentation of a resource efficiency proposal?
6. What evidence is there that certainty is a key barrier to resource efficiency? How do issues of certainty influence the approach taken when seeking commitment for a resource efficiency programme?
7. Is the Jevons Effect significant? To what extent can it affect the goals or outcomes of our programme?

5 A Framework

Over two decades and hundreds of projects I have been fortunate to encounter some of the best, and worst, examples of resource efficiency programmes. My own observations are that around a third of projects have done very well, another third delivered some value but have not been sustained in the long-run and the final third fell short of expectations from early on. This success rate is typical of most change management or business process re-engineering projects,[344] so we should not single out resource efficiency programmes for particular criticism.

The emphasis on difficulties is not intended to dishearten organizations or individuals who are about to set out on the path of greater resource efficiency – for the journey can be stimulating and rewarding. I raise these difficulties to emphasize the need for a method or process that can increase the likelihood of success, which can help the reader avoid obstacles on their journey to greater efficiency, and provide a process that is more enjoyable and value-generating.

> *One-third of energy and resource efficiency programmes do well, one-third deliver some value but are not sustained in the long run and one- third get into trouble from early on. This is* ***typical*** *of most change programmes.*

In the previous chapter, we saw that there are many barriers which act against resource efficiency. General barriers such as the relative scale or fragmented nature of opportunities make resource efficiency appear of little benefit or difficult. Structural barriers such as split incentives stand in the way of an agreement to change. Psychological and behavioural barriers create resistance and tie us into inefficient standards and routines. Even if we overcome all these challenges, the lack of availability of technologies, people or money may block progress.

The good news is that there are solutions to all these challenges. The preceding chapters have given us many suggestions for the approach that we could adopt to increase the probability of success.

We know that energy and resource efficiency programmes have continual improvement at their heart. We are aware that management focus is needed and that a wide range of activities within organization influence resource use. We understand that there are many sources of value, so we need to engage many teams in realizing the full potential for our organization. We know that informed decision-making at all levels is necessary and that this goes beyond mere technical fixes. The following pages bring together all these strands in a structured way so that the reader can anticipate challenges and plot the best route forward for their organization.

5.1 The Framework

This Framework sets out energy and resource efficiency as a structured process with many discrete activities. An organization may choose to implement all these elements or they may focus on specific elements for guidance on how to overcome specific barriers and achieve success.

The Framework specifically addresses the common causes of failure of many energy and resource efficiency programmes, as well as the barriers to their adoption in the first place.

To address the challenges and complexities of resource efficiency, we need a Framework to guide our programme design. This Framework, illustrated opposite, breaks down the resource efficiency process into a sequence of activities. It specifically addresses the common causes of failure of many energy and resource efficiency programmes, as well as the barriers to their adoption in the first place. Using the Framework does not guarantee success, but it can increase the probability of a successful outcome, in other words, it will help in selling the notion of a programme in the first place, achieving the programme goals and sustaining the programme in the long-term.

At the centre of the Framework is Value. Value describes the monetary worth of private companies, reflected in the interlinked elements of profit, equity price or asset valuation. Value also includes the notion of service effectiveness for public sector or not-for-profit institutions, which may mean simply increasing the capacity for service provision by reducing costs, or may mean delivering a better quality of service or meeting the wider needs of users. In essence, Value is shorthand for the principal measure of attainment of the organization's core purpose.

The Framework set out here is based on experience gained in the implementation of many hundreds of programmes in industrial, commercial and public sector organizations over a period of 25 years. It is intended as a *profoundly practical* guide to help design and implement a successful resource efficiency programme. In keeping with the notion of resource efficiency as a process, the Framework is task-oriented, outlining a natural sequence of activities which will help ensure success. The Framework can be applied at any scale, from a single facility through to a global operation and is suitable for all types of organizations, public or private, profit or socially driven, service or manufacturing.

The Framework is conceptual rather than prescriptive: that is to say organizations are free – in fact encouraged - to adopt different terminology or use different methods for individual elements of the Framework. The Framework describes many interlocking components of a successful programme - such as management commitment - which should ideally be present in any approach to energy or resource efficiency. If these elements are already in place or can be achieved using different methods, then that is fine. At its simplest, the Framework is a checklist of best practice as well as a tool to diagnose – and hopefully remedy – factors that are impeding success.

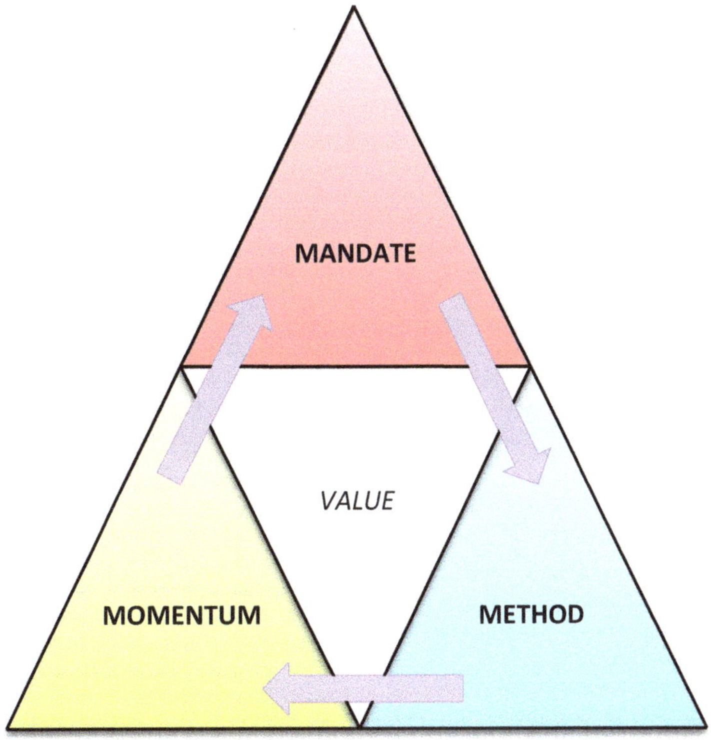

5.1 **The energy and resource efficiency Framework has three components, centred around the idea of increasing Value**
Source: Niall Enright

The techniques and processes that form the Framework have been honed and refined in the field, in hundreds of sites worldwide; this is not a theoretical toolset, but rather a distillation of the most successful techniques in use today. For each element of the Framework, I will explicitly examine why a particular approach increases Value or overcomes obstacles, so that the critical thinking behind the design is understood, and can be applied to shape your organization's programme.

This is not a theoretical toolset, but rather a distillation of the most successful approaches in use today, observed in hundreds of organizations worldwide.

Above all, the Framework is intended to offer a common sense approach to resource efficiency. I start off with the proposition that we need three fundamental things to make our programme a success. We need a Mandate for action, we need a Method or process for engaging our organization in the necessary changes and we need sufficient Momentum so that the programme persists and has the potential to achieve fundamental change over the long term.

Although each of these may require markedly different levels of effort, no programme will prosper unless it has all three. The absence of a mandate means that the programme is unlikely to receive the necessary resources or attention. The lack of a method means that an outcome is likely to be more expensive and less effective than expected, leading to disappointing results. Poor momentum means that the programme is likely to decline when the next initiative comes along or the impetus for more substantial change wanes following initial success.

Real World: ***Too rosy a picture***

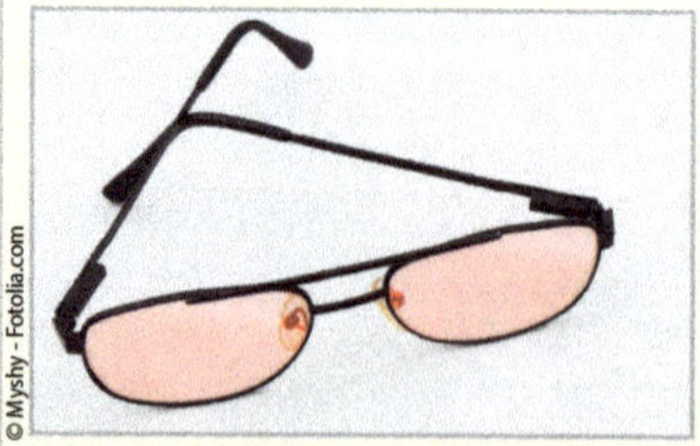

© Myshy - Fotolia.com

If we consider what many organizations say publicly about their resource efficiency programmes, we can only find stories of success. Few or no mention anywhere of challenges, disappointments, steps backwards.

Consultants, like me, are also silent on the true level of success of resource efficiency programmes, because we are either bound by client confidentiality or we quite simply don't want to advertise our involvement in failed projects – it is not good for business! There is a similar silence from agencies promoting efficiency that are anxious not to make the process seem difficult.

We have seen in Chapter 3 the considerable savings that many large organizations *have* achieved as a result of closer attention to energy and other resources. However, those of us on the inside track of these programmes know that they seldom go smoothly and often fall short of their real potential. Indeed, the quoted results are often exaggerated - after all, reputations are at stake.

This is borne out by research that suggests that there remains a huge amount of further improvement possible (see Figure 4.1). If energy and resource efficiency were easy, then this value would already have been captured. So, for those embarking on resource efficiency, my advice is: do not always believe the rosy picture you are given.

The three parts of the Framework are largely implemented in sequence - Mandate leading to Method leading to Momentum, as shown in Figure 5.1. However, while each step needs to anticipate the next, one can step backwards in sequence to change or modify a process based on learning. For example, the Mandate may need to be renewed on a regular basis, to address the next series of opportunities; hence we return to the beginning of the cycle.

Placing Value at the heart of the process reminds us that, at every stage of our programme, we need to demonstrate the benefits that the resource efficiency efforts are producing. Later, we shall consider the importance of aligning the resource efficiency programme with the primary objectives of the business, but it is worth noting now that the word Value is flexible. In not-for-profit institutions, we could replace *Value* with a term that reflects the mission of the organization, such as *Learning Outcomes* or *Service* or *Patient Care*. For-profit businesses may also want to apply terminology that matches the corporate imperative: *Earnings per Share*, *Profit* or *Brand*.

Each of the three high-level components of our resource efficiency programme – the *"three Ms"* – has a range of additional activities within it. These, coincidentally, each break down into three elements, illustrated opposite in Figure 5.2. Within each of the three high-level components of our Framework - Mandate, Method and Momentum - there is an element indicated in underlined bold text – Leaders, People and Stakeholders. These are the usual initiators of our resource efficiency programme – for example, the impetus could arise because of the needs of a particularly influential stakeholder, e.g. Walmart, requiring change from its suppliers. Or it could come about because a particular group of employees have identified a value-creating opportunity, such as the marketing department in L'Oreal. Or because a member of the leadership team, such as the chairman, CFO or CEO, believe that resource efficiency is a valid focus for the organization. Regardless of whether the case for resource efficiency originates inside or outside the leadership team, the sequence of Mandate, followed by Method and Momentum is recommended; thus, early engagement of Leadership to craft a Mandate is always desirable.

Of course, organizations are complicated things, and it is not expected that every resource efficiency programme will start with an all-encompassing mandate. In many cases, organizations are simply not set up for that type of approach – examples would include conglomerates where businesses are run as separate entities or organizations where the facilities are effectively in competition with each other (as is often the case in automotive plants). It is quite common that a resource efficiency programme is limited in scope to a business unit, division, branch or even a single site. In this abbreviated scope, the same cycle of Mandate, Method and Momentum will apply, but oriented to the Leadership, Staff and Stakeholders at that operating level. Thus, we should think of the Framework as being capable of being replicated at multiple levels of an organization: a corporate programme could, in turn, have a number of a related site programmes with their own Frameworks, in a sort of nested, or *Russian Doll*, structure.

5.2 **The Complete Energy and Resource Efficiency Framework**
The three high-level components, Mandate, Method and Momentum, each in turn have three further interlinked elements. A successful programme will address all nine elements during its evolution. Each element is described in further detail in the following chapters.
Source: Niall Enright. Available in the companion file pack, along with a poster version incorporating the key tasks at each stage.

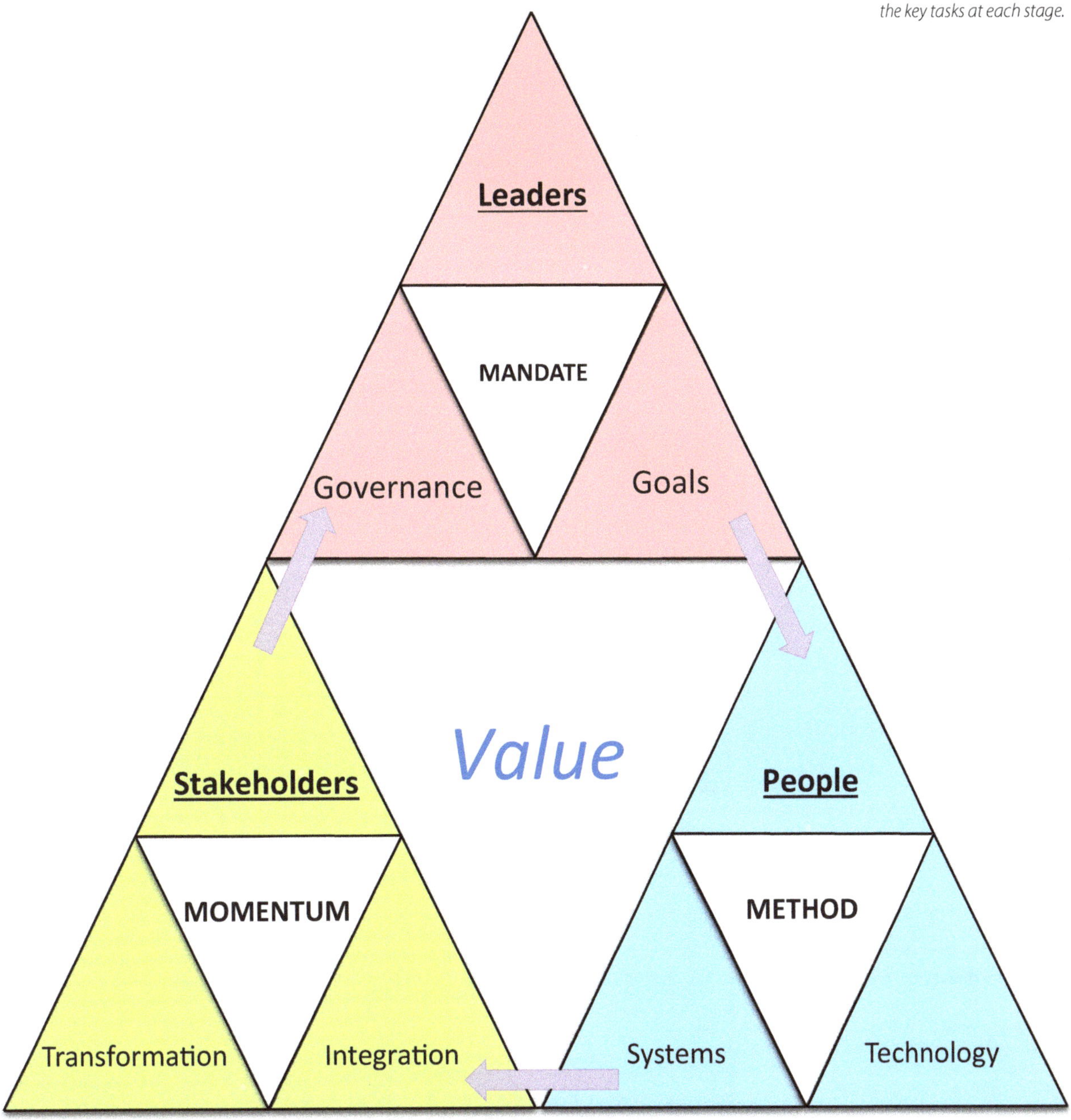

Real World: *Perfect is the enemy of good*

This Framework unapologetically sets out an ideal. However, the real world is imperfect. As I write this piece, I am working with a global organization for whom resource efficiency is simply not a big enough priority for them to implement a programme of the scale and ambition to have a material impact, even though they estimate savings from such a programme would be worth US$0.5 billion (yes that is half a billion dollars!). With one fatality in the week I write this and a perception that stakeholder pressure is declining, this particular client is choosing instead to focus heath, safety and environment efforts on developing a stronger safety culture, which is quite clearly needed given the tragedy involved in a workplace death.

However, the section *Why now?* (page 209) argues that we need to stamp out the notion that resource efficiency is an either/or proposition. There is absolutely no reason why an energy and resource efficiency programme cannot be run at the same time as a safety culture programme – these are not *mutually exclusive* activities. However, perceptions are what they are, and it is perceptions that matter.

So in the real word, few organizations wholeheartedly embrace the notion of optimum resource efficiency. The truth is that the vast majority advance in much smaller baby-steps. Sometimes they may even retreat.

Thus, we often have smaller programmes focusing on parts of the organization or perhaps on specific resources such as energy. Incremental improvement may come about simply through plant replacement programmes accessing more efficient technologies. Legislation may raise the *"floor"* on our performance, with little management involvement. There may be episodic drives on efficiency when costs rise, but soon the organization's efforts will turn to other priorities and other initiatives may displace efforts on resource efficiency. Many programmes will treat superficial aspects of resource use, rather than consider the fundamental design of products and processes.

The techniques and approaches set out in this Framework are equally applicable to these piecemeal programmes. The Framework is a problem-solving resource, a set of ingredients to enable practitioners to successfully deliver whatever resource efficiency dish they are preparing now, rather than as a one-size-fits-all recipe for that perfect banquet which the organization simply has not ordered! While we may wish for an all-encompassing programme, the timing may not be right for many reasons, and I would urge all advocates of resource efficiency not to hold out for perfection if doing so sacrifices or delays progress towards a good programme. Perfection is great but rare and, in my experience, it is better to move forward slowly than not to make progress at all.

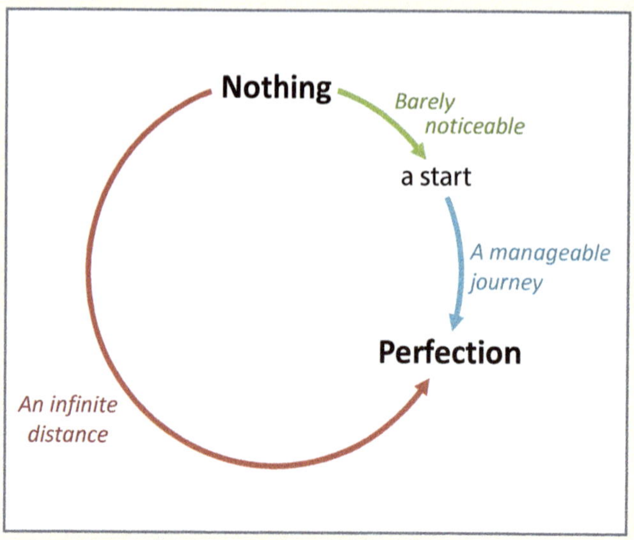

5.3 **The direct path from Nothing to Perfection is infinite**
It is often a better strategy to simply get started with a process of improvement than to hold out for the perfect approach.
Source Niall Enright, available in the companion file pack

If I had £1 for every time that an organization has said that it needs to fully install its sub-metering before it can initiate an improvement programme, I would be a rich man (I am not)! Each day of procrastination means that potential savings are lost (see *Why resource efficiency is like the hotel business* on page 333).

Sometimes the well-meaning ambitions of those who are championing energy and resource efficiency impede progress; by demanding that a perfect approach is adopted, they ask more than the decision-makers are willing to give.

As the illustration to the left shows it is often faster and more effective to start small and build from there. Once the value of the process has been demonstrated then the potential to scale up is greater. Initial learning can also feed into the design of subsequent phases. As long as there is appropriate preparation, a low-key start to a programme has many merits.

5.2 Quality systems

There are a number of mature business improvement processes, such as Lean, Six Sigma, Total Productive Maintenance etc., which are concerned with efficiency. Standards such as ISO 14001 and ISO 50001 set out how elements of a resource efficiency programme should be designed. Here we demonstrate that the Framework is entirely compatible with these.

"Continuous improvement is better than delayed perfection."

- Attributed to Mark Twain

Your organization may have already have implemented business improvement processes such as Six Sigma, TPM or Lean Manufacturing, as well as environmental standards such as ISO 14001. You could already have capital allocation processes which consider some non-financial criteria. Management and employee incentive schemes may already reward improved resource use. You will almost certainly have formal procedures in place to ensure environmental compliance.

All these system are highly relevant to an energy and resource efficiency programme. This Framework is not conceived as an *alternative* or *replacement* for the existing processes used to run an organization. Rather, the Framework provides a complementary set of tools and techniques which can usually be introduced *within* the current systems to drive improvement.

The Mandate elements of the Framework will help us develop the case for improvement, to put in place an effective organization and set appropriate goals. These probably exist already in your organization. What the Framework offers is inspiration, trouble-shooting, best practice and techniques which may bolster your existing processes' capability to address energy and resources.

Within the Method, there are two tools, an Opportunities Database and Monitoring and Targeting (M&T), which are continual improvement tools specific to energy and resource efficiency. Ideally, if they are not already in place, these tools can be used as extensions of the existing business processes.

The activities under the heading Momentum are about ensuring that our programme is fully integrated within our organization. We want to avoid a separate *cottage industry* of effort, remote from other business activities, with the risk that it will be discarded at some point in the future because of its lack of connection with what is perceived to the organization's core purpose.

The intention is that you, the reader, selectively incorporate the processes, behaviours and tools described in this Framework into your organization's "*everyday way we do businesses*". If there are no equivalent systems in place, or they need a radical overhaul, then you can implement this Framework as a *complete* improvement process. However, most organizations will cherry-pick those aspects that are most useful. This book has been written with this approach very much in mind. Certainly, I make no claim to superiority over other systems, which have been designed for other purposes.

Standards: ***ISO 50001:2015***

There is a chapter on the ISO 50001 Energy Management Systems standard later (page 717). This international standard, and the Framework set out here, are entirely complementary.

5.3 Change Management

The Framework presented in this book is a change management process designed to change organizational systems and culture. It has many of the components of long-established change management methodologies, as well as features specific to resource efficiency.

Organizations are complex systems. They consist of many component parts (human resources, sales, design, operations, finance, etc.), all of which contribute to the achievement of the core goals. These systems are maintained through the underlying aspirations, thought patterns, recognition and reward systems, training and skills, measurement and communications processes which determine the decision-making and behaviours of individuals. Thus, to achieve lasting change in resource use, it is not sufficient to only implement technical fixes; one needs to make changes to these *systems*.

The term change management is used to describe processes that aim to change organizations' cultures and systems.

<table>
<tr><th colspan="2" rowspan="2">This Framework</th><th colspan="2">Leading Change Management Processes</th></tr>
<tr><th>J. Kotter's 8 Steps[448]</th><th>GE CAP process[523]</th></tr>
<tr><td rowspan="3">Mandate</td><td>Leaders</td><td>1. Create urgency</td><td>1. Leading change</td></tr>
<tr><td>Governance</td><td>2. Form a powerful coalition</td><td>2. Creating a shared need</td></tr>
<tr><td>Goals</td><td>3. Create a vision</td><td>3. Shaping a vision</td></tr>
<tr><td rowspan="3">Method</td><td>People</td><td rowspan="3">4. Communicate the vision
5. Empower others to act
6. Create short-term wins</td><td>4. Mobilize commitment
5. Making change last</td></tr>
<tr><td>Systems</td><td>6. Monitoring progress</td></tr>
<tr><td>Technology</td><td></td></tr>
<tr><td rowspan="3">Momentum</td><td>Stakeholders</td><td></td><td></td></tr>
<tr><td>Integration</td><td>7. Consolidate Improvements and produce more change</td><td></td></tr>
<tr><td>Transformation</td><td>8. Institutionalize new processes</td><td>7. Changing systems and structures</td></tr>
</table>

5.4 **Comparing the Framework with two leading change management methods**
Source: Niall Enright

Those familiar with change management methodologies will probably spot the similarities between some of the leading models of change and the elements in the Framework. John Kotter was one of the earliest people to systematically study why change management succeeded for some organizations and why it failed for others. Kotter's *8-Steps to Transforming Your Organization*[448] was a milestone paper, which summarizes the necessary ingredients for success. Companies like GE's *Change Acceleration Process*[523] have similar clear steps to success.

Real World: *Why now?*

If one were to ask 100 senior managers what they thought of resource efficiency they would probably all agree with the proposition it is a *"Good Thing"*, like *"apple pie and motherhood"*.

But the data shows us that, in the majority of organizations, there remains a significant untapped potential to improve efficiency. It seems that resource efficiency gets stuck on the organizational "to do" list, in limbo, its value unrealized or only part-realized.

In many cases, the challenge is not only to convince management about the benefits of resource efficiency but why implementing a programme now is desirable. We must accept that there is always some other corporate imperative that can claim greater priority: the big acquisition; the reorganization; addressing a downturn; the new management system; the expansion; or the new service development.

But how on earth did we get to the assumption that resource efficiency is an either/or proposition? *"Either I achieve corporate objective A, or I implement a resource efficiency programme"*. We need to mercilessly stamp this notion out wherever it appears because it is plain *wrong*. In fact, if we listed corporate objectives A, B, C and so forth, resource efficiency will most likely enhance all of these core objectives directly or indirectly.

So the key is not to delay. Energy and resource efficiency are not an either-or proposition, and other objectives do not need to be sacrificed to progress in this area. Every day that passes is a day where savings are lost and can never be recovered.

Exploration: *Rolling the DICE to predict the outcome*

© krasyuk - Fotolia.com

Back in 1992, a team from Boston Consulting Group set out to establish if they could determine what factors could be used to indicate the success or failure of change management programmes in general.[665] They concluded that four "hard factors" were highly effective at predicting the outcome of the programme. These factors and the questions they asked were as follows:

- The Duration of the programme, and specifically the amount of time between reviews of progress. *Question: How often do formal project reviews take place?* Score: 1 for less than two months, 2 for 2-4 months, 3 for 4-8 months and 4 if more than 8 months apart.
- The Integrity or effectiveness of the individual or team leading the change, which includes aspects of skills, motivation and time dedicated to the task. *Question: How capable is the leader and how strong are the team's skills and motivation? Do they all have enough time to dedicate to the change initiative?* Score: 1 if all these criteria are met and 4 if the team are lacking in all areas. Score 2 or 3 for capabilities in between.
- The Commitment of two groups to the project: the senior management, C1, and the local employees, C2, affected by the change. *Question 1: How often, consistently and urgently do senior managers communicate the need for change and have they allocated enough resources to the programme?* Score 1 for clear commitment, 2 or 3 for neutral approach and 4 for resistance or reluctance. *Question 2: Do the employees involved in the change understand why it is happening and are they enthusiastic and supportive or anxious and resisting?* Score 1 for an eager adoption of the change, 2 for just willing or neutral, and 3 or 4 for resisting change, depending on the level of reluctance.
- The Effort required from employees over and above their existing workload. *Question: How much extra effort must employees make to deliver the change and does this come on top of a heavy workload?* Score: 1 if less than 10% incremental effort is required, 2 for 10-20%, 3 for 20-40% and 4 if over 40%.

These four hard measures can be easily recalled as they form the word DICE.

The Boston Consulting team then scored 225 change management initiatives in a range of organizations using the questions set out above. By carrying out regression analysis on this data they were able to determine the relative importance of each factor in terms of the impact on the success of the programme. They concluded that Integrity, I, and management commitment C1 were particularly strong and so their scores should be doubled when calculating a total DICE score, which is expressed as:

$$DICE = D + 2(I) + 2(C1) + C2 + E$$

Since each value can be between 1 and 4, the best (lowest) possible DICE score is 7 and the worst (highest) 28. The chart on the next page plots the distribution of these 225 scores against the degree of success of their change initiative. As can be seen, low scores correlate with successful outcomes, while high scores are associated with poor success.

DICE Score

Increasing Degree of Success	7	8	9	10	11	12	13	14	15	16	17	18	19	20	21	22	23	24	25	26	27	28
9	1		3	5	10	5	7	6	6	3	1	1										
8					1		1		1		1											
7					1	2	5	9	14	9	11	5	2									
6									1		3	1		1								
5									4	1	9	3	7	2	2	2	1					
4														3	4	1	2					
3										1		1	6	6	6	4	5			1		
2																1	1	1	1	1		
1												1	2	3	5	4	1		2			1
	Win Zone								Worry Zone						Woe Zone							

5.5 **The Boston Consulting analysis of 225 change management initiatives plotting the DICE scores in relation to the degree of success of the project**
Source: Adapted by the author from The Hard Side of Change Management, Sirkin et al, Harvard Business Review 2005.[665]

The numbers in the table above show the number of projects with a particular success score (vertical axis), and the DICE score (scores rounded up to the nearest integer).

The Boston Consulting team observed that projects with a score of 14 or less are likely to achieve the majority of their intended outcomes (the *Win Zone*), while those with a score of more than 20 were very unlikely to make the desired change (The *Woe Zone*). Scores between 15 and 20 show a broad range of outcomes, and so executives need to assess these projects carefully (the *Worry Zone*). Over time, the Boston Group have decreased the lower boundary of the Woe Zone to 18 points, providing an earlier indication that the project outcome is no longer predictable.

In their paper, the authors of this scoring system provide some practical examples of its application, and they make the observation that DICE scores can be applied to individual projects or initiatives within a larger programme. Thus site participation in a resource efficiency programme can be evaluated using this method as well as individual projects (involving technology as well as behaviour).

The beauty of this approach is that it is simple and focuses attention on many of the practical aspects of change that will influence the outcome of a resource efficiency programme. The authors stress, however, that there are many "soft" aspects of the programme, such as culture, communication and attitudes, that are also important – it is just that these are not so consistently relevant or easy to measure. Another key benefit of this performance metric is that it is a leading indicator or a predictor of the programme outcome, so can be applied early in a project to determine if corrective actions may be needed.

The resource efficiency Framework set out on these pages will lead to a low DICE score if properly implemented, as it:

- Encourages regular measurement and reporting (D)
- Ensures that the skills of the team are adequate (I)
- Emphasizes the commitment of leaders and participants (C)
- Does not demand unrealistic effort from all involved (E)

Summary:

1. As few as one-third of energy and resource efficiency programmes achieve their objectives, but this is rarely acknowledged in literature on the subject.

2. A structured methodology or Framework can help avoid the most common causes of failure. The Framework presented in this book is not theoretical - it draws on the lessons taken from hundreds of energy and resource efficiency programmes.

3. Techniques from change management are useful in resource efficiency programmes since we often need to make changes to our organization's underlying culture and systems.

4. Another source of inspiration are quality systems and standards, such as Lean, Six Sigma, TPM, ISO 14001 and ISO 50001. The Framework complements these methods, although we should note that they lack some specialist tools usually needed for energy and resource efficiency since they are designed for different purposes.

5. The Framework is not prescriptive. You are encouraged to change the terminology and adapt elements of the Framework to meet your organization's needs.

6. We should express Value in terms of our organization's core mission.

7. It is better to make a start than to hold out for the perfect programme. As long as the foundations that are laid support continual improvement, the effort will not be wasted.

8. Energy and resource efficiency should be considered a core requirement, not an optional extra. It is perfectly possible to initiate an improvement process that does not impede other organizational priorities.

9. Ideally, we will develop a resource efficiency process which is embedded in the "ordinary way we do business". Thus, it is usually much better to adapt existing systems and processes than to replace them. Integration is the key idea here.

10. The DICE methodology from Boston Consulting Group can help assess whether a programme is likely to succeed.

Further Reading:

1. John Kotter, *Leading change - Why transformation efforts fail*, Harvard business review 73 (1995).[448] This paper was ground-breaking because it was based on the analysis of actual change programmes. From this analysis some very useful insights into the role of leaders and processes to drive change emerged.

2. *The Effective Change Managers Handbook* ed. Smith, King, Sidhu and Skelsey (2015).[668] This title provides a great range of ideas, tools and techniques for change management. Chapters are authored by experts in each field, so the content feels authoritative. There are good references and didactic materials.

3. Paul Gibbon's *The Science of Successful Organization Change* is a thought-provoking title, which challenges many of the assumptions about how change in organizations takes place. I very much like the emphasis on the need for scientific and academic evidence for specific techniques, as well as the acknowledgement that change programmes often fail. Excellent!

4. *Managing Change*[495] is a great resource focusing on change management to deliver energy and resource efficiency in organizations. Produced by The Government Office for the Southwest & Envirowise back in 2007, its content is still highly relevant. And it is free!

Questions:

1. According to the DICE methodology, what factors influence the outcomes of change management processes? Do you agree that some factors are more important than others? Why is this?

2. Why do we believe that energy and resource efficiency requires change management?

3. Describe the nine elements of the Framework and how they might apply in your organization.

4. Why is Value placed at the heart of the Framework?

5. Consider the names given to different part of the Framework (Value, Method, Mandate, Momentum, and the sub-elements Leaders, Governance, Goals, People, Systems, Technology, Stakeholders, Integration and Transformation). Determine if alternative words would better suit your organization and explain why this is.

6 Mandate

The first part of the resource efficiency Framework is the Mandate. This is shown at the top of the Framework, as this element is the one that drives and directs the necessary change. In my experience, the single most important determinant of the success of an energy and resource efficiency programme is the strength of the Mandate.

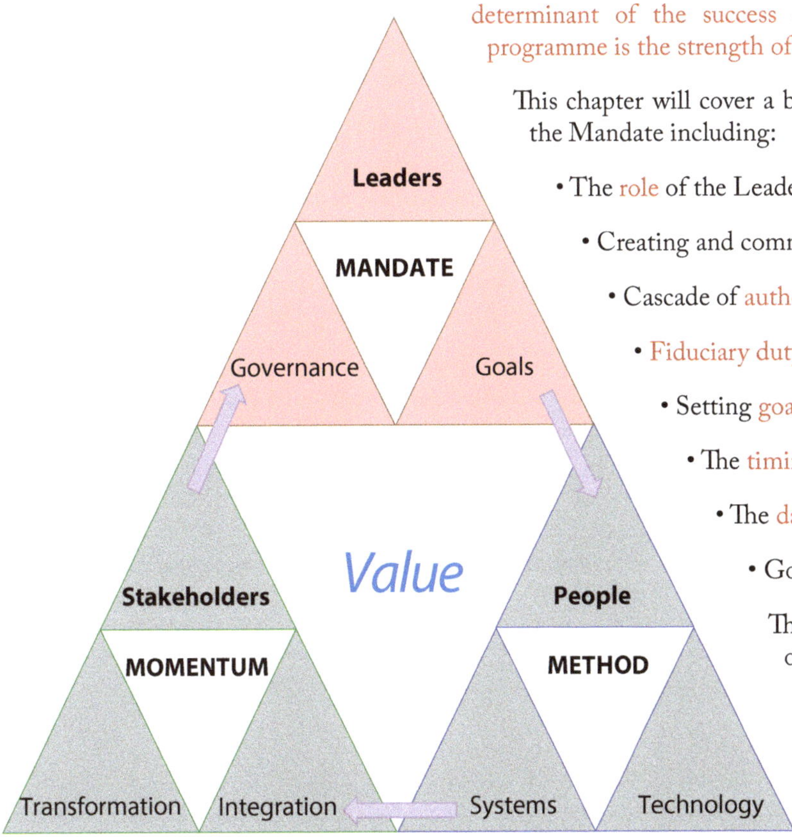

This chapter will cover a broad range of different issues related to the Mandate including:

- The role of the Leader
- Creating and communicating a vision
- Cascade of authority and feedback
- Fiduciary duty
- Setting goals
- The timing of change
- The danger of a focus on immediate savings
- Governance models.

These Framework chapters provide an overview of the key elements in a resource efficiency programme. For each of these elements there are accompanying sections in the techniques chapters which provide much more hands-on advice about each topic.

6.1 The importance of leaders

It goes without saying that Leaders are important. However, for them to make the best contribution to the programme we need to be clear about their role and the tasks we want them to complete.

The importance of Leadership is borne out by the elevated score given the management commitment in the DICE methodology from Boston Consulting - see *Rolling the DICE to predict the outcome* (page 209).

There are four reasons why Leadership commitment is so important.

- Resource efficiency is a change management process. If we are to take our organization to the optimum, then we will inevitably need to change the dominant culture and systems, both of which require strong leadership.
- Efficient use of resources can depend on many different functions within an organization working effectively together, from the end-users of the energy through to maintenance, facilities managers, engineering, through to design and finance. Leadership is often needed to overcome the silos that would impede this collaboration.
- Excessive resource use often does not have immediate consequences so it is common for resource efficiency to be considered "*nice to have*" rather "*must have*". Leadership is required to create the compelling case for action and, if necessary, to align rewards and incentives with the programme.
- More substantial changes to the existing business model, such as a move to products-as-services or remanufacturing, require leadership support.

These four Leadership tasks are not limited to the start of the programme but are ongoing. These expectations of the Leader reinforce the observation that our Mandate involves more than passive approval for our programme from the Leader. In developing our Mandate, we should secure continuing active involvement of the Leader in our programme, which is why requesting a definite commitment of time, with a specific description of what is needed in that time, is recommended. Lack of management commitment to the programme is the most common cause of failure. However, much of the blame for this should not be placed on management, but on those who have designed the programme, for not having articulated the role and expectation of the Leader effectively.

In developing our Mandate we should secure continuing active involvement of the Leader in our programme.

The nature of the Leadership commitment depends on the scope of the programme. If the focus is on one site or facility, then the management team at that level need to be engaged. The Leadership referred to in the Framework

Leaders are not free agents; they need to **justify** their actions to shareholders, boards and other stakeholders.

encompasses all the intermediate business, departmental and functional managers who need to be actively supporting the programme. There is no point having the full backing of distant executives if local managers on the ground are contradicting the call to action. For a case study on this, see *Real World: Spanning the intent gap – enManage™ at BP* (page 239).

Of course, it is entirely feasible for resource efficiency to happen without significant Leadership commitment. Perhaps this is achieved as a normal part of cost or inventory control. Maybe the efficiency arises as a happy side-effect of process improvements and capital equipment replacement. Alternatively, it could be as a result of efficiencies external to the organizations, such as the reduction in carbon per unit electricity in a national grid as renewable production increases. Possibly, as total production volumes increase, a decrease in energy use per unit occurs, due to the *baseload effect* discussed in the chapter on goals. Perhaps a reformulation or "lightweighting" of our products leads to a concomitant reduction in resource requirements.

All these sources of improvement, if real, are to be welcomed. Indeed, in situations where it is hard to obtain Leadership commitment at the outset, proponents of resource efficiency may well find themselves having no choice but to drive improvement activities *below the radar*. However, no organization is likely to achieve or maintain its optimum resource use without there being some form of explicit programme with a mandate from Leaders. The nature of continual improvement demands that a long-term focus is maintained with systems of measurement, review and revision. This clearly requires the participation of Leadership. It is senior management who are best equipped to initiate this internal shift in thinking and to create the requirement for action.

"The **first responsibility** of a leader is to **define reality**."

- Max DePree

The Leadership, in particular, the chief executive, chairman and other members of the board (or their equivalents in public institutions) are responsible to the stakeholders of the business: they have a fiduciary duty to look after their interest. In many organizations the Leaders cannot just act on a whim, they need to be able to justify their actions to shareholders and boards or other more senior managers. However passionate they are about the change they seek, Leaders need to carefully prepare the ground internally and externally, to create a rationale in support of the change, to anticipate the response of markets and customers, to assess the impact on performance. This constraint applies just as much to public-service institutions as it does to the private sector – although listed companies may be under particularly intense scrutiny with analysts reviewing performance at least every quarter. This is a reason why the commitment of resources to the programme in the initial proposal is often framed as a *"business case"* – because the pros and cons need to be articulated and weighed up, expectations managed, and the groundwork laid to convince first the board and then external stakeholders. Leaders too will be limited in the number of issues they can focus on. There may well be *"bigger fish to catch"*, and so it is important that the demands that we make of them are proportionate to the benefits.

6.2 Urgency and forgiveness

Articulating the reason for action is one of the most important jobs of the Leader. They also need to help people move beyond past mistakes and avoid the defensiveness that can arise when we discover imperfections in the way we operate.

Of course, Leadership is about much more than overcoming initial barriers. Once a decision to proceed has been arrived at, and if the programme is of sufficient profile that it needs to be shared with stakeholders, it will be either the chief executive or chairman who will convey the essence of the programme. This communication is usually in the form of their letters to shareholders which outline the current vision and goals of the organization.

Often the leadership team need to initiate a change in perception of stakeholders – to articulate why the resource efficiency programme will add value to the organization, or enhance its mission in a fundamental way. Some examples of these letters are shown opposite. A further advantage of the Leader articulating the programme goals to shareholders is that this gives them a stake in the success of the programme. 3M's aims, opposite, are clear and ambitious and, having communicated these goals personally, then we can expect CEO Inge Thulin to make available the resources necessary to deliver.

John Kotter, a change management expert, believes that the key task of the Leader is to establish and maintain a sense of urgency (see *The importance of urgency when driving change* on page 321). The Leader needs to convey the message that action *now* is essential, not because of a distant future threat of climate change, but because there is value here at this moment, today, that the organization needs to realize. Impatience, eagerness, enthusiasm and excitement better describe the essence of the desired message than dissatisfaction. The Leader's call to action needs to overcome the complacency that many organizations have around resource use which we described earlier in the piece *Why Now?* on page 209.

"Management must demonstrate its clear and durable intent to progressively improve the energy performance of the entire organization".

Christopher Russell

In his outstanding book[635] on managing energy in industry, Christopher Russell sets out "7 *Steps to Successful Energy Cost Control*". His first step, as in our Framework, is to align leadership: "*Top management must demonstrate its clear and durable intent to progressively improve the energy performance of the entire organization. Management should also declare amnesty and hold individuals blameless for past choices that caused energy waste.*"

Few people have more experience and have better articulated the challenges of industrial energy efficiency than Christopher Russell, and it is good to see that he is emphasizing the nuances of the Leadership message to staff – in this case, a no-blame culture – that help the programme to succeed. ⇒page 223.

Letters to shareholders or stakeholders indicate the importance that an organization puts on sustainability and resource efficiency.

Real World: ***Leaders' letters to shareholders***

Some examples of leadership statements around resource efficiency and sustainability are shown below.

3M:

"Sustainability is a real-world business initiative that is woven into the culture of 3M.

We are building on our years of success to further reduce our environmental footprint. Our goal is to reduce volatile air emissions 15% by 2015, reduce solid waste 10% and improve energy efficiency by 25%, all from a 2010 base.

Our sustainability goals also measure progress on social responsibility and economic issues. For example, we aim to increase sales of 3M products that offer environmental advantages and we are developing and implementing water conservation plans in areas with scarce or stressed water resources."

Inge Thulin
Chairman, President and CEO
(2012 Sustainability Report[2] p7)

DuPont:

"Over the six years of my tenure as Chair and CEO... We have dramatically reshaped and retooled our company to ensure we can solve for the needs of the world's growing population during a new age of resource scarcity—and can translate that directly into compelling and sustainable value for shareholders.

This programme of change has taken direct aim at DuPont's productivity and shifted the portfolio to centre on the highest-potential opportunities where DuPont's science and engineering can deliver the greatest value. We have already driven more than US$2 billion of cost productivity and expect to deliver additional annual run-rate savings of at least US$1.3 billion by the end of 2017."

Ellen Kullman
Chair of the Board & Chief Executive Officer
5 May 2015"[232]

Kingfisher:

"We also remain committed to being a truly sustainable company and we continue to make good progress in this area. As we develop our unified ranges we are improving sustainability performance in areas like materials, manufacturing and transport.

Sales of sustainable home products were £2.9 billion in 2015/16, accounting for 28% of sales across the year. Through sales of energy efficient products and services we have helped customers save 10.8TWh of energy since 2011/12. We also aim to have a positive impact on the communities we work in and source from. For example, to increase the impact of our community investment, we updated our communities strategy in 2015, identifying three areas where we can have a positive impact both at a local and global level. These are homes, skills and forests.".

Daniel Bernard
Chairman of the Board, Kingfisher
(Annual Report 2016)[441]

The directors of many companies and shareholding institutions use the concept of their ***fiduciary duty*** *as an excuse for failing to address issues such as resource efficiency.*

Exploration: ***Why fiduciary duty and shareholder rights need to be reinterpreted***

A fiduciary is someone who is entrusted with another person's money, rights or assets. The word is derived from the Latin fiducia for *"trust"* and reflects the owner's expectation that the custodian will act exclusively with the owner's interests in mind.

It is a lovely concept. It describes how one person, perhaps vulnerable, in good faith places reliance and trust in another, and how that second person in good conscience, and bearing in mind the trust placed in them, acts in such a way to benefit and support the first person. A fiduciary does not allow their personal interests to interfere with their obligations to the first person, nor will they profit from the relationship.

Microsoft Office Clipart Gallery: MP900387776.JPG

This model of fiduciary duty has a legal basis in jurisdictions founded on English common law. A company has a fiduciary duty to the shareholder, in other words, the directors of a corporation are trustees for the stockholders and as such they have an obligation to act in their interests at all times, not to profit without the shareholders' consent and to avoid any conflict of interest. Fiduciary duties extend across many types of relationships: that between doctor and patient, for example; educator and student; estate agent and seller; trustees and beneficiaries. There is a comprehensive body of law that sets out the expectation of these different relationships, and the remedies and penalties for failing fiduciaries. In many of these fiduciary relationships, there is an overriding sense of the stronger, abler, better-informed person exercising their power justly and not exploiting their position.

In private corporations fiduciary duty has become horribly confused with another key issue of governance, the notion that the shareholder has primacy in the affairs of the corporation. The assumptions behind shareholder primacy are that since the shareholders 1) own the corporation they are 2) entitled to any residual assets the corporation owns - after meeting any existing obligations - and therefore 3) shareholders are the principals who employ the company directors to act as their agents.

In her book, *The Shareholder Value Myth,*[686] Professor Lynn Stout at Cornell Law School demolishes each of these notions in turn. First, shareholders do not own corporations; they own stock in corporations which - in very simple terms - confers privileges not much different to those of a debt-holder. Stock ownership is a contractual relationship between the company and the stockholder. Indeed, owning shares in Apple does not give one a right to walk into an Apple store and help yourself to goods. On the second point, it is the boards which in effect determine the dividend (i.e. payments to shareholders) since these are only payable from profits and boards determine the level of profits through their control of expenditure. Finally, in the US at least (unlike the UK), shareholders have no rights to call an extraordinary meeting, appoint CEOs or determine dividends. In other words directors do not work for shareholders, they work for corporations

Indeed, we can see that many public institutions, which do not have shareholders, still have boards of directors, or their equivalents, whose individual duties are essentially indistinguishable from those in private corporations. These Directors are perfectly able to discharge their functions on the basis that they are employed by and serve the needs of the institution.

The current dominance of the notion of shareholder primacy is fairly recent and seems to have originated in the neoliberal free-market Chicago School of economics in the 1970s. One of the movement's principal thinkers, Nobel Prize-winning economist Milton Friedman, wrote a paper in 1970 entitled *The Social Responsibility of Business is to Increase its Profits*[295] in which he stated that:

> *"In a free-enterprise, private-property system, a corporate executive is an employee of the owners of the business".* and concludes *"there is one and only one social responsibility of business--to use its resources and engage in activities designed to increase its profits so long as it stays within the rules of the game, which is to say, engages in open and free competition without deception or fraud."*

This statement is however not a legal opinion but rather a cultural or ideological position, written in the context where the dominant view at the time was that corporation and executives do have wider duties to society. US law has never enshrined an obligation on directors to maximize the profit of the corporation (or the share value for that matter) above all other considerations.

In the corporate law context, fiduciary duty is a rather limited duty setting out the director's obligations not to defraud the shareholder or seize their assets. Critically, it is not an exclusive duty. Under US law, directors enjoy an enormous amount of autonomy: they can choose to improve the welfare of employees if they so wish, they can give corporate money to charity, they can invest in environmental protection measures. All these things decrease shareholder wealth, yet these are perfectly legal because it is the directors who determine what is right for the corporation.

Where shareholders have challenged the decisions of directors, the US courts have often stated that directors owe a duty to both the *"corporation and the shareholders"* as two distinct considerations. Indeed, there have been some key judgements that have explicitly stated that directors can look beyond shareholder wealth. As Stout observes:

We need to go beyond saying that assessing the impact of resource efficiency is merely ***permissible*** *to a position that this is* ***obligatory****, with the consequence that a failure to do so is a breach of fiduciary duty, creating a legal and moral liability.*

Framework

"There is evidence that many companies have profitable opportunities to improve energy efficiency but do not pursue them because of lack of management focus or other internal obstacles. In these cases, the role of the board is to ensure that the organization has a governance structure that ensures action on climate change receives sufficient priority"

- Investor Expectations

> *"They* [US judges] *uniformly refuse to actually impose legal sanctions on directors or executives for failing to pursue one objective or another. In particular, courts refuse to hold directors of public corporations legally accountable for failing to maximize shareholder wealth".*

The essential principle here is that courts and corporate law recognize that executives need to exercise their *"business judgement"*. This judgement is what makes it acceptable for the directors to do things that may appear to be against the short-term interest of the shareholders. The autonomy of the directors clearly makes sense in the modern shareholding context in which hundreds if not thousands of individuals own the shares in the corporation, do not have the necessary proximity to the business to make informed decisions about how it should be run, or indeed are highly unlikely to all share a common view on what those decisions should be.

So, how has it come to the situation today that many directors believe and indeed assert that they have a legal requirement to act exclusively on shareholders' interests, regardless of the wider needs of the organization. To some extent, this can be understood in the context of the cultural shift towards the neoliberal thinking of the Chicago School, where the rationale of shareholder primacy seemed logical and was adopted, despite any formal legal basis, at least in the US, simply because people wanted it to be so. A more cynical view is that this focus on share value served the interests of senior executives who could claim that they would focus better on the share price if they received generous, if not exorbitant, share packages in addition to their normal salary. Thus, the devotion to the share price by directors became self-fulfilling once shares became their dominant form of compensation.

In its extreme form, this phenomenon has been referred to as *"employee capture"*, the notion that some organizations have been taken over by their senior executives who act entirely to serve their own interests. A prime example is the top 10 investment banks where, over a period of almost 10 years, between 1999 and 2008, staff took out four times as much as shareholders did in profits (*Leader, The Economist, Oct. 2009*), often in the form of share packages or bonuses related to short-term earnings.

The problem with this focus on share price above all else is that there is no evidence at all that it serves the interests of shareholders in the long run. Corporate law in the UK more closely enshrines the legal duty of directors to meet the needs of shareholders than in any other major jurisdiction - if this were to lead to superior returns we would expect UK companies to consistently out-perform their competitors in the US or Europe, but this is not the case. Indeed, some countries with longer-term, more pluralist forms of governance, such as Germany, have an equally good track record in the creation of world-class firms.

The reason that this matters from a resource efficiency point of view is that the idea of shareholder primacy acts as a brake on corporate responses to longer-term resource and environmental issues. Friedman's notion that the *"only social responsibility"* of business is to create profit *excludes* the idea that there could be other social responsibilities.

Company directors could correctly point out, in their defence, that regardless of where their prime duty lies, shareholders are simply not strongly articulating a need for better performance on resources (or other social and environmental issues). After all, the notion of primacy says it is up to the shareholder to advise the board of their priorities, and shareholders consistently send the message that share price is the key priority.

However, herein lies a second-order effect of this distorted notion of fiduciary duty arising from the fact that most shareholders are institutions, not individuals. Just as in the case of the company directors, many of the individuals making investment decisions for these institutions also use narrow interpretations of their fiduciary duty as a reason for inaction. They, possibly with greater justification, presume that their sole objective is to increase the short-term value, or share price, of the assets that they hold in trust for others and thus they fail to signal to the companies they "own" that they should consider longer-term aspects of value.

Some initiatives are seeking to address this investor behaviour and, as a minimum, develop the view that considering sustainability is permissible if not yet a requirement. The United Nations Environment Programme Finance Initiative (UNEPFI) has set about providing a robust legal examination of fiduciary duty to ensure that investors take proper consideration of environmental, social or governance (ESG) issues. Its comprehensive 90-page study[732] makes for fascinating reading and one quote from lawyers Arnold and Porter reinforces the view that considering resource efficiency is OK if it relates to share price:

> *'...it does not appear that current US law forbids integrating ESG considerations into an asset manager's decision-making process, so long as the focus is always on the value inuring to the beneficiaries and not on achieving unrelated objectives—even if positive collateral benefits result.'*

In other words, investors or asset manager can consider issues such as resource efficiency where they can demonstrate a link to value. This link with value is important since in many jurisdictions legislation, such as the Employee Retirement Income Security Act (ERISA) of 1974 in the US,[678] limit fund managers from considering social or environmental factors.

It is the historical disconnect between ESG issues and asset performance which lies at the heart of the fiduciary duty problem. Company directors have said "*since the investors and shareholders don't value this we can't address this in our business*", while institutional investors state "*since companies are not demonstrating how this affects share value we cannot invest on this basis*". While both parties accept that ESG issues, including resource efficiency, are material to asset value, their behaviours demonstrate an unwillingness to move first on the topic, in part from a lack of tools to quantify the value (share price) impacts of ESG considerations and also from an innate conservatism.

The CFA Institute, home of the chartered financial analysts designation, states in its *ESG Manual for Investors*[134] that:

> *"Successful investing is dependent on one's ability to discern the factors that influence the market's valuation of a company and then judge the accuracy of that valuation. Analysts are generally well versed in using financial metrics to understand those drivers of corporate value and lend skilled interpretation to what is often highly detailed accounting data. In recent years, however, non-financial factors—including environmental, social, and governance factors—have figured ever more prominently in the value of corporations.'*

Also, it goes on to state that:

> *"A company that incorporates ESG exposures into its long-term strategic planning and adequately communicates these factors and strategies to investors will provide a more complete picture of that company's prospective value. Strategically incorporating ESG analysis may also position companies to better anticipate future operating environments, including potential costs or burdens to their existing business model."*

Current and future resource efficiency (unlike some other environmental, social and governance issues) can easily translate into future cash flows so there is every reason to include this in investment or share price appraisal.

So we have two misunderstandings of fiduciary duty that need to be addressed:

1. Company directors should accept that shareholder primacy is a myth and that they have responsibilities to the corporation that include addressing risks related to resource use, which they are empowered to act upon regardless of the short-term impact on shareholder value; and
2. Investors must take into consideration environmental and other issues when formulating investment strategies as these can be material to value.

The proper understanding of the fiduciary role of directors and investors is an essential foundation for action on resource efficiency. Corporations are very much the solution to resource efficiency challenges. The power of capitalism to efficiently deliver innovation and drive change is not in dispute. The problem is that this potential engine of transformation is, and will continue to be, hampered if the directors of corporations focus myopically on short-term earnings, or if they are denied capital by investors who discourage innovation by focusing excessively at quarterly earnings.

The initiatives mentioned above seek to show that consideration of resource efficiency is *compatible* with the notion of shareholder primacy and fiduciary duty. These efforts try to demonstrate that it is *permissible* for directors and investors to consider these issues on the basis that they affect shareholder value.

A bolder, and more honest, approach would be to reject the notion of shareholder primacy altogether. There is a compelling argument, that rather than being simply permissible, that action on resource efficiency is *obligatory*, and that a failure to do so is a breach of fiduciary duty, creating a legal and moral liability for directors, trustees and analysts, etc.

This suggestion is not diminishing the obligation of directors towards shareholders, in fact quite the opposite. It recognizes that the current approach causes many companies to focus on activities that are reckless, irresponsible and endanger investors and shareholders alike. BP's disaster at the Macondo well in the Gulf of Mexico proved the inadequacy of short-termism as an effort to save US$1 million a day on well operations led to safety lapses that destroyed US$100 billion of shareholder value.[686] VW's attempted short-cut to regulatory approval on diesel engine emissions will likely prove very costly too.

Indeed, investors and fund managers should also recognize that sentiment on this matter is changing. The recent Arch Coal and Peabody Energy case in the US[626] alleged that the directors of these corporate pension schemes failed in their fiduciary duties by not considering financial risks driven in part by climate change.

Real World: ***Institutional investors***

Forward-thinking investors are beginning to interpret their fiduciary duty to include a consideration of climate change in investment decisions.

In *Investor Expectations*,[394] three investor organizations, the *European Institutional Investors Group on Climate Change (IIGCC)*, the US-based *Investor Network on Climate Risk (INCR)* and the *Investors Group on Climate Change (IGCC)* in Australia and New Zealand, representing €21 trillion of investments, recognize that climate change represents a risk to investors. Their justification for action reads:

> *"In order to fulfil our fiduciary duty to safeguard the long-term interests of our clients and beneficiaries we believe that it is essential to take action now that will result in substantial reductions in global greenhouse gas emissions within a time frame that minimizes the risk of serious impact."*

6.3 Cascading authority

The Leader acts through others. As the Mandate cascades down, the Leader needs to ensure that there is an corresponding upwards reporting mechanism that can track progress and inform them of the true performance of the programme.

A resource efficiency programme can be initiated at any scale, so it may be the case that the Leadership input is coming from a site manager, business unit head, or regional manager. These managers, too, need to craft a Mandate that is relevant to their local operations, regardless of whether the CEO has instigated the programme or whether it is local in nature. Leaders of operational units do not write open letters to the shareholders, but their visible and continual commitment to the resource efficiency process is just as important.

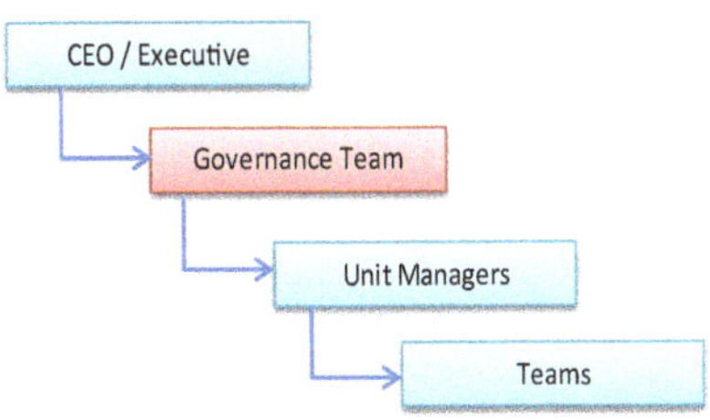

6.1 **The Mandate from the topmost sponsoring Leader cascades down the organization, being variously repeated or reinterpreted in the light of the local situation** The topmost level may be the whole organization or it may be a single division or facility and there could be more or fewer levels in this hierarchy. *Source: Niall Enright*

It is important to appreciate that the Mandate cascades from the highest level Leaders sponsoring the programme – whether that is at the top organizational level or a single facility – down to the heads of individual functions and individual teams. At each level, the Mandate needs to be reiterated (i.e. repeated) or reinterpreted (i.e. translated into the language, priorities and context of the local unit). This process of repetition and reinterpretation is the role of leadership at each level and the manner in which they accomplish this is crucial to the outcome of the programme.

This point cannot be stressed enough – it is my experience in many larger corporate resource efficiency programmes that the battle is won or lost at the middle-management level, as shown in the BP case study on page 239. By and large employees on the shop floor, production lines, or delivering service, are well disposed to the idea of better environmental performance and reduced waste. Similarly, most senior executives "get" resource efficiency. It is at the level of middle management that the greatest resistance can occur.

One immediate cause of resistance in management is concern about the impact of the initiative on existing targets, deliverables and workloads. All organizations today are driven by a plethora of KPIs, targets, performance indicators, service charters, etc., and it is fair to say that many managers feel that they are fully stretched meeting these imperatives. How often have I heard this *cri de cœur* from middle management on learning about an energy or resource efficiency programme: "*Oh no, not another initiative*"! To overcome this reaction, I try to be explicit about their personal commitment to the process from the outset.

We have seen that the call to action needs to be clear about the precise input required from each part of the organization and to communicate that this reasonable input is non-negotiable.

Changed decision-making implies changes to authority, to the flows of information and to the allocation of resources in an organization – a reason why clear leadership is important.

At the same time, we have said that our Mandate needs to be empowering. This delegation is necessary because the use of resources in an organization is not something that can be determined by a single decision – or even by a single department. The sources of value are spread throughout the organization. Resource consumption is affected *upstream* by leaders and planners who respond to customers and set the overall level of activity and direction for the organization. Product or service designers who fix the resources intrinsic to products have an impact. Also, the marketing departments which address the supply chain and stakeholder expectations around resources. Or *downstream* finance and procurement folks who choose between alternative raw materials or capital plant to meet the service or product needs. Engineering departments and maintenance teams influence resource use when they determine the design and efficiency of facilities and equipment. The teams manufacturing or distributing the goods, or providing the service, can operate efficiently or wastefully. It is inconceivable that the chief executive or sponsor of the resource efficiency programme will make the decisions for all these specialist functions – hence the emphasis on empowering informed decision-making.

Changed decision-making implies changes to authority, to the flows of information and the allocation of resources in an organization – another reason why clear leadership is essential. For example, when we make end-users of energy in a manufacturing process more accountable for the energy that they consume, then they are likely to put pressure on the maintenance team to focus on energy waste – e.g. compressed air leaks – which will in turn influence the effort allocated to this task compared to other maintenance tasks. It may be that the maintenance team don't like this change and so leadership will be needed to resolve the potential conflict. The process of delegation to the point of optimum control is a characteristic of the most successful resource efficiency programmes – it is no longer the utilities department on the supply side which is accountable for energy but production units on the demand side. In a college it is the department heads and lecturers who use the energy or create the waste paper who are made accountable – not just the facilities manager. Energy budgets may no longer be allocated at the site level but are paid for by individual departments. Responsibility for waste shifts from procurement and environmental services to the product design and operations teams. These changes in authority, and possibly budget responsibility, are a potential cause of resistance in organizations – resistance that can be overcome by strong leadership at every level as well as an effective governance structure, as we shall see shortly.

Of course, empowerment is not the same as abdication. We mustn't simply say "*You're the engineer – you figure it out*", rather "*You're the engineer – you figure it out and let me know what we need to do differently*". In the Framework approach, to resource efficiency, there is always upwards accountability back to the originator of the Mandate. These feedback loops are critical to the long-term success of the programme – at the end of the day, it is the fact that *the boss* is interested in the process and measuring how we perform that keeps every level of the organization honest and engaged. In addition to providing the

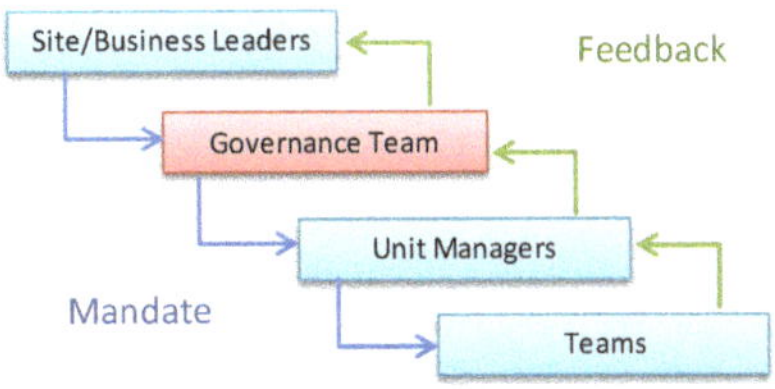

6.2 **Leaders at every level need to ensure that they give and receive feedback on the progress of the project**
There could be many more or fewer levels in this feedback model depending on the complexity of the organization.
Source: Niall Enright

original Mandate, it is the role of leadership is to ensure that these feedback processes are working well – and it is this commitment to regular review that lies at the heart of the*The "15 minutes is all" pitch (page 195)*. The two-way flow of information is also needed to identify cases where there is an *"intent gap"* in which conflicting objectives cause middle management to resist the programme.

In Volume II, we shall explore the types of measurements and metrics that can be used to ensure that the feedback at every level of the organization provides a clear indication of progress, even when senior managers are quite distant from the operating units where the resource efficiency changes are taking place. Wherever possible, the feedback needs to be informed by hard quantifiable indicators – ideally metrics that are linked to the core measures of Value - profit, operating costs, service delivery, etc. - that drive the organization. The feedback, in other words, needs to be quantitative not just qualitative and put a figure on the value that the programme is delivering.

It is important to note that the two-way feedback systems illustrated here do not imply the development of a vast *cottage industry* of reporting, working in parallel to the existing business systems and processes. This is absolutely not the intention – what we want is for the feedback to take place within the existing business decision-making forums: at the Monday morning team meeting, at the weekly department head meeting or the quarterly sustainability steering group. Nor do we want our programme metrics to be generated by a parallel data-processing effort – ideally these metrics will be incorporated into the existing key performance management systems. In this way, energy and resource efficiency become *"part of the normal way we do business"*, not a bolt-on initiative that can be sacrificed the minute our organization encounters turbulence of some form. ⇒ page 228.

6.3 **Listening is the most important skill of a Leader or Champion**
The folks at the *coal face* usually understand better than anyone else where improvements can be made. The quality of the feedback we receive will greatly influence the outcome of our programme.
Source: photo: © biker3, Fotolia.com

Real World: *The impetus for change at Interface*

Ray Anderson was the founder, chairman and CEO of Interface Carpets, a US company that today has over US$1 billion of global sales of carpet tiles. Having started the company in 1973, Anderson had a profound insight in 1994 which led to a compelling Mandate for change at Interface. The story below is largely from Anderson's highly readable autobiography, "*Confessions of a Radical Industrialist*",[25] published shortly before his death in 2011.

6.4 **Ray Anderson on the shop floor at Interface Carpets. He summarized his vision as "Mission Zero" – for his business to create no waste, use no non-renewable inputs and create no pollution.** *Source: reproduced with kind permission from the Ray C. Anderson Foundation, photo by Karin Koser*

The story of Interface's sustainability – or more accurately resource efficiency - has been quoted many times as an inspiring example of how you can "*build a successful business without destroying the environment*". For this transformation, Ray Anderson and Interface received many accolades: Time Magazine called him a "*Hero of the Environment*"; US News & Business Report said that he was "*America's Greenest CEO*"; Fortune listed Interface as one of "*100 Best Companies to Work For*". The GlobeScan 2007 survey of sustainability experts – those in the know – put Interface at the top of the list of companies with the greatest commitment to sustainability, ahead of Toyota, GE and DuPont. The honours are impressive for a business whose core activity can be considered dirty and resource-intensive, which is to take large amounts of petrochemical-derived molecules such as nylon, energy and water and combine them into a carpet square. What Interface achieved over 12 years between 1996 and 2008 was to:

- Reduce GHG emissions in absolute terms by 71% (and in relative terms by 82%);
- Reduce the waste sent to landfill by 78%. The scrap is now recycled back into the products – saving valuable organic molecules;
- Saved a total of US$405 million of avoided costs;
- Increased turnover by two-thirds while doubling profits.

Ray Anderson said that Interface would not stop making carpet tiles, they would not give up any orders or relinquish market share.

What they would do is to run the business as a good steward of the planet, and that this in turn would be good for the business.

The results prove this.

According to Anderson, the change process started in a small way with a customer enquiry. *"A short memo was dropped on my desk in the summer of 1994. It was a handwritten note passed along by Jim Hartzfeld, then an associate with our research division. It has been sent to him by a sales associate out on the West Coast. On it was a simple question: 'Some customers would like to know what Interface is doing for the environment. How should we answer?'"*

Apparently the customer who had raised the question was quite an important prospect for Interface, and they had previously told the sales associate that Interface *"doesn't get it"* when it comes to the environment. So this latest note spurred Anderson to try to figure out what the "it" was that concerned the customer. At the time, Anderson admits not to having a clear answer as to what answer he should give about Interface's actions on the environment. He could talk about being 100% compliant with regulation, but he felt that was not inspirational.

The note's author, Jim Hartzfeld, suggested that an environmental task force consisting of representatives from all the divisions around the world should be convened to consider a response, to which Anderson readily agreed. Hartzfeld also asked Anderson to kick off the meeting with an *"environmental vision"*. It was this need to find something to say other than *"we comply with the law"* that caused Anderson to dig deeper into the topic and so come across a book that was to transform his thinking: *The Ecology of Commerce* by Paul Hawken.[359] In his autobiography, Anderson recalls his shock at reading a chapter of the book called *The Death of Birth* about the decline of our environment and the extinction of species (never to be born again). The culprit was business, industry and corporations just like Interface. However, if business was to blame so, according to Hawken, business was also the solution. It was up to business to put right the damage – not government, not education, not the church – business is the only institution wealthy enough, pervasive enough and powerful enough.

On the 31st August 1994, Ray Anderson made his speech to the environmental task force. In front of an astonished audience, he recalled his epiphany on the decline of the planet and committed Interface to *"Mission Zero"*, in other words, no waste, no non-renewable raw inputs, zero pollution.

This commitment was a classic example of a mandate established by an influential and visionary leader. As the founder of the business, the chairman and CEO, Ray Anderson had the power to initiate the resource efficiency programme. However, he also created in his workforce a sense of urgency about the change that was expected – he said that they would not stop making carpet tiles, they would not give up any orders or relinquish market share. What they would do is to run the business as a good steward of the planet, and that this in turn would be good for the business.

Ray Anderson's first great achievement was his ability to create a mandate that fired up his organization. However, his other achievement was to engage with his team and people to ensure that the vision was turned into reality. Revolutionary recycling initiatives, employee engagement programmes, company-wide suggestions schemes, strong governance and his continued personal leadership were all essential to delivering the success of Mission Zero. To this day, the Interface business remains fired up by the vision and continues to reap the huge business benefits and competitive advantage that arise from the programme.

6.4 Strategy

Strategy is the responsibility of the Leader. Others may be involved in formulating the strategy, but they must ultimately buy into it. Here we explore whether an explicit resource efficiency strategy is advisable and what such a strategy may contain.

The earlier chapters on climate change, resource depletion and value have covered the overriding "*Why?*" of our resource efficiency programme.

There is, however much more that needs to be considered. We need to describe "*What?*" we want to achieve - the overall goal. We need to establish "*How?*" the goal is going to be met; which in turn tells us "*Who?*" needs to be engaged; we need to know "*When?*" we get folks involved and "*Which?*" systems and processes need to change in order to initiate and sustain the improvement, and finally we need to establish a process that will let us know "*Where?*" we are in the journey towards our goal.

6.5 **The strategy and tactics of a resource efficiency programme can be clarified by asking a series of fundamental questions.**
Source: Niall Enright

Question	Objective	Type
Why?	Create dissatisfaction - the overriding reason why the status-quo is not acceptable	Strategic
What?	Craft a vision of *what* the aims and benefits of the programme will be and how these will be measured - define the goals	Strategic
How?	Determine *how* the change will come about – what process will be put in place to drive the improvement	Strategic
Who?	Establish *who* will be involved in making the changes to achieve the goals and empower/ equip them to act	Tactical
When?	Ensure that there is a clear timetable for *when* action is needed and convey a sense of urgency	Tactical
Which	Establish *which* systems, technologies and processes need to change to create continual improvement	Tactical
Where?	Establish feedback mechanisms to tell us *where* the programme is on an ongoing basis.	Tactical

The answers to these questions collectively form the strategy for the programme. A resource efficiency strategy is essentially a definition of the goals of the programme, and the plans and actions that will enable these to be achieved. It may be that many of these questions will have been addressed in the original pitch for the resource efficiency programme – particularly if it was a detailed proposal. The Leader may have also expressed some views on aspects of the programme that will have provided some additional clarification. Because this is a leadership task, strategy is considered in this part of the Mandate, but

*There are some strong arguments **against** creating a separate resource efficiency strategy.*

it cannot be divorced from the other elements of the Mandate: Goals and Governance.

Although the Leader will review and approve the strategy, it is important, that the final detailed plan has input and ownership from a sufficiently senior team across the business, particularly from the departments most affected, which can refine and validate the strategy and tactics and put the programme in motion. This need for diverse input is why it often falls on a Governance team, described next, to refine, or at least approve, the strategy.

Before we leap into discussions about the strategy, we should note that there are some strong arguments against creating a separate resource efficiency strategy. We need to recognize that a specific strategy has the potential to set resource efficiency apart from other activities our organization is engaged in, which works against our objective of integrating resource efficiency into the day-to-day decision-making processes. This school of thought would maintain that the best approach is to ensure that existing strategies are updated to reflect our resource efficiency goals. The argument goes that to do otherwise runs the risk of misalignment with other business strategies, potentially leads to diminished ownership and divorces the resource efficiency plan from other strategies, which increases the possibility that it may be abandoned at a later date because it is separate from core business activities.

These are extremely compelling reasons for us to approach the topic of resource efficiency strategy with care, and in particular to think carefully about the form our plan takes and how it links with wider organizational decision-making. If the value case for our programme has been effectively aligned with the core objectives of the organization, then it is logical that resource efficiency should feature in the core strategy, which will reinforce the mandate for change. In these circumstances, it is quite reasonable to develop a plan for how the resource efficiency programme will support the overall strategy. For example, at L'Oreal we have seen that resource efficiency is aligned to the core strategy to diversify into new markets by developing consumer trust in the brands in order to recruit a billion new customers, which will be achieved "*through the strategic management of raw materials and the optimization of intangible value-drivers*". In these circumstances, a separate strategy for resource efficiency is not really called for, but a plan to ensure how resource efficiency, executed and communicated effectively to underpin that customer growth, does make sense and will reinforce the business case for this activity.

Of course, different organizations approach strategy in a variety of ways, so it is not possible, nor desirable, to prescribe a one-size-fits-all approach. The goals of the programme are what determine the strategies to be employed, and these vary hugely. We shall see how an ambitious goal can stimulate more radical thinking, but can also lead to greater resistance. Timescales, too, influence the potential for a programme – thus, longer-term objectives can leverage the full asset investment cycle of an organization, while short-term objective will dictate a strategy based around optimizing existing operations.

Real World: *A total optimization strategy vs a resource efficiency strategy*

Another very compelling argument against having a separate resource efficiency strategy is put forward by "*total optimization*" advocates.

The thinking goes that the objective of any organization is to optimize the overall process of delivering the product or service rather than a focus on just one narrow set of inputs and outputs. A focus on water, for example, may lead to increased energy use due to the extra pumps, etc. that treatment systems require. Every process has a broad range of inputs, some not material – such as people's time and money – which together contribute the value added by the activity. Thus, the argument goes, to single out one set of resources for special treatment may disturb the equilibrium of the system and lead to inefficiencies elsewhere.

One memorable occasion when I came across this situation for myself, was at a BP petrochemicals site in Köln, Germany. Köln was a site that was upper quartile in the Solomon Energy Intensity Index, i.e. very good, and I was excited to learn how they achieved their high levels of efficiency.

When we sat down to discuss the potential to run an energy management programme at the site we were quickly advised that the idea was a non-starter. This was because the way that this site was operated was that each night the plant staff would enter the following day's desired production quantities for a range of different chemicals into a mainframe computer model. This model would then run for several hours, eventually producing the optimum recipes, operating schedule and conditions to maximize the financial return of the entire system. The production planning programme optimized the whole site and also took into consideration a number of tightly integrated partners which would take steam, chemical feedstock and power from the BP facility.

What the site engineers had assumed was that this production planning process meant that *separate* consideration of energy was inappropriate. Actually, this meant nothing of the sort, for two very fundamental reasons:

- BP had made specific corporate commitments around emissions and so managing energy use to specifically meet these commitments was entirely valid;
- The production planning process optimized the system as a whole, but it could not guarantee that *individual* items of process plant were operating efficiently. Heat exchangers could be fouled, steam traps leak, control sensors read incorrectly. Correcting these inefficiencies would not conflict in any with the optimization of the system as a whole.

Thus putting in place an energy efficiency strategy was entirely appropriate at Köln, despite the existing total process optimization programme. My colleagues and I advised that the site should determine if the range of emissions forecast from the production plan were acceptable to BP. If the emissions were too high, we proposed that they should incorporate a constraint into the production planning programme that would limit the emissions – just as there were multiple other constraints in the model limiting things like tank fill for safety reasons. The second part of the recommended strategy was that the site should ensure that there was some dedicated resource to look specifically at the equipment and unit energy efficiency issues, which had been neglected in the *total optimization* regime, and set some priorities around these.

Unfortunately, this advice was not taken up as it ran counter to the belief that the system was operating optimally.

A similar paradigm clash can occur with other forms of improvement programme. Many quality systems such as Statistical Process Control (SPC), Total Productive Maintenance (TPM), Lean and Six Sigma have a methodology to optimize sites, units and processes. Facilities using these systems are often resistant to the idea of introducing an additional optimization programme focused on resources, especially as these quality systems may have been expensive to implement and require considerable effort. In these circumstances, I tend to work on the basis that these quality systems can, with appropriate modification, incorporate resource efficiency within them. However, it is important that the additional tools and techniques required by resource efficiency are adopted and that the limitations of some of the key methods of the quality systems are understood. For example, I have often seen metrics such as Operational Equipment Effectiveness (OEE) mislead users into believing that general optimization equals resource optimization. Of course running equipment for fewer hours and greater throughput is good, but *so* is running the equipment at optimal efficiency.

The bottom line is that these are not "*either/or*" decisions. These quality systems support resource efficiency, but the latter has tools and techniques which deliver *additional* efficiencies. It is always advisable to create a specific focus on resources as well as any other improvement processes taking place, as this invariably will lead to even greater overall system effectiveness.

Challenging an organization that is doing **the minimum** around resource efficiency is especially difficult.

Strategies will be as numerous as there are organizations undertaking resource efficiency. As a consequence, I will not attempt to set out a particular strategy for resource efficiency, but to describe what such a strategy might include and the sorts of tools and techniques that are available to inform and communicate the strategy. The emphasis will not be on the many changes that will reduce resource use, but rather on the design of the resource efficiency programme which will enable such changes to take place.

We also need to be very pragmatic when we think about our programme strategy. If our organization is straightforward, there is a clear mandate in place and the emphasis is on achieving a rapid impact, strategy development can be simple and direct, as demonstrated by the case study, on the next page, from MediaCityUK.

Another key challenge in gaining a commitment for resource efficiency lies in the situation where an organization thinks that it is already doing resource efficiency, when the reality is that its efforts are small in comparison to the value potential. In these circumstances there are some strategies to engage the organization such as:

- Following an incremental approach whereby the existing mandate is developed to extend the programme to "*its next logical stage*" and so span the gap in opportunity, essentially working on the notion that "*if it is not broken, don't fix it*";
- Or undertaking a more fundamental re-examination of the existing programme, seeking a renewed management commitment and mandate. In this case, benchmark comparisons with other organizations in the sector can sometimes provide a helpful argument for action.

Challenging an organization that is doing the minimum around resource efficiency is particularly challenging. Management in these organizations may believe that their response is adequate or even good. They may also feel that the incremental effort needed to take improvements to the next level are disproportionate to the benefits. Vested interests may be reinforcing this view as functional managers caution against efforts on resource efficiency which may create additional workload or illuminate mediocre practice.

In this situation, an external consultant can help, as long as they are appropriately experienced and willing to be honest (not all are). I have on many occasions been employed by operational staff who are dissatisfied with their organization's response to resource efficiency to provide an impartial, objective, honest and independent assessment of the opportunity presented by energy or resource efficiency. It is not a process of criticism, but rather an evaluation of the scale of opportunity, backed by example and analysis, that is required. Sometimes the fact that this comes from someone external makes all the difference.

⇒page 236.

In My Experience: *Lewis McIntyre – MediaCityUK – "just do it!"*

Lewis McIntyre is Finance Director at MediaCityUK.

Here he shares his own experience of a successful and, at times, challenging, energy efficiency programme.

This case study shows how resource efficiency programmes can be started and led by non-technical staff, and reinforces the value of senior management commitment throughout.

MediaCityUK is a vibrant and sustainable development in Manchester's historic waterfront at Salford Quays. It has 780,000 sq. ft. of office accommodation; a 250,000 sq. ft. high-definition film studio, one of the most advanced in Europe; 378 apartments; over 60,000 sq. ft. of retail space; a 218-bed hotel as well as extensive public spaces and a multi-storey car park. The tenants include major British broadcasters such as BBC North and ITV, as well as the University of Salford and over 80 small businesses, most of which have been drawn to the huge potential that these innovative facilities offer the creative industries.

6.6 **MediaCityUK transformed industrial docks at the end of the Manchester Ship Canal into a European centre of excellence for the creative industries**
Phase 1 occupies 36 acres with a potential to develop a further 200 acres over the next few years.
Source: reproduced with kind permission from Peel Media Ltd.

MediaCityUK was designed with sustainability as a key objective. It was one of two projects that piloted the BREEAM Communities sustainability rating system – the other was the Olympic Park for the London 2012 Olympics. A trigeneration plant that uses free cooling from the Manchester Ship Canal provides low carbon electricity, heat and cooling to many of the tenants on the site.

As most property owners will know, it is one thing to get prizes at the design and post-construction stage and another to ensure sustainable operations once tenants have moved in. The basic assumptions that are made in the design stage are often quite different from the actual needs of the occupiers when the building is delivered. Controls are modified, tenants install additional equipment and the physical performance of the building can change significantly. Landlords' commissioning of the buildings usually simply takes the form of confirming that systems are operating as designed, to identify defects within the guarantee period, rather than to ensure the performance is ideal.

The team and I at MediaCityUK had other ideas. With a strong mandate from the Managing Director, Stephen Wild, I set up an *Energy Focus Group* to drive cost improvement. This brought together the customer-facing property managers, with in-house engineering, finance, the energy suppliers to the site and the external facilities management contractors. This group, which I chair, meets monthly and acts as the Governance team for the energy cost-reduction programme. We then issue a monthly progress report to the board to keep them involved. Niall Enright was our external adviser and helped me to establish the programme and also participates in the ongoing monthly energy focus group meetings.

The approach that I adopted was incredibly simple: *"just get on and do it!"*. The programme had no quantitative goals as such, other than to reduce operating costs for MediaCityUK and our tenants (who pay for the energy costs in common areas through the service charges). I suspected that there were plenty of opportunities for improvement available. For me, the key was to organize the programme to implement these opportunities quickly while putting in place the foundations for a continual improvement programme.

The programme kicked off in January 2012 and a web-based Opportunities Database, using Verco's Carbon Desktop™ tool, was put in place. Within the first month over 47 projects were identified and 13 completed delivering £70,000 of annual savings. The project benefited from a hugely enthusiastic Champion, the MediaCityUK Technical Services Manager, Derek Elliott.

Of course, things don't always go smoothly, and one of our first challenges was to get the facilities management (FM) contractor fully onboard with the energy-savings programme. As is common in FM contracts, energy efficiency took a back seat to reliability and maintenance. Fortunately, as I launched the energy programme, the FM contract was up for renewal so Derek Elliott could put in place clear requirements around energy efficiency in the new contract. This part of the process was time-consuming, and at times it felt like the MediaCityUK team had to educate our potential suppliers. However, the result was the successful bidder, a company called Engie, was clear about the importance we placed on energy efficiency, and this was built into our contract. In fact, Engie provided a dedicated team member, Phil Harris, with a background in sustainable buildings, to work with Derek on identifying and delivering projects, which has been extremely useful.

The diagram below is the closest document there is to a strategy for the programme. We started with four activities:

1. Connect existing metering to Carbon Desktop™ and examine the day/night/weekend profiles to decide if use is excessive or variable, and remove the excess or variance.
2. Assess what metering is missing and fill the gaps. Over time, as the buildings are fully occupied, to set up relationships with weather and other variables that can give further insights into performance.
3. Capture all ideas for improvement, and track their stage of implementation as well as the outcome as evidenced from direct measurement of consumption. This is the "Opportunities Database".
4. The fourth activity was to provide a monthly board report to keep the MD and senior team engaged and to celebrate success when it occurred In order to motivate all the participants.

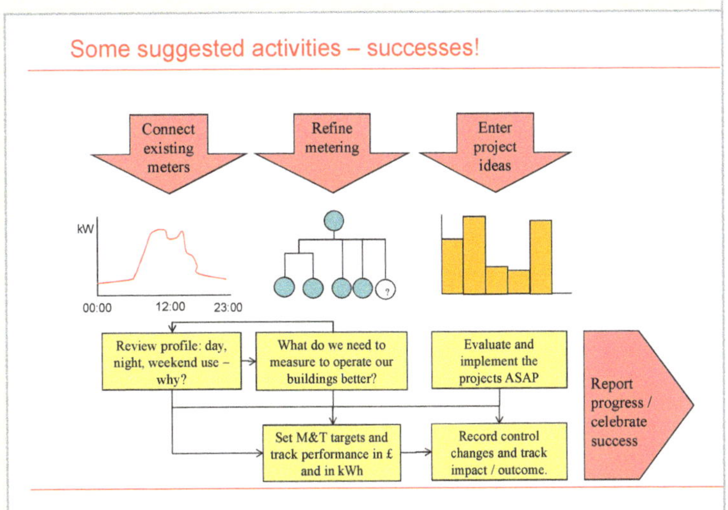

6.7 **The MediaCityUK energy efficiency strategy summarized in one slide**
Source: Nlall Enright, reproduced with kind permission from Peel Media Ltd

Rather than tackling the whole estate from day one, the 430+ meters were configured in Carbon Desktop™ building by building, so the team were rapidly able to assess the first building's performance and start on the savings initiatives. The instruction from Derek and I was that *"no or low-cost"* projects should be implemented immediately, without the need for further approval. Projects that needed greater investment or could involve tenants would come to the monthly focus group meetings for an instant decision. This rapid decision-making was one of the key aspects of the programme, in my mind.

MediaCityUK

ENERGY STAR

THIS CERTIFIES THAT

GEOFF MANN

HAS BEEN RECOGNISED AS A MEDIACITYUK ENERGY STAR FOR HIS OUTSTANDING CONTRIBUTION TO ENERGY MANAGEMENT DURING

MAY 2012

STEVE WILD, MD

LEWIS MCINTYRE, FD

2012 Energy Star

6.8 **Celebrating success was an important part of establishing early momentum in the MediaCityUK energy efficiency programme**
Source: NIall Enright, reproduced with kind permission from Peel Media Ltd

An important feature of the MediaCityUK approach was our emphasis on performance measurement right from the outset. As they say: *"you can't manage what you don't measure"*. The measurements gave me confidence in the claimed savings, allowed Derek to assess the efforts of the FM contractors and gave the team the chance to celebrate success very early on. Roughly every month the focus group selects an *"Energy Star"* based on their contribution to the programme. This Energy Star is not necessarily a member of staff – they could be a supplier or a tenant. They receive a certificate signed by Stephen Wild and me, as well as the opportunity to take a partner or colleague out for a meal for up to £100. Celebrating success not only provides a positive impetus for a programme, but it is also very enjoyable!

This programme has one full-time equivalent focusing on technical issues (Derek Elliot and the dedicated Engie engineer also had other jobs to do as well). The Carbon Desktop™ consultant, Caroline Robertson-Brown, spends one to two days a week on the software and reporting. Other team members' contributions were no more than just a few hours a month. The FM contractor's workload did not increase but was better focused. So for around 1½ full-time equivalents the programme has delivered, as of November 2012, £500,000 of annual savings, at a cost under £100,000 largely on small CAPEX and metering. There are over 100 further projects at various stages of evaluation or implementation! From my perspective, this has been an outstanding success. Overall energy use has been almost halved, despite increases in tenancies.

Data quality and progress reporting have been two of the key challenges of the project. At times, I was frustrated by the lack of visibility of what is happening on the ground – a feature of busy people and sometimes unreliable data.

Our next challenge is to ensure that MediaCityUK can maintain and build on the improvements made. It would be easy to declare victory and treat the programme as an outstanding example of post-occupancy building commissioning – which it is - rather than as the foundation for a process of continual improvement. Having halved energy use in some areas, halving energy use again will be much more difficult, time-consuming and costly. Just as exponential growth impossible, so too is exponential reduction – at some point any facility must reach its theoretical optimum!

The team recognize this and are evolving the programme. Having initially had no specific objective for the programme – other than "save money" - we are now setting individual targets for each building. Formalized audits (based on the CIBSE TM22 methodology) are driving out more opportunities. In addition Monitoring and Targeting is being deployed using weather data variables, to ensure that seasonal operational variability is reduced. In particular the ongoing motivation of the team and celebration of success will be the key to consolidating the gains made – many of which are reversible operational improvements.

We decided in early 2013 to see if we could enter our programme for some awards. This was in order to get an external assessment of our efforts, give the team recognition for their work and possibly achieve a positive marketing message that would help demonstrate our credentials to existing and prospective tenants. The particular awards we chose to enter were those established by professional institutions and so would have the added value of scrutiny by experienced practitioners who would be able to differentiate between "green gloss" and real underlying technical excellence and innovation.

Much to our surprise and pleasure we were awarded:

- The British Institute of Facilities Management (BIFM) Award for Sustainability and Environmental Impact 2013. We were up against some very strong competitors including Google, Napier University and the International Monetary Fund.

  The judges kindly said of our work: "*Sustainability and environmental management have been designed in from the very start of this project and right across the facilities spectrum of activity; from power and utilities generation and distribution through to day-to-day maintenance and operational management.*"

- The Premises and Facilities Management awards where we and our FM contractors, Engie, were voted the Partners in Expert Services 2013 as well as receiving the Overall Award 2013 for the best project submission.

As well as motivating the team, the success of the programme has had another benefit for MediaCityUK. One of the largest tenants has asked us to provide the same energy management programme in their building, providing some additional revenue and further enhancing customer relations and the distinctiveness of MediaCityUK as a location.

For me the key lessons are:

- A resource efficiency programme doesn't have to be hard or complicated.
- Rapid impetus can be established by focusing on specific areas, empowering people to act and making quick decisions on investment.
- Data can be a problem. It is important not to let the lack of perfect information get in the way of action.
- Getting the contract with FM providers right can be difficult but will reap rewards in the long run.
- The correct tools are essential. Without an Opportunities Database, proper metering and data analysis the programme would not have been successful.

On the basis of this hard work we were successfully certified under the prestigious and rigorous ISO 50001:2011 standard in July 2015, which has helped us to embed the best practice we have developed over the last couple of years. In 2016, the team were awarded a Platinum Award for Carbon Literacy across the estate.

Not only did the MediaCityUK team "*Just do it!*" but they did it *right*.

6.5 Goals and targets

Confusion between high-level programme goals and the specific targets that drive performance at a lower level in an organization is a common problem which undermines energy and resource efficiency programmes.

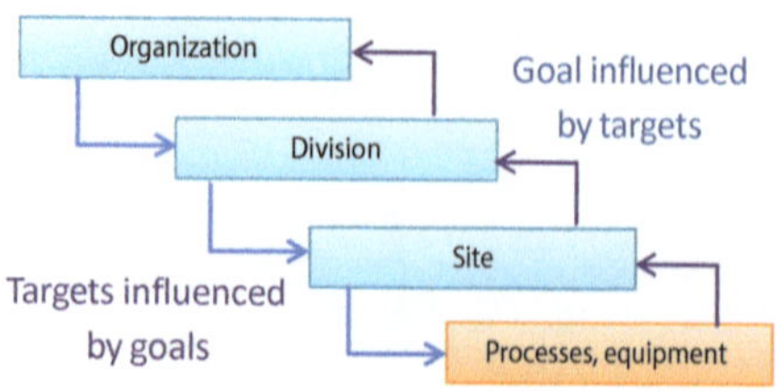

6.9 **Goals and targets are usually different but complementary measures** The former describes the overall aspirations of a resource efficiency programme while the latter are specific performance objectives set at the level of individual processes, teams or items of equipment. *Source: Niall Enright*

It is important to note the distinction made in this book between goals, which apply at the high level, and targets, which apply at a very low level of an organization to assess the performance of specific processes or items of equipment. I know that many organizations refer to their high-level objectives as targets, and this is perfectly reasonable; however, for clarity, the distinction between these two is made in this book. This chapter is all about the higher-level goal-setting – although many of the principles discussed here apply to target-setting as well.

Targets can vary within organizations. This variation may arise because there are differences in potential for improvement in different divisions or locations because the value to be gained from resource efficiency differs across the organization, or perhaps because such matters are devolved to local management who simply have diverging aspirations or priorities. In the process of cascading the mandate down an organization, the process of reinterpretation to align the goals with specific needs at the lower level can often lead to a restatement of the objectives in different, more relevant, terms.

While it is common for efficiency programmes to be driven by relevant local targets, most organizations still describe an overarching programme goal which provides a powerful call to action and allows for external communication. Targets, on the other hand, are never uniform because different items of equipment or processes will have different inherent capacities for improvement. Thus, two different boilers may have a target of a 5% and a 3% reduction in energy use per unit steam produced, respectively, while a refrigeration plant's target may be for a 25% increase in efficiency. All these targets may contribute to an overall programme goal of, say, 20% absolute emissions reduction, but are determined by the inherent potential of the equipment or process to be further improved. Techniques to set meaningful targets for equipment and processes in such a way that those responsible consider them to be fair, honest and achievable, are discussed later.

Normally goal-setting precedes target-setting as part of the mandate development process, but in some cases, particularly in smaller organizations, the top-level goal is simply the sum of the improvement targets set at the lower level. It is important early on in a resource efficiency programme to ensure that the lower-level goals and targets will combine to deliver the overall goals of the programme, especially when goals may be set in absolute terms, whereas

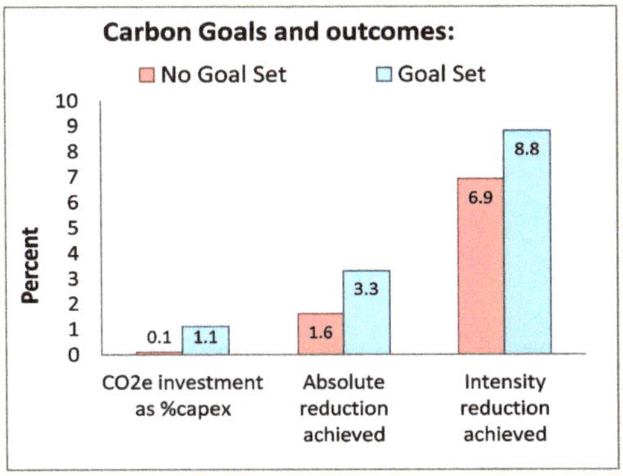

Real World: *Goals do influence results*

There is evidence that setting a programme goal, at least in terms of emissions reductions, leads to better results, as shown in the analysis by the CDP and Bloomberg, left. This link is further supported by recent research among building owners.[397]

We can see those organizations with a goal invest 10x more CAPEX than those without – although the figure, at 1.1%, is a small proportion of total CAPEX spend. Thus, it seems that setting an explicit goal is a key to unlocking CAPEX for emissions-reduction investments. The existence of a goal also doubled the absolute emissions reduction achieved to 3.3% compared to organizations without a target, while intensity reductions were 27% higher in the group of companies with a goal.

6.10 **Carbon Goals and Outcomes**
Source: CDP and Bloomberg,[127] adapted by Niall Enright, available in the companion file pack

targets are more often expressed relative to some measure of activity. Goals and targets may also emerge from a formal strategy development process as described later.

Another important consideration about goal-setting is the degree to which a goal will stretch an organization. It is helpful to have a goal which people feel is achievable, and which provides positive feedback when accomplished but if the objective is too easy it could mean that the focus may be on changes around the periphery of resource use, without addressing more fundamental opportunities. On the other hand, too hard an objective means that the value case for resource efficiency may be compromised and the basis for the programme questioned (particularly if the organization encounters financial difficulties). Challenging objectives may lead to greater resistance and so require a stronger Mandate to succeed.

A further factor when setting a goal is the overall profile that is being given to the resource efficiency efforts. In organizations where there are already many changes taking place – where there is *initiative overload* - it makes sense not to set the programme up as an *initiative* at all but rather to position it as simply *business as usual*. Another circumstance where we want our programme to *fly below the radar* would be where we don't have a Mandate from the senior team but feel compelled nevertheless to act on resource use. In these situations, it is common to set objectives that are not overly challenging or even not to set goals at all but perhaps to put in place criteria for investment (such as a payback threshold) and simply measure the outcome as the programme gains momentum. In circumstances where a business is struggling financially or where environmental issues are not prioritized, it might make much more sense to set a goal based on cost savings (rather than a reduction in resource use) – it translates into the same actions. This is what we would call an indirect goal.

Effective goals:
1. Link to the core purpose
2. Are stretching
3. Have the right visibility
4. Are owned
5. Are simple and relevant

It is also important that the goal is owned by the people that will deliver the programme, that is the Leader or Sponsor and the Governance team who will be tasked with achieving the goal. If these folks haven't been involved in the goal-setting process and don't have ownership, then it is going to be tough to get them motivated. So an obvious starting point in goal-setting is to either get the Leader to set the objective or to ask the Governance team to make recommendations on the goal for approval by the Leader. The reader is advised to look at the next section on the Governance team roles when considering how to set a resource efficiency programme goal.

Another fundamental principle which applies to just about any organizational initiative is to keep the goal simple and relevant. The goal needs to be communicated widely and underpin a call to action. Objectives that are easy to understand include a simple percentage reduction in resource use or emissions; a reduction per (global and simple) unit of activity; or an absolute cost reduction. More complex goals would include relative performance in indices or benchmarks (e.g. being in the top three in the Dow Jones Sustainability Index – an AkzoNobel goal) where it's hard to know exactly what actions are needed to achieve the target score unless one is an expert in the Index. It is also helpful if the goal is something to which the individuals who have to deliver the improvement can relate – so a goal may be translated into a more meaningful figure such as forests preserved, homes electrified, jobs created. This context is particularly relevant to lower-level objective-setting. We should recognize that there is a tension between the creating a simple, but possibly misleading, communication and the reality that resource efficiency is usually more complex and diverse in reality.

Working against ownership and simplicity is the proliferation of goals within organizations. There are many reasons - not least regulatory, customer and investor pressure - which cause organizations to have to set a large number of interlocking objectives. A case in point is the later example of Ford (page 362) where there are at least eight areas of high concern related to resource efficiency, encompassing energy, water and raw materials. In these circumstances, where the organization is advancing on a broad front, there will be challenges relating to focus, project interactions (e.g. on-site water purification may reduce water use but increase energy demand), accountability and performance measurement.

6.11 **A helpful mnemonic for goal-setting.**
Source: © nasakid12, Fotolia.com

On the other hand, a broad and multi-faceted approach may reflect the true complexity of an organizations and the risks and opportunities posed by energy and resource needs. It may also encourage more "systems" thinking to bring about holistic changes that impact on many resources. Multiple objectives are perfectly achievable as long as those responsible for them have thought through the implications and heed some of the suggestions in these pages.

It is useful to remind ourselves of the mnemonic SMART, which tells us how to set objectives. It stands for Simple, Measurable, Achievable, Realistic (to those who are being set the target) and Timely (i.e. it is clear over which time frame we are expected to achieve the objective). Some people add ER at the end, to remind themselves that a goal should also be Ethical and Reasonable.

⇒page 242.

It is perfectly acceptable, and often necessary, to have different **targets** at an operating level.

Real World: *Spanning the intent gap – enManage™ at BP*

One of the largest corporate energy efficiency programmes I have worked on was with BP, from around 2003 to 2004. This programme was seeking to drive carbon reductions in the refining and petrochemical units in Europe and North America. My role was project director for the consultancy Enviros (now part of Jacobs) and we provided a large consulting team and some data management tools to help the internal BP team spot and drive energy savings. I helped to develop the programme with Kevin Ball, the Director of Energy Efficiency at BP, and his colleagues, Mark Siddle, Brian Turner and Tim Sullivan, who led BP's internal consulting team, as well as Chris Stubbs at Enviros.

To ensure we had an effective programme, we adapted a methodology called *enManage*™. This methodology is a structured approach to energy management, based on Monitoring and Targeting, which provided an early version of the Framework set out in this book.

At the time, BP had already set itself apart from other oil majors regarding its reaction to climate change. Back in 1997 John Browne (later Lord Browne) made a speech at Stanford University where he declared: *"The time to consider the policy dimensions of climate change is not when the link between greenhouse gases and climate is conclusively proven but when the possibility cannot be discounted. We in BP have reached that point."*[91] It is fair to say that this speech created somewhat of a shockwave within BP and the wider community of oil and gas businesses.

The speech was followed up with swift action. A goal of a 10% reduction in GHG emissions by 2010 per unit production compared to 1990 levels was established. An innovative carbon trading scheme was piloted in 1999 and then rolled out globally in 2000 and 2001. All business unit leaders (BULs – the people who run the BP sites) had an element of their variable pay (bonus) linked to performance in emissions reductions. And the very strong, clear and unequivocal messages from the very top of the organization were frequently repeated so that no one was in doubt that emissions reductions was something that the organization was going to achieve. It was a textbook example of good leadership.

Unfortunately, these actions alone failed to achieve the objectives intended. After an initial decline from 1999 to 2001, emissions rose in 2002 and 2003, particularly in the big complex refinery and petrochemicals businesses.[79] At the time, the BP management culture was regarded as among the most effective in the industry, and John Browne appeared on Management Today's list of most admired CEOs for five years in a row. The superior management culture – generally described as *"management by objectives"* – was seen to give the organization improved prospects, one of the reasons why BP's share price was trading at a premium to the other oil majors. This programme was also not long after BP launched itself into the global super-league by audaciously acquiring Amoco in 1998. However, as with all such acquisitions, it was taking time to instil the BP culture and way of working.

So we wind forward to 2003 when the executive team first learnt of the increases in emissions and were puzzled about why the objectives that had been set at the BUL levels were not being delivered. Their response was to provide a much more focused *"hit team"* to go and work with the downstream petrochemicals and refining businesses to establish why the emissions had started to rise again and provide some external technical assistance to help the local engineers kick their programme off. This was the team that Chris Stubbs, Kevin Ball, Tim Sullivan and I

enManage™ is a trademark of Jacobs

6.12 **BP's Coryton Refinery on the Thames estuary.** This was the pilot site for the enManage™ programme, and proved that a systematic, well structured, energy efficiency programme based on monitoring and operational focus could deliver savings even in very complex facilities. In the pilot programme 40,000 tonnes of CO_2 were saved at no cost. *Source: BP Sustainability Report 2003,[78] p25*

helped to deploy, and we rapidly engaged with some of BP's largest sites including Texas City, Cherry Point, Decatur, Chocolate Bayou and Whiting in the US, as well as Coryton (where the approach had been piloted in 2002), Lavera and Köln facilities in Europe.

What we found when we talked to BULs was that they all felt that the GHG targets were reasonable and essential, but that they had other more important priorities. For example for a refinery manager, availability in a refinery is a key figure and small percentage variance in output for businesses with a low operating margin and high fixed costs can make the difference between profit or loss. According to publicly available data at the Chemicals Safety Board in the US,[439] the Texas City refinery was running at 95% availability in 2002 compared to a target of around 97% - a number the management team were very focused on improving because they knew that their careers depended on this.

When the *enManage™* teams arrived at the BP sites around 2003, it became clear that energy efficiency was perceived to undermine availability. This is because in the refineries (and also petrochemicals sites) the operators, fearing the slightest interruption in production, would run secondary systems such as pumps and compressors permanently in standby mode just in case the primary system failed. They were worried that we would target these *"obvious"* wastes leaving them more vulnerable to unplanned shut-downs. The was also a belief that a focus on energy would distract management efforts from keeping the somewhat under-invested equipment functioning. Budgets at the time were being squeezed hard and managers at all levels were concerned about resources for their core activities, let alone finding people for new initiatives. No wonder then that emissions targets were missed!

As Chris Mottershead, a Senior Adviser with BP, put it in Esty and Wilson's book *Green to Gold*[268] (page 243):

> *"The stumble was that we told refineries to chase volume and to produce cleaner fuels and lower greenhouse gases. Those three goals are not independent – you drive one and is has real consequences for the others."*

These conflicting goals are what Esty and Wilson called the *"middle management squeeze"* and were one of 13 reasons they give for the failure of environmental initiatives in general. What Mottershead and the BULs had incorrectly assumed was that the BP goal was for absolute emissions reductions - a real challenge in a rising production environment. In fact, John Browne had set a goal of 10% fewer emissions *per unit production* compared to 1990, which was completely doable (and BP were later to claim that they had achieved this goal in 2001, nine years early,[91] although that was not clear at the time).

The solution for the *enManage™* team was to stop talking about emissions altogether and start talking to the site teams in terms that mattered to them. First of all, we established the notion that tracking the energy consumption of their major plant would increase availability because tracking energy use is a great form of condition monitoring since a decline in energy efficiency can signal imminent equipment failure. Thus, the programme would increase reliability. Also, because some of the systems were capacity constrained we could potentially squeeze more from them through efficiency and so have a positive effect on output.

We also made it clear that we wouldn't tamper with the standby systems unless we could clearly demonstrate that we would not affect reliability. Secondly, we focused the programme on increasing the energy efficiency per unit production, particularly by delivering immediate low-hanging US$ savings, thus releasing funds to support other core activities. Lastly, we made it clear that the extra effort to deliver savings would be modest (see *The 15-minute pitch* earlier on page 195) and the *enManage™* team itself was offering an additional dedicated energy manager resource to the site to pull together the projects as well as external support for setting up systems and identifying opportunities. All of the efficiency activities had the happy effect of helping the BULs achieve their corporate emissions targets (and hence bonus) but this was not the main reason articulated for the programme. See also *The power of pairwise comparison (page 167).*

As far as the BULs were concerned, their people had cost-reduction targets related to their energy use, which brought with them availability and reliability benefits arising from a better understanding of the equipment's performance.

This reinterpretation of the programme goals was about making sure that it aligned with the priorities at the site level. The repackaging of the programme was successful and at most sites the *enManage™* programme was given a warm welcome. This was in part because, as results started to come in, it became easier to *"sell"* the next site the concept since the BULs clearly spoke to one another and would learn that the process did not undermine their principal targets. Overall, the programme contributed to a fall in absolute emissions in 2004 and 2005, and lowered energy costs by US$33m for an investment of US$8.8m, as shown below:

6.13 **Savings achieved over the first three years of the BP enManage™ programme in European and North American refineries**
Source: Niall Enright.

$=US$	Investment	New savings each year	Continuing savings	Total saved
Year 1	$1.3m	$3m		$3m
Year 2	$3.5m	$7m	$3m	$10m
Year 3	$4.0m	$10m	$10m	$20m
Total	**$8.8m**			**$33m**

Lack of middle-senior management buy-in to change is a frequent obstacle where energy and resource efficiency are concerned. This lack of support arises, in my experience, not because folks in these roles are naturally obstructive, but because they have conflicting objectives. Ensuring that communications, incentives, performance indicators and targets align with our programme is essential. It is not enough to assume that this alignment will happen when the chief executive articulates their support for the programme - we must test this.

6.6 Opportunities and timescales

Opportunities for improvement have a range of lead times to implement. Understanding these ensures goals are achievable.

So far we have discussed the "what" of our goal, but equally important is the "when". In this section, we shall see the value of incorporating longer-term change into our programme right from the beginning, because different opportunities for improvement require different lead times.

If I want to turn down the thermostat to save gas in my office, then I can do that immediately. On the other hand, if I want to replace my office boiler for a more efficient model it may take up to a year or more to get approval for the investment, find the right model, procure the replacement, install and commission it. Finally, if I want to undertake an extensive retrofit to my office, it could take many years until a vacant or void period arises, and the existing fabric and equipment becomes sufficiently dilapidated to justify the change.

Economists refer to the idea of a capital replacement cycle for major assets such as infrastructure, water, waste and transportation systems. The notion is that different categories of assets have different useful lives and that when this date is reached organizations will replace these assets. My experience – confirmed by research conducted by the Pew Centre[470] among others – is that asset replacement decisions are not made in this clear-cut way. It seems that many capital assets will continue to operate well beyond their design life and organizations will avoid making any capital upgrades until they absolutely have to. As a consequence, some assets have remarkably long lives. For example, most US power plants are at least 20 years old and over a third are older than 50 years. This longevity of assets clearly has real implications for organization's ability to make fundamental changes in energy and resource efficiency.

There are other practical timing constraints on making major equipment and process changes in facilities. For example, many chemical plants and refineries are operated continuously – with very expensive and complicated shut-down and start-up sequences needed to make significant changes. As a result, these turnarounds are often scheduled years ahead and it is essential that the resource efficiency plan understands when these are and can take advantage of them to seize the window of opportunity for improvement.

Because of the costs associated with lost production during a turnaround, there is usually limited time available and the programme of work can be hotly contested The resource efficiency elements may need to be discussed months

Real World: ***Chemical phases***

This is how a well-known global chemicals company sees its resource efficiency opportunities.

It categorizes opportunities as falling into three types: operational improvements, medium-term process redesign and very long-term fundamental changes in technology. It recognizes that meeting their full potential means addressing opportunities over different timescales.

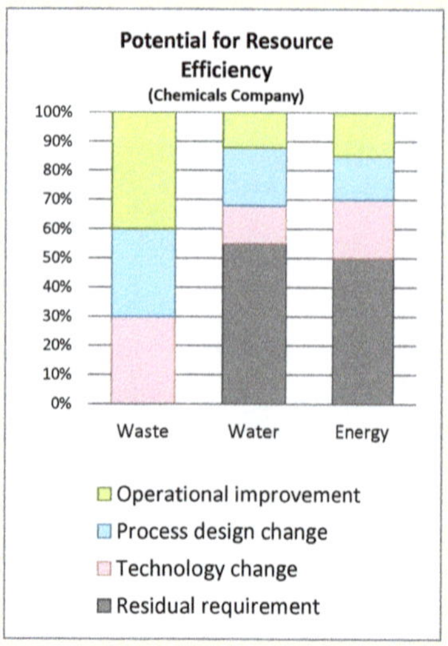

6.14 **The internal assessment of the potential for organization-wide reduction in waste, water and energy demand in a global speciality chemicals business**
Source: taken from real client data modified to preserve confidentiality, Niall Enright. Available in the companion file pack.

There are broadly three types of opportunities with different time frames: **Optimize**, **Modify** and **Transform**.

It is highly desirable to work on all three right from the start of the programme.

or years before the turnaround takes place if they are to have any hope of being incorporated into the plans.

In some regulated industries, there are other windows of opportunity. In the UK, for example, the water industry operates five-year asset management plans (AMPs) in which they agree on capital and operations spending plans with their regulators. Getting resource efficiency investments (such as CHP/cogeneration power plants) into these plans is essential if the change is to be made within the capital allocation cycle.

Thinking of energy and resource efficiency not just in terms of technology or assets, but in terms of systems, we should be aware that some forms of change can occur rapidly and others take longer to bring about. For example getting a simple behaviour change for folks to separate their waste paper into different bins for recycling should be relatively quick and easy if the right incentives are in place. At a systems level, it will take longer to change the procurement process to support waste reduction – for example, by introducing supplier take-back on packaging. Finally, more fundamental changes to the business model such as a move towards a paperless organization could take significantly longer and require many process and behaviour changes. Culture change activities also fall into this longer-term category.

Thus, almost every organization is bound to have some short, medium and long-term opportunities for improvement, depending on the various time constraints mentioned above: the remaining operating life of the assets that they have, the decision-making processes that exist or the degree of systems change that they want to bring about.

The first step in a resource efficiency programme is usually to improve the operation of existing equipment, systems and processes without making massive changes. The second phase involves some CAPEX expenditure or systems changes – modifying the equipment or technology supporting the current business model to make it more efficient. Finally, more radical and strategic changes done over a longer time frame which may enable more substantial improvements to be delivered by changing the business model fundamentally - such as a transition to a circular economy model.

In order to avoid the impression of a fixed time frame, and to encourage all three modes of improvement to start simultaneously, I will refer to these as Optimize, Modify and Transform opportunities.

In the Optimize category, the opportunities are about making the *existing* systems and technologies work better. Here, the emphasis is on delivering rapid value by putting in place the governance processes, metrics and tools to start to drive down resource use. High levels of discovery characterize this programme of work; that is to say that considerable efforts may be invested in mapping out resource flows, establishing internal and external drivers, establishing measurement and incentive systems, capturing and evaluating opportunities and generally working the resource efficiency pyramid hard. The Optimize opportunities often involve delivering a lot of *low-hanging fruit*, in other words, changes to operations and behaviours which don't require significant expenditures, so the financial returns from these are usually very attractive.

However, while we do want to demonstrate early value, it is crucial that our programme does not focus exclusively on immediate savings at this stage, but puts in place the continual improvement systems necessary to maintain the savings in the future and to begin to address longer-term and higher-cost improvement measures.

If organizations take the view that the goal of the resource efficiency programme is simply to optimize resource use in existing operations, managers will insist on a project-by-project justification with very rapid paybacks, leaving no room for the implementation of overarching continual improvement systems, or the investigation of more significant changes to existing business models.

In some organizations, I have seen the programme managers themselves get so carried away with early success and praise they receive from it that they become totally preoccupied with delivering the short-term savings as rapidly as possible and so don't put in the effort to lay the foundations for longer-term change. Weak incentives and performance indicators can also be to blame.

In either case, there is a grave risk that at some point the low-hanging fruit will be exhausted and the organization will assume that there is little more to be done within the existing operations, and so premature victory will be declared and the programme will come to an end. Because no groundwork has been done on the Modify and Transform opportunities the effort now needed, as well as the time delay for the benefits to be delivered, makes building a case to realize these opportunities tough, if not impossible.

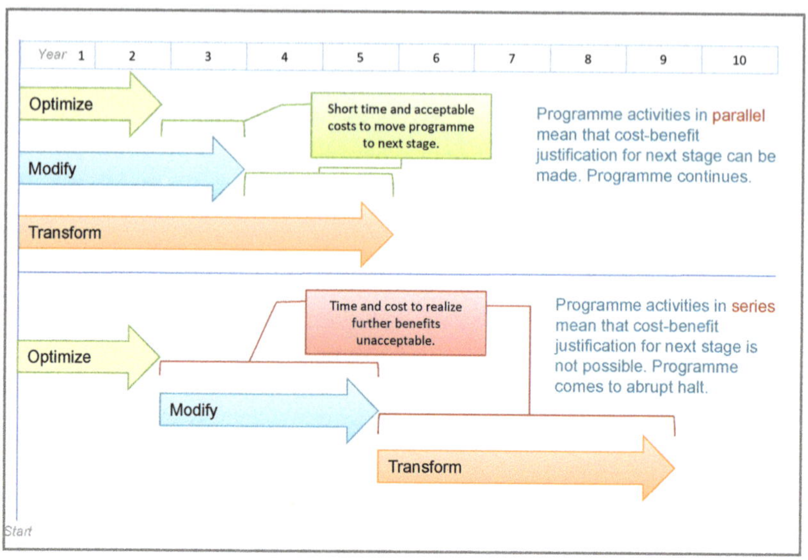

6.15 **Failure to drive medium and longer-term opportunities from the outset is one of the most common reasons energy and resource efficiency programmes are not sustained in the long run. The key is to take a parallel rather than a serial approach.**
Source: Niall Enright.
This illustration is available in the companion file pack, including an A3 poster version.

It is essential that ***medium and long-term improvement actions are initiated sufficiently early*** *in the programme to deliver value when the low-cost savings are exhausted.*

Unfortunately, this outcome is surprisingly common, especially in programmes where the primary emphasis has been on cost reduction rather than on wider aspects of competitiveness, brand, risk or stakeholder engagement. In declaring victory, these types of *flash in the pan* programmes send an unwitting signal internally that resource efficiency is now complete and so accelerate the return to the *bad old ways* as behaviours revert and set off a cycle of value destruction.

It is thus desirable, when setting programme objectives, and in the execution of the Optimize opportunities, to initiate the subsequent phases of the resource efficiency programme. These simultaneous efforts ensure that the investments in continual improvement systems are realized and the process of more radical change begins immediately.

This is called the parallel approach to resource efficiency in contrast to the serial process where each type of saving is considered one by one, as illustrated opposite. It is important to note that while savings may initially flow more slowly from a parallel process than from a serial one, the superior financial case for this approach is clear. Quite simply, programmes with continual improvement and parallel approaches, provide better returns than those without, as illustrated in the panel *"The Hare and the Tortoise"* on page 248.

The Modify phase of energy and resource efficiency involves upgrading equipment to the latest, more efficient models, or changing existing systems by introducing new ways of working, new standards, additional controls and so forth. This type of investment does not aim to change the process fundamentally, but rather bring it to its optimum potential using best available technology not entailing excessive costs (BATNEEC).

Transform opportunities are where a more significant change to the existing processes or business models are proposed. Examples would be Interface Carpet's move to leasing carpet rather than selling it, so that it can introduce a closed-loop recycling process by taking back its product and recycling it at the end of its useful life. Changes arising from new disruptive technologies would also fall into this category, such as the introduction of all-electric vehicles into public transportation fleets or the decision by Amazon to sell eBooks which eliminated demand for paper, and systems changes, such a move by some electricity companies away from the conventional pricing model that rewards high usage with lower costs, towards a rising-tier pricing model which charges more per unit the more electricity the customer uses. All of these are examples of Transform opportunities, which may take some time to realize but which bring about large change and deliver considerable value to the organization. In some cases, Transform opportunities depend on a fundamental technological breakthrough which may be outside the control of an organization. In other cases, the only barrier to Transform opportunities is vision and lateral thinking

The key is to understand that activities of all types can (should!) take place at the same time within an energy or resource efficiency programme even though the border between one type and the other are blurry.

6.16 **Companies' production waste intensity improvements (i.e. per unit of output)**

This report has similar rate-duration curves for energy, transport and greenhouse gases. *Source: Inspired by "Food and Beverage Sector Non-Labour Resource Efficiency", Lavery, Pennell and Evans,[593] adapted by Niall Enright. There is a spreadsheet model for plotting "isosavings" charts available in the companion file pack.*

Real World: *Ambition, time and effort*

It is evident that the cumulative improvement that an organization achieves is the average rate of saving multiplied by the number of years of its programme has been operating.

A report by resource efficiency consultants Lavery/Pennell, the 2Degrees network and the Institute for Manufacturing at Cambridge plotted waste improvements from a number of companies on a rate-duration chart.

In the chart below, Xerox and Interface have achieved roughly the same cumulative improvement, only Xerox has done so rapidly with a compound annual rate of reduction of 8.3% over four years while Interface has achieved 2% a year improvement over 18 years. This places both organizations between the 25%, yellow, and the 50%, pale green, *isosavings* lines, in the chart below.

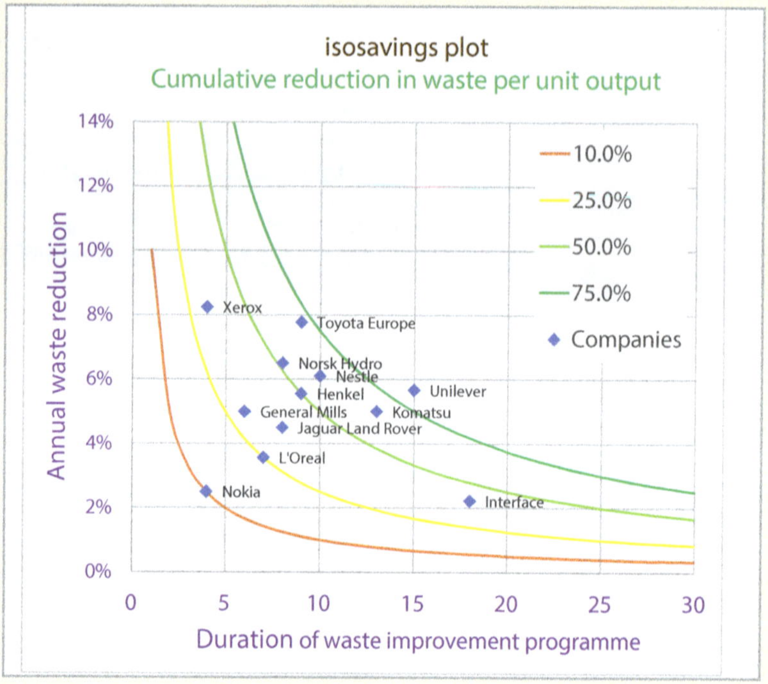

A study by Accenture, Marks & Spencer and Business in the Community, *Fortune Favours the Brave,*[6] argues that the days of incremental thinking and low aspiration are over. This report set out five areas of innovation that will drive huge material and productivity gains: *"Shared values"*, *"More with less"*, *"Circular economy"*, *"New consumption models"* and *"Transparency and customer engagement"*. The potential value for the UK economy is over £100 billion per year, much of which comes from resource efficiency and material productivity gains.

It is important to note that these benefits include not only cost savings but also the new business opportunities that a low-carbon economy offers. In setting resource efficiency goals, we have seen companies like AkzoNobel express aspirations regarding additional sales of eco-premium products.

6.17 **Most resource efficiency programmes can be thought of as falling into one of three categories**
Note that while the maximum timescales may vary, opportunities of every category can sometimes be implemented immediately.
Source: Niall Enright

Category	Technology View	Systems View	Timescale
Optimize	Operational and small CAPEX improvements available immediately within existing processes.	Develop an understanding of resource use through discovery, measurement, incentives which change attitudes and behaviours.	1-5 years
Modify	Changes in existing processes requiring greater funding commitment or technical change. Often an "upgrade" of existing equipment to the latest model.	Improvement requiring some modification in existing processes or decision-making systems.	1-10 years
Transform	New large-scale process using best available technology to create a step-change improvement. May involve abandoning old technology and writing off assets.	Improvements requiring a fundamental rethink of a business model – e.g. move from product to service model, or relinquishment of some service or product lines altogether.	1-25 years

One other important distinction between the different categories of projects is that they often have different approval processes. Operational savings can often be agreed at a low level if they meet certain payback/financial criteria or if they result in net in-year savings Design changes in existing facilities may need a strong project/technical input within the CAPEX budgeting cycle. Major new plant, systems or technology decisions, on the other hand, may well involve very significant scenario planning, risk assessment and approval at the main board. Thus, there will often be a different group of people who need to be involved and engaged for the three categories of improvements.

An ambitious goal with a 5-10 year time frame can challenge conventional thinking, but it should be accompanied by annual objectives which cannot be deferred.

Recognizing that a programme should incorporate all opportunities for improvement from the outset does not mean that we should necessarily set goals for multiple time frames. Too many goals can be confusing, so it tends to be advisable to have one clear overarching goal but to put in place other measures (such as an Opportunities Database, discussed in the next chapter) to provide some quantitative confirmation that Modify and Transform opportunities are being developed.

Immediate goals tend to focus efforts on optimization, rather than contemplating more profound changes in resource use, or the larger capital investments that an organization may make. On the other hand, too distant an objective can lead to complacency and postponement of effort.

In my experience, setting a more ambitious goal with a time horizon of 5-10 years seems to work well in terms of stimulating a more profound examination of the organization's resource requirements, so long as this goal is broken down into near-term objectives which cannot be deferred. For example, an organization may set a goal to reduce waste by half in 10 years, which is translated as a compound annual 6.7% reduction (see page 594 for how this is calculated).

⇒ page 251.

Continual **improvement**, *far from being an unnecessary cost, substantially enhances the* **value** *of the resource efficiency programme.*

Exploration: *Hare Corp. and the Tortoise plc*

One of the most common errors that folks make in the delivery of energy and resource efficiency programmes is to focus exclusively on maximizing the short-term returns from the programme. The consequence of this can be disastrous.

We can illustrate this with a simple scenario comparing two different approaches to resource efficiency. Let us start by examining our first company, Hare Corp., which has undertaken some initial discovery activities and established that there are some excellent opportunities to reduce operating costs through resource efficiency. Many of these £20 million savings will come from no/low-cost behavioural and maintenance changes, so the initial payback is less than six months on an investment of £10 million, creating a net saving of £10 million in year one and winning praise for all involved. In year two, the new savings are somewhat lower, £15 million for an investment of £15 million, giving a nevertheless attractive payback of one year. However, as time passes, the paybacks get longer and longer, fewer projects get funding and the programme profile starts to decline, so that by year 10 our programme is now much reduced, adding only a net £0.5 million of savings. To all intents and purposes, the programme has lost momentum. At Hare Corp., the programme has been driven to accelerate financial returns. Every project has been expected to meet a minimum payback so there has been little investment in sub-metering to monitor behavioural savings, and so only 60% of savings in any given year are maintained in the following year (if anything an optimistic figure in these circumstances). There has also been no investment in additional audits and studies to identify further projects, as this is seen to be an overhead that distracts effort from implementing the existing opportunities.

The contrast at Tortoise plc could not have been more different. Here, the programme managers have decided to put in place the systems to underpin continual improvement from the outset. In particular, they have argued that the first-year savings should all be reinvested into the resource efficiency programme. As a result, £10 million is available to spend on appropriate sub-metering to provide feedback to users and teams on performance, which means that 80% of savings identified in any given year are carried forward to the next year, rather than Hare Corp's 60%. Money was also spent on an Opportunities Database to capture project opportunities systematically and for external assistance to help evaluate these projects. The result of this was that each year Tortoise plc had 25% more projects to invest in than Hare Corp at the same payback rate. Of course, all this additional effort around continual improvement has a cost, so Hare Corp has added 20% of the direct project investment each year from year two onwards into a "systems investment" pot to maintain the continual improvement systems.

The table opposite shows the 10-year returns for each organization. Despite the extra cost entailed by the investment in continual improvement systems (£34.5 million), Tortoise plc has done remarkably well compared to Hare Corp. We can see that at year 10, when Hare's programme is gasping its last breath, Tortoise plc is delivering net annual savings of £18.7 million. Tortoise's cumulative savings to this point are more than double Hare's and the overall payback is more attractive. On every measure, Tortoise is delivering more value than Hare.

Continual improvement, far from being an unnecessary cost, increases the value of the resource efficiency programme through three mechanisms.

Hare

£m YEAR	New Savings	60% Carried over	Total Saved	Investment Direct	Systems Investment	Total Spent	Net Return	Cumulative Saving	Cumulative Cost	New Savings Payback (y)	Cum. Programme Payback (y)
1	20.00	-	**20.00**	10.00	-	**10.00**	**10.00**	**10.00**	**10.00**	0.50	1.00
2	15.00	12.00	**27.00**	15.00	-	**15.00**	**12.00**	**22.00**	**25.00**	1.00	1.14
3	10.00	16.20	**26.20**	15.00	-	**15.00**	**11.20**	**33.20**	**40.00**	1.50	1.20
4	8.00	15.72	**23.72**	16.00	-	**16.00**	**7.72**	**40.92**	**56.00**	2.00	1.37
5	5.00	14.23	**19.23**	15.00	-	**15.00**	**4.23**	**45.15**	**71.00**	3.00	1.57
6	3.00	11.54	**14.54**	12.00	-	**12.00**	**2.54**	**47.69**	**83.00**	4.00	1.74
7	2.00	8.72	**10.72**	9.00	-	**9.00**	**1.72**	**49.41**	**92.00**	4.50	1.86
8	1.50	6.43	**7.93**	7.50	-	**7.50**	**0.43**	**49.85**	**99.50**	5.00	2.00
9	1.00	4.76	**5.76**	5.50	-	**5.50**	**0.26**	**50.11**	**105.00**	5.50	2.10
10	0.50	3.46	**3.96**	3.00	-	**3.00**	**0.96**	**51.07**	**108.00**	6.00	2.11
			159.07			**108.00**	**51.07**				0.68

Tortoise

£m YEAR	New Savings	80% Carried over	Total Saved	Investment Direct	Systems Investment	Total Spent	Net Return	Cumulative Saving	Cumulative Cost	New Savings Payback (y)	Cum. Programme Payback (y)
1	20.00	-	**20.00**	10.00	10.00	**20.00**	**-**	**-**	**20.00**	0.50	
2	18.75	16.00	**34.75**	18.75	3.75	**22.50**	**12.25**	**12.25**	**42.50**	1.00	3.47
3	12.50	27.80	**40.30**	18.75	3.75	**22.50**	**17.80**	**30.05**	**65.00**	1.50	2.16
4	10.00	32.24	**42.24**	20.00	4.00	**24.00**	**18.24**	**48.29**	**89.00**	2.00	1.84
5	6.25	33.79	**40.04**	18.75	3.75	**22.50**	**17.54**	**65.83**	**111.50**	3.00	1.69
6	3.75	32.03	**35.78**	15.00	3.00	**18.00**	**17.78**	**83.62**	**129.50**	4.00	1.55
7	2.50	28.63	**31.13**	11.25	2.25	**13.50**	**17.63**	**101.24**	**143.00**	4.50	1.41
8	1.88	24.90	**26.78**	9.38	1.88	**11.25**	**15.53**	**116.77**	**154.25**	5.00	1.32
9	1.25	21.42	**22.67**	6.88	1.38	**8.25**	**14.42**	**131.19**	**162.50**	5.50	1.24
10	0.63	18.14	**18.76**	3.75	0.75	**4.50**	**14.26**	**145.45**	**167.00**	6.00	1.15
			312.45		34.50	**167.00**	**145.45**				0.53

Scenario: Tortoise carries over 80% of savings year on year compared to Hare's 60%
Tortoise identifies 25% more new savings each year compared to Hare
Tortoise doubles expenditure in year one to contribute to systems investment, thereafter spends 25% more per annum

6.18 **The programme results for Hare Corp. and Tortoise plc**
Source: Niall Enright. Spreadsheet available in the companion file pack.

- Opportunity availability: An improvement programme ultimately depends on the availability of feasible interventions. Ideas for projects do not arise spontaneously. If there are no processes to replenish and evaluate these ideas, then the programme will run out of steam. A best practice resource efficiency programme continually engages with resource users and experts to capture their ideas and to systematically evaluate them, usually using a tool called an Opportunities Database. In the scenario above, Tortoise plc has invested time and effort to encourage new ideas and to investigate these, and so has 25% more project opportunities available in each year than Hare Corp., which leads to a significant cumulative increase in investments and hence savings.
- Opportunity return: The next source of additional value from continual improvement approaches to resource efficiency lies in an increase in no/low-cost opportunity identification. These are opportunities that arise from behaviour or control changes and are often only identified when resource users become engaged in eliminating sources of operational variability in resource use through continual improvement techniques like monitoring and targeting. The result is typically a large number of relatively small projects whose aggregate savings are highly attractive and which decrease the overall payback of the programme.

*If there are **more, cheaper** savings which will be maintained for **longer**, the net return will improve.*

- Savings conservation. Very few resource efficiency interventions (particularly the lower cost ones) are irreversible. Stepping down a thermostat in a building by 1° C will typically deliver 8% energy savings at no cost. However, it is entirely possible for people to reverse this change and raise the temperature setting, and so increase the energy use. Continual improvement processes put measurement and reporting systems in place to ensure that once an intervention is made, the change persists. In the scenario above Hare Corp. only sustained 60% of savings from one year to the next, while Tortoise plc was able to maintain 80% of savings year on year. This small difference cumulated to a big variance for the programme.

It is important to note that the three effects above, opportunity availability, opportunity return and savings conservation, reinforce each other. That is stating the obvious fact that if there are more, cheaper savings which will be maintained for longer, then the net return will improve. Small improvements in these three metrics can compound together to produce a large improvement in the overall programme performance and will usually offset the additional costs associated with implementing the continual improvement systems many times over, as shown by the results of Tortoise plc.

In addition to the benefits of substantial additional cost savings, incorporating continual improvement methods into a resource efficiency programme prepares the organization to move beyond the optimization of existing processes. It enables the organization to start to contemplate how to transition to new business models which will dramatically transform the organization: in other words, to realize Modify and Transform opportunities.

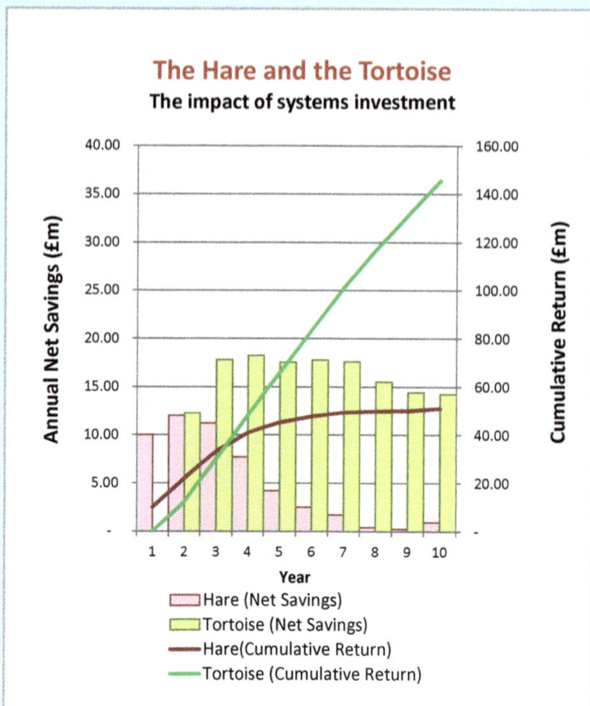

6.19 **The end results for Hare Corp. and Tortoise plc could not be more different**
Source: Niall Enright, spreadsheet available in the companion file pack.

6.7 Governance team

Experience shows that the composition of the team leading the resource efficiency programme is very important. Here, we discuss these roles and responsibilities.

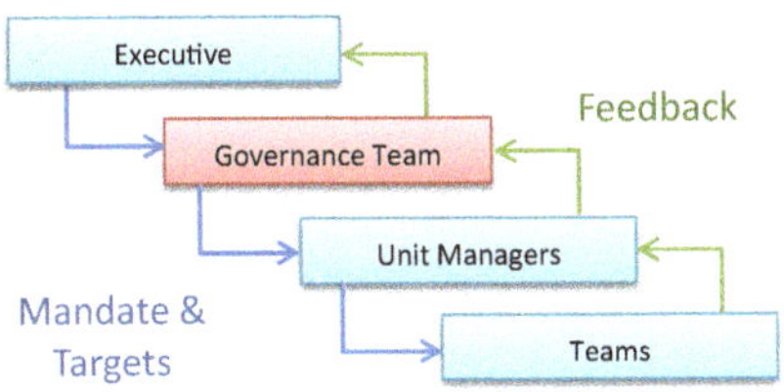

6.20 **The Governance team should ideally operate within the existing management structures in the organization**
Source: Niall Enright

The most common way that organizations manage resource efficiency programmes is to bring together a team to take on the job of delivering the objectives. Because there are so many different parts of the organization that influence resource use, it is common for this team to draw on a broad range of expertise. In fact, the creation of a team offers a golden opportunity to overcome a major barrier to resource efficiency which is compartmentalized or "silo" thinking. For our purposes, we will call this the Governance team, but many different names are given to this team, such as steering group, programme committee or board, supervisory panel, etc. etc. etc. While the name is not particularly important – although names do have meanings that are organization-specific – the composition of the Governance team is critical to the success of the programme.

I am often asked if the Governance team should not simply be the executive management team of the organization. The executive clearly has the authority and cross-functional representation which are essential to success, so their participation in the programme is to be applauded, particularly given the earlier observation that management commitment is the #1 requirement for success. In smaller organizations I have worked with, typically with one facility and a flat management structure, I have found that an executive management team can fulfil the role of the Governance team. However, in larger and more complex organizations, the executive usually lacks some ingredients to take this task on for themselves, particularly the time to dedicate to the programme and technical expertise in resource issues. In these circumstances, I would tend to try to get the best of both worlds – I would maintain the involvement and support of the executive by reporting regular progress and involving them in key decisions, while also establishing a separate Governance team to drive the day-to-day participation of the business. The essential point is that the programme needs to well-connected into the existing management of the business.

Framework

In the illustration, left, we can see that the Governance team is integrated within the existing management structures of the organization. It reports to the executive of the organization, which is encouraged to have a short agenda item on progress at every meeting. The individual teams, units or divisions which are participating in the programme should drive the process through their existing line management structures and have similar agenda items on progress at their own meetings. In this way, at unit and team level there will

Real World: *A dedicated team?*

One of the questions that I am often posed is whether it is not a more effective use of existing management teams to drive the resource efficiency programme, rather than to create a separate Governance team. We will see later under Momentum that I strongly favour adapting existing organizational processes to drive resource efficiency instead of creating new structures, which can be easily jettisoned if business conditions get tough.

However, there is a strong argument for the formation of a dedicated Governance team at the top level to drive the overall programme. This is because where decision-making is subsumed within an existing executive or management committee a number of disadvantages can arise:

- There may not be enough time dedicated to the programme.
- The composition of the current executive committee may not encompass the right functions or departments who are in a position to make informed decisions about resource efficiency.
- It may be difficult to involve the Champion or advisors in the programme governance if they do not usually sit on the management body.

At a lower level of the organization, such as in a site or business unit, the argument for a separate Governance team is much less clear-cut. Here, I have seen many successful programmes where the local management team have been able to drive the programme within existing meetings. Clearly, a Champion to drive the programme forward is desirable, even if there is no Governance team.

be clarity about the Mandate and sufficient feedback on performance to the management meetings to take corrective action. At this point, the *15 minute commitment*, set out earlier on page 195, can be negotiated with leadership at all levels! Thus, although the role of the Governance team is to shake up the organizational silos, if they exist, and to challenge existing behaviours and systems where they impede resource efficiency, the action and content for the programme are ideally *delivered* in the existing management structures. As we shall see later in the section on Momentum, this integration of resource efficiency into the day-to-day management processes is what ensures that the programme is not just a *flash in the pan*, but is sustained for the long term.

To be clear, the responsibility for delivery of the improvement actions to achieve greater resource efficiency should lie at unit or department level, in other words, within the existing organization structures. The Governance team is responsible for the overall programme design, facilitation and monitoring of the programme acting on the Mandate and reporting to another existing management team, the executive. Clearly, if the programme is only within a division or facility, the same principles would apply, only with the local management structures.

Ideally, the Governance team is chaired by the Leader who initiated the programme, or a very senior member of the management team, acting with the authority of the Leader, whom we will refer to as a Sponsor. We then need to have decision-makers from each of the functional or operational parts of our organization where we need to drive change. We will almost certainly have a programme Champion, who may also have been the initiator of the original proposal for the resource efficiency programme. Another useful role is that of an external adviser who might be a stakeholder representative, a participant from an NGO, a seasoned resource efficiency practitioner from another organization or a consultant.

The intention is that the members Governance team can make decisions, on behalf of the departments represented, about the programme and allocate resources (money and people) to achieve the targets. It should not normally be necessary for decisions taken in the Governance team to be referred elsewhere for approval – the exception may be an annual budget proposal, or where there is a deadlock within the Governance team, or where an action has a big impact and so consultation is appropriate. The participation of cross-functional departments on the Governance team is essential because we want to be in a position where no one can pass the buck: engineering can't blame operations; operations can't blame maintenance; maintenance can't blame procurement; procurement can't blame finance. Everyone in the Governance team is collectively accountable for the delivery of the resource efficiency target and if a particular department or function is not supportive, then the Governance team is in a position to examine why that department is struggling and provide constructive solutions. The multifunctional composition of the Governance team is critical to overcoming the barriers that *siloed* organization structures create and to get everyone working together towards the resource efficiency goal. In the example above, we have looked at involving different functions, such as engineering, but if it is clear that other organizational silos, such as regions, business units or services are a bigger barrier, then the Governance team may have representatives from different divisions and businesses rather than from functions.

The other considerations that will determine who to involve in the Governance team are the strategy for delivering the improvement, the impact on resource a particular department can have, as well as the effectiveness and commitment of the individual representative. For example, if whole life costing is important to our strategy, then we would involve the procurement or finance functions.

Over time there is no reason why the membership of the Governance team can't change, although it will be necessary to ensure that new participants take ownership for the goals set by their predecessors. There is also merit in inviting *"guests"* to participate in the meetings, either because they can provide expertise in a particular area or perhaps because they help to keep the meetings fresh and interesting. These outsiders could fill a role akin to non-executive directors on a board, who can challenge an organization simply because they are more distant from the day-to-day delivery activities.

Decision-making in the Governance team will tend to follow the organizational convention, that is to say, there may be a formal vote or decisions may be taken by consensus. Whatever the process, I would strongly recommend that the Governance team meetings are formally documented with actions, those responsible and due dates recorded. In many teams I have seen the Champion act as the secretary at the meeting, keeping the minutes; however, where possible, I would recommend that this task is undertaken by a dedicated secretary, freeing up the Champion to participate fully in discussions.

6.21 **Typical roles in a Governance team**
Source: Niall Enright.

Role	Description
Senior Sponsor	If at all possible, this will be the Leader who approved the programme. Alternatively, this could be someone who reports to the Leader and is their nominee to chair the Governance team programme. They need to be a senior individual with broad authority over the departments which are participating. I have seen finance directors effective in this role.
Department Representatives	These are senior figures from each of the participating departments who have the authority to commit their departments to actions. It is not uncommon for there to be half a dozen or more departments involved in a programme with the key ones being the departments that use resources (operations), the departments that operate equipment and buildings (engineering, facilities or maintenance), the departments that allocate investment (finance) and the departments that design products or services (e.g. design or marketing). Every department must be represented at every meeting so that if one of the representatives is unavailable they should send a delegate who can participate and take decisions with their full authority. Commitments made by departments need to be reported back to the management team of that department so that the desired action is driven by the existing line management processes. Similarly, the departmental representative will take the intentions, needs and priorities from the departmental management team back to the Governance team. It cannot be overstated how important this two-way flow of actions is – and in many cases it is only achieved where the departmental representative also sits on the departmental management team. In essence, the departmental management team need to take ownership for the programme decisions that relate to them and feel bound to deliver on their part of the process.

Role	Description
Champion	The programme Champion is the individual with the day-to-day responsibility for driving the programme forward. This may be a full-time responsibility or simply another duty depending on the size of the programme. The Champion will be responsible for administering the programme – that is to say for providing the reporting and measurement that the Governance team needs to ensure progress. They will work with departments on individual projects and continually drive forward the message of change and value that underpins the original Mandate. The Champion is a change agent, so their communications skills need to be first class. They need to be diplomatic, persuasive and persistent. They should be enthusiastic and have lots of energy. They should be able to take setbacks in their stride. They should be open-minded and willing to accept new ideas if existing actions are not succeeding. They should be intelligent and able to assimilate the many technical aspects of resource efficiency. They need to be a team player and generous with praise of others. In short, they need to be Superwoman or Superman! Ideally, the Champion has been with the organization for some time so that they know how things work, and have established a certain level of personal credibility. On the other hand, they should not have been with the organization so long that they are afraid to challenge existing practices or to rock the boat. In many large multinational corporations, careers are made in a series of five-year roles and the culture can often be very risk-adverse: *"to survive the current assignment without any goofs"* and so move up the corporate ladder. In this situation, it is difficult to find a Champion who can challenge the status quo effectively and so there may be little choice other than to bring in an external Champion on a contract – although that can create resistance, too. We should note that the Champion role is as an enabler and facilitator. They will rarely be responsible for the actions or changes that deliver the improvements. Their role is to coordinate and support, not to instruct or command. Some organizations which have Lean, Six Sigma or a similar programme will combine the resource efficiency Champion role with the Six Sigma Black Belt role. While this is an excellent idea, so that these processes are integrated and reinforce each other, it is crucial that the differences between these methodologies are clearly understood. For example, Six Sigma is very project-based, while the approach to resource efficiency set out here is based more on a systems approach focusing on continual improvement and behaviour change. Lean, TPM and Six Sigma have very formalized methodologies that can sometimes constrain the options for resource efficiency. In larger organizations, complementing the programme Champion, there may be a similar Champion role at a business unit or facility level. In these cases another important role of the programme Champion is to ensure that these other Champions are properly networked together, learning from shared experience and motivated to fulfil their roles as change agents.
External Adviser	This role should really be called external challenger, but that is not a title that sits easily, so I have used the term adviser. Change almost always means challenging existing ways of thinking and of working. Sometimes this is difficult to do if one has been in an organization for a long time, or one is afraid to *"ask stupid questions"* or to imply that a process that others take for granted needs to change. This is why I often recommend that an external advisor is put in place on the Governance team to fulfil two essential functions. First, to support and mentor the Champion (it is critical that they both understand that this is a supportive role). Second, to act as an honest adviser to the Leader or Sponsor (if necessary to tell them if their programme is effective or not or if the Leader needs to do more). The external adviser needs to be able to *"speak truth to power"*. Ideally, the external adviser will bring lots of experience of resource efficiency from other organizations. They should have gravitas, credibility and track record to lend authority to their contribution. They should not be arrogant or opinionated and should see their role as helping the team as a whole to succeed. In particular, they should help the team to generate new ideas when the programme appears to be reaching a plateau without taking away ownership for those ideas. External consultants can fill this role – I have served as the external adviser on several programmes - or perhaps by an NGO, or by a Champion from another organization or expert from academia or a trade association, or maybe an external member of the organization's HSE committee. The sources of external advisers are plentiful, but it is important to be clear about their role and necessary qualities in advance of appointing them.
Secretary	Sometimes the chore of keeping minutes and recording actions can be fulfilled by the Champion, but if this task impedes their ability to contribute fully to steering group meetings it makes sense to bring in someone to act as the secretary. This person does not necessarily need to have an administrative background – they could be an assistant to the Champion or a member of the corporate HSE or sustainability function for example. The Champion is bound to be very busy, so any help we can provide them with is desirable.

6.8 The site Champion

The site Champion's role is hugely satisfying but there are a number of challenges they normally face.

Almost all large organizations will identify an individual within each targeted business or facility to act as the local coordinator and contact point for the resource efficiency programme. This is the site Champion. Obviously, if this is a very large facility, such as an oil refinery, this local Champion may be dedicated full-time to the programme, but in most cases, it is a part-time commitment.

The profile of the site Champion should be the same as that set out in Table 6.21 – we are looking for an enthusiastic, intelligent, energetic, collaborative individual with excellent communications skills who can drive change. This is easier said than done, of course, because it is unlikely that someone will be hired specifically to do this role and so they will be selected from the existing workforce. Unfortunately, individuals with the attributes I have suggested are already likely to be in high demand so releasing them to Champion energy and resource efficiency may mean that some other important contribution needs to be foregone. However, it is important to hold out for the best candidate as the quality of the site Champion can make a difference between success or failure for the programme at the site level.

Even if they meet the high specification set out here, the local Champion will face some particular challenges, especially if their role is a part-time one. The first challenge will be to create separation from their *day job* – that is to say to be able to carve out enough time to fulfil the duties of Champion and to be able to operate outside their particular departmental agenda. I have frequently seen Champions frustrated by their immediate line managers dragging them back to address crises in their usual role so that the efficiency work is squeezed out.

We are looking for an enthusiastic, intelligent, energetic, collaborative individual with great communications skills who can drive change.

Another problem is the issue of site politics where the line manager may not support particular initiatives by the Champion because they conflict with their own agenda. For example, an engineering team may favour capital investment projects which will deliver shiny new kit, over behavioural change programmes which will undermine the case for investment in equipment. If the Champion's day job is in the engineering team, they may be unable or unwilling to put forward behavioural approaches or may even be biased against these themselves. A further challenge will be to be able to work in relative isolation and to form effective support networks with other local Champions and the overall programme Champion so that shared lessons can be learnt and systems and processes are not being constantly reinvented. It is a key task of the programme Champion to put in place the appropriate network to support these site Champions. The Champion should monitor the time the local team are dedicating to the programme and enlist the appropriate senior manager in the Governance team to intervene if the local management is constraining the effectiveness of the local team.

Another issue for all Champions is the matter of expertise. We shall see shortly that the resource efficiency method usually involves some specialist tools and techniques like Monitoring and Targeting, an Opportunities Database or whole life costing. The resource efficiency programme may emphasize activities that are outside the current experience of the Champion, such as running motivation and awareness campaigns to drive behavioural change. Thus the Champions will almost certainly need initial support and training to use these new tools and implement the unfamiliar processes.

As if this is not enough, the Champions will be under pressure to deliver early successes. In fact, one of the key objectives of a resource efficiency programme is demonstrating value by delivering quick wins which create positive momentum for the programme. Achieving these early successes will require the ability to focus and prioritize effort from the outset and to balance activities with an immediate impact with getting the longer-term foundations right.

Furthermore, Champions have to achieve this change without direct control of the departments or functions that need to implement the improvements. The site Champion has to rely on their powers of persuasion and the authority of the Governance team to drive change.

Without sounding like a sales pitch for consultants like myself, I should observe that many organizations can provide an effective support programme to overcome these challenges for site Champions, by engaging external consultants to make available some additional capacity and deliver the necessary training and mentoring. Sometimes this external consultancy is from a corporate team and at other times it may be from outside the organization.

One justification for the use of consultants is that the level of support needed by the local Champions will initially be high as they require training and the tools and measurement systems are being implemented, but then declines markedly. Another is that consultants may be more willing to communicate some of the difficult messages that may be necessary at the outset of the programme - they may be able to challenge senior managers at the site without the anxiety the site Champion may have about whether this would be career-limiting.

Despite the challenges set out here, the role of site Champion can be hugely rewarding. It can bring an individual into contact with many aspects of their organization. The role can provide an excellent showcase of their skills and exposure to senior managers, and offer very interesting and varied challenges and opportunities for growth. It provides fulfilment that comes from helping to address an important global issue. I have seen many people thrive in this role.

Real World: ***Winding down***

A mistake that organizations sometimes make is to assume that the role of Champion can be filled by someone who may be towards the end of their career. The role can seem like a good transition position as the individual approaches retirement.

That is not to say that every employee at the end of their working life is ineffective or uncommitted - far from it! The extensive network and credibility of a long-serving employee can be very beneficial. But they do need to have *all* the attributes of a Champion as well.

Unfortunately, I have seen the role of programme Champion treated as a solution to an HR problem on too many occasions not to flag up the dangers that this can pose to our resource efficiency programme.

The role is most definitely not suitable for an individual with whom the organization is having difficulties. It should never be treated as a final swan song, a way to ease an employee gently out of their current role. Far too much value is at stake.

© matthias21 - Fotolia.com

6.9 Collective responsibility

The Governance team need to be collectively responsible for achieving the programme goals. As a consequence they must offer an interface to the existing management structures in the organization through which improvements will be delivered.

The Governance team will act together to achieve the resource efficiency goal. They should take joint ownership for developing or approving the strategy and tactics to reach the goal. If members of the Governance team differ sharply, they should nevertheless accept the decisions of the team and show unanimity outside the meeting. If the goal is achieved, they will all get the credit, if the goal is missed then they will all be accountable.

In practice, because of their seniority, the Leader or Sponsor will have an overriding say on many decisions, which will ensure that the Governance team do not make a choice that is to the detriment of other organizational objectives. This authority can be used to make sure that decisions are driven by the needs of the programme rather than the local interests of one department or another. However, the Leader or Sponsor needs to use this casting vote with care, as it can reduce the ownership by the Governance team.

A more subtle form of domination occurs when the Governance team members sit on the fence and wait for the Sponsor's view before taking a position that aligns with the Sponsor. This passivity is especially prevalent when there is a significant gap in the seniority of the Sponsor and the other members of the team, and is something that the Sponsor needs to be mindful of. If the Governance team turn into a committee of *"yes men"* then they will be much less effective – it is the role of the external adviser to check that this is not happening and to raise the issue with the Sponsor if there are hints that this is the case.

*Past errors need to be forgiven if they are to be identified and corrected. The Governance team need to create a "**no blame**" culture.*

Another important aspect of the dynamic of the Governance team is that they should operate a no-blame culture. The first objective of the resource efficiency programme is to find and eliminate waste, and it is entirely counter-productive if the departmental representatives are taken to tasks for any waste identified. Past errors need to be forgiven if they are to be identified and remedied. While I would not go so far as to suggest that inefficiencies are something to be celebrated, they are nevertheless the basis for improvement, and any obstacles to their identification need to be removed.

Another important motivational duty of the Governance team is to ensure that success is recognized and celebrated. By acknowledging the benefits that the programme is bringing and singling out individuals or teams for their efforts a positive perception of the programme can be created.

In a nutshell, the role of a Governance team is:

- To agree on the objectives of the programme and take collective responsibility for ensuring that these objectives are achieved;
- To develop or approve the plan for the resource efficiency programme and take collective ownership for delivering the plan;
- To ensure the appropriate measurement systems are in place and used to assess progress being achieved;
- To provide a decision-making forum where teams wish to cascade up proposals which require senior approval or where decisions relate to organization-wide changes that individual teams cannot approve;
- To regularly review progress and provide support and encouragement (and instruction, if appropriate) to those making changes to improve resource use. To celebrate success;
- To help overcome the many barriers to resource efficiency, in particular, the barriers that can exist between functions, departments or businesses, by working collaboratively to achieve the goal and communicating a shared vision externally;
- To provide the Champion with an effective Mandate and resources to deliver the plan.

The Governance team thus acts as the programme management team, setting strategy, approving resources and ensuring that individual business units, functions or departments are making progress against the plan and targets. Except in the smallest of organizations, or for the largest of investments, it is not anticipated that the Governance team will issue detailed instructions on specific resource conservation measures. Rather the Governance team will tend to empower teams to come up with strategies and actions for themselves, thereby ensuring that ownership for the opportunities is achieved locally. The Champion is the day-to-day coordinator of this effort and acts on behalf of and with the authority of the Governance team, facilitating change within teams.

In the vast majority of programmes, it is desirable that the delivery of the improvement measures is carried out through existing management structures. That is to say that the Governance team establish the change or action that is needed, but that this becomes the responsibility of the operational businesses or functions. In this way, the programme will not be a separate *"bolt-on"* but will become part of the ordinary way the organization carries out its mission. By making the improvement, the responsibility of the existing line management their knowledge is harnessed and ownership increased. It is the responsibility of the divisional, business or functional representatives on the Governance team to communicate to their line management the requirements of the programme and to ensure their participation - hence why they need to be relatively senior.

Real World: ***Specialist sub-groups***

It is important that the Governance team do not attempt to micro-manage the programme. They should try to avoid getting distracted by the detail and processes.

This is one reason why it is common in larger efficiency programmes for sub-groups to be established to deal with specific issues.

For example, within Peel Land & Property Group - where I act as the external adviser - the Governance team (called the Sustainability Steering Group) set up a specialist Carbon Management Group (CMG). This group drew in engineering, procurement and utilities experts to consider the technical aspects around energy and make recommendations to the Steering Group. The CMG acted within the overall mandate and policies set by the Sustainability Steering Group, so their activities were aligned with the high-level sustainability goals.

It often makes sense if a member of the Governance team sits on the sub-group to provide a direct link between the two bodies. Sub-groups might have a specific task (e.g. to establish a new process or standard) and may be disbanded once that task is completed. Indeed the CMG morphed into a wider Facilities Management Group over time when the ISO 50001 standard was fully embedded in Peel.

6.10 Multi-tier governance

In large organizations, Governance will take place at many levels. Creating the appropriate accountabilities between these teams is important.

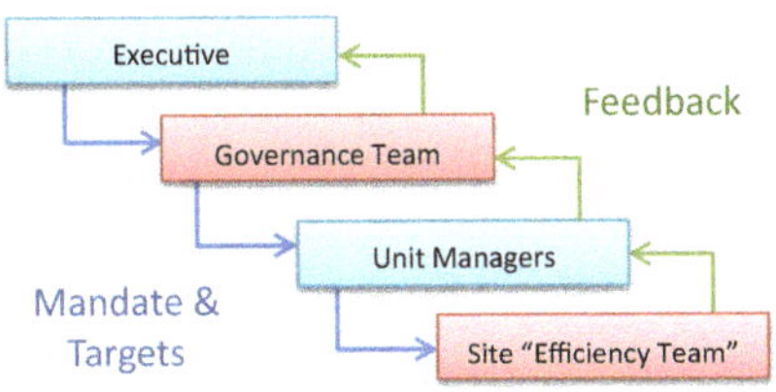

6.22 **It is not uncommon for there to be multiple Governance teams at different levels of an organization, as shown here in red.**
Source: Niall Enright

The notion of bringing together the different functions that can influence resource use applies just as much to a single facility as it does at the corporate level. Thus, many programmes in large and complex organizations will spawn further Governance teams to enable the cross-functional, focused effort that is so often the key to success.

Getting the balance right between the higher-level Governance teams and those at the coalface can be tricky. Certainly, the lower-level teams will receive their Mandate from above, but they should also have a degree of autonomy to enable them to identify the most effective strategies to apply to their own part of the organization. In this way, they take ownership of the problem and any credit for successes they achieve.

On the other hand, too much autonomy can be inefficient and lead to constant *reinvention of the wheel.* The top-level Governance team do need to be assured that the actions at the site level will, in the aggregate, deliver the desired goals. So the feedback mechanism is quite important as is some means of direct engagement between the groups.

In the illustration on the next page, I have shown what a top-level corporate Governance team might look like in a large, multinational organization. In big organizations, we have already stated that it is important to cascade down and reinterpret the Mandate to lower levels and to provide upward feedback on progress from these levels. It is often sensible if these lower-level parts of our organizations, in turn, implement their own Governance team to provide the same coordination and decision-making functions close to the operations of the organization where the resource savings will be made. The scope of decision-making of these teams will be limited to their site or business unit, but in all other respects, they will operate like the corporate Governance team. The role of the Leader - to frequently and visibly articulate the need for change - remains critical. It should also be clear that the Mandate and authority flow from the Leader through the Governance team to the operations shown in green, which is where the resource efficiency improvements take place.

If we turn to the structure of the site Governance team, illustrated overleaf, in Figure 6.5, we can see that it is almost identical to the high-level Governance team. The roles are similar, with the same participation of folks drawn from across functional teams and departments.

6.23 **An illustrative top-level Governance team in a multinational organization**
Source: Niall Enright, available in the companion file pack.

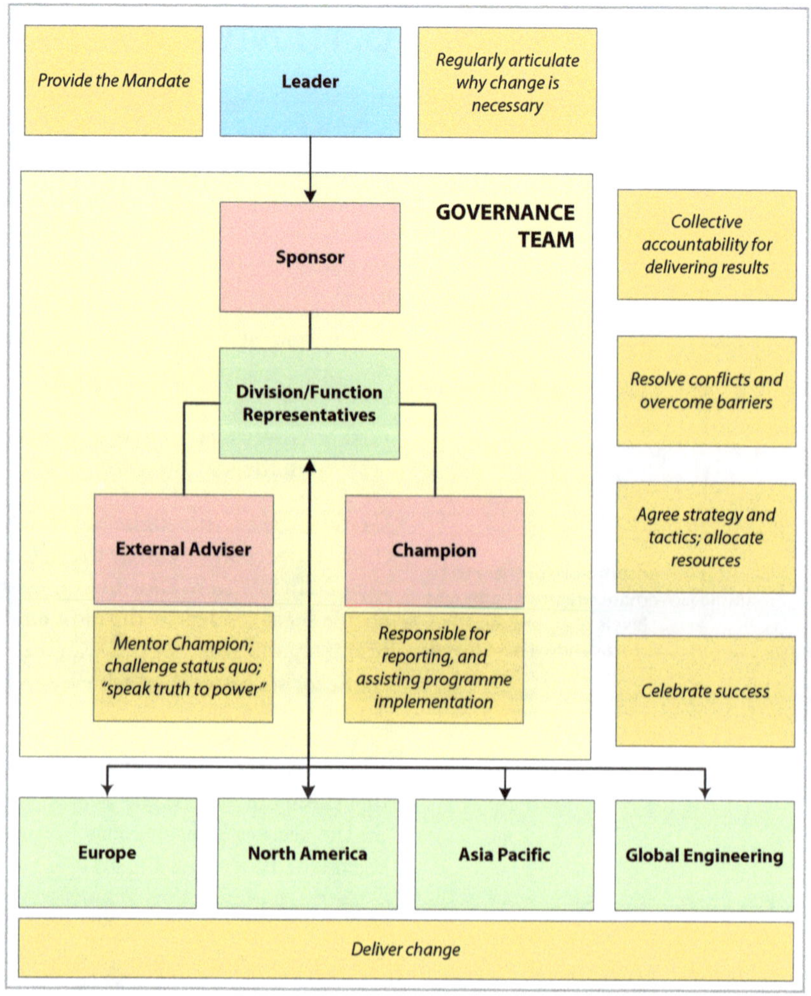

When we refer to Governance, we need to understand that this means all the management processes that underpin the resource efficiency programme. It can't be emphasized enough that the vast majority of the management of our resource efficiency programme happens outside the dedicated Governance team. It occurs within the ordinary management systems in our organization.

> The vast majority of the **management** of our resource efficiency programme happens outside the dedicated Governance team.

There are many examples of how resource efficiency can be incorporated into these normal management processes of an organization. For example, it could be that the previous week's energy use is covered in the *Monday morning meeting* that each shop floor team in a factory hold. Maybe the monthly *New products review committee* is expected to consider the embedded carbon in the materials used and the waste that the products will generate in manufacture. The *investment approval process* could be based on a whole life cost assessment rather than just a capital cost. Employee inductions can incorporate a session on expectations around resource efficiency. The maintenance team could be

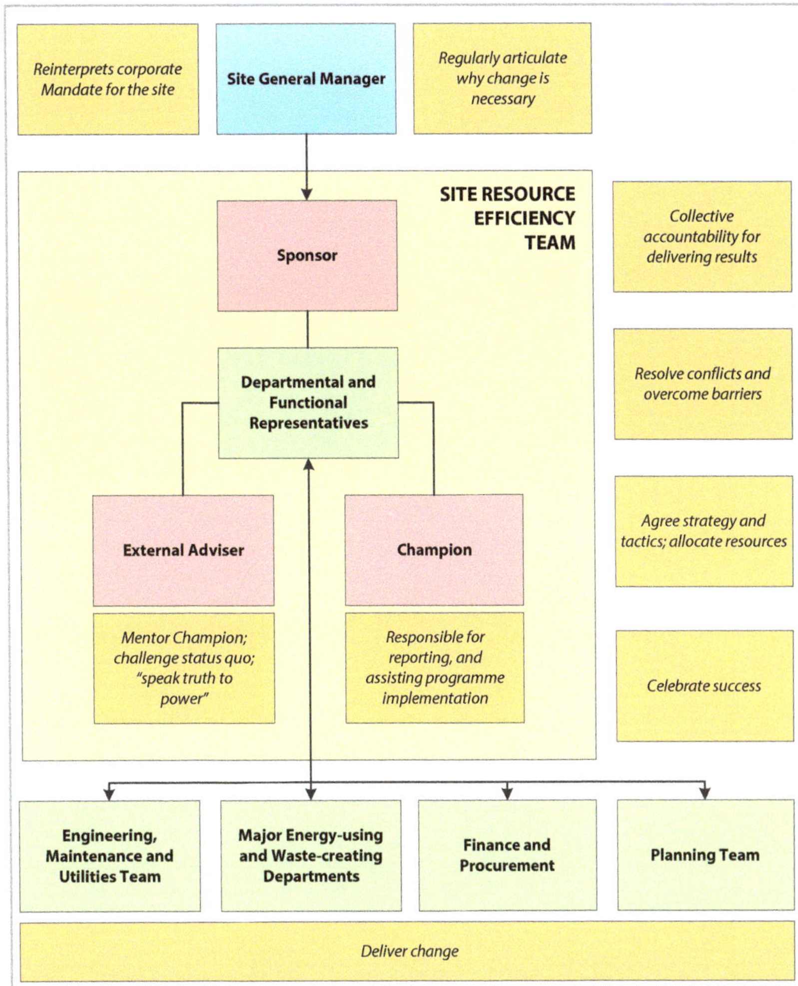

6.24 **An illustrative site-level Governance team, engaging many different functions within the site**
Source: Niall Enright, available in the companion file pack.

set specific targets around condensate return (steam trap maintenance) and compressed air leaks, which are reviewed monthly. The property team could establish a minimum Energy Performance Certificate rating for all new office accommodation in the EU, and an Energy Star Rating for North America, and report every quarter on the scores achieved. It is these management processes that drive the *continual* aspects of our improvement process, and which are vital for the long-term sustainability of our programme.

The core of the resource efficiency programme is enabling teams to create these changes in operations, investment or processes. Having made the changes they need to be integrated into the *existing* management systems, along with measurement, so that progress and feedback can be cascaded up the management structure. In the next chapter, we will explore our Method for energy and resource efficiency which will put some structure around the techniques that will deliver the desired change.

Real World: *Clear accountabilities at BP*

The relationship between corporate Governance teams, site team, corporate and site Champions, departments and consultants can become very complex. In the BP enManage™ programme, which I helped design and lead while at SKM Enviros, we needed to clearly communicate the Mandate to the sites, i.e. the instruction to act and the goals that needed to be achieved. We also needed to make a distinction between those who had accountability, i.e. were responsible to others to deliver the goal and those with support roles in the programme.

At all the BP sites - Texas City, Lavera, Chocolate Bayou, Whiting, Coryton, etc. - the programme would provide a dedicated enManager from the BP corporate team. Four to six energy efficiency experts, some from SKM, some from BP, some from our US engineering partners, Veritech (Mike Rutkowsky and Bruce Pretty), would spend a month on a site audit from which a list of energy savings opportunities would be created. Without this initial external input, the programme would not have been able to identify sufficient opportunities to achieve the goals.

Although there was significant outside help, it was clear that the goal and accountability for selecting and implementing the energy-saving measures were the responsibility of the individual departments. The business unit Leader was held accountable by the BP corporate Governance team. At the site level, the role of the site Champion was to support (not instruct) departments in achieving the goals, and the Champion was accountable to the Governance team for the quality of support provided. In turn, once on site, the external consultants had a clear chain of accountability through the site Champion to the site Governance team. The consultants could not instruct departments what to do. All opportunities identified were verified by the department and credit given to them for every saving measure implemented. A no-blame culture ensured that some poor - even shocking - practices could be openly discussed and remedied without fear of penalty. The various obligations are shown in the diagram below.

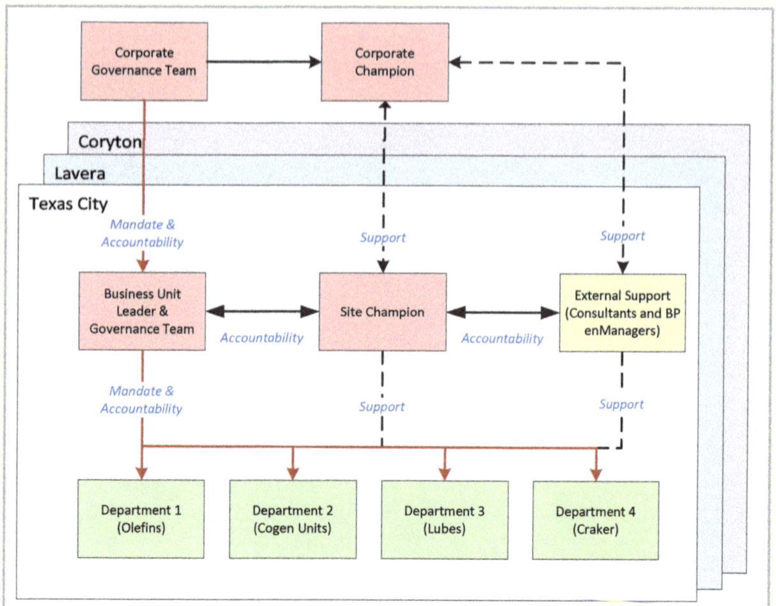

6.25 **Organization of the BP enManage© programme. The red line indicated the clear chain of responsibility from the corporate Governance team through the site management to the departmental heads.** Only four departments are shown in this illustration, a typical site could have 20 or more. *Source: Niall Enright*

Summary on Leadership

1. The role of Leadership is:

   - To express the need for change *by communicating dissatisfaction with the status quo, primarily, focusing on the positive benefits that change will bring;*

   - To mandate others to act *by making it clear that inaction is not an option; by communicating a sense of urgency so that action happens now; by ensuring that sufficient resources (time, expertise or money) are made available so that change can be achieved;*

   - To change the way the organization behaves *by empowering informed decision-making at all levels;*

   - To define or approve the goals and strategy of the programme. *To ensure that appropriate feedback loops are in place to measure progress against the goals;* and

   - To celebrate success.

2. Leaders are not free agents. They need to be able to *justify* action.

3. In order to improve, we need to create a culture in which mistakes can be acknowledged openly and lessons learned. It is the job of a Leader to ensure that this is understood.

4. Outdated concepts of fiduciary duty need to be revised to meet the needs of an organization in the 21st century.

5. Letters to shareholders can often provide some powerful reasons why a resource efficiency programme should go ahead.

6. Goals are the high-level aims of the programme; targets are the specific performance that individual items of equipment, teams or departments have to achieve. Goals can cascade down the organization, and it is perfectly OK to reinterpret the goal in terms that are most relevant at that particular level of the business.

7. Don't assume that a resource efficiency strategy is needed. In some cases, treating efficiency as a separate issue could lead to it being compartmentalized and not integrated with broader business objectives.

8. Improvement actions can be categorized by the time needed to implement these. Immediate opportunities fall into the Optimize category, longer-term changes of equipment fall into the Modify category, while Transform actions involve a more fundamental change to a system or business model.

9. Cherry-picking opportunities can boost short-term success, but usually only at the price of longer-term outcomes.

10. Sometimes the greatest resistance to projects come from middle management who have multiple, potentially conflicting, objectives.

11. A Governance team should ideally include representatives from all the major resource-consuming teams. They should take collective responsibility for the outcomes of the project.

12. The critical role in an energy and resource efficiency programme is the Champion. It is important that they are selected carefully and have the appropriate tools and sup[port that they will need.

Further Reading:

Richard Lynch. *Strategic Management*. Financial Times/Prentice Hall; 6th edition (9 Dec 2011) Pub. ISBN-13: 978-0273750925. *This highly recommended book is aimed at mainstream strategic planning and management but many of the principles apply to the development of resource efficiency.*

Marc j Epstein. *Making Sustainability Work*, Greenleaf Publishing. ISBN-13:978-1906093051. *A treasure trove of practical advice for those working in organizations to drive sustainability. Chapter 2 discusses leadership and strategy of corporate sustainability.*

Lynn Stout. *The Shareholder Value Myth*, Berrett-Koehler Publishers. ISBN-60509-813-5. *A superb exposition of how putting shareholders first harms investors, corporations and the public.*

Questions:

1. What is the role of a Leader in a resource efficiency programme and why?
2. All a Leader must do is to "*empower informed decision-making*". Discuss in the context of energy and resource efficiency.
3. What are the pros and cons of integrating resource efficiency strategy into existing businesses strategy versus developing a stand-alone strategy?
4. Consider the current interpretation of fiduciary duty. Is this a manifestation of market short-termism and a barrier to action on resource efficiency or does it increase value by ensuring an organization is not distracted from its core objectives. For supporting material you may consider *Fiduciary Responsibility,*[732] *Sustainable Capitalism*[309] and *The Short Long.*[343]
5. Why was Ray Anderson successful in transforming Interface?
6. What are the characteristics of the ideal programme Champion?
7. Consider the composition of the Governance team for resource efficiency in your organization. Who would you include and why?
8. Why is it important to start working on longer-term Modify and Transform opportunities from the outset of a programme?
9. What sorts of time frame constraints could one encounter in a resource efficiency programme, and how could these be overcome?
10. Consider the savings model from The Hare and the Tortoise - what factors lead to the latter achieving so much better results?

7 Method

In this chapter, we will put some structure around the tools and techniques that will ensure our energy and resource efficiency programme is successful. Collectively, these tools and techniques are referred to as the resource efficiency Method.

The Method described here is focused on delivering the continual improvement processes that will provide immediate value from the programme, and lay the foundations for a longer-term, more radical transformation of the organization.

Most of the techniques described in this chapter apply to the Optimize and Modify phases of the resource efficiency programme, where we are taking an *existing* organization and business model and making it as resource-efficient as possible. Longer-term, more profound transformational activities are discussed in the next part of the Framework: Momentum.

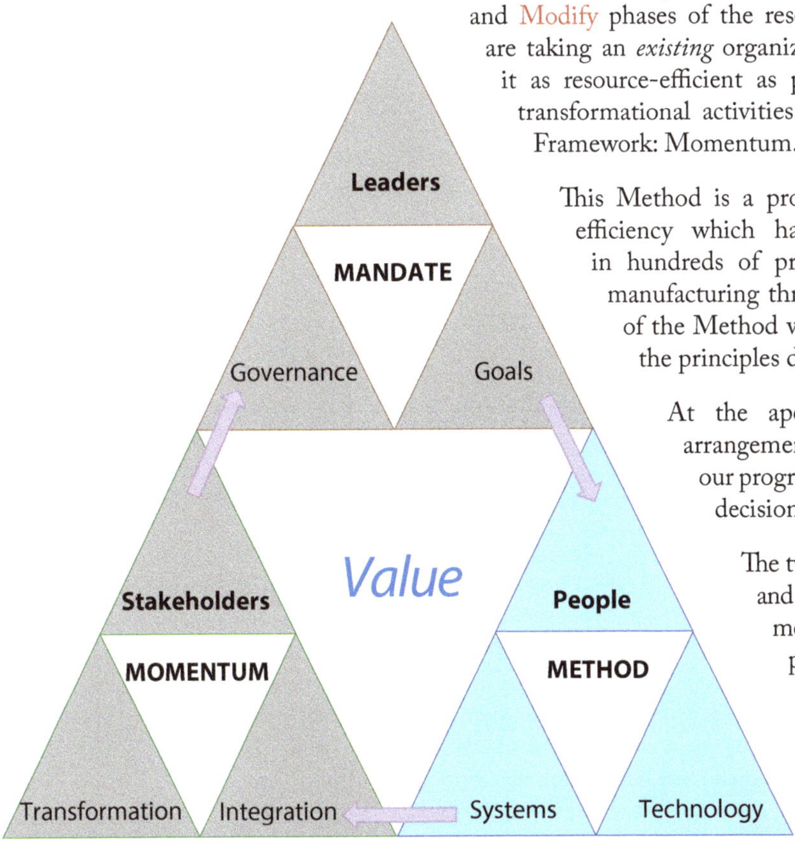

This Method is a proven approach to energy and resource efficiency which has been developed over many years in hundreds of projects worldwide in all sectors, from manufacturing through to public services. Not all aspects of the Method will be suitable for all organizations, but the principles described here are fairly universal.

At the apex of our Method is People. This arrangement reflects the fact that the initiators of our programme are people and it is their informed decisions which bring about change.

The two other parts of our Method are Systems and Technologies. These represent the other mechanisms for change that organizations possess. Our Method is about getting these three powerful levers for improvement in balance and working effectively together to put in place a continuous improvement process which will deliver value and endure.

7.1 Achieving balance

Organizations have three tools to achieve change: their people, their technology and the systems which inform and control. It is important that our efficiency programme achieves the right balance of efforts in each of these areas.

Real World: ***People, Systems, Technology***

There are three routes to change in an organization:

- People: their behaviours and decisions;
- Systems: the rules, information and feedback processes that govern choices;
- Technology: the physical equipment that convey, transform or convert resources.

Some folks have argued for finance to be considered the fourth mechanism for change. On examination, money cannot act alone - it achieves its results through the other three mechanisms.

Our Method for resource efficiency categorizes our activities under these three headings.

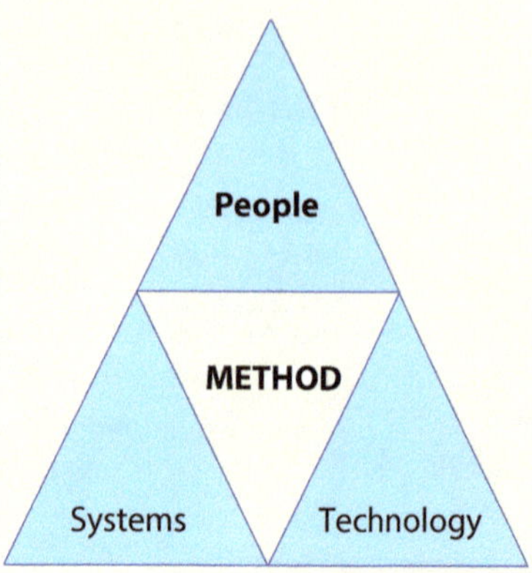

7.1 **The three parts of the Method, with People shown at the apex**
Source: Niall Enright

In the illustration left, People are shown at the apex of our pyramid. This is fitting, because it is only individuals who can initiate the resource efficiency programme and because changes to systems and technologies are actually made by people. The success or failure of our resource efficiency method will be primarily down to the people element.

Technology refers to equipment and processes that consume or produce resources. A piece of technology could be a humble flow-limiting valve on a water hose, though to a complex gas turbine generator, or waste recovery plant. These technologies can be operated efficiently, or not, usually due to decisions taken by people. At the same time, these technologies may also have limits to their efficiency, which can only be overcome by upgrading or replacing the equipment.

Finally, we have Systems which is sometimes used to describe the data, software and controls systems (closed-loop or with human intervention) that influence the performance of People and Technology. In this Method, a broader definition is used, encompassing much wider decision-support processes, such as standards, incentives, policies, processes and structure, all of which have a critical part to play in resource use. Indeed, organizations can be described as *"a collection of People and Technologies whose activities are coordinated or directed by Systems"*.

These three elements, People, Systems and Technology, represent the three levers that any organization has to achieve improve resource efficiency. One of the key messages to organizations embarking on a resource efficiency programme is that these three aspects need to be appropriately balanced if the benefits of the programme are to be sustained in the long-term.

For example, I have encountered dozens of cases where there has been a sustained motivation and awareness programme around energy efficiency – the "*switch it off*" campaigns with posters, etc. – where people have nevertheless reverted to their original behaviour several months after the campaign has come to an end. This reversion arises because maintaining behaviour change requires systems of measurement and management to provide regular feedback on the improvement and to identify and correct any deviation back to the original poor performance. Savings through People are not sustained without Systems, as illustrated in the first part of Figure 7.2, left.

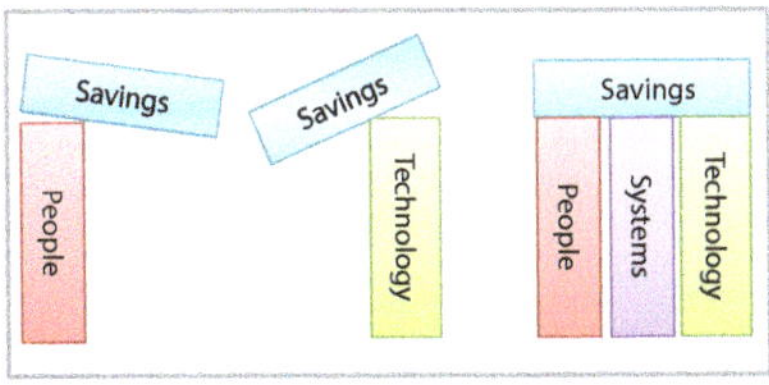

7.2 **Focusing on People or Technology in isolation will not lead to long-term sustained savings**
Only by combining People, Systems and Technology will improvements be sustained.
Source: Niall Enright

In a similar fashion, it is remarkable how often Technology improvements can be undone through the actions of People. There are many examples: windows left open in buildings which undermine the space-conditioning heating, ventilation and air conditioning (HVAC) systems, or the industrial plant switched from *auto* to *manual* mode because the operators don't fully understand or accept the built-in control logic, or the water conservation spray nozzles in hoses which are stolen by employees to use at home and so no longer fulfil their flow-limiting function. Equipment improvements will not succeed unless people are fully engaged in *why* this has been done and *how* the equipment should be operated. Savings through Technology are not sustained without People.

It may be tempting to think that we can "*engineer out*" the human altogether. For example, as is quite common in modern offices, putting dummy thermostats in rooms gives occupants the impression that they are controlling the temperature, whereas, in reality, the setpoint cannot be overridden. However, who is to say that the centrally set temperature and timing of the HVAC system will remain optimum as the building's use evolves? Who will ensure that when the office is closed over Christmas that the heat is turned down? People, of course. While Systems can lead to savings on their own, they ultimately rely on the operating logic set by People and the underlying efficiency of the Technology they control. People control Systems, not the other way around. I recently came across an HVAC system in a prestige office building that had been using three times the energy it should have done for many years simply because the air replacement rate was still set for the days when indoor smoking was allowed. That was a People, not a System error.

It may be tempting to think that we can "engineer out" the human altogether. However, **People control Systems, not the other way around.**

Of course, the degree of effort and opportunity in the People, Systems and Technology areas will vary from organization to organization, site to site and even across the different programme phases of Optimize, Modify and Transform. Where an organization already has a strong culture and awareness of resource issues, perhaps because they have implemented a Monitoring and Targeting (M&T) programme in the past, then less work will need to be done on employee engagement. Where an organization is resource-intense, then improvements are more likely to come from technology and control systems to optimize the process at a greater, near real-time, frequency than is possible with People-based initiatives alone. However, in all these cases there needs to be an appropriate balance between the three elements and no programme is likely to succeed without some activity in each area.

We shall start our exploration with People. I will examine the importance of knowledge, motivation and capacity to act, touching on the latest lessons from psychology and social sciences. We shall see that there are many roles which need to be aligned in order to address resource efficiency in organizations, and that systems of measurement and feedback are critical to success. Hopefully, those who see engaging people as "*heavy lifting*" and would prefer "*fire-and-forget*" technology solutions, will be persuaded of the benefits of engagement and be reassured that it is not as difficult as at first presumed.

7.2 Fundamentals

To deliver change we need to do things differently. This Method is designed to provide the tools and motivation to permanently alter decision-making in our organization.

This Method is designed to support informed decision-making, to change People and Systems, not just Technologies, to deliver short-term while thinking long-term, to ***change "the way we do things around here".***

We have seen in previous chapters that resource efficiency is about continual improvement; that management commitment is essential; that it is, in essence, a change process; that we need to look to optimize immediate performance but at the same time lay down the foundations for future transformation.

We have seen that there are many barriers to making the necessary changes, even though there has been abundant evidence of the benefits that this will bring to our organizations.

So what are the key ingredients of our Method that will maximize the probability of success for our programme? In fact, there are just five essential elements:

- A strong Mandate
- Engaged People
- Continual measurement, feedback and valuation
- Celebration
- Constant change

Underpinning these elements of our Method are two important tools:

- Monitoring and Targeting (M&T)
- Opportunities Database

The precise names given to these programme elements or tools do not really matter. What is important however is, that we structure a process with the right characteristics. Successful programmes share certain common factors that we want to replicate; we want to create multiple layers of accountability, to support informed decision-making, to change People and Systems, not just Technologies, to deliver short-term while thinking long-term, to change "*the way we do things around here*".

In setting out these objectives, we need to be realistic about the aspirations and capabilities of our organizations. Few programmes achieve perfection from the outset (see *Perfect is the enemy of good* on page 206). This Method recognizes that the foundations often take some time to develop. As a consequence, the

METHOD: Operational Excellence and the Foundations of Continuous Improvement

	PEOPLE	SYSTEMS	TECHNOLOGY
Discovery	See Baseline & Technical Audit		
Scope	Communicate the mandate, provide the resources, launch the programme, set the tone	Establish Governance . Align with quality Systems if needed. Set hurdle rates & approval criteria	
Prepare	Provide training for system operator (Champion)	Install M&T System Opportunities Database Make Systems changes	Fit additional metering and data capture Enter all existing ideas into Database
Deploy	Work with users to set honest, fair and achievable targets, and populate the Opportunities Database	Establish reporting cycle at appropriate levels (top management, units, teams), create networks for sharing and learning	Eliminate poor, repeat good variances Find and implement feasible projects GEMBA
Celebrate	Celebrate success with individuals and teams	Track value and relate this to core objectives of stakeholders. Management align incentives and rewards CUSUM	Celebrate and exploit better understanding. Develop aggregate projects, lead/lag, best practices, case studies etc.
Change	Culture of positive challenge to change *Systems* not just Technology	Management process to understand and rectify common *Systems* causes of inefficiency. ROOT CAUSE ANALYSIS	Deploy standards and best practices. Examine future Technology developments in this light

7.3 **The resource efficiency Method.**
This Method sets out a series of activities under the headings People, Systems and Technology. *Source: Niall Enright*

The key to change is ***informed decision-making****. When people have a full understanding of the impacts of the alternatives available, they will make the right choice.*

focus is on putting in place Systems that will deliver immediate value and at the same time serve the organization when it chooses to embark on the longer-term Modify and Transform processes.

The Method is illustrated on the previous page. It may well be that initial audits have been completed or it may be that we need to start our process with some form of discovery, as shown in the top box. Having identified the areas of opportunity and gained a Mandate to proceed, we then need to prepare our programme, focusing primarily on the People side of things, and their capacity, so that we are clear on who we are going to involve, and how we are going to organize, support and oversee the process. Here we will also install and prime the meters and software we need. Once installed, with an appropriately trained Champion, we will deploy these tools, engaging with the folks at the site or organization, educating them on the programme, and putting in place the reporting processes. Our approach is to win "*hearts and minds*". Here, we want to achieve some "*early wins*" so that we can celebrate success and develop a positive context for the programme. This chapter expands on these steps.

Although the Method appears linear, the processes we have implemented involve a continual improvement cycle of measurement, analysis, response and review. Thus, the discover, deploy and celebrate stages are repeated continually. Indeed the whole process may be regularly re-initiated if we have chosen to implement our programme in a staged fashion to align with our capabilities and other organizational objectives.

Once we have demonstrated that the continual improvement process is working well, and we have developed some learning from the programme, we will naturally - inevitably - start to identify and consider bigger, more fundamental and longer-term changes in our organization's use of resources. Here we need to constantly beware of the trap of just thinking in terms of Technology *quick fixes* and to recognize the transformational potential in the Systems that underpin the use of resources and the decision-making that People make.

Informed decision-making is at the heart of this Method. It seeks to make waste visible, to measure and assess resource use, and to make People accountable for delivering improvement. We all make countless decisions every day that affect resource use. Most of these decisions are made in an automatic fashion, by habit or custom, often using mental shortcuts or *heuristics*. What this Method does is to provide the reason and data for People to step outside the customary decision-making processes around resource use, and to put aside conventional assumptions.

When presented with a true understanding of the implications of their decisions on the value of the organization (in the widest sense), the presumption is that People will make the right choices. Thus, this is essentially an optimistic, positive and affirming Method, which is why celebration is such an important element of this approach to reinforce the value delivered and thus the value inherent in continuing to develop and extend the process further.

⇒ page 274.

Exploration: *Informed decision-making*

Several years ago I was demonstrating an energy management software tool to a plant manager at an Abbott Laboratories facility in Ohio and, at the conclusion of the presentation, he summarized the purpose of the tool in a powerful way:

> *"I get it; what you've got here is a way for my folks to make better-informed decisions."*

Following the meeting, I just couldn't get this description out of my mind. The more I thought about it, the more it seemed to me to sum up the change at the heart of resource efficiency and sustainability. In fact, today when I am asked to define sustainability, I rarely come out with the standard Brundtland Commission[92 p15] description of *"development which meets the needs of current generations without compromising the ability of future generations to meet their own needs"*, or reiterating sustainability as *"balancing economic, environmental and social needs".* Instead, I say *"sustainability is a process of ensuring fully informed decision-making".* What I am trying to do is to emphasize the change that is needed to achieve the goal, not the end in itself.

An informed decision is one where the costs and benefits of each alternative (including the *"do nothing"* choice) are made available to the decision-maker. In simple terms, this book is all about the Mandate that drives people to consider the alternatives and the tools and techniques to then inform their choices.

The rationale for this viewpoint is that change happens when decisions are made. The choices organizations make may lock in consequences for decades – for example, a power station may have an operating life of 60 years, a building will stand for 50 years or more, a chiller plant will be maintained for 20 years, a car design will be marketed for over five years and hundreds of thousands of units produced.

It is self-evident that these types of choices should be sufficiently informed in order to protect future value. Just ask the shareholders in General Motors about their 1999 investment in the Hummer SUV brand and manufacturing. Ten years later this investment was completely wiped out by the collapse in sales arising from increasing petrol prices, which were entirely foreseeable. Companies like Ford and Toyota, in particular, who saw that the *status quo* was likely to change, profited – rising resource costs have created an opportunity for them. Recent falls in gasoline prices have not reversed the trend.

This brings us to one of the harder aspects of responding to the resource limits we face – the difficulty in challenging the *status quo*. If a business model is working today, delivering profit or service, then it is difficult, or even inappropriate, to change – particularly if it is a company operating in a market that is driven by the need to deliver increasing profits and growth quarter on quarter. The key is to recognize that maintaining the current course is *an explicit choice*, and so we must thoroughly consider the pros and cons of the *status quo* on an equal footing with the alternatives (do something v. do nothing).

We have seen that economies, and hence organizations, are likely to be strongly impacted by resource issues in the future. even if the effects are not big today, it is sensible that the organization should devote effort to anticipating the potential future impact of resource scarcity or regulatory limits on the current business model and preparing for the changes that might arise. This might be by researching its vulnerabilities across the supply chain and product/service life cycle, or by starting the design of less-resource intensive alternatives, or by putting in place the continual improvement processes needed to manage resource use, which should in any event reduce operating costs today.

Another particular challenge for organizational strategists is that sources of information about future trends are inconsistent in their treatment of resource limits. A case in point is Peter Dicken's otherwise excellent, and highly influential, exposition of the key global trends that are driving businesses, *Global Shift*,[211] which barely touches on the subject of resource efficiency. Take too HSBC's perspective of *The World in 2050*,[772] just one of many typical business analyses:

> *"On the road to 2050, we expect what are currently 'off balance sheet' costs – whether in terms of carbon emissions or biodiversity loss – to be brought more formally into economic decision- making. This will reward the corporations and countries that make resource productivity a key element of long-term strategy."*

So far, so good. However, in the final chapter *What might go wrong?*, which sums up the risks to businesses, it is trade protectionism, war, natural disasters, economic cycles and reduced demand that are featured, not natural limits. This tendency to acknowledge environmental limits, without recognizing the implications on growth and value, is not helpful, as it diminishes the importance of resource considerations in business strategy.

It is the contention of this book that resource efficiency is one of the most significant potential creators or destroyers of business value in the next decades and should be close to or at the top of the agenda of business strategists. While organizations may find it difficult to influence war, natural disasters or economic cycles – they certainly can affect the resource-intensity of their operations, and managing this will be highly material to their future success.

Of course, the simplest way to save resources is to shut our organizations down. What sustainability requires of us is much more challenging - that we keep the factories, schools, shops, offices open while we address the social and environmental impacts that arise directly through these operations, or indirectly upstream or downstream in the supply chain.

In order to keep our organizations open in a resource-constrained - aka resource-expensive – world, not only do we need a robust, high-level strategy but we also need to reach down and influence the myriad choices made by employees at every level of the organization every day. Choices such as the selection of replacement lamps; the scheduling of a production run; whether contracts for raw materials should include a supplier *"take back"* on the empty drums; the decision to set a two-year payback hurdle for energy efficiency investments. Or the company car policy; the temperature setting on the office thermostat; the choice between bottled and tap water at training sessions; setting the printers to output duplex by default. Our organizations' policy on home-working; the incentives to encourage efficiency; the type of maintenance regime we operate; the feedback and communications that employees receive about sustainability. All will determine our use of resources.

On their own, these operational decisions may seem insignificant, but in their aggregate, they can have a very large impact – and the process of informing those decisions and putting in place appropriate rewards and sanctions is key to transforming organizational culture to recognize the value inherent in efficient resource use.

Informed decision-making is not just limited to organizations and their employees. Regulators are also making increasingly informed decisions about the environmental challenges we face, and are willing to act on this information. As mentioned earlier, the UK government has set itself a legally binding target to decarbonize the economy by 80% by 2050, which is leading to a whole raft of policies and regulations to achieve this goal. We have seen how some powerful intermediaries between the public and manufacturers, such as Walmart, are beginning to apply pressure on their suppliers to deliver products which are more resource-efficient. Manufacturers are starting to make decisions about the ingredients in their products based on their embedded carbon content.

Consumers too are decision-makers. While generalizations are risky given the vast diversity of people and cultures around the world, we can certainly discern some long-term patterns in the increase in ethical and environmental consumerism. Significantly, surveys put consumers in the newly emerging markets such as China as among the most likely to have purchasing decisions influenced by environmental performance.[698] These are the very customers on which many of the world's leading brands are pinning their hopes of future growth.

Most consumers today are unaware, or possibly in denial, of the actual scale of the environmental problems we are facing. But the evidence of harm is becoming ever more real in people's lives. It will possibly only take a few more extreme weather events, coupled with some high-profile environmental crises such as a series of massive plankton blooms, or the extinction of a popular species, for the penny to drop and for people to understand that we – they, us - are facing a crisis. There are many such potential turning points in people's attitudes, the collapse of more fisheries, coupled with the rising costs of the basic means of survival of energy, food and water. At that point – assuming that they are following a Kübler-Ross model[454] of *Denial, Anger, Bargaining, Depression and Acceptance* – consumers will move swiftly from denial to anger and want to blame someone else. That *"someone else"* is likely to include companies and institutions that are perceived to be profligate with

natural resources. In this scenario, there is a consumer backlash which has the potential to rapidly undermine the value of brands and corporations that have been built up over decades. What, for example, happens to the bottled water industry if shipping water halfway around the world becomes *"uncool"*; or to the car industry if owning two cars is seen as selfish or attracts a huge tax penalty; or to the holiday industry if stay-at-home *"staycationing"* becomes the norm rather than the exception; or to golf businesses if the huge amounts of water needed to irrigate the courses become prohibitively expensive or offensive to public perception (just consider today's water scarcity in California for a real world example).

Just as there are tipping points in the environment, so it is the case that tipping points in societies occur. These are points where the established patterns of thought are overturned and replaced by a different mindset. The natural limits that we face are potentially significant enough to bring about a paradigm shift – which is a fundamental change in the underlying world view brought about by the accumulation of incontrovertible evidence that demonstrates the inadequacy of the current model. Thomas Kuhn, in his ground-breaking study of scientific progress, *The Structure of Scientific Revolutions,*[455] says that a paradigm shift is not an overnight event, but a long process, in which, at some point, it is evident that the collective view has moved from one worldview to a better worldview that more adequately integrates the new evidence:

> *"A new scientific truth does not triumph by convincing its opponents and making them see the light, but rather because its opponents eventually die, and a new generation grows up that is familiar with it."*

Kuhn is quoting the great physicist Max Planck. Paul Gilding, in his book *The Great Disruption,*[313] likens the current situation to a *"dam about to burst"*. According to him, once the dam of public denial (or ignorance?) bursts, the nature and scale of the climate challenge will involve a mobilization akin to the Second World War, with acceptance of the need for a similar collective effort. If this type of response were to emerge, then it could be a game changer for many organizations. In the Second World War, for example, petrol was rationed to four gallons of fuel a week and the speed limit was set at 35 miles per hour – something that was accepted by the public, given their recognition of the dire threat that they collectively faced.

Lest we dismiss this idea out of hand, we should note that a more sophisticated form of rationing is being actively developed: personal carbon allowances. This rationing is the ultimate form of informed decision-making, where individuals are fully aware of the carbon content of the resources they consume and have the option to reduce their emissions (and thus sell their surplus allowances) or to emit more which will require them to go to the market to purchase allowances from others. These ideas are not fanciful – in the UK, for example, there has been considerable work done in the area of tradable energy quotas that provide individuals with quotas for carbon or energy. According to one design, set out in an All Party Parliamentary Group report in 2011,[284] 40% of the UK's quotas would be issued free of charge to every adult citizen on an equal basis. The remaining 60% would be auctioned to all other energy users. Irene Lorenzoni and her colleagues see personal carbon trading as a key driver to informed decision-making, calling the ability to integrate carbon impacts into decisions *"carbon capability"*,[480] akin to financial capability.

While Gilding's *One Degree War* is not certain and consumer and regulator responses to climate change are by no means predictable, we should not dismiss the vision he and many other commentators offer. This imagination of people's response to climate change represents a carefully considered and self-consistent view of the changes that may occur should the public and governments take the view that climate change is an overwhelming threat to our wellbeing. It is easy to dismiss these visions as fanciful, speculative, not grounded in reality, biased against capital. It is easy to dismiss the notion of a parallel economy based on personal carbon allowances as impractical. This would, however, do our organizations a disservice. The nature of the current environmental challenges and the incompatibility of the current perpetual growth paradigm with the accumulating evidence to the contrary, mean that it is entirely possible that a fundamental shift in world view may take place, driving entirely new behaviours from shareholders, markets, consumers and regulators alike.

Informed decision-making is the key to achieving linking the future impact of resource availability and risk with Value. By creating measurement and feedback systems, by developing the knowledge of our people, by connecting more firmly with the needs of stakeholders we should position our organizations in the best possible place to address the changes and opportunities that will *inevitably* arise in a resource-constrained world.

7.3 Scope

Before installing any tools or systems we should consider who within our organizations needs to be involved in the programme. Some folks may have formal roles, while others may be in a position to influence the outcomes we are seeking. Once identified we need to communicate the Mandate to these colleagues and get their support.

"Success has many fathers, but failure is an orphan."

- attributed to Tacitus

When we have the Mandate to start our programme of resource efficiency, an obvious starting point is to establish who needs to be involved.

There are many functions within an organization which can influence resource use. These functions will change depending on the stage of our programme and the specific activities that we want to complete.

One simple approach to mapping out the folks that one wants to involve is to create a grid with the main functions and departments in our organization across the top and the key tasks we want to accomplish on the left. We can then place a mark against those functions that have some influence or a significant impact on that specific activity. A fairly elaborate example is shown on the next page.

We will also need to determine who is going to take a formal role in the programme, both in delivering the processes and in the Governance. The table on page 253 provides a summary of some of the specific roles we might consider.

In most programmes, one does not launch the resource efficiency efforts simultaneously across the whole organization or even across a whole facility. So, we also need to decide where the initial focus of our efforts will be. That is "*where, precisely, to begin?*"

One way to consider this question is to work backwards from where we want to be. I am referring specifically to the notion of gaining *quick wins* for the programme, of being able to celebrate success early on. In my experience, nothing leads to success like success. There is the old saying "*success has many fathers, but failure is an orphan*" (which appears in many variants going all the way back to the Roman Tacitus). When things are going well there is a natural tendency for humans to want to be seen to have contributed to the success.

Thus, in establishing our initial focus one piece of advice is to start where we feel that we can gain early success. We might start, for example, in a department or facility where we have an enthusiastic management team, where there is good data and where there is sufficient variability in our operations that quick changes can deliver rapid improvement. Clearly, the Leader of the programme and the local management team will have their views, but the potential for early success is an important consideration.

Real World: *Where to start?*

These are the five essential characteristics that I look for when determining where to begin a resource efficiency programme in a large organization.

1. Look for enthusiasm above all. There is nothing more positive than an engaged management team and workforce and nothing more stifling than an uncooperative team.
2. Data is important because it provides clues for improvement ideas and the basis for verifying success. Also, readily available data avoids delays in installing measurement equipment so you can hit the ground running. So, good sub-metering and records are helpful.
3. If there is a lot of variability in our operations and if resource use is spread over a large number of process steps or items or plant, then we will be more likely to identify quick wins. For example, in a steel mill I wouldn't start in the blast-furnace but in the peripheral energy-users.
4. If the initial area is considered representative of several other facilities, or perhaps is seen as a better performer, then the improvement delivered will be considered to be replicable more widely rather than being dismissed as a "special case".
5. Timing is an issue. I want an initial focus to be in a department or facility which is not going to undergo a major disruption in the future.

These rules have served me well over many years. It is unlikely that you will see all five features, but you certainly don't want none to be present!

Having established where we are going to focus our efforts, it is essential that we gain the full support of the team in that part of our organization. In particular, effort needs to be put into obtaining the support of the local management - department heads (e.g. the site financial controller), function heads (e.g. the local facilities, engineering or maintenance managers), service leaders (e.g. the faculty head at a university), and so forth. This point cannot be stressed enough – it is my experience in many larger corporate resource efficiency programmes that the battle is won or lost at the middle-management level.

By and large employees on the shop floor, production lines, or delivering service, are generally well disposed to the idea of better environmental performance and reduced waste. Similarly, most senior executives *get* resource efficiency. It is at the level of middle-management that the greatest resistance can occur. For a real world example, see *Spanning the intent gap – enManage™ at BP* (page 239), which illustrates how local priorities can be at odds with the overall programme. This potential for a misalignment of objectives justifies spending some time and effort reinterpreting the overall goal in terms of the needs, priorities and aspirations of the local team.

One way to ensure that the local management are on board is to ask the Leader to engage with them and communicate the importance of the programme. Another way is to give the local management a formal role on the Governance team. Thus, they will be at the heart of the collective decisions about the programme, rather than on the sidelines throwing rocks.

We should now know precisely who is involved in the programme. Hopefully, many of the key folks will have been involved in developing the strategy, in audits or in goal-setting and so should feel ownership for the findings. For those who are new to the programme, the initial briefing is critical. *You only get one chance to make a first impression.* Some of the communications techniques set out in *Chapter 9 Creating a Mandate (page 307)* will be helpful at this point.

Moving from the People to the Systems preparations, it is advisable at this point to hold an initial meeting of the Governance team. If not already defined, it would be useful for the team to clarify just what the economic return (payback or internal rate of return) is acceptable for the programme so that efforts can be focused and expectations managed.

The Governance team should also clarify how it will track progress and ensure that adequate reporting systems are in place. Other considerations for the Governance team will be how to align the resource efficiency initiative with any other overlapping or competing initiatives or processes in the organization. It is not unusual for the Governance team to meet several times in advance of the programme launch to ensure that these initial issues are properly considered.

Before we can proceed to launch the programme on the shop floor we need to put in place some tools and competencies to support our efforts in the chosen location. This is what is referred to as the prepare stage. ⇒ page 278.

7.4 **The number of people and roles involved at each stage of a resource efficiency programme can be quite staggering** Below are the phases of the programme and opposite an indication of the functions involved.
Source: Niall Enright. This is available in the companion file pack, including a poster version, and an editable model in Excel.

People and Activities in a Resource Efficiency Programme

	Processes and Activities	Phases: Initiatiate	Optimize	Modify	Transform
Mandate	**Leadership**: get management commitment to resource efficiency	x			
	Governance: define wider roles and accountability	x			
	Goals: set programme goals, scope and measurement	x			
Discovery	**Audit:** quantify potential value from resource efficiency	x	x	x	
Operational Excellence Method	**Prepare:** communicate the mandate, establish governance	x	x	x	
	Initiate: implement software tools and measurement systems		x	x	
	Deploy: engage users, embed in management, drive improvement		x	x	
	Celebrate success: quantify value, optimize programme, set standards		x	x	
	Change: culture and root cause analysis drive *systems* changes		x	x	
Transition Existing Model	**Integrate:** redesign capital allocation CAPEX in existing facilities		x	x	x
	Integrate: incorporate resource efficiency in product and service design			x	x
Transition Existing Model / Transform by Organizational Redesign	**Integrate**: harness CAPEX allocation in new facilities			x	x
	Integrate: redesign other decision-making processes & business metrics			x	x
	Transform: supply chain and end-user engagement/co-design			x	x
Transform by Organizational Redesign	**Transform:** redefine the vision and purpose			x	x
	Transform: redefine brand, disclosure and reputation management			x	x
	Stakeholders: engage and involve stakeholders in co-design			x	x
	Transform: test and deploy new business models				x
Transform by Organizational Redesign / Transform with Technology	**Transform:** consider fundamental service or product re-design				x
Transform with Technology	**Transform:** harness the power of new technology to shift the paradigm				x

Functions

Stage	Operations	Engineering	Health, Safety and Environment	Maintenance	Property	Procurement	Finance	Marketing and Sales	Design	Logistics	Human Resources	IT	Business Management	Legal	Corporate Affairs & Comms.	Senior Management
Mandate	○	○									○		●		○	●
Mandate	●	●					●				○	○	●		●	●
Mandate	○	○	●				○			○		○	●	○	○	●
Method	●	●		●	○		○			●		○	●			
Method	●	●	○	○	○	○	●	○			○		●		○	●
Method	○	●		●	○					○		●	○			
Method	●	●		●	○		○		○	○	○	●	●			
Method	●	●	○	○		●	○		○	○	●		●		○	●
Method	●	○		○	○	○	○		○	○		○	●		○	○
Method	●	●		○	○	○	●						●			●
Momentum	●	●	○				○		●	○			●	○	○	●
Momentum	●	●	○	○	○	○	●	●	●	○		○	●	○	○	●
Momentum	○	○			○		●	○				●	●			
Momentum	○	○				●	○	●	●	○			●			
Momentum	○	○	○				○	●	○	○	○		●		○	●
Momentum			○			○	○	●	●			○		○	●	●
Momentum								●	●				●		○	●
Momentum	○					○	●	●	○				●			●
Momentum	●	●						●	●	○			○			
Momentum	○	●	○	○	○	●		○	●	○			○		○	○

Framework → Mandate → Method → Momentum

Audit can precede Mandate

● = *Key role* ○ = *Some involvement*

7.4 Prepare

Right from the start of our programme we want people to engage in the process of questioning current resource use and generating ideas for improvement. In this phase of our project we will make sure that the key tools to support this are in place and ready to support this change.

The core concept in our resource efficiency Method is the notion that, given the right information, and accountability, people will identify waste and develop strategies to eliminate the waste.

Solution identification is not achieved by simply throwing data at resource-users. To support this process, we need to put in place some clever tools and systems. These tools are going to achieve two purposes. The first objective of the tools we are going to implement will be to produce improved inputs into decision-making processes.

Our first such tool, used to inform decisions around the operation of existing equipment and processes, is M&T. This tool's objective is convert data (which is often plentiful) to information (much rarer). This definition from *BusinessDictionary.com* describes the distinction:

> *Information is: data that is (1) accurate and timely, (2) specific and organized for a purpose, (3) presented within a context that gives it meaning and relevance, and (4) can lead to an increase in understanding and decrease in uncertainty.*
>
> *Information is valuable because it can affect behaviour, a decision, or an outcome. For example, if a manager is told their company's net profit decreased in the past month, they may use this information as a reason to cut financial spending for the next month. A piece of information is considered valueless if, after receiving it, things remain unchanged.*

"Success depends on upon previous preparation, and without such preparation there is sure to be failure."

\- Confucius

What M&T does is to take direct resource use or waste measurements and relate these to key drivers such as activity, weather or production to provide a meaningful indication of performance for a given day/week/month for items of equipment, processes or facilities. M&T is the tool to convert metered data into information in a systematic fashion.

But that is the *Monitoring* part of M&T. The second part, *Targeting*, is a process where individuals analyse the performance information and set (ideally themselves) an improvement target. M&T analysis can be carried out in a spreadsheet or using one of a half-dozen or so commercially available software packages (a system specification can be found in the companion file pack). Volume II has information on the M&T data analysis techniques (Chapter 14, page 431) and the improvement processes (Chapter 20, page 707).

*Because of its quick impact and good financial return, **M&T** is almost always the first tool to be deployed in a resource efficiency programme.*

In the prepare phase, we will select our M&T tool and populate it with the appropriate resource and activity data. This M&T system configuration can be quite time-consuming if we have to go through a formal specification and procurement process, so we need to get this activity moving as quickly as possible. Indeed, the training of our Champion and site team in the use of the tool cannot proceed until the selection has been finalized.

Because of the effort needed to install M&T systems, it can seem tempting to bypass this altogether and instead to focus immediately on any technology changes identified in our audit phase. Technical *quick fixes*, such as lighting upgrades, waste separation or leak reduction, may indeed help us reach an immediate target. The problem is that once these ideas are exhausted, we then need to embark on another discovery process to find the next set of opportunities.

M&T, on the other hand, helps us improve the operation of existing equipment, processes and buildings while generating a continual stream of new projects. M&T offers quick improvements often at no or low cost, by identifying and eliminating variability in operations. Its rapid impact and good financial return means that M&T is almost always the first tool to be deployed in a resource efficiency programme.

Because M&T is measurement-based it helps build confidence in the resource efficiency programme as a process that can be properly evaluated by management, even in dynamic organizations where processes or activity change frequently. This ability to discern performance is especially attractive where improvements arise from behaviour changes, which are considered more difficult to quantify than simple hardware upgrades.

One trap to avoid in the installation of an M&T system is the temptation to hold back in deploying the system until all the metering of the resources at a facility is complete. I have seen many programmes held back years (with the resultant loss of accumulated savings) as the Champions on-site push for elaborate automated sub-metering systems to feed data to the M&T tool. In reality, M&T can be effective with manually read metering and with a significant proportion of the resource being *"unaccounted"*. It is invariably better to start the improvement process and then incrementally enhance the data systems, than to defer the process awaiting a perfect metering and data acquisition infrastructure. Installing additional metering is common in the prepare phase but should not prevent progress using the metering already available.

A final characteristic of M&T which reinforces its role as the foundation tool for resource efficiency, is the way that it can stimulate new insight and curiosity about the operation of systems. When the folks that manage these processes are given the data on performance and have an improvement target, they will start to question the status quo. They will ask themselves where the variability in resource use compared to activity comes from, they will question the fixed loads that are not influenced by activity and they will want to know why apparently similar equipment behave differently. These questions may lead to further investigations and deeper insights into the specific equipment and processes. Out of this will then arise changes in operation, maintenance or even the business case to upgrade or replace the equipment. Although M&T is the key Optimize phase tool, it is a fantastic supplier of opportunities into the Modify and Transform phase of our programme, where knowledge of the pinch-points and inefficiencies of our processes can be addressed in the capital-replacement cycle or expansion plans of the organization.

While most organizations' resource efficiency efforts focus on the use of resources in operations, this is not the only area where we may want to modify decision-making. In Figure 7.3, illustrating the Method, there is a box in Prepare titled Make System Changes. Here, I am referring to many different tools, techniques and standards which can help influence decision-making across various activities. For example, we could introduce a minimum Energy Star rating target for our buildings which will influence all future construction and retrofit projects. This is a system change where we are using an existing standard to influence the choices designers and engineers are making.

7.5 **Just some of the Systems within an organization that may be modified to achieve greater resource efficiency**
Source: Niall Enright.

System	Sector	Decision Support Tool or Standard
Operation	All	Departmental cost allocation
Design	Buildings	Building codes
Design	Buildings	Building ratings (e.g. Energy Performance Certificates)
Design	Buildings	Equipment efficiency (e.g, Energy Star, Energy Technology List)
Design	Buildings	Sustainability rating (e.g. LEED or BREEAM)
Maintenance	All	Planned preventative maintenance and condition monitoring
Maintenance	All	Reduce standardization (don't oversize!)
Procurement	All	Financial appraisal (use internal rate of return not payback)
Procurement	All	Whole Life costing
Energy Use	All	Carbon prices and internal emissions trading
Waste	All	Allocate disposal costs to waste-generator
Transport	Logistics	Route optimization

The table above illustrates just a few of the systems that exist to support decision-making. As our resource efficiency programme matures and develops, we are likely to modify many (if not all) of these systems to ensure that they are aligned with our improvement goals.

Although it is wise not to bite off too many changes at one time, there are a couple of areas of wider decision-making which I often tackle in the prepare phase of my resource efficiency programme:

- Departmental cost allocation is a powerful incentive for end-users of energy and resources to use less. It shifts accountability from a centralized engineering function to the shop floor.

- I would try to move the basis for approving investments away from a payback to an internal rate of return (or other more mainstream method of cost evaluation), ideally using whole life costing, as this can significantly increase the availability of funds for resource efficiency projects.

We can see a direct connection between these tools and the three basic improvement types described in Table 6.15 on page 244. Our M&T tool is driving the first type of opportunity, Optimize, improved operation of existing processes, while the focus on systems will support the third category of improvement, Transform. Our next tool, although broad in its focus, is particularly suited to driving Modify improvements through equipment upgrades and replacement.

The Opportunities Database is the other foundation tool for resource efficiency. Its purpose is to capture all ideas for improvement and ensure that these progress through six stages: study needed; study begun; feasible; funding requested; in progress; achieved. There are two other categories to deal with projects which are not practicable, infeasible, or which are not being considered at present, archived.

By tracking the movement of projects from one stage to the next the organization can ensure that there is a continual flow of new ideas, and that there are no bottlenecks in the evaluation, approval and implementation process. Thus, if there is a large number of projects at the funding requested stage one can question whether sufficient resources are being made available. Alternatively, if the bottleneck is at the study required stage, then we know we need to put more resources behind assessing the viability of the ideas.

One of the characteristics of initial phases of energy and resource efficiency programmes is that they can throw up a large number of relatively small projects, which individually may not be seen to be material, but which in the aggregate offer a very dramatic improvement with a highly attractive return. Projects falling into this category would include things like presence detectors on restroom lights, flow restrictions on hoses, low-resistance filters on air handling units and so forth. These kinds of low-value projects often fail to progress because they are perceived to be so small, but when they are added together and the value appreciated, there is suddenly a big push to get them to happen.

There is much more information about the Opportunities Database in the Techniques section. Once again, this tool usually takes the form of a software application, either a spreadsheet, database or commercial energy management application. Many organizations have some existing project capture and management systems which may fulfil the requirements of the Opportunities Database. If this is the case, then the advice is always to use existing systems, as long as the key functionality of the Opportunities Database to track project status is available.

At the end of the prepare stage, the organization will have an M&T system and an Opportunities Database in place and will be ready to train the Champion and other users before deploying these tools and processes.

Real World: *Skills and effort*

Earlier, the DICE analysis on page 284 identified the ability of our people to deliver changes as a key success key factor.

An important part of our prepare activities must be to assess two key elements in our people's capacity:

- The knowledge and skills they have and will need to succeed;
- Their available time to engage in the programme.

People at the operating level of the organization, as a minimum, will need to have training in the M&T process and the Opportunities Database. They will need to be briefed on the reports that will be produced and given help on how to identify and deliver improvement opportunities.

We also need to allow time for them to contemplate the programme and to provide input into its focus, the targets that will apply to their area of operation and to discuss the implications for them. Without this contribution to the programme design, we will not gain ownership.

Deploying tools, changing systems and engaging with and training folks will require an initial peak of effort (as shown in Figure 4.15 on page 193). It is essential that the planning process quantifies the nature of this effort and ensures that resources and business demands are appropriate.

It may well be the case that some external consultancy support may be needed in order to get over this initial peak. These external contributions will need to be deployed carefully so that they do not compromise the long-term objective of embedding resource efficiency into day-to-day operations.

7.5 The two core tools compared

Our two core resource efficiency tools offer us complementary approaches to improvement. Because they underpin different sources of improvement, they should almost always be implemented together.

Only by addressing all sources of improvement - better operation and better equipment - can we achieve the best possible resource efficiency.

M&T and the Opportunities Databases are different but complementary tools.

Let us consider a simple energy-consuming system - lighting. We can manage improvements, i.e. achieve a reduction in the electricity consumption, in one of two ways. We could monitor the electrical use by comparing it to the daylight hours to ensure that the lights are not left on unnecessarily. This is an example of operational control - we are running existing equipment better, which M&T achieves

Alternatively, we could carry out a study to see if we can replace the existing lamps with more efficient models, such as using light emitting diodes (LEDs), which give the same amount of light for a lower electrical consumption. This is an example of a project-based saving - we are making discrete, one-off improvements. Here, the improvement idea would be managed with an Opportunities Database.

An important difference between these projects is the funding basis for the improvements. The monitoring of lighting energy use on an ongoing basis would tend to be paid for as an ordinary operating cost of the business and so come from operational expenditure, OPEX (although the metering and software needed for M&T, may be considered capital expenditure, or CAPEX, items). See page 189 for these definitions.

A project to replace the lamps with LED models will almost certainly require capital expenditure or CAPEX. The distinction between CAPEX and OPEX is important because, in most organizations, there are different approval and planning processes for these funds.

There are other contrasts. M&T requires data analysis and change management skills, to engage the users in the process so that they respond to the variances. Project opportunities usually need strong technical and financial assessment skills. Many engineers are comfortable working on discrete projects so they are at ease with implementing an Opportunities Database and like the fact that, once completed, the project savings will be secured and they can move on to the next opportunity. In fact, the training of engineers tends to be very project-centric. On the other hand, the people engagement aspects of driving operational savings are far less familiar to project engineers, so M&T is sometimes perceived as "*heavy lifting*" by engineering or process folk.

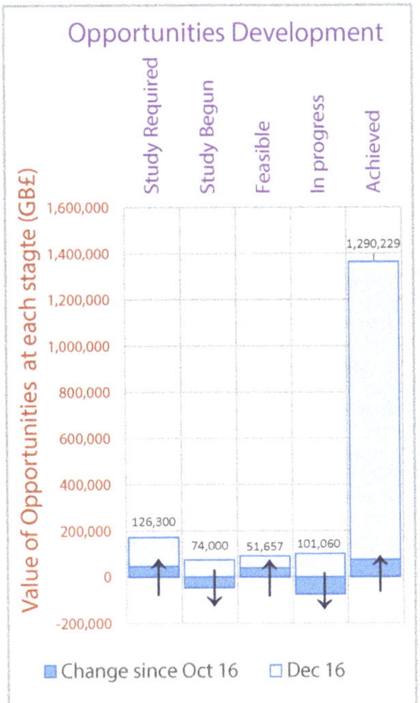

7.6 **Opportunities development**
The Opportunities Database should report on the movement of ideas from origination to completion. We are especially interested in seeing the change in value or number of projects at each stage of development. Unfortunately, at the time of writing, none of the commercial software packages provide this analysis, so I have created a prototype, using Peel Land & Property Group data, shown above.
Source: Niall Enright
Spreadsheet available in the companion file pack

The table below emphasizes some of the differences between programmes focused on operational improvements, using M&T, and those focusing on capital investments, using an Opportunities Database. When deploying our programme, it is important to understand the differences between these approaches.

Just because all CAPEX projects would be managed through an Opportunities Database, we should not fall into the trap of thinking that *only* capital projects can be tracked in this way. We could have an idea for a one-off process change - a real world example would be changing the temperature to cook baked beans, which would involve no capital cost at all but needs a rigorous feasibility assessment before it can be implemented. Thus, operational changes can also be driven by the Opportunities Database, as can system changes such as the introduction of new standards and procedures. Likewise, M&T, although focused on operational decisions, will inevitably produce ideas for one-off capital investments.

Improvements in almost all organizations require both strategies - operating equipment optimally and updating equipment to the most efficient models - so there are very few circumstances in which M&T and an Opportunities Database would not be deployed together. Particular caution needs to be taken to avoid taking a project-centric, opportunities-only approach, as this can bypass the cheapest 10-15% improvements and leave the organization devoid of systems and processes to sustain savings in the long run, even if substantial equipment improvements have been made.

7.7 **Comparison of OPEX and CAPEX resource efficiency efforts**
Although a resource efficiency programme should almost always have strategies focusing on both OPEX and CAPEX, understanding the differences between these approaches can help ensure we get the most from our programme.
Source: Niall Enright

	Operational Costs (OPEX)	*Capital Costs (CAPEX)*
Main Improvement Tool	Monitoring and Targeting	Opportunities Database
Description	Savings through behaviour change	Savings through equipment change (upgrade/replace)
Speed to deploy	Quick(er)	Slow(er)
Return on investment	Higher (typically < 1 year payback)	Lower (Typically > 2 years payback)
Primary Skills	Change Management	Engineering
Perceived effort	"Heavy Lifting"	"Green Plug"; "Fire and Forget"
Savings Potential	Reliably, 10%-15% ; up to 25%	c20%-60% reduction possible
Primary Emphasis	Demand-side (use of resource)	Supply-side (resource conversion)
Verification	Metering: on-going M&T	Engineering calculations
Strengths	Brings people on board so more sustainable	Doesn't require people so may be easier to implement
Weaknesses	Needs people's time in order to succeed	Behaviour can undermine the technology

7.5 The two core tools compared

7.6 Deploy

Rolling out our programme on-site is about ensuring that there is sufficient high-quality two-way communication to inform, reassure, and excite the folks who will be driving the improvement. We need to be receptive to their inputs in order to create ownership for the process.

"You only get one chance to make a good first impression." - various

The key personnel are engaged. Roles are clear. The tools have been installed. Governance systems are working. Training resources are in place. It's Monday morning and we need to roll out our programme. Where to begin?

We can be guided by the fact that our programme is a change management process. This suggests that our first objective must be to win over the hearts and minds of the folks at the facility where we are rolling our programme out. We want to be able to demonstrate to people that this process:

- does not require excessive additional effort;
- will put people in a position to gain recognition and positive feedback;
- supports existing priorities;
- is urgent and is not optional;
- requires their active participation; but
- does not threaten.

There is a sort of "*good cop, bad cop*" feel to many of the programme launch activities I have been involved in. The bad cop is the Leader, who must articulate the compelling need for change, the urgency to achieve change and the fact that participation is not optional. Balancing this strong message is the good cop, usually the programme Champion, who will reassure people that the process will be manageable, supported by appropriate tools and resources, will provide positive rewards and (definitely!) be fun. There is a tightrope to walk here; we want to ensure that there is a strong requirement to act without, at the same time, making people feel powerless to engage with the specifics over which they have control (see the piece on *The importance of urgency when driving change* on page 321).

For example, we could launch the programme with a presentation to staff covering the reason to act, the scope, goals and people involved. The message should engage at a practical and an emotional level. The programme launch could be accompanied with the announcement of a suggestion scheme and the gathering of initial ideas, which creates some ownership over the solutions. If at all possible one or more of these initial ideas should be immediately put into action (or at least approved). In this way, the message will be sent: "*This*

Real World: *The initial target-setting*

No one likes targets.

People often have a load of performance targets to meet. So when I appear on the scene and start to talk to them about setting additional energy or resource efficiency targets their initial negative reaction is entirely understandable.

The only way we are going to get acceptance of a target in these circumstances is if we can demonstrate that it is honest, fair and achievable.

The first of these means that the target is set using an appropriate method - for example, by using the historical performance of a process or activity to set the objective rather than the theoretical design efficiency. The method must be able to distinguish good from bad, rather than simply describe a goal.

For a target to be seen to be fair, the effort and improvement sought must be considered comparable with the effort that others are expected to make. That is not to say that every target should be the same - squeezing a small improvement out of a well-controlled system is more challenging than getting a larger improvement from a poorer system.

Finally, we must be able to show we can achieve the target. One way to do this is to demonstrate that we are already meeting the target regularly (e.g. on at least half the occasions).

Using the best-fit line from regression we can create targets that people accept as *fair, honest and achievable.* The key is that the targets are explained well and they are involved in choosing the variables and the baseline periods. More on this fascinating topic in the section on Data Analysis on page 460.

programme is about results." The quicker we can provide evidence to people that we mean business, the faster they are likely to become involved.

As soon as practical after the launch, ideally a matter of hours or days, staff in the key resource-consuming functions should have a one-to-one session with the programme Champion, where their resource use is analysed over the previous year or so and a baseline established (using production or other key variables). They should be invited to set their own targets (the line of best fit being the obvious candidate). The message should be clear that the aim is to *repeat the good, eliminate the bad* through *honest and achievable targets* that they themselves control.

This point about ownership cannot be emphasized enough. There may be an overall programme goal of a 10% improvement in resource use, but that does not mean that every production line or building should have a 10% target. Some may have a bigger target based on the observed pattern of resource use and the opportunities already identified, and some may have a lower target. In fact, it is usually better to have a modest target which is seen as realistic than an ambitious target which folks do not believe in. Targets can always be changed over time, once people have achieved the first objective they will be more inclined to accept a new target, whereas nothing is more demotivating than chasing an unattainable target.

The choice of people to focus on early should be based on their influence over resource use, the degree to which they may be perceived as role models and the enthusiasm with which they would be expected to embrace the programme. Representatives from unions or work councils, if present, would usually be involved in the process from the outset.

The key message throughout should be "it is your programme". The aim is to authorize people to take action. Where people express doubts over their ability to influence aspects of their resource use, these concerns should not be dismissed. It may well be the case that the engineering team feel that the operations team are the people who have the greatest control, while the operations team may feel that the maintenance team are the only people who can actually change the equipment, while engineering and maintenance complain of not having the budget to replace old equipment. Every such objection and barrier should be recorded and, critically, the person raising the objection should be asked to propose a solution. The sooner folks appreciate that this is their programme, the sooner they will start to engage.

When we meet with the resource users, we should not forget to ask them for improvement ideas which can go into the Opportunities Database. We must make a significant effort to get at least some of these ideas assessed as quickly as possible and, where feasible, implemented. Nothing will be a better signal that this rapid response to their contribution.

In our Systems column for the deploy stage (see Figure 7.3 on page 269) the key activity is to implement effective reporting. This reporting involves a bottom-up process: first, we will start to report on the specific targets for each piece of equipment, process stage, building or activity which have been agreed with our resource users. Around each of these performance targets, we will almost certainly set a *traffic light* process of reporting by exception - for example, a green symbol indicates a significantly better performance than expected, a red symbol a worse than expected performance and a blue symbol a performance in line with expectation.

Media City ISO:50001

Readings | Cost | CO2e

Default Meter Target Report: 01/06/2015 - 30/06/2015 / (Reading)

Meter	Target Ranges	Unit	Consumption	Target	Variance	% Variance
Blue Tower - Main Utilities						
V PML Blue		kWh	50,353	49,350	1,003	2.033
Orange Tower - Landlord Services						
V PML Orange		kWh	90,253	80,700	9,553	11.84
The Garage - Main Utilities						
A MSCP LV Switchboard (SCC/FAREBROTHER)		kWh	40,278	52,770	-12,492	-23.67
The Pie Factory Limited - Main Utilities						
M Pie Factory Gas		kWh	0.00	10,067	-10,067	-100.0
A Pie Factory HH Electricity		kWh	83,135	104,070	-20,935	-20.12
White Tower - Landlord Services						
V PML White		kWh	11,116	11,700	-584.4	-4.995
		Grand Total:	**275,134**	**308,657**	**-33,522**	**-10.86**

7.8 **A simple, but effective, target report**
This report shows the performance of key buildings at MediaCityUK against targets that use a range of factors such as cooling degree days, lighting or occupancy. Simple traffic lights give an instant indication of performance. The software tool used to produce this report is Verco's *CarbonDesktop*,™ The author assisted in implementing the energy management, M&T and ISO 50001 systems at this site over a number of years. See the case study on page 232 for more information.
Source: reproduced with kind permission from Peel Media Ltd

The next reporting requirement is to aggregate the performance up at the various levels of management, such as at facility, department or shift. Since we may be tracking many resources that are non-additive (e.g. we can't add a kWh of electricity to a litre of water saved), these aggregate reports typically use the cost savings or losses for the resources, which can be added together to show a total. Since we have an expected use and an actual use we can also have an expected and actual costs and we can provide percentage improvements (bearing in mind that we need to be cautious in circumstances where resource costs are volatile as this can mask underlying consumption changes).

In a similar fashion, we can provide reports on the number of new projects identified in the Opportunities Database by department, the number and value of the projects implemented, the rate of return achieved and so forth.

The deploy phase is all about **engaging people positively**, creating visibility and getting some **quick wins**.

The aim of these intermediary-level reports is to provide management at every level, all the way to the top, with a summary of the resource efficiency programme performance. These tiers of scrutiny are a key part of maintaining the momentum for the process as questions will be raised if reporting ceases.

Generally speaking, it is better if we integrate the reporting within the existing reporting and management processes. We want our report to be part of the Monday morning briefing and planning pack, or part of our monthly departmental reports and the quarterly board reports. Thus, the more that we make the resource management *"part of the ordinary way we do things"* from the start, the more this will prove true in the long run.

Having put our reporting processes in place, it would be a good idea for the programme Champion to attend each of the management meetings in order to introduce the reports. Ideally, the senior manager at the meeting will have been briefed to ask questions about the performance of specific items, teams, or processes. They should have been briefed to focus on any especially good performance (*"green exceptions"*), to congratulate the team and ask what brought this about and how it can be repeated. Where a team is struggling (*"red exceptions"*), the emphasis is on understanding the cause of the variation and providing help to eliminate it (*"what resources do you need to investigate or fix this?"*). Ideally, the discussion on resource use will be short, focused and solution-oriented. While organizational cultures and management styles vary widely, my personal experience has been that programmes that focus on praising and repeating good performance tend to achieve greater support than those which just focus on criticizing and eliminating bad performance. After all, if you see that you will be praised for doing well then you will have a big incentive to understand and eliminate poor performance as well as repeat good performance.

It is important that everyone who receives the reports or has access to the M&T software has been trained in interpreting the reports. They should understand how the information produced by these systems can point them in the direction of an improvement or better decision.

As well as engaging our organization's vertical management structures in our programme, it is worth considering if some additional cross-cutting learning networks are appropriate. It may well be desirable to have a lead person for the programme in each department or facility and to host a regular meeting for them to share experience, receive training, work across the organization and feel part of the bigger programme.

The last part of our deploy phase, illustrated under the Technology column, is the act of eliminating bad performance and repeating good, and finding and implementing feasible projects. This is listed under Technology as the actual change will generally impact on resource-consuming equipment or processes, but improvement can also consider the wider Systems within our organizations which might be contributing to poor performance.

7.7 Celebrate success

Once people are engaged and settling in to the day-to-day operation of the programme, we need to keep them motivated and informed of progress.

7.9 **Focusing on Success**
When considering the scope of our resource efficiency programme it is worth thinking in terms of the early success that we can create. Everyone naturally wants to be involved in a programme that is going well.
Source: © michaeljung Fotolia

The emphasis on celebration in the Method is quite deliberate. It has not come about from a "*happy clappy*" enthusiasms for resource efficiency, the promotion of a "*team hug*" culture, or a preference for levity and frivolity over hard organizational realities.

This Method stresses positive messages because they play a powerful role in achieving the results that we are seeking. Celebration:

- enables us to change the feedback loops that govern people's decision-making;
- creates a positive context for the programme and reinforces desirable behaviours;
- encourages folks to verify results more systematically, to justify and articulate the gains openly and transparently;
- can create a healthy competition between teams;
- underpins continual improvement rather than one-off quick fixes;
- supports the Leader in defending the programme to stakeholders;
- correctly designed, can encourage some risk and innovation;
- may enable our resource efficiency programme to out-compete other organizational initiatives in terms of mindshare and adoption.

The People part of our Method specifies that we will *celebrate success with individuals and teams*, which includes, as a minimum, those folks at the front-line of our programme, their managers and the Governance team.

The Systems heading for the celebrate phase of our Method states that we will *track value and relate this to the core objectives of our stakeholders*. This reminds us that we should always be able to place our efforts in the context of the core organizational objectives. That is to say, we should be able to quantify the Value, capital V, that the programme is delivering, whether that is in terms of lower cost, increased profit, enhanced service delivery or greater asset value. So the processes to translate results into this measure of Value should be in place from the start of our programme.

Real World: *What is success?*

On the face of it, success is simple to define - it is the outcome of our resource efficiency projects.

This is the Value described in Chapter 3, a *lagging* indicator, which includes:

- The number of projects implemented
- Reduced operating costs
- Greater profits
- Improved share price
- Competitive advantage
- Greater asset valuation
- Lower risk or improved reliability
- Greater product throughput
- Improved brand awareness
- Greater market share
- Enhanced service delivery
- Rating, certification or other scoring systems measures

Alternatively, we could define success is terms of the desired behaviours from people or the processes we want to put in place.

These are the *leading* indicators of our programme:

- Amount invested
- Suggestion scheme ideas raised
- % of resource use metered
- Examples of risk-taking, innovation or radical thinking
- Training hours delivered

Our Leader may have a view on what success looks like, but the folks at the operating level may see this differently, and their views should be canvassed. There is no reason why we can't report different measures to different audiences if that is feasible and will engage them.

However, as the box to the left shows, there are many other measures of success that we may wish to celebrate, other than the Value delivered. At the start of our programme we may want to focus on leading indicators, such as the number of improvement suggestions received, instead of the overall results, because it is too early to quantify the outcomes.

There are also subtle judgements to be made about whether formal rewards and incentives should accompany our celebration of success. We need to be aware that some teams may be able to achieve rapid success, not because they are particularly able or committed but simply because they happen to be in a function that is a large resource consumer (and maybe one where resources are not well-controlled). Singling these out for praise may be demotivating to other teams, which may have to work much harder to achieve less. While absolute improvement is good for headline-grabbing, where we are linking effort into reward schemes, we tend to do this through the attainment of targets or goals that take into account factors that adjust for the circumstances in each team. There is more on motivation and reward systems on page 676.

Celebration and rewards encourage engagement and participation. We should take full advantage of this enthusiasm to develop our people's appreciation of the wider aspects of resource use. Through our communications, mentoring, coaching and one-to-one sessions we want to enable teams to:

- Learn to identify the real causes of variance (both good and bad), through root cause analysis, and to take ownership for managing this variance;
- Understand how to develop the most compelling business cases for action and to bundle projects together to have good opportunities support the case for weaker ones.

Despite our focus on the positive, and our careful preparations for success, we should anticipate that things may not, despite the best intentions, turn out as expected. Where initiatives fail or projects disappoint, we need to address these in a positive fashion. Just as good feedback encourages participation, disapproval is a huge turn-off. If something goes wrong, it is essential that the response is not to blame but to learn lessons, to acknowledge (and if appropriate praise) the effort, to encourage the team involved to try again. We should be problem-solvers, not blame-givers.

One of our biggest barriers to innovation will be fear of failure. If folks feel frightened to take a risk or to propose radical change or see that failure leads to punishment, then our ability to take our programme forward beyond a few obvious *low-hanging fruit* will be severely damaged. Indeed, we need to celebrate risk-taking (small r), and radical ideas, not just the measured outcomes, in order to support this innovation. People will be led by what we give visibility to and to what we praise. This is another reason why it is important to celebrate leading indicators at the outset of our programme. While results may be difficult to predict, we know that the number of projects identified is likely to be a product of effort and so well within our control.

7.8 Change

Chasing early success will create the initial momentum needed to embed our programme. However, we must not lose sight of the longer-term, more substantial changes that our organization may have to achieve to truly transform its resource use.

This is the point, when things seem to be going very well, that many resource efficiency programmes lose their way and set up the conditions for their ***ultimate failure.***

Let's imagine that we are a few months into our programme. By now we should have deployed our M&T and Opportunities Database tools, which are being used to produce regular reports of progress and engage teams across our facilities. We will probably have quite an extensive list of actions and investments which we believe will deliver efficiency gains at acceptable financial returns.

At this point, it is tempting to sit back and let the process continue. It is probable that we have a considerable pipeline of opportunities to work through, so there will be a lot to do. The initial success may lead to facilities which are not part of the programme asking to participate, creating yet more work.

Unfortunately, it is usually now, when things seem to be going very well, that many resource efficiency programmes lose their way and set up the conditions for their ultimate failure. The mistake that most programmes make is to focus exclusively on Optimize projects - that is to say, *quick win* projects that deliver value by improving existing systems and processes.

There is an additional category of improvements, called Modify, that involves larger capital expenditures to upgrade equipment. Modify opportunities may also require changing established Systems, something that can be time-consuming and meet resistance within organizations.

More radically, we have Transform opportunities, which involve a fundamental change in the organization's processes or business model which can result in a dramatic step-change in resource used (and, paradoxically, render much of the Optimize and Modify changes irrelevant, causing us to revisit these types of opportunities).

Because Modify and Transform improvements don't offer the same return on investment or effort as the Optimize improvement, they are often "*parked*" for action in the future. The problem is that, by the time these are revisited, the Optimize opportunities are coming to an end and the long timescale and cost to get even Modify, let alone Transform, opportunities off the ground now appears prohibitive. The only way that these opportunities have a hope of success is if they are considered right from the outset of the resource efficiency programme, where the initial success of Optimize (and the resultant celebration) can release resources to focus on the longer-term projects.

Exploration: ***"How many engineers does it take to change a light bulb?"***

- ***None.*** *That's an electrician's job.*
- ***One.*** *Once they gather requirements, obtain financial approval, get signatures in triplicate from various departments, raise purchase orders for the bulb, procure the bulb and complete end-user acceptance tests.*
- ***Two.*** *One always leaves in the middle of the project.*
- ***Four.*** *One to design the change, one to implement it, one to document it and one to maintain it afterwards.*

Corny, unfunny, perhaps a little close to the bone, these are just a few of the *"How many engineers..."* jokes on the internet. But do any of these ring true for your organization? If you were to select the key"functions" in your organization (doctors, salespeople, production managers, accountants), how would you characterize *their* ability to change resource use?

Asking ourselves *"How many"* people, or hours of time, consultant fees, approvals, etc., are needed to change inefficient equipment or systems can be very illuminating, as it can help us shine a spotlight on the systems barriers to improvement (*ok - no more lighting puns, promise!*).

You could consider posing the question *"How many {xxxxxxx} does it take to change a light bulb?"* as a tool to engage employees in suggestions for improvement.

[Apologies to any engineers offended by the incorrect use of the word "bulb", when of course we know that we are replacing a "lamp". Bulbs grow in the garden, whereas lamps provide light!]

The change phase of this Method ensures that we incorporate longer-term projects into our programme from the outset. This phase is about harnessing some of the initial enthusiasm and goodwill towards the project (which ideally will have been generated by our early success) into the more difficult task of Modify and Transform improvements. Change is designed to create the parallel improvement programme shown earlier in Figure 6.15 on page 244.

Our first task, under the heading of People, is to create a "*culture of positive challenge to change systems, not just technology*". A bit wordy, but what it is saying is that we are encouraging folks to dare to challenge the way things are done. For example, we might look at our capital cost allocation process and establish if whole life costing or marginal costing (techniques described later in the finance section) are used. It may take some time to introduce these changes, but the impact could be great in terms of investment decision-making. So what we want to do is to encourage people across all the organization to reassess the way things are done in the light of resource efficiency.

In parallel to this, we want to introduce "*root cause analysis*" as a technique to encourage people to assess why certain inefficiencies exist in their organization. Root cause analysis enables us to look beyond the superficial reason why equipment or processes are inefficient into the system's causes. For example, we may observe that our steam production uses more gas than expected; this might be caused by the age of the boiler being used, which in turn may reflect the low investment in new equipment. The reason for this may be a shortage of capital for equipment upgrades, which in turn is due to prioritization of investments away from efficiency measures. Understanding that the root cause is an investment issue, rather than a technology issue, means that we can then set about resolving the problem. We can either increase the available capital (see Volume II for topics such as third-party funding) or change the organization's priorities. Making more money available may require the Leader to act on our behalf - yet another reminder that management commitment is critical. Even if the root cause is not immediately rectifiable (and this is rarely the case), it should be identified and recorded as an area for future action.

On the Technology front, we should be looking from the outset to systematize and embed the improvements we have identified into standards, specifications and best-practice methods. While equipment recommissioning, maintenance, repair or upgrades may have dealt with the source of inefficiency, we want to ensure that our business processes do not enable the problem to *recur*. In other words, we want to put in place systems around our choice and operation of equipment that fully incorporates the learning we have achieved into our ongoing processes. There will often be a temptation to skip these activities - after all, we may have *solved* the problem and the benefit of this additional effort may appear to be low - but unless we put in place the necessary preventative measures to avoid repetition we cannot truly claim to have fixed the problem.

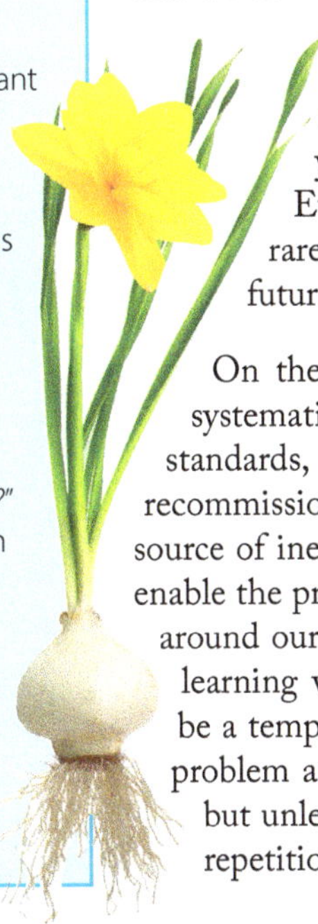

Photo © Valentina R - Fotolia

7.9 The right mindset

In our enthusiasm to recognise progress achieved, we need to avoid the trap of becoming too backward-looking. Looking ahead and anticipating the outcome of decisions are characteristics that the best programmes share.

Our M&T tool is incredibly powerful in helping us to interpret *past* performance. It helps us to understand whether yesterday, last week or last month were periods of good and bad performance, which gives those who are close to the resource use the insight needed to *repeat the good and eliminate the bad.*

However, by its nature, this is a backward-looking process, so I want to close this chapter by emphasizing the importance of a future-facing *mindset* among those empowered to act on resource use. I want to introduce what I call anticipatory thinking (which is I have discovered, is a term used in academia and systems thinking, where it has a slightly different meaning).

For our resource efficiency programme, it is really helpful if we can get folks to think *forward* in terms of their resource use. "*Next week I plan to produce x tonnes and will be running lines 4 and 5, so I need to watch out for the daily shut-down process, deal with the leak on the compressed air line to 4 and take the opportunity to experiment with minimizing the material losses on the cleaning in process on these lines*".

Anticipatory thinking goes beyond mere **prediction**.

By encouraging people to consider a range of possibilities we are supporting innovation and building a greater capacity to respond to the unexpected.

Anticipatory thinking goes beyond mere *prediction* of resource use (which can be derived from the regression-based targets for each activity once the variable terms are known), and considers and prepares for a range of scenarios, not simply the most likely one. Anticipatory thinking will help teams to respond to unexpected changes (perhaps by putting additional measurement or focusing extra attention on possible problem areas). In fact, in anticipatory thinking, you *mustn't* assume everything will go as predicted.

Anticipatory thinking is necessary because one can never change the past. All change takes place in the present using knowledge and data about the past and the future. The right mindset will question the meaning of data; experiment with small/test changes to derive new data; forecast the expected results and check against these as they arrive; recognize potential problems before they arise; and put in place appropriate observation or contingencies.

This thinking does not occur spontaneously. It needs to be nurtured and taught across all the activities described here. The Leader and Champion can encourage this thinking by posing as many future-facing questions: "*What if...*", "*How will...*", "*What's needed...*", "*When should....*", as backward-facing questions, such as "*Why did...*", "*What happened...*", "*Who was...*".

Exploration: *How this Method came about*

This Method is not an abstract concept, but a proven approach to resource efficiency that has been developed over many years in hundreds of projects worldwide in all sectors, from resource extraction through to manufacturing and public services.

The foundations of the methodology are in the work of the UK's Energy Efficiency Office in the mid-1980s, working with sector trade associations to promote "*Computerized Energy Monitoring and Targeting*". With the arrival of the personal computer in the early 1980s, it was becoming practicable to use statistical techniques, drawn from statistical process control (SPC), to manage energy data. In particular, the techniques of regression analysis and CUSUM could identify otherwise undetectable deviations in energy use, which could be used to set performance targets. The basic concept of M&T was operational excellence – "*repeat good variances and eliminate bad*".

I became heavily involved in this methodology when I joined March Consulting Group (now part of Jacobs Engineering) in 1991 and took responsibility for an M&T software application called Montage™, developed by Jane Galloway and Ray Gluckman. Montage™ went on to become the leading global M&T tool for manufacturing industry, selling many hundreds of licences worldwide.

In the mid-1990s, colleagues at March Consulting Group and I were unhappy about the surprisingly high drop-out rate for the M&T programmes. To address these issues, Keith Webster and I went on to develop the enManage™ methodology (standing for energy or environmental management) which incorporated much more formalized Governance processes. The main aim was to reduce the drop-out rate by getting the programme to dedicate as much effort on the People aspects of the process as on the Technology side. The methodology was essentially a compilation of what the consultants saw as best practice from the many M&T programmes that March Consulting Group had implemented at that time. The methodology came about because of the realization that:

1. Energy management is as much about change management as it is about project identification; and
2. A key weakness in previous approaches was the lack of management commitment within the client organizations.

What emerged from these discussions was empirical and practical, not theoretical. For example, an important innovation was to introduce an Opportunities Database into the process alongside the statistical tools of M&T, so that discrete efficiency measures could be captured and driven to completion. Thus, enManage™ catered for both an operational improvement and a projects-driven approach to energy efficiency. Around that time, Keith Webster saw that the process could be usefully communicated under the three headings "*People, Systems and Technology*", and this formed the structure that we used to categorize tasks. Coinciding with the development of enManage™, there was a strong push for waste minimization in UK industry. Furthermore, March Consulting Group was also working heavily in Eastern Europe in an EU-funded energy efficiency technology transfer programme. This enabled the enManage™ approach to be tested, refined and proven for resources other than energy, such as water, wastewater, raw materials and solid wastes, as well as in a wide range of different countries and cultures. An extensive programme of work in North America and in extremely large and complex facilities, such as BP's refineries, allowed for further development of the methodology into the early 2000s, with key inputs from Chris Stubbs, Brian Turner, Kevin Ball and Tim Sullivan. Collaborations with North American Utilities, through pioneering work with Donald Gilligan at Predicate, tested the approach in other US manufacturing sites.

Just as enManage™ borrowed heavily from M&T, the QUEST™ methodology I went on to develop later with global consultancy ERM, around 2010, also incorporated M&T concepts. QUEST introduced additional techniques to the process, notably the cost determination and reduction workshop developed by Peter Fink and Martin Hess at ERM's Frankfurt office, which had its origins in Six Sigma project identification techniques. Quest™ (which stands for "Quick Energy Savings Technique") has been successfully implemented by ERM in a number of large energy and resource efficiency projects in Central Europe, China and South Africa, and is also a key service of Human Element Consulting led by Arne Springorum in the Czech Republic, while enManage™ continues to be a core offering from Jacobs. At the same time, many other consultancies and software providers are offering similar methodologies based on the foundations of M&T.

This history is presented so that the practical origins of the methodology outlined here are understood, and the reader has confidence that this is a proven approach. The Method described here represents a synthesis of these various best-practice approaches which continue to demonstrate their effectiveness today. The Framework takes these successful strategies one step further by emphasizing the initial Mandate for action and the systems changes that will ensure long-term success. Many individuals and organizations deserve credit for developing these techniques over many years, not just those mentioned here. They are the real heroes of resource efficiency.

Summary:

1. The levers for change in any organization are its People, Technology and Systems. A successful programme will balance these.
2. People are able to undermine Systems and Technology, however good these are, so make sure you carry folks with you.
3. Most resource efficiency programmes involve a wide range of people and functions, Mapping these roles and understanding when they should be engaged is important to developing a successful plan.
4. The core of our Method is empowering people to make better-informed decisions. Success will come from focusing on those activities, systems, tools and processes that will support better decision-making at all levels. Judge your plan in this light.
5. There are two critical tools that you will need in most programmes: Monitoring and Targeting to drive operational improvement, and an Opportunities Database to drive projects and investments. These are tools to better inform decisions and so create visibility and accountability for resource use. Even if your organization uses other systems or metrics, such as SixSigma, OEE, Lean or TPM, you will need these resource-specific tools.
6. Strike a careful balance between delivering early success (low-hanging fruit) through Optimize projects and initiating the longer-term Modify and Transform initiatives which will have big impacts in the future.
7. When you deploy your programme on the shop floor make sure that it is not a *fait accompli* and that folks feel that they can influence their targets and make suggestions on the priorities of the programme. Their ownership of the process is critical.
8. Give people permission to challenge the current systems, to innovate without fear of failing and to anticipate the effect of their actions and choices.
9. Characteristics such as enthusiasm, good data, large resource use, relevance and stability are desirable in our initial programme facilities.
10. Looking ahead is key to being able to influence resource use. Encouraging anticipatory thinking is an important part of the resource efficiency programme.

Questions:

1. Do you agree that people are key to the success of a resource efficiency programme? Why?
2. How are decisions around resource use made in your organization? What is the input data? What systems support these decisions and how could they be improved?
3. Consider Figure 7.4 on page 276 showing the impact of a range of functions on different resource efficiency activities. Design a similar matrix for your organization (see the companion file resources). How does it differ from the illustration in this Framework?
4. What motivates people in your organization? Don't just think in terms of explicit rewards and incentives, but also of the hidden drivers, such as curiosity, discovery, variety, recognition and so forth. How would you go about harnessing these drivers in support of a resource efficiency programme?

8 Momentum

In this chapter, we will consider what is needed to ensure that our resource efficiency programme is successful in the long run.

Momentum builds on the solid foundations of our day-to-day Method and the continual improvement systems put in place in that phase. It is the success of these earlier activities that will carry us forward to the next stage. Although we should start to focus effort on Momentum activities during our Method phase, we should not do this at the expense of the success of the Method activities. Indeed, if the Method activities are struggling it may be desirable to take a step backwards to see if further work around the Mandate might resolve the issue.

As with the previous stages of our Framework, Momentum has three separate elements, which will need to be aligned to the specific nature and needs of our organization.

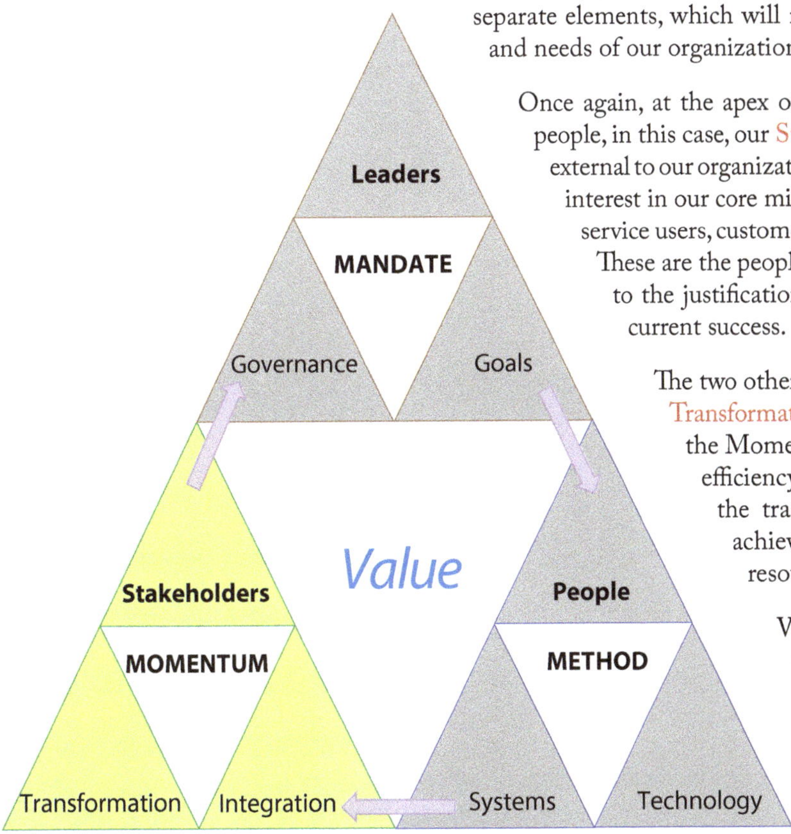

Once again, at the apex of this phase, and in bold, are a group of people, in this case, our Stakeholders. These are the folks which are external to our organization who have the greatest influence on and interest in our core mission. I am thinking here of shareholders, service users, customers, regulators, media or campaign groups. These are the people who will keep us honest and contribute to the justification for our organization going beyond our current success.

The two other parts of our Method are Integration and Transformation. These are taken to mean *outcomes* from the Momentum phase - the integration of resource efficiency into day-to-day decision-making, and the transformation of our business models to achieve even more radical improvements in resource use (and Value).

While it may appear that Momentum is a defensive phase (designed to mitigate premature declarations of success or changes in management/fashion/ sentiment), this is actually one of the most creative, varied and exciting phases of our programme.

8.1 Stakeholders

Our organization does not operate in a vacuum. As we seek to extend the value of resource efficiency we will find that engaging with external stakeholders will bring considerable benefits.

The Momentum phase is ***not an end point;*** *rather it is the* ***stepping-off point for a new Mandate.***

Stakeholders' engagement is the essential ingredient for the long-term success of our resource efficiency programme. This is for three reasons.

1. Much of the Value associated with our programme will only be realized through effective Stakeholder communications.
2. Ongoing feedback to Stakeholders about the programme will significantly reduce the possibility of a premature end to the programme arising from a change in fashion, organizational focus or leadership, or as a result of easy wins being exhausted.
3. The organization, or Leader, is likely to need Stakeholder support to gain a Mandate for more radical transformations in resource use that may require significant investments or a change in business model.

The good news is that the involvement of Stakeholders is almost inevitable once a programme starts to achieve concrete results in terms of cost savings and environmental impact. This is because, as Chapter 3 on Value has amply demonstrated, the benefits in terms of competitiveness, brand, reputation, compliance, risk and asset valuation that resource efficiency can deliver are largely a product of communication with Stakeholders.

Brand value arises from the perception of customers, which is a product of marketing and public relations. Additional asset value for owners is created through a premium associated with the future cash flow our better-performing assets will deliver. Value related to lower risk arises from an appreciation of the decreased exposure to resource-related harm by insurers or investors. Corporate communications can position our efforts positively with regulators or planning authorities and so improve our licence to operate. Better staff recruitment and retention can only come about if we effectively communicate a distinctive and positive vision of our organization's contribution. Competitive advantage arises from quantifying and communicating the superior efficiency of our product or service (e.g. a lower carbon content) up the supply chain.

Most organizations these days understand the importance of good stakeholder engagement and may have many functions, such as corporate affairs, marketing, legal and investor relations which will seek to control the messages to Stakeholders. These functions are often far removed from the activities that have driven the efficiency programme in the Method phase, such as operations, engineering, maintenance and so forth. A specific programme of engagement with the *stakeholder management functions* will almost certainly need to be developed by the Governance team or Champion.

It is not just the internal interlocutors with stakeholders who will be embraced in this phase of the programme. As we shall see in the next section, many other functions within our organization will be engaged as we seek to integrate our programme into the wider range of decision-making systems. Finance, procurement, design and business planning are just some of the teams which can have a large influence on resource use.

One way of looking at the Momentum phase is to see it as a marketing campaign to expand our programme into new constituencies, internal and external. For this campaign to succeed it needs to have several characteristics:

- It should be carefully planned;
- It should be based on proven success (i.e. it builds on the facts created by the programme in the Method phase);
- It should have strong elements of co-design. Just as we need to be open to the depth of experience of those on the shop floor as we deploy our programme and create ownership, we also need to recognize that the specialist teams we are about to engage have a unique body of knowledge to offer.

Many of the pitching techniques described later in Volume II Chapter 9, Creating a Mandate, are equally applicable here. For example, we may be able to more effectively gain a team's support if the proposal comes via a person or function that they trust. We need to sell the *benefits, not the features* and understand the "*pebble in the shoe*". Simple techniques such as pairwise comparison can be used here to map support or resistance.

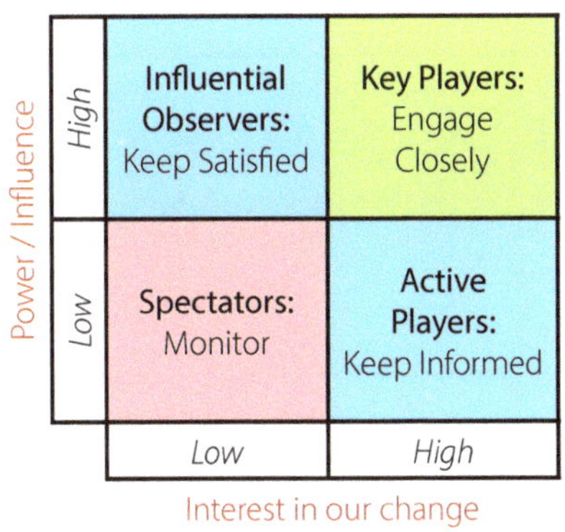

8.1 **An influence/interest matrix**
Source: Adapted from Patrick Mayfield, "Stakeholder Strategy" in The Effective Change Managers Handbook[668]

The activities in this Momentum phase are illustrated on the next page. Because of the varied nature of organizations and their ambition for change, the process is necessarily generic and only some steps may be relevant. Nevertheless, it provides a helpful reminder that the more carefully we assess and plan our expansion of the programme from the current core, the better the outcome.

One useful planning technique is to place our stakeholders and internal functions on an influence/interest matrix, illustrated left. Not only can this help us establish just who we need to target but also to differentiate between those we want merely to keep informed and those we need to invite to become active participants in our programme.

Tasks such as prioritizing functions and stakeholders, form part of the scope and prepare activities in this phase. It is anticipated that the Governance team will play a major role in the planning process. At this point, there may be a number of changes we want to develop - more straightforward systems changes such as project approval processes, and more profound changes such as

MOMENTUM: Transformation based on an established track record			
	PEOPLE	SYSTEMS	TECHNOLOGY
Scope	Develop and gain support for an engagement plan Influence/Interest Matrix	Identify key Systems to change, e.g. CAPEX allocation, KPIs	
Prepare	Develop the pitch, test and then engage	Co-develop Systems and communications changes	Identify break-through technologies Technology Learning Curves
Deploy	Communicate the change, establish SMART targets, update Governance to reflect wider participation	Deploy the Systems changes, and implement reporting against targets	Monitor, experiment, test and evaluate break through technologies
Celebrate	Celebrate success with individuals and teams	Track and report value CUSUM	
Change	Reassess Mandate in light of wider stakeholder and functional engagement	Integrate change into the core mission of the organization	Further innovation

8.2 **Activities to achieve Momentum**
There are a number of activities that will drive our programme beyond simple optimization of the existing system to a more radical transformation. These are categorized under the headings People, Systems and Technology.
Source: Niall Enright, a Visio version is available in the companion file pack as well as an A3 poster version describing all the tasks in the Framework.

an investment in technological innovation or engagement of key Stakeholders in some aspect of policy or goal-setting for the next stage of our programme.

It is probable that there will be many different initiatives or projects within this Momentum phase, each of which will potentially follow the steps set out in Figure 8.2, opposite. Some changes will be quick to implement, others may take years of patient work or may require certain preconditions before they can be executed.

A general assumption in moving into an active Momentum phase, is that we have achieved some success in our Optimize activities in the Method phase. If our programme has been unable to achieve traction on the *low-hanging fruit*, we need to question whether we have a sufficient Mandate to move into the more challenging (in terms of costs and risks) Momentum activities. This is, of course, a generalization, and there may be situations that favour profound systems changes from the outset (for example, see *Rio Tinto and CAPEX on page 569*). One reason for wanting there to be solid results before engaging wider functions in our organization and external Stakeholders is the importance of gaining trust in many of these interactions. If we can focus our discussion on verifiable benefits, using objective data, our arguments become considerably more persuasive.

This issue of trust is of particular importance if one wishes to engage in partnership with non-government organizations (NGOs) or regulators. These stakeholders may well start from a position of distrust and may fear that the proposed association is intended to merely "*greenwash*" our image. Indeed, engagement may bring about a negative response from the external body's own staff or supporters. I know how sensitive this issue can be because, many years ago, I was head of corporate fundraising for Oxfam, a leading development charity, and it was essential to make sure that the integrity of the organization was not compromised by relationships with donors who were behaving in a manner incompatible with Oxfam's objectives and values.

Aside from the Value benefits arising from better communications, the active involvement of Stakeholders will make it more difficult for our organization to end the resource efficiency programme prematurely. Indeed, active Stakeholders, together with the wider group of engaged internal functions, will very likely cause the existing Mandate to be re-examined. This, in turn, may empower the Leader to drive the programme further, with new goals and a reinvigorated Governance team, which will create a new Mandate for action. This Mandate will be delivered using the now-established Method systems and processes coupled with the newly enrolled functions and Stakeholders. And so the cycle will begin again, with a refreshed and updated vision and an appetite to tackle more profound change. Thus, the Momentum phase is not an end point; rather, it is the stepping off point for a new Mandate. The final set of Change activities illustrated opposite are about explicitly reassessing and reinvigorating our Mandate in the light of the learning and success to date and the needs of the latest participants.

Real World: ***B&Q and FSC***

Establishing an NGO to support a resource efficiency goal may seem a little radical. But that is exactly what the progressive UK home improvements retailer B&Q did in the early 1990s, when it helped to set up a certification and labelling system for sustainably procured wood, the Forestry Stewardship Council.

Now, almost 25 years later, in a recent seminar by the Carbon Disclosure Programme,[128] Ian Cheshire, CEO of the Kingfisher group, which owns B&Q, explained that timber remains a very real issue for the company:

> *"Most businesses have a set of resources which are particularly critical to them. In our case at Kingfisher we really have timber as a big issue because about a third of our products, about 16,000 product lines in the average store, contain some form of timber. And we started on a journey which was a) where does it come from and b) will it be around in the future and it is much more serious because the available supply of certified wood is coming under pressure...*
>
> *So, for us, we have a clear issue and a clear long-term business issue as well as wanting to have sustainable products. So we are focusing very hard on that issue. For Nestle, it might be water. But every business will have one or two critical resource issues which they need to start addressing."*

Ian Cheshire is articulating the idea that external stakeholders can help solve resource challenges and that an active partnership can benefit business and the environment.

8.2 Integration

For our resource efficiency programme to be sustained we want it to be integrated into the day-to-day processes in our organization. We can measure integration in terms of the depth (i.e. quality, impact or persistence) and the breadth (i.e. scope or reach) of the changes we have achieved.

One way of thinking about our organization is as one huge system designed to deliver our particular service or business objective. This system is made up of many processes or subsystems. A hospital has processes that deliver treatment to patients in line with guidelines and standards of clinical practice; there are processes to recruit, train and supervise staff; processes to ensure that the environment of the hospital is appropriate, that the nutrition of patients and staff is adequate; other processes that record patient data and recover costs from funders; there are reporting, strategy and planning processes that enable the hospital to respond to change over time; there are also processes for communicating, informing and motivating staff and patients alike. While not all processes have a bearing on resource efficiency, it is often surprising just how many of them do.

We integrate our resource efficiency improvements into the day-to-day operation of our organization by changing existing decision-making processes and modifying the overall system to support the change. We want the more efficient approach to become "*the way we do things around here*".

Some changes that we may make to our system are more profound and longer-lasting than others. For example, setting a new standard for an item of equipment (e.g. specifying high-efficiency motors) will be beneficial, but is potentially reversible (people can choose to ignore the standard, or modify it to buy the cheapest motors at all times, rather than the most efficient). On the other hand, publicly setting a goal to be the most resource-efficient organization in our sector in a decade, and reporting regularly against this aspiration, is a much more profound change (although BP's climate change goals show that these types of pronouncements are not necessarily set in stone either).

8.3 **System leverage points**
There are numerous forms of systems changes, some of which have greater impact and longevity.
Source: Bob Doppelt, "Leading change towards sustainability"[222] modified by Niall Enright.

#	Description	Leverage Points
1	Change the dominant mindset out of which the current system arose	Vision, business model and purpose
2	Rearrange the parts of the system	Organization and structure
3	Alter the goals of the systems	Goals
4	Restructure the rules of engagement of the system	Workflow, decisions and authority
5	Shift the flows of information and communication in the system	Understanding and collaboration
6	Correct the feedback loops of the system	Performance indicators, incentives
7	Adjust the parameters of the system	Standards and targets

Reflecting on this issue of the effectiveness of systems changes, Bob Doppelt has taken the work of Donella Meadows[516] (see page 764) and John Kotter[448] and devised a hierarchy of seven types of system change, in descending order of effectiveness, illustrated opposite. In this model, the least significant change that can be made is to change a parameter of the system while the most effective modify the core mission of the organization.

Let us consider the activities in our Method phase. Here, we introduced two new tools or processes into our organization, Monitoring and Targeting (M&T) and an Opportunities Database. These tools enable us to set targets (in the one case operationally and in the other in terms of projects) and adjust the parameters of our system. The question we can ask ourselves is *"how can we increase the integration these tools?"* If we look at the figure below, we can imagine changes at each point in the system to increase the effectiveness of these tools:

Leverage point	Description of the change	Impact
Parameters	Our M&T system and Opportunities Database are providing new parameters (targets) for operational performance.	
Feedback	A method of reporting the targets across the various groups is implemented, along with incentives to act on the reports.	
Information	Training, technical assistance and networks to collaborate are put in place. People now have the knowledge needed to make improvements based on the reports they receive.	
Rules	People are given greater authority to act on the improvement and allocate resources for the fix. (e.g. they have a credit card to enable low-cost improvements to be fast-tracked).	Low High
Goals	The business unit's overall goals are adjusted to incorporate the outcomes our processes can deliver (e.g. these are firmly set in annual plans). Management are now closely involved.	
Structure	Accountability is aligned with our processes, (e.g. energy costs are allocated at the operating level rather than being treated as an overhead. Maintenance, Engineering and Operations are now jointly responsible for resource use rather than Finance who treat this as a fixed overhead).	
Purpose	We define one of our business objectives as being the most resource-efficient producer in our sector and so our new tools are critical to the organization.	

8.4 **System leverage points for method tools**
We can achieve greater value and integration from our Monitoring and Targeting and Opportunities Database tools by understanding how these can be leveraged in our existing systems.
Source: Niall Enright

As well as being a useful way to identify additional changes to our system, we can also use the System leverage points model to assess the impact of changes we have put in place already and how likely these are to endure in the long run by categorizing the changes into the seven different levels. This represents a measure of the depth, or quality, of the change made.

Real World: *CAPEX allocation at Peel*

I have worked for the last nine years with a main board director at Peel Land & Property Group, David Glover. He is chair of Peel's Sustainability Group but also leads a team of project directors who deliver Peel's ambitious development programme (£50 billion spend, in 50 projects, over 50 years).

With construction costs of up to £500 million per project, you can imagine that the project directors are very skilled at meeting the three core aims of any development: on time, on budget and in specification.

In thinking how we could introduce further consideration of energy efficiency and sustainability into the construction process, it became clear that making a change to the three objectives above was unnecessary and undesirable. Nor was adding a fourth criterion *"is sustainable"* the way to go - all that would happen is that it would be secondary to the other three established objectives.

The solution was remarkably simple and effective. We built sustainability into two of the three criteria. We started to include whole life costing for large energy-consuming equipment. The second change was to aim for BREEAM Excellent as the target specification for every major building (Very Good as minimum).

This is a great example of how a system can be transformed by simply changing the inputs to a decision-making process rather than the process itself. The rules did not change: project directors were still tasked to drive the lowest costs that will achieve the specification, on time. Only this cost now incorporates data on the in-use phase, not just the initial cost, and the specification incorporates asset value-adding sustainability features.

It is often helpful also to assess the extent or the breadth of our changes. At one level this is a simple measure of the reach of our new tools. Thus, if all our major production units are using the new M&T and Opportunities Database tools, then this has much greater breadth than if only one or two sites have adopted these processes. But breadth can also be measured in more subtle ways - for example, if our finance team are relying on the relationships between production volumes and energy use to procure energy most cost-effectively, we have created an additional *customer* for our M&T system data. If our corporate communications team are using the verified savings from our Opportunities Database to gain positive PR with our investors, then the impact of our system is reaching further into our organization. The previous description of Stakeholder engagement is largely about activities which extend the breadth of our programme.

Another way to think about how we integrate resource efficiency into our organization is not to think about the activities that we have put in place and how these can be better tied in, but rather to look at the systems that already exists and to consider how they impact on resource use. We could, for example, use the illustration of the functions and processes in our organization in Figure 7.4 on page 276, as a starting point to consider the potential for that function or process to affect resource use. Perhaps we could put these roles and processes in an influence/interest matrix, as shown in Figure 8.1 on page 297.

Having determined which processes or functions are likely to have some impact, then we can go about assessing how decisions are currently made and how consideration of resources can be incorporated. Again, we can use the Systems leverage points to brainstorm possible points where change can be made.

In an ideal world, we will aim for a change at the top of the list of effective leverage points, but this can in practice be pretty difficult to achieve. Resource efficiency is not the only priority for an organization and fundamental changes of purpose, structures, goals - with the disruption they can bring - are often simply out of the question. We are likely to find that changes are limited to setting targets or standards, putting feedback processes in place and increasing knowledge, but are not able to touch on authority, organization or power.

It is important that we acknowledge the limits of the changes that we can introduce. While parameter changes may be theoretically the weakest form of intervention in a system, in many cases these are the only possible change we can make. For example, there is one function, finance, and one process, capital allocation, that are almost always seen to have a significant impact on resource use. However, the processes around finance are often very resistant to change (in part because they touch on the power structures of our organization), and so the strategy that I most often use in these circumstances, is not to change the way in which decisions are made but to change the data or inputs into the decision. The real-world case left, demonstrates that changing the rules of a system is sometimes unnecessary and counter-productive, and that there may be other ways of achieving the same objective.

As with previous interventions, it is important when we are considering system changes that we involve those affected in co-developing the change. One philosophy that I tend to use is that of "*lightening the load*" wherever possible. This was an expression that I learnt from folks working for a global mining company in Australia who said that the effect of sustainability initiatives from HQ was to add to the "*deadweight*" that they had to carry simply in order to do their job.

The Liberal Democrat and Conservative coalition government in the UK introduced in 2013 the idea of "*one in, two out*" for almost all new regulations.[371] This would not be a bad principle to follow when considering the changes we want our resource efficiency programme to make.

The reason that folks often fear and resist changes to systems is the presumption is that there will be additional effort associated with these changes. Resistance occurs because compliance processes govern most systems: a mix of requirements, rules, regulations, standards, measurements and policies which people need to understand and must follow. Adherence to these compliance processes is often achieved by some form of undesirable sanction.

8.5 **People fear compliance, not change**
Resistance to new systems and processes often arises because people fear the compliance burden that these will bring, not because they object to the aims of the change.
Source: image: © cacaroot, Fotolia.com

More generally we can think of two basic approaches: principles vs rules. Often it is more effective to set out some overarching principles of resource use rather than defined rules. This approach has numerous advantages: there is likely to be greater acceptance; the guidance can be applied in a wide range of circumstances, as they arise; it reduces the risk that our changes will become merely a "*box-ticking*" exercise; it empowers people to take responsibility and apply their own professional judgement. In my experience, in an organization where there are rewards, recognition, support and celebration around success, a principles-based approach can be a much more effective way to integrate resource efficiency than a rules and sanctions regime.

8.3 Transformation

Simply optimizing our operations and introducing new equipment with more efficient models at the end of its lifetime will get us so far in terms of resource efficiency. To go further we need to identify and deliver more profound changes. This involves innovation, design, and planning over a longer time frame.

While integration is about consolidating the existing changes in our resource efficiency process, Transformation is about taking our organization to the next level. The boundary between these is naturally blurred, but as our Framework illustrates, they are both strongly dictated by the needs of Stakeholders.

Although our initial mandate may not allow us to change our business model in a fundamental way, we need to recognize that resource consumption is likely to remain a long-term challenge for our organization - an issue which may well become acuter in the future, and which may demand more significant interventions.

It is, therefore, prudent to embark on work on identifying Transformation opportunities at the earliest possible moment in our resource efficiency programme. Our analysis of Transformation opportunities is likely to produce a set of projects which fall under the following headings.

1. Viable and value-adding interventions that can be initiated now but which may take a long time to complete.

2. A Transformation that may become feasible in the event of some change in conditions in the future, which we can anticipate now. For example, some opportunities may only be viable when major plant reaches its end of life, or when there are factory shutdowns, or when a technology price-point reaches a certain level (see page 358 on *Technology Learning Curves*). These are projects for which a lead-in time can be estimated, and which make sense from a Value perspective, and so should be planned for.

3. Changes which are not currently viable and which are not anticipated to be value-adding given current predictions (e.g. of utility costs, policy. customer demand, etc), but which may need to be implemented to meet regulatory or other objectives.

These Transforming opportunities may be identified as part of a strategy development process, or may just emerge as we encourage people to think more profoundly about improvements. What is important is that these are not simply set aside for a future date or plan, but become part of the programme of work in our organization (they can be entered as Opportunities in our Opportunities Database like any others). We must avoid at all cost deferring action on these ideas until after the Optimize savings are delivered, as this will run a risk that the timescales and costs are too great for these changes to receive support at that point.

We should also recognize that, as our focus changes from the existing operations where Optimize and Modify opportunities dominate, to the Transform projects, we will be likely to engage new functions in our organization.

A particularly influential function is design, in its broadest sense: those who design products or services, and those who design facilities and buildings, or those who design supply or logistics systems. The great industrial designer, Victor Papanek, accused designers of creating useless, unnecessary and unsafe products; of wastefully propagating product obsolescence; of creating *"stuff-lust"* that promoted materialistic lifestyles. If designers do have such a power for evil, then the converse can be said to be true – that product and service designers can achieve very significant improvements in product efficiency.

The new ideas around the circular economy should challenge designers of goods and services to think more radically. In Volume II topics such as Design for the Environment (e.g. screwing things together rather than gluing them so that they can be disassembled), page 754, and Life Cycle Management, page 440, are providing the tools to understand where in a design resource impacts are greatest. We have seen, for example, that the design of a detergent product's biggest impact is on the in-use phase of the product's life because the energy to heat water is much more significant than the energy to make the detergent in the first place.

Although the Momentum phase of our programme has been illustrated in the same People, Systems, Technology columns using the same scope, prepare, deploy, celebrate and change rows described in our Method phase, the reality is that the Transformation activities are the most variable and unpredictable stage. Whereas I can present a series of tasks that will enable the Optimize opportunities to be delivered in the Method phase, the Transform phase cannot be reduced to a simple formula.

Organizations that succeed in Transforming resource use, will do so because they can anticipate change better than others, because they can innovate more effectively, because they can scale changes quickly, because they understand the value proposition better, possibly because they are more willing to take risks.

It is sobering to think that the organizations that exhibit the above characteristics are most often disruptive new market entrants, rather than established businesses and institutions. Many large organizations have such a large *sunk cost* in the present way of doing things that they are unwilling or incapable of writing these assets off and moving forward with a radical new approach. Recognizing this reality, it might be appropriate for our organization to experiment with radical change in an *"incubator"* businesses. Perhaps our organization needs its own resource efficiency *"Skunk Works"*, as Lockheed-Martin calls their separate innovation division, where heretical, disruptive, frightening ideas can be imagined, tested and developed.

Real World: ***Rules for innovation***

The *Harvard Business Review*[554] summarized some simple rules that organizations can follow to innovate around sustainability.

- Don't start from the present. It is always better to gain consensus on the imagined future and then understand what steps will get there, than to project the present forward which usually leads to overoptimistic assumptions.
- Ensure that learning precedes investment. Smart companies take small steps, learn fast and then scale rapidly.
- Stay focused on the goal while always adjusting tactics. It is impossible to anticipate every event, so flexibility is important.
- Build collaborative capacity. Few innovations can be developed by one organization alone.
- Use a global presence to experiment. Organizations that operate in many markets may find that innovation is easier in emerging economies where there are fewer regulations and fewer entrenched ideas.

These basic rules will help most organizations develop Transformation opportunities.

Real World: *UPS and right turns*

It is important to recognize that energy and resource efficiency are not just about implementing technical projects but also about modifying systems.

One of my favourite examples is from the US, where vehicles drive on the right side of the road and can often make right turns at a junction even if the traffic light is red (so long as they have stopped and nothing is coming from the left).

Back in 2004 the major US delivery company UPS decided to design all its delivery routes to get rid of left turns. Within a couple of years around 90% of turns on UPS delivery routes were right turns!

So why are right turns preferable? Where a left-hand turn is required, the vehicle may have to wait for the light to turn green, thus wasting petrol and time.

That little system change has led to savings of 3 million US gallons of fuel in 2006,[629] and avoided sending out 1,100 trucks onto the road. Neat!

8.6 **A simple change can have a profound effect**
Transformation is about seeing the potential for these types of changes
Image: Vprisivko, Wikipedia CC3 licence[174]

Summary:

1. Engaging Stakeholders in our programme can significantly increase the Value from resource efficiency. It can reduce the risk of prematurely ending the programme and can strengthen the Mandate for change.

2. It is very important that we work in parallel on Optimize, Modify and Transform opportunities from the start. A sequential approach runs the risk that we never get to the more significant Modify and Transform projects because of the timescale and cost remaining to implement them.

3. Change is about much more than technology. Systems have many leverage points for change. Understanding and changing the highest of these points will help our programme to endure in the long term.

4. Transform opportunities may require us to abandon old, proven ways of working. We should see this not as a risk, but as an opportunity to move our organizations forward to the next phase of their development.

5. Design has a huge impact on resource use, and the effects of some design decisions (such as buildings) can endure for decades.

Questions:

1. What external organizations or stakeholders would have an interest or ability to influence your organization's resource efficiency? Place these in an influence/interest matrix, shown in Figure 8.1. How could they be engaged to support improvement? Are there risks in doing so?

2. Describe an aspect of design in your organization or with which you are familiar. How does this influence resource use? How could the designers approach their decisions differently to reduce resource use?

3. Is your organization or industry capable of a radical change in its business model, or will it require a new entrant to create disruptive change? Discuss.

4. Why is it important to consider longer-term and higher-cost opportunities from the outset of an energy and resource efficiency programme? Illustrate your arguments with some examples of Optimize, Modify and Transform opportunities.

5. Describe some energy or resource efficiency opportunities. For each, quantify their degree of reversibility and the measures that you could take to reduce the probability that the improvement will be undone.

9 Creating a Mandate

The Framework uses the word Mandate to convey the importance of authority and obligation at the heart of a resource efficiency programme.

The Mandate is often interpreted as the instruction to the organization, usually emanating from a Leader, which sets in motion the improvement process. However, there is a lot more to a Mandate than a diktat. There is the rationale that justifies the programme, the process whereby Leaders are engaged; there are the Goals of the programme, as well as the Governance structures. Altogether, these things form the Mandate and make the programme *happen* in our organization.

Of course, this type of Mandate is not unique to energy and resource efficiency, all types of change require similar foundations. Where energy and resource efficiency differ, is that, to maximize the results, they depend on the participation of an unusually broad range of people. This wide engagement means the Mandate often needs to be exceptionally strong.

In developing our Mandate, we would be wise to start from the assumption that there will almost certainly be some resistance to the changes proposed – particularly when the existing systems and processes are felt to be working well. In Volume I, we have seen that deep psychological factors, such as status quo bias and loss aversion are at work, as well as the usual problems that arise when existing power structures in an organization feel threatened. Some barriers, such as split incentives, require senior managers to reconcile the conflicting demands being made on staff.

Only a strong Mandate, with the visible support of the organization's Leaders, can overcome much of the resistance. Thus, the starting point for our programme, and the focus of the following pages, is how we can engage our Leaders sufficiently in the resource efficiency programme to provide a sufficiently strong Mandate.

If the initiator of the programme is a Leader, then they are already persuaded of the merits of resource efficiency, which is wonderful. However, most initiators are lower down the organization and so they need to develop a persuasive argument for change. US president Dwight Eisenhower once said: "*Leadership is the art of getting someone else to do something you want done because he wants to do it.*" The Champion or proponent for change needs to do just that. Here's how.

9.1 The proponent and the pitch

Change is about following a course of action. The proponent is the individual who defines and puts forward that course of action. They are the person who introduces and sells the notion of change.

Real World: ***It is a sale***

Getting commitment to a resource efficiency programme is a sales process. It may be disguised as a budget discussion or a technical assessment, but it is a sale nevertheless.

This idea can be quite frightening to many proponents. The world of sales may be alien to many advocates of change - maybe they are engineers, managers, service providers, EHS executives. They are possibly more comfortable with numbers than with *unique selling propositions*. They will often not have had any sales training and may even view selling in a negative light.

In reality, recognizing that our initial pitch is a sale opens us up to a whole raft of tools and techniques that can help us get to *"yes!"*

I am not advocating the hard-selling of forcing an unwanted product on a trusting victim. Quite the opposite. The sales techniques set out here will enable proponents to identify the key business or organizational need that their resource efficiency programme will address. They will help present the proposal in a more compelling fashion, and align people around the decision-maker to support our recommendations.

Selling is simply the art of listening, engaging and then persuading another person of the merits of your solution.

While energy and resource efficiency may already be established as a goal within an organization, or may have been initiated by external factors of stakeholders, it is the proponent who grasps the opportunity and develops a rationale for further or new action.

Sometimes, they are proposing change because it is their job to do so: perhaps they have an environmental, sustainability or corporate affairs role within their organization. Alternatively, they may be concerned with strategy, competitiveness or profitability and see how resource efficiency can deliver these. Possibly they do not have a formal role in setting out their organization's direction, but the need for change compels them to act.

In all these cases, the function of the proponent is the same. Their job is to sum up the problem that needs to be solved, or the opportunity that exists, in a way that engages decision-makers sufficiently for there to be a Mandate for change. It is a tough (but very rewarding) task.

As proponents vary hugely in terms of seniority, experience, age and influence, it is evident that there is no *one-size-fits-all* approach to winning management support. The culture of organizations, their willingness to change, the resources at their disposal, all differ. The scale and nature of the change being proposed also vary considerably - perhaps the first step is a *no-brainer* and will readily be accepted by the organization; on the other hand, maybe the proponent is working against a backdrop of a previous failure in this area and so has a real challenge of persuasion.

It may be that management has formally requested a proposal, or it may be the case that the proponent is taking the initiative themselves. If the latter is the case, the proponent may choose to go to management without a polished programme proposal, but instead with a request that they engage in developing such a proposal.

What the initiator of a resource efficiency programme needs to do is to sell the idea. They need to create the most compelling argument in support of the resource efficiency programme possible; in other words, they need to craft a sales pitch, *the statement and promise that make someone buy something*. Although we cannot prescribe the content of the pitch we certainly can suggest a roadmap for its development, opposite, which will increase the probability of success.

MANDATE:

(verb)
1. give (someone) authority to act in a certain way; and
2. require (something) to be done; make mandatory

(noun)
1. an official order or commission to do something; and
2. the authority to carry out a policy or course of action

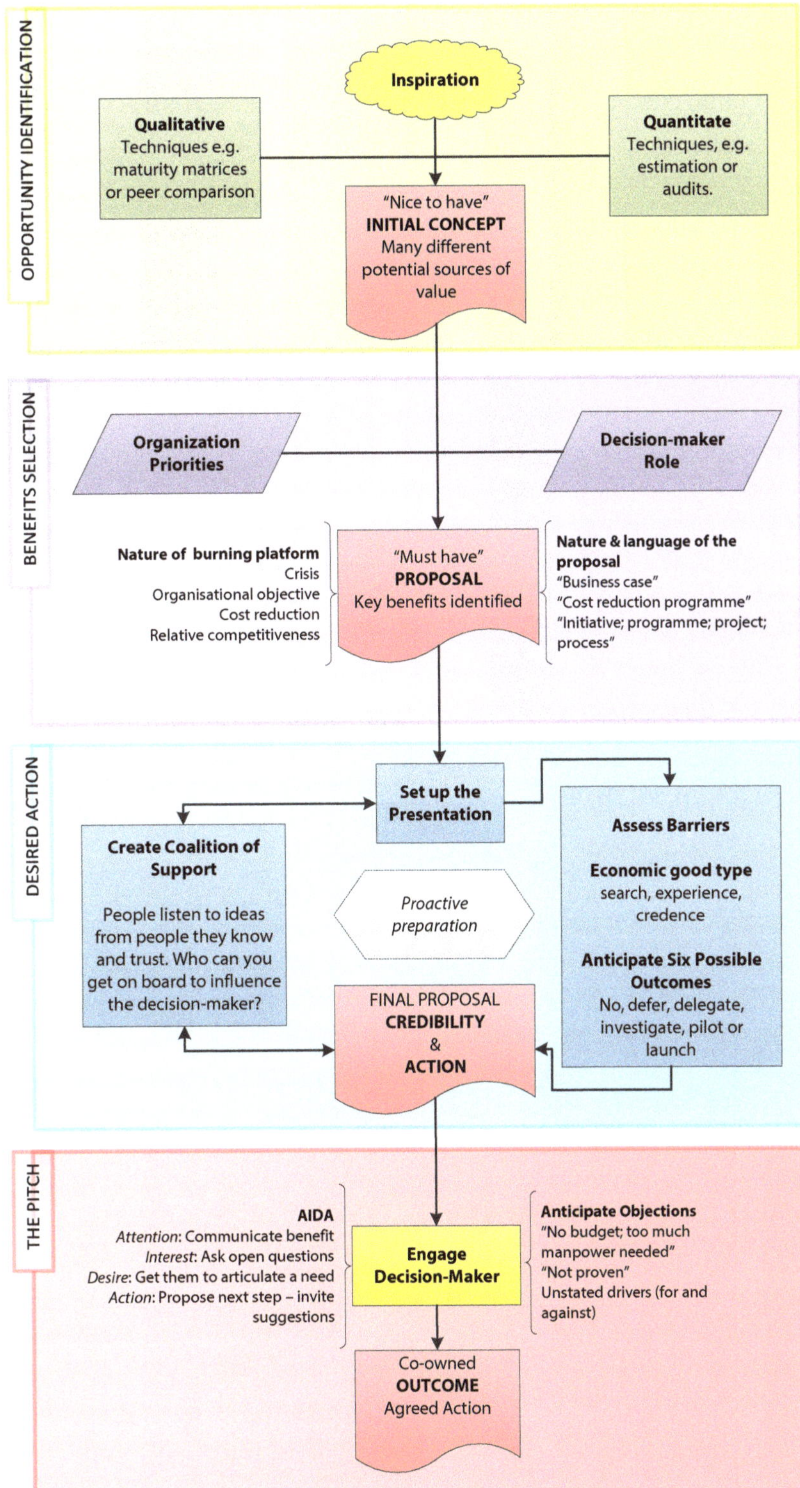

9.1 **Getting from an idea to a pitch the four stages of selling resource efficiency to management**
Preparation is a key to success, especially so for a proponent bottom-up selling the notion of a resource efficiency programme to a senior decision-maker. The activities illustrated here provide a useful check-list when preparing a proposal and a pitch.
Source: Niall Enright, available in the companion file pack

9.2 The proposal type

The first thing the proponent needs to decide is whether to pitch a detailed proposal or a conceptual proposal. The former increases the certainty in the decision-maker's mind but involves more effort, while the latter affords more opportunity to get the decision-makers to co-develop the programme.

The plan for a resource efficiency programme rarely emerges fully-formed in the mind of the proponent. The first task is, inevitably, to translate an aspiration to move our organization forward into an initial concept or proposal where some fundamental aspects of the programme have been determined.

Questions we are likely to be asking at this stage are:

- What is the scope of the programme? Will it be focused on our own operations or will it involve our suppliers or maybe our customers as well? Will the whole organization be involved or perhaps just a region, division, facility or function?
- What sorts of changes can we make to our existing activities which will impact positively on resource use? Will they save money or cost money? How do we compare to peers and others in our sector? What are our strengths and weaknesses?
- What are the trends in resource costs, regulation, risk and so forth which could influence our programme? What are the views of customers and other stakeholders?
- What are the potential benefits of the programme, both direct and indirect? How reliably can these be quantified, and what evidence will be needed to prove the benefits?

The most common mistake is for proposals to go into ***huge amount of detail on the technologies*** *to achieve the improvement but* ***far too little on the quantification of the financial and other benefits.***

Here, we need to make a fundamental decision – whether we want to obtain quantitative data to support the proposal or if we are happy to proceed with a more conceptual approach. The earlier section on how certainty influences the sales process, on page 185, illustrates the importance of selecting the right proposal type for the available evidence. If we choose to go down the route of a detailed quantitative proposal, backed with data, we need to make balanced judgements:

- about what level of supporting evidence is needed to provide the most credible quantification of the benefits; and
- about the degree of detail of the proposed programme. Too little detail will reduce the certainty of the outcome, but too much detail will reduce our ability to get the decision-makers to co-develop the programme and so take personal ownership for the process.

One could be forgiven for thinking that selling a conceptual approach will be less effective than putting forward a quantified and detailed proposal. However, there are some merits to this strategy which should not be dismissed.

Mandate

Generally, in a full cost-benefits proposal, it is more important to provide evidence for the benefits than set out the process in excessive detail. One of the most common mistakes is for the proposal to go into huge amounts of detail on the technologies and methods to achieve the improvement but far too little on demonstrating the financial or other benefits that will arise.

The alternative to a proposal based on quantitative data, is to approach the decision-maker with a high-level rationale for a programme. This justification might be based on what peers have achieved, or some simple estimation techniques to extrapolate to our own situation (see *Real World: Estimation using "the 10% rule" and the "rule of thirds"* on page 314). In this situation, the request made to the decision-maker is unlikely to involve a full-blown commitment to a programme, due to the lack of quantification of costs and the estimation of benefits. It is much more likely that we will introduce an intermediate step - maybe to set up a team to come up with recommendations, or to approve a series of audits or to participate in a workshop to define the programme focus. This is what we call a qualitative or conceptual proposal; we are selling the idea of embarking on an energy and resource efficiency journey, without specifying the final destination, but just one or more intermediate stops.

Because the final decision on the nature of the programme is usually deferred in a conceptual proposal, one could be forgiven for thinking that this strategy is not ideal – however, the approach has a number of merits:

- The initial request is relatively modest and so more likely to be approved;
- The amount of work that needs to be done in advance of the initial approach to the decision-maker is less and so the organizational appetite for resource efficiency can be tested early. It also means we can move forward quickly;
- Because we are not presenting a detailed programme or budget, there is much more opportunity for the decision-maker to participate in co-developing the programme, which increases their commitment and ownership;

I have seen many conceptual proposals succeed with just an initial high-level pitch. This is particularly the case where the organization is already broadly committed to efficiency or sustainability. These proposals often rely on introducing a high-level estimate of cost and benefits, to set the expectation of decision-makers and to test the availability of resource. In my experience, it is comparatively easy to quantify the potential sources of value from a resource efficiency programme. For example, if a competitor boasts of a 15% reduction in energy per unit product we could see what the same improvement would mean for our own organization. Given the amount of information in the public domain as well as some of the industry sector data earlier in this book and the organization's own knowledge, getting a credible estimate of potential savings for a resource efficiency programme should not be too taxing. The area where data is usually much more challenging is in estimating the costs (of time and money) to achieve the savings, as this is rarely in the public domain and can vary widely with location.

9.3 Opportunity identification

Quantitative proposals rely on the identification of specific opportunities for improvement in a process commonly referred to as an "Audit".

Although widely used, the term audit does not do justice to the plethora of tools and techniques available to uncover value in an organization. It also has negative connotations associated with financial audits. "Audit" somehow implies an external inspection of performance with resulting criticism for any deficiencies found. An audit is typically something that is done *to you* – not something that you embrace and look forward to. The word also implies a relatively static quantification of flows or resources – *"device x uses y resources a year"*, *"a total of z resource is used a year"* – much the way that a balance sheet or profit and loss statement in a financial audit illuminate what has happened, not what could happen.

I much prefer terms like "*ideas generation*", "*opportunity identification*", "*business case review*", "*resource analysis*" or "*value discovery*" which are much more descriptive of the collaborative, energizing, interesting and rewarding experience that an investigation into resource efficiency truly is! However, in the absence of a simple alternative, I will stick with the word audit here despite its disadvantages, simply because it is so widely used to describe the quantification of benefits.

It is important to emphasize from the outset that the most successful audits are always team efforts, engaging a range of expertise to collectively find changes that will reduce resource use. Key participants are the end-users of the resource; the finance team; logistics, engineering, maintenance and facilities professionals; as well as the local management team. As a consultant, I have often been in situations where the auditor is a very experienced external engineer from my own business who comes to a client site to conduct the audit and who invariably can rapidly identify the key focus areas and come up with a credible list of potential opportunities. But delivering a list of opportunities is not the purpose of the audit – the purpose is to create a list of opportunities that the site personnel believe are feasible, which they feel are worthwhile to implement and which they have ownership of long after the auditor has left. No matter how experienced the external auditor is, unless they understand why particular processes and systems operate as they do, what the rationale is for historic decisions, what the cultural, financial or other barriers to change are at a site, their recommendations are unlikely to be adopted. In essence, the auditor, whether internal or external, is marshalling the creativity knowledge and problem-solving skills of a wide range of individuals to create a set of recommendations that are collectively *owned*. This last point is too often overlooked.

An audit is typically something that is done to you – not something that you embrace and look forward to.

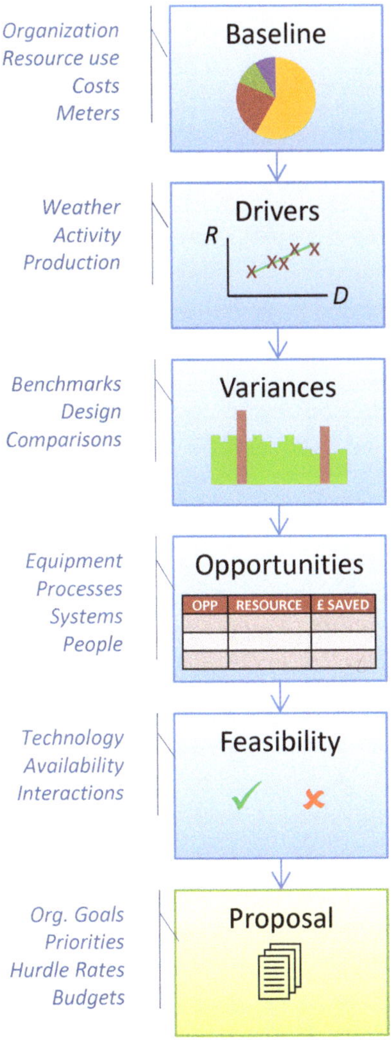

9.2 **Simplified audit process (shown in blue)**
Most audit processes start with baseline data collection. By comparison with drivers such as weather, activity or production, variances from best practice or design performance can be identified. This then highlights the resources, processes or equipment which merit detailed investigation. Performance assessment against design, benchmarks or similar items gives rise to an improvement opportunity. The "longlist" of opportunities is then assessed to determine which are feasible.
Source Niall Enright

An audit usually starts with a mundane quantification of the precise resource inputs into each part of the organization and the true (marginal) costs of these. This process is sometimes called baseline data collection. Where the resource efficiency programme will cover water and waste, a comparable assessment of the volume and costs for outflows of resources/wastes is also made. From this high-level mass balance, it is possible to determine which resources have the largest costs or variability and in which parts of the organization.

From this starting point, a more detailed investigation of selected resources will take place. First of all, the team will establish what work has been done historically to improve the efficiency of the target resources, whether there have been previous studies conducted and what recommendations have been made. They will seek to obtain much more detailed data such as half-hourly, daily, weekly or monthly measurements, as well as driver data on activity, operating hours, weather, production, square footage which could all help to understand current performance. If the audit is looking at the supply chain embedded carbon, data from suppliers or materials databases will also be used to provide a baseline. The auditor may also ask for information on the current supply contracts to assess if there is room for improvement in the pricing. This data will also help to determine how the resource savings should be valued – this should usually be at the marginal cost rather than the average costs of the resource (see page 576).

The next stage is to identify unusual patterns of resource use or variances. This can involve comparing the observed resource consumption with the design consumption or benchmark consumption. Statistical techniques like regression analysis can provide an accurate and reliable assessment of the scope for behavioural change improvement. This process of variance investigation can then quantify potential areas for improvement, called opportunities.

It is important that this stage of the audit is highly collaborative, with site personnel being able to contribute ideas and involve themselves in the investigation of opportunities. In the first place, the knowledge of the local resource users and managers is invaluable in identifying opportunities and explaining the cause of any variances. Secondly, unless these users are involved in the process they won't take ownership for the results, and so the programme is much more likely to fail.

Once the longlist of opportunities has been completed it is now necessary to determine which are viable or not. It should be noted here that feasibility is not limited to a technical assessment, but a whole range of practical considerations such as the impact on production, the expected lifetime of the measure, availability of staff to implement the change and opportunity costs that could arise. The feasibility assessment needs to consider the combinatorial effects of multiple changes on the same resources. For example, if we simultaneously replace lights for more efficient models and, at the same time, reduce the hours of operation of the lighting, the combined savings will not be the same as the sum of each project executed in isolation.

Real World: *Estimation using "the 10% rule" and the "rule of thirds"*

When making a conceptual pitch, the proponent often needs to have some rough estimate of the scale of the efficiency programme. There is some evidence of what other organizations have achieved (see Figure 3.7 on page 94) and the theoretical potential for various industries (Figure 3.4 on page 91), which can form the basis for an estimate. Where these data are not relevant, an alternative approach is to use *"the 10% rule"*.

This is a simple rule of thumb that I have used on many occasions, which states - for energy efficiency at least - that most organizations new to a continual improvement approach to resource efficiency can comfortably achieve 10% savings with a basket of projects which together have a one-year aggregate payback.

"Aggregate" here means a collection of projects, some of which might have an 18-month payback and some a six-month payback, which *collectively* produce a one-year payback or less.

This rule consistently applies to all but the most energy-intense manufacturing industries (such as metals like aluminium and steel, where energy is a big input and so is usually well controlled), as well as commercial, retail and public sectors. Where this rule applies, it is self-evident that the initial budget needed to achieve the one-year payback for the programme is 10% of the annual energy costs of the organization. A related rule is the *"rule of thirds"*, which states that this 10% cost can usually be broken down into three different headings: approximately a third on metering and small capital expenditure (CAPEX), a third on manpower to start the programme (internal or external) and a third on software tools to drive the programme.

I recently worked with an international tobacco company whose global utility costs (electricity, gas and water) were around US$120 million a year. I wanted to estimate if they had allocated sufficiency resource to their energy management programme to achieve a material impact (10% or more) on their utility consumption. Assuming that the 10% rule applied, it meant they should budget to spend at least US$12 million. In fact, this organization released a corporate fund of US$5 million for metering and software, and made additional funding for manpower and small projects available at each factory. From this I quickly concluded that the resources dedicated to the programme were in the right order of magnitude to have the desired impact, especially since there had been little work on energy efficiency historically, so there should be plenty of *low hanging fruit*.

Although the 10% may in practice be 8% or 12% or 20%, this rule of thumb nevertheless enables us to have a sense of the level of resources that may be needed to make our programme a success. In this way, even if we are putting forward a high-level conceptual programme, we can *"roll the wicket"* with the decision-maker to test whether they feel this level of funding could be made possible. Setting expectations is very important, and being able to provide an indication of the order of magnitude of the programme benefits and costs can elevate the seriousness with which our proposal is greeted. Our proposition could be: *"It is common for a resource efficiency programme to break even in the first year and achieve a 10% improvement. Thus, we could confidently expect to permanently reduce our operating costs by US$12 million per annum, for an initial investment of around US$12 million. Before committing to this budget, I am seeking your support for a series of audits to confirm the scale of the opportunity, which will only cost US$200,000 and will in themselves almost certainly lead to savings greater than the audit cost."*

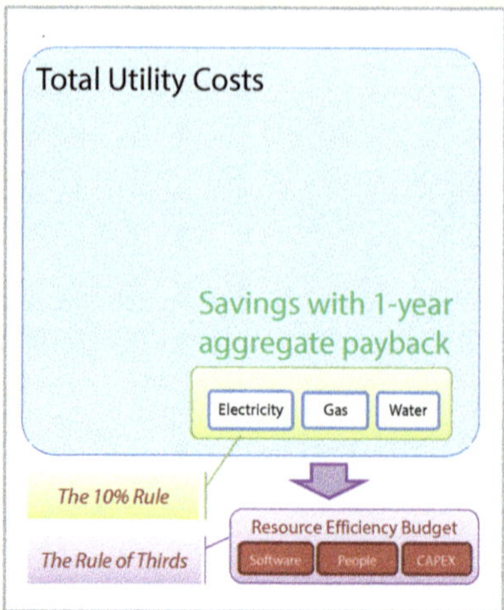

9.3 **Simple but effective rules of thumb**
Even a crude estimate of the scale of effort for a programme can help set expectations.
Source Niall Enright

This 10% rule is supported by much better evidence than merely my own long experience in the field of energy efficiency. Remember the study[328] quoted earlier by McKinsey & Company, which showed that efficiency savings of US$1.2 trillion were available for an investment of US$770 billion (US$520 billion on the projects and up to US$150 billion on programme costs)? Economy-wide that is less than eight months payback for a reduction of energy use of 23%. If anything, there is plenty of evidence to say that 10% is an underestimate of the potential savings with a one-year payback, as shown in a recent UK study covering waste, water and energy.[373] Finally, we need to take to heart Amory Lovins' message[483] that savings are more likely to rise over time than to fall, as resource costs increase and technology costs decrease.

9.4 Qualitative proposal scope

Mandate

Every resource efficiency proposal needs to have some definition of scope. While the focus of quantitative proposals usually flows from the most cost-effective opportunities, there are some tools that can be used to define their scope.

One of the most useful qualitative tools is what is rather grandly called a maturity matrix, which is simply a check-list of the stage of development of resource efficiency in an organization. Maturity matrices do not provide help with scaling our programme; for that we can use estimation techniques such as sample audits or the 10% rule (opposite). They do, however, help with scoping our programme, i.e. defining just what we want to focus on.

I have been using variants of a maturity matrix called the Energy Management Matrix very effectively for over 25 years. This matrix helps to engage folks in a dialogue about the *qualitative* nature of their energy management processes. The matrix identifies current versus ideal performance in six key areas and so helps to define the scope and focus of our programme, as well as to communicate the notion that energy management is about much more than technology. It is also useful in determining if there are any gaps in perception between different folks, for example between the management team and shop floor. See page 346 for a much more sophisticated example of this engagement process.

9.4 **An energy management matrix**
This example is taken from The Carbon Trust[177] in the UK. This matrix has been around for many years and variants can be found in the UK (where it originated) the US, Canada, Australia and many other countries.
Source: Carbon Trust

Score	Policy	Organizing	Training	Performance Measurement	Communicating	Investment
4	Energy policy action plan and regular review have active commitment of top management	Fully integrated into management structure with clear accountability for energy consumption	Appropriate and comprehensive staff training tailored to identified needs, with evaluation	Comprehensive performance measurement against targets with effective management reporting	Extensive communication of energy issues within and outside organization	Resources routinely committed to energy efficiency in support of business objectives
3	Formal policy but not active commitment from top	Clear line management accountability for consumption and responsibility for improvements	Energy training targeted at major users following training needs analysis	Weekly performance measurement for each process, unit or building	Regular staff briefings, performance reporting and energy promotion	Same appraisal criteria used as for other cost reduction projects
2	Unadopted policy	Some delegation of responsibility but line management and authority unclear	Ad-hoc internal training for selected people as required	Monthly monitoring by fuel type	Some use of company communication mechanisms to promote energy efficiency	Low or medium cost measures considered if short payback period
1	Unwritten set of guidelines	Informal mainly focused on energy supply	Technical staff occasionally attend specialist courses	Invoice checking only	Used to promote energy efficiency	Only low or no-cost measures taken
0	No explicit energy policy	No delegation or responsibility for managing energy	No energy-related staff training provided	No measurement of energy costs of consumption	No communication or promotion of energy issues	No investment in improving energy efficiency

9.5 Benefits selection

Once the scope and potential for the resource efficiency programme have been quantified it is essential to describe the benefits in terms that will justify action by the decision-maker. This involves much more than simply stating the savings - it requires aligning the programme with the core mission of the organization.

Now we should have the outline of an initial proposal for resource efficiency with a range of possible benefits, financial and non-financial, as well as some idea of which decision-makers must be involved in order to gain approval. At this point, we need objectivity to determine which aspects of the resource efficiency programme are critical to the success of our proposal. We need to set aside our own views about what our organization *should* be doing and think in terms of why they *must* do resource efficiency. This distinction between *should* and *must* is important.

We have seen that there are many reasons, not least our survival, why we *should* become more resource-efficient. This notion that resource efficiency is the *right thing to do* is what drives me personally – the sense that it is good and worthy and positive and fulfilling. I, and most readers no doubt, passionately believe that organizations *should* all adopt vigorous, transformational resource efficiency programmes at the earliest possible time. However, when I meet with those very same organizations I don't pitch it this way – after all, that is just *my opinion*. What I seek to do is to find the compelling reason why they must become more resource-efficient, *why it is essential for their organization at this time*.

Thus, for a resource efficiency proposal to be justifiable it needs to be:

- Aligned with existing business priorities; *and*
- Appropriate to the role and responsibilities of the decision-maker.

To enable an individual to take action on resource efficiency our pitch should ideally provide them with the means to justify the decision to their bosses (e.g. the board) or stakeholders (e.g. shareholders). We could, of course, convince them of the moral case, for example, and then let them find their own justification – but the probability of success is likely to be greater if we provide them with some basis for this justification in the first place.

Failure to find a compelling reason for action, in the context of the organization's mission and the individual's role, is termed the justification constraint. It is a common, largely avoidable, reason why otherwise seemingly strong arguments for resource efficiency are unsuccessful. The rejection of the proposal, in these circumstances, is usually due to poor benefits selection and presentation by the proponent.

Real World: ***Adapting the message***

In his book *The Great Disruption,*[313] Paul Gilding makes an interesting observation on the importance of aligning messages with the priorities of an audience.

He recalls that for decades as an environmental activist, even during the period where he had the authority of the role of head of Greenpeace International, companies listened politely to him, for the most part agreed, but did little to change their behaviour.

It was only when he linked his arguments to the issues that mattered to the senior management, that he felt he could drive change. In his book, he explains how his new approach is working in his recent series of talks to executives at Cambridge University:

> *"I no longer argued that this was about the destruction of ecological systems or the arrogance of humanity's disrespect for nature; rather, I warned my listeners that the global economy was at risk of a sudden collapse and with it, their pension funds, their personal wealth and their companies. The level of engagement in response was a quantum leap from what I had seen previously."*

The fact is, people identify more strongly with tangible, immediate risks, rather than abstract global ones.

Given the alternative between a moral call to arms which fails and a business case which succeeds, it is the latter approach which surely has the greatest ethical and environmental integrity.

Real World: *The limits of authority.*

Every individual in an organization, regardless of seniority, has responsibilities and duties which mean that they are not acting as free agents, but rather as representatives of shareholders and other stakeholders and within a defined role and authority.

Within organizations, decisions are rarely taken because of personal sentiment but as a result of the current objectives, norms, values, culture, priorities and governance structure of the organization. In order to make a decision to support a resource efficiency programme, any individual - including the CEO - will need to justify their decision to others.

So, whatever its merits at an personal level, an argument for resource efficiency that is not based on the benefits that it will bring the organization is likely to fail. Global environmental issues do not resonate with CEOs in the private sector, not because they are heartless people, but because the environment is seen as largely the responsibility of others, such as governments, which should legislate if an issue becomes important to the nation. Put quite simply, issues such as biodiversity are not in the job description of the CEO, whereas maintaining the intangible value of the brand, which underpins a significant proportion of the share price, most definitely is, so the CEO can justifiably act on the latter but not on the former.

Whether this disconnect is a moral failing of capitalism is arguable, and whether this detachment is in the ultimate interest of shareholders is also debatable (see the section on fiduciary duty on page 218), but it is as it is, and so we need to work within these constraints to succeed.

The same need to focus the pitch on justifiable benefits applies to public institutions. For example, I worked with a UK university whose carbon programme was approved only when it became clear that it would enhance the reputation of the institution and so help attract more foreign fee-paying students. In the numerous hospitals I have provided consultancy support for resource efficiency, the case for the programme has almost always been about releasing additional funds for patient care, a justifiable effort aligned to the core mission of the institution.

We can construct a more effective pitch to an organization if we present resource efficiency as a means to achieving one or more current organizational priorities, not as an end unto itself.

For those who advocate resource efficiency from personal conviction and passion, this need to align with corporate goals can seem to cheapen the basis for the programme. However, given the alternative between a moral call to arms which fails and a business case which succeeds, it is the latter approach which surely has the greatest ethical and environmental integrity.

Finding legitimate reasons to undertake resource efficiency should not be too difficult because there are numerous direct links between resource efficiency and Value, as shown in Volume I, Chapter 3. In fact, we are spoilt for choice, and the challenge is more often to find the *right* source of value for the particular individual we are pitching to. Here, an examination of the decision-maker's goals is helpful. For example, a site manager's priority may be to maximize output, rather than any notion of brand value, so we can talk to them about how the resource efficiency programme will contribute to that specific goal, for instance, by reducing re-work, which increases productivity and adds to increased output.

We should try to avoid a proposal with a scattergun blast of dozens of alternative justifications for our programme, since we will be most effective where we target one or two top priorities relevant to the role of the individual we are trying to sell to and focus on providing compelling evidence to support those benefits. For me, this process of aligning the resource efficiency opportunity to the needs and responsibilities of the decision-maker we are looking to convince is a particularly enjoyable part of developing the case for a resource efficiency programme. It is a bit like a working out the motive in a detective story. The process has a lot of science – researching the business needs and thinking of which priorities we should focus on – and a bit of art, getting the decision-maker or people around them to open up about what is really important to them.

Another neat consequence of this mission-focused approach is that we avoid one of the most common pitfalls of environmental evangelism, which is to present an argument for a *balance* between our environmental impacts and our business priorities. This approach can unwittingly create the impression that we need to forsake business performance to reduce environmental impact. Our proposal should be founded on the direct beneficial impact of resource efficiency on our business or organization's core objectives (profit, service delivery, etc.). If we achieve this, then we can quite happily introduce wider notions of benefit to the environment without appearing to be "*green for green's sake*" or that we have a false *either/or* choice between our core objective and the planet.

By emphasizing the mission of the organization, we can shift the proposal from *should* to *must*. This link to the core objectives can make our proposal totally compelling: a *no-brainer* rather than a *nice to have*. One reason our proposal needs to be compelling is that senior managers often don't see the

9.5 **Justifying action is not easy**
Contrary to most people's assumptions, managers in organizations need to have a compelling case to act on external issues such as climate change by identifying the internal benefits that action will bring.
Source: Niall Enright. Drawn using Pixton. Available in the companion file pack.

need to change, or even if they do, they may put a low priority on it. Our argument needs to be extraordinarily powerful if it is to succeed and overcome the many barriers to resource efficiency mentioned earlier (see Volume I, Chapter 4). So to get the decision-maker's approval, we need to create what is called a "burning platform" in the sales jargon - because when you are standing on a burning platform, you need to jump!

Here are some examples of the compelling arguments used to justify resource efficiency programmes, which I have recently encountered or used:

1. The first, and most powerful, driver for resource efficiency is a crisis, which confirms that an organization needs to change its approach. For Evian, a major French bottled water manufacturer, resource efficiency is now a *must* as they aim to cut CO_2 emissions to Net Zero by 2020, in order to blunt a powerful environmental backlash against bottled water. For Ford and other US automotive manufacturers, it is recent oil prices at US$4 a gallon that make the production shift from gas-guzzlers to fuel-economic models a shrewd move – here, it is not so much crisis as opportunity. For Dupont, it was the Toxics Release Inventory in the US in the late 1980s that spurned their radical transformation from polluter to possible paragon. For a Catalan food company client of mine, it was a sudden reduction in their borehole water abstraction licence that caused them to put water conservation centre-field. Clearly, where there is an urgent or mandatory requirement to improve resource use this becomes a *must*. However, we should bear in mind that compliance drivers can be a double-edged sword, as it can lead to a *quick-fix* mindset where there is no real long-term commitment to continual improvement once the initial problem has been solved.

2. The second most compelling argument occurs when resource efficiency is aligned closely to the current organizational priorities. For example, L'Oreal argued that boosting its environmental credentials through resource efficiency had to be done to help it win over the next one billion customers because it found that consumers in emerging economies are particularly keen on environmental performance. In a well-known and trendy internet business I worked with, it was the need to be seen to be green so that they could continue to recruit the brightest people (in 2007, Monster.com found that 92% of the undergraduates it surveyed wanted to work for a green company[533]). The key argument for a global titanium dioxide manufacturer was that its largest customer insisted that it reduce the carbon-intensity of its TiO_2 to retain business in the competitive paints ingredients sector. A large European property portfolio felt it must reduce energy and water consumption in its prestige offices in order to remain competitive with new buildings and so avoid *accelerated depreciation*. A Latin American oil and gas business I advised some years ago, had to reduce its energy consumption to keep *lifting costs*, which are closely tracked by investors, as low as possible as the fields matured.

Real World: *The "pebble in the shoe"*

Sales people talk of every individual having a *"pebble in their shoe"* – an immediate challenge that they are facing (in fact, some unfortunate individuals have many *"pebbles"*).

If we can identify this pebble for our decision-makers, and demonstrate how resource efficiency will help deal with this, then we have gone some considerable way towards gaining their support.

Another important driver for decision-makers is how they are rewarded in their role. Do they get a bonus, for example, by achieving sales, managing costs or by exceeding production volumes? Understanding this could dramatically change the outcomes that we select in our pitch to them.

Finally, let us not forget that there are also unstated but quite powerful personal motivators for decision-makers to act. For example, an individual may be quite ambitious and could believe that association with a new programme that generates substantial value for the organization would be career-enhancing. In these circumstances, a pitch that emphasized the innovation, leadership and visibility of the programme – as well as the low risk of failure – would hit the right *"hot buttons"*.

Institutional politics also play a role. I have been able to enlist the support of many environment health and safety (EHS) directors for resource efficiency programmes on the basis that the programme can be seen as a value-adding initiative from the EHS department. This perception of the EHS function as value-adding rather than cost-generating is something than many EHS directors would be keen to convey.

3. There is also the financial argument. Cost savings are possibly the most common justifications for resource efficiency – although it is third on the list of things to consider because organizations invariably have numerous *alternative* ways to save money. For the property team at Peel Land & Property Group in the UK, energy management is a must-do as part of a focus on keeping tenancy costs for customers as low as possible and thus reducing vacancies. For Walmart, resource efficiency in its operation and supply chain is all about saving money for customers (and the company) and improving the environment to keep its promise: *"Save Money, Live Better"*. For numerous public service organizations I have worked with, the driver has been cost savings which could be diverted into more services. Similarly, for government departments, the requirement is to show leadership for political reasons, while reducing overheads. I have worked with several major electricity utilities in the UK, Ireland, Canada and the US, where the biggest driver is to reduce power consumption by end-users to avoid making large capital investments in generation and distribution to meet growing demand (so-called *demand-side management*). Ideally, the cost saving argument is used in conjunction with a crisis or priority driver – so that we can say, "*we must become more resource efficient because of X but the good news is that we can cut US$Y a year off our operating cost doing so*".

4. Another strong argument is relative competitiveness. Here one can point to benchmark data from other organizations in the same sector to highlight a shortcoming or potential opportunity. For example, at several BP refineries in the US, the fact that they were third quartile (i.e. below average) in the Solomon Energy Intensity Index was the motivation to improve. Numerous brewery and dairy clients in the UK were driven to improve by benchmark GJ/Ml data from the Energy Best Practice Programme, which showed them lagging behind their competitors.

There are four main arguments for action:

1. a crisis;

2. existing priorities;

3. financial benefits;

4. competitiveness.

This notion of a burning platform corresponds well with the message from change management gurus like John Kotter, who advises that without urgency, change is difficult to achieve (see piece opposite). The bigger the change desired, the greater the urgency required.

However, this idea can frighten many proponents of resource efficiency. This is because they may feel that they are not setting out to achieve a fundamental transformation of their organization, but rather to embark on a simpler, practical programme that can be accommodated within the organization's existing activities. In other words, they may not be setting out to *rock the boat*. Indeed, where the initial proposal is conceptual, the evidence and rationale necessary to create the burning platform may simply not be available.

Most proposals for resource efficiency focus on savings and so **miss the other compelling reasons to act.**

Nevertheless, the concept of urgency is useful. That urgency may not need to be stated in apocalyptic terms. Rather it may rationalize why action *now* by the decision-maker and all the participants in the resource efficiency programme is important. It is the antidote to procrastination.

The notion of a burning platform corresponds well with the message from change management gurus.

Exploration: *The importance of urgency when driving change*

A giant of change management thinking, John P Kotter was one of the first people to study the causes of failure in organizational change in a systematic way. After reviewing over 100 organizations, he concluded, in a paper published by the Harvard Business Review,[448] that the #1 cause of failure was "*not establishing a great enough sense of urgency*". He stated that the first, most essential, step in any change programme is "*establishing a sense of urgency. Examining market and competitive realities. Identifying and discussing crises, potential crises, or major opportunities*". "*Without motivation,*" he said "*people won't help and the effort goes nowhere. Compared with other steps in the change process, phase one can sound easy. It is not. Well over 50% of the companies I have watched fail in this first phase.*" Kotter goes on to say who is responsible for this sense of urgency: "*If the renewal target is the entire company, the CEO is key. If change is needed in a division, the division general manager is key. When these individuals are not new leaders, great leaders, or change champions, phase one can be a huge challenge*". According to Kotter, failure in any one of the eight transformation steps he outlined (see Figure 5.4 on page 208) can lead to failure of the change programme – but it seems that this first step is the most difficult. "*Sometimes executives underestimate how hard it can be to drive people out of their comfort zones. Sometimes they grossly overestimate how successful they have already been in increasing urgency. Sometimes they lack patience: 'Enough with the preliminaries; let's get on with it.' In many cases, executives become paralysed by the downside possibilities.*"

As the sustainability change practitioner, Bob Doppelt says: "*I have seen very few efforts that were initiated without convincing people that the status quo was an accident about to happen.*"[222] Bert Spector put it a different way in his 1989 paper in the Sloan Management Review,[671] which examined six different companies embarking on change programmes and concluded that the role of the leader is first to recognize the need for change themselves and then to communicate that to their organization. Spector notes: "*While leaders may be convinced of the need to change based on their own dissatisfaction with the status quo, that dissatisfaction is not enough. They must find ways of sharing it with the members of the organization who will actually institute new ways of thinking and acting. This distinction between a dissatisfied leader and a leader who diffuses dissatisfaction throughout the organization is more than a simple refinement of the existing theory of organizational change. Overlooking the diffusion step can be (and often is) profoundly debilitating. When leaders jump directly from being dissatisfied to imposing new operating models, they fail to generate any real commitment to change.*" Spector goes on to examine the various strategies for diffusing dissatisfaction and concludes that the least effective is mandating the change. In other words, directly demanding the participation of subordinates in the change programme destroys ownership and fosters silent resistance: "*Top-down commands and threats violate the notion of free-choice; doubters don't feel they 'own' the choice to adopt new patterns of behaviour.*" This confirms my earlier observation that our Mandate should be about communicating urgency and then empowering people to respond with the changes they feel most appropriate. It is entirely reasonable to set a goal and put in place systems to measure performance, but real commitment will flow when people can apply their own knowledge, talents and experience to achieving the objective.

Creating dissatisfaction with the current model and articulating a *burning platform* for change is needed whatever the scale of the efficiency programme.

"You never want a serious crisis to go to waste." - Rahm Emanuel

9.6 Language

Creating a compelling proposition means that we should be conscious of the language we use. We also need to understand the differences between benefits and features.

Real World: ***The mission***

If we are pitching at a very senior level, then it can be helpful to establish how the overall organizational priorities are described in annual reports and letters to investors:

Amazon: *The power of invention*
3M: *Inspired Innovation*
Ford: *Profitable Growth for All*
Toyota: *Rewarded with a Smile by Exceeding Your Expectations*
CEMEX: *A rigorous transformation to capture untapped value*
HSBC: *Connecting customers to opportunities*
AkzoNobel: *Building global leadership in core markets*

While resource efficiency may seem distant from some of these aspirations, we can often align our pitch to the sentiment of the CEO. For example, inspiring innovation at Amazon or 3M can be supported by energizing and retaining staff through a resource efficiency programme. Ford, AkzoNobel and CEMEX's focus on financial performance will be helped by reducing resource costs. The Toyota and HSBC emphasis on customers will be reinforced if the brand is associated with superior environmental performance.

It will do no harm to use language that resonates with or acknowledges the higher aspirations of the organization, particularly if our pitch is to the executives who set that vision.

If our case for action does pass the justification and urgency test, there is still work that needs to be done to increase the probability that it receives approval.

Our proposal will not be complete until we have thought carefully about the language that it uses. This is because the words that we use in presenting our programme can bring baggage. For example, the word *audit* often has negative connotations. On several occasions, for example, I have been surprised to hear senior executives strongly condemn the word *sustainability* for a variety of reasons such as vagueness, or overuse, or being old-fashioned or simply because it is *"not business-like"*. It appears that this word can elicit quite negative reactions among some people – so we would clearly want to avoid it if we need to get approval for a programme.

The term *sustainable development* might be associated with the narrow field of poverty alleviation or global institutions like the UN. Another word that seems to cause difficulties in some sectors is the word *"environment"*, which is associated in some people's eyes with compliance and additional cost which do not add value. As mentioned above, the word *audit* may be most commonly associated with an external hit-squad of experts who descend on a facility and dig through their financial performance data in an excruciatingly painful way. *Resource efficiency* may suggest manpower reductions. For one client, an oil major, I was told not to use the words *climate change* (see real world items on page 180). No kidding!

The label that we give to our proposal is also important - I tend to use a business-like title which conveys the seriousness that I expect the proposal to receive. By calling the pitch and associated documents a *"business case"*, I can establish that this is a carefully considered proposal focused on Value. Alternative phrases I have used include: *"investment proposal"*; *"brand value programme"*; *"yield improvement process"*; *"asset value proposal"*; *"recommendations for portfolio enhancement"* and *"margin maximization programme"* and there are countless variants. If it has been quantified at the pitch stage, I also tend to put the primary benefit in the title such as: "*A business case to reduce operating costs by £1.6 million per annum*" or "*Least-cost route to legal compliance on waste*". We can also use this summary when we are describing why we want to pitch: "*I'd like to discuss an investment proposal to increase the output of the site by 3% while reducing annual operating costs by US$500,000.*" That should get attention!

We also need to be conscious that the name we give to our programme also has implications:

- an initiative may suggest that we are proposing an additional effort, over and above business as normal;
- the word programme might imply too large a scale effort;
- most projects or campaigns have definitive start and end dates so may undermine the notion of continual improvement;
- a scheme sounds optional;
- the words process or methodology come across as somewhat formulaic or procedural.

If an organization has a large number of change programmes in place, and so may be suffering from *initiative overload*, it may be desirable not to give our programme proposal a distinct identity at all. We may simply present our programme as a "*proposal for a more efficient allocation of energy and waste budgets*".

If there are doubts about the acceptability of certain expressions we need to test the language we use in advance of presenting a proposal for resource efficiency. Later, in the chapter on People (page 653), we shall see how other aspects of language, such as comparisons (norms) and numbers (anchoring), used in our proposal, can also influence perceptions and outcomes.

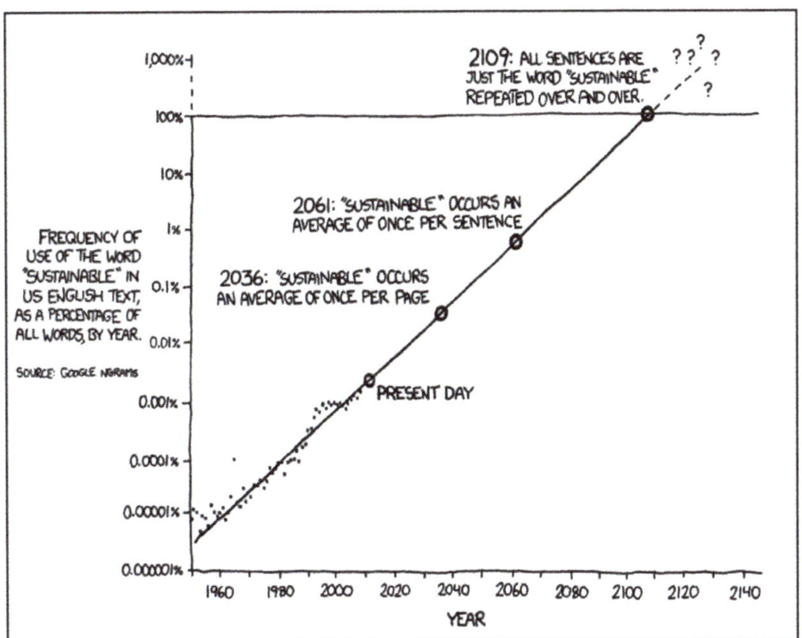

THE WORD "SUSTAINABLE" IS UNSUSTAINABLE.

9.6 **An amusing take on the increasing use of the word *"sustainable"* in the English language, based on Google NGrams which shows the distribution of words in published books.**
Source: xkcd, reproduced under a creative commons 2.5 attribution non-commercial licence. https://xkcd.com/1007/

Mandate

9.7 Desired action

Every proposal is simply a request for an action. In order to succeed, our request needs to be as easy as possible to approve.

Having developed a very compelling justifiable case for proceeding and considered the format and language of the presentation, we now need to decide what precisely we are going to ask the decision-maker to do. This is what sales people refer to as "*the call to action*".

Just as they say *"you don't get married on a first date"*, so our pitch will be more likely to succeed if we ask for the right level of commitment at the outset. The initial proposal doesn't have to be "*let's run a resource efficiency programme across the whole organization*". Instead, it can be "*let's run a pilot at site X*"; or "*let's do one/several audits to establish the savings we can make*"; or "*let's do a review of our competitors to see how we stand in comparison*"; or "*let's get together a team from across the business to figure out how we respond to this challenge*". The important thing is to start the process. If we are asking for an intermediary step on the way to a full programme, then it makes sense to signal the next steps: "*Let's do a pilot first and assuming that gets us the target 8% savings we can roll this out across our XYZ operations.*"

One of the key determinants of what we ask for is the nature of our proposal: whether it is a high-level concept we are pitching or a more detailed business case based on data specific to our organization. One of the main weaknesses in a conceptual proposal for resource efficiency is that, despite the vast amount of evidence in the public domain that there is substantial financial value available, most organizations are simply unwilling to commit to significant investments on the basis of this general data. The crux of the issue is the level of certainty that the decision-maker has in the outcome of the resource efficiency programme (see *Why certainty drives the resource efficiency proposal* on page 185). One of the few ways that we can gain credibility for such a generic proposal is to win endorsement by peers or experts in which the decision-maker has high levels of trust. Even with a strong endorsement, because of the lack of directly supporting evidence from within the organization, most high-level, conceptual proposals don't seek outright approval for a programme. Instead, they seek agreement to quantify the overall savings and costs in more detail or to establish a group of trusted experts (such as a steering group) to develop formal proposals, or perhaps to carry out a limited pilot to demonstrate the concept works. Each of these steps is designed to increase the certainty of the outcome of the subsequent action while enabling the decision-maker to feel in control of the process to the extent that they can abort if they feel that it is not achieving the desired result.

Real World: ***Starting points***

Different projects I have been involved in have had many different starting points: in Evian's case, the first requirement of their programme was to carry out a life cycle assessment on the bottled water so that a reduction target could be established. For L'Oreal's programme, the starting point was a series of audits at manufacturing facilities, to determine where savings in energy, waste and water could be made. In BP's case, the initial proposal was to pilot the resource efficiency programme at one refinery, Coryton in the UK, in order to develop the methodology and prove that it would work in such a complex facility. For Skoda cars in the Czech Republic, the energy department asked for a test in just one manufacturing hall at the huge Mlada Boleslav site to show how savings could be achieved by engaging the shop floor and maintenance teams. At Knorr-Bremse, the key was the connection of the programme to the ambitious CO_2 emissions reduction goals already in place. Having established the link, the programme involved a series or *"rolling audits"* of global manufacturing facilities.

Each of these requests was relatively low risk and well within the authority levels and budgets approval of the decision-makers involved. Successful proposals are usually the ones which are easy to approve.

🔑 A "call to action" must **precisely describe what we want the decision-maker to approve,** and we should ensure that this is well within their **authority and budget approval levels.**

If we believe that only a detailed assessment of cost-benefits will enable our programme to gain acceptance, then we need to develop a detailed business case based on data specific to our organization before submitting a proposal. It is quite common for functional specialists – such as environment directors or operations managers – to commission external consultants to carry out detailed audits across a representative sample of the organization to be able to present a credible business case for the value of resource efficiency. In this case, the aim is to gather evidence that will reassure the decision-maker of the benefits of the programme.

Even where there are strong data to support the outcome, it may be sensible to embark on an initial pilot or rolling deployment so that the organization can develop and refine its approach in an evolutionary fashion. Indeed, if there are already some corporate programmes underway, there may be strong arguments for a phased approach where the resource efficiency programme is not portrayed as an initiative at all, but rather as merely a series of incremental improvements on the way we currently do business. In any case, it is important that any suggestion of a staged approach to the programme is not interpreted as a lack of confidence in the outcome or process, but rather as a means of improving the programme's effectiveness. Where a staged approach is proposed, I would recommend that the subsequent steps are understood and mapped out in outline in the proposal. This is to ensure that, at the end of the initial activity, we don't come to a full stop, but proceed on the proposed course, assuming of course that the initial objectives were achieved.

Counter to the idea of a simple initial request is the notion of materiality. Here, we need to be conscious that senior decision-makers may be looking for results that have a visible impact on their organization. If our programme delivers a 5% reduction in energy costs which are 1/10th of operating costs, then we are offering a 0.5% improvement in operating costs. This may not be enough to capture the support of the top management. Thus, every proposal needs to ensure that the scale of benefit aligns to the goals of the decision-maker and is clearly signalled.

As well as the degree of certainty in our proposal, we should also review our pitch and proposal in relation to the wider barriers to resource efficiency that are set out in Volume I, Chapter 4. This is not intended to cast a shadow of doom over the proposal but simply to help tweak what is already almost certainly by now a very compelling proposition. The barriers we should consider include psychological factors such as status quo bias, loss aversion, underestimation and bounded rationality; economic factors such as hidden and missing costs; and financial factors such the use of the correct appraisal techniques or impact of limited access to capital. Considering these issues will help to strengthen our programme, improve the way that we sell it to decision-makers and clarify the specific action we want approval for. I often prepare by writing down the possible objections and solutions: e.g. "*not enough staff time*" can be countered by "*can use external support to boost the team*", or "*no funds available*" can be addressed by "*third-party funding would be possible*".

9.8 Decisions

The response to our proposal need not be either "yes" or "no". There are a total of six possible decisions that can be made and we need to prepare for them all.

As well as considering the barriers and possible objections to our programme in advance, it also makes sense to think about the decisions our decision-maker could arrive at. Anticipating their response is particularly important if we have entered into an open dialogue where we are asking them for their thoughts on how to proceed. There are six kinds of decisions that they could take:

- A decision not to proceed – i.e. *"no thanks"*;
- A decision to defer – i.e. *"not now"*;
- A decision to delegate (with or without a target being agreed);
- A decision to investigate or qualify;
- A decision to pilot - i.e. a *"yes"*;
- A decision to launch - i.e. a *"yes"*.

One approach to the "call to action" is to **engage the decision-maker in selecting among a number of alternatives.** This involves an open dialogue and requires careful preparation, to anticipate the possible decision.

Obviously any one of these decision-types - apart from *"no thanks"* - could be the desired outcome we are seeking from the decision-maker. It makes sense in preparing our pitch to anticipate what each of the other decisions would mean for our programme. Thus, if we are asking to launch our programme but the decision-maker indicates that they would prefer to proceed more slowly then we might switch the discussion to a pilot – having already formulated some thoughts on what such a pilot might look like, and which facilities or departments it might involve. In some cases, we might explicitly offer the decision-maker a couple of options, as this engages them in the process.

A decision to investigate may be the most appropriate outcome where the decision-maker is expressing some uncertainties around the benefits of the programme. Here, we should have some ideas ready, such as where we might undertake a series of audits or surveys to understand the potential and focus for our programme. We might be going in with the hope that we get a decision to launch or pilot – but this can be one of our fall-back options. Thus, if there is a lack of confidence that the opportunity is real an audit might be the only realistic action our decision-maker could approve. If this is the case, we should try to determine which aspects of the proposal the decision-maker is uncertain about and what kind of evidence they would like to come out of the audits or further investigations to address that uncertainty.

Real World: *Co-developing*

Photo © Monkey Business, Fotolia.com

When we make the request of our decision-maker, whether for a full programme or an intermediary step, it is often wise to create an opportunity for them to respond to the proposed action and so co-develop the solution. In other words, we do not have to set our request in such a way that the only response is a *"yes or no"* decision. Instead we can have a discussion about *alternatives*.

This is sensible because we have already stated that it is their perception of risk that will influence the approval and this is not something we can definitively predict in advance. If our decision-maker is more confident than anticipated of the need for action they may upscale the initial commitment, and if they are less certain than we expect there is still the opportunity to salvage some progress forwards from our proposal. Thus our call to action might be: *"I propose that we pilot this in three facilities, one in the US, one in Europe and one in the Far East. What are your thoughts on this as a way of proving the benefits of the programme?"*

This open question has both a clear action for approval while giving the decision-maker an opportunity to offer an alternative approach which they are comfortable with. The other benefit for inviting input is that this can create ownership in the decision-maker. On the other hand, it can lead to the proponent's programme design being compromised.

A decision to delegate often involves forming some sort of task torce or steering group, or perhaps mandating a member of the leadership team to make recommendations to the decision-maker or board on whether a programme should be initiated and what it should look like. Sometimes the Mandate from the decision-maker to such a group is limited to *"tell me how we should do this"* - in other words, a decision to proceed has been arrived at - rather than *"tell me if we should do this"*. Clearly, the former option is better than the latter, and in some cases, the decision-maker will go as far as to agree or propose a target for the programme. Often the team/individual with the initial recommendation role will remain in place to oversee the programme going forward, so it is important to anticipate this kind of decision and be ready to put forward possible participants on the panel if this is the way the discussion goes.

Sometimes a decision-maker will approve a full-blown programme even where this is not something we have asked for. Perhaps the decision-maker is already convinced of the benefits and does not want to delay the process, or perhaps there are some strategic reasons we are not aware of that make an accelerated programme desirable. In any event, this has happened to me on a few occasions, and I have been pleased that I had previously considered what the maximum realistic rate of deployment of the programme might be and the resources that would be needed.

Obviously, if we anticipate a decision not to proceed at this point of our preparation, it means that our proposal is not ready for approval and we need to revisit some of the previous steps in the proposal development process. For the sake of completeness here we will cover what we should do if, having proceeded, the decision-maker says *"no"*. If, in the course of the presentation of our proposal, we feel that this is going to be the answer we should try to do a couple of things. We should try to understand why the arguments we have set out have failed to persuade (or rather what the objections to the programme are) and we should do our best to gracefully change the *"no"* to a *"not now"*. Often the truth is not that we have structured our case poorly, but that there are other competing demands in the organization for management time or there are other alternative methods of achieving the desired business goal – e.g. reduce operating costs. Perhaps the decision-maker knows something we do not which would influence a programme, such as impending reorganization or business divestment. In any case, we should prepare for this eventuality by working out what we might say to leave the door open for a future return to the subject when circumstances change.

Possibly the most frustrating and least desirable decision is a decision to defer. At least where there is a *"no"* we can explore the objections and get down to modifying our approach. If deferral is unavoidable, it is essential to get the decision-maker to provide a timescale in which the programme can be revisited in order to avoid the project being left in limbo. If they are unable to indicate a time when the project might be revisited, then this can suggest that the decision to defer is a *"no"*, masking as a polite *"later"*.

People buy ***benefits****, not features.*

Exploration: ***Benefits, not features***

An important lesson from sales is always to focus on the benefits, not the process. A car salesperson will talk about the smoothness and comfort of the ride, not the fabrication of the shock-absorbers for a good reason: people buy benefits, not features.

The sales trainer Elmer Wheeler used to say: *"Sell the sizzle not the steak."* In other words, it is the outcomes of the resource efficiency programme for the organization that matter in the initial sale, not the mechanics of the process. No one is particularly interested in the precise cut or the packaging of the steak, but how it will taste.

In other words, our pitch will be more successful if the decision-maker understands the benefits that the resource efficiency programme will bring in meeting organizational objectives, rather than if they get bogged down in the mechanics. Of course, we may need to communicate some detail of the process and methodology that we are proposing, in order to build confidence and understanding – but we should relate these back to a core benefit. *"Because we can run the resource efficiency programme through our existing Six Sigma teams* [a feature] *we will accelerate the rate at which realize the projected cost savings of US$1 million a year* [a core benefit]". This *"feature x means benefit y"* approach is a good way of keeping a more detailed presentation engaging and avoiding feature overload.

Some attributes or features of the programme we are presenting, such as measurement, can change the perception of the risk associated with a proposal (a key issue that was explored in the earlier section, *Why certainty drives the resource efficiency proposal* on page 185). In these circumstances it is appropriate to include reference to these features, e.g. *"because impartial measurement is built into the programme from day one* [a feature] *there is a negligible risk that we will approve projects below the target of 25% internal rate of return* [a benefit]".

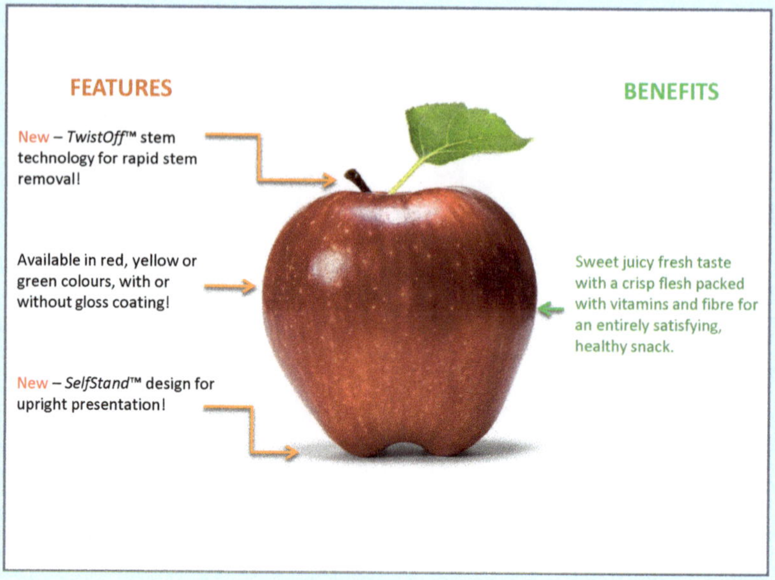

9.7 **Features do not create the same desire for a product or service as benefits**
What sells the apple is the anticipation of the taste and the pleasure in eating. When presenting our resource efficiency programme, we need to remain focused on the benefits that will arise, not the features of the programme itself. *Source: Niall Enright, apple image © Mariusz Blach / Fotolia*

9.9 Testing and endorsement

In advance of our pitch, we should try to gather a coalition of support for the proposal. This will give us an opportunity to test our case for action and identify potential objections.

Another aspect of preparation is to build up a coalition of support for the resource efficiency proposal in advance of the final pitch to the decision-maker.

We might want to get some functional teams on board, such as HSE or engineering, or perhaps enrol the management of a prospective pilot site first. Here, we ideally want to focus on people who can help actively move the idea forward by, for example, helping to refine the proposal and pitch, or identifying possible objections and reasons not to proceed. Our supporters may also have better contacts with the decision-maker which they can put to use in setting up the meeting in the first place.

Because a comprehensive resource efficiency programme tends to involve many functions in an organization – from finance, procurement, operations, engineering and maintenance, to name just a few – this multi-level selling is not uncommon. In fact, the need to engage many individuals and teams is the chief contributor to the long sales cycles, usually months, sometimes years, which I have often encounter when selling resource efficiency services to organizations. This process of garnering support may increase the likelihood of a decision to delegate by the decision-maker, and so we need to consider the pros and cons of this outcome.

Creating a coalition of support and, in particular, gaining the endorsement of someone whom the decision-maker views as authoritative, is especially important where we have identified that our proposal can be characterized as a *credence good* (for a full explanation of this, see *Why certainty drives the resource efficiency proposal* on page 185).

*Wherever there are intermediaries involved between the initiator and the final decision-maker, **never let the intermediary make the sale.***

Wherever there are intermediaries involved between the initiator and the final decision-maker, my advice is never to let the intermediary make the sale. It may well be the case that an intermediary says "*I am real excited about this, leave this with me and I will raise it at the next executive meeting*". In this case, the outcome is now entirely in the intermediary's hands, and they may not have the same passion, commitment and detailed knowledge as the proponent. A much better outcome in this situation is for the proponent to make the presentation at the next level - alongside the intermediary - in the knowledge that the supporter will endorse the idea. Of course, preventing the intermediary from taking the lead is not always possible, but the proponent may be able to reduce the likelihood of this by being clear from the outset that they are looking for an introduction and a sponsor, rather than an emissary.

9.10 AIDA

The most effective way to get the support of decision-makers is to engage in a structured conversation rather than to have a one-way monologue leading to a simple "yes" or "no" response. This is an example of how a formal discussion or presentation might be structured.

At this point, the key meeting or presentation with the decision-maker(s) is set up, and we have prepared our business case and perhaps validated this with a coalition of supporters. Now we need to think about the face-to-face encounter.

Let's see what the world of sales can offer by way of advice. Well one of the first things I ever learnt in selling was the mnemonic "AIDA", which I no longer associate with a famous opera of the same name by Verdi, but with the key steps in any sales encounter.

- Attention: to engage someone you first must get their attention and start a two-way flow of communication.
- Interest: once you have a dialogue flowing you need to get their interest very quickly, usually by getting them to discuss their needs or problems related to energy and resource efficiency.
- Desire: having revealed need, the next step is to explore a vision and programme that might address that need. Before moving on to the next step we need to check if there is *desire* to act on the need.
- Action: finally, you need to be clear about the action you want the decision-maker to take, e.g. approve a budget for a study, set up a team to work on a resource efficiency strategy, or agree to an organization-wide programme. Bear in mind that you will have considered alternative actions as part of your preparation process and be ready to change your request depending on the feedback you are receiving.

Of course, one does not follow this model in a mechanical fashion. It is simply a reminder of some good common sense things that will help us get to the decision we want. Nor should these techniques be reserved for the final pitch, they are useful in any discussion of the proposal, however long or short, from shop floor to boardroom.

Our fundamental aim is to create a conversation around the proposal. Although we may have some very carefully developed and detailed suggestions, we need to recognize that the decision-makers are probably approaching the subject with little previous knowledge. Within the time available they need to understand the benefits fully, conclude that these address an organizational

Real World: *Narrative*

Image © AKS, Fotolia.com

Every conversation has a narrative theme:

- Positive or Negative: e.g. emphasizing different drivers such as opportunities and benefits or compliance and risk;
- Defensive or Offensive: e.g. pitching a necessary action to protect the status quo or a chance to gain a significant competitive advantage – here, knowledge of what others are doing is key;
- Incremental or Transformational: e.g. from a low-key set of improvements to business as usual to a large-scale change in the way things are done;
- Safe or Innovative: we need to establish if what we are proposing has been done before. If the programme is innovative, then we need to articulate why "first-mover" advantage could be beneficial.

An important contribution from our supporters could be to advise on the appropriate narrative style for our discussions with the decision-makers.

Thinking more broadly about narrative, we need to recognize that stories are a powerful way to convey complex meaning and persuasive messages. It is beyond the scope of this book to provide guidance on how to craft stories, but I have included a useful starting point on this in the Further Reading list at the end of this chapter.

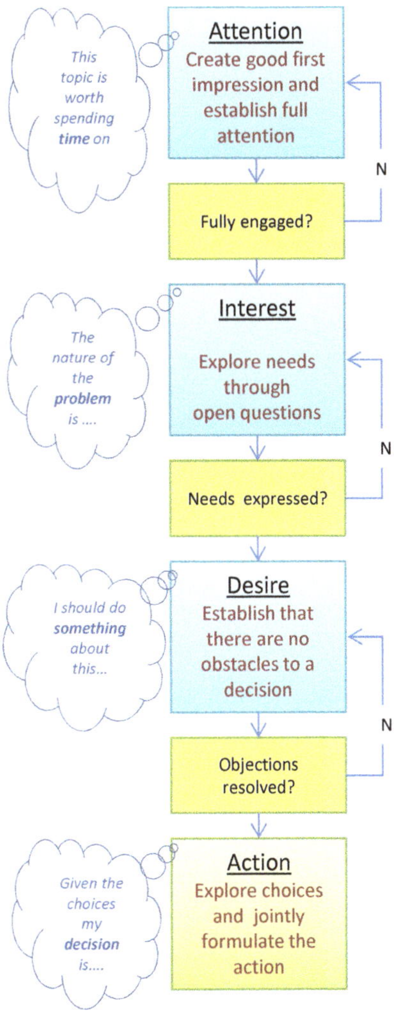

9.8 **The sales technique known as AIDA can help develop a structured conversation with decision-makers that will result in action**
Source: illustration by Niall Enright

need, confirm that they would be justified in acting on the proposal, address any concerns they may have personally and then finally decide the course of action that they want to take. It is important that if a PowerPoint presentation is used, that it is structured to enable a conversation to take place.

The first part, attention, is about making the initial impression. Here, we want to set the tone, look professional, maintain eye contact and project confidence. We might start the conversation/presentation by summarizing the theme we want to explore: "*I believe that there are some considerable business benefits if we approach resource efficiency in a more structured way. My initial assessment shows that we could improve profit margins by at least 5% within 18-24 months, for a relatively modest investment.*" Here, we are putting the value case clearly and succinctly – based on the research we have done before and the feedback we have had in practice. We might even employ some subtle psychological techniques like anchoring (see *Perception is everything* on page 178 and *Using framing techniques 19.3 (page 658)*). Hopefully, we understand how resource efficiency will impact on the role of the decision-maker we are pitching to, and the opening statement will resonate.

However, the key is not to leap into a full-blown pitch at this point, but to get the conversation flowing as soon as possible by following up with open questions which get the decision-maker talking about the benefits our programme can bring. Open questions end with an invitation to share ideas "*What are your thoughts about...*", "*Have you considered...*" or "*I would value your direction on...*". If the question can be answered with a word or with "*Yes*" or "*No*" then it is not an open question. Examples might be: "*You might be interested to hear that Competitor X's programme has saved them US$YYYY and I wanted your views whether this is a useful comparison for the savings we could achieve*", or "*I believe that we could enrol a further 150 students if we deliver the savings from our operating costs through a resource efficiency programme, but I wanted your advice on this.*" or "*I wanted to talk to you about Walmart's sustainability questionnaire. In particular, I wanted your thoughts on whether this will influence the volumes they take from us and if so how we should respond?*". The example open questions above have focused the discussion around the benefits that are a key to the success of the pitch: cost reduction, service provision and sales. Preparing for our pitch could involve preparing a number of questions such as the ones above, so that if the conversation drifts away from the central theme, we might gently nudge it back on track.

Whatever we do, we should avoid making the rookie mistake of leaping in with a one-way sales pitch at this stage; unless the decision-maker is sharing their perspective, we do not have their attention. We might have eye contact, but we will be speaking *at* them, lecturing them rather having a conversation with them. If we are using PowerPoint, now would be a good time to pause. What we want to do is to get the decision-maker to recognize that there is a problem or an opportunity and to tell us in their words why they think resource efficiency is worth looking at. We want them to feel that the need for resource efficiency is their idea. This is what salespeople call "*pull*". If the other person has articulated a problem, or need, then they are much more likely to be receptive to the possible solution that we have to offer. If, on the other hand, we try to offer a solution to a problem that they don't think they have, then the likelihood of success is pretty much zero.

So we now have attention, because the other person is speaking to us and they are fully engaged. Hopefully, because of the way we framed our initial question, they have articulated a need to improve resource efficiency in some way, or perhaps they are asking us for more information at this point. We are having a conversation. The crucial next step is to gain their interest by exploring the area for improvement they raised, or perhaps by providing the information that they requested. Again, we want to finish with open questions such as: "*Yes, mandatory reporting is coming in for FTSE listed companies from next April and it looks like carbon will be the main issue. How do you think our shareholders will react to such a report?*" One useful technique here is to build trust by repeating back the need that the leader articulated: "*So you're saying that keeping Walmart happy is important, but that you would like to know where we might make improvements before we make any promises. It seems to me that the best way of achieving this would be to do some audits across the business which will tell us what we can deliver. What do you think?*" Not only does this tell them that we are listening to their needs, but also enables us to confirm what that need is.

We have interest because we are discussing their problem at this point. We are exploring possible solutions to the needs they have told us about, not telling the decision-maker what they should do. There may be several ways to kick off a mandate for resource efficiency, and we should be prepared to discuss several alternatives approaches, rather than have a fixed solution in mind. This is where our preparation is so important – we should have anticipated a range of possible decisions and approaches. Maybe the decision-maker doesn't have the information or expertise to craft the solution in which case it makes sense to offer some alternatives: "*We could perhaps start by doing some audits across the group which would give us some base data, but possibly the quickest way to demonstrate the process is to simply pick a facility and run a pilot. What do you think?*"

> It is worth bearing in mind the old adage of a salesperson having one mouth and two ears and needing to use them in that ratio.

Desire should arise when we demonstrate that we can meet the stated need cost effectively, reliably and easily. At this point, the issue of resource efficiency will be in the open, and the parameters and appetite for a possible solution can be examined. A vital part of this stage of the pitch is to surface any objections or concerns that the decision-maker may have about the programme and address each of these in turn.

The desire stage is about the decision-maker wanting to act on the opportunity in some way. They will only do so if they feel that all their concerns are addressed. We therefore, need to ensure that all the objections are brought to the surface by asking open questions. For example: "*Although it will save US$1 million with a two-year payback, I wonder if you can anticipate any challenges in running a programme focusing on energy, water and waste across our manufacturing units?*" We have now introduced the potential benefit, around a US$1 million a year,

> **Exploration:** *Why resource efficiency is like the hotel business*
>
> 
>
>
> © Natalia Bratslavsky, Fotolia.com
>
> There are many businesses where the principal product becomes worthless after a certain point in time. An empty hotel room cannot be sold the next day again; theatre tickets have no market once the curtain goes up; and an unoccupied airline seat is valueless, once the plane has taken off.
>
> Resource efficiency is the same. The potential savings in any day can never be realized again at a later date - they are gone, lost forever. Needless money has been spent; it has flowed out of the organization, never to be seen again. Meantime, the emissions, waste or other environmental impacts are accumulating, becoming more difficult, and expensive, to undo.
>
> We need to inject some of the same urgency into our resource efficiency programmes. We should say, positively: *"Every day/week/month/year that passes we can add US$x to the bottom line of our business, equivalent to US$z in sales, and also prevent y damage to the environment… so we need to act now, not tomorrow/next week/next month/next year."*
>
> The emphasis on eliminating something *now* that is a pointless drain on our performance can be very powerful.
>
> Change must always challenge the inertia that maintains the status quo.

and discussed a payback threshold and a scope of water, waste and energy in manufacturing, but also created the space for any concerns to be expressed. We are always testing for the level of desire to proceed. It is at this stage of the conversation where I typically seek to establish a vision for the programme and create ownership for that vision in the decision-maker.

It is a common error to try to get to the action step before the decision to proceed has been fully made or all concerns addressed. The decision-maker will usually need to conclude that *something* needs to be done before deciding *what*. In our relief at having made the pitch, it is natural to want to get the approval quickly, but rushing this can have a negative effect. If we *assume* that the decision has been made, for example by asking "*So who do you think I should involve?*" it looks like pressurized selling. This is the old car salesperson's trick of closing the sale by getting the conversation going around "*which colour do you prefer?*" Our senior audience is likely to be much too sophisticated to fall for this kind of pressure, so what we really need to do, before we proceed to the discussion of possible actions, is to double-check that the decision to proceed really has been made.

What we want to do is to understand if any hidden objections or concerns could prevent us moving forward, and get these on the table so that we can address them. So our final question might be "*It seems to me that there is some agreement that we should tackle this subject, but are there any issues that we should be thinking about?*" If we find we make a proposal for action and we get push-back then it is probably because there is a hidden objection and so we need to circle back to the interest and desire stages again. Once again, our preparation for possible objections should prove invaluable as we will hopefully have a response for some or most of the points that will be raised.

When we have gained acceptance of the need to proceed in some way, we can move to the next step, action, where we have a discussion around what the next step should be. Here, too, our preparation will prove useful as we will have one or more suggestions that we believe are practical, low risk and doable. Just to remind ourselves, there are six kinds of decisions that we can arrive at:

- A decision not to proceed – i.e. "*no thanks*";
- A decision to defer – i.e. "*not now*";
- A decision to delegate (with or without a target being agreed);
- A decision to investigate or qualify;
- A decision to pilot - i.e. a "*yes*";
- A decision to launch - i.e. a "*yes*".

Once one of these decisions has been taken, we should ensure that we have properly captured the outcome and next steps, which will usually be followed up in written form following the meeting.

Exploration: ***Credo***

In describing the importance of resource efficiency, the emphasis in this book is on Value, and building a case for action that is specific to the organization. This is a deliberate choice, as senior decision-makers have to frame and justify their actions in terms of their organization's core purpose and competing objectives.

There are, however, a series of over-arching arguments that complement or underpin our call to action, which run like a thread through both volumes of this book and which are hopefully adequately supported by evidence and references. These ideas provide a broader rationale for why our organization should act now to address resource use and are set out below.

1. Significant change in our use of natural resources is inevitable. We have passed a number of important boundaries in our planet's ability to support us, which puts our future wellbeing and our economy at grave risk. The change we face can be either because we have decided to do things differently, or because natural limits force this on us. One thing is clear, change is inevitable. We need to accept this as fact.

2. We need to get moving now to work our way through the implications. We have many of the technologies, the finance and human capital in place to restore the balance, but we need to act urgently. Although we don't have all the evidence for the extent of harm we face, the grave nature of the risks means that we should make preparations now. The big incentive for us is that early action will result in greater value or lower cost to our organizations, and potentially give us a significant competitive advantage.

3. All organizations must make a contribution to the problem, so we cannot sit on the sidelines. Every country and every part of society, from individuals to governments, through to businesses and public sector institutions, will need to play a role. The same could be said for all parts of our own organization, from finance to procurement, from design to operations. If we want to ensure that others are playing a meaningful role, we should engage and challenge other organizations and institutions to step up to the plate. We can only do this by showing leadership ourselves.

4. We need to increase our effort markedly, but not lose sight of the fact that the task is achievable. Our response to date has not reduced our risk sufficiently, so we need to do more. If we break down what we need to do across the economy and over time into discrete steps and interim goals, and determine what this means for different part of our organization, we can see that our contribution should be affordable and achievable.

5. The outcome will enhance value and so action is in the interest of shareholders and stakeholder. Although the changes may be large, so too will be the opportunities for organizations. In some cases, our response will be defensive, to protect value or redesign business models that underpin our organization's success; in other cases, our response will be offensive to try to capture competitive advantage. What is certain, is that inaction will be value-destroying.

In putting together a rationale for a resource efficiency programme, it may be helpful to test and articulate each of the five points above in the context of your own organization's missions, vision and values.

Mandate

Summary:

1. Getting acceptance for an efficiency programme is a *sales* process. Do not be put off by this; selling is a form of conversation.

2. Whenever you are discussing the resource efficiency programme, you should avoid excessive detail on technologies and features and instead concentrate on the benefits, which should be as closely aligned to the core objectives of the organization as possible.

3. People buy benefits, not features.

4. Decision-makers are not free agents. They will usually need to be able to justify their support for energy and resource efficiency efforts. The more you support their role and help them justify their decision, the more likely approval is to be forthcoming.

5. A conceptual proposal, which usually involves further investigation or pilots, has some advantages in terms of the initial effort and time required to get started. However, it may lack the key element of *certainty* that is needed to mobilize a larger effort.

6. The language used and the narrative style really does matter. It is sensible to get feedback on these before presenting formally.

7. In discussing a proposal, effort should be taken to elicit any concerns in the mind of decision-makers. Unless these concerns are tackled head-on, they can lead to an unexplained rejection of a seemingly attractive project.

8. Being aware of the barriers to resource efficiency can help anticipate and overcome possible objections from decision-makers. There are broadly six possible decisions to any proposal: no; defer the decision for another day; delegate the decision for others to take; investigate the potential; proceed at a smaller scale (pilot); proceed as proposed.

9. If you can develop a *coalition of support* around your proposal, you are much more likely to get approval.

10. Do not accept the idea that energy and resource efficiency is an *"either/or"* proposition; this is almost never the case.

11. The knowledge, beliefs and biases of decision-makers are critical. Anyone bringing forward a proposal for investment will benefit from understanding these.

12. Pitching to decision-makers should ideally be a conversation. The AIDA sales technique can offer a way of keeping that conversation on track and engaging the decision-makers effectively.

Further Reading:

1. *The NEXT Sustainability Wave, Building Boardroom Buy-in,*[791] Bob Willard. A structured list of arguments for sustainability (many relevant to energy and resource efficiency), which includes an excellent primer on objection-handling in Chapter 5 with numerous examples.

2. *Story Telling with Intent - the sustainability story that makes your point wins* by Tim Hedgeland,[362] has a good introduction to using the power of stories to persuade and has good further references on this subject.

3. Chapter 5 of *The Effective Change Manager's Handbook*[668] by Ranjit Sidhu has a good summary of AIDA and other useful communications techniques.

Questions:

1. Consider the differences between a conceptual and a quantitative proposal for a resource efficiency programme. What are the pros and cons of each approach, and in which situations should they be used?

2. Why is urgency an important aspect of an energy and resource efficiency proposal?

3. What is the difference between a feature and a benefit? Imagine that you want to convince someone to buy a ballpoint pen. List five features and five benefits that you could use in the sale.

4. What does AIDA stand for? What do we need to achieve before moving from each stage in the conversation to the next?

5. What are the six basic decisions that can be made in response to a resource efficiency proposal? How would you respond in each case and why?

6. List *at least* 20 reasons why a decision-maker will turn down a proposal. Can these objections be eliminated or minimized and if so, how?

7. Score your organization using the energy management matrix (see Figure 9.4). Can you suggest improvements or changes to the matrix and why?

8. Write down the narrative that you might use to convince your organization to adopt an energy and resource efficiency programme. Describe the characteristics of the narrative: is it positive or negative, defensive or offensive, incremental or transformational, safe or innovative? Explain why you have used the chosen approach.

9. Thinking about *credo* (page 334), do you believe that any of these arguments could be used to support the pitch for resource efficiency in your organization? If so which ones and why, or why not?

10 Developing a Strategy

Illustration © Rassco, Fotolia.com

Energy and resource efficiency strategy development is something that few organizations do properly. What most organizations call *"strategies"* are, in my experience, actually *"plans"*. They are plans because they arise from relatively narrow considerations, such as how to achieve a particular goal, or how to comply with regulations, or as a response to an audit or a study focused on a limited range of activities or locations. There is nothing wrong with a plan; we just need to be aware that we can step back, and take a more strategic approach.

True strategies are much broader and more rigorous assessments of internal and external context, trends, capabilities, expectations, risks and opportunities, from which a series of *insights* arise. A strategy can appear, at first glance, to be the same as a plan, which is a list of actions. However, the actions arising from strategy are always connected to the strategic insights, and often involve many more functions in our organization, sometimes over a long timescale.

A strategy should start with a *"clean sheet of paper"* and take care to avoid *sunk cost fallacies* (see page 572), confirmation bias (see page 179) and other distortions. Strategy development might use techniques like *"backcasting"* (see page 703) to foster innovation and *out of the box* thinking. Strategy development is one area where external consultants can be especially helpful.

In the following pages, I will explore the main approaches to corporate strategy development and the specific areas that an energy and resource efficiency strategy might include. I will share some tools that I have developed to align corporate and site strategies, and touch upon some common strategy errors.

Even if your organization already has a strategy or plan, this chapter should provide ideas to refresh these.

10.1 Types of strategy

Depending on whether a quantitative or qualitative mandate has been obtained we are likely to have a prescriptive or an emergent strategy. There are pros and cons to each.

In his classic book, *Strategic Management*,[489] Richard Lynch describes two approaches to strategy. Prescriptive strategy, where the three core areas of strategy (analysis, development and selection of options and implementation) are sequential, and the emergent approach where the analysis is refined on an ongoing basis and options selection that is not necessarily fixed from the outset, as feedback from the implementation provides additional learning.

The strategy approach that is selected depends in part on the organization's approach to strategy in general, and also whether the initial proposal for the resource efficiency programme is quantitative (which suggests a prescriptive strategy) or qualitative, which implies an emergent strategy.

While the processes of corporate and resource efficiency strategy development are broadly comparable, the latter has unique features that we need to take into consideration. In particular, the analysis of the potential for resource efficiency to add value involves some specialist techniques whose importance is not fully conveyed by the broader headings "*analysis of the environment*" and "*analysis of resources*" (not resources in our sense, but the capacity of the organization) used by Lynch. For this reason, I have included an additional area of investigation called "*analysis of opportunities*". This activity involves a number of techniques, described in Discovery (page 385), which are designed to determine the potential for resource efficiency in the three programme phases:

- The potential to optimize existing operations, in other words, improvements possible without large capital investment or product or service redesign. This is the Optimize phase potential.
- The potential to reduce resource demand in existing processes/services either through the use of capital to improve existing equipment, material flows and processes (either by harnessing existing planned capital allocation or releasing new capital) or through the modification of systems, standards and policies. This is the Modify phase potential.
- The potential to radically change the business/service model to achieve a step change in resource efficiency. This will probably involve a significant investment and require a more profound change to the organization. This is the Transform phase potential.

There are two types of strategies we can choose from: ***prescriptive*** *or* ***emergent****.*

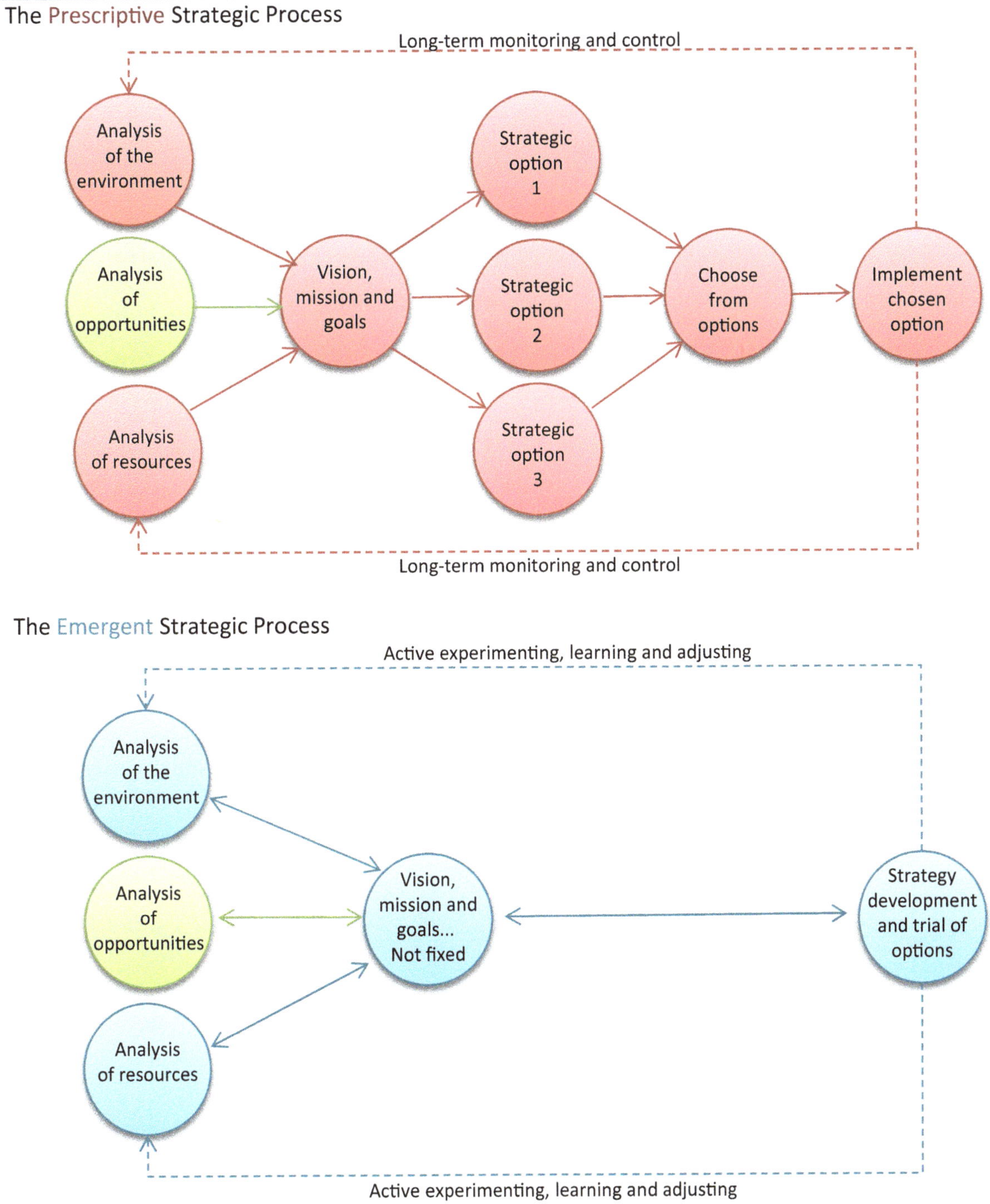

10.1 **Two main approaches to strategy formulation, the prescriptive and emergent processes**
Source: Adapted from Richard Lynch, Strategic Management. Pearson[489]

It is quite common to move from an emergent strategy approach to a prescriptive one, but less so to go from a prescriptive to an emergent strategy.

The "*analysis of opportunities*" activity should establish the scope of economically and technically feasible projects that can deliver improvements, bearing in mind that the timescale of the strategy may rule out one or more of the phases described above. If we are close to theoretical efficiency limits in a process (e.g. as per the Gibbs Energy for a chemical reaction, see page 367), then we need to consider alternative approaches to improvement, looking lower down the resource hierarchy. The intention in separating out the analysis of opportunities from the analysis of environment and resources is to emphasize that resource efficiency strategy analysis should be based on a hard appraisal of opportunity, costs and benefits, rather than just a qualitative approach. The resource efficiency goals will fundamentally depend on the value justification that is available through this analysis of opportunities.

The MediaCityUK case study (page 232) is an example of an emergent strategy for resource efficiency. Here, there was a clear vision to reduce costs, but no specific target. This approach was the right one at the time, given the wide variety of opportunities which were being constantly re-evaluated in the light of the learning made by all the participants in the process. With the introduction of formalized audits and building targets, the strategy is changing from an emergent to a prescriptive approach. This fits in with the greater knowledge of the team and the increasing investment needed to deliver further savings. It is quite common to move from an emergent strategy approach to a prescriptive one, but less so to go from a prescriptive to an emergent strategy.

Different organizations will have different theories of strategy, and it is generally advisable that our resource efficiency strategy is developed using the same approach. For example, organizations may view strategy as a process of adapting to the external factors, to better realize the primary goal. This is called the environmental theory of strategy, because of its emphasis on external issues (environment here meaning the milieu in which the organization operates, rather than anything to do with sustainability).

For prescriptive approaches to strategy, Lynch describes:

- The environmental theory, which emphasizes the organization's response to external factors as the principal source of competitive advantage;
- Resource-based theories, which state that it is the organization's resources (cash, brand, reputation, patents, assets, people) that provide the source of advantage and differentiation;
- Game-based theories, which seek to consider the implications of a strategy on the strategy of other influential competitors, suppliers and stakeholders, and then use the anticipated response to modify the initial strategy;
- Cooperation and network-based approaches to strategy, which see an organization being at the centre of a range of informal and formal relationships which could enable them to forge forms of collaboration which allow outcomes that they could not otherwise achieve alone.

There are also some theories of strategy that are prevalent in the emergent approach:

- Survival-based theories of strategy which believe that in very competitive environments, inefficient organizations are being regularly weeded out. Here, the most appropriate strategy is to focus on running very efficient operations which are highly responsive to competitive threats;
- Uncertainty-based theories of strategy state that the degree of unpredictability in the major drivers for organizational performance is such, that strategy is more likely to emerge from a dramatic shift arising from a severe challenge, followed by a period of stability. Essentially this is a strategy for anticipating and rapidly responding to risk;
- Human-resource-based theories of strategy believe that the motivation, capacity, culture and effort by individuals drives performance and see strategy as primarily a process of identifying how change in these aspects can take place;
- Innovation and knowledge-based theories of strategy focus on the organization's ability to use its resources in innovative ways, as well as to develop new products and services.

Strategy development for resource efficiency is likely to follow the prevailing theory of strategy in the organization, although this is by no means obligatory. The complexity and uncertainty, which can characterize resource supply and policy, suggest that a more dynamic strategy, following the emergent approach, may be more appropriate than prescriptive approaches. A strategy can, of course, incorporate elements of more than one theory. The key is to recognize when a simple *"just do it"* strategy is applicable, and when not. More formal strategy development is advisable where an organization is complex, or where there are many functions within the organization which can impact resource use or where the goals of the programme are particularly ambitious.

The **complexity and uncertainty,** which characterize resource supply and policy, suggest that a more dynamic strategy, following **the emergent approach, may often be more appropriate** than prescriptive approaches.

The strategy can, of course, vary from unit to unit, so that local conditions, culture and opportunities can be addressed more effectively. This was the approach of BP, which had a well-regarded *management by objectives* process where the centre stated the goals that needed to be met, but the periphery could set out their own strategy to achieve those aims. This is a form of emergent strategy that relies on the capability and knowledge of the local operations to deliver the goals in the most efficient way. In this approach, the centre must still ensure that there is sufficient urgency conveyed in respect of the goals, that the goals do not conflict with other business objectives, that rewards and incentives are aligned to the goals and that the resources needed to achieve the goals are made available.

A useful tool for determining the role of individual functions within an overall strategy is the illustration of teams and stages of a resource efficiency programme on page 276.

10.2 Strategic analysis

For a formal strategy, it is quite possible to analyse dozens of different issues. Here is a list of some of the main areas of analysis which draws upon real strategy development work with a global corporation.

What separates a strategy from a plan is the scope of the analysis undertaken. A proper strategy will take into consideration many internal and external factors that can influence resource use directly or indirectly, whereas a plan is much more limited in scope and is usually focused on achieving a specific goal.

It is helpful to have a model or checklist to guide the strategy development to ensure that all possible aspects have been considered. The example below is taken from a strategy development assignment I undertook recently for a global corporation. The analysis for this organization examined 20 different aspects of resource use. I have placed these under the same three headings as in the modified strategy development illustration on page 339.

My suggestion in terms of analysis is *keep it simple*. There are bound to be many external factors which cannot be fully quantified and effort should be focused on those aspects that have the greatest bearing on the key benefits and/or goals identified for the programme. Where factors can be confidently quantified and forecast, then a prescriptive strategy may be appropriate, otherwise an emergent approach may be better.

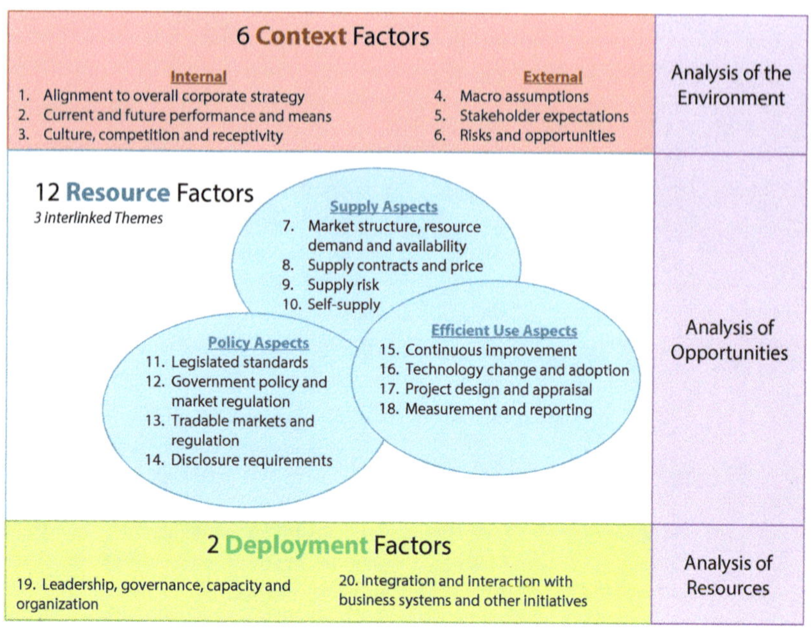

10.2 **An example of the breadth of topics that one might include in a resource efficiency strategy analysis**
Here, the analysis encompassed 20 different factors: six related to the broader context in which the business was operating; 12 factors examined specific resource issues in three linked themes; and two aspects of deployment. More details on each of these factors are set out in the table overleaf.
Source: Niall Enright, adapted from a real strategy development process undertaken recently for a global multinational corporation. Available in the companion file pack.

A challenge for large organizations is **resistance to top-down interference,** which can make strategies difficult to develop and implement.

Mandate

The example strategy illustrated in Figure 10.2 is one of the most comprehensive strategic analyses around resource efficiency that I have undertaken. Any one factor, such as a commercial supply/procurement strategy at a global level, can involve many different considerations and potentially dozens of different alternative options. Clearly, the level of detail one needs to go into in the analysis phase will depend on the impact of resources in the organization. The example above is taken from a very diverse organization with a total energy bill of well over US$1 billion and which is exposed to a wide range of regulatory regimes, and so investing in a thorough analysis is worthwhile. On the next page, I have described each of the 20 factors in more detail.

For many multinational organizations, it can be impossible to develop a single corporate resource efficiency strategy, simply because of the very large degree of variability in the internal and external environments at a site or national level. Water management approaches that work in Australia may be completely irrelevant in North America. Cultural attitudes to water in Europe and Africa are entirely different. Costs and risks related to water also differ greatly. That does not mean to say that a *free for all* approach to strategy development is advisable in these circumstances. In the first place, it may be that the organization wishes to have an overall goal for resource efficiency, even if that goal needs to be reinterpreted at a local level to make it meaningful and achievable. Clearly, then, the local strategies need to relate to the global objectives. Secondly, the allocation of internal resources (money and people) to resource efficiency cannot be optimized unless there is some form of consistent appraisal of opportunities across the facilities or units in the scope of the programme, such that those operations where savings can be delivered most cheaply can be identified and prioritized.

Another challenge for multinational, large or diverse organizations is resistance to top-down interference in local operations. In many cases, this resistance is the cause of the failure of resource efficiency programmes that are mandated from above. This resistance creates a real dilemma for proponents of resource efficiency. If they compel their sites to certain actions, they know that these are more likely to fail because the sites don't buy into the need for or the effectiveness of the measures proposed. However, if they don't mandate specific actions, then there is a risk that the sites may either select ineffective approaches or take no action at all.

An approach that seems to address the problems of resistance and quality is to require a process of consistent strategy development at a local level. In this approach, the sites are asked to develop their own strategy but following a globally consistent template – covering the same factors in every facility and business, but enabling the sites to set out their individual priorities under each heading. This approach has the benefit that the strategy is owned by the local team and relevant to their circumstances, with reassurance for the corporate function that all the key aspects of resource use will have been considered and documented in a consistent way with the desired quality. The case study on page 346 is an example of how I successfully applied this approach. ⇒ page 349.

Factor	Description
1. Alignment to corporate strategy	The importance of resource efficiency is set against other key organizational objectives: whether this is profit generation; service delivery; market or product development. This is essentially an analysis of the materiality of resource efficiency.
2. Current and future business performance	This analysis considers how the organization is meeting current objectives and is forecast to perform in the future, which will impact on the means (time, money, people etc.) that are available for our programme and the hurdle rates required for investments. Here, we would also seek to benchmark our performance with competitors or peers.
3. Culture, competition and receptivity	Here, we will assess the readiness of the organization to embark on a resource efficiency programme. Are people sensitive to the reasons why resource efficiency is important, or are we likely to see resistance? Are there other major corporate initiatives that could compete for attention and resources – is our organization suffering from *"initiative overload"*? Alternatively, is there a change in the pipeline that offers a one-off opportunity to elevate resource efficiency? Are we short of specific skills? This analysis is often omitted because it is difficult, but factors in this category are among those that have the highest impact on programme success.
4. Macro assumptions	Under this heading, we would provide a baseline analysis of aspects of the organizations' markets, demographics, technology development, and other macroeconomic trends such as globalization. Peter Dickens' *Global Shift*[211] is an excellent source of themes under this heading.
5. Stakeholder expectations	Stakeholders fall into the following broad categories: owners, shareholders, commissioning bodies; customers or service users; government, regulators and trade bodies; opinion-formers such as NGOs and the media. We have seen how in the example of Interface Carpets that it was a customer that provided the initial impetus for change and how for L'Oreal data on consumer attitudes in emerging markets is evidence in support of their approach to resource use.
6. Risks and opportunities	Here, we will consider broader aspects risks and opportunities that issues around resources could present – such as an opportunity to show our organization as a leader in the field of resource efficiency and the resulting branding benefits that would arise. We might consider broader impacts of adaptation, social upheaval and market sentiment from the environmental consequences of resource use, not just for our organization, but in society as a whole. Depending on the time frame of the strategy this analysis can be quite revolutionary, contemplating fundamental changes to business and service delivery models. One technique that is particularly useful for this step is "backcasting", which is discussed later.
7. Market structure, resource demand and availability	In terms of resource-specific analysis, one obvious starting point is to look at our forecast demand for resources and understand how the market will be able to respond to that requirement in the future. Who are the players who control the resources we require and how are these likely to change?
8. Supply contracts and price	Obviously, if one could accurately forecast future prices for resources, then one should probably be a commodity trader, not a resource efficiency practitioner! Nevertheless, broad price trends can be understood, and for some resources, such as energy, there are numerous sources of data on likely future price expectations, such as national government energy agency data. In our analysis, we should examine if our organization has in place contracts that may fix the price of resources for a period, or which may even have a *"take or pay"* clause that can undermine the economic value from a resource efficiency programme. Another consideration here might be the availability of third-party partners who would be willing to enter into EPCs (energy performance contract), where they provide the investment in the resource efficiency technology. All of these can affect the marginal "price" that we use in determining the cost/benefits of the resource efficiency programme.
9. Supply risk	Allied to the contractual and price analysis are supply risk considerations. For some organizations this might be the direct physical risk to supply. For example, at Rio Tinto's Diavik diamond mine in the Canadian Arctic, the availability of diesel fuel is constrained by the two months of the year when the ice-road to the mine is frozen (see *Diavik, diamonds and diesel* on page 113). Climate change may reduce this time and so introduce a future supply risk. In the case of other resources, such as rare earth minerals, the risk might relate to physical scarcity, or regulation may make the resource unobtainable, for example, as fresh water abstraction licences are revoked in many parts of the world.
10. Self-supply	Energy and water are resources where an organization may have options for self-supply either through on-site generation or using closed-loop recycling systems. Within these self-supply options, there may often be opportunities to develop profitable business by providing supply to third parties – for example, it is common for mining companies to develop generation to serve local communities as well as their own operations.
11. Legislated standards	No analysis of resource use would be complete without considering current and emerging regulations. Choice-editing may disallow some forms of resource use, such as prohibiting flaring of natural gas from oil extraction, requiring complicated and expensive methane recovery or reinjection. In the buildings sector, the trend to *"zero carbon"* building is driven by ever-increasing building standards which are limiting the use of non-renewable energy sources.

Factor	Description
12. Government policy and market regulation	This is a huge topic in resource efficiency terms, as there are clear signals that governments worldwide are turning increasingly to regulation to force organizations to mitigate their environmental impacts. In some cases, the interventions are direct, such introducing carbon taxation for vehicles and, in other cases, the policy impact is indirect such as market incentives for renewable energy generation, or land set-aside, whose cost is borne by all consumers. There are some on-line databases that provide useful lists of regulation for energy[808] and renewable generation.[617]
13. Tradable markets and regulations	Related to policy are a range of emerging mandatory and voluntary markets for carbon, water quality and biodiversity, alongside established markets in quotas for resource use such as fishing and water abstraction. In strategic terms, these market mechanisms will provide additional incentives for resource efficiency throughout the supply chain. Even if an organization is not involved in these markets they can have an impact via the supply chain.
14. Disclosure requirements	At an international level, there is already considerable pressure on global businesses to report on their carbon and water impacts through the Carbon Disclosure and Water Disclosure programmes. In the UK, there is a requirement to include carbon reporting in the annual reports and accounts of some companies listed on the main market of the London Stock Exchange. Where an organization has a brand to maintain or has stakeholder expectations around emissions, these disclosure requirements could act as a powerful driver for resource efficiency, particularly if the organization does not rate well compared to its peers.
15. Continual improvement	Here, we should examine the readiness and system available to implement continual improvement. If the organization uses Lean or Six Sigma, or similar, this may provide some useful structures to drive resource efficiency. We would wish to incorporate the findings of quantitative studies (e.g. using regression analysis) which can provide a measure of the improvements available from behaviour change. We could also do a gap analysis for ISO 50001(page 722).
16. Technology change and adoption	It is useful to consider the evolution of technology in a resource efficiency strategy because this is an area of rapid progress. In particular, the strategy should bear in mind that the costs of some technologies are decreasing quickly and so should incorporate regular reappraisal of the cost/benefits of these technologies in the final programme. There should also be a consideration of existing capital allocation plans within the organization so that the adoption of resource efficiency technologies can be considered on a marginal cost basis – i.e. where the business case is developed using the incremental cost of the efficient option over and above the committed capital cost of the standard technology (page 567). Where a technology might not be currently cost-effective, e.g. solar panels, the strategy can nevertheless future-proof project to allow for later adoption, e.g. by selecting a roof that can take panels in the future. The converse of this analysis is an understanding of the risks associated with inappropriate technology selection, which can lead to *"stranded assets"* which depreciate at a faster than anticipated rate. In this section, evidence from the analysis of opportunities should provide hard data on improvements available by focusing on equipment.
17. Project design and appraisal	An analysis of the existing processes around design and innovation (projects, services or products) will provide the foundation for understanding the potential for longer-term transformational changes in resource use, in quantitative and qualitative terms. It may be the case that changes to financial appraisal processes, such as the use of whole life costing, would align the financial and resource objectives of the organization.
18. Measurement and reporting	Here, an appraisal of the sources of data to evaluate progress in resource use is critical in ensuring that the goals and expectations from the programme can be adequately tracked. There is no point in setting a target that cannot be measured. Of course, measurement is not just about passive *"score keeping"*; it is a process that drives change in itself and the applicability of techniques like Monitoring and Targeting to deliver improvement should be analysed. Other considerations in this analysis might be the way that budgets are allocated within an organization. A move to allocating the costs of resources down the end-users – rather than treating these as a corporate or site overhead – has a big impact on motivation to reduce consumption.
19. Leadership, governance, organization, capacity	Here, our analysis should consider aspects of the internal organization which will hinder or favour resource efficiency, in particular, the possible form of the Mandate for the programme which can address these organizational findings. We also need to assess the capacity (money, people, skills, etc.) of the organization to realize the opportunities identified.
20. Integration into business systems and other initiatives	This is a very broad field of analysis, which encompasses key decision-making processes within the organization and how these might impact on resource efficiency. It is not just the systems but the incentives within our organization that have an impact. For example, the finance team may be incentivized to minimize capital spend on projects, leading to the cheapest, less-efficient technologies being used. Techniques like pairwise comparison, (discussed in *The power of pairwise comparison* on page 167) can help to highlight some of these conflicting incentives. We also need to consider if other initiatives, current and future, will favour or hinder our programme.

10.3 **20 Strategic considerations for a resource efficiency programme** *Source: Niall Enright, adapted from a real strategy development process*

Real World: ***Consistent site strategy development at a global resources business***

I have worked on a recent project where a maturity matrix has been the central tool to catalyze energy efficiency strategy development in a global business. This particular organization faced a number of very common challenges in constructing a global energy efficiency programme: the operating units were very diverse and varied hugely in their approach to energy efficiency; the corporate team did not want to be seen to be imposing an external process on the sites (which rarely works); and, as usual, there was often a perception at sites that energy efficiency was just about technology.

What emerged as a solution was an advanced form of maturity matrix, which could provide the basis for a full-day workshop that brought together a number of departmental heads and specialists to define a prioritized plan for their own site. The maturity matrix was a souped-up Excel spreadsheet with six themes: *Leadership and Context*, *Measurement*, *Opportunity Assessment*, *Project Implementation*, *Continual Improvement* and *KPI's and Communication* on different tabs.

Within each theme, there were between six and 11 topics, each exploring a particular aspect within that theme. For example under *Leadership and Context* the first question asks if there is an energy efficiency policy and plan. There then follow five statements from *No policy* through to *Site-wide EE policy and action plan, fully integrated with site business plan*, and the respondents are asked to select which of these statements most closely represents the current situation at the site.

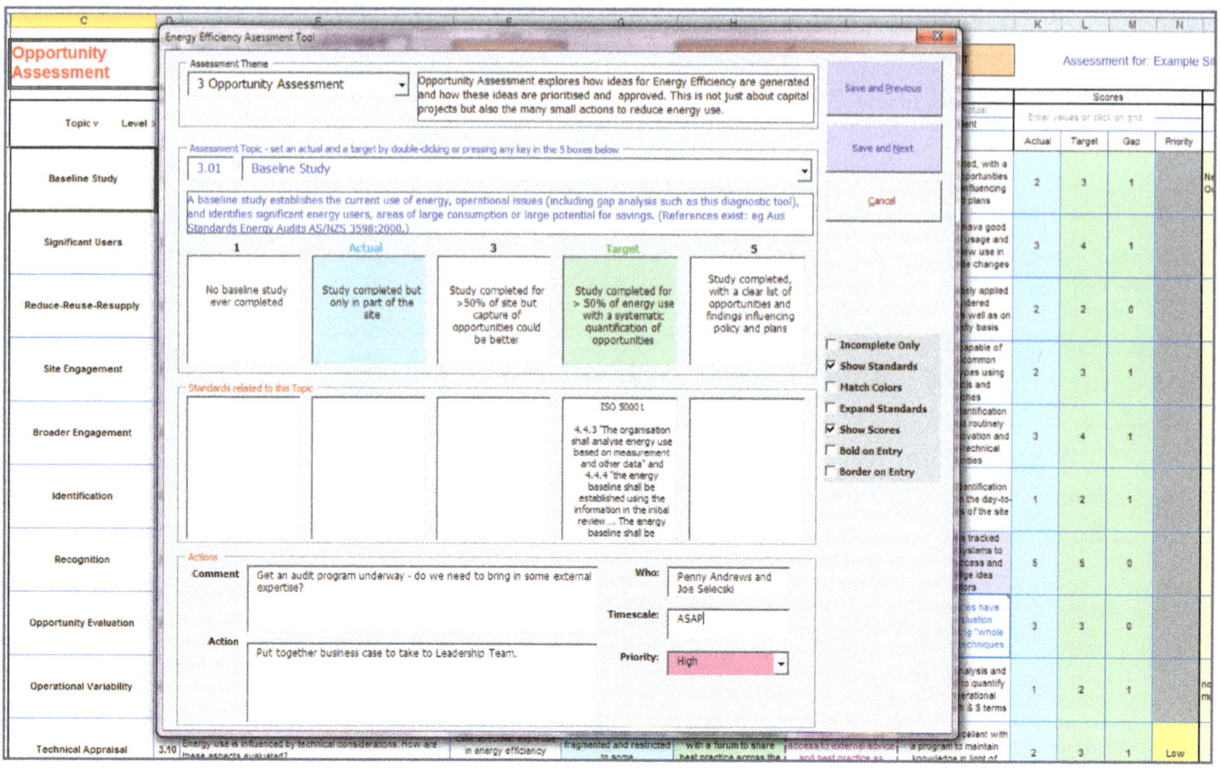

10.4 **The energy efficiency assessment and planning Tool**
This Excel-based tool, developed by the author, was used by a global resources business to enable sites to create strategies and plans which they owned at a local level, but which considered a consistent range of factors.
Source: SustainSuccess Ltd

This tool gave ownership of the strategy to the sites, while ensuring that it was comprehensive and consistent.

Mandate

In effect, the matrix provides a structured analysis for energy efficiency, giving a consistent approach to the assessment across all the sites.

Our approach incorporated a number of innovations, compared to conventional maturity matrices. First of all, the site team were asked not just to rank their actual performance but to state what their target would be. To help set the target, the tool displayed the score needed to achieve certain standards (both the corporation's internal requirements and, as appropriate, the requirement of the Australian Energy Efficiency Opportunities Act and ISO 50001). There were also features to run multiple assessments and aggregate these, as well as to instantly create PowerPoint presentations that summarized the key conclusions. The input was designed for a "live" interaction, with opportunities to capture comments and actions and to set tasks and priorities, through an easy to use data-entry form (illustrated opposite).

The tool could also link to additional documents relevant to specific questions, or capture additional information – thus the question on "*Does the site have a designated Energy Efficiency Champion?*" could allow the name and contact detail of the individual to be recorded or provide a link to the Champion's role-description.

The way in which the responses were obtained was important. Initial communication with the site would establish the key players around energy efficiency and then a multidisciplinary workshop would be organized to bring these folks together. At the workshop a facilitator from the corporate team would take the participants through each topic one by one, enabling a discussion on the subject and ensuring that a consensus was reached on the actual and target ratings as well as the priority of the issue and actions identified.

One reason for the success of this approach was that the corporate team who ran the workshops were excellent at managing the discussion and knew about the site's process and culture. With 46 topics or questions, it was important to allow space for debate, but then get to the decision around the actual and target rankings. Here language is critical – one finding is that the wording of the possible responses to each topic should be neutral, and carry no implications of poor or inferior performance. We also thought long and hard about the scores themselves – should they be 0 to 4 (as per the Energy Management Matrix on page 315), 1 star to 5 star, or words? We settled for the numbers 1-5 and the words "*Rudimentary*", "*Transitional*", "*Progressing*", "*Mature*" and "*Excellent*" as being fairly neutral. In running the workshop, the corporate facilitators ensured that a *no blame* approach was in place, to make sure that people felt open to give themselves realistic scores and articulate actions that would genuinely deliver improvements.

So, the skills of the teams running the workshops, and the care and attention taken in the wording of the topics, were both important to the success of the tool. A third, equally important, aspect was that the project was not perceived to be about outside assessment or imposition of targets – rather it was about the site's *own* perception of their strengths and weaknesses and their *own* priorities. The process was seen to bring value because it was something around which the various functions at the site level could unite to create a plan they had developed and owned. Here, the selection of the site's participants at the workshops was also crucial, to ensure that their conclusions carried weight and that all departments that could influence energy use were involved: not just engineering but operations, finance and maintenance.

10.5 **The energy efficiency assessment and planning tool provided a range of charts and reports that could be used to summarize the assessment process**
Here we can see that Measurement is one of the areas where there is there is a desire for the largest improvement. These results could be sent automatically to PowerPoint for a rapid presentation to the site management.
Source: SustainSuccess Ltd

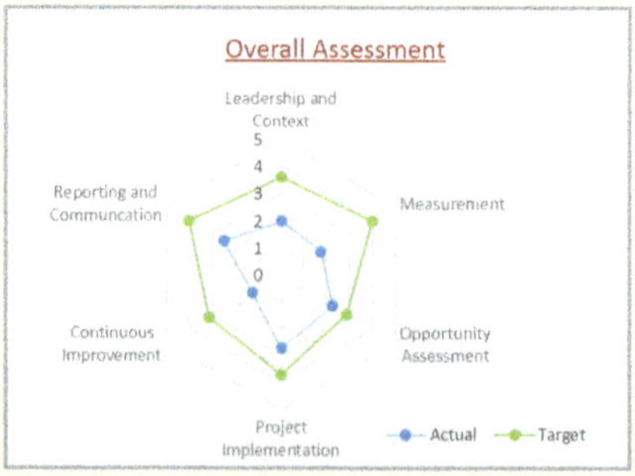

For the corporate team, the process brought many benefits. First and foremost the workshops delivered a strategy and action plan at the site level, which the site owned and would implement.

Secondly, the structured process enabled folks at the site to appreciate that there are many more aspects to resource efficiency than just spending money on technology; that aspects such as procurement, incentives, metering and management all have an important role. The process also enables the sites to see what is considered best practice under each heading.

Another important outcome for the corporate team was the emergence of common priority areas and challenges across multiple sites. Issues emerged, such as the lack of specification for energy metering and the need for more technical guidance on the operation of compressed air systems, which resulted in several best practice notes being developed by the corporate team which the sites found extremely useful and avoided *reinventing the wheel* at each facility.

Undoubtedly, the biggest benefit for the corporate team was that this was a form of collaborative engagement with the sites, which the sites perceived added value and so embraced. For the organization as a whole, there was a much more consistent approach to strategy development and a clear set of objectives at each site, against which progress could be measured.

It is important to note that there are some possible shortcomings with this approach, which need to be managed to achieve the desired outcome. The first issue is around honesty of reporting – we found that there is a natural tendency for folks to err on the positive side when evaluating their performance. This bias could be, to some extent, overcome by undertaking the assessment as a group where there was less likelihood of individual agendas dominating. Capturing data to support the score awarded also helped keep folks honest. For example, on the topic about whether there was an Energy Champion, if the response was *"yes"* we would ask for the person's name and a document setting out their role. Making the questions *neutral* and ensuring that there is no negative feedback for a low score in any area were also important.

Finally, the matrix tool provides a qualitative assessment of energy management at the facility, and so needed to be accompanied by a formal audit of the site to provide a quantitative assessment of the available opportunities for energy efficiency. The former shows the potential improvements to systems to achieve best practice, while the latter provides the business case for action and highlights the focus areas of the programme.

As with all maturity matrices, such as the energy management matrix (see page 315), the tool offers the opportunity to revisit the scores at a later date to assess how energy management has matured, i.e. improved, over time. Early suggestions that this type of analysis would be incorporated into the ISO 14001 and ISO 50001 standards on environmental and energy management proved false, but indicates how highly these tools are viewed in continuous improvement circles.

10.3 Options selection

Strategies throw out alternatives. There are a number of methods for selecting which of these alternatives are relevant and should be considered in more detail or built into the final strategy.

Once the analysis phase of strategy development has been completed, there is a process of options selection. In the prescriptive approach, this will tend to be a series of binary "yes" or "no" decisions from which a clear plan and timetable for action arise. In the emergent approach, the choice will be around selecting initial series of activities and determining how the outcomes of these will be evaluated so that the strategy can be refined.

Several criteria are typically used to select from alternative strategic options.

- The cost-benefits assessment of each option in comparison with the alternatives. Clearly, those options that deliver the greatest benefit for the lowest cost are likely to be favoured.
- The materiality and alignment of the option in relation to the primary purpose of the organization. Thus, in a private company, choices that lead to increases in profits are more likely to be approved, whereas in a public organization choices that enhance the quality of service may be more important. Clearly, if an option addresses a *"burning platform"* it is more likely to be advanced than one that is merely *"nice to have"*. Similarly, an action that has little impact on the overall goal is less likely to be approved than one which can have a sizeable effect. In essence, the greater the benefit the organization perceives from a particular choice, the more desirable that choice will be.
- The degree of stakeholder approval that the option will elicit. Every organization's existence depends on acceptance by a range of different stakeholders, such as shareholders, employees, managers, local communities and regulators. An example would be the choice to discharge wastewater rather than treating it - this may gain favour with shareholders, who would see costs reduced, but would attract the ire of regulators, communities and, possibly, employees. Thus, every choice is a sensitive balance between the needs of these stakeholders, and options that best achieve this balance are more likely to move forward.
- The credibility and validity of the assumptions underpinning the option. We have already seen that the more we can confirm the outcomes of a choice, the more likely it is to receive approval. Thus, well-documented and evidenced options are more likely to be favoured than those where there is uncertainty. Choices supported by existing case studies or best

Formulating a great strategy is about **correct analysis** *to identify and value alternatives, followed by* **correct selection of options.**

practice evidence, in particular from internal pilots or respected peers, are more favoured – especially where the organization is being asked to do something new.

- The practicality and feasibility of the option. Clearly, an option which is seen as relatively easy to carry out is more likely to be implemented than one which is not, all other things being equal. If the activity requires a high level of internal resources (time, money, management effort, etc.), then that may also make it impractical.

- The risks and limitations that will arise in the future as a result of exercising the option. This is because making a choice has the potential to constrain future action. For example, betting heavily on biofuels could lead to exposure to higher costs if demand for the fuel increases faster than supply. A similar problem arises where a low-mass roofing structure is selected, which is efficient in terms of the materials used in construction, but which could preclude the more important future use of the roof for solar PV generation, which requires a heavier structure for support. A choice, once implemented, can reduce the future options and, given the relatively fast-changing issues around resource use, flexible solutions will probably tend to be favoured by most organizations over those with obvious constraints.

Many of these decisions are formalized by organizations using a materiality matrix to determine the priority areas for environmental, social and labour issues. They typically engage with internal and external stakeholders to determine which are the priorities for the interested parties and which areas have the biggest impact on the organization. See *Focusing on the right areas* (page 362) in chapter 11, on Goals, for more on materiality matrices.

It should be noted that this options selection process usually occurs in an incremental, iterative manner. The first set of choices may be made on only one of the criteria above, typically by considering the cost/benefit for each option, ruling out those alternatives which are simply too expensive. One of the most common tools to quickly rank the financial return from a basket of projects is a marginal abatement cost curve, discussed on page 610.

The initial strategy analysis and the first pass at options selection may have been delegated to specialist staff in finance, environment, engineering or planning functions. However, as the options are reduced, and the strategic choices become clearer and more material, it is very common, and desirable, for this initial development team to engage with the Leader or decision-makers, such as the Governance team, to whittle down the choices further. These more senior folks are more likely to be able to make the subtle choices around stakeholder approval and the role of resource efficiency within other organizational objectives than the specialists. Furthermore, the participation of the senior team in the final options selection means that they have greater ownership over the strategy.

10.4 Common strategy errors

Some strategy errors around energy and resource efficiency occur with depressing regularity. Here is a list of the most common mistakes and how these can be avoided.

Strategy errors fall into two types: a failure of analysis and a failure of choice. In the first instance, incorrect assumptions or scope in the analysis leads to an incomplete set of option, while in the second, the wrong choices are made. There are a number of very common analysis errors.

- Failure to adequately consider business as usual (BAU). Doing nothing is a choice, with cost, reputation and other implications that can be assessed just like any other options. Doing nothing not only commits the organization to the current level of resource costs, but also potentially locks in future increases in price or supply risks, or could lead to a poor image in comparison to peers. As the resource efficiency options are always compared to the BAU case, quantifying this accurately is essential to the analysis. The most common mistake is to assume that BAU means that costs etc. are unchanged.

- Failure to consider planned capital expenditure. It is remarkable how often resource efficiency strategies are divorced from organizational capital investment plans. Here, it is critical to consider the marginal cost associated with resource efficiency investments; in other words, how incremental capital spend over and above that committed, including for maintenance and similar tasks, can deliver highly cost-effective improvement (see page 567).

- Failure to consider system changes. Much strategy analysis for resource efficiency starts from the viewpoint that this is a technical issue and so focuses on discrete projects or equipment changes, rather than the more profound way organizations reach decisions. System changes, such as the use of whole life costing in capital allocation, can have an enormous impact on resource efficiency and can be implemented relatively quickly. Other examples of system changes include introducing a company vehicle policy based on emissions – a form of internal choice-editing.

"Well, now we know what not to do."

- Failure to evaluate non-technical issues. The success or failure of a resource efficiency programme is not just about the technical opportunities for improvement, but about the organization, culture, incentives and motivation of people. Unless people factors are taken into account

Analysis errors fall into two categories. First there are errors of scope, such as omitting types of improvement like system changes. Then there are errors of calculation, such as failure to take into account opportunity interactions or underestimation of costs.

in our programme, technology solutions are likely to perform less well than expected, and we are unlikely to treat the process as one of change management leading to continual improvement.

- Failure to appreciate that improvements are often non-additive. For example, combining a project to reduce lighting hours with one to replace the lamps for more efficient units will not lead to the sum of the savings that each project would deliver individually.
- Underestimation of costs. The actual costs of implementing a technology can sometimes be more than double the purchase cost for the equipment alone, once one takes into consideration the specification, procurement, installation and commissioning costs. Also, we sometimes need to consider opportunity costs, such as the interruption of production that can occur when the technology is installed.
- Inadequate sensitivity analysis. The cost-effectiveness of many resource efficiency activities depends on the future price of resources, which may be very unpredictable. Sensitivity analysis will test the option with a range of future costs scenarios and so will help provide some indication of the probability that a project will achieve the desired financial return. In many cases, choices are made on a single "central" assumption about prices which may, in fact, lead to apparently more attractive but riskier actions being approved ahead of choices with a slightly lower return but which are less influenced by price (see page 593).
- A very common error is to assume a lack of available workforce and money. We shall see later that there are plenty of sources of external expertise and financing that can compensate for the lack of internal resource. Far too often, rather than going out to find these resources, a strategy is developed on this presumed limitation, rather than on the true potential for improvement.

Given a sound analysis, it follows that the selection of the options that will deliver our resource efficiency goals should represent the optimum balance between cost and reward, depending on the decision-maker's view of risk and stakeholder acceptance. Unfortunately, even with perfect data from the analysis, decisions around resource efficiency options are often irrational for a number of reasons, such as the psychological status quo bias and loss aversion, the preference of options where the outcome is certain and underestimation effects when considering benefits.

By far the most common error in choice selection, which I have observed over the years, is to rush to new capital expenditure too early. In this situation, organizations leap-frog no and low-cost operational savings and go straight to technical fixes to meet their resource efficiency challenges. Thus the behavioural aspects of resource efficiency are bypassed – with the unintended consequence that the technology fixes often fail to deliver the anticipated improvement because the resource users are not involved or engaged. Not only does this lead to worse financial returns for the programme, but the project-focused nature of the technical solutions undermines efforts to put in place continual improvement processes. Throwing money at a problem will almost always solve a problem, but it is rarely the most cost-effective way to do so.

There are many reasons why organizations exhibit this tendency to rush to technology for a fix. The plethora of self-congratulatory case studies that give the impression that resource efficiency is straightforward is certainly unhelpful, as are technology vendors' often over-optimistic predictions for the impact of their products – look at the claims for voltage optimization, for example. Often, the individual tasked with developing the strategy simply is not experienced enough in resource efficiency programmes. It is very common for engineers, who are mainly technology and project-centric, and behavioural change adverse, to be asked to develop the detailed programme. Is it any surprise then that they will inevitably provide a technically focused solution rather than an approach incorporating changes in behaviours or systems? It is in the nature of their training to take a project-centric view. For many folks, a *"fire and forget"* technical fix is much more preferable than the *"heavy lifting"* perceived to be involved in engaging resource users in a continual improvement process. Developing an end-of-pipe fix enables the resource efficiency box to be ticked quickly and the individuals to move on to other projects. And, of course, there is the bias towards more *"credible"* technical fixes – search goods in the parlance of economics - rather than more cost-effective but less predictable operational improvements which fall into the less desirable category of credence goods (see *Why certainty drives the resource efficiency proposal* on page 185).

Choice errors include biases towards choices with the greatest certainty and hence capital investment projects, rather than behavioural or systems changes.

A more recent phenomenon leading to bigger and earlier capital expenditure on resource efficiency is the tendency for organizations to set much more ambitious goals, well above the level of improvement that can be achieved by operational efficiencies alone. Assuming that typically 10% emissions/energy improvement may be delivered with little or no capital expenditure, it follows that organizations with greater improvement goals will need to invest capital in technology to achieve their target. The problem is that some of these organizations, recognizing the scale of improvement needed, skip the initial Optimize phase and leap straight into the Modify or Transform stages in a resource efficiency programme, as described earlier in Section 6.6 on page 242. Not only does this lead to a more expensive process, which is slower to deliver, but the failure to engage folks in continual improvement can impact the long-term sustainability of the programme, as well as leaving much *low-hanging fruit* unharnessed. Having said that a rush to early capital expenditure is often a mistake, there can be situations where the exact opposite is the case (see *Rio Tinto and CAPEX on page 569*).

Until resource efficiency is universally understood to be a change management activity, there will naturally be a tendency to focus on equipment rather than people, systems, processes and design, which can all have a more profound and long-lasting impact on resource use. ⇒ page 357.

Real World: *Strategic challenges in combining resource efficiency and energy supply*

In the late 1990s, the North American electricity generation sector was undergoing very significant change. Deregulation of utilities was occurring across broad swathes of the country with individual states and public utilities commissions seeking to separate the generation and distribution side of the business from the retail side. In simple terms, large electricity consumers would be able to purchase their electricity from any supplier rather than being obligated to buy from the local utility company, opening the market up to competition. In a similar fashion, generators would need to compete to provide electricity to the utilities that had the supply contracts. This led to a wave of consolidation as companies tried to reduce their costs to remain competitive and gain market share.

One such company was FirstEnergy, formed in 1997 from the merger of power companies in Ohio and Pennsylvania. Like other utilities, FirstEnergy was being forced to make a choice between focusing on the wholesale generation business or competing directly for large customers, within and outside its historical supply territories. FirstEnergy chose the latter route and started to consider how it might differentiate itself from other utilities in the north-east of the US. The conclusion it came to was that providing a range of "*energy services*" would help with the customer engagement and thus support electricity sales, and so it created a new subsidiary company, The E Group, to deliver these services.

Within the FirstEnergy territory, there were many manufacturing businesses which are large consumers of electricity and The E Group felt it needed some innovative industrial energy efficiency propositions to begin a dialogue with these customers. At the time I was with Enviros (now part of Jacobs), promoting the enManage™ methodology in North America, which focuses on delivering no/low-cost energy, water and waste savings through behaviour change. In 1999, Enviros entered into a deal with The E Group to supply the enManage service and accompanying software Montage™ into 13 states in the central northern region of the US. The attraction of enManage as a proposition to FirstEnergy was that it offered a very high return on investment compared to the typical offerings of other energy services companies (ESCOs). This is because enManage focuses on no/low-cost energy savings with a much better return than the traditional motors, lighting and boiler programmes.

Unfortunately, this deal was not a success. The problem lay in three mistaken assumptions that underpinned the strategy:

- That the large manufacturing customers would recognize the value inherent in a behaviour-based approach to energy savings;
- That the large manufacturing companies would see The E Group as a partner for these services;
- That the dialogue would cross over into opportunities for electricity sales.

It seems that the manufacturers were certainly interested to hear about this approach to energy management focused on continual improvement, not least because of the excellent rates of return that appeared to be on offer. In fact, over the first six months in 2000, the proposition was presented at over 40 sales calls with very large customers such as Ford, DuPont, BF Goodrich, General Mills, SmithKlineBeecham and so forth, many of which I participated in. However, despite the positive engagement and warm welcome we received, we were

Mandate

simply unable to overcome two basic objections: *"It wouldn't work here in the US"* and *"I can't understand why an electricity provider would want to help me save energy"*. The core problem in this strategy, which was poorly understood by us at the time, was that we were essentially offering a credence good, which therefore relied on high levels of customer trust in the outcome. The absence of local evidence as to the effectiveness of enManage, and the lack of credibility of a utility as a supplier of energy efficiency services, both undermined customer confidence in the benefits.

The final nail in the coffin of the strategy was the fact that big electricity users will always buy energy on a commodity basis – the quantities are so large and the costs so high that the only factor determining who gets the supply is price, down to the sixth decimal point in the unit charge if necessary. It doesn't matter at all if a supplier has helped reduce usage, the concept of loyalty simply does not apply in a commodity market, where the product, electricity, was identical from all suppliers.

Of course, hindsight is perfect. FirstEnergy and The E Group were to be applauded for innovating and bringing new services to the market in an effort to differentiate their offering. Today I am pleased to see that The E Group is thriving, and was recently awarded the accolade by Energy Star of *"Partner of the Year – Service Product Provider"* for the second consecutive year running. By focusing on more conventional services such as energy monitoring, Energy Star Building Certification and energy efficiency rebates, largely to less complex commercial and public sector customers, The E Group has been successful.

By contrast, a similar partnership with Enviros was in place in Canada at around the same time with an Ontario utility, Enbridge Gas, which was to have a very different outcome. Here the service being promoted was a more generic M&T programme - in actuality enManage™ - aimed at industrial gas users as part of Enbridge's wider demand reduction programme.

These gas users were not cynical about Enbridge's motives in the way FirstEnergy customers were, because the programme had been in place some years and they were aware that demand reduction was a key requirement underpinning the price agreement between Enbridge and regulators. The M&T approach was also just one of several improvement packages on offer through the programme, and many of the companies approached had benefited from rebates related to gas savings using other technologies. The second feature that enabled the Enbridge programme to succeed is that they overcame the credence good objection by providing a significant part of the scoping study costs for the M&T programme. This scoping study was essentially a site audit that would be conducted by an Enviros expert who would identify the very specific actions and projects that would deliver the no and low-cost energy savings, thus creating a much greater degree of certainty for the programme. Perhaps if The E Group had offered a similar commitment, they might have been able to overcome the credence hurdle, signed up some initial facilities and so put in place a local case study to prove that the approach does work in the US. This scoping study turned the credence good proposition into a search good.

Energy is always brought on a commodity basis by large users. As a result, bundling other services will have little effect on the user's buying decision unless the energy price is competitive.

The Enbridge programme was so successful that it still has a case study from 2002 on its website,[250] entitled *"Unilever's Monitoring and Targeting Programme Captures $1 million in Savings"*. This describes how the continual improvement approach to energy management at the Unilever Rexdale edible oils (margarine) facility delivered savings through systematic improvements in operation and technical improvements.

10.6 **Left to right: Bing Sitahal, Unilever Canada; R. John Efford Minister of Natural Resources, Canada, and Doug Dittburner, Chief Engineer, Unilever Canada, receiving an Honourable Mention at Canada's Energy Efficiency Awards 2004 presented by The Canadian Industry Programme for Energy Conservation (CIPEC).**
Source: Niall Enright from correspondence with Doug Dittburner, May 2004

The Unilever Rexdale programme began in a modest way in 1999 under the banner of *"Watt Watchers Energy Teams"* where shop floor employees at this 170-strong manufacturing facility were encouraged to become actively involved in energy savings. The information flow was strong, with display boards charting progress and celebrating success. In the following year, the team embraced the M&T approach, with the support of Enbridge and my colleagues at Enviros. This enabled Unilever to systematize the measurement of savings to reduce operational variability in energy and water use, as well as implementing an Opportunities Database to capture all the project ideas.

By 2004, cumulative savings (credited for the first year only) had exceeded C$3 million for an outlay of C$1.5 million, and there were some very dramatic reductions in the intensity ratios of key utilities as shown in the table, left.

Utility	Decrease per lb production
Natural Gas	39%
Electricity	24%
Steam	50%
Compressed Air	27%
Water	52%

10.7 **Improvement in specific ratios for Unilever Rexdale from 1999-2003 arising from the Watt Watcher continual improvement programme.**
Source: Niall Enright from correspondence with Doug Dittburner, May 2004

One would be forgiven for thinking that by this point the savings opportunities at the site had been exhausted. But Doug Dittburner and the team at Rexdale had different ideas: *"After all we've done to improve energy efficiency, we still aren't close to finished. Think your plant is running efficiently? Think again."* was typical of the culture of continual improvement. Each year new projects would be brought forwards, new ideas explored and new tools deployed. The energy savings programme was aligned to the Total Productive Maintenance (TPM) programme which provided additional tools such as the *"TPM Loss Tree"*, enabling savings to rise to a total of C$5m through 123 discrete projects by 2007.[336]

The contrasting outcomes in these two utility-led resource efficiency programmes indicate just how important careful analysis is in developing a strategy. In both cases the end-user proposition was very similar, but both the credibility of the seller and the nature of the initial commitment differed enough to lead to quite different results in the end. Perhaps the fundamental difference was that Enbridge needed the gas saving to be achieved, whereas The E Group were more equivocal once they saw that energy efficiency was not a channel to delivering more electricity sales. In the former case resource efficiency was aligned to the core business objectives and in the latter not.

10.5 Modelling and strategy

Models allow strategy analysis to remain "live". Assumptions can be updated in light of changed circumstances and plans adjusted. Where a resource efficiency programme is complex or goals are set some time into the future, modelling is highly recommended.

If our strategy is intended to operate over a reasonable time frame, say several years, it may make sense to develop a dynamic model to support our decision-making. In the emergent strategies approach, the model would be used to continually inform decisions, while, in a prescriptive approach, the model may be run once or twice a year to confirm if the strategy is still valid. The idea of the dynamic model is that it allows resource prices, compliance costs, mitigation options and risks factors to be adjusted as circumstances change, enabling the most cost-effective choices to be clearly seen at any time.

The illustration below shows one such model. The model takes cost and emissions projections and relates these to the total potential for emissions reductions from many different projects and project types to determine the most cost-effective carbon neutral strategy. The model also takes into consideration approaches that may not be acceptable from a reputational perspective, such as ruling out "*green energy purchasing*" (discussed in the next chapter).

These types of decision support tools are common when the resource efficiency programme involves substantial costs and risk to an organization. They are particularly useful in undertaking sensitivity analysis at the outset of the programme, where the impact of changing one parameter, such as the price of carbon or gas or electricity, can be fully understood.

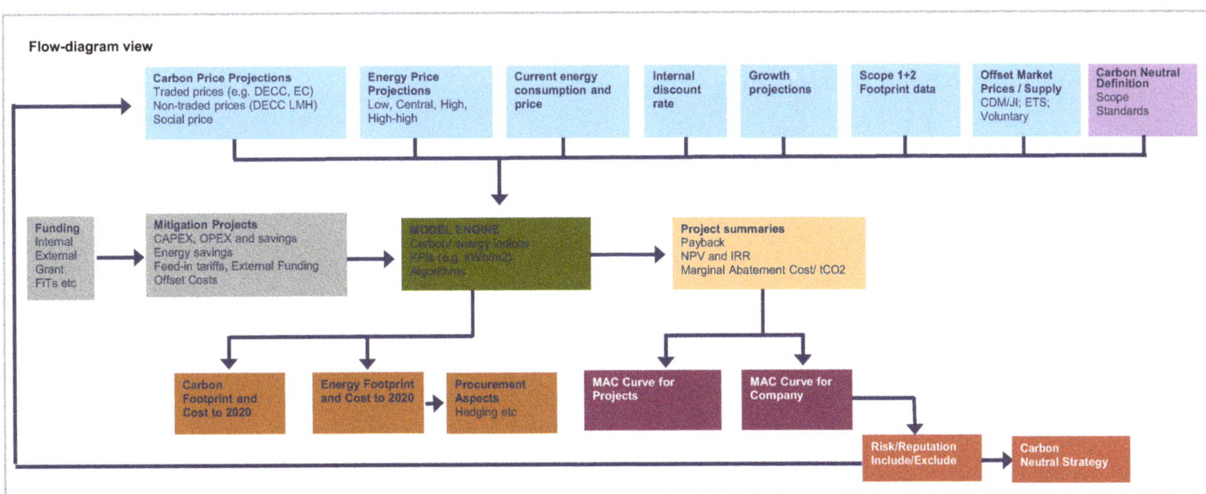

10.8 **A dynamic model for a global corporation seeking to achieve carbon neutrality by 2020**
The outputs include a cost-benefit analysis, a carbon price required to achieve the desired change and a prioritized list of projects that will achieve the desired outcome in the form of a marginal abatement cost curve.
Source: Niall Enright

10.6 Technology learning curves

The rapid decrease in costs for many energy technologies means that opportunities that are infeasible economically today could be very attractive in just a year or two. Our strategy needs to take this into account if we are to reach our full potential for improvement.

One aspect that we need to bear in mind when developing our energy and resource efficiency strategies is the rapid pace of developments in this field, both in terms of regulation and technological advances.

We often hear about *economies of scale* leading to reduced cost of technologies. Economists refer to this as a *learning curve*. As can be seen from the charts below, quite dramatic cost reductions have been observed in many new energy technologies, which have been matched by rapid increases in deployment. Wind energy generation costs in the US have decreased by a factor of three between 1984 and 2014, as shown in the chart left. There was a small increase after 2000 due to increasing raw material costs, but the trend is once again downwards. For solar PV, the decrease in costs between 1976 and 2014 has been almost 100-fold. When considering the return on investment on technologies, it is vital that the strategy includes an assessment of the rate of decrease of cost. For example, LED lighting, which was prohibitively expensive just a few years ago, now offers some great returns on investment.

10.9 **Learning curves for wind and photovoltaic technologies (note that the axes are logarithmic)**
Source: Adapted from JP Morgan report "A Brave New World".[429]
Original data, chart left NERL, GWEC 2014, chart right Bloomberg, IEA 2014.
Available in the companion file pack.

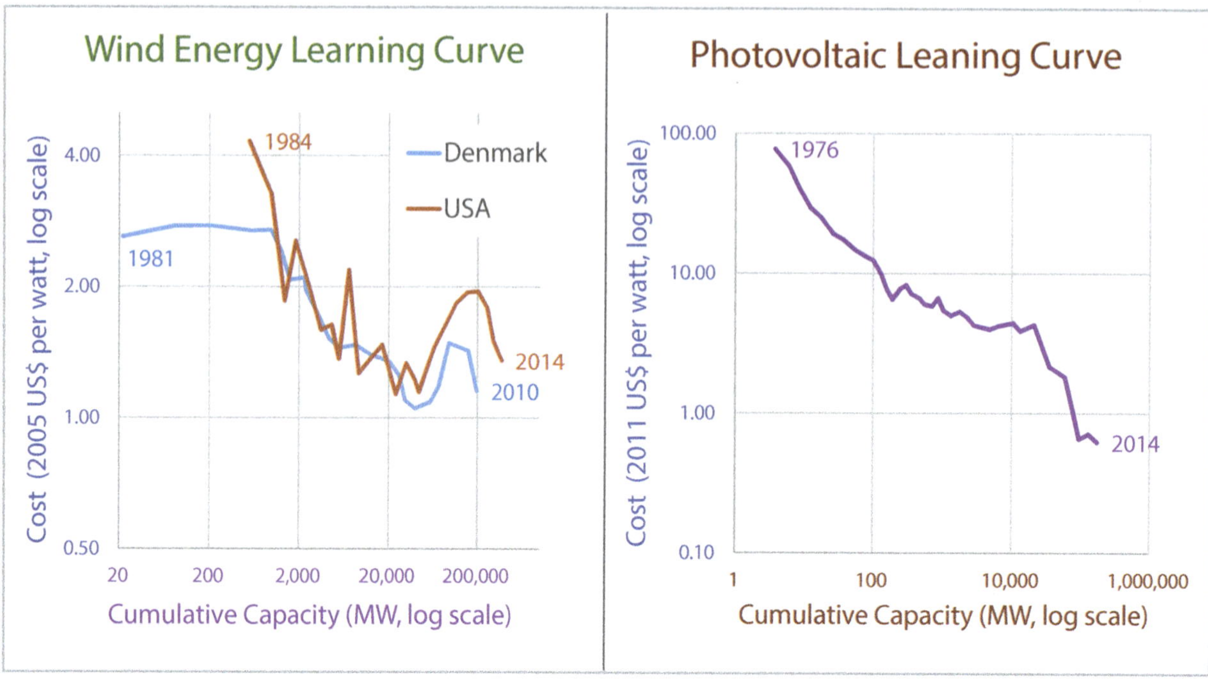

Mandate

Summary:

1. Consider carefully whether a separate resource efficiency strategy is needed – it may be better that resource efficiency becomes a fundamental part of *other* organizational strategies.

2. Understand that the strategy you get depends on who you ask: where possible, involve a wide cross-section of folks in its development.

3. The most effective strategies approach resource efficiency from a change management perspective. Reviewing the barriers to energy and resource efficiency in Volume I Chapter 4 will also help design a more effective strategy.

4. Recognize that resource efficiency is a continual improvement process until "zero use" is achieved. Don't plan to stop.

5. Beware the many common pitfalls in the analysis phase. As they say: *"Garbage in, garbage out."*

6. Analysis developed with a particular goal in mind leads to a plan. A strategy should start with a blank piece of paper and identify what is truly possible, which can then be compared with an existing goal, if required.

7. Always strive for improvements at the top of the resource efficiency hierarchy (see page 57).

8. A strategy does not have to be set in stone; you can use an emergent process to create a dynamic programme that adapts and responds to change as lessons are learnt.

9. Try to keep the strategy simple. There are many aspects of resource efficiency and it is easy to lose sight of what is important if you are not careful.

10. Strategy development involves analysing opportunities for improvement and then selecting from the alternatives. Be aware of the possible errors of analysis and choice that can occur and take steps to avoid these.

11. Just because a technology is not feasible today does not mean that it should be permanently ruled out. The learning curve for many technologies means that they could be cost-effective in the near future.

12. No matter how clever or sophisticated it is, a strategy will not succeed unless there is ownership of it. Try to involve all the folks, from leaders to the shop floor, who you will need to deliver the strategy so that they feel that it is *their* strategy.

Further Reading:

Richard Lynch. *Strategic Management*. Pearson, 7th edition (21 Jan 2015) Pub. ISBN-13: 978-1292064666. *This highly recommended book is aimed at mainstream strategic planning and management, but many of the principles apply to the development of resource efficiency.*

Questions:

1. What is the difference between an emergent and a prescriptive strategy? What are the pros and cons of each?

2. What sorts of challenges do large, complex, multinational organizations face in developing a resource efficiency strategy and how can these be overcome?

3. How is the choice of strategy influenced by whether a proposal is quantitative or conceptual? (See also Chapter 9 Creating a Mandate on page 307.)

4. What is the difference between a strategy and a plan?

5. Does your organization or department have a strategy? What kind of strategy is it (emergent or prescriptive)? Does the existing overarching strategy provide any guidance on how a resource efficiency strategy might be formulated?

6. What are the most common energy and resource efficiency strategy errors and how can these be avoided? Use some real examples for your own organization, if available.

7. Why is it important to understand the learning curves for different efficiency technologies when formulating a strategy?

8. What is the role of modelling in strategy?

9. Consider the 20 aspects of strategy listed on page 345. Which of these are most important for your organization and why?

10. Who would you include in a strategy analysis in your organization and why?

11 Goals

An effective Mandate for resource efficiency requires that there are clear goals for the programme against which resources can be allocated, progress can be measured and expectations managed.

This chapter will cover a wide range of different issues related to goals including:

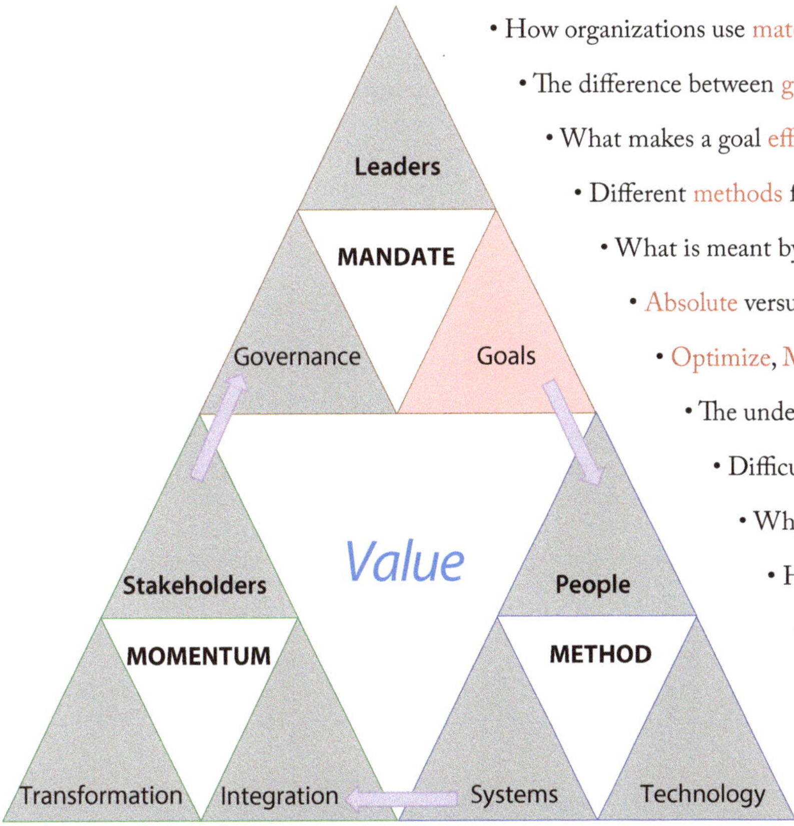

- How organizations use materiality to decide on their goals.
- The difference between goals and targets.
- What makes a goal effective.
- Different methods for setting a goal.
- What is meant by a sustainable goal.
- Absolute versus relative goals and targets.
- Optimize, Modify and Transform opportunities.
- The underlying weakness with specific ratios.
- Difficulties with renewable electricity.
- Why additionality is important.
- How to calculate rates of growth.
- How goals can be articulated.
- Leading and lagging indicators.

At the end of this chapter the reader will be comfortable with the key aspects of goal-setting.

11.1 Focusing on the right areas

Goals normally flow from the Leaders who define the Mandate. However, there are a number of tools available to help us to prioritize goals.

Most organizations publicly describe their energy and resource efficiency objectives in terms of their environmental outcomes, not the business benefits that this will bring. However, when these goals were set, there will have been a judgement of the cost and benefits of the changes, so there will also be an internal goal or value proposition for the resource efficiency programme.

As we have seen earlier, there can be many drivers: a crisis (often a compliance or licence to operate issue), cost reduction, improved competitive advantage, or simply because *"it is the right thing to do"*. If the main aim is cost reduction, then the internal goal will be based on the potential savings that can be achieved with a desired rate of return. On the other hand, if the primary objective is an enhanced brand, then the goal may be influenced by what competitors are doing (no one sets out to be mediocre in comparison to their peers). For those organizations whose primary concern is the environment, like Interface (page 226), then the objective might be *zero waste*, and a much more fundamental examination of the business model will be required. Even though the public goals are expressed in terms of the environmental outcomes, the programme manager will also be expected to measure performance in relation to the internal goals. The principles set out in this chapter cover both the public and internal programme goals.

A tool that has grown in popularity recently is the materiality matrix, which can help determine the business, environmental or sustainability issues likely to have the greatest impact on the organization and its stakeholders. This matrix consists of a simple two-axis chart in which issues are rated by the level of concern to stakeholders (customers, staff, regulators, etc.) and by the degree of impact on the organization, as shown below. Priority issues for the organization are in the top right-hand corner of the matrix.

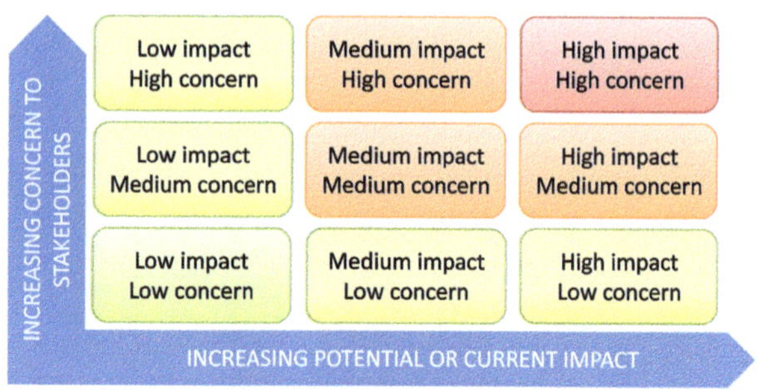

Real World: *Materiality at Ford*

As energy and resource efficiency become a mainstream concern in organizations (usually as part of their duty to manage risk) the development of goals in this area has become more systematic and objective.

Ford identified over 500 material business issues which have the capability to influence the future. It whittled these down to 57. Of these, 14 are in the category *"high impact, high concern"* and eight, over half, are to do with resource efficiency.

These are the 14 most material issues identified by Ford (with resource efficiency issues highlighted):

- Low carbon strategy
- Vehicle GHG emissions
- Fuel economy
- Electrification strategy
- GHG/fuel economy regulation
- Water strategy
- Product competitiveness
- Risk and cost management
- Sustainable mobility
- Vehicle safety
- Supplier relationships
- Supply chain env. sustainability
- Sustainable raw materials

Against most of these issues, there are specific goals for improvement so that progress can be reported annually. In some areas, like climate change, there may be several goals. The materiality matrix itself is regularly reviewed and changes in priority reported.

11.2 Goal-setting methods

There are more choices in goal-setting than most people imagine! Choosing the right goal-setting method will increase the probability of success. In particular understanding your organization's "sustainable goal" is key to reducing long-term risk.

Two or more goal-setting methods can be used at the same time: for example, we may have a fact- based **empirical** goal which uses **indirect** cost savings, or a goal which is both **guesstimate** and **aspirational**.

There are many different ways that organizations go about setting efficiency goals. This variety reflects the diversity of organizations, motivations and sources of value. Table 11.1, overleaf, summarizes the main techniques used.

Perhaps the most common approach is empirical goal-setting where an organization is using information from their operations to set the overall programme objectives. This is the popular *business case* approach to resource efficiency programmes in which audits, often carried out by independent third parties, have established the scope of savings and the necessary investments. The advantage of this approach is that the degree of certainty for the resource efficiency programme is elevated, and so the *leap of faith* required approving the programme is diminished (see the item on page 185). In some jurisdictions, such as the Netherlands, authorities oblige some sectors to complete energy audits and require organizations to implement all projects which fall below a threshold payback, making this approach mandatory in this case. Other jurisdictions require audits, without mandating the implementation of the opportunities identified. Empirical objective-setting is a *bottom-up* process. A business case for resource efficiency will most likely be based on a list of opportunities that will deliver the desired improvement, and so the programme will be designed to implement those opportunities which meet the desired rate of return within the required time frame, and the goal will naturally flow from this.

In a business case approach, the internal goal is usually described in terms of net cost savings, not resource use. This is an indirect goal. In many organizations, this type of target can form a very motivational *headline* objective for a resource efficiency programme. For example, a programme "*to reduce operating costs by £1 million per annum through improved resource efficiency with a rate of return equal or superior to our standard investment criteria*" has a very clear and desirable outcome. Almost all resource efficiency targets based on business cases start with the scope of work being defined by the cost-benefit analysis and then, if desired, can be translated into a more direct goal – e.g. to reduce energy use by 20% per annum or to reduce emissions by 15%. It is also perfectly acceptable to articulate both a direct and indirect objective: "*we aim to reduce waste by 10% and so cut raw material costs by US$2.5 million*".

There is a common pitfall associated with an empirical, project list approach to resource efficiency goal-setting: that of properly quantifying the benefits

Method	Description	Advantages	Disadvantages
Empirical	A goal based on measurement, observation and data, not theory. Typically would involve some form of value discovery process or audit.	Basis in data gives greater certainty (search good argument see page 185). Clear justification for executive action through business case. Ability to focus investment in ascending order of return and to bundle projects into a basket which meets the hurdle rate.	Cost and time to establish opportunity. Two sources of error: tendency to bias towards technology fixes rather than behavioural change, leading to underestimation of potential; on the other hand failure to assess project overlaps can lead to overestimation of savings. Risk of "project-centric" approach to resource efficiency.
Comparative	A goal based on a benchmark from a comparable organization. An example would be the Solomon Energy Intensity Index for refineries, or the CIBSE or ASHRAE benchmark kWh/m^2 for different types of buildings.	Can provide a very simple and clear indication of opportunity, especially where a sector is relatively homogeneous e.g. refining, steel, milk. Helpful to understand comparative position where improved brand value is an objective of the programme.	Depends on the quality of the benchmark data and the degree of similarity of organizations. Often distrusted: *"we're different!"* Rarely applicable internationally. More often used to create a *hit list* of worst and best performers rather than as a goal per se (see page 386).
Theoretical	A goal derived from a theoretical understanding or model of a process, building or equipment. An example would be EPC certification of buildings whose recommendations are based on theoretical potential.	Can provide an upper boundary for savings potential for a process or building. Works better at the level of simple equipment – e.g. lighting or pumps, rather than in complex systems like a whole building.	Often relies on *expert advice* so may have less ownership at operating level. Equipment and people rarely operate at theoretical efficiency, so often overstates savings.
Qualitative	A non-numeric goal, for example, to achieve the *Energy Management ISO 50001 Standard* or *Carbon Trust Standard Accreditation.*	It is a simple goal, as it will be clear if the objective has been met. A good starting point if it is difficult to quantify or estimate benefits at the outset. ISO 50001 incorporates continual improvement.	Potentially too focused on process – obtaining the certification can get in the way of reducing resource use. Not particularly inspiring. Without quantifying benefits, the case for the programme may be poor.
Guesstimate	A goal based on a *"gut instinct"* of potential. Almost always set at 5% savings but the *"10% rule"* is probably a better *"random"* choice (except in energy-intense industries).	Any goal is better than no goal! As long as progress is monitored and the goal can be adjusted, it might mean that the programme can get started more rapidly.	Depends on who is setting the goal, but it will most likely have little basis in reality. Lack of ownership. Tendency to err on the side of safety and so understate potential and lead to reduced value.
Indirect	A goal based on an indirect measure influenced by resource efficiency. A common indirect goal is operating cost savings, but other goals could include measures of productivity such as overall equipment effectiveness (OEE) as used in Lean or even brand reputation metrics.	Can ensure that resource efficiency is integrated into wider business objectives such as cost reduction, increased profitability, market share, or within programmes like Lean. Where there is a strong link to value or other key business objectives, support for the programme is elevated.	If there are multiple means of achieving the goal, resource efficiency may be pushed to the back of the queue. Can lead to misaligned incentives e.g. reduced investment may help deliver lower operating cost in the short term but can lead to greater inefficiency in the long run.
Aspirational	A stretch goal designed to challenge the organization. For example Interface Carpets' *"Mission Zero"* goal for zero waste.	Can bring about re-examination of the core business model and hence may result in deeper reductions and greater value than would otherwise be possible. Forces fundamental change rather than superficial quick fixes.	Resistance may be even higher than normal so a very strong Mandate is needed to overcome objections. Lack of ownership is an issue. Unless the goal can be broken down into interim steps can lead to disillusion or difficulty tracking progress.
Sustainable or science-based	A goal based on an assessment of external sustainability needs. E.g. a goal to reduce emissions may be based on the UK government goal of 80% reduction by 2050; or based on perceived future availability of resources or natural services.	An approach that can be characterized as *"responsible"* and based on science. There is a rational basis to justify the goal. May be very appealing to stakeholders and customers as organization is meeting its obligations to society.	Sometimes difficult to quantify as sustainability often depends on several parties acting in concert – e.g. for decarbonization we need changes to supply side (utility companies) and within organizations (demand side). Investors may see this as an ideological *"green"* goal rather than one based on value.

11.1 (Left) **The advantages and disadvantages of common goal-setting methods** *Source: Niall Enright*

of multiple projects which influence the same resource streams. We shall see later (page 607) that if we make multiple savings in the same resource flow these savings percentages must be multiplied, not added together, a mistake that is still made far too often in business cases today. I would recommend that, in putting together an empirical business case for resource efficiency based on multiple projects, one should err on the side of overstating costs and understating benefits. An important part of getting the project assumption right is to get the people who are close to the resource use in each project to confirm that they are happy with the estimation.

The second approach that I would like to dwell on is that of sustainable objective-setting, which organizations are increasingly willing to contemplate. The objective of sustainable goal-setting is to align the organization's objectives with what science and economics are telling us are the reductions needed to bring our use of resources back into balance with nature.

Understanding what the sustainable goal (aka science-based goal) is for an organization, and tracking performance against this, is recommended even if this is not the objective that is eventually selected or publicized. This insight will enable the organization to assess its contribution to society and so better assess risks and opportunities in this area. Performance against sustainable goals will become ever more important as organizations face increasing environmental disclosure, in particular, carbon reporting. It is entirely possible that this disclosure could result in a backlash from customers, stakeholders and NGOs against those organizations that are considered *free riders*; in other words, who are not pulling their weight in the overall goal of decarbonization. Thus, sustainable goals weigh heavily on brand and reputation and so it is prudent to take these into consideration, even if internally at first.

The process of breaking down a sustainable goal into its parts, as described overleaf, makes the objective much more achievable. If we had said to the senior executives at the supermarket, *"you need to reduce emissions by 22% by 2027"*, they would most likely have responded that this was not possible. On the other hand, stating that the objective is to maintain absolute emissions steady in line with growth and achieve an additional 1.1% reduction of direct emissions each year sounds much more achievable. In fact, the same could be said for the power sector whose headline goal of a 46% reduction in 15 years becomes much more achievable if one reduces it to a compound goal each year – a 2.6% year per annum reduction. This annualized target is an example where compound growth works in our favour – so long as we take early action.

Even if it is never published, understanding your organization's **sustainable goal** is highly recommended.

It is important to note that, by definition, sustainable goals should always be expressed in absolute terms, not in relation to production or any other variable. Absolute goals are necessary because most natural systems we depend on have absolute thresholds beyond which we are no longer sustainable. Thus, an objective to reduce emissions by 80% per unit GDP by 2050 is not a sustainable goal because GDP is expected to increase three-fold over this period – largely cancelling a large part of the improvement.

Supermarket			Power Sector
Scope 1	Scope 1	Scope 2	Scope 1
Gas	HFC refriger-ants	Electricity demand	Fossil fuels for electricity
UK government 2012-2027 projections (net -22% emissions across all sectors)			
-11%[1]	-62%[1]	+18%[2]	-46%[1]
Annual equivalent (2012-2027)			
-0.8%	-6.3%	+1.1%	-5%
Supermarket sustainable goal 2027 (absolute annual improvement vs 2012 *including growth*)			
-0.8%	-6.3%	+1.1%	

11.2 **Emissions reductions are shared between the supermarket group and power generators**

While the UK, as a whole, has a target of a 22% overall improvement between 2012 and 2027, the sustainable goal for the supermarket is considerably lower. What appears at first to be a daunting target is actually much more manageable when we understand the relative contributions of the parties. This sustainable goal-setting methodology is based on identifying the relative contributions needed by different sector of the economy derived from national policy objectives or scientific data. Organizations that wish to adopt this approach are recommended to look at the new science-based targets methodologies, which are currently being developed (http://sciencebasedtargets.org/).

Sources: 1[194] 2[195] and Niall Enright

Real World: ***The sustainable (aka science-based) goal at a leading supermarket***

Several years ago I was consulting for a major UK supermarket group when the subject of what emissions reductions the supermarket should make over the next 15 years (for the period 2012 to 2027). This goal is not in the public domain so the organization will remain anonymous.

The question I was asked is: "*Should we use the UK emissions trajectories to achieve 80% reduction as our targets?*" The answer is "*Yes, at a high level, and no, not in practice.*" To explain why this is, we need to be aware that responsibilities for emissions are shared. The Greenhouse Gas Protocol (GHG Protocol)[613] is the nearest thing we have to a global standard for reporting organizational emissions. This protocol defines emissions as falling into one of three types. Scope 1 emissions are direct emissions from sources that are owned and controlled by the company such as vehicles, fuel use, refrigerant leakages, etc. Indirect emissions from purchased electricity, heat and steam are Scope 2, and finally Scope 3 indirect emissions are from all other sources, such as employee travel or arising from the use of the goods that the organization produces.

For the supermarket, its major impacts are Scope 1, its direct emissions mainly from gas for heating and baking, as well as from refrigerant leakage, and Scope 2, from the electricity it purchased. Scope 3 was not material. Furthermore, a chunk of the emissions reductions due to the Scope 2 electricity usage was expected to be as a result of separate actions taken by the power generators, not by anything the supermarket did.

Fortunately, the UK government has published data on the expected emissions by sector which will ensure the overall 2050 goal is met.[191, p191] The data shows the business sector – which includes our supermarket - achieving a reduction in direct emissions of CO_2 from 79.9 $MtCO_2e$ in 2012 to 71.2 $MtCO_2e$ in 2027 – i.e. a decrease of a little under 11% (equivalent to 0.8% a year compounded). For the supermarket, this is their sustainable target for gas. The similar data for indirect emissions from refrigerants is shown in Table 5.2.5 in the same document.

For the electricity demand, we also need to know what estimates the UK government has made in respect of demand growth in the supermarket's sector. In this case, we can turn to another table of data which projects demand for electricity growing by 18% over the relevant period. This equates to an annual growth rate of 1.1%. Over the same timescale, the energy supply sector would decrease its emissions from 185 $MtCO_2e$ to 100 $MtCO_2e$, a decline of 46% (4% per year compounded). This substantial decrease at the supply side explains why the supermarket's own demand-side target did not need to be set at the 22% overall figure for emissions reductions 2012 to 2017.

Thus, the supermarket's sustainable emissions goals for 2012-2027 are: for each year:

- A 0.8% per annum decrease in absolute gas usage;
- A 6.3% per annum reduction in absolute refrigerants emissions;
- No more than a 1.1% per annum increase in absolute electricity demand.

Taking into account anticipated growth of 2% per annum, the above figures will change to -2.8%, -8.3% and -0.9% respectively, but the emphasis on absolute reductions ensures this is included. I am glad to see that this approach has been taken up more widely - search for science-based targets.

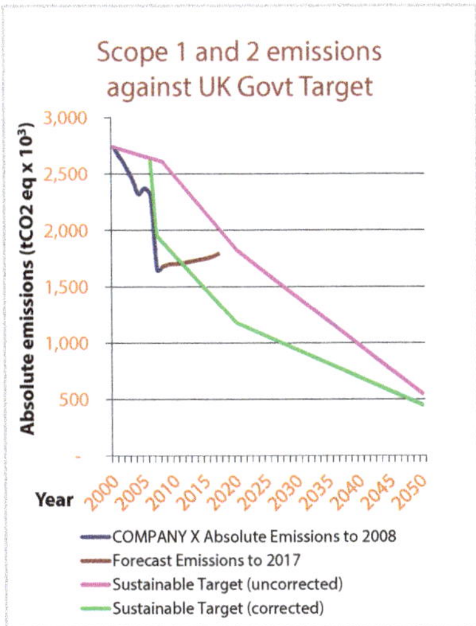

11.3 **The sustainable goal for this global corporation (pink line) needed to be adjusted when they outsourced their transportation (green line). The blue line indicated progress to date, while the brown line was the forecast to 2017, which shows the corrected goal being missed.** For an explanation of Scope 1 and Scope 2 emissions, see the box opposite. *Source: Niall Enright, based on a real organization's data*

The chart left, shows an example of a real organization that has been tracking its absolute emissions in comparison with UK government goals which are shown by the pink line. The organization details have been anonymised as this data is not in the public domain; however, it provides a helpful illustration of the challenges that can occur with sustainable goals. The dark blue and brown lines represented the company's actual and forecast emissions, which appears to show that despite an anticipated increase in emissions the company was going to remain below its initial sustainable goal in pink. However, the large drop in actual consumption in 2007 was due to the outsourcing of the organization's distribution to a third party, rather than a reduction in emissions per se, which meant that the emissions were now Scope 3 and thus no longer reported. In these circumstances, the only way a like-for-like comparison can be made is by correcting the sustainable goal to take into account the change in the treatment of these emissions, as shown in the green line in the chart. Looking at the corrected goal, the company's emissions, far from being comfortably below its sustainable goal, were now forecast to exceed the goal quite soon.

Thus, as well as being expressed in absolute terms, a sustainable goal needs to be adjusted whenever there is a significant change in the boundary of the organization. Failure to recalibrate the goal against these changes can lead to a false sense of security as over time the original goal becomes less and less relevant to the current organization. Absolute goals are discussed in greater depth in the next section.

One type of goal which is fairly infrequently used is a theoretical goal. In my experience, these goals are usually helpful only in terms of setting an upper limit to the savings potential. For example, the Gibbs Energy for a chemical reaction helps us to understand the absolute minimum energy requirement of that process. From this, we can calculate the absolute minimum amount of energy needed to make steel from iron ore, say, which is 6.7GJ/tonne or to make aluminium from ore, 29.5 GJ/tonne. The best technology available for each process today uses just over double this amount of energy, so we can conclude that efficiency gains over 50% for these technologies are not available.[22, p102] In practice, theoretical goals are much better applied to specific items of equipment or processes, where they can provide insights that help identify specific opportunities, rather than at an organization-wide level where complexity gets in the way of creating an effective model. There are some analytical techniques which are helpful, such as Pinch Technology or Water Pinch, which can optimize flows of a resource (heat or water in this case) across a complex series of processes, but these are much more relevant in target-setting rather than goal-setting.

So, we can see that there are many different approaches to setting goals. Selecting the appropriate one for an organization requires great care and an understanding of the pros and cons of each method. A goal can lock an organization into behaviours for many years, and it is important to reflect on how this goal will work in the future, not just at the outset of a programme.

11.3 Absolute vs relative goals

There are some very real pitfalls to beware of in both absolute and relative goals. In particular, specific ratios or intensities should be used with great care.

The notion of absolute goals is well illustrated by considering how organizations report emissions. When it comes to climate change, there is only one measure that matters, the total quantity of greenhouse gases (GHGs) in the atmosphere. That is why policymakers tend to favour absolute measures of emissions. The UK Climate Change Act goals set out a legal obligation to reduce absolute emissions of *"carbon units"* by 80% by 2050 compared to 1990 levels. Given progress to date, this equates to around 3% per annum reduction every year for the next 40 years.

However, despite their environmental credentials, it does not take long to spot some glaring weaknesses with applying absolute emissions goals to organizations or individuals. The first issue is that of economic efficiency – some organizations can achieve emissions reductions much more cheaply than others, and so setting a uniform 3% per annum objective across the economy, or within an organization, would not be the most efficient way to achieve a given emissions reduction. Another flaw with absolute goals is the issue of boundaries. Whereas a national frontier does not change, organizational boundaries are rarely static; acquisitions, disposals or outsourcing can give the appearance of good/poor performance in absolute terms without any indication of true underlying improvement, unless there is a readjustment of baseline data to account for the changes. To compound the boundary issue, there are aspects of equity ownership or control which make absolute emissions reporting at an organizational level even more complex.

These problems with absolute emissions reporting explain why emissions intensity, in which emissions are divided by some form of activity measure – such as service delivered, floor area, production, or turnover – has gained widespread acceptance. Broadly called specific ratios, metrics based on intensity are all around us, from the gCO_2 per kilometre we see quoted for motor vehicles, through to gCO_2 per pack used in product labelling, or the tCO_2/M£ turnover in the UK's CRC reporting, and many companies use these ratios to describe their emissions goals. The widespread use of these intensity ratios reflects their simplicity and the ease with which they can be understood by the public.

Intensity goals, *i.e. resource use per unit of activity, are widely used but have a* **critical weakness.**

An underlying attractiveness of specific ratios is that they permit comparisons to be made. Knowing the g/km for a single vehicle is of relatively little value, but knowing it for a range of vehicles can inform choice by the consumer, help manufacturers establish their relative performance compared to others and permit policymakers to set performance goals for different sectors of the economy. These types of specific ratios often rely on a defined scenario to set the benchmark – for example, the UK vehicle emissions are quoted for the *"typical journey cycle"*. More complex industry benchmarks of energy and emissions, such as the Solomon Energy Intensity Index in the refining sector have hugely complex scenarios to enable comparisons to be made.

Specific ratios can also be used for internal benchmarking, typically tracking changes in product intensity over a number of years, and many organizations quote these goals rather than absolute goals as they feel it will overcome the negative effect growth would have on absolute emissions. However, there is a huge hidden risk in placing reliance on specific ratios as a measure of emissions performance: the baseload effect. Many organizations are oblivious to this effect, which is described in more detail in the box below.

Because most emissions data are a product of an energy conversion factor, there is an additional challenge in their interpretation, given the regional and temporal variations in these factors. If a product is manufactured in a country with a low electricity conversion factor, such as France which has a

Real World: ***Marketing trucks in Canada***

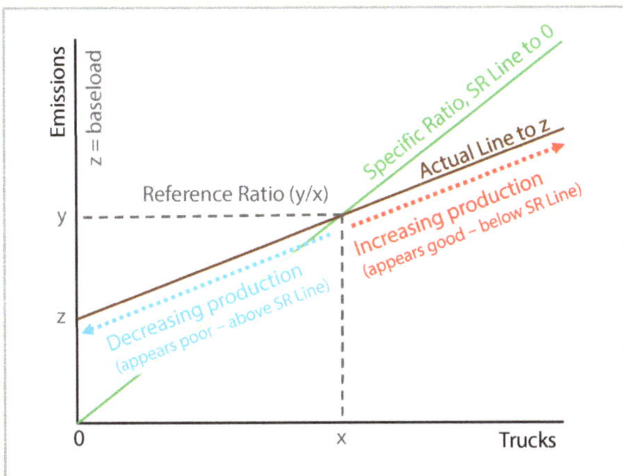

11.4 **Specific ratios have a major weakness: they ignore the baseload effect** *Source: Niall Enright, available in companion file pack*

Almost a decade ago I met a senior executive for a major Canadian automotive manufacturer who proudly boasted that over the previous five years they had continually reduced the energy intensity, hence emissions, of the manufacture of their trucks, even as volumes increased. In fact, these great emissions figures formed part of the marketing of the company concerned.

I sketched out the diagram shown left on a napkin to explain to that executive why I felt that publishing their performance in the form of a specific ratio could entail considerable risk.

What the truck manufactured did to establish their intensity was to take the total emissions in a particular year and divide them by the production that year, as shown in the figure above, y tonnes for x trucks. However, the specific ratio suggests that there is a linear relationship back to zero as shown by the green line – in other words, if there is no production then there will be no emissions.

The reality is quite different, as can be seen by the brown *"Actual"* line. Car factories in Canada, like everywhere, use a lot of energy, and so produce emissions, completely independently of production. Examples are the lighting and heating of the assembly halls, offices, canteen and so forth. In fact, it is not unheard of for the fixed load in an automotive plant to approach 50% of the overall demand. Taking the point of our initial ratio y/x back to the fixed baseload, z, gives us the actual line. So what does that mean for our Canadian example? Well, as production rose, the fixed baseload was divided over more vehicles, and so the intensity improved irrespective of the underlying performance of the plant. In effect, they tracked along the actual line following the red arrow to the right of the reference ratio, which is lower than the specific ratio. The risk lay in the opposite effect – if the production fell below that of the reference year, then the plant would track along the actual line the direction of the blue arrow, above the specific ratio line, as the baseload is divided by an ever-decreasing volume of cars, and performance would appear bad. Unsurprisingly, the improvement quoted by the executive was quickly dropped in a subsequent downturn.

Real World: *A real reduction?*

A particularly dramatic example of how specific ratios can confuse can be seen in the published environmental performance of a major UK water plc.

In 2006, it reported a specific ratio reduction of over 18% of their emissions per megalitre, compared to 1990, which sounds impressive. Well, we already know from the previous example that when the volume of production increases the baseload effect tends to improve the intensity, so some of the improvement could have been as a result of the baseload effect, and in fact, water production did increase over this period.

However, there was another, bigger effect at work here: energy conversion factors. In 1990, the conversion from a kWh of electricity to a kg of CO_2 using UK grid-average factors, was 0.77 kg CO_2/kWh, while it was 0.52 kg CO_2/kWh in 2006/07 - a reduction of 32% in the period.

Thus, rather than falling, the water company's energy use per unit must have been rising, as otherwise, everything else being equal, the improvement should have been 32% (electricity is the major source of emissions)!

The fact that the specific ratio of the company had improved is technically correct; however, the *implication* that this reflects underlying performance improvement, rather than the decarbonization of the electricity supply in the UK, is not.

In fact, the water company was a free rider on the efforts of utility companies – something that I suspect the senior executives were unaware of, but which could have potentially negative impacts on perception if it came to light.

lot of nuclear power in the generation mix, it will have a lower emission per unit. This does not mean to say that French factories are more efficient than those in higher conversion factor countries. However, from an environmental perspective, the comparison is perfectly valid, and a French product is better in emissions terms than, say, the same product manufactured in the UK. Another issue with conversion factors is that they tend to change over time, which again can mask true performance as the real world example to the left describes.

If the use of intensity goals is problematic at a company level, it becomes doubly so as a means to measure operational performance, where different equipment and processes are driven by different variables: not just production volume, but weather, etc. To overcome this baseload effect, the most common operational modelling technique employs regression analysis to derive a formula in the form $y=mx+c$ where y is the expected emission, m is the emissions intensity per unit activity (production, or weather etc.) and c is the baseload. The effectiveness of this form of modelling has been proven in thousands of M&T programmes where this technique is employed to set targets to drive operational improvement. M&T targets differ from intensity goals in that they are rarely used for benchmarking, e.g. comparing different buildings, but to provide a comparison with the historical performance, against which corrective action can be taken.

M&T targets tend to work best at a very low level, such as individual plant or equipment, and so a single facility can have many different targets using a variety of different variables, which can be complex and difficult for outsiders to understand. Luckily, the performance can be aggregated up from these individual models using techniques such as CUSUM, which are discussed later, and which make M&T appropriate to track and report organizational-level performance.

If we examine the three main interlocking schemes to manage UK emissions we can see how hybrid approaches involving both absolute and intensity metrics are emerging in practice. As long as they get their absolute *"caps"*, policymakers are willing to incorporate more flexible intensity ratios in the *"trade"* element of many emissions reduction schemes. In the EUT-ETS Phase 3, emissions permits were allocated across 50+ product categories using the average emissions intensity of the best 10% of producers. If you are in the top 10%, then you will receive all or most of the emissions allocations you need – whereas if you are a poor performer according to the benchmark you will have to buy more allocations in the market, or undertake your own emissions abatement activities to compensate. The total emissions allowances remain capped – but their distribution is based on specific ratios. In the Climate Change Levy, and associated Climate Change Agreements, most sites have opted for an intensity-like, product-mix algorithm to relate their target emissions reductions to the production output(s) of their facilities. Even in the Carbon Reduction Commitment League Table (now discontinued), there was some concession towards growth, in that 25% of the score was based on the emissions intensity change intensity per £ turnover of the participant.

11.5 **A comparison of different types of metrics used in goal-setting**
Source: Niall Enright

Type of Goal	Example	Typical Application	Strength	Weakness
Absolute	tCO_2 or MtC	National or corporate reporting	Most relevant environmental assessment of impact. Basis for sustainable goals	Difficult to apply where boundary changes are frequent or boundary is unclear
Intensity or specific ratio (SR) (Sometimes called a specific energy ratio or specific emissions ratio (SER))	g/km for cars	Most often applied to product benchmarking. Also used in corporate emission goals	Easy to understand and can be applied across a product category	Comparison often requires a reference scenario. No account of underlying factors driving usage or of baseload effects
Operational (regression)	A formula in the form $y=mx+c$ or a more complex product-mix algorithm	Applied within facilities or operations at a low level to track performance	Most accurate measure of performance. Usually focused on the underlying resource so no conversion-factor issues. CUSUM can aggregate performance at higher levels	Can be complex, as multiple variables influence different resources. More applicable to historic benchmarking than comparative benchmarking across facilities or products

Mandate

In fact, the fundamental global vision for equitable emissions allocation combines an absolute goal, contraction (of total emissions) with an intensity goal, convergence (on per capita emissions).

Absolute, intensity and operational metrics all have their pros and cons when it comes to managing resources, and it is not unusual to see these used together at both a national and corporate level. Where the resource being managed is emissions, there are particular complications associated with the use of conversion factors to calculate emissions from different energy sources.

Setting **absolute goals** has the highest environmental integrity.

The form of reporting with the highest environmental integrity must remain absolute reporting. But, like a balance sheet in a company, this form of reporting rarely provides enough insight to inform day-to-day performance management, and for that, we need the *"management accounts"* of resource use – intensity or operational metrics, which provide many advantages as long as we recognize and avoid potential pitfalls. The use of regression to establish operating level performance metrics is covered later (see page 460).

11.4 Rate vs quantity targets

It is important to understand the difference between a target based on a rate of resource use (or emission or waste) and a quantity of resource use. Quantity-based, or cumulative, targets almost always require a larger improvement than rate-based targets.

In the previous pages, we have seen that absolute goals, i.e. targets expressed in the quantity of resource use, have the highest credibility. Here, we will examine aspects of absolute goals that can cause confusion.

Most organizations that set absolute reduction goals for resources do so in the form of a target rate of resource use or pollution. A rate-based goal would be, for example, "*to reduce absolute annual energy consumption by 20% over five years*". However, there is a problem with this type of goal. For example, we could keep our energy consumption at the present level for the first four years and only make an improvement in year 5. If we did this, then we would only achieve a reduction in our energy use over the period of 4% (a 20% improvement spread over five years is a decrease of just 4% a year). In fact, energy use could rise in the early years and the target still be met, despite the overall increase in energy use over the period. I refer to this as trajectory uncertainty.

There are a couple of ways of overcoming this uncertainty:

- by setting intermediate or milestone targets which make the trajectory clear;
- by expressing the goal in terms of a cumulative total or absolute quantity of resource use over the target period.

The vast majority of climate change goals established at a policy level are cumulative or quantity-based goals. They are designed to ensure that the level of greenhouse gases in the atmosphere does not exceed given concentrations. Supporting these goals, policymakers put in place market mechanisms such as carbon-trading in the EU Emissions Trading Scheme, which is based on fixed and declining quantities of allowances being available each year.

*A cumulative goal averaged over time translates into a **greater** rate goal at the end of the period as improvements usually ramp up.*

Most private sector organizations set themselves rate-based targets as this aligns with most performance goals (e.g. businesses set an annual profit target not a cumulative profit goal). However, as the example opposite shows, it is not simply a question of taking the *headline* percentage of a cumulative goal as the basis for a rate-based target. This is because, assuming that improvements ramp up, the improvements over the final years need to be higher than the headline percentage to make up for below-average improvements in the early years. However, a cumulative goal has the advantage that *overachievement* in early years contributes to the overall target - avoiding postponement of actions because the current year's target has been met.

Real World: *Two very different 20% improvement goals*

The easiest way to explore the difference between a rate goal and a cumulative goal is to look at some numbers. The table below illustrates a company which wishes to reduce its current emissions of 100 tonnes of CO_2e a year. In the first case, it sets itself a target to achieve a 20% reduction in the *rate* of emissions over eight years. In the second case, it sets itself a target of a 20% reduction in total, *cumulative* emissions over the same period.

In column 3 we can see the rate target in orange. Here we have assumed that the reductions will be continuous until the company reaches the emissions goal of 80 units a year. This requires an annual reduction of 2.8% in each of the 8 years. To the right we can see, in green, the cumulative target of 640 units of emission (20% less than the business as usual emissions of 800 units). To achieve this target will require a constant 5.0% reduction in emissions every year, and the rate of emissions in year 8 is 66 tonnes of CO_2 a year compared to the 80 tonnes for the rate target, a 34% rate improvement. Clearly, these 20% targets are not the same!

	Business As Usual (BAU)	**Rate Target (20% *rate* improvement)**			**Cumulative Target (20% *cumulative* improvement)**		
	For this example assume 100 tonnes CO_2e a year in baseline Year 1.	Start rate (emissions/year):		100	BAU (Cum. emissions):		800
		End target (emissions/year):		80	Target (Cum. emissions):		640
		Annual reduction (%):		-2.75%	Annual reduction (%):		-4.98%
		Reduction over period:		11.62%	Reduction over period:		20.00%
Year	Cumulative	Addition	Change	Cum. Total	Addition	Change	Cum. Total
1	100	97.25	2.8%	97	95.02	5.0%	95
2	200	94.57	2.8%	192	90.28	5.0%	185
3	300	91.97	2.8%	284	85.78	5.0%	271
4	400	89.44	2.8%	373	81.51	5.0%	353
5	500	86.98	2.8%	460	77.45	5.0%	430
6	600	84.59	2.8%	545	73.59	5.0%	504
7	700	82.26	2.8%	627	69.92	5.0%	574
8	800	80.00	2.8%	707	66.44	5.0%	640

11.6 **Two 20% improvement targets, one stated on the basis of the final rate of resource use and the other on the cumulative quantity of resource use**
In the latter case, the improvement needed is almost twice the former case.
Source: Niall Enright, example spreadsheet model available in the companion file pack

These differences do matter in the real world. For example, the EU has a 20% improvement target, 2013-2020 under the Kyoto Protocol, as well as a separate target set for the 27 EU Members of a 20% improvement by 2020. The former target is for average annual emissions throughout the period, i.e. a cumulative target, while the latter is for the rate of emissions to be reached in 2020.[192] In practice, the reductions achieved since 1990 (the base year for both commitments), as well as differences in scope and base year measurements for each or the goals, means that both these objectives are broadly compatible. Nevertheless, it is critical that organizations, which usually wish to use simpler rate-based targets, should be aware that international goals, such as those set through the UNFCC negotiation process, are usually cumulative and thus matching these sustainable targets may mean a higher eventual rate reduction than the simple headlines suggest.

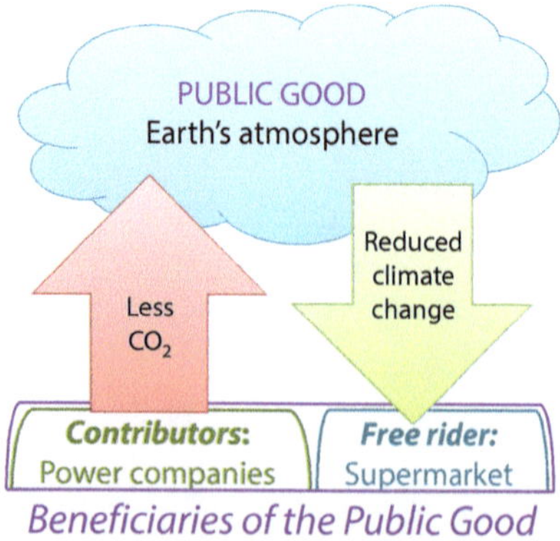

11.7 **In economics, *"free riders"* are organizations or individuals who benefit from a public good without having contributed to it.** This simplistic example shows a supermarket group benefiting from the efforts of the power companies. If the supermarket were to incorporate the efforts of the power companies into its own resource efficiency goals, then not only would this be misleading, but the supermarket would have little control over meeting its objectives. *Source: Niall Enright. Image available in the companion file pack*

11.5 Additionality and free riders

Energy and resource efficiency can involve several parties working together - claiming the efforts of others can lead to criticism.

A key sustainability idea, widely applied to carbon markets, is additionality. It centres on whether a particular intervention that an organization makes to improve sustainability delivers a change that would *not otherwise have occurred.*

While it is clear from the previous section that the measure of resource efficiency that has greatest environmental integrity is the measure of the absolute resource use in relation to the sustainable capacity of the planet, it does not necessarily follow that all improvements an organization makes in absolute resource use can be recognized towards the organization's *own* efficiency goal.

The reason for this lies in whether the improvement has occurred because of actions attributable to the organization or whether this improvement would have taken place regardless. For example, we have seen earlier (page 366) that our supermarket group will reduce its absolute emissions by 46% if the UK power generation sector achieves its goal of decarbonization by 2027. Thus, if the supermarket were to set a goal of a 40% reduction in emissions over the same period, this would be misleading, as all of the improvement would have come from the efforts of others and would have happened regardless of any action taken by the supermarket. In this case, additionality cannot be demonstrated and the supermarket would be a free rider because it is not contributing to the improvement.

It does not necessarily follow that all improvements an organization makes in resource use can be recognized towards the organization's own efficiency goal.

The term *free rider* has a wider meaning than just claiming the efforts of others as your own. In economics, there are resources known as public goods, like the atmosphere's capacity to absorb CO_2 or the ocean's ability to provide fish. The problem that arises is that some activities, such as burning fossil fuels, or industrial-scale sea-bottom trawling, damage those public goods, but the organizations responsible do not pay the cost of restoring those goods to their original state. In these cases, those organizations can also be said to be free riders on the wider society which bears the cost to restore these public goods.

An area where organizations often inadvertently set themselves resource efficiency goals which are not truly additional is when they incorporate the purchase of green electricity - i.e. energy from renewable and low-carbon sources such as wind, hydro or solar - into their emissions goals. This is because the purchase price paid for the green energy often does not cover the cost or provide the incentive to generate the green energy in the first place, which is funded from the public purse or through the bills of all electricity users.

The lesson here is that emissions performance claims are especially problematic, as organizations are open to criticisms if these objectives are achieved through the efforts of others and additionality cannot be proven. The PAS 2050:2011 specification for the assessment of the life cycle greenhouse gas emissions of goods and services[95] can provide very useful guidelines on how to deal with issues like renewable energy (allowed only if you can prove you used the energy and no one else also claims an emissions factor reduction from the same generation) or emissions offsets such as tree-planting (not allowed). ⇒ page 378.

Real World: ***Additionality and renewable electricity***

Many organizations buy *"green electricity"* and report lower emissions as a result. These claims have been legitimized by a major revision to the Greenhouse Gas Protocol[670] which introduced the concept of the market method for accounting for emissions, which allows a supplier-specific emissions factor to be used to convert electricity to emissions (in the case of *green* or renewable power, the factor is usually zero). This use of supply-specific emissions factors contrasts with the alternative location method, which stipulates national or regional average grid factors. This rule change has a huge - in my mind, negative - effect on emissions reporting, given that the GHG Protocol is the global standard.

There are many problems with this treatment of *green electricity*. First of all is the question of whether the claim of zero emissions is justifiable. The principle of additionality states you can only claim an improvement if it can be demonstrated that it is as a direct result of your actions. In the UK, the vast majority of organizations claiming reduced emissions as a result of the purchase of *green electricity* are, in fact, *free riders*. This arises because the extra costs associated with generating renewable electricity compared to that from conventional sources are mostly met by a market mechanism called the Renewables Obligation Certificates (ROC), which is worth some £42 per MWh (depending on the renewable technology). This payment compares to the average cost of the electricity of around £30 MWh, so ROCs are an essential subsidy that the renewable power producer receives, *without which the electricity would not have been produced*. The cost of these ROC subsidies is, in turn, incorporated into the overall cost of all electricity supplies and paid for by *all* electricity users. Although the *green electricity* may pass a *"chain of custody"* test to confirm it is from a renewable source, via Guarantees of Origin (GOs) in the UK or Renewable Electricity Certificates (RECs) in the US and Canada, *green electricity* fails the additionality test because the small premium the buyer has paid is usually nowhere enough to fund equivalent additional renewable generation.

A second problem is that of double-counting, as the UK average grid factor for all electricity declines in line with the increasing levels of renewable generation. Other electricity consumers will be reporting improvement in their emissions using the location method, which is based on the *same green electricity* that the buyers have claimed for themselves, using the market method. This overlap makes the carbon neutrality claims incompatible with PAS 2050 (see 7.9.4.1 of the Standard[95]).

My third, and biggest, concern with *green electricity* and the market method of reporting is that these undermine the case for investment in *reducing energy use*. By offering organizations a relatively simple, painless and lower-cost route to improvement, *green electricity* helps them avoid making the substantial reductions in energy use that we need them to achieve.

For example, a UK corporation states that it has achieved, in 2016, over 80% emissions reductions since 1997 by buying 100% green electricity and reporting this under the market method. However, using the location method, the emissions reduction over this time is a less flattering 19%, much of which, arguably, could be credited to an overall decarbonization of supplies rather than the organization's own efforts. The problem for this business is that, if the public comes to perceive these claims of improvement as *"greenwash"*, this could have a big negative reputational impact. Those consumers who are increasingly willing to make personal sacrifices to address climate change, or have been affected by impacts such as flooding, will not be favourably disposed to organizations that appear to be *gaming* the system, regardless of whether *"the rules"* permit this.

For these various reasons, my advice is: avoid *green electricity* emissions goals or claims, unless you can genuinely show you have paid fully for the generation of that electricity (in which case *well done!*). In fact, if your organization feels strongly about us all working together to reduce emissions, you could go further, and explicitly state that you *"do not claim reduced emissions from green electricity which has been subsided by others"* in your reporting, and so help educate people about these issues.

The biggest challenge for the net positive approach is that **many forms of natural capital cannot be substituted by man-made capital.**

Exploration: *Net Positive*

A recent development in the world of corporate social responsibility is the notion of "net positive".

The concept is simple. An organization can be said to be net positive when it has quantified all its material social, economic and environmental impacts and can demonstrate an overall positive contribution to society and the environment.

Some leading companies such as ATT, AMD, CapGemini, Dell, The Crown Estate, Dow, IKEA, Kimberley Clark, Kingfisher and other have joined to support the Net Positive Project.

This project team, together with the sustainability advocates, Forum for the Future, The Climate Group and WWF, established the following 12 principles that organizations need to follow to be able to claim a net positive impact.[292]

1. *The organization aims to make a positive impact in its key material areas.*
2. *The positive impact is clearly demonstrable, if not measurable.*
3. *As well as aiming to have a positive impact in its key material areas, the organization also shows best practice in corporate responsibility and sustainability across the spectrum of social, environmental and economic impact areas, in line with globally accepted standards.*
4. *The organization invests in innovation in products and services, enters new markets, works across the value chain, and in some cases, challenges the very business model it relies on.*
5. *A net positive impact often requires a big shift in approach and outcomes, and cannot be achieved by business as usual.*
6. *Reporting on progress is transparent, consistent, authentic and independently verified where possible. Boundaries and scope are clearly defined and take account of both positive and negative impacts. Any trade-offs are explained.*
7. *Net positive is delivered in a robust way and no aspect of a net positive approach compensates for unacceptable or irreplaceable natural losses, or ill treatment of individuals and communities.*
8. *Organizations enter into wider partnerships and networks to create bigger positive impacts.*
9. *Every opportunity is used to deliver positive impacts across value chains, sectors, systems, and throughput to the natural world and society.*
10. *Organizations publicly engage in influencing policy for positive change.*
11. *Where key material areas are ecological, robust environmentally restorative and socially inclusive methods are applied.*
12. *An inclusive approach is adopted at every opportunity, ensuring affected communities are involved in the process of creating positive social and/or environmental impacts.*

These are fascinating principles. They speak of a much more profound change in approach than the simple emissions reduction goals that have been described in this chapter so far. The net positive collaborators admit that the project is still in its infancy, but they are signalling that progressively tougher goals for improvement in single areas, such as GHG emissions, may not be the best way to drive real transformation in organizations.

In practical terms, some big challenges need to be overcome before organizations should consider a net positive approach to goal-setting.

1. There are several net positive *"brands"* and organizations out there, which can lead to confusion.
2. There is no commonly accepted methodology for measuring net positivity. While standards like the Greenhouse Gas Protocol are well established for some impacts, how do you equate a negative environmental impact on water, say, with a good social gain on jobs? While some emerging methods, such as the Natural Capital Protocol,[544] measure environmental impacts and assets in economic and social terms, these are still rudimentary.

   Also, many forms of natural capital simply cannot be substituted by man-made capital (see page 42). Irremediable harm, by definition, cannot be cancelled out or offset by positive action elsewhere.

   Where the concept does have some integrity is where *equivalences* exist. Thus project A which emits x tCO_2, can be offset by project B, which decreases methane emissions by the equivalent of 2x tCO_{2e}, leading to a net positive outcome. These impacts can be equated because greenhouse gases are distributed uniformly in the atmosphere and because the equivalences between them are (broadly) understood. Here, the locations of the impacts are not particularly important - the effects are said to be homogeneous. On the other hand, a project which conserves 2 x units of fresh water in a water-rich environment like Scandinavia cannot in any way be said to compensate for a project with reduces fresh water by x in a water-scarce environment like the Sahel in Africa. Here, clearly, location is highly relevant and the impacts are said to be heterogeneous. The case study on page 602 touches on the notion of homogeneous and heterogeneous impacts incorporated into project assessment using financial proxies for impacts.
3. The issue of additionality and double-counting is problematic. Organizations like BT and Dell are aiming for multiples of their emissions (three times and 10 times, respectively) to be reduced by their customers through the use of their products and services. However, proving the savings is problematic. Take, for example, BT's claim that its provision of broadband leads to dematerialization and reduced consumer travel:

   "Broadband enables a range of technologies that remove or replace the use of carbon-intensive products or processes (dematerialization). We measure the effect of dematerialization and reduced consumer travel enabled by online news, music streaming, online banking, online retail, online education, digital photos and email."[101]

   However, the banks, music streaming companies, email providers, etc., etc. could equally claim that it is *their* services which are resulting in the changed behaviours which lead to the reduced emissions. Indeed, broadband provision in rural areas of the UK is significantly funded by government.
4. Implied in the notion of net positive is that the good outweighs the bad. However, there are some impacts on natural capital or society which cannot be mitigated by a positive action elsewhere. In the examples above, BT's or Dell's CO_2 emissions will still contribute to global harm, even though their customers are reducing their own emissions by using the company's products.

   The fact is that the organization *and* the customers may both need to make a drastic reduction if we are to avert disaster. While the organization should be applauded for being *"part of the solution"* that should not absolve them from actions on their own emissions.

   To be fair to the Net Positive Project, they recognize this problem and are very clear that *"covering unacceptable negative impacts with positive impacts"* is a form of greenwash which is incompatible with good reporting.

The net positive concept is worth monitoring closely. If the limitations can be addressed and a methodology standardized, it has the potential to provide a compelling and useful framework to help organizations take into account the wider impact of their actions. What this approach must not do, however, is diminish the responsibility for action by organizations, for example, by enabling them to outsource their obligations. To do so would forego the value available and reduce the overall improvements needed by society.

11.6 Leading and lagging indicators

Some goals are backward-looking while others signal future outcomes. A combination of both of these types of goals can help anticipate the future progress of the programme.

Without leading indicators at the start of the programme, those responsible would be flying completely ***blind.***

Financial services providers often use the expression "*past performance is not an indicator of future results*". The same warning could be applied to many published energy and resource efficiency goals.

The fact is that the goals we have described here are all about the outcomes achieved, not the underlying changes in the organization that drive the improvement. These goals are reported in terms of results: cost savings, emissions reductions or waste elimination. These objectives are admirable - they speak of the real, tangible change that an organization desires to achieve.

There is, however, a problem with these goals. They are backward-looking. They track what the organization has done in the most recent reporting period. They do not tell us what is likely to be achieved in the next reporting period. These types of measures are called lagging indicators because they reflect the past, they *lag* behind the underlying changes that are driving the improvement.

A leading indicator, on the other hand, tells us how our resource efficiency may change in the future. There are numerous leading indicators in energy and resource efficiency, which will help predict future results, such as:

- The number of improvement projects in the pipeline or audits completed;
- Hours of staff training on efficiency;
- Current and future budget made available for efficiency investments;
- The number of ideas generated by employee suggestion scheme;
- Staff employed on efficiency measures;
- Percentage of total utilities metered and reported;
- The number of sites certified to a standard like ISO 50001;
- The ratio of predictive to reactive maintenance spend;
- The reinvestment ratio (proportion of savings reinvested vs taken to profit);
- The value of future capital procurement, using a whole life costing approach.

Each of these measures is predictive of the following period's programme. If these numbers are declining then one would expect the programme outcomes to worsen; if they are increasing, then one would expect the results to get better.

That is not to say that we should set our organization goals based on leading indicators - there are often much too detailed and nuanced for general consumption. Rarely does any single leading indicator on its own predict the outcome.

Whenever I have led a large resource efficiency programme, I have always wanted to get a good range of leading indicators established so that I may assess early on whether the programme is on track. Typically these would touch on several themes: Discovery: e.g. How many sq ft of the property portfolio has been audited? What proportion of the energy/water/waste is measured and reported daily/weekly/monthly? How many opportunities are there in the Opportunities Database?; Engagement: How many senior managers have been briefed on the programme?, What incentives are in place for staff?, How strong is the leadership commitment? Investment: What funds are committed to the programme? How many staff are allocated?

The DICE methodology, described on page 209, is an example of a formalized assessment of four key leading indicators: the frequency of programme reviews, the effectiveness of the team, the level of resistance to change and the incremental effort needed to achieve the desired outcome. It serves as a remarkably good measure of likely programme success or failure. Leading indicators may not form the goal itself, but they are the best tool for establishing if the goal will be achieved or not.

This brings us to a key point about leading indicators, which is that they are particularly important at the outset of a programme. This is the period when the programme has been launched and before any progress towards the overall goals has been measured. Without leading indicators at the start of the programme, those responsible for it would be flying completely blind.

Thinking about leading indicators forces us to think about what changes are needed to achieve our desired goal, and in particular, how these changes can be measured. The question of which leading indicators to use can often be put to the programme's steering group to get them to think about the underlying *levers of change* at their disposal.

Leading indicators may not form the goal itself, but they are the best tool for establishing if the goal will be achieved or not.

While I would not usually advocate using leading indicators as the basis for the resource efficiency goals of an organization, I do often recommend that organizations consider putting this "*how*" into their overall goal: "*we will achieve x improvement (lagging indicator) by investing an additional US$y amount into our manufacturing operations (leading indicator)*". The benefit of doing this is that the required resource or activity that will bring about the change is clearly set out for all to see. In other words, by incorporating a leading indicator in our overall goal we may arrive at a stronger mandate for the programme, which not only sets out where we want to get to, but also how we intend to get there.

11.7 Reviewing the goals

When goal-setting, the folks responsible for delivering the change should have the opportunity to confirm that it is achievable. This process must be genuine; in other words, the input of the participants must be able to refine the desired outcomes, activities or resources.

We have seen in Section 11.2 that there are many goal-setting approaches. Perhaps we have a goal that is aspirational - that comes from a leader that wants to see a big transformation. Sometimes a goal may be a *"guesstimate"* where senior folks choose an objective that seems worthwhile but also feels *"doable"* (by the way, this most often results in an improvement target of 5%!). Maybe we have an empirical goal, built up from detailed research and analysis.

It may be that the goal has been developed from an extensive study of the organization's capacity for improvement involving all those who would be expected to deliver the necessary change. In these circumstances, a review of the goals may not be necessary.

However, it is possible that the goals may have originated from a proposal from a programme Champion or a small group of senior executives, or as a result of recommendations from external consultants. It may be that there is no solid evidence base as to the viability of the goal. In these circumstances, it makes sense that the organization confirms its ability to *deliver* the goal before committing irrevocably to it.

The best folks to be involved in the validation process are those who will be responsible for delivering the goals. If this is a large group, then representatives from the group can be involved. Ideally, we would also include senior executives and sponsors in the validation process, as well as any project Champions who will facilitate the programme.

It is important that the review process should genuinely permit the goal to change. It must not be seen as the first step in cascading the goal to the rest of the organization. Clearly, if there are minimum expectations (perhaps because, below a certain level of performance, the organization is not legally compliant or because there is a minimum level of return that is financially justifiable), these should be clearly spelt out. It should also be made clear that the review can lead to increase in the goal just as much as it can to a decrease.

It is important that **the review genuinely allows the goals to change.** It should be possible to increase as well as decrease the goals.

The important thing here is to ensure that we take the opportunity to create *ownership*. Here the senior leaders need to be very aware of their influence. There is a risk that everyone participating in the review process *appears* enthusiastic and confident, but in reality, no one wants to be seen to disagree with senior management. Organizations that have established Kaizen or Lean approaches should be well-versed in open and unbiased discourse; for others, external facilitation may be necessary to ensure effective contributions from all.

A technique that can be helpful in reviewing the goals starts by considering the practical steps that will be needed to deliver the improvement. I have adapted this technique from a policy matrix used in Lean, as illustrated below.

We start by setting out the strategic aims and objectives of the programme and the specific goals that have been set. Then we need to consider the changes or projects that will deliver the goals. The individuals who initially proposed the goal will undoubtedly have some idea as to the projects or changes that are needed to achieve the goals. That is a good starting point, but new participants in the validation process should also be free to propose alternative actions.

Once one or more actions have been identified, the next step is to work out what resources these actions will require and over what time frame. Finally, having set a chronology and scale of the projects, the results can be estimated.

At this point, we will see that the results match, exceed or fall short of the desired goal (ideally, they will be a little above). If there is a shortfall, we can either add new projects, deliver the projects at a greater scale or adjust the high-level goal downwards (it is not advisable to change the results to fit with the goals without altering the resources, as this often leads to unrealistic targets).

What this process does is to test if there is a likely set of projects and changes that can deliver the goals. It tests if there is sufficient capacity to achieve the desired scale of change. The aim should be not only to confirm that the goals are achievable but also to identify if the organization can do more than it has initially proposed. It also ensures ownership by those who have to achieve these results, by involving them in defining the changes that they have to deliver.

11.8 **Policy matrix**
Source: Niall Enright, adapted from The Lean Toolbox, Bicheno and Holweg[70]

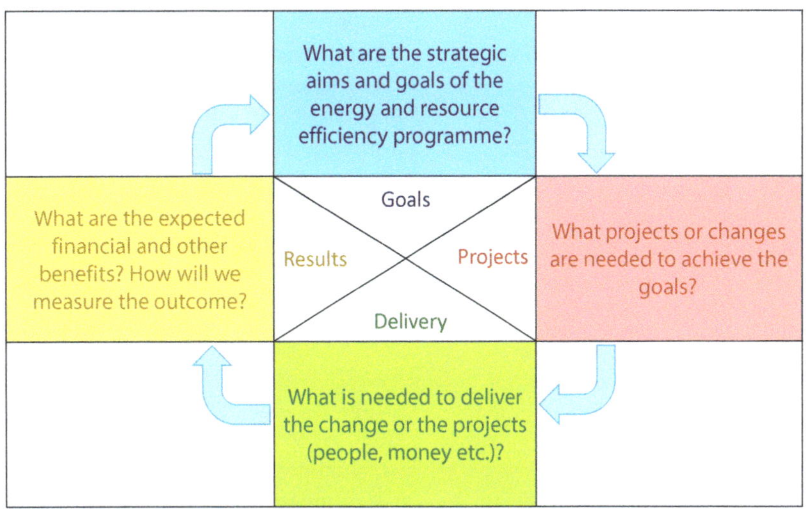

11.8 Articulating goals

The way a goal is articulated is important. Sometimes they may ambiguously phrased. Where goals affect performance, promotion or incentives it is human nature to "game" these.

> **Real World:** *When exactly?*
>
> One of the most common areas of confusions around goals relates to how the target date is described.
>
> Consider an organization that says: *"By 2015 we will achieve XXX".* To me, this statement means that the objective will be completed by *31 December 2014*, but some people interpret this as meaning *during 2015* or *by the end of 2015*.

When defining a multi-year goal, we need to take care in how this is articulated. I have come across several cases in organizations where employees, having achieved a given year's goal, defer further action so that it can contribute to the next year's goal. This problem is common where there are financial incentives linked to achieving the goal. Paradoxically, those well-intentioned, multi-year goals become the defining point at which employees believe efforts should stop!

One solution is to make the goals cumulative – that is to say, to set a goal of 10% reduction by the end of year 1, 20% by the end of year 2, 30% by the end of year 3, and so forth. Thus, early action is not penalized – to the contrary, early action now helps employees feel that they are *on top* of the improvements required and so can reduce the pressure. Where financial incentives exist, it makes sense to increment these linearly with outcomes and not to have an effective *cap* on the reward, once the goal has been attained, although I appreciate that this easier said than done. Finally, to overcome the semantic implication that a goal implies a cessation of effort, I recommend the use of the words "*minimum*" or "*at least*", as in: "*Our goal is to have reduced waste sent to landfill by a minimum of 10% for each year between now and 2020.*" Again, note the careful use of words *"for each year"*, rather than *"in each year"*, which makes it clear that the goal is related to the number of years elapsed, not the outcome in the year.

Another pitfall to watch out for in setting goals is in defining the way that improvements are credited. For reasons I have never been able to fathom fully, many organizations will record the value of resource efficiency improvements for just one year, wiping the slate clean in the following year. If we apply this approach to the BP *enManage™* programme described earlier (see Table 6.13 on page 241), the actual programme returns from the investment of US$8.8 million will be reduced from US$33 million to US$20 million, as continuing savings from previous years are not added to the totals. Not only does this approach significantly understate the real benefits of the programme, but it has an even more undesirable consequence, for it ensures that efforts will be entirely devoted to driving new improvements, rather than putting in place systems and processes to *maintain* existing improvements. This might be an acceptable approach if every improvement were a technical fix which, once implemented, would remain in place forever – but the reality is that many of the opportunities for resource efficiency are *reversible*, especially those that rely on behaviour change. I would therefore strongly caution against any goal with the words "*new*" in it, such as "*x new savings a year*".

Summary:

1. Any goal is usually better than no goal at all.
2. Goals should focus on the most important aspects of resource efficiency. A materiality matrix can help with this.
3. Goals should be SMART: specific, measurable, achievable, realistic and timely (or SMARTER, adding ethical and reasonable to the list)
4. The folks who have to deliver the goals should own them. If they have not been involved in setting the goals then a goal review process is suggested.
5. Goals should align with the key mission and preoccupation of your organization.
6. With items 3-5 in mind, it is entirely acceptable for goals to be reinterpreted at different levels of an organization to align the programme with local priorities.
7. Targets represent the goals for machines, processes and business units. These should reflect what is achievable at this level of the operation and reconcile upwards to exceed the high-level goal. Organizations have goals while individuals are responsible for delivering targets.
8. Absolute sustainable goals have the greatest environmental integrity. Every organization should be aware of its sustainable goals, even if these are not publicly stated, because any gap between these and the current objectives represents risk.
9. Headline targets from policymakers (such as the UNFCC) may be for cumulative improvements so the final rate of improvement needed to match these objectives is likely to be considerably greater than the headline figure if, as in most cases, improvements "ramp up".
10. A resource efficiency programme is like a marathon, not a sprint. Goals should not seek to maximize short-term results at the cost of long-term value.
11. Goals which assume that improvements are irreversible or which discourage continual improvement should be avoided, as these will usually destroy value in the long run.
12. Work on Optimize, Modify and Transform opportunities should take place at the same time. In other words, don't put off starting big changes until tomorrow, even if there is plenty of *low-hanging fruit* today.
13. If a goal is relative to a measure of activity (e.g. turnover, service delivery or production) beware of the *baseload effect*. Intensity or specific ratios can be very misleading.
14. Do not be tempted to set goals which depend on *free riding* on the efforts of others. Folks are becoming far too sophisticated to fall for these and they will eventually backfire.
15. Ensure the integrity of the goals. For example, adjust the baseline of absolute goals in line with boundary changes or in line with changes outside your control such as emissions factors.
16. Be aware that the timescale over which the goal is to be achieved greatly influences the nature of the resulting programme.
17. Goals are usually lagging indicators, so it is important to track leading indicators to be able to forecast how the programme will perform and make adjustments if necessary.

Real World: *Some random goals*

Dow: identify and implement business-driven project alternatives that will best enhance nature and deliver US$1 billion in net present value.[225]

JX Nippon Mining and Metal Corporation: Domestic CO_2 emissions: cumulative emissions of less than 3.17 million tonnes for fiscal 2013 to 2015.[430]

Ford: Reduce waste sent to landfill by 40% on a per-vehicle basis between 2011 and 2016 globally.[291]

Cargill: Improve our freshwater efficiency 5% by 2015.[118]

AkzoNobel: aims to restrict its absolute scope 1 & 2 greenhouse gas emissions (based on its current business portfolio) to no higher than 2008 levels, by offsetting organic growth with energy efficiency and fuel mix improvements.[14]

NHS England: a 10% carbon reduction by 2015 compared with 2007 levels.[553]

Interface: Our goal is to eliminate waste by 2020.[398]

Harvard University: has committed to a GHG reduction goal of a 30% reduction from fiscal year 2006 levels by 2016, including growth.[356]

Anheuser-Busch InBev: reduce global water usage to a leading-edge 3.2 hectoliters of water per hectoliter of production.[4]

UK Parliament: both Houses have agreed the following targets (based on 2008/09 baselines): a) To reduce absolute carbon emissions by 34% by 2020/21; b) To reduce water consumption by 50% by 2020/21; c) To reduce the weight of waste generated by 30% by 2020/21; d) To recycle 75% of waste generated by weight by 2020/21.[726]

Further Reading:

1. *Marc j Epstein. Making Sustainability Work, Greenleaf Publishing. ISBN-13-978-1906093051.* Chapter 5 on performance evaluation and reward systems.

2. *Leading Change Towards Sustainability*, 2nd Edition, Bob Doppelt. ISBN 978-1-90609-334-1. Chapters 7 and 8 touch on how changing the purpose and goals of an organization can catalyze fundamental change. See my review of this book and a link to purchase it.

Questions:

1. Consider the goals in the box left. Which are absolute and which relative? What are the strengths and weaknesses of these? Can you foresee any pitfalls in the way they are described?

2. Some commentators argue that the only goal of a private company is to make returns to shareholders, and so setting any other objective is not appropriate. Can and should all resource efficiency goals be expressed in terms of shareholder value? (You may want to refer to the item on fiduciary duty on page 218)

3. In a similar vein to the question above, if the primary purpose of a public body is to deliver service to its stakeholders (e.g. hospitals to their patients, schools to their students) how could this influence goal-setting around resource efficiency. Is this reflected in the real world goals of these types of organizations?

4. What are the characteristics of effective resource efficiency goals and why?

5. What steps should organizations take to avoid being accused of *"greenwash"* in the goals that they set?

6. What are the problems associated with goals based on intensity ratios and how can these be overcome?

7. Examine the resource efficiency objectives of one or more of the organizations in the box left. Suggest what leading indicators could be developed to forecast performance ahead, and explain why these are suitable.

8. Consider the efficiency or sustainability goals in your organization. Are these clearly described? Would you suggest any changes and why? Are there any leading indicators that you would propose and why?

9. How does the language used influence the nature of a goal? What kinds of pitfalls exist and how can these be overcome?

12 Discovery

Discovery is seeing what everybody else has seen, and thinking what nobody else has thought.

- Albert Szent-Gyorgyi

The processes of identifying and prioritizing opportunities for resource efficiency which add value to our organizations is called discovery.

In this chapter, we will explore the traditional process of identifying, at a facility or process level, the scope for improvement in our resource use, an audit. We have previously discussed the shortcomings of the word audit in Section 9.3, but concluded that its wide usage means that it cannot be easily replaced.

The methodology set out in the following pages is by no means the only approach to discovery. There are many other ways to generate ideas for improvement other than a formal audit, such as suggestion schemes, workshops, vendor investigations. No matter how the ideas are generated, a number of key principles are described here which apply to all discovery efforts. Nor should we see discovery as a one-off activity - in a continual improvement process we are always identifying further changes which will lead to reduced resource use.

Within the overall Framework, discovery activities form part of the Method. However, organizations frequently undertake audits to understand the business benefits of energy and resource efficiency, to set achievable goals and so help develop their Mandate. Because audits are often the first activities in our programme, it is doubly important that we get these right.

It is also not uncommon for discovery to *follow* programme launch. In these circumstances, we have a Mandate in principle and are undertaking discovery activities to provide a focus for the programme. For example, we may be rolling out a programme to a large organization, and so we need to develop a plan of action that is specific for each facility that is participating in the programme.

Discovery is one of the most exhilarating and rewarding aspects of energy and resource efficiency. It involves careful detective work, the accumulation of data and facts, observation of behaviours and processes and the ability to challenge current thinking. It is essentially a team effort, where the most important participants are not necessarily those with the greatest technical knowledge, but rather the people who need to implement the solutions at the end of the day. Without their understanding and commitment, the opportunities are unlikely to be realized. Discovery is about seeing business goals, technologies, systems and processes with new eyes. It involves asking some apparently very dumb questions, usually preceded by the word "*why*", in order to arrive at some profound new insights.

12.1 Site selection for audit

If an organization is large and complex, it can be quite challenging to choose where to carry out initial audits. There are some characteristics of sites that will tend to lead to better audit outcomes. Using these we can arrive at a technique for site selection.

If we are initially just going to audit a proportion of our organization in order to establish the potential for resource efficiency, then we need to take care with our choice, as the audit results will determine the scope of our programme and set the expectations of our sponsors. Whether we are looking at a sample of facilities, sites or even some departments or buildings within a larger campus or facility, there are some fundamental questions that we can ask which will help us with the decision where to start.

- Does the scale of resource use merit a detailed investigation? Clearly, we want to create a compelling business case for action, and if the audit costs are disproportionate, this might undermine our objectives. This would not apply in an organization where there are many similar small operations, such as a fast-food chain, where the initial audit aims to identify widely repeatable projects to be implemented across the business.
- Is the location typical of the organization's operations? It may be sensible to focus on demonstrating value in core activities as a way to gain support for a broad resource efficiency programme, rather than have the findings dismissed as not applicable to the organization because the site is a unique case.
- Is the site or facility well instrumented? Many opportunities for improvement can be identified through data analysis, so clearly the ready availability of historic resource and activity (e.g. production) data down to sub-meter level will enable the audit to achieve a better analysis.
- Is there an indication of an outlying performance? If the site is among the largest resource users per unit of activity, there may be greater opportunities for improvement and so it makes sense to focus efforts here.
- Is there good management commitment? If the local management team are behind the process and are open to suggestions, then the audit is more likely to be successful.
- Is there an impending major change? Where the facility is going to undergo a significant change, it may not be a good candidate for an audit because the findings may become irrelevant or the site personnel are too preoccupied with the change to provide adequate support. This example could include the implementation of a major new management system, such as SAP, as much as a physical change to the facility.

When selecting a sample of sites to audit, an important consideration is what these sites can tell us about the whole organization.

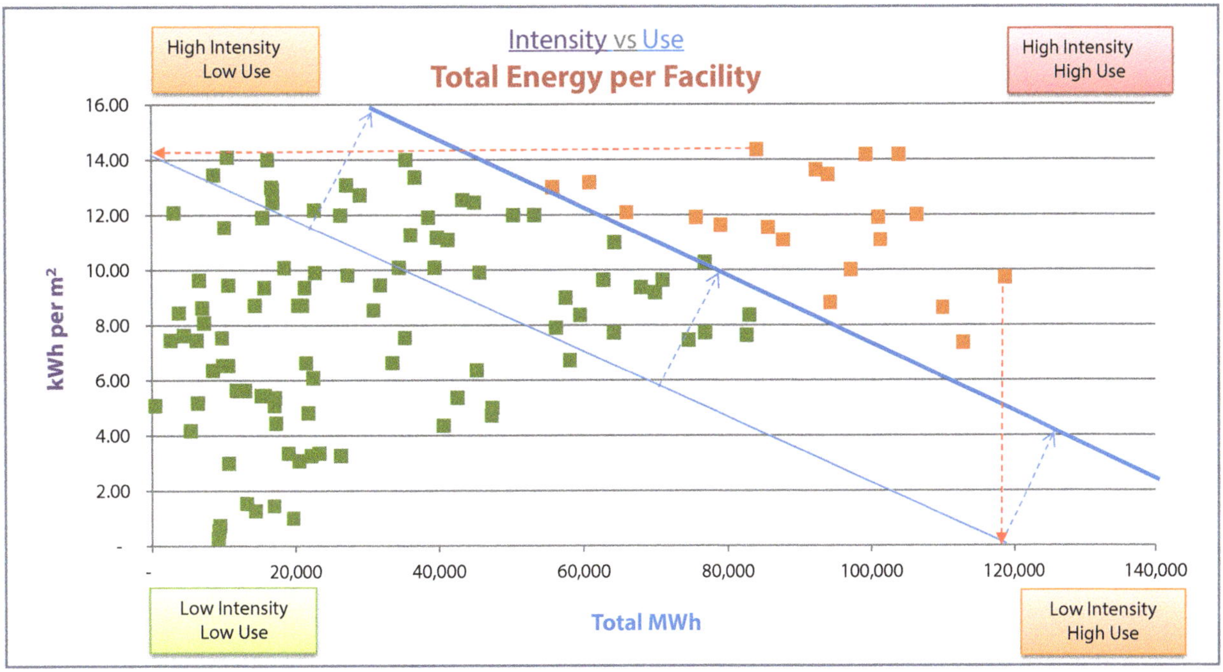

12.1 **Intensity-use charts can help prioritize sites for audit**
The slope of the line determines whether the focus is on intensity or use, as shown below
Source: Niall Enright, available in companion file pack

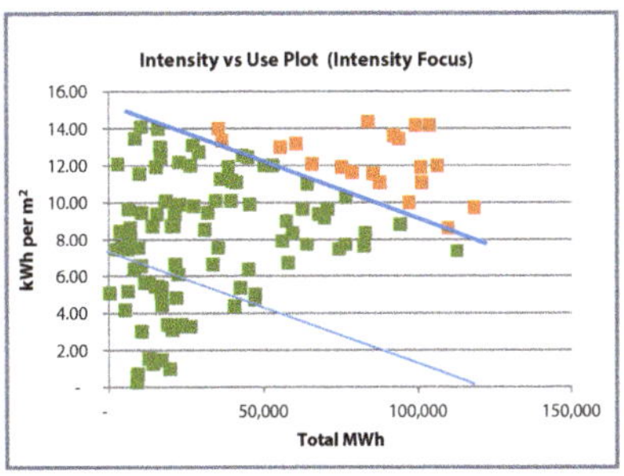

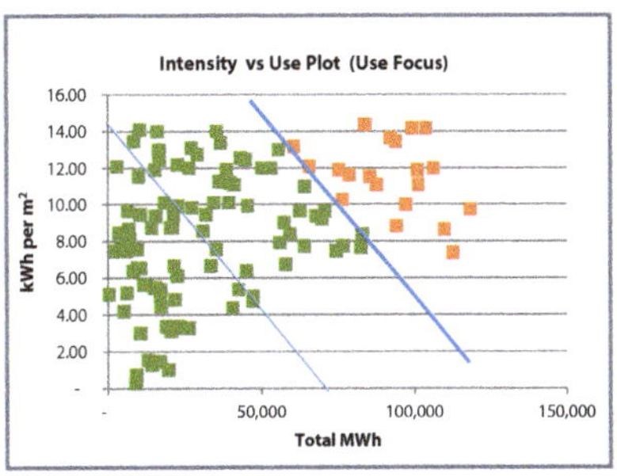

When considering a range of locations, as in a supermarket chain, simple benchmarks, such as kWh per m2 floor area are often used to develop the audit list, usually after categorizing the stores into subgroups such as *out-of-town 24 hr. superstores*, *metropolitan convenience stores* or *mid-size high-street supermarkets*. Typically, the largest users in each category are targeted for audits, on the assumption that their higher use arises from greater inefficiencies, although this may lead to an overestimation of the total potential.

A similar approach to site selection involves plotting the sites on an intensity-use chart. In the example above, a manufacturer of electronic components has a large number of facilities but initially only wants to carry out an audit programme at 20 sites. To select the sites to be audited, the manufacturer plotted an intensity (in terms of kWh total energy use per m^2 conditioned factory floor space) vs use (total MWh) chart, as shown above.

The manufacturer then drew a line between the highest value for the intensity on the vertical axis and the highest value for use on the horizontal axis. This produced the thin blue line joining the points indicated by the red dashed arrows. As there were more than 20 sites of interest above this line, they then moved this line in parallel away from the origin (shown by the dashed blue lines) until 20 facilities remained, as shown by the heavier blue line, with the selected sites shown in orange. If there were fewer than 20 facilities above the initial thin blue line, then the line would have been moved in the opposite direction, towards the origin, until the desired number of facilities was reached. If needed, the emphasis can be placed on either intensity or use as a site selection criterion, as shown in the smaller charts, left.

12.2 Audit baseline

As with many things in life, the outcome of a discovery or audit process often depends on the quality of preparation. Here, we set the right expectations in the facility we are investigating and gather some baseline information which will help gain insight into resource use.

An audit is concerned with the potential to improve resource efficiency, and so must be centred on the physical operations of the organization. Typically, we would wish to collect baseline data about our energy and resource-consuming systems prior to the costs and effort of visiting the facility. We also want to be sure that the folks that we will be interacting with at the site, from the senior management team, through our contact person, and people in the operations, understand the purpose of the visit. In the briefing for our hosts about the scope and details of for the audit, such as the resources to be investigated, facilities, cost and acceptable return on investment, I would also emphasize:

- the collaborative nature of the process;
- the fact that the effort required from staff will be proportional to the benefits; and that
- all opportunities identified will be reviewed by and credited to the site team.

This approach is not only designed to reduce resistance to the study but to place ownership for the findings with the facility. Key to the design of engagement is the notion that the purpose of an audit to create the basis action.

Our initial, usually pre-visit, activities are summarized in the illustration opposite, where tasks are separated into three columns: People, Systems and Technology. There are four activities in this first phase of the discovery process. Tasks are shown in the blue boxes and tools in the red.

In prepare, the site team are briefed as above. We ready the facility for the visit and the data that we will need. We also ensure that the logistics for the audit are in place. Typically, a formal data request is sent to the organization, an example of which is shown in Figure 12.3 on page 390. The baseline data tasks consist of gathering the basic numeric data on the flow of resources (e.g. utility bills) and an inventory of major equipment.

Next, we want to understand how the facility measures, performance. In other words, we want to know what are the key business drivers for the people at the facility. We need to know the goals that are set for the facility and how these are measured. We need to know how investments are approved and also what has happened to previous resource efficiency studies and requests for funding.

Real World: ***People, Systems, Technology***

The techniques in this chapter expand on the Method stage of our energy and resource efficiency Framework, described in Volume I.

The activities are therefore categorized under the same three headings: People, Systems and Technology.

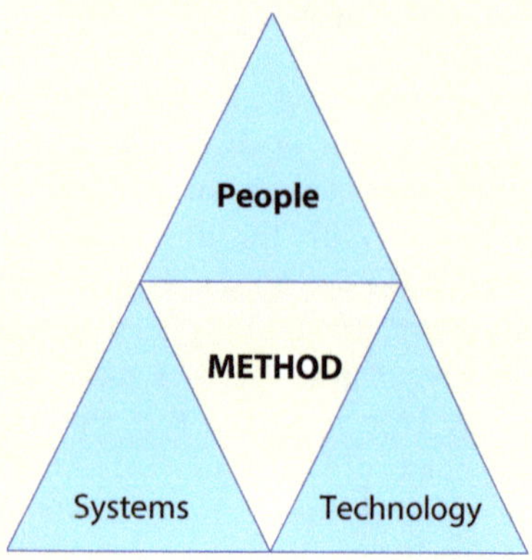

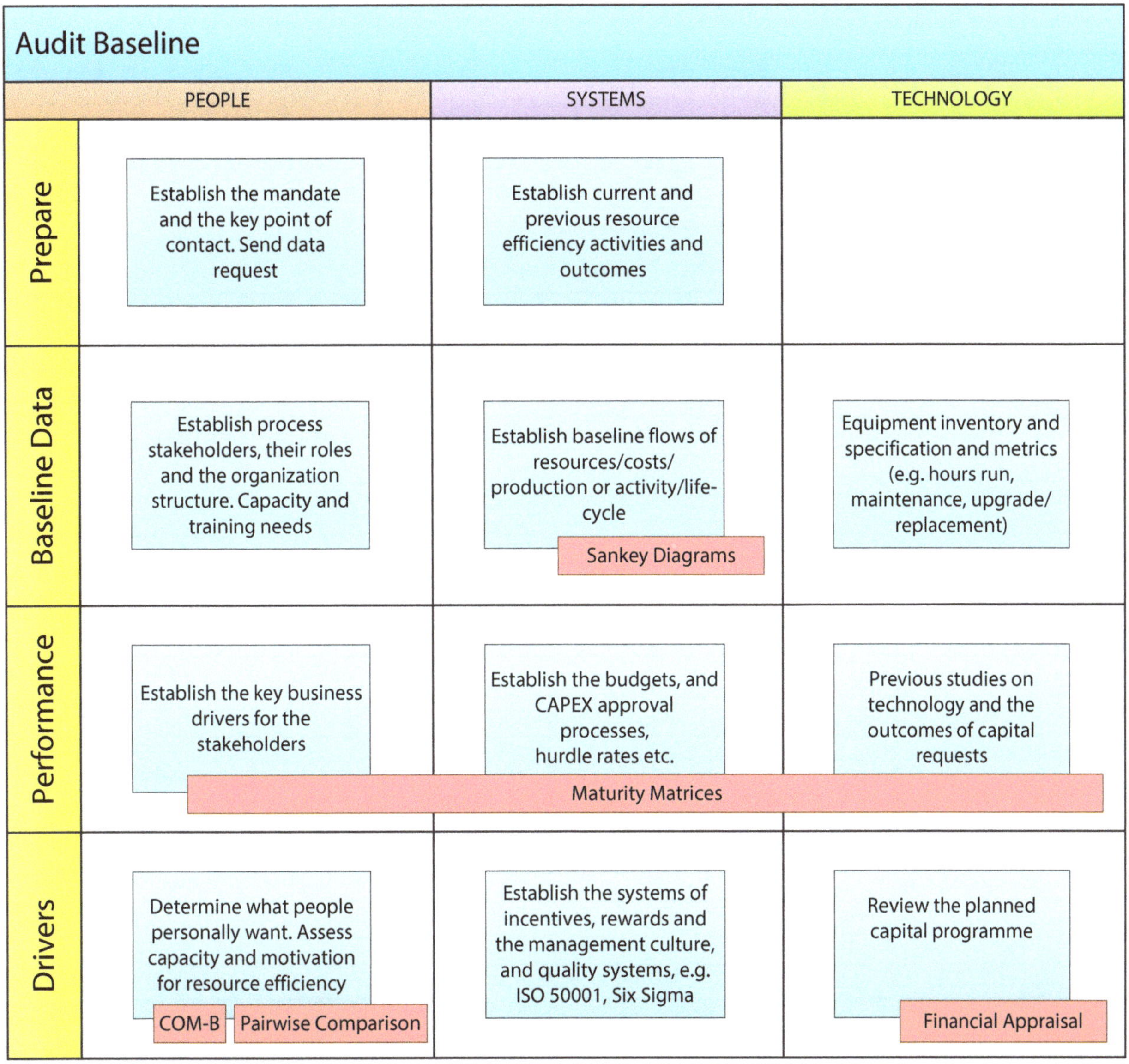

12.2 **Baseline activities**
These baseline activities form part of our Method for energy and resource efficiency. They are the first stage of the audit process, undertaken prior to a visit to the site or facility. Not all activities are relevant to all facilities. Tools and techniques described in later chapters are shown in red boxes.
Source: Niall Enright, available in the companion file pack

Finally, we want to explore the drivers and barriers, for action which could influence our recommendations. For example, if there is a bonus scheme in place we could propose aligning it with efficiency improvements. We want to know if there are formal quality systems in place. Are there other initiatives underway or planned which could compete for resources or management attention? If there is an existing capital expenditure programme, we can see how we can tap into this to deliver investment. People's motivation and capability to achieve the necessary changes should also be assessed at this point, as these are critical to the overall programme success.

The table below lists the baseline data that would be collated in advance of an energy and resource efficiency audit at a facility.

	BASELINE Information Request
Overview	Basic location information for the site or facility
	A narrative description of the site or business: maps, plans, schematics, process flow diagrams
	Description of how the facility fits into the wider organization (if relevant). Is it typical?
	Key contacts for the audits, their availability
	Health and safety details and procedures to be followed during the visit. Personal protective equipment (PPE) required
	Logistics for the visit: travel, accommodation, security, on-site facilities required (e.g. room, desk, phone, network connection)
	Details of existing quality systems like ISO 50001, Lean, Six Sigma or TPM
Operations	Patterns of activity e.g. hours and days of operations, shifts, closure\shutdowns. Periods of abnormal activity (e.g. Christmas for a retailer)
	Detail of management systems, meetings and processes. What are the primary objectives and key performance indicators for the site?
	Relevant details of employee training, performance management, targets, incentives and rewards in respect of resource efficiency
	Details of current accountability for resource use. Who pays the bills? How are improvements initiated and approved
Resource use	Relevant distribution diagrams (e.g. water, effluent, steam, electricity), location of waste collection bins etc.
	Metering and sub-metering schedules and schematics. Information about data collection systems
	Details of major raw materials volumes, storage and handling on site
	Details of historic and current studies, proposals, CAPEX requests and initiatives related to energy and resource use
	Benchmarking data that may help us assess the site's performance, such as the Solomon Energy Intensity Index
	Details of legal or compliance requirements related to resource use
Resource Data	Three years' data of resource use down to sub-meter level: monthly weekly or daily values depending on volume/cost of resource and data
	Three years' data on waste arising, treatment and disposal to department or unit level at same frequency as resource data, if possible
	Three years' activity data (e.g. production, sales, # patients, meals served etc.) down to department/unit/equipment level at same frequency as resource data, i.e. monthly, weekly or daily, if possible
	Details of resource/waste costs and rate structures over the period. May be appropriate to provide resource/utility/waste treatment bills
	For life cycle assessments we will need to have schedules of the component parts, source, packaging and transportation of different inputs to our facility or the types/destinations of finished products over several years
Plant	Inventory of major resource-consuming or waste related plant and equipment. Details include name, description, manufacturer, model number, purchase date, planned replacement date, location, maintenance approach, annual hours run and rating or throughput data
Organization	Organization chart for management team, and a summary of key management meetings and reports. Details of key performance KPIs and also of any incentive schemes or bonuses related to improvement
	Details of individuals responsible for resource issues: e.g. procurement, engineering or maintenance, operations, finance
	Site directory with telephone and email contact details
	Information about the site's budgets, approval hurdle rates and expenditure sign-off levels

12.3 (above) **Typical baseline data request**
Source: Niall Enright

12.4 (right) **A report is NOT the objective of an audit**
Image Source: Microsoft Clipart library

Real World: *Challenging the conventional notion of an audit*

Over three decades, I have taken part in many hundreds of energy efficiency and waste minimization audits, as part of a team of external consultants working on behalf of a wide range of industrial, commercial and public sector clients. From this experience, I have come to the conclusion that the objective of most audits is fundamentally wrong.

From my observations, it is clear that most consultants and clients mistakenly see the aim of an audit being the production of the audit report, detailing a range of recommendations along with a cost-benefit analysis. Everything about the assignment reinforces this viewpoint. The consulting firms each have their proprietary reporting format, which they jealously protect. Their proposal documents detail the specific content and time frames for delivery of the report. The idea of the report as the product is further reinforced by the payment schedule, where final fees become due once the draft report has been reviewed and approved.

This notion is crazy, since a report alone achieves nothing. Goodness knows I have come across hundreds of efficiency audit reports gathering dust at the back of cupboards.

The aim of the audit should be to gain commitment to improvements, not to produce paper. The process should be outcomes-driven rather than report-driven. With this insight, the nature of the audit process changes dramatically. In particular:

- The business case analysis must be strengthened considerably and aligned much more closely to the client's own methodology so that the additional effort to get from report to investment case and, hopefully, approval is eliminated.
- The feasibility and cost/benefits associated with the opportunities identified should be rigorous and validated through discussion with the site team so that there is strong internal support and ownership for the opportunities put forward.
- Decision-makers must be much more explicitly involved in the audit process. Thus, we must request an initial discussion with the site manager and finance director, and a closing meeting at the end where the *"investment opportunities"* are *"selected"* (by implication the closing meeting is about approving actions, not receiving data).
- The structure of the audit outputs (reports) must be oriented to support decision-making. Dense technical details should be relegated to appendices and the main body of the document made much shorter and more cost/benefits focused. In some cases, reports can be dispensed with altogether, replaced by a PowerPoint presentation and an Excel Opportunities Database sheet setting out the opportunities.

Because the audit is a merely a waypoint, consultants are no longer under pressure to write massive tomes and can focus more on the impact, practical implications and effectiveness of the recommendations rather than the number of pages.

These changes will help to prevent the audit report being seen as *the end of a process*. We need to achieve a smooth transition from opportunity recommendation to implementation. The way the site responds to the audit has much to do with how we set expectations in the process. Thus, in conversations with decision-makers, we should refer to the end-of-audit meeting as their *"opportunity selection"* session where we would be looking for decisions, not passively conveying information. Clearly, this places a greater burden on the audit team and the site stakeholders to agree on some recommendations then and there. However, it also creates much higher levels of satisfaction among the audit team, who can see their ideas being adopted, and delivers much greater value for money from the audit process to the client.

In moving towards this much more outcomes-driven approach to audit, a significant change must take place; the audit team should work with the site to implement any feasible no/low-cost improvement opportunities during the audit process itself, giving full credit for the savings to the site team.

Thus, the audit is a now process where change is seen to be delivered. Because some opportunities have been realized immediately (and hopefully the audit costs more than covered), it is natural for the next step to be organized around deciding what *further* opportunities should be implemented. The journey towards improvement has already begun.

If needed, the consultants should remain available to help the client deliver the audit recommendations. Not only can this help the customer to overcome any bottleneck in staff availability to deliver the improvements, but it has the important additional benefit of ensuring that the consultant's findings are honest and realistic (after all, they may be asked to deliver them).

12.3 Audit overview

The objective of an audit is to identify opportunities for improvement and prepare the organization for their implementation. The process involves close collaboration with the site teams and should be seen as the beginning of a journey, not as a one-off activity.

I have used the term "technical audit" here to refer to the investigative work that is carried out to identify resource efficiency opportunities involving equipment and processes. This work includes the baseline activities described in the previous section, which may have to wait until the site visit to take place. A generic technical audit process is summarized opposite. Again, the activities fall into three columns based on primary focus: People, Systems or Technology.

The first step in the audit is to understand which metrics and data are available to drive improvement. We will assess how data is used in the facility, what key performance indicators (KPIs) are available and possibly put in place some temporary measurement to fill gaps in the existing data. Lastly, we need to analyse the performance of equipment using standard calculation methods, such as coefficient of performance (COP) for refrigeration plant.

Having consolidated our data, we next determine the sources and causes of variance. The objective is to separate out periods of good and bad performance for the major resource-consuming equipment and processes. This assessment can be done through workshops with staff, statistical analysis of data, or detailed technical investigation of items of equipment. From these activities, we will generate many improvement opportunities.

With a longlist of opportunities identified, we then need to carry out some cost-benefit analysis. This will be done at the level of individual equipment upgrades or replacement and also at the level of an overall programme. We should price the metering and other systems to provide feedback on performance which will drive and sustain behaviour change. We may also need to develop an understanding of the effort and manpower cost of our proposed resource efficiency programme so that this can be properly included in our recommendations.

The final step is to gain the local Mandate for the proposed opportunities. For this to happen, we need to have close collaboration with the site team so that they have full ownership of the recommendations. While these activities are taking place we should implement any immediate improvement measures that we identify and the site team approve. This will reinforce the notion that the audit is about making things happen, not about pieces of paper (see the previous page on *"Challenging the conventional notion of an audit"*). This will provide us and the site team with an early opportunity to celebrate success.

Real World: ***Opportunity identification***

This chapter focuses on formal audits, but there are many other ways to generate and evaluate ideas for improvement.

Discovery Techniques
Customer or staff suggestions
Workshops
Tariff, invoice or bill Analysis
Technology vendor's suggestions
Rating methods (e.g. LEED\BREEAM, Energy Performance Certificates etc.)
Life cycle assessment
Benchmarking
Visualization (e.g. Thermographic analysis or smart leak detection and reduction)
Process integration and pinch analysis
Statistical analysis, such as regression analysis comparing resource use with drivers such as weather.
Gemba walks
Activity-based costing

Once savings ideas are identified, it is important that they are captured and then assessed. This is the purpose of the Opportunities Database tool. This database (which could take the form of a spreadsheet) should record all ideas, however trivial, and to ensure that these are driven through a process of assessment and, where feasible, implementation.

Technical Audit

	PEOPLE	SYSTEMS	TECHNOLOGY
Baseline	See Baseline		
Metrics & Data	Establish dissemination of information and KPIs.	Data assessment: Systems review (software and metering), KPIs. Gaps and temporary measurement	Specialist data for equipment performance, e.g. coefficient of performance, boiler efficiency
Variance or *Muda*	Discovery workshops and one-on-one conversations CDR – Six Sigma	Statistical data analysis and benchmarking Regression Analysis	Detailed investigation of equipment and processes Demand-led Audits
Consolidate	See Consolidate		
Cost-benefit	Cost estimation for program manpower (internal and external) including training	Estimation for systems (e.g. measurement, Monitoring and Targeting and Opportunities Database) Variability Analysis	Technical project cost estimation (all elements, not just capital) Financial Appraisal
Mandate	Business case, next steps, Mandate for programme MACC	Defined scope, goals, resources and Governance	

12.5 **Audit activities**
Source: Niall Enright.
Available in the companion file pack

12.4 It's not just technology

Audits are usually carried out by engineers. As a result, they are often stronger on the opportunities for improvement through technology, than through behaviour or systems changes. It makes sense to include folks with management experience in the audit team to work on these important additional sources of improvement.

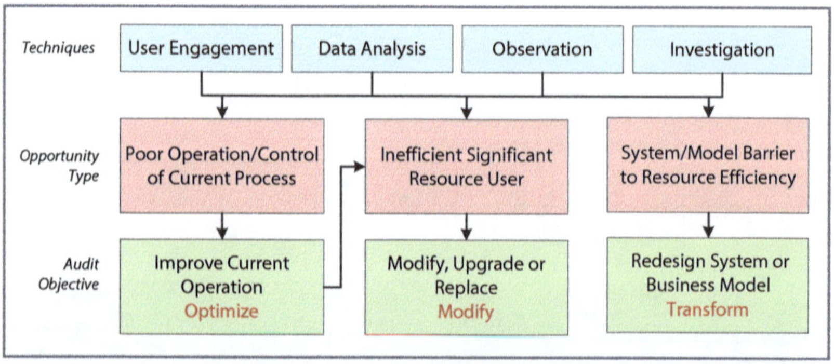

12.6 **Resource efficiency audits can identify three types of opportunities for improvement. These correspond broadly to the Optimize, Modify and Transform activities described earlier.** The first two areas, better operation and upgrades to significant resource-using equipment, are what traditional technical audits focus on. However, the most important opportunities may actually lie in the Systems that drive decisions within the organization, or the current design of products or services (the business model). Some Systems changes can be quick to implement and can have a rapid impact (e.g. introducing whole life costing into investment appraisal) *Source: Niall Enright, image available in companion file pack*

There are broadly three sets of recommendations that an audit process can make:

1. How to operate existing equipment and processes better to more closely match its design efficiency, usually by fixing faults, improving control or improving yield; or

2. How to modify or upgrade existing equipment, processes, products, services, to achieve a step change in the underlying design efficiency; and

3. How to change systems in order to change decisions around resource use, for example, by modifying the reward systems or strengthening procurement standards or introducing whole life costing into financial appraisal. In this category of improvements, we might also consider the business model of the organization, although this is rarely done in a traditional audit.

The first objective in most resource efficiency audits is to identify all instances of poor control/operation, as these can be very quick and cheap to rectify. Indeed, these operational improvements are the kinds of changes that can be implemented during the audit, and so send the important signal that the audit is about change, not reports.

Next, the team will consider equipment, process, product or service changes, but will usually only investigate significant resource users, and will rapidly exclude those modifications that fall outside the site's required return on investment.

The last category of improvement, system changes, is most often neglected by audit teams in the pursuit of technical improvement. This is unfortunate

Audits should not just consider **technology,** but also **systems and designs.**

because most often the greatest barriers to resource efficiency are issues that fall under this heading, such as: split incentives (i.e. utility budgets are not allocated to end-users, so there is no incentive to reduce use); higher hurdle rates for resource efficiency projects (usually arising from use of payback as an indicator of return) than for other investments; a culture of procuring the cheapest capital cost equipment rather than the cheapest whole life cost; insufficient attention to resource use in management meetings and KPIs; *siloed* operation of teams who need to collaborate to bring about savings.

Often the expertise, and confidence, of the audit team do not extend to assessing these systems issues adequately during the audit. The ability to dispassionately assess the cultural, procedural and organizational barriers at a site is a compelling argument for including an external expert as part of the audit team. Indeed, the site personnel may not feel able to challenge the existing conventions and paradigms as they have to work within the system and cannot afford to *burn any bridges*, particularly with their superiors.

Insight

The evaluation of the scope for Systems improvements is not formalized in the technical audit method set out in Figure 12.5, as this is much less easily described in terms of tasks or phases. However, a consideration of this aspect is built into the consolidate phase of the audit, described later.

Real World: *Self-auditing at Unilever*

The stereotype of an audit has the *expert* auditor examining equipment and processes and, like some Sherlock Holmes character, coming up with fundamental changes that no one has previously identified and which will instantly transform the situation. Of course, this is complete fantasy, but this view nevertheless colours the expectations of the sites as well as the behaviour of some auditors, who feel under pressure to *pull a rabbit from the hat*. In fact, the idea of the external *"expert"* coming to tell them how to do their jobs is one of the biggest turn-offs for site personnel involved in audits.

Auditing should, in fact, be a much more collaborative process where the deep experience of the site team can be augmented by external consultants who bring knowledge of similar processes in other locations and an *"ability to ask dumb questions"*.

An example of this good practice was an extensive programme for Unilever's food operations in Europe supported by my former employers, now Jacobs. Here, the site audit tasks were deliberately structured as a training activity – whereby very experienced energy auditors from Enviros, Jane Galloway and Peter Cohen, came to the site and worked with the site engineering and maintenance staff in a structured programme of opportunity identification. The auditor would, for example, provide some training on the basics of efficiency in steam distribution systems and then the site personnel would survey their steam system and identify issues such as poor steam traps or low condensate return and go on to calculate the associated savings potential with the auditor's help.

From what I could see at the time, this approach was very effective in terms of getting site ownership for the opportunities that were identified, although I did get the impression there were fewer and less complex opportunities identified compared to the more independent audit approach previously used.

In practice, a hybrid approach ended up being developed – where some technologies were addressed in the training activities, and others relied on the auditor themselves carrying out the assessment in its entirety. This was particularly the case when it came to refrigeration equipment, where there is a very high level of specialist knowledge required to interpret performance and spot opportunities. Sometimes detailed equipment data was needed from the manufacturer, which introduced delays so that the analysis could not be completed during the site visit, thus effectively eliminating any training potential.

Local ownership for an audit process differentiates good audits from mediocre ones as these are the occasions where the opportunities identified are more likely to be implemented.

12.5 Start with demand

Understanding the demand for resources is central to a successful audit. If we know "why" we use a certain quantity of a resource in a certain form, then we can begin to establish "how" to achieve the desired result with less.

All audits should start by understanding the **demand** for energy and resources.

The waste hierarchy described in Chapter 2 states that the first strategy to adopt in an energy and resource efficiency programme is remove - in other words, the elimination of the demand for the resource altogether. Only once we have reduced the demand for the resource should we consider minimizing the usage further by making the supply side more efficient.

In follows naturally from this that all audits should start by considering the demand for energy and resources, before investigating the supply. This may sound self-evident, but it is remarkable just how often audit teams head directly to large items of equipment at a facility – the boiler, chillers, heating and ventilation systems, air compressors, dryers, wastewater treatment plants, and so forth. This arises from the tendency for audits to be conducted by engineers, whose primary training is focused on equipment, and by the dominant belief that technology is the solution to resource issues.

There is a very good financial reason for pursuing demand first. This is because demand improvements tend to be cheaper and faster to implement per unit saving, than equipment retrofits or replacement. In fact, demand improvements are at the heart of our initial no and low-cost opportunities for improvement.

The vast majority of audits, instead of focusing on demand, concentrate first on the supply side, as illustrated in the figure below. These audits, often carried out by equipment manufacturers or energy services companies, will start by looking at the big items of equipment and ask themselves what upgrades can be made to this plant to achieve efficiency improvements *given the current demand.* Thus, the equipment-focused auditor will consider whether the water treatment plant can be fitted with newer technology, rather than ask themselves the more fundamental question about the source of the wastewater and how that can be reduced. In this approach, supply is the focus and demand an afterthought.

12.7 **An audit should begin with the demand for a resource before examining the supply-side equipment servicing that demand, as shown by the green arrow**
Source: Niall Enright, in companion file pack

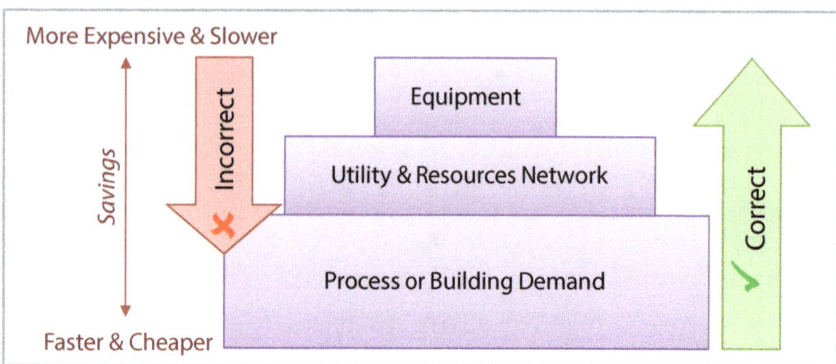

This focus is understandable because the demand-side picture is often more complicated and time-consuming to consider than the performance of individual items of

Real World: *Three steps to interpreting performance*

The objective of an audit is to initiate a process of improvement in resource use. Developing strategies and recommendations for improvement requires an analysis of current performance. The quality of this analysis will determine the effectiveness of the audit, i.e. how credible and actionable the recommendations are.

Far too many audits fail in this respect, providing a large volume of evaluation-free information, leaving the recipients of the audit to interpret the data and develop for themselves the interventions that can lead to improvement.

For an audit to be effective, three analytical techniques are required to assess resource use at a facility. These take place in the steps "baseline", "metrics and data" and "variance" in the audit process set out Figure 12.5.

1. Starting from the demand, we quantify the quantity, location and flow of the resource use. Pie charts or Sankey diagrams are helpful presentation tools.

Remarkably, some audits end their analysis at this point, only providing a detailed snapshot of resource use. What is needed is for the auditor to interpret this information on behalf of the decision-maker, by, for example, discussing distribution losses, conversion efficiencies, bottlenecks or costs.

2. The next key piece of analysis is to assess the evolution of the resource use or flows over time. Here, the auditor will look at longer-term trends (ideally over several years) to identify the direction of resource use or seasonal effects. For resources that are used continuously, much shorter timescales data such as time-of-day and day-of-week data is essential to identify different patterns of resource use (e.g. weekends vs weekdays or night vs day).

Time-series data certainly takes the resource-use narrative beyond the instantaneous snapshot described previous, It also provides the basis for investigating historical events - *"what happened then?"* - that can lead to opportunities for improvement.

However, we need to avoid an over-reliance on time-series data in the interpretation of resource performance. Trends of a value of time, while pervasive in our lives, are often very poor tools for the interpretation of performance.

3. To understand how resources are used, we need to relate the trend of use to influencing variables, such as activity, output, weather, the physical qualities of materials and so forth.

It is obvious when we think about it. The amount of natural gas I use to heat my house will depend on the ambient temperature because a thermostat controls it, so I use more gas in the cold winter than in the hot summer. In fact, just about every resource that an organization uses is driven by one or more of these variables. Interpreting resource use without comparison to variables is impossible - yet many auditors fail to do this and so arrive at incorrect conclusions.

There is a range of techniques that can be used to establish the influence of the variables, the most important of which is called regression analysis, which is described on page 460. This technique has a number of strengths, such as the ability to separate the fixed or baseload resource use from that which is influenced by the variable (in the example above, the fixed load use of natural gas in my home would be to heat water - a year-around activity) as well as the ability to put a number, the correlation coefficient, to the impact to the variable.

*Interpreting trends of resource use without comparison to **variables** is impossible - yet many auditors do just this and so arrive at incorrect conclusions.*

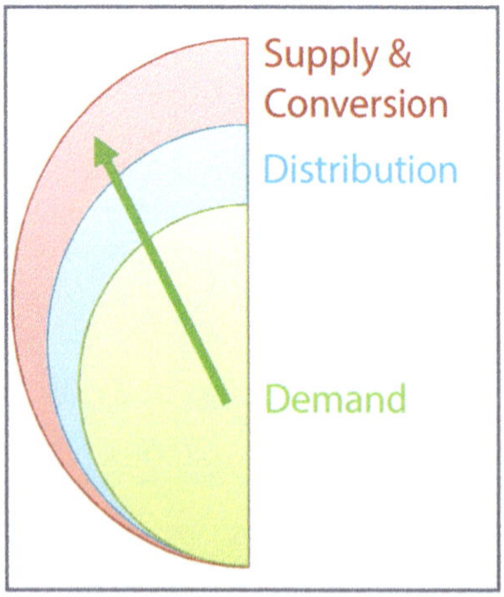

12.8 **Resource efficiency audits are sometimes likened to an onion: you should start at the centre with the core demand for the resource and then work outwards following the distribution system back to the point of supply**
Source: Niall Enright, available in companion file pack

plant and equipment, and many auditors have not been trained on the basic skills around using regression analysis to understand operational variability, so lack the means to understand the scope for demand reduction through better control or behaviour change. For many technically minded people, reducing environmental impacts is an end-of-pipe solution rather than a change management activity. It may also be a reflection of the nature of the business that the auditor is in – equipment manufacturers usually don't sell more kit by pointing out behavioural savings and ESCOs are usually interested only in large capital investments, as that is the way their business model works.

Justifications aside, it is clear that a good audit should first establish what can be done to reduce the demand for a resource at the point of use. This demand may be a measure of volume or mass of a resource (e.g. water or packaging material) or energy (e.g. the volume and temperature of a chilled water supply, or the volume and pressure of a compressed air circuit, or the voltage of an electrical supply). The auditor should be asking: *"Why do you need this much water in this part of the process?", "What quality does it need to be?", "Why at this temperature or pressure?", "Why at these times?"*. The auditor will consider what happens to the resource in the process: *"Is waste heat discarded which could be used elsewhere?", "What are the sources of yield losses?"*. They will use regression analysis (described later) to understand the fixed losses, or baseload, that occur in a building or process regardless of activity, and how this can be reduced. They will identify the variability which, when managed out, can result in lower usage. Only when they have considered these aspects will they have an understanding of the underlying *demand* that needs to be met by the *supply*.

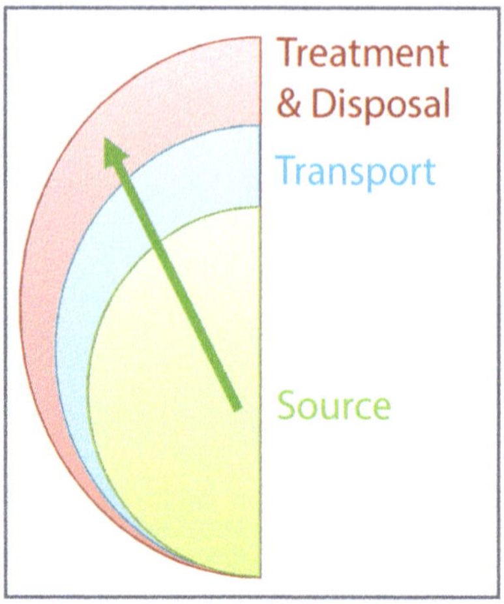

12.9 **For waste systems the focus is again on the root cause or source of the waste, working outwards following the distribution and transportation systems back to the point of treatment or disposal**
Source: Niall Enright, available in companion file pack

The next focus of attention is on the distribution systems that bring the resource to the point of use to meet the process demand. Here, we are referring to pipework for products, steam, compressed air, heating and ventilation systems, but we can also think of conveyors and forklift trucks, as well as the systems that take away the waste to the point of treatment. Distribution networks are often the source of considerable potential savings, as they may be incorrectly sized, inefficient (e.g. leaky or uninsulated), operated to the wrong setpoint (pressure, temperature, etc.) or otherwise poorly controlled (e.g. not isolated when there is no demand).

Only when we have taken into account the lowest true process demand and the avoidable distribution losses should we examine the supply-side equipment. The phrase supply side here is used in its widest meaning to include both the equipment that carries out primary conversion in order to supply a secondary utility (such as an air compressor which converts electricity to compressed air, or a boiler which converts gas to steam), as well as the plant that supplies the downstream treatment requirements demanded by our process (e.g. the conversion by our wastewater treatment plant of process wastewater to sludge and effluent, which can be safely discharged). It is often the case that the true demand when operational and distribution inefficiencies are eliminated is 10%, 20% or even 30% lower than the current level. Also, the characteristics of the supply (e.g. temperature, quality or pressure) may also be changed

*Understanding the **true demand** transforms our analysis of distribution networks and supply systems.*

significantly. For example, I have seen many compressed air systems over-pressurized to compensate for leakage, which once corrected can lead to reducing the outlet pressure and considerable savings in electricity.

Understanding the true demand fundamentally changes our analysis of the supply or treatment plant. We may have new options to shut units down, or to change the operating conditions, which will yield additional savings. If replacement of the equipment is advisable, the size of the replacement plant should be smaller and the capital cost significantly reduced by having first focused on the demand and distribution.

Insight

The investigation of larger items of resource-consuming or conversion equipment involves two tasks: establishing if there is operational variance, as described earlier, and also a comparison with design operation. Here, we need to issue a word of caution: the absence of variability does not necessarily mean that there are no problems. Consistent performance does not indicate efficient performance, so comparison with the design conditions, and also similar equipment or benchmarks (if available), is a critical second stage in establishing if current performance is optimal. ⇒ page 402.

Real World: *Design data*

Design data tells us how equipment should perform, given certain conditions, and provides insight into a range of improvement opportunities.

- Maintenance and faults: Comparison with design performance will point to maintenance issues or faults which may be quickly resolved, such as fouled heat exchanger surfaces.
- Sizing: It may well be that we can see from the design data that a piece of equipment is oversized for a particular function, so replacing it for a smaller, more efficient, unit will lead to savings.
- Control and optimization systems: The design data may also alert us to the presence of systems intended to improve the efficient operation of equipment, such as economizers or automatic blow-down in boilers, which may or may not be operational.

Unfortunately, reliance on design data for specialist equipment is one of the most significant sources of delay in an audit, and it can sometimes be difficult to anticipate what design data is desirable in advance of the visit to site. This problem can be exacerbated when the site doesn't have the data to hand and the manufacturer has to be approached for the information.

At this point, it should also be noted that few older plant can realistically be expected to operate at original design conditions, as the passage of time will have introduced inefficiencies that no amount of maintenance can have completely prevented. The job of the auditor is, therefore, to assess whether current performance can be improved, not to assume that the current operations can be made to match the original design.

Nothing is more likely to upset operators of old plant more than a blind assumption that the original design performance continues to be possible.

Real World: *Root cause analysis*

In the 1990s, colleagues and I did a lot of work on energy management in the dairy sector in the UK. I recall an audit where there was a finding that the biological oxygen demand (BOD) of the wastewater was high, leading to additional effluent disposal costs.

A *siloed* solution to this problem would be to improve the wastewater treatment to bring down the BOD, which would certainly have some economic benefits for the dairy. This is a typical end-of-pipe solution to a waste problem.

However, by going back to the source of the waste milk, the team discovered that the excessive loss of milk was mainly due to poor cleaning-in-place (CIP) with unnecessarily frequent, over-long flushes of tanks and pipework.

In fact, the BOD charges were only the tip of the iceberg in terms of losses. The poor CIP was leading to considerable further costs from the waste of product (excess milk in the wastewater); from the excess use of hot water for the cleaning (wasted water, detergents and wasted energy to heat and pump the water); and additional demand on the effluent system (yet more energy, treatment chemicals and waste).

Taking the root cause analysis a step further, the reason the CIP process was poor was mainly due to human factors (poor supervision, staff training and monitoring of the cleaning activities). These root causes could be addressed at a fraction of the cost of the technical improvement to the effluent plant to *fix* the BOD issue and delivered a host of additional savings in terms of milk, energy and chemicals.

Exploration: *Audit variability*

Building 101 was originally a marines barracks at The Navy Yard in Philadelphia, US, built in 1911. Following the departure of the Navy in 1996, the Navy Yard is now a commercial development covering 1,200 acres run by the Philadelphia Industrial Development Corporation. One of the distinguishing features of this campus is its dedication to energy innovation and sustainability.

When the Navy moved out, the building was completely refurbished in 1998, with all the major mechanical equipment being updated at that time. The building covers 61,700 square feet (5,600 m^2) over four floors (a basement and three floors, all of which are tenanted. The building is in the shape of a T with three wings, and it has offices, a canteen, workshops, mechanical spaces and a lobby/atrium located in the centre of the building.

With the support of federal funds, the Consortium for Building Energy Innovation (CBEI) was formed and set about making Building 101 one of the most instrumented buildings worldwide.[696] 500 sensing points capture more than 1,000 values a minute, including factors such as CO_2, volatile organic compounds, air flows, thermal comfort, temperatures and humidities, occupancy, lighting quality, acoustics, as well as detailed energy use information.

One of the experiments carried out by CBEI was to invite three firms of consultants to come and audit Building 101. The consultants all had access to the same data, and so the researchers at CBEI expected their findings to be broadly similar.

When the results of the three audits were compared, there was a remarkable variation in the projects that they identified.[424] In fact, only three opportunities, shown in the centre of the Venn diagram opposite, were identified by all three consultancies.

Consultants A	Consultants B	Consultants C
9 opportunities	9 opportunities	12 opportunities
US$60,200 savings	US$22,495 savings	US$34,000 savings
US$497,000 cost	US$150,000 cost	US$138,000 cost
38% cost reduction	14.5 % cost reduction	24% cost reduction
8.3 year payback	6.8 Year payback	4.1 year payback

Some of the variances could be attributable to the different assumptions that the auditors had made about the acceptable payback of projects. Company A, for example, included a *"big ticket"* item, to replace the building management system, which significantly impacted the payback period, while Company C suggested measures with a much shorter overall payback.

12.10 **The improvement measures identified by the three consultancies in the Building 101 study**

Source: Niall Enright, based on the CBEI report "Variation in Energy Audits"[424]

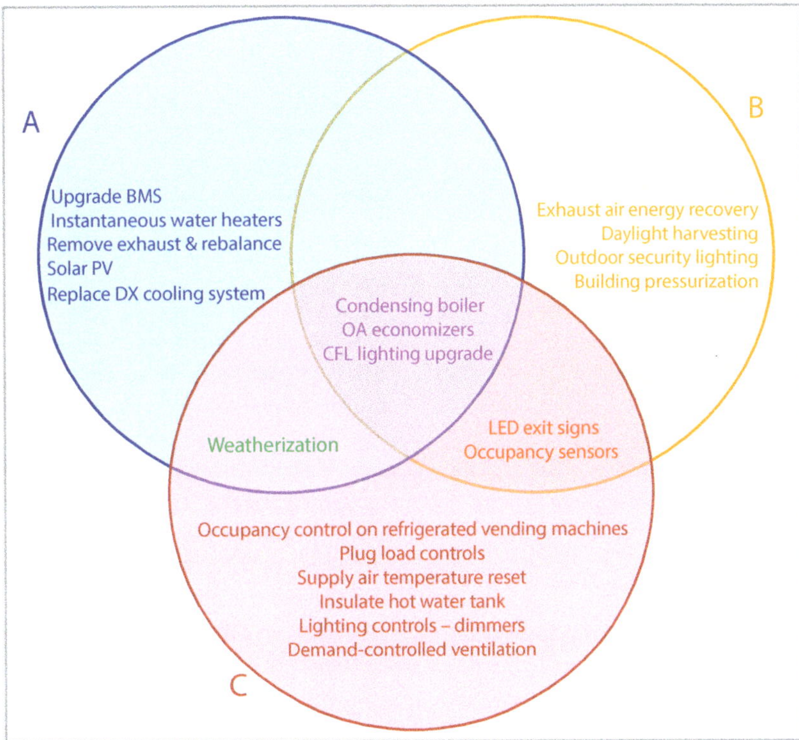

The study does not state if all the consultants received the same briefing. They did, however, all have access to the same data, although it appears that the analysis techniques used did vary somewhat. Company B, for example, used a building simulation to test a series of *"what if"* analyses to develop their recommendations.

Another striking feature of the study is that no behaviour-based improvement opportunities were identified by any of the consultants. This contrasts with the UK where techniques such as M&T are well-established methods to drive improvement through better operational control.

The CBEI concluded from this study that a much greater standardization of audit methods is required, in particular, the ASHRAE Level 2 audit involving a site walk-through and outline recommendations. There was an especially large variation in the language used by the auditors, the way they categorized and aggregated opportunities and their estimation of the implementation costs of measures.

While there is merit in greater standardization of audits, we need to recognize that energy systems in buildings are complex and highly varied, as well as the knowledge and expertise of individual consultants on which the audit process depends. Greater standardization, while welcome, is not a panacea.

Other factors may also be important. For example, more time may be needed to reveal the widest range of opportunities. Perhaps audits would benefit from a broader range of skills amongst the audit team (e.g. in technical and behavioural improvement techniques). Indeed, we may need to stop thinking of audits as one-off activities, and move to a process of continuous project identification, as suggested by the Opportunities Database tool that is part of this framework.

12.6 Gemba

In Japanese there is a word for where waste occurs: "gemba" which means "the real place". We in the west might call this "the frontline" or (no pun intended) the "coalface". The key concept in gemba is that opportunities can only be identified by going to the location where the activity is taking place.

Complementing the focus on demand is the notion of *gemba*. Understanding this concept will help make our audit process more effective. For a business, gemba is the place where the products or services are formed, the manufacturing line, the oil well, the field of crops. For a service business like a hotel, it is everywhere - it is the reception desk, the kitchen, the bar and the guest's room. For a public sector organization, it is the teaching auditorium, the laboratories, the university library, the hospital ward. It is at gemba where the needs of the customer are met, depending on which the organization will either prosper or fail, and it is at gemba that we need to focus our audit efforts. A manager's role is to be in gemba, not their office since it is only here that value can be added or destroyed.

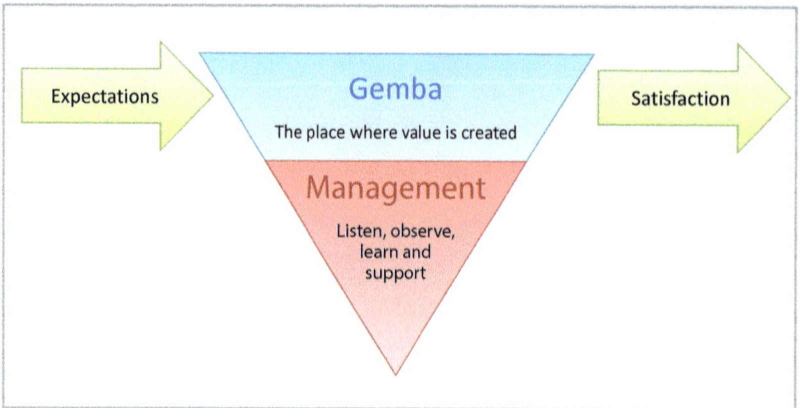

12.11 **An inverted pyramid provides a useful metaphor for the role of auditors and managers**
The aim is to help those who create value to better satisfy the expectations of stakeholders and so deliver greater value to the organization. The audit and management response are *enabling* activities.
Source: Niall Enright, inspired by "Gemba Kaisen" by Masaaki Imai[502]

Gemba is not just about being at the correct place to diagnose waste, but also about the flow of information. The identification of opportunities through data analysis can be carried out remotely. However, no conclusion can be reliably arrived at until the auditors have familiarized themselves with the process in question and have spoken with the operators or personnel at that location to get their input into the interpretation and validity of the data.

This input requires that there is an honest and open two-way flow of information between the auditors and the shop floor. If an operator believes that a particular opportunity is challenging or difficult, the audit team must genuinely listen to these reasons so that the opportunities that are put forward are firmly grounded in reality. My earlier example of the huge brewery chillers running on manual control (page 182) is a perfect illustration of the importance of gemba. If the engineering managers came to the control room at nights they would have seen the switch turned to manual control and if the shift supervisors felt better able to communicate their concerns about the automatic control system then their misgivings could have been addressed. In either case, a problem would have been resolved and waste avoided.

It is the role of the audit team to create an open flow of information by:

- Actively soliciting ideas and contributions from people using resources;
- Acknowledging and capturing all ideas, large or small;
- Sharing thinking that relates to the resource use;
- Not being critical, remaining matter of fact;
- Giving the credit for ideas and improvements to the site team.

Insight

When they are on the shop floor, the audit team need to convey this positive message to the local personnel. It is quite common that the sight of strangers walking around with clipboards and taking notes leads to anxiety about the purpose of the investigation. Often the immediate assumption of shop floor workers is that the organization is rating their work or looking to weed out poor performers. Thus wearing the right apparel (that usually means no suits or neckties on the shop floor!), projecting a friendly demeanour and explaining the objectives of the audit are essential. "*Hi, my name is Niall, I'm working with the team here to see if we can reduce your energy bills*" is the kind of introduction I will make many times a day. I will often ask the shop floor folks if there is any area that they would like us to focus on or any aspect of their energy systems that were causing them specific concern. I will ask them about the start-of-day and end-of-day processes, the kinds of management meetings they attend, what targets they have or what other initiatives are going on at their site at the moment. All of this is very useful information and helps to break the ice. If I intend to review some specific equipment or observe a process, then I will let them know that I am going to see how their area works and to please come and speak with me as I wander around if they have any questions.

It is important to know who to engage with; for example, in many food factories which employ seasonal contract labour, the most informed folks are the shift supervisors or team leaders who are full-time employees rather than hourly contractors who may have limited experience of a process.

Credit for **all improvements** must be given to the people on the frontline.

Observation is critical. On a familiarization walk-through of a manufacturing facility, an experienced auditor will *hear* the compressed air leaks. They will be looking up at the lighting and down at the drains. They will *see* if the warehouse doors are left open. They will *feel* if a room is warm or cold. They will look for steam vapour leaking out of a valve. They will observe the number and contents of waste collection bins. They will look at the condition of insulation on pipework. They will check out the contents of the noticeboards scattered over the shop floor. They will see if water hoses are left running and will feel if the water flowing out is hot or cold. They will *smell* the ammonia leak from the chiller plant. It is quite remarkable just how much a seasoned auditor can pick up from even the briefest site tour. This experience and the ability to see a facility through fresh eyes are some of the benefits of including an external expert in the audit team. ⇒ page 407.

12.12 **Peter Fink, from ERM's Frankfurt office, shown facilitating a CDR workshop**
Source: Photo courtesy of ERM

Real World: ***Cost determination and reduction workshops – the TRW success story***

An excellent example of a formal method to engage site personnel in value discovery around resource efficiency is a process developed by former colleagues of mine, Peter Fink and Martin Hess at the global sustainability consultancy, ERM. This process has a great title, *Cost Determination and Reduction* (CDR), which makes it clear right at the outset that it is all about delivering financial value. The process was refined with the input of ERM's client, TRW Automotive, under the leadership of Thomas Koening, VP Global Heath Safety and Environment (HSE).

In the CDR process, a multidisciplinary team of 6-8 people from a facility are brought together in a two to three-day workshop to drive health, safety and environmental costs down. It is important to note that these teams comprise resource users and engineering staff, but also non-technical staff from functions like HSE, finance and procurement. ERM provides the specialists to support the process, with expertise in the areas for improvement being addressed. The workshop methodology is based on the DMAIC techniques in the Six Sigma quality system:

- **D**efine (the problem to be addressed)
- **M**easure (the amount of waste or cost)
- **A**nalyse (discover the cause of waste)
- **I**mprove (remove the cause of waste)
- **C**ontrol (to make sure the causes do not recur)

The CDR workshop begins with a process of prioritization. Here, the participants are invited to look at a high level at the categories of resource used and to prioritize these based on three factors: overall cost impact, potential for improvement and ease of implementation. From this initial Define process, one or two resource streams are selected for detailed further investigation, usually by illustrating the flows and costs associated with that particular stream. The Measure phase of the workshop is much easier if the organization involved has some form of activity-based Costing in place in which resources are allocated at functional, process stage or departmental level.

Once an area for improvement has been identified and the basic data collated, the Analyse phase involves the workshop participants, possibly in one or two teams, going to the shop floor to investigate the resource use, to speak with operators and to gather more information. This Analyse phase is the point in the workshop where the specialist is particularly helpful because they can, without taking over the process, point the team at the specific areas to investigate and data to collect.

The team then return to their classroom and "*work the problem*", refining their understanding and defining one or more discrete opportunities that will Improve the resource use, as well as the Control strategies to ensure that this improvement is sustained in the long-term.

Where the opportunity is no or low-cost and can be implemented rapidly, the necessary change is made there and then. If it requires more detailed investigation or development or higher level approval, then it is handed over to the appropriate team, such as engineering or operations, to execute.

If this process sounds simple and straightforward, that is because it is just common sense. The rationale behind CDR is that you do not need to be an expert to find improvement; you just need to look at your resource use systematically. The presence of the non-technical team members is important because they often have no preconceptions about a particular production activity and so can ask the *"dumb questions"* that can uncover unexpected solutions. The role of the specialist in the workshop is to help guide the team towards the areas of greatest opportunity, in part by bringing best practice from other sites to the workshop and in part because of their training in relevant technologies. However, the specialist should not dominate the process or prescribe the outcome – they are there to offer expert support when the workshop participants are unclear about a particular issue or hit a brick wall in interpreting data. The specialist will provide ongoing assurance to the team that their methods and conclusions are correct and avoid the process going down a blind alley.

In most CDR engagements, ERM provides the external specialist consultant to support the workshops and bring expert knowledge of the client's target resource (e.g. energy, waste or water) and industry (e.g. manufacturing, automotive, food, etc.). Also, ERM often provide a second consultant facilitator who simply manages the overall process, ensures that time-keeping is maintained, introduces the tools used such as prioritization grids, brainstorming, flow-charting, and records the ideas generated. However, either of these roles can be provided by the host organization and the participation of an external consultancy is not a requirement of CDR, as demonstrated by the manufacturer TRW Automotive. In 2002, TRW worked closely with ERM to incorporate the CDR methodology as the central resource efficiency process within the TRW HSE management system.

TRW Automotive is a leading automotive components manufacturer headquartered in the US, employing 60,000 people in 190 facilities in 26 countries, with sales in 2011 of US$16.2 billion. It is now part of ZF Friedrichshafen. Like many businesses in the automotive sector, TRW has a long track record of applying formal quality systems such as Operational Excellence and Six Sigma in its production activities, to improve quality and reduce cost. What attracted TRW to the ERM CDR process was the combination of a known and proven technique, DMAIC, with an emphasis on a new category of costs – health, safety and the environment. The HS&E costs that could be *"determined and reduced"* using the ERM approach included process environmental resource costs such as energy, emissions, noise, water and waste, and HS&E system costs associated with managing incidents, dealing with HS&E related absences, providing training. etc. In all, TRW identified 13 categories of cost that would be managed via the CDR programme, shown in the purple boxes in the illustration, left.

12.13 **The ERM CDR process, as adapted by TRW Automotive, focuses on 13 categories of costs**
Source: Reproduced from TRW's 2010 Sustainability Report[717]

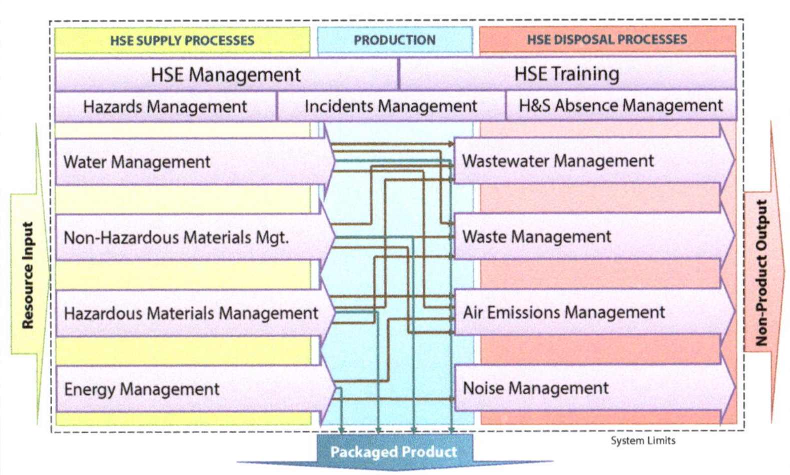

Each year, every large TRW facility was expected to undertake at least one CDR workshop focusing on two of the 13 cost categories. By 2011, over 300 workshops had been completed, with over US$23 million validated annual savings achieved through 2,100 measures. The mix of measures, in terms of numbers, has been dominated

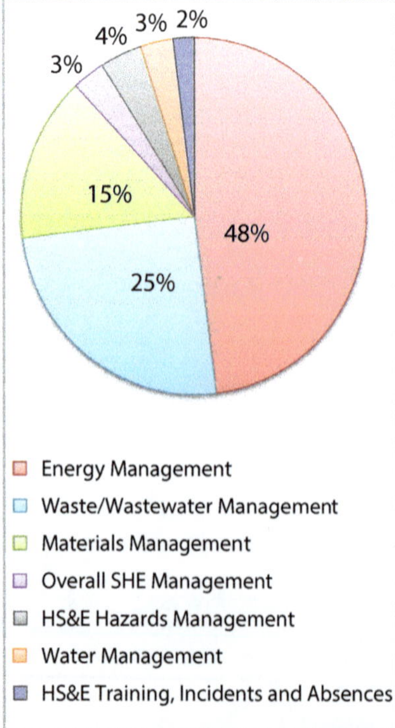

12.14 **The TRW CDR opportunities mix by process element, 2006-2011**
Source: Reproduced from TRW Automotive 2011 Annual Report for Health, Safety, Environment and Sustainability[718]

by process improvements, as shown in the chart left, with over 90% focused on resource efficiency, waste reduction, and energy conservation.

Although public reporting on the CDR programme ceased following the acquisition of the TWR business in 2015, it is clear that this was a very successful resource-efficiency initiative which was sustained for well over a decade. The last three years on public record, 2009 to 2011, produced the largest ever number of measures and verified savings.[718] However, private correspondence with Thomas Koening indicates that, by December 2016, the savings had risen to US$ 57.9 million, which also brought other HSE benefits: less waste, less material, less risk, less CO_2, etc.

That is not to say that CDR doesn't have some limitations, which are recognized by ERM and TRW. For example, the number of opportunities that can be identified is limited by the time available for the investigations and the level of expert knowledge in the team conducting the workshop. The workshops are episodic, and with an average of seven measures identified, the savings made represents just a small number of the total opportunities that would be available in a complex production facility. This is confirmed by the fact that subsequent CDR workshops undertaken at the same facility in TRW did discover further savings opportunities.

Addressing some of these weaknesses, in 2010 TRW initiated its *Energy Efficiency Programme*, again lead by Thomas Koening. The programme continued to use CDR as a key opportunity identification tool, but introduced two other opportunity identification processes to provide additional expert focus on energy at a facility level and capture some of the potential savings missed by CDR:

- Energy efficiency workshops, where employees from several facilities receive in-depth technical training from external experts and go on to identify a further raft of opportunities; and
- Detailed energy audits by third parties to find more complex opportunities.

Alongside the additional opportunity identification activities, TRW introduced more extensive sub-metering of energy as well as programmes aimed at specific technologies, such as motors or compressed air. Management systems to reduce energy included targets, benchmark data and the incorporation of energy programme performance indicators into key business measures. The TWR energy programme demonstrated considerable depth and breadth, encompassing supply as well as demand-side initiatives, all based on solid measurement. In effect, it represented an ideal mix of *People, Systems and Technology*.

CDR shows that resource efficiency can be a process that everyone in an organization can contribute towards, particularly if there is an appropriate structure to work within. It is important to note that while TRW had extensive experience in Six Sigma techniques, this was not essential for the success of CDR, which could stand completely alone as a process. According to TRW:

> *"CDR works because it draws on the knowledge and experience of employees at every level of the organization to reveal areas that work well and those that need improvement."*

While CDR alone will not uncover all opportunities for improvement, as a means of engaging site personnel in a process where they have full ownership of the savings, it is highly effective. With the external expertise and facilitation from a team such as ERM, this method is highly recommended for both new efficiency initiatives or as a means of reinvigorating an existing programme.

12.7 Consolidation

The words "consolidate" has two meanings: on the one hand it means to "bring together" and on the other to "strengthen". At this stage of an audit we need to take a step back from individual opportunities for improvement and look at the big picture.

Insight

There is a critical stage in an audit when the audit team take a step back and consider the results of their analysis. Usually, audits consider one resource at a time (e.g. water, electricity, gas, raw materials, waste) and then drill down into specific instances of use (items of equipment, or processes). Various specialist techniques of measurement, data analysis, benchmarking and financial appraisal, relevant to the technology in hand, then throw up opportunities for improvement in that particular resource.

The training of auditors, oriented as it is to different systems such as HVAC, lighting, water and wastewater, reinforces the tendency to consider opportunities discretely. Resource-consuming processes and systems do not, however, operate in isolated silos in the real world. Resource users are usually highly interconnected.

Very little is written about this critical stage of an audit, so I have set out in the illustration 12.15, on the next page, a three-stage process to consolidate the findings into an actionable proposition. In practice, these three stages may not be sequential and may merge as insights gained through one step causes others to be revisited.

In the consolidation process, one task is to understand the interactions between the projects and the performance of the system as a whole. There are quite a few useful tools that can support this task. Root cause analysis can draw the auditor towards the centre of the *"onion"*, creating a focus on demand for resources of the source of wastes. Integrative design involves a multidisciplinary team working holistically to develop a solution. Portfolio appraisal provides the tools to quantify resource and financial savings from overlapping projects.

There is a very important stage in an audit when the audit team take a step back and consider the results of their analysis

Perhaps the most important interactions are how the recommendations from the audit relate to the goals of the organization and the aspirations of individuals. Unless there is a clear alignment of the project recommendations with these drivers, the impetus for action will be poor. Here, the auditors will investigate the strategic objectives of the site and articulate how the proposed activities will support those objectives. Non-financial benefits such as recognition, stakeholder engagement and the desire to do good should be understood. Anything that can be done at this stage to ensure the site personnel take ownership and are emotionally engaged with the proposed

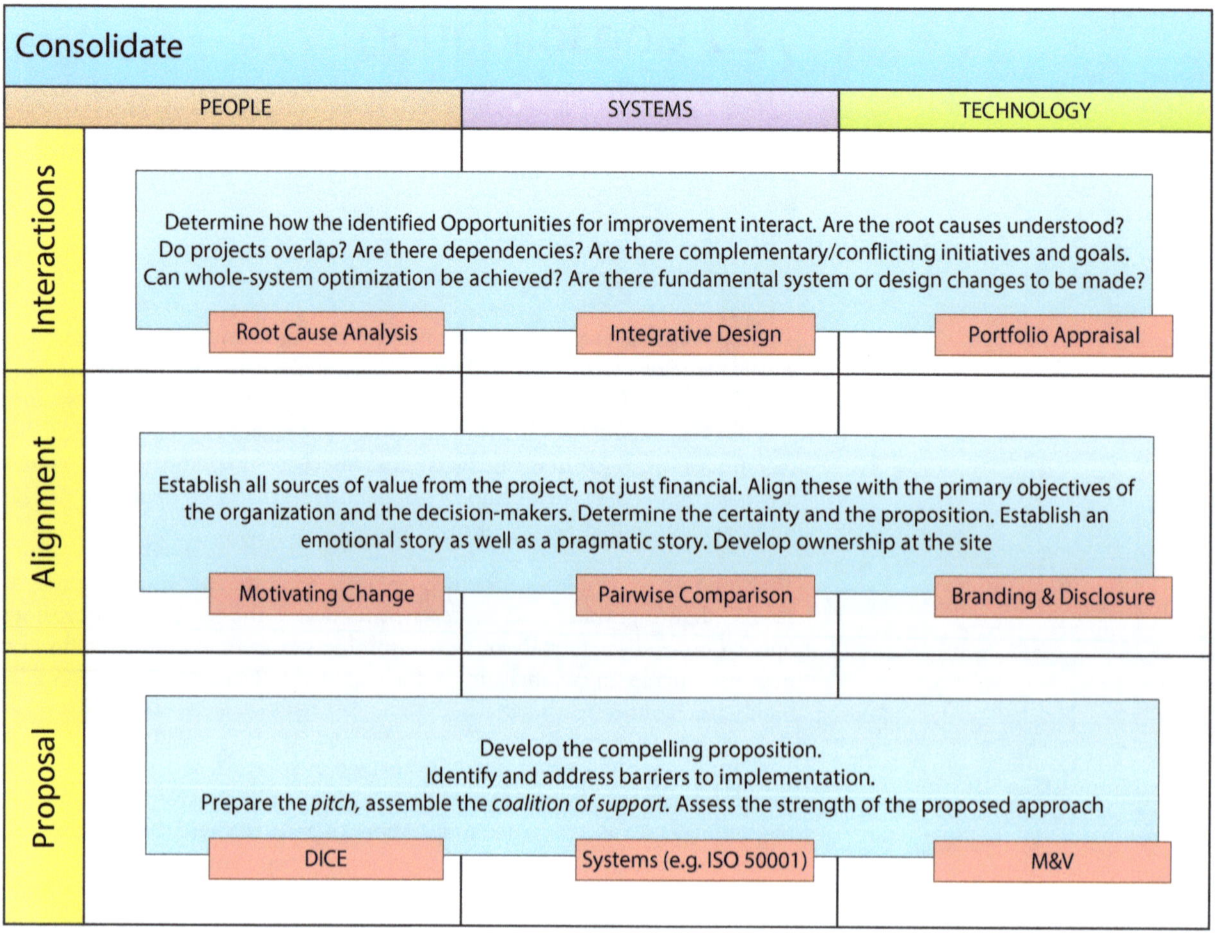

12.15 **An audit report should be more than a list of projects and their financial return. To identify the full potential for improvement and create the most compelling call to action we need to take a step back.**
Source: Niall Enright

programme of work is to be encouraged, and a number of motivation techniques are described later, in Chapter 19 on People. Ideally, the audit will be seen as a joint product of the site team and the external audits, and the site personnel become strong advocates for the recommendations that emerge from the audit. Techniques here include pairwise comparison which helps to understand whether the folks at the site share a common improvement goal.

Although not part of a traditional audit, in these interaction and alignment activities we might want to consider more profound system and business model changes that could be driven by efficiency considerations. These potential changes could arise from branding and disclosure considerations, or from a deeper strategic analysis, such as that described in Chapter 10.

In the final proposal step, the audit team will refine the recommendations to ensure that they are clear and compelling, as well as practical and actionable. Here, it is important to understand what other initiatives the organization has in place which could either support or hinder the resource efficiency opportunities.

Timing is also important. If the future of the facility is in question then long-term opportunities are unlikely to gain support. Where there are plans for process shut-downs, then these may offer a unique window when changes, such as installing metering, which otherwise would not be possible because of the process interruption, can be made. Planned capital investments, operational and major process changes should be examined so that incremental or marginal investment opportunities (see page 567) can be considered.

If the focus is on longer-term change, then the team may choose to use a methodology like DICE (see *Rolling the DICE to predict the outcome* on page 209) to provide an objective rating of the probability of success of the proposed programme and adjust accordingly.

It is important to note that in the consolidation process we will have taken fully into account the cost-benefits of the individual recommendation using our portfolio appraisal tools. But the cost will have just been one of the considerations: practicality, ownership and alignment with other goals and projects should have been addressed. Thus, the next audit step, cost benefit analysis is about putting together the cost of the *whole programme of work* set out in the proposal. If there are a number of changes that will rely on human factors, such as behaviour change, then the cost of the training can be spread. In this way, we avoid presenting a list of individually-priced projects (which can be *cherry-picked*), but rather put together a comprehensively costed programme of work where opportunities reinforce each other and more can be achieved for the investment proposed.

12.16 **An ironic take on the energy audit**
Source: Reproduced with permission from a series of articles entitled "How to Waste Energy" by Vilnis Vesma. See www.vesma.com.

Real World: ***How to waste energy - No. 9: audits***

For the keen energy waster faced with demands to have an energy audit, it is vital to employ an incompetent assessor - one who can be expected to follow these principles:

1. Just turn up at site with a clipboard and start counting light fittings.
2. Never analyse historical data to identify anomalies that you could productively focus on during site visits.
3. Base your report on a previous one for a different client. A good trick is to use 'find and replace' to change the name in the body of the text, but overlook where it appears in headers and footers.
4. Always make at least ten recommendations, even if there is only one substantial worthwhile measure.
5. Always include recommendations for LED lighting and voltage reduction.
6. Over-estimate the savings expected from each recommendation.
7. Ignore any possibility of interactions between recommended measures.
8. Never obtain actual installation costs. Reverse-engineer them: take the annual savings and multiply by an assumed payback period.

Of course as a client, the keen energy waster has their own part to play in making the audit a futile exercise. Here are some tips:

1. Do not let anybody in the organization know about the audit visit.
2. Render all relevant data and drawings inaccessible.
3. When you receive the report, ignore it.

Standards in Chapter 12: ***Energy audits***

Energy auditing is a mature field. As a consequence, there are several well-established standards and certifications available to guide those wishing to undertake audits.

There are British and European Standards for energy audits. BS EN 16247-1:2012 sets out the general requirements for an audit. Additional standards in the series provide further requirements for buildings, BS EN 16247-2, in processes, BS EN 16247-3 and in transport, BS EN 16247-4.

The International Standard ISO 50002:2014, is based on these standards and so is almost identical in content but goes a step further in providing a flow chart of the audit process, shown below.

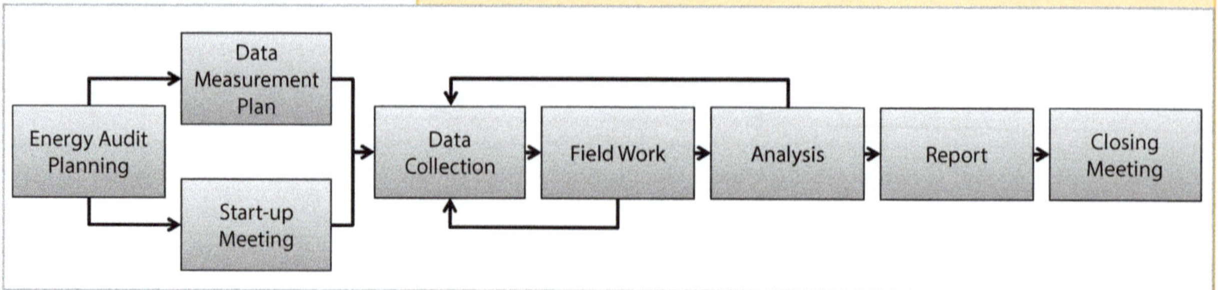

An accompanying document in this series describes a methodology for energy efficiency benchmarking: BS EN 16231:2012.

These standards, as currently written, provide very little of value to anyone other than a complete novice auditor as they only describe the audit process in the most general terms. This poor value is exacerbated by the high-cost associated with purchasing these documents.

Although not standards per se, the American Society of Heating, Refrigerating and Air-Conditioning Engineers (ASHRAE) describes three levels of investigation which are widely used:

ASHRAE Level 1 – An initial one day walk-through analysis/preliminary audit, focusing on low-cost operational savings.

ASHRAE Level 2 - A standard audit, which builds on the Level 1 audit but has a more detailed analysis of equipment and operations, to provide a set of recommendations and a financial analysis of the savings potential.

ASHRAE Level 3 - Detailed analysis of capital-intensive modification. Here, variables are used to assess the year-round energy performance of the building. In many cases, additional measurements are taken, and the economic analysis will be more detailed than in Level 2.

A further category of audit is what is sometimes referred to as an investment grade audit, which goes beyond the Level 3 audit, and is described by the Association of Energy Engineers as an audit that includes weighted financial risk in the economic calculations. This type of audit often includes further analysis techniques such as computer simulation and whole life costing, and can be used to obtain funding for the projects identified.[708]

cea CERTIFIED ENERGY AUDITOR ®

12.17 **Not only may we wish to ensure that our audit meets a recognized standard, but we may also want to ensure that the auditor is competent to do the work.** *The Association of Energy Engineers can certify individuals as being competent to carry out energy audits. Source: logo © Association of Energy Engineers. See* www.aecenter.org.

Summary:

1. Avoid the word *"audit"* if it has negative associations.
2. An audit is a highly collaborative process which usually benefits from objective external expertise as well as local knowledge and commitment. The absence of any of these will diminish the outcome.
3. A good auditor (internal or external) will use their ears and mouth in the appropriate ratio (2:1). Sometimes inefficient behaviours have good reasons, sometimes they are due to an entrenched false logic and sometimes because the system is flawed. Only by listening can the root cause be properly understood and the correct solution proposed.
4. All ideas for improvement should be credited to the site team. Even if they did not conceive the idea, they will need to implement it and so they need to own the solution. Auditors need to leave their ego behind.
5. The desired outcome of an audit is change. It is not a report. One of the best ways to demonstrate value and the possibility of change is for the audit team to deliver improvements during the audit. In this way, the audit has initiated the improvement process.
6. All audits, without fail, should start with the demand for the resource in the first place (or the source of the waste, if that is what is being examined). Only when the potential to reduce the demand has been fully examined can the distribution and conversion processes be properly analysed.
7. No analysis of resource use is complete unless the influence of external factors (weather, production, activity, etc.) has been undertaken and the impact determined. Regression analysis is the key technique for this, as it can separate the fixed load from the variable load and provide profound insights into the usage.
8. An audit recommendation should not be a list of projects. Interactions between opportunities need to be understood, the practicality assessed and the recommendations need to be placed into a wider context of what is going on in the facility or organization.
9. There are only three sources of improvement: People, Systems and Technology. An audit cannot be considered comprehensive unless it has considered all three, no matter how extensive the technology review.
10. The lowest-cost, most rapid and most profound improvements usually come from changes in the People or Systems element, not Technology.
11. Every audit should go to *gemba*, the heart of the organization's service or activity.
12. Seek to understand and address the root cause of inefficiency. Thus the reason a chiller is inefficient may be because the financial approval systems in the organization are biased against equipment upgrades compared to other business investments.
13. Audit recommendations need to actionable. The only criterion for the effectiveness of the audit is if it acted upon.
14. The emphasis of the audit should be the rationale for the treatment proposed rather than the diagnosis. To provide a diagnosis alone is unforgivable - if further investigations are needed these need to be described and rationalized.
15. The extent of the audit needs to be described fully. If resources, facilities or processes have not been investigated there will be an assumption that these are operating efficiently unless they are explicitly ruled out of the scope.
16. Remember that audits should be fun, creative and collaborative.

Further Reading:

1. *Handbook of Energy Audits,*[708] Thumann, Niehus and Younger. This book is organized primarily by technology and its focus is almost entirely on equipment and fabric. It is a very useful reference but the reader should be aware that it uses US Imperial units rather than SI units of measure.

2. *Energy Audits Undone, How to find 101 simple ways to slash your utility bills,* Zaitsau, Viktar.[821] Again there is a focus on equipment, but here the author is providing a broad overview of each technology and the major areas where savings can be found rather than giving calculation methodologies, so this title would suit the less-experienced auditor. There are some helpful comments on the importance of trust, the mindset of the auditor and the need to focus on demand: "*Replacing the boilers in the energy centre, without doing any works in the buildings themselves, is a sin.*"

3. *Energy Audits - A Workbook for Energy Management in Buildings*[17] by Tarik Al-Shemmer, provides a good foundation in the calculations to quantify energy use and identify improvement opportunities (it uses metric/SI measures).

Questions:

1. What are the characteristics that differentiate a good audit from a poor one and why? What are the qualities of a good auditor? Explain.

2. How can the true demand for resources be understood and why is this important?

3. What human factors influence resource use and how can these be assessed by an audit team?

4. Review an existing energy or resource audit report. Describe the criteria that you will use and why? Evaluate the report against these criteria and suggest how/if you would make any changes. Were the recommendations in the report acted upon and if so why, or why not?

Resources:

The Canadian Industry Programme for Energy Conservation has a useful free *Energy Auditing Manual and Toolkit.*[110] Although the Australian Energy Efficiency Opportunities Programme was repealed in 2014, the free *Energy Efficiency Opportunities Assessment Handbook*[227] has a wealth of practical techniques for audits and opportunity identification, with a focus on industrial and natural resources businesses.

In the UK, the CIBSE *TM22 Energy Assessment and Reporting Methodology*[147] provides a benchmark-driven methodology to audit a wide range of buildings. It includes a spreadsheet for the analysis of performance. There is a modest charge for this.

Another paid-for resource is ASHRAE's *Procedures For Commercial Building Energy Audits*[38] which provides very useful guidance on best practices. Again there is a charge for this, but it is modest given the coverage.

13 Meters

"Messen ist Wissen"
(Measurement is knowledge)

Georg Simon Ohm (1789-1854)

Photo: © Alex Yeung. Fotolia.com

This chapter provides guidance on obtaining data for the Method phase of our resource efficiency programme, which concentrates on Optimising existing operations. Other activities in our programme, such as strategy development or audits may also require meter data, but in these cases, it may be more appropriate to use temporary metering devices than to install permanent meters.

In this chapter, we will look at the mechanics of what to meter and how often, different types of data, substitutes for meters and "virtual metering", designing a metering network and standards related to metering. We will also cover topics such as how to estimate the budget available for metering and how to work out the resource use needed to justify any given meter.

Before you start make sure you ***understand the purpose of the data.***

Metering may be installed for several reasons, such as for billing or as feedback for a control system. This chapter presumes the purpose of the metering is to provide data to support a resource efficiency programme. This chapter should be read in conjunction with the next chapter on analysing meter and variable data. By understanding what we are going to do with the data we get, we can make better-informed decisions about what and how to meter resource or activities.

13.1 Choosing what to measure

The aim of our Optimize activities is to provide individuals who are taking decisions about resource use with the information that they need in order to make informed decisions. There are essentially two types of data that we will require: resource use data (often simply called meter data) and activity data (called variable data).

The Method phase of our resource efficiency programme is based on the notion that giving people appropriate performance information will enable them to make better decisions in their day-to-day activities and so Optimize the existing operation.

Data is the raw ingredient for this process. However, excessive data can overwhelm people. Raw data without context can be meaningless and data arriving too late can be useless.

In fact, what we need our data to do is to provide information (see page 278). This information is data that is organized and processed for a particular purpose, in our case to inform us of the performance of resource-using activities.

The most common problems one finds with metering systems is that they have been developed for purposes other than resource efficiency in mind.

1. For accounting, billing or cost allocation. These tend to divide a facility into physical units (e.g. buildings) rather than functional units (processes).
2. For real-time process control. These tend to focus on a few high-consuming activities with much a higher frequency of measurement than is typically used in Monitoring and Targeting (M&T).
3. For power quality and electrical distribution purposes. These tend to be over-specified meters in many cases located at power distribution points, rather that at the point of consumption.

Although the original purpose of many meter systems may not have been to support resource efficiency, keeping in mind a few basic principles will help us to adapt these systems to our needs.

The first thing to understand is that, in order to assess performance, we need to compare our resource use with something. In effect, we will usually be working with two values - first a meter value, which is the quantity of resource used, and second a variable value, which is a quantification of an activity which we expect to influence resource use. In fact, variables are often measured using meters - these are simply the naming conventions that we use in M&T. Examples of these paired measurements are shown in the table opposite.

Real World: ***Making use visible***

Let us consider two expensive resources that we almost all use: electricity and petrol ("gas" in the US).

When we drive our car, we have a constant indication, the fuel gauge, of how much petrol we have used. If the tank runs low, a warning light comes on. When we fill up the tank, we get an immediate signal of the cost involved. We all know that to use less fuel, we need to reduce our journeys and moderate our speed,

Compare this to electricity. Here, we have no equivalent to a fuel gauge and we often have little awareness of how much power different items of equipment in our homes use. We are connected to an apparently infinite supply of electricity, so we have no warning light. Furthermore, when the bill arrives it is often for usage weeks, if not months, ago, and has a very complex rate structure which makes understanding the consumption even more difficult.

The resource efficiency Method and techniques are designed to provide visibility around the resources we need to manage. We want people to have information about:

- how much they use
- what affects usage
- what it costs

With this information and support, people will reduce usage.

13.1 **Examples of meter/variable combinations**
Source: Niall Enright

Meter	Variable
Lighting Electricity	Hours of Darkness
Process Steam	Total Production
Waste to Landfill	Total Production
Water Use	m^2 Irrigated
Diesel Used	km Travelled
Chiller Electricity	Temperature
Conveyor Motors Electricity	Tons of Grain Moved
Paint Sprayed	Cars Produced

The statistical techniques available comparing a meter's value and a variable's value are covered in detail in the next chapter.

However, it is important here to reiterate the comparative nature of our analysis. We are seeking to understand if we use more or less resource *given a particular level of activity*. In the table above we have listed some uses of electricity for activities such as lighting, chilling and conveyors. Each of these uses has a different "activity" variable for comparison.

This brings us to the first rule of metering for resource efficiency. We want our metering, insofar as is practical, to separate out our resource use by function. Thus we want to be able to measure the electricity to our chillers (whose usage is governed by how hot it is) separately from the electricity used by our lighting systems (whose usage is governed by the number of daylight hours).

In fact, as it tends to get lighter in the summer, our lighting electricity use will tend to decrease, while it also tends to get hotter (in the northern hemisphere), so our cooling demand is likely to go up. If we just measured the total electricity use, it is possible that these effects would counteract each other and so we would apparently see little variation in use. By separating out the functional use we will (hopefully) see much clearer seasonal effects and be able to establish a performance indicator *for each function* using the related variable.

We need to **separate use according to function.** For example, we want to separate electricity used for cooling, lighting and production.

Do bear in mind that the same practical applies to metering for the variable. If we want to set performance targets for different units of productions (e.g. different assembly lines), then we will need to have the measurement of output down to each assembly line. The meter and variable pairings should refer to the same scope of activity (you wouldn't assess your own car's petrol use by reference to someone else's vehicles' mileage).

It is this functional separation that is the most common challenge when faced with metering systems designed for billing or other purposes. For example, it is common to find electrical sub-metering in an office building to separate the use by floor, but nothing to separate use of chillers, lighting or lifts. Metering in substations or distribution panels tends to be at a circuit level, not necessarily at a functional level.

13.2 The meter hierarchy

We tend to think of a meter as a grey box with dials on. In fact there are many alternative sources of data for to use, which can help reduce the cost of data collection in our resource efficiency programme.

Direct metering, the measurement of the value we require through a dedicated instrument permanently in place, is clearly the preferred source of data. Assuming that the meter is accurate, we can rely on it to produce a regular set of data for our resource management. However, there are many circumstances where it would be too costly, impractical or otherwise impossible to install a dedicated metering device. In these cases, we have some alternative strategies available to us.

For devices which have an "on-off" operating state (i.e. where the load or resource use is constant), we can use hours run as an alternative to direct measurement of the energy used. Examples of such systems would be motors (without variable-speed drives), conveyor motors, HVAC system fans and many

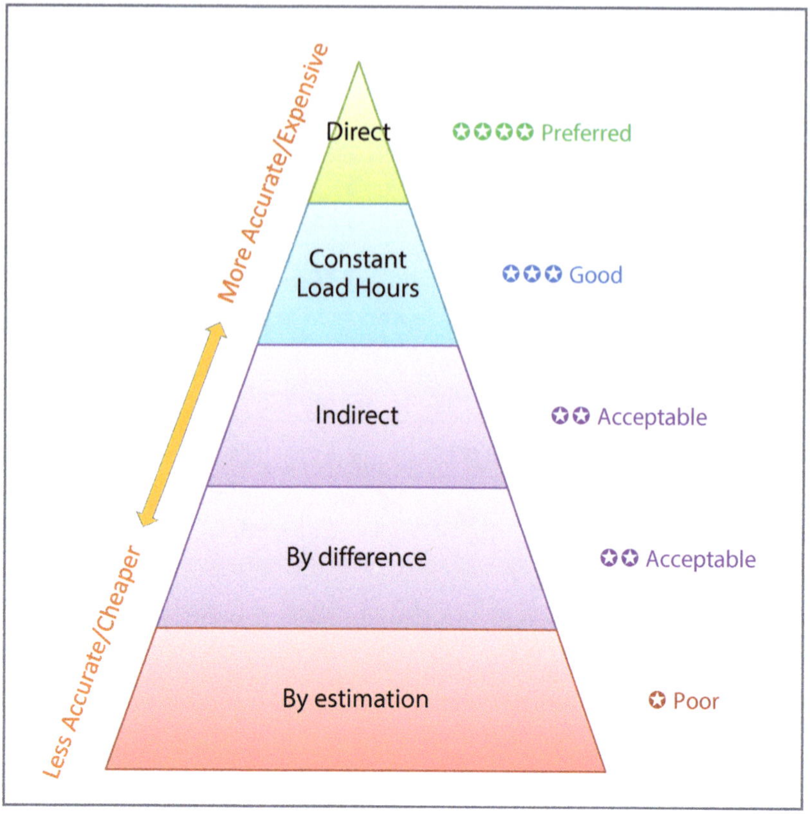

13.2 **The meter hierarchy**
Source: Niall Enright based on Carbon Trust[179] and others. Image available in the companion file pack.

A meter doesn't have to be a meter. There are many alternative methods to calculate a value.

lighting systems. Once the load in kW is understood (ideally by measurement rather than by reference to the "nameplate" rating), the use in kWh simply is that kW demand multiplied by the hours of operation, which is usually available from a building management system or process control system.

Another widely used technique is indirect metering, where a resource use is calculated from one or more data. A typical example is heat metering, which will have a flow meter as a primary measurement device which, combined with a flow and return temperature, enables the heat supplied to be calculated to an acceptable accuracy. Similar calculations are used to measure coolth delivered by cooling systems.

The next metering method, similar to indirect measurement, is by difference. Here, the resource use is calculated by reference to a network of meters from which the balance of usage can be determined. This metering technique is widely used in M&T, where many sub-meters may be desired to separate functional use of resources at a very low level. These meters are often referred to as virtual meters.

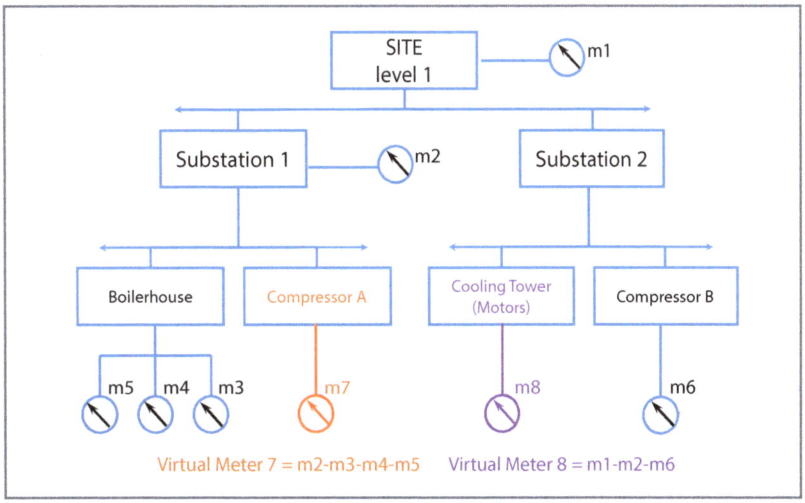

13.3 **Virtual meters are meters that are calculated "by difference"**
In the example here, we can calculate two meters, compressor A (m7) and cooling tower (m8), using consumption data from other meters in our meter network. Note that this technique should not be used where a small consumption value is being determined from much larger metered values, as the variation in accuracy of the larger meter may exceed the value being calculated, due to the meter resolution (discussed later).
Source: Niall Enright.
Image available in the companion file pack.

Our last proxy for meter data is by estimation. Estimation is the least accurate method and should be avoided if possible. A common example of metering by estimation is waste collected from a site; many waste collection companies only measure the number of bins emptied or even just the number of collections. Even if they weigh the waste, the actual waste content is likely to be estimated by reference to standard refuse types.

Generally speaking, if a resource is valuable enough to manage, we can justify some form of measurement that is superior to estimation. The one time that I regularly use estimation is to eliminate small power loads, for example, in a cooled data suite, I may want to subtract the fairly *small, constant* power load of the ancillary systems (e.g. lights) from the load to the local chiller units.

13.3 Frequency

The subject of data collection frequency or monitoring periods is often contested. Clearly, modern technology and cheap storage makes it possible to collect data more often, but over-collection may undermine the objective of providing simple, effective and timely information.

Few topics around resource metering are as hotly debated as that of metering frequency. On the one hand, we have the *real-time* aficionados, usually supported by metering and data systems vendors, who want a larger investment. On the other hand, we have conservatives, like myself, who argue for collecting fewer data and investing more in the processes of understanding the data.

Just to be clear, as we shall see from the next chapter, time-series data is helpful. Here, I would recommend collecting data at a frequency up to the usual billing interval data for *some* resource (e.g. this is half-hourly in the UK for electricity, 15-minutely in many other countries). This is because this data can be useful for checking tariffs as well as for understanding consumption. However, the M&T process itself is a management process, where we want the performance data to be integrated into the normal management cycles of the business, so the frequency of our M&T performance indicator is, at most, likely to be by shift, if not daily or weekly. If we don't manage money (or production or patients treated) in our organization minute-by-minute, then it doesn't make sense to do so for resources.

Also, before we rush out and invest in metering to collect high-frequency data, we need to recall that we will generally be analysing data in pairs, a meter value for our resource use and a variable value for the measure of activity or the external driver. In my experience, it is usually the frequency of the variable data that determines the overall monitoring period or frequency, as this production or activity data is often only available at a daily or weekly frequency.

There are some very practical considerations around the subject of monitoring periods or data collection frequency. If we look at the illustration below, collecting values for one meter every minute means that we will have over half a million records in a year, more than five hundred times as much if we collect the data three times a day at each change of shift.

13.4 **Frequency and data volumes**
More frequent data collection means a greater volume of readings to store and analyse.
Source: Niall Enright

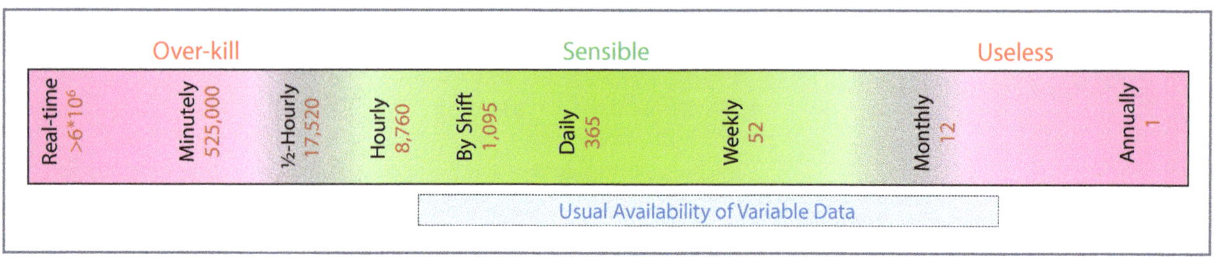

Collect data at an appropriate frequency. *This should align with variable data and management reporting.*

Not only does a lot of data cost more to store and analyse, but it usually requires much more expensive data collection systems to gather in the first place. It is not unreasonable to ask a security guard to read a water meter at the start of a night shift, but completely impractical to ask them to do so every half-hour.

The other problem with high-frequency data is that it changes the emphasis of our analysis. If you are overwhelmed with too much detail you *"can't see the forest for the trees"*, as they say, and you will run the risk of focusing on time-series trends, rather than investigating relationships between resource use and operational decisions.

Insight

Real World: *Manual meter readings*

Although costs associated with data collection using network, radio or GSM (cellphone) technologies have dropped considerably, and many meters now provide "smart" data-logging capabilities, almost every M&T system will have some manually read meters. This might be because the meters are difficult to connect, they are old and do not have a pulsed output, or simply because of cost.

Indeed, it may well be the case that, at the start of our programme, many meters that are capable of being automated are not connected. In these circumstances, I urge folks to get on with the M&T process using manual readings, to demonstrate the effectiveness of the approach and not to delay savings. Once the value of the data is appreciated, then automation of data collection can be readily justified.

When undertaking manual meter readings, I would recommend that the data is collected electronically if at all possible. This could involve entering the data into a spreadsheet on a tablet or using a dedicated hand-held device (such as a Zebra Workabout). The aim is:

1. To allow the data entry to follow a set route, which will help make sure meters aren't missed.
2. To enable the reading to be automatically time- as well as date- stamped so that if they are late/early the value can be adjusted (normalized) if appropriate.
3. For the value to be validated at the point of entry. Errors to be checked for are range errors (value higher or lower than expected), digit errors (too few or too many digits) and missing errors (meter missed). The intention is that the meter reader corrects errors in front of the meter.
4. To reduce the effort, and the potential for error, in re-entering the data into the M&T analysis tool.

A significant additional benefit of using one of these devices is that it elevates the status of the meter-reading task and so can be positively perceived by the people who take the readings.

One piece of advice that I give regardless of whether the readings are captured on a clipboard or an electronic device is to tell the meter-reader to enter every digit that they see, regardless of red or black numbers and regardless of decimal points. The M&T tool can easily apply a meter multiplication factor to adjust the value entered, but this simple instruction makes it much easier for someone unfamiliar with the meters to collect the data, should that be required.

13.5 **Hand-held devices are recommended for manual meter readings**
Some devices even incorporate a barcode scanner to enable the meter to be quickly selected.
Image: © Bacho Foto, Fotolia

13.4 Meter data types

Meter data comes in various forms, which may need to be manipulated in different ways in order to get a useful value.

Correctly treat the different meter data types: *incrementing, consumption and absolute.*

The meter type that we may be most familiar with at home is an incrementing meter. This is one whose value rises steadily, for which we need to subtract the previous reading from the current reading to get a value (unless, of course, the meter goes "around the clock", in which case calculating the value is more difficult). Examples of these meters are standard domestic electricity and gas meters.

There is another type of meter that is widely available, which is a consumption meter. This is a meter that gives the use for a specified period of time. Therefore all we need to do to calculate the use for a period is to add up the interval consumptions. When we get half-hourly demand data from an electricity supplier, this is usually in the form of the kWh used in each half-hour period.

The final type of meter is called an absolute meter. An example of an absolute meter is a thermometer, which provides an instantaneous value. The data from absolute meters cannot be added up in the same way as incrementing or consumption meters.

In a resource efficiency programme, we are almost certain to be using a mix of all these meter data types. They each have advantages and disadvantages. In particular, when it comes to the two basic manipulations of resource data, normalization and estimation, each of these data types needs to be handled differently (see box opposite).

By far the most challenging data to work with is absolute data. The most common absolute value that we work with is temperature, and there is a special series of calculations needed to turn this data into a value, called a degree day that is suitable for M&T. This is covered on page 476.

13.6 **There are three basic data types** Each of these has advantages and disadvantages. *Source: Niall Enright.*

	Incrementing	Consumption	Absolute
Application	Common for manually read meters	Common for automatically read meters	Used for sensors
Description	Gives a "reading" which increases each time	Gives consumption over a fixed interval (or since last read)	Gives an "instantaneous" value
Advantage	Easier and more accurate to deal with missing readings	Data is simpler to manipulate	Suitable for its purpose. Can increase accuracy with # of reads
Disadvantage	Meter *"lap"* conditions when the meter goes *"around the clock"*	Less accurate to estimate missing reading values	Value must never be normalized (averages OK)

Read the meter at the start of every monitoring period.

Insight

In Numbers: *Normalization and estimation*

Two data transformations are commonly used in M&T.

The first is normalization. This is the process of adjusting a measured value to fit into a given time frame.

Usually, when we establish an M&T system, one of the first things we will do is select the frequency of our system; for example, we may define this as daily, weekly or monthly. The system will then report performance at this frequency. The M&T monitoring frequency is normally determined by the availability of variable or activity data. In a hotel operation, for example, activity data on occupancy, restaurant sales, laundry volume, weather, etc., may only be available weekly and we will, therefore, set our monitoring frequency as weekly.

The basic rule is you should not set a monitoring frequency that is for a shorter period (greater frequency) than the availability of most variable data. Thus, if activity data is only available weekly, then we wouldn't set a monitoring frequency as daily, but if activity data is available daily, then it is perfectly ok to choose to set the monitoring frequency as weekly.

The reason for this is that normalizing data to a lower frequency (e.g. from daily to weekly) is very accurate as the process just involves summing the higher frequency reading together. On the other hand, normalizing to a higher frequency (e.g. going from weekly to daily) will lead to poor approximations, as we have to spread the consumption over a fixed number of periods, usually through some form of pro-rata allocation.

To avoid normalizing our data, we need to read our meters at the start of every monitoring period. So, if we have a monitoring period of weekly starting each Monday at 00:00 AM, then that is when we want our meter readings to be taken. Gathering *additional* meter data at regular intervals during the monitoring period is fine, as long as it doesn't affect the alignment of the reading times with the monitoring period start.

Estimation is the process of calculating the value when a meter reading is missing. There are a couple of common techniques for estimation:

- Pro-rata allocation: this involves taking a recent, reliable rate of consumption and scaling it to the missing period. *Incrementing* meter readings have the advantage since we always know the total consumption between two dates. Thus, if I have a daily system and I failed to read the meter on a Saturday, I can take Sunday's reading and divide it by 2. I will know that the sum of Saturday and Sunday is correct, but not necessary the breakdown between the days. Pro-rata allocation can be used for missing *consumption* meter values, but without the same confidence that the total over the period is accurate.
- Direct comparison: this is like pro-rating, but involves selecting a period which is similar to the missing period. For example, for my missing Saturday value, I could choose to use the value from the previous Saturday.

We will never normalize the value of an absolute meter. If I read the temperature a day late in a weekly system, then I can't simply pro-rate it (divide it by eight to get a daily rate and then multiply by seven to get back to the weekly estimate). The best I can do is to use the value given, take an average value or use a value from a comparable period.

13.5 Trueness and precision

Accuracy refers to the overall correctness of a single meter reading. Trueness is how close a group of readings are to the real values, while precision relates to the similarity between the readings. For M&T, we need precision above all. For billing, on the other hand, trueness is critical.

One of the considerations when working with meter data is the accuracy of measurement of the device. Manufacturers often specify meter accuracy in two ways, as ± a percentage or absolute value for any given reading, or ± a value at the full-scale reading of the instrument. The latter tends to mean that the actual percentage error is greater at lower reading ranges. The manufacturer's accuracy statement represents the best performance of the meter since the assessment is usually carried out under very controlled conditions.

In the "real-world" setting, we can expect the meter accuracy to be worse than the manufacturer's statement or even the accuracy defined by the appropriate standard. This reduction in accuracy may be because the device has been installed incorrectly, or because it is measuring at one extreme of the range (flow meters are especially sensitive to being matched to the correct flow), or because of mechanical wear or degradation due to contamination.

While maintaining and calibrating meters regularly will get us close to a scientifically accurate *"true"* value, we need to apply some common sense to

13.7 **Typical meter accuracies**
Within almost all meter categories, accuracy is a product of careful meter selection for the operating conditions, along with willingness to pay.
Source: Niall Enright. Some data from DRET: Energy Savings Measurement Guide.[229]

Resource	Notes	Cost	Typical Quoted Accuracy
Electricity	kWh meters simple and cheap. Additional features such as power quality can add to the cost and have little impact on resource efficiency.	Low	±0.05% to 1%
Gas	May want a meter with full compensation for temperature, pressure and calorific value to get a kWh value rather than m².	Low	±0.5% to 1%
Combustion	Combustion analysers will measure a range of gases such as O2, CO, and temperature in boilers and combustion equipment.	Medium	±0.1% to 5%
Steam	Need temperature and pressure correction, also the sizing is very important as accuracy varies with flow. High maintenance costs.	High	±1% to 5%
Water	Small (domestic size) meters can be quite cheap and accurate if they are fitted to standard size pipes. Larger meters more problematic.	Low to Medium	±1% to 5%
Oil Meters	Density (i.e. temperature) compensation needed.	Medium	±2% to 4%
Compressed Air	Temperature and pressure compensation is required. These can also be quite expensive to maintain and to calibrate.	High	±1% to 2%
Other Fluids	Flow meters exist for wide range of fluids. As with water meters, sizing is critical to accuracy.	High	±1% to 5%
Weight	Scales are relatively cheap measurement devices. Accuracy can vary greatly with the type of scale used.	Low	±0.05% to 5%

Be sure you understand whether you need a **true meter** or a **precise meter** or both.

the subject. The purpose of metering in the M&T programme is to measure resource use and to compare this with appropriate measures of activity. The savings come from the differentiation between good or poor periods. In practice, this distinction can be made with an *"untrue"* meter, so long as the measurements are precise. If our key performance indicator (such as a single linear regression relationship) was made with the meter over or under-recording by a given value, as long as this variance remains consistent, the distinction between good and poor usually remains valid.

Trueness represents the degree to which the measured value matches the known value. It is the correctness of the reading. Precision, on the other hand, is the degree to which the measurement of a value remains consistent over a large number of repetitions. Accuracy (often mistaken for trueness) is the combination of these two factors and describes the closeness between an *individual* measurement and the true value.

Insight

Because of the statistical approach employed, M&T depends primarily on achieving precision (systematic error is less of a concern than random error in techniques like regression analysis). M&T is thus a process which does not like meter drift, where a meter's precision reduces, as this can simulate a gradual improvement/worsening of resource use compared to the initial conditions. Whenever we see a gradual change in performance, we should consider deterioration in precision as a possible cause.

There will be plenty of situations where both trueness and precision are required in our measurement. This is especially true when we are taking a *"spot"* or instantaneous value, such as the oxygen concentration in a boiler exhaust. Since this value will indicate the combustion efficiency of the equipment and may rule in or rule out certain strategies for improving efficiency, it is important that we get this measurement right. Similarly, if the purpose of the measurement is to benchmark two similar activities, then we need both measurements to be accurate. However, by understanding the distinction between trueness and precision we can avoid situations where we pay a premium for a highly accurate device, when in reality what we need is a cheaper but precise instrument.

13.8 **Trueness vs Precision** (below and right)

Measure	TRUENESS	PRECISION	ACCURACY of Single Measure
BLUE	*GOOD*	*POOR*	*LOW*
RED	*POOR*	*GOOD*	*LOW*
PURPLE	*POOR*	*POOR*	*LOW*
GREEN	*GOOD*	*GOOD*	*GOOD*

The blue star-shaped readings are true because they are clustered around the central, correct value for the measurement, but the red diamond-shaped values are more precise because they are closely grouped i.e. they are more consistent. The purple circle-shaped values are neither true nor precise. The dark green cross-shaped readings are both true and precise. *Source: Niall Enright. Image available in the companion file pack.*

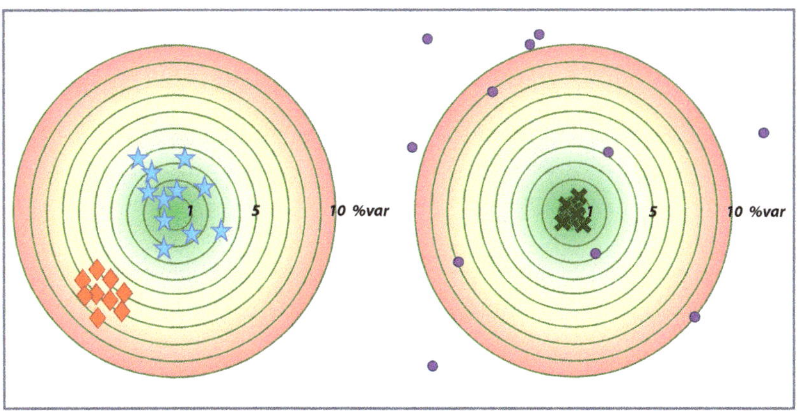

Real World: *A handy discovery*

A long time ago, I was involved in a waste minimization project at a bakery and pie factory. This plant was very small, with no separation of the waste streams from each production line. Undeterred, we used a common trick, which was to measure the inputs for each activity (i.e. the quantity of ingredients used) and compare these to the outputs (the weight of pies produced), on the basis that the difference in the mass balance would represent waste.

When we came to analyse the data we saw an intriguing result - on the same three days each week, there was an approximately 10% greater input of the pie fillings' ingredients, compared to the other production days.

Puzzled by this, we set about observing how the pie line worked. As the pie bases came along the production line, these would be manually filled by a worker, who scooped the filling from a bin and packed it into the pie. The simple fact of the matter was that one operator, *Mark*, had much bigger hands than the other, female, operator, and so he was consistently over-filling the pies on the days he was on shift!

Although this was not a waste issue, it still represented a *resource loss* for the factory, which they were grateful we identified. This story reminds us that accuracy is not just a matter for analysis data, but also in the control systems that govern resource use. In this case, Mark's *"hand"* measurement device was not *"true"*.

Closely related to the notion of precision is the resolution (sometimes called sensitivity) of a meter. If you imagine that I have a meter ruler whose scale is in centimetres, then the resolution of the ruler is 1 cm. Any point between, say 5.5 cm and 6.5 cm will be interpreted as 6 cm. This is quite different from the question of whether the ruler is calibrated correctly (the scale markings are in the right place) or the accuracy with which I can *read* the scale when measuring.

A meter manufacturer may express the accuracy of the instrument as a ± value (in percent or absolute terms), but in reality, the instrument can never be more accurate than its resolution (actually expressed as ± half the resolution). Scale resolution is important because, taking the ruler example again, if I have measured a length of 5 cm with my ruler and the maximum accuracy is ± 0.5 cm, then the accuracy of the measurement is ± 5%, meaning I could be 10% out. Whereas if I used the same meter ruler to measure an object that is 80 cm long, then the potential error is still ± 0.5 cm which is much lower in percent terms, ± 0.625%. Sometimes, this scale accuracy is explicitly stated in an instrument's data sheet as two figures, e.g. *"±5%, ± 1kWh"*, but is often omitted.

Standards: *International standards and nomenclature*

In this description of accuracy, I have followed the standard ISO 5725-1:1994[411] and the 2008 issue of the *BIPM International Vocabulary of Metrology*[415] These documents have established the terms *"trueness"* and *"precision"*, used here.

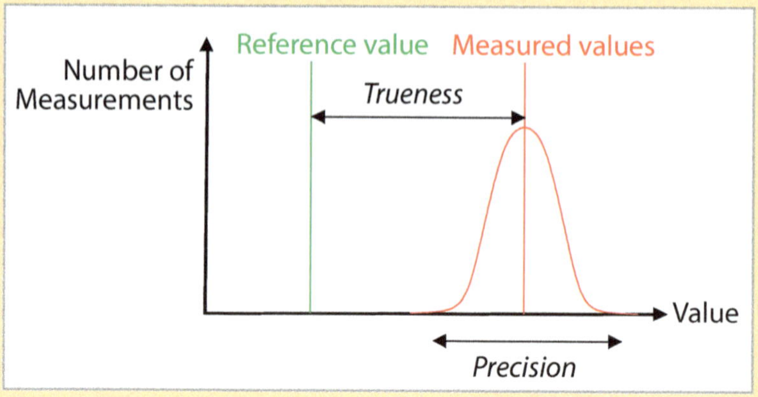

The illustration above shows a true or reference value in green and our measured values in red as a range of readings in red. For a single measurement, there is no precision component, so the accuracy is the same as the trueness. In statistics, the term systematic error is used to describe trueness, random error describes precision, and uncertainty is used to describe accuracy. There are statistical methods (described later) to calculate random error from the distribution of the measurement, whereas the systematic error is much harder to calculate because all of the data is off in the same direction (either too high or too low). The only real way to determine the accuracy is, therefore, to compare the instrument's measurement with a known reference value, a process called calibration.

Exploration: *Measurement systems analysis*

Given how much of our modern society depends on measurement, there is a wealth of information on how to assess measurement systems. There are many aspects to consider, four of which we have already touched on.

- Trueness: This is the difference between what the measurement system reports and the true value. The opposite of trueness is called bias.
- Precision: This is the measure of the consistency of measurements, i.e. whether the bias of the system is consistent or changes.
- Accuracy is the bias and precision observed over a large number of measurements (usually expressed as a ± value around the true value).
- Resolution: This is the maximum accuracy of the meter based on its units of measurement.

But there are other factors to think about.

- Sensitivity or discrimination: This is the smallest resolution of the measurement device. If our measurement system cannot distinguish meaningful changes, then it is not suitable for its intended purpose.
- Stability: This is the ability of the measurement tool to maintain the same measurement over time. This is sometimes called measurement drift. Mechanical metering systems tend to demonstrate greater levels of drift than electronic or solid-state systems.
- Linearity: This is the degree of bias throughout the scale of measurement. For example, flow meters are notoriously inaccurate at low flows. If the bias is the same over the whole scale, then the meter is said to be linear.

Calibration of a meter or measurement instrument involves the comparison of the reading derived from the instrument with a known accuracy, called the standard. The purpose of calibration is to remove bias by adjusting the instrument so that it corresponds with the standard. A general rule of thumb is that one should allow 10% per annum of the original meter costs for calibration. There are many standards that govern calibration.

Where we have manual readings, we may find that operator effects come into play (examples would be taking thermometer readings or dip-stick tank volume measures, etc.). Six Sigma has a technique to assess the validity of these measurements: Gauge Repeatability & Reproducibility (abbreviated to Gauge R&R).

- Repeatability: This is the ability of the measurement system to produce the same result on the same measurement under the same conditions. It reflects the inherent variability of the measurement system (a reflection of precision and linearity). This is referred to as equipment variation.
- Reproducibility: This describes the operator's ability to produce the same result if all other factors are unchanged. This can be measured by asking several operators to measure the same item and comparing them. This is also called appraiser variation.

The details for calculating Gauge R&R are beyond this book - good references (such as The Six Sigma Performance Handbook)[340] will describe this.

13.6 Investing in meters

Virtually every resource efficiency programme which adopts M&T needs to invest in additional metering. These are a few guidelines to help assess the investment.

Metering is very rarely ideal at the outset of an M&T programme. That is why the availability of data is one of the recommended criteria for the selection of sites in the first place.

Thus it is almost certain that we will need to invest in meters at some point in our programme. To specify, procure, install and commission metering systems can sometimes take a long time. I strongly urge folks not to wait for the completion of their metering system before embarking on M&T, even if only part of a facility can be covered at the start. I have countless examples of organizations which have delayed savings for years waiting for *perfection* in their metering.

Another principle when investing in meters, is that every meter needs to justify its existence. The BEST formula opposite helps us to determine an overall budget for metering. However, at the individual meter level, we can use the flow formula below to calculate the minimum value of resource per meter.

$$\text{Required Flow}_{\$} = \frac{\text{Meter Installed Cost}_{\$}}{\text{Resource percent saving in payback period}_{\%}}$$

For example, assuming that an electricity meter costs US$600 installed and will save a compound 6% of electricity over the desired two-year payback period (3% each year), I can justify one electricity meter for every US$10,000 of expenditure (US$600/0.06). Please note the sensitivity of this analysis to the installed cost of the meter. From an M&T perspective, we are looking for "*cheap and cheerful*" metering devices. Many vendors, particularly of electricity meters, can propose hugely over-specified meters able to measure all sorts of aspects of power quality, etc. For M&T we only need a kWh measure.

Don't wait for **metering to be perfect** *before you start your M&T programme.*

Ideally, we will include in our metering cost some form of data acquisition system. Again, these can be simple and effective, or complex and expensive. Here my recommendation is always to go with open systems rather than proprietary ones, but the choice in practice is largely governed by the existing systems at a facility. It may well be impractical to connect every meter to the data acquisition system, but I would nevertheless specify some form of standard pulsed output on the meter so that the option to connect in future remains open.

*Every meter needs to **justify its place in** the M&T programme.*

In Numbers: ***The BEST Equation for calculating the justifiable investment in M&T***

There is an easy-to-remember formula that we can use to calculate the budget available for investment in M&T:

$$B_{\$} = E_{\$} * S_{\%} * T_{y}$$

Where B is the available total budget in US$, £, €, etc., E is the annual expenditure on the resource metered (in the same currency). S is the savings percent, and T is the time in years in which we want to see the meters paid back.

Resource	S = typical M&T Savings %
General Electricity	*3%*
Chillers Electricity	*10%*
Gas	*5%*
Oil	*5%*
Steam	*5%*
Compressed Air	*10%*
Hot Water	*7%*
Cold Water	*7%*
Materials	*10%*
Transport Fuels	*5%*

We may have determined ourselves a value for the savings, but if not we can use the table left, which shows typical savings that we can expect from M&T when we apply the technique to different resources.

For example, if I spend US$1 million a year on general electricity and need the programme to pay for itself in two years, then I would be able to justify an investment of US$60,000 calculated from

US$1,000,000 * 0.03 * 2= US$60,000

I can then calculate the budget for each resource in my programme. So if I spend another US$1 million on gas I could justify a US$100,000 investment for a two-year payback. For electricity used in chillers, a conservative estimate of annual savings is 10%, so a US$1 million expenditure here would justify an additional US$200,000 spend on M&T.

From the "rule of thirds" (page 314), we can estimate that a third of the available M&T budget will be spent on metering, $Bm_{\$}$, which, in the above case, is a third of US$360,000 or US$120,000). Although this is an approximation, it is very helpful in establishing if the metering budget is sufficient for the programme.

Resource	$M_{\$}$ = typical installed meter cost US$
Electricity	*$200-$900*
Gas	*$300-$2,400*
Oil	*$1,200-$1,600*
Steam	*$6,000-$15,000*
Comp. Air	*$4,000-$15,000*
Water	*$400-$1,500*

Given some prices for metering, we can then calculate the actual number of meters we could afford for each type of resource, using the formula:

$$\#Meters = \frac{Bm_{\$}}{M_{\$}}$$

Where $Bm_{\$}$ is the budget for metering and $M_{\$}$ is the average meter cost. Since I have US$20,000 available for general electricity metering (a third of the US$60,000 M&T investment budget), and the average installed cost per meter is US$200, I could justify installing 100 meters in support of the resource efficiency programme.

Insight

13.7 Metering structure

Meter network design is as much an art as a science. If we take into account the "rules of metering" described in these pages, and apply some common-sense, we can create very effective metering strategies.

The reason I have put the topic on investing in metering before determining the meter structure is that, in reality, people often establish how much they have to spend and then work out where to put the meters to get the maximum value from the spend.

I know that this sounds like a backwards way of approaching the design of metering system, but in practice it helps us to provide a comparatively quick top-down ceiling for expenditure. It provides a good estimation of the overall cost, which allows us to set expectations and get outline budget approval.

Having done this top-down work, we then need to do the more time-consuming bottom-up assessment of where, given existing metering and a budget ceiling, we should place new meters. There are few hard and fast rules for metering network design. The important issues that we should take into account have already been covered:

- The need to separate resource metering by function.
- The placement of the existing meters so that virtual metering can be used instead of physical metering, which can reduce costs.
- The availability of proxy measures such as hours run.
- The required flow for a meter to justify its existence in terms of savings.
- The specification of the meters, so that we focus on *"cheap and cheerful"* meters rather than *"gold-plated"* meters.

As well as meter-specific factors, we need to consider broader issues. For example, it is not sensible to place new meters in a facility or process that is likely to change dramatically. If we cannot find a corresponding variable value to use in our analysis, it may not make sense to install a meter. It may not be possible to install meters in certain processes unless these are shut down, which may mean that we have to delay installation until there is a "turn-around" or periodic closure, e.g. for holidays.

Don't aim for completeness; "unaccounted" is OK.

It is perfectly acceptable, indeed common, for a significant proportion of the usage of a resource at a facility to be unaccounted. This is not a sign of poor metering design, quite the opposite - it reflects on care and attention being taken to align metering with potential for improvement.

Real World: *Justifying resource metering in a global corporation*

Not long ago, I was appointed by a global manufacturer based in Switzerland, with over 50 sites, to assist them in developing a business case for a considerable investment in utility metering. The purpose of the meters was to help reduce energy use further to respond to supply chain pressures on carbon emissions. A systematic assessment of every site was out of the question in cost and time terms, so we needed a top-down analysis which would be seen to be credible.

This manufacturer spends just over US$120 million on three utilities: electricity, gas and water. 40% of the power is used to generate compressed air, while half the gas is used to produce steam. Savings in compressed air or steam will result in a cost reduction in electricity and gas, but these resources have different savings potentials, so they need to be assessed separately.

The target payback for the utility-saving programme investment was under two years. Metering was assumed to be around half the programme costs (with 10% on central software and the remainder on staff training, consultants and communications). Thus it was decided that the metering needed to pay for itself in a year.

The next step in the business case development was to agree the potential savings for each resource. Luckily there were a number of audits, pilot projects and case studies which gave a high degree of confidence in these figures. A feature of the organization is that its product and production process was virtually identical at every site, which gave a high degree of confidence in the scalability of pilot cases. The final step, again supported by evidence from pilots, was to determine the typical meter costs for each resource type. These costs were rounded up to create credibility in the calculations.

Given the data above I then constructed the table below, designed to estimate the number of meters of each type that could be justified by the programme. This is not necessarily the number of meters required, but rather an estimate of the maximum expenditure, in this case, US$7.4 million, that could be justified. Taking into account the existing metering, the client presented a proposal for US$5 million investment to the board, which was accepted.

Although the project was a success from a budget approval perspective, there was one aspect that did not go well. The client asked a central engineering team to design the metering systems at each facility. The individuals concerned were determined to achieve 100% utility metering, to the extent that they insisted that the shower facilities and washrooms should be separately equipped with expensive water meters. There was absolutely no way that these meters would ever pay for themselves. Hard as I tried, I could not convince these folks that it is ok to have unaccounted usage.

13.9 **Meter budget calculations**
Source: Niall Enright, based on a real client example

Utility ($=US$)	Percent Value	Total Input Cost	Resource Value	Est. Savings %, $	Cost per Meter	# Meters Justified
Electricity	50%	$60m	$36m	~5% = $1.8m	$640	2,800
Compressed Air			$24m	~10% = $2.4m	$6,600	400
Gas	40%	$48m	$24m	~5% = $1.2m	$1,500	800
Steam			$24m	~5% = $1.2m	$4,700	255
Water	10%	$12m	$12	~7% = $0.840m	$3,200	262
	100%	*$120m*	*$120m*	*(=6.2%) = $7.44m*		*4,517*

Summary: *The 10 rules of metering*

1. Be clear about the purpose of the metering. If it is for billing or automatic control, its specification may be different to that for resource management.

2. Meter to separate use according to function. When deciding what to meter, consider the complementary variable that you will use to assess performance.

3. A meter does not have to be a meter. There are many proxies for meter data and virtual meters which can reduce metering investments significantly.

4. You need to collect the meter data at a frequency equal to or greater than the variable data frequency.

5. There are three meter types, incrementing, consumption and absolute, which each have different implications for how they can be normalized or estimated.

6. Read all meters at the start of the monitoring period.

7. There is a difference between trueness and precision. Don't pay a lot for trueness when what you really need is precision.

8. Don't wait for the metering network to be perfect before you start resource management.

9. Every meter has to justify its place in the programme. The BEST formula can provide an overall budget, while the Flow formula tells you how much of a resource you need to measure in order to pay for the meter.

10. Don't aim for complete coverage: unaccounted use is perfectly OK.

Further Reading:

Carbon Trust. *Metering - Introducing the techniques and technology for energy data management.*[179] A UK-focused guide with good advice on how to use the data available from existing utility suppliers.

Jones, Phil. *Metering energy use in new non-domestic buildings - a guide to Part L2 compliance,* Energy Efficiency Best Practice Programme (2002).[428] Although a little old this contains some useful flow charts on how to create a metering strategy. Of interest not just to UK practitioners.

The Institution of Engineering and Technology (IET) produces the *Guide to Metering Systems: Specification, Installation and Use,*[392] ISBN 978-1-78561-059-2.
I was a minor contributor to this publication, which offers more detailed technical guidance than the other suggested documents, albeit at a cost.

14 Analysing Data

Illustration © Rassco, Fotolia.com

Exploration: *A missing toolkit*

It is remarkable how little formal advice on analysis is provided by management systems such as ISO 50001 on Energy and ISO 14001 on the Environment. Their emphasis is on describing soft qualitative activities rather than hard quantitative measures.

In fact, there is a very well-established set of analysis techniques, with broad applicability, which is essential to all energy and resource efficiency programmes.

It is these generic techniques that will be explored in this chapter. I would go as far as to state that the analysis methods described here are essential. Without using most of these methods of understanding resource use, we cannot claim to have an effective resource efficiency programme in place.

Of course, complementing our core analysis tools, there are many other specialist analytical techniques for a myriad of specific types of equipment or process, which may be useful in our programme.

These specialist techniques are so numerous that they have been omitted from this book, but are well covered in other publications.

In this chapter, we will explore how we can convert raw data to information: a process of analysis. The test of the value of any given data analysis technique is the extent to which it will provide *insight* into the resource use. The source of this insight can take many forms.

- We may gain an understanding of the quantum and location of resource use. Sometimes simply knowing how much we use and where is enough to tell us where we need to focus our efforts. If we use more than expected (e.g. in comparison to a benchmark or an identical piece of equipment) we have a potential improvement opportunity.
- We can identify trends in our use over time. Information about seasonal, shift, or time of day patterns can provide powerful clues about the degree of control we may have over resource use.
- We can associate resource use with influencing variables such as weather or activity. From this relationship, we can differentiate between good and bad performance and begin to understand what causes this.

The first techniques presented in the following pages are primarily useful in the discovery or audit phase of a programme. These are quantification tools, locating resources in time or space.

Later, as we look more deeply into understanding time series data, we will discover techniques which also have a bearing on the control of resources in our the day-to-day optimization efforts. Many of these techniques are essential to Monitoring and Targeting (M&T), which is a process which, hopefully, the majority of readers will be using in their efficiency programmes.

I will present important statistical techniques such as regression analysis, a powerful way to understand resource use in relation to activity. Please don't be put off by the word statistics - these are easy and common sense ways to look at resource use, and the maths involved requires little more than addition or multiplication. The supplied MS Excel™ formulae (and the many spreadsheets models in the companion file pack) will make applying this knowledge easy.

14.1 Inventories

Our first data analysis technique sets out to establish the quantity of resources used in a period of time. Although inventories appear simple we shall see that there are a number of fairly subtle considerations to be taken into account. Where inventories are formalized these are often called footprints.

A snapshot of resource use over a fixed period is called an inventory. There are some preconditions for creating an inventory.

- We need to know the scope or boundary for the inventory.

  There are usually two broad inventory types: a product inventory and an organization inventory. For products, the boundary needs to be stated in terms of the life cycle (usually cradle to grave, but sometimes cradle to gate). There should be a clear designation of the product itself (e.g. is it just the soap powder, or the soap powder and packaging). These inventories are usually expressed in terms of the mass/units of product.

  In an organization inventory, there must be a definition of which facilities, processes, business units, etc. are included in the inventory.

- If we are considering an organization's inventory, then we also need to set the time frame for the inventory. The most common time frame used is a year, often aligned with an organization's financial year or a calendar year. Where inventory comparisons are being undertaken, we need to be clear about the periods involved (for example, emissions are often compared to the base year of 1990).

One of the valuable aspects of inventories is that they require us to classify or categorize our resources in specific ways. Because of this that we can use inventories to make valid comparisons between organizations and products.

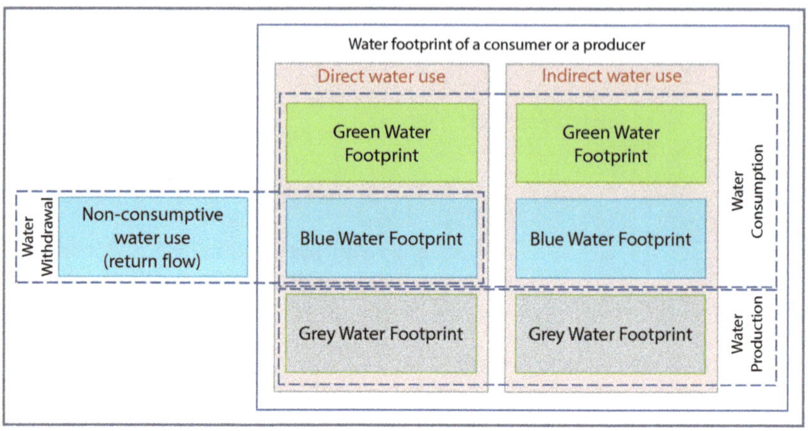

14.1 **Most inventory definitions provide very specific guidance on how to categorize resources**
This example is from the Water Footprint Assessment Manual, which categorizes water into Green (rainwater, soil water and transpiration losses from plant mass), Blue (surface or ground water, e.g. rivers and aquifers) and Grey (which is a measure of the fresh water needed to assimilate the wastes of the organization)
Source: The Water Footprint Assessment Manual[33]

One of the largest collections of organization inventories is maintained by the Carbon Disclosure Project (CDP), which has over 5,000 companies reporting their emissions using GHG Protocol categories. The CDP sets out to create peer-to-peer comparison opportunities, which are intended to encourage laggards to catch up with leaders. This *network* effect is an important factor when considering the comparison benefits that an inventory can offer.

There are, however, some challenges that we will face if we wish to track our resource consumption over the long-term using inventories.

First of all, there will be problems associated with methodology changes that are introduced from time to time in the standards. In these cases, we would normally expect to go back and recast previous results in the light of the latest methodology to be able to make a like-for-like comparison. Problems occur, too, when we have changes of emissions factors, for example. A decline in the national emissions factor for electricity would give us an apparent improvement in our carbon footprint emissions, which is genuine, but which might incorrectly be attributed to our own actions.

Insight

The second challenge in terms of consistency is changes to scope. Organizations are rarely still and will tend to acquire and dispose of assets and products. Once again, it is important to reassess previous inventories in light of the new boundary if we intend to make comparisons.

For this reason, inventories are usually poor tools to manage resource use on a day-by-day basis and tend to be used no more frequently than annually.

Possibly the most significant concern about the use of inventories to understand resource use, is the potential to treat this form of analysis as a purely accounting or categorization exercise. In reality, the value from the process comes only comes from questioning the impacts of particular resource use, the reasons that it arises, trends in usage, and why our organization is better or worse than its peers. If these questions are posed, *hot spots* are identified and opportunities for improvement produced, footprints are a valuable addition to the toolkit of any resource efficiency practitioner.

14.2 **Some standards**
Inventories add greater value if they are based on standard methodologies. These are just some of the more widely used systems.
Source: Niall Enright

Inventory	Standards and Guidance	Document References (bold for standard)
Carbon Reporting	Organization emissions arising from energy use. Standard is the **GHG Protocol**. Three types: Scope 1, or direct emissions; Scope 2, arising from imported electricity use; and Scope 3, indirect. The **Carbon Disclosure Project** ensures that a large number of organizations report emissions following this methodology.	178, **613, 670**
Product Carbon Footprinting	These standards provide guidance on aspects that should or should not be included, including standard factors for land use changes etc. The relatively new **ISO /TS 14067:2013** brings together many aspects of life cycle assessment and carbon accounting, including the notion of "product category rules" which ensure consistency within products. **PAS 2050:2011** is a somewhat more detailed standard than the ISO standard.	**95**, 412
Water Footprinting	The new **ISO 14046:2014** standard for water product footprints brings this in line with the carbon standards mentioned above. The Water Footprinting Assessment Manual provides a more descriptive methodology.	33, **410**

14.2 Benchmarks

A benchmark is a single figure which can be used for comparison purposes. Benchmarks can apply at all levels, from individual items of equipment to buildings, plants and organizations.

When presented with resource use data, most users appreciate that the absolute value alone has little meaning. If 9,000 m^3 of water is consumed in an apartment block in a year, it is difficult to understand whether this is good or bad, the figure is *data* rather than *information*. The search for meaning in the data leads to benchmarking.

If I know that there are 100 occupants in the whole apartment block, then I can calculate that the consumption is, on average, 90 m^3 per person per year. I can then compare this with the typical use for homes with two people, which is 110 m^3 per person per year,[162] and conclude that my building is performing better than the *benchmark*.

This simple illustration provides some valuable insights into using benchmarks.

1. First of all, there are the input values. In practice, the numerator (i.e. the resource use) is usually relatively well established (after all, it is usually a value that we have measured ourselves). The real challenge with most benchmarks is the denominator. In the example above, this is the number of people in my apartment block, which (unless I have some form of security or occupancy sensor) is likely to be an approximation.

   Even where we talk about apparently straight-forward denominators such as the sq ft of a building we get into subtleties such as *"what sq ft"* we are referring to: is it the gross floor area (the area withing the dimensions of the building); the gross internal area (the actual floor space available based on the internal dimensions); the net internal areas (the gross internal area less common areas such as lobbies, lifts etc.); or the conditioned area (i.e. just that space which is heated or cooled, which excludes unconditioned space such as garages)?

2. Then there is the variability that exists within the benchmark. Unless we know the range of the benchmark, we are not really in a position to interpret what 90 m^3 per person per year really *means*. We are aware that it is 18% less than benchmark, but is that *good*? According to the data on water use for households with two people, 55 m^3 per person per year is low use (136 m^3 per person per year is high). In that context, we can see that there is still significant potential for further improvement and 18% under benchmark may not be as good as we think.

Real World: *The rise, fall and rise again of benchmarking*

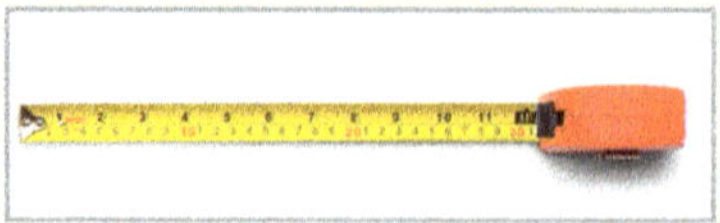

At the beginning of my resource efficiency career in the early 1990s, UK industry benchmarks for energy efficiency were relatively common. Some studies openly compared and published similar facilities' performance data. Others, such as the UK's *Best Practice Programme* provided aggregate data on the performance of different sectors. A brewery, for example, would know the best, average and worst values for MJ of energy per hl total production.

Around the mid-1990s, fears of collusion between organizations led competition authorities to restrict firms from sharing operating data, since this information could be used to fix prices between participants. As a result, energy managers participating in industry groups were forbidden from revealing any information about their organization's performance, which had a chilling effect on knowledge-sharing.

It is now accepted by antitrust bodies that benchmarking is ok, as long as the data cannot be attributed to specific organizations. Current practice usually requires that at least four companies are included in any comparison group. Since an organization could extract their own data from the comparison set, this would still leave three firm's data, from which it would be difficult to establish who's data was whose.[631]

There is, of course, the caveat that none of the comparison group are so exceptional that their identity can be determined from the data itself.

A benchmark should be a) **relevant**, b) based on **reliable and available data**, c) **adjusted** in ways you understand and d) able to **separate good from bad** performance.

In Numbers: *Quartiles*

Many benchmarks, such as the Solomon Energy Intensity Index for refineries, use the notion of *quartiles* to categorize performance.

Organizations are divided into four groups. The *first quartile* (aka *top quartile*) is the best, in other words, has the lowest energy use; the *second quartile* is above average, while the fourth quartile is the worst category.

The key to benchmarking different plants in this way, is to use weighted averages to calculate the energy use. The boundaries of the quartiles can be computed using the average ū and the standard deviation σ (discussed later), as shown below.

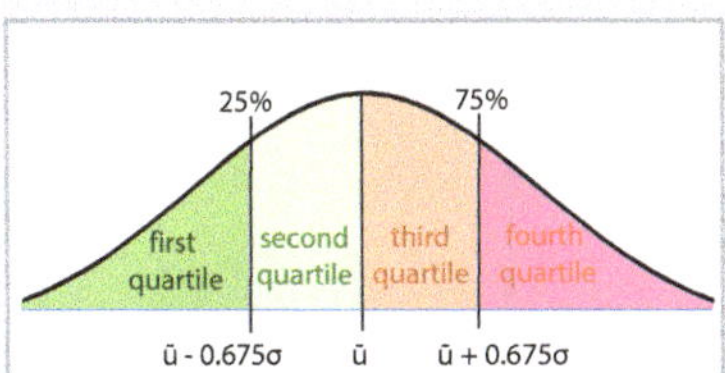

14.3 **Quartiles** *Source: N. Enright*

3. Finally, there is normalization of the benchmark. In the example above, we have used an average occupancy for our apartments, but it is possible that we have a mix of one-person and three-person flats, which each have different benchmark figures. To assess our performance, we would need to calculate the weighted benchmark using the number of each type of flat.

   Similar normalization occurs in the UK's energy in buildings benchmarks provided by CIBSE,[149] which are widely used to estimate building performance and form part of the government's methodology to rate buildings in operational terms via Display Energy Certificates.

   These benchmarks allow for two normalizations or adjustments: weather adjustments to compensate for mean temperature differences across the UK; and occupancy adjustments to make up for the fact that a building may be in use for more hours than the benchmark states.

From the above examples, we can see that benchmarking based on external criteria can be much more complex than at first appears to be the case. On examination, many published benchmarks are of limited value because they only have a single figure against which performance can be compared, or at best they simply provide different countries' averages. Another set of benchmarks to be treated with caution are equipment design performance data, which may be impossible to match as they are achieved in idealized conditions.

Given these challenges I would propose the following test of the potential effectiveness of external benchmarking.

1. How closely aligned is the benchmark to your organization (i.e. are the benchmarks in your sector, geography and reasonably recent)?
2. Are accurate input data (numerator and denominator) available?
3. Are adjustments (if any) to the benchmark understood and appropriate?
4. Does the benchmark offer you a means of interpreting performance (i.e. is there a distinction between good, average and poor performance)? Will these ranges differentiate your performance sufficiently? If every benchmark gives the same result, will it be useful?

The alternative to external benchmarking is to create internal benchmarks. These serve to differentiate between similar activities in the organization and tend to have fewer of the drawbacks or complexities of external benchmarks designed to work across organizations.

The fundamental problem with all benchmarks, is the potential for people to say "*yes, but...*" and to point out some characteristic of the benchmarked item that makes it different from all the others... "*the benchmark is for homes and we are flats so we don't water our gardens...*". That is why later in this chapter the focus is on *intrinsic* measures of performance where the comparison is made with an item's past performance, rather than the performance of others.

14.3 Flows

Almost all resource efficiency programmes start with some form of mass balance. Sometimes these are simple process flow diagrams which can be used to understand a process; at other times, they can provide fundamental insights about where opportunities for improvement might lie.

A fundamental data analysis technique for resource efficiency is a mass balance. What a mass balance does is to compare one set of quantity data at one physical point in a system with the amounts of the same resource (or its derivatives) at another stage in the same flow (upstream or downstream, so as to speak).

A resource efficiency programme focusing on energy tends to use an energy distribution diagram as its basic roadmap, while a programme working on materials will use a process diagram to visualize and understand usage. That is why these items of information are high on the list of data requested when an audit is carried out in a facility.

The US Manufacturing Energy Footprints[219] are simple energy balances for different manufacturing sectors in the US. An example of one of these energy flow diagrams is shown below, for the plastics sector. In this case, the flows,

14.4 **US plastics energy and carbon flow** The energy flows are indicated by the arrows and the indirect and direct carbon emissions are shown in the brown boxes below each step. The flows in these steps are then broken down further in a second illustration (not shown here). *Source: US Department of Energy.*[220]

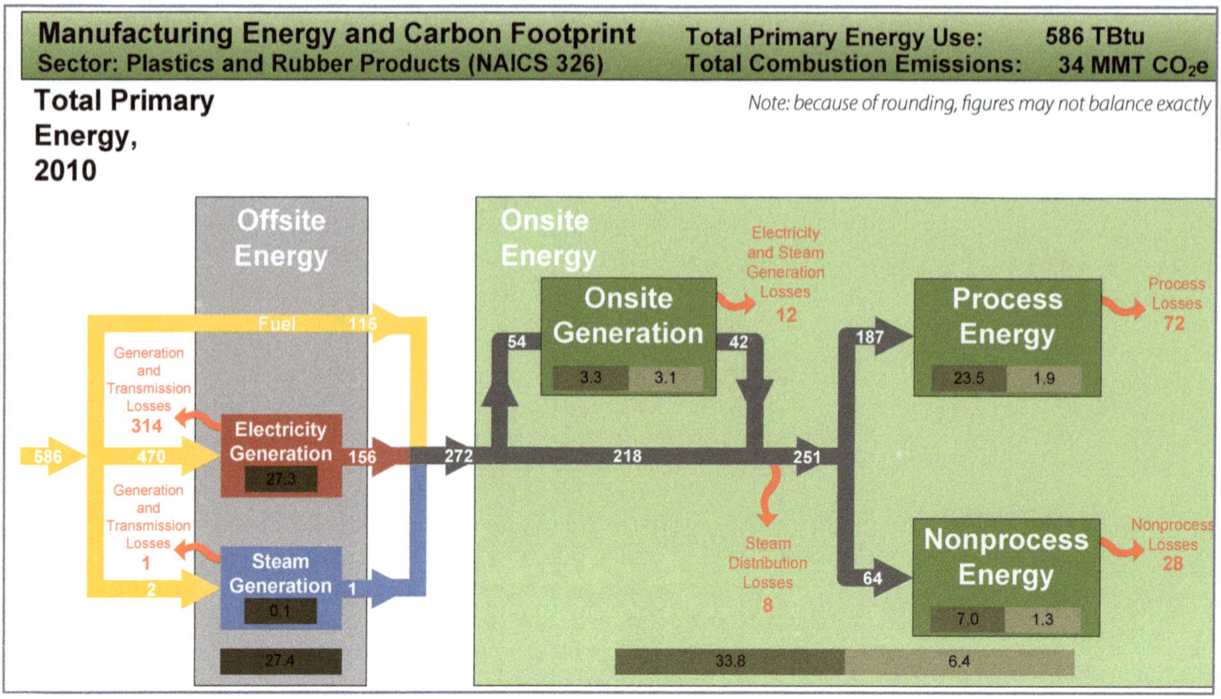

Don't automatically assume the ***largest users of resource*** *offer the greatest savings potential.*

and the losses at each step are shown in absolute terms, but these figures could just as easily have been presented as a percentage value.

Most folks looking at this kind of diagram will tend to focus on identifying the largest users, assuming the Pareto principle that states that 80% of use will be located in 20% of equipment. However, I would caution against automatically assuming that a focus on large consumers is best because this equipment is most likely to be optimized already. I found this out first hand from energy surveys in steel mini-mills in Ohio where the electric arc furnace consumed the lion's share of power and was controlled in quasi-real time and where all the big opportunities lay in comparatively neglected peripheral systems.

In the illustration opposite, of the total of 272 units of energy input into the facilities, I can see that 64 units are for non-process uses, which is 24%. It may well be the case that a lot of effort goes into optimizing the production activity, so a good focus could be this quarter of the energy use that is not adding value in the sense that it is not directly going into the final product.

Typically, these types of mass-flow diagrams can only be created for just a few resources at a facility level. Adding more resources makes the diagrams increasingly complicated to draw and interpret. For processes involving chemical reactions, we may need to know the *stoichiometry* of the process, i.e. the reaction involved and the chemical composition of the various components so that we can determine the flow of elements at each point in the process. This is particularly important where one stream (e.g. a gaseous wastes stream) cannot be measured fully and has to be calculated.

Determining the mass balance in a facility that produces one consistent product is relatively straightforward. If the product is produced in batches, then we could use the batch totals for our analysis, whereas if it is a continuous process, such as in a refinery, then we would need to set a specific time frame for the analysis of the flows based on the available data. It is much harder to assess the materials balance in a multi-product situation, e.g. a speciality chemicals plant which produces many different products by combining reactors and processes in unique ways on each production run. Here, our materials balances are likely to focus on the unit or equipment level, rather than the plant as a whole, or be developed for individual products.

In all batch processes - that is processes where there is a start and stop - whether single or multi-product, we should be aware that it is often the *"turnarounds"* that can offer the greatest number of improvement opportunities. At the changeover from one product to another, there will be shut-down and start-up processes, or sub-optimal periods of operation, or cleaning activities, or out-of-spec product, that can lead to considerable waste. Thus, we should not just look at the *"in production"* mass flows for inspiration, but also these periods of change. We should therefore not be over-reliant on the mass balance alone.

In the chapter on presenting data we see some further flow illustrations, such as Sankey diagrams, that show the amount of a flow using the size of the line.

14.4 Intensity use

Categorizing resource consumers in terms of their relative intensity, i.e. their resource-demand, or relative use (i.e. their utilization-rate) can help identify strategies for improvement.

We have already seen how intensity-use analysis can help us determine which sites to audit (page 386), but this technique can also help in determining our resource efficiency priorities and strategy within a facility or business.

Intensity use takes the previous flow analysis one step further by separating these flows into their two elements, intensity (in terms of resource use per time period) and use (the number of periods of operation over a year or percentage of total time the process is running). These components might also be referred to as demand and utilization.

Whatever the naming convention, the analysis usually involves plotting the flows on a chart with the intensity vertically and the use horizontally. Having plotted all the flows (which might be items of equipment or process steps), the chart is usually divided into four quadrants by the average of each element, as shown below.

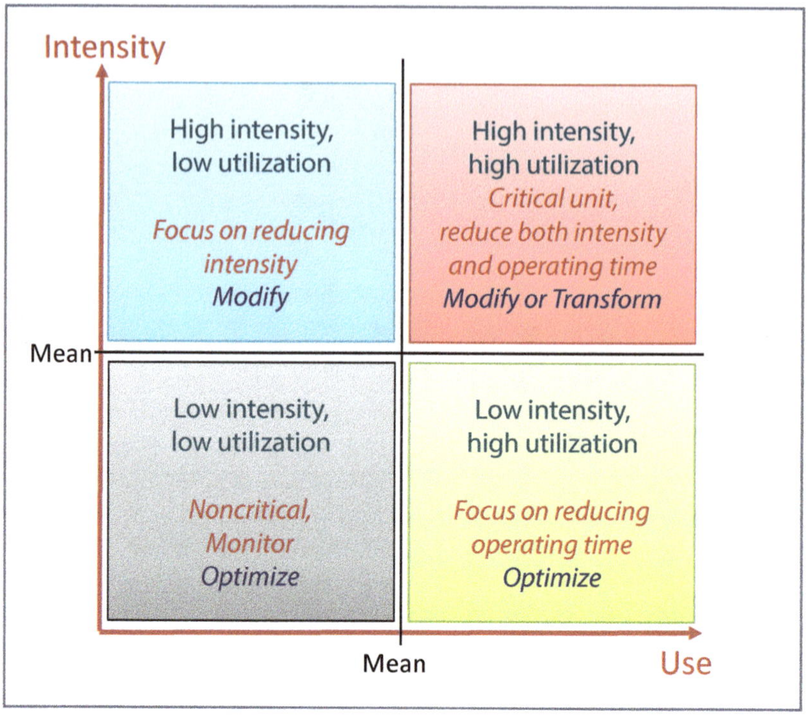

14.5 **Intensity-use plot**
The bottom-right quadrant is shown in green. This is because low-intensity, high-use resource consumers tend to offer good opportunities for rapid savings, typically using control strategies to reduce the hours of operation or switch off equipment and processes when not required.
Source: Niall Enright based on Thiede et. al.[704]
Image available in the companion file pack.

Use reduction is usually cheaper as it involves better control or housekeeping, compared to intensity reduction, which often requires equipment changes.

Although the illustration shows each quadrant as being the same size, we will use the mean of intensity and use (the average), rather than the median (the midpoint). Thus, the upper right quadrant is likely to be larger in area than the lower left one.

For each quadrant we will have different strategies in our resource efficiency programme.

- The High U - High I quadrant, shown in red, will have our critical resource-consuming units. Typically, there will be a relatively low number of units in this quadrant but they use a large proportion of the resources. Here, we will aim to reduce both intensity and use. Here, the opportunities for change will fall mainly in the Modify or Transform categories.
- The High U - Low I or Low U - High I quadrants, in green and blue, will tend to have strategies to reduce use or intensity respectively. It is in these quadrants that we are likely to find our opportunities for improvement, since the High-High quadrant equipment may (should) already be highly controlled and optimized.
- The final quadrant represents the non-critical equipment, which is likely to have little effort invested in improvements as the impact is relatively small.

Insight

All resource efficiency programmes need some form of prioritization method. Typically this is done using Pareto, a simple ranking based on total consumption. The intensity-use analysis takes this ranking a step further and provides insight into the improvement focus.

There is a caveat. What this analysis does not do is to recognize that the cost inherent in intensity reduction (which usually involves changing equipment) tends to be much greater than the cost associated with use reduction (which usually involves better control or housekeeping). To use a lighting analogy, it is cheaper to switch off lamps than to make them more efficient. Thus, our opportunities tend to skew towards use reduction first, whatever quadrant they are in. This is why this quadrant is shown in green.

Exploration: ***Energy intensity vs energy efficiency***

Energy intensity, such as MJ primary energy/US$ GDP, is the main measure favoured by economists and policymakers to explain trends in energy efficiency. The key policy objective is to decouple energy from economic growth, i.e. to reduce the additional energy use as GDP rises. In the US, for example, the economy grew by 84% between 1990 and 2015, but energy use only increased by 17%. Thus, the energy intensity has improved (decreased) by 56% since 1990, 13% since 2007 and 2.3% between 2014 and 2015, largely as a result of energy efficiency measures.[75] Sometimes we see this data expressed as energy productivity, which is the inverse of energy intensity, with US$ GDP as the numerator and primary energy the dominator.

Energy intensity (or productivity) is the best measure we have for energy efficiency within a large organization or an economy, but it is not without its limitations. This is because it is only helpful at a very high level - we can't use this to compare different sectors or facilities with each other. Also, although this is assumed to measure efficiency, there could be structural changes, which influence the result. Examples of these structural changes are a shift in the economy away from manufacturing to service industries, or an ageing population who like to keep their homes warmer, or protracted periods of unusual weather.

14.5 Life cycle assessment

Life cycle assessment formalizes the basic process of mass balances and material flows into a much more rigorous, standards-dominated field of data analysis. There are many reasons, not least understanding resource risks, why organizations may wish to invest in this powerful technique.

Nowadays the term life cycle assessment (LCA) has fallen somewhat out of fashion, in favour of life cycle management. The move away from the notion of passive quantification to a process of active management clearly makes sense. Here, though, our focus is on the LCA process - what data is needed and the insight that it can bring to our resource efficiency programme.

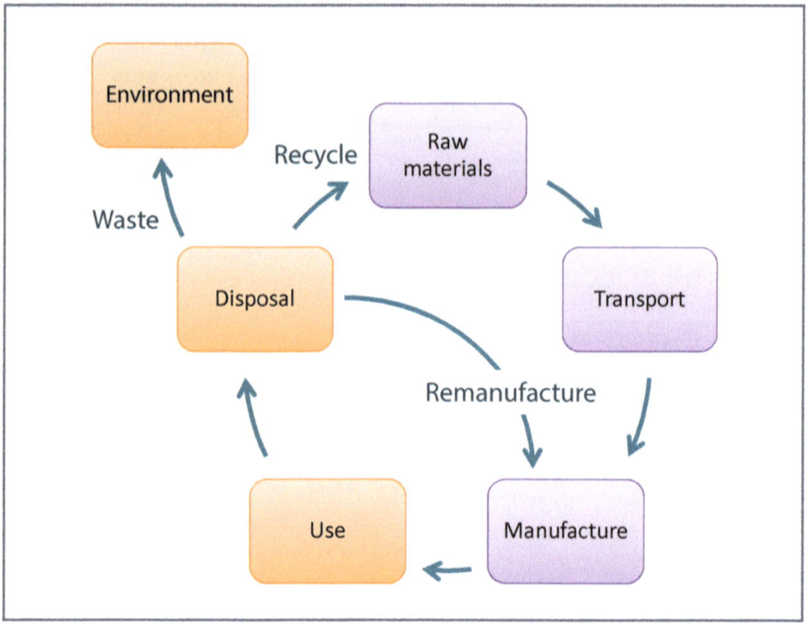

14.6 **Simple product life cycle**
This diagram is idealized; in practice there are many steps in the manufacture of modern products, as supply chains often cross continents and products may have many hundreds of individual components.
Source: Niall Enright.
Image available in the companion file pack.

A product life cycle starts with the definition of the scope and boundary. Commonly we might look at a "*cradle to gate*" scope shown in purple above. In practice, this scope can be much more complicated as there may be multiple raw materials, several intermediate processing stages as well as component manufacture, various transport stages all feeding into our manufacture and packaging to the final factory gate product.

If we include the whole of the life cycle, then we need to consider a use phase and then the disposal of the product, with or without some degree of recycling or remanufacture. This cycle, shown by the purple and orange steps above, is called "*cradle to grave*" where the materials are disposed of or "*cradle to cradle*" where they are largely recycled or reused.

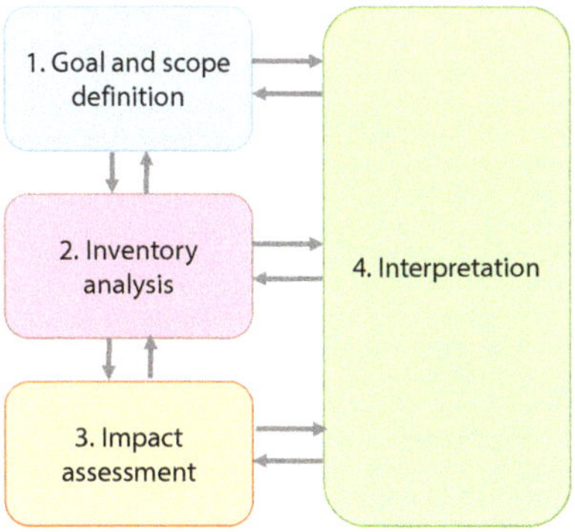

14.7 **Steps in life cycle analysis**
Source: Niall Enright, based on ISO 14040. Image available in the companion file pack.

The ISO 14040:2006 standard, shown left sets out the formal processes in an LCA. The first step is to describe the Goals and Scope, in terms of the life cycle stages and the externalities (of which there are many, such as water extraction, climate change emissions, acidification, eutrophication and human toxicity) which are going to be assessed.

Once the Goals and Scope have been defined, the next step is to create the inventory of the product(s), which quantifies the various material flows at each step of the life cycle in scope. From this inventory, we can then determine the impacts in terms of the externalities in scope. As you can imagine the data-collection aspects of an LCA are not inconsiderable, and most practitioners use dedicated software such as SimaPro, Ecobelan TEAM, Umberto or OpenLCA. The data is drawn from a myriad of free and commercial databases which provide standardized data for materials and compounds (GaBI, Sweden's SPIN, EcoInvent, NERL's LCI database and PlasticsEurope are among the better-known sources).

Interpretation involves a number of data analysis stages, such as normalization, where we adjust the impact to population size; weighing, where we may combine multiple impacts into a single score; uncertainty analysis where the variability and accuracy of the data is reviewed; and sensitivity analysis, where the impact of changing parameters of the study or constituents of the product are assessed.

The complexity of LCAs means that the results that they provide are highly sensitive to the input assumptions. "S*mall differences in assumptions related to system boundaries or valuation techniques can lead to radically disparate results*", according to a 2009 Deloitte study.[204] That does not mean that LCAs don't offer considerable benefits, especially to a business dominated by a few basic core products or variants, such as dairies, cement manufacturers, soda or water bottlers.

For a successful life cycle assessment, it is important that ***the business objectives are clear*** *and* ***up to date standards are followed.***

Despite, or possibly because of, the challenges, a systematic framework for the development of LCAs is coming into place. The foundations are based on product category rules (PCRs) which define the rules for different product categories (a recent visit to a PCR database[403] shows that these exist for an eclectic mix of products from commercial planes, through kiwi fruit, concrete, to leather). There is an ISO standard, too, for PCRs: ISO 14025. Once a PCR is in place any manufacturer can create an environmental product declaration (EPD) following the appropriate ISO standards, with a reasonable degree of confidence that this will be comparable to other manufacturer's EPDs. While these declarations are not mandatory, the EU is undertaking some pilot project on product environmental footprinting and organization environmental footprinting, which could see mandatory requirements in this area in future. Until then, it is important that those who set out to use LCAs to understand their resource use and risks, should do so with clear objectives, such as cost reduction, environmental performance or customer engagement in mind. We should also ensure we use existing standards, so that the results are future-proofed as far as possible. ⇒ page 444.

Real World: *Where happiness equals lower emissions!*

Life cycle assessment can provide some very useful information to assist our resource efficiency programmes. We saw earlier how knowledge of a detergent's use phase enabled the manufacturer to reformulate the product to work as well at lower temperatures. This real-world example focuses on how LCA data can be used to inform real business decisions.

Typically the use phase is the most difficult to inventory and assess in an LCA and is something which a manufacturer cannot easily control. That is why a lot of organizations take a practical approach and focus on *cradle to gate* - on the basis that they can influence their supply chain (both for data or for changes in their process) and can measure and control what goes on within their own organization.

Together with a former ERM colleague Zomo Fisher, now at Accenture, I was involved in an interesting assignment to assess if a grocery chain could offset its direct emissions by investing in emissions reductions at other points in the cradle to gate flow of products. *Gate* here was not the manufacturer's gate, but the grocer's shelves. The rationale was that the organization had already done a great deal to reduce energy use and further emissions reductions were becoming increasingly costly. If some categories of products had *low-hanging fruit* in terms of resource efficiency, then this might with help the overall emissions goal, and also potentially lower production costs to the benefit of the supplier, grocer and end-customer.

One of the categories we looked at is milk. In this case, the retailer sourced milk from low to medium-intensity dairy farms. There was already significant data available from ERM's extensive work on carbon footprinting, from questionnaires provided by the grocer's milk processors, and from academic studies. This data allowed an accurate breakdown of the emissions at each step to be created, so we didn't actually need a full-blown LCA to arrive at the chart below.

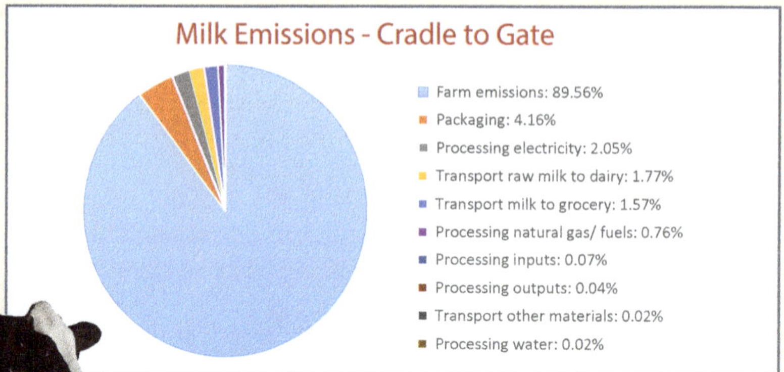

Of the average 1.3 kg of CO_2 equivalent emissions per litre to get milk from the farm to the shelf, around 90% occur at the farm (over half of this is from the high global warming potential gas methane, produced by enteric digestion in cows and from manure storage).

Armed with this information, Zomo examined opportunities to reduce emissions at each stage of the milk production process. For example, wind turbines at a farm

14.8 **(below) The largest CO_2e source** Keeping her happy will reduce emissions. *Photo: ©Eric Isselée, Fotolia.com*

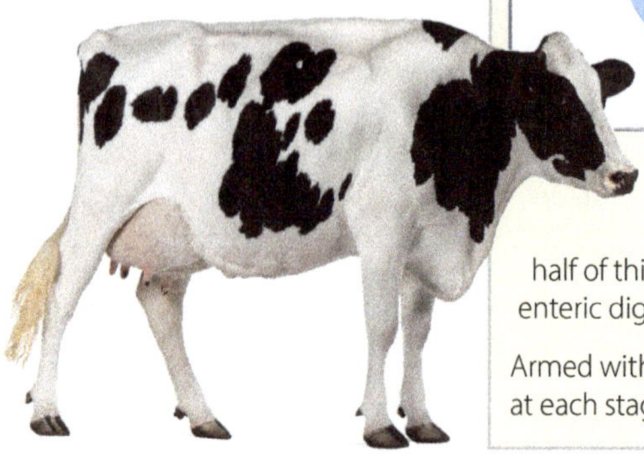

would have a very high emissions reduction per dollar invested, but since farms use little electricity, the overall potential impact on total emissions was very low. We got similar results for other non-farm interventions, such as changing the product packaging, using renewable energy at the processors and improving transport efficiencies. These all had a positive return per dollar invested but had a comparatively small impact overall.

Clearly, the solution, if any, was to be found at the farm. After considerable analysis of the factors that influence milk production, such as herd stocking levels and replacement rates, supplementary feeding, water and fertilizer use we came to the conclusion that the key factor was the cows themselves. It seems that there is quite a lot of variability in the amount of milk each cow will produce (the yield). Since the amount of emissions per cow is relatively constant regardless of the milk production, the greatest reduction in emissions *per litre of milk* could be achieved by improving the average yield of each cow in a herd. In fact, this had the potential to reduce emissions at the farm by 25% per litre, and to do so at the lowest investment dollars per tonne CO_2 equivalent.

One of the easiest ways to improve the yield is to reduce the incidence of mastitis, an infection that affects the udders of the cows. Thus, an opportunity was identified by the grocer's dairy buyer to work in close collaboration with suppliers to assist in education and veterinary support to reduce mastitis. Indeed, the topic of cow welfare became critical, as another yield-increasing strategy, highly intensive, indoor rearing, could be seen by some as impacting negatively on cow health.

Transparency about impacts at each point in the life cycle is behind the idea of EPDs to help inform decision-making. As we are on the subject of milk, I have reproduced below just one table from the comprehensive, fascinating EPD from Italian milk producer Granarolo.[329] This extract shows that their emissions at 1.45kg CO_{2e} per litre, are similar to the UK's, and also gives data on the water and land-use (ecological footprint) per litre.

14.9 **Milk EPD**

Source: Granarolo.[329]

This declaration reinforces the fact that the most environmental impacts from milk production occur on the farm, and so it is here that improvement efforts are best focused.

ENVIRONMENTAL FOOTPRINTS (1.5 litres bottle)	Milk production	Packaging production	Other materials	Granarolo process	Transport	TOTAL	Conservation	End of life of packaging
ECOLOGICAL FOOTPRINT	3.6	0.3	0.1	0.3	0.1	4.4 global m^2/l	0.4	<0.1
CARBON FOOTPRINT	1.15[6]	0.10	0.03	0.13	0.04	1.45 kg CO_2 eq/l	0.14	0.01
WATER FOOTPRINT	1.350	<10	<10	<10	<10	1.350 litri/l	<10	<10

14.6 Understanding variation

Only by understanding why our resource use changes can we hope to improve. The majority of the variation is usually due to external "drivers" such as weather, called common factors. There is another source of variation, due to exceptional events, which also provides a source of improvement.

*Understanding **exceptional variation** is the key to resource optimization. From this we can separate good and bad performance and **repeat the good and eliminate the bad.***

A lot of data analysis involves comparing data values. For example, we might be comparing values in a time series (e.g. how one facility's resource use changes over time) or in a population (e.g. how a facility compares with others in the group). In this type of analysis, our principal aim is to understand the variation present, what makes one value different from another.

We can categorize our variation into several types. The first kind of variation is called common variation because it exists in the samples we measure *all the time*. We can divide common variation into two further types.

- There is measurement variation, which reflects the accuracy and precision of our metering equipment, or perhaps the consistency with which we can take measurements, or maybe due to the gauge of the instrument or our skill. We covered error in the last chapter on metering.
- There is common factor variation in every resource-consuming activity or process, due to the influence of external factors. The factors are things that affect the value measured. For example, the humidity of a tobacco leaf affects its drying time, hence the input gas to the dryers; the external temperature will affect the electrical consumption of a cooling system; the number of guests in a hotel affects the measured water use. These are common variations because the source of the variation is always present. It is sometimes referred to as "*random or uncontrollable*" variation - in the sense that we can't change these factors - the temperature will be the temperature, the number of guests will be what it will be. These factors that influence our measurement have a special name in M&T: variables. Thus humidity, weather and guests are all variables. We have already come across these variables in benchmarking, where they are used to normalize data in population or cross-sectional studies.

As well as the common factor variation, we also have exceptional variation, called *"attributable variation"* in Six Sigma and Statistical Process Control (SPC), that is to say, variation in our measurement that can be linked to a specific events or conditions which are *not always present*. Examples of these are equipment failure or abnormal operation (such as leaving equipment running when it shouldn't be, or getting a product recipe wrong and having to discard a whole batch of product). This variation is sometimes referred to as *"controllable variation"*, as we can usually take steps to control it (even if

14.10 **Variation of gas use in a building**
There are three sources of variation in resource use: measurement error, common factor variation (e.g. due to weather) and exceptional variation due to unusual events. The key to improvement is being able to separate out the effect of each of these. For example, a below average gas consumption may be due to error, because the temperature is mild or because someone temporarily turned down the thermostat.
Source: Niall Enright, Image available in the companion file pack.

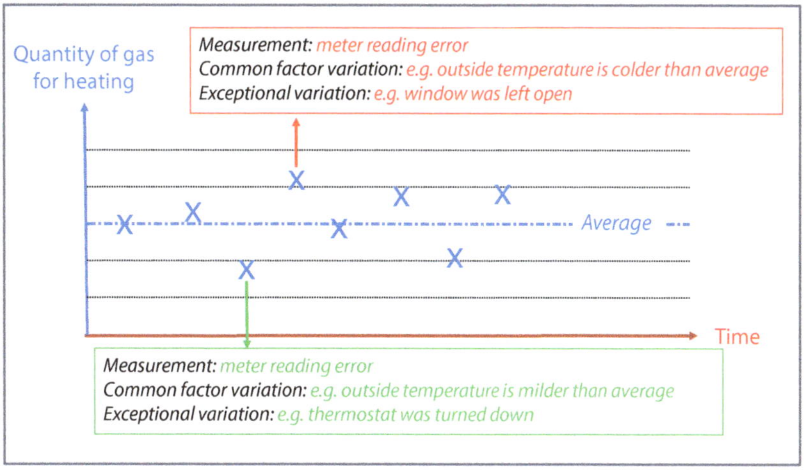

it appears to be an accident that causes it). Improvement opportunities and control strategies come from creating a model that can separate common factor variation and exceptional variation in our resource use.

The way we highlight exceptional variation is by creating a model (i.e. a formula) that can predict our common factor variance due to variable(s). If the model has correctly adjusted for the variable(s), any remaining difference, or variance, between the actual resource use and the prediction is due to either error (which hopefully is small) of as a result of some unusual event (i.e. the exceptional variation).

If there are no significant variables driving resource use (or the values of the variables are constant), we can use simple statistics like the average (mean) to model the expected resource use. If, on the other hand, we have identified one or more significant variables that do change and drive the resource use, the statistical analysis technique we use to model the influence of the variable is called simple regression (for one variable) or multiple regression (for more than one variable). These surprisingly easy but powerful modelling techniques will be described shortly.

Although **common factor variation**, such as that due to the weather, may appear "uncontrollable", our response **is modifiable.**

Let's consider a real example, illustrated above, gas consumption in a building, which is influenced by the outside temperature (our variable which produces our common variation). We would use regression analysis to model the effect of the outside temperature. Having eliminated the effect of temperature, we can then discern exceptional variation due to unusual events, such as people leaving windows open (bad variance, to be eliminated) or turning down the thermostat (good variance, to be repeated).

That is not to say that we shouldn't also look to improve the common factor variance as well. Although we can't directly control the weather, we can modify the way our building responds to weather. For example, a key strategy may be to insulate the building against the cold better, since this will change the underlying heating requirement in response to the weather.

The purpose of the **model** is to **isolate the influence of the common factor variable** which normally masks performance, **so that we can detect the presence of exceptional variation** and take corrective action.

14.7 Creating a model

There is a fairly standard sequence of events when analysing resource-use data. First we check for data errors, then determine if the use is material, before establishing a model that takes into account common factor variables, or is based on a simple statistic such as an average, if no variable is available.

In the context of resource efficiency, a model is a grand name we give to the formula we use to predict resource use. The model might be as simple as the average resource consumption, or it may be a more complicated linear regression equation.

The first thing that we will do when we start to develop a model is to establish if the data is *"clean"*, in other words, if there are any errors and if the data is for consistent time periods. Once we know the quantity and timing of the resource use are accurate, we can make a decision as to whether further analysis and monitoring are necessary. This is usually a very obvious decision, but if needed, we can apply intensity/use categorization and cost analysis to inform the decision.

Having decided that it is worthwhile to analyse a particular resource use, we then look at its values in isolation (for example, by plotting trends). The word *look* is used advisedly as all data analysis should involve a visual inspection of the data series in a chart form to spot issues such as outliers or non-linear relationships, which may not be easily evident from statistics alone.

The next stage in our analysis will be to establish if one or more variables influence the resource use. This requires some understanding of the resource-using equipment. For example, the purpose of the air blowers in a wastewater treatment plant is to oxygenate the water to feed the bugs that are treating the waste. The number and activity of the bugs depend on the water temperature and the level of waste (measured by the biological oxygen demand, BOD) as well as other factors. In my analysis, I would want to obtain the data for the same time periods for these variables. I would then use a range of statistical techniques, described shortly, to determine if I can create a model that incorporates one or more of these variables.

If a driving variable cannot be identified, then it is possible that there is no significant common factor variable (or the variable influence is constant), and so the average, or some other measure, can be used to model the resource use. Alternatively, we may need to go back to first principles and identify additional potential variables or carry out more detailed investigations that will enable a model to be created.

All these steps and the related techniques are described in greater detail in the following pages.

Exploration: *A note on models*

Models are a fancy name given to a mathematical representation of a real world process or activity.

The models we are interested in are formulas that predict the resource we use based on one or more variables that influence the use. For example, I could have a formula that says that my petrol consumption p depends on the distance I travel d. This model, predicts p will change with d in a linear way, but in fact there will be differences due to other factors, such as my speed. Understanding this exceptional variation helps me better manage my petrol consumption, e.g. by driving slower.

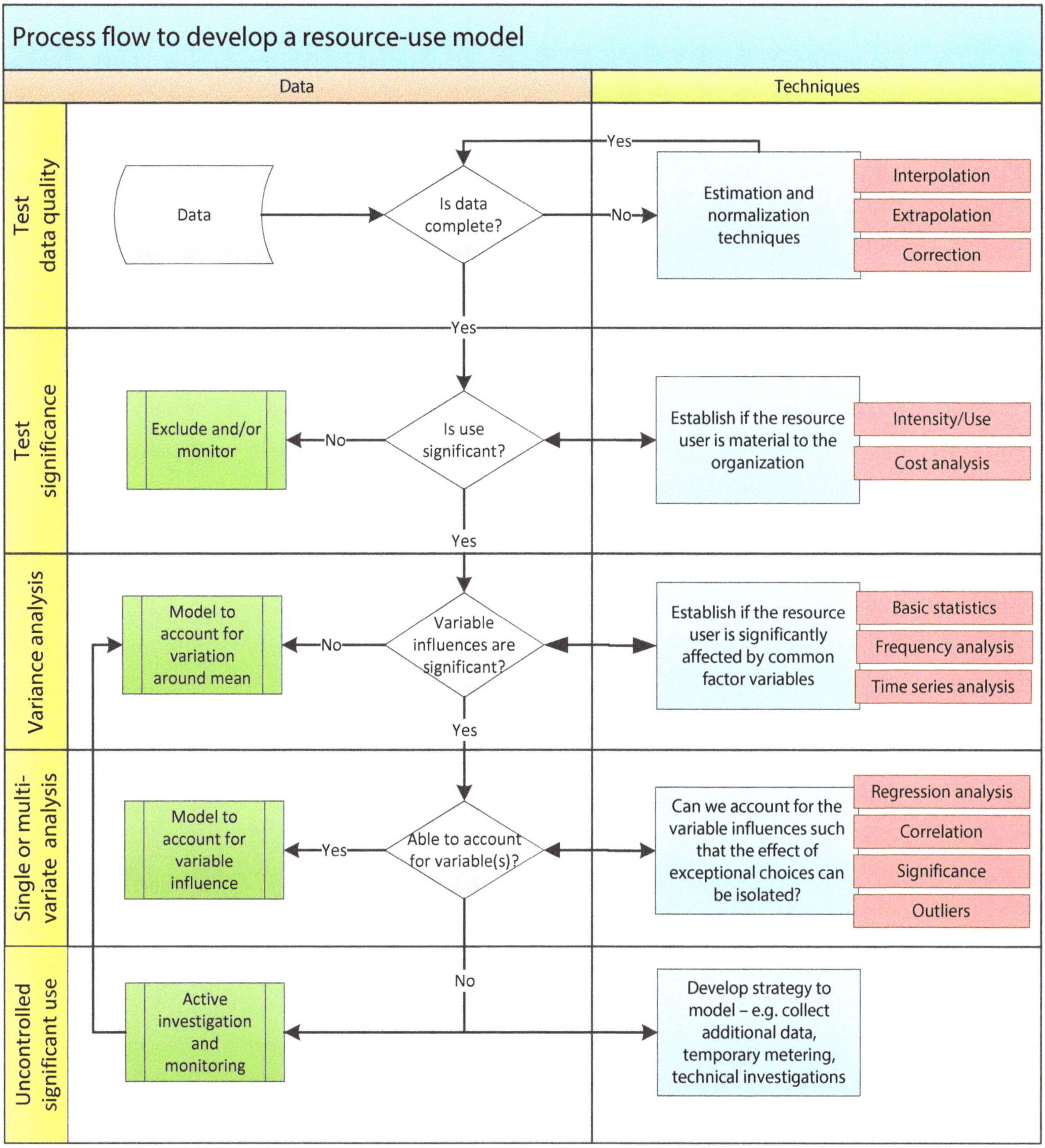

14.11 **Process flow for analysing resource data to identify improvement opportunities or control strategies**
The key to using data to enhance control or identify improvement projects is to understand why variations in use occur. There are a number of techniques available, but these should be used with care as they may not always be suitable.
Source: Niall Enright, image available in the companion file pack.

14.8 Correcting data

When using resource data we always need to be on the alert for potential errors. In the event that we do encounter a genuine error, as opposed to an inconvenient value, the best approach is to exclude that data point from the analysis or use other techniques to correct, or fill in, the data if needed.

In order to assess time series data we need to be sure that the values are at ***uniform intervals.***

One of the key things we need to do when analysing resource-use data is to establish if there are any errors. In time series data there are two common mistakes:

- an incorrect measurement or reading, e.g. due to instrument error; and
- a correct reading that is taken at an incorrect interval.

The most common form of incorrect measurement in time series data is a missing reading. Some data collection systems will record a missing reading as if it is the value zero, so wherever this value occurs we need to test if it is genuine or not. Other measurement errors will be due to the accuracy of the instrument, and it is important not to draw conclusions from variances which are less that the meter accuracy. Tests such as variance from the mean, or z-scores (explained later) in normally distributed data, can help us to pick out exceptional readings that need to be investigated for error.

If we are going to undertake statistical analysis on time series data we first need to establish whether the periods of time for each value are uniform (e.g. half-hourly, daily, weekly) or irregular (e.g. monthly). It is surprising how often people are looking at small variances in month-to-month data as if these are significant, where in fact, these are merely a reflection of the different lengths of time involved. A 30 vs 31-day reading period is a 3% difference in time, while the 28-day February period is 10% less than a 31-day month.

Particular attention needs to be made to ***zero values*** *as these may be genuine values or artefacts due to missing data.*

If our time series data does not have consistent intervals, then this will need to be normalized. This normalization can be through shifting, where we are adjusting existing reading pairs, interpolation, where we are determining the normalized value between two existing values, or extrapolation, where we are determining a value that falls outside (either at the end or beginning) of our data series. Because we have data either side, interpolation tends to be more accurate than extrapolation. Aspects of normalization and meter types were covered earlier on page 421. Techniques for correcting data are referred to as estimation methods, and we need to be aware of any specific rules that apply to specific schemes, such as the Carbon Reduction Commitment in the UK.

Where a time series value is missing, we should consider letting this be a blank or null entry (if necessary deleting zeroes in the data), as most statistical functions in Excel and most M&T software packages will simply ignore blanks, and this signals that the value is unknown rather than measured as zero.

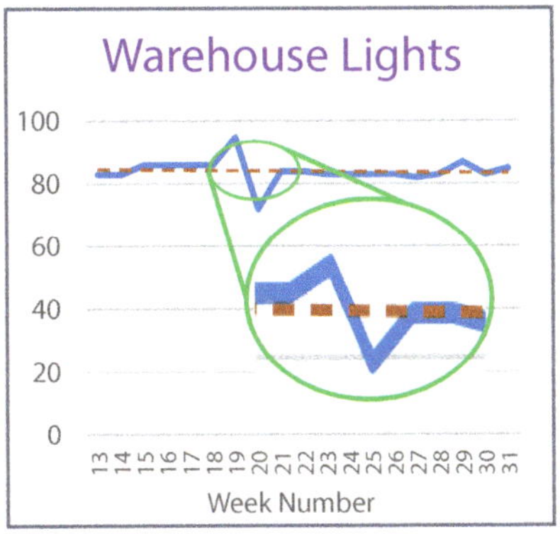

14.12 **Sawtooth data**
A sawtooth pattern can arise when a reading in a uniform series is taken early or late.
Source: Niall Enright.
Illustration available in the companion file pack.

In Numbers: ***Data correction***

When looking at time series data, I tend to start off with a simple line graph, such as the one shown left. Here, we can see the weekly energy consumption of warehouse lighting. One of the key patterns to look for in this data is a sawtooth pair, which has been highlighted in the example. This pair of readings, in weeks 19 and 20, consists of an unusually high reading followed by a low reading. This shape suggests that the first reading was possibly read late (e.g. it could be over eight days' use rather than seven) and the second reading, read on the correct date, contains a shorter period of data than usual (e.g. six days rather than seven). Similarly, a sawtooth with a low value followed by a high value indicates that the first reading was possibly read early.

If I had confirmed that the reading was taken a day late, I would use shifting to correct this value, since the error lies between existing data. I would divide the consumption by the eight days, and then subtract this value from the consumption in week 19 and add this value to the consumption in week 20.

For missing values, we can interpolate a value, where the missing value falls within our series or we can extrapolate a value where the missing value falls at the start or end of the series.

Interpolation, shown in orange in the chart below, involves drawing a straight line between the consumptions preceding and following the missing reading and calculating the missing value. For one missing value, we can simply add the previous and following values and divide by two to get our new value. However, the generic formula for interpolating one or more values is as follows:

$$V_i = \frac{V_f - V_p}{nperiods} + V_{i-1}$$

Where V_i is our missing value, Vp is the previous value preceding our gap, V_f is the following value after our gap, and V_{i-1} is the value before the current interpolated values (which may be V_p for the first value). There is an identical formula for extrapolation, only *nperiods* is 1, so this can be left out, (purple in the chart below):

$$V_i = (V_p - V_{p-1}) + V_{i-1}$$

Where V_i is our missing value, V_p is the value before our gap and V_{p-1} is the value preceding this one and V_{i-1} is the value immediately preceding the interpolated value (which may be V_p for the first value).

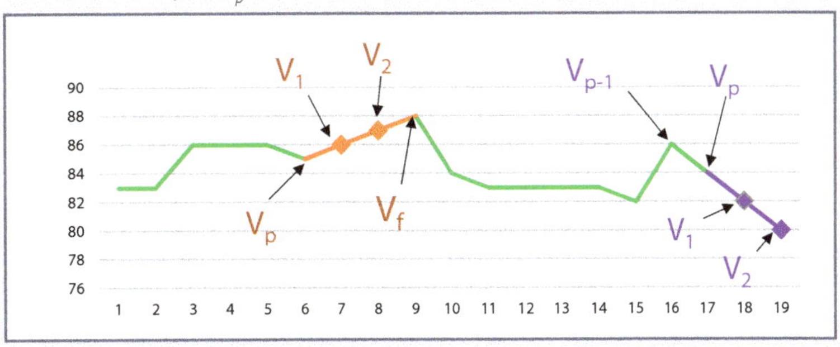

14.13 **Interpolation and extrapolation**
These techniques can be used to fill in for one or more missing data points.
Source: Niall Enright.
Illustration available in the companion file pack.

Insight

14.9 Simple statistics

Many people find the idea of statistics off-putting. In reality we use statistics all the time, and a number of very simple statistical values can give us profound insight into data.

The statistics we will use are usually ***descriptive values*** *that tell us something about our data. If we understand how to interpret these we can gain really useful* ***insights.***

Basic statistics can provide insights into collections of numbers, called a data series, such as electricity meter readings, waste values or production data. Statistics provide scientific methods to interpret these data series and to test assumptions about the data. Probably the most common statistical calculation that we use for series data with one value (univariate) is an average. This should better be referred to as the arithmetic mean (or usually just mean), which is the total of all the values in the series, divided by the number of values.

Not to be confused with the mean, we have the median, which is the middle value in the series. If we have an odd number of data points in the series then the median is the value with the same number of observations above and below, if we have an even number of points the median is the average of the two middle points). Then we have the mode, which is the value that occurs most often in our series (there can be more than one mode).

Where we have calculated the mean of a data series, we have a statistic which enables us to model the data. Another statistic about our data is the sample size, i.e. the number of values we have in our series. Sometimes statistics can tell us something with certainty about our data, for example, with both the sample size and the mean we can calculate the sum of all the values in our series, without knowing any actual values. Some statistics give a firm conclusion about your data; others are descriptive and require *"rules of thumb"* to interpret. When thinking about the statistics in the following pages, it is worth remembering that *in every case*:

Data = Statistical Model + Error

While statistical models may *represent* a data series, we also need to understand the error that is present. Unless every value is the same as the mean, data in our series will differ from the mean. Some values will be above and some values will be below. The difference between a model value and a measured value is called a deviation, residual or variance (we will tend to use the latter). Note that *"error"* here does not mean *mistake* or *meter error;* it means *difference.*

Taken together, these variances represent the *error* in our model, so we need a way of calculating what these are, in other words, what the distribution of values is around the mean. The formula we use is called the standard error and is explained in the box on the right. A large standard error means that our data is widely dispersed, while a small one indicates that it follows the mean closely.

Real World: ***The meaning of "error"***

Statistical words do not always have the same meaning as in common parlance. Error does not mean mistake, but is simply another word for variance around the model prediction.

	Value	Mean	Error	Error^2
	8	7.6	0.4	0.16
	7	7.6	-0.6	0.36
	6	7.6	-1.6	2.56
	8	7.6	0.4	0.16
	9	7.6	1.4	1.96
Sum	**38**	**38**	**0**	**5.2**

14.14 **A data series and statistics**
The five values on the left are simple data series from which a number of statistics can be calculated.
Source: Niall Enright

***In Numbers:** Series statistics*

Let's take a very simple data series: the hours of lighting for a store in a five-day working week:

8	7	6	8	9

The mean of these values is:

$$\frac{8+7+6+8+9}{5}=\frac{38}{5}=7.6$$

The table left summarizes the data. From this table we can see that the sum of the errors (or residuals) in column 3 is zero. The reason for this is that the negative and the positive variances cancel each other out, so this is not useful as a measure of the distribution of the errors around the mean.

The way to get rid of the negative values is to square the errors, as shown in column 4. We can then add these up to get the total of 5.2. If we then divide these by the number of values in our series and take the square root (to reverse the earlier squaring), we get the standard error (also called population standard deviation) which is 1.02.

$$SE=\sqrt{\frac{5.2}{5}}=\sqrt{1.04}=1.02$$

Insight

***Exploration:** Statistical formulae and notation*

Statistical calculations are very simple if one understands the notation that is used. For example, the equation for the arithmetic mean of a sample of values is:

$$\bar{x}=\frac{\sum x_i}{n}$$

Where the letter x_i represents *all* the data values in our data series ($x_1, x_2, x_3 ... x_n$) from the first value (x_1) to the n-th value (x_n), n is the number of values, and Σ (the Greek upper case letter Sigma) means the sum of all the values. So simply said, the first formula is *"take the sum of all the values of x and divide by n"*. The arithmetic mean is indicated by a bar above the variable name, $\bar{x}$ (called *"x-bar"*).

The formula for the standard error, or population standard deviation, described above, is:

$$\sigma=\sqrt{\sigma^2}=\sqrt{\frac{\sum(x_i-\bar{x})^2}{N}}$$

In simple terms, the formula says: *"take each value and subtract the mean"*, denoted by $x_i-\bar{x}$, *"which gives the variance (aka residual)"*. Then *"square each variance"*, $()^2$, *"and sum these squares together"*, $\Sigma()$. Finally, *"divide the result by the number of values in the sample"*, N, *"and then take the square root"*, $\sqrt{}$. The result is usually denoted by the symbol sigma, σ, which we saw earlier when discussing quartiles (see page 435). Sigma is a very, very useful statistic, as we shall see next.

Statistical calculations are very simple, once you understand the notation. The maths is usually uncomplicated.

14.10 Using probability

Where data falls into a known distribution curve, such as a normal distribution, then it is possible to use probability to assess if a value is due to exceptional variation or simply random influences.

Exploration: *Sample vs population*

There is a subtle difference when the numbers we are analysing are a sample of a larger population.

With the whole population, i.e. N values, we can be confident that our statistics are most accurate, but with a sample of n values, we lose accuracy, as our sample may be slightly biased *compared to* the population values.

To reflect this greater uncertainty, when we calculate variance or standard deviation from a sample we divide by **n-1**, rather than **N**, which has the effect of *increasing* the variance or standard deviation. The mean, however, is calculated by dividing by n or N respectively, so will be the same for both types of data.

As our sample size approaches the size of the population, then n gets closer to N and the differences between the sample and population statistics decrease.

Sample and population statistics use different symbols, as illustrated in the table below. In most cases, we are using sample data rather than population data for our resource efficiency analysis.

	Sample	Population
# Values	N	n
Mean	$\bar{x}$	μ
Variance	s^2	σ^2
Std deviation	s	σ

Probabilistic statistics are critical in many quality systems, such as Six Sigma or SPC. Probability statistics describe the likelihood of a given result or value occurring, from some very basic statistics about the readings, namely the mean and the standard deviation.

This type of analysis is used in quality systems like Six Sigma and SPC because these disciplines use what is called normally distributed data, which allows univariate statistics to be calculated to understand performance. Examples of normally distributed data are the diameter of a ball bearing, the number of errors in a customer survey, or the number of damaged parts per thousand.

A normal distribution is a distinctive, bell-shaped *probability distribution* which occurs when the variance in a data set is random, due to measurement errors or from common factors with a relatively small impact. The width of the bell curve is related to the standard error or standard deviation (same thing), which we learnt about in the previous section, and which describes how far our data varies from the mean. The highest probability of a value occurring is around the mean and the lowest probability is at either of the tail-ends of the distribution (this is the central limits theorem if you want to find out more).

So, the standard deviation describes the shape of our bell curve (peaky or flat), and it is also a measure of the probability of any given variance from the mean occurring. In a normal distribution, we can use the 68-95-99.7 rule to remind us of the likelihood of a random value falling within one, two or three standard deviations, denoted by σ for population standard deviation or s for the sample standard deviation. Thus the probability of a value being within $\pm 1\sigma$ from the mean is 68%. Within $\pm 2\sigma$ it is 95% and $\pm 3\sigma$ it is 99.7%.

This probability rule applies to all normal distributions (of either sample or population data), so that, for example, only 0.3% (three in a thousand) ball-bearings measured will vary from the mean by more than 3σ. The name Six Sigma refers to achieving better than $\pm 6\sigma$ or 99.99966% or just 3.4 defects per million. It follows from this that if we calculate the number of σs from the mean for a measurement, we can determine its probability. The variance expressed in the number of standard deviations is called the z-score and is calculated by dividing the value's variance from the mean by σ (or s). Knowing the z-score can tell us if a value is likely to be an exception, either due to error or an exceptional variation. The example opposite shows how this works in the real world.

Z-scores provide an easy to use and powerful test whether an outlying value is due to chance, or exceptional variation.

Real World: *Understanding frequency data*

A lot of our data analysis work is designed to tell us if a particular value we measure is significant or not. One useful tool is to determine the distribution of the data available, which can indicate whether an individual value is within the expected range. This technique involves plotting a frequency chart of our data series, although the statistics involved can be produced without charting the data.

Let's look at a real-life example taken from a home and personal care products factory making shampoos. This facility uses the same mixing and bottling equipment for different product variants, and in between each batch they rinse the system out. The volume of lost product is calculated by subtracting the volume of rinse water put into the system from the amount that comes out. The full data series is available in the companion file pack.

In this particular example the trend of the data is pretty flat - that is to say that there is not a great variation from batch to batch. So a simple trend chart does not tell us a lot about the losses.

One way of visualizing the data more clearly is to plot a frequency histogram, such as the one shown below. This is not difficult to do - we simply create a number of bands (or "bins") of volume losses, in this case in 2 litre-steps, e.g. 80-82 litres, 82-84 litres, etc., making sure we have covered the full range of values. We then simply count the number of occasions our data falls into each band, and plot a bar chart as shown.

14.15 **Frequency histogram**
This type of chart helps us to visualize if a distribution is approximately normal.
Source: Niall Enright.
The image and source spreadsheet are available in the companion file pack.

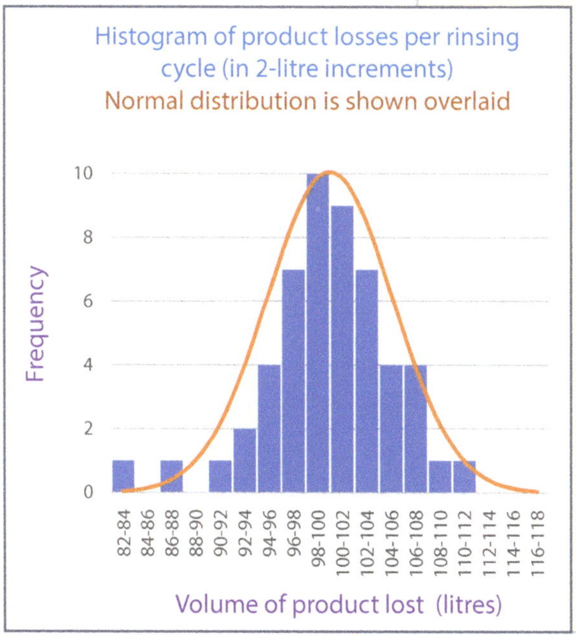

As you can see from the chart, the data cluster around a central value of 96-98 litres (the mean is actually 99.97 litres), and then rapidly decline so that by the time we get to the outermost frequency bands, 80-82 litres and 114-116 litres, there are few or no readings. In fact, these values follow a bell curve or normal distribution as shown by the orange line. The standard deviation, s, which described how the values in a normally distributed data set are dispersed from the mean, is 5.13litres which gives us a fairly *"peaky"* curve.

So we can look at any of the values in our data series and calculate its z-score, which is simply the number of standard deviations or s's, it varies from the mean. The two loss values on the left of the distribution are 83.49 litres and 86.0 litres, which have z-scores of -3.21 and -2.72 respectively (e.g. (83.49-99.97)/5.13 = -16.48/5.13 = -3.21).

Using the reference table z-Table 3 on page 779 we can see how likely a particular z-value is to occur. The frequency of the z-score ±3.21 is once in 757 readings and for z-score ±2.72 it is once every 153 readings. As we have taken only 52 readings, this suggests that these occasions are not due to random variation and they merit further investigation for exceptional variation.

As well as learning that the values on the left of our distribution are exceptional, we have also seen that the data on losses follows a broadly normal distribution, which means that we can confidently use the variance from the mean as a performance indicator. We could even apply some of the more advanced control chart techniques from SPC to identify early any underlying changes in the losses.

Insight

14.11 Analysing time series data

The analysis of resource data over time aims to discern periods of good and bad performance, from which we can identify the root cause and so improve. The approaches for the longer-term trends and detailed profiles tend to differ slightly.

By far the most common way for organizations to analyse resource use is to compare usage over time.

This approach is reinforced by the way that we measure all sorts of organization metrics: we may track sales on a weekly basis; track variances against budget on a monthly basis; report to investors quarterly; and produce our annual profits statement. In manufacturing and process environments, the data resolution moves to the other extreme – with important parameters being tracked in *real time* and any deviation from expected performance corrected rapidly through automatic feedback mechanisms. All this provides us with a potentially large volume of date and time-stamped resource data.

While we have a lot of data to play with, there tends to be practical differences in our analysis at each end of the spectrum: time of year and time of day.

Let's consider the long-term view of resource use, where we are looking at daily, weekly or monthly consumption data taken over one or more years to assess patterns of consumption. This is often called trend analysis, because here we may well discern seasonal patterns or trends. For example, the electricity for air-conditioning will rise in the summer and fall in the winter (in the Northern Hemisphere), opposite to the gas used for space heating, which rises in winter and decreases in summer.

The challenge of longer-term ***trend analysis*** *is to separate out the effect of* ***common factor variables*** *from the influence of* ***manageable factors****.*

The fundamental objective of time series data analysis is to determine if any particular period of consumption is good or bad. To do this we need to separate out two influences:

- The effect of a common factor, or variable, such as weather, which we cannot change easily (if it is hot we will need more cooling);
- The effect of intermittent, controllable factors (such as leaving a window open in winter), which can lead to good or poor performance, i.e. exceptional variation.

It is customary for people analysing consumptions that vary seasonally to use year-on-year usage as a measure of performance. This is because a comparison with the same period in the previous year is likely to be more meaningful than with the last month. In effect, they are looking to eliminate the influence of any seasonal variables, by choosing periods when this will be approximately

In high-frequency, short time frame, profile analysis, we are usually looking for clues about the quality of control of our resource users.

the same, in order to discern if other, controllable, factors that affect use are good or poor. Sometimes, where we can't formally assess the impact of the variable in order to exclude it more scientifically (covered in the next section), we have no alternative but to use these *"similar"* periods of operation as a means to get to an understanding of performance.

In these circumstances, we should take care to select a reasonably *long* comparison period as this may even out any inherent variability in the variable. Thus, comparing a single day's gas use with the same day a year ago may be unreliable, as temperatures can vary day to day. But comparing a whole, similar, month should give less variation, and therefore provide a more reliable comparison. Indeed, a day exactly a year ago may fall on a weekend, but is on a weekday in the current year and if - like most building - we have a different pattern of operation at weekends, our comparison would be invalid.

Insight

At the other end of the spectrum, the analysis of resource use data over short time frames using high-frequency data is called profile analysis. Indeed, this kind of chart for electricity is known as an electricity *profile* (or *demand profile* if the y-axis shows kW rather than kWh). The name reflects the fact that the analysis tends to focus on the *shape* of the consumption as it changes over a day, or from batch to batch, shift to shift, or weekday to weekend. When comparing profiles, the same principle applies as with trends - one wants to look at periods of time where the external common factor variables are broadly consistent, allowing us to hone in on the effects of controllable factors.

The study of profiles often ties in closely with an assessment of the effectiveness of the control systems we have around a resource use. If we see that we are using resources at unexpected times, then we would question the time settings of the control system. Alternatively, if we see that we are exceeding desired parameters (such as temperature), we would question the setpoint of the control system or the accuracy of the feedback instrument (i.e. the thermostat).

Interpreting profiles usually requires greater experience than trend analysis. For example, we may only spot that night-time use is higher than expected if we have seen the profile of many other similar facilities and even then we may not be able to arrive at a judgement unless we know the activities that are taking place at the site. The example on the next page shows that subtle, easy to overlook, aspects of a profile can show opportunities for improvement.

14.16 **"Before and after" electricity profiles (following pages)**
These profile charts show the reduction in electricity consumption at the Peel Land & Property Group head office in Manchester, resulting from energy management measures taken between 2014 and 2015.
Source: Courtesy of Peel Land & Property Group. The images and source spreadsheet are available in the companion file pack.

In my own work, I tend to use trend analysis only to understand the general direction of use and cost, rather than as a means of identifying improvement opportunities. For this frequency of data, I favour the much more accurate regression analysis techniques, described shortly. However, the lack of variable data at short time intervals, or the number of influences that come to bear over a short timescale, means that regression analysis is rarely possible for high-frequency data. If this high-frequency data is available, I will almost always use *profile* analysis to look into specifics, like the effectiveness of control systems, or to spot unusual patterns of use which merit further investigation.

⇒ page 458.

Real World: *Profiles in action*

Profile data is helpful at picking out what is happening to resource use at different times of the day. The most widely used form of profile data is electricity, not only because many electricity tariffs have a time of day component and so the data is often available from utility companies, but also because electricity sub-metering and data-logging are relatively cheap.

At Peel Land & Property Group, the Energy Champion Chris Foran, Engie FM contractor John Cardus and data analyst Caroline Robertson-Brown have been paying close attention to the electricity demand profiles from the many office buildings in the investments portfolio.

We can step through an example of the analysis that has been done to get an appreciation of the benefits of profiling. Let us take a typical week's demand profile for Peel's head offices over a week in May 2014, when the data was first reviewed. This profile is shown below.

The first thing we will note is that there is a clear night and day separation; the weekday night-time use is around 15-20 kWh per half-hour (i.e. a demand of 30-40 kW), or around 25% of the daytime use of around 65kWh (which is what we would generally expect in an office with ordinary operating hours and without a data centre or other constant load). Indeed, we also see, as expected, that the daytime load for Saturday 10 May and Sunday 11 May, on the right of the chart, is less than the weekdays, as the offices are empty.

But, as was mentioned earlier, the purpose of the analysis is to identify potential performance improvements. Two distinct patterns caught the attention of the team.

- Just after midnight every day there is a "shoulder" where demand rises from around 00:00-00:30 until 05:30, when other systems come on to prepare for people's arrival at work. We can estimate the electricity consumption this represents by looking at the chart - the lows (19:00 to 24:00) are around 15-20 kWh per half-hour and the "shoulder" (ca. 00:30 to 05:30) is around 30 kWh, so we have an increase of, say, 12 kWh per half-hour for about five hours, which is 120 kWh per weekday night, which adds up to a possible saving of 600 kWh per working week.
- The second observation is that overnight at the weekend we are not achieving the low energy consumption we can attain overnight on a weekday. The observed use is around 30 kWh per half-hour, which could potentially be 15-20 kWh per half-hour if the weekday minimum could be achieved. If we assume that the decrease could occur between 19:00 and 05:00 then we have a savings potential of 12 kWh per half-hour for 10 hours or 240 kWh per day, 480 kWh per weekend.

This total 1080 kWh per week potential savings, 600+480, is a not inconsiderable amount of electricity, £108 in cost terms at 10 pence per kWh, per week - over £5,600 per year - and about 8% of the weekly total. The message here is that seemingly small and subtle deviations from expected consumption over a day can add up to a significant opportunity for improvement.

Having identified a possible improvement, this was immediately logged as an opportunity in the Peel Energy Management Opportunities Database. John Cardus was now tasked with devising a programme of work to investigate the pattern of use and assess if a reduction was feasible from a cost and technical perspective. Two obvious aspects needed to be investigated, the

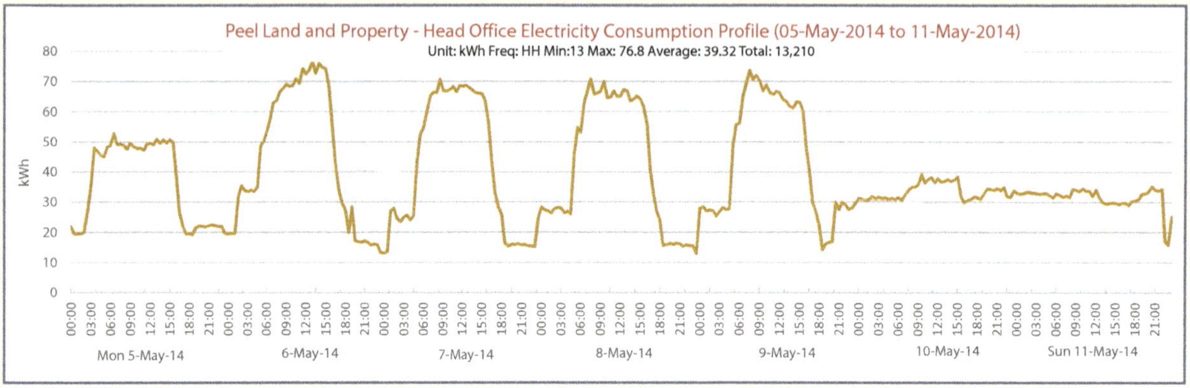

buildings management system (BMS) and the heating and ventilation and air conditioning (HVAC) plant. As a result, an engineer from the controls company TREND came on site and assessed the equipment and the controls strategy in place at the time.

The engineer's study identified many issues with the control of the HVAC plant: timing of operation of some plant; less efficient equipment being the *"lead"*; controls on some extractor fans in the washrooms had been bypassed; and variable speed drives on some of the air-handling units were operating at full load (which is not efficient). Many of these problems could be remedied straight away. Control of the plant was centralized and individual room controls deactivated, which would have the effect of preventing cleaning staff switching on equipment at night, which was believed to cause some of the night-time *"shoulder"*.

The impact of all these small changes is dramatic. If we draw 2015 year's profile for the same week in May, shown below, we can see that the overall consumption in the week has decreased from 13,210 in 2014 to 10,449 in 2015, a drop of over 20%. On an annual basis this saving would be worth over £14,000, for an investment of a few hundred pounds in the analysis work.

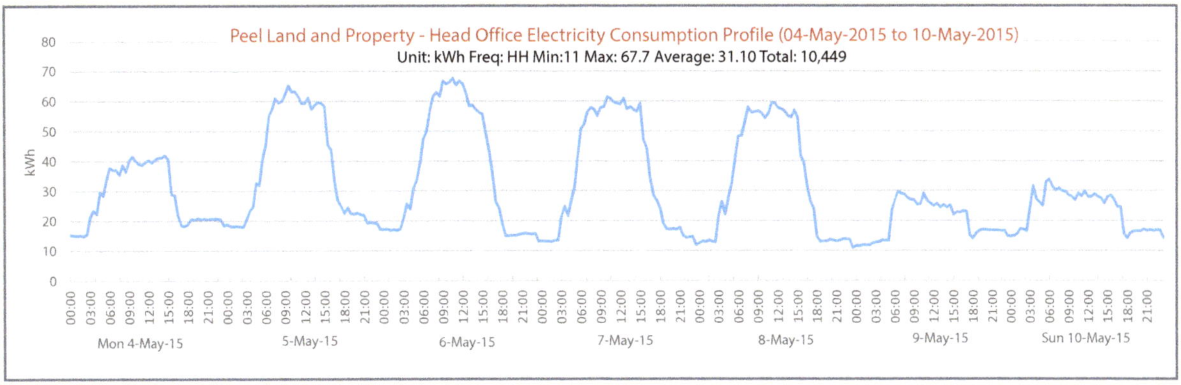

At first glance, the two profiles (above and left) seem very similar. However, by overlaying the two periods on one chart, as shown below, the differences are more obvious. The "shoulder" at night has disappeared, the weekend night-time use is now the same as the weekday night-time use, both of which were identified as key issues initially. Additional improvements can also be seen. For example, the daytime maximum use has decreased significantly (despite staff number growing over the year) and equipment seems to be coming on slightly later and switching off slightly earlier.

For those who doubt the ability of organizations to make significant reductions in resource use, this example provides confirmation that the process of looking closely at a resource use profile can lead to great savings. In a situation like this, where the building is being heated and cooled fine, there is a widespread tendency to fall into an assumption that everything is working well whereas an examination of the profiles will often reveal many easy-to-correct inefficiencies.

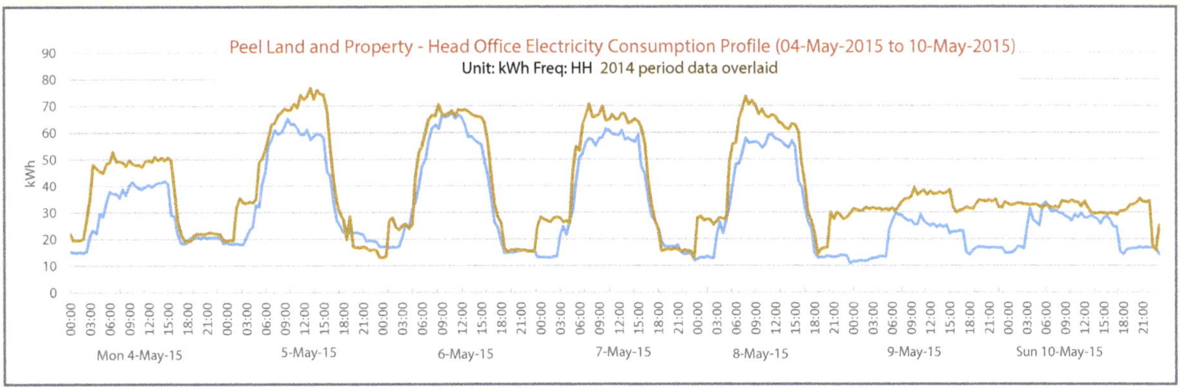

14.12 Using specific ratios

Specific ratios compensate for the effect of common factor by dividing the resource use by the factor. While this goes some way towards separating good and bad performance, there are a number of underlying issues with ratios which we need to be aware of when using them.

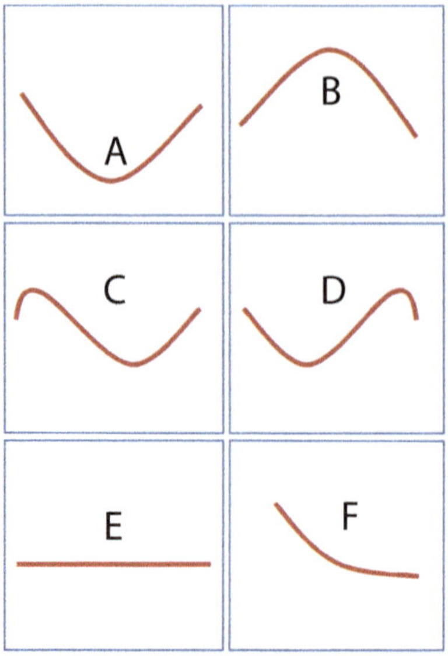

14.17 **Differential specific energy consumptions plots (SEC against activity)**
Source: Niall Enright, adapted from Kit Oung, Energy Management in Business.[581]
Illustration available in the companion file pack.

One way that people seek to take into account a common factor variable is just to divide the resource use by the variable, to create a specific ratio which adjusts for the variable so that the exceptional variation can be more readily detected. Thus, we could report our lighting energy use in an office per sq ft, water use in a golf course per acre, waste in a factory per tonne of product.

We covered specific ratios earlier in the chapter on Goals (page 368) and have seen that these ratios are widely used to benchmark (see page 434) where ratios are used to *normalize* data.

In benchmarking, we are comparing different things using the same metric, called a cross-sectional study in statistics. An example would be to compare the energy use across a portfolio of supermarkets using the kWh/m^2 sales area. There is another use of specific ratios, as a tool to compare the same resource user over different periods of time, in other words, as a measure of change in time series data, or a longitudinal study.

Kit Oung, in his excellent book *Energy Management in Business,*[581] describes an interesting application of specific ratios based on longitudinal data. Here, the specific energy consumption, SEC (i.e. the specific ratio of energy use per unit production, e.g. kWh per product), is plotted against total output (or some other measure of activity), as shown in the illustration left.

Thus, we can see in case A the *"sweet spot"* for operating our factory is mid-way through the plant capability. In example B, we probably want to run our facility at full load (since we won't be making much at very low load when the ratio is as good). More complex plants can exhibit several peaks or troughs as shown in C and D. According to Oung, these patterns are due to a range of rotating machinery acting on the production. Then, rarely, one may get a plant where the specific ratio is unchanged through the production range, as shown in E. Assuming we can control for other influences and we can get data for a range of production, this analysis adds valuable insight to time series data. The analysis above would, presumably, only be valid if our production operations were not changing fundamentally from one measurement to another.

The chart 14.18 above right, show the SEC of UK domestic households from 1990 to 2014 (adjusted to 1990=100), in green. On the face of it, the chart is very good news - the SEC has declined from 100 to 62.2. The problem with this, however, is that energy use (the top part or numerator in the ratio),

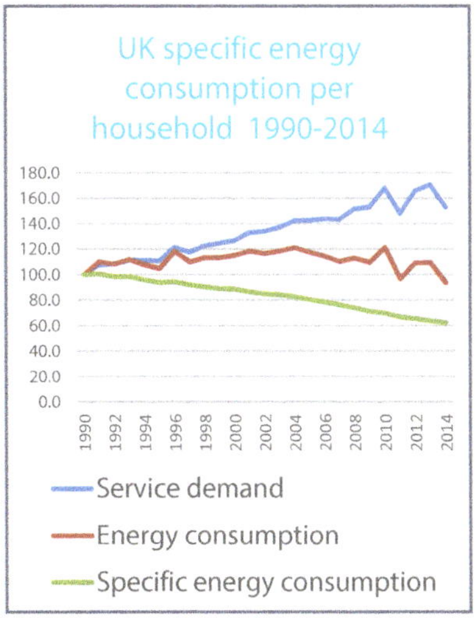

14.18 **SEC of UK domestic energy consumption 2009-2014**
Source: DECC.[189]

shown in red, has only decreased to 93 (it was 109 in 2013). The apparent improvement is entirely down to the increase in the denominator (the bottom part or divisor) of the ratio, the service demand which is shown in blue on the chart which has risen to 152 in 2014. Service demand has increased as people have wanted warmer homes, they use more hot water and have more gadgets. Clearly, it is a good thing that we have been able to meet this 50%+ rise in service demand while holding energy roughly constant, through better efficiency, etc., but nevertheless, the ratio has masked the fact that absolute energy use has not changed. This is important because, as we saw in the introductory chapters to this book, the impact of many resources, such as emissions, are related to the absolute value.

Insight

The key shortcoming of specific ratios as a resource analysis and management tool for time series data is that a ratio does not distinguish between the effect of a *desirable reduction in the numerator* (i.e. the quantity of resource used) or *an increase in the denominator* (i.e. the driving variable). Specific ratios take our eyes off the sustainable goal of reduced absolute resource use. Specific ratios are also prone to adjustment over time of the definition of either the numerator or the denominator value, making like-for-like comparisons difficult.

Another flaw with specific ratios was mentioned in the earlier piece on goal-setting. Here, we pointed out that the ratio dictates that when the denominator is zero then the numerator will also be zero. This is a very dangerous assumption to make. In reality, most resource-consuming systems exhibit a baseload component which is unchanged in respect of the variable. For example, a building will use some gas for hot water even though the heating requirement ceases in the summer; a car factory will use quite a lot of energy for non-productive purposes (such as running a staff canteen) so, at zero vehicle production, we will see some energy consumption; a bank will use electricity for security lighting that is not related to their opening hours.

Not only do we see baseload effects in longitudinal data, but also in cross-sectional data. In a supermarket group, electricity use will be strongly related to the sales area of each facility, but the relationship will not approach zero as the floor area decreases as there will be other energy use in the mix, such as electricity for car parks or external lighting, which represent consumption unrelated to the floor area. This baseload tends to flatter the larger users at the expense of the smaller one (see *Marketing Trucks in Canada* on page 369). Often a specific ratio is quite simply an incorrect model with which to differentiate good and poor performance. Returning to Oung's differential SECs (illustrated in Figure 14.17 opposite), the most common pattern observed is a decrease of the SEC with activity, shown in Box F, as the fixed baseload is divided by an ever-increasing number of units of activity (see page 486).

Be careful using specific ratios as they miss baseload effects.

For this reason, we need a more sophisticated approach to establishing relationships between resource use and common factors, called regression analysis, which takes into account the baseload resource use and which is explained in the following pages.

14.13 Simple linear regression

Regression analysis is an easy technique to enable us to take into account the effect of common factors, or variables, on resource use. This permits us to develop models that can highlight the exceptional variation.

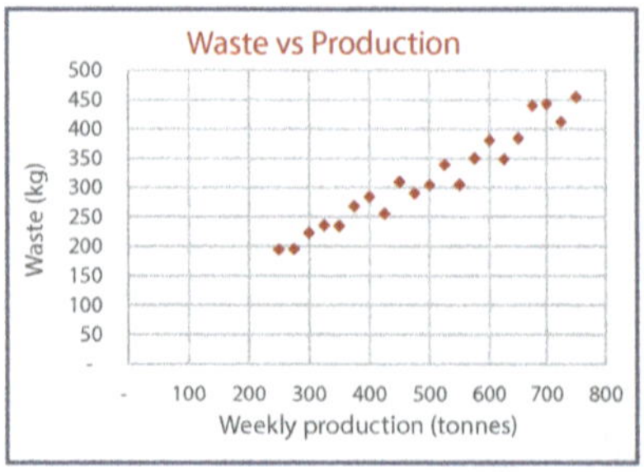

14.19 **Scatter plot**
The convention is that resource use is always shown on the vertical axis and the variable on the horizontal axis
Source: Niall Enright.
All the illustrations shown here and the source spreadsheet data are available in the companion file pack.

If instead of plotting time series data by date, we plot resource use on the vertical (y) axis and we plot the variable that we believe influences our resource consumption on the horizontal (x) axis, we will get what is called a scatter plot. This chart is shown in the figure, left.

We can see that resource use rises as the variable increases (e.g. as our weekly production increases then the quantity of waste also increases). We can reinforce this relationship by drawing a linear regression line (or best-fit line) through these scatter points. This can be fairly easily done by eye, but software like Excel will also draw the line on a scatter chart (see instructions opposite). The regression line is shown in green in the second chart, below left.

The regression line provides a relationship between waste and production. In fact, there is a very simple formula that we can use to describe the relationship:

$$y = mx + c$$

We can say that the waste produced rises with production by the factor m, which is the *slope* of the linear regression line, and that there is an additional fixed amount of waste *each week*, c, which is the intercept of the regression line with the y-axis. This intercept is commonly referred to as the *baseload*.

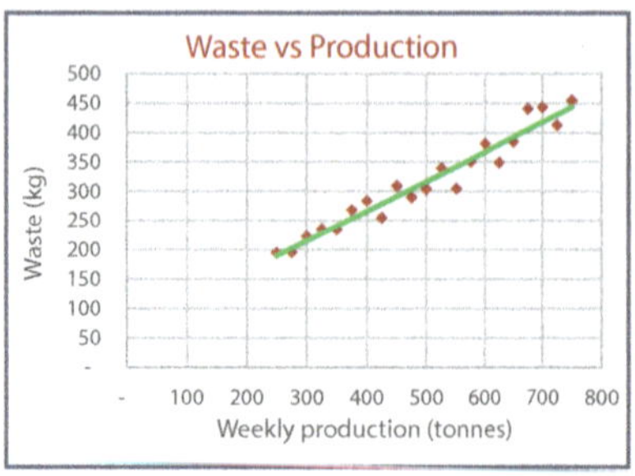

14.20 **(above) Scatter with regression line**
The convention is that the Regression line only extends horizontally for the range of the variable.
Source: Niall Enright

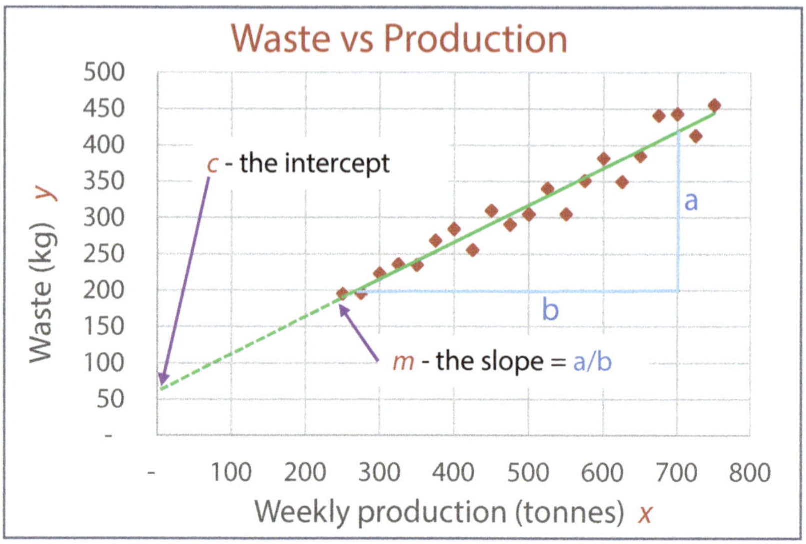

14.21 **(right) Elements of a regression line**
Illustrating the y = mx+c formula elements
Source: Niall Enright

In Numbers: Linear regression and Excel

Week	Production	Waste	Calculated values		
Equation->	x	y	x^2	y^2	xy
1	675	441	455,625	194,216	297,472
2	450	310	202,500	96,020	139,442
3	325	236	105,625	55,698	76,702
4	350	235	122,500	55,306	82,310
5	275	196	75,625	38,388	53,880
6	300	223	90,000	49,877	66,999
7	700	443	490,000	196,404	310,222
8	575	351	330,625	123,096	201,739
9	600	382	360,000	145,770	229,079
10	425	256	180,625	65,393	108,681
11	725	413	525,625	170,696	299,536
12	625	349	390,625	122,128	218,418
13	650	385	422,500	148,376	250,378
14	525	340	275,625	115,683	178,564
15	550	305	302,500	93,083	167,802
16	250	195	62,500	38,105	48,801
17	750	455	562,500	207,388	341,549
18	475	290	225,625	84,381	137,980
19	500	305	250,000	92,821	152,333
20	375	269	140,625	72,150	100,728
21	400	284	160,000	80,894	113,767
SUM()	10,500	6,664	5,731,250	2,245,873	3,576,383
Equation	Σx	Σy	Σx^2	Σy^2	Σxy
AVERAGE()	500	317	number of values n =		21
Equation	$\bar{x}$	$\bar{y}$		$n\bar{x}\bar{y}$	3,332,182

The table to the left has the same production and waste for 21 weeks as in the charts opposite. There are three columns of calculated data, the squares of x and y and the product of x and y. At the bottom of each column we have calculated the sums, and the averages for x and y.

Given this data, we can calculate the slope, m, of our regression line as follows:

$$m=\frac{\sum x_i y_i - n(\bar{x}\,\bar{y})}{\sum x_i^2 - n(\bar{x})^2}$$

Substituting the values we get the result:

$$m = \frac{3,576,383 - 21(500)(317)}{5,731,250 - 21\,(500)^2} = 0.5074$$

The equation for the intercept is given by rearranging our $y=mx+c$ formula:

$$c = \bar{y} - m\bar{x}$$

Substituting the values gives us the intercept:

$$c = 317 - 0.5074\,(500) = 63.635$$

We don't need to remember these formulae because tools such as Excel make it easy to draw scatter charts, add regression lines and display the equation of the line. To draw the scatter plot put your data into two columns, making sure that the first column has the x-values (variable) and the second column the y-values (resource). Highlight the two columns and choose **Insert**, **Chart** and then select the "scatter chart". You should see a plot like Fig 14.19. To add the line of best fit, right-click on the series (the points in the scatter) and choose **Add Trendline** from the pop-up menu. This displays the **Format Trendline** panel (in Excel versions prior to 2013 this is shown in a dialogue box). Choose **Linear** for a linear regression line. You can also tick the option "**Display Equation on the chart**" and the equation $y = 0.5074x + 63.635$ is added to the chart.

Excel has some functions that make calculating these values from data easy.

Statistic	Excel Function	Result
Slope, m	=SLOPE(known_ys, known_xs)	0.507431044
Intercept, c	=INTERCEPT(known_ys, known_xs)	63.63514164

Note that the Excel functions above take the y column (our variable) first and then the x column (the observed resource use). When creating a regression graph the order is reversed, with x being the first column of data, followed by the y-values.

*Linear regression is used to create "**fair, honest and achievable" targets,** as half the variance will be below the line (good) and half above (bad).*

14.14 Scatter and correlation

We instinctively know that if our data is widely scattered then the effect of the variable is weak. Now we will see how we can express this formally using correlation coefficients.

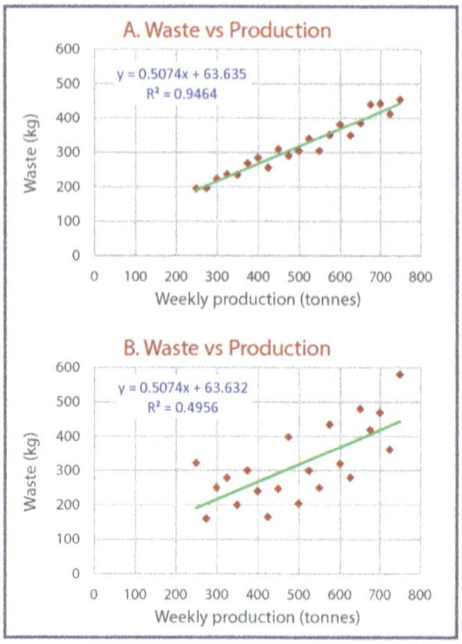

14.22 **Two charts with identical Linear regression equations**
The top chart is the same data as was used in our explanation of linear regression, the bottom chart has data points which are much more widely scattered.
Source: Niall Enright.
All the illustrations shown here and the source spreadsheet data are available in the companion file pack.

In the simple linear regression example of waste vs production on page 460, we determined that the formula to predict the amount of waste produced (following the $y=mx+c$ equation) was:

$$\text{Waste} = (0.5074 * \text{Production}) + 63.635$$

This formula was calculated using the data from the chart A, left. It may be surprising to learn that the data shown in chart B, below, also produces the same equation. The slope m and intercept c have the same value for both sets of data, even thought the points in chart B are more widely scattered.

Clearly, there is another aspect of our relationship between waste and production that we need to understand. This is commonly referred to as the scatter, which is a measure of how well the two values, waste and production, are related. If we imagine all the data points fall close to or exactly on our line of best fit, as in chart A, then we could say that the linear regression equation *strongly predicts* the amount of waste for a given level of production. On the other hand, if the data points are much more widely scattered around the line, such as in the chart B, then we would say that the relationship between waste and production is weak.

Fortunately, rather than relying on our visual senses, there is a statistic, developed by Karl Pearson, to measure this scatter. This statistic is called the Pearson correlation coefficient (from *"co"* for more than one and *"relation"* for relatedness, and *"coefficient"* because we have one ratio that expresses the overall relationship). The correlation coefficient is shown with the symbol r_{xy}. The correlation coefficient for chart A is 0.972, and the correlation coefficient for chart B is 0.704. Thus, the larger the correlation, the closer the points are to the linear regression line (the best-fit line).

If every point is on the best-fit line, then we would end up with a perfect correlation of 1 or -1, depending on whether the line of best fit slopes upwards, as in our examples, or downwards. In practice, we tend to use an alternative measure of correlation, called the coefficient of determination, symbolized as R^2, which describes the proportion of the variance in the waste created (x) that can be explained by the production (y). Because R^2 is the square of r_{xy}, it will always be positive and between 0 and 1. The R^2 value for chart A is 0.9464 (0.972*0.972), and for chart B it is 0.4956 (0.704*0.704). Thus we can say in chart B that only half of the variance in waste is production-related and so there must be other significant factors at play.

Understanding this relationship transforms our ability to influence the resource use. In the following pages, there is more on interpreting whether a correlation is significant and on errors or choices that may affect the results. ⇒ page 466.

***In Numbers:** Calculating correlations*

Week	Production	Waste	Calculated Values				
Equation->	x	y	$x - \bar{x}$	$y - \bar{y}$	$(x - \bar{x})(y - \bar{x})$	$(x - \bar{x})^2$	$(y - \bar{y})^2$
1	675	441	175	123	21,586	30,625	15,215
2	450	310	- 50	- 7	374	2,500	56
3	325	236	- 175	- 81	14,236	30,625	6,617
4	350	235	- 150	- 82	12,327	22,500	6,753
5	275	196	- 225	- 121	27,320	50,625	14,744
6	300	223	- 200	- 94	18,804	40,000	8,840
7	700	443	200	126	25,165	40,000	15,832
8	575	351	75	33	2,512	5,625	1,122
9	600	382	100	64	6,445	10,000	4,153
10	425	256	- 75	- 62	4,622	5,625	3,798
11	725	413	225	96	21,556	50,625	9,178
12	625	349	125	32	4,015	15,625	1,032
13	650	385	150	68	10,177	22,500	4,603
14	525	340	25	23	569	625	519
15	550	305	50	- 12	- 613	2,500	150
16	250	195	- 250	- 122	30,537	62,500	14,920
17	750	455	250	138	34,512	62,500	19,057
18	475	290	- 25	- 27	672	625	722
19	500	305	0	- 13	0	0	161
20	375	269	- 125	- 49	6,093	15,625	2,376
21	400	284	- 100	- 33	3,293	10,000	1,085
SUM()	10,500	6,664			244,201	481,250	130,932
Equation	Σx	Σy			$\Sigma(x - \bar{x})(y - \bar{y})$	$\Sigma(x - \bar{x})^2$	$\Sigma(y - \bar{y})^2$
AVERAGE()	500	317			correlation coefficient	r =	0.9728
Equation	$\bar{x}$	$\bar{y}$			coefficient of determination	R^2=	0.9464

The table to the left has the data for production and waste for 21 weeks in green cells.

The additional calculated columns provide the "residuals" or variances of x and y from their means, the product of the residuals and the squares of the residuals. If we plug these number into the formula:

$$\frac{\sum(x_i - \bar{x})(y_i - \bar{y})}{\sqrt{\sum(x_i - \bar{x})^2 \sum(y_i - \bar{y})^2}}$$

we get the result:

$$r_{xy} = \frac{244{,}201}{\sqrt{(481{,}250)(130{,}932)}} = 0.9728$$

The equation for the coefficient of determination R^2 is given by:

r_{xy}^2 thus $R^2 = (0.9728)^2 = 0.9464$

Excel will provide us with the coefficient of determination if we follow the steps on page 461, and choose "**Display R-squared value on chart**".

If we simply want to place these values in cells, Excel has the following functions.

Statistic	Excel Function	Note
Correlation, r_{xy}	=CORREL(Array1, Array2)	Array 1 and Array 2 are the cell ranges with y and x values
Coefficient of determination, R^2	=RSQ(known_ys, known_xs)	Note that the y's and x's need to be in the right order

Exploration: *Why regression is such an important analysis tool*

The top chart shows a gas consumption trend of an office building over a period of almost 2.5 years. It is real data which I use in a workshop on energy efficiency.

The question is: can you work out from the chart which were good months and which were poor months in terms of gas use? The course participants were asked to select between six and eight months in total.

In the workshop, folks initially seem to think that this is a fairly straightforward question to answer using Chart 1, the trend. There is usually agreement that the big peak in the middle of the chart, for Jan 2012, is much higher than normal and so this is a poor month. Others will point to the fact that the lowest month's use is July 2012 so that must be very good.

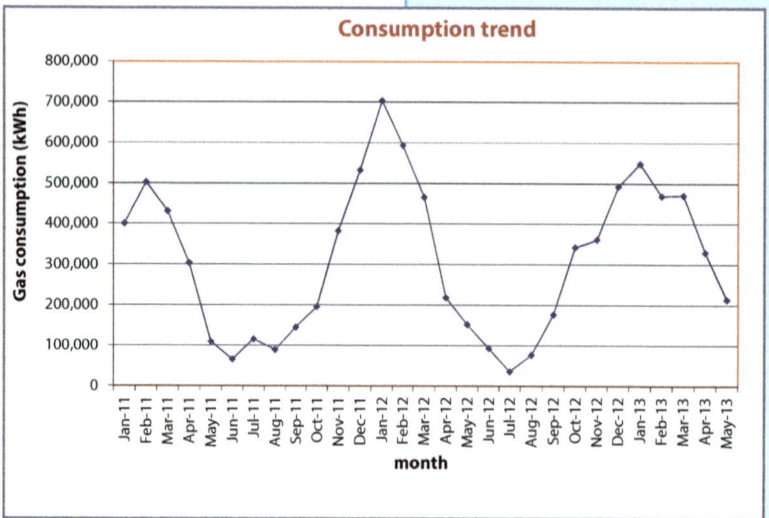

The seasonal influence (where gas consumption rises in winter and falls in summer) is obvious, and there then follows a process where folks use their rulers and try to compare like months with like.

At this point, much to everyone's relief, I will introduce Chart 2, below, which shows the same data, only overlaid as a year-on-year comparison. This new chart sometimes causes people to change their mind about whether particular months are good or bad. Clearly, overlaying the data makes differences easier to spot.

In the exercise, I ask folks to work in small groups and to complete an answers table with a column marked good and one marked bad for each month for each chart.

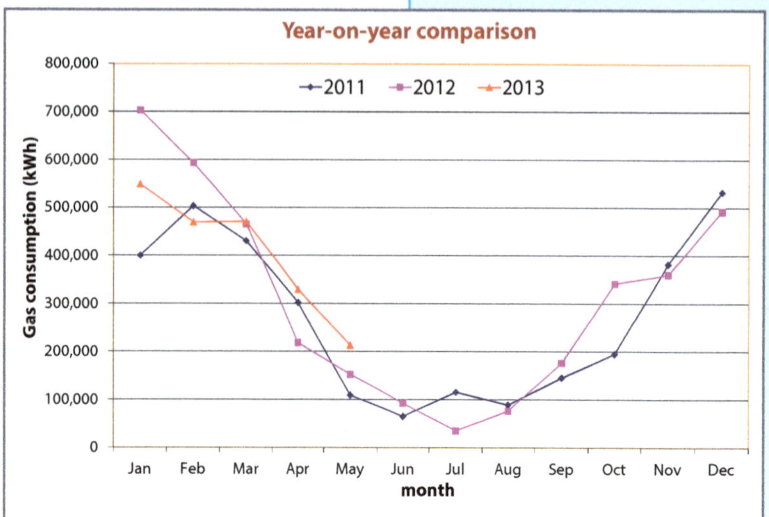

It is as the participants fill in the answer table, that there tends to be a debate - how much of a difference constitutes good or bad? Is a higher use a bad sign anyway - could it just have been that the weather was colder that month?

Uncertainty and doubt creep into the discussion, particular when the groups compare notes and they see that they have different answers.

The reality, of course, and the learning topic of the exercise, is that it is impossible to establish good or bad performance from these charts. Yet this method - trend analysis - is often the only method that organizations use to assess performance and manage resource use.

At this point of the task I introduce the technique of simple linear regression, and in a few quick clicks of their mouses, the teams have created a scatter chart and overlaid the line of best fit and equations.

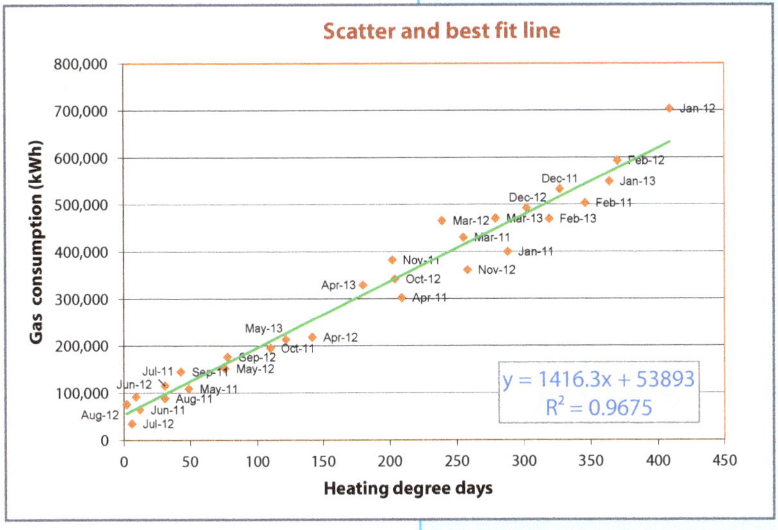

14.23 **Exercise to interpret performance**
These charts compare how easy it is to discern energy use performance in three different types of charts: trend, period-on-period and simple linear regression.
Source: Niall Enright, images and data are available in the companion file pack.

[Hint: to label the points on the scatter chart, I will right-click on the data series and choose **Add Data Labels**, then right-click again and choose **Value from Cells** and I select the date column. Finally, I deselect the **y-values** checkbox so that only the date is displayed.]

OK, so now the teams have the chart as shown on the left. Now they can begin to discern the points that are furthest from the line; those above the line are using more gas than predicted and so are bad, and those below the line are using less than predicted so are good.

The important thing to note is that this regression incorporates a variable, called heating degree days (more on degree days shortly), which is a measure of how cold it is and thus how much gas we would expect to use. The R^2 figure displayed on the chart shows that this variable accounts for 96% of the variability in gas use, so it is clearly a very powerful predictor.

	Trend		Year-on-Year		Regression	
Month	**Good**	**Bad**	**Good**	**Bad**	**Good**	**Bad**
Jan-11			✓		✓	
Feb-11						
Mar-11						
Apr-11					✓	
May-11			✓			
Jun-11						
Jul-11		×		×		
Aug-11						
Sep-11						
Oct-11	✓					
Nov-11						×
Dec-11						
Jan-12		×		×		×
Feb-12		×		×		
Mar-12						×
Apr-12			✓		✓	
May-12						
Jun-12						
Jul-12	✓		✓		✓	
Aug-12						
Sep-12						
Oct-12			✓			
Nov-12					✓	
Dec-12						
Jan-13						
Feb-13						
Mar-13						
Apr-13		×				
May-13						

Now the team are able to agree on a specific distance from the line that represents good and bad and to complete the final columns of the results table shown left. The fascinating thing is that the regression has identified four occasions (in purple) where the gas consumption was abnormal which were totally invisible in the trend data. Indeed, many of the supposedly abnormal periods in the trend data were in fact perfectly normal, and on only two occasions (in orange) did all three charts agree.

Regression analysis is clearly a much superior technique for identifying exceptional variation. By correcting for the weather (heating degree days) the true performance has become visible.

Regression compensates for that which we can't manage, the common variation, weather in this case, highlighting other sources of variation, exceptional variation, such as leaving windows open, which we can repeat if good or eliminate if bad.

14.15 Interpreting regressions

Each part of the linear regressions equation provides a great insight into resource use.

Real World: *Jaguar's real baseload*

Many years ago, I worked with my colleague Gary Armstrong for Peter Dipple and the team at Jaguar Cars in Coventry on energy M&T.

One of the key pieces of information that Peter was uncertain about was the factory's baseload of electricity use - that is to say the quantity of electricity that was production-related compared to that which was fixed regardless of the number of cars produced. Historically, this figure was assumed to be very low because during the *"shut-down fortnight"* the load fell to a very small proportion of the total during production. To check this Gary and Peter identified the regression equations for all the major electricity users, based on the most relevant production variables. They then added up the intercepts (the c figures in our equations, which is perfectly ok to do for the same resource), to get the overall plant electrical baseload at times of normal production.

The result was staggering. Almost half the site electricity use in normal conditions was unrelated to car numbers or other activity data. This was vital information about where savings could be made; not only should production equipment efficiency be targeted, but also things that lead to the high baseload such as lines running empty, HVAC, lighting and non-productive demand.

Our linear regression equation can be described in three numbers represented by the letters m, for the slope of the line; c for the intercept; and R^2 for the coefficient of determination. Each one of these numbers provides a fantastic insight into our resource use. In the hands of someone knowledgeable about the resource user, I cannot overemphasize how powerful this information is.

First, let's consider the slope m. The sign of m tells us if our resource use increases or decreases with the variable. Inverse or negative correlations can be seen, for example, when one relates street lighting to daylight hours, as the daylight increases, the lighting electricity use decreases. Although most slopes are positive, no special significance should be read into a negative slope, unless it is unexpected, of course.

We can directly compare the slope of similar items of equipment. Thus, if we have two boilers and one has a steeper slope in terms of steam raised per unit of gas consumed, we can state that it is less efficient overall and investigate the root cause of the difference. Please note that, as in all efforts to interpret variance, we should not just focus on what makes the one boiler bad, but equally focus on what makes the other boiler good.

The intercept c of our equation is very informative, too. It tells us what proportion of the resource use is independent of the variable. For example in Figure 14.23, on the previous page, my building's gas use was very closely related to the variable heating degree days, but there is a baseload, given by the intercept c, of 53,800 kWh per month which is not heating-related but is due to hot water demand which is present throughout the year. This ability to differentiate fixed load from variable load is truly remarkable because we can employ different strategies to reduce each aspect of the resource use.

Thus, to reduce the variable-dependent element, I am aiming to reduce the slope, which I can do by increasing the efficiency of my heating system, by setting the thermostat to a lower temperature or by switching on the heating system later in the year. On the other hand, to reduce the baseload, I will look at reducing the intercept, for example, by putting flow restrictors on the showers and taps or by lowering the hot water temperature setpoint. Not only can the intercept and slope tell me where to focus my efforts but, following changes, I can recalculate the regression line and see how m and c changed and so separate the effect of my actions on the activity-related versus the baseload part.

The intercept and slope together allow us to predict resource use given a value for the variable i.e. to forecast our resource use. Since our variable is often a value such as production for which we have future plans, or degree days for which long-term averages a published, we can simply plug this variable number into our linear regression equation and predict resource use. This is immensely useful for some resources, such as electricity, where understanding future use enables better procurement and demand-management strategies to be implemented. Because we typically model many key resource users at a facility, by plugging data for the specific variables that apply to them, we get a much more powerful and accurate bottom-up forecast than that which can be achieved by a simple top-down model. It is important to note that the intercept c is defined by the time period of our regression analysis. If our data is monthly, then c is the *monthly* fixed load. So if we are forecasting on a yearly basis we would have to multiply c by 12.

The last of our trio of eye-opening statistics is the coefficient of determination, R^2. This tells us the degree to which the change in our x-value, or resource use, is due to the change in the y-value or variable. R^2 can range from 0 to 1, and the closer it is to one, the greater the one variable influences the other.

The most common assessment of R^2 is usually: *"Is this value as high as I would expect?"* For example, if a gas consumption shows little relation to heating degree days in a building, I may be concerned that the systems that are meant to control the gas use (such as the thermostat) are not working. Alternatively, the low R^2 value could indicate that there is another factor affecting the gas use (such as a restaurant whose use is driven by the number of meals served), which I need to capture in the form of a multiple variable regression or there could simply be data errors such as outliers, which are influencing the result. There is more on multiple regression and outliers later.

14.24 **Different types of correlation**
Note that the value for R^2 is positive regardless of the direction of the slope.
Source: Niall Enright. Illustration and data are available in the companion file pack.

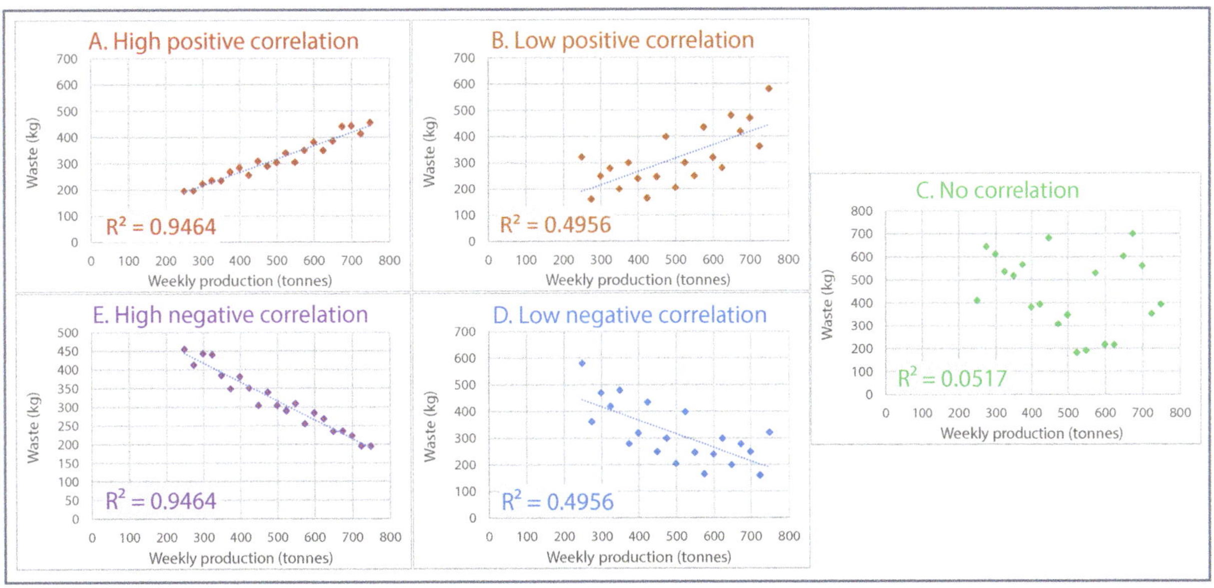

14.16 How significant is R^2

It is possible for a correlation to be very strong, that is to say that R^2 is close to 1, yet not very significant. On the other hand, it is also possible for a correlation to be very weak, yet the result be highly significant. It all has to do with sample size.

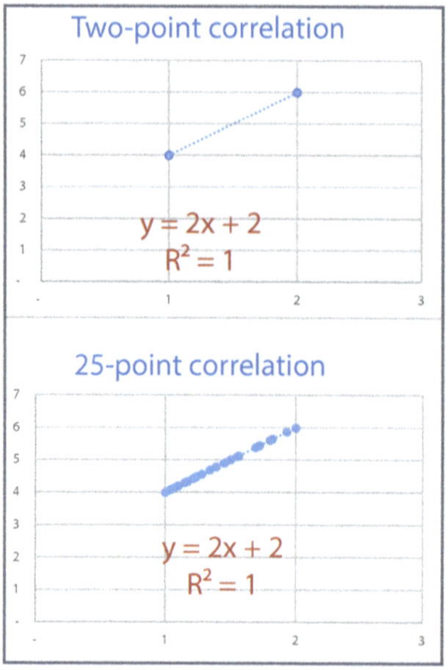

14.25 **Same result, different # points**
The number of data points will not affect the linear regression equation or correlation, but it does impact on significance.
Source: Niall Enright. Image and data available in the companion file pack.

Real World: ***Significant/significance***

Another example where statistical words do not always have the same meaning as in common use is significance. For example, in statistics, because something is significant, it does not mean it is important. It simply means that, according to a statistical test, it is unlikely to have happened by chance. Something can be significant, but entirely trivial or even irrelevant, if its effect on resource use is minimal.

One of the aspects of linear regression that confuses people is the difference between correlation and significance.

Consider the two charts on the left. The top chart has just two data points so my line of best fit is bound to go through both the values. In the second chart I have 25 data points, every one one of which falls on my line of best fit. Which of these data series is more convincing that the linear regression equation ($y = 2x+2$) is valid? Clearly, the second data series is more persuasive, or to use the correct term, it is more significant.

From this we can conclude that significance is related to the number of data points we have in our data series. The more data, the more significant a given result is, the more firmly we can accept the conclusion. Thus, if we have a very large series of data that shows that the correlation between gas use and heating degree days in a building is low, we are more likely to accept that this is a true reflection of the situation, and there really is genuinely little relationship between the two factors (and an opportunity to improve!).

As one would expect, statistics comes to the rescue here, with a formal way to determine the significance of a correlation. We've had sigma, x, y, r, m, c and z so far; so our next important statistic will be denoted by the letter t. We can calculate the t-value if we know our correlation coefficient and the number of data points in our series (see the statistics reference on page 770). For a correlation to be significant, the t-value needs to be greater than the value given in the critical values table shown in the table on page 780.

While the t-test is a widely applicable test of significance, in practice there is an easier way to confirm that our regression is meaningful. That is to use the r_{xy} critical values or the R^2 critical values tables on page 782. For example, taking the earlier waste and production plots, I obtained a correlation of 0.972 for Chart A with the points closely lined up with the line of best fit and 0.704 for Chart B where the points were more scattered. Since I have 21 data points in this data set I will look up row 19 (n-2) on the critical values table and see that for a significance of 1% the r_{xy} value needs to be 0.549. Both the correlations are *greater* than this critical value, so I can say that the probability of these correlations being due to random chance is considerably less that 1%. As this quick test is so useful, a simplified version of the table is shown in Figure 14.26 opposite.

Number of data points	Minimum coefficient of determination, R^2	
	α = 0.05 (5%)	α = 0.01 (1%)
10	0.399	0.585
12	0.332	0.501
15	0.264	0.411
20	0.197	0.315
24	0.163	0.265
25	0.157	0.255
30	0.130	0.214
35	0.112	0.185
36	0.108	0.174
40	0.097	0.162
45	0.086	0.144
50	0.078	0.130

14.26 **Simplified R^2 critical values table**
The minimum required value for R^2 is given for different data series sizes at two levels of significance: the 95% confidence level and the 99% confidence level. (Note that the number of data points in the table above takes into account the degrees of freedom, values for 12, 24 and 36 points are provided for monthly data).
Source: Data from NIST/SEMATECH e-Handbook of Statistical Methods[555] and Turner[721] modified by Niall Enright. There is a fuller version of this table in the statistics reference section on page 782.

I am often asked: *"How many data points do I need to do a regression?"* To answer this question I need to establish what the coefficient of determination, R^2, is and the degree of confidence that I want in the regression. If I have an R^2 of 0.5, and I want to be confident to ±1% that my result is not due to chance, I will look up 0.5 in the right-hand column of the table 14.26, left (or the complete table in the Statistics Reference section on page 783). I can see that if I have 12 data points, then R^2 needs to be greater than 0.501 for the regression to be significant to a level of confidence of 1%, so the answer is approximately 12 points (13 points should be enough).

So we now have a way to assess if our correlation is due to randomness or if it is significant. Don't, however, fall into the trap of assuming that correlation equals causation, however significant the correlation is.

Just because something is correlated with something else, no matter how perfectly, it does not mean that it is caused by it! For example, the number of jellyfish stings in Australia is strongly correlated with ice cream sales. That does not mean that eating ice creams makes jellyfish attack you - in fact, these are both due to the fact that when it is hot, more people are in the water and ice cream consumption is also high.

If A and B are correlated, we cannot conclude that B causes A. It could be that a third factor C causes both B and A (as in the jellyfish/ice cream example). It could be that the relationship between the two events is a complete coincidence. In his book, *Spurious Correlations,*[762] Tyler Vigen has collected some amusing and purely coincidental correlations - such as the 99.3% correlation between total US wind generation capacity and the total number of Facebook users. These are unconnected values that happen to have grown at approximately the same rate between 2005 and 2013.

The key message in this and the previous topic on interpreting regression is that there are statistics which are easy to apply and which can give us a great insight into our resource use, making visible what would otherwise be invisible. That does not mean to say that we should abandon our common-sense altogether and simply take the statistics at face value: we still need to apply judgement and intelligence to assess if the statistics are true, as the next section on dealing with outliers will reinforce.

14.27 **xkcd's amusing take on correlation and causality**
Source: xkcd, reproduced under a Creative Commons 2.5 attribution non-commercial licence. http://xkcd.com/552/.

14.17 Removing outliers

Since we need to understand the normal relationship between resource use and a common factor variable, it is necessary to exclude influential data points where exceptional variation or error have occurred.

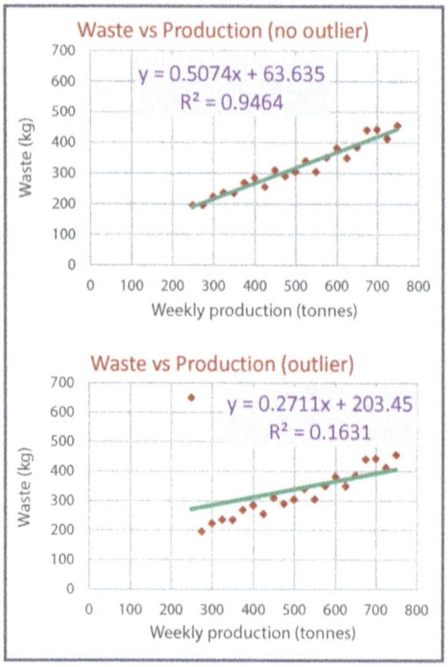

14.28 **Effect on an outlier**
Illustration of the effect of one outlier (bottom chart) on a linear regression.
Source: Niall Enright, data and illustration available in the companion file pack.

Exploration: *Topiary*

I recently heard of the practice of data cleansing referred to as topiary, the art of clipping a plant into a desired ornamental shape. Needless to say, I am not encouraging the reader to exclude outliers in the data to support a predetermined conclusion! As with the observation earlier on correcting errors (page 448), one should only exclude an outlier if there is evidence that this is due to error or exceptional variation.

Consider the two charts left. On the top is the data series that we previously used to explore simple linear regression, which shows a high coefficient of determination R^2 of 0.9464, an intercept value of 63 kg waste per week and a slope of 0.5 kg waste per tonne of production.

The second chart is based on the same series with the first waste value changed from 195 to 650. As can be seen from the green line of best fit, this data point is now skewing the line upwards with the effect that the intercept is now over three times as large, at 203 kg, the slope more shallow, 0.27 kg waste per tonne of production. Furthermore, the value for R^2 has decreased to 0.1631 which means, from the discussion in the previous section, that the R^2 has gone from being highly significant to no longer meeting the usual 95% confidence level (which would require our R^2 to exceed 0.187 for 21 data points).

It is quite remarkable how one single point, called an outlier, can so dramatically affect the results. In fact, this is a weakness of Pearson's correlation and one which we need to have a strategy for.

Clearly, we need to understand the possible sources of the unusual value or outlier. From the earlier discussion, we know that there are three types of variation. First of all, the outlier could be due to an error (usually due to measurement). If an error can be proven, then the data point can be quite reasonably excluded from our analysis. Second, the deviation from the expected line could be due to common variation, i.e. the value is simply a random value we would expect from the relationship between waste and production. Or, third, it could be due to exceptional variation, e.g. a batch went wrong, and a much larger volume of waste was created than would be normally expected.

The important thing to remember here is that we want our linear regression to only model the effect of common factor variation, not exceptional variation (e.g. the influence of production, not the effect of batch errors). That is the purpose of our model in both day-to-day operational control and one-off analysis - to account for the common factor variables like production and temperature which are "*givens*" which we can't change, so that the effects of other sources of variability (such as mistakes with batches) become visible and hence *controllable*.

Given that the *purpose* of our linear regression is to model only the common variation, it follows that we must exclude outliers if they are due to error or if they are due to exceptional variation as they distort our model.

We **must** exclude outliers if they are due to **error** or if they are due to **exceptional variation.**

The question now is, how do we know the source of the variation is due to error or exceptional variation? Fortunately, there is one statistical feature of our data that can help us greatly. This is the fact that the variance, i.e. the difference between the actual resource use and the line of best fit (called the residuals in statistics), should follow a normal distribution. We would expect the values to cluster around zero (remember the variances all add up to zero when we create the line of best fit). This normal distribution means that if we calculate the variance in units of standard deviation, we get a z-score for each value (which Excel calls the standard residual). We learnt earlier that the z-scores indicate the probability of a particular value occurring (page 452). Thus, 68% of values should have a z-score between 0 and ±1, 95% a z-score ±2 and 99.7% a z-score ±3. The outlying data point in the bottom chart has a z-score of 3.97 which means that the probability of this variance (or residual) occurring in our model is 0.00008 (Table 24.8 on page 778), i.e. once every 12,500 readings (Table 24.9 on page 779). On this basis, we can confidently remove the data point, as it is highly unlikely to occur within the 21 data points.

Insight

We should be aware that the standard deviation we used to calculate the z-scores above (and which Excel will use to calculate standard residuals) *includes* the outlier. This will have the effect of making the standard deviation larger than expected. If we recalculate the standard deviation excluding the outlier, we can then use this s value to calculate the deleted z-score of the outlier. This deleted z-score is 9.4 and occurs so rarely than its probability does not appear in any of the published tables and online calculators state that it happens once in infinity! For more on deleted z-scores, see *Applying regression and correlation: A guide for students and researchers*[526] page 75 onwards.

When a data point has a substantial impact on the regression results, it is called influential. Unfortunately, there are no capabilities in Excel that help us identify influential observations, and I am not aware of any M&T packages that offer this facility.

One characteristic of an influential data point is that either the x-value or the y-value, or both, are likely to be at the extreme of the range of x's and y's in the data set. In the example we have been following, the y-value of 650 is much higher than expected. Thus we can use a simple z-score on just the x's and y's as a means of identifying possible influential data points. This item-by-item test is especially useful when we are looking at influential values for a multiple regression which cannot be plotted on a chart, so we do not have the opportunity to visualize the data and spot outliers. It follows from our definition that we can verify that a point is influential if the regression results change dramatically when the point is removed, and this is often the simple method that is used to confirm that the value is influential.

Variation is the change in resource use over time; **variance** is the difference between actual and predicted resource use.

Excluding data points, as long as it is done with consideration (rather than to get the desired result or make a weak correlation strong) is not only desirable but essential, if we are going to see the wood, not the trees. The next section reinforces, yet again, the importance of looking at the data.

14.18 The shape of the data

By looking at the shape of our data in a chart, we can tell if our use of linear regression is appropriate and establish the range of values over which our regression equation holds true.

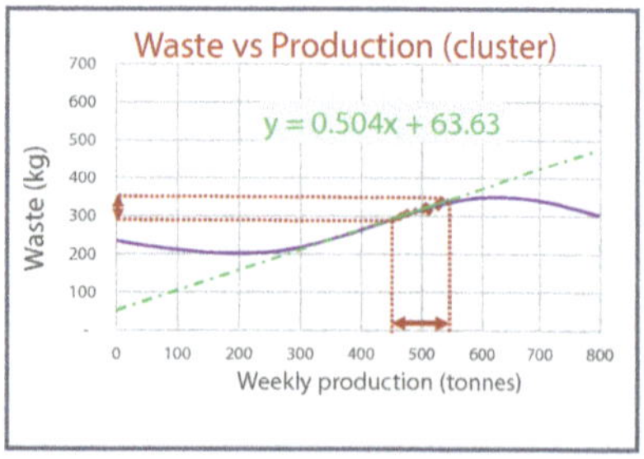

14.29 **Clustered values**
It is unwise to make predictions for values outside the actual range of values that have been modelled.
Source: Niall Enright, Image and data available in the companion file pack.

Statistics are descriptive values from which we can draw inferences and form conclusions. However, they can rarely be relied on as absolute evidence of a particular hypothesis since they are usually based on probabilities. The outlier in the previous section was extremely unlikely to occur by chance, but it could still, however improbably, reflect some aspect of the relationship between waste and production which does not manifest itself often.

One key rule of regression (and modelling in general) is to never draw a conclusion outside of the range of x and y that has been modelled. If you look at the chart left, you can see that our data points are clustered around a narrow range of x (450-550) and y (290-350), from which we calculate the same linear regression equation as in the previous examples. The linear regression equation has been extended in the dashed green line to meet the y-axis at the intercept value and to the right beyond the highest value of x. However, it is quite possible that the actual relationship between waste and production over the full range of values is modelled by an entirely different, non-linear equation, such as the purple line on the chart. The linear regression formula that we have developed represents just a small part of the true model, for which we happen to have data, and it is only valid for this range of values.

It is not uncommon for industrial processes to be clustered around a high value for production. That does not mean to say that, for the range studied, we don't get an excellent model that enables us to predict resource use effectively and to differentiate between the common factor variance and the exceptional variation *at high levels of production*. It is just that we must not draw conclusions about performance at low levels of production until we have sufficient data points at this lower mode of operation. The bottom line is that we can only infer from a model over the range of data we have analysed.

Consider Anscombe's Quartet opposite. To the human eye, these four data sets are obviously different, even though many of the statistics are near-identical. If we had chosen to calculate some other statistical properties such as z-scores, which we have already discussed, or skew and kurtosis (which are beyond the scope of this book), we would be able to differentiate between these data sets. Statistics are funny; sometimes the more you have, the more you want.

We can only infer within the range of data we have available.

Non-linear relationships between resource use and measures of activity are covered in a later section.

Exploration: *Anscombe's Quartet*

Named after the American statistician Francis Anscombe, these four sets of data were developed in 1973 to produce the same line of best fit from radically different sets of data. The purpose of these data sets is to reinforce the point that one should review a plot of data prior to drawing conclusions, rather than rely just on its statistical properties.

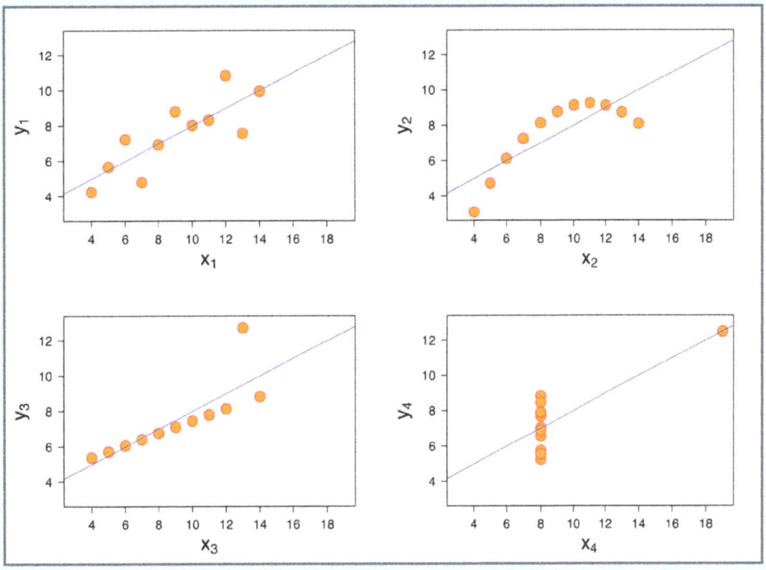

14.30 **Anscombe's Quartet**
These four data sets have very similar statistical properties but are clearly quite different.
Source: Schultz reproduced under Creative Commons 3.0 Licence.[174]

The first chart looks very much like our standard scatter plot. The next plot (upper right) is clearly non-linear. In the third chart, bottom left, the relationship is clearly linear, but one outlier has skewed the line of best fit. The fourth chart shows how one single outlier can determine the line of best fit. In his paper, Anscombe published the full statistical characteristics of the data sets to demonstrate that it was not just the linear regression line that was the same.

Property in each case	Correspondence of value in the data sets
Mean of x	9 (exactly)
Sample variance of x	11 (exactly)
Mean of y	7.5 (to 2 decimal places)
Sample variance of y	4.122 or 4.127 (to three decimal places)
Correlation of x versus y	0.816 (to three decimal places)
Linear regression	$y = 0.500x + 3.00$ (2 /3 decimal places respectively)
R^2	0.67 (to 2 decimal places)

This goes to show that we need to look carefully at our data and apply common sense when interpreting linear regression statistics. If the data we are plotting is not a normal, linear scatter plot, then we should use different techniques to analyse it.

All data analysis should involve a ***visual inspection*** *of the data series in a chart form, if possible.*

14.19 Choosing variables

Selecting the appropriate variables for a model is usually fairly straightforward. However, there are basic principles that will make this more effective, for example variables must be independent, not simply a disguised measure of the resource use.

A variable in resource-use models, such as linear regression equations, is a *driver* of the resource use. In many cases, this is an obvious value. The lighting in an exterior car park should be governed by lighting hours. The fuel used in a truck should be determined by the tonnes carried and the km travelled. The hotel laundry's water use should be related to the number of guests checked in.

The word *should* is used advisedly because the process of variable selection starts with an assessment of all the potential drivers for activity, which is then followed up by testing the correlation in practice. Sometimes the seemingly obvious variable is not the best one to help us identify variance due to exceptional factors, as the box right shows.

When analysing a facility's resource use for the first time, I tend to request access to all the variable data available. This is because I may need to compare several alternative candidates before I establish the most reliable indicator of resource use. Indeed, I may want to create a model that uses multiple variables (a process described later in this chapter). These are a few rules of thumb I would propose when modelling.

- Try to use variables that are as *"local"* or close to the resource use as possible. For example, if I had a measure for total factory production and another for production in a specific hall, then I would want to model the hall resource using the hall production.
- Think. It is quite possible for a variable to show a good correlation with resource use by chance when, in fact, there is no relationship. These spurious correlations are more prevalent with smaller data series so be especially alert when the number of values you have is low (say <30).
- Conversely, it may be the case that a variable that *should* be related to the resource shows little correlation when, in fact, this is a signal that the feedback influence between variable and resource is defective (e.g. the earlier case of gas being unrelated to heating degree days due to a disconnected thermostat). In these circumstances, don't dismiss an unexpectedly low correlation as it may be a goldmine for improvement opportunities.
- Be open-minded and objective. Some M&T applications (such as the version of Montage, which I wrote many years ago), will *scan* through all variables and indicate which ones show the greatest correlation with

Real World: *Airbus' activity value*

Photo: A380 at Toulouse © Airbus [13]

Regression modelling can be challenging when the variables you are working with don't change much. I encountered one such measure at aircraft-maker Airbus in Toulouse, where production is a monthly integer *"number of aircraft completed"*, which is usually around 30. This was not helpful for modelling daily energy use.

Fortunately, the major energy users such as lighting and HVAC at Toulouse are more significantly influenced by factors other than production, such as daylight hours and degree days for which daily data was available. Furthermore, the process management quality systems at Toulouse had many daily activity measures (such as daily man-hours worked) that were good alternative measures of production activity.

This real world example reinforces the general principle that activity can be measured either as an output measure (which most people automatically select) or as an input measure (such as labour hours in this case), which may often be a better choice (see also the story opposite).

Real World: *Jaguar's paintshop*

Another example from the work at Jaguar Cars, in Browns Lane, Coventry, recalls the efforts to establish a target for the weekly electricity use in the paintshop.

The obvious variable to use was the number of cars through the paintshop. When we did the analysis, however, this showed an unexpectedly low value for R^2. Although it was a significant result, so was statistically valid as the basis for target-setting, it was clear that car numbers alone did not account for the electricity use.

So naturally, we investigated this further and discovered - as the data had hinted - that not all cars are equal. Although there was only one model of car through the paintshop (the S-Type shown below), those vehicles receiving metallic paint finishes had more coats of paint applied, so the process of painting these was more energy-intensive.

Armed with this information the team identified another variable, volume of paint used, which had a much stronger relationship to the electricity use. We now had a model which could more reliably differentiate between variance due to the common activity (painting cars) and the exceptional variation due to other factors (primarily leaving equipment running idle) which needed to be identified and managed. The confusing effect of the car finish was now eliminated.

Image: Rudolf Stricker, Wikipedia, CC[74]

a given resource use. Although mechanistic, this kind of brute-force analysis can throw up unexpected relationships that provide new insights and so opportunities for improvement and control.

- Look and check. Always seek to visualize the data in chart form where possible. This will help identify the errors, outlier effects and non-linear artefacts discussed previously. A particularly common problem in modelling is offset timing in activity data and resource data, which may purport to be for the same time frame (e.g. week or day) but in practice is collected at different times (e.g. production data at the end of a shift, energy data at midnight).

Sometimes a variable can also be a resource. For example, I may model the gas consumption for a boiler against the variable total steam raised. This gives me an indication of whether the boiler is being operated efficiently. On the other hand, I may also want to model the resource total steam raised against production as an indicator of the efficient use of steam in my process and to enable non-production factors (such as steam leaks and venting) to be picked up. In the one case I am modelling resource-supply or conversion, and in the other I am modelling resource demand or use. When modelling fluids such as steam we need to take into account the measurement accuracy of our instruments and bear in mind that these tend to be less accurate at lower flow rates.

One fundamental requirement in all variables is that they are independent. That is to say that the variable is not a disguised measure of the resource use. The table below sets out some possible variables for an airport.

Resource	Variable	Quality
Baggage hall conveyors electricity	Hours run	Bad (not independent)
	Number of passengers	OK (independent)
	Number of bags	Best (independent, local)
Baggage hall heating gas	Temperature setpoint	Bad (not independent)
	Number of passengers	OK (independent but weak?)
	Heating degree days	Best (independent)

The hours run for the conveyors is simply another measure of the energy use (in fact, we discussed using hours run on motors as an alternative to fitting meters in the previous chapter). Similarly, the temperature setpoint directly determines the response of the hall to the weather, so is not independent of the gas use.

Identifying candidate variables for resource use and exploring the insights from our analysis is one of the most enjoyable and satisfying aspects of resource efficiency. It is here that we can test our theories and draw conclusions which may well be new insights in the resource use. There are lots of *eureka* moments to be had in this process.

14.20 Using degree days

Degree days is the name given to a measure of ambient temperature which we can use to assess the performance of systems responsible for heating and cooling, as well as other temperature-driven systems.

Almost 40% of all energy consumption in commercial properties in the US is for space heating or cooling.[218] All this energy use is driven by the ambient temperature around the buildings. Unfortunately, temperature is quite a tricky thing to pin down, as it rises and falls during the day. As a result, two measures, called heating degree days and cooling degree days (HDD and CDD) have been developed to quantify the amount of heating and cooling a building may need. These are a measure of the difference of the outdoor or ambient temperature *at a particular location* in comparison to a baseline temperature. HDD reflects how far and for how long the ambient temperature is *below* the baseline (because heating is needed when it is colder than the baseline) and CDD measures how far and for how long the ambient temperature is *above* the baseline (because cooling is needed when it is hotter than the baseline).

For HDD, different countries use different baseline temperatures. In the UK we use 15.5° C. The rationale for this is that most buildings are heated to an internal temperature of 18° C, but since many buildings have additional sources of heating, the heating systems don't need to come on until the external temperature falls below 15.5° C. The additional sources of heating fall into two categories: internal gains of heat from lighting, computers and even people; and external gains, such as passive heating from solar irradiance. There is an exception to this baseline in the UK for hospital HDD, where the baseline is 18.5° C recognizing that the desired internal temperature in a hospital is closer to 21° C. In the US, the HDD baseline temperature is 65° F (18.3° C), while in Germany it is 20° C.

Whereas there are well-established baselines for HDD, the situation for CDD is more complex. Building owners and operators don't cool their building to the same temperature and internal heat gains can add a substantial cooling demand that influences the building's response to ambient temperature. Thus a building with a data centre will require much more cooling than a simple office. Although we have standard CDD, in the UK based on a 22° C baseline (in the US these are based on 65° F /18.3° C, the same as HDD), it is usually better to establish the specific cooling baseline temperature for each building, as described shortly on page 480.

Degree days reflect how far and for how long the ambient temperature is from a given baseline.

When we use degree day data in linear regression, we may find that we get negative intercepts, as illustrated in Figure 14.31. This usually occurs when we have not determined the building's heating balance point and are simply

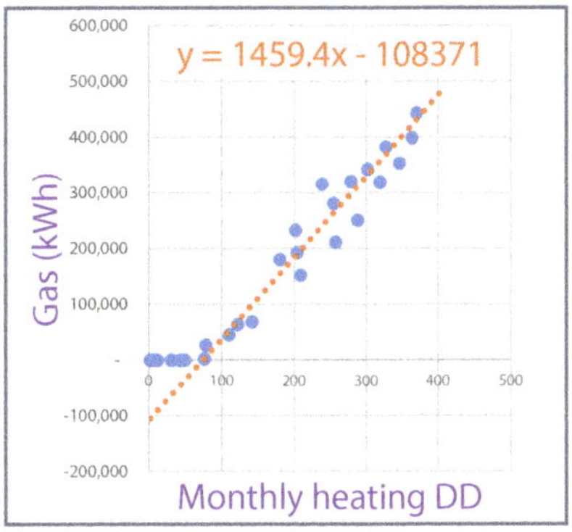

14.31 **Negative intercept**
This is not an uncommon result when using linear regression with degree day data.
Source: Niall Enright

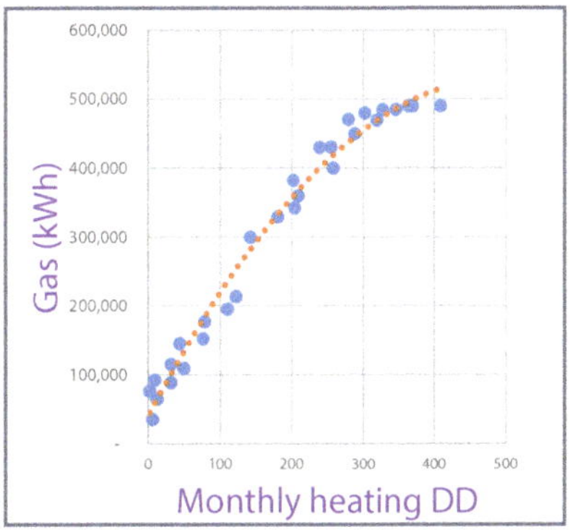

14.32 **Levelling off**
Where the heating systems longer responds to the stimulus of a greater heating demand, we should consider the possibility of a capacity constraint in the system.
Source: Niall Enright

Real World: ***Degree days per day***

Earlier, we observed that there are 9% fewer days in February than in January. This can be a problem with the baseload calculation in particular, as the assumption is that this is constant for each period.

One way to overcome this is to plot degree days *per day* against energy use *per day*. This effectively adjusts for different length months and should improve the correlation achieved.

using HDD based on 15.5° C. We will see this effect negative intercept value if the building has a comparatively small baseload energy use (for example, if we only measure the space heating energy, so domestic hot water is out of the equation, literally) and we don't initiate our heating at 15.5° C but a lower temperature. This situation could occur, for example, because we have large passive gains or our indoor thermostat is set to a lower temperature than the usual 18° C, or because the building is simply highly efficient (i.e. well insulated). Clearly, we are not generating 108,000 kWh of energy out of thin air each month as the linear regression equation implies, it is just that the assumed HDD baseline temperature is too high for this particular building. In these circumstances, we should ideally establish the true baseline temperature for the building (as described shortly), and create a model with HDD based on this temperature. If that is not feasible, we should at least ensure in any performance assessment that we set predicted energy use to zero if the HDD are below the point at which the line crosses the x-axis (this is calculated by the formula +c /m which is 74 HDD in the example given).

Another pattern we should recognize when using HDD or CDD is when a best-fit line appears to level off. This often reflects the fact that the heating (or cooling) system in the building is reaching its maximum capacity and so no longer responds to increasing heating demand, which means that the energy use simply stops rising. This is a very helpful diagnostic (not just when using degree days but in terms of any relationship between activity and resource use). In the example, left, we can see energy use levelling off above 300 HDD per month, and no doubt the building occupants are complaining of feeling cold in these especially cold months.

A third pattern to recognize is the *sawtooth* pattern that was described in Figure 14.12. If we have one point below the best-fit line and then another above, or vice-versa, for consecutive periods, it is possible that either the resource data or the degree data were read early or late.

Degree days are location-specific. Before the advent of the internet, data tables would be published for 18 specific degree day regions in the UK,[114] reflecting the fact that the weather is very different in northern Scotland (2,873 HDD per year, on average), compared to south-western England (1,947 HDD). These tables, however, are only crude approximations and so the tendency today is to use data from local weather stations (usually airfields which have good records) or, ideally, from temperature readings taken in the building's own building management system.

Clearly, climate can change considerably from one year to another, so 20-year average degree days are often used as a means of forecasting heating and cooling energy consumption in a building over the following year. Here, too, we need to show some caution as the climate warms - for example, in the EU-27 the average annual HDD declined by a staggering 16% between 1980 and 2009.[270]

⇒ page 482.

In Numbers: *How degree days are calculated*

The best way to calculate degree days is from degree hours, which is the difference between the mean temperature and the baseline temperature for each hour of the day. The heating degree hours are shown in the chart right by the green columns and the cooling degree hours by the red columns. Thus, on Day 1, there are 144 heating degree hours, which when divided by 24 gives us exactly six degree days. On Day 2 there are 73 heating degree hours which means that we have just 3.04 heating degree days. On Day 3 we have 19.5 heating degree hours which is 0.81 heating degree days, and on Day 4 there are no heating degree hours, hence the heating degree days is zero. Please note that it is possible to have both heating and cooling degree days in the same day, as illustrated in Day 3 on the chart opposite.

ASHRAE in the US[36] uses an average daily temperature method of calculation, subtracting the mean temperature (if below the baseline temperature) from the baseline temperature of 65° F (18.33° C). The disadvantage of this method is that a day such as Day 3 in the chart opposite would be recorded as having zero heating degree days or cooling degree days, when there was, in fact, a small heating and cooling requirement. The counter-argument is that a building would be unlikely to be heated or cooled in a day where the average temperature fell between both baseline temperatures. Germany is among other countries to use the daily average approach.

The UK Meteorological Office uses a maximum and minimum temperatures method to calculate degree days, according to the following table.

14.33 **UK Met Office DD calculation**
The table right sets out the conditions and calculations used by the UK Meteorological Office, summarized in the illustrations below.
Source: Niall Enright based on Met. Office[725]

Case	Heating Condition	Heating DD Calculation
1	$T_{max} \leqslant T_{baseline}$	$T_{baseline} - T_{mean}$
2	$T_{mean} \leqslant T_{baseline}$ & $T_{max} > T_{baseline}$	$0.5(T_{baseline} - T_{min}) - 0.25(T_{max} - T_{baseline})$
3	$T_{mean} > T_{baseline}$ & $T_{min} < T_{baseline}$	$0.25(T_{baseline} - T_{min})$
4	$T_{min} \geqslant T_{baseline}$	0
Case	**Cooling Condition**	**Cooling DD Calculation**
1	$T_{min} \geqslant T_{baseline}$	$T_{mean} - T_{baseline}$
2	$T_{mean} \geqslant T_{baseline}$ & $T_{min} < T_{baseline}$	$0.5(T_{max} - T_{baseline}) - 0.25(T_{baseline} - T_{min})$
3	$T_{mean} < T_{baseline}$ & $T_{max} > T_{baseline}$	$0.25(T_{max} - T_{baseline})$
4	$T_{max} \leqslant T_{baseline}$	0

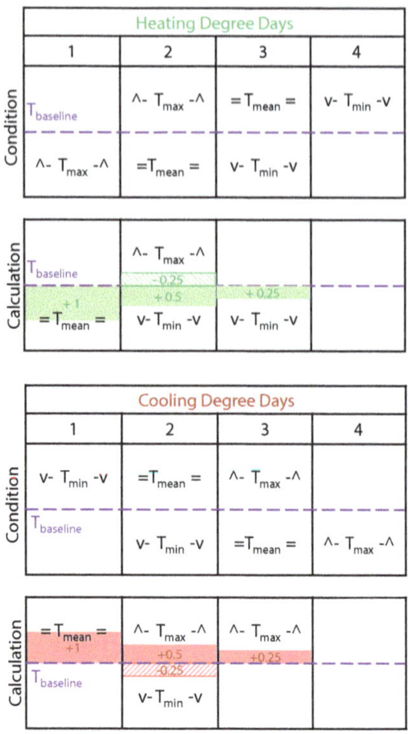

Where T is the symbol for temperature and $T_{baseline}$ is the baseline temperature, T_{max} and T_{min} are the maximum and minimum temperatures on one day. For example, let us consider Day 3 in the chart. Heating Case 3 applies since T_{mean} (on this day 19° C) is greater than $T_{baseline}$ (15.5° C) ***and*** ($T_{min} < T_{baseline}$) (12 < 15.5), so the calculated heating degree days is given by the formula ¼ ($T_{baseline} - T_{min}$), which equals ¼ (15.5 - 12) = ¼ (3.5) = 0.88 HDD.

Cooling Case 3 also applies on Day 3 since T_{mean} (19° C) is less than $T_{baseline}$ (22.0° C) ***and*** ($T_{max} > T_{baseline}$) (30 > -22), so the calculated cooling degree days is given by the formula ¼ ($T_{max} - T_{baseline}$), which equals ¼ (30-22) = ¼ (8) = 2.0 CDD.

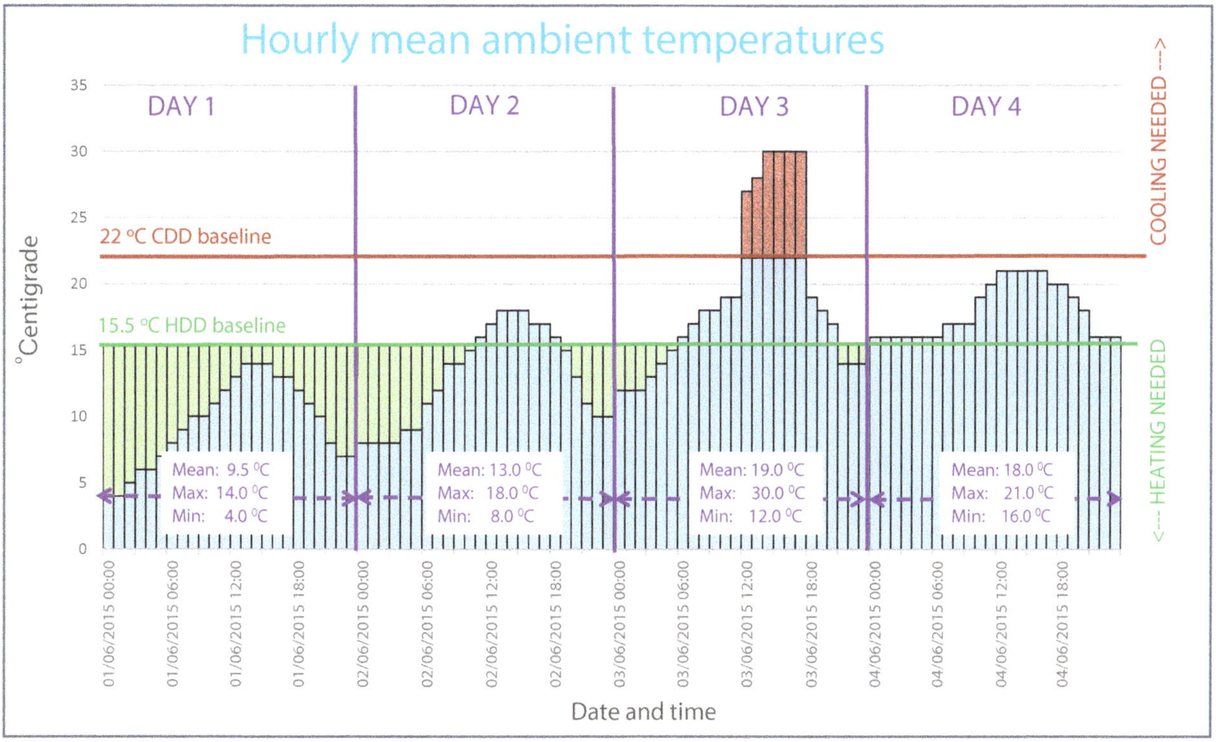

14.34 **Degree hours plotted over four days**

Source: Niall Enright. Images, data and calculation available in the companion file pack.

Below we have a table of results comparing the various calculation methods using the data in the chart above.

Calculation method	Day 1	Day 2	Day 3	Day 4
	Heating degree days			
Hourly degrees	6.00	3.04	0.81	0.00
Daily mean temp.	6.00	2.50	0.00	0.00
UK Met. Office	6.00	3.13	0.88	0.00
	Cooling degree days			
Hourly degrees	0.00	0.00	1.79	0.00
Daily mean temp.	0.00	0.00	0.00	0.00
UK Met. Office	0.00	0.00	2.00	0.00

In practice, degree day data rarely needs to be calculated as it is obtained from data tables, or more commonly, from internet providers such as *www.degreedays.net* which can derive degree days from temperature records from weather stations worldwide (using an integration method similar to the hourly method).

Because of the high degree of variability in the methods used, and the differing baseline temperatures, it is important not mix to degree day data from different sources in the same model.

14.35 **Simplified building energy signature** This chart shows both heating and cooling balance points to illustrate the concept of an energy signature. In practice there tends to be one chart for heating and one for cooling. Sometimes it is necessary to separate out the data points representing weekday energy use from weekend energy use in order to develop a clear chart as the heating and cooling systems may operate in fundamentally different control modes at these times. The steeper slope of the cooling line per degree C, indicates the greater energy intensity for cooling compared to heating. *Source: Niall Enright. Data and illustration available in the companion file pack.*

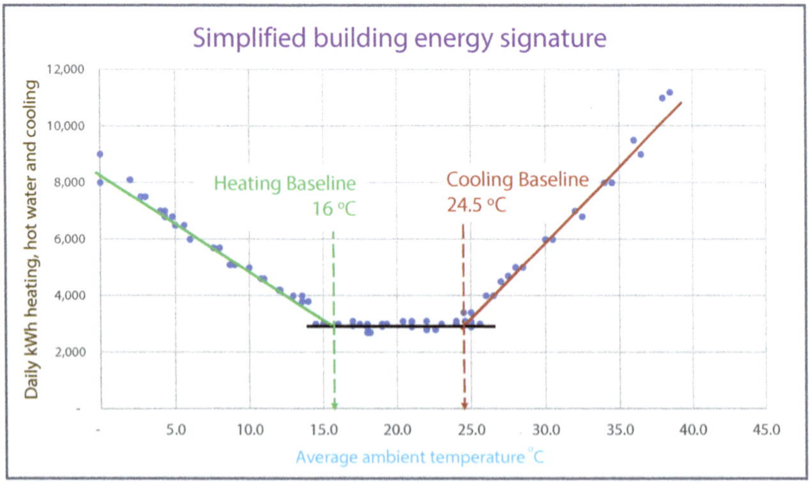

*The **baseline** temperature is not to be confused with the **baseload** of the building. The first is **the temperature** at which the building begins to need heating or cooling; the latter is the **amount of energy** the building uses for non-heating or cooling purposes each period.*

In Numbers: ***Baseline temperatures and building balance points***

There are three methods to determine the heating and cooling baseline temperature of a building. The first is referred to as the energy signature method, which is illustrated above. In this approach we plot the energy used for heating or cooling against ambient temperature and identify the point at which the requirement ceases, i.e. where the energy use levels off with temperature. This "inflexion point" may be zero or greater depending on whether there is additional energy used by other systems such as for hot water or air-handling. The points where the energy uses level off are heating or cooling balance points, which are the temperatures that should be used for baselines in degree day calculations.

The disadvantage of the energy signature method is that it requires a lot of data over a large range of ambient temperatures, typically daily values over a full heating and cooling season. An alternative technique is the performance line method, which plots our usual weekly or monthly degree days against energy. In this case, however, we plot multiple degree day data series based on different baselines around our expected building balance point, as illustrated in Figure 14.36, opposite. We also add a polynomial line (as simple as choosing **polynomial, order=2** in Excel, rather than linear, when adding a trend line). If we ask for the equation of the line it will look somewhat like $y = 0.5534x^2 + 1016.2x - 23124$ (this is the polynomial equation for the workshop example on page 465).

In this method, we are interested in the first term of the equation, the polynomial one with x^2. If this is positive (e.g. 0.5534) it suggests that our baseline should be *lower* for heating, *higher* for cooling, whereas if this is negative we should have a higher heating baseline or a lower cooling one. There is much more on this method of calculation in Section 5.3 of the excellent *CIBSE TM41 Guide: Degree days: theory and application.*[148] This guide points out the potential pitfalls of this approach, which may give unusual results such as where the heating or cooling systems are limited in their capacity, which is in itself useful information.

The intercept method uses data taken from the degreedays.net website[201] which handily offers an *"include base temperatures nearby"* option so that the technique can be easily applied. In this method the slope(), intercept() and RSQ() functions in Excel can be applied to multiple series of DD data determine the balance point.

14.36 **Performance lines**
This chart shows second-order polynomial plots for a building's gas use against three heating degree day series, one at 12.5° C, one at 16.5° C and one at 18.5° C. The line with the least curvature (indicated by the smallest value for the x^2 of the polynomial equation) is the green series at 16.5° C. This suggests that this is close to the baseline temperature of the building. Note that a negative x^2 element suggests the DD temperature is too low (for a heating series), or too high (for a cooling series) and a positive figure indicates that it is too high (or too low for cooling).
Source: Niall Enright. The spreadsheet and illustrations are available in the companion file pack.

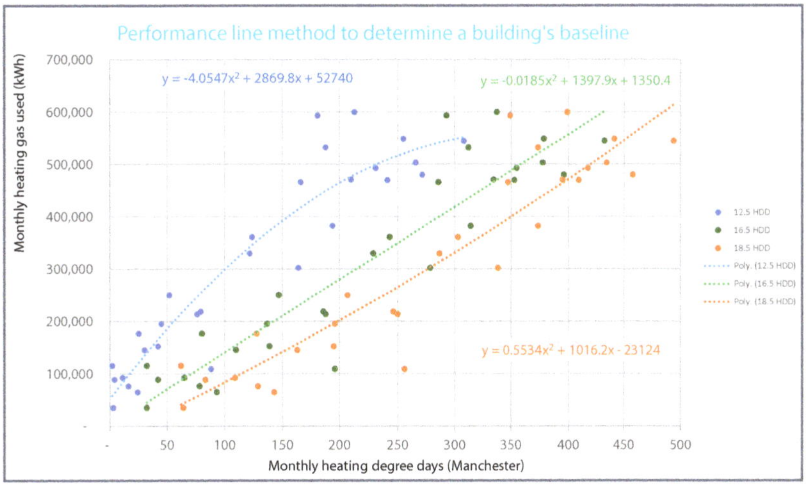

These calculations are shown in the table below. Here, assuming we have a building whose energy is only for space heating, we are looking for the series with the intercept value closest to zero (or closest to the calculated non-heating load if the energy has other uses). Looking along the slope calculation, we can see that this is at 16.5° C (shown in bold). We would also expect the coefficient of determination to be reasonably high at this balance point although this is not the key determinant.

A lower than usual expected baseline temperature for a heated building (or higher temperature when cooling) indicates an energy efficient building, which is likely to be well insulated and have reduced levels of air leakage. The reverse is true, indicating inefficiency that may offer opportunities for improvement.

14.37 **Zero intercept method**
The 16.5° C series shows the closest intercept to zero (in red) so the balance point is near to this temperature. This data is also used for the Performance Lines chart above.
Source: Niall Enright, using "nearby temperatures" data from www.degreedays.net. Excel sheet in the companion file pack.

	Manchester Heating (Degree Days in °C)													
	12.5	13	13.5	14	14.5	15	15.5	16	16.5	17	17.5	18	18.5	Gas
January-2013	272	287	303	318	334	349	365	380	396	411	427	442	458	480,000
February-2013	266	280	294	308	322	336	350	364	378	392	406	420	434	503,247
March-2013	308	323	339	354	370	385	401	416	432	447	462	478	494	545,000
April-2013	164	178	192	205	220	234	249	264	279	294	308	324	338	302,496
May-2013	88	99	112	124	138	152	166	181	196	210	225	240	256	109,235
June-2013	24	29	36	43	52	60	70	81	93	104	117	130	143	65,213
July-2013	2	4	6	8	12	15	20	25	32	38	45	53	62	115,568
August-2013	4	6	9	12	17	21	27	34	42	51	61	71	83	88,959
September-2013	30	37	45	54	64	74	85	97	110	122	136	149	163	145,482
October-2013	45	54	64	74	86	97	110	123	137	151	166	181	196	195,256
November-2013	194	209	224	239	254	269	284	299	314	329	344	359	374	382,548
December-2013	188	203	218	234	250	265	280	296	312	327	342	358	374	532,689
January-2014	213	228	244	259	274	290	306	321	337	352	368	383	399	600,000
February-2014	181	195	209	223	237	251	265	279	293	307	321	335	349	593,458
March-2014	166	180	195	210	225	240	255	270	286	301	316	332	347	465,870
April-2014	79	90	103	116	130	143	157	172	186	201	216	231	246	218,550
May-2014	42	52	63	74	86	98	112	125	139	152	166	181	195	152,400
June-2014	11	15	20	24	31	38	46	55	65	74	86	97	109	92,556
July-2014	3	5	8	10	13	17	21	26	32	39	47	54	64	35,455
August-2014	16	20	26	32	39	46	56	66	78	90	102	115	129	76,523
September-2014	25	29	35	40	47	54	62	70	80	91	103	115	128	176,889
October-2014	52	60	71	81	93	106	119	132	147	161	176	191	207	250,000
November-2014	124	138	153	168	183	198	213	228	243	258	273	288	303	361,258
December-2014	231	246	262	277	293	308	324	339	355	370	386	401	417	492,550
January-2015	255	270	286	301	317	332	348	363	379	394	410	425	441	548,997
February-2015	241	255	269	283	297	311	325	339	353	367	381	395	409	469,589
March-2015	210	226	241	256	272	287	303	318	334	349	364	380	395	470,556
April-2015	122	134	147	160	173	187	201	214	229	243	258	272	287	329,840
May-2015	76	88	101	115	129	143	158	173	188	204	219	234	250	213,559
Slope	1,761	1,692	1,631	1,576	1,529	1,486	1,448	1,417	1,390	1,366	1,347	1,328	1,313	
Intercept	90,252	81,004	70,363	60,728	49,476	39,007	27,294	15,303	1,960	-10,868	-25,067	-38,714	-53,776	
R-squared	0.856	0.860	0.862	0.865	0.865	0.865	0.864	0.862	0.860	0.859	0.857	0.855	0.853	

14.21 Using lighting hours

Another widely employed variable based on ambient conditions is daylight hours (often converted to lighting hours).

The amount of daylight available is due to the position in the sky of the sun, which depends on the latitude (the distance north or south of the equator), and the season. The amount of daylight, daylight hours, has a significant influence on how much lighting we use. Ideally, street lights should turn off when the sun rises and turn on again when the sun sets. Indeed, public buildings and offices may benefit from natural light and so their interior lighting should be modelled, in part, against lighting hours.

When calculating lighting hours from daylight data, it is not just a question of subtracting daylight hours from 24, since many lighting systems will run for a certain length of time before sunset and after sunrise. It should also be noted that daylight is not binary, on or off, since we may want to consider the level of natural light. In practice, lighting is also affected by weather, so that dark clouds can substantially reduce the daylight available. For this reason data from light sensors attached to building management systems are more accurate than location-based calculations, such as the tool described left.

14.38 **Lighting hours tool**
Given their usefulness, obtaining lighting hours is surprisingly difficult. I have had to resort to writing my own *"mash-up"* Excel tool which uses the US National Oceanic and Atmospheric Administration routines to calculate daylight hours, coupled with Google Maps to look up latitude and longitude and daylight savings tables from Wikipedia. The options panel for this tool is shown below, and the complete tool is available in this book's companion file pack. The sinusoidal pattern of lighting hours over a year, in the chart below right, is lovely to see.
Source: Niall Enright

Options

Location	
Type here for Geolocation search ->	Liverpool
Location matched:	Liverpool, Merseyside, UK
Latitude:	53.4083714
Longitude:	-2.9915726
Multiple Time Zones (see comment)	0.00

If you do not have an internet connection or wish to manually enter longitude, latitude and time zone data you can do so in the blue area above.
If you manually enter a country in the "Location matched" blue cell the tool will seek to match the time zone and daylight savings data.

Lighting Pattern	
Hrs after Sunset (decimal):	1.0
Hrs before Sunrise (decimal):	1.0
Exclude Specific Days:	
Use Summer Time or Daylight Saving	Yes
Summer Time Starts:	L,Sunday,March
Summer Time Ends:	L,Sunday,October

If performance is slow, reduce the length of the data requested to 1 year.

Output	
Initial Date and format for output	03/01/15
Frequency:	Daily
Weekly data begins/ends on a:	Saturday
Length of Data:	1 Year

Sat

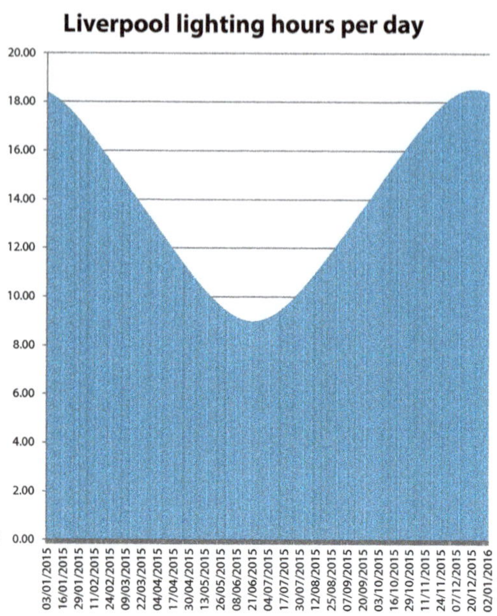

Real World: ***Using lighting hours to calculate savings at MediaCityUK***

Derek Elliott (the technical services manager for Peel Media) and Phil Harris (an energy engineer with the site FM contractors, Engie) set about an ambitious programme to reduce the car park energy consumption in MediaCityUK in late 2013. This is a significant electricity user, consuming just over 1,000,000 kWh (worth £100,000) in calendar year 2013.

The first thing Derek and Phil did was to understand what was driving the lighting use. Their analysis of 2013 showed a very good correlation with lighting hours. Although lights are on all day in the central core of the car park where natural light is low, since the lamps on the periphery can be switched off on a bright day, lighting hours is still the significant driver (aka common factor variable) for electricity use. Lighting hours were used to establish a baseline model using the 2013 data, to predict future use. The fit between actual use (in blue) and predicted use (in red) for 2013, was very good.

Having established the baseline model, the team then went about making the existing system more efficient: by automating the peripheral lights with a daylight sensor; by shutting several levels of the car park at night; and by reducing lighting in stairwells during the day. This was the operational improvement phase.

Once the system was operating optimally using the existing hardware, a proper analysis of the business case for replacing the T5 fluorescent lighting system with LED fittings could be made using these lower use patterns. The business case was approved, so a lighting replacement programme began in July 2014 and continued until late August.

The effects of the initial operational savings programme, which started around December 2013, can be clearly seen, as the blue electricity consumption begins to fall below the values predicted by the lighting hours in red. There is then another sharp decrease as the LED replacement programme started in July 2014. The model using lighting hours provided a seasonally independent performance indicator, against which all the improvements delivered could be measured. During the operational control improvement phase, Dec 2013 to July 2014, savings of 69,000 kWh were made (equivalent over a full year to 120,000 kWh or 12% of the total). The variance from baseline for the LED replacement phase demonstrated an *additional* saving of 45% against the 2013 use.

Insight

14.39 **A baseline using lighting hours**
The baseline established for the multi-storey car park (MSCP) at MediaCityUK provided a highly credible means of tracking improvements made by the engineering team, and gave the finance folks confidence that the improvement process was well managed and the benefits would be transparent to all. The baseline also supported the ISO 50001 energy management programme at MediaCityUK.
Source: Niall Enright. Data reproduced with the kind permission of Peel Media Ltd.

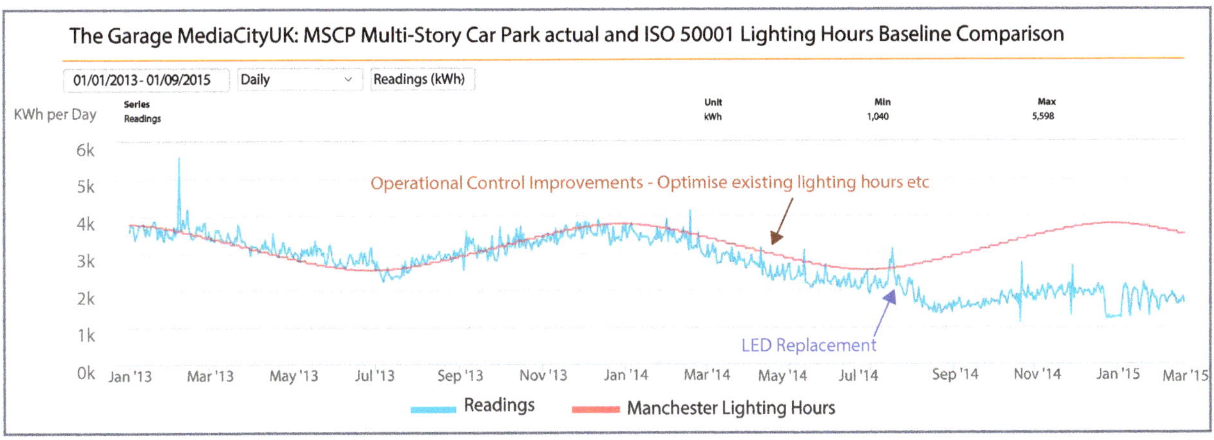

14.22 Using multiple regression

Multiple regression is the same a simple linear regression but with more than one variable. It is not uncommon for resource-using systems to have multiple influences and this technique enables us to establish which variables have the greatest influence.

Exploration: ***Predictor and response***

The y-value in our x-y scatter plot (usually the resource we are trying to control), is known as the response variable, whereas the x-values (degree days, lighting hours, production, etc.) are called the predictor variables.

All the regression analysis techniques described here are univariate, which means there is just one response variable and one (simple linear regression) or more (multiple linear regression) predictor variables.

We should not confuse multiple linear regression with multivariate regressions where there is more than one response variable. Multivariate regression is beyond the scope of this book, but could occasionally be useful where we have a system that influences more than one resource (e.g. a dual-fuel boiler that can operate with two different input resources, gas or oil). Techniques such as multivariate analysis of variance (MANOVA), can pin down the relationships between the variables. However, it is usually much easier to see if a proxy measure that incorporates both variables into one measure can be calculated. In our boiler example, we could use the volumetric calorific values of each fuel to arrive at a single variable "input energy" and then correlate this against the predictor variable value (e.g. heating degree days).

In a typical energy M&T system, around a third of the performance targets will use regression equations which have more than one independent variables.

There are two main reasons why multiple variables are needed.

- We have not been able to separate out the resource use into a single function or activity. For example, we only have data for the electricity use in the whole building, and this delivers two services (chilling and lighting), so we need to use cooling DD and lighting hours in our model.
- The measure of activity may change. For example, we may be producing two products on the same production line over the monitoring period, so we need to incorporate both of these into our model.

In the first case, we should consider installing additional metering to separate out the functional use of the resource. A new meter on the chillers would allow us to obtain the balance of electricity for lighting by subtraction from the main meter. The decision on whether install additional metering can be guided by the coefficient of determination, R^2, that we obtain when we relate the electricity to both variables. If the coefficient of determination is low, we may have no choice but to add sub-meters, so that we can model the electricity use using each variable separately.

In the second case, we may wish to consider developing an aggregate variable, which may permit us to use a single number to model the activity. For example, we could create a variable called total production, the sum of product A and product B. When creating this type of measure, we may have a choice between various units, such as the number, volume or mass of the products, and the selection of the value to use should reflect which of these attributes of the variable affect the resource use more. We may also have data on the resource-intensity of products A and B, and so we may be able to incorporate these into our equation so that our aggregate variable might be:

Production = (Quantity A * Intensity A) + (Quantity B * Intensity B).

Single measures of activity that can incorporate several production variables have become commonplace in UK industries which have climate change agreements, since the CCA performance measure often requires a

Real World: *Aggregate variables*

It can sometimes be tough to determine which variables to use.

In 2011, I participated in a QUEST™ audit of energy, waste and water at a speciality chemicals facility, ImproChem, an AECI subsidiary in Durban, South Africa. The ERM team led by Massimo Bettanin, with the support of Richard Wise, David Baudains, Tim Price and myself, found that the site manufactures a large number of products in batches taking different lengths of time. There were far too many products for a multiple regression to capture their impact and our analysis proved this.

When presented with this challenge, the very enthusiastic and involved site team, Molebatse Letswalo, Gravin Phyfer and Nonhlanhla Ciliza, suggested that we aggregate the products into classes such as acrylates and monomers, which had broadly similar production paths.

Subsequent analysis showed that these variables provided a sufficiently good indication of activity to understand the performance of the plant and estimate the potential savings that might be achieved through M&T.

However, this analysis had to be done using monthly data in order to *"dilute"* some of the additional variability caused by batch starts and ends not coinciding precisely with the periods when the water, energy and waste data was collected. Over a month, the impact of this was much smaller than over a weekly or shorter time frame.

This example reinforces two key points made earlier: the importance of the site team being an integral part of any data analysis work, and the fact that looking at the same data over a longer time frame can compensate for timing issues.

product mix algorithm which combines the output of the facility into a single value or formula.

A regression equation using multiple variables looks similar to the simple linear regression equations described earlier:

$$y = m_1x_1 + m_2x_2 + m_3x_3.... + c$$

Where x_1, x_2 and x_3 are the variables we are using, and m_1, m_2, m_3 are the coefficients for each variable, while c is the intercept. Note that I haven't used the word slope to refer to the coefficients m. This is because in multiple regression equations, there is more than one value for m, and the data cannot, therefore, be charted. Although we can't visualize the multiple linear relationship on a graph, we can still think of the m coefficients as reflecting the amount that our resource use y will increase for each unit change in x, the same as in simple linear regression. The c coefficient remains the amount of the resource that is used regardless of the value of the variables x_1, x_2 and x_3.

Most M&T software packages will readily produce the equation of the line for multiple variables. The method for Excel involves the Analysis Toolpak, which is described in the reference section of this book on page 774. Please note that the Excel functions CORREL(), SLOPE() and INTERCEPT() will not work with x and y ranges which are different in size, and therefore will not work with more than one variable.

If using Excel to undertake multiple linear regression analysis, it is worth noting that one of the statistics that Excel provides is called the adjusted R square, which is also referred to as the adjusted coefficient of determination, symbolized as R^2_a. This statistic is used because of the tendency of the *"normal"* coefficient of determination, R^2, to increase when even very weakly significant variables are added to the multiple regression. To avoid simply adding lots of variables to chase an ever more perfect (but complex) correlation, many analysts favour using the adjusted coefficient of determination which takes into account the number of variables. Thus, if one adds a variable to a regression, and R^2_a does not increase, then this variable should be omitted.

Excel, and most M&T packages, will provide the correlation for each variable used (as well as an indicator of significance such as the t-statistic). These should be used to determine the relative influence of each variable on the results. If a variable has a low correlation and a low significance and its inclusion does not increase the adjusted coefficient of determination, then the variable should be omitted from the model. Tuning a multiple regression model requires patience, as experimentation with several candidate variables is usually needed.

Since we can't graph the output of the multiple regression, we must take care to ensure that we do not have outliers (by looking at the residuals or plotting single regressions for the dominant variables).

14.23 Non-linear relationships

Sometimes a relationship between resource use and a variable does not fit a straight line. Here, we explore how these non-linear relationships arise and how we can interpret the data.

The efficiency of some types of equipment, such as electric motors or boilers, does not vary in a straight line with load or activity, but instead, exhibits an efficiency curve. In these cases, we can use a tool such as Excel's line fitting (see page 772) to deliver an equation which we can use to model the common factor variation due to the performance characteristic of the equipment.

14.40 **Specific ratio plots lead to curves**
Wherever the relationship between a resource and activity has a fixed or baseload element, any effort to plot the specific ratio of the resource against activity will lead to a curve, as shown below. The upper chart plots the ratio, the lower one is the conventional scatter plot of the same data.
Source: Niall Enright.
The data is available in the companion file pack.

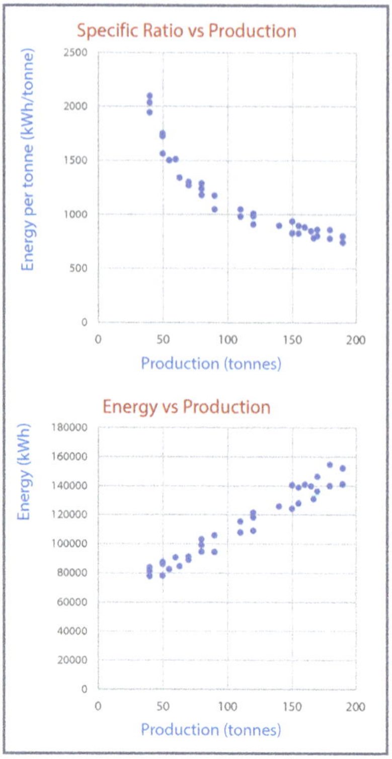

In my experience, however, non-linear models are relatively rare when accounting for common factor variation in resource use. It is not that these effects do not exist; e.g. an air compressor will be more efficient in the winter because the ambient temperature is cooler (typically, there is a 0.25% increase in electricity use for each 1° C increase in the inlet air temperature).

This temperature effect will modify the linear relationship of power against compressed air production, but *in practice*, the impact of production is such a dominant effect that the non-linear temperature effects are swamped. With few exceptions, extrinsic activity factors are much more influential than intrinsic performance characteristics of individual items of equipment.

Although uncommon, there are some circumstances where non-linear relationships can be found. For example, when there is more than one item of equipment in the resource-using system, merit order effects come into play as the items of equipment are unlikely to exhibit the same efficiency. Capacity constraints can also lead to non-linear relationships, as already described in the topic on degree days (illustrated by the office example, opposite). Finally, additional inputs can result in non-linear patterns, in some cases where these are deliberate (as in the example of the shaft furnace, opposite), or in other situations where these are accidental (as in the example of heat stratification).

It is also possible to create a curved relationship through an incorrect choice of the resource measurement. For example, if you plot a chart using the specific ratio of a resource (e.g. kWh/tonne of product) against the production variable (e.g. tonnes of product) then, as long as there is a baseload, we will get a chart as shown above left. This is easy to explain, for as production increases the fixed element (shown by the intercept in the normal linear regression line) is divided over more units of production. If the Intercept is positive, this means the that the chart curves downwards as illustrated; if it is is negative, then it curves upwards. If there is no baseload, the chart of the specific ratio will be roughly horizontal, around the value of m, the slope of the regression.

Real World: *Examples of real-world non-linear energy use relationships*

Here, we see the energy consumption rises as production increases. This is typical of a situation where there is a merit order of equipment, with the most efficient being used first and the less efficient being deployed as production rises. Clearly, an opposite curve could suggest less-efficient equipment is being used first.

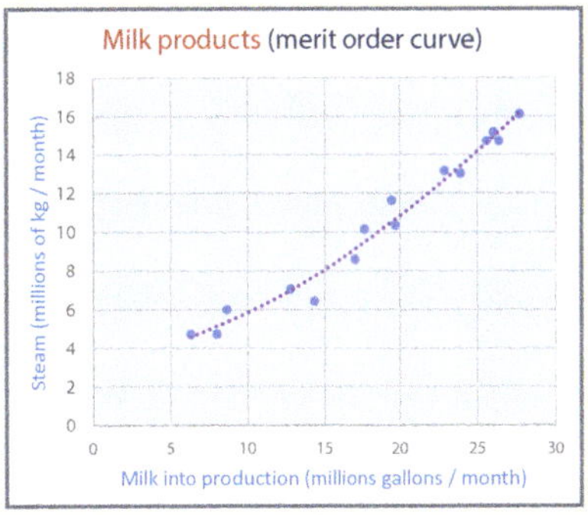

This data is from an aluminium production in a shaft furnace, which includes heat recovery from the exhaust to preheat the incoming material. At low production throughput, the heat recovery system is not very effective, hence the greater energy input indicated by the steeper slope.

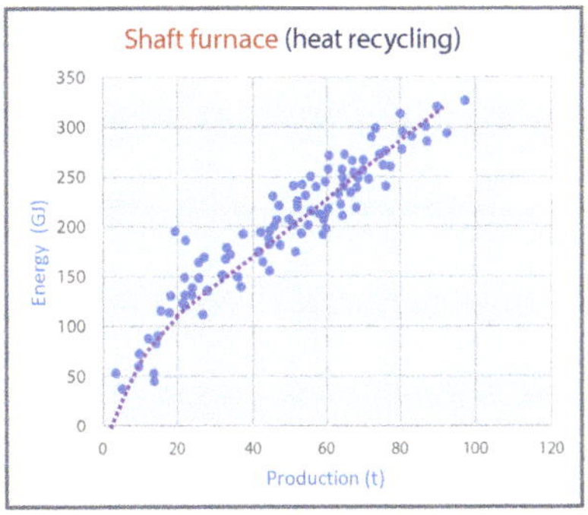

As described earlier, a building whose heating system is constrained will show a curve that levels off at greater demand as the system fails to cope with the additional requirement for heat.

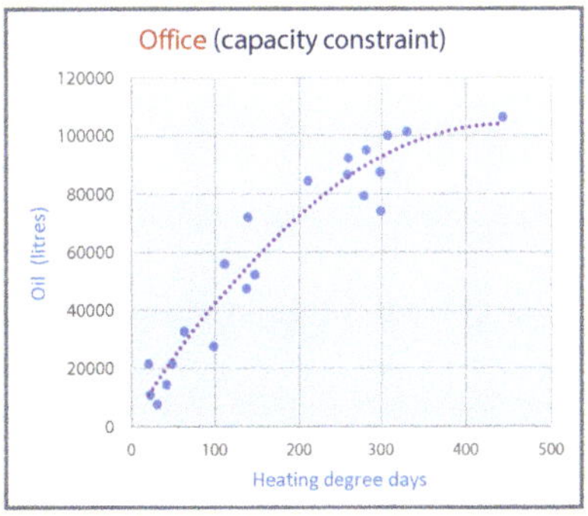

In this example, we can see the effect of heat stratification in a warehouse. As the temperature falls (heating degree days increase) more cold air enters the warehouse, forcing warm air towards the roof. This additional effect of ambient temperature is greatest at lower temperatures and causes the gas use to increase more rapidly than expected.

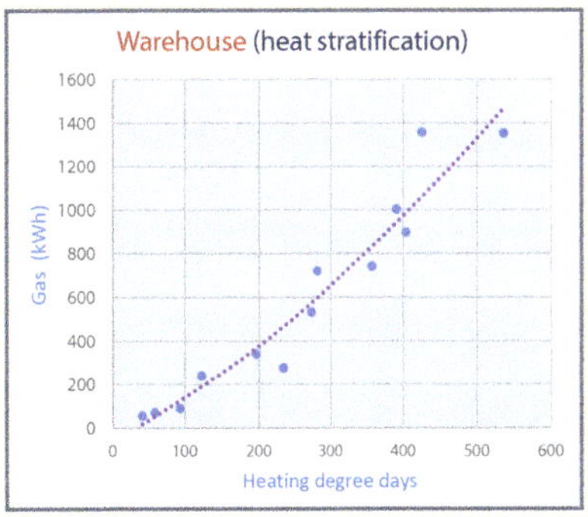

14.41 **Non-linear energy use cases**
Although relatively uncommon there are some cases where energy use follows a non-linear relationship with activity or demand.
Source: Niall Enright, taken from examples in the Energy Good Practice Guide 112, Monitoring and Targeting[257] *originally written by Peter Harris © Crown Copyright.*
Images and data in the companion file pack.

14.24 Model and variance

Our model provides clues about how the constant driving factors such as weather and production determine the resources we use, and hence offers insight into how these relationships can be changed. Variance from the predicted use given by the model will highlight exceptional factors, which also provides opportunities for improvement.

W. Edwards Deming, the great quality systems guru, once described variation as ***evil.*** *Nothing could be further from the truth.* ***Variation represents opportunity.***

We have seen that we can model the relationship between resource use and influencing factors in a variety of ways. Usually, a simple linear relationship describes the way the resource responds to changes in the common factor, but sometimes we need more sophisticated combinations of factors or non-linear relationships to describe the relationship adequately.

The purpose of the models is to explain as fully as possible the effect of the common factors such as weather or activity on our resource use. These common factors are known as the predictor variables because we can use them to forecast the amount of resource we expect to use, our response variable.

There are two sources of insight that these models provide, from which we will generate opportunities for improvement.

1. The model itself contains information about the resource use from which we can identify strategies to reduce consumption. If there is more scatter than expected, then we may need to improve the feedback loop between the factor and the resource by changing the control inputs. If the baseload is high, then we may have to look at resource use unrelated to the factor. If the slope is steeper than expected, then we may have cause to question the efficiency or maintenance of the resource-consuming equipment. We should never simply accept our model as the *status quo*, however great the correlation. Consistency does not equal efficiency.

2. The second source of insight is the variance of actual resource use from that expected (or predicted) by the variable. This variance can be due to the random distribution of the resource use around the mean predicted by the factor. Or it can be due to meter error or due to exceptional variation. Managing this variance from prediction (called a residual in statistics) is central to M&T. Eliminating variance is often regarded as the *low-hanging fruit* of the Optimize phase of a resource efficiency programme. If the system was able to operate at a particularly good level of performance, then it should be able to do so again. By eliminating (or reducing) the periods of bad performance and increasing the periods of good performance, *"free"* savings can be achieved.

The first type of analysis, an examination of what the model is telling us, often requires help from the Champion leading our efficiency programme, external consultants and folks with a technical background. This is in part because improving the underlying resource/activity relationship usually requires a fundamental change to equipment, process or control, which could be beyond the capabilities of the operators of the equipment and may involve greater risk and cost. To fully understand what the model is telling us, we often need to make comparisons, using the model's features (scatter, significance, intercept, slope) against reference data (similar equipment, design specifications, etc.), which may require considerable technical expertise and access to data.

Although the interpretation of the model fundamentals is often a more complex process, it will undoubtedly benefit from the participation of the operators of the equipment. They will, at the very least, be able to advise on the control of the equipment and process, but also need to understand and support any more fundamental opportunities for change that are identified.

The reverse is true when it comes to the interpretation of variance. This, ideally, should be led by the folks who operate the equipment. Since they control the resource use on a day-to-day basis, it is they who will be able to find out what patterns of behaviour lead to better or worse performance. Creating ownership among operators for the operational improvement is critical, and this is one of the main reasons why we want to make the underlying model as easy as possible to understand (see page 514 on the trade-off between complexity and ownership). It is also important that we present the variance in a clear and unambiguous way, as described on the following pages. ⇒ page 492.

14.42 **A dairy baseload**
Understanding the baseload in a resource-consuming process can give real insight into improvement opportunities.
Source: Niall Enright. Images and data available in the companion file pack.

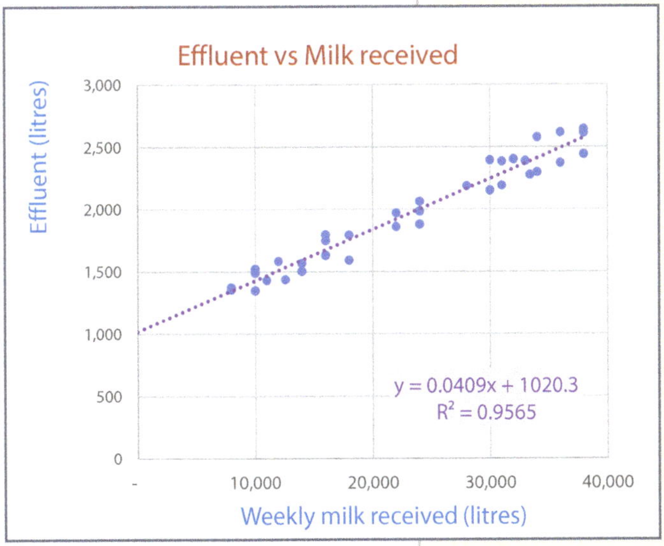

Real World: *A flash of insight at a creamery*

One of the most enjoyable aspects of working in resource efficiency is the pleasure one can get from discovering something fundamental about the way that resources are being used. I have lots of *eureka* moments, but one that I would like to share is from a creamery that I visited in the early 1990s.

Look carefully at the chart left. What is it telling us? We can see that on the left-hand, vertical axis we have the amount of effluent through the wastewater treatment plant (this is the expensive-to-treat waste that we want to reduce). The driving variable is milk received (in litres).

The point that caused real excitement during the presentation of the findings is the baseload of 1,020 litres of effluent, which is not production-related. During normal production, wastewater arises from cleaning activities, as expected, so this mysterious large volume of liquid through the effluent plant when not cleaning was very worrying. This waste was not being created out of thin air and only two explanations came to mind. First, the unexpected treatment flow could be leaking milk from holding tanks (highly undesirable). Second, it was rainwater which should go to drain being sent to the wastewater treatment (leading to unnecessary effluent discharge costs, so this, too, was equally undesirable).

Exploration: *Managing variance*

Earlier (page 464), we saw how we can take real data for gas use in a building and create a regression model using heating degree days. Below I have plotted this predicted gas consumption, in orange, against our actual gas use, in blue, as a time series chart (left, top). We can see that the model is remarkably good at accounting for the seasonal change in gas consumption due to external temperatures. Because the coefficient of determination, R^2, is 0.967, we can say that changes in the heating degree day variable accounts for over 96% of the variation in gas use (with a few caveats around significance and causation).

14.43 **Variance plots (below and opposite)**
Source: Niall Enright. Available in file pack.

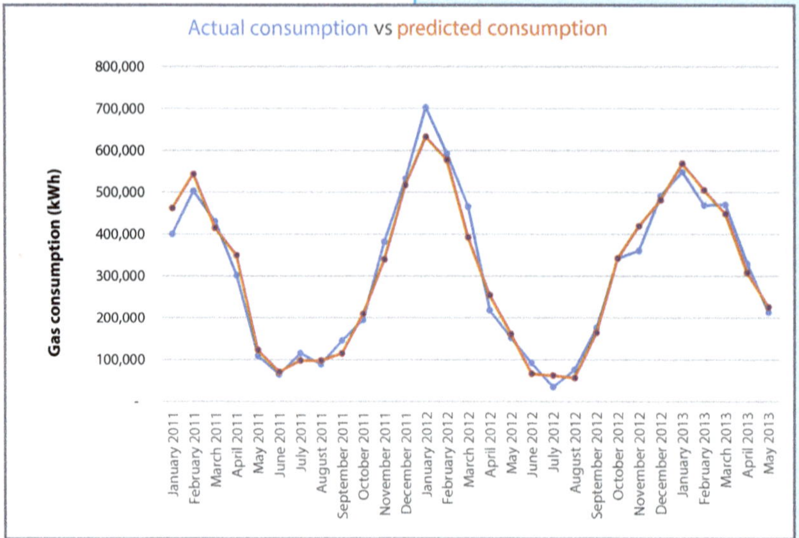

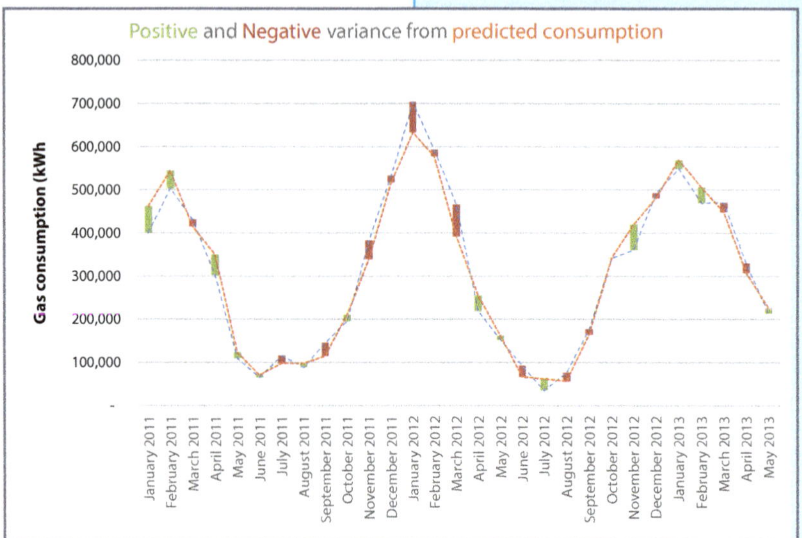

The people who use the energy in the building generally don't understand the meaning of the statistical values, so we rely on using chart data to communicate the fit of the model. In this case, they will have little doubt that the model is a fair representation of the gas usage in the building and, crucially, they are likely to agree that the model will be a good predictor of future gas use and so can be used for ongoing control.

We can be satisfied that we have now captured the dominant influence on gas use of the uncontrollable common factor variance due to temperature.

Of course, *uncontrollable* does not mean fixed. We can set about changing our building's response to ambient temperature, by, for example, installing insulation, draught-proofing, changing our boilers for more efficient models, or lowering the heating setpoint in our control system. These are fundamental changes that modify the resource response to the common factor driving stimulus.

There is a second set of opportunities that are available to us as a result of this analysis. Although weather seems to account for 96% of variation there is some 3-4% variation in use due to other factors, as seen in the time series chart (left, below). Sometimes we use more gas than predicted, the red columns, and sometimes less, green.

In the time series trend chart these differences can be difficult to see, so they are more often plotted as a variance from predicted chart, as shown opposite top, which has the same data.

One piece of advice when showing folks these variance charts derived from a best-fit line, is that the sum of negative (good) variances and the sum of positive (bad) variances balance out over the analysis period. Put simply the chart is *objectively* separating good and bad periods in equal measure. It is an *"honest and fair"* analysis.

To further reinforce the positives, I always start the discussion not by leaping on the periods of excess consumption compared to prediction, but on the good periods, when we used less resource than predicted. *"That was a great month, why do you think that was?"*. After all, M&T is about *"repeating the good and eliminating the bad"*, usually in that order!

Put a different way, the variance chart is not a theoretical model that I have devised using some design data of how the heating system should work, it is an empirical model based on the observed variation in actual gas use driven by weather over a period of time. The challenge for folks is, given a set of good months and a set of bad months, to work out what was different on those occasions that led to the higher or lower than predicted use. Some of these differences may be controllable in the future and can lead to operational improvement (e.g. windows left open, timers overridden at weekends, hot water taps left on). By monitoring actual use each month against predicted use, we have established continual control of the resource use.

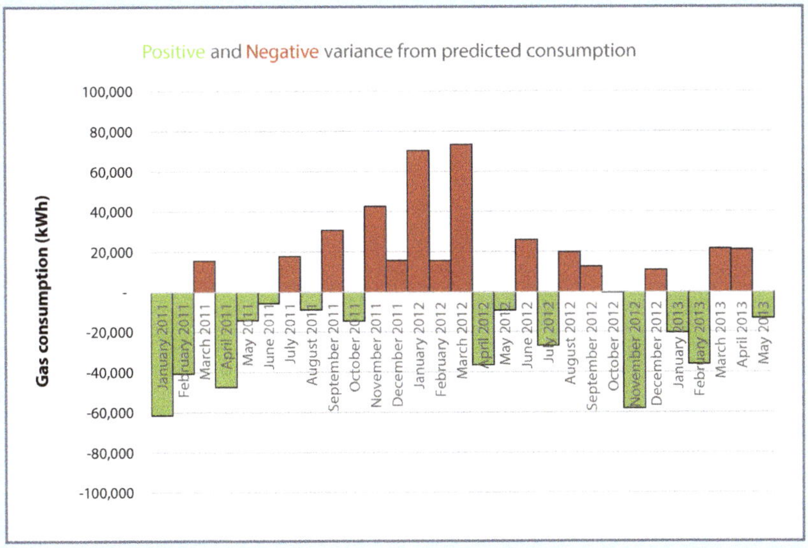

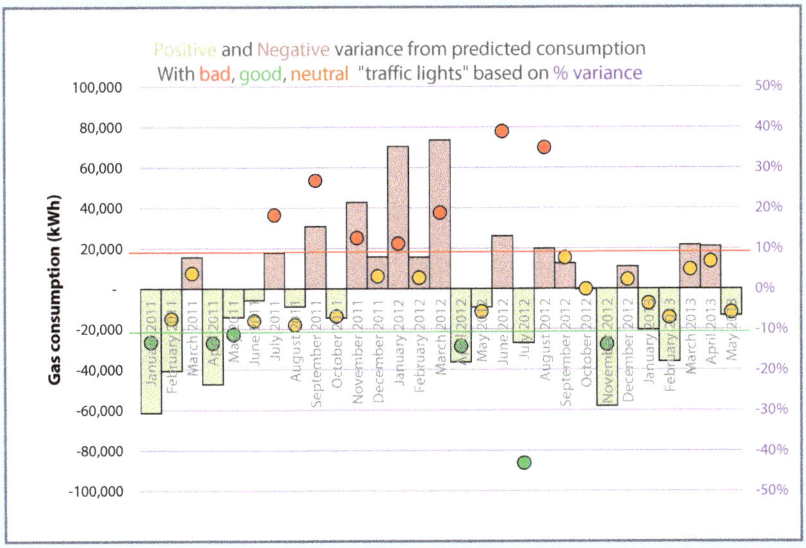

In this control process, we would typically define a level of variation that is acceptable. As a minimum, this acceptable range may reflect the metering accuracy of our gas and heating degree day data, but may also involve other considerations such as the age of the equipment, the cost of gas, the importance of this resource compared to others.

The acceptable boundary may be expressed as a percentage variance (as shown by the circular *traffic lights* plotted against the right-hand scale on the chart left, bottom) or as an absolute value. In some cases, a cost or quantity threshold can also be incorporated to exclude variances which are not materially significant.

However, the boundaries are set, the red and green traffic lights are called exceptions in M&T. Note good exceptions are expected to be analysed and responded to at least as vigorously as bad exceptions.

Exception reporting lessens the workload of the M&T programme and makes the process more acceptable by focusing limited people and effort where it matters most. Periods with yellow traffic lights do not usually merit additional investigation.

14.25 How to compare variance

Sometimes we may want to compare variances across several different resource users or we may want to combine a variety of indicators of variance into a composite score. In this section I will describe the statistically correct way to do carry out these comparisons.

Z-scores allow us to ***compare improvement on an equal basis*** *across completely different systems and processes.*

Performance reporting based on percentage variance from predicted is widely employed in M&T because the underlying metric,

((Actual-Predicted)/Predicted) *100

is easy to explain and easy to calculate. This calculation underpins exception reporting, where a threshold variance around the predicted value is defined, above or below which an exception occurs, requiring the cause of the variance to be investigated. Note that it is just as important to investigate negative (good or lower-than-expected use) exceptions as it is to investigate positive (bad or higher than expected) ones.

In the practical world of M&T programmes, a lot of effort goes into refining these exception conditions to avoid false positives. Thus we can introduce further criteria such as a dollar value threshold below which an exception will be suppressed.

The more sophisticated users will also recognize that a 5% variance in energy use on a boiler, which is a large user and should be tightly controlled, is much more critical than a 5% variance in, say, the electricity use in the PCs in the offices. So different items of equipment will, sensibly, have different thresholds of acceptable performance around which they are controlled.

If we were to look at the data for the boiler and the PCs, the level of control being achieved would be reflected in the shape of the distribution of the data.

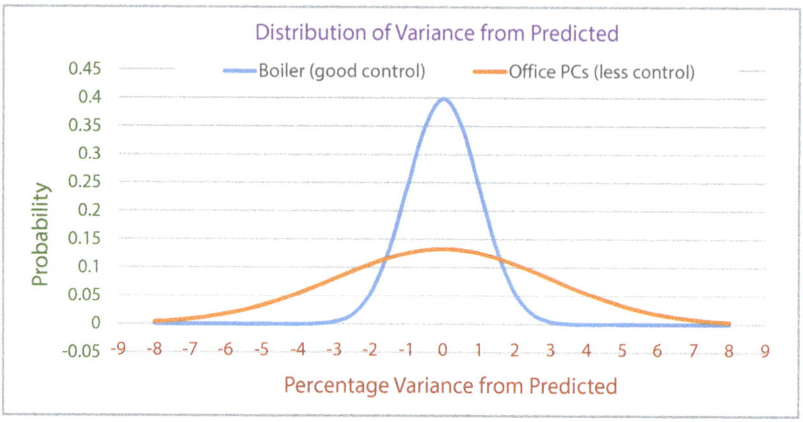

14.44 **Distribution of variance in percent actual versus predicted resource use for a boiler and office PCs**
As the boiler exhibits much less variation (i.e. it is better controlled) than the electricity use in office PCs, it will be more difficult to deliver operational improvements. Also, simply reporting the headline % improvement is not a fair comparison. *Source: Niall Enright. Illustration and data available in the companion file pack.*

As we can see the boiler (the blue line) is well controlled within a ±3% range. On the other hand, the distribution of the variance of PC electricity use in the offices (the orange line) is much broader.

Going back to our earlier work on probability (see page 452), we know that 68% of values will fall within one standard deviation of the mean. This gives a good starting point as to where we might set our percentage exception thresholds. For the boiler, the one standard deviation in terms of percentage variance is ±1.6% and for the PCs it is ±4.5%.

You will also recall from the probability statistics that we can describe our variances as a z-score, which is simply the variance divided by the standard deviation. Thus an improvement of -1% for our boiler would have a z-score of 1/1.6 = 0.625, whereas an improvement in the PC electricity of -2% would have a z-score 2/4.5 = 0.444. The z-scores enable us to compare improvement on an equal basis across completely different systems and processes. Thus a 1% improvement in the boiler consumption is more significant than a 2% improvement in the PC electricity since it is much less likely to occur randomly.

While comparing A with B is relatively easy, comparing the performance of multiple resource users, such as one manufacturing site with another, is much more challenging. This is because these comparisons rely on composite measures of performance. Where we are improving our use of a range of different resources, such as water, gas, electricity and so forth, we cannot roll the improvement up into one single unit. Nor can we take the mean of the percentage savings (or the z-scores), as these take no account of the relative scale of different resource use.

The most common solution to this is to work with dollar savings rather than unit savings. We can do this by calculating the units saved or lost for each resource compared to prediction, converting this to a dollar saving (or loss), using the appropriate cost of the resource. This can then be added up at a department or site level. If we then wanted to normalize this data to compare sites, we would typically take this cost and divide it by the total resource costs to get the % cost saving achieved.

> *For improvement comparison purposes, we would want to apply the same units costs to all our site, but for investment appraisal purposes we want the costs to reflect local prices.*

By and large, we tend to use this data as is: if site X has reduced its resource costs by 3% and site Y by 5%, we assume Y has done a better job. To be really objective we should take this variance one step further and calculate at a site level the standard deviation in the cost variance from predicted and so calculate the z-score for costs for each site and make the comparison using the z-score, although I have never seen this metric used in practice.

There are some challenges when working with costs as a tool for comparison - for example, for improvement comparison purposes we would want to use the same unit costs across all sites, but in terms of appraising investments, we would want to calculate the real savings made using the local costs. On balance, I tend to favour using local costs as this adds credibility to the reporting of improvements, despite diminishing the ability to make objective comparisons.

14.26 Control charts

Control charts are widely used in manufacturing industry to maintain consistency and quality. Here, we describe one type of control chart that can be applied to the analysis of variation in resource efficiency.

Control charts were first developed by Walter Shewhart working at a US power plant in 1931[660] and form an essential part of SPC. Shewhart thought that variation was the enemy of quality and the control chart techniques that he developed are designed to identify and eliminate variation. For example, if you are producing ball-bearings which should have a diameter of 2 cm to a tolerance of ±0.1mm, you would use control charts to help reduce the incidence of non-conforming bearings.

Because of the emphasis on consistency, one industry that has embraced SPC is the automotive industry, which relies on large numbers of components meeting very finely defined tolerances. Indeed, the US Automotive Industry Action Group produced my main reference for SPC back in 1991,[47] a gift from Peter Dipple at Jaguar Cars who was the first person I met applying these charts to variation in energy use. In its later incarnation as part of Six Sigma, the SPC control charts described here are widely used in many different industries and even in some service sectors.

The difference in control strategies in SPC/Six Sigma and resource efficiency is that the former aim to maintain ***consistency*** *while the latter seeks to achieve* ***continual improvement.***

Shewhart designed four kinds of control chart, but the only type that is appropriate for resource efficiency is called an individual and moving range (X-MR) chart, as the others involve multiple samples or measurements (our resource use data has just one reading at a time). An example of this chart, using the same buildings' gas consumption data as used in earlier examples, is shown opposite. The control X-MR chart is the area shown shaded in yellow, but I have added, in blue, the same *variance from predicted* chart described on page 491.

The control chart consists of two trend graphs (individual value above, moving range below) as well as a table of data (with values in this case shown as #### due to lack of space). The elements and calculation that make up the charts and data table are shown at the top of the illustration. The upper control chart is essentially a standard variance chart, with the addition of an upper and lower control limit, placed three standard deviations either side of the mean.

Although widely used in industry, control charts are rarely applied to energy and resource efficiency (and are not formally part of traditional M&T) because they are designed to achieve different objectives. Control charts are fundamentally about maintaining consistency, whereas a resource efficiency programme is about achieving continual improvement. In other words,

14.45 **X-MR control chart**
The control chart is illustrated in the area shaded yellow. The individual elements of the chart are described in the table.
Source: Niall Enright, based on The Fundamentals of Statistical Process Control, AIAG.[48]
Worksheet in the companion file pack.

resource efficiency wants to change the *status quo* by eliminating the bad variation and repeating the good variation. By contrast, in SPC all variation around the mean is considered bad. There have been some discussions about the applicability of these charts to environmental indicators (Corbett and Pan[169] and the evidence is that techniques like CUSUM (discussed shortly) are more suited to this type of data than control charts.

Legend (colour matches charts)	Description	Calculation
Measurement/Value (X)	Variance from predicted	Input data
Upper control limit of X (UCL_x)	Upper control limit (~equal to 3 sigma)	$\overline{X} + (2.66 * \overline{MR})$
Grand average ($\overline{X}$)	Average of the variances (0 for the linear regression period)	Average(X:X)
Lower control limit of X (LCL_x)	Lower control limit (~equal to 3 sigma)	$\overline{X} - (2.66 * \overline{MR})$
Moving range (MR)	Absolute difference between the current (C) and previous (P))variance	Max(C:P)-MIN(C:P)
Upper control limit of MR (UCL_{MR})	Upper control Limit for the moving range (there is no lower control limit)	$3.267 * \overline{MR}$
Average range ($\overline{MR}$)	Average of the moving ranges	Average(MR:MR)

14.27 Interpreting control charts

Despite some questions about the suitability of control charts for managing energy and resources, we need to understand how these charts can be interpreted in the event that they are used to assess variability in our resource use.

In the conventional quality-driven application of control charts, the range between the upper and lower control limits is called the system *"capability"*. The capability ratio is the ratio between the actual performance and the capability using the formula (USL-LSL)/6s, where USL is the upper specification limit or tolerance, LSL is the lower specification limit or tolerance and 6s is six standard deviations (i.e. UCLx - LCLx).

From our earlier work on probability (see page 452) and Table 24.8, we know that 99.73% of values have a z-score of ±3 (i.e. a fall in a range of six standard deviations). This z-score means that the defect rate of a process with a capability ratio of 1 is 0.27% (also expressed as 2,700 defects per million). Small changes in capability ratio have a big impact on quality: a capability ratio of 1.33 (±4 sigma) has a defect rate of just 0.0063% (63 parts per million).

A lot of effort is put into SPC processes to ensure that they achieve the desired capability ratio in order to meet an acceptable (low) failure rate. Such a "tuned" system is said to be in statistical control or stable and the purpose of the control chart is to maintain that stability by alerting the operator to potential significant variation to the system caused by what we refer to as exceptional variation, but is called special cause or assignable variation in SPC. The key point is that control charts are for systems that are in statistical control, i.e. where there is no exceptional variation in the data used in the charts.

Where the organization is **experienced** using control charts, these may also be applied to energy and resource use **as long as other techniques, such as CUSUM, are also used.**

The control chart of the gas use in our building in Figure 14.45 does not fail any of the tests opposite. In fact, with this monthly data, we would expect only to see a point outside the UCL or LCL once every 31 years (0.27% of the time)! This is a commendably low rate of *"false positives"*, but not a practical proposition in terms of helping to identify potential areas of improvement. The fact is that there is still considerable variation in actual gas use compared to predicted use (up to ±40%), due to exceptional variation over the analysis period. By setting the control thresholds using this data, we have effectively hidden exceptional variation from view. This is a clear demonstration of why control charts are generally unsuited to resource efficiency: the systems we are trying to improve are not in statistical control, *the control data available includes occasions where exceptional variation exists* and measurements are comparatively infrequent.

There is also some broader debate around the statistical validity of the control chart approach. An example of the discussion can be found in a paper by

William Woodhal,[802] *Controversies and contradictions in statistical process control*. Many of the discussions centre around the black-and-white "*in control*" and "*out of control*" rules, as well as whether the data used is actually normal. Furthermore, the X-MR chart is considered the weakest of the control charts by SPC professionals in terms of analysing variation, with questions raised in particular about the moving range chart.[779]

Even if we have a system under statistical control, the interpretation of control charts takes real skill and, unless the operators are already soundly grounded in this, and for the reasons above, I would not generally recommend the use of control charts as a technique in energy and resource efficiency. However, where the organization is habitually using control charts, and we have systems where the resource use is in statistical control, these charts may also be applied to energy and resource use as long as other techniques such as CUSUM, described next, are also used.

14.46 **Control chart tests for special causes** There are a number of tests that indicate that a process is being influenced by special causes and so is no longer under control. Note that the bands A,B,C are each one standard deviation. *Source: Niall Enright*

Rule	Example	Interpretation
Point above the UCL_x or below the LCL_x.	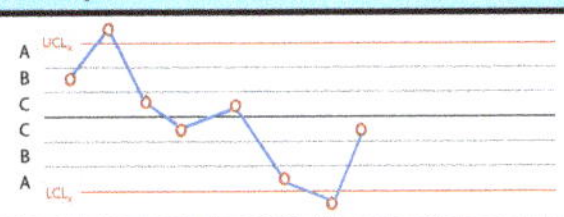	Special cause variation (p <0.27%), a permanent change may have occurred which will shift the mean or control limits.
6 or more consecutive points increasing or decreasing.	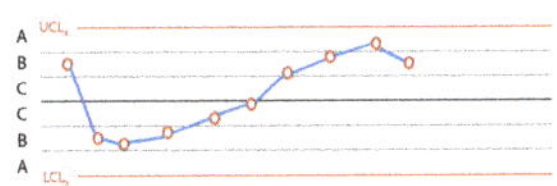	Something has changed in the process to cause the increase/decrease. This is usually caused by a shift in the process mean.
9 or more consecutive points on the same side of the median.	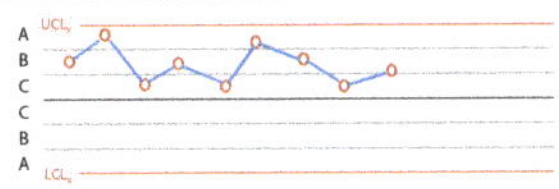	There has been a shift in the process mean. Identify what has changed at this time.
14 or more points alternating up and down (sawtooth pattern).	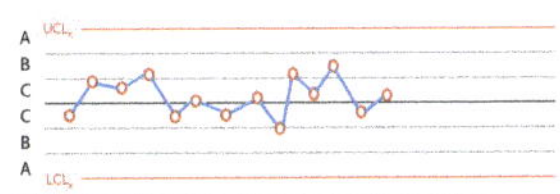	There are systematic differences in the process, e.g. two alternately used lines or different shifts.
2 of 3 points in a row beyond two standard deviations (i.e. in zone A or beyond),	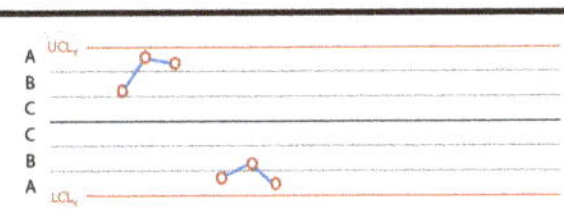	There has been a shift in the process mean or a change in the standard deviation.
4 of 5 points in a row beyond one standard deviations (i.e. in zone A or B or beyond).	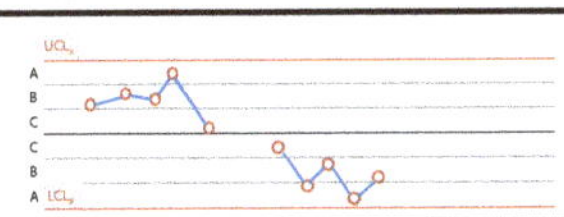	There has been a shift in the process mean or a change in the standard deviation.
15 points in a row within one standard deviation.	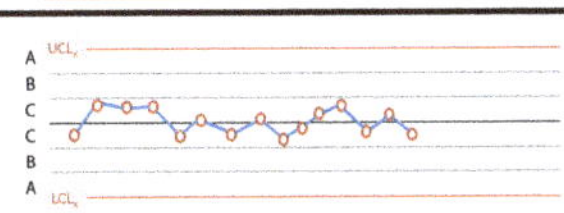	We may be measuring a sub-group - i.e. weekend performance values only instead of a combination of weekday/weekend.
8 points in a row on both sides of the centre line within two standard deviations.	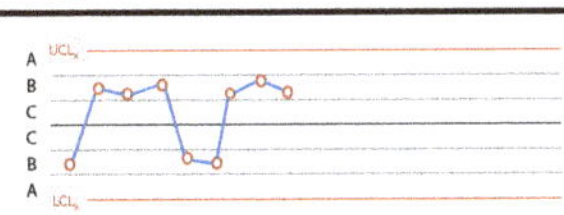	We may be measuring a sub-group - i.e. weekday performance values only instead of a combination of weekday/weekend.
Count of runs (groups of one or more points on each side of the mean) that fall outside the lower and upper limits in Table 24.14.	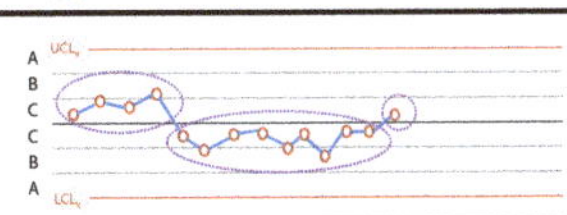	If runs are too low, as in this case, there may be an underlying cycle in the process. If too high there may systematic difference as per sawtooth.

14.28 Using CUSUM

The single most important analytical tool for energy and resource efficiency is the cumulative sum of variance (CUSUM) chart. No other technique is as powerful in highlighting underlying changes in resource-consuming systems, and as such this is central to both audit and control.

The cumulative sum of variance (CUSUM), is a charting technique which helps us to see if there are underlying trends in resource use by examining the variance from predicted values. Because it is a technique to analyse variance, CUSUM starts with a model of resource use (usually linear or multiple regression, but it could be any of the models described earlier) and the variance is our actual use less our model predicted use. In this example, a positive variance indicates that we have used more resource than expected and a negative variance means we have used less (but see the box opposite). The period of data used to create the model is usually called the baseline period (also known as the *"training period"* by some statisticians), and the model on which the CUSUM variance is based is called the original equation in M&T.

14.47 **Simple CUSUM table**
This simplified table of 20 daily pump electricity consumptions omits the first column with the dates. We can see that the CUSUM is simply calculated by adding the current variance to the previous CUSUM. *Source: Niall Enright. The image and data are available in companion file pack.*

Pump Daily Electricity Use (kWh)			
Actual	Predicted	Variance	CUSUM
85	95	-10	-10
109	113	-4	-14
97	105	-8	-22
102	110	-8	-30
97	105	-8	-38
106	111	-5	-43
92	96	-4	-47
92	113	-21	-68
103	102	1	-67
94	94	0	-67
104	101	3	-64
96	94	2	-62
111	113	-2	-64
115	106	9	-55
108	101	7	-48
122	114	8	-40
100	93	7	-33
125	119	6	-27
130	120	10	-17
120	117	3	-14

Constructing a CUSUM series is simple. We just take the current period variance and add this to the previous period *cumulative* variance. Looking at the Table 14.47 left, on Day 1 the electricity use by the pump is 10 kWh less than predicted, so the first value for the CUSUM is -10 kWh (as it is the first day of my analysis, we don't have a previous total to add). On Day 2 my use is 4kWh less than predicted, a variance of -4kWh, which gives us a CUSUM of -14 kWh. On Day 3 my use is 8 kWh less than expected, so the CUSUM is now -22 kWh, taking into account the previous total of -14kWh.

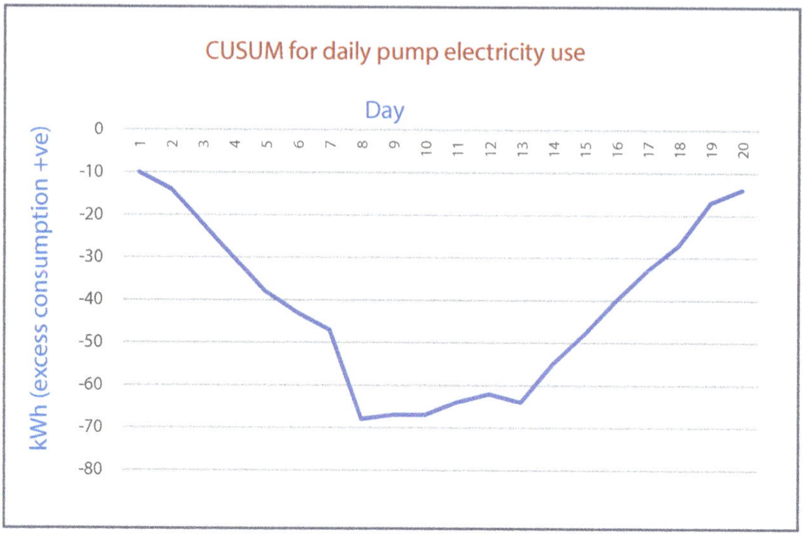

Real World: *Beware of the basis for the CUSUM - cost or saving*

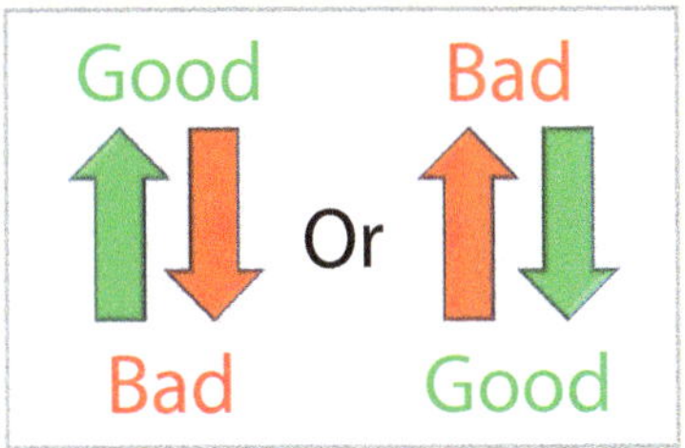

In the example left, we have plotted our CUSUM based on variance being *actual - predicted* use i.e. positive indicates we have used more resource than expected and negative less. Here, a downward slope is good while an upward slope is bad. This is often called a CUSUM of consumption or, in money terms, a CUSUM of cost (as costs falling is good and costs rising is bad).

Some commercial M&T packages plot CUSUM in the opposite way. That is to say, that they show the amount of resource saved, so an upwards slope is good and a downward slope is bad. In many ways, this makes sense as people like to see improvement as a positive value. It is a matter of preference.

However, it can sometimes be difficult to establish which is the basis for measurement in the CUSUM. In the M&T packages I have designed, I explicitly have a system setting to allow users to set the approach system-wide. I then use the words *"cost"* or *"consumption"* to indicate that we are looking at the actual variance from predicted use and *"savings"* in the axes labels to indicate we are seeing predicted variance from actual. For consistency, this cost or savings basis for plotting should extend to the variance trend chart, with lower than predicted use shown as a negative costs or consumption, or as a positive saving.

If we then plot the CUSUM values, in the chart opposite, we can see that there has been a dramatic change in the underlying efficiency of the pump. From Day 1 to Day 8 the slope of the line is downwards; that means that each day the pump has used less electricity than predicted, and so is more efficient than the model (original equation) predicts. Because the line is reasonably straight, the level of improved efficiency is relatively constant, and so we can conclude that the pump is in a steady operating condition. Then, on Days 9-13, there is a clear change in the pump's performance. Now the negative variances previously recorded turn to small positive variances as the pump starts to perform slightly worse than predicted (the line is almost horizontal), a pattern that worsens dramatically between Days 13 and 20 where the pump consistently uses more electricity than predicted by the model (original equation).

An early enthusiast for the use of CUSUM, Peter Harris, described five ways that the chart could be interpreted, back in 1989.[354]

1. Straight lines correspond to periods of constant energy performance.
2. Sustained changes in a CUSUM slope signify a change has occurred. The slope inflexion point indicates when the change happened (or became noticeable).
3. Horizontal lines correspond to the energy performance of the model training or "baseline" period.
4. Jumps in the CUSUM plot correspond to one-time or brief energy performance changes.
5. The chart y-position shows *"Cumulative savings/loss since the beginning of the period"*.

The reason that a CUSUM chart is so important is that it quantifies and highlights change. Since the chart can pinpoint the precise dates when changes happen, it is an invaluable tool to diagnose the root cause of good and bad performance. We can monitor the CUSUM and react quickly (within one or two measurements) to any change in the direction of our chart.

CUSUM is often considered the *"scorekeeper"* of any resource efficiency programme, because it can unambiguously report the underlying saving made between any two dates (usually from the start of the programme, or from the beginning of the current financial year, depending on the reporting requirements). If we express the CUSUMs of all the resources we are managing in cost units, these can be added together to summate the savings or losses made at department, site or even whole-programme level.

It also has the advantage of being able to tell us if the models we are using to predict performance are valid. If we run a CUSUM for the baseline period which we used to model our performance, the value of the CUSUM should return to zero at the end of the period. This is because the variances on either

Real World: *Two models*

Image© kchung. Fotolia.com

Having introduced the notion of an original equation, it is helpful to expand on how this is used.

In M&T (and also in the energy management standard, ISO 50001) we actually have *two performance models* for each resource user.

The original equation (the *energy baseline* in 50001), is used to measure changes in a system compared to the starting situation. This model provides the variance for CUSUM and *remains unchanged unless there is a fundamental change in the system* (e.g. equipment is replaced). It is usually the best-fit line obtained at the start of the energy or resource efficiency programme.

The second model is the target equation (*energy performance indicator* or EnPI, in ISO 50001). The target incorporates the desired improvement goal and is regularly updated - typically once a year when the improvement plans and budgets for the year are agreed.

So, at any time, we can report on where we are compared to the current target, using the target equation and we can measure our overall improvement, using the original equation.

So we can celebrate success and aspire to do better, at the same time! We can look backwards and forwards.

side of our best-fit line (or other models) should cancel each other out if the model is valid, i.e. the sum of the positive variances should match the sum of the negative ones.

There are a some practical warnings that we need to bear in mind when using CUSUM. The first is obvious but is often overlooked, which is that changes in CUSUM charts of costs are only meaningful where the cost has remained constant. If there are price changes during the CUSUM period, the underlying efficiency may be constant, but we would see the same sorts of changes in slope that would falsely cause us to think that there has been a shift in the system efficiency. Thus the interpretation of CUSUM charts should usually be done in original measured units (such as kWh, litres, etc.), not in secondary values such as costs or emissions.

Another, rarely discussed, issue with CUSUM charts is the rate effect. That is to say that the first rule of interpretation of a CUSUM, that a straight line CUSUM represents a new operating state, is true only *where the throughput of the system does not vary greatly*. In fact, Harris' Rules 1 and 2 are more effective at identifying baseload changes than changes in the relationship between the driving variable and the resource use.

By way of explanation, let us imagine a system operating at 10% efficiency improvement over the model prediction. If my predicted use is 10 units, then we will measure a -1 unit variance in period 1. If on the subsequent occasion the throughput is much higher, and my predicted use is 100 units, I would see a -10 unit variance. Finally, in the third period, the predicted use is 20 units, so I will see a -2 unit variance. My CUSUM will be -1,-11,-13, clearly not a straight line, even though the efficiency improvement, in percentage terms, is steady.

In systems where the rate of resource use changes greatly, such as the gas use in our example office, which varies from 35,000 kWh in summer to 700,000 kWh in winter, changes in the CUSUM plot should be interpreted with caution. Indeed, for this kind of changeable resource, a chart of percentage variance from predicted use (as shown on the bottom of page 491) may well be a more valuable analysis tool.

A further caveat to Rules 1 and 2, is that large baseload changes can have a dramatic effect on the slope of the CUSUM, and subsequently mask any changes in common factor variation or exceptional variation that is taking place. The solution for this is to develop a new original equation when the baseload changes, taking into account this new baseload, but this has the disadvantage that it could take several weeks or months to have sufficient data points in the new mode of operation to provide a valid equation.

Despite these cautions about the interpretation of CUSUM charts, they remain a vital tool to quantify programme savings and to verify the validity of models. As long as we recognize the pitfalls, the example opposite shows that, for systems with fairly consistent levels of resource use, CUSUM can provide new and precise insight into events that may otherwise be difficult to see.

14.48 **CUSUM data from a hotel boiler**
This is real data without a site name in order to preserve confidentiality.
Source: Niall Enright

In systems **where the rate of resource use varies greatly,** *changes in* **the CUSUM plot should be interpreted cautiously.**

Real World: *A hotel CUSUM interpreted*

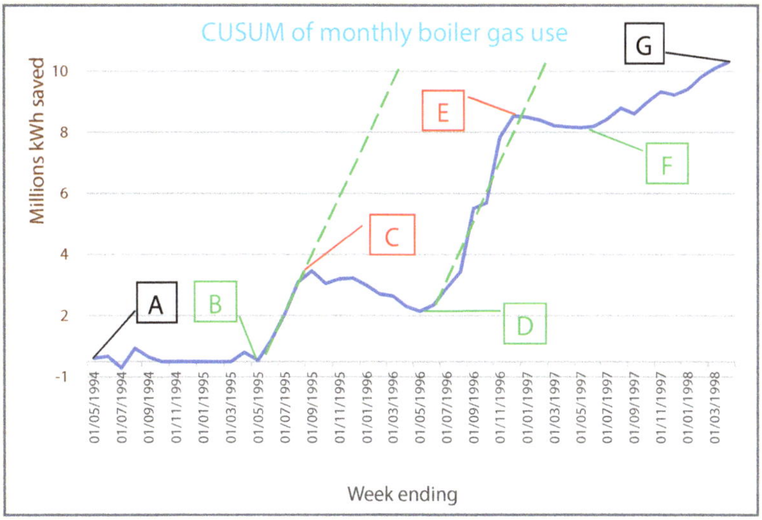

Above is a CUSUM chart taken from an energy efficiency programme for Hilton International. I worked on this project in the late 1990s with a colleague Dave Covell doing the technical audit and Simone LeRoy and I analysing the data.

The chart is for four years' monthly gas use in a hotel boiler, which we analysed during the initial audit of the hotel. First, we have to note that this is a CUSUM of savings chart, so an upwards-sloping line is good, indicating that we are using less gas than expected. Secondly, there was a fairly constant gas use throughout the year as the steam produced was used in the laundry and for hot water rather than for heating.

The first year was set as the baseline period for the performance, and the data was used in a linear regression to model the predicted gas use. This can be clearly seen as the period A-B when, as expected, the CUSUM returns to zero. It is from this original equation model that the variance was calculated.

This analysis picked up five occasions, labelled on the chart above, where there was a fundamental change in the efficiency of the boiler: the points in green show an improvement and those in red a decrease in efficiency. Since we had definite dates for the events, we could work with the hotel engineers to establish what had happened on those occasions. Luckily, the boiler logbooks were very well kept and we could see that these changes coincided with the dates when the external maintenance company came into the hotel to service the boilers. Another piece of information the chart provides is an indication of the efficiency possible from this boiler. On two occasions, labelled B-C and D-E, the boiler achieved almost identical high levels of efficiency (indicated by the green dashed line). While it was clear that overall the hotel had saved almost 10M kWh in three years, it was also clear the current boiler efficiency F-G did not match the best possible. Armed with this information, the maintenance contractors were swift to ensure that high levels of efficiency were achieved from that time forward - representing a significant and "no-cost" improvement. Needless to say, the subsequent CUSUM was used to confirm this.

14.29 Estimating "best"

Behaviour-led resource efficiency efforts often struggle to gain support because the savings potential is less tangible. There is a remarkably simple and credible analysis technique to quantify the savings potential and so gain support and set realistic objectives.

Linear regression provides a powerful and credible way to estimate the savings that can be achieved by better operational control.

A line of best fit for a data series (whether produced through linear regression or a curve-fitting technique) has the effect of separating our resource use data into two populations: above the line are the periods of poor performance and below the line are the periods of good performance.

We can exploit the fact that we have these two groups to develop a hypothesis: *"What would the resource use be if we succeed in repeating the good and eliminating the bad?"* There are two subtly different ways we can answer this question: one called the optimistic method and the second the conservative method.

The first method can be thought of as assuming we "*repeat the good*" on all those occasions where performance has been *"bad"*. Here, we ignore those bad points and simply plot a new line of best fit, using just the good data points below the purple original line. This line is shown in solid green in our fourth chart, lower right, in the illustration opposite. Now that we have a new line of best fit, calculating the difference between the actual resource use and that predicted by the new line is straightforward, and in the example in the charts opposite, the optimistic method predicts a reduction of 19.9% in gas use.

This approach has the advantage of being simple to explain - "*We will repeat our best performance during those occasions where we are currently poor.*" - and simple to calculate. This method is still sometimes called the *"EEO method"* after the original UK Energy Efficiency Office guidance in which it was first published, so we also have the added credibility that this is the *"official"* method to calculate savings.

In my experience, though, this method does tend to lead to some very optimistic savings estimates or targets. If we imagine that we achieve good performance half the time, and then divide this into two again, which is what we are doing with our new line of best fit, we would expect, at the current level of performance, to be better than the new line of best fit about one quarter of the time, and worse three quarters of the time. That level of negative feedback may not be desirable when we start our programme, particularly as it makes some currently good periods now appear bad.

As a result of this, I have developed an alternative model to determine savings, called the conservative method. It is very similar to the previous method but can be considered to be more focused on "*eliminating the bad*". Here, we work on the assumption that all bad periods of consumption can be brought to the average performance, represented by the purple best-fit line. This is illustrated in the lower left chart, below. The bad points that were above the line (represented by diamond-shaped markers) have now been brought to the purple original line of best fit. A new line of best fit, shown in solid green, can then be calculated using the new distribution of the data points. This new line predicts how our system could perform if *"we bring our periods of poor performance to the current average level of performance"*. Again it is relatively easy to calculate the savings that would arise if we follow the level of performance predicted by the new best-fit line, and in this example the, seemingly small, difference between the purple original line of best fit and the new green line of best-fit is a reduction of 8.7% in gas use.

The choice of which method to use is entirely up to you. This may be determined by the ease with which the method can be explained or the level ambition for change. The optimistic approach assumes that good performance can get even better, while the conservative approach acknowledges the reality that folks are more likely to focus efforts on improving currently bad performance.

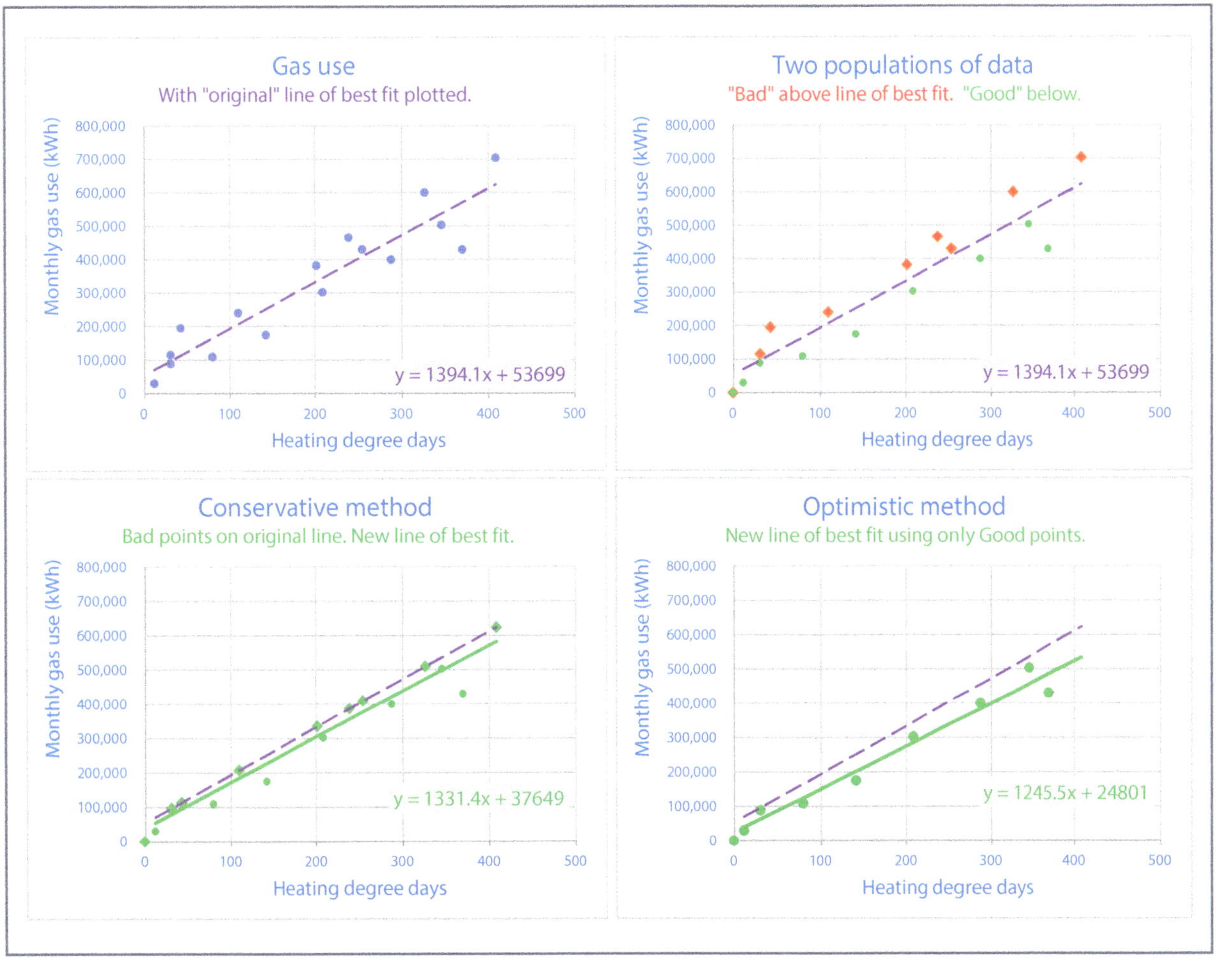

14.49 **Alternative methods to assess savings potential**
One of the benefits of statistical analysis of resource use data is that it offers an objective and credible method of evaluating the savings potential that can be realized by simply repeating the best performance already achieved. Since this level of efficiency is already being delivered, the presumption is that this can be done with the existing equipment or systems and so does not require large capital investment. The charts below illustrate the two methods that can be used to calculate the savings. There is a third variant, which was briefly promoted but is now discredited, which set the savings potential as the line joining the two points furthest below the line of best fit - which I might call the "exceptional" method.
Source: Niall Enright. The example spreadsheet is available in the companion file pack.

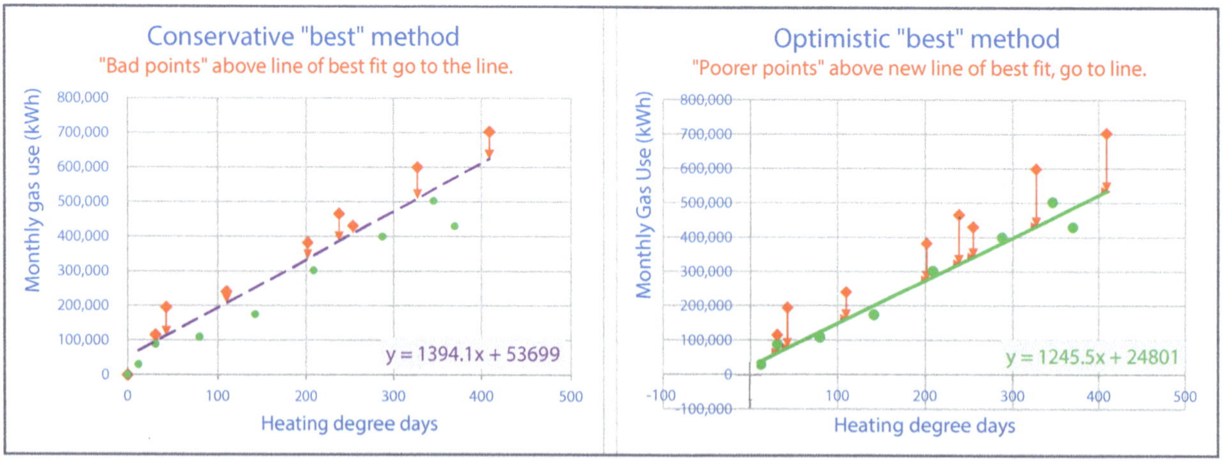

14.50 **Estimating saving potential**
Whatever definition for best performance is used, the savings potential is estimated by summing up the difference (or residual) value for each point above the best-fit line and the value on the best-fit line.
Source: Niall Enright. The example spreadsheet is available in the companion file pack, and includes the array formula methods described opposite.

The estimation of savings using these methods is straightforward, albeit a little fiddly. The conservative method is shown in the graph, above left, and the optimistic method in the graph above right. Both charts show the same data.

Savings are calculated by determining which points are above the standard regression line or the new optimistic regression line and calculating the effect of bringing these points down to the chosen line of best fit, as shown by the red arrows in the charts above. Clearly the optimistic method, right, shows the greater savings.

Given an Excel spreadsheet with a column of data for our resource (the gas data in this case) and another column of data for the driving variable (heating degree days in this case), it is not too difficult to create a third column with the calculated value using the regression formula for the line. We can then determine which values are above the predicted value and sum these differences up to calculate the savings. There is an example of this approach in the companion file pack.

However, this requires some additional columns to carry out the calculations, and so a method of calculating these values using single formula is shown in the box opposite.

Please note that I have used the word *"estimate"* for the savings potential. There is a degree of uncertainty in our calculations, as we are looking back at historic performance and assessing what this would mean if we had modified our operation to repeat good performance and eliminate bad performance. We are extrapolating future performance from past data.

Some uncertainty about the savings available arises because our data sample may not be large, or because the values are scattered around the best-fit line (in other words, less predictable), or because there may be instrumentation or metering inaccuracies. In the next section, we will look at how we can describe this uncertainty, and so increase confidence in our savings estimate.

In Numbers: ***Estimating the conservative and optimistic savings potential using Excel***

The method for estimating the savings potential from a "best" analysis described opposite requires that we examine each of our historic values in turn to determine if they are above our selected best-fit line and then add these up to create a total. It would be helpful if we could calculate the savings with a single formula.

One Excel capability we can make use of to achieve this is the TREND function. This takes the form of TREND(X_values, Y_values, [NewY_values]), where X_values is the range with the energy or resource, Y_values is the range with the variable (or variables). The TREND function returns the values of X which fall on the best-fit line, whose coefficients and intercept were determined on the basis of X_Values and Y_Values. If we wanted the output to be based on an alternative set of variable data, then we would enter a value for the NewY_values range. TREND offers a quick way of calculating the points on a line of best fit.

The second feature of Excel that we are going to use is an array formula, which can carry out multiple calculations within a single formula. Imagine the following four cells in Excel: A1=1, A2=2, B1=3, B2=4. If I want to work out what the maximum value for the sum of each row is, I would normally need to add a column to add A1+ B1 and so forth and then work out the largest value in the column. Using an array formula, {MAX(A1:A2+B1:B2)}, I can get the result in a single step. The + in the array formula instructs Excel to take *each item* in the two ranges, A1:A2 and B1:B2, and add them together (i.e A1+B1, and then A2+B2), and the MAX() will return the largest value in the resulting array (in this case the value 6). To tell Excel the formula needs to be treated as an array formula, you press CTRL-SHIFT-ENTER as you save the formula.

Our conservative "best" savings calculation also uses ranges, which I have colour-coded in the formula below, as follows, RngResource and RngVariable.

This is an Excel array formula. When you type in the formula, you press CTRL-SHIFT-ENTER together and the brackets "{" and "}" will appear. You DON'T type the brackets in yourself.

This is the Excel range with our resource (e.g. "A1:A20")   This is the Excel range with our variable (e.g. "B1:B20")

{=SUMPRODUCT(RngResource-TREND(RngResource, RngVariable),--(RngResource-TREND(RngResource, RngVariable)>0))}

The residuals: The first part of the formula subtracts the best fit consumption from the actual resource use using the TREND function (which returns all the points on the best fit line using the X and Y ranges supplied). The residual is positive for values above the line and negative for values below.

Return 1 if it is positive, 0 if not: The second part of the formula is similar to the first, but just checks if the result is > 0, in other words if we have a positive value above the best fit line. This returns a TRUE or FALSE, which we convert to 1 and 0 using the "--" double-negation function.

Now add up the positive residuals: The SUMPRODUCT multiplies both sets of numbers and adds up the totals. The positive values, above the line, are multiplied by the 1 and the negative values are multiplied by the 0, so we only include the positive residuals in our result.

The formula above will return the *conservative* savings estimate for the period in question. It can be used for multiple as well as single regression ranges (in which case rngVariable would have more than one column of variable data). In many cases, we would prefer to have a percentage saving. This is very easy to produce, by simply taking the formula above and dividing the output value by the total resource used in the period in question using SUM(RngResource) as shown below.

{=SUMPRODUCT(RngResource-TREND(RngResource, RngVariable),--(RngResource-TREND(RngResource, RngVariable)>0))/SUM(RngResource)}

It is possible to construct an array formula to calculate the optimistic savings (as shown in the spreadsheet model provided in the companion file pack); however, it is considerably more complex than the example above and so is potentially more error-prone. As a result, for conservative calculations, I suggest using the step-by-step process of creating columns with intermediate values.

14.30 Expressing uncertainty

Accuracy describes the degree to which a measurement or value approaches the true value. Uncertainty is a wider concept which seeks to include sources of error other than measurement. These basic techniques will help us understand how to estimate uncertainty from statistical data, or how to combine values with known uncertainties.

In the previous chapter on metering, we noted that there is no such thing as an absolutely correct measurement. Every value has a possibility of error. When I record that my consumption of electricity in a month is, say, 1,000 kWh, I should indicate the uncertainty in the measurement (which can arise from random variation or measurement error).

To determine the uncertainty, I should start with the resolution of my electrical meter. If it has a resolution of 1 kWh I know that there is a potential error of at least ±1 kWh (the maximum sensitivity). Furthermore, if the meter manufacturer has stated that the accuracy of the instrument is ±3% then I have a further uncertainty of ±30 kWh (3% of the reading of 1,000 kWh). I can state my electricity consumption was:

1,0000 kWh ±31 kWh or 1,0000 kWh ±3.1% or 969 kWh - 1031 kWh

I favour the middle description, with a percentage error, because variance from the target is often indicated as a percentage. Thus, if I know my metering error is 3.1%, I would be foolish to respond to a 2% variance in the energy consumption (unless of course the variation is repeated over many readings).

When describing accuracy, it is customary to present the value no more than one decimal point greater than the measured value. Thus a reading of 10.5 kWh with an accuracy of 3% would be shown as 10.5 kWh ± 0.32 kWh (not ± 0.315 kWh) or 10.18 kWh to 10.82 kWh rather than 10.185 kWh to 10.815. Significant digits are explored further in the next section.

When working with regression analysis results, we have already seen that the coefficients (the slope and the intercept) that are produced are merely descriptions of the data, which themselves have uncertainty about them. In Table 14.26 on page 469, the notion of a confidence interval was introduced. This describes the probability of statistical results not being due to chance - a 95% confidence interval means that the results are less than 5% likely to be due to error and a 99% confidence interval means that it is less than 1% likely to be due to error.

On the previous page, I described how we can quantify the savings potential using the conservative method. This estimate depends on the sum of the gap for points above the regression line and the regression line (expressed in statistical terms as the *sum of the positive residuals*). Now you will recall that the

In Numbers: ***Absolute precision***

Our confidence in the accuracy of a measurement increases with the number of measurements we take.

Earlier we described the t-value as a test of significance (page 468). It is also useful to calculate the absolute precision of a series of readings.

absolute precision = t * standard error

The value for t is looked up using the table on page 780. If I have 10 meter readings and I want the t value with a 95% confidence level, I would look up the value on the row labelled 9 (because I subtract 1 from the number of readings to allow for the *degree of freedom* of 1). Then I would choose the column with 0.05 (=95% confidence) on the two-tailed heading (because my reading can be ±, it is two-tailed). The t-value is 2.262.

If the average of my 10 readings is 205 kWh and the standard error (or standard deviation) is 9.6 kWh, the absolute precision of our readings is:

2.262 * 9.6 kWh = 21.7 kWh

The relative precision is simply the absolute precision divided by the mean

i.e. 21.7/205 = 10.6%

So I would state my reading as being 205 kWh ±21.7 kWh or 205 kWh ±10.6% with a 95% confidence level.

For a greater confidence level, say 99%, the value of t goes up to 3.25, and the absolute precision is ±31.2 kWh).

residuals of a regression are normally distributed (see Section 14.17 on page 470). We are interested in the values on just one side of the distribution since it is just the positive variances from which we have calculated our savings, so we have a *one-tail distribution*. Using the one-tail normal distribution table on page 777, we know that 50% of values are below the best-fit line, so to get to 95% probability we need to look up 45% (0.450) in the table, which gives us a z-score of 1.645. Thus 95% of the residuals in our regression should be less than +1.645 standard errors. In the companion file pack, there is an Excel example of this calculation. The results of the example savings estimate, rounded, can be expressed as:

450,000 kWh ±114,000 kWh at a 95% confidence level or
8.7 % ±2.2% (at a 95% confidence level) or
between 6.5% and 10.9% savings (at a 95% confidence level)

Note that the 2.2% is the 114,000, divided by the total resource use in the regression, not 114,000 kWh divided by the savings, 450,000 kWh. The standard error can be found using the Excel Data Analysis ToolPak Regression Tool (described in Section 24.4 on page 774), or you can use the Excel function STEYX(RngResource,RngVariable).

In Numbers: *Combining accuracies*

Where we are adding together several values, $E_1 \ldots E_n$, with a known uncertainty we use the following formula to calculate the error:

$$SE = \sqrt{SE(E_1)^2 + SE(E_2)^2 + SE(E_3)^2}$$

So, if my total savings are calculated by adding together two projects which save on electricity, the first of which saves 100 kWh and is accurate to ±20 kWh and the second which saves 200 kWh and is accurate to ±30 kWh, then the overall uncertainty is $\pm\sqrt{((20)^2+(30)^2)} = \pm\sqrt{((400)+(900))} = \pm\sqrt{(1300)} = \pm 36.05$ KWh. The total saving is thus 300 kWh ±36.05 KWh. Note that the combined uncertainty will always be greater than the largest single uncertainty value.

We cannot plug percentages directly into this formula. We need to multiply them by each value (in other words, the percentages are weighted according to the proportion to the total), using the following formula:

$$SE = \sqrt{(\%E_1 * E_1)^2 + (\%E_2 * E_2)^2 + (\%E_3 * E_3)^2}$$

Where my savings are calculated as a multiplication or product, P, of several numbers, $Y_1 \ldots Y_n$, then we would use the formula:

$$\frac{SE}{P} = \sqrt{\left(\frac{SE(Y_1)}{Y_1}\right)^2 + \left(\frac{SE(Y_2)}{Y_2}\right)^2 + \left(\frac{SE(Y_3)}{Y_3}\right)^2}$$

Taking a very simple example, if I am calculating the area of a rectangle of width 10 cm and length 20 cm and my measurements are accurate to ±1 cm then the accuracy of the area calculation is $\pm\sqrt{((1/10)^2 + (1/20)^2} = \pm\sqrt{((0.01)+(0.0025))} = \pm\sqrt{(0.0125)} = \pm 0.112*200\ cm^2 = \pm 22.4\ cm^2$.

You will note that SE(Y)/Y is the uncertainty expressed as a percentage. I could have completed the formula with the percentage which gives a combined uncertainty of ±11.2%, which when multiplied by our area of 200 cm^2 gives us the same absolute value of 22.4 cm^2. In all these formulae the units of the uncertainty need to be the same, and they also need to have the same confidence level.

14.31 Significant digits

Presenting results correctly involves more than making a statement of the accuracy and confidence levels. We also need to present the result with the correct number of digits. Here, we explore how to do this.

Anyone who has worked for the person responsible for giving me my first job in energy management, Ray Gluckman, will have had occasion to hear him rant (in the nicest possible way) about "*spurious precision*". Heaven help you if you put a report in front of him to review which gave over-precise savings from an opportunity, e.g. 15,347.84 kWh a month, or suchlike! Given a typical electrical meter accuracy of ±1.5% this will be ±230 kWh (at a 95% confidence level or whatever the manufacturer states), so what on earth is the final .84 kWh doing?

In addition to the aspects of statistical accuracy in our data, described previously, we also need to consider the issue of significant digits (sometimes called significant figures) in our calculation, which will affect how we express the results of any calculations. The notion of significant digits is quite easy to understand, and centres around how zeroes in a number are treated.

The following are the rules to work out the number of significant digits we have in a value:

1. All zeroes placed between non-zero numbers are significant digits. Thus the value 1003 and 10.03 have four significant digits; and
2. All zeroes after a decimal place are significant. Thus 10.0 has three significant digits, 10.010 has five significant figures; and
3. All zeroes at the beginning of a number are not significant. Thus the value 0.045 has only two significant digits; and
4. Similarly, all zeroes at the end of a number without a decimal point are not significant. Thus the figure 45600 has only three significant digits.

In practice, we only need to remember rules 1 and 2 (which explain which zeroes are significant) or 3 and 4 (which explains which zeroes are not significant).

It should be noted that there is some debate about rule 4, as there is ambiguity as to whether the trailing zeroes are present because they represent a real value that coincidentally ends in zero or whether they are placeholders to give a scale to the value and so represent an unknown quantity. This is one reason why scientific notation is recommended, because it eliminates the ambiguity. The value 45600 could be shown as 4.56×10^4 (three significant digits) or 4.5600

Real World: *Uncertainty is often large*

We should not underestimate the potential uncertainty in our calculations.

For example, take two values which are read to ±10% accuracy.

The test below is taken from the CMVP examination[42] by the Association of Energy Engineers (see page 529):

> *The baseline power requirement of a circuit is measured to be 100 kW, with a meter rated at ±10% of reading. After retrofit the same meter measures power as 80 kW. What is the uncertainty in the demand reduction?*
>
> *a) 10%*
>
> *b) 14%*
>
> *c) 20%*
>
> *d) 64%*

Plugging in the actual kW accuracies for the two measurements into the formula on the previous page we get $\sqrt{((10\% * 100)^2+(10\% \times 80)^2)} = \sqrt{(10^2+8^2)} = \sqrt{100+64} = \sqrt{164} =$ ±12.8kW. This can be changed back into a percentage by dividing by the demand reduction as follows: 12.8/(100-80) = 12.8/20 = 0.640312 = 64%.

Thus, these two readings combine to give us a much higher level of uncertainty than we would intuitively expect. Our savings could be much greater than measured or much less.

14.51 **Misleading outputs**
Hand held calculators and spreadsheets like Excel can present results with a large number of digits. In reality, these are often meaningless as the input values did not share the same level of precision.
Source: photo © Kurhan, fotolia.com

x 10^4 (which has five significant digits), as in scientific notation all the digits in the mantissa (the part before the X^x exponent, e.g. 4.56) are significant. For my version of rule 4, I have used the most common convention here, which is to treat the trailing zeroes as ambiguous. Please note that if the value was entered with a decimal place 45600. then it would have five significant digits.

We should be aware that mathematical constants used in formulae are regarded by convention as having an infinite number of significant digits, so if we are calculating the area of a circle from the formula πr^2, the only significant figure we are concerned about is for r, the radius because π is a constant.

Multiplication, division, trig functions, exponentiation. Here, the number of significant digits in the result is the same as the least number of significant digits in the input values. So if I am calculating the volume of a water tank using length x width x height and my input values are 0.56 m (two significant figures) x 4.01 m (three significant figures) x 1.234 m (four significant figures), my result is 2.7710704 m^3, which expressed with two significant figures becomes 2.8 m^3. Note that it is not the same as the end result expressed using the least number of decimal points in the input values, two, which would be 2.77.

Insight

Addition and subtraction. Here, our answer will match the least number of decimal places of our input values (not the least number of significant digits). Thus, if I am adding four electrical meter readings - 123.56 kWh and 11.2 kWh and 1.15 kWh and 0.43 kWh - the result is 136.34 kWh which to one decimal place is 136.3 kWh (rather than to the least number of significant digits in the inputs, two, which would have been 140 kWh).

The reason that the methods differ in multiplication and addition is that the uncertainty that arises when numbers are multiplied is greater than when addition takes place.

Note that the first result was rounded up and the second rounded down based on whether the first discarded digit in the answer is 5 or greater (rounded up) or less than 5 (rounded down). In order to avoid creating errors by rounding, if one is carrying out multi-step calculation one should carry at least one more significant digit or decimal point in the intermediate steps.

The two sins when presenting results are:

1. Writing more digits in an answer (intermediate or final) than justified by the number of digits in the data. (Ray's rule!)
2. Rounding-off, say, to two digits in an intermediate answer, and then writing three digits in the final answer.

The data we are using includes the stated accuracy of a value. For the result mentioned at the start of this discussion, 15,347.84 kWh (7 significant digits) * 0.015 (2 significant figures) the calculated saving is ±230 kWh (two significant figures, no decimal places) and because this is a ± the overall results can be stated in the same decimal places i.e. 15,347 kWh ±230 kWh.

14.32 Working with exergy

Not all energy is equal! If we save a kWh of electricity (high quality energy) this is worth more than saving a kWh of heat in hot water (low quality energy). Although rarely used in practice, understanding how exergy can be calculated will help us differentiate between the quality of energy that we are saving.

If one were to consider the first law of thermodynamics, aka *the law of conservation of energy*, we would never need to worry about energy efficiency as energy is never lost - it can change form, but the quantity of energy in a closed system always stays the same.

The second law of thermodynamics states that as energy is transferred within a system there is a tendency towards greater disorder (entropy). The second law is really cool. It makes the notion of a perpetual motion machine impossible. The tendency to go from ordered state to a less ordered state is what drives *"time's arrow"*. This law reflects the fact that energy always flows from the less possible state (hot, disordered) to a more possible state (cold, ordered) leading to an overall increase in entropy. You never get heat flowing from cold places to hot places without external work being put in, as in a fridge. Similarly, if I mix a hot gas with a cold gas I get a warm gas, but a warm gas never spontaneously separates itself back again into separate regions of cold and hot gas.

A brilliant French engineer, Sadi Carnot published a book in 1824, at the age of 28, which revolutionized people's understanding of steam engines. Before the laws of thermodynamics were really understood or accepted, Carnot developed the notion of a "*heat engine*", an idealized and simplified form of a steam engine, from which a deeper understanding of the workings of engines could be developed. What Carnot did was to prove that the work of a steam engine arises from the flow of heat from a high-temperature reservoir to a lower temperature one. Without this temperature difference, there is no flow of heat and so no ability to do work (e.g. move a piston).

From these insights, Carnot was able to calculate the maximum efficiency possible of a heat engine for any given pairs of temperatures (hot and cold). The equation is shown right. The greater the temperature difference, the greater the efficiency of the machine. So two equal quantities of steam (in energy KJ or kWh terms) discharging to ambient, one at 180° C and one at 120° C do not have the same ability to do useful work. The usable energy of the steam (that which can be converted to work) is called exergy. Unlike energy, exergy can be reduced when energy is converted from one form to another. Instead of energy efficiency, we should focus on exergy efficiency. For example, we should not use high-temperature heat for low-grade purposes (e.g. space heating) as this rapidly reduces exergy.

Real World: *Exergy in practice*

Considering exergy efficiency rather than energy efficiency gives us insights into energy systems.

For example, in the table opposite the boiler is by far the most energy-efficient system producing 90 units of heat energy for 100 units of gas energy in. However, the output of the boiler is (relatively) low-temperature hot water at 70° C (343K) which can't do much useful work (it cannot drive a piston engine or a turbine, for example). So, it is quite inefficient in exergy terms and so are the units of greenhouse gases per unit of exergy.

We are getting less work potential from 100 units of energy going into a boiler than into the CHP plant. Indeed, although it appears less energy-efficient, focusing on the power output of the CHP gives us a greater output of exergy.

At a national level, some startling conclusions can be drawn by looking at trends in exergy efficiency. For example, exergy efficiency in the UK improved from 9% to 15% from 1960 to 2010 but in the US it has stagnated at around 11%.[89]

It appears that in the US energy efficiency gains at an equipment level have been offset by an increase in the proportion of lower exergy-efficient processes (such as using electricity for air-conditioning).

In Numbers: *Calculating exergy*

The basic formula for exergy is:

$$\text{Exergy } \psi = \text{Energy} * \text{Quality Factor}$$

The quality factor reflects the fact that not all the energy in our energy source can be converted to useful work. For electricity the quality factor is 1, in other words, 100% of the electrical energy can be converted to useful work.

Similarly, when we consider the chemical energy inherent in solid, liquid and gaseous fuels (coal, oil, natural gas etc.) we can assume that the exergy is close to one (with a small deduction for the fact that some of the heat released is absorbed evaporating the water that is a product of the combustion reaction). In simple terms, we can consider exergy is net calorific value divided by gross calorific value.[646] For natural gas, this would give us a quality factor of 0.99.

Where the quality factor becomes a critical issue is when we are considering energy stored as heat in a medium (e.g. hot water or steam). From the work of Carnot, we know that this heat can do useful work for us if its temperature is higher than the temperature of the reservoir to which it flows (typically ambient). In this case, the quality factor is given by the Carnot efficiency η for the two temperatures, ° Kelvin, T_H (Hot reservoir) T_C (Cold):

$$\eta_I = 1 - \frac{T_C}{T_H}$$

The example below shows how we would calculate the exergy in a combined heat and power plant. Here, energy quantities are shown in MWh. The electrical output has a multiplication factor of 1, so its exergy is the same as the energy. However, the useful work that the heat produced can do depends on its temperature, so we use the Carnot efficiency formula to calculate the quality, which is 0.422. When we multiply this by 40MWh, we get 16.8 MWh of exergy in the heat (assuming the ambient temperature is 0° C or 273K). Note that the exergy efficiency of the plant (52.8%) is considerably lower than the energy efficiency (76%).

Figure 24.1 on page 771 provides some example exergy values for several energy sources.

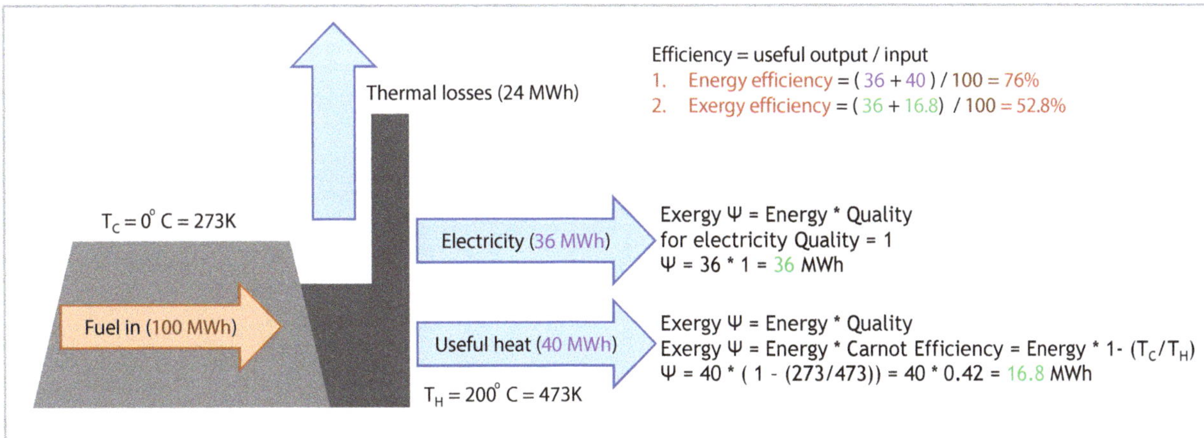

14.52 **Exergy of various power plant types** Assuming all plants have 100 units energy input and the varying quantities of heat and electricity lead to different efficiency in exergy. The most energy efficient plant (boiler) has the highest global warming potential per unit of exergy. *Source: Adapted from Schuller et al.*[649]

System	Elec Out	Heat Out	Heat Temp	Energy Eff.	Exergy Eff.	GWP/ψ
Boiler	0	90	343	90%	18%	5.5
Power Plant	50	0		50%	50%	2.0
CHP (heat led)	20	60	473	80%	45%	2.2
CHP (power led)	50	12	473	62%	55%	1.8

14.33 OEE

Overall equipment effectiveness is a measure that is fairly widely used in manufacturing industry to achieve the maximum throughput from equipment. Here, we will explore how it is calculated and why it is not a substitute for measurements of efficiency.

Overall equipment effectiveness (OEE) is a measure of performance that is used for single items of equipment. OEE is the primary metric in a productivity process called total productive maintenance (TPM). The formula for this is:

$$OEE = Availability * Performance * Quality$$

Each of these figures is entered as a percentage value. Where:

Availability = actual hours run / scheduled hours run

Performance = actual throughput / max design throughput

Quality = percentage yield or (units - defective unit) / units

Availability is sometimes simplified as "*time*" and Performance as "*speed*". Sometimes the metric loading = scheduled hours/total hours is also used to indicate that a machine's scheduled hours is less than the maximum hours available. In practice, scheduled hours represents the overall operation of the factory (five days per week, no weekend) so the loading is academic as one would not run one machine alone while the rest of the production process is idle.

It is important to understand the purpose of OEE. This metric is intended to apply to equipment that is a bottleneck in a production process and to provide guidance for the operators on the nature of the events that are leading to a bottleneck. Thus if a piece of equipment has an OEE of 35% *and* is a bottleneck, then increasing the OEE to, say, 50% may be a key priority. However, it is possible that the OEE of 35% is perfectly acceptable as the machine can comfortably meet the required throughput of the plant as a whole (called the "takt time") and there is no point at all beating folks up to improve this value.

14.53 **The six big losses**
When OEE is not achieving the desired level the causes can often be traced back to one of these six most common causes
Source: Niall Enright

Availability
Breakdowns
Start-up/shut-down (aka planned stops)
Performance (speed)
Minor stops
Reduced throughput
Quality
Start-up rejects
Production rejects

It is widely held that an OEE of 85% is world-class and thus there is a common misconception that this is the level of performance that one should achieve for every piece of equipment. In reality, the idea of an OEE target is somewhat misleading in the sense that one identifies the item of equipment that is the primary bottleneck in a production process, calculates the OEE and then sets about improving it. At some point, this equipment will no longer be the production bottleneck, and we record the OEE as the minimum we need to ensure that the equipment is not the limiting step. We then move on to the next bottleneck item of equipment and repeat the process, again recording what the minimum OEE needed is. This cycle repeats itself, possibly returning to earlier items of equipment to further optimization, until the required OEEs are understood, and the overall production throughput is achieved. In this process, the manufacturing team will have developed a good understanding of the critical items of equipment in their factory and will have used other TPM techniques to ensure that this equipment operates effectively.

Even the most ardent fans recognize that OEE has its limitations. One criticism is that OEE applies equal weight to production losses due to availability (equipment stoppage) compared to quality (units rejected due to defects). In practice, the cost implications of these may be quite different. We can get an OEE of 20% (assuming performance is the same) in an item of equipment with 20% availability and 100% quality or an item of equipment with 100% availability and only 20% quality. In reality, these are very different operating situations with very different solutions. Another criticism of OEE is that it is quite easy to "*fudge*" the figures, for example, by treating downtime due to maintenance as outside scheduled run hours, and so effectively removing maintenance time from scrutiny.

A focus on OEE is one of those whole system optimization activities (see page 230) which can actually impede efforts on resource efficiency (see left). This should not be the case, as OEE and resource efficiency are actually separate but complementary activities - one usually supports the other. For example, the most difficult term to understand and manage in the OEE equation is performance, yet this is an aspect for which resource efficiency brings additional tools to bear. A real-world example is some breweries in which I have worked, which are limited by the availability of coolth, i.e. the chillers are the bottleneck. Measures of the chiller's energy efficiency, for example, using cooling degree day targets, would highlight issues such as heat exchanger fouling which can be corrected to maintain the performance component of the OEE metric for the refrigeration equipment. Conversely, the OEE metric quality highlights material waste in a process arising from discarding or reworking product. Making sure that we maximize the yield within tolerance clearly is good for resource efficiency. The obvious relationship between OEE and energy efficiency has been demonstrated statistically.[133] But because these are related does not mean they are the same. Effectiveness does not equal efficiency. Thus, I can maintain the OEE of an item of equipment by increasing availability while decreasing performance. But we know that most equipment has a fixed baseload, so the efficiency will usually decrease in this case.

In practice, integrating OEE in a resource efficiency programme is more than a calculation issue. The challenge lies in folks on the shop floor being able to apply multiple measures of performance to their work. All the good practices that TPM brings, such as greater autonomy, tidy workplaces, training and knowledge, will improve efficiency but are designed for another purpose. It is not a case of "*either/or*" but "*both*". ⇒ page 516.

Real World: ***The right tools***

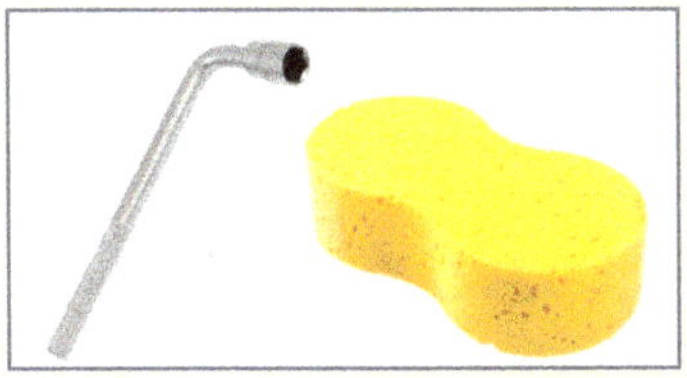

©Marek Walika/NorGal . Fotolia.com

Martin Hess, Richard Wise and I carried out a resource efficiency review at a tea bag producer in the UK (see Figure 16.18 on page 547).

Following some analysis, we presented the results to the engineering manager at the site.

The stumbling block we encountered was that the engineering manager was focusing on deploying TPM at the site and so OEE measurement was his primary performance indicator. As a consequence, the energy efficiency opportunities were regarded as secondary. A few *low-hanging fruit* projects would be implemented, but the more systematic continual improvement process we proposed was rejected.

The problem was that OEE was seen as a measure of efficiency, which it clearly is not.

OEE is a tool to de-bottleneck an item of equipment or a process. Resource efficiency is a tool to help maximize the return on investment of the same item of equipment.

Just as you wouldn't use a wrench to wash the windshield of your truck, you wouldn't use a sponge to loosen the wheel nuts. They are different tools for different purposes.

OEE keeps the truck on the road for the desired time (availability) at the desired speed (performance) carrying the required load (quality). Resource efficiency lowers the fuel consumption.

Exploration: *Putting it into perspective - ownership vs perfection*

Data analysis and statistical techniques can tell us a great deal about the performance of our resource-consuming systems. We have seen that by using these techniques we can reveal opportunities that would otherwise be invisible.

We must, however, avoid the temptation to get carried away with the power and beauty of these techniques, to the detriment of our resource efficiency programme. We must always remember that it is not the number on the spreadsheet that leads to change, but the decisions and actions of people.

Photo © Monkey Business, Fotolia.com

A key element of our resource efficiency method is enabling folks to make better-informed decisions (page 271). An informed decision implies a decision based on *understanding*. This is where our search for ever greater correlation needs to be balanced against the increased complexity that this statistical rigour can bring.

In my view, it is much better to have a simple model that users understand and which accounts for the *majority* of the variability, than a complex model in which the significant variables are poorly understood, but which achieves a higher correlation and so accounts for a greater proportion of the variability.

We need to remember that the purpose of our analysis is to inform and motivate action. If our conclusions are incomprehensible without an advanced understanding of statistics, then we will very likely have difficulty in communicating these. This has the effect of making our recommendation precisely what we don't want them to be: a "credence good", i.e. a leap of faith - "*Trust me because I know what the numbers are saying.*" (see page 185).

We also need to recognize that we are usually working with complex systems with multiple influences. We are unlikely to ever have a completely *"clean"* dataset, with only common factor variation and no exceptional variation. So any effort to overanalyse the data is misdirected, given that it is not pristine, aka *"normal"* in statistical terms.

Paradoxically, if our organization is already focusing effort on improvement, this may also make it harder to create effective performance models. This is because our models depend on a baseline period (what statisticians sometimes call the learning period), which is an initial period where only the common factor variable influences the resource. Improvements cloud this ability to develop a sound baseline, as overlapping changes create ambiguous model inputs.

On many occasions, I have encountered clients who recognize the problem of setting a baseline while delivering improvement and so deliberately hold back from implementing immediate savings opportunities. The intention is to establish a sufficiently long baseline period against which they can subsequently demonstrate savings. This is understandable psychology, but not good in a programme where we may want to achieve early success and momentum.

Another common danger with our data analysis is to fall into the trap of spurious accuracy where the model we have created is sensitive to changes smaller than the precision of the meters - in effect leading to false positives and wasted effort.

I would like to think that the techniques described here enable us to go beyond being passive interpreters of past performance and instead become active ***experimenters.***

Efforts have been made to address some of the weaknesses of tools such as CUSUM. For example, Antony Hilliard, Greg A. Jamieson and colleagues at the University of Toronto, Canada, have recently proposed alternative statistical techniques, using recursive estimate (RE) charts, as a potential alternative to CUSUM[369] This technique has the advantage of separately assessing changes in the baseline or variables in our model, enabling users to more rapidly home in on the root causes of change. However, as the authors point out, RE charts themselves have a number weaknesses, and *"will only be useful if colleagues can understand what model parameters represent"*.

In practice, the analysis described in these pages can be undertaken in several contexts. On the one hand, it can form the basis for a decentralized energy management or control process, where individuals and teams use the models as the basis for the M&T of resource use on an ongoing basis. Here, the constant challenge is to get managers to accept the basis for the performance objectives suggested by the analysis, and so an approach of *"simple is best"* is usually advisable.

Alternatively, we may have an ongoing resource management and control objective, but the analysis and response process may be centralized. Here, experts do the analysis, interpret the causes of variation and initiate the appropriate corrective or repetitive actions (depending on whether the change is bad or good). In these circumstances, a much more sophisticated analytical approach may be merited.

The analysis of data in audit usually falls halfway between the extremes above. On the one hand, the auditor is usually well-versed in data analysis, but on the other hand, they are not empowered to make changes, so need to explain the rationale for a particular change to others who have a less detailed understanding of the data analysis processes. Clearly, if the auditor is highly credible or the systems being analysed are particularly complex, a more sophisticated statistical approach may be suitable and more readily accepted.

Underlying these tensions about rigour versus simplicity is the fundamental challenge of interpreting variation: is it caused by error, by a change in a system's response to the common factors that we have modelled, or is it due to a new exceptional factor coming into play? This is the challenge for everyone working with empirical *"real world"* data, not just those analysing resource efficiency.

So what do scientists do when presented with uncertain and ambiguous results? Well, they design an experiment to resolve the uncertainty, of course.

I would like to think that the techniques described here enable us to go beyond being passive interpreters of past performance, and instead become active experimenters. Where changes are simple to make (because they do not require significant investment or detailed review), then we should be encouraging folks to *"give it a go"*, to try out new modes of operation. The metering and analysis we have put in give us a safety net that allows us to experiment much more freely as we can see quickly if a particular change has delivered benefits or not.

Instead of simply waiting for data to inform us of periods of good operation that we want to repeat (and bad we want to eliminate), we should be setting out to *experiment to create the good performance*. That way we will learn and innovate our way to a much-improved performance.

14.34 Science and art of analysis

The techniques set out in these pages can appear dry and theoretical. However, when applying these to our own organizations using our own data, the insights that they provide can be truly exhilarating.

To **gain insight** you need to think about what the data is telling you about the **real world**.

In the preceding sections, we have discovered some neat techniques that can bring new insight into our use of resources. Some of these tools, such as linear regression and CUSUM, are foundation-stones for any programme, while some of the other tools may only be helpful in very specific circumstances. However, all the techniques explored here are deliberately generic in nature - being applicable to a diverse range of resources and organizations.

We haven't touched upon a huge body of equipment-, resource- or sector-specific analysis that is also an essential part of any resource efficiency programme. There are thousands of specific techniques such as heat and water pinch analysis, refrigeration COPs, boiler efficiencies, logistic network analysis, compressed air and steam system pressure modelling, for which there is simply not enough space in this book (and which are much better covered in other, specialist books on these subjects).

The emphasis in this chapter on the mathematical and statistical aspects of data analysis should not blind us to the fact that there is also an art involved. This is because our analysis *has a purpose*. In many cases, the aim is to initiate change through greater understanding, and so it is often not *our* understanding that needs to change but that of *others*, such as the folks who operate our systems on a day-to-day basis, or the folks who allocate finance. Thus, the *way* we do the analysis can be just as important as the conclusions of the analysis - people will take ownership of discoveries that they make. This is the art in the process.

We also need to approach the analysis with the right state of mind. Data does not always conform to the neat compartments we would like it to fall in. There may be many competing common factor variables that can explain our resource use; spurious correlation can lead us down the wrong track. We truly need to wipe our minds clear of any assumptions and try to understand what the data is *telling* us. Consider the last example, opposite, taken from a very recent project I supported in the Czech Republic. As you can see from the narrative, the team were able to come up with a model that could account for the two modes of operation of this production line. But in fact, clever though it is, the lines in this model are not the key. The elephant in the room spotted by Arne and his colleagues are all those points on the vertical axis of the chart, showing electricity used by each line when there is zero production. If that could be reduced, the savings would be enormous. This is the kind of *eureka!* moment that makes this job such great fun.

14.54 **Production scatter charts**
This chart, and the one below, show a manufacturing process which has two modes of operation.
Source: Niall Enright

To truly understand what data is telling us we must put aside our assumptions. In this example it is not the lines of best fit that provide the greatest insight.

Real World: *Multiple modes of operation in the Czech Republic*

It is not just curves that we see when we begin to plot resource use against activity. Sometimes more exotic patterns can emerge, as shown by the chart below. This is taken from a real client's data at a site in the Czech Republic, where Arne Springorum of HEC is delivering a QUEST energy efficiency programme. The vertical axis shows the resource, in this case, daily electricity use, and the horizontal axis shows the activity, in this case, number of parts produced per day.

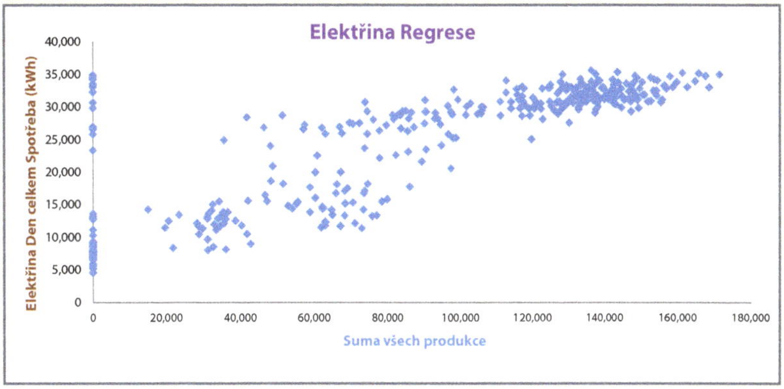

There seems to be a clear linear pattern of electricity vs production above about 100,000 units of production. Similarly, at lower levels of production there appears to be a different, but possibly linear relationship. Further investigations by Arne confirmed that there are indeed two modes of operation in the department, with either one or two production lines running. Sometimes the whole day would run in one mode, and at other times there would be one or more changes in operation, leading to quite a low correlation (i.e. a lot of scatter). By separating out the periods with one or two production lines, Arne and the team were able to construct the following model:

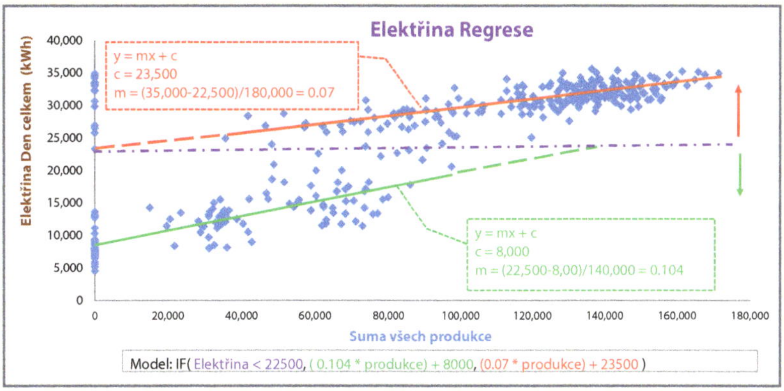

The model was described by the equation shown at the bottom of the chart where, based on the electricity use, one or other of the linear regression equations would apply. The key is that folks could have a clear indication, at least on a day when the operating mode was constant, whether performance was good or bad.

Note: *Companion files*

Method 1: from the data

Calculating the Slope

$$m=\frac{\sum x_i y_i - n(\bar{x}\,\bar{y})}{\sum x_i^2 - n(\bar{x})^2}$$

m= (3,576,383-3,332,182) / (5,731,250 - (21*500)^2

m= 0.50743104

Calculating the Intercept

$$c = \bar{y} - m\bar{x}$$

c= 317-(0.5074*500)

c= 63.6351416

The Microsoft Excel™ spreadsheet models in the companion file pack provide much more than the data to create the illustrations in this chapter.

These workbooks expand on the methods described in the text, with further explanation and examples. The spreadsheets are designed as teaching and learning aids to add to the information in this chapter.

For example, in the Excel spreadsheet model "*14.20 Calculating the slope and intercept of a linear regression. xlsx*", we look at how the slope m and intercept c of a best-fit line can be calculated (see page 460). It:

1. Shows how the values can be *calculated from first principles* using the raw data and the equations in the reference at the back of this book;
2. Shows how the values for *m* and *c* can be placed on a chart;
3. Shows how the *Data Analysis tool* in Excel can be used to create a table with these values (among others); and
4. *Uses the Excel functions* SLOPE() *and* INTERCEPT() to calculate m and c respectively.

These spreadsheets are available free of charge to buyers of the print edition of this book and for a modest fee for those using the free PDF version. See www.sustainsuccess.co.uk/iwik for further details.

Further Reading:

Linear regression techniques are well described in the Carbon Trust's *Monitoring and Targeting guide CTG077*.[116] A more detailed examination is provided in ASHRAE's *Inverse Modelling Toolkit*[443] which focuses on building energy use.

Detailed information about degree days can be found in CIBSE's *Degree days: theory and application*[148] but if this 98-page guide is too daunting, then the Carbon Trust has a simpler guide: *Degree days for energy management - a practical introduction*.[114]

Many excellent titles cover statistics, but I will focus on those that also provide guidance on using Microsoft Excel®. If you are after a textbook-style publication, then the very expensive *Essentials of Modern Business Statistics with Microsoft Excel* by Anderson et al.,[26] or much more economical *Statistics for Managers Using Microsoft Excel* by Levine et al.,[473] are both excellent guides, loaded with practical examples. For those wanting to get into detail on the statistical aspects, while still keeping a connection with Excel, there are three complementary titles by written by Conrad Carlberg for QUE, of which I think *Statistical Analysis: Microsoft Excel 2013*[109] provides a very good general foundation and P*redictive Analytics: Microsoft Excel*[108] covers the ground well for the kind of analysis we do in resource efficiency.

For those interested in SPC and Six Sigma, there are numerous publications available, but the little *Lean Six Sigma Pocket Toolbox*, by Michael George et al.[310] can't be beaten for brevity, scope and value for money if you are new to this field.

Stephanie Bell's. *Measurement Good Practice Guide No. 11*[65] is an excellent introduction to measurement accuracy and uncertainty.

We haven't discussed any of the technology, sector or equipment specific data analysis techniques in this book. For details on the appropriate data analysis techniques, the best source is an appropriate engineering handbook covering the specific resource type. Turner and Doty's *Energy Management Handbook* is probably the leading publication for energy, albeit it is in imperial units, which makes it less accessible to those more familiar with SI units. Eastop and Croft's *Energy Efficiency: For Engineers and Technologists* although last published in 1990 is still relevant and is in SI units. For water systems, *Water Loss Control* by Thornton et al.[707] provides a wealth of practical examples, albeit with a focus on public water distribution networks.

On energy efficiency, many of the titles in the Further Reading section in the chapter on Discovery also offer more specific analytical methods (see page 412).

Resources:

Most of the original spreadsheets used to develop the explanatory examples in this chapter are available in the companion file pack, see box, left. The lighting hours tool is also available in the companion file pack.

15 Measuring and Verifying Savings

15.1 **Proving savings achieved is important**
Not only is it good practice to confirm that an efficiency measure has achieved the desired results but it can also be an essential requirement to obtaining funding.
Source: Photo by naypong, Fotolia.com

Insight

An important field of data analysis involves the verification of savings. There are many situations where the impact of a resource efficiency measure needs to be objectively determined.

On the one hand, we have verification of fact, which is confirmation that a measure has been implemented. Then at the other extreme, we have verification of outcomes, which is a structured assessment of the direct impact that a project has had.

Some processes, such as the energy management standard ISO 50001, require verification of savings but are very relaxed about how this is done. It is up to the organization to describe how savings are to be measured, and as long as the approach is reasonable, there is no right way or wrong way to do this.

In other situations, we can have much more formalized verification requirements. This is usually where there is some form of contractual commitment from a third party around the outcome of a resource efficiency measure, such as in an Energy Performance Contract (EPC). In the EU, in 2013, for example, 52% of EPCs have a guaranteed savings component[588] (where the third party commits to a certain outcome), and the remainder is on shared savings or combined basis. In the US, the annual revenues from EPCs in 2011 were US$4.4 billion.[677] All of these activities require a rigorous and objective measurement of the savings delivered.

Our previous chapter on data analysis has shown that this is not always as straightforward as it might appear. First of all, there is the issue of natural variability in energy use to consider, due to weather, occupancy levels or other activity effects. Then there is the fact that there may be multiple conservation measures being implemented at the same time.

Fortunately, several methodologies have been developed that offer a standardized approach to the verification of savings. These are from the Efficiency Valuation Organization®, the American Society of Heating, Refrigerating and Air-Conditioning Engineers (ASHRAE), the Investor Confidence Project and branches of the US federal government. It is not the intention in this chapter to reproduce these methods but to describe the key principles, applicability and suitability of the techniques in order to signpost the reader to the appropriate standards and documents. Although initially US-centric, these standards are now largely globally accepted.

15.1 Fundamentals of verification

Being able to prove the return from an investment reduces risk and builds confidence. Not only does this help us gain internal support for projects, but it can also open the door to third-party financing based on the savings achieved.

There is the old saying: "*If you can measure it, you can manage it.*" Perhaps we might consider a variation on this, "*if you can verify it, you can monetize it*", reflecting the important enabling role that verification plays both in obtaining third-party funding for energy efficiency and in developing internal decision-maker confidence that the benefits promised have been delivered. The role of verification in obtaining funding is explored in more detail in chapter 18 on page 621; here, we focus on the verification techniques available.

Measurement is a key prerequisite for verification, which is emphasized in the collective name given to the various methodologies, Measurement and Verification (M&V), not to be confused with Monitoring and Targeting (M&T).

In terms of measurement, we first record the resource use before we implement a project, the so-called baseline use (again not to be confused with the similar-sounding *baseload* or non-variable related use, or the temperature *baseline* of a building, both referred to in the previous chapter). The M&V baseline use is the same as our original equation in M&T.

Once we have implemented our improvement, called in M&V an energy conservation measure (ECM), we can continue to measure our energy use to get our post-retrofit use or reporting period use. "*Retrofit*" is a buildings-

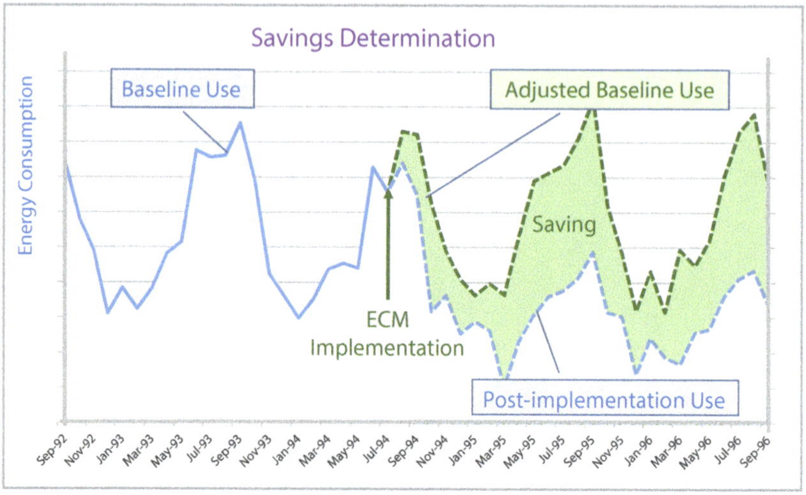

15.2 **Illustration of the challenges in determining of savings based on actual and predicted energy use**
The energy consumption illustrated in this chart changes quite markedly over the year. Once the energy conservation measure (ECM) has been installed we continue to measure the actual energy consumption. Savings are determined by the difference between the predicted use (labelled Adjusted Baseline Use) and the actual use (labelled Post-implementation Use), giving us the shaded green area on the chart.
Source: Based on Guideline 14, "Measurement of Energy and Demand Savings" from ASHRAE.[37]

Real World: *Justifying the cost*

Image: © pogonici. Fotolia.com

A factor in choosing the correct M&V approach is the costs. It would make no sense at all for the savings to be reduced substantially because of the verification overhead.

The US government reports that, for Federal EPC projects, average annual M&V costs range from about 2%–5% of annual project cost savings[277]

Some more complex ECMs may warrant greater M&V costs, but the overall M&V costs for these projects should be balanced by other ECMs that do not require significant annual costs.

Selecting the correct M&V strategy very much depends on the costs and benefits involved.

Let us imagine we have an ECM which saves US$500,000 a year and which has an uncertainty level of ±20%, meaning that the savings *"at risk"* are US$100,000 a year. Now we need to select from two M&V options: A which will reduce uncertainty from ±20% to ±10% at an annual cost of US$25,000; and B, which will reduce uncertainty from ±20% to ±5% at a cost of US$50,000.

Well, the increased certainty/cost ratio for M&V option A is US$50,000/US$25,000, i.e. 2.0; while the same ratio for B is US$75,000/US$50,000 i.e. 1.5, so A appears the better choice. We can also use a total savings/M&V cost ratio to ensure that the impact on the overall savings is reasonable (the ratio for option A is 20, i.e. costs are 5% of the savings).

related term meaning the addition of something post-construction. A better term is *"reporting period use"* or *"post-implementation use"*, as some ECMs, such as behaviour change or modification of existing control strategies, don't involve installing equipment.

All M&V methods exist to predict what would otherwise have happened if an ECM had not been implemented. In the chart opposite, this is the dashed green line, labelled the adjusted baseline use. From this predicted use, we calculate the solid green region, the savings delivered by the ECM, simply by subtracting the actual use shown in the dashed blue line.

As can be seen from the chart, the pattern of energy use may be complex and influenced by a range of controllable or uncontrollable variables. A key decision will be which M&V method is appropriate to the circumstances and budget available (see left). Not only do we need to consider the core methods, but also anticipate any adjustments that may need to be applied if the circumstances in which the ECM was implemented change in any fundamental way.

The nearest we have to a global standard for M&V is the *International Performance Measurement and Verification Protocol®*. The IPMVP® sets out the following fundamental principles for any M&V process:[406]

- Accurate: the M&V method must be as accurate as can be justified by the project budget. *"Consideration of all reasonable factors that affect accuracy is a guiding principle of IPMVP."*[406]
- Complete: the M&V should endeavour to measure the key influences on energy use and estimate others.
- Conservative: where there is uncertainty or estimation, we need to ensure that savings are not overstated.
- Consistent: not in the sense that M&V methods used by an organization should always be identical, but the M&V approach should be comparable and similar across project types, individuals involved, time frames, and across both new supply and demand reduction measures.
- Relevant: the parameters that most strongly influence the energy use must be measured, whereas those with little influence can be estimated.
- Transparent: the methods used should be fully disclosed and documented. In some cases, there will be formal M&V plans and regular M&V reports.

In the next pages, we will look in more detail at the published M&V methods which all have their genesis in the US Department of Energy's *North American Energy Measurement and Verification Protocol, 1996*. Because of their similar origins, the IPMVP,[406] ASHRAE's *Guideline 14-2014*[37] and the US Federal Energy Management Program's *M&V Guidelines: Measurement and Verification for Performance-Based Contracts Version 4.0*[277] all share the same basic approach, which we shall explore next.

15.2 Basic approaches to M&V

M&V is a relatively mature aspect of resource efficiency, and as such there are well-established methods to quantify the effect of an efficiency measure. Here, we list the four M&V options and describe some of their features.

Measurement and Verification techniques can be divided into two general types based on the measurements involved. The first approach is based on isolating and measuring the ECM and the second is based on measuring the whole facility. Within each of these, there are two sub-types or categories, giving a total of four methods, labelled Option A to Option D.

The most rigorous M&V method is Option B, where the ECM's energy use is separately measured, along with any other key parameters which may influence the energy consumption (such as activity, load, run hours, etc.). Option A is similar; the major influences on energy use are measured, but some aspects such as hours run may be estimated or stipulated (agreed by all the parties to be a particular value).

Option C is a whole facility measurement, where energy flows to several items of equipment, not just the ECM. Here, the effect of the energy use is normalized using regression analysis to model the influence of key variables, as described in the previous chapter. This option is not advised where the effects of the ECM on the energy use are low (e.g. $<10\%$), as the other uses will mask the improvement. In federal EPC this method is rarely used because of the effort needed to itemize the other equipment in the facility to exclude its effect.[277]

The final option, D, is to create a simulation of the building (this sounds grand, but could actually be in Excel rather than a commercial building modelling package). The key is that the model is *calibrated* against the building's actual data so that the outputs of the model match the actual measurements of the energy use. When the model is complete, it represents the *baseline* against which savings can be determined - either by comparing the actual meter reading or comparing the simulation output without the ECM and the simulation output with the ECM modelled as well.

The table opposite gives a number of scenarios where one or other of the M&V options would be favoured. However, it is important to note that there is no hard and fast rule and it is entirely possible to use these methods in different ways. For example, simulations Option D or regression analysis Option C can be used on individual isolated ECMs instead of Option A. This table and the flowchart on the next page illustrate the most common approaches, not the only ones, to be taken.

15.3 **The four basic M&V options (opposite)**
The table opposite describes the approaches, their coverage within the common standards documents and if the approaches are used in the Investor Confidence Project (ICP)
Source: Niall Enright, adapted from FEMP v4.0[277] and from IPMVP EVO 10000 - 1:2012[405]
Illustration available in the companion file pack.

		ECM or Retrofit Isolation Approach		Whole Facility Approach	
Illustration		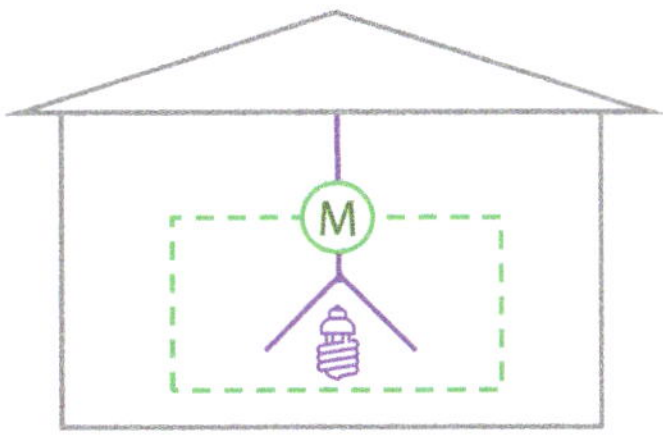		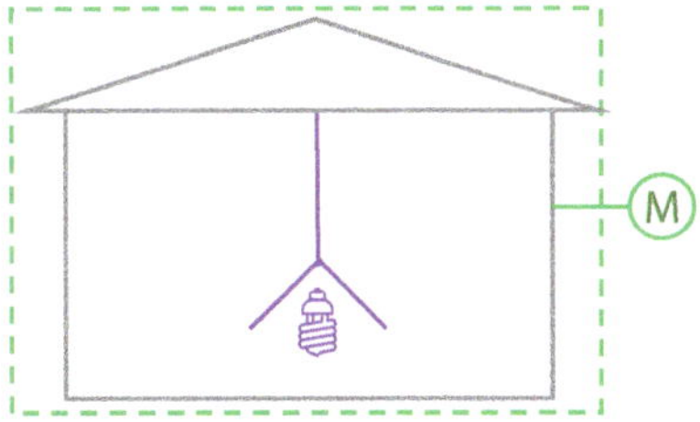	
Category		Option A	Option B	Option C	Option D
Name		Isolation with key parameter measurement	Isolation with all parameter measurement	Whole facility measurement	Calibrated computer simulation
Description		A combination of measured and estimated factors	All the key factors are measured	Measurement of the whole facility and regression with key variables	A computer simulation is used, and calibrated with actual energy use
Measurement Frequency		Short-term, periodic or continuous	Short-term, periodic or continuous	Continuous (needs a fairly long term baseline). > 9 data points	Use simulation data before/after or simulation before/ actual use after
Example		Lighting, power is measured, hours of operation are estimated	Variable speed drive, where the power use is measured directly	Replacing a gas boiler using a heating degree days baseline	Major refurbishment of a building with many overlapping ECMs
Application & Limitation		Appropriate for simple ECMs	Appropriate for more complex ECMs	Rarely used in federal projects	Complex and needs engineering knowledge
Common choice for different situation	Need a focus on Individual ECMs	Yes	Yes		Yes
	Need to assess the whole site			Yes	Yes
	Savings expected to be <10%	Yes	Yes		Yes
	Baseline unavailable/new build				Yes
	Significance of variables not clear		Yes	Yes	Yes
	Changes within boundary possible	Yes			Yes
	Project interacts with others			Yes	Yes
	Understandable reports needed	Yes	Yes	Yes	
	Metering skills needed	Yes	Yes		
	Computer modelling skills needed				Yes
	Regression skills needed			Yes	
Standards	IPMVP[406]	Yes	Yes	Yes	Yes
	FEMP M&V Guidelines v4.0[277]	Yes	Yes	Yes	Yes
	ASHRAE Guideline 14-2014[37]	No	Yes	Prescriptive/performance	Yes
	UK ERD Pilot M&V Guide[242]	Yes	Yes	Yes	Deemed
	Investor Confidence Project[387]	Yes	Yes	Yes	No

15.3 Choosing the M&V method

Choosing the right M&V solution is not always easy. Here we set out some of the reasons one would choose one particular option over another.

Choosing an M&V method depends on many factors. The first consideration is the purpose of the M&V. If there is no specific requirement for a method, e.g. for an item in an ISO 50001 Action Plan, then the user is free to choose whatever method is most convenient, including a very simple "*it has been done*" confirmation such as a photo (Option E in the illustration opposite).

If the M&V underpins a legal contract such as an EPC, then a specified M&V methodology will certainly be required, depending on the contract form. In the unlikely event that M&V is not proposed, I would certainly recommend it. For example, projects labelled as an *Investor Ready Energy Efficiency™* under the Investor Confidence Project must follow one of the IPMVP Options A, B or C. Similarly, projects submitted for the UK's Electricity Demand Reduction programme will follow IPMVP Options A, B or C, or can be *"deemed"* to deliver given savings if the equipment is listed. This latter model is a variation on Option D, Simulation, where the savings are calculated on a spreadsheet based on agreed parameters. The US Federal Energy Management Programme guidelines also follow the IPMVP standards but have additional provisos described in their *M&V Guidelines v4.0.*

Once any specific requirements are understood, the project partners will then select the M&V method that is most cost-effective. Clearly, where existing metering is good, the more that one can isolate an ECM, the more reliable the performance data will be, with Option B usually offering the greatest accuracy. However, where the project is not sufficiently large to merit additional investment in metering, one often has to fall back to a whole building approach. Here we need to note that an ECM which has a negligible effect on the total building's energy use (say <10%), is too small to be picked up from the other variation in energy use and so a simulation approach will be the fallback.

15.4 **Procedure to select an M&V method (opposite)**
The illustration opposite describes some of the considerations that will help in the selection of the best M&V method for a given situation. There are many other factors that could influence any decision.
Source: Niall Enright.
Image and poster available in the companion file pack

Simulations are usually the least desirable M&V method because they are complex and expensive to set up and often difficult for non-experts to understand. Nevertheless, where there are many ECMs, where there is no baseline data because the building is new, or where significant changes in the building are anticipated, a simulation may be the only realistic M&V option. In this case, there is a useful listing of some simulation software by the Northwest Energy Efficiency Alliance,[452] although most verification professionals will recommend a package they are familiar with or the parties may already have a specific requirement as part of a standard contract.

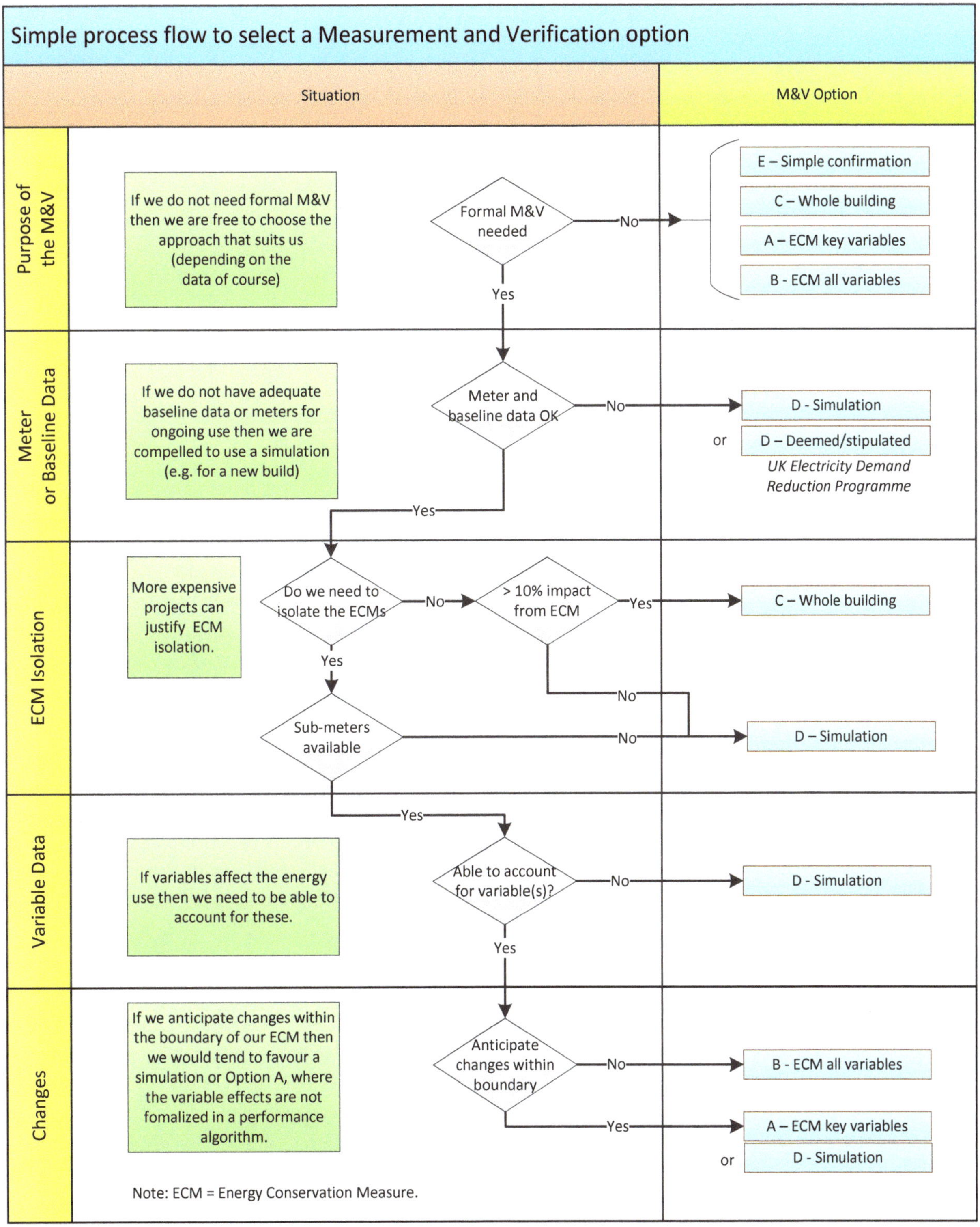
Simple process flow to select a Measurement and Verification option
Situation
M&V Option
Purpose of the M&V
If we do not need formal M&V then we are free to choose the approach that suits us (depending on the data of course)
Formal M&V needed
No
Yes
E – Simple confirmation
C – Whole building
A – ECM key variables
B - ECM all variables
Meter or Baseline Data
If we do not have adequate baseline data or meters for ongoing use then we are compelled to use a simulation (e.g. for a new build)
Meter and baseline data OK
No
Yes
D - Simulation
or
D – Deemed/stipulated
UK Electricity Demand Reduction Programme
ECM Isolation
More expensive projects can justify ECM isolation.
Do we need to isolate the ECMs
No
Yes
> 10% impact from ECM
Yes
No
C – Whole building
Sub-meters available
No
Yes
D – Simulation
Variable Data
If variables affect the energy use then we need to be able to account for these.
Able to account for variable(s)?
No
Yes
D - Simulation
Changes
If we anticipate changes within the boundary of our ECM then we would tend to favour a simulation or Option A, where the variable effects are not fomalized in a performance algorithm.
Anticipate changes within boundary
No
Yes
B - ECM all variables
A – ECM key variables
or
D - Simulation
Note: ECM = Energy Conservation Measure.

Insight

15.4 The M&V plan

When M&V is used to underpin contractual arrangements between parties, it is essential that a formal M&V plan is developed. This is not just about measuring the current situation but also about anticipating changes that could occur in the future.

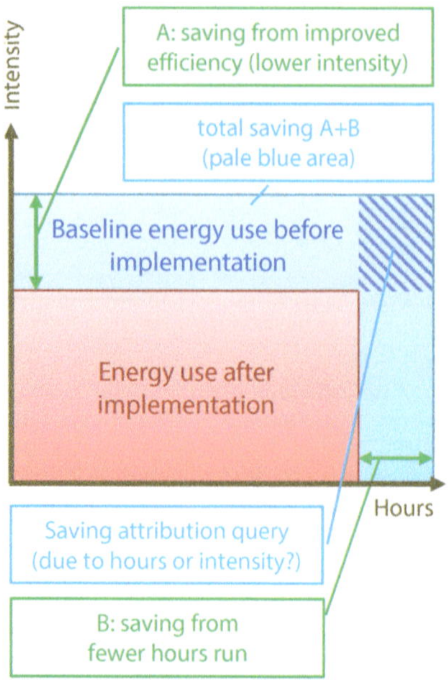

15.5 **Attributing savings is important**
The illustration above shows savings from an energy efficiency measure arising from the reduced energy intensity per item output (vertical axis) and reduced operating hours of the equipment (horizontal axis). One source of saving could be attributed to the investment in new equipment by a third party and another to the better operation by the end-users. It should be noted too, that the combined savings, A+B, above, are not equal to the reduced hours * original energy use plus original hours * reduction in energy use, as this will lead to double-counting the hashed area. Failing to describe how these two sources of improvement are measured and allocated in an EPC can be a source of conflict between the parties. A well-designed M&V plan will anticipate these types of issues.
Source: Niall Enright

As mentioned earlier, the crucial aspect of M&V is getting the baseline right. The baseline is what determines what would otherwise have happened absent the ECM, and so it forms the basis for the calculation of savings.

There are two approaches to the "*what would have otherwise happened*" model of savings. The first involves creating a model (like the linear regression model described in the previous chapter) that describes how the baseline energy changed in relation to a variable such as heating degree days (HDD). Then the M&V process records the actual use, and the baseline is adjusted based on the HDD observed in the reporting period, and the savings are the difference. This approach is called an avoided energy use calculation. Although the calculation may have predicted the savings expected, say by using average HDD, the actual saving will be different (as the actual HDD are fed into the adjusted baseline equation to get the revised predicted use for the conditions).

The alternative method is to use a normalized energy use calculation. Here, the baseline energy is not changed, as a fixed relationship between the key variables and energy use is established. For example, we may determine that the baseline lighting for an office is x kWh per day. Savings are then determined by subtracting the measured daily energy use from the baseline daily use. One consideration with this type of calculation is that the baseline may involve several modes of operation (e.g. weekday, weekend and holiday) and the ECM may need to operate for a full cycle before savings can be determined.

Where a formal M&V process is required for one or more ECMs, such as when a third party provides funding as part of an EPC, there should be an M&V plan. The IPMVP[406] (and FEMP[277]) guidelines provide excellent advice on what the plans should include, such as:

- A description of the ECMs and a top-level summary of any key assumptions and risks associated with the project.
- The M&V option selected and the measurement boundary.
- The baseline energy use, which usually includes information about:

  a. the baseline period

  b. the baseline energy consumption and data

Standards: *ISO 50015:2015*

This standard, *Energy management systems - Measurement and verification of energy performance of organizations - General principles and guidance*, complements the ISO 50001 standard on energy management (see page 717) in that it applies the same terminology and structure, although it can be used as a stand-alone method to undertake M&V.

In this standard, ECMs are called Energy Performance Improvement Actions (EPIA).

As with the existing M&V methods, the standard sets out the first requirement as an M&V plan. Although the M&V plan in ISO 50015 is a little more detailed in terms of the content, the underlying principles are the same.

Where the ISO standard differs from the other M&V methods set out here, is that it does not refer to the four approaches to M&V. Instead, there is a generic *baseline* Energy Performance Indicator (EnPI) against which the influence of variables and static factors are characterized in the M&V plan. This EnPI baseline could be an absolute energy value, a specific ratio, a regressions formula or a more sophisticated algorithm (such as that produced by a simulation), so any of the M&V approaches are acceptable within the ISO standard. The plan also describes the routine and non-routine adjustments expected, similar to the existing M&V methods.

I would view the ISO standard as a reaffirmation of the existing best-practices and an updating of the terminology in line with ISO 50001. The current IPMVP guidance offers much more detailed guidance for specific applications of M&V, so ISO 50015 should not be seen as a replacement of these well-established approaches.

c. related variables and data

d. presumed static/estimated factors (e.g. floor area, hours of operation)

e. operating conditions (e.g. temperature setpoints)

f. current equipment details (an inventory, nameplate details, location, maintenance schedules, problems, etc.)

- The reporting period (that is to say the time frame for which the savings will be proved - this could be continuous or fixed).
- The type of adjustments to be made to the baseline. This will depend on whether we are using an avoided or a normalized savings calculation.
- The algorithm used (e.g. our regression formula) and how this was developed (including if any data was excluded).
- Details of energy costs and how changes in price will affect the savings calculation or contract terms. Factors such as time-of-use tariff changes may need to be considered.
- Metering details. This should include information about the maintenance, calibration, data connections and how to deal with missing data. If we are using a simulation (Option D), we will provide information here about the software to be used, inputs and outputs, calibration specifics, etc.
- Data collection responsibilities and processes. It is important that there is rigour around the definition of the variables and how these are to be measured. If there are presumed static or estimated factors, the plan needs to describe how these will be monitored in case they change materially. If we are using Option A (ECM isolation with key parameter measurements), we need to justify why some parameters can be estimated and the effect of the estimates on the energy use (for example with a sensitivity analysis).
- The accuracy of the M&V (see *Expressing uncertainty* on page 506).
- The source of savings. It may be that some savings are not energy-related (e.g. in lighting projects there may be large savings from reduced manpower replacing lamps). How these savings are calculated and allocated needs to be spelt out.
- The budget and resources required for the M&V, the report format (possibly with a sample report) and any specific standards or quality assurance processes that will be followed in carrying out the M&V. There also needs to be a section on how disputes will be resolved.

A good M&V plan should not just describe the situation today, but will anticipate changes and adjustments that may be needed in the future. These are discussed in more detail in the next section.

15.5 Compensating for changes

The conditions in which our efficiency measures are implemented are very unlikely to remain static. It is essential, therefore, that we define how and when our calculated savings can be adjusted.

The most common method of assessing operational performance in resource use is an **avoided energy use** calculation, which reflects real-world improvements. In some cases **normalized use** calculations are favoured by investors as they provide a guaranteed return.

M&V is about quantifying the change from the baseline period resource use compared to the reporting period after a conservation measure has been implemented. However, improvements do not occur in a vacuum, as other changes may influence the energy use positively or negatively.

The general formula for calculating the savings is:

savings = baseline period use - reporting period use ± adjustments

The adjustment term represents a modification to the calculation to compensate for changes in the conditions before and after the implementation of the conservation measure. These adjustments can take various forms:

First, let's consider routine adjustments to compensate for influencing variables. These are the common factor variables introduced in the previous chapter which influence resource use. Examples are environmental factors such as weather or lighting hours or measures of activity such as production, occupancy or service provided. Routine adjustments for common factor variables usually involve an avoided energy use calculation, where routine adjustments are made to the baseline to reflect conditions in the reporting period, using techniques such as linear regression. The formula takes the form:

avoided energy use = adjusted baseline use - reporting period use
± non-routine adjustments

More rarely, the adjustment for the influencing factor might be made to the reporting period use, for example, when the data quality is better in the reporting period. In this case, the formula would be:

avoided energy use = baseline use - adjusted reporting period use
± non-routine adjustments

This is sometimes called *backcasting* as the current energy use is recalculated to match conditions in the past.

Avoided energy savings are influenced by the reporting period conditions (e.g. the weather), so this may make this approach unsuitable for some third-party investors who are expecting to see a guaranteed return. For example, an investment in a chiller could show a smaller saving than expected not because the technology is not as efficient as expected, but simply because there was

Real World: *CMVP professionals*

Image: © AEE and EVO

The US Association of Energy Engineers (AEE®) and the Efficiency Valuation Organization (EVO®) run a certification programme for Certified Measurement and Verification Professionals (CMVP®).

The certification is based on the IPMVP standards. Candidates need to meet strict eligibility criteria, attend a mandatory seminar and complete a four-hour examination.

Over 3,700 CMVPs are listed in the AEE certified professionals directory.[41] The US and Canada account for a third of those listed, with registrations from a total of 59 countries.

As we shall see later, the availability of independent M&V professionals is an important enabler of third-party EPC funding for resource efficiency and it is very encouraging to see the extent of the CMVP programme.

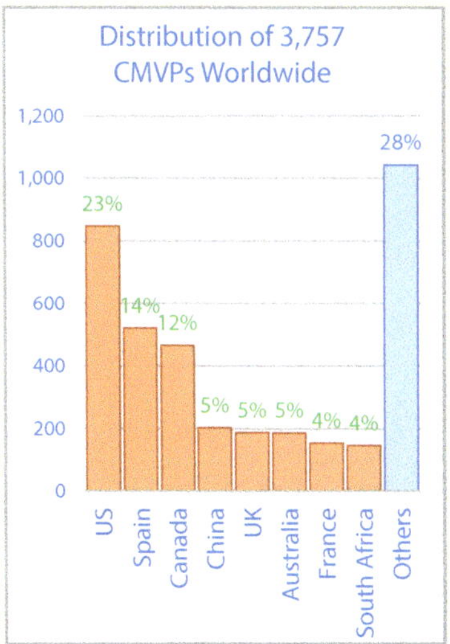

less demand for cooling because the weather was milder than usual. Another disadvantage of this approach to calculating savings is that it is hard to compare savings achieved across several conservation measures, as the conditions in which each operates are different.

These disadvantages are eliminated when we are using a normalized use calculation, where a fixed set of conditions is used to calculate the savings. The conditions that describe this *"normal"* operation may not necessarily have been present in the baseline period and so it is possible that the calculated savings may involve an adjustment of the baseline as well as the reporting period use to reflect the norm. Here the formula is:

normalized use = (baseline use ± routine adjustments to fixed conditions ± non-routine adjustments to fixed conditions) - (reporting use ± routine adjustments to fixed conditions ± non-routine adjustments to fixed conditions)

The advantage of a normalized approach is that the calculated savings are unaffected by the reporting period conditions because they are based on a fixed set of conditions. This can provide investors with more confidence of the returns - although it can also lead to situations where a saving is calculated when, in fact, given the actual conditions, none was made. Another disadvantage of a normalized approach is that a lot of data over a full cycle of operating conditions (e.g. winter/summer) may be needed before the basis for the normal (or typical) performance can be calculated.

In all the formulae above, we have a non-routine adjustment term. This reflects adjustments made to compensate for changes in supposedly static aspects, such as the conditioned floor area of a building, or the shift pattern, or fault conditions in equipment. It is expected that these adjustments would only be made occasionally, and if these factors are found to change frequently, then it is recommended that these are built into the routine adjustments, for example, as a variable in a multiple regression.

The IPMVP standard allows for operational verification measures that can complement the four standard M&V options and strengthen the basis for the savings calculations. These verification actions can include visual inspection, spot measurements, short-term performance measurement, control systems reviews, or on-off testing where the measure is easily switched on or off.

Where the calculation of savings forms the basis for a contractual arrangement between a resource user and a third-party investor or equipment manufacturer, it is highly recommended that an independent third party with a CMVP qualification (see left) undertakes the verification. This chapter has only highlighted the key principles in M&V and the reader is strongly advised to consult the detailed guidance in the IPMVP and other standards referenced here.

Standards: *Related standards & tools*

Although the IPMVP itself does not stipulate the data to be used for M&V, there are several other standards and guidelines that can ensure our approach is robust.

The Investor Confidence Project's *Project Development Specification*[387] recommends that building data should follow ASTM-E2797-15 Building Energy Performance Assessment,[43] which has a list of data commonly collected in section 10.3.

The ASHRAE Inverse Modelling Toolkit[443] sets out some building-related numeric modelling techniques such as linear regression and baseline temperature adjustments. These methods have been described in depth in the previous chapter, but this paper provides a helpful overview.

For ECM isolation M&V in buildings and industry, more specialist techniques may be needed. The Bonneville Power Administration provided a very helpful guide Verification by Equipment or End-Use Metering Protocol[682] for equipment operating in one of four modes:

- constant load, timed schedule (e.g. lighting under timing control);
- variable load, timed schedule (e.g. wastewater treatment plant air blasters maintaining dissolved O_2 levels);
- constant load, variable schedule (e.g. lifts); and
- variable load, variable schedule (e.g. industrial air compressor with VFD compressor).

For calibrating building simulations, ASHRAE paper RP-1051[615] by Agamy Reddy and colleagues provides some helpful methods.

Further Reading:

Essential reading for anyone contemplating formal M&V is the International Performance Measurement and Verification Protocol (IPMVP). The IPMVP *Core Concepts,*[406] reference EVO 10000 - 1:2016, was updated in October 2016. This describes the terminology used in M&V, the core principles, the four options (A to D), the M&V plan and how to adhere to the IPMVP guidance.

The Efficiency Valuation Office also publishes a more detailed IPMVP reference, the *Concepts and Options for Determining Energy and Water Savings, Volume 1,*[405] reference EVO 10000 - 1:2012, which provides greater detail on the Core Concepts material and also discusses some of the challenges in M&V and gives many worked examples. This document also provides country-specific guidance for IPMVP in the US, France, Spain, Catalonia, Romania, Bulgaria, The Czech Republic, Croatia and Poland.

The detailed IPMVP guidance sets out the statistical basis for modelling savings. The previous chapter of this book describes the same techniques, albeit there are naming differences. The IPMVP calls Standard Error the root mean squared error (RMSE) and the Mean Square Regression is referred to as the mean bias error (MBE). IPMVP also introduces the coefficient of variation of the root mean squared error - CV(RMSE), which is simply the Standard Error divided by the average resource use ($\bar{y}$). These three statistics are particularly important when calibrating a simulation model in Option D.

Intended for those involved in undertaking M&V in US public, state or federal institutions and facilities, the 2015 *M&V Guidelines: Measurement and Verification for Performance-Based Contracts Version 4.0*[277] produced by the Federal Energy Management Program (FEMP) sets out additional examples of best practice in M&V. This guidance is fully compliant with the IPMVP, so should not be considered as an alternative methodology, but rather as a complementary reference, which is helpful to all M&V applications. There is extensive guidance on the application of M&V to specific conservation measures such as lighting, motors, boilers, steam traps, variable air volume (VAV) systems, building envelopes, control systems, pumping, air compressors and many others. Water conservation measures are included, demonstrating that M&V can apply to resources other than energy.

In Australia, the New South Wales government has published an excellent *Measurement and Verification Operational Guide*[552] to implementing IPMVP, which has lots of practical examples in its 100+ pages and is highly recommended (this an supersedes earlier guide).[10]

Finally, the Investor Confidence Project's *Project Development Specification*[387] version 1.0, 2014, sets out the quality processes to meet certification to the ICP Investor Ready Energy Efficiency (IREE) methods, which will facilitate investment by third-parties. The M&V elements of the ICP framework are, again, all based on the IPMVP standard, but this framework provides further advice on setting a baseline and calculating savings. The document also covers aspects such as design, construction and operations and maintenance, which fall outside the formal M&V process, but can nevertheless strongly influence the outcome of a conservation measure.

16 Presenting Data

We have seen from the previous chapter that there are many analytical tools at the disposal of resource efficiency practitioners. However, success may depend not only on our insights, but also on the way we communicate our results and engage decision-makers with our findings. As they say, "*a picture is worth a thousand words*" and choosing the right presentation method can have a big effect on our ability to convince people to act.

Here we set out some different data presentation techniques to stimulate ideas. This list is by no means an exhaustive catalogue of data presentation or visualization techniques (indeed, there are some fine books entirely dedicated to this subject), but rather a summary of some of those which are particularly relevant to resource efficiency practitioners.

Further aspects of data presentation are covered in other parts of this book. Information on how to develop marginal abatement cost curves is covered in the chapter on financial analysis since this is chart is a way of presenting financial information. In the chapter on data analysis, we have already explored some methods of charting data that can help us to visualize good and bad performance.

16.1 **Get the basics right**
No wonder the woman on the right is looking slightly puzzled. None of the charts are labelled! Although this is just a stock photo, it is surprising how often in the real world basic items of information such as scope, date range, units, axes labels, or the legend are missing from charts. I am sure that no readers of this book would ever commit these elementary errors!
Source: Photo © alotofpeople, available at Fotolia.com

16.1 Fundamentals of presentation

There are many presentation techniques that can help make sense of our resource data. Choosing the right approach requires us to understand the purpose of the presentation.

Data presentation techniques, or visualizations, are used to convey information efficiently. In simple terms, every visualization can be reduced to two basic types: comparing different values at a fixed point in time or comparing values over time. Note the use of the word compare. Every visualization uses a scale to convey a value, such as width or height (a bar chart or line chart), volume or area (a bubble chart), angle (a pie chart), colour (a heat map) or coordinates/position (a scatter plot). Where necessary, the visual depiction of quantity using the scale can be complemented by labels or legends that convey additional information. For example, we can have a label providing a percentage of the total, or we can have a colour that categorizes values.

Selecting the correct visualization requires a clear goal. It may be that the purpose of the visualization is to convey one key piece of information for several objects. At Peel Land & Property Group, as part of the ISO 50001 programme, for example, colleagues need to know if they have achieved the performance target in the buildings they manage. In this case, a very simple *"glance and go"* type of visualization such as a traffic light indicator for each building is ideal, as shown below. With this type of presentation, there is no ambiguity about the meaning and the team can quickly home in on the exceptions (good as well as bad, remember). Once the user knows which buildings are of interest,

16.2 **A meter target report**
This reports shows the performance of a number of buildings at MediaCityUK, as part of an ISO 50001 certified energy management programme.
Source: Reproduced with kind permission from Peel Media Ltd., Report produced by Carbon Desktop™, from Verco[111]

Media City ISO:50001

Readings | Cost | CO2e

Default Meter Target Report: 25/09/2015 - 24/10/2015 / (Reading)

Meter	Target Ranges	Unit	Consumption	Target	Variance	% Variance
Blue Tower - Main Utilities						
V PML Blue	●	kWh	47,360	49,350	-1,990	-4.033
Orange Tower - Landlord Services						
V PML Orange	●	kWh	80,554	80,700	-145.6	-0.1804
The Garage - Main Utilities						
A MSCP LV Switchboard (SCC/FAREBROTHER)	●	kWh	42,526	52,770	-10,244	-19.41
The Pie Factory Limited - Main Utilities						
M Pie Factory Gas	●	kWh	0.00	10,067	-10,067	-100.0
A Pie Factory HH Electricity	●	kWh	70,818	104,070	-33,252	-31.95
White Tower - Landlord Services						
V PML White	●	kWh	12,457	11,700	756.5	6.466
		Grand Total:	**253,714**	**308,657**	**-54,942**	**-17.80**

The most effective presentations work because they are communicating the ***required information*** *at the appropriate* ***level of detail*** *to the* ***correct audience.***

the report provides the recipient with additional information about the consumption, target and variance, without overwhelming the user with data.

These type of reports are sometimes called monitoring reports since they focus on the status of one parameter (in this case, the variance from the target). At the other end of the spectrum are presentation techniques designed to support data exploration and to help in the interpretation of large volumes of information. A scatter plot would be an example of this kind of visualization. Here, the presentation requires the user to examine the visualization closely to extract the greatest meaning from the data. For a scatter plot, the user interprets the intercept, the scatter of the points, the scales of the axes, whether a linear or curved relationship between the values is evident, as well as the presence of outliers. This interpretation may be an entirely unconscious analysis by an experienced resource champion or may require a step-by-step approach by a less experienced user.

As they say: *"Every picture tells a story."* The most effective presentation techniques work because they are communicating the required information at the desired level of detail to the correct audience. Far too often I see poor charts or other visualizations chosen just because they are readily available from the analysis tool, because they are pretty, or because that is the usual way a particular value is presented.

When choosing a data presentation method, you might consider if it is:

1. Valid: do you have the correct data without errors or noise?;
2. Appropriate: is the information you want to convey relevant to and actionable by the recipients (in other words, will they do something when they get it)?;
3. Comprehensible: can the information be easily understood?; Is the comparison scale clear and capable of differentiating values (e.g. good vs bad performance)?; Are units provided?; Are colours or other indicators labelled correctly?; Is the date range or scope of the data communicated?.

This chapter covers *static* presentation techniques, that is to say, visualizations used in documents or presentations. Most modern information management systems provide interactivity so that users can drill down into data. For example, when the Carbon Desktop *Meter Target Report*, opposite, is displayed live, the user can simply click on the building name and display a time series chart of actual use versus target consumption, with a CUSUM overlaid, which enables the user to determine the point in time of any exceptional variation. From the trend display the user can then drill into the target definition, the raw data, or change the output (e.g. to CO_2 or cost). These interactivity features are incredibly useful and should form an important consideration when selecting an information management tool to support a resource efficiency programme, as should the capability to export data to other tools such as Excel, should further analysis be required.

16.2 Basic charts

Many data presentation techniques rely on our ability to interpret dimensions to convey information. Being aware of the limitations of our senses can help us to choose the correct presentation method.

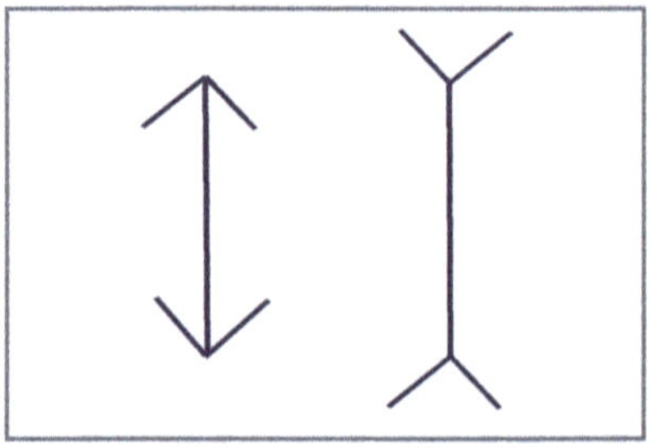

16.3 **Judging height**
Consider the shapes above, which vertical line is tallest? Although both lines are the same, the one on the right can appear longer.
Source: Niall Enright.

There are some things that we do well as humans and some that we do not. When it comes to data visualization, understanding how we interpret size and shape, where these are intended to convey meaning, can help us select the right methods.

One of the most often misused forms of presentation is the pie-chart. Take for example the chart below, left. By default, this chart in Excel does not have any labels so although we can assess the relative size of different resource savings we have no idea of what the scale is. Making comparisons using pie charts is more difficult than it needs to be because the human eye is not particularly good at judging angles. For example can you tell from this chart if the diesel savings are greater or less than gas?

The second chart in the illustration shows the same data in a stacked bar. In this case, at least, we have some units shown on the left. However, to assess the value of any given resource we now need to judge its height, which involves subtracting the preceding resource's value of the stack from the current one.

The third column chart, below right, shows the same data in the forms of separate columns, which makes establishing the values for each resource very much easier. However, there are still some challenges in interpreting the data as the values are not sorted and we need to refer to a legend to know which resource is in each column.

16.4 **Three chart types**
These charts are all based on the same data. There are a number of flaws with each of these, which are drawn in Excel using standard default settings.
Source: Niall Enright.
Spreadsheet and images available on website.

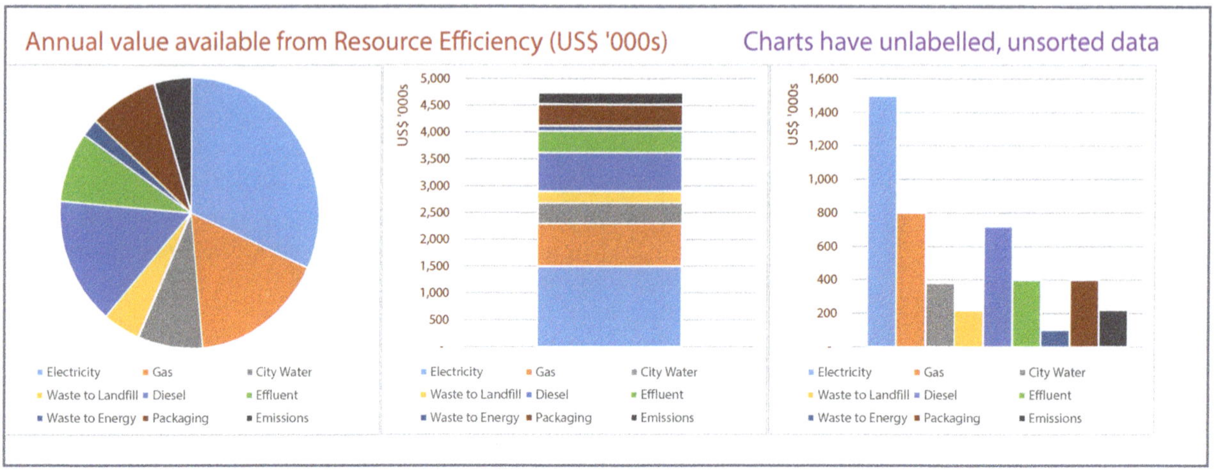

If we turn our attention to the alternative versions of the same charts shown below, we can see that we have made some improvements. First of all the data have been labelled in two of the charts that need it: the pie chart, and the stacked bar chart. This labelling removes the work needed to interpret the value shown.

A second enhancement to these charts is sorting the data in order of value (in this case descending), which means that the values decrease as we go clockwise around the pie chart, or upwards on the stacked bar. People often overlook sorting when presenting data, but it has a big impact on users' ability to understand the relationships between the values being compared - we are often looking for biggest/smallest values in a data set and so we should make it easier for the user when this is something they are considering.

	A	B	C

radius	1.5	1.06	0.47
area	7	3.5	0.7
ratio C	10	5.0	
ratio B	2		

16.5 **Judging area**
Consider the circles above. The area of A is twice that of B although it can appear bigger. C is one fifth the size of B but seems smaller. Judging areas is quite imprecise.
Source: Niall Enright

16.6 **Three chart types revisited**
These charts are all based on the same data and are drawn using Excel. Some of the deficiencies with each chart type have been addressed.
Source: Niall Enright

One problem with both the pie chart and the stacked bar is that we need to refer to a legend to determine the resource. Working out the resource involves glancing at a value on the chart, committing the colour to memory, then examining the legend to find the matching colour. Not a difficult process, but one that nevertheless takes a discrete, though small, amount of time. While the legends are colourful, they are not especially practical and also impede the ability of colour-blind people to interpret the charts and act as a barrier to reproducing the charts in black and white, which typically uses less energy and ink/toner and so should be encouraged. The charts below work in black and white and for colour-blind people since they dispense with legends (the pie chart, left, and column chart, right) or use patterns (stacked bar, centre).

Given this data, the best chart is the column chart below, right. This chart is simple, does not require a legend, can be reproduced in black and white and has a label associated with each value so that the user does not even need to scan along to the left and read the value on the vertical axis. The data is sorted so that the user can quickly discern the top and bottom resources. Although it is less colourful, this chart communicates the information about our savings potential more effectively than any other.

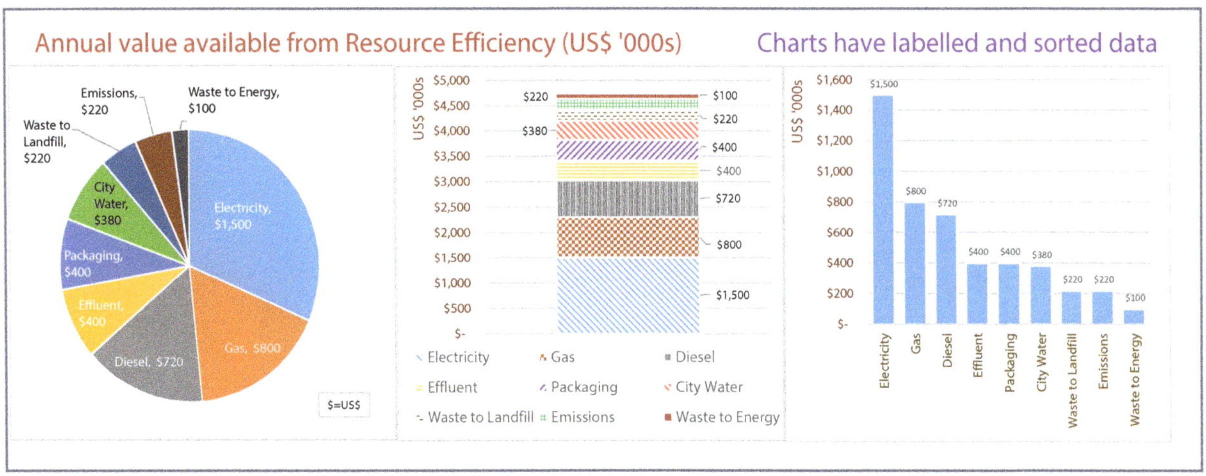

16.3 Block diagrams

Although they appear simple, block diagrams can convey a very great deal of information efficiently.

Often simplicity is the key to effective communication. Take, for example the illustration below setting out the scope of a 2011 study of public sector emissions reduction potential by Gill Bryan, Robert Cohen and Paul Stepan at the energy management and sustainability consultancy Verco. This simple block diagram clearly sets out what categories of emissions are included in the study and, equally important, which are not. If the Verco team had to use just text to describe the scope, it would undoubtedly have been much harder for the reader to grasp the scope.

I reproduced this diagram very quickly and easily using an Excel spreadsheet. The columns and rows could be easily resized to fit the text (note that line breaks can be forced using ALT-Enter in the text in the formula bar).

Block diagrams don't just have to convey dimensionless data. The illustration on the opposite page shows the breakdown of water distribution in Dalton Utilities, in Georgia in the US. Here, we can see seven stages or classifications in the distribution of potable water, as set out in the enhanced AWWA/IWA water balance calculations (I have labelled these at the bottom of the chart

16.7 **An illustration of a programme scope**
This very effective illustration makes it clear just which Kyoto gases, countries, emissions types and organizations are included in a report on Public Sector Emissions in England
Source: Reproduced with kind permission from Verco,[93] redrawn by Niall Enright

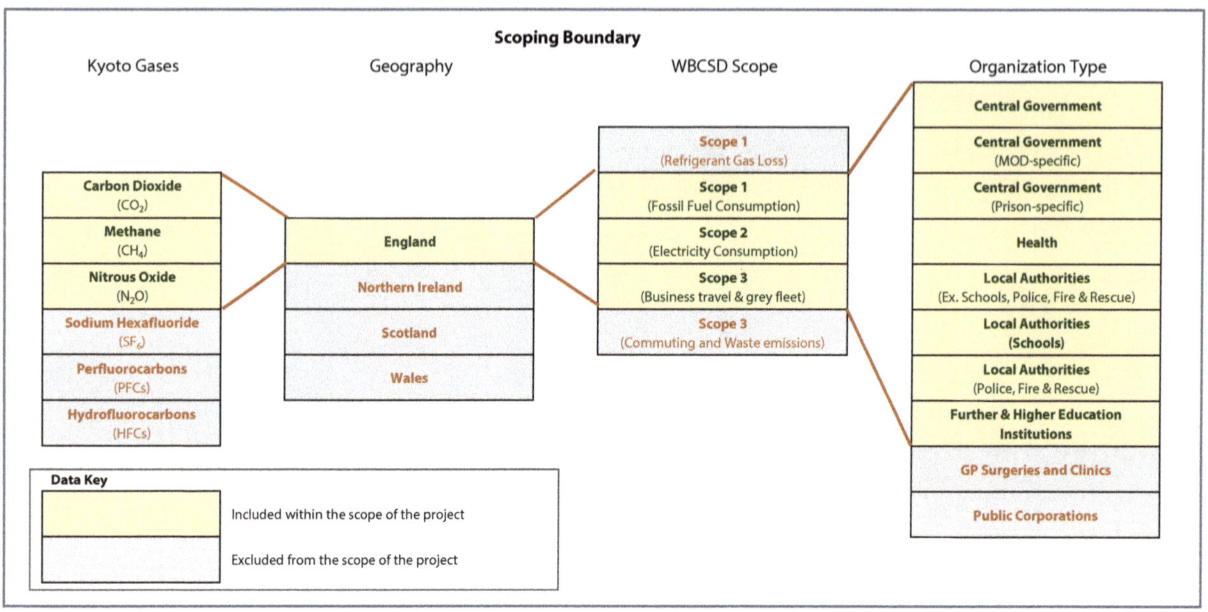

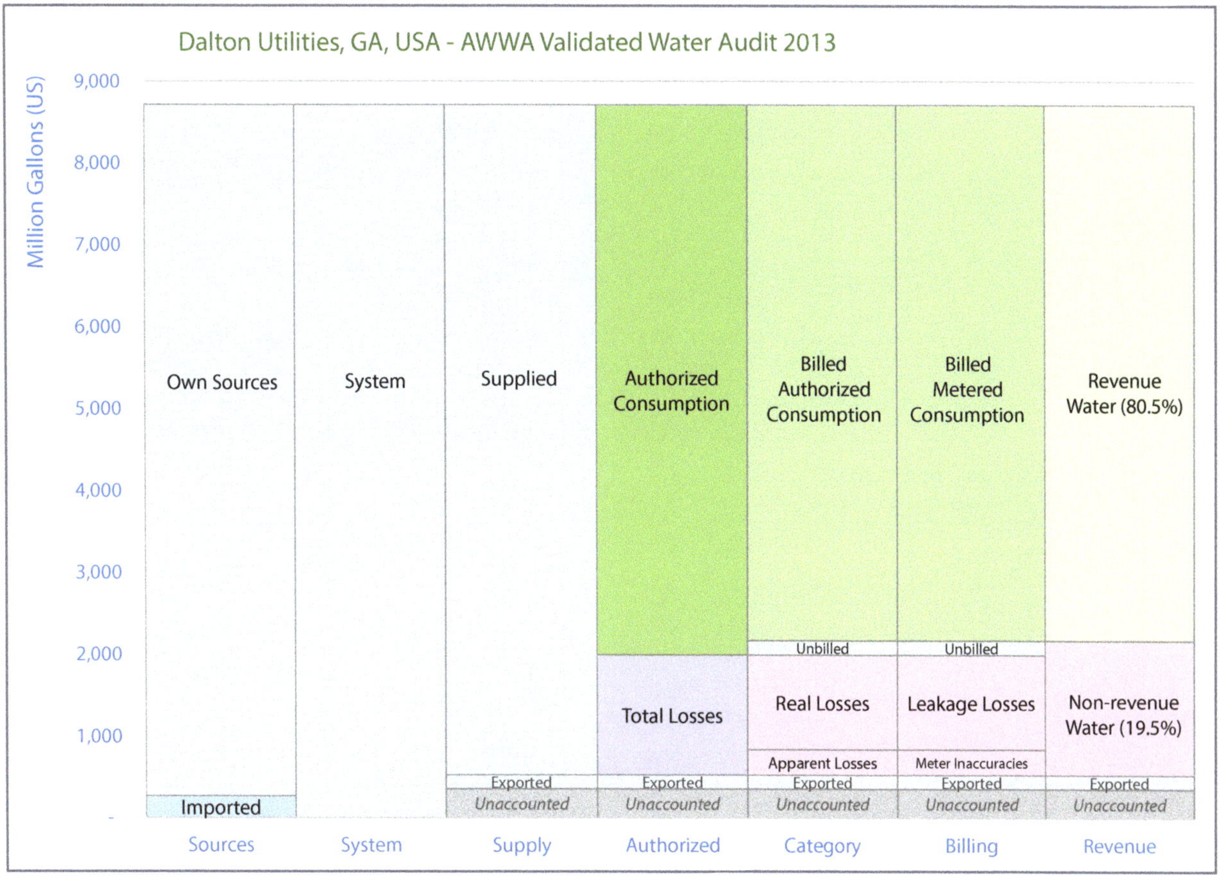

16.8 **Water audit for Dalton Utilities, GA**
This chart illustrates the relative volume of water in different stages of the potable water distribution for this US utility.
Source: Data from the American Water Works Association Water Audit Data Initiative[24] charted in Excel by Niall Enright. The spreadsheet is in the companion file pack.

with my own category names). The illustration shows the movement of water from the abstraction of the water on the left, to the final use and billing on the right. I have added an *"Unaccounted"* category to the data in so that the total volume of water at each stage is the same (this is not in the AWWA/IWA standard). As you can see in this diagram, the vertical axis has meaning, representing the volume of water in each category. If the area, i.e. the height and width, was used then the diagram could be referred to as a tree chart.

This illustration was easily constructed using an Excel stacked bar chart with each of the seven steps in a column (e.g. *"Sources"*) and the different categories (e.g. *"Imported"*) in rows. To ensure a consistent colour scheme, the individual data points had to be shaded manually and labelled with the series name (taken from the 19 different rows that the chart required). Please note that Excel places the series in *reverse* order (i.e. bottom up); thus, the first row of the data range was *"Imported"* followed by *"Own Sources"*, both with data in the first column, then *"System Input Volume"* with data in the second column, etc. The series overlap was set to 100% and the gap width to 0% in order to achieve the effect of the columns being one single block. The Excel 2016 tree chart type can also create these charts.

16.4 Bubble charts

If we have three pieces of information to communicate, a bubble chart may be the ideal solution. They are sometimes known as Boston charts after Boston Consulting which popularized these in the 1970s to separate products into "stars, cash cows, question marks and dogs". With the right data these charts can be a very effective presentation tool.

A variation of an X-Y chart, where two characteristics are portrayed by the X and Y axes, and the third characteristic is shown by the size of the marker, is called a bubble chart in excel. The example below uses data from an annual review of efficiency opportunities in a UK chain of convenience stores.

16.9 **Classic bubble chart**
This chart illustrates three items of information: total payback, percentage savings and annual savings potential for a chain of convenience stores in the UK.
Source: Based on client data charted in Excel by Niall Enright. The spreadsheet is in the companion file pack

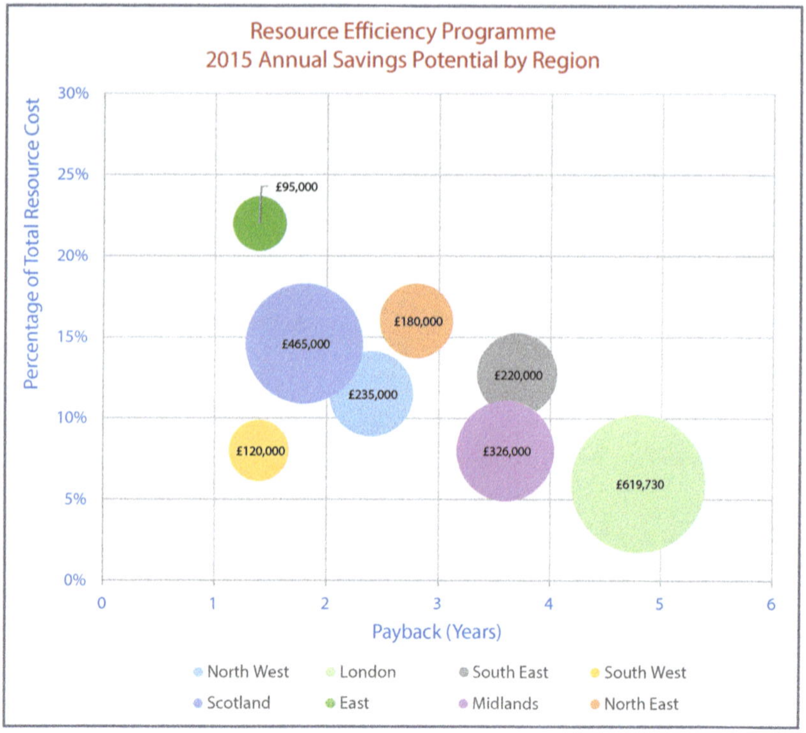

16.10 **Dynamic bubble charts**
For a superb illustration of dynamic bubble charts which change over time, visit Hans Rosling's Gapminder.com. The chart below shows how CO_2 emissions have grown as income per capita has risen.
Source: Gapminder.com.[304]

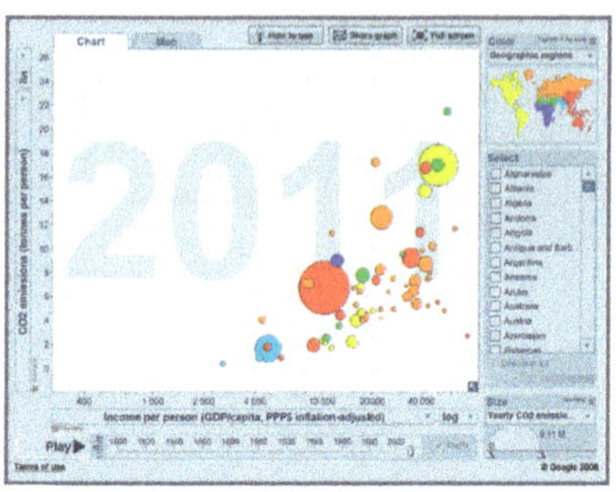

Each bubble represents the total value of savings available in each of the eight operating regions. We can see that the London regions has the greatest value of savings because its bubble is largest. It is easy to see at a glance that the biggest percentage of savings is available from the Eastern region, as shown in the vertical axis, while the shortest overall payback is in the South Western region, as indicated on the horizontal axis.

When creating these charts in Excel, I will use bubble size to represent the quantity characteristic (e.g. money, emissions etc.). Because there is no axis scale, I always set the bubble label to display the quantity. I then often place the next most important factor on the x-axis simply because people in western countries tend to read from left to right, and so this value is the most prominent. In the example above, I have placed payback on this axis because it is probably the key determinant of which activities are funded in what order. Rather than add a second label I have chosen to indicate the regions with a legend.

We should be aware that bubble charts can sometimes be impractical when the values closely cluster and so the bubbles overlap, even with the scaling turned down. The problem of overlap can, to some extent, be mitigated by using semi-transparent fill colours, but the labels can also interfere with each other and become difficult to read. The only way to determine this is to plot the data so I would not recommend the routine use of this type of chart, e.g. as part of a standard reporting template.

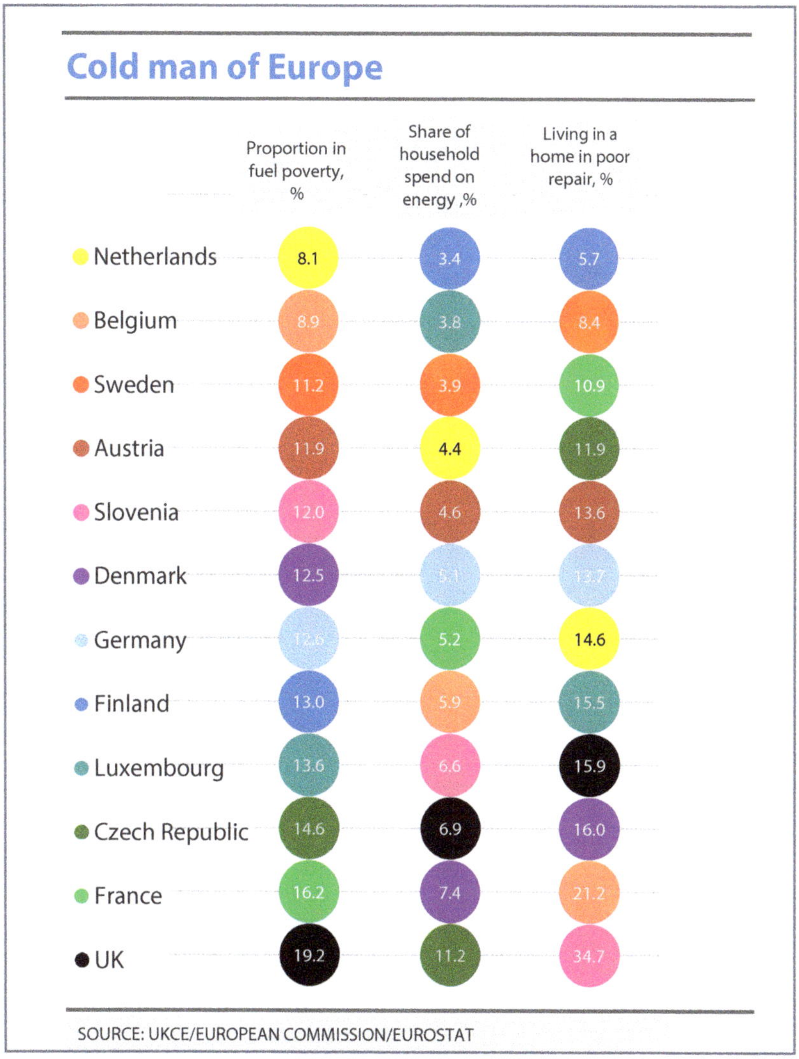

16.11 **Ranking chart**
I had quite a bit of fun constructing the chart right, which can best be described as a ranking chart. This has been drawn as an Excel bubble chart and here each bubble has the same size (a value of 1 which was then adjusted using Excel's "scale bubble size"). There were 12 series in the chart, one for each country, with a Y-value of 1-12 indicating its position vertically and an X-value of 1-3 indicating the metric is being plotted. A separate data range provided the labels.
Source: Inspiration taken from a Guardian newspaper item on energy efficiency,[120] charted in Excel by Niall Enright. The original source had lines connecting the values and the legend text naming the countries matched the series colour, which is not possible in Excel. The spreadsheet is in the companion file pack.

16.5 Waterfall charts

Waterfall charts are useful for two general presentation tasks: to set out a path from a current position to a future one and to describe a sequence of changes that affect a single resource as it flows through a system.

A waterfall chart is a modified bar chart in which we have a starting point on the left-hand side and a finish point on the right. In between these two points, we illustrate sequential increases or decreases in our resource by a vertical bar, connected to the previous bar. These intermediate steps seem to hang in mid-air, hence the other name that is given to these charts - flying brick charts. Usually, downward changes are indicated in one colour and upwards in another.

The example of a waterfall chart below is from a real project in the US where a team of auditors were establishing if a home and personal care products factory could get to zero emissions. It is a slightly unusual chart in that I have illustrated two possible finish positions - one with a conventional cogeneration plant and another with trigeneration. The chart reads from left to right. On the left-hand side is the initial status of the system, 100% of base

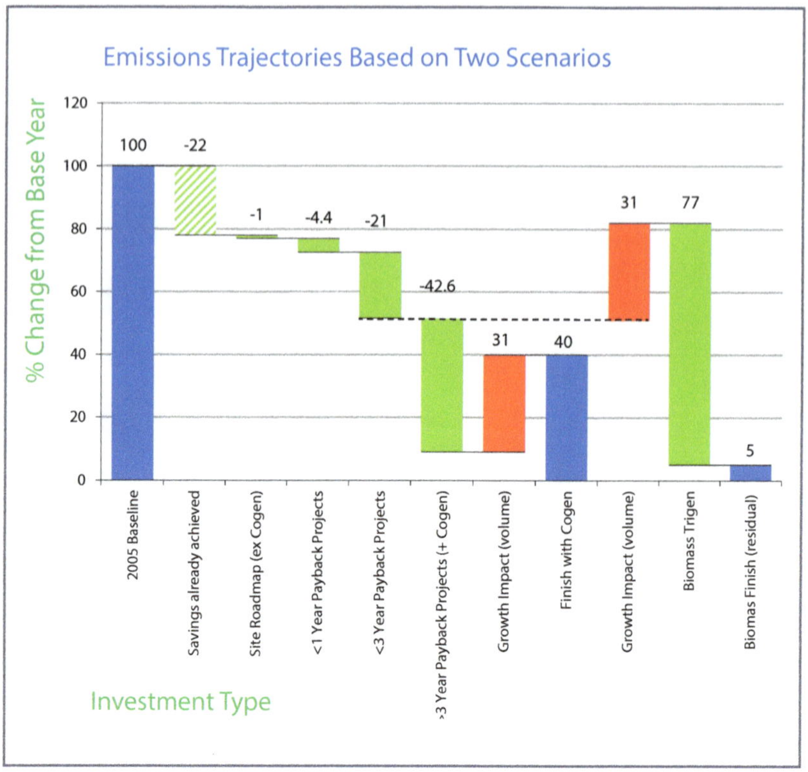

16.12 **A waterfall chart illustrating a trajectory towards a desired goal**
In this unusual example, two different end results are illustrated. This is based on real data, anonymized to preserve client confidentiality.
Source: Niall Enright. Drawn using Peltier Charts 3.0. Waterfall Chart, with some manual changes to allow two end points. Example sheet is available in the companion file pack (note requires Peltier Charts).

year emissions. The first step indicates that the plant has already achieved an impressive 22% emissions reduction (because this was completed, we used a hashed pattern to differentiate this from the other improvements identified). The next downwards bar shows that the plant has a further 1% emissions reduction in its plant roadmap. There then follow the savings identified from the audit, broken down into under one year, under three year and over three-year paybacks. Although that gets the plant to about 10% of the original emissions, the next bar in red indicates that production growth is expected to increase emissions by 31%, leading to an end-position of 40% of current emissions. On the other hand if, instead of implementing the >three-year payback projects and cogeneration system, the organization implements a biomass trigeneration plant, the emissions would be reduced to 5%.

Excel 2016 for Windows has just introduced a waterfall chart option. Before this, we had to create the chart using a series of bar charts with X-Y error bars providing the connecting lines. It is very fiddly to do this manually and so I would recommend an Excel Add-in, Peltier Tech Charts for Excel 3.0.[592]

The example of a waterfall chart, opposite, shows the trajectory towards a goal. We could include in such a chart a *"gap"* to highlight improvement not yet identified. Waterfall charts can also provide a kind of simplified Sankey diagram for a single resource flow, as illustrated below, where I have used the same data as in Figure 2.8 on page 53.

16.13 **A waterfall chart of a flow of a resource**
This example has been draw using an Excel 2016 waterfall chart.
Source: Niall Enright, data adapted from Amory Lovins.[485]
The spreadsheet is in the companion file pack.

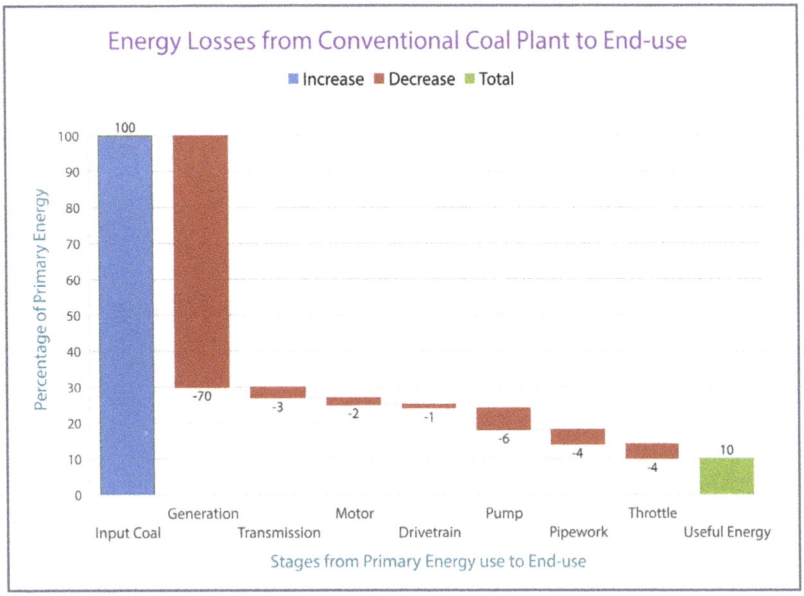

The Excel 2016 version of waterfall charts has a few drawbacks, for example, the connecting lines between the steps are very faint and the legend text cannot be edited (I would have preferred the legend to read *"Start, Losses and End"* rather than *"Increase, Decrease, Total"*). If this proves a problem, the Peltier Charts do work well in Excel 2016.

16.6 Thumbnails and sparklines

In resource efficiency, trend data is universal. As we have seen from the chapter on data analysis it can be difficult to present this data in a meaningful way since apparently small variations from period to period can represent quite large improvements.

There are many different ways of presenting time series or trend data. We have already seen that profile data can be very helpful in identifying opportunities for improvement (see *Profiles in action* on page 456). The weaknesses of longer-term trend data have also been examined (see *Why regression is such an important analysis tool* on page 464).

One widely employed data presentation technique is the use of miniature daily profiles, often arranged in the form of a calendar to easily compare like days of the week with each other. The figure below shows a simple example of this type of report, with 31 small profile charts showing electricity use for an office on each day in May 2015. The key to these profiles is that they should all share the same scale on the vertical and horizontal axes.

With the advent of the sparklines feature in Microsoft Excel, a raft of new possibilities to enhance the presentation of trend data is available. To demonstrate some of the capabilities, we have used the same data as shown in the mini profiles below in a more tabular format report, shown above, opposite.

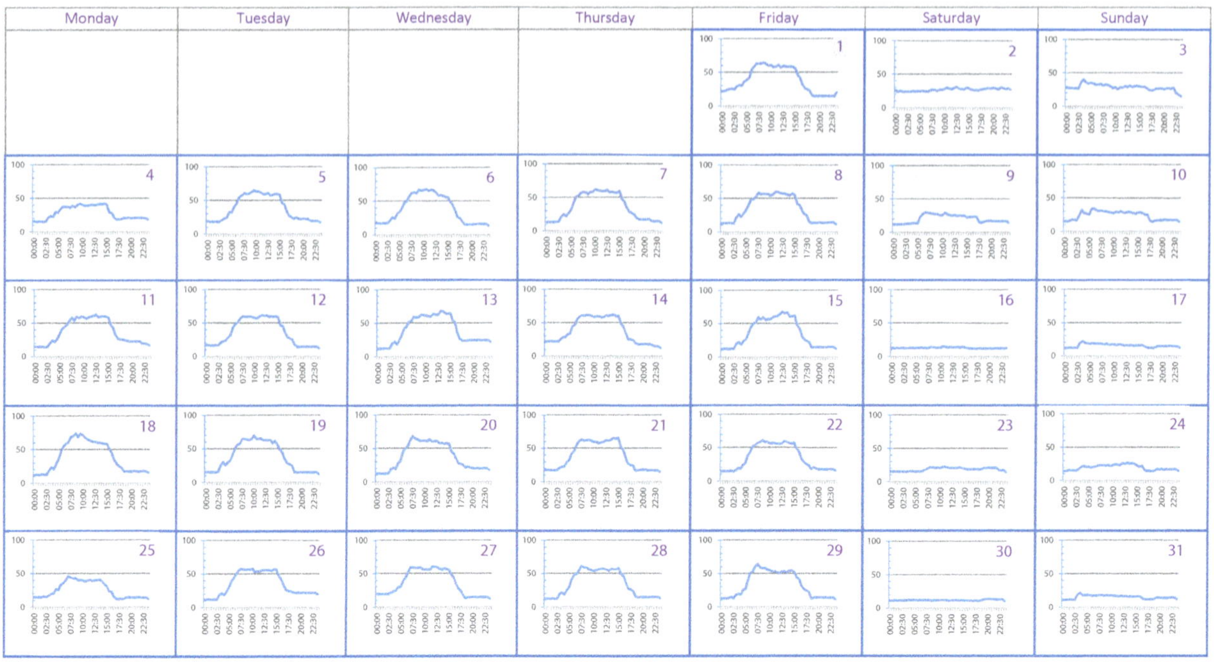

16.14 **Mini profiles in a calendar**
Source: Niall Enright.
Based on real data courtesy of Peel Land & Property Group Ltd. Drawn using Excel.
The spreadsheet is in the companion file pack.

16.15 **Mini profiles in a calendar**

Source: Niall Enright. Based on real data courtesy of Peel Land & Property Group Ltd. Drawn using Excel sparklines. The spreadsheet and image are available in the companion file pack. Please note how the "traffic light" icons at the right of the report have a cross (bad) and a tick (good). This is to enable the report to be understood if reproduced in black and white or if the recipient is colour-blind. The other coloured elements (like the high average and low bars) are differentiated by their position, so no special proviso is needed for these.

Date	Sparkline	High	Average	Low	HAL Bars	Total	±5%
01-May-15 (Friday)		64.3	38.2	14.2		1,833	
02-May-15 (Saturday)		30.8	26.7	23.2		1,282	
03-May-15 (Sunday)		40.2	28.6	15.0		1,372	
04-May-15 (Monday)		42.0	28.3	14.9		1,358	
05-May-15 (Tuesday)		65.3	38.3	17.2		1,838	
06-May-15 (Wednesday)		67.7	37.5	13.2		1,800	
07-May-15 (Thursday)		61.5	36.0	12.0		1,727	
08-May-15 (Friday)		59.4	33.9	11.0		1,629	
09-May-15 (Saturday)		29.8	20.1	11.6		967	
10-May-15 (Sunday)		33.9	23.6	14.3		1,131	
11-May-15 (Monday)		62.2	37.0	14.2		1,774	
12-May-15 (Tuesday)		61.7	36.4	12.2		1,749	
13-May-15 (Wednesday)		67.9	38.7	11.9		1,858	
14-May-15 (Thursday)		62.1	38.2	12.4		1,832	
15-May-15 (Friday)		67.5	35.2	11.9		1,689	
16-May-15 (Saturday)		14.9	12.7	11.5		609	
17-May-15 (Sunday)		21.8	15.5	11.8		743	
18-May-15 (Monday)		74.1	38.7	12.0		1,859	
19-May-15 (Tuesday)		70.1	38.0	12.4		1,826	
20-May-15 (Wednesday)		68.2	37.5	12.3		1,801	
21-May-15 (Thursday)		65.4	38.0	15.1		1,825	
22-May-15 (Friday)		61.3	35.4	14.6		1,701	
23-May-15 (Saturday)		22.2	18.2	14.3		876	
24-May-15 (Sunday)		26.8	19.8	14.3		952	
25-May-15 (Monday)		45.7	25.7	12.0		1,236	
26-May-15 (Tuesday)		57.6	35.5	11.8		1,704	
27-May-15 (Wednesday)		60.0	35.9	12.6		1,722	
28-May-15 (Thursday)		61.6	34.8	12.3		1,668	
29-May-15 (Friday)		64.6	33.5	11.2		1,610	
30-May-15 (Saturday)		14.0	12.0	11.0		576	
31-May-15 (Sunday)		21.7	15.3	11.3		736	

Although widely available in commercial software packages, these **small time series charts should be used with caution** as the variation we are looking for is often hidden at this small scale.

The tabular report above incorporate two sparkline features, one showing a very simple profile (which could also take the form of a line, but works better as column), and the other repeating the high, average and low values. As with the mini profiles, it is essential that the sparkline axes' minimum and maximum values are set to *"Same for all sparklines"* so that visual comparisons are possible. On the right of the tabular report are some *"traffic light"* icons, which can be placed using the Excel conditional formatting option. Here, the thresholds are set to ± 5% from the average (with weekdays and weekends having different thresholds).

These two illustrations using the same data reinforce the diversity of tools available to help make meaning of our resource-use data. Given the choice, it is important that we select presentations that help us understand the true performance of our systems, rather than those which are most visually appealing. While I am a great fan of traffic light indicators, I am less convinced by sparklines and mini profiles. My reservations lie with the very small size of the charts compared to the quantum of variation we want to be able to identify. As a result, I think that these small charts should be employed with caution, only where we anticipate large variation in use, even though many of the commercial energy management tools seem to have variations of these.

16.7 Heat maps

Heat maps are a great way to spot patterns in large quantities of time series data.

16.16 **Heat map (opposite)**
Source: Niall Enright.
Based on real data courtesy of Peel Land & Property Group Ltd.
Drawn using Excel. The spreadsheet is in the companion file pack and shows an alternative row-per-day heat map. Please note that, for comparison purposes, I have used the same data as in illustration 16.14 and 16.15. This kind of presentation can be problematic for people with colour-blindness, although the inclusion of the actual values in the cells can help with the interpretation of the chart. The companion file pack has a "black and white" version of the heat map to illustrate the problem differentiating between the red and green cells by the darkness of the shading alone.

A heat map chart uses a colour range to indicate intensity. Usually green or blue represent low values and orange or red high values. This chart is very good at highlighting patterns in half-hourly data.

Heat maps can be created relatively easily in Excel, using the conditional formatting colour scales facility. The example opposite uses the same data as shown on the previous two figures with a red-yellow-green colour scale applied. The data covers the last four weeks of May, with each day shown as a column and each hour as a row. Some people prefer to see days as rows and time in columns, and there is no correct approach. For four weeks of data, the rows-as-hours approach creates a portrait format chart while the columns-as-hours format tends to a wider landscape appearance.

In the example data opposite, there are a total of 1,344 data points, but thanks to the powerful pattern recognition of the human brain, we can very quickly pick out some clear trends. For example, we can see the warm red regions correspond to the daytime energy use starting at around 5:30 AM GMT on weekdays through to about 4:30 PM GMT. Since the month of May is in British Summer Time the actual peak usage is an hour later, from 6:30 AM to 5:30 PM. As an aside, daylight saving time is one of the traps to look out for when using any time series data, as many data systems time-stamp using unadjusted time all year round.

A few immediate questions arise from the heat chart. For example, why were the second and fourth weekend (16/17 May and 30/31 May) electricity use considerably lower than the first and third weekends (09/10 May and 23/24 May)? Have fewer people come into the office on those days? On some nights, electricity consumption does not fall as much as usual (see 13, 20 and 26 May as examples) - is a security guard or someone else leaving the lights on? Monday 25 May had a lower electricity use compared to other Mondays as this was a holiday, but should the consumption have been even lower?

One thing to be aware of when using these heat charts is whether the colour scale is absolute or relative. In Excel, for example, the range from green to red depends on the actual values in the data, so a particular hue of red may represent a different value in one chart to the next. Also, a single outlier value in Excel can force all the remaining values into a narrow range of colours at the other range of the spectrum, making the heat map ineffective.

	04-May-15 (Monday)	05-May-15 (Tuesday)	06-May-15 (Wednesday)	07-May-15 (Thursday)	08-May-15 (Friday)	09-May-15 (Saturday)	10-May-15 (Sunday)	11-May-15 (Monday)	12-May-15 (Tuesday)	13-May-15 (Wednesday)	14-May-15 (Thursday)	15-May-15 (Friday)	16-May-15 (Saturday)	17-May-15 (Sunday)	18-May-15 (Monday)	19-May-15 (Tuesday)	20-May-15 (Wednesday)	21-May-15 (Thursday)	22-May-15 (Friday)	23-May-15 (Saturday)	24-May-15 (Sunday)	25-May-15 (Monday)	26-May-15 (Tuesday)	27-May-15 (Wednesday)	28-May-15 (Thursday)	29-May-15 (Friday)	30-May-15 (Saturday)	31-May-15 (Sunday)
00:00	15.3	18.8	17.2	13.3	12.5	11.7	14.8	14.3	16.7	12.3	22.0	12.2	12.7	11.9	12.3	15.2	12.4	17.4	14.8	14.9	14.3	14.8	11.8	19.7	12.8	12.6	11.3	11.3
00:30	15.1	18.2	17.3	13.2	13.2	11.6	15.0	14.4	16.8	11.9	22.3	12.4	12.3	12.3	12.5	15.4	12.6	17.8	14.9	14.8	14.5	14.6	12.1	19.4	12.6	12.8	11.2	11.3
01:00	15.0	18.2	16.9	13.2	13.0	12.0	15.6	14.6	16.5	12.0	22.5	12.4	12.6	12.3	12.0	15.2	12.5	17.3	14.7	14.9	16.0	14.8	12.0	19.5	12.3	12.6	11.7	11.3
01:30	15.2	18.3	17.1	13.1	13.6	11.9	17.3	14.5	16.6	12.4	22.1	11.9	12.2	12.1	13.0	15.3	12.3	17.4	14.8	14.7	15.5	14.9	11.9	19.5	12.7	12.9	11.3	11.9
02:00	14.9	18.1	17.0	13.5	13.1	11.9	17.2	14.2	17.1	12.2	22.2	12.5	12.5	12.3	12.8	15.3	12.6	17.5	15.2	14.9	15.5	16.2	12.0	19.6	12.4	15.0	11.6	11.3
02:30	15.5	18.1	17.3	13.6	12.9	12.5	16.5	14.4	16.8	12.1	22.0	12.4	12.5	12.3	12.5	15.4	12.9	17.4	14.6	14.8	15.6	15.0	12.0	19.6	12.4	12.7	11.4	11.6
03:00	21.3	20.4	21.0	21.4	21.6	12.9	24.5	19.8	20.7	19.1	24.7	20.4	12.3	19.6	18.5	24.0	19.3	21.0	16.1	14.6	17.8	17.2	19.2	22.0	20.0	17.8	11.4	18.9
03:30	23.4	23.4	25.7	24.9	26.4	13.0	31.6	24.0	22.7	22.1	28.2	22.1	12.4	21.8	23.5	30.5	21.1	22.5	19.1	15.0	20.7	19.6	21.2	23.3	21.6	19.6	11.6	21.7
04:00	22.2	24.6	24.0	21.7	22.2	13.6	27.3	22.0	23.9	18.2	28.2	18.7	12.7	19.3	20.6	25.5	19.0	23.0	18.7	15.4	22.3	20.2	18.9	24.8	18.5	18.6	11.9	18.3
04:30	29.6	32.7	31.1	26.6	27.6	13.5	26.2	25.9	28.6	24.3	32.8	23.7	12.6	18.9	28.5	29.8	27.7	29.4	24.0	15.6	20.4	26.8	24.4	30.0	25.7	25.7	11.8	18.1
05:00	28.5	32.1	34.0	31.3	32.4	13.4	25.1	33.9	32.2	31.8	33.5	26.6	12.6	18.4	36.2	32.4	31.2	32.4	27.0	14.8	21.0	25.8	30.6	30.5	29.9	26.8	12.0	18.3
05:30	34.0	40.6	39.7	41.7	40.2	23.5	33.0	41.2	43.8	41.6	44.4	36.0	12.4	18.7	47.7	42.1	44.4	41.6	38.3	15.6	19.8	33.5	40.0	39.6	41.6	40.8	11.7	17.9
06:00	37.8	45.5	47.5	50.8	48.3	26.9	33.9	44.3	48.3	44.9	46.3	44.6	12.9	18.2	54.5	50.7	50.1	46.2	44.6	14.8	20.2	35.9	43.9	44.2	47.9	51.3	11.8	17.3
06:30	37.1	55.1	50.2	52.2	48.5	29.8	31.4	47.2	51.4	51.5	55.6	49.6	12.4	18.6	58.6	56.4	53.9	54.9	51.9	16.1	21.3	40.0	49.6	52.1	49.2	53.2	11.6	17.4
07:00	37.1	57.2	57.1	56.4	52.7	29.1	30.2	53.7	57.5	53.6	58.7	51.6	12.6	18.3	61.0	57.4	62.3	60.5	53.4	18.1	22.5	45.7	55.6	59.2	54.5	61.2	12.1	17.3
07:30	35.5	61.0	61.7	57.9	58.0	28.9	30.8	58.2	60.1	59.3	61.1	59.4	12.5	17.8	68.6	64.7	68.2	62.8	57.3	19.6	23.3	44.1	57.6	59.2	61.6	64.6	11.5	17.8
08:00	38.7	59.5	63.0	57.3	56.2	27.7	29.9	53.8	59.4	58.3	60.4	55.9	13.9	17.4	70.0	64.9	64.5	61.6	58.8	21.2	23.3	42.5	56.8	58.7	59.3	59.0	11.8	17.9
08:30	36.5	60.0	61.6	55.1	56.4	27.0	29.7	59.2	59.3	58.8	61.5	59.1	13.7	18.0	74.1	64.0	63.5	62.1	61.3	21.4	22.7	43.3	56.8	58.8	59.1	58.0	11.3	17.3
09:00	40.2	62.3	66.7	57.9	56.7	27.0	28.6	57.3	59.5	62.3	61.2	55.5	13.2	16.6	68.6	63.5	60.7	62.5	58.7	20.2	23.5	40.2	56.7	59.1	56.1	58.2	11.6	17.9
09:30	41.6	65.3	65.8	58.2	56.1	25.4	28.0	59.8	59.1	61.9	60.2	56.5	13.2	17.2	73.8	64.6	61.5	61.6	59.1	20.8	22.3	40.2	56.9	56.3	55.1	55.7	11.6	17.5
10:00	40.2	63.2	66.3	61.5	54.4	25.6	26.9	58.5	57.7	61.2	58.7	56.2	13.2	16.7	71.2	70.1	61.2	59.6	56.3	20.3	23.3	40.0	57.5	56.0	54.0	55.2	11.7	17.6
10:30	39.1	63.3	67.7	61.0	55.7	29.2	29.2	56.4	57.0	60.5	59.4	59.4	14.9	16.1	67.7	65.1	60.4	59.6	57.3	21.5	24.5	38.0	53.4	56.3	55.1	53.2	12.2	17.2
11:00	38.8	61.7	65.4	59.7	59.3	26.8	28.2	58.8	59.9	59.7	59.2	60.4	14.3	17.0	64.8	64.8	63.5	57.5	56.8	22.2	24.0	39.5	53.9	56.0	56.0	53.5	12.0	16.8
11:30	39.6	59.3	66.7	59.3	59.4	25.8	29.8	59.4	61.7	63.9	57.8	61.1	13.4	16.7	63.2	65.6	60.0	58.5	56.3	21.0	23.3	40.7	53.8	60.0	57.5	51.3	11.9	16.6
12:00	40.4	59.4	65.9	59.1	57.8	25.0	28.0	60.9	60.7	62.6	58.3	64.3	14.2	15.6	61.6	62.8	60.7	59.3	55.9	20.2	26.3	39.0	55.8	59.9	57.3	50.8	11.4	16.9
12:30	39.6	61.2	63.0	61.0	57.4	25.7	28.0	62.2	61.3	62.3	60.7	67.5	13.5	15.8	61.6	62.8	60.7	61.8	56.6	20.3	25.2	40.1	55.7	59.4	55.9	53.0	11.6	16.4
13:00	40.6	57.4	58.5	57.4	56.7	24.5	28.9	58.5	59.1	67.9	60.0	65.4	13.6	15.6	60.3	63.0	57.8	61.5	60.0	20.3	26.8	39.8	55.6	57.2	54.4	53.2	11.9	16.6
13:30	41.2	58.7	58.7	58.2	55.2	25.3	28.1	59.8	60.2	67.5	61.2	66.5	14.1	16.1	60.0	63.0	58.3	62.5	58.9	20.2	25.6	40.8	55.8	55.7	57.6	52.1	11.7	16.2
14:00	41.2	59.7	57.4	57.2	54.5	24.5	27.5	59.9	59.2	64.4	62.1	60.3	13.4	15.2	59.2	58.2	56.7	65.3	57.0	21.4	26.3	38.3	55.0	57.6	56.9	54.4	11.9	16.2
14:30	42.0	59.3	56.3	56.6	57.0	25.3	25.8	59.8	60.1	63.7	59.4	60.8	14.2	15.4	58.5	61.8	57.4	64.6	56.1	20.1	25.0	34.1	56.5	56.7	56.8	54.6	11.9	16.0
15:00	40.8	58.3	55.7	59.3	54.8	22.1	28.0	58.3	58.1	64.8	60.3	61.6	13.0	15.6	57.9	56.2	57.2	65.4	57.6	19.1	22.7	31.0	56.7	55.1	58.5	52.2	11.9	16.1
15:30	29.0	45.4	49.3	47.3	41.7	22.9	28.5	46.7	46.3	53.4	48.7	47.5	12.7	15.9	50.2	48.4	45.0	52.6	48.3	18.1	21.0	23.1	46.7	47.9	49.2	44.6	11.7	16.6
16:00	28.6	43.8	43.7	44.3	39.7	22.8	27.0	43.3	45.3	51.6	43.7	42.6	11.9	15.3	44.2	44.9	42.0	47.5	44.1	18.3	22.1	22.6	41.3	44.1	46.2	41.6	11.4	15.7
16:30	21.9	33.6	35.8	34.8	31.6	23.3	24.7	32.2	35.4	35.1	33.7	31.1	11.7	13.3	33.6	34.8	34.8	35.4	32.9	18.5	16.7	14.7	32.8	34.7	36.6	31.0	11.4	12.6
17:00	18.5	27.0	26.3	28.6	26.4	23.1	24.6	26.6	28.5	26.1	24.8	26.0	11.5	11.9	28.2	28.6	27.6	27.2	28.4	18.6	14.5	12.0	26.0	27.4	26.9	26.3	11.0	11.7
17:30	18.3	25.0	24.3	26.9	24.2	15.2	15.5	25.6	26.7	23.6	25.1	23.8	11.7	12.0	25.0	26.7	26.5	25.2	25.1	18.2	14.7	12.0	24.5	24.5	26.5	24.0	11.6	11.7
18:00	18.8	22.5	19.0	24.3	14.9	14.2	14.3	25.0	25.1	23.4	23.7	22.0	12.0	11.8	22.5	25.3	25.2	24.6	23.2	18.1	14.5	12.1	23.6	18.5	23.2	22.1	11.3	12.0
18:30	20.6	24.4	15.0	19.2	13.1	15.8	16.0	24.4	17.3	23.6	19.2	14.3	11.8	13.0	17.3	15.2	21.3	16.9	17.9	18.3	14.7	12.3	22.3	13.1	15.8	12.3	11.7	11.8
19:00	20.4	22.4	15.0	17.3	13.3	16.8	16.5	23.3	14.2	24.6	17.9	14.6	11.8	14.4	17.6	14.5	22.7	17.4	20.1	20.5	17.1	14.0	21.6	14.7	14.7	13.5	13.1	13.4
19:30	20.8	22.3	15.2	17.2	13.2	17.1	16.6	22.7	14.4	24.4	17.9	14.7	11.7	14.4	17.3	14.9	21.2	17.9	18.5	20.4	16.9	14.3	22.1	14.8	15.0	13.9	13.8	14.4
20:00	20.5	22.6	15.1	17.3	13.8	17.0	16.6	22.5	14.5	24.7	18.3	14.6	12.0	14.4	17.2	15.0	20.1	18.1	17.2	20.8	17.1	14.2	21.8	15.0	14.9	14.0	13.9	14.2
20:30	20.7	22.1	15.5	17.2	13.5	16.9	17.2	22.6	14.4	24.3	18.2	14.7	11.7	14.3	17.8	14.8	20.0	18.5	17.4	20.6	16.6	14.3	22.2	15.1	14.9	14.0	13.8	14.1
21:00	20.5	21.9	15.7	17.8	13.3	16.8	16.7	22.5	14.8	24.5	18.0	14.6	12.0	14.3	17.4	15.0	20.2	17.1	17.0	20.6	17.1	14.4	21.9	14.8	14.9	13.8	14.0	13.8
21:30	20.7	19.3	15.9	15.1	13.5	16.8	17.0	22.6	14.4	24.1	17.9	14.5	12.3	14.9	17.3	14.8	19.8	17.3	17.1	20.8	16.8	14.3	21.9	15.2	15.0	14.1	13.6	14.0
22:00	20.7	19.5	15.7	14.4	14.0	16.8	16.7	19.6	14.4	24.6	16.1	15.1	11.9	14.4	17.4	14.8	19.9	17.4	17.4	17.3	17.1	14.4	21.9	14.9	15.4	13.8	13.3	14.3
22:30	20.7	19.3	15.6	14.6	14.0	16.7	17.0	19.3	14.6	24.3	15.0	15.0	12.3	14.5	17.8	14.8	20.2	17.2	17.3	16.6	17.2	14.2	21.7	14.9	15.2	13.7	13.5	14.5
23:00	20.5	19.1	15.8	14.7	13.8	16.8	16.9	18.9	14.4	24.5	14.8	14.7	11.9	14.2	17.7	14.9	20.1	17.3	17.3	17.1	17.2	14.1	22.0	15.1	15.2	13.8	13.7	14.4
23:30	18.4	17.2	13.2	12.0	11.0	14.9	14.3	16.7	12.2	21.8	12.4	12.3	12.4	12.2	15.5	12.4	17.8	15.1	15.4	14.3	14.7	12.2	19.7	12.6	13.1	11.2	11.0	11.9

Min 11.0 Colour Scale (kWh) 74.1 Max

16.8 Sankey diagrams

Sankey diagrams are a fancy name for an illustration of the flow of materials in a system. They can convey a very large amount of information in one image; however, specialist software is required to created these.

One of the most versatile and information-dense presentations of resource flow data is in the form of a Sankey diagram. These are named after a Canadian Captain Matthew Henry Phineas Riall Sankey, who is said to be the first person to use these flow charts in 1898 to compare the energy flow diagram in two steam engines (a real one and an "ideal" one).

The key feature of a Sankey diagram is that, although the placement of the various steps in a system is usually arranged to create a close approximation to the actual equipment or process, the relative thickness of the arrows joining the parts indicates the flow of materials between the steps.

In his interesting introduction to the history of the Sankey diagrams,[648] Mario Schmidt describes these as "*the main tool for vizualising industrial metabolism*" and these diagrams are now widespread in areas such as life cycle assessment. However, it is debatable whether Sankey should get the credit for this form of illustration as Charles Minard's diagram of Napoleon's Russian Campaign

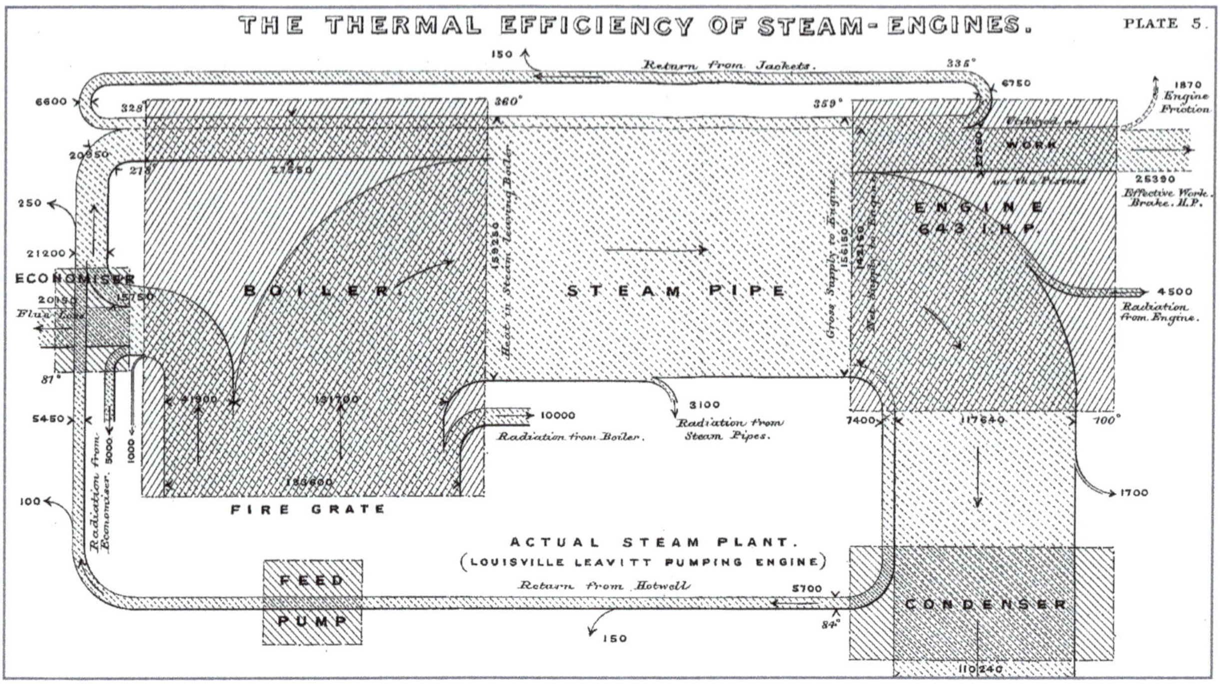

16.17 **Diagram illustrating the thermal efficiency of steam engines**
The original Sankey diagram showed an actual steam engine (illustrated) and an ideal steam engine (not shown) and represented the flow of steam through the engines by the thickness of the lines, as well as adding temperatures at various points.
Source: Capt. MH Sankey. Minutes of Proceedings of The Institution of Civil Engineers. Vol. CXXXIV, Session 1897-98. Part IV. Public domain.

of 1812, drawn in 1869 predates Sankey's, but has the same key features of a spatial representation and flows with a quantity represented by the thickness of the line. Many illustrators consider Minard's diagram[787] to be the finest ever illustration. It displays six items of data in just two dimensions: the number of Napoleon's troops; the distance they travelled; temperature; latitude and longitude; the direction of travel; and location at specific dates.

The Sankey charts below show the current and proposed waste flows in a tea-bag factory, following a waste audit by my colleague Martin Hess at ERM, which I contributed to in 2011. Like Sankey's original illustrations opposite, I have used the Sankeys to compare states plotted a *"before"* and *"after"* scenario. The yellow arrows on the right-hand diagram show changes to the flows. The plan calls for greater segregation of the waste streams to divert valuable waste like cardboard from the mixed waste stream, and for an additional Pelletizer to handle some of the currently discarded burnable waste such as Filter Dust.

One additional benefit of most of the professional Sankey drawing tools is that they will check on the mass balance of the system that is being illustrated, ensuring that the inputs and outputs of the various elements are balanced. Although the diagram below illustrates solid waste, it is quite possible in some of the Sankey tools to draw multiple streams between system components such as water and energy, and to show these selectively at the click of a mouse. Some tools offer inputs from Excel, creating dynamic diagrams.

While Sankey diagrams are commonly used to represent the physical flow of a resource through a facility, organization, or supply chain, I often will use a Sankey chart to summarize the flow of money, or opportunities following an audit, as shown on the next page.

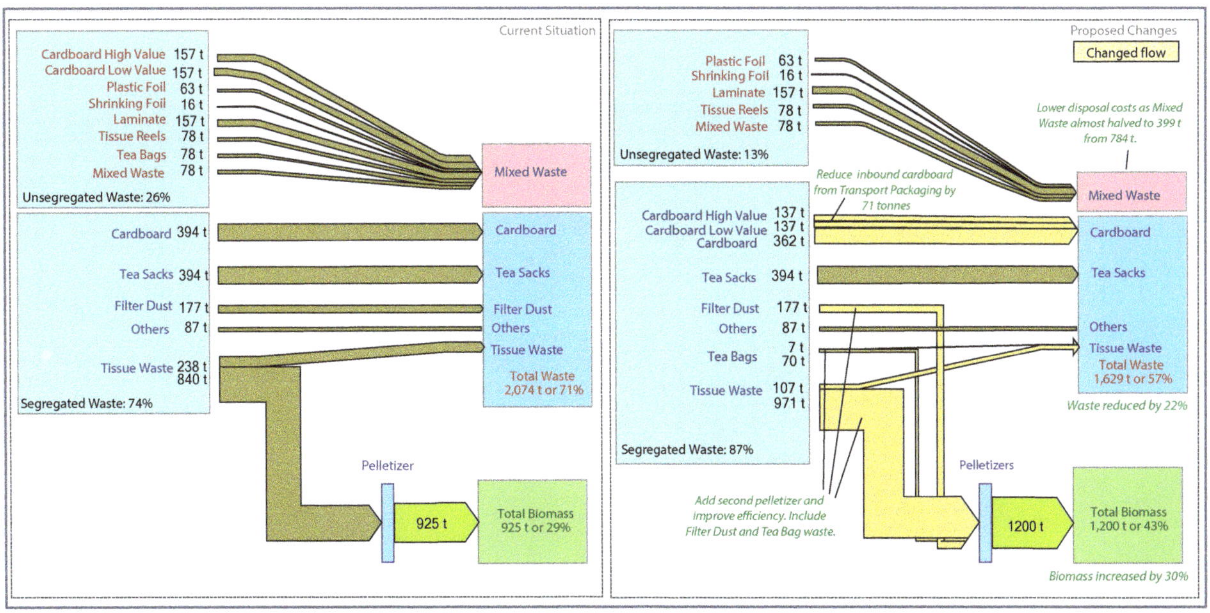

16.18 **Sankey diagram illustrating the flow of solid wastes in a tea bag manufacturer**
The current situation is shown left and the proposed changes resulting in a reduction of waste by 22% on the right. Summaries of the key actions and benefits (in green) have been added to the illustration, which provides a very accessible overview of the proposed resource efficiency improvements.
Source: Niall Enright. Based on real client data. Drawn using eSankey!! software from ifu.

16.9 Creating Sankey diagrams

There are a number of software tools available for drawing Sankey diagrams. Here, we examine some of the capabilities of these tools using a mass flow and a financial Sankey diagram.

16.19 **Sankey diagram illustrating the proposed flow of solid wastes in a tea bag manufacturer**
Source: Niall Enright. Based on real client data. Drawn using eSankey! software from ifu.

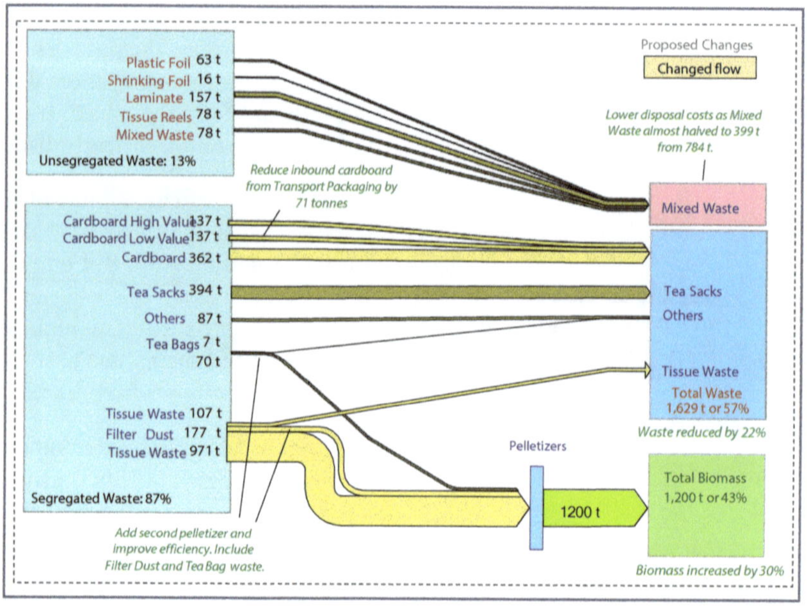

16.20 **Sankey diagram illustrating the proposed flow of solid wastes in a tea bag manufacturer**
Source: Niall Enright. Based on real client data. Drawn using SankeyMATIC.

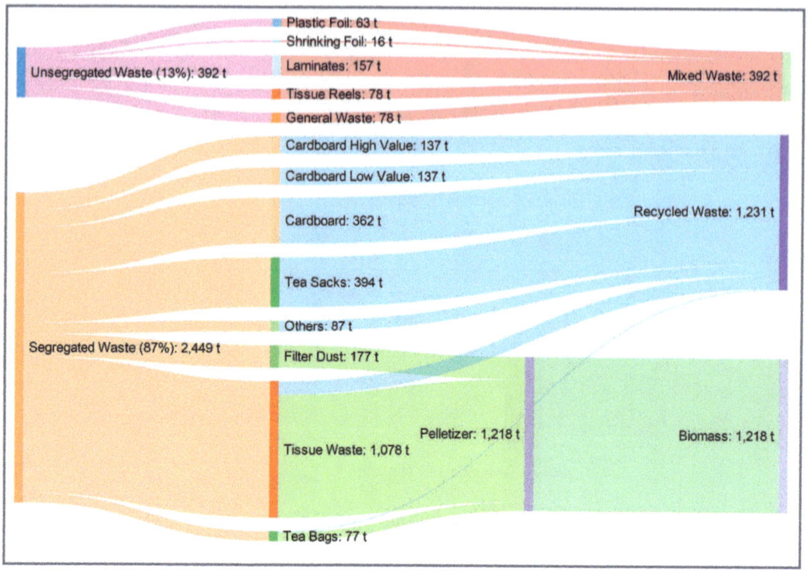

Broadly speaking, I would divide Sankey software into two general types.

1. *"Free-flowing"* drawing packages that may have additional features such as illustrating multiple resource streams, using images in the charts, connecting "live" to Excel spreadsheets, tabulating data flows, etc. Examples of these packages are eSankey! and SDraw, both of which are fairly expensive but very competent.

2. *"Block drawing"* packages that usually take a simple tabular input and connect the nodes automatically. There are several free of charge websites that employ the open-source Java library d3.js. Examples of these sites are SankeyMATIC, WikiBudgets and Eco-data, which offer varying degrees of flexibility in the design of the chart and some remarkable interactivity in the web interface.

The illustrations opposite show the results that are possible with each type of tool: above eSankey! (version 3.0) and, below, SankeyMATIC. I have chosen to illustrate roughly the same flows in each package, the desired waste strategy at our tea bag factory. The eSankey! model tool took just under two hours to produce while the SankeyMATIC model took about 30 minutes. eSankey! clearly has much greater flexibility in terms of the positioning of elements, the inclusion of additional text and the characteristics of each flow. However, SankeyMATIC also has some neat features such as the ability to set the colours for each flow, checks on the balances at each node and the ability to move the nodes around before producing an output file. The Sankey diagrams website[642] has a very comprehensive list of Sankey software.

The Sankey chart below shows the flow of costs in a small dairy and is designed to emphasize the potential savings, with low capital expenditure avoidable costs shown at the top of the chart.

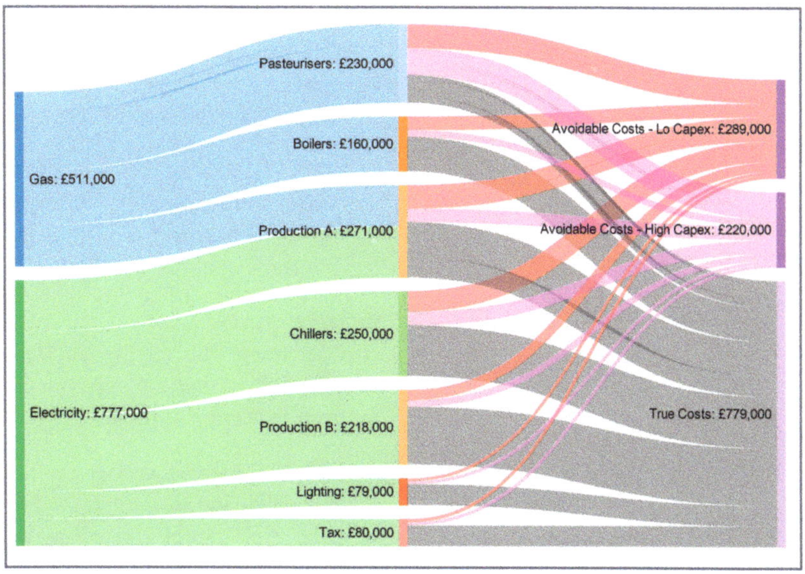

16.21 **Sankey diagram illustrating the savings potential for electricity and gas in various departments in a small dairy**
Source: Niall Enright. Based on real client data. Drawn using SankeyMATIC.

16.10 Sunburst charts

Sunburst charts are especially useful at conveying information where data is organized in a hierarchical manner.

In Excel 2016 Microsoft has introduced some new charting options. We have already seen that the waterfall chart provides a useful illustration of a path towards a given efficiency goal. Another chart that may provide an effective overview of resource use is a sunburst, chart, so called because of the data radiates outwards from a central point.

The chart opposite illustrates the resource breakdown in a small components factory. I like the fact that we can show the resource use in the process at the centre, with the distribution and supply at the edges (see page 396).

When using a sunburst chart you should consider the following:

1. Units: All values need to be in the same unit. In practice, this means that resource quantities are usually presented as costs.

2. Plot order: The charting order is determined by the total value of the each entry in the first column of data (plotted as the innermost circle, in the example opposite that is "production"). The values for the second column of data are shown again in ascending order (in this case "electricity"). There is no way to change this order. Up to four levels can be shown.

3. Labels: For anything other than the simplest sunburst charts, each segment will need to be labelled with its name. That leaves little room to show values as well. For very small segments, Excel will simply leave the segment unlabelled, which is not ideal.

4. Scale: There is no scale provided, but in the example opposite I have added a scale as a simple table. For degrees, I have taken the total costs and divided by 365 (I have included radians, where I have divided the total cost by 2π, but these units are not likely to be generally understood).

Sunburst charts have the same challenge as pie charts in that it is difficult to differentiate the relative angles of different segment. There is a further complication in that the apparent area of each segment increases as we move out from the centre of the chart.

Nevertheless, these charts can add to the presentation of resource data, for example, following an audit, if used with caution. For those without Excel 2016, the same effect can be created by overlaying several pie charts, but the process can be laborious.

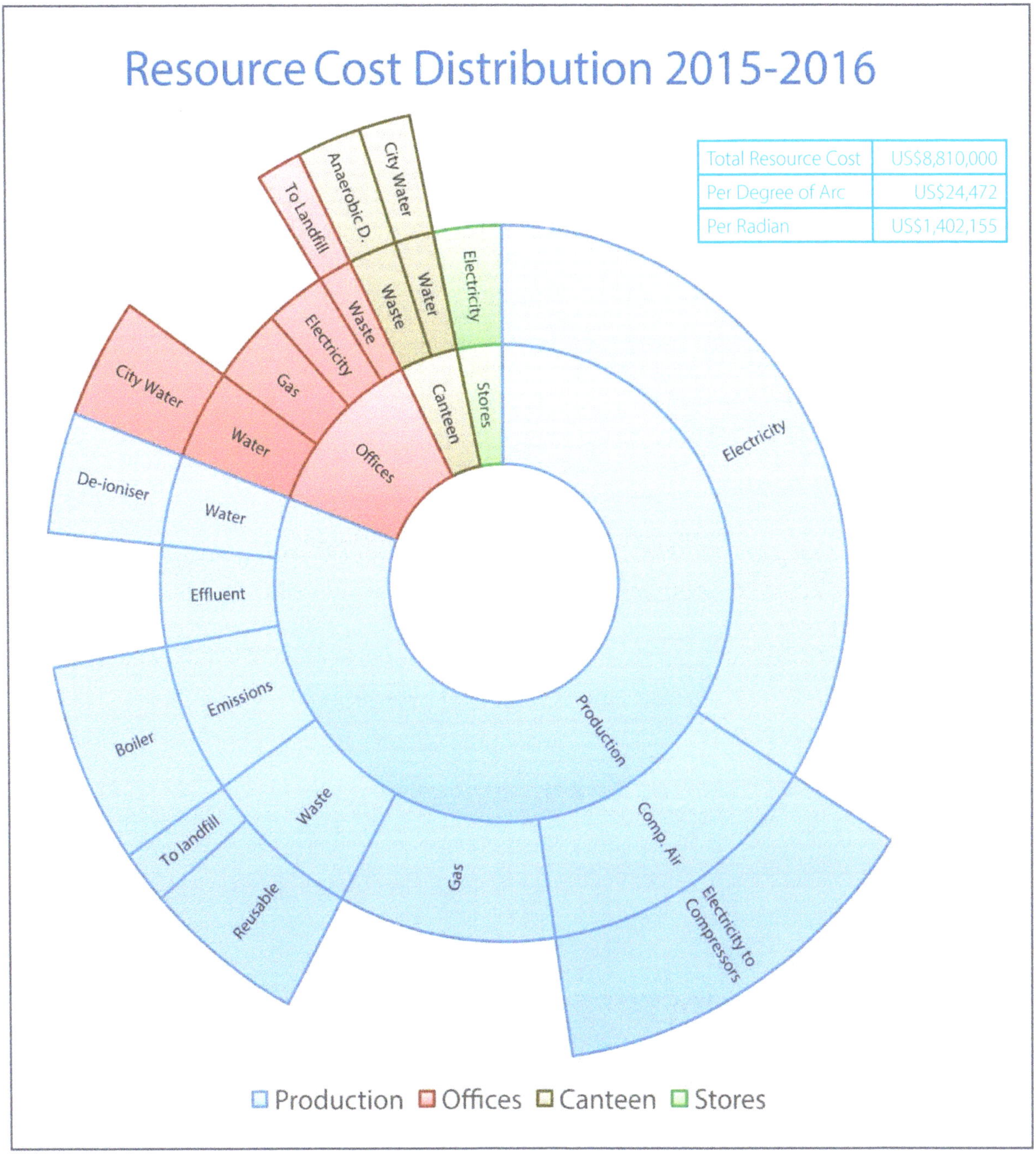

16.22 **Sunburst chart showing resource costs for a small manufacturing facility**

Source: Niall Enright. Based on real client data. Drawn using Excel 2016. Note that the summary table has been added manually. The spreadsheet is in the companion file pack.

16.11 Fishbone diagrams etc.

An important presentation tool is the fishbone diagram. This is not only a way to visualize opportunities for improvement, but offers a way to engage people in a structured root cause analysis.

A particularly valuable presentation tool in energy and resource efficiency is the fishbone or Ishikawa diagram, named after the shape of the diagram or its Japanese originator respectively. Although not strictly a data presentation tool, this illustration is nevertheless a powerful way to encapsulate a large amount of information in one image.

The fishbone diagram is used in root cause analysis to help move beyond the symptoms of resource inefficiency to potential causes, which can then be corrected. The diagram starts with a problem statement, as illustrated below on the right-hand side. Possible causes of the problem are placed in one of the six main categories. In manufacturing, these categories have traditionally

16.23 **Fishbone chart**
In this example, the causes of excessive hot water use in a dairy are examined. The six categories proposed are intended to focus on systems and behaviour aspects as well as equipment.
Source: Niall Enright. Image and spreadsheet template are in the companion file pack.

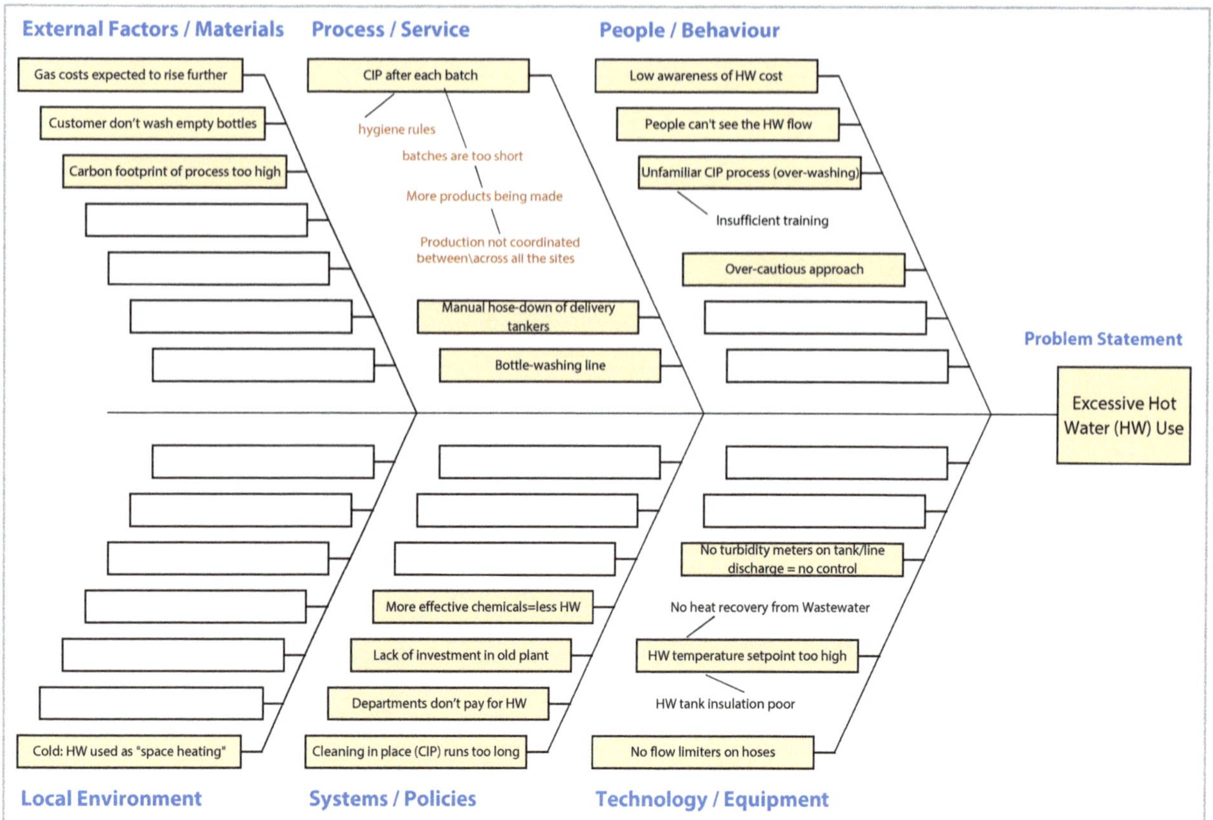

been: machine; method; material; people; and, optionally, measurement and environment. In keeping with the emphasis in this Framework, I have proposed the following categories: people/behaviour; systems/policies; technology/equipment; process/service; local environment; and external factors/materials. These categories should comfortably cover manufacturing and service organizations, but there is no requirement to use these headings if you feel they don't suit your needs.

Once a problem has been identified, the process of populating the diagram usually involves a workshop/brainstorm activity. Here we would use a technique like the "*5 whys?*" to keep probing for deeper level insights. Each possible cause is entered on one of the main category branches but can have other sub-causes branching off. In the example opposite, one cause of the excessive hot water use, was in the process, *"CIP [cleaning in place] after each batch"*. This, in turn, is caused by the *"hygiene rules"*, or by the fact that the *"batches are too short"*, which in turn is due to *"more products being made"*, which in turn could be caused by *"production not coordinated between\across sites"*. All the usual caveats about brainstorming apply (no criticism, etc., described later on page 672) and, as with most team problem-solving efforts, it is important that an appropriately diverse, experienced and creative team is assembled. I usually use Post-it notes during the brainstorming stage as this allows causes to be moved around and easily linked together, usually on a whiteboard with a large fishbone diagram template.

Insight

When we have completed the fishbone diagram, the next step is to prioritize the most significant, lowest-level, root causes and seek to change these, rather than addressing the intermediary symptoms. Note that this process is good at spinning off multiple solutions to a problem and so is particularly useful at highlighting interactions between people, systems and technology. Furthermore, by asking folks to consider *"People/Behaviour"* and *"Systems/Policies"*, we get away from a technology-focused mindset. Reminding folks to think about what causes the demand for hot water will also help them to understand the drivers for hot water consumption, some of which may be controllable.

Many other specialist presentation techniques can support our energy and resource efficiency programme. One big category involves spatial location so that we can visualize our resource use both in place and time. Examples of these tools range from very simple (but effective) mimic diagrams or dashboards offered by most control systems, through to sophisticated geographical information systems (GIS), which can provide maps of resource use overlaid with a wide range of other physical or environmental data.

Fishbone diagrams are a great tool to engage folks in **establishing the root causes of inefficiencies.**

While space is too limited to allow us to go into detail on these other visualization tools, the same principles apply as in any other presentation tool. In other words, the output must be valid (i.e. correct), appropriate (relevant and actionable) and comprehensible (understandable).

Summary:

1. Get the fundamentals right. Are the units clearly set out? Are values or series labelled? Can differences or scales be accurately discerned? Can the information be conveyed in black and white?

2. Make sure that colour-blind people can understand what is being communicated. There are various forms of colour-blindness, so it is important to choose a palette that works for the particular situation. Especial care needs to be taken with red-yellow-green "traffic light" indicators and with legends which rely on colour alone to indicate which series is which. While the use of colour can enhance the attractiveness of a presentation, it is important that the information conveyed by the colour is communicated in other ways (e.g. the earlier heat map has actual values in the cells, as well as a colour). Care should also go into other aspects of the presentation such as the use of red laser-pointers to highlight information on the screen. If a presentation works when reproduced in black and white, then it will usually be fine in all circumstances.

3. When choosing a data presentation method, make sure that it is valid, appropriate and comprehensible. This requires that the *purpose* of the presentation and *the capabilities and needs of the recipients* are understood.

4. Ask yourself: *"Will the way I present the information support the action I want the receiver to carry out?"* If not, change it.

Further Reading:

There are many titles on the use of *infographics* to convey information effectively. One of my favourites is *Information is Beautiful* by David McCandless (ISBN 978-0007492893), partly because it has some very neat images illustrating environmental issues.

For those seeking to improve their Excel presentation skills, *Excel Dashboards and Reports* by Michael Alexander and John Walkenbach (ISBN 978-1118490426) is a very comprehensive introduction to the presentation possibilities in Excel and many basic *dos and don'ts.*

I also recommend Charley Kyd's excellent www.exceluser.com website, which has lots of practical advice on creating very professional-looking outputs using Excel. There are dozens of downloadable files.

Finally, also for Excel users, is Jon Peltier's website, peltiertech.com/ which covers all sorts of exotic charting techniques. You can also find the Peltier Tech Charts 3.0 add-in for Excel.

17 Financial Analysis

Photo © Kitti Kornkasame , Fotolia.com

In this chapter, we will focus on quantifying the value that resource efficiency provides. We will build on the sources of value first described in Chapter 3 by introducing techniques to enable us to express this value in monetary terms. This quantification is designed to help us to justify effort around resource efficiency, to create compelling business cases for action.

First, we will consider how cost savings and resource savings are not necessarily the same thing. Then we will look at what an investment business case needs to include. The core notion of a future cash flow using a base case will be explored. From this, the important notion of marginal costing will emerge. We will also discover that the most widely used measure of investment quality in efficiency, payback, is entirely unsuitable for its intended purpose.

Next, we will consider the timing of money: why a dollar now is worth more than a dollar in a year's time. This will introduce us to the common financial measures used in corporate finance - net present value (NPV) and internal rate of return (IRR) - with a view to levelling the playing field for efficiency investment compared to other corporate investments. The emphasis throughout will be on whole life costing. We then extend this analysis to consider how externalities can be incorporate into financial decision-making.

We will consider how project interact and how a portfolio of projects can be described using marginal abatement cost curves (MACCs). We will explore the weaknesses of MACCs and look at alternatives.

Although this chapter describes the formulae to be used to calculate value, in practice the financial analysis techniques described here could not be easier, as the functions are built into most spreadsheet tools. In fact, the most challenging issue is to select the right financial function for the case in hand and to ensure that the cash flows are correctly described. The techniques described here are in many ways far easier than some of the statistical tools described earlier.

The companion file pack for this book has a large number of spreadsheet models to rapidly create very credible, compelling and attractive business cases. In the reference section, there is a summary of the key financial functions and useful tables to accompany this chapter. Finally, this chapter is about calculating and presenting the business case for investment. The later chapter on obtaining funding complements this one and speaks about the sources of investment when the value case is made.

17.1 Delivering financial efficiency

Resource efficiency is focused on achieving lower costs by consuming fewer resources. However, the resource efficiency practitioner can also deliver resource cost savings by procuring resources more cost-effectively or by changing the pattern of resource use. These sources of value are an important part of any resource efficiency programme.

Resource efficiency and financial efficiency are different, but usually, complementary, objectives.

Resource efficiency practitioners often end up delivering cost savings without reducing the quantity of resource use. Although this may seem a distraction from our pursuit of greater efficiency, in practice this value can be very important. Part of the role of the resource efficiency practitioner must be to support the financial efficiency of resource use.

A classic example of cost reduction without consumption reduction is load shifting, also called demand response. Here we are preoccupied with the time of day, or day of the week, of our electricity consumption. Many electricity utilities charge different amounts for electricity consumed during the various times of day and raise their prices during peak demand periods such as morning and evening. By adjusting our pattern of activity, we can modify our consumption profile so that we use less electricity during these peak periods, which can significantly reduce costs. This process is sometimes called *peak shifting* and our ability to take advantage of this depends entirely on the cost structures, or tariffs, in place and the flexibility of our electrical demand.

Although moving our consumption around to reduce costs does not modify our use or energy, it is nevertheless an important part of a resource efficiency practitioner's role. On closer examination, shifting electrical demand usually does have some usefulness in respect of resource consumption, albeit indirectly. This is because of the nature of electrical power as a distant voltage potential, created in a power station, which can bring about the movement of electrons, i.e. a current which does useful work at our location. Put simply, electricity generation has to match demand moment by moment. Thus, the utilities need to put in place sufficient generation, and transmission, infrastructure to meet the moment of highest demand - even though this infrastructure can lay idle much of the time. This is further complicated by the different electricity generation technologies having different response times. The output of solar and wind, for example, cannot be significantly increased on demand, while nuclear power stations take a long time to start up and shut down. Because of this, it often falls to plants such as gas-fired turbines which can be rapidly dispatched, to provide the most flexible generation capacity which can be ramped up or down quickly to match demand changes. The high price of electricity at peak periods is intended to reflect the significant additional costs maintaining this peak capacity. If all electricity users could manage their consumption so that the demand was constant, we would need a lot less power generation capacity and so would reduce the quantity resources needed

Real World: *Part of the job*

Photo © casnowe, Fotolia.com

Resource efficiency practitioners see cost reduction as a core part of their role. For example, my colleague at ERM Richard Wise carried out an energy audit of a relatively newly built components plant in an East European country. Here, the context is important as this country has previously experienced disruptions in its natural gas supply from Russia via Ukraine.

Because of the supply problems, the plant had been designed with a liquefied petroleum gas (LPG) supply which was in use during the audit.

What Richard spotted was that less than 30 metres from the factory fence there was a connection to a natural gas pipeline. Not only was this connection very close to the plant but it was even fitted with a meter.

Since the factory had been built, the natural gas supplies had become very much cheaper and more reliable than LPG. Richard worried about how the news of his discovery would go down with the management team at the factory. In fact, rather than get into recriminations about why this had been overlooked, the team rapidly and enthusiastically set about getting the connection established in a record-breaking three weeks after they learned of the alternative supply. Although not a reduction in gas *use*, this project saved the plant €150,00 a year of costs, which could help justify the audit expenditure.

to build all the additional generation capacity. Since much flexible supply is currently fossil-fuel based then, by lowering our peak demand, we will help to reduce emissions. Participating in demand response, where we voluntarily reduce electricity use as a consequence of spikes in the system, is usually good for the environment as well as for the wallet.

A big part of the financial efficiency of resource use involves procurement. Take, for example, Peel Land & Property Group, a large and complex infrastructure company I work with, headquartered in north west England. Here, Dale Mullane, the senior procurement manager, working with a third-party brokerage, LG Energy, has aggregated the demand of all the Peel companies, which exceeds £20 million per annum of electricity use. This has enabled Peel to buy its energy on the wholesale market and to *lock in* the price months ahead when the price seems most attractive. This sophisticated procurement strategy has led to considerable savings in the electricity cost - in the order of 12% for winter 2015 power, for example, based on the average wholesale price. In fact, the saving is even greater compared to the regular *retail* prices Peel would have had to pay if it had not aggregated its use sufficiently to go into the wholesale market. This gives Peel a competitive advantage compared to organizations which do not have the same buying power or expertise since these lower energy costs translate to lower costs for tenants and customers.

While Dale Mullane works very closely with the energy Champions in Peel, it is often the case that there is a separation between those tasked with buying resources and those tasked with managing them. This separation can cause problems where the procurement folks make buying decisions that do not take into account the lower volumes that efficiency can bring, or where the efficiency folks do not have an understanding of the marginal costs of resources, so do not apply the correct valuation to the resources that are being saved (see the Skoda example on page 576).

Thus, the very first activity in any financial assessment of resource efficiency opportunities is to obtain a detailed understanding of resource pricing. This cost is usually much more than an average US$ per unit cost. The resource practitioner needs to understand how the costs/tariffs are structured.

- Are there seasonal or time of day cost elements?
- Are there volume elements (paradoxically using more can lead to a lower unit cost, which runs counter to an efficiency objective)? Sometimes organizations have *take or pay* contracts in place where there are penalties associated with using fewer resources.
- If resource use is reduced, what is the actual value of each unit saved?

It is probable that future resource costs will become more complex and unpredictable as a result of changes such as renewable and distributed generation, smart networks and decreasing demand. Assessing how pricing is likely to change is critical to appraising long-term efficiency investments.

17.2 What is an investment?

An investment is an allocation of money to support an organization's core objective. The investment usually adds value to the organization (in the broadest terms) or supports its licence to operate.

17.1 **Categorization of investment decisions by finance directors**
Based on a 2011 study of 35 Geneva canton firms consuming more than 1GWh of electricity per year, 19 in the manufacturing sector, 16 in the commercial sector (e.g. retail, conference centres, parking lots). We can see that there are many reasons why businesses make investment decisions: to increase production, to meet legal obligations, to carry out research.
Source: Catherine Cooremans.[168]

Investment Categories	# of citations
To maintain or renew existing production capacities	14
To increase productivity of existing means of production	13
To improve production process	9
To reduce energy consumption	9
Legal conformity of equipment (pollution, etc.)	9
Equipment replacement	9
Marketing of new products	8
Research	7
Product quality improvement	7
In-house development of new products	6
Working conditions improvement (beyond legal obligations)	5
To increase productivity of support functions	5
Internal communication	1
External communication	1
Others	0

There are two basic reasons why organizations spend money.

- Because they have no choice. The expenditure is a necessary part of doing business, e.g. they have to pay taxes or ensure legal compliance. This is expenditure to support the organization's licence to operate; and/or
- Because the expenditure will add value to the business. This is expenditure to increase the organization's capacity to achieve its core objective (increase shareholder returns, treat more patients, educate more students, etc).

Before we go on to talk about value, we should note that the compliance expenditure may also offer opportunities for energy and resource efficiency. In some ways, there may be an advantage as the spending does not need to demonstrate value to proceed, but on the other hand, compliance expenditure is seen as a non-productive overhead, and so there is often a desire a to minimize this costs if possible.

Even though most environmental compliance is expected to involve BATNEEC (best available technology not entailing excessive cost) the reality is more often CATNIP (cheapest available technology not incurring prosecution). If we have a culture of spending the minimum on compliance, then it is important that we carry out a marginal cost analysis (see page 567) on the additional cost to deliver a more efficient alternative than the cheapest compliance solution.

Most expenditure on energy and resource efficiency is optional, that is to say, that it is not required in order to support the organization's licence to operate. Optional expenditure that is intended to increase value is called an investment.

To get approval for investment, we usually have to produce a business case. These can vary greatly, but they usually boil down to two simple elements:

- Quantifying the risk associated with the investment; and
- Quantifying the expenditure and the value that the investment will produce, i.e. the return on investment;

All things being equal, decision-makers will favour investments that produce the greatest value in relation to the expenditure, that is to say, the greatest return and the lowest risk (or the highest certainty).

17.3 The right level of accuracy

In reality, there are a number of stages of financial analysis with increasing levels of accuracy required.

A business case needs to address two aspects of an investment: risk and return.

The financial case for efficiency involves a degree of estimation. That is because we are usually looking at a future cash flow or asset value which can be influenced by many factors (such as the cost of resources, the *"learning curve"* of technologies, the level of activity in our organization or even the weather).

There are many possible sources of inaccuracy in a business case. First of all, let us consider the accuracy of the costs and savings that we use to calculate the return on investment. One cause can be due to data errors; for example, we have already seen that meters have a given accuracy due to bias or limits to precision. Another source of inaccuracy is the fact that many aspects of the business case, such as equipment cost or operating characteristics have to be estimated at the time the business case is being developed.

The US Department of Energy, and many other organizations, use a five-class definition for cost estimation, which acknowledges that financial assessment is undertaken for different purposes. Thus, for a high-level concept screening exercise, a much less detailed level of definition of the elements of the project is acceptable and so the accuracy of the analysis will be low - in the range of -50% to +100%. This is a Class 5 - order of magnitude - estimate which is usually used to screen out infeasible projects quickly.

17.2 **Estimate classes and typical accuracy**
These estimation classes are widely established and reflect the fact that different types of decisions are made with varying levels of detail.
Source: Niall Enright, based on US DOE.[740]

Estimate Class	Name(s)	Degree of Project Definition	Purpose of Estimate	Accuracy Range
Class 5	Concept or Order of Magnitude	0%- 2%	Concept Screening	L: -20% to -50% H:+30% to 100%
Class 4	Opportunity or Intermediate	1%-15%	Study or Feasibility	L: -15% to -30% H:+20% to +50%
Class 3	Preliminary or Pre-design	10%-40%	Budget Authorization	L: -10% to -20% H:+10%to +30%
Class 2	Substantive or Detailed Design	30%-70%	Bid or Tender	L: -5% to -15% H:+5% to +20%
Class 1	Definitive or Implementa-tion	70%-100%	Check Bid or Tender	L: -3% to -10% H:+3% to +15%

The level of accuracy I would expect from a typical energy efficiency audit would be Class 4 - that is to say an estimate in the range -30% to +50%. Note that these classes have an off-centre range of accuracy; the presumption is that we are more likely to be higher than the estimate than lower than the estimate, reflecting the observed pattern of estimation in the real world. There are usually greater uncertainties on the downside than on the upside.

One of the problems with most energy efficiency audits is that the level of certainty around the value case for investment is often insufficiently detailed for budget authorization. This doubt means we cannot go straight to approval for opportunities without further definition of the project, and more detailed quantification of the costs and benefits.

The investment analysis must ***meet the required level of accuracy.***

In most resource efficiency audits, quantifying the benefits is easier than quantifying the costs, although we shall see later that most audits simply think in terms of direct resource cost savings rather than the many co-benefits that can substantially enhance the value.

If we are exploring an opportunity that involves capital expenditure (CAPEX), getting a realistic price with only a sketchy outline of the equipment specification can be difficult. As soon as you enter into a dialogue with vendors, they will often require a lot more information than can realistically be provided at the time of the audit. Fortunately, there is a rule of thumb that can help us. The six-tenths rule says that if you double the size of a piece of equipment, then its cost will rise by 0.6 or 60%. Thus we can scale the known cost of similar equipment to estimate the cost for our business case. If the reference equipment was purchased sometime in the past, we could use cost indices such as the Chemical Plant Cost Index[783] (or similar relevant indices) to adjust to today's prices.

If all I have is a capital cost for our equipment and I need an approximation of the installed cost, then I tend to double the purchase price, to account for additional costs such as procurement, installation, disruption, commissioning and training. In complex industries or for larger installations, this multiplier is even greater. In the refining and petrochemicals industry, for example, Lang Factors (or more recently Guthrie Factors) are used to estimate the cost of projects based on equipment costs. For example, in a solids processing plant, the total cost is 3.1 times the equipment cost; for a predominantly fluids processing plant the cost is 4.7 times the equipment cost. These multipliers are for a whole factory or process, so they are almost certainly too high when estimating the total installed cost of single items of equipment within an existing process.

In Australia, the Energy Efficiency Opportunities Act 2006 (EEOA) represents one of the most comprehensive mandated programmes for energy efficiency anywhere in the world. Although since repealed, the detailed guidance for these previously mandatory audits provides an excellent resource for those wanting to take a more systematic approach to evaluating opportunities or energy (and resource) efficiency. One aspect that is especially interesting is the emphasis on accuracy. The EEOA introduced the concept of estimation accuracy – all projects with a payback of fewer than four years are required to be assessed to an accuracy of ±30%, i.e. Class 3 estimation.

We covered the subject of uncertainty in previous chapters on metering (see page 422) and on data analysis (see page 506). In those chapters, the causes of uncertainty were primarily related to measurement. However, the biggest source of uncertainty in most business cases isn't measurements but poor assumptions that underpin business models (such as overconfidence about savings or underestimation of costs).

A recurring theme in this chapter will be how to ensure that assumptions about a future cash flow are credible and realistic.

Real World: *How much time is needed to analyse opportunities?*

The guidance accompanying the Australian Energy Efficiency Opportunities Act (EEOA) gives an indication of the level of effort that would be recommended so that energy efficiency opportunities are adequately assessed. The EEOA effort level is based on the assumption that the necessary data is readily available, that staff are knowledgeable, and that the projects are relatively low complexity.

Number of **hours** effort required to bring forward opportunity recommendations based on the AUD ($) cost of the opportunity.	Saving <$10k p.a.	Saving $10k–100k p.a.	Saving $100k–$1M p.a.	Saving >$1M+ p.a.
Opportunity estimation. Analyse how the opportunity impacts site energy use. Develop clear assumptions on the impact of future changes. Use the above to estimate energy savings. Confirm accuracy of savings. *(Estimation Class 3)*	1-3	3 – 16	16 – 50	50 – 200
Business case. Forecast energy saving. Calculate simple payback. Develop assumptions and calculations. Make implementation recommendations. Additional time. *(Estimation Class 4)*	1	2 – 8	8 – 20	> 20
Total	**2-4**	**5-24**	**24-70**	**70-220**

Is this level of effort realistic? When I compare the actual hours spent by the audit teams identifying and evaluating opportunities in five recent energy efficiency audits I was involved in, we can see that the real world falls far short of the EEOA ideal.

Type of Organization	# Opportunities	Minimum EEOA suggested hours (above)	Real world actual audit hours
Chemicals (South Africa)	41	135	112
Automotive (Czech Rep.)	20	135	75
Tea (UK)	34	115	90
Cosmetics (US)	63	978	105
Airport (UK)	27	95	21
TOTALS	**185**	**1458**	**403**
AVERAGE (hours per opportunity)		**7.8**	**2.2**

In defence of the authors of the EEOA guidance, the actual audit time above did not take all the projects identified to a Class 3 level of assessment. Some of the opportunities would have been quantified at a much lower level of accuracy or even left open; in other words, requiring further investigation to get an estimate.

The key lesson is that the evaluation of opportunities is time-consuming. However, I don't buy the idea that the time is related to the savings value - a low impact project can be just as complex as a big-ticket item - but in the real world one naturally spends more time on bigger value projects, resulting in these being more accurately defined than lower value projects.

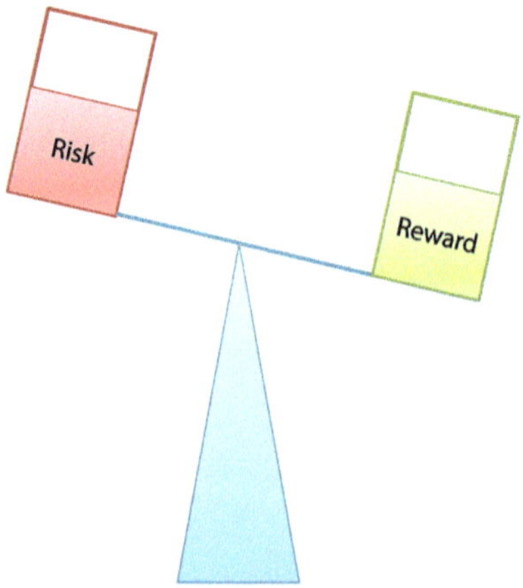

17.3 **A business case needs to properly account for risk and return**
To be credible, our cash flow doesn't just need to be structured correctly, the financial figures we present need to reflect an impartial assessment of risk and reward.
Source: Niall Enright

Exploration: *Just how accurate are project cost estimates?*

There is a large gap between the potential for resource efficiency and what is being achieved. This chapter seeks to address one of the reasons for this, the poor quality of business cases that are put forward, where costs and benefits presented are not credible or compelling.

Lest we assume that poor business cases are confined to resource efficiency practitioners, an interesting paper[374] by John Hollmann for the American Association of Cost Engineering sets out a pretty woeful picture of cost estimation accuracy across the board (at least in mining, process and infrastructure projects in the US).

Hollmann states that the cost of big projects tends to be significantly understated through a failure to consider all sources of risk and from wishful thinking on behalf of developers or asset owners who want the project to go ahead. We can all think of expensive projects that have come in several times over budget.

Conversely, the paper finds that the cost of small projects (typical of many Optimize phase investments in resource efficiency) tend to be biased towards overestimation of cost. This overpricing has many adverse effects; the projects may not be approved, or, if approved, it will lead to project under-runs, which is also damaging, since without pressure on costs the "fat" in the projects tends to be spent unnecessarily. Finally, where cost have been overestimated, investment resources are unnecessarily tied up in the inflated project budget.

According to Hollmann, the cause of the overestimation of cost for small projects is that the system is geared to rewarding delivering projects under budget. From my experience, overestimation of cost for small projects may also reflect a reluctance on the part of the project manager to implement these. Small projects with small returns are seen to be a hassle and much less glamorous than large projects with big impacts.

"Small" projects here is used in a relative way - this could be a project requiring thousands of pounds investment in an organization with a considerable resource spend, which would be considered large in smaller organizations.

The interesting thing about these observations is that the root cause of the poor business cases is not a lack of technical expertise about how to construct a cash flow, but largely about behaviours and attitudes relating to risk. However rigorous our methods, pricing decisions are often a matter of judgement or choice and it is critical that these choices are balanced and objective.

As resource efficiency practitioners, we often encounter biases against investment. In order to give our projects the best chance of approval, it is important that we understand how to create credible business cases, based on realistic cash flows where all the assumptions and risks (or uncertainties) are clearly set out. In particular, where these risks are material to the business case, we should acknowledge them and describe how these risks can be managed.

Every organization has a different expectation and culture around investment selection and risk. A big part of the *"art"* of the most successful resource efficiency practitioners is understanding these expectations and delivering superior proposals for investments within this cultural frame of reference. There is no *right* way, although some of the techniques set out here should help.

17.4 Simple payback

The most common method for evaluating energy and resource efficiency investments is using payback. Unfortunately, there are significant problems with this method, and its use should be limited to the assessment of risk rather than the evaluation of return.

Simple payback, often referred to just as payback, describes the length of time needed for the savings generated by a resource efficiency project to return the initial investment made. It is calculated using the formula:

$$\text{Payback}_{\text{years}} = \frac{\text{Initial Investment}}{\text{Annualized Savings}}$$

Thus, an investment of US$1,000 that yields regular annual savings of US$500 has a payback of two years. The annualized saving could be determined by a number of uneven cash flows, e.g. US$250 in year 1 and US$750 in year 2; here, the payback period is determined by adding up successive savings in the cash flow until they match the investment. If we are interested in the payback in months, then we would multiply the annual payback by 12.

A fractional payback is only strictly correct where our savings are continuous throughout the year (e.g. if I install new energy-efficient lamps the saving occurs every day the lighting is used). On the other hand, if the saving results from a discrete payment at the end of the year (e.g. a waste collection charge that is levied annually), then the payback period will be a precise number of years, even if the final saving exceeds the amount needed to cover the investments. Thus, if I have invested US$1,000 and this results in savings of US$600 at the end of year 1 and US$600 at the end of year 2, strictly speaking, the payback is two years, not 1.7 years as it won't be until year 2 that the investment is repaid.

The saving is a net figure, that is to say, interest or other finance charges are included, unlike most other most methods of financial appraisal. We should also allow for tax in our savings. In other words, if these savings result in an increase in profit which is taxed, we would usually use the lower after-tax cash flows that are created by the investment. Alternatively, if we have a tax benefit - such as an enhanced capital allowances or depreciation - as a result of our investment, the net savings should be increased by the tax benefit.

Payback is a measure of **risk**, not of return.

Despite its widespread use, payback is a very poor method of evaluating investments in energy or resource efficiency projects and is only really suitable as a high-level indicator of project risk rather than as a method of financial appraisal. Payback may be helpful when we have a return on investment of less than a year, where other measures of financial performance such as IRR are unsuitable, but should otherwise be avoided.

17.5 Accounting rate of return

Accounting rate of return is a variant of payback which shares many of its disadvantages and so has limited value in assessing the financial performance of alternative investments.

The accounting rate of return is the inverse of the payback calculation, with the savings as the numerator and the investment cost as the denominator:

$$\text{Accounting Rate of Return}_{\%} = \frac{\text{Annualized Savings}}{\text{Initial Investment}}$$

Thus, an investment of US$1,000 that yields regular annual savings of US$500 has an ARR of 50% (equivalent to a simple payback of two years).

The annualized saving could be determined by a number of uneven cash flows, e.g. US$250 in year 1 and US$750 in year 2; here, the annualized savings are determined by adding up successive savings and then dividing them by the number of years over which they have occurred.

As with payback, the saving is a net figure, that is to say, interest or other finance charges are included in the figures. We should also allow for tax in determining our savings.

ARR has the same basic flaws as payback since it does not take into consideration the timing of the cash flow or the duration of the project. As a consequence, ARR cannot be recommended as a tool for the financial appraisal of alternative efficiency investments.

17.4 **Payback does not measure the return on investment, only the speed of return.** *Source Niall Enright, drawn using Pixton. Image available in the companion file pack.*

17.5 **Illustration of two projects with the same payback**
Although the payback for both projects is identical, Project B is more attractive than project A.
Source: Niall Enright. This cash flow model and those on the following pages are available in a spreadsheet in the companion file pack.

Payback is the single most commonly used investment appraisal technique in energy and resource efficiency, and yet it is **manifestly inappropriate** for this purpose.

Exploration: ***What's wrong with payback (and accounting rate of return)?***

Consider the two projects below. They both have identical payback periods of five years (and the same ARR of 20%). Clearly, project B is much more desirable as it makes significantly greater savings earlier on, whereas project A defers almost all the savings until much later on. Payback is unsuitable for comparing investment options because it fails to consider the timing of the savings within the payback period.

£	Project A	Project B
Initial Investment	£1,000,000	£1,000,000
Year 1 Savings	£1,000	£934,000
Year 2 Savings	£5,000	£50,000
Year 3 Savings	£10,000	£10,000
Year 4 Savings	£50,000	£5,000
Year 5 Savings	£934,000	£1,000

Payback (years)	5	5
Accounting Rate of Return	20%	20%

Now consider the two alternative investments below:

£	PROJECT A	PROJECT B
Initial Investment	£1,000,000	£1,000,000
Annual Saving	£200,000	£250,000

Duration (years)	25	4
Payback (years)	5	4

It would appear that project B with a shorter payback of four years is more attractive than project A. However, project B has a duration of just four years, so it adds no value at all to the organization as it only just covers its costs. Project A, on the other hand, continues to deliver savings for 20 years after it has paid for itself. Clearly, Project A is the better investment choice, even though it has a longer payback. This example forms the basis of the cartoon opposite.

Payback is unsuitable for comparing investment options because it fails to consider the duration of the savings beyond the payback period.

However, the payback does provide a measure of the risk of a project. Clearly, a project which pays for itself quickly is has less uncertainty around it than one which takes a long time to return the initial investment. Given two projects whose financial performance is similar, then payback can help us make the final decision.

Don't just take my word for it. According to *Capital Budgeting:*[187]

> *"Payback is a very unsophisticated and misleading measure and it is not recommended for accepting or rejecting projects."*

There are always ***two cash flows, business as usual*** *and* ***investment.***

17.6 Working with cash flows

Expenditure that is intended to add to the future value of the organization is called an investment. Evaluating an investment involves an assessment of the future value that is gained and the probability (or risk) that this value will materialize. This involves creating cash flows.

An investment acquires an asset or modifies a process which provides a future cash flow (such as the rental income on a house) or appreciates in value (such as a painting or precious materials). Most business cases for efficiency investments are determined by savings (a future cash flow) from reducing operating costs. The business case for a lighting upgrade, for example, is based on the lower volume of electricity used *and* the reduced effort replacing lamps.

When we talk about a cash flow assessment of the business case for investment, we are actually considering the incremental cash flow which is the difference between the predicted cash flow in the existing, do nothing, business as usual base case and the modified cash flow that arises if the investment takes place.

US$	2016	2017	2018	2019	2020
Current cash flow (BAU)	100,000	100,000	100,000	100,000	100,000
Cash flow with investment	320,000	20,000	20,000	20,000	20,000
Incremental cash flow	-220,000	80,000	80,000	80,000	80,000

In the simple example above, we have a process which uses US$100,000 of resources each year. We can invest US$300,000 to increase the efficiency by 80% and reduce our annual cost to US$20,000. In this example, the investment and the savings are assumed to both occur in the first year, so we spend US$300,000 on the equipment and only US$20,000 on resources.

The incremental cash flow is shown in the third row, which is the cash flow without the investment less the cash flow following the investment. So our business proposition is whether we invest US$220,000 *net* in order to achieve annual savings of US$80,000 each year for four years at least.

We will often see a business case with just the incremental cash flow stated. This is fine, as long as this is *genuinely* the incremental cash flow, that is to say, that BAU and investment cash flows have been properly considered.

Sometimes there is no obvious BAU cash flow, say, when we are investing in building a new factory. Assuming the broad investment case is made, our resource efficiency business case should be made on the basis of the *additional* value that higher-specified equipment will offer compared to the *committed* lowest-cost equipment using the marginal expenditure basis described next.

Real World: *Business as usual*

Every business case is based on two cash flows:

- The business as usual (BAU) cash flow if the investment did not take place; and
- The Investment cash flow that comes about when the investment is made.
- The business case is based on the incremental cash flow, that is to say, the BAU cash flow *less* the Investment cash flow.

Typically, an incremental cash flow will have a big negative value at the beginning (representing the cost of the initial investment), but will then have positive values for subsequent years, representing the savings that are helping to pay back the investment.

17.7 Marginal return on investment

An absolutely critical part of a resource efficiency practitioner's investment armoury is the ability to leverage existing investments that the organization has planned. Where there is an existing commitment this becomes the BAU case against which higher efficiency systems should be evaluated.

Some expenditure is inevitable. For example, we may be replacing failed equipment or we may need to make an investment to satisfy our licence to operate (e.g. install some equipment to achieve environmental compliance), or we may already have decided to build a new factory for strategic reasons.

A business case is always a comparison between a BAU case, which is the future cash flow to which the organization *is already committed*, and the altermative cash flow which arises as a consequence of the investment. Thus the base case cash flow should always include all committed expenditures.

Let us imagine that we need to replace a failed electric motor. The cheapest model that delivers the required performance costs £10,000. However, we have an alternative of buying a more efficient motor for £12,000. If we did so, we would save £1,000 per year in electricity costs due to the lower electrical consumption of the more efficient motor. The question is how long will it take to recover our investment?

Funding

	BASIC MOTOR BAU Case	EFFICIENT MOTOR Marginal case
Motor Installed Cost £	£10,000	£12,000
Efficient Motor Additional Cost £	n/a	£2,000
Annual Saving £	n/a	£1,000
Full Cost Payback (years) - high risk	n/a	12
Marginal Payback (years) - low risk	n/a	2

If expenditure is **committed,** always assess the return on investment on **a marginal expenditure basis.**

Far too often the answer given is 12 years; the full cost of the motor is divided by the annual saving. In this case, the investment in the efficient motor appears to have a high level of risk. However, this is not the correct way to assess the investment return of our high-efficiency motor. The incremental cost over and above the BAU case is only £2,000 and so the investment will pay for itself in only two years. Our decision is not "*should we buy a £12,000 high-efficiency motor*" but rather "*should we invest an **additional** £2,000 in order to generate a cash flow of £1,000 a year?*"

Organizations often have large amounts of committed expenditure and the most cost-effective efficiency improvements frequently lie in making marginal investments on the back of this planned expenditure.

Most investment decisions about energy and resource efficiency involve making a choice between efficient and inefficient options within ***investments that are being made for other purposes.***

17.6 **CAPEX=Opportunity**
Source Niall Enright, drawn using Pixton. Image available in the companion file pack.

Real World: ***Marginal funding works!***

A great example of the use of marginal investment cases is the famous refurbishment of the Empire State Building in New York managed by Jones Lang LaSalle, with Johnson Controls providing the technical expertise.

Here, the allocated capital was increased if it could be demonstrated that there was a good return on the additional investment.

The final table in the 2010 article by Dana Schneider and Paul Rode shows how an incremental investment of US$13.2 million provided additional annual savings of US$4.4 million.

There are several challenges with developing a marginal case for investment on the back of planned capital expenditure. First of all, is the fact that projects have budgets. Project managers and financial decision-makers may simply refuse to increase these, no matter how attractive the marginal investment case is. Indeed, the opposite effect often happens when projects come under financial pressure; *value engineering* results in non-critical expenditure being reduced, resulting in cheaper, less efficient equipment being installed.

Another problem is that of split incentives. In some areas, such as property, the developer of the asset sells it on when it is completed. Because they don't retain the asset, the developer will not gain the benefits of the greater efficiency and so there is simply no marginal business case for investment (unless the increased efficiency can be translated into a greater asset attractiveness and so a greater asset valuation, for example, by obtaining a good Energy Star, LEED or BREEAM rating). Where a building has been pre-sold, or pre-let and these split incentives exist, it is often worthwhile for the developer to engage directly with the buyer or tenant, who may choose to make the additional investment themselves. I have seen this on several occasions at Peel Land & Property Group, for example, on a major construction project for the healthcare group Bupa.

Real World: *Rio Tinto and CAPEX*

An interesting and rewarding recent project was with the global resources giant Rio Tinto. It was looking to provide some corporate support for energy efficiency and I helped the team pilot an approach based around using maturity matrices as a tool to enable facilities to develop its energy efficiency strategies. This was to result in a programme proposal for several US$ million to roll the process out, along with some standards for key technologies etc.

However, during this assignment, I became aware that Rio Tinto's approved CAPEX budget, at the time, was US$37,000,000,000 (yes, US$37 billion!)[622]

This is an example where the CAPEX budget commitment of the organization is so large that an immediate focus on marginal investment opportunities is called for.

Thus, my recommendation to the Rio Tinto energy efficiency team was that the priority should be to address the decisions being made around this allocated CAPEX. If not properly informed, these decisions could lock in inefficiencies for years to come, whereas the current operational savings opportunities could be addressed at any time. This approach was advisable because the CAPEX was already approved.

One of the first actions in every resource efficiency programme is to consider committed CAPEX. Changes to the design and procurement process to incorporate concepts such as whole life costing and marginal investments can unlock considerable sums for resource efficiency - sums which can sometimes dwarf those available for Optimize savings efforts. So don't assume every programme needs to start with Optimize!

The key message here is that most investment decisions about energy and resource efficiency involve making a choice between efficient and inefficient options within investments that are being made for other purposes.

While the programme Champion and the resource efficiency team may be dashing around developing proposals for investment in efficiency opportunities, it is important to understand all the other investments in plant, processes and assets that the organization is making. It is often the case that the decisions around these investments may be much more significant in efficiency terms than the opportunities for dedicated investments the teams are developing (as the case study from Rio Tinto, left, describes).

Indeed, many efficiency technologies, such as heat recovery systems, are much more costly to retrofit to equipment than to include in the initial construction of a plant or building. Failure to consider these options in the initial design may rule out the opportunity in the future, as the retrofit costs are too great.

It is my experience that detailed decision-making around these larger investments can be jealously guarded by those involved. The suspicion by the decision-makers is that the core objective of the investment - whether that is to increase production, to deliver greater services, or to enhance competitiveness - will be diluted by other considerations such as resource efficiency. This defensive attitude can be quite harmful to the organization, for we have seen that resource efficiency is usually material to future value.

There are a number of solutions to this problem. We could:

1. Require a whole life cost approach to the selection of design alternatives. This ensures that decisions are not just based on the up-front capital costs, but on the operating expenses over the whole life of the project, where more efficient solutions will perform better.
2. Introduce a rule that for significant resource-consuming equipment alternative efficiency options are considered (a high-efficiency model and the cheapest model, as a minimum). This is a requirement of ISO 50001.
3. Provide an additional budget for marginal investments over and above the agreed project budget, where the investment exceeds an agreed return on investment or hurdle rate.

The whole life cost need only be considered for those items where there is a significant efficiency effect. Thus, the selection of marble paving in an airport concourse can probably be done on the basis of up-front cost alone, while the selection of the motors powering the baggage conveyors should be done on the basis of the whole life cost (an average motor uses its capital cost of electricity in just 10 weeks!). It should be emphasized that whole life costing, is the only method that properly captures the value of an investment because it fully analyses the resulting cash flow. It should be the norm, rather than the exception.

17.8 The three parts of a cash flow

A whole life cash flow can have three elements. First, there is the investment, then there is the operating cash flow and, finally, we may have the terminal cash flow when the asset is disposed of. These may not always be required in our business case and are not always clearly separated in any event.

	£	2020 (Y0)	2021 (Y1)	2022 (Y2)	2023 (Y3)	2024 (Y4)	2025 (Y5)	2026 (Y6)	2027 (Y7)
Investment	Capital investment (fully installed cost)	-300,000							
	Indirect costs (50%) (treated as OPEX)	-150,000							
	Total Investment (A)	-450,000							
Operating	Current cash flow (BAU)	100,000	100,000	105,000	100,000	100,000	100,000	100,000	100000
	Cash flow with investment	100,000	20,000	20,000	20,000	20,000	20,000	20,000	20,000
	Net operating cash flow (B)	0	80,000	80,000	80,000	80,000	80,000	80,000	80,000
Terminal	Proceeds from selling old plant								100,000
	Dismantling and other sales costs								-50,000
	Net terminal cash flow (C)								50,000
	Incremental flow (A+B+C)	**-450,000**	**80,000**	**80,000**	**80,000**	**80,000**	**80,000**	**80,000**	**130,000**

17.7 **The three parts of a cash flow**
Source Niall Enright, spreadsheet available in the companion file pack.

The analysis above illustrates the three parts of a whole life cash flow. There are the costs associated with the initial investment. Then there is the net operating cash flow (that is to say the difference between the BAU cash flow and the post-investment cash flow). Finally, there is the terminal cash flow in the final year of the project. As it takes a project from *"cradle to grave"*, this form of investment appraisal is also called whole life costing.

Key to the success of our business case is how credible it is. In the following pages, we shall examine each of these three parts of the cash flow with the intention of making them as plausible and accurate as possible. In the example above, the investment case incorporates a line for indirect costs which is meant to reflect the overhead and disruption procuring and installing a new piece of equipment. In the terminal cash flow, we haven't just included the sale price of the equipment but the costs associated with the sale.

Use whole life costing for all but the most trivial of investments.

It is often the case that the terminal value is not considered in resource-efficiency cash flows at all, simply because the equipment has zero end-of-life value and the costs associated with disposing of it are incorporated in the business case for the replacement. Another reason why terminal values may be ignored is when the cash flows are discounted and the gain from disposal is so far in the future that the discounted value is negligible.

Standards: *Life cycle costing*

The nearest we have to a standard for whole life costing is a life cycle costing (LCC) guide for buildings and assets: BS ISO15686-5:2008. The differences between whole life costing and LCC are illustrated in the following diagram:

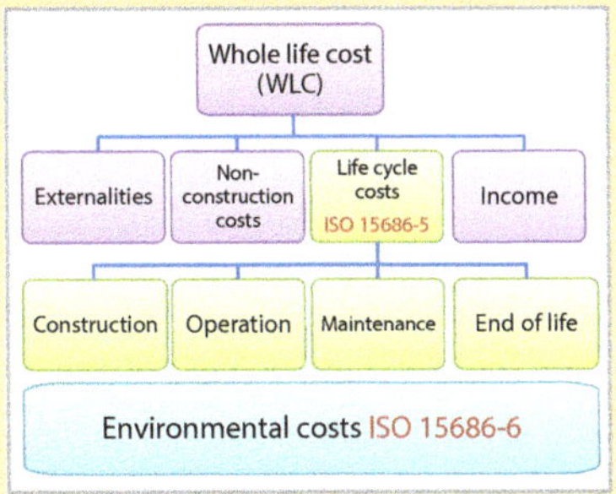

Whole life costing takes into account the costs of externalities, income, and non-construction costs while LCC is focused on the physical assets. The aspects covered by LCC are shown in green, above.

If you follow the methods set out in this chapter as required, (such distinguishing the three phases of a project life, discounting, NPV, IRR, annual equivalent cost and sensitivity analysis), your financial analysis will fully conform to the ISO standard. Indeed, the standard is not particularly prescriptive as to *when* each technique should be employed.

One useful aspect of the standard is a checklist of the major cost items that may need to be considered under the headings construction, operation, maintenance and end of life.

In practice, though, an LCC is not usually carried out on all aspects of a building, but rather to compare the relative merits of alternative solutions for major items of equipment in the building. In this case, guides like the BSRIA guide[142] or the US FEMP guide[298] may be more useful.

Exploration: *Whole life costing*

Whole life costing is simply common sense. For example, as an individual you would not buy a car without thinking about what it costs to run and what you might get when you come to resell it. The importance of whole life costing is summed up by the 1:5:200 rule of thumb, which is taken from a Royal Academy of Engineering report[269] on the relative cost of construction, maintenance and operation of buildings over their lifetime. Indeed, we can add 0.1 at the front for the design cost. The message here is that the construction costs of buildings pale into insignificance with the operational costs (although these precise figures have been challenged,[382] the overall message remains true).

And yet many investments - and I am thinking in particular in buildings and infrastructure assets - end up being made on the basis of the lowest initial cost, regardless of operational efficiency. This narrow capital budgeting viewpoint arises because once an outline business case is made for the asset and an initial budget decided, all efforts are focused on getting this cost down as low as possible, regardless of the longer-term consequences in terms of operational costs. This is exacerbated by the split incentives that arise when a building is to be sold immediately on completion so the cost of inefficiency is borne by someone else.

In this framework, whole life costing is the only methodology proposed. Where this is not the current approach in your organization, then this is one of the system changes that should be explored as a matter of urgency.

It should be emphasized that whole life costing is a methodology for the consideration of alternatives. We have already stated that one alternative is BAU, but we should be looking at multiple alternatives where we have to make choices about large resource-consuming equipment or processes. For example, we could heat a space with a boiler (of which there may be several designs available), or use heat pumps or solar hot water. The methodology adds greatest value where there are multiple options available.

In practice, many whole life cost assessments of technical alternative such as the ones described above are actually life cycle cost assessments (see left). BSRIA's excellent guide to LCC[142] by David Churcher and Peter Tse, illustrates the highly dynamic nature of the process, with a constant reassessment of data and assumptions throughout. In the previous (2008) version of the guide, Step 4 was labelled "sensitivity analysis", to emphasize the importance of this technique.

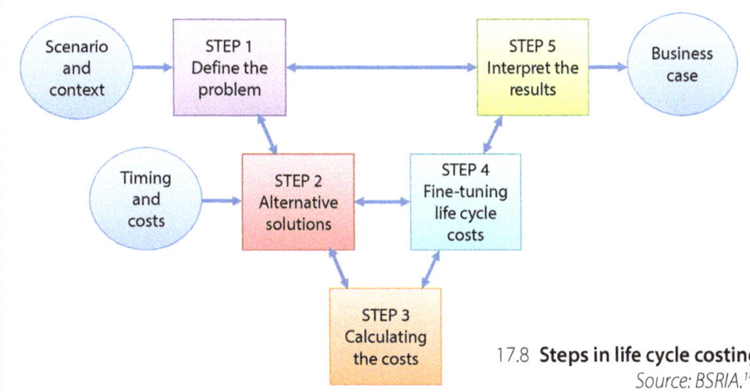

17.8 **Steps in life cycle costing**
Source: BSRIA.[142]

17.9 The investment cash flow

The first part of our business case is to assess the investment needed. Here we need to ensure that we have properly quantified the fully installed costs as well as indirect costs, such as those due to disruption. If this is an option, we will need to decide if we book savings to the same year as the investment. Finally, there is a note of caution about the treatment of past cost.

Real World: *Optimism bias in the UK*

A study by Mott McDonald for the UK Treasury[536] examined the procurement of major public projects in the UK.

It concluded that optimism bias, the tendency for a project's costs and duration to be underestimated and/or benefits to be overestimated, is widespread.

% Bias (expressed as an overrun)			
Duration	*CAPEX*	*OPEX*	*Benefits*
17	47	41	-2

While this study included notoriously badly managed areas of procurement such as defence equipment, it should be noted that even straightforward projects such as "standard buildings" and "standard civil engineering" exhibited 24% and 44% average levels of optimism bias in respect of the CAPEX (i.e. the investment cash flow). In the case of building projects, the largest contributor to the bias arose from inadequate business cases and the second largest effect was due to disputes on the contract (this was the biggest issue for civil engineering projects).

The study points out that these figures have improved as a result of better procurement practices. It demonstrates that the quality of business cases is important, as is the ability to anticipate risk (e.g. disputes on contracts).

An important way to gain credibility for an investment case is to take into account all the hidden and missing costs which often undermine the proposal. The most common mistake in pricing an efficiency investment is to use the capital or purchase cost of the equipment as the sole basis of the cost. In this case, a reasonable estimate of the total investment cost is double the capital cost (or more if the equipment is complex). Even if we have been given a *fully installed* cost, the indirect costs of procurement, disruption, engineering, commissioning and training can add another 50% or so to the installed cost (see Figure 4.9 on page 174).

When thinking about the investment case at any moment in time, it is important not to let historical investments influence your judgement. The presence of past sunk costs can lead to irrational decisions. First, having committed to an investment decision, those responsible tend to develop a more optimistic outlook on the likely success of their decision.[34] This danger applies to our programme Champions, too - having espoused a proposal we should recognize the risk that we *"talk it up"* excessively and so diminish our own credibility in the long run if the benefits are less than we promoted. Second, having made an investment, this personal responsibility towards the idea or projects can lead decision-makers to exhibit poor judgement in respect of future decisions. What results is an over-commitment or bias towards existing projects, a form loss aversion, the sense that "*we have gone too far to turn back*", which can result in irrational decisions that lead to "*throwing good money after bad*".

Sunk costs are a very real barrier to energy and resource efficiency investments as they extend the capital replacement cycle for equipment and infrastructure. By rejecting economically sensible choices in areas where we have invested heavily in the past, our opportunities for improvement are diminished.

Classical economics states that there should be no weight given at all to previous investments. If last year I invested £1 million on a lighting upgrade project and then this year a new lighting investment opportunity comes along which delivers an acceptable return on investment, then theory says I should make that investment even if it means that I have to rip out the almost-new system. So why do people behave *irrationally* and reject the investment? In the first place, there is the not insignificant matter of the embarrassment that would be caused by such a decision. Preston McAfee and his colleagues point

Understand which costs are treated as capital and which not.

out[504] that, as well as the reputational considerations just mentioned, there are other reasons why considering sunk costs may actually be rational. Firstly, they argue, the information content around the sunk cost makes it more attractive than the alternative. Also, financial and time constraints faced by individuals cannot be ignored. Harking back to our earlier discussion on certainty and the decision to support a proposal (page 185), the sunk cost project may represent a "*better the devil you know*" choice, which is known to be imperfect, but has fewer uncertainties than the new investment.

In its report[606] on barriers to energy efficient lighting, accounting firm PWC also contradicts the established notion that sunk costs should be ignored. It argues that, although the cost was in the past and nothing can change that, the original investment was made with the intention that these costs would be recovered over time and so the outstanding balance of the sunk costs should be factored into the future decision. Not to do so would be to ignore the obligation on the part of the lighting operators to pay for the historical costs they incurred. Also, a failure to recover the sunk costs would call into question future investments, as this signals that the recovery of the new investment, too, could be abandoned at some point. This rationale is very much a financier's view of sunk costs, rather than an economist's, but one that I have encountered on several occasions.

Where this view exists, we may have no other choice than to add a *recovery charge* to our initial Investment or to the operating cash flow for the purposes of determining if the new investment still meets the organization's hurdle rate, even though this is an entirely imaginary cost and one which the literature say we should be ignoring.

Finally, we need to ensure that CAPEX, is treated correctly. Earlier, in Section 4.8 on page 189, we spoke about a lack of availability of CAPEX as a barrier to investment. There are often special considerations in respect of CAPEX, such as the balance sheet effects, depreciation and tax treatments that can materially affect a business case (see later discussion on page 578). For this reason, it is important that expenditure is correctly allocated; for example, some organizations may allow software to be capitalized but not consultancy costs setting it up. It may be the case that our initial investment cash flow will have two lines, one for the capital costs and another for other, non-capital, costs.

If the investment cash flow alone is considered, then the analysis may be called a capital appraisal. This form of investment decision-making usually occurs in the context of a project with a defined budget (e.g. constructing a building or factory) where the overall business rationale has been made previously and the objective now is to deliver ahead of time, on specification and under budget. In this situation, it still pays to take decisions on resource-consuming equipment on a whole life basis, if at all possible, and so develop the operating cash flow as described next, which will help justify investment in efficiency over and above the lowest-cost option meeting the specification.

17.10 Operating cash flow

Following the initial investment, we need to develop the operational cash flows from which the benefits of the investment can be assessed. There are a number of important choices to be made which could affect the variability of the investment and the credibility of the business case.

Now that the investment cash flow has been quantified, we need to consider two operating cash flows, the BAU cash flow and the cash flow following the investment.

The first important decision that will need to be taken is over how many years this cash flow will be considered.

The five principal aspects that can influence the term or length of a whole life costing assessment are:

- The operating lifetime of the equipment or asset (e.g. the anticipated capital replacement cycle for the asset).
- The commitment period of the organization to the activity (for example, a building may only have a certain number of years left on its lease, in which case the business case will be based on the shorter of the operating lifetime and the commitment period).
- The time frame (known as the tenor) of any financing being made available. In this case, the business case needs to analyse the risk and rewards over the same period to ensure that the financing is viable.
- The discount rate being used. We will discuss this in more detail later, but it is important to note here that a large discount rate gives much greater weight to the cash flow in the short term than in the long term, which may shorten the length of time over which the analysis need be undertaken.
- The period needed to make comparisons valid. Some alternative technologies have unequal lives. In this case, a replacement chain for the projects may be developed such that the lowest common multiple of the alternatives is achieved. For example, I have an option, A, with a two-year life and another, B, with a three-year life, I would compare these using a six-year cash flow which included a replacement reinvestment of project A on three occasions and project B on two occasions.

Select the **period of the analysis** with care.

Where the length of the whole life assessment is smaller than the operating lifetime of the equipment, the issue of a residual (or salvage) value in the terminal cash flow naturally becomes more significant.

In Numbers: *Annual equivalent costs*

Sometimes we may want to treat the initial investment as if it was part of operational cash flow, to get to an annual equivalent cost, for example, as an alternative to a replacement chain when making comparisons.

Imagine that I want to assess two building designs with components with different lives. The building costs US$50 million and lasts 50 years, the furnace has a life of 25 years and costs US$10 million, and a furnace lining costing US$1 million needs to be replaced every four years.

If I am not discounting my cash flow, I divide the costs by the lifetime of the item of equipment so that the annual equivalent cost of the building is US$1 million per year (US$50 million / 50 years), the furnace is US$400,000 a year and the furnace lining US$250,000 a year. The total annual equivalent cost is thus US$1,650,000, which I could use to compare another building with different equipment and different lifespans.

As we shall see shortly, it is common to discount future cash flows. If we are doing a discounted analysis, I can use the table on page 790 to look up the capital recovery factor (CRF) by choosing the appropriate discount rate.

If I assume the discount rate is 8% then the CRF is 0.0817 for 50 years, so my annual equivalent cost of the building is US$50,000,000 * 0.0817 = US$4,085,000 a year. Similarly, the cost of the furnace is US$10,000,000 * 0.0937 = US$937,000 and the cost of the furnace linings is US$1,000,000 * 0.3019 = US$319,000.

These costs are also called the levelized cost of capital, not to be confused with levelized cost, the price of energy needed to repay the initial investment in the plant.

Another important decision regarding the timescales of the financial analysis is whether savings are booked in the same year as the initial investment. Traditionally capital investment is treated as a separate Year Zero cost. If a choice on this is possible (many organizations have standard investment models where this is fixed), it needs to be made with care as this decision can make the difference between a viable investment case and an infeasible one.

The example below is the traditional Year Zero, denoted by Y_0, investment model, where the investment has been assumed to be at the end of Y_0. Do not be confused by the naming of "*Year*" Zero - it doesn't have to be a year, it is there to convey an indeterminate amount of time in advance of the first year of operation of project. In this model, savings have no impact on the initial investment costs in Y_0, as they are said to occur at the *end* of each year, starting in Y_1.

	£	2020 (Y0)	2021 (Y1)	2022 (Y2)	2023 (Y3)
Investment	Capital investment (fully installed cost)	-300,000			
	Indirect costs (50%) (treated as OPEX)	-150,000			
	Total investment (A)	-450,000			
Operating	Current cash flow (BAU)	100,000	100,000	105,000	100,000
	Cash flow with investment	100,000	20,000	20,000	20,000
	Net operating cash flow (B)	0	80,000	80,000	80,000
	Incremental flow (A+B)	**-450,000**	**80,000**	**80,000**	**80,000**

This separation of the initial cost and the benefits of the resource efficiency investment may hold true for many significant investments where the lead time from decision to completion of the project could well be a year or more. However, for many smaller projects, such as lighting upgrades, this is not realistic and deferring the benefits can undermine the attractiveness of the investment. In this case, it may be appropriate, indeed correct, to model the investment and savings in the same year, Y_0, as illustrated below. In the example below, the Y_0 savings shown are the full £80,000 per year but there is no reason why I couldn't adjust this to represent a saving of, say, £40,000 to incorporate six months of savings in Y_0 by making the cash flow with investment £60,000, not £20,000.

	£	2020 (Y0)	2021 (Y1)	2022 (Y2)	2023 (Y3)
Investment	Capital investment (fully installed cost)	-300,000			
	Indirect costs (50%) (treated as OPEX)	-150,000			
	Total investment (A)	-450,000			
Operating	Current cash flow (BAU)	100,000	100,000	105,000	100,000
	Cash flow with investment	20,000	20,000	20,000	20,000
	Net operating cash flow (B)	80,000	80,000	80,000	80,000
	Incremental flow (A+B)	**-370,000**	**80,000**	**80,000**	**80,000**

Real World: *Marginal cost*

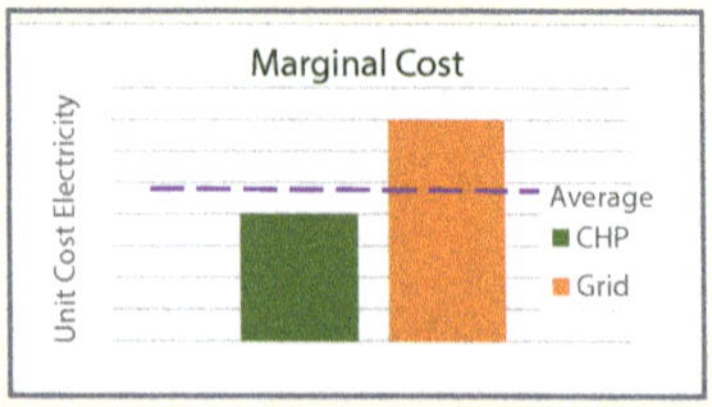

My colleague Arne Springorum and the team at HEC are leading an energy efficiency programme at ŠKODA AUTO. The main production site at Mlada Boleslav in the Czech Republic has its own power station which provides cheap electricity, which has the advantage of being lower carbon since the waste heat from the power plant is also used.

This power plant does not meet the entire needs of the site, however, so additional electricity is purchased from the grid at the standard rates for large industrial firms.

When valuing the electrical savings, the price that ŠKODA AUTO should use is *not* the average price of the electricity paid for by the site, but rather the higher price of the grid electricity that is imported (in orange above).

Any reduction in demand for electricity from the energy efficiency projects will *first* lead to a reduction in imports from the grid since CHP is the preferred supply. Hence the marginal cost is the grid import price, not the average cost.

This is a reason why declining tier cost - i.e. cheaper rates for higher volumes - for electricity use discourage efficiency measures. The marginal cost of the electricity is lower than the average cost, which makes a justification for investment harder. See *"Rising block tiers (page 164)* for how some regulators are addressing this barrier.

Thus, showing savings in Y_0 can be perfectly reasonable. In some cases it may be necessary to align the implementation of the project to match the start of an organization's financial year to maximize these Y_0 savings and so make a viable and more credible business case. Because of the need to consider Y_0 savings, especially for smaller projects, all the business case models provided in the companion file pack all have a *Return in Year 0* option.

Having established the timing and number of years over which we are going to develop the cash flow model, we now need to incorporate the individual cash flow items.

Since many business cases for resource efficiency depend on the cost savings arising from reduced resource use, a key area to get right in these calculations is the pricing of the savings. All cash flows should value the resource saving using the marginal cost of the resource, i.e. the actual value of the resources saved. Often the marginal price is higher than the average price paid, as illustrated by the example, left, so it is important to get this detail right.

The next aspect that we need to consider is the additional benefits of the investment, over and above the resource savings. These co-benefits can sometimes exceed the resource savings in terms of value and often make the difference between success and failure in a business case. An example of co-benefits that are commonly missed is in light emitting diode (LED) lighting upgrades which can more than double the lamp replacement cycle time, which reduces the cost of lamps and cost of labour in fitting the replacement lamps, which when combined usually exceed the value of electricity savings.

This point cannot be emphasized enough. If we look at the figure opposite, we can see the many co-benefits that arise from energy efficiency over and above the direct resource costs. Sometimes, these co-benefits can be quantified financially, such as savings on emissions or savings on electricity network upgrades to meet increased demand, in which case they should appear in the cash flow. In other cases, these co-benefits have broader brand and societal impacts which, while not being quantified in cash flow terms, are important elements in the narrative of the business case.

There are numerous cases where these co-benefits can become the overriding reason for the investment. For example, the *raison d'être* of many utility-led energy efficiency programmes is demand-side management. By reducing demand, they don't have to invest in supply and save money because the marginal cost of providing another watt of capacity is greater than the cost of delivering a *"Negawatt"* of demand reduction. In the earlier example of the BP energy efficiency programme, the rationale for the business unit leaders crystallized around greater reliability through a form of condition-monitoring delivered by a better understanding of energy use. The carpet maker, Interface, like many other organizations, has based its unique selling proposition on being the most efficient and sustainable producer in its category. Peel Land & Property Group drives its energy efficiency programme because reducing energy costs makes Peel's properties much more competitive to discerning tenants.

Always use the true or **marginal cost** of resources saved.

There are some expenditures which should *not* appear in the cash flow analysis:

- Financing costs, such as interest payable on debts to implement the project, are usually omitted, This is because the purpose of the whole life cost assessment is to determine the intrinsic return of the investment, compared to other possible investments. That is not to say that financing will not affect the ultimate decision to invest, far from it, but there are so many aspects to this decision that it should be kept separate from the project appraisal.

- Overhead costs, such as an allocation of central staff time to manage the project, are usually omitted as these are already committed costs which will be borne regardless of the decision to proceed. On the other hand, if new people need to be employed as a direct consequence of the investment then these costs would be included.

Increases in working capital may be included in the cash flow. For example, if I switch from a gas boiler to biomass boiler and I need to have a stock of fuel on hand, then the cost of this stock can be considered in the cash flow. The broad principle as to whether items are included or not is if they represent an opportunity cost - in other words, are resources diverted from other productive uses as a result of the investment decision.

One area that often creates confusion is taxation and the related area of depreciation. This will be covered next.

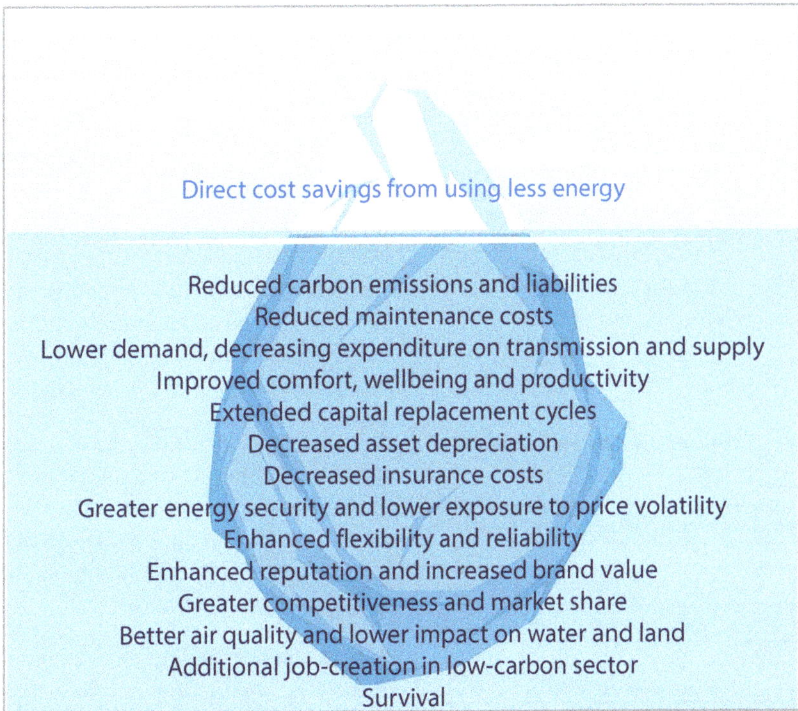

17.9 **The hidden value of energy efficiency**
There are many sources of value from energy efficiency, over and above the direct costs savings from lower energy use.
Source: Niall Enright, iceberg image © Oceloti - Fotolia.com.

17.11 Should tax be included?

Taxes can be complex. For many smaller investments it is simply not worth the effort to consider taxation implications, but where the investment is considerable and tax has a material effect, the credibility of our analysis can be enhanced by considering the after-tax cash flow.

	£	2020 (Y0)	2021 (Y1)	2022 (Y2)	2023 (Y3)	2024 (Y4)	2025 (Y5)	2026 (Y6)	2027 (Y7)
Investment	Capital investment (fully installed cost)	-300,000							
	Indirect costs (50%)	-150,000							
	Total investment (I)	-450,000							
Operating	Current cash flow (BAU)	100,000	100,000	105,000	100,000	100,000	100,000	100,000	100000
	Cash flow with investment	100,000	20,000	20,000	20,000	20,000	20,000	20,000	20,000
	Net operating cash flow (O)	0	80,000	80,000	80,000	80,000	80,000	80,000	80,000
Terminal	Proceeds from selling old plant								100,000
	Dismantling and other sales costs								-50,000
	Net terminal cash flow (T)								50,000
A	**Incremental flow (I+O+T)**	**-450,000**	**80,000**	**80,000**	**80,000**	**80,000**	**80,000**	**80,000**	**130,000**
Tax	Balance of capital investment (X)	*300,000*	*240,000*	*192,000*	*153,600*	*122,880*	*98,304*	*78,643*	*62,915*
	Depreciation (20% on balance) (X*20%) Y	-60,000	-48,000	-38,400	-30,720	-24,576	-19,661	-15,729	-12,583
	Taxable income (O+T+Y)	-60,000	32,000	41,600	49,280	55,424	60,339	64,271	117,417
B	**Tax (@20% of profits)(positive= tax credit)**	**12,000**	**-6,400**	**-8,320**	**-9,856**	**-11,085**	**-12,068**	**-12,854**	**-23,483**
	Incremental cash flow after tax (A-B)	**-438,000**	**73,600**	**71,680**	**70,144**	**68,915**	**67,932**	**67,146**	**106,517**

17.10 **Cash flows including tax (UK)**
Source Niall Enright, spreadsheet available in the companion file pack.

The cash flow model above incorporates depreciation and tax effects in a separate block below the incremental cash flow. This model includes a calculation of the tax on the additional profit that will be made because of the investment, to determine the post-tax incremental cash flow on the bottom line.

17.11 **Cash flow (Y_1) including tax shield (US)**
Source: Niall Enright

A	**Incremental flow**	**80,000**
B	Tax (@20% of profits)	-16,000
C	Tax shield	12,000
	Cash flow after tax (A-B+C)	**76,000**

If an organization is not making a profit or if it is exempt from taxation (for example, because it is a public sector body), these tax calculations may not be relevant. To complicate matters further, even where there are material tax effects, the project approval processes in some organizations will exclude these tax effects from consideration on the grounds that they are too complex for most investment decisions (just as financing aspects are often excluded).

If we are taking tax into consideration, then we need to understand what the tax rate is (the organization may have to pay less than headline rate of tax because of

Include tax *if there is a material effect.*

previous years' losses carried forward, for example). In the example, I have used the UK corporation tax rate of 20%, but this varies from country to country.

The second aspect of taxation that we need to understand is the treatment of depreciation of capital expenditure. Depreciation is a reduction of the value of an asset over time, in this case our newly purchased equipment's initial capital cost of £300,000. In the UK, you are typically allowed to depreciate equipment at 20% of the reducing balance each year and show this in your profit and loss account. Please note that this is not a *real* cash flow, but rather a notional entry for bookkeeping purposes when calculating taxable profits.

So, in the example given, the taxable income is reduced by the depreciation of the capital cost £300,000 at 20% of the remaining balance each year, i.e. £60,000 in Y_0. In the US, this calculation is slightly different: there would be a tax shield, which is simply depreciation times tax rate (which varies according to the type of capital equipment) which would be added back to the after-tax income, as shown in Figure 17.10, opposite. In many jurisdictions, incentives for investment in resource-efficient equipment are provided though advantageous tax treatments, in which case considering tax is unavoidable/desirable.

Real World: ***Enhanced capital allowances***

An example of a tax incentive for investment in energy-efficient equipment in the UK is enhanced capital allowances (ECAs). The purpose of this incentive is to encourage buyers to purchase higher-efficiency equipment, by allowing the capital cost to depreciated (i.e. be written off) over one year, rather than several years.

		£	2020 (Y0)	2021 (Y1)
A	**Incremental flow (BAU less investment)**		**-450,000**	**80,000**
Tax	Balance of capital investment		300,000	-
Tax	Depreciation (100% on eligible investment)		300,000	-
Tax	Taxable income			80,000
B	**Tax (@20% of profits)(positive= tax credit)**		**60,000**	**-16,000**
	Incremental cash flow after tax (A-B)		**-390,000**	**64,000**

17.12 **Cash flows including tax (UK)**
In this extract from the cash flow opposite, the effect of an enhanced capital allowance has been modelled, demonstrating that the initial investment appears more attractive because of the tax credit available.
Source Niall Enright, spreadsheet available in the companion file pack.

Taking the example opposite, if the equipment benefited from ECAs then the depreciation line would show the full capital costs of £300,000, as illustrated in the extract of the cash flow, for Y_0 and Y_1, shown left.

This sum can be set against profits, which will result in a tax saving (at 20% corporation tax) of £60,000, shown as a tax credit. This credit can be incorporated into the final, incremental cash flow, which means that the Y_0 costs are £390,000 compared with £438,000 without the ECA.

Note that in Y_1, the incremental cash flow after tax is lower that when the ECA was not available. This is because all the depreciation has taken place in Y_0 and there is no residual balance to be depreciated in the subsequent years. The ECA does not change the long-term cash flow, it just changes the timing of the tax benefits available, so it is essentially *neutral* overall, but is important when we come to consider the timings of cash flows.

In order to qualify for an ECA, the equipment purchased needs to be on the Energy Technology List[256] or meet Energy Technology Criteria. To be on the list, a technology or item of equipment needs be among the most efficient in its category. This brings us to another benefit of the ECA system - it provides an objective list of the most efficient equipment. As a consequence, many organizations, such as Peel Land & Property Group, specify that suppliers must state if their equipment is on the list and, if it is not, provide a justification for the lower-efficiency solution.

17.12 The terminal cash flow

Where an investment is being made to increase an asset value, then the terminal cash flow is the key part of the business case. The terminal cash flow is not just about additional value, but can sometimes represent a significant cost to the project (such as the decommissioning costs of a nuclear power station).

Most energy and resource efficiency business cases ignore the terminal cash flow as it is assumed that the equipment has no residual value and there are no special measures to be taken when it comes to the end of its life. Indeed, it is often the case that the costs associated with disposing of the existing equipment are incorporated into the financial analysis of the replacement project.

For most mechanical equipment the residual value is taken to be the scrap or resale value, although after a few years of use the resale value is usually nil. For buildings, the residual value may well be the value of the land less the cost of demolition of the building and remediation of any contamination. Building components such as roofing or paving may have a residual value in proportion to the remaining lifetime of the component. Thus, a roof which initially cost £1 million and has a serviceable life of 30 years, if sold after 20 years should be worth 1/3 of its original cost, i.e. £333,000.

We saw earlier that the remanufacturing businesses, such as Caterpillar, are working hard to ensure that the residual value associated with equipment at the end of the serviceable life remains attractive enough to encourage the owner to return the *"core"* to them (see page 62). In this case, the mechanism is a refundable deposit available when the parts are returned to Caterpillar for remanufacturing, which should exceed the scrap value of the part.

One particular area of technology where residual values are especially important is renewable energy technologies and in particular solar photovoltaic (PV) panels. Here, the serviceable life of the panels can be in the range of 25-30 years, and it is possible that the building or roof on which they are fitted needs to be redeveloped before the panels come to the end of their life, or the building is sold on. There are three ways that the residual value can be estimated.[505] An income basis looks at how much the panels can earn in the future: an approach which can give a superior result if there are existing subsidies linked to the panels and if electricity prices have risen. A replacement value basis considers how much would it cost to replace these panels with new ones with the same generation capabilities, which can reduce the apparent value since the cost per W of PV may reduce significantly over time. Finally, a market basis sees what the actual price is that buyers have been willing to pay. Despite these established methods, uncertainty around residual values has been a significant barrier to renewable investments.

Real World: ***Residual value of solar PV***

For some technologies, the question of a residual value is highly uncertain. Take, for example, solar photovoltaic (PV) panels.

I have 3.3 kW capacity PV fixed to my house (see photo above), but will I get my investment back if I sell the property? Will the buyer appreciate the PV panels? Will it translate into an increase in the sale price of the house? The jury is out on this at the moment.

You could argue that the point is moot since I could simply remove the panels when I sell and take them elsewhere. However, the PV installation works because the cost of the panels is subsidized through a feed-in tariff, which would be lost (or at least greatly reduced to the much lower current level of subsidy) if the panels were moved from this property to another.

The key decision that I, and other PV investors, have to make is: will I move from the property before I repay the investment (in this case seven years)? If the answer is yes, then there is a risk of making a loss on the PV system.

Exploration: *Liquidity, reversibility and risk*

Photo © R Timages Fotolia.com

Some investments, like gold, bonds or stocks and shares, that can be easily bought and sold are "liquid". By contrast, investments in efficiency usually involve technologies which are embedded in buildings or processes. Investments such as insulation or glazed windows or light fittings or heat exchangers are costly to remove and may have little or no resale value.[694] Because these investments cannot be easily sold once implemented, they are said to be illiquid or irreversible.

Clearly, the more liquid an investment, the lower risk that it represents to the investor. One reason that is often cited for the bias against energy efficiency investments is that they are illiquid. However, this argument does not fully stack up as it does not explain why efficiency investments are treated less favourably than similarly illiquid investments in areas such as production which similarly involve investments in equipment and processes.[669 p 57] The different treatment of production-enhancing investment and cost-reduction investment is more related to the value aspects of the investment, as discussed next.

Although our business case cash flows will be based on an anticipated project end date and the terminal cash flow at that time, we do need to address the liquidity of our investment as part of the discussion of risk for the investment.

We could, for example, identify the potential loss if the project comes to an end earlier than anticipated. Here, I would highlight the payback period of the investment and then estimate the likelihood of a termination event during this period, since this is when a loss could be incurred. In my analysis, I would state what the events are, how they could be avoided and what the residual cash flows might be available under the circumstances.

One of the common problems with manufacturing industry is that production can only reasonably be assumed for a relatively short time frame. An example is when I was working with Roche Vitamins on energy services companies (ESCO) financing for a number of large energy efficiency upgrades in its US production sites in the early 2000s. Here, there was a recurring stumbling block around the issue that the factories could only foresee production for a decade at most, more often only five years, yet the ESCO investors needed 20 years of operation for some of the equipment to deliver the required return on investment.

Another challenge created by irreversible investments is that of commitment delay. This issue is particularly pronounced in technologies such as LED lighting where the costs have been dropping rapidly (and the quality also increasing fast). These factors can lead to a deferral of investment on the basis that the price is likely to improve further (akin to what happens to general expenditure in an economy in a *deflationary spiral*).

By contrast, some energy efficiency investments, such as projects under the auspices of the Clean Development Mechanism, are designed to create highly liquid assets in the form of certified emissions reductions (CERs), which are tradeable emissions allowances. In this case, the investment decisions will take into account a whole range of factors such as the volatility, volume and price of the CERs, factors normally associated with more liquid asset classes such as shares.

It may be helpful to include an assessment of the risk associated with premature project termination, where the investment is considered illiquid.

Real World: *What discount rate?*

The discount rate represents the value - or cost - of money over time. At its simplest, we can think of the discount rate as the same as the rate of interest we might obtain if we put money in a safe investment such as a bank or government bonds, rather than invested in a project.

Another way of thinking about a discount rate is what money costs to our organization. This cost of money is a blend of the cost of borrowing and the expected returns for shareholders, who, by initially buying shares, provide capital. This cost of money is called the weighted average cost of capital (WACC).

The discount rate used in capital investment decisions may be greater than the WACC, because there are risks and overheads in implementing any project. The actual discount rate used is sometimes referred to as the risk-adjusted discount rate (RADR).

So what is a reasonable discount rate to use? Well, it varies greatly from sector to sector and between public and private companies. Here are some typical values.

Sector	Rate (%)
UK public sector[714]	3.5
US property (average)[125]	10
Global private sector[184]	4-12

Aswath Damodaran of NY Stern University provides some estimated cost of capital[184] for different sector and regions in the world, which can give a good starting estimate of the discount rate to use.

Public bodies and Institutions tend to use lower discount rates which result in longer-term investments being given a greater present value, i.e. being more favourable, than in the private sector.

17.13 Future and present value

Intuitively, we recognize that a dollar now, in our hand, is worth more than a promise of a dollar in five years' time. But how much more? Is it worth more than US$2 in five years? Here, we explore why the future value of money is less than the current value and how we can calculate that future value in our business cases.

We have already seen from the earlier examples on payback that two projects can have an identical payback even though the return on investment for one project is heavily front-loaded and the return for another is back-loaded. Clearly, getting the money we invested back early is more attractive as these funds can then be reinvested or distributed to shareholders. The reason why the longer money is tied up in a project, the less it is worth is because of the opportunity costs that this represents. If I didn't have my money tied up in an investment, then I could use it elsewhere, for example, by depositing that money in a bank and getting interest on it.

If my bank offers an interest rate of 10%, then if I invest US$1.00 today it will be worth US$1.10 in a year. If I reinvest that US$1.10, it will be worth US$1.21 at the end of year two and then US$1.33 at the end of year three. This is called the future value of money. The future value, or FV, of US$1.00 at 10% interest per year in three years is US$1.33.

Another way of looking at this is that a promise of US$1.33 in three years' time is worth the same as US$1.00 today, assuming an interest rate of 10% per year. We can say that the present value of US$1.33 delivered in three years at a discount rate of 10% is US$1.00 today. We call the interest rate used in present value calculations a *discount* rate because it is reducing or discounting the future cash flow to get back to today's value.

Just about every organization uses this notion that the value of money reflects the time when it is received when making significant financial decisions. Business choices are not based on the face value of the future cash flow stream but on the discounted cash flow. By discounting back to the present value the organization can compare alternative investments which may have very different timings of income and expenditure.

This simple notion that money in the distant future is worth less than in the present is true whenever there is a positive interest rate for funds deposited in banks or loaned, which is usually the case.

We shall consider the effect of inflation separately; for the moment we will work on what is called the nominal cash flow, that is to the actual dollar amounts without any adjustment for inflation. When calculating future value or present value we need to compound the effect of the growth or discount rate over several periods, as shown opposite.

It is present value that is of concern in a financial appraisal of an investment. as we are evaluating the worth of a future cash flow. Looking again at the simple payback for the two projects in Figure 17.4 on page 564, we can now apply a discount factor to get a discounted value as illustrated below.

£ using a discount rate of 10%	Discount Factor	Project A	Present Value A	Project B	Present Value B
Investment	*1*	-1,000,000	*-1,000,000*	-1,000,000	*-1,000,000*
Year 1 Savings	*0.9091*	1,000	*909*	934,000	*849,091*
Year 2 Savings	*0.8264*	5,000	*4,132*	50,000	*41,322*
Year 3 Savings	*0.7513*	10,000	*7,513*	10,000	*7,513*
Year 4 Savings	*0.6830*	50,000	*34,151*	5,000	*3,415*
Year 5 Savings	*0.6209*	934,000	*579,941*	1,000	*621*
Sum of Present Values		0	*-373,354*	0	*-98,038*

17.13 **The present value of two investment cash flows compared**
These cash flows give an identical payback but their present values are very different. *Source Niall Enright, spreadsheet available in the companion file pack, with this and the following cash flow models and a number of additional examples of the time value of money.*

In simple terms, the preferred investment is the one with the greatest present value, in other words, the one that is worth more. We can see that the sum of the present value of Project A, at -£373,354 is much less attractive than that of Project B at -£98,038. We shall explore the power of net present value next.

In Numbers: *Present and future value*

The formula for future value is:

$$\text{Future Value, } FV_t = PV_0\ (1 + r)^{t}$$

Where PV_0 represents the initial amount, FV_t the future amount after t periods, r is the interest rate per period as a decimal figure and t is the number of periods (usually years) over which the investment has been made. Thus US$1.00 * (1.1)^3 = US$1.00 * 1.331 = US$1.33.

Similarly, the present value is calculated by compounding the discount rate for the number of periods in the future the cash flow occurs.

$$\text{Present Value, } PV_0 = \frac{FV_t}{(1 + r)^{t}} = FV_t\ (1 + r)^{-t}$$

Again, PV_0 represents the current amount in Year 0, FV_t the future amount after t periods, r is the interest rate per period as a decimal figure and t is the number of periods (usually years) over which the investment has been made. The expression $(1+r)^{-t}$ is called the discount factor since this is the amount by which the future sum is discounted. The discount factor at 10% discount rate over three years is $(1.1)^{-3}$ = 0.7513, so US$1.33 in three years time is worth US$1.33 * 0.7513 = US$1.00 today.

Instead of applying the formulas above we can use a discount factor table to look up the factor for a given discount rate and number of years. There is a discount factor table on page 788. Excel methods to calculate present and future value are shown on the next page.

Funding

17.14 Net present value

The sum of the present value of a cash flow is the net present value. This is one of the most powerful financial appraisal tools as long as we are aware of some of the practical issues in applying it.

	A	B	C
1	Symbol	Value	Description
2	FV_n	$1.33	Future Value
3	r	10%	Discount Rate
4	t	3	Periods (years)
5			
6	Method	Result	PV Formula
7	Excel	PV=PV(r,n,,-FV)	
8	Function	=PV(B3,B4,,-B2)	
9	Result	$1.00	
10			
11	Linear	PV=FV*((1+r)^-t)	
12	Function	=B2*((1+B3)^-B4)	
13	Result	$1.00	

	A	B	C
1	Symbol	Value	Description
2	PV	$1.00	Present Value
3	r	10%	Discount Rate
4	t	3	Periods (years)
5			
6	Method	Result	PV Formula
7	Excel	FV=FV(r,n,,-PV)	
8	Function	=FV(B3,B4,,-B2)	
9	Result	$1.33	
10			
11	Linear	FV=PV*((1+r)^t)	
12	Function	=B2*((1+B3)^B4)	
13	Result	$1.33	

17.14 **Methods to calculate present value (top) and future value (above)** The Excel functions FV and PV can be used respectively. In this case, be aware that the input values of PV and FV used are the fourth parameter and are shown as negative values. Alternatively FV and PV can be calculated from first principle using the formulae provided. *Source Niall Enright, spreadsheet available in the companion file pack.*

The sum of the present values of a cash flow is termed the net present value, which is abbreviated to NPV. The word net emphasizes that it is the sum of *all* the values, including the initial investment.

Typically, the NPV is calculated after the final cash flow has been determined so that there is only one series of numbers to work with. Although it is feasible to add together multiple discounted cash flows, the discounted values would not be as intuitive to work with and so this approach is not recommended.

The previous example had us obtaining a discount factor for each year of our cash flow from a table such as that in Section 24.9. Excel provides a function, NPV, which takes two inputs, the discount rate and the cash flow, and provides a single figure for the NPV of an investment. Details of how to use this function are given below.

In Numbers: *NPV in Excel*

The Excel formula NPV(rate, values) provides a method to calculate the net preset value of an undiscounted cash flow in the range, values, given a discount rate.

However, there is one quirk that can seriously affect its results. That is that Excel assumes that the initial value in the cash flow occurs at the end of one period (ignore the misleading help reference that labels this Y_0), whereas the convention in this book (and much more widely), is that the initial investment in Year Zero is not discounted.

As a consequence, we need to modify the inputs to the Excel function to achieve the desired result. There are two ways that this can be achieved:

$$\text{NPV} = \text{NPV (discount rate, cash flow } Y_1\text{-}Y_t) + Y_0 \text{ investment}$$

That is to say that we take the investment cost in Y_0 out of the values range and add it back at the end of the equation (it should remain as a negative value). Another approach that achieves the same result is to compensate for the unintended discounting of the investment. Placing all the values in the NPV() function leads to a result that is understated by a factor equivalent to the discount rate (as one more discounting took place than intended). Knowing this we can adjust the formula to read:

$$\text{NPV} = \text{NPV (discount rate, cash flow } Y_0\text{-}Y_t) * (1\text{+discount rate})$$

Personally, I prefer the first method as this makes it clear which values are being discounted and which not, and it also simplifies the profitability index calculation.

Let us consider how we can use the NPV of a project. Examine the three projects A, B and C shown in the table below. Projects A and B have the same initial investment, while Project C has an initial investment twice the size of the others. The comparison between Projects A and B is straightforward: given equal investments we should favour the project with the larger NPV, in this case, Project B.

However, the question arises about which of Projects B or C is favourable. Project C has a much greater NPV, but also a larger initial investment.

Net Cash Flow	Discount rate = 10% £	2020 (Y0)	2021 (Y1)	2022 (Y2)	2023 (Y3)	2024 (Y4)	2025 (Y5)	NPV	Profitability Index
	Project A	-1,000,000	50,000	50,000	50,000	50,000	50,000	-810,461	0.19
	Project B	-1,000,000	300,000	300,000	300,000	300,000	300,000	137,236	1.14
	Project C	-2,000,000	600,000	770,000	400,000	250,000	950,000	242,973	1.12

17.15 **Using the profitability index to differentiate between investments**
Because NPV is influenced by the size of the investment, as well as the timing, the profitability index is used to compare projects with different investment amounts.
Source Niall Enright, spreadsheet available in the companion file pack.

To compare projects with different initial costs, we use the profitability index, which is shown on the right-most column of the table above. The profitability index is calculated using the following formula:

$$\text{Profitability Index} = \frac{\text{NPV}\,(Y_0 \ldots Y_t)}{\text{Cost}(-Y_0)}$$

Where the numerator is the NPV of the cash flow from Y_0 to Y_t and the denominator is the initial investment expressed as a positive value. We can use the profitability index to compare projects whose initial investment is not the same. The project with the highest profitability index represents the best investment, in this case Project B, by a whisker. A profitability index less than one means that, taking into account discounting, the project costs have not been recovered - i.e. we have lost money, as in the example of Project A, above.

This bring us to the main criticism of NPV, which is that it is not possible to compare projects by simply comparing the NPV value, as this is an absolute financial amount. If I invest US$10,000,000 for an overall gain - NPV - of US$10,000 then that is much less attractive than if I invest US$1,000 and gain the same NPV of US$10,000. The NPV is meaningless on its own.

Use the ***profitability index*** *to compare the present value of projects with different investment amounts.*

Another criticism of the NPV method is the sensitivity of a project to the discount rate used. It is possible to make an attractive project unattractive by raising the discount rate and vice versa, and also to change the relative attractiveness of projects using the same rate. If we go on to change the discount rate for individual projects, say because we feel one is riskier than another, then we can run the accusation of "*getting the answer we want*" by manipulating the discount rates.

For these reasons, NPV analysis tends to be combined with another form of analysis called internal rate of return (IRR), covered next.

17.15 Internal rate of return

The internal rate of return works on the undiscounted cash flow to give the equivalent of the "interest rate" earned by the investment. It is a very useful financial indicator because projects with dissimilar initial investments can be compared, although there are some other weaknesses to consider.

Project A Cash Flow		
Y0	Y1	Y2
-120	20	130

Discount Rate	NPV Project A
0%	30
5%	17
10%	6
20%	-13
IRR	12.7%

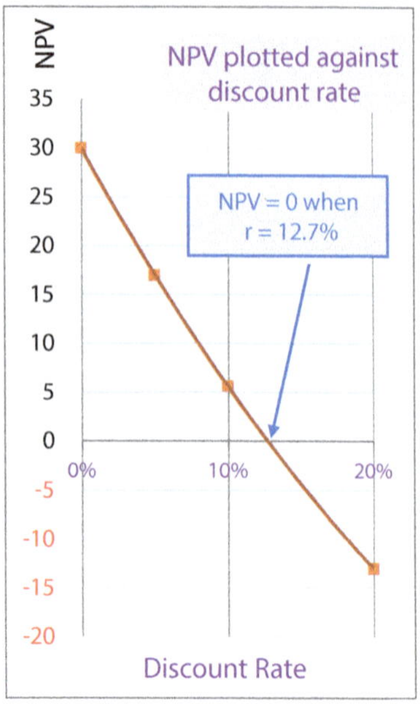

17.16 **Illustration of the internal rate of return**
The top table has a cash flow for Project A. Below this are the NPVs for the cash flow at different discount rates from 0% to 20%. When these NPVs are plotted against the discount rate we get the chart, below. *Source Niall Enright, spreadsheet available in the companion file pack.*

A very commonly used financial analysis metric is called the internal rate of return (IRR). We can think of this as an "interest rate" that the project is earning as a result of the undiscounted cash flow.

The IRR is the discount rate where the net present value is zero. This rate is much easier to understand when we look the simple cash flow, left, and the linked chart. In this example, Project A has an investment of -120 in Y_0 and then a cash flow of 20 in Y_1 and 130 in Y_2 as illustrated in the top table.

In the table below, we have calculated the NPV of Project A, for different discount rates, 0%, 5%, 10% and 20% using our Excel NPV function. So, at a discount rate of 0% the NPV of the project is +30 (-120+20+130 as would be expected); at 5% it is +17; at 10% it is +6 and at a 20% discount rate the NPV is -13.

We can then plot these values in an X-Y chart as shown below. The interesting section of the chart is when the line joining our points crosses the horizontal x-axis where the NPV is zero. We can see that this is somewhere around the discount rate of 13% (actually 12.7%). Since the definition of the IRR is the discount rate where the net present value is zero, we can conclude that the IRR for Project A is 12.7%.

In order to obtain an IRR, we don't have to plot a chart. The Excel worksheet IRR function will provide the IRR for a cash flow, as described opposite.

IRR has a significant advantage over the NPV, which is to say that IRR is scale-independent. We can quite happily compare two mutually exclusive projects with different initial investments and select the one with the highest IRR knowing that it offers the greatest return. That does not mean to say that we aren't interested in the amount of the initial investment, just that the IRR can be compared across projects with different investments, whereas NPV cannot.

There are a few disadvantages to IRR that we should note. In the first place, it may sometimes be impossible to compute an IRR. Take, for example, the cash flow $Y_0$200; $Y_1$450; Y_2-255 which gives an NPV of -5 at 0% discount rate, -2 at 10% and back to -5 at 30% discount rate. In effect, the orange line in the chart, left, is an inverted U-shape which never crosses the x-axis. This chart, and one with two correct IRR values, are in the companion file pack. Changes in sign in the cash flow are an indication that the IRR *may* be incorrect (see opposite).

	Y0	Y1	Y2
A	-120	20	130
B	-120	110	31

Discount Rate	Net Present Value: Project A	Project B
0%	30	21
5%	17	13
10%	6	6
20%	-13	-7
IRR	12.7%	14.4%

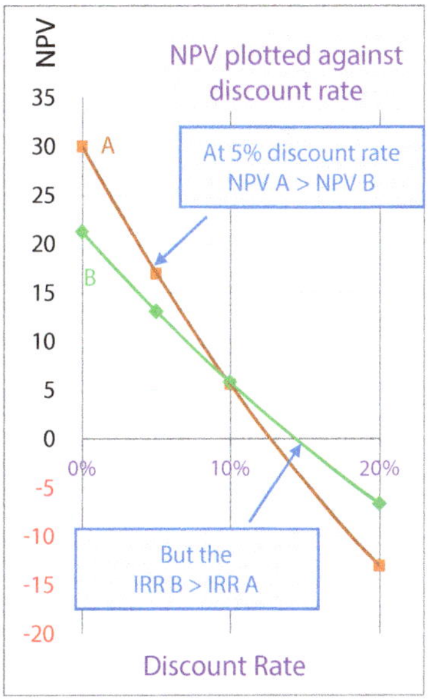

17.17 **Illustration of the relationship between IRR and NPV for two projects** The top table shows the cash flow for Project A and Project B, both with the same initial investment. Below this are the calculated NPVs for the projects using different discount rates from 0% to 20%. When these NPVs are plotted against the discount rate we get the chart, bottom. The point where the lines cross the x-axis equate to the IRR for each project, which has also been calculated. Note how the two lines cross, indicating the discount rate where the NPV of Project A ceases to be as favourable as the NPV of Project B. This crossing point is around the 10% discount rate. *Source Niall Enright, spreadsheet available in the companion file pack.*

Another issue with IRR that we should be aware of it that the IRR ranking can sometimes conflict with the NPV ranking. Take the example of the two projects, A and B, shown left. We can see the IRR calculations give Project A an IRR of 12.7% and Project B an IRR of 14.4%, so B should be favoured over A.

However, at a low discount rate, below 10%, the NPV of project A has a greater NPV than B, whereas above 10% the NPV of B is better. The discount rate where this change happens is where the two lines cross. Now if my organization's WACC is 8% and I used this discount rate to evaluate the two alternatives, then Project A would rank higher than B, but if I used IRR, I would get a different answer. This is all to do with the fact that greater discounting devalues the big slug of late income in Project A in Y_2 in comparison to the small income in Y_2 for Project B.

The reason that IRR tends to be used in preference to NPV is that decision-makers are accustomed to the notion of "rate of return" and because of its superiority in comparing projects with different initial investments. However, NPV is a generally the better method to evaluate returns as it represents the value added by the project to our organization.

In Numbers: ***IRR in Excel***

The Excel formula IRR(values[, guess]) gives the IRR of a cash flow in the range values, which in this case should be the entire cash flow. The optional second parameter guess is available if Excel struggles to compute the IRR (Excel uses an iterative method to approximate closer and closer to the IRR value and providing a guess can reduce the number of calculations needed. The default for guess is 0.1 or 10%). So for our cash flow, we would use the function as follows:

$$IRR_{\%} = IRR\ (cash\ flow\ Y_0\text{-}Y_t)$$

Please note that it is possible that Excel cannot compute an IRR, in which case the result will be #NUM. This is definitely the case if all the values in the cash flow have the same sign. It is also possible that a given cash flow may result in one or more values for the IRR calculated by Excel (according to ***Descartes' rule of sign*** there are as many positive answers as there are changes of sign in the cash flow) and Excel only returns the first correct answer. In practice, problems with computing IRR are greatest where there are additional negative values in the cash flow $Y_1..Y_t$.

In the examples given, we have assumed that the project cash flows occur at regular periods. If this is not the case, the Excel function XIRR(values, dates[, guess]) is available, where dates is a series of dates for the values given.

The IRR function in Excel makes an assumption about interest rates which may not be valid. Our cash flow has periods of negative cash, where we might need to borrow money and periods of positive cash where we have money available to reinvest. The standard IRR function in Excel assumes that both of these rates are the same. If this is not the case (and it is unlikely to be so), then there is the modified IRR function MIRR(values, finance rate, reinvest rate), where the finance rate is what we are charged to borrow money, and the reinvest rate is what our money earns when deposited.

17.16 How to deal with inflation

Incorporating the effect of inflation into our financial analysis need not be difficult. It is all a question of selecting the right discount rate for our calculations.

In our financial analysis so far, we have created incremental cash flows based on the predicted incoming and outgoing cash flows that arise in the BAU case and the investment case. These are likely to be expressed in terms of today's prices, in other words, unadjusted for inflation. This unaltered cash flow is what we call the nominal cash flow.

This nominal cash flow *should* take into account prices rises that are anticipated in the future, for example, expected increases in the cost of electricity and gas or employment costs, or known office rent increases. In this way, the cash flow remains the very best estimate of the *actual* dollars (or pounds or yen) coming into or going out of our organization as a result of the investment.

However, over a number of years, the underlying value of each unit of currency will change as a result of general inflation. For example, US$100.00 in 2016 is the equivalent of US$74.00 in 2000 because inflation has decreased the purchasing power of the dollar over time (put another way, I would need to spend US$135.00 in 2016 to buy what US$100.00 in 2000 could buy).[738]

So, if I am comparing my organization's financial performance over time, I need to convert the nominal income or profit into an equivalent cash flow at *"constant"* dollars. This cash flow is called the real cash flow. It allows the value of the cash flow each year to be considered in terms of what it can really buy. The letter R before the currency symbol is sometimes used to indicate a real value.

Just as we create a discount factor from a discount rate, so can we create an inflation factor using the same formula, where p is the rate of inflation and t the number of periods:

$$\text{Inflation Factor Year}_t = (1+p)^t$$

If inflation is 3% then in Y_1 the inflation factor is 1.03, in Y_2 it is 1.061 and in Y_3 it is 1.093, as inflation is compounded. Converting a nominal cash flow to a real cash flow involves *dividing* each value of the nominal cash flow by the inflation factor for the year to get the real cash flow. This division can confuse folks, but we need to remember that *positive* inflation makes money worth *less* in real terms, so we *deflate* the nominal values to get the real values by *dividing* by the inflation factor. In fact, you sometimes see a deflation factor, $(1+p)^{-t}$, with which the nominal cash flow can be multiplied to get the real cash flow.

The nominal cash flow should take into account anticipated price rises, so that it is the very best estimate of the **actual** cash flow.

Going the opposite way from real to nominal involves multiplying by the inflation factor (or dividing by the deflation factor). The example below shows a nominal cash flow on the first line, the inflation factor based on an annual inflation rate of 3% and the resulting real cash flow on the bottom line.

US$	2020 (Y0)	2021 (Y1)	2022 (Y2)	2023 (Y3)	2024 (Y4)	2025 (Y5)
Nominal Cash Flow	-1,000,000	300,000	300,000	300,000	300,000	300,000
Inflation Factor	1	1.030	1.061	1.093	1.126	1.159
Real Cash Flow [2020$]	-1,000,000	291,262	282,779	274,542	266,546	258,783

Discount rate		NPV $	IRR
Nominal	10%	137,236	15.24%
Nominal	10%	47,491	11.88%
Real	6.796%	137,236	11.88%

17.18 Converting a nominal cash flow to a real cash flow
The top row is a nominal cash flow representing the actual expected inflows and outflows. Below that is the inflation factor, based on an inflation rate of 3% per annum. The bottom row, in green, is the real cash flow at constant 2020 prices. If we apply the nominal discount rate to the NPV calculation for this real cash flow, we get a decrease in the NPV of the project, compared to the nominal cash flow NPV. On the other hand, if we use the real discount rate, we get the same NPV.
Source Niall Enright, spreadsheet available in the companion file pack.

Note that my real cash flow is lower than the nominal cash flow as the values have been depreciated. As a consequence, the various financial measures such as NPV and IRR give a smaller result (see the middle row in the table on the right, where the NPV is US$47,491 and the IRR 11.88%).

The solution to the NPV calculation is very easy - it just involves changing the discount rate. The discount rates we have used so far are the nominal discount rates which we apply to our nominal cash flows, 10% in the example above. So if I have a real cash flow then I need to use the real discount rate.

The equation that links real and nominal rates is called the Fisher equation, where n is the nominal discount rate, r is the real discount rate and p is inflation:

$$(1+n)=(1+r)*(1+p)$$

Rearranging the formula above the real discount rate can be derived:

$$r=\frac{(1+n)}{(1+p)}-1$$

Assuming that the inflation rate is positive, we can see that the real rate will be divided by a value greater than one, that is to say, that the real rate will be less than the nominal discount rate. In the example above, the nominal discount rate used in the NPV calculation was 10% and the inflation rate is 3%, so the real discount rate will be (1.1)/(1.03)-1 = 1.06796 - 1 = 0.06796 = 6.796%. When I calculate the NPV using this real discount rate (bottom row), then I get the same net present value for my project, US$137,236 as in the nominal case.

Use the appropriate discount rate or hurdle rate for **nominal** or **real cash flows.**

The IRR cannot be adjusted in the same way. An IRR represents a percentage increase in value for a given cash flow; in this case, the 11.88% figure is the *real* internal rate of return. The way that IRR is used when making investment decisions is to establish if the IRR exceeds an internal minimum IRR or hurdle rate. Just as we have a nominal discount rate and a real discount rate, then organizations will have a nominal IRR target rate and a real IRR target rate. The next section will look more closely at this notion of a hurdle rate.

17.17 Hurdle rates

The required return from an investment will vary from organization to organization depending on many factors. The first thing one should do when carrying out an audit is to establish what this hurdle rate is. Sometimes the hurdle rate for efficiency projects is higher than for other investments and this may need to be challenged.

Seek to use the **same financial metric** for all investment decisions.

The threshold for financial approval of an investment is called the hurdle rate. It can be typically be expressed as a minimum payback period or a minimum IRR. So one organization may have a hurdle rate of 15% IRR and another a payback of three years.

Not only is payback an inappropriate measure of return on investment, for the reasons stated earlier, but where payback is used, it is often set at a much greater level than the IRR. Time after time, surveys show that payback hurdles are tougher to meet than IRR ones. In Prindle's survey[605] of manufacturing energy efficiency in the US, the median IRR hurdle rate was 15%, which is equivalent to a greater than six-year payback (1/IRR = 1/0.15 = 6.667), whereas those companies that used payback hurdles had a requirement that projects should achieve under two to three years'[605] payback. Although there were examples of very high IRR requirements in the study (35% in one case) as well as some long payback terms (five years), the difference between the two groups showed a significant disadvantage for those who use payback as the hurdle rate.

A very common frustration among resource efficiency practitioners is the tendency for many organizations to define the hurdle rate for efficiency investments in terms of payback while using IRR (or NPV/profitability index etc.) for other investments, which disadvantages resource efficiency investments compared to other investments.

In a rational world, investments would be decided purely on their rate of return and their risk. However, we do not live in an economically perfect world; in many organizations, capital is either explicitly rationed (e.g. sites have fixed capital expenditure budgets) or capital allocation is biased towards value-enhancing activities rather than cost-reduction (see page 598 on core value).

It is true that efficiency investments may legitimately be considered a higher risk that other investments due to irreversibility, the *"credence good"* nature of the decision to invest and the other factors mentioned previously. However, this risk is not several times greater. Two consequences flow from this:

- Organizations should adopt the same financial metric for all investment decisions so that they can be compared on an equal basis; and
- Resource efficiency practitioners need to acknowledge and better quantify and mitigate risk in their investment proposals, as described next.

17.19 **Challenging payback (opposite)**
Many organizations use payback to set the hurdle rate for energy or resource efficiency investments. Not only is payback unsuitable as a tool to evaluate investments (as we have seen earlier) but the approval threshold is, in effect, much higher.
Source: Niall Enright, Image available in the companion file pack.

Real World: *Cost of saved energy and other alternative hurdle rates*

Sometimes we want to easily compare capital expenditure and operational expenditure. One way of doing this used by the American Council for an Energy Efficient Economy (ACEEE), is the cost of saved energy, which uses the capital recovery factor, (CRF) to annualize the costs of a one-off capital payment for a project whose life is t years.

$$\text{CSE (US\$/kWh)} = \frac{\textit{Cost} \ * \ \textit{CRF}}{\textit{kWh Saved}} = \frac{\textit{Cost} \ * \ \dfrac{r\,(1+r)^t}{(1+r)^t - 1}}{\textit{kWh Saved}} = \frac{\textit{Cost} * \dfrac{t}{1-(1+r)^t}}{\textit{kWh Saved}}$$

CSE is useful to compare different programme types such as those focusing on behaviour change and capital investment. It also provides a simple metric to estimate the overall cost for a given outcome. There is a CRF table on page 790 of this book.

It is not uncommon for organizations to have their own alternative metrics for financial appraisal. One example that I and colleagues have come across when evaluation efficiency investments for the UK retailer Tesco is cash return on investment (CROI), which is simply the net annual savings (i.e. earnings before interest, tax, depreciation and amortization) divided by capital invested as a percentage. It is essentially payback as a percentage figure, with all the disadvantages that we have seen that this measure has, such as not taking into account the life of the project or timing of the savings. However, this is the standard metric for return on investment used by Tesco for everything from investor presentation to individual projects, so it at least compares efficiency projects in the same way as all other investment decisions, for which they should be applauded.

17.18 Quantifying risk

A weakness of many resource efficiency proposals is their failure to address risk. Risk can be managed, so it should not be ignored. In fact, there are whole industries such as the insurance business, or energy services companies (ESCOs), that are based on risk management and can bring solutions to the table.

Exploration: ***Methods for scenarios***

Where possible, our cash flow predictions should be based on hard numeric data drawn from historical or market sources.

However, there may be situations where the data is less reliable, perhaps because our investment is in a relatively new technology or because there are a large number of variables that could influence the results.

There are some established processes that we can use to arrive at the *"best guess"* values for our cash flows in these circumstances. These involve a structured process of getting experts to provide estimates (usually without conferring in the first instance and sometimes in complete anonymity in order not to skew answers), followed by a review process where the experts can elaborate on their guess and revise it based on the inputs of the other panellists.

A leading example of these techniques is called Delphi which was developed by the Rand Corporation in the 1950s. There is, however, a great deal of variation within this technique.[187 Ch. 4]

Academic research by Rowe and Wright[633, 634] provides some advice on how the Dephi Technique can achieve the best results and compares it to other similar group methods of estimation.

Risk falls into two broad categories. There is the intrinsic or model risk related to the items in the cash flow calculations, and there is the extrinsic risk of the more general assumptions in the proposal. The first, model risks, relates to the accuracy of the numbers presented, such as whether all hidden and missing costs have been properly quantified. A significant source of model uncertainty for resource efficiency cash flows relates to assumptions about the future costs of resources (on which many business cases stand or fall). In these cases, where the investment is large, a sensitivity analysis such as the one shown opposite, is recommended. Such an analysis can give decision-makers comfort that the probability of project losses, for example, is low.

The second category of risk arises from the external factors that influence a project, but which may be less easily quantified, especially if they fall into the high-impact/indeterminate probability category. We saw earlier that contract disputes were a big source of overruns in UK government investments. These are the big risks John Hollmann was referring to earlier (page 562).

One starting point with this kind of project risk is to ensure these worst-case scenarios are identified, using some of the methods illustrated in the box, left. Clearly, there is a limit on the type of scenarios that can or should be considered (meteor impact from outer space would not be on the list!). But where there is a significant risk that is relevant to the investment, then it should be addressed in the business case. Once the risk is described, methods to monitor and respond to this can be developed. For example, we can test the risk in a controlled fashion (run a pilot project), we can mitigate the risk and diminish its impact (e.g. take out insurance), or we can offload the risk (e.g. involve a third party or get the manufacturer to guarantee performance).

The key thing to note is that we should confront risk in our projects where it has a material effect. Doing this may be as simple as acknowledging that the risk exists, which reassures decision-makers it has been considered, or using sensitivity analysis to assess the probability of losses, through to explicitly designing the proposal with risk mitigation at the core.

If the investment is small, we may consider bundling the project into a portfolio where under performance by one project may be offset by over performance by another and the overall value attained. This approach is most appropriate where we have projects acting on different resource streams in various locations.

Always ***describe risk*** *where it is material to the business case.*

Real World: *Sensitivity analysis*

A sensitivity analysis is just a test of the influence of a particular value (or sets of values) on the results of a calculation. This technique is used to assess the risk of an investment if the values differ from the assumed values.

Consider the example below, taken from a real organization. Here an investment of €2.2 million results in a decrease in annual electricity consumption of 5 GWh per year. The business case has been formulated using the current electricity price of €0.12 per kWh, as shown in the pale green row, which gives a net present value for the investment of €1.8 million over 10 years using an 8% discount rate.

However, this €0.12 per kWh is an estimate of the average future electricity price. By modelling the savings using a range of electricity rates, we can see to what degree our result is sensitive to this cost - hence *"sensitivity"* analysis.

The analysis carries out the same NPV calculations using different assumed future electricity prices, in this case from €0.04 per kWh to €0.20 per kWh. The results are plotted on the chart at the bottom. We can see that the net present value becomes zero at an electricity price of around €0.065 per kWh, which is just over half the business case assumption.

Although electricity prices are clearly material to the business case, the organization concluded from this analysis that there was relatively low risk associated with electricity pricing.

Sometimes, we may want to carry out a sensitivity analysis on an aspect of our business case to prove that the factor is not material. We could, for example, look at the impact of varying the initial capital cost, or on labour rates and so forth.

17.20 **Example of a sensitivity analysis for electricity unit costs**
Source Niall Enright, spreadsheet available in the companion file pack.

Business Case Sensitivity Analysis
Electricity costs ± €0.08

Initial Investment	€ 2,200,000
Duration (years)	10
Discount Rate	8%

€ / kWh	Base Case (kWh) 10,000,000	Investment Case (kWh) 5,000,000	Annual Saving	Net Present Value
€ 0.04	€ 400,000	€ 200,000	€ 200,000	-€ 857,984
€ 0.06	€ 600,000	€ 300,000	€ 300,000	-€ 186,976
€ 0.08	€ 800,000	€ 400,000	€ 400,000	€ 484,033
€ 0.12	€ 1,200,000	€ 600,000	€ 600,000	€ 1,826,049
€ 0.16	€ 1,600,000	€ 800,000	€ 800,000	€ 3,168,065
€ 0.20	€ 2,000,000	€ 1,000,000	€ 1,000,000	€ 4,510,081

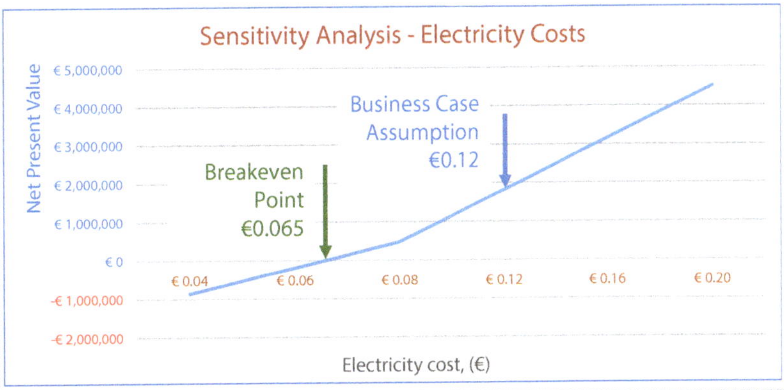

	A	B	C
1	Symbol	Value	Description
2	PV_0	100	Present Value
3	FV_n	54	Future Value
4	R	46%	Total % change
5	t	15	Periods (years)
6			
7	Method	Result	PV Formula
8	EXCEL	r=RATE(n,0,-PV,FV)	
9	n, PV, FV	=RATE(B5,0,-B2,B3)	
10	Result	-4.02%	
11			
12	Method 2	r=(1+R)^(1/t)-1	
13	R, n	=B2*((1-B3)^-B4)	
14	Result	-4.02%	
15			
16	Method 3	r=(FV/PV)^(1/t)-1	
17	PV,FV,p	=(B3/B2)^(1/B5)-1	
18	Result	-4.02%	

17.21 **Different methods in Excel to calculate r using PV, FV and t or R and t**
Source Niall Enright, spreadsheet available in the companion file pack.

	A	B	C
1	Symbol	Value	Description
2	r	-2%	Annual Rate
3	t	10	Periods (years)
4			
5	Method	Result	PV Formula
6	Formula	R=(1+r)^t-1	
7	n, PV, FV	=(1+B2)^B3-1	
8	Result	-18.29%	

17.22 **Excel calculation for the total change R after t periods at r rate.**
Source Niall Enright, spreadsheet available in the companion file pack.

In Numbers: ***Working with growth***

Like discounting, the maths of growth also involves compounding values. Here, it is often helpful to calculate the annual rate of improvement r which will be needed to achieve a goal of R% over t periods. This calculation is not the same as dividing R by t because the savings are compounded. The formula we should use is:

$$r = \sqrt[t]{(1+R)} - 1$$

This formula can be written in a linear fashion as r=(1+R)^(1/t)-1, which is helpful in spreadsheet versions of the calculation, such as those shown, left.

Using our earlier example from the supermarket sustainable goal calculations (see page 366), we know that the power generation sector is expected to decarbonize by 46% over 15 years, but may want to know what that means in practice in terms of the annual improvement needed. We can work this out by substituting into the above formula: (1-0.46)^(1/15)-1=-4.02%. Note the use of the negative sign before the 0.46 figure to denote that our goal is a *decrease* of 46%.

If we know a particular future value, FV, then we would use the following formula which includes the present value PV:

$$r = \sqrt[t]{\frac{FV}{PV}} - 1$$

Again, there is a linear version which is r = (FV/PV)^(1/t) -1. Remember, earlier we had an organization which saved 19% of its emissions between 1997 and 2016 (page 375). We can say that emissions in 1997 were 100 units (PV) and in 2016, 81 units (FV). Substituting into our equation we get an annualized rate of savings of (81/100)^(1/(2016-1997))-1 = -1.12% compound growth each year, sustained for 19 years.

As seen earlier, we can work out the future value FV_t from a given annual growth or reduction rate r over n periods starting with a present value PV_0, using the formula:

The linear form of this equation is FV = PV (1+r)^t.

$$\text{Future Value, } FV_t = PV_0\ (1 + r)^t$$

Finally, if we want to see what the total percentage change R is, given t periods at r rate, we don't need the values for PV or FV, just the formula:

$$R = (1+r)^t - 1$$

In linear form, this is R=(1+r)^t-1.

Rule of thumb: ***The rule of 72 for doubling /halving***

If we have a positive compound growth rate of r then this is equivalent to a *doubling* after 72/r periods. So a growth rate of +8% a year would lead to a doubling in nine years. Similarly, a negative growth rate of -8% would give rise to a *halving* in nine years. Thus, if we hear that China has a GDP growth rate of 12%, then we could quickly work out that its GDP would double in 72/12 = six years.

17.19 Investment models

We have considered the inputs to our financial models item by item, as well as the outputs such as NPV, IRR etc. Although many organizations already have well-developed models for financial appraisal, in this section we will consider what the inputs and outputs of a financial model might look like.

The cash flow elements and key parameters such as the discount rate used in a business case are the inputs to an investment appraisal model. The outputs typically include at least one metric such as payback, NPV and IRR. More sophisticated models may produce cash flows from simple inputs, create depreciation schedules and provide charts.

An investment model should not be confused with the business case. Although the model will determine if the opportunity achieves the required hurdle rate, there are many other details that the decision-maker will need in order to approve the projects.

Within most organizations, the investment models used for budget approval are quite prescriptive. Many of the key inputs, such as the discount rates used, as well as the output metrics may be fixed in advance. There will also be policies in place, for example, whether taxation or depreciation are included in the cash flow data.

Funding

An example of an investment appraisal model developed in Microsoft Excel is shown in the following pages. This investment appraisal model is provided in order to illustrate the many of the practical issues described in this chapter.

The model has a large number of inputs, shown in the image at the top of the next page. The mandatory inputs are shown in pale blue and include key parameters such as the discount rate used, the project term, asset life, treatment of Year 0 savings, and resource units costs (shown under the Key Assumptions section). The model creates a cash flow from an initial equipment cost, annual operational savings or annual resource savings (shown under Project Cost and Savings. From these eight inputs, a full cash flow is created and the key financial parameters determined. The cash flow is visible to the user and if a more complex timing of costs is required the user can override the calculated values by inputting directly into the appropriate cells.

Two further entries are needed in order to produce the desired output. These are a currency symbol (which bypasses Excel's currency formatting to allow the user to assign a currency symbol such as £, $ or GBP, USD) and a resource name for the first resource listed (up to three resources can be included in the model and, if named, the entry fields appear automatically). These entries are shown in the Settings section (not illustrated).

KEY ASSUMPTIONS

Project Fundamentals:		
Base Year	2016	*Year 0 year*
Project Term	5	*Years (max asset life or 25, d=5)*
Asset Life	5	*Years the asset will last (d=5)*
Return in Year 0?	N	*Y if savings appear in year 0, d=Y*
Project Discount Rate	9%	*(typical range is 8% to 12%)*
Inflation	0%	*(d=0, i.e no inflation)*
Utilities Cost Inflation	0%	*(+ve is increase)*
Tax and Depreciation:		
Year 0 Depreciation	2,000	*£ (Optional)*
Corporation Tax Rate	22%	*If the business is not profitable put 0 here*
Depreciation Rate	20%	*Straight Line = years e.g. 20% = 5 years.*
Depreciation Method	Straight Line	*Blank, "Straight Line" or "Decreasing"*
CAPEX Grant £	-	*CAPITAL grants towards the equipment*
Costs and Emissions:		
Emissions Value	16.90	*£/tonne CO_2*
Electricity Rate	0.100	*£/kWh*
Electricity CO2 Factor	0.496	*kgCO2/kWh*

PROJECT COSTS AND SAVINGS

These values will create a cash flow below. For more complex models value can be entered into the cash flow directly in the columns with a heading in blue

Investment Cash Flow:			
Design and Planning	0	*£. An OPEX cost such as Consulting costs*	
Equipment	-8,000	*£. The CAPEX for the project*	
Installation	0	*£. This is an OPEX cost*	
Project Management	0	*£. Internal OPEX costs*	
Initial Costs	-8,000		
Operating Cash Flow:			
Operating Saving	0	*£ /year. +ve value is a saving, -ve a cost*	
Maintenance Saving	2,000	*£ /year. +ve value is a saving, -ve a cost*	
Operational Savings	2,000	*£ /year. +ve value is a saving, -ve a cost*	
Electricty Saving	500,000	*kWh/year*	50,000 £
Other Savings		*£(year). Negative is a saving*	
Resource Savings	50,000		
Terminal Cash Flow:			
Residual Value	1,000	*£ total. +ve value is a saving, -ve a cost*	
Residual Cost	0	*£ total. +ve value is a saving, -ve a cost*	
Terminal Value	1,000		

17.23 **Investment appraisal inputs**
These are the inputs for the investment appraisal model. The user will enter a number of mandatory (pale blue) or optional (grey) details about the investment. The values illustrated above correspond with the outputs shown opposite.
Source Niall Enright, spreadsheet available in the companion file pack.

As well as the mandatory inputs, the model offers many optional entries to include factors such as depreciation (using a straight line or decreasing approach), emissions factors and costs, as well as inflation effects (with resource cost inflation and general inflation separated out). These optional entries are shown in the grey shaded cells. Not shown are further options to enter the accuracy of the initial costs, operating costs and terminal costs, from which a ± value is calculated, and a text indication of the estimation class generated.

The results of the model are shown opposite. This output is designed as a one-page overview to be included in a business case. The financial metrics are prominently placed below the summary text. Note that there is space beside these for the user to enter the additional benefits (or co-benefits) provided by the project.

A section on risk assessment includes the project payback, as well as the text representation of the estimation class (the user can modify this text in the model settings). If one or more sensitivity analyses have been undertaken, the filed would be changed to "*Yes*" to show this.

Probably the most compelling part of the investment appraisal is the cumulative discounted cash flow chart on the bottom left of the page. This chart will not only add up to the net present value for the project, but will demonstrate the speed at which the project moves from a negative to positive cumulative return.

The investment model describes the direct effects of the investment in terms of cost, risk and value. However, in order to succeed, our business case needs to provide more than financial metrics and describe how the investment supports fundamental organization objectives, as described next.

17.24 **Investment appraisal summary (opposite)**
This is the summary output sheet for an investment (using the values shown above). The objective is provide the key information about the investment in a simple but compelling fashion.
Source Niall Enright, spreadsheet available in the companion file pack.

INVESTMENT APPRAISAL

Project Title:	Baggage Conveyor Motors
Project Type:	Motors Upgrade
Business Unit:	Aiports
Owner:	Susan Smith, Engineering.
Brief Description:	The existing HVAC controls are not maintaining the correct air flow in the terminal and do not offer sufficient flexibility in terms of zoning. An audit has established that we can save 500,000 kWh of electricity per year by moving to a more flexible and open system.

FINANCIAL BENEFIT SUMMARY

	TOTAL	CAPEX	
Costs	£ -8,000	£ -8,000	
Annual Saving	£ 50,000	±£ 20,016	/year

Discount Rate Used	9%
NPV	£ 211,226
IRR	702%

ADDITIONAL BENEFIT SUMMARY

	£ equiv. per year
Comfort - ↑customer satisfaction	£10,000
Supports Climate Change Goals	N/A

These benefits are not included in the figures, left.

RISK ASSESSMENT SUMMARY

Payback	1.2	*Years*
Estimation Class	Initial Approval	
Sensitivity Analysis	N/A	

CASH FLOW SUMMARY (All figures in £)

Year	CAPEX	*Discounted* CASHFLOW	*Undiscounted* P&L
2016	-8,000	-8,000	-7,296
2017		51,554	56,458
2018		47,298	56,458
2019		43,392	56,458
2020		39,809	56,458
2021		37,172	57,194
2022			
2023			
2024			
2025			
2026			
2027			
2028			
2029			
2030			
2031			
2032			
2033			
2034			
2035			
2036			
2037			
2038			
2039			
2040			
2041			

CUMULATIVE DISCOUNTED CASHFLOWS

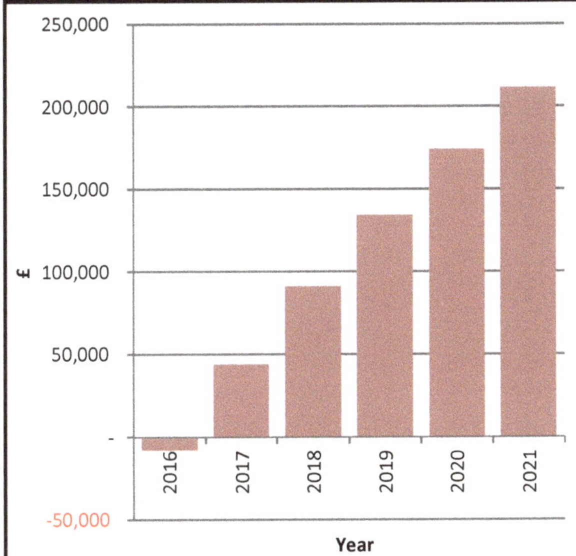

Funding

17.20 Emphasize core value

However compelling our financial appraisal seems, we need to place special emphasis on the effects of the project on the core organizational objectives. Here, we digress to examine how we can enhance the investment case beyond simple cost reduction.

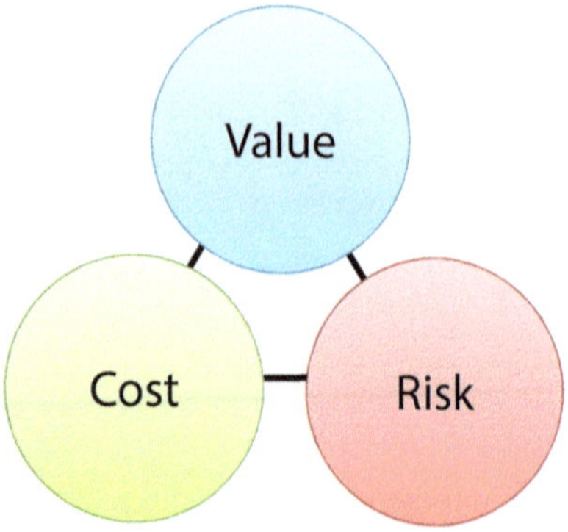

17.25 **The elements that financial decision-makers consider when evaluating investments**
These components are not equally weighted when decisions are taken, with the value element having greater prominence. Most resource efficiency business cases rely on a cost rationale for investment and so may struggle against alternative uses of funds which appear more strategic.
Source: "Investment in energy efficiency: do the characteristics of investments matter?" by Catherine Cooremans.[168]

In the preceding sections, we have explored what the investment, operational and terminal cash flows should contain. There has been an emphasis on ensuring that these are complete and transparent, with hidden and missing costs properly considered, but also with hidden benefits highlighted.

While the financial appraisal is about creating a credible cash flow, that is not the only factor that influences decision-makers. Recent research by Catharine Cooremans[168] examined energy efficiency investment decision-making in 35 firms in Switzerland and asked finance decision-makers to rate the importance of energy efficiency, in terms of strategic value to the business, cost saving and the risk reduction. Out of 5, the average score for strategic value was 1.9, for cost saving 4.0 and for risk reduction, 2.7. The low score of 1.9 for value, in Cooremans' study, indicates that finance directors see energy efficiency as not particularly important in terms of strategic impact.

The key thing to note is that these characteristics are not equal in weight: value-enhancing projects receive much higher priority. No matter how compelling the cost-saving case, efficiency investment will have a low priority where energy costs have a low impact or low visibility.

A similar study[203] of 135 Dutch sites in 1999 found that the third most significant barrier to investment was that "*energy costs were not sufficiently important*" to the organization. This is not new; in his 1986 study,[630] Marc Ross found that around a third of firms rationed capital for smaller projects by requiring much greater financial returns (usually expressed as payback) than the return needed for investments in production improvements. One of the possible reasons given for this is:

> *"Many energy-conservation projects' lack of impact on production capacity, product quality, and product flexibility is perceived as a reason for giving them a low priority. Cost-cutting projects can be postponed, it is thought, without losing much of the opportunity, whereas market opportunities associated with new or improved products or increased production may be altogether fleeting."*

Where possible, describe the benefits delivered in core value terms.

So energy and resource efficiency projects are in competition for scarce resources and are often seen as peripheral when they lack a connection with the core objectives of the organization. This competition means that our business case should not just rely on the cost saving, but should also include value and risk benefits, in financial terms if possible, in narrative form if not.

Real World: *20 basis points*

As I was putting this chapter together, David Glover, of Peel Land & Property Group, and I took part in an interesting call with Lloyds Bank.

This call was to discuss the introduction of a £1 billion Green Lending Fund for Commercial Property by Lloyds.

Property owners will be able to access the fund, which is the first of its kind, after undergoing a test to assess how much energy efficiency they can achieve. For loans over £10 million in value, with the maximum potential for energy efficiency, the loan could be a much as 20 basis points (0.2%) below the standard cost of borrowing.

The intention is to incentivize the bank's customers to implement sustainability measures in their properties and capture growing investor demand for green and sustainable fixed-income products.

The actual discount offered will be determined by a benchmark tool developed by the consultancy Trucost, which will help set the KPIs that borrowers have to meet to maintain the discount.

For anyone in the commercial property sector, an up to 20 basis points discount for commercial loans is very significant. This is the kind of initiative that places resource efficiency firmly into the value category of investment returns and gets the attention of the finance director.

So how do we go about identifying the core value benefits? Here are some of the possible approaches.

- Look for direct effects on core objectives. For example, the decreased cost delivered by an efficiency programme can lead to increased profit, which can be translated into a share price increase or release funds sufficient to provide an additional N patient treatments a year. This type of value tends to be the easiest to quantify. Numerical techniques for translating saving to value were explained on page 108.

- Look for equivalence. This approach is where we translate a purely financial saving to its equivalent expressed in terms of the core mission of the organization. At its simplest level, I could state that a saving of US$100,000 a year is equivalent to US$2,000,000 of sales a year in a sales-oriented organization.

- Look for complementarity. Here, we want to know how the investment complements and supports a stated core objective of the organization or business unit. For example, energy efficiency performance is important to a brand's position on climate change (see page 114). Although there may not be a direct or quantifiable relationship between the financial benefits from the resource efficiency investments and these core values, that does not mean that this link is unimportant. Financial decision-makers are quite capable of integrating qualitative as well as quantitative data into their decision-making.

- Look for enabling effects. Sometimes a resource efficiency investment is a prerequisite for a value-enhancing activity of some sort (as the example left shows). This is often the case where resource efficiency meets supply chain requirements or where it provides a licence to operate. Other very common situations occur where greater efficiency in manufacturing or process sites can improve the overall throughput of services or equipment. An example would be how improving the efficiency of the chillers at a brewery can lead to greater output at the site. Here, we are thinking about efficiency as opening otherwise closed doors.

In terms of risk, our business case should not only reflect any real risks that could undermine the return expected from the investment, using techniques such as sensitivity analysis, if needed, but it should also articulate what the investment might do to reduce risk. We saw earlier that decreasing the risk associated with diesel supply to the Diavik diamond mine (page 113) in Canada is a powerful argument for investment in its own right.

So far, our financial analysis techniques have been applied to traditional assets (balance sheet items) and cash flows (profit and loss account items). There is a growing school of thought that says we need to extend financial analysis to a wider range of forms of capital, such as natural and human capital, and so integrate sustainability in a much deeper way into our decision-making processes, as we shall see next.

17.21 Valuing sustainability

There is an argument that says sustainability should be incorporated better into everyday decision-making. One technique involves identifying the monetary value of decisions that impact on services provided by nature, although this is not without some controversy.

Real World: ***The value of ecosystems***

Even the briefest of examination of *The Economics of Ecosystems and Biodiversity*[691] cannot fail to impress the reader with the enormous value of the services provided to us by nature:

- Conserving forests avoids global emissions worth US$3.7 trillion;
- Bees in Switzerland provide pollination services worth US$213 million (five times the direct value of bee products like honey);
- It is five times cheaper for New York Water Authorities to pay Catskill farmers to improve water run-off, than to build new water treatment plants;
- A major Australian mining project was abandoned when it became clear that the WTP value of the forest (see box opposite) was four times greater than the net present value of the mine.

The benefit of these valuations is that they belie the presumption in favour of economic development that there is of little value in leaving land or resources unexploited.

In some cases, such as the reduction of emissions from deforestation and degradation (REDD), this approach has led to real payments flowing to the preservation of natural resources in poorer countries.

In this book, I have defined value as the expression of our organization's ability to achieve its prime objective, whether that is to make money or to deliver a social service. The application of resources creates value. Indeed, we saw in the introductory Chapter 2 in the work of Herman Daly, that all value ultimately depends on the natural capital of our planet.

The desire to incorporate forms of value beyond money into decision-making is very strong. The first step involves quantifying value itself. Take, for example, the UK National Health Service. Here, health has been quantified into a unit called a QALY (quality-adjusted life year), in which 1 represents one year of full health, and 0 represents death. The QALY of various forms of ill-health is derived using techniques such as getting respondents to rate the severity of particular conditions, or to make trade-offs involving time and risk, or using objective questionnaires of life quality.[789] Once the QALY been quantified, it is then possible to compare the cost of alternative treatments which will improve QALY. Today the UK's health service sets a financial limit of between £20,000 and £30,000 for a QALY.[213] That is how much they are prepared to spend on a treatment that extends good quality life by a year.

This example is deliberately chosen to demonstrate some of the controversial aspects of monetization, that is applying economic valuation to non-financial social and environmental benefits.

First, we have the challenge of quantifying "value". The *Natural Capital Protocol*[544] describes value as the importance, worth or usefulness of something. So when we are defining the broader value of resource efficiency, we need to understand that the frame of reference is very important. A resource such as a river, for example, may be a great facility for local people to fish or for businesses as a source of tourism revenues, as a flood defence for homeowners or as a source of energy to drive a turbine for a power company. Which of these is more valuable? Whose competing needs are most important? There is clearly a challenge comparing different forms of value: the financial value of a dam across the river can be calculated but what about the lost beauty of the natural landscape itself, or the climate change benefits from low(er) carbon electricity?

Although we are primarily concerned here with monetary aspects of value we need to accept that for some aspects, e.g. historical, cultural or spiritual, there may be no absolute measure, just a qualitative indication of relative importance.

Exploration: *Contingent valuation*

Photo © Dmitry, Fotolia.com

Survey techniques have been used since the 1970s to attempt to put a monetary value on environmental resources. Willingness to pay (WTP - how much people are willing to pay for a particular environmental service or resource) and willingness to allow (WTA - how much compensation they would require for the loss or an environmental service or resource), are two of the most common methods.

Paradoxically these approaches do not give the same result. It seems that WTA can be much larger than WTP when there are few substitutes for the resource[346] (people only have a finite amount they can pay, whereas they can accept an infinite amount of compensation). There have also been suggestions that loss aversion may be involved (there needs to be a greater amount of money to accept a loss than to pay for a similar gain).

In the design of these surveys, one can generalize and say that if people already have a legal, moral or assumed right to a resource, then we should ask them about their WTA to reduce the availability of the resource. Where they do not have any rights, it is more reasonable to ask them their WTP for an increase in the resource.[347]

Although these surveys can be greatly influenced by the design of the questions and behaviours such as anchoring (page 178) can affect the outcome, there is a large body of academic research available to help in the design of the surveys so that the results can usefully inform decision-making.

In other circumstances, we may stop at a numerical quantitative measure, such as a QALY, because folks are unwilling to take this measure one step further and convert it to a monetary value. Monetary values can be difficult to accept for moral or ethical reasons, or simply be hard to calculate or interpret.

Quantification of value is almost always a necessary step to calculating monetary equivalents. Here, aspects such as the scope (in terms of location and time) need to be clearly defined. The Natural Capital Protocol defines the three key impacts that should be considered (I have rephrased these slightly).

- Impacts on your organization (resulting from your impact on the resource)
- Your impacts on society (as a result of your effects on the resource)
- Your dependencies on the resource (the benefits it bring your organization)

Resource here can be conventional resources (e.g. minerals), natural capital (e.g. a river or food supply), social capital (e.g. the value of an educated and supportive workforce) or intangibles (such as brand value). The key principle is that unless these three components, covering all the internal effects and the external effects (aka externalities) of the resource use, are fully considered, the results will be limited. In the example of the river, the various competing values would need to be defined for an accurate valuation of each alternative to be produced - any one decision affects all the others (including *"do nothing"*).

A variety of well-established methods is available to help with monetization. The most common methods are based on stated preferences (see left), but there are other techniques based on revealed preferences (such as the travel cost method which originated in the US to value national parks), production functions based on the resource's direct effect on goods and services with known costs, market prices and replacement costs methods. More recently risk analysis techniques from the insurance world have successfully built on climate change adaptation cost assessments to price carbon sequestration services. It is beyond the scope of this book to cover these, but most good texts on environmental economics will provide a good foundation (some are listed at the end of this chapter). Robert Ayres' paper *Sustainability Economics, where do we stand?*[349] provides a wide-ranging primer.

Some readers may consider the notion of applying a monetary value to everything is abhorrent - reflecting a utilitarian view of nature solely in the service of humankind. Others believe the lack of objectivity in the methods makes them little more than an academic exercise. Nor should we assume that, because we may measure natural capital and financial capital using the same units, that they are interchangeable. For much of our natural capital, there is simply no substitute. Despite these reservations, large-scale natural capital accounting has proved itself a useful tool within national and regional government departments and agencies to help formulate policy. These techniques are beginning to move into mainstream investment assessment by organizations, as we shall see next. ⇒ page 604.

Real World: *Decision-making incorporating externalities*

Back in 2010, I was involved in a very interesting project for a global manufacturer (which will remain anonymous). This client had (and still has) an outstanding commitment to sustainability and was concerned that its traditional methods of investment appraisal were not differentiating adequately between projects which had similar financial returns but which might have radically different environmental consequences.

My role in the project was as the *"finance guru"* and expertise in matters related to environmental valuation was provided by James Spurgeon, now of Sustain Value, who is a recognized leader on environmental economics.[672, 673] The project manager was Erik Wijlhuizen, who is now at another consultancy, Sustainable Endeavour in the Netherlands.

The objective of the project was to establish if a few key parameters from real investments could adequately separate *"good"* and *"poor"* projects and if this would have any effect on decision-making.

The model we created is illustrated opposite. We wanted to keep things simple, so we used nine Dow Jones Sustainability Index (DJSI) parameters as our environmental measures. For each of these measures, we arrived at two prices. The proxy market price was the average market prices that had been established in a market somewhere in the world for each of the nine parameters. So for CO_2, we had the EU emissions trading scheme price (which was a real financial cost to the Barcelona site), for NOx and SOx we had prices established in the US emissions trading markets. These are tangible costs which the organization is paying today or could be paying in the future, for these DJSI impacts.

Then we had the societal costs, which were based on numerous studies on the health consequences of air emissions (e.g. for EU countries ExternE, CAFE, IMPACT, NEEDS or GRACE). These societal costs are much greater than the market prices because they take into account externalities such as, for CO_2, the costs of adaptation to climate change, impacts on production, property, food and mortality arising from severe weather, etc. The table with the cells in orange, lower left in the model, lists all the costs we used in the model.

In this project, we found that a key issue was acceptance of the costs, especially the high societal costs. We advised that the only way to get decision-makers to understand and accept these costs, and take ownership for decisions using these, would be to run a workshop with them where they could be involved in evaluating the sources of price information and choosing an appropriate price themselves.

Here, it is important to note that some of the costs, such a fresh water intake, are heterogeneous because the value varies considerably depending on location (a m^3 of fresh water in the Sahara is more precious in environmental terms than one in the Amazon). So, for some of the measures we developed a cost based on location.

Once accepted, these costs and the data for each project enabled the team to calculate a straight financial analysis (the conventional approach), one based on proxy market prices or one based on societal costs and benefits, as illustrated in the pale blue cells, top right opposite.

Looking at this project, we can see that in financial terms alone, the IRR of 17% is less than the organization's hurdle rate of 18%, so the investment would have

Decision-makers needed to be involved in selecting the costs of externalities if these are to be ***accepted.***

17.26 **An investment model including the monetization of externalities**
This is based on a proof-of-concept developed for a leading global manufacturer.
Source: Confidential client

Project Name: Barcelona Plastics Plant Process Fuel Switch

Location: SPAIN

Model Parameter Costs -ve Savings +ve	Unit	Increase in unit per year	Decrease in unit per year	Financial cost in euro per unit	Financial costs in euro per year	Additional proxy market price	Additional social cost/ benefit
Energy	TJ		10	-33,334	333,336	333,336	333,336
CO2	ton		917	15	13,750	13,750	68,751
VOC	ton	0				-	-
NOx	ton	0	20			20,000	69,820
SOx	ton	0	4			2,000	20,800
COD	ton					-	-
waste (sludge from WWT)	ton	0				-	-
fresh water intake	m3	0	0			-	-
Labor cost	€						
Capital cost	€	-1,000,000					
Equipment Life	years	10					
Annual Operational Cost	€	40,000			40,000	40,000	40,000
Discount rate	%	8%					
Depreciation rate (10%)	%	10%			-100,000	- 100,000	- 100,000
Cost of capital (7%)	%	7%			-70,000	- 70,000	- 70,000

Economic Appraisals			
Total annual cost or benefit	217,086	239,086	362,707
Net present value (10 yrs @ 8%)	422,838	559,525	1,327,584
IRR	17%	20%	34%
Externalities Valuation			
Externalities (-ve) +ve / year	13,750	35,750	159,371
Environmental ROI	1.4%	3.6%	15.9%

Dow Jones Sustainability Index Parameter	Unit	Proxy market price (EURO €)	Societal cost (EURO €)
Energy purchased	TJ		
Energy from site	TJ		
CO2 equiv	ton	15	75
VOC	ton	100	2100
NOx	ton	1000	3491
SOx	ton	500	5200
COD	ton	650	2814
Total waste	ton	50	77
Fresh Water Intake	m3	1	2

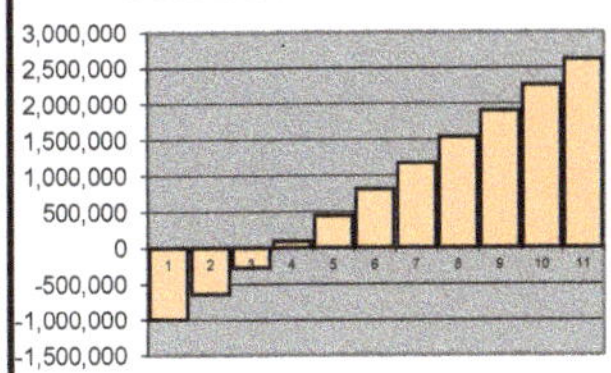

been rejected. On the other hand, if we include the costs of externalities we get an IRR of 20% using proxy market prices, or 34% using social costs.

There are several indicators that decision-makers could use. They could consider the € total of the project in terms of cost or benefit. This indicator is the simplest, but does not take into account timing, so an alternative would be to consider the net present value (NPV). If NPV were to be used, one could separate this into a *cash NPV* and an *environmental NPV* both using the same initial investment cost, but one with the real financial savings and the other with the proxy market or social benefits for the DJSI factors. These values could be added together. However, as NPV does not take into account the scale of the initial investment, an IRR figure may be preferred.

In addition to NPV and IRR, which are both well understood in the client organization, we looked at another simple measure which we called environmental return on investment, which is simply the net value of the externalities divided by the initial investment costs, expressed as a percentage.

This example shows just how seriously innovative and committed organizations like our client are considering sustainability in investment appraisal.

Funding

17.22 Value-added approaches

Establishing absolute monetary values for resource impacts can be difficult. An alternative technique compares an investment or organization with an appropriate reference (e.g. peers or a national economy) and establishes if the relative performance is better or worse across a range of environmental impacts.

In the preceding examples, we have seen that the process of applying a monetary value to externalities is problematic. Take, for example, CO_2 emissions - there is the advantage that this is a homogeneous externality, that is to say, that its impact does not vary from location to location - the CO_2 has the same contribution to global warming no matter what its source or where it is emitted. Another advantage for CO_2, compared with other impacts, is that there are several well-established emissions trading schemes which will give us a market price (although the prices do vary considerably across both the regulated and the voluntary markets, so there is still a decision to be taken about which price to use). However, the markets do not factor in the full costs to society, so we then have a societal cost of carbon, which considers all the harm that climate change can produce. This cost is much higher and subject to much greater uncertainties. In 2015 the US government put the figure at US$37/tonne, the Stern Review of The Economics of Climate Change[680] put the figure at US$86, while a recent study[535] suggests that these are underestimates and the real figure is closer to US$220/tonne. By comparison, a 2013 survey[129] from the Carbon Disclosure Project lists 30 or so companies which have adopted internal carbon prices using prices ranging from US$6 (Microsoft Corporation) to US$60 (ExxonMobil Corporation).

An alternative approach to monetizing an organization's sustainability impact has been proposed by Frank Figge and Tobias Hann.[279] They suggest combining financial and impact data in such a way that a measure of performance, labelled "sustainable value added", can be calculated.

The technique builds on conventional finance concepts. For example, there is the notion that an organization is adding value if it uses capital more efficiently than the markets. The average cost of capital in a market such as the UK could be described as the total value created in the UK (i.e. net domestic product) divided by the amount of capital employed to get that value (total net wealth). Both of these figures are readily available from national statistics organizations.

A company operating in the UK is adding value if it employs capital more efficiently that the country (and conversely destroying value if it is less efficient). The equivalent inputs are the organization's profit and its non-financial assets (calculated by subtracting all shares, which are accounted for in other companies' balance sheets, from the total assets).

Exploration: ***Nothing is new***

I am still amazed and thrilled when I discover that the ideas that we have around resource efficiency today are not new.

Here is an extract from a paper[331] by David Green from 1894, quoted by Figge and Hann.[279]

> *"Not only time and strength, but commodities, capital, and many of the free gifts of nature, such as mineral deposits and the use of fruitful land, must be economized if we are to act reasonably.*
>
> *Before devoting any one of these resources to a particular use, we must consider the other uses from which it will be withheld by our action; and the most advantageous opportunity which we deliberately forego constitutes a sacrifice for which we must expect at least an equivalent return."*

Green is reinforcing the tenet of investment appraisal, that all the alternative uses of our resources (including *"do nothing"*) must be considered. Where the choice is to develop an opportunity of a lesser value than the best on offer, we should expect to be compensated in some way. This is at the root of schemes like biodiversity offsetting, where damage in one area is offset by improvements in other, although we should recognize that some resources cannot be substituted for.

The difference between the ratio of value to capital employed at a national level and the ratio at organization level is called the value spread. If this is negative, it means that the organization's return on the total capital employed is inferior to the country's, and so value is being destroyed.

Once these basic concepts have been understood, we can then extend the idea to include the return on other forms of capital, not just total capital employed. For example, we can consider CO_2, methane (CH_4), particulates ($PM_{[10]}$), or even accident rates as forms of natural capital that are employed to generate value. Again, these figures are readily available for many national accounts and individual organizations. Where multiple forms of sustainable capital are considered, the sustainable value added is the average of the contribution made by these forms of capital (i.e. the total divided by the number of items).

We can better visualize the steps in the process by looking at the example provided in the paper, which considers the performance of BP in 2001.

Factor	UK 2001 [£ million] A	BP 2001 [£ million] B	BP Return on Capital C		UK Opportunity Cost of Capital D		Value Spread C-D E	Value Created by BP [£M] B*E, F	Sustainable Cost of Capital B*D, G
Value Created *	884,718	15,563							
Capital Applied ‡	4,375,200	69,885	0.2227	-	0.2022	→	0.0205	1,431	14,132
CO_2 [t]	572,500,000	73,420,000	0.0002	-	0.0015	→	-0.0013	-97,897	113,460
CH_4 [t], Methane	2,195,238	367,201	0.0424	-	0.4030	→	-0.3606	-132,425	147,988
SO_2 [t]	1,125,000	224,541	0.0693	-	0.7864	→	-0.7171	-161,020	176,583
NO_X [t]	1,680,000	266,133	0.0585	-	0.5266	→	-0.4681	-124,587	140,150
CO [t]	3,966,500	124,584	0.1249	-	0.2230	→	-0.0981	-12,225	27,788
Work accidents [number]	132,696	83	187.5060	-	6.6673	→	180.8388	15,010	553
PM_{10} [t]	178,000	16,666	0.9338	-	4.9703	→	-4.0365	-67,272	82,835

Average	**-72,373**	**87,936**
Sustainability Efficiency		0.7947

* UK: Net domestic product [£ million], BP:net value added [£]

‡ UK: Total net wealth [£ million] from National Balance Sheet, BP: Non-financial assets [£ million]

17.27 **Calculation of BP's sustainable value for 2001**
This method compares an organization with its national economy, but a similar approach could be to compare an investment to the average value added by the organization to date. This will clarify if the investment is better performing, i.e. value adding in sustainability terms, compared to the current situation. Where targets for improvement exist, these could also be easily integrated into this methodology, by adjusting the inputs or the value spreads.
Source: Frank Figge and Tobias Hahn. "The Cost of Sustainability capital and the Creation of Sustainable Value by Companies" (Ecological Economics, 2004). The spreadsheet model is available in the companion file pack.

The table above shows the steps involved in the calculation of the total sustainable value created for BP (column F). In only two categories, return on capital applied and work accidents, did BP make a positive contribution compared to the mean of the UK economy. On methane emissions, for example, BP would have had to create £132,425 million more value (profit) to match the average £/emissions of the UK economy as a whole. The average of £-72,373 can be considered BP's overall sustainability value created, a negative indicating that it has a worse impact than the average UK organization. By looking at which factors are worse, BP can identify where it under performs in sustainability terms. Figge and Hann propose another metric, the sustainability efficiency, which is the actual capital applied divided by the sustainable cost of capital. In this example it comes to £69,885 million/£87,936 million, i.e. 0.7947 (I have modified the paper's method to ensure that if this figure is unity, it means that the organization is delivering as much value as the comparator - e.g. the UK economy). The paper goes into more details, but this intrinsic approach may offer a more objective method of comparing a company with its peers or appraising an investment.

17.23 Working with portfolios

Most assessment of improvement involve more than one change. In this situation, the overall benefit of the combined changes may be less than the sum of the individual opportunities. In essence, implementing one opportunity changes the BAU cash flow for a subsequent opportunity.

Bundle projects to increase the total savings achieved at a given hurdle rate.

While individual investments will be assessed in isolation, it is important to note that the saving for a portfolio of projects is often not the sum of the individual opportunity savings. Some projects, for example, a lighting reduction project and a water recycling opportunity, are genuinely independent in that they are acting on completely different resource streams. In this case, the overall cost and benefits are the simple sum of each opportunity and the cash flow of one does not affect the other. However, many opportunities are acting on the same resource stream. Take, for example, the projects in the table below left: I have the possibility to reduce the interior temperature of a building by 3 °C which saves, say, 24% on a heating bill and a second opportunity to reduce the heating time from 16 hours to 14 hours, which saves 12%. Both of these projects act on the same resource stream, the gas I use in my boiler, so these are called compound projects. If I implement both projects the savings will be 33%, less than the simple sum of the savings, which is 36%.

There are other forms of opportunity interdependence that need to be considered when building a business case. For example, two projects might be mutually exclusive – implementing one prevents the implementation of the other. These projects are also sometimes referred to as substitute projects and can be a particular challenge when conducting an audit. Problems arise because most audits are limited in the amount of time and effort available to identify and assess opportunities and so time can be wasted investigating projects for which there turns out to be a better alternative. For this reason, many auditors will try to identify substitute projects early on in an assessment, using some high-level screening criteria such as cost or risk, and then focus on the strongest candidate for detailed investigation, reverting to the next strongest alternative only if that candidate proves infeasible for some reason. The ability to triage opportunities with incomplete data is a key audit skill.

The opposite effect is seen where a project is contingent on another. We can have mutual contingency, in the sense that either both projects go ahead or neither, or one-way contingency where a second opportunity depends on the first, but not vice versa. If two or more projects are mutually contingent, especially if they are compound opportunities, i.e. they influence the same resource stream, they are often merged into a single overarching investment for the financial assessment to reduce complexity in the subsequent business case calculations.

17.28 **Illustration of two projects acting on the same resource**
These are called compound projects and the savings achieved cannot be simply added together.
Source: Niall Enright

Opportunity	*Start kWh*	*Saving %*	*Saved kWh*	*End kWh*
↓ temp by 3 °C	1000	24%	240	760
↓ heating hours	760	12%	91	669
Combined Saving		*33%*	*331*	

In Numbers: *Compound savings*

Let us define S_1 as our starting quantity of resource used for project 1 (1000 kWh in the example opposite) and r_1 is the savings (i.e. 24% expressed as 0.24) and E_1 is the end quantity of resource after applying the savings rate for the first project (760 in the example). For just one project the end amount used is determined by the formula:

$$E_1 = S_1 * (1 - r_1)$$

If savings are expressed as a negative number, the minus sign above becomes a plus. Feeding the actual numbers into the equation we get: $E_1 = 1000*(1-0.24) = 1000*0.76 = 760$

We can generalize this first equation for multiple projects, 1 to n, in effect decreasing the original consumption S_1 by the percentage savings rate of each project one by one until project n where we reach E_n:

$$E_n = S_1*(1-r_1) * (1-r_2)\ldots * (1-r_n)$$

Again, substituting in our example numbers for the two projects above: $E_n = 1000*(1-0.24)*(1-0.12) = 1000*0.76*0.88 = 668.80$

It is often more useful to have a figure for the total savings percentage from the basket of projects. In the example above, we have saved $S_1 - E_n$ kWh and the percentage saving of the combined projects, r_n, would be described with the formula:

$$r_n = (S_1 - E_n)/S_1.$$

We can substitute our expression for E_n into our equation to get the general formula:

$$r_n = (1 - (S_1*(1-r_1)*(1-r_2)\ldots*(1-r_n)))/S_1$$

Simplified, the compound savings rate r_n of n projects acting on the same resource stream becomes:

$$r_n = 1 - ((1-r_1) * (1-r_2)\ldots * (1-r_n))$$

The picture of opportunity interactions is further complicated by the fact that some opportunities may influence several resource streams. Many resources are fundamentally and inextricably bound, such as water, wastewater and energy. Reducing water demand reduces the energy used in the distribution, conditioning and treatment of the water. It may also lessen the requirement for chemical inputs and sludge disposal. These second and third-order benefits can have a very significant impact on the business case for improvement – for example, energy is usually a much more expensive input than water, so the business case for reduced water use may be driven by the associated energy costs rather than the water reduction. We have already seen that for many lighting projects it is the reduced parts and labour costs associated with replacing lamps that drive LED investment decisions, rather than just the energy savings. These co-benefits can also exhibit compound effects if they arise from savings on the same resource stream.

The final recommendations of an audit team usually set out a number of opportunities. The team will, therefore, need to take into account the interactions and interdependencies of these projects as described above, to arrive at a credible business case. Where the organization has established criteria for investment, then clearly the portfolio of projects needs to meet this hurdle rate (unless there are other overriding non-financial drivers for improvement, such as brand, compliance or ethical objectives). Assuming that meeting the financial hurdle rate is critical, this then becomes the primary basis for opportunity selection into the portfolio. Where there are mutually exclusive projects or opportunities with broadly similar returns, then the selection process tends to take into account factors such as ease of implementation, site acceptance or perceived risk.

Bundling multiple projects together into a portfolio has several advantages.

- The portfolio can include projects which would individually fail to meet a financial hurdle rate as they are balanced by projects with a better return.
- Programme cost for continual improvement and management can be more readily justified and allocated.
- Risk associated with uncertainty about individual project outcomes should be balanced as some will do better than expected and some worse.
- The problem of *cherry-picking* the best projects is reduced.

The biggest benefit of a request for a single budget covering multiple projects is the ability to get approval for projects above the investment threshold by combining them with projects that fall well below the threshold.

A key strategy for any energy Champion should be to gain approval for a *programme* meeting the investment hurdle rate, rather than for individual projects. We shall see next that compound projects are also especially sensitive to the sequence or order of implementation.

17.24 Portfolio sequence

In putting together a portfolio of projects which have to meet a required rate of return, we will often come across a situation where we may be forced to choose between the highest absolute saving or the highest return on investment.

Properly assess **project interactions** and **don't assume cheapest is best.**

A strong theme in our Discovery process (see Section 12.5 on page 396) has been the benefit of starting with demand. By taking this approach we avoid the tendency to leapfrog no/low-cost opportunities and go straight to equipment upgrades or replacement.

However, an emphasis on no/low-cost savings can run counter to the objective of achieving the maximum possible resource efficiency. The challenge is that, as better operation and control reduce demand, the financial case for the equipment upgrades becomes weaker, potentially pushing the larger capital investments off the table altogether as they no longer meet the hurdle rate in the new more operationally efficient context. While the strong emphasis on demand reduction set out here is logical, cost-effective and will lead to more rapid savings, the unintended result can be a lower total improvement.

The table below left is based on a real case in a UK car park. Here we have two compound opportunities, i.e. they are opportunities that are not mutually exclusive and act on the same resource: lighting electricity. The first opportunity, O_1 is a project to control the car park lighting so that half the lights are switched off when ambient daylight is sufficient high and it requires a very modest investment in a daylight sensor to automate the control and achieve a 15% reduction in electricity use. The second opportunity O_2 is based on a quotation for the complete re-lamping of the car park replacing T5 fluorescent lamps with LED replacements, which will reduce lighting electricity use by 50%.

17.29 **Illustration of the sequencing of opportunities and payback**
By implementing the no/low-cost Opportunity 1 first, the larger Opportunity 2 falls outside the required hurdle rate.
Source Niall Enright, spreadsheet available in the companion file pack.

Opportunity	*Start Cost £p.a.*	*Saving %*	*Saving £p.a.*	*Invested £*	*Payback (months)*
Treating these as isolated single opportunities					
O_1 Automate Lighting	205,000	15%	30,750	2,937	1.1
O_2 Change to LED lamps	205,000	50%	102,500	190,000	22.2
Sequencing - best return first					
O_1 Automate Lighting	205,000	15%	30,750	2,937	1.1
O_2 Change to LED lamps	174,250	50%	87,125	190,000	26.2
Sequencing - biggest first					
O_2 Change to LED lamps	205,000	50%	102,500	190,000	22.2
O_1 Automate Lighting	102,500	15%	15,375	2,937	2.3
Portfolio - combining both projects together					
O_3 Combined: O_1 and O_2	205,000	58%	117,875	192,937	19.6

Given an annual cost of £205,000 on electricity, both opportunities in isolation meet this organization's two-year or 24-month payback hurdle rate. The combination of both projects into a single opportunity O_3 "*Improve Car Park Lighting*", also falls within the required two-year payback with a 19.6-month payback, as shown in the bottom row of the table. However, a problem arises if opportunity O_2 were to be brought forward separately for funding *after* opportunity O_1 had been implemented, as the saving already achieved through the daylight sensor reduces the saving possible through the LED re-lamping, pushing the second opportunity payback to 26.2 months, above the desired hurdle rate. Thus, by first implementing a 15% saving on the demand side, we have potentially forsaken an initially viable 50% saving on the supply side.

Opportunities which cease to be financially viable as a result of other interventions are called displaced opportunities. Paradoxically, the order in which the projects are implemented will determine whether they will both meet the necessary return on investment or not. As we can see from the table left, the control opportunity O_1 remains financially attractive after the LED change opportunity O_2 has been implemented; although the payback has risen from 1.1 months to 2.3 months, it still provides a highly attractive return.

One of the judgements that the audit team will need to make in constructing their portfolio of projects is to assess whether early action on some opportunities will displace others. It is advisable to assess the incremental payback from projects as a matter of course when reviewing a portfolio of opportunities. In the event that some opportunities will be displaced, the decision on how to progress is far from clear-cut, and depends largely on the funds available and the organization's overall objectives:

- If there is no shortage of funds, i.e. the portfolio of projects will be fully funded so long as it meets the required hurdle rate, I will incorporate all the projects in the portfolio and get approval on the basis that the portfolio as a whole meets the required return.
- If there is a shortage of funds, or a fixed budget, or a fixed target for improvement, which means that some projects which are closer to the hurdle rate will not be funded, I will tend to try to include any projects that could be displaced in the initial list of recommendations to be implemented. This may mean that these opportunities are given preference over some opportunities with a better return. The intention is that the projects with a better return are merely delayed and can still be approved at a later date knowing that they will still meet the hurdle rate.
- An alternative strategy is to assess if the opportunity causing the displacement and the potentially displaced opportunity are mutually contingent and merge them into one single new opportunity. Thus I could have an opportunity O_3 *"Improve Car Park Lighting"*, which is the combination of opportunities O_1 and O_2, with a payback of 19.6 months. However, I would need to be careful that this project will pass the approval hurdle else I risk throwing out the *"no-brainer"* opportunity O_1 of the mix taken forward.

There is a very powerful presentation technique that can be used to illustrate and rank a portfolio of opportunities, called MACCs, which we will consider next.

Exploration: ***Starting with expensive***

Photo © iQoncept, Fotolia.com

It makes common sense to implement the lowest-cost improvement measures first, right?

Well not necessarily. We have seen that implementing a cheap solution first can, in fact, lock in inefficiency by making the more profound changes no longer viable financially.

This conclusion is articulated well in two papers[764,765] by Adrien Vogt-Schilb and Stephane Hallegatte at the World Bank. They also present the argument that *"starting with the most expensive option makes sense"* because deeper and more expensive solutions may be the only means to achieve long-term goals. If that is the case, then these expensive projects may need to start early is because they have a long lead time to deliver improvements.

A similar point was made strongly in our earlier discussion of opportunity timescales (see page 244), where the high return from low-cost projects can support efforts developing the longer-term, more expensive solutions.

There is also an important argument that many opportunities can only be implemented at specific times. Thus, some projects will need a plant shutdown to take place, and other technologies are only cost-effective to install at the initial construction stage, marginal returns on investment need to *"piggy-back"* on planned capital expenditure. Thus, we need to apply common sense in sequencing our investments and not just cherry-pick the low-hanging fruit.

17.25 Creating MACCs

MACC charts offer an at-a-glance overview of a complex portfolio of projects. Used with care, they can greatly assist in the interpretation of the financial and environmental performance of projects.

***In Numbers:** MACC calculations*

The vertical or y-axis axis value in a MACC chart is the net present value (NPV) of the opportunity divided by the annual emissions reduction (for variable emissions reductions we would use the average over the project lifetime).

The NPV will be calculated in the usual way, as shown earlier, using the appropriate real or nominal discount rate for the cash flow concerned.

There is a debate among economists whether the denominator, in this case, tonnes of carbon dioxide equivalents, should be discounted. On the one hand, it is true to say that we want emissions reductions sooner rather than later (damage is cumulative) but on the other hand, the discount rate is based on the weighted cost of capital, not on any environmental datum for a compounding effect. Having debated this extensively with colleagues at the World Bank working on *MACTool* (see later), I have settled on a definition where the denominator is not discounted. The IPCC[453] took this approach in their reports on climate change. Thus the MAC costs is:

$$\frac{NPV(Costs)}{Emissions\ Reduction^{-y}}$$

The MACC abatement is simply:

$$Emissions\ Reduction^{-y}$$

Marginal Abatement Cost Curves (MACCs) are widely used in carbon emissions management programmes as a means of ranking a portfolio of investments. The name gives an impression of complexity, but in practice, a MACC is simply a fancy chart which can help visualize two key attributes for multiple opportunities at a glance: costs and tonnes of CO_2 reduced. The word *"marginal"* means additional, *"abatement"* means reduction and *"cost"* is self-explanatory, while *"curve"* means a chart or graph.

The illustration opposite shows a MACC chart. On the vertical axis we have the cost, in this case €/tCO_2e, and on the horizontal axis, we have the cumulative total tCO_2e abated. Each project is one block, and the projects are sorted in ascending order by the cost. The first project illustrated, on the left of the chart, is "*Lighting Control*", which will save 107 tCO_2e per annum for a cost of -€85 per tCO_2e (the figures are also given in the table below the chart). Since the cost is negative, this project saves money over its lifetime.

The projects are ranked by cost so the ones that deliver greatest savings are always on the left of the chart. The second project "*Comp Plant UG*" saves a little less, -€70 per tCO_2e, but delivers a greater annual emissions reduction of 366 tCO_2e, as indicated by the width of the block. As we move to the right, the emissions abatement becomes increasingly expensive to the point where we see the "*HE Motors*" project has a net cost of €10 per tCO_2e saved and so is shown above the horizontal axis.

Policymakers like this chart because it shows the total emissions reduction that can be achieved at no cost (in this case 1,488 tCO_2e per year). It also indicates that, if a carbon price of €20 per tCO_2e is available, as shown by the dashed line on the chart, the "*HE Motors*" project is cost effective to implement.

One of the problems with MACC charts is that Excel cannot produce variable-width column charts. There are some workarounds to this, such as using the error bars in an XY scatter plot to draw the exterior of the individual blocks.

The chart opposite has been drawn using MACC Builder Pro, an Excel MACC charting tool which I developed several years ago. This tool is used purely to draw MACCs, using five very basic inputs about the projects: the project name, the capital cost, annual emissions abatement and project life. Some system settings such as currency symbol and discount rate complete the inputs. You will also notice that the MACC chart has each block coloured depending on the technology category of the opportunity. MACC Builder Pro supports user-definable categories based on labels or numeric values.

MACC Builder Pro offers a broad range of customization options. You can add a CO_2 price line to the chart. Labelling can be via callouts (as shown) or numbers/labels linked to a data table. There is an ability to include values in the table (below I have chosen to include the tCO_2e abatement and the marginal abatement cost). Statistics can be added to the table (I have included total CO_2e and the *"intercept"* CO_2e, which represents the sum of all projects with a negative cost, i.e. a saving; as well as axis and full control over colour and font styles for all the chart elements. ⇒ page 614.

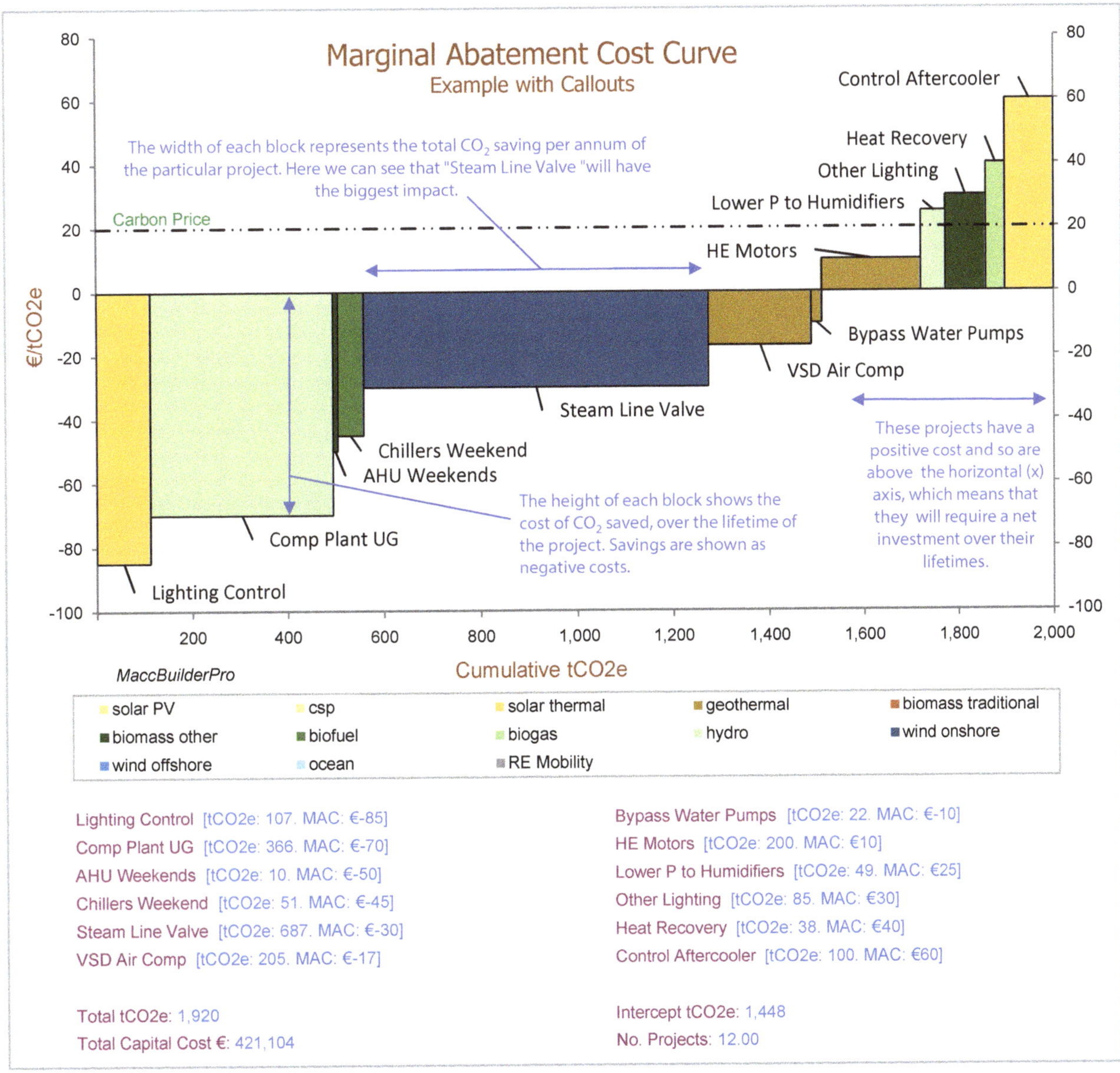

17.30 **MACC chart**
The chart is illustrated with explanatory text overlaid in blue.
Source: Niall Enright using MACC Builder Pro software, available in the companion file pack

As one would expect from a very knowledgeable team of economists at the World Bank, the flexibility and capabilities of MACTool are truly staggering.

Real World: *Advanced MACC production, the World Bank /ESMAP's MACCTool*

In the last year, I have been involved as the lead Excel developer in a fascinating project to create a powerful MACC analysis package for the Energy Sector Management Assistance Program (ESMAP) of the World Bank. The lead on the client side was Victor Loksha, a senior energy economist, with input from Martina Bosi, Perre Audinet, Grzegorz Peszko, Adrien Vogt-Schilb and many others. The project was led by ERM's Peter Rawlings and Braulio Pikman, supported by Sandra Seastream, Wairimu Mwangi and additional Excel development by James Joyce and Calvin lost. This project was a major redesign, building on an earlier version of MACTool, developed by Andreas Mastle and led by Christophe de Gouvello at the World Bank.

MACTool is a remarkably powerful MACC creation tool. Unlike MACC Builder Pro, which simply draws MACC charts, MACTool provides inputs for multiple cash flows in a project and will calculate the marginal abatement cost and emissions abatement values given these inputs. Features provided by MACTool include:

- Separate cash flows for a baseline and low carbon case. These cash flows can be created from templates designed for different sectors, so that a generation technology, for example, can have a revenue cash flow from the sale of electricity. The templates can be modified by users and reused.
- An intuitive interface that takes away the complexity of the underlying model and allows the user to manage many different projects using a pop-up navigator.
- The ability to categorize projects by sector/technology, SIC code, or region.
- MACC charts can be set to show an overview of each category which can then be *"drilled down"* to show individual projects within the category.
- Complete flexibility over the discount rates used so that these can apply at the model level, by sector or globally. There is the option to discount the abatement term, if desired (see the discussion on the previous page), using a separate discount factor.
- A lot of attention has gone into bringing the key assumptions for all the projects into a single table which allows for rapid validation and error checking. Models can have an *"in progress"* status, which means they are not included in totals until they are marked *"complete"*.
- The ability to chart investment costs, investment intensity and other financial metrics, not just the marginal abatement cost.

As one would expect from a very knowledgeable team of economists at the World Bank, the flexibility and capabilities of MAC Tool are truly staggering, with the ability to use multiple currencies and to run cases also built into the system.

One metric that appears as standard in the templates that ship with MACCTool is the breakeven carbon price. This is based on the notion that while there may be an overall market discount rate set which takes into account the cost of capital in an economy, private investors will usually expect a higher rate of return. Thus the present value of a project can be worked out using the desired rate of return rate, such as 20%. If the project PV is negative it means that it has made a loss, so MACTool will then calculate the *additional* incentive in US$/tonne CO_2 that is needed to deliver the percent target return required by the investor. Neat!

Another innovation in MACTool is the wedge-MACC chart. An example of this chart is shown below. On the left is a normal wedge chart showing the emissions abatement per year from a range of different projects/technologies. Because these are added together the height of the chart is the cumulative total. Well, we know that a MACC chart also shows a cumulative total along its horizontal axis. So, if we rotate the MACC vertically, as shown on the right-hand side of the image, we can see the MAC for each opportunity in ascending order from best return to worst return, with the height matching the total saving at the end of the analysis period. This combined wedge-MACC chart was first described in a paper[766] by Adrien Vogt-Schilb, Stephane Hallegatte and Christophe de Gouvello at the World Bank.

The team at ESMAP and the World Bank have made MACTool freely available[267] and would welcome feedback from users. There is a handy guide on how to set up your first model, which users are strongly advised to work through before leaping into the tool. It is also worth mentioning that I have worked with Pedzi Makumbe from ESMAP and Braulio Pikman on another ERM project to deliver a different Excel-based tool called TRACE which looks at the potential for municipalities to reduce energy use by selecting relevant opportunities across numerous sectors.

17.31 **A wedge-MACC chart in MACTool**
The chart below is taken from an analysis of Costa Rica's emissions abatement opportunities prepared in advance of the Paris UNFCC Summit in 2015 (COP21). The models were prepared by local experts Francisco Sancho, Luis Rivera and German Obando working with Martina Bosi and Marcos Castro from the World Bank and Braulio Pikman from ERM.
Source: World Bank/ESMAP MACTool. The tool is available on the ESMAP website.[267]

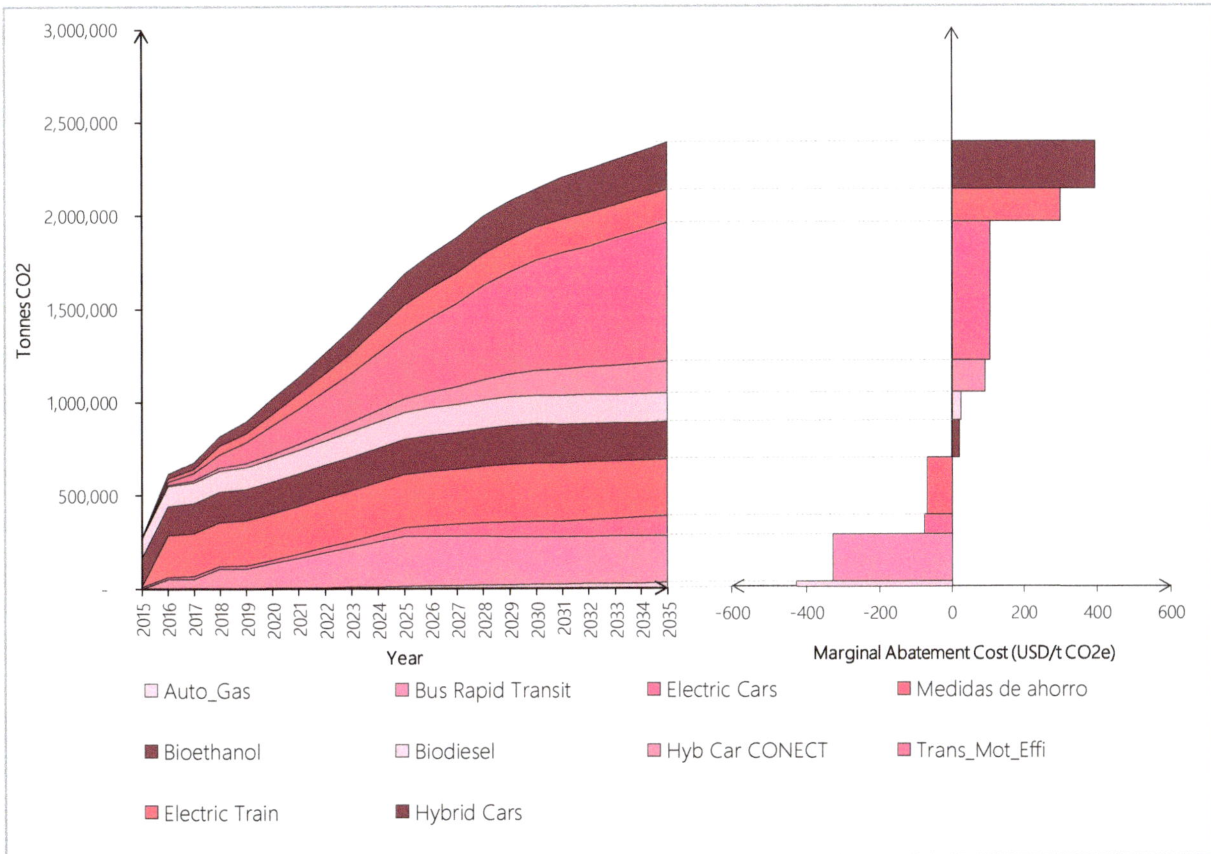

17.26 MACC variants

MACC charts can illustrate resources other than emissions. Here, we look at some variants that show electricity savings.

Traditionally, MACC charts have been used to illustrate abatement of CO_2 emissions. However, there is no reason why another resource cannot be illustrated, both as the denominator on the y-axis cost calculation and as the cumulative figure on the x-axis. In the two examples on this page, we have the MACC based on MWh of electricity. In fact, MACCs, also called conservation curves, were first developed for electricity consumption ($/kWh) by Alan Meier[521] following the oil shock of the 1970s.

MACC Builder Pro also permits MACCs to be drawn based on IRR (in which case the y-axis order is set in reverse, from negative at the top to positive at the bottom, so that the projects with the largest positive IRR are shown in the bottom left of the chart, maintaining the convention of MACCs). As a result of customer request, despite the issues with this measure, MACC Builder Pro also produces MACCs based on payback, where projects are ranked from those with the shortest payback to the longest. In both IRR and payback charting, the absolute value can be shown or the variance from a threshold (thus the x-axis can be set to match the hurdle rate so that the projects below the axis are those that meet the requirement).

Real World: ***Energy MACCs in UK office sector***

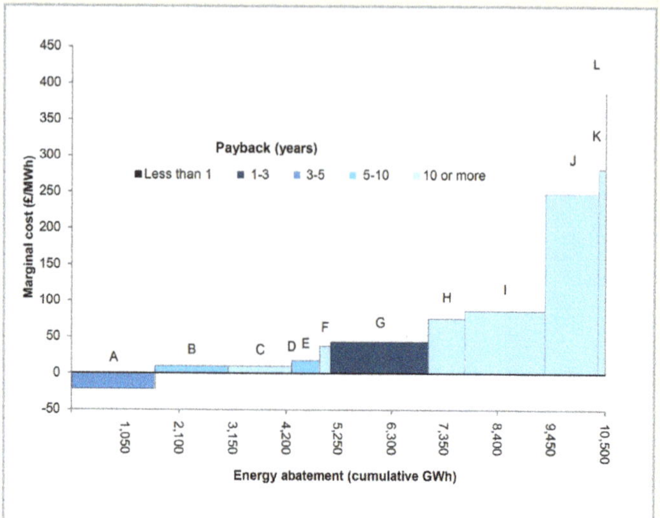

17.32 **MACC illustrating energy abatement costs and opportunities in UK offices** This chart is taken from the sector reports in the Building Energy Efficiency Survey[205] carried out by consultants Verco and GFK on behalf of the Department for Business, Energy & Industrial Strategy, and published in November 2016. The MACC chart shown here illustrates MWh of energy rather than CO_2. The MACCs were drawn using MACC Builder Pro (MBP). Please note that MBP has calculated the payback of each item, and shaded the MACC accordingly. Note, too, that the interactions between abatement opportunities have not been considered, so the total reduction in energy use achievable may be less than that shown in the chart. *Source: "BEES Sector Report Offices", UK Department of Business Energy and Industrial Strategy, 2016. Reproduced under the terms of the Open Goverment Licence.*[724]

The abatement costs were calculated using the UK Government's guidelines on the valuation of energy, which takes into account the capital expenditure, operational expenditure, social cost of energy, air quality impacts, and value of emissions, all discounted at the social discount rate.

KEY to chart:

A Lighting [MAC: £-21 per MWh. GWh: 1,650]
B Building instrumentation and control [MAC: £10 per MWh. GWh: 1,430]
C Building fabric [MAC: £10 per MWh. GWh: 1,260]
D Cooled storage [MAC: £14 per MWh. GWh: 10]
E Air conditioning and cooling [MAC: £18 per MWh. GWh: 570]
F Hot water [MAC: £38 per MWh. GWh: 220]
G Carbon and energy management [MAC: £44 per MWh. GWh: 1,950]
H Ventilation [MAC: £76 per MWh. GWh: 720]
I Space heating [MAC: £86 per MWh. GWh: 1,580]
J Small appliances [MAC: £248 per MWh. GWh: 1,030]
K Building services distribution systems [MAC: £281 per MWh. GWh: 140]
L Humidification [MAC: £388 per MWh. GWh: 1]

17.33 **MACC illustrating carbon emissions saved at the Reef HQ Aquarium**
When the team at the Reef HQ Aquarium use the results of their energy efficiency programme as an education tool, they chose a MACC as a highly effective way to communicate the improvements achieved. This chart was drawn with MACC Builder Pro, and can incorporate different types of data and factors, depending on the audience. *Source: Sascha Thyer, Technical Operation Manager, Reef HQ Aquarium, Great Barrier Reef Marine Park Authority, Queensland, Australia.*

Real World: *Communicating savings using a MACC chart*

Reef HQ Aquarium is the Australian National Education Centre for the Great Barrier Reef, a major tourist attraction in tropical north Queensland. In eight years, Sascha Thyer and the team at the Aquarium achieved an outstanding 50% reduction in grid electricity use by embarking on a comprehensive range of infrastructure upgrades designed to reduce the facility's power costs, and greenhouse gas emissions.

Reef HQ Aquarium demonstrated that environmental sustainability in business can be greatly increased, whilst still growing and improving their business. Key initiatives included operational measures, installing energy efficient equipment and a large photovoltaic (PV) solar system. The Aquarium is now registered as a solar power station (206 kW) and has achieved payback for all measures adopted since 2007.

With a vast array of measures undertaken, a tool was required to help visitors to the Aquarium and other stakeholders to easily understand the main outcomes. A key driver for the Aquarium is to assist individuals and businesses to decrease their environmental footprint, which helps to protect the Great Barrier Reef. The MACC chart summarises the large volume of data gathered over 8 years into a simple visual comparison which is used as an education tool to engage visitors in discussions about business sustainability and the Great Barrier Reef. The MACC chart allows comparison between common measures requiring large capital investment (like solar power) with low cost operational changes such as adjustments to indoor temperature or minimising building air leaks.

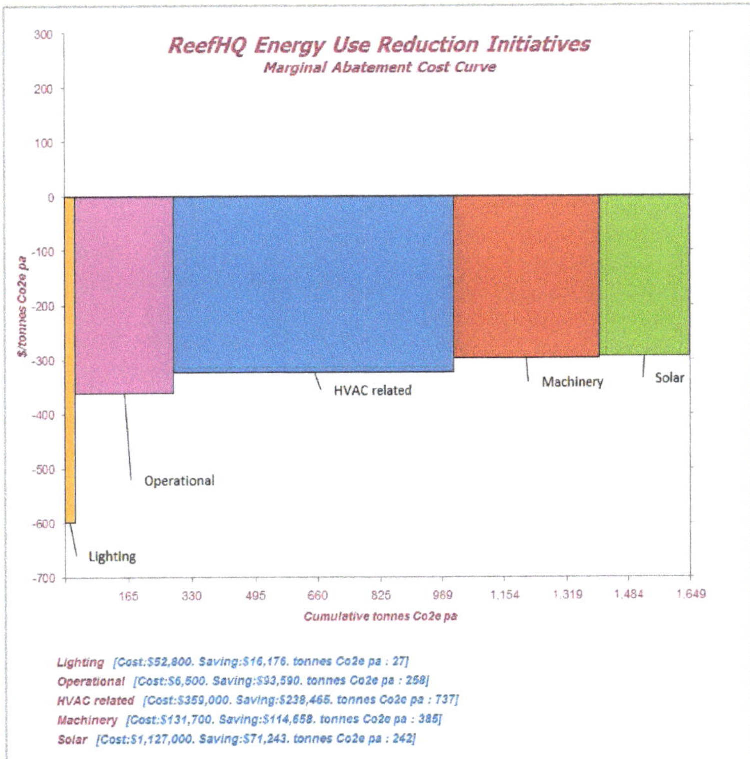

Funding

17.27 MACC problems

MACC charts are not without their challenges. In particular, placing the desirable feature - abatement - in the denominator of the calculation can lead to some unexpected results.

The global consultancy McKinsey & Co has been credited with popularizing MACC charts through a series of international studies on emissions abatement potential in 2007,[259] later updated[260] to reflect the effects of the global economic crisis. Although McKinsey published the data used for the charts, the assumptions behind the calculations were not made clear, leading some critics of the models to argue that the wider benefits of greenhouse gas abatement, such as improved health due to lower air pollution, were left out of the cost analysis.[247]

Another problem related to the cost calculations is that, since MACC costs are expressed as the net present value, the choice of the discount rate used has a huge impact on the results. Low discount rates, as used in the public sector and studies such as the Stern Report on the Economics of Climate Change, make abatement opportunities appear *"no-brainers"*, but may not reflect *"real world"* returns expected in the private sector. This ties in with the idea that a negative cost should simply not be possible in a well-functioning economy[8] (see page 154).

A further weakness of MACCs is that they cumulate the total abatement achieved along the x-axis[440] whereas we have seen that projects may have interactions that mean that you cannot simply sum them to get a total. Thus, MACCs may *overstate* the emissions abatement potential for any given carbon price, their primary function in policy-making.

Then there is the practical problem that MACCs cannot easily be drawn in Excel, which requires some inelegant workarounds or third-party tools. Even when drawn, MACCs can only practically show up to 100 projects, beyond which labelling and display width become real issues (many of the McKinsey MACC charts only have a few of the projects' names, leaving the user to guess what the missing ones are).

Beyond the theoretical and production problems with MACC charts is an even more serious shortcoming. By placing the emissions saved on the denominator term, MACC prioritizes projects with a high-cost saving and a low emissions reduction. In fact, people intuitively prioritize high emissions savings for low-cost, the precise opposite of what a MACC chart illustrates. The case study left and box right explore the counterintuitive nature of the MACCs prioritization of abatement opportunities.

Real World: ***The devil is in the detail***

A while ago I had a very interesting email exchange with the Worldwatch Institute, which is putting together sustainable energy roadmaps for small island states in the Caribbean.

The data they were using was:

- Costs - oil: \$220.5, natural gas: \$137.6, hydro: \$52.7
- Emissions - oil: 0.66 tCO_2e, natural gas: 0.40 tCO_2e, hydro: 0 tCO_2e

The McKinsey approach (see Exhibit A.III.2)[545] defines the marginal abatement cost as:

$$Abatement = \frac{Project\$ - Ref\$}{RefCO_2 - ProjectCO_2}$$

That is the difference between the cost and emissions of the project and that of the reference emitter. Using oil as the reference emitter, we thus get to the following calculation of marginal abatement cost:

- Natural gas: (137.6 - 220.5)/(0.66 - 0.40) = -\$318.8/tCO2e
- Hydro: (52.7 - 220.5)/(0.66 - 0) = -\$254.2/tCO2e

Using this method, we get to a situation where, counterintuitively, Hydro appear less attractive from an emissions perspective than natural gas. But the figures are correct! See opposite for an explanation.

Exploration: *The counterintuitive nature of MACC rankings*

Photo © Zerophoto, Fotolia.com

A MACC curve is a prioritization tool (sometimes called a merit order tool) to enable a portfolio of emissions reduction projects to be ranked. It is designed to highlight projects in order of net present \$ cost per tCO_2 saved, which is shown on the vertical axis. The secondary datum, the absolute emissions reduction for the project, is shown by the horizontal axis and is not used to rank projects.

In selecting emissions reduction projects the merit order in which most people would place projects could be described as: *"to achieve the greatest emissions reductions at the lowest possible costs"*.

A MACC merit order is actually doing something subtly different. It is showing projects in order of the lowest possible cost per unit of emissions reduction. This value is the ratio we see on the MACC chart, $\$/tCO_2e$, which determines the left-to-right ranking of the projects. This subtle difference is at the heart of the query posed in the real world example, opposite. Keeping the numbers simple, let us consider three projects, which will help to illustrate the difference:

- A delivers US\$100 savings and 4 tonnes reductions in emissions
- B delivers US\$50 savings and 2 tonnes reduction in emissions
- C delivers US\$30 savings and 1 tonnes reduction in emissions

Comparing A and B, we would instinctively say that project A is the most attractive because we save more money and more emissions than project B. We also might conclude that project A is four times more effective that B – it delivers twice the savings and twice the emissions reductions. Clearly, if we had the choice of just one project we would choose A rather than B. However, from the MACC perspective both projects have the same ranking because the marginal abatement cost (MAC) of both projects is -US\$25 per tonne of CO_2 abated (-US\$100/4 or -US\$50/2). That is to say, from a financial perspective they are indistinguishable in terms of cost to deliver a tonne of emissions reduction.

In a similar way, we might also instinctively feel that project A and B are more attractive than C because both the savings and the emissions reduction delivered by A and B are larger than C. However, in a MACC approach, project C would be prioritized over project A because it has a MAC of -US\$30 per tonne, (which is 20% better than A). A MACC is ranking purely on the basis of lowest cost per unit of emissions reduction. The MAC is telling us that, from a financial perspective, project C is the cheapest as it delivers the greatest cost saving per tonne of CO_2 reduced. Thus project C is first in the merit order.

We perceive the results for hydro and gas in the case opposite as counterintuitive, because we instinctively expect the MACC to show the emissions reduction per dollar spend, whereas it really shows the dollar spend per emissions reduction. Hydro beats gas in both the emissions saved and the dollars saved compared to oil, so we intuitively rank it higher. When I got the email, I also thought that *"this can't be right!"* However, the reality is that the marginal abatement cost of the gas project is better, at -US\$318 per tonne, than that for hydro at -US\$254 per tonne and so the MACC chart would place it to the left of the hydro project, a higher ranking. Gas ranks higher than hydro, because the ratio of the saving of dollars to the savings of emissions is greater for gas than for hydro. A project with a large cost saving divided by a small emissions reduction gives a more negative MACC, than a project with a large cost saving and a large emissions saving.

To illustrate the point: project A, above, delivers US\$100 saving and 4 tonnes of emissions reduction and has a MAC of -US\$25. Now imagine project D, which achieves US\$100 saving and an emissions reduction of 2 tonnes, so its MAC is -US\$50. Project D is ranked higher than A despite having half the emissions abatement for the same cost. Instinctively we want both cost savings and emissions savings to be high, but placing the emissions savings on the denominator means that the MACC does not do that. MACCs actually prioritize projects with large cost savings and small emissions reductions. That's simply the way the maths works - if we want to prioritize both we should multiply the numbers together to get a *"compound savings product"*, or similar.

$$\textit{Marginal Abatement Cost} = \frac{\textit{Cost}}{\textit{Emisions Reduction}},\ MAC_A = \frac{-\$100}{4\ tCO_2e} = -\$25/t,\ MAC_D = \frac{-\$100}{2\ tCO_2e} = -\$50/t$$

17.28 ERICs

An alternative to the MACC chart has been proposed, based on the internal rate of return. The emissions return on investment curve can illustrate the return on a portfolio of projects and has the added benefit of encouraging bundling of opportunities.

Recognizing the limitations of MACC, curves a number of alternatives have been put forward by resource efficiency practitioners. One such is the emissions reduction investment curve (ERIC). This chart was proposed[465] by Dr Greg Lavery, one of the directors of Lavery/Pennell, a consultancy specializing in resource efficiency and the circular economy.

An ERIC shows projects in descending order of the internal rate of return (IRR). As with a MACC, the most desirable projects are shown on the left of the chart and the horizontal axis shows the cumulative emissions abatement achieved. An example of an ERIC is shown on the page opposite. The key feature of the chart is the red line which plots the cumulative IRR for the basket of projects from left to right. The advantages of the ERIC chart are:

- It presents the merit order of projects clearly, maintaining the same left-to-right presentation as a MACC. Also more positive = better, which is intuitive.
- The cumulative abatement on the x-axis, although suffering from the same problem as MACCs in that projects may not simply sum up due to interactions, nevertheless gives an indication of which investments are needed to achieve a particular goal.
- The ERIC line encourages bundling of projects for approval. We can see a hurdle rate of 25% IRR overlaid on the chart. Individually, projects G-J would not receive funding, but in a basket with the other projects, the ERIC shows that they exceed the required hurdle rate.
- Because IRRs are used, there is no need to worry about whether the appropriate discount rate has been used, or for that matter about whether all projects should have the same discount rate.
- IRR is understood by finance decision-makers, whereas MACC may take some time to explain.

One day ERICs may be as familiar in the dictionary of resource efficiency practitioners as MACCs.

There are a few, relatively minor, disadvantages that come to mind. Hopefully, these won't impede a wider take-up of this form of presentation.

- We have already noted that the project with the best IRR does not necessary represent the project with the best net present value given a specific discount rate.

17.34 **ERIC chart of a portfolio of emissions reduction projects**
This chart shows the IRR of a basket of projects on the vertical axis and their cumulative emissions abatement on the horizontal axis. Overlaid in red is the ERIC line showing the cumulative IRR of all the projects to that point, and in blue is the hurdle rate of 25% IRR.
Source: Niall Enright, using a beta version of the new MACC Builder Pro. Inspired by Dr. Greg Lavery.[465]

- The same charting problem of variable-width columns means that creating these charts in Excel is not particularly easy and may require specialist tools.
- Given a particular set of costs and annual savings, the red ERIC line may inflect upwards, which is counterintuitive and potentially confusing.

Finally, I think that the name ERIC should be changed to make this a generic resource efficiency portfolio tool - we could call it an "*efficiency return on investment curve*", so that it keeps the memorable abbreviation, ERIC.

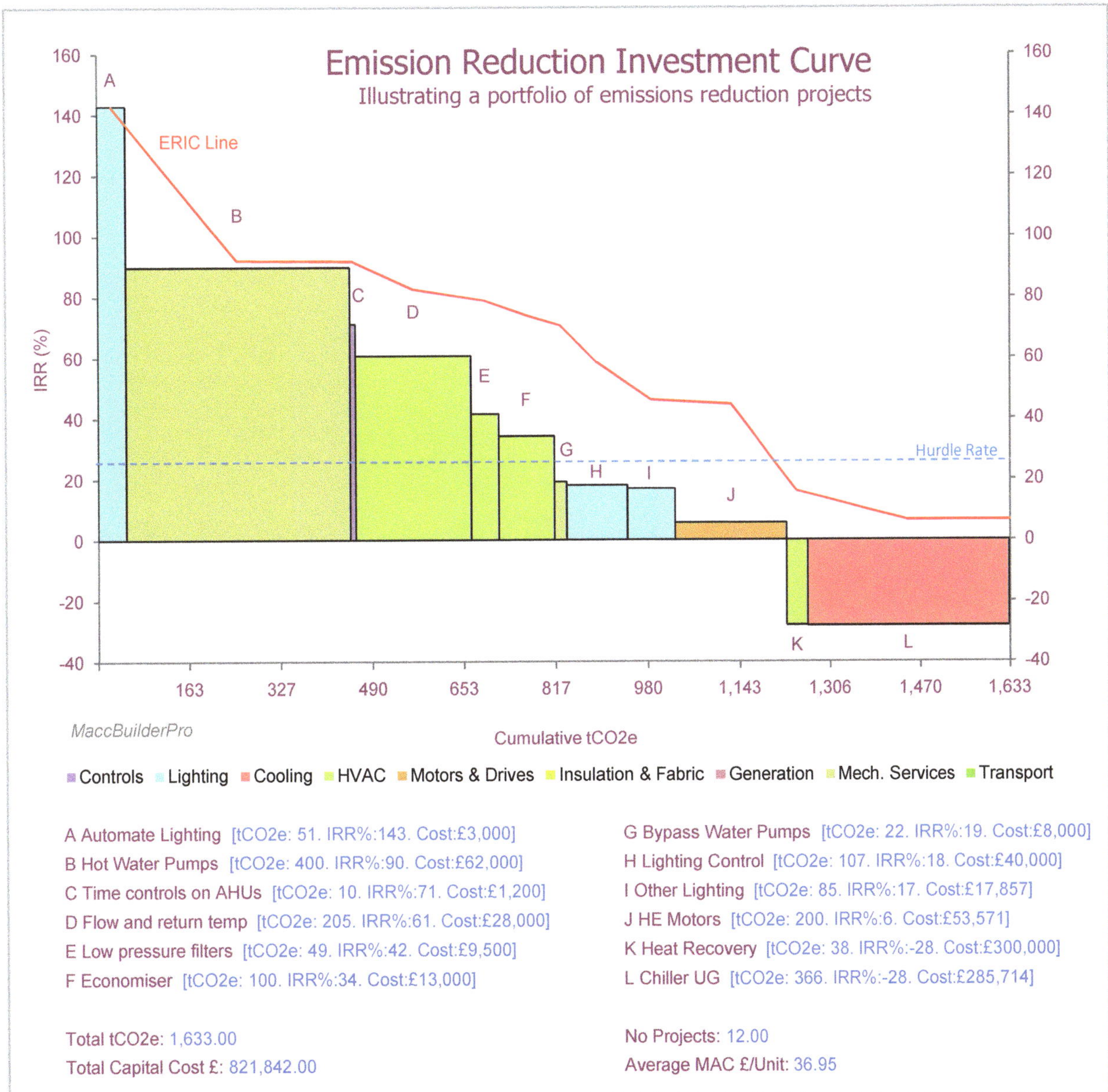

Note: *Companion files*

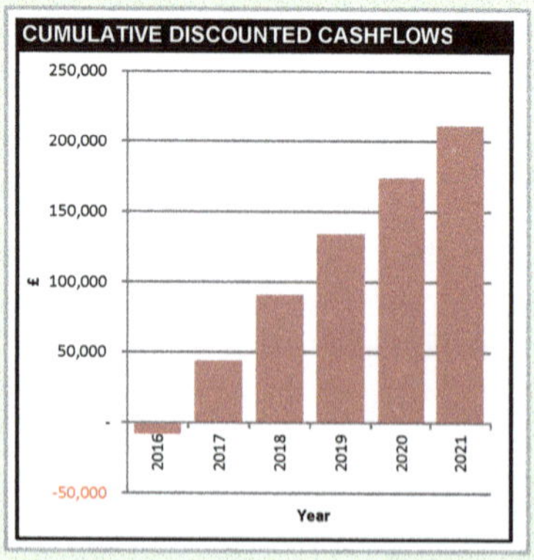

The Microsoft Excel™ spreadsheet models in the companion file pack provide further examples of the financial appraisal calculations in this chapter.

These workbooks expand on the methods described in the text, with additional explanation and examples. The spreadsheets are designed as teaching and learning aids.

There are also additional tools for resource efficiency practitioners.

1. There are five full financial case models: for a "generic" project, lighting, motors, Monitoring and Targeting, and wind projects.
2. There is a full working copy of MACC Builder Pro which previously was sold for £85/US$100.

These resources are available free of charge to buyers of the print edition of this book and for a modest fee for those using the free PDF version.

There are a few limitations on the resale, copying and commercial use of these tools, as well as limited warranties.

See www.sustainsuccess.co.uk/iwik for details on these limitations and how to download these resources.

Further Reading:

Bierman and Smidt, *The Capital Budgeting Decistion*. Routledge. ISBN-13:978-0-415-40004-6. *This is one of my main references for financial analysis of investments.*

Don Dayananda and colleagues. *Capital Budgeting, Cambridge*. ISBN-13:978-0-521-52098-3. *An excellent explanation of all aspects of financial appraisal of projects.*

Questions:

1. What are the characteristics of a good investment and why?
2. What influences the credibility of a business case and what can be done to improve this?
3. What are the strengths and weaknesses of the following financial metrics: payback, net present value and internal rate of return?
4. Should we always invest in the cheapest projects first? If not, why not?
5. Consider some recent investment business cases made in your organization. For each identify if there are any core benefits that were not presented in the original business case and how these might be included in future.
6. Take a sample business case from your own organization. Considering the concepts set out in this chapter, is there anything you would change and if so why?
7. In what ways are MACC charts counterintuitive? Is it possible that less favourable opportunities could be prioritized higher than more favourable ones? Discuss.
8. What do you think makes for a good business case?
9. Why should we consider the cost of externalities in our business cases? What are the challenges you would anticipate adopting this approach in your organization and how do you think these could be overcome?
10. Review the key concepts set out in this chapter (in green) and sort these in order of importance. Explain why you have made this choice.
11. What effect does risk have on an investment business case? What kinds of risk exist? How can risk be quantified and managed?
12. How can timing or sequencing of projects lead to different investment decisions?

18 Funding for Improvement

Photo © Oleksandr Dibrova, Fotolia.com

It would be easy to assume that, having made a compelling business case which spells out the great value that our efficiency investment provides, funding would be readily forthcoming.

The reality is quite different, as we discussed in the Chapter 4 commentary on barriers (page 151). Within organizations, there are structural barriers, split incentives, psychological factors, perceptions and behaviours which act against investments in resource efficiency. All too often these are summed up by the phrase, "*we don't have the money*". In survey[605] after survey,[397] the lack of funds is cited as a major constraint to achieving energy efficiency, although this is also used as an excuse for a broader disinclination to support investment.

The availability of funds for resource efficiency is not just an organizational challenge; it is a global one. The International Energy Agency's *World Energy Outlook 2012*[390] reported that for their "*Efficient World Scenario*" (page 297):

> *"Additional investment of US$11.8 trillion in more efficient end-use technologies is needed, but is more than offset by a US$17.5 trillion reduction in fuel expenditures and US$5.9 trillion lower supply-side investment."*

More recently, the New Climate Economy estimated that the necessary expenditure on sustainable infrastructure is US$6 trillion *per year.*[549]

In this chapter, we shall explore techniques to help us unlock funding for deserving projects. First, we will consider the different sources of financing, the importance of budgets and budget-setting processes within organizations. Continuing on the theme of internal funding for investments, we will then look at centralizing funding in order to achieve a better return on investment. We will also consider how a revolving fund can sustain investment in the long-term.

If internal funding is not available, we can turn to third-party finance. We will discuss the various forms that this can take and the practical aspects of obtaining external support. Here, we will consider debt, lease financing, bonds and energy performance contracting. We will also see how financing can be attached to a property and recovered through local taxes. Finally, market support mechanisms are explored. We will look at carbon markets and other forms of incentives as well as direct investments by utilties arising from demand side reduction or public policy requirements.

18.1 Budgets and ad-hoc funding

Understanding how budgets are allocated is a key skill of any resource efficiency practitioner. If conventional budgeting processes are not helpful there are other approaches we can adopt, ranging from simple ad-hoc requests for support through to sophisticated revolving funds.

The money for investments almost always exists. It lies in the existing resource budgets, as **waste**.

In the next chapter, we will see that many of the decisions that people take are based on unconscious "*fast brain*" routines. Organizations, too, have routines that are designed to simplify decision-making. One such habit is the use of simple payback to evaluate investments. Another powerful set of routines exists around budgets, as described by Paul Stern and Elliot Aronson in their 1984 paper, *Energy use: The human dimension:*[681]

> *"The most obvious organizational rules are those associated with budgets, which represent plans and agreements. Organizations use budgets and rules about them to manage expenditures. They can be made somewhat flexible and somewhat contingent on uncertain future events, such as revenues, and they may be renegotiated to some degree, but budgets function as routines for delegating expenditure authority. They are rarely underspent, relatively rarely overspent by much.*
>
> *Expenditures that can be fit into a current budget require less organizational consultation and approval than those that cannot. As a result, the real availability of funds for a project in an organization depends on such things as the stage of the budget cycle, the departmental location of the project, and the amount of slack in the budget. Many organizational investments relevant to energy consumption involve asking whether there is money in the budget for the project, rather than what its return on investment or payback period might be."*

Budget development in larger organizations is as much about power and prestige as it is about maximizing value to the organization. Managers tend to be reluctant to decrease the allocation of resources to their function or to have that allocation directed in ways that they do not control. Budgets are also barriers to change, as people base current budgets on previous budgets.

Which brings us to a fundamental conundrum in resource efficiency; the money for investments almost always exists. It lies in the existing resource budgets as waste. Our challenge is to redirect that allocation of money elsewhere.

In essence, we have our energy/utilities/materials budget lines and our facilities/maintenance/engineering/operations budget lines - by moving money from the former to the latter, we can release funds for improvement. Sounds simple? In reality, this is far from easy as we have first to prove the quantum of waste (i.e. the savings we will make) and secondly we need to time the expenditures so that the investment has the desired effect within the budget year if the costs and savings are to balance. Moreover, some of the investments in efficiency are liable to be on capital equipment whereas the resource budgets are operational costs and, as we shall see later, these are rarely interchangeable.

This is where our Mandate comes in. If our leader is committed to resource efficiency, then they should intervene on our behalf to ensure the redistribution of expenditure across our budgets. A good starting point, if there is no other evidence, is to assume the "10% rule" applies (page 314), and redistribute 10% of the resource costs in the target areas for resource efficiency investments. If we have an ISO 50001 programme, we can take advantage of the requirement in the standard to make available the necessary resources to achieve the continuous improvement goals to support our funding requests.

To ensure that we get budget allocation for efficiency projects, we need to understand the timing and people involved in the budget-setting process, and work well ahead of time to engage with these folks so that they are primed to submit/support/approve the projects, depending on their role.

The alternative to budget allocation for resource efficiency is to go for ad-hoc, case-by-case approval. In some organizations, deserving projects may receive funding, regardless of whether they have been budgeted for. There are advantages and disadvantages to this approach as shown by the table below.

Budget		Ad-hoc	
Pros	Cons	Pros	Cons
There is a clearly defined pot of money.	*Effort - need to predict savings and costs in some detail.*	*Easy to administer - no budget hoops.*	*A tendency to focus on in-year funding.*
There is a tendency to take a more strategic approach to prioritizing projects.	*Pot may be limited - once exhausted there may be no more money.*	*May be able to access more funding than in a budgeted approach.*	*Projects are not necessarily prioritized - less effective projects may be funded first.*
Costs are matched by a requirement to save - so projects have to deliver.	*Location of budgets may not match accountability (split incentives).*	*No certainty - effort may be wasted on projects that are turned down.*	*Since there is no budget- there is no target.*

One funding technique that I have seen employed very effectively is to carve out a *revolving fund* for efficiency investments. Here, the organization makes an initial financial commitment to resource efficiency which is invested in cost-saving measures. The savings from these measures then accumulate in a fund for further projects, sometimes after initial investment has been repaid or sometimes straight away.

It is quite common for efficiency programmes to work on an ad-hoc or case-by-case basis if fundamental changes are needed to budgets before efficiency investments can be properly allocated. This approach is usually because costs are not correctly allocated, as we shall see next.

Real World: *The Harvard Green Fund*

HARVARD UNIVERSITY | Sustainability

We saw earlier that Harvard University had set itself a goal of reducing CO_2 emission by 30% by 2016. One of the key programmes to achieve this goal is the Green Revolving Fund.[358]

This programme has a fund of US$12 million and will accept projects with a payback of up to 11 years. Over 200 projects have been funded so far.

To gain support, the projects must have a positive NPV and are prioritized by greenhouse gas reductions (apart from 25% which demonstrate innovation).[357] They must also have a Measurement and Verification plan in place to confirm the project results.

Although the projects can have very long paybacks, the overall fund has achieved an annual return on investment of 29%,[40] delivering savings of US$4.8 million a year in 2013.[286] That is an outstanding return on investment and further demonstrates the great value that resource efficiency at scale can bring.

Participating schools and departments don't get to keep the savings, which are returned to the fund. However, their incentive is to upgrade the efficiency, comfort and functionality of their facilities at no cost. We shall see later that, in the US in particular, funding for facility upgrades through energy savings is a well-established process.

The Harvard fund was one of the first established in 2001[656] and today the Association for the Advancement of Sustainability in Higher Education lists 85 revolving funds with US$118 million to spend serving 81 higher education institutions.[39]

18.2 Departmental cost allocation

In larger organizations, central cost allocation can be a significant impediment to engaging resource users in reducing consumption. Here, we explore how a simple cost allocation system can be made to work to drive funding for improvement.

Real World: *Budgeting losses*

© Drobot Dean, Fotolia.com

Departmental cost allocation is a very useful system change to create visibility and awareness of resource use, but what about creating awareness of the resource waste?

One technique that I have employed a few times is not only to allocate costs, but also to allocate losses explicitly. This allocation can only be done where there has been a reliable audit or assessment of efficiency. Thus the budget would look something like:

Budget:

Useful electricity	£35,000
Wasted electricity	£18,500
Total electricity	**£53,500**

I must confess that this approach has provoked some very strong reactions. Imagine sitting down once a year budgeting for the amount you plan to waste in the following year!

This technique has proven to be particularly useful in gaining management support for investment in improvements.

One of the key problems with budgets is that they compartmentalize organizations, which can lead to irrational behaviours. It is often the case that the cost of resources are borne centrally, whereas the expenditure to deliver improvement needs to be allocated elsewhere, for example, in the budgets of the operations, engineering or maintenance teams. If these teams are not paying for the resources (and so will not see any benefit in a reduced consumption), then they have little incentive to reduce waste - especially if this involves cost and effort on their behalf and they have other targets to meet. Furthermore, by not breaking the costs down to the location in which they are borne, it is not clear where efficiency efforts might best be focused.

For these reasons, departmental or account centre cost allocation is almost always a prerequisite for a sustained continuous improvement programme. Where possible, this cost allocation should follow the management structure of the organization, so that costs can be discussed in regular review meetings, and incorporated into the usual budget-setting processes. In short, line managers need to be responsible and accountable for their resource use and improvement targets in their areas of the business. Energy *"Managers"* are rarely managers in the real sense, as they do not have the power that line managers have over the deployment of money and people.

First, we should note that the procurement team have an important role to play in this process. The buyers need to ensure that the contract for the resources means that reduced consumption will result in reduced costs. In some cases *"take or pay"* or minimum volume contracts mean that a reduction in unit use does not lead to a saving. In this case, allocating resource costs may be a waste of effort.

Assuming that the resource costs are variable, by and large, the best place to locate these costs is as close as possible to the end-users of the resource. The awareness of these costs, will drive the first category of savings, the Optimize improvements driven by reduced demand.

This allocation to end-users can sometimes cause problems. Take, as an example, the engineering teams at a factory. They are a cost centre in their own right. They operate the boilers to raise steam for the process and, because of their technical expertise, they may consider themselves better qualified to assess the opportunities for improvement in steam use. As a result, they may

Making **losses visible** at the point of resource use is a key strategy to release funding for improvement.

feel that there is a good case to be made that both the cost and investment budgets should lie with them. Moreover, if all the gas costs are devolved to the operations team, then the incentive for the engineering team to operate the boilers and distribution systems efficiently is diminished. On the other hand, if none of the gas costs are allocated to end-users, then they can be profligate in their use of steam.

This division of responsibility means that a more nuanced allocation of costs might be called for. End-users are charged for steam on the basis of the current conversion and distribution efficiency (so they want to use less). At the same time, the engineering team have to pay for the gas which means they are motivated to operate the boilers and steam network as efficiently as possible to make a profit from their steam recharge to operations. Importantly, now that each team has a cost associated with the resource, they can go about making a business case to reduce the cost, in the same way as they would for any other operational investment. This kind of cost allocation challenge exists for lots of other resources, such as compressed air or coolth, that provide multiple opportunities for improvement on the demand side, distribution and conversion/generation within one facility.

18.1 **The Monday morning meeting**
Successful cost allocation leads to discussion or resource use in management meetings at every level of the organization.
Source: © Monkey Business, Fotolia.com

A warning should be given here about the tendency for these budget allocations to create conflict. In reality, both the operations and the engineering (as well as procurement and who knows what else) need to work together to drive improvement. This is why this Framework has emphasized that multidisciplinary teams need to work on, and be accountable for, delivering the overall site-level goals. The budget allocations are proposed to facilitate funding and must under no circumstances be allowed to impede collaboration. Where we have multiple resources going through many conversions and processes, then we need a more rigorous, standards-based, cost allocation process, as described next.

Real World: ***Best practices in departmental cost allocation schemes***

We need to remember that departmental cost allocation has a specific goal of encouraging and empowering business units to drive improvement. Thus, whatever system we design should be proportionate to the value that it will produce. It is quite acceptable to have a system that only traces a proportion of the resource use or which does not perfectly balance inputs and outputs. Poor examples of cost allocation involve excessive effort to allocate minuscule consumptions where there is little potential for improvement. Remember, burdensome systems tend not to persist long.

Over many years working with organizations that have implemented a range of cost allocation systems, I have identified a number of general principles that appear to be common to most successful systems:

1. They allocate costs as close to the point of resource use as possible (to drive demand reduction).
2. They allocate costs within existing management structures and the costs are shown in existing management reports.
3. They are based on real metered quantities of the resource rather than proxies (sq ft. floor area, production, etc.).
4. They may separately record *upstream* costs (to drive distribution and conversion efficiency).
5. They take account of the full value of the resource. For example, they use the marginal cost per unit (i.e. the real value of savings) not the standard cost, include the value of emissions and the savings from lower waste handling.
6. These costs are treated as controllable cost, not fixed overheads.

18.3 Material flow cost accounting

In large organizations, such as manufacturing sites, which consume many resources, quantifying the true value of waste can help release funding for improvement measures.

We have already discussed the value of material flows in assessing resource use (page 436); however, here I would like to discuss their importance in releasing funding for investment in efficiency. Material flow cost accounting (MFCA) is a form of mass balance analysis which provides an unambiguous cost of waste at significant points of an organization's service or production process. Traditional accounting will only value waste in terms of the input costs of the non-productive material with any direct waste handling costs added. In the example opposite, the 30 kg of waste would be accounted for as costing US$300 for the wasted raw materials and US$80 to dispose of: a total of US$380. Using MFCA we see that this waste has had inputs of energy and labour prior to being disposed of, so the true cost is US$635, or 67% greater.

Although it can be complex to implement, those organizations seeking to drive excellence in resource use should contemplate MFCA. I would not recommend this as a first step because there may be many other more rapid opportunities available to improve resource use. However, as a medium-term activity, it is highly recommended as the information gained can not only support further Optimize activities but also Modify and Transform.

Standards: ISO 14051:2011 - Material flow cost accounting

This international standard sets out the formal framework for materials (i.e. resources) cost allocation systems. The general principles set out in the standard applies equally to manufacturing, extractive and service organizations, and the numbering in the 14000 series indicates that this is an Environmental Management Standard. First developed in Germany in the 1980s, this became a huge success in Japan where, by 2010, up to 300 companies had applied this approach, with strong support from the Japanese government.[786]

The core concept is a quantity centre (called an accounting centre in Monitoring and Targeting). Here, a materials balance is undertaken, looking at the inputs, outputs, initial stock and final stock of the quantity centre.

Three basic costs are considered: materials costs, systems costs and waste costs, although energy may be treated separately. Systems costs need to be explained; these are the costs (other than energy, waste and materials) that are incurred in handling a material within a quantity centre, such as labour costs, cost of depreciation of the equipment, maintenance costs, transport costs. By capturing the systems costs, we obtain an accurate price of defective products - the system costs often far exceed the material, energy, and waste disposal costs and are usually not recoverable (for example, materials can be recycled, waste heat can be reused). The outputs from one quantity centre (with all the cumulated costs) become the inputs to another, so the gas-to-steam cost allocation issue described in the item on department cost allocation is neatly addressed. The standard is more than just an accounting methodology - there is a "Plan, Do, Check, Act" continuous improvement process that should drive reduction in losses. A complementary standard ISO 14052 is under development and will cover the application of material flow cost accounting in supply chains.

In Numbers: *A worked example of material flow cost accounting*

An MFCA assessment starts by breaking down the organization into a series of discrete quantity centres (QCs). These QCs are significant resource users where we will quantify the physical and monetary material flows. QCs are places where materials are stored, transported or transformed. They could, for example, be items of equipment, such as a boiler, or stages in a production process. Ideally, the boundaries of the QCs will align with the management and budgeting boundaries of the facility or organization so that accountability for the QCs can be readily assigned.

The next step is to analyse the flows within a QC over a fixed time period - ideally, this will correspond to a typical reporting period for other forms of management accounting in the organization, such as a year, quarter, month or week. The reason for choosing a longer time frame is to allow for seasonal and other factors to be incorporated into the totals. The input and output material flows are determined. Inputs should match outputs on a mass and monetary basis, unless there is an inventory of material within the QC which also needs to be accounted for.

QCs have energy costs, which should usually be the marginal cost of the energy (i.e. the real value of the energy it saved) rather than the average costs. There is a system cost, which represents the labour costs to operate the QC, equipment depreciation and maintenance costs and other overheads deemed appropriate. Finally, there may be a waste management cost which represents the additional handling costs to treat or dispose of the waste. The input material costs will be determined by the purchasing or marginal costs of the material if it comes straight from the supplier. If not, it is the output costs of the materials from a preceding QC. An example of these costs, taken from the ISO 14051 Standard, is shown below.

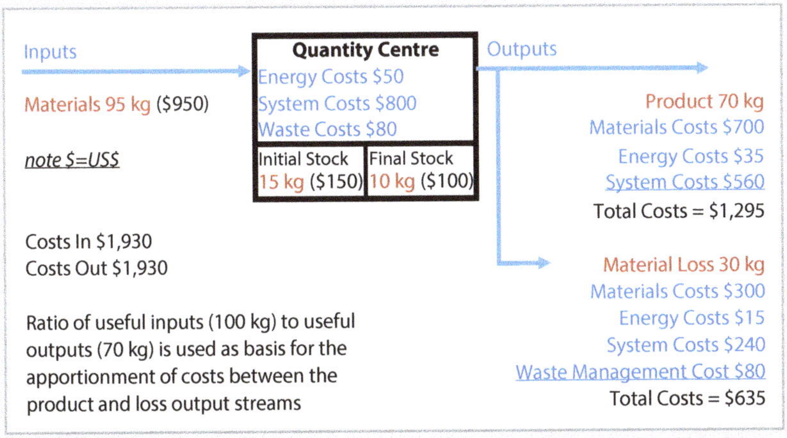

The allocation of the QC energy and system costs in this example is based on the ratio of useful products out compared to the losses (70:30). The ISO 14051 Standard allows another basis for allocation of these costs, for example, it may be known that the useful energy consumption is lower than the 70:30 ratio, as the machinery has a start-up and shut-down process that consumes additional energy, or there are known faults with the equipment.

Where the material flows involve a large volume of a non-product material, such as rinsing water, then the proportion of material assigned to the losses stream may be much greater than the product output, thus inflating the apparent costs associated with the losses, which does not help management decision-making. Appendix B of the standard sets out some alternative approaches, although an organization is free to define whatever method suits them. The waste costs are always assigned to the waste stream.

It is important to note that the material costs "cascade", thus the input costs of the material into the next stage of the process is US$1,295, i.e. US$18.5 per kg (US$1,295/70 kg), considerably higher than the initial material input cost of US$10 per kg. reflecting the embedded energy, labour and other costs associated with the output material. If a material is recycled, then it is treated as just another input stream with all the embedded costs that the material has accrued. Although this may appear to discourage recycling by making the recycled materials seem more costly than virgin ingredients, the fact that the material has to be recycled in the first place is reflected in the losses from the system, and this should act as a strong disincentive to the generation of the waste in the first place.

Although the example above appears relatively straightforward, MFCA can become incredibly complex in processes where there are multiple material streams, such as bulk mixing processes, complex assembly or food preparation. In these cases, the standard allows complex materials to be treated as one simpler intermediate material whose cost is the sum cost of the components divided by the mass. As with many of the ISO standards, a best-practice framework is provided, within which the organization can define the most appropriate approach for its needs.

18.4 CAPEX and OPEX revisited

As we consider third-party funding for our efficiency investments, we need to understand that it is an inability to take on debt to finance improvements that holds back many organizations. Being able to treat these investments as operating expenses can make all the difference.

Money is a resource like any other. Compared to the physical resources of energy, water and materials, money usually has very active management and control processes placed around it, as failure to report an organization's finances correctly can lead to penalties. Understanding how this resource is managed can help us unlock it for our improvement projects.

Earlier we explored the difference between CAPEX (capital expenditure) and OPEX (operating expenses): the day-to-day running costs of an organization (page 189). CAPEX is shown in the balance sheet, which records what the organization owns (its cash in the bank, equipment, buildings, brands, and suchlike) as well as its liabilities (long-term debts).

Investment can come from either OPEX or CAPEX. If I invest in buying a piece of equipment with many years of life, such as a new boiler, then this is a CAPEX cost which affects the balance sheet. I may finance this out of my own funds, in which case I am exchanging one asset on the balance sheet (cash) for another (equipment). On the other hand, if I finance this with debt then I am adding to my assets (equipment) *and* my liabilities (debt). In the first example, the source of funds is shareholders' equity (retained profits) and in the latter, the source of funds is debt (see box opposite for more on this). In either case, ignoring depreciation, the overall balance sheet total should remain unchanged, but the ratio of assets to liabilities does differ.

By contrast, if I invest in an awareness campaign to influence resource use, then this is OPEX and does not appear on the balance sheet. Instead, this cost will appear in the profit and loss statement (P&L), which shows all the income and expenses in the current year, and records a profit (if income exceeds expenses) or a loss (if expenses exceed income).

CAPEX funds are usually scarce because there are limits on the debt that organizations can incur, because cash on the balance sheet only increases as profits are made and because raising additional shareholder equity is difficult. For this reason, when projects are competing for limited CAPEX, many efficiency investments can only proceed if they are *"off-balance sheet"*, where the cost of the equipment, for example, can be converted to some form of OPEX, such as a lease repayment. Many of the techniques we shall explore later in this chapter are about understanding how to structure funding for efficiency investments as an operating expense.

Real World: *BP's CAPEX fund*

Following on from the success of BP's energy management programme (described on page 239), Kevin Ball and his colleagues were able to secure a US$100 million CAPEX fund earmarked for energy efficiency investments in the refining and petrochemicals sites.

This fund was not new money, but a small proportion of the committed CAPEX in these facilities, now earmarked for efficiency. Although no new money was available, some significant advantages arose from establishing this fund:

- The facilities saw that energy efficiency was being supported.
- Funds were earmarked for improving efficiency.
- By centralizing the fund, all sites could bid for support, allowing the team to select those projects with the highest impact per dollar invested.

A constant challenge for resource efficiency programmes in large organizations is the tendency for the better-managed sites to attract funding due to their greater competence in developing investment cases. Unfortunately, this can result in funds being allocated to better-performing sites, while good opportunities among a *"silent"* minority are missed.

A centralized pot, such as that at BP, if allocated based on project outcomes, can encourage the weaker sites to submit proposals. This is because the weaker sites can see the probability of support is higher than average, and they are not competing at their site level for limited funds which they may need to address other priorities. This same fund approach can work within a site, to focus effort on the weaker departments.

The difference between CAPEX and OPEX is ***not a matter of choice****: there are very clear rules on how expenditure should be treated and severe penalties for getting these wrong.*

Exploration: *How corporate finance works*

© kevers, Fotolia

The reason why private organizations seek funding is to invest in activities which will generate a greater return than the cost of the funds and so increase the value of the firm over time, which will benefit the owners of the company.

All private companies, no matter how complex, are funded by a mixture of owners' funds (equity, i.e. the funds from stock and shares issued, or retained earnings) and borrowed money (debt). This funding comes at a cost. The debt (in a variety of forms such as loans or bonds) will usually have a rate of interest (or other charges in Islamic systems), while the equity holders (who own shares or common stock) will expect a dividend (a distribution of the profits of the organization).

Debt finance tends to be cheaper than equity finance, because debt interest payments are usually tax-deductible, whereas dividend payments and interest earned on retained earnings are not. The cost of the debt, i.e. the interest, will depend on the central bank interest rates at the time, the amount borrowed, the risks and security offered, and the tenor (duration) of the loan.

Reflecting the fact that funding of companies usually comes from a mixture of sources, the overall cost is called the weighted average cost of capital (WACC). As long as the investment returns over the long term exceed the WACC, then the investment will increase the value of the firm, and it is sensible to proceed.

However, that does not mean to say that all possible investments with a greater return than the WACC can or should be funded. Companies have limits to the amount that they can borrow, as loans are often secured against the finite assets of the business. Another brake on borrowing is that shareholders expect the ratio of debt and equity to be appropriate to the industry that the organization is operating in (called the gearing) and will discount the share price of companies which are too highly indebted (since they represent a greater than average risk). Some institutions, such as banks, are also highly regulated and are expected to maintain certain levels of reserves which also limits funding available.

Corporate finance is the name that is given to the process of obtaining and managing funding. Corporate finance is usually a complex activity since there are many competing interests. Shareholders want to extract as much value from the shares as possible. Lenders want to ensure their loan is repaid. Employees are often more concerned with their own remuneration than shareholder value. Finally, customers want the firm to sell its goods at the lowest possible price.

For companies that are publicly listed, there may be pressures to invest in areas that achieve quick results as the share price is tracked daily and performance updates are expected quarterly. This may disadvantage longer-term investment choices.

Although they do not typically have shareholders, similar finance issues apply in public sector institutions, who need to balance funds from their institutional stakeholders (e.g. local or national government) and their ability to raise finance directly. Public institutions may not be allowed to retain excess cash, so there may be incentives to convert money into other assets or service enhancements.

18.5 Loans or share issues

Borrowing is a common method used by organizations to raise funds for investment. Private companies can also raise funds by issuing and selling more shares. In all the cases here we are investing in equipment, which is a capital expenditure and so appears on the balance sheet.

OK, so we have a situation where there genuinely "*is no money*"; what can we do? The obvious answer is to find the money from an external source. Before we leap to this conclusion, we should note that if the funds we require are OPEX, then we probably *do* have the funds, but these are simply hidden in our other operating budgets in the form of waste.

Where a real scarcity of funds exists, it is usually in the form of CAPEX. Here we can ask ourselves "*why don't we simply borrow the funds*"? Indeed, there are many banks and lending institutions which are willing to provide project finance. In some countries, there may even be institutions specifically set up to invest in energy efficiency (such as Salix Finance or the Green Investment Bank here in the UK). In practice, however, using project finance or other forms of bank or corporate loans is far from straightforward.

- The commercial and contractual considerations can increase the administrative burden, to the point at which the effort is not worthwhile. This overhead may be further exacerbated by the lack of expertise in evaluating resource efficiency measures among lenders.
- Because of the administrative effort, projects need to be above a certain value before they become attractive to lenders (we may be talking about minimum loan levels of £1m+).
- There can be challenges securing the loan. Often lenders will expect the loan to be guaranteed against the property involved and restrictive covenants or existing mortgages preclude additional debt being placed on these assets. Alternatively, the loan may be secured against the equipment being funded, which is fine if a fleet of vehicles has been purchased as these are easy for the lender to repossess in the case of a default, but much more tricky for light fittings or wastewater system improvements.
- The loan will appear as a debt on the balance sheet of the organization, thus impairing its ability to borrow for other investment or affecting key ratios that impinge on the share valuation.

Public sector organizations face similar problems in accessing external funds, In many cases, they may simply be forbidden from taking on debt or may lack sufficient creditworthiness (in the sense that they may not have a guaranteed future revenue stream to assure repayments).

Real World: ***Green bonds***

The Climate Bonds Initiative tracks *"climate aligned"* bonds (i.e. whose purpose is to finance low carbon and climate-resilient infrastructure) and puts the overall market value of these at US$694 billion[153]

Within this category of bonds are green bonds, which explicitly signal environmentally responsible investments to the market, which have raised US$118 billion to 2016.

With pension funds and other investors seeking sustainable investments and anticipating the potential grown in efficient goods and services, the demand for green bonds is very strong, and these often sell out rapidly.

A case in point is Toyota's first green bond, used to provide finance for customers wishing to buy lower emissions vehicles, such as the Prius. When first issued in 2014 the issue was set at US$1.25 billion but quickly rose to US$1.75 billion due to customer demand.[47]

Apple Inc. has raised US$1.5 billion for its green bond, directly targeting measures to move to 100% renewable energy, installation of more energy efficient heating and cooling systems and an increase in the company's use of biodegradable materials.[534]

Green bonds are finance tools for larger investments, with the majority being in the range of US$10 million to US$100 million in value[153] and usually have a higher investment grade and shorter term than other climate-aligned bonds. They are not just restricted to corporations, as municipalities, states and international institutions also access these markets. Green bonds should be certified to comply with the Green Bonds Principles[401]

Risk	Explanation
Credit	Risk that the borrower will not repay the loan.
Performance	Risk that the efficiency measures do not deliver expected savings.
Debt	Risk that debt will rise or impair the borrower.
Market	Risks from energy price changes.

18.2 **Categories of risk in efficiency finance**
There are four broad types of risk associated with financing efficiency investments. Different finance methods allocate these risks in different ways among the parties involved.
Source: Niall Enright

Another form of external finance is to issue bonds. A bond commits the organization to pay a fixed interest rate for the term of the bond and then repay the face value at the end. For private sector organizations, raising corporate bonds for resource efficiency projects rarely makes sense as the bonds often need to be sizeable (tens or hundreds of millions of dollars) to attract investors. In the US, some "investment grade" government agencies can also issue bonds and, where the interest earned on some government bonds is tax-free, the bonds can offer a lower interest rate and still compete with commercial bonds. Overall, bonds have many of the disadvantages of conventional loans.

If an organization can borrow (via a loan or a bond) then this has some benefits compared to our last source of external finance, issuing stock. When a company issues more stock (i.e shares), it has committed itself to a future payment, in the form of dividends. These dividends are not tax-deductible, unlike interest. On the other hand, dividend payments are more flexible, as they are not guaranteed, unlike debt repayments. Selling stock also dilutes existing shareholders and affects the capital structure (debt-to-equity ratios and so forth), which may or may not be desirable. If the savings produced by the investment do not exceed the price-to-earnings ratio of the organization's shares, then value is reduced. In practice, selling stock is not something that an organization would do to support resource efficiency per se - it is something that is usually done to fund core strategic development of the business. Selling stock is something that most public institutions can't do as they do not have share capital.

As we can see, all these forms of external funding have some challenges. In the next section, we shall examine how we can release funds using our retained earnings, or existing cash.

18.3 **Funding models and risk**
The illustrations below and on the following pages are available in an A3 poster in the companion file pack
Source: Niall Enright

Loan (all types and sizes of organizations)

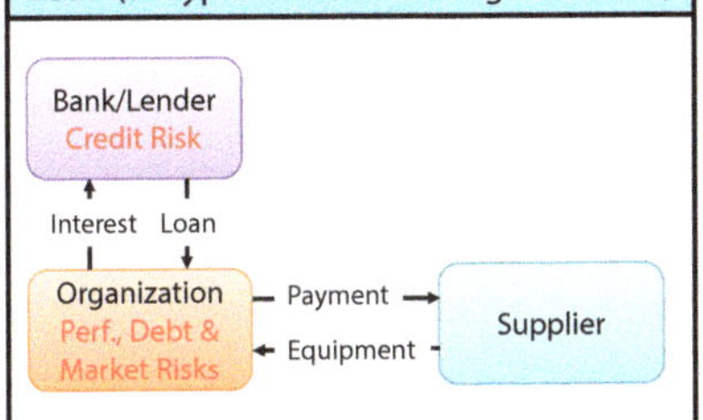

Pros	Cons
Process of obtaining a loan is understood.	*Some administrative effort needed to apply for loan.*
All savings from the investment are retained by the organization.	*No risk-sharing with lender, if investment does not pay organization loses.*
Depreciation and interest payments are both tax-deductible.	*Increases debt; lender may require collateral or a security against an asset.*

Bond (larger companies or agencies)

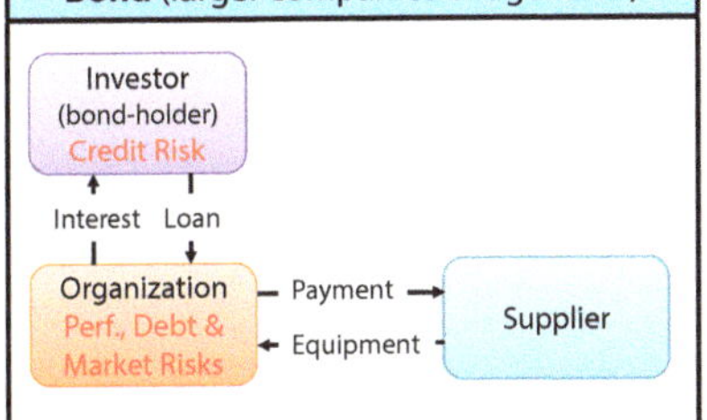

Pros	Cons
US govt bonds may be interest-free, lowering rates.	*More complex to administer than a loan.*
All savings from the investment are retained by the organization.	*No risk-sharing with lender, if investment does not pay organization loses.*
Depreciation and interest payments are both tax-deductible.	*Increases debt for the organization.*

Stock or Share Sale (companies)

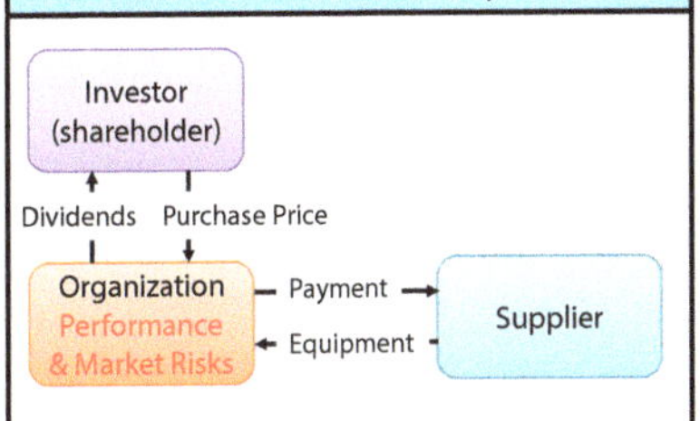

Pros	Cons
Issuing shares does not increase debt. Dividends not fixed.	*In practice, shares are not usually issued to fund one purchase.*
All savings from the investment are retained by the organization.	*No risk-sharing with lender, if investment does not pay organi-zation loses..*
The depreciation of the equipment is tax-deductible.	*Dilutes existing shareholders. Dividends not tax-deductible.*

18.6 Purchasing or leasing

Because of the tax treatment of interest payments, outright purchase of equipment may not be the most cost-effective approach. Where we have identified that we can divert "wasted" cash flow towards our investment, then we can use a lease to align the savings with the repayments.

The easiest way to fund a viable resource efficiency investment is from the organization's retained earnings, that is to say from money the organization already has. In practice, the retained earnings don't necessarily need to be in the bank to be available, as the investment could be paid for by switching funds from one budgeted operating cash flow (e.g. the energy costs budget) to another (an equipment lease payment).

Paying outright for an investment may be less financially efficient than borrowing money to fund the investment. If the capital cannot be written off as quickly as a loan, the net savings will be higher and thus the tax payable will be higher, at least in the short term. Loans are favoured, because the interest is immediately tax-deductible, whereas depreciation of CAPEX can take longer.

Leases come in two different flavours. First, there is what is referred to as a true lease, also known as an operating lease, which is essentially a "*rental agreement*". Here the organization, the lessee, simply rents equipment from a supplier, or intermediate leasing organization, called the lessor. At the end of the lease term, the equipment remains the property of the lessor - the lessee has simply been paying for the use of the equipment. There are several benefits from this kind of lease. The payments made by the lessee are usually fully tax-deductible as an ordinary business expense. The lessor may contribute to the maintenance of the equipment and at the end of the lease the lessee is free to enter into a new lease and so obtain the latest, most up-to-date equipment.

True leases have the big advantage that they are not treated as CAPEX payments and do not affect the balance sheet.

Following the Sarbanes-Oxley Act in 2002 in the US, there has been a general clamp-down on accounting practices by corporations, especially those activities that disguise ownership. Updated International Financial and Reporting Standards (IFRS) and the US Generally Accepted Accounting Principles (GAAP) have had a considerable impact on the treatment of leases in general, reducing the ability to report "*off-balance sheet*" in particular.

In the US, recent guidance (ASU 2016-02 / ASC 842) from the Federal Accounting Standards Board (FASB) retains the notion of an operating lease, but requires that these are shown on the balance sheet of the organization if they have a duration of over 12 months - although the tax on repayments is unchanged, the "*off-balance sheet*" benefit is lost. By contrast, the International Accounting Standards Board has chosen to treat all leases in the same manner, essentially as capital leases (IFRS 16), discussed next.

Most finance options add cost compared to outright purchase, as the third party also needs to make a return on the money they have employed.

The second form of lease is a capital lease. This lease is comparable to a *hire-purchase* agreement. The key attribute of this type of lease is that the equipment reverts to the lessee's ownership at the end of the lease or there is some form of "*bargain purchase option*" set out at the beginning of the lease (e.g. the asset reverts to the lessee for a US$1 payment). If the present value of the lease payments exceed 90% of the initial value of the equipment, or the lease is for over 75% of the equipment life, then this also makes the lease a capital lease in the eyes of the Internal Revenue Service (IRS).

A capital lease has the advantage of spreading payments for equipment, but unlike a true lease, only the interest element of the repayment is usually tax-deductible. Furthermore, the equipment leased is treated as if the organization owns it, i.e. it is "*on balance sheet*", which brings the advantage that the organization can depreciate it. On the other hand, the fact that the equipment is on the balance sheet may be undesirable, as this affects some key performance measures of the organization, such as the capital-to-earnings ratios. Also, the outstanding lease amounts may be viewed as a debt, which we have already noted tends to be rationed in organizations and affects the ability to borrow more funds.

A true lease is usually suitable for shorter-term projects where the equipment ownership at the end of the lease period is not especially important (e.g. a fuel-efficient vehicle employed for a particular project with a fixed timescale). Capital leases are best suited to longer-term equipment which the organization wishes to retain (e.g. a wastewater treatment plant which it would be impractical to dismantle).

Outright Purchase - Retained Earnings

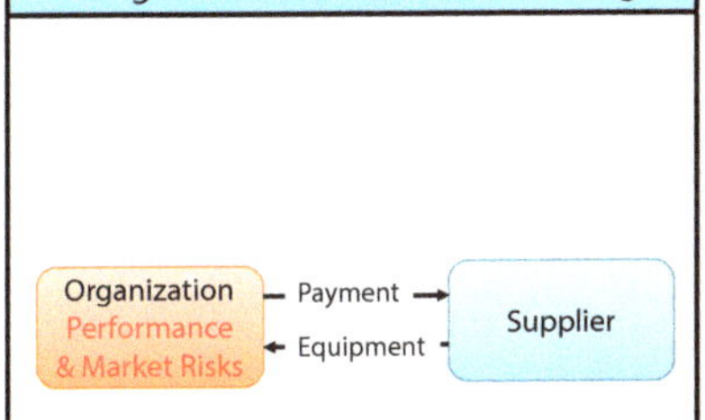

Pros	Cons
Simple.	*Procurement may be time consuming if this is an unfamiliar product category.*
All savings from the investment are retained by the organization.	*No risk-sharing with lender, if investment does not pay organization loses.*
No interest payments.	*Because interest is tax-deductible, this may be a less efficient form of funding.*

Lease (aka a True or Operating Lease)

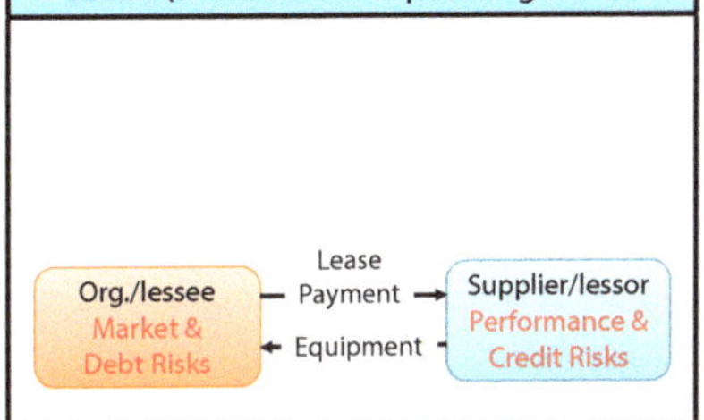

Pros	Cons
Usually relatively straightforward to administer.	*May be more expensive than purchase - depending on tax treatment.*
All savings from the investment are retained by the organization.	*No risk-sharing with lender, if investment does not pay organization loses.*
Lease payments are fully deductible. Lessor may pay some maintenance costs.	*Lessee cannot depreciate the equipment.*

Capital Lease

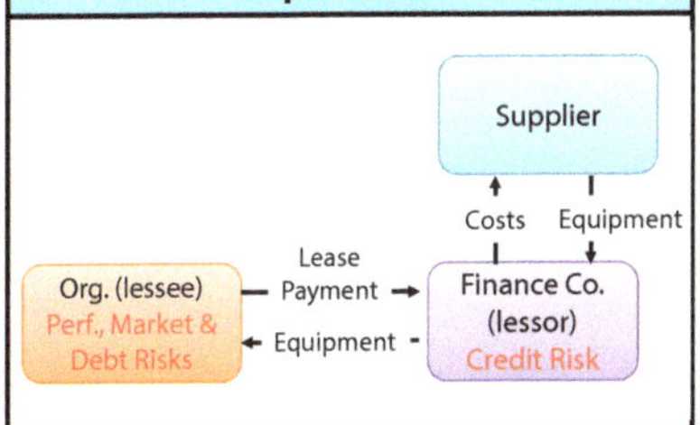

Pros	Cons
Usually relatively straightforward to administer.	*May be more expensive than purchase - depending on tax treatment.*
All savings from the investment are retained by the organization.	*No risk-sharing with lender, if investment does not pay organization loses.*
Can depreciate the equipment. Ownership at the end of the lease.	*Treated as CAPEX. Can only expense the interest element, not capital costs.*

18.7 Energy performance contracts

Rather than switch funds from our resources budget to our equipment budget, we can enter into a contract for resources which incorporates a payment for investments in greater efficiency. Today, there is a huge market offering a diverse range of energy (and water) performance contracts.

Switching funding around from our utilities, waste and maintenance budgets to our equipment budgets to fund efficiency measures is not always as easy as we would like, particularly when the investment involves CAPEX. Even if we can overcome the CAPEX issues, the cost and return are rarely aligned, equipment must be paid for up-front, but the savings take several years to cover the investment. We may need to use debt or leasing to align the costs and savings, so as to maintain profitability. Sometimes top management simply fail to understand that financing efficiency out of avoided resource cost does not require new budget allocations.

Energy services companies bring much more than finance to a partnership. They offer expertise in identifying, procuring, installing and operating a wide range of energy-consuming equipment and will back that expertise up with a performance guarantee.

Where some organizations see a difficulty, others see an opportunity. Over the years, many guaranteed savings schemes have been developed by creative energy services companies (ESCOs) to enable organizations to direct their previously wasted energy and maintenance expenditure (old equipment usually costs more to maintain than new equipment) to pay for improvements. These schemes are generically termed energy performance contracts (EPCs) because the ESCO provides some form of guarantee as to the energy savings that will arise from their investments.

Often the client organization makes an immediate saving when they enter into an EPC, as the balance of the costs of energy post-implementation and the ESCO finance costs are lower than the pre-implementation energy costs. Sometimes the client agrees to pay the same amount as they would have done before the EPC; in other words, there may be no net saving, as they are more interested in maximizing improvements to their buildings than cost reduction and are happy to settle for a financially neutral outcome. For example, a school board may look for improved insulation, a new roof and upgraded lighting leading to lower maintenance, greater comfort and learner productivity. The ESCO may even be willing to incorporate improvements that are not directly energy-related, as part of a wider bundle of upgrades funded by the overall energy savings.

As our history of EPCs shows, (see Exploration on page 640), savings guarantees are now typically expressed in units of energy, rather than in dollar terms, which means that the ESCO and client are usually sharing the risks around utility price movements. The big danger here is that if energy prices drop significantly, the value of the units of energy saved reduces and the period for the recovery of the investment costs increases.

Real World: *A deal for every occasion*

Photo: Fotolia © Tomasz Zajda

In these pages, we have discussed EPCs where the ESCO provides equipment upgrades funded by energy savings in return for payment.

An alternative model involves the ESCO providing power as well as equipment (perhaps because they are operating generation plant). This arrangement is called an energy services performance contract (ESPC).

If the only payment to the ESCO is for energy (e.g. because they have installed a new solar PV system), then the agreement may be structured as a power purchase agreement (PPA). In a PPA (sometimes called a design-build-own-operate project) the expectation is that the cost of the energy will be lower than the cost staying with the incumbent utility. Since the client does not take ownership of the equipment, PPAs are usually off-balance sheet.

Managed energy services agreements (MESAs) are similar to PPAs in that the client pays the ESCO based on their historical energy use, and the ESCO pays the energy bills directly and uses the savings to fund improvements. This deal is suitable where a landlord wants to upgrade a property but can only charge tenants for energy, not capital, as the MESA charges are treated like the old utility bills and are a service-charge item.

There are many variants of these contract forms so that almost every conceivable situation can be catered for.

Properly structured EPCs can be off the balance sheet of the client organization. They are essentially operating lease-type arrangements where the ESCO provides much more than single items of equipment. The additional input that an ESCO can provide is a very important aspect of the scheme.

Although at its heart an EPC is a financing mechanism, a modern ESCO brings a great deal more than money to the client. They will invest in identifying feasible project (often at their own risk). They will develop a portfolio of projects and technologies to suit the needs of the client. They will procure, install and commission the agreed improvements; they operate and maintain the equipment on-site; they will maximize the savings wherever possible, to include grants, demand response or emissions payments if available. They will commission independent Measurement and Verification (M&V) to confirm the savings achieved. If necessary, they will set up Special Purpose Entity companies to ensure that the capital expenditure is off-balance sheet and there is transparency around the contractual and savings arrangements. The value of the expertise that the ESCO brings, their purchasing ability and operational skills should not be underestimated, as many clients will not have these capabilities in-house. Indeed, some EPCs, like that for the famous refurbishment of the Empire State Building in New York[352] were funded by the client with no external finance - the driver was the expertise of the ESCO.

Where financing is involved, an ESCO is usually offering a form of operational lease. This lease approach means that the ESCO is incentivized to maximize the capital expenditure in the EPC, so there could be a potential for split incentives between them and the client. On the other hand, because the ESCO is guaranteeing the performance of the equipment - they are taking on the technology risk in the projects - they will have to be cautious about overstating the savings or over-investing, if they are to make a return.

Because of the need for certainty around savings and the emphasis on equipment, ESCOs rarely incorporate purely behaviour-based savings into their portfolio of projects. Indeed, one of the key things to be aware of in the arrangement with the ESCO is the form of M&V that will be undertaken. Whole-facility approaches may mean that other, independent initiatives that a client may take, such as awareness and motivation programmes, could end up being tied into the EPC savings rather than being *"new"* savings in their own right. My own experience working with Don Gilligan in attracting ESCO interest in programmes such as Monitoring and Targeting (M&T) in the US around 2000, demonstrated the difficulties of incorporating *"soft"* improvement measures, which are harder to quantify, into *"hard"* contracts built on performance guarantees. Everyone accepted that M&T was a proven approach to delivering significant savings and that the return on investment was highly attractive, but the challenge in quantifying the savings ahead of time meant that it was too risky to incorporate into contracts.

It is important to note that EPCs are not for everyone. In the most developed market, the US, EPCs have been dominated by the *"MUSH market"*

Standards: ***Energy services financing***

The only formal standard concerning energy services is BS EN 15900:2010, which is of little value to practitioners, as it merely provides some general definitions and the sketchiest of processes.

In 2009 the Efficiency Valuation Organization published the *International Energy Efficiency Financing Protocol*[402] which sets out some standard concepts and terminology. It is a conceptual document which focuses on the types of financial models that can be developed (with a US-centric focus). The intended audience is the finance community as much as the client.

The EU has a *Code of Conduct on Energy Performance Contracts*[713] which sets out a brief set of principles of good governance and fairness around EPCs, but this is a very high-level *"apple pie and motherhood"* list which lacks detail on what a contract should contain.

Of much greater value is the *DECC's, Guide to Energy Performance Contracting Best Practices*[197] which sets out the steps that an organization may go through to secure an EPC. Important activities covered include the definition of the objectives and scope of the EPC, the selection of an ESCO, the investment grade audit and M&V plan that should form the basis of the ESCO offer. As well as the guidance notes, the DECC has also provided a fully annotated *EPC Model Contract*[196] with guidance, which is an invaluable resource for any organization embarking on its first EPC. These resources have come out of the RE:FIT[618] programme improving public sector buildings in London through a partnership with 15 ESCOs. The Sustainable Energy Authority of Ireland also has excellent guidance on EPCs.[654]

(municipalities, universities, schools and hospitals), which together with federal and public housing markets, accounted for over 88% of ESCO revenues in 2011,[677] a figure relatively unchanged in the National Association of Energy Service Companies (NAESCO) 2014 survey.[314] There are many reasons why some significant sectors such as retail, manufacturing and industry have not embraced EPCs. Key issues are:

- Most EPCs involve long-term contracts, often 10,15 or even 20 years and many private sector organizations cannot guarantee that their businesses or facilities will remain sufficiently unchanged for this duration;
- ESCOs bring a large body of expertise to their clients but private sector organizations, especially in industry, may well have greater capabilities in-house (indeed, in-house staff may see ESCOs as a threat to their roles);
- The cost of finance for a large corporation may be significantly less than the cost of finance that an ESCO can achieve, so an EPC does not make sense.
- There is a threshold of around US$500,000 project value, below which large ESCOs find it difficult to fund economically and lenders are reluctant to underwrite loans.[314]

ESCOs have balance sheets like other companies. Because their ability to borrow would decline rapidly if they were to finance all their projects directly, the vast majority of ESCO contracts involve the client entering into a finance agreement with a third-party lender, as shown on the left, below. These *"guaranteed savings"* deals are more complex than simpler single *"shared savings"* arrangement, below right, where the equipment is on the ESCO balance sheet.

Guaranteed Savings EPC (90% of US EPCs)

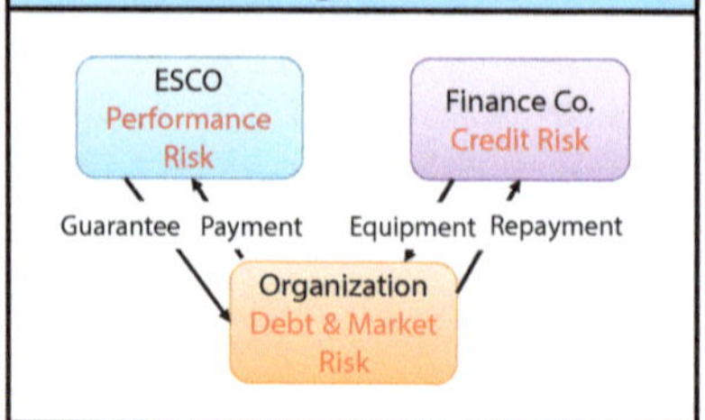

Pros	Cons
Well-established model. Particularly appropriate for long-term projects.	*Administratively complex and time-consuming to set up.*
ESCO organizes finance and there usually is a simple fixed repayment.	*Debt may not be off-balance sheet. Separate contract with finance co.*
ESCO brings skills and guarantees performance of installed technology.	*Organization has debt and market (energy price) risks.*

Shared Savings EPC

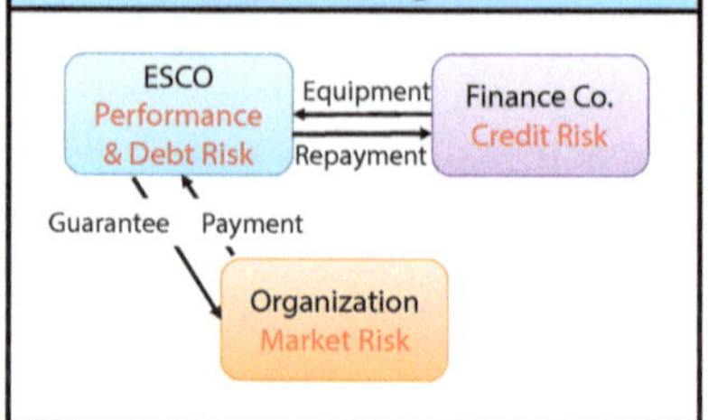

Pros	Cons
Well-established model. Particularly appropriate for long-term projects.	*Administratively complex and time-consuming to set up.*
ESCO gives performance guarantee and finance. Single payment.	*Organization has risks associated with movements in energy prices.*
Simpler contractual relationship between the parties.	*May be more expensive as ESCO taking on greater risk.*

In the illustrations, there are three categories of risk identified. The first is the performance risk, which we have already said falls on the shoulders of the ESCO. A further risk is the market risk. This risk is that the energy costs may rise or fall, potentially reducing the value of the EPC to the organization. Since the mid-1980s, when a sudden drop in oil prices almost wiped out the ESCO industry (see the Exploration piece on page 640), these risks are now typically borne by the client. Finally, there is a debt risk, i.e. an obligation to make a series of future payments, which is usually borne by the customer but may be taken on by the ESCO in a shared-savings scheme. If there is a third-party finance company providing the cash for the equipment (and there usually is, as few ESCOs are large enough to fund their projects from their own reserves), then the finance company bears a credit risk.

The typical risk allocation between the parties is further broken down in the table below:

Risk	Finance Co.	ESCO	Shared	Organization
Credit risk of default by the organization	✓			
Identification of opportunities		✓		
Engineering design		✓		
Procurement and capital costs overrun		✓		
Installation and commissioning		✓		
Failure of opportunities to deliver savings		✓		
Failure during expected life of equipment		✓		
Equipment replacement/maintenance costs		✓		
Insurances for equipment		✓		
Operating and maintenance costs		✓		
Operating efficiency of the equipment		✓		
Health and safety (related to the equipment)		✓		
Damage to equipment or property			✓	
Legal or regulatory changes			✓	
Force majeure event			✓	
Non-payment by tenants (if any), debt risk				✓
Actions by organization's own staff				✓
Energy price changes				✓
Changes in the pattern of energy use/demand				✓

18.4 **Risk and parties to an EPC**
The majority of the risks in an EPC fall on the ESCO, which is guaranteeing the performance of the projects that it has recommended. Other parties bear fewer, but still significant, risks which need to be clearly understood for the EPC to work.
Source: Niall Enright

This table demonstrates the range of commitments that the ESCO will make to underpin the performance guarantee. However, the ESCO cannot be held liable for external factors outside its control and so the client will need to be satisfied that the arrangements will work for a range of energy prices and/or patterns of energy use or demand. If both parties agree to these reciprocal obligations in an open and collaborative way, then the EPC should be a great success. ⇒ page 642.

18.5 **Accreditation builds customer confidence in the capabilities of an ESCO**
To some extent, selecting an ESCO partner is an *"experience good"* (see page 185); that is to say that the range and complexity of the skills and expertise required by an ESCO make these attributes very difficult to assess prior to project initiation. Accreditation by a body such as NAESCO, which can objectively assess the capabilities of the ESCO, helps clients in the pre-qualification process when procuring these services. NAESCO has three categories of certification: energy service companies which provide performance contracts; energy service providers which, in addition to ESCO services also supply energy; and energy efficiency contractors, which tend to offer a limited set of service, typically focused on one or more technologies or types of service.
Source: image © NAESCO

In My Experience: *Energy performance contracting: is your facility a good candidate?*

Donald Gilligan is President of the National Association of Energy Service Companies (NAESCO) which has 70 members with aggregate revenues of US$6 billion a year. Although NAESCO is based in the US, its members operate internationally.

Don understood over 40 years ago that finance is a key barrier for organizations seeking to improve efficiency and ever since has been very influential in bringing together regulators, institutions, services and clients to address this problem. Here he shares his insight on what makes a good facility for an EPC.

Let's begin with a simple working definition of EPC as the re-purposing of money you are currently spending on energy that is wasted and maintenance expenses that are necessary to keep your obsolete energy equipment running, into a payment stream for capital improvements financed over a long term, typically 10-20 years (in the US). Energy performance contracting (EPC) is typically comprehensive, that is it blends quick payback measures (e.g., lighting and controls upgrades) with long-payback items (e.g., boilers or chillers) into a single project that addresses the full scope of the facility's needs.

EPC projects are delivered by specialized ESCOs, which provide turnkey services (energy audits, design engineering, construction management, providing or arranging project financing, monitoring and verification of project savings, and maintenance of the equipment installed in the project). EPC projects can include the full range of facility energy efficiency and renewable energy capital improvements and usually guarantee that the project savings will be sufficient to repay its capital cost.

Most property managers and financial managers of organizations that own substantial facilities have heard of EPC, but many are not sure if their facilities are good candidates for EPC. So here is a quick list of the characteristics of organizations that are good candidates for EPC.

Does your facility need critical improvements now?

Do you have older facilities that have suffered from years of inadequate maintenance and capital improvement budgets? Is critical equipment, such as lighting, heating and control systems obsolete and requiring constant, even extraordinary, efforts to keep it functioning? Is there a major facility need, such as a new roof, that you cannot see any way to finance? Are some or all of these needs urgent, as in you cannot hope to get through next winter without a new boiler?

Are you realistic about what you can, and are willing to, pay for?

A good EPC candidate makes sure that the ESCO understands its financial situation at the outset. Some facilities have a target list of improvements, realize that the cash flow from energy and maintenance savings is not sufficient to finance their full cost, and have some funds available to buy down part of the cost. Other facilities have no funds available, and expect the ESCO to understand and prioritize the project measures to meet the needs of the facility within the limits of the project savings cash flow. All facilities have to be willing to commit to a long-term financing deal, usually in the form of an equipment lease or other debt instrument that puts a liability on the customer balance sheet, or carries a premium cost to keep the liability on the balance sheet of the ESCO or the project financier.

"Good EPC candidates realize that an EPC is a business deal, not an engineering exercise."

Donald Gilligan, President, NAESCO

Are you looking for help?

Good EPC candidates realize that energy technologies and the procurement of energy supplies in today's commodity markets are not their core competencies, and are willing to turn over key functions that have been handled by in-house staff to an ESCO that is an expert. Staff have to be willing to share responsibility for the procurement and operation of energy equipment and cannot be threatened by talk of job elimination through outsourcing.

Do you want a partner or a vendor?

A typical EPC in the US has a term of 10-20 years, during which the ESCO has a substantial interest in the operation of the facility, insofar as it is guaranteeing the performance of the project equipment. The EPC specifies the maintenance and operations responsibilities, and is premised on a cooperative, rather than an adversarial, relationship of the ESCO and the facility personnel. An EPC is not a zero-sum game. Both parties are working to minimize utility costs and maximize facility comfort and productivity over the long term and should, over the life of the contract, deliver more benefits to the facility than were contemplated, or guaranteed, in the project contract.

Are you a "best value" buyer?

Good EPC candidates realize that procuring an EPC is different from many other procurements, because an EPC is designed to provide the lowest life cycle cost, rather than the lowest first cost, of energy equipment. Good EPC candidates fully value all of the attributes of energy efficiency and renewable energy technologies – reduced energy and maintenance costs, reduced environmental emissions, reduced water and sewer usage, etc. – and are willing to pay for that value.

Are your decision-makers willing to be involved from the beginning?

Good EPC candidates realize that an EPC is a business deal, not an engineering exercise. Successful ESCOs have learned, usually through painful experience, that project development is the riskiest part of their business and that engineering work rarely solves business problems, so they work to keep the business and technical discussions in phase.

Expect the ESCO to quickly develop and present the outlines of the project in the form of a preliminary proposal: if I bring you a project with this list of measures, and which costs roughly US$5 million with a 10-year simple payback, will you buy the project? The ESCO expects to present this outline to, and get a conditional acceptance (or preliminary close) from a decision-maker. This may seem presumptuous to an organization that is used to paying for project development engineering and specifications, and then walking a fully designed project through its internal approval processes.

So, bear in mind that most ESCOs are interested in implementing projects, not in earning project energy audit and/or engineering fees, which they view as a lost opportunity for their specialized staff. ESCOs understand that between the preliminary close and the project contract lies 6-12 months of engineering and financial development work, but they have learned that a customer that does not approve the preliminary proposal is unlikely to buy the fully developed project. So they will work to surface and resolve the objections to the business deal, before they spend the time and effort to flesh out the project engineering and financial details.

Characteristics of organizations that are not good candidates for EPC

As you might expect, the characteristics of organizations that are bad candidates for EPC are the opposite of the characteristics of good candidates. Bad candidates are *"cream skimmers"* (interested only in quick payback measures), have unrealistic expectations of what the savings available in their facilities can pay for, have no perceived urgency in implementing a project, are leery of long-term debt, are low first-cost buyers, and are not interested in a long-term partnership with an ESCO.

If you believe you are a good candidate

Should you feel that an EPC deal is right for your organization, then one way you can proceed is to check out the list of NAESCO member companies. You can get details of these companies at the NAESCO website (http://www.naesco.org/members). Many of the member companies operate internationally, and there is a resume of their services and contact details. You can also find a wealth of case studies across a range of sectors and technologies. Bear in mind that a successful partnership at one facility can turn into a long-term relationship over many sites, so it pays to find an organization with a good fit with your own. There is an equivalent list of EU ESCOs at EUESCO (http://euesco.org/members/index.html).

18.6 **Boulton and Watt**
It is absolutely fantastic to see that Energy Performance Contracting is celebrated in the UK's highest value banknote! The text under Matthew Boulton, on the left, says: *"I sell here, Sir, what all the world desires to have - POWER"*, and under James Watt, right, it says: *"I can think of nothing else but this machine"*. There are 192 million Boulton & Watt £50 banknotes in circulation, worth £9.6 billion.
Source: Niall Enright

***Exploration:** The fascinating history of shared savings*

The notion of paying for savings through a utility bill is not new. In the 18th century, the entrepreneur Matthew Boulton and engineer James Watt offered to finance the installation of their more efficient steam-powered water pumping engines in Cornish tin mines on the basis that the mine-owners would pay for the improvement by passing on some of the savings.in fuel.

> *"There was some local resistance in Cornwall, where the new engines were certain to save costs in pumping out water from the tin mines, ….., the 'no cure, no pay' terms offered by Boulton and Watt – based on one-third of the savings in fuel over a period of 25 years – saved the day."*
>
> From: *Thomas Crump, The Age of Steam, p58, London, Constable and Robinson, 2007*, quoted in *Energy Efficiency* by Dr Steve Fawkes.[274]

There are several characteristics of this deal that are present in modern-day EPCs: the contract is long-term (25 years); the performance is guaranteed (that is to say that the mine-owners will pay nothing if no savings take place); the client makes a saving and obtains superior equipment; the partners (Boulton and Watt) bring financing and expertise to the table. Note that this is a *shared saving* rather than a *guaranteed saving* as the payment relates to the actual cost saving, not reduced fuel use.

The next developments originated in France and elsewhere in Europe, in the early 20th century through *"chauffage"* contracts where heat was supplied by companies such as GDF-Suez for a single *all-in* price which covered the capital costs for boilers and distribution systems, operation and maintenance costs and the fuel costs. Although these contracts encouraged the heat producer to be as efficient as possible, there was still an incentive to maximize the consumption by the end-user, as this helped the supplier recoup their costs and make a profit.

Chauffage contracts have persisted to this day in industrial facilities where the energy supply has been outsourced to a third party. In the UK this evolved into contract energy management (CEM), where much of the financial saving for the client was derived from the de-manning of outsourced boiler-houses and reduction of maintenance costs that the updated equipment offered, rather than fuel efficiency improvements per se.

One of the things to be wary of in chauffage contracts is a *"take or pay"* obligation on the client to use a minimum quantity of the resource (heat, steam, electricity, compressed air, treated water etc.). This minimum requirement can impede energy or water efficiency efforts, as reductions beyond the minimum may not lead to financial savings.

The modern-day EPC market started to take off in the 1970s when Scallop Thermal, a division of Royal Dutch Shell, introduced the concept of providing building owners with *"conditioned space"* for 90% of their current utility costs. Their key innovation was to extend the chauffage and CEM supply-side improvement to the demand side, which enabled Scallop to reduce the energy used to well below the 90% level and so repay the equipment purchases and make a handsome return.

According to Shirley Hansen's history of performance contracts,[350] both the UK and US governments were initially opposed to the concept of shared savings. Future British prime minister, John Major, when a government accounting officer, was responsible for enabling the language used in UK regulations to accommodate this approach. In the US, on the other hand, the embryonic EPC industry had to get laws passed state by state to enable their business model to work. We shall see that regulations remain a key influence on the ability of the ESCO industry to innovate and deliver finance to this day.

The EPC shared savings sales took off in the US in the early 1980s. However, the energy price falls in the mid-1980s led to the near collapse of the ESCO sector. With lower prices, it took much longer for ESCOs to recover the costs of the investments they had made, and some investments could never be repaid. A few ESCOs folded, others reneged on their commitments to their clients and lawsuits proliferated. The notion of *shared savings*, where the ESCO revenues depended on energy prices, died, and out of its ashes arose the *guaranteed savings* model where the promise was centred on the amount of energy that could be saved. In the new deals, the customer committed to paying a certain, generally fixed, amount to the ESCO for their services, usually less than the cost of energy at the time of the deal. The customer themselves took on the risk that if energy prices fell they would be stuck with higher payments than they would otherwise have had.

Indeed, one of the criticisms of the early deals was that shared savings could lead to excessive payments by the clients if prices rose. Thus if an organization was funding US$4 million of investment by paying US$500,000 a year (say, 70% of their original energy bill) then, because the payment was linked to energy prices, a doubling in the energy cost would mean that their payment could rise to US$1 million, at which point the deal is rather less attractive.[804] Indeed, in the early days, some contracts were quite unfair in that some ESCOs benefited from the upside on contracts but were protected by a "*floor price*" against falling energy prices, which led to a reputational problem for the sector.

From the 1990s onwards, the move away from contracts based on energy prices eliminated the confrontational nature of client-ESCO relationships. Greater competition as major equipment manufacturers entered the sector, new standards (especially around M&V), a trade association (NAESCO), greater end-user education and stricter regulations all helped put an end to cowboy practices and misunderstandings and restored the sector's reputation.

At the same time as the US sector was professionalizing, there was an active promotion of ESCOs internationally as governments and institutions saw the benefits of the combination of finance and expertise as a means of unlocking untapped energy efficiency opportunities. Between 2003 and 2010, the China Energy Conservation Project of the World Bank aimed to enable ESCOs to access finance through regular commercial banking channels, recognizing the important role ESCOs can play in capturing China's considerable commercially viable, energy-saving potential.[811] Article 18 of the EU's Energy Efficiency Directive (2012/27/EU)[271] explicitly requires member states to adopt measures to support the development of ESCO markets.

ESCO services are highly tailored to local markets, depending on the nature of the demand, local financial regulations and the nature of the delivery organizations. In Germany, for example, many EPC providers are public bodies (such as the technical departments of a municipality) which provide services to other public institutions and are funded by cross-payments within the overall budgets of the municipalities. These arrangements are known as public internal contracting (PICO).

Today, the international ESCO sector is in robust health. In 2013 surveys,[81168] there were over 500 ESCOs in Germany, 300 in France, 20 in the Czech Republic and 30 in the UK. In 2016, there were 38 ESCO members of NAESCO in the US, several of which are large multinational operations and equipment businesses such as Honeywell, Johnson Controls and Schneider Electric. This market is growing quickly; between 2009 and 2011, for example, the US ESCO market grew at 9% per year, despite the effects of the financial crisis, although more recently sales have levelled off.[689] The chauffage-type arrangements still dominate European contracts, but there are many new and innovative models emerging in different markets.[68]

There remain some challenges on the horizon. The US EPC market faces further difficulties from the continuing tightening of regulations[314] which require ESCOs that arrange financing to be registered with the Securities and Exchange Commission (SEC), which puts a much greater onus on the customer's attorneys and advisers to understand and package the loans, leases, tax credits, etc. Furthermore, new accounting rules, mentioned earlier, have substantially eliminated operating leases, so the asset liability has to appear on the ESCO or the customer's balance sheet, unless more complex Special Purpose Entities (separate companies) are established, with all the complexity that entails.

What this history tells us is that the ESCO sector has proven very adaptable and innovative and no doubt will overcome these challenges in time. With the strong backing of institutions, increasing support from finance providers, standardization of contracts and verification, a growing recognition of the importance of facilitators (e.g. national energy efficiency agencies), greater education of the market and the underlying need for change, the prospects for ESCOs look very strong.

18.8 Commercial property finance

Funding for energy efficiency investments in the commercial property market poses a number of challenges, primarily related to the ability to finance improvements by taking on debt. Changes in regulations and the treatment of efficiency investments may offer a solution.

One especially challenging sector for efficiency finance is property. We have seen how US ESCOs are providing excellent support to the federal, state and municipal sectors, but there are real difficulties in these funding models for the private residential, industrial and commercial property sectors. As buildings account for around 40% of global CO_2 emissions, it is critical to release investment for all types of property.

Efficiency investments in commercial buildings face particular challenges:

- There is the landlord-tenant split incentive in that the former has to pay for major improvements while the latter gets the benefits;
- There is the problem that most mortgages will not allow further debt to be secured against the building;
- The timescale for a return on many larger investments, such as solar photovoltaics (PV), 7-15 years, exceeds the period during which the landlord can be certain of retaining the property, so there is concern over the residual value of the investment (will the next owner value it?).

There have been many efforts to circumvent the barriers to investment in private or commercial buildings. One of the simplest is to undertake what is called on-bill financing which, like the MESA previously described, repays efficiency investments through an addition to the utility bill.

There are basically two forms that these schemes take. The first is a loan scheme where the debt is usually tied to the individual but may be required to be paid off in full if the property is sold. This has the advantage of needing fewer regulatory approvals, but is not especially popular with utilities which consider loans to be outside their normal business model.

The second mechanism of on-bill financing involves linking the repayment to the tariff for electricity through a given meter. This approach has the advantage that the loan stays with the property so that the person repaying the investment is the one who has had the benefit (the additional cost should have been offset by the units of energy saved). There are, however, legal implications, as almost all energy market tariffs are heavily regulated. California and New York have enacted legislation that obliges investor-owned-utilities to offer on-bill financing schemes, which should significantly increase the market.[274]

On-Bill Financing (linked to tariff)

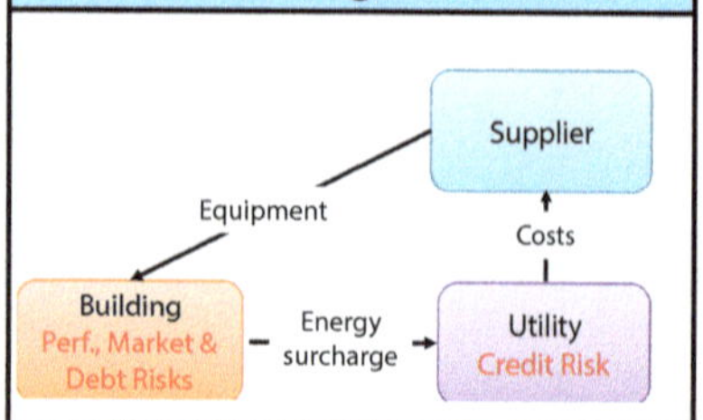

Pros	Cons
Collecting payments is simple.	*Utilities do not see this as a core business activity and so may resist this facility.*
Does not conflict with mortgage debt, landlord can recover in service charge.	*Only really works if final bill is lower than initial bill.*
Householder only pays for benefit while they own the property. Obligation attaches to the meter.	*New owner may not necessarily value the benefit in the same way so could affect property price at sale.*

The fundamental challenge for financing efficiency investments in property is that loans secured against the building are cheaper, but existing mortgage lenders are not willing to have additional debt in place.

An alternative mechanism, available in the US, for tying investments to a property, rather than an individual is called property assessed clean energy (PACE). This funding vehicle involves a property owner (residential or commercial) taking out a loan from a PACE provider (usually a local municipality or a body established by a municipality). This loan can provide 100% finance for energy efficiency or renewable energy investments and, because the term can be long, it is usually possible to ensure that the net repayment is less than the savings made at the time of the loan. The repayment of the loan becomes part of the property taxes levied on the building and will remain attached to the property even if it is sold.

Proponents of PACE point to several advantages of the scheme. First of all, because the funding is coming from a municipal bond and is based on local tax collection powers, it is seen as a relatively low risk and so the interest rate is low. Secondly, the treatment of the loan repayment as an additional property tax does not constitute an additional *debt* on the property and so is not prohibited in the existing mortgage. Critically, the repayment of the initial loan remains attached to the property, so when it is sold the remaining cost and benefits of the efficiency measures will transfer to the new owner.

Together these features have led to PACE being described as a revolution in finance to address climate change. An article in Harvard Business Review named PACE as one of 10 *"Breakthrough Ideas for 2010"*.[355] Unfortunately, as with many complex finance mechanisms, the reality has been a more modest success rather than an overnight revolution.

In part, this slow uptake was because PACE depends on individual states creating enabling regulations and raising the finance so that the availability of PACE depends on whether the local municipalities have chosen to embrace the idea. Another major stumbling block for residential EPCs was a ruling by the Federal Housing Finance Agency (FHFA) in 2010 to prohibit PACE in the residential sector. The concerns revolved around the priority of *"liens"* against a property - that is to say that, in the event of a default by the property owner, the repayment of taxes has a higher priority over mortgage debt and so PACE repayment obligations subordinate the mortgage lender's loan. This decision affected a large part of the private mortgage market since two semi-public institutions under the supervision of the FHFA, Fanny Mae and Freddie Mac, provide liquidity to a large proportion of the US mortgage market.

From a mortgage issuer's perspective, it is easy to see why a new debt against a property with a higher repayment priority is undesirable. Indeed, there are other criticisms of PACE,[173] but this is by far the most important one. Two things followed the FHFA ruling. First, states took the FHFA to court to overturn the decision and, second, the states addressed the concern of mortgage lenders by putting in place mechanisms to ensure that defaults on PACE obligations would be funded by a form of insurance, rather than from the money raised by liquidating of the property. For example, in California Governor Jerry Brown established a US$10 million loan-loss reserve fund

Property Assessed Clean Energy (PACE)

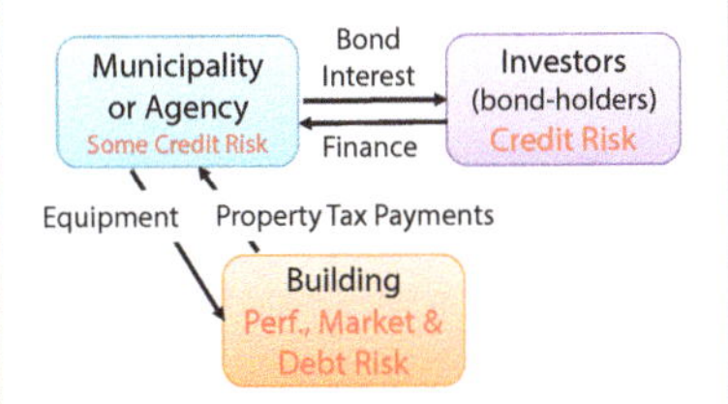

Pros	Cons
Well-established model which avoids debt being added to property.	*Requires changes to laws in each municipality to allow PACE liens.*
Lien on property is superior to a mortgage, so may have low risk and lower cost.	*Opposed by mortgage lenders, so municipality may need to take some credit risk.*
Householder only pays for benefit while they own the property.	*New owner may not value the benefit in the same way, so could affect price.*

to ameliorate mortgage lenders' default concerns. To date, 2016, that fund has never had to be tapped. As a consequence of these provisions, the FHFA recently decided to accept mortgages with a PACE obligation against the property, and so residential PACE is set to take off.

The PACENation website in October 2016 provided the following statistics about commercial property PACE programmes in the US: US$280 million had been invested in 790 commercial properties; 43 states have PACE programmes in place; 49% of the projects supported were for energy efficiency, 37% for renewable energy and the rest mixed; 30% of projects were under US$75,000 and 55% from US$75,000 to US$750,000, which means that PACE seems to be able to support projects below the typical ESCO EPC size.[583] The scale of the commercial PACE programmes is very modest in comparison to the market size, although the doubling rate of 18 months is an encouraging sign. Residential PACE, in contrast, is much larger:[584] in October 2016 over US$2,200 million was invested in 104,000 homes; 58% of projects were for energy efficiency, 37% for renewable energy and 4% for water conservation.

In the UK, an on-bill financing scheme called the Green Deal was launched in 2013, but the scheme proved too complex and poor value (both for the homeowner and the taxpayer) and it was closed down two years later (see the case study opposite).

Another form of residential property-linked finance in the US is an energy efficient mortgage (EEM). This is provided by the property's existing mortgage lender and there is usually a ceiling on the amount of the building's current value that can be borrowed. EEMs are recognized by the FHFA as they do not affect the priority of the debt on the property. To apply for an EEM, the homeowner must submit the house or their construction plans to an energy audit by a certified auditor. The lender will then determine the mix of investments they will support, on the basis, like the Green Deal Golden Rule, that the overall savings will exceed the additional repayment costs.

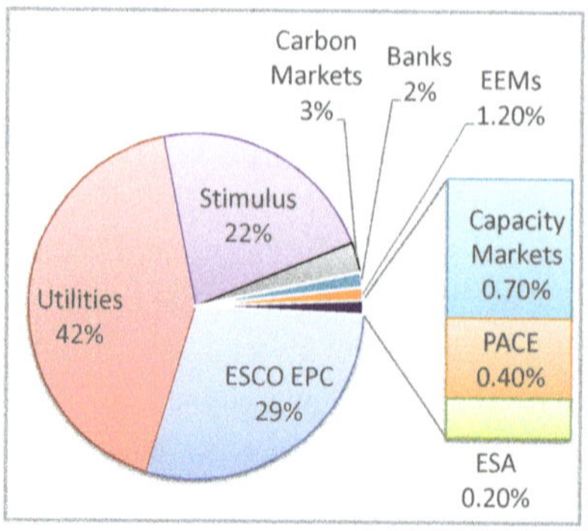

18.7 **US energy efficiency expenditure 2010**
The largest source of funding is utility operated programmes, which essentially are paid for by energy end-users. This is followed by ESCO EPC contracts, then direct stimulus funds on energy efficiency retrofits which are taxpayer-funded. Other programmes represent a small proportion of the total.
Source: Hesser, "Energy Efficiency Finance, A Silver Bullet among the buckshot", in "Energy Efficiency: towards the end of demand growth".[664]

The emphasis in the previous sections has been on the lessons from the US on property financing, where the data is good and innovation has led to a range of different solutions for different customer groups. Although a little dated now, we can see from the chart, left, that in 2010 only 31% or so of the total finance for energy efficiency (US$14.4 billion), was in the form of debt, primarily via ESCO EPCs with some bank lending, EEMs and PACE investments, too. Although PACE markets have increased substantially (to around US$2.4 billion, as reported above), ESCO markets were flat between 2011 and 2014[689] at about US$5.3 billion. If we estimate the market by rounding up these figures to US$10 billion of debt finance for US energy efficiency measures today, this total is a drop in the ocean (0.07%) compared to the outstanding debt in the US mortgage market of US$13.9 trillion.[276]

In the next chapter, we shall explore some of the non-debt sources of finance illustrated in the chart, such as direct investment by utility companies and carbon markets.

18.8 **Utility-led programme funds efficiency improvement**

The images above are of the rear of my semi-detached home. My house is on the right half. As can be seen from the infrared image, top, the cooler yellow colour shows the effectiveness of the cavity wall and roof insulation we have installed, compared to the neighbour's uninsulated property on the left, which is radiating considerably more heat, shown in red. We received a grant that covered some of the cost of the cavity wall insulation under an Energy Company Obligation, although the roof insulation was well above the standard specification, so we funded that entirely ourselves. In the UK there were around 12 million homes lacking wall insulation in 2015 (that is cavity-walled and solid-walled homes that could be insulated). These measures could save up to half the heat lost by homes, and so improving these homes is a key government objective

Source: Niall Enright

The cost per tonne of CO_2 avoided of the Green Deal was a staggering £94!

Real World: *Hard lessons from the Green Deal*

They say that you learn most from your mistakes. In that case, the UK's ambitious Green Deal should teach us a very great deal about how to design an on-bill finance mechanism.

Launched in January 2013, the scheme was intended to boost energy efficiency investment in residential property, with the specific objectives of stimulating private investment, improving harder-to-treat properties (e.g. those requiring solid wall insulation) and addressing fuel poverty by providing measures that would reduce the energy bills of poorer customers.

The scheme involved a complex set of participants. Certified Green Deal auditors would carry out independent assessments of improvement opportunities. Approved Green Deal providers would then be able to bid against each other to perform the work (which can be subcontracted to Green Deal installers) and provide the finance based on the assessor's reports. The provider, who is not necessarily the auditor, would submit a Green Deal plan to the occupier, which sets out the measures to be implemented and the costs. Central to this is the Golden Rule, which means that the utility bill post-installation should be lower than the original utility bill. The plan also had to be approved by the Green Deal Finance Company, which would provide the finance to the provider (at an average rate of 8.3% APR).[253] Of interest is the fact that the occupier could initiate the audit process and a landlord was obliged to allow any approved plan to be carried out.

In hindsight, it is easy to understand why this scheme got into trouble. Its complexity was an immediate turn-off. Consumers did not appreciate that the loans were unsecured and felt that the rates of interest were far higher than was available from high street banks for secured borrowing (i.e. loans against an asset such as a mortgage). The much-lauded Golden Rule was based on average energy consumption for dwellings, and so those in fuel poverty could find themselves paying more post-installation if their consumption was lower than average. The warning signs came early on, with the BBC reporting[85] that in the first six months of the scheme there had been 38,000 assessments, but only four Green Deal loans.

In a bid to save the scheme, the government relaunched it in June 2014 with fixed grants being provided instead of, or to supplement, the loans,[84] thus enabling some of the measures that had been identified in the audits to be completed. This change, too, was criticized as there was no means test involved and the grants were seen to favour middle-class households rather than poorer homes. Recognizing that the scheme would have to be completely redesigned, the government closed the Green Deal Fund to new loan applications in July 2015.

The National Audit Office carried out an investigation[543] of the Green Deal and a related scheme, the Energy Company Obligation (ECO), which obliged larger energy suppliers to fund efficiency improvement measures for households. It found that only 14,000 homes had received loans under the Green Deal (compared to 1.4 million households helped under ECO). The total expenditure by the Department of Energy and Climate Change on the Green Deal (including the grants to stimulate demand) was £240 million and the cost per tonne of CO_2 reduced was a staggering £94, compared to £34 for the previous schemes. There are many lessons from this scheme, not least that demand is not just about finance but about consumer motivation and behaviour.

18.9 Incentives and subsidies

Because finance markets work imperfectly when it comes to efficiency investments, many governments offer financial or fiscal support, either directly or through third parties such as utilities.

The ECO programme in the UK, mentioned in the case study on the previous page, has financed improvements in 1.4m British households (my own included - see previous page). Similar programmes exist in the US (see Energy Efficiency Resource Standards)[23] and in many other countries, where utility companies are obliged to deliver certain level of efficiency improvement among their customers, and will be penalized if they fail to hit these objectives. These programmes, called energy savings obligations, can range from single-focus activities, such as distributing compact fluorescent lamps, to much more tailored engagements starting with energy audits. For the customer of the utility, these interventions are either free or highly subsidized. In some markets, there may be a means test or some form of qualification for the support - e.g. in the UK the programmes have generally been targeted at less affluent customers.

Why would utilities undertake to provide this assistance? Well, in the first place they may be mandated to do so by their regulator, and since they are able to recover the costs through the charges they make to all customers, there may be little overhead for them. Secondly, utilities may also wish to decrease electricity consumption as part of a demand side management (DSM) programme (sometimes referred to as capacity market programme). Here, the intention is to avoid building new electricity generation plant through the cheaper option of reducing electricity demand. Where DSM is involved, the utility company may be putting its own money on the table. In some countries, organizations can offer to reduce demand and if their bids are accepted they will be paid for each kW of demand that they are able to shed when requested to do so during peak electricity use periods, called demand response payments.

In addition to energy savings obligations, another mechanism to support resource efficiency are renewable energy obligations (REOs). These place a requirement on a utility to produce a certain proportion, usually rising each year, of their electricity from renewable energy sources. This can be as simple as a specific obligation for each utility or a complex market where the obligations can be traded and so delivered in the most cost-effective manner. In either event, these obligations are key to funding many renewable energy technologies, through the subsidies that they produce. Again in my own case, I benefit from a feed-in tariff for the electricity I generate from solar panels on my own home, towards which all electricity users are contributing to in their bills. Without subsidy, the PV panels would not have been economic to install.

Utility Incentive Programmes

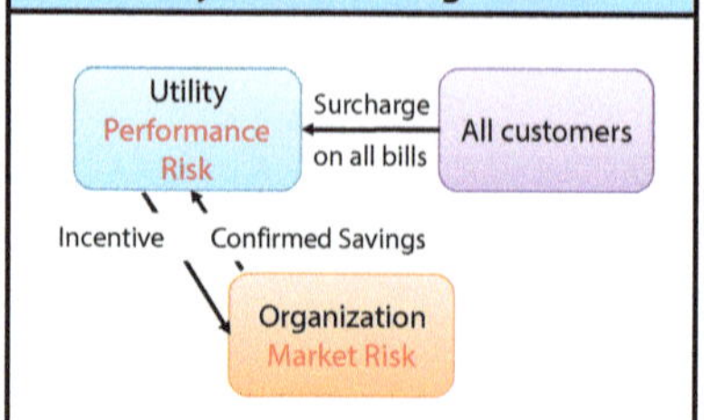

Pros	Cons
Simple proposition for recipient of the incentive.	*Adds to the energy cost of all customers. "Giving away stuff for free" is harder than it appears.*
Incentives overcome market failings.	*Utility is committed to certain savings and so takes on performance risk.*
No debt is usually involved (although some incentives are "soft loans")	*Effort may be focused on more expensive but scaleable solutions.*

Technology subsidies are dominated by lighting, solar heating, renewable and transport. An alternative to a direct subsidy is a *"soft loan"*, where credit is guaranteed by the government but accessed through banks, which helps to overcome the initial purchase price of the equipment. In Germany, a public bank, KfW, supports the refurbishment of existing buildings by offering home owners low-interest loans with a range of subsidies linked to the energy efficiency performance of the improvements. Other examples of financial incentives are subsidies for energy efficiency audits, which can help uptake in countries where these are not mandatory.

Direct funding for efficiency is available in many countries or regions. These may be administered by organizations such as the World Bank, Global Environment Facility, UN Development Programme, regional development banks or national agencies. In the UK, for example, the Energy Savings Trust funded a boiler *"scappage"* scheme to enable householders to replace old, inefficient boilers with newer, more efficient models.

One drawback of these financial incentives for energy-efficient products is that they can distort competition if the number of products meeting the requirement is low, as vendors can increase their prices in anticipation of the additional incentive given to the consumer.

In addition to the financial incentives described above, other policy tools to support resource efficiency are fiscal incentives. These consist of measures that relate to the taxation of investments in efficiency. One example we have already encountered is enhanced capital allowances for energy-efficient equipment in the UK (see page 579), a form of accelerated depreciation. Other examples would include lower value added tax (VAT, a sales tax) on efficient goods or on labour costs related to efficiency investments. Reduced import duties for more efficient vehicles is yet another example of a fiscal measure. In Germany, companies that adopt ISO 50001 receive tax concessions, while in the UK industrial firms with climate change agreements have a reduced climate change levy charge.

Instead of providing lower taxes on efficient products, there is an alternative of increasing taxes on inefficient ones. Road congestion funding can be seen as an example of this, as most schemes price the road use depending on the emissions of the vehicle.

Obviously, measures based on taxation will tend to work best in economies where there is a high tax collection rate and will have a low impact elsewhere. Fiscal measures need to be targeted carefully depending on the market issue that they are designed to address. For example changes to VAT or import duties would be helpful where the initial purchase prices for efficient goods is a concern, whereas accelerated depreciation would not. Carbon taxation is another area where there is an indirect fiscal incentive for energy efficiency as the additional cost of the emissions increases the value that efficiency projects provide. Examples of carbon taxes include the Climate Change Levy and the Carbon Reduction Commitment in the UK.

Real World: ***Programme to market***

Although national, regional and municipal incentives (whether directly delivered or driven through utilities) remain the dominant form of external support for energy efficiency, these have a number of problems related to their programmatic nature, because each incentive requires unique rules, delivery mechanisms and so forth. There is now a trend towards single market-based approaches.

Leading this innovation is, unsurprisingly, California, where in 2015 the legislature passed a bill[643] to require that all future incentives must be based on *"normalized metered savings"* and that utilities are authorized to support customer efficiency initiatives on a *"pay for performance basis"*.

In effect, the utilities will still have state-imposed demand-reduction targets or self-imposed demand-side reduction objectives to reduce investment in new plant. If they find that they are struggling to achieve these targets, then, they will have to increase the payment to the customers until they see that sufficient customers respond to meet the target.

This technology-agnostic approach will encourage customers to focus on low-cost behavioural improvements as well as equipment. Indeed, there is a great deal of interest in the scheme from sophisticated technology firms which can provide apps to help customers maximize their payments and can potentially act as *"aggregators"* of improvement.

Although it is early days yet, this approach may truly open up an efficiency market in its widest sense, where investment returns follow measured improvements.

Funding

18.10 Investor confidence

There are a number of initiatives underway to attract investments to the efficiency sector by packaging the projects in standardized ways that the investment community understands.

*"Although the lack of finance is often cited as a major barrier to investing in energy efficiency, the real problem is not lack of finance per se, but **a lack of structures that address investor concerns.**"*

Dr Steven Fawkes, "Energy Efficiency".[274]

We have seen that there are many problems on the *demand* side of efficiency funding - from management simply being disinterested in this investment, through to poor quality business cases and biases such as artificially high hurdle rates. But even if we create a compelling case, it still "*takes two to tango*" and we are dependent on the sentiments of the *supply* side funders.

In the commercial building retrofit market, a key barrier is collateral. Traditionally, a loan is secured against an asset and a maximum loan to value ratio of 70% to 80% is on offer. In energy efficiency projects, the problem is that there is rarely little collateral value in the equipment supplied once it has been installed, and the loan amount sought may be 100%+ of the equipment value, to pay for all the related costs of procurement, installation, commissioning and so forth. The investment is also illiquid; that is to say, there is no secondary market for this kind of loan, where the investor can sell the investment. A secondary market would require there to be a high degree of standardization of the contracts and the involvement of the rating agencies in assessing the risk profile of the loans (e.g. AA).

Many commercial building mortgages have restrictions on further debt being placed against the building. With some efficiency measures having a long tenor, i.e. needing 10, 15 or even 20 years to deliver a return, it is quite possible that there will be a change of ownership of the building before the investment has repaid itself (US commercial mortgages are usually 5-7 years). The Basel III banking regulations have also made it more difficult for banks to make longer-term loans. To cap it all, most investments in property, even for large property owners with extensive portfolios, are simply too small in total value to be able to access funding from green bonds or pensions funds.

There are several initiatives designed to address these investor concerns. One of these, started by the Environment Defence Fund, but now part of the wider certification body, Green Business Certification, Inc., is the aptly named Investor Confidence Project (ICP). This project seeks to standardize investment opportunities in property by reducing the "*friction*" associated with transactions:

- A series of protocols provide clear guidelines on the pre- and post-installation data requirements, measurements, design construction, operation and maintenance and verification of savings for six project types (see the table opposite).

Large Commercial	Standard Commercial	Targeted Commercial
Whole building *> $1M project costs* *Savings > 15% energy* *IPMVP Option C*	*Whole building* *< $1M project costs* *Any level of savings* *IPMVP Option A and/or B*	*One or a few ECMs* *Limited interactivity* *< $1M project cost* *IPMVP Option A and/or B*
Large Multifamily	**Standard Multifamily**	**Targeted Multifamily**
Whole building *> $1M project costs* *Savings > 15% energy* *IPMVP Option C*	*Whole building* *< $1M project costs* *Any level of savings* *IPMVP Option A and/or B*	*One or a few ECMs* *Limited interactivity* *< $1M project cost* *IPMVP Option A and/or B*

18.9 **Investor Confidence Project protocols** To date, six protocols have been developed. Here, *"multifamily"* refers to residential structures with five or more units. The Large protocol is for projects where the investment is significant and so greater effort is justified. The Standard protocols are used where the investment is smaller, but there are a number of interventions, or where the analysis can only practically be applied at a whole building level. The Targeted protocols are for single measures or a number of measure with small overlap, where a limited part of the building is affected and a more streamlined or focused approach is merited.
Source: The Investor Confidence Project.[416]

Investor Ready Energy Efficiency™

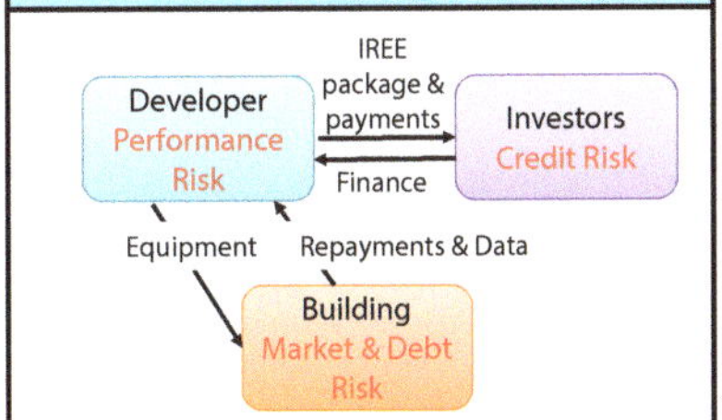

Pros	Cons
Standardization lowers costs and "friction" for investors.	*Depends on investor education to develop market at sufficient scale.*
Independent M&V and assurance underpins performance expectation.	*More expensive than simple purchase due to overheads and packaging as investment.*
Secondary market can develop which would make investments "liquid".	*Market is not nearly large enough yet. Ratings agencies not involved yet.*

- The project development specification provides an overall roadmap to meet the requirements of the protocols. For a project to meet the specification, the developer must be an ICP-certified project developer, and the software tools must also be appropriately certified. Once a project has been developed, it is independently certified by an ICP-certified QA provider, of which there are over 20.
- If the certification is passed, then the project can be certified as an Investor Ready Energy Efficiency™ (IREE) project. This certification should enable investors to more easily make investments and, equally important, trade in these investments because of the standardized data, contracts, best practices and the rigorous independent review.

The ICP is not a source of financing itself. It is rather a series of processes intended to make financing easier by: a) reducing due diligence costs for investors/lenders; b) reducing performance risk; c) enabling aggregation through standardization; d) enabling banks to build teams around standard processes; and e) baking in ongoing M&V, which will be essential for a future green bond market. Whether this is sufficient for investors to make funds available at scale, or if they will revert to bespoke EPC arrangements which are supported by the customer's credit rating or collateral, remains to be seen.

An initiative that uses ICP is the European Sustainable Energy Asset Evaluation and Optimization Framework (SEAF)[653] which is designed as a standardized database of demand response, energy efficiency and distributed renewable generation projects which targets projects under €1.5m. The plan is to unlock investment using the IREE methodology, together with performance-based energy efficiency insurance from Hartford Steam Boiler (a major player in equipment insurance and part of Munich Re) and a software platform to bring together projects and investors. There are two aims:

- They are making packaging the investments in ways that the investors understand; and
- They strive to make the investment tradeable. That is to say that the cash flow that it generates can be priced effectively and can be sold on to other parties in secondary markets in due course.

18.11 Carbon markets

Carbon markets are designed to provide funds for energy efficiency programmes. The requirement of additionality, location-specific credit-types and low allowance prices all make this a relatively impractical source of funds for most organizations.

In our discussions on value, we touched on the notion of carbon markets (page 131). Here, we want to explore how an organization might access these markets to fund energy efficiency programmes.

Most of the carbon markets are designed to produce tradeable certificates representing a tonne of CO_2 emissions reduction, sometimes called carbon credits. There are several types of compliance market carbon credits.

- European Union Emissions Trading System (EU-ETS) and Emissions Allowances (EUAs). Organizations will either have been issued with these free or charge or have purchased these at auction. These cannot be generated through carbon-reduction projects, so their only relevance to efficiency investments is if the project reduces emissions in an ETS installation (essentially power generation plants or large heat systems), then the value of EUAs saved can be included in the investment case.
- Certified emissions reductions (CERs) created under the Clean Development Mechanism (CDM). From a project developer perspective, the projects must be in a developing country (called *"non-Annexe I countries"* by the United Nations).[734] The country must also have an operating designated national authority (DNA) with which the project can be registered.
- Emissions reductions units (ERUs) created in a developed country and used to meet emissions obligations in another developed country. In essence these projects are in Annexe I countries, except Belarus and Turkey.

The demand for these certificates is driven by the EU-ETS, within which organizations emitting CO_2 from installations need to surrender credits (aka allowances) equal to the volume of emissions each year. There are mechanisms that permit CERs and ERUs to be used instead of EUAs. However, the over-allocation of EUAs, together with stagnating demand due to low growth in the EU, means that the interlinked prices for all three of these carbon credits have fallen to very low levels (around €6 per tonne).

For those organizations that wish to create CERs or ERUs, the process involves first obtaining the consent of the DNA for the project, which will involve demonstrating that the project will contribute to sustainable development.

Carbon Credits

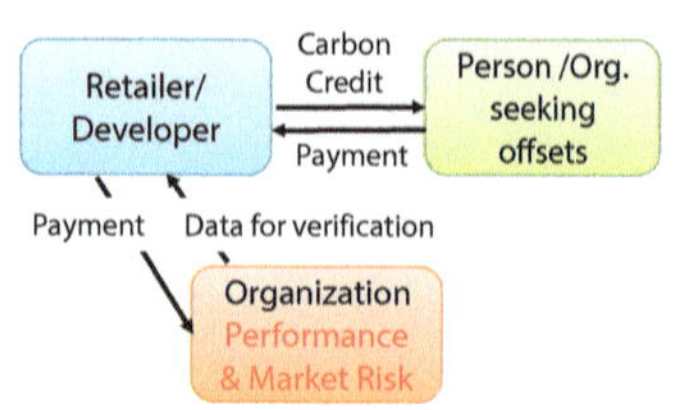

Pros	Cons
Additional source of funds which can complement other sources.	*Need to demonstrate additionality.*
Supporting sustainable development as well as generating funds.	*Early projects had reputational problems due to gaming of system.*
Credible certification bodies in place will assure buyers of credits that these are legitimate.	*At present there are more sellers than buyers in the market so prices are very low.*

Funding from carbon markets should only ever be sought where **additionality** can be demonstrated. That is to say, the project must depend on the carbon funds in order to go ahead.

If this hurdle is passed, the project developer then needs to identify the appropriate project design document (PDD) to follow. The PDD sets out the methodology for accounting for the emissions - there are PDDs for a wide range of technologies from hydro-electric power through to reduction of industrial gases. The process of documenting the project and the baseline emissions against which the CERs are issued is undertaken by an independent party, a designated operational entity (DOE). The DOE submits the project (and changes to the methodology) to the CDM Executive Committee for approval and must verify that the project offers additionality (i.e. without CERs being issued it would not be viable, see page 374). This process makes the pursuit of carbon credits impractical for smaller projects.

Some of the CERs that are produced are not used in the compliance market but instead are used for *carbon offsetting* in the voluntary market. This market is relatively small, at 85 $MtCO_{2e}$ emissions in 2015,[240] just 4% of the approximately 2,000 million allowances surrendered annually in the EU-ETS.

Voluntary allowances are referred to as verified emissions reductions (VERs), sometimes called voluntary emissions reductions as they are not underpinned by a mandatory emissions trading scheme. There are a number of standards in place that govern the issuance of VERs, the largest of which is the Verified Carbon Standard[761] (VCS), which accounted for 49% of the market in 2015,[240] followed by the Gold Standard[322] at 24% and the Climate Action Reserve[152] at 22%, which focuses on the North American market. The VCS is based on the CDM and requires that VCS credits must be real (have happened), are additional (beyond BAU activities), are measurable, permanent (not just temporarily displaced emissions), independently verified and unique (not used more than once to offset emissions).

VERs can be produced by a range of project types. The most common are wind followed by reduced emissions from deforestation and forest degradation (REDD). However, carbon prices are at a historical low at around US\$3.30 per tCO_{2e}[240] in 2015, a 14% decrease compared to 2014. This low price has prompted the development of the Fairtrade Carbon Standard (a collaboration between the Gold Standard and Fairtrade International), to ensure that prices are set at levels that support sustainable development in poorer countries (currently €13 for forest projects, €8.20 for energy efficiency projects and €8.10 for renewable energy projects).

Voluntary emissions are used by organizations and individuals to offset carbon emissions. For example, for the last five years, for all the flights that I have taken, I have offset the emissions by purchasing the appropriate number of carbon credits. I choose to offset my emissions with the World Land Trust,[812] whose offsetting price is considerably higher than average at £15 per tCO_{2e} (€17 or US\$19) but where I feel that the payment will have a more direct impact on a carbon-reduction programme, rather than contributing to the profits of an emissions credit retailer. Although these offsets are not yet fully certified, I have made the judgement that they are high quality.

Summary:

1. A good business case does not mean that a project will automatically be funded. There are many constraints on capital and competition for resources within organizations.

2. Internal funding is almost always the cheapest and most rapid solution. Where funding is limited, then there are a number of third-party sources of finance.

3. External financing is largely about risk and the accounting treatment of debt.

4. Many of the financing models presented in this chapter can be *internalised*. That is to say, a large organization could create its own internal ESCO to finance projects and provide specialist expertise. Indeed, some companies have even experimented with internal carbon markets to focus investments on the most effective projects.

Further Reading:

For those considering working with an ESCO to raise finance then the UK's *Guide to Energy Performance Contracting Best Practices*[197] is an excellent "generic" methodology.

Eric Woodroof's and Albert Thumann's *How to Finance Energy Management Projects - Solving the "Lack of capital" problem*[804] covers much of the material in this chapter, but in more detail, and with a US focus.

***Exploration:** Global financial institutions respond to climate change*

In 2015, the value of the capital markets (stocks, bonds and loans) was estimated at US$294 trillion.[796] That is equivalent to over three years (378%) of global GDP. Indeed, the return on capital is now so low that some major banks are demanding that depositors pay them for holding their money, through negative interest rates. At the same time, there is a huge need for investment in energy and resource efficiency and in renewable energy: US$6 trillion a year,[549] as we mentioned at the start of this chapter. In that context, the current level of green bond provision, for example, appears insignificant.

There is clearly a mismatch between the global need for capital for sustainable infrastructure and the ability of the markets to respond to that need. The good news is that this problem is recognized.

According to Andrew Voysey, Director, Finance Sector Platforms, Cambridge Institute for Sustainability Leadership and Academic Visitor, Bank of England, speaking at a recent conference organized by Lloyds Bank,[476] there are three specific responses that will have a material effect on the provision of finance.

- There will be much greater disclosure as financial regulators believe that climate change presents a financial stability risk to the markets.[280] This increased disclosure is in part because of the direct risk of climate change, but also because of the risks associated with an aggressive decarbonization of the economy (essentially *"stranded asset risks"*, see page 111). If the markets are primed with data about climate risk, then investors will be sensitized to take this into account in their decision-making and, by implication, will attach value to low-risk, climate-friendly investments. This work is being supervised by The Task Force on Climate-related Financial Disclosures (TCFD), chaired by Michael Bloomberg under the auspices of the Financial Stability Board.
- Regulators will build climate risk into their market supervision.[679] A recent example was the Dutch Central Bank, which carried out a review of the financial sector's exposure to climate risk. The result of this is that the loans on the balance sheets of commercial banks will be assessed for their exposure to climate risk and the banks may be asked to hold greater reserves to offset the risks.
- Central banks will put in place measures to improve the availability of finance for climate-related investments. In a recent speech,[119] Mark Carney the Governor of the Bank of England, stated that to achieve the Paris Agreement target of no more than 2° C warming would require an investment of €45 trillion. One way of *"mainstreaming green finance"* to *"reach escape velocity"* is for the green bond market to tap into the €100 trillion fixed-income pool of private capital. This involves standardizing the terms and conditions for green bonds globally; harmonizing and ensuring certification and validation of the green credentials; integrating the certification and risk into the credit ratings; developing green bond indices to enable passive investors to support *"tracker-type"* products.

While these developments may seem distant to the lowly programme Champion seeking to fund discrete projects, they should lead to greater investment and market innovation around resource efficiency.

19 People

The recurring theme in the preceding chapters is the importance of people in a successful energy and resource efficiency programme. Indeed, people, in the form of leaders, employees, stakeholders, consumers and regulators, can be the chief driving force or a major obstacle to the changes we need to achieve.

People are complicated. Unlike technology or data, there is rarely one specific technique that will apply in all cases. In some situations, all we may need to get people to act is to instruct them, as our organization's culture and our authority can mandate the change. In other cases, we may need to persuade people to make the change of their own volition. Thus, the drive to change can be extrinsic or intrinsic and can vary widely from individual to individual. Motivation, of course, is only part of the picture. We also need to consider people's ability to act; do they possess the skills, resources and access needed to make the desired change? Another issue is that people simply do not behave in a rational manner. We have already seen in Chapter 4 that many aspects of human psychology influence decision-making and can lead to counterintuitive behaviours.

This chapter focuses on techniques that I feel are useful when engaging people in energy and resource efficiency. Where you consider a technique may be helpful in your circumstances, you are strongly advised to supplement this content using the many references provided.

First I will explore the difference between motivation (the willingness to act) and capability (the ability to act) and the schools of thought around driving change. Then I have set out a process to design a change programme which, if followed, will help avoid most of the major pitfalls.

The following sections will consider many practical engagement techniques. How to frame a message for maximum impact. How to disseminate a message. Using social norms to support change. The pros and cons of engaging in a small way to prime the desired behaviour. How commitments and goals have a big influence. Engaging people through incentives and suggestion schemes. How games are expanding the way people get involved. Then we will look at how to deal with resistance and climate change denial.

Motivation and capability *are both needed for people to change.*

I will close with some advice on capacity building, learning and training and finally with a more profound look at imagination, empathy and emotion in the context of our change programmes.

19.1 **Presenting *"Does Not Equal"***
The material on this page was presented at the Dubai Sustainable Cities Summit, 2015, hosted by the Dubai Land Department. The panellists shown above are: (L-R)), Romily Madew, Chief Executive, Green Building Council of Australia; Niall Enright; Dr Abdulla Al Karam, Chairman of the Board of Directors and Director General of KHDA; Giovanni Schiuma, Vice Mayor of Matera (European Cultural Capital 2019); Professor Mohamed Al Shami, Founder and CEO of Brain Group.
Source: Photo © DSCS

Behaviour change is not simple. People often behave counterintuitively and many decisions are made **unconsciously**.

Exploration: ***Does not equal*** $\neq$

At a recent sustainability conference, I had just five minutes to present some thoughts on behaviour change. I chose to emphasize the danger of making simple assumptions around drivers for change, using the *"does not equal"* symbol $\neq$:

Awareness $\neq$ behaviour change: The Canadian One Ton Challenge campaign in 2004 spent C$37 million to encourage people to reduce their emissions by one tonne or around 20%. While the programme was successful in raising awareness of climate change from 6% to 51%, a review[715] concluded that few people reduced their emissions: awareness alone is insufficient to change behaviour.

Information $\neq$ behaviour change: Of all the interventions designed to change behaviour, information campaigns appear to be the least effective.[171] In 2012, Dan Kahan published a paper[431] in Nature Climate Change which came to the remarkable conclusion that:

> *"Members of the public with the highest degrees of science literacy and technical reasoning capacity were not the most concerned about climate change."*

An alternative interpretation of the awareness/knowledge school of thought on behaviour change states that motivation comes from attitude, feeling and beliefs.

Positive Attitude $\neq$ behaviour change: A positive attitude towards something has also been shown not to lead to behaviour change. In one classic study, by Leonard Bickman, in 1972,[71] 94% of 500 people interviewed agreed that it was their personal responsibility to pick up litter, yet as they left the meeting only 2% of them actually picked up litter that had been planted by the researchers.

Feedback $\neq$ behaviour change: Another school of thought states that feedback is the key ingredient to drive behaviour, in particular, *normative* feedback where an individual is compared with their peers. However, recent studies[650] have shown that this can fail spectacularly with individuals told that they have lower emissions than their peers actually increasing emissions as a result. This boomerang effect is discussed more extensively in the section on norms. Clearly, feedback may lead to change - but of the unintended type.

Yet another widespread belief among change practitioners is that people are essentially driven by self interest. It is thus assumed that all that is needed to change behaviour is to highlight the benefits to the individual.

Cost savings $\neq$ behaviour change: Anyone who has worked for any length of time on energy efficiency will know that cost savings are not a sufficient incentive to change. The weak effect of costs on motivation is borne out by the evidence, shown in Figure 4.1 on page 153, that there is a huge unrealized potential for energy efficiency improvement with a positive financial return. Clearly, showing people cost savings alone is not enough.

Behaviour change is not simple. People often behave counterintuitively. We make many decisions unconsciously - our *"fast brain"* makes rapid decisions using simple rules called heuristics and so we do not consider many choices. We are creatures of habits which are difficult to change. There are also many psychological factors (confirmation bias, anchoring, loss-aversion, the Hawthorne effect, to name just a few) that distort our judgement. It is possible to design a programme that will get most people to change behaviour, but only if we understand the complexity of people and incorporate the lessons from previous campaigns and research.

19.1 Capability and motivation

For change to happen, people and teams need two things: the capability to carry out the required action and the desire to do so. Capability reflects many aspects such as knowledge and skills, as well as resources like time and money. Motivation can be intrinsic due to beliefs and attitudes or extrinsic due to instructions, incentives, penalties or social norms.

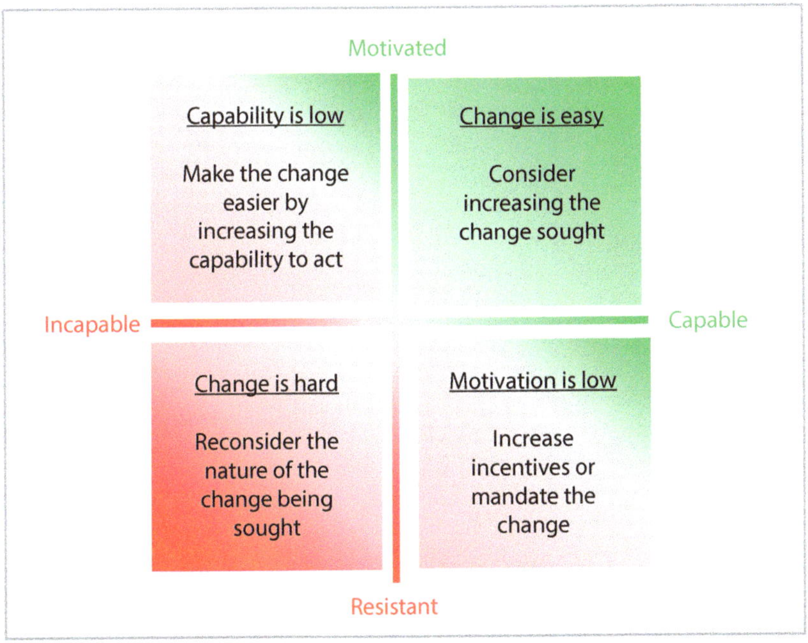

19.2 **The capability motivation matrix**
Please note that we are interested in the capability and motivation of individuals and of teams.
Source: Niall Enright

Capability and motivation are related. An easy-to-implement behaviour change (i.e. one for where the existing capability to act is high) will need much less motivation than a challenging behaviour change (i.e. one which requires lots of resources, time, effort, risk-taking, knowledge, etc).

A useful first step of our behaviour change process is to define the precise action we want people to perform and assess the level of motivation and capability needed to carry out that action. Having understood the requirements, we can then determine the actual levels of motivation and capability within the target group and develop strategies to bridge any gaps.

A lot of the literature on behaviour change focuses on changing the intrinsic motives of individuals by, for example, engaging them on an emotional level on the implications of climate change. While this is certainly desirable, as strong beliefs can lead to more sustainable behaviours in the long run, it is also more difficult achieve. When working on motivation, extrinsic factors such as incentives, penalties or social pressure can be easier to put in place but have a drawback in that the behaviour change can be harder to maintain once the external stimulus is removed.

I tend to start my planning by focusing on reducing the capability and opportunity barriers to change rather than on motivation. For example, I consider how to make the change easy and convenient to implement, what tools and training are needed, and how to ensure that the necessary time and money are available. Once this aspect is maximized, I then consider what messages and motivation are needed to prompt people to act. By ensuring a high capability to act I also reduce the negative emotions and frustration that can arise if motivated people subsequently encounter barriers to action.

Behaviour change requires proper analysis and preparation.

19.2 How to design a programme

Because people - either individually or in teams - can behave counterintuitively and because driving change is complex, we need a methodology that will help us design a winning strategy.

Real World: *The COM-B model*

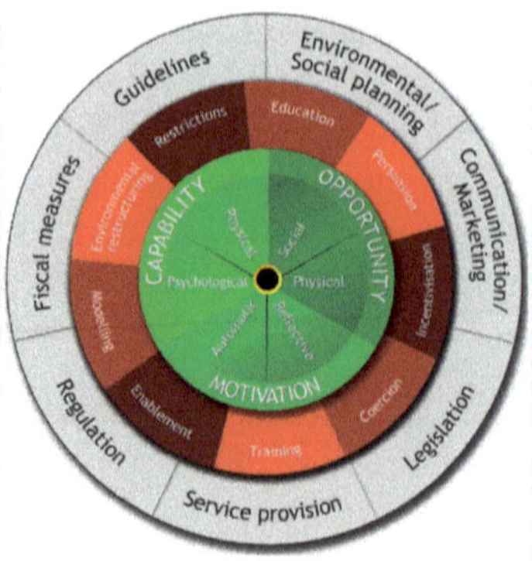

© Mitchie et al.

A recent paper by Susan Michie and colleagues[525] specifically set out to examine how the *analysis* of behaviour change interventions could be improved.

The study considered behaviour as three concentric wheels, illustrated above. In the inner circle, what I have referred to as capability is represented by two distinct factors, capability and opportunity, alongside motivation.

These sources of the behaviour are shown in green, above. There are then nine intervention functions that can influence the sources - such as training or incentives. Finally, we have a number of policy categories which in turn can enable or support the intervention functions.

This model is used by top behaviour change consultants such as STRIDE[685] to help design effective programmes.

The method illustrated opposite provides an objective approach to designing a behaviour change programme. The first step in this process is to define, in a very detailed way, the change we want to see. This definition should be very precise - not "*save energy*" or even "*switch from incandescent to compact fluorescent lighting (CFL)*", but with the intermediate actions "*obtain CFL*" and final actions "*fit CFL*" that are desired. A rookie mistake is to assume that an intermediate action will automatically lead to the final action. This assumption explains why some programmes which have distributed free CFLs to consumers have not achieved the desired effect - the consumers simply put the lamps in their drawers rather than fitting them. Doug McKenzie-Mohr quotes a solution to this problem from Queensland, Australia, where the homeowner had to give up an equivalent number of incandescent lamps when receiving the free CFLs.[510 page 15] Another reason why clarity about the precise behaviour is needed is that apparently similar actions, such as installing insulation, have quite different intermediate and final steps when examined closely (e.g. draught-proofing can be carried by the homeowner, cavity wall insulation requires a specialist).

Having defined the action we are seeking, we enter the analysis phase. First of all, we need to check that the final action will achieve a sufficient degree of improvement. The impact is the result of the action on the level of resource used (e.g. the reduction in kWh per CFL used per annum) and the uptake (e.g. the number of replaceable incandescent lamps in our target audience times the proportion changed for CFLs). Here, it is important to understand the existing penetration of the technology and to be realistic about the uptake rate.

Assuming our action will have a sufficient impact, we then examine it in terms of capability and motivation as described in the previous section or using the more advanced COM-B model (left). An alternative or complementary approach is to look at barriers and incentives. In either case, we are seeking to understand the accelerators and inhibitors of the desired action and what we can do about them. Having identified the characteristics of our action (e.g. if it is one-off or repetitive, whether capability or motivation need to be developed, etc.), we should carry out a literature and case study review to see what has and what has not worked in the past, where similar actions are being promoted. The further reading at the end of this chapter provides some good starting points, and the papers quoted also have extensive cross-references which offer insights into key aspects of the actions we are analysing.

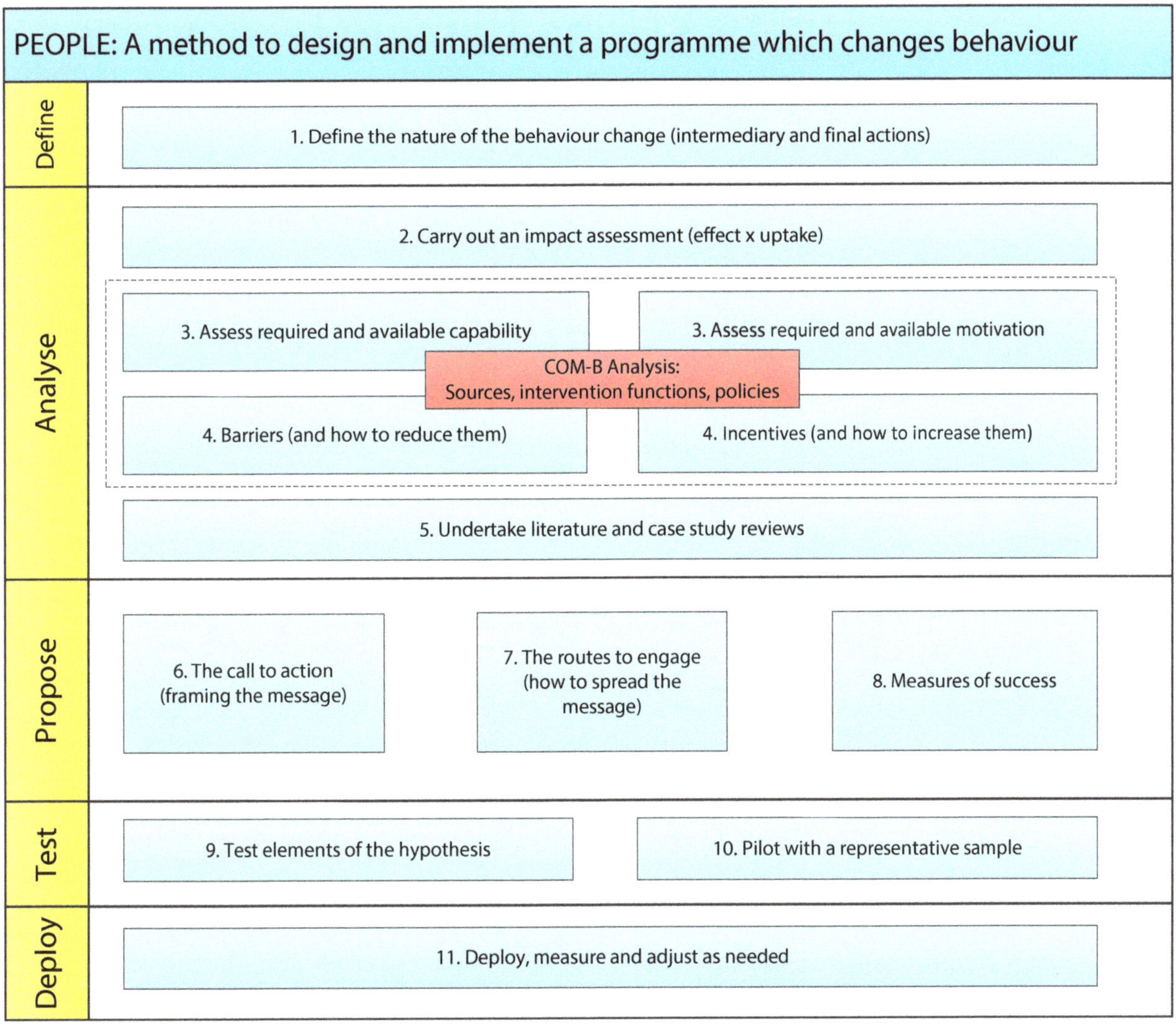

19.3 **Change framework** We cannot trust simple assumptions when designing and deploying a behaviour change programme. By following the steps illustrated above, we can take a more objective approach which reduces the risk of unexpected outcomes. This model shares many similarities with the five-step process developed by the UK government.[159] *Source: Niall Enright. Image available in the companion file pack*

From the analysis, we develop a hypothesis: the proposed messages, routes of engagement and measurement techniques that will support the required behaviour change. The proposed approach may, at this point, contain some options or uncertainties, so we may test these separately (e.g. through focus groups or surveys) before we pilot the approach as a whole on a representative sample of the target audience. If the pilot is a success, we can deploy the programme, ensuring that we are measuring results and adjusting as needed.

The process described above is not linear, and we may well backtrack at several points if we encounter problems. In practice, we may want to drive several different actions and behaviours at the same time to several audiences, and so our strategy could involve several applications of this process running simultaneously, with some features in common and some quite separate.

19.3 Using framing techniques

Framing is a general term for the way that the phrasing of a call to action can influence willingness to act. Understanding framing enables us to craft messages which have greater impact. While these psychological techniques appear to be somewhat manipulative, the truth is that their effects can be considerable.

We have already encountered the concept of framing when we considered the effect of anchoring on our estimation of energy savings potential (page 178) and loss-aversion where again it is valued less than an avoided loss of the same size (page 181).

Framing arises because humans have to use decision-making shortcuts, called heuristics, without which we simply could not function. Imagine what it would be like if we had to consider every decision we ever take in detail - we would be paralysed. It has been proposed[432] that we operate two thinking systems: a *"fast brain"* (System 1) for most routine decisions, and a *"slow brain"* (System 2) for decisions that need careful consideration.

The primary reason why information campaigns are poor at modifying behaviour is that we use System 1 decision-making the majority of the time, particularly for habitual decisions, and this system relies on instinct and prior responses instead of information. We are indeed creatures of habits, and habits are difficult to break. There will be more on the topic of habits later in this chapter when we think about the effect of rewards on behaviour (see page 680). Where information does influence decisions is when we have to make non-habitual decisions, such as when purchasing a fridge, but even here we can see that framing plays an important role as demonstrated by the example left.

Framing is important because - particularly in System 1 thinking - the shortcuts that we use lead to cognitive biases. That is to say that when identical information is presented in a different way, our choices are different. For example, patients told that the success rate of an operation was 96% were more likely to proceed with the surgery than those told that the failure rate was 4%. In both cases, the facts are the same but the way they are framed has biased the decision one way or the other.

Resource efficiency practitioners are making framing decisions every time they communicate the benefits of change. The decision is whether to frame the benefit in terms of a gain (positive framing) or a loss (negative framing): *"If you adopt these recommendations you will save US$50 per day"* or *"Unless you adopt these recommendations you will lose US$50 a day"*. The commonest approach used by resource efficiency practitioners is to communicate the gain, but there is lots of evidence that negative or loss-framing leads to a greater intention to act.[185]

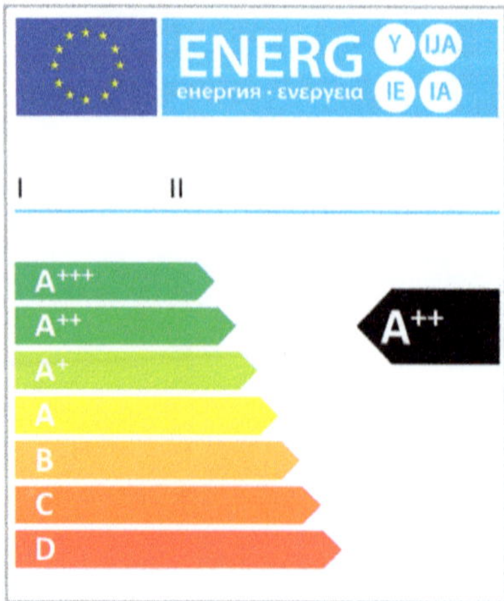

Real World: *Anchoring on A*

If you look at the figure above, you can see that the ratings are labelled from A+++ to D, compared to the earlier version of the label show on page 141 which is labelled from A to G.

A recent study[364] shows that German consumer selection of the top-rating appliance fell from 33% when a scale of A-G is used to just 7% with A+++ to D. The change in the letter range represents a 4.5-fold decrease in the selection of the most efficient option.

The explanation is that consumers tend to anchor on the A rating, and so see A, A+, A++ and A+++ as roughly the same. Loss aversion may be a factor too, as going from A to B may be perceived to be a greater loss than going from A+++ to A++.

Real World: *Default choices*

Photo: Fotolia © slaved[13]

A common application of framing occurs when we consider what default choices are offered. For example, a study in Denmark[574] offered homeowners two options for smart meter installation:

> *[ ] YES I would like to have a smart meter with remote control installed in my home*

or

> *[ ] NO I would not like to have a smart meter with remote control installed in my home*

Above these choices were the instructions, which the researchers designed to take the form of either an opt-in or an opt-out.

> Opt-in: *"Tick the box below if you accept installation. If not, please continue to the next question."*
> **Result:** 60% of people - ticked YES to accept installation. (i.e. a 60% acceptance rate resulted).

> Opt-out: *"Tick the box below if you do not accept installation. If you accept installation, please continue to the next question".*
> **Result:** 22% of people - ticked NO to reject installation (i.e. 78% acceptance rate for the default).

It seems that we are biased (or simply lazy) to accept what appears to be the default or *"do nothing"* choice. If our resource programme depends on consumers installing smart meters, then this is very pertinent.

The effect of loss-framing appears to be influenced by the level of prior knowledge of the subject. In cases where the salience of the issue is low (in other words, where the recipient is unaware that a problem or opportunity exists), then a negative message is most effective,[566] whereas when the problem is recognized, then a positive message is more effective. This has been described as *"the sick baby/well baby"* alternatives.[566] If one is unaware of an issue, then a stark presentation of a problem can spurn action, whereas if one is already aware of the problem, then action is enhanced by focusing on the positive impact of the solution. Because of this salience effect, different aspects of resource efficiency may be best approached with loss or gain framing. The study by Obermiller[325] suggested loss-framing was more effective when encouraging energy conservation, while gain-framing works better for recycling (possibly because awareness of the latter subject was greater at the time of the study or because waste is more visible).

Context also has a bearing. For example, I was involved with an energy efficiency programme at a Molson brewery in Toronto many years ago. The brewery was scheduled for closure, so framing the message in terms of a positive gain for the company (saving money) had little appeal to employees who were going to lose their jobs at some point in the future. In this case, the call to action was framed strongly in environmental terms, which still resonated with the employees.

Where numbers are used in a message, it is important to understand the anchoring effect. Make the first savings figure mentioned high, not low, to increase the perceived impact of the action (or if we are talking about costs make the first figure low!). We should also remember that people tend to overestimate resource use when it is visible (e.g. waste) and underestimate it when invisible (e.g. energy).

Comparison is another area where efficiency practitioners would benefit from understanding the biases people exhibit. It appears that framing has a big influence on comparisons, as well as the order of the comparison. The Exploration piece on the next page goes into this in more detail but, in short, we should, to increase the intention to act, select a poorer performing peer (i.e make a positive comparison) when comparing ourselves to others. Conversely, we should choose a better-performing peer when comparing them to ourselves (a negative comparison which makes us look worse). Likewise, if we are comparing ourselves with a period in the past, then choosing a better period of performance (i.e. there is a negative comparison) increases the intention to act.

In a similar vein, a study by Peggy Sue Loroz[481] concluded that when self-referencing in a call to action (i.e. describing how one's actions will affect oneself) loss-framing has a greater effect. On the other hand, when self-other referencing (i.e. setting out a call to action to affect oneself and others) then gain-frames appear more effective. These variations in the framing effects further reinforce the value of testing each alternative approach before full rollout.

Exploration: *The psychology of framing comparisons*

© Simone Capozzi . Fotolia.com

Comparison is widely used as a tool to engage and motivate people on environmental issues. It is common, for example, for energy efficiency programmes to have league tables ranking a building/facility/department with other buildings/facilities/departments within an organization.

The challenge for resource efficiency practitioners is to get these comparisons right, to ensure that they motivate action rather than act as a disincentive. A recent study[611] by Anna Rabinovich, Thomas Morton and Christopher Duke provides some objective evidence of how we can best approach comparison in our messaging.

Let's take the example above: comparing two business units. Here, we have what is called an intergroup comparison. "We", the *ingroup*, are comparing ourselves with "Them" the *outgroup*. To test these types of comparisons, the researchers asked a sample of British people to compare Britain and Sweden (a negative comparison as most people consider Sweden to be ahead of Britain on many issues) and Britain and the US (a positive comparison given that many British people feel the US performs worse). Following these comparisons, the respondents were asked to rate their environmental intentions (note that the environment had not been explicitly mentioned in the earlier comparison task).

In an echo of the psychological trait of anchoring, it seems that a positive comparison (i.e. comparing themselves with a worse performer) led participants to rate Britain's environmental performance higher and state a greater intention to act personally on environmental issues, than those participants who had previously compared Britain with Sweden (a negative comparison).

Furthermore, the order of the comparison seems to be important. When we compare something with something else, it is the second item that is usually the constant against which the first is compared (it is what is called the *normative* item). Thus, the second item is seen as the standard for the intergroup comparison. When the researchers asked participants to make a direct comparison between the US and Britain, and Britain and the US, the reported intention to behave sustainably was noticeably greater following the first comparison. With Britain as the norm, we have a tendency to emphasize the differences of the others from our norm, whereas with the US set as the norm we have a tendency to want to stress the similarities of ourselves with the norm (we subconsciously want to conform to the norm).

As an alternative to external comparisons, resource efficiency programmes often compare aspects of our own current performance with our previous performance. This self-comparison is what is called an intragroup comparison. The question is whether intragroup comparison follows the same general principle of intergroup comparison, that is to say, is a positive comparison more motivational than a negative one, especially if we compare a worse performer to our performance?

Fortunately, Rabinovich and the team also tested this question. What they found was that negative feedback (i.e. an unflattering comparison of the present with the past) led to a slightly greater intention to act than a positive comparison.

In summary, when making a comparison with something external (an *outgroup*) a greater intention to act will follow:

- If the comparison is with a worse performer external to us (i.e. there is a comparison which puts us in a positive light);
- If we compare the external performer with ourselves, rather than ourselves with the external performance. In other words, if we frame our performance as the *second* term or the norm.

However, when making a comparison with our own performance (an *ingroup* comparison) there is a greater intention to act:

- If the comparison is with our own better past performance (i.e. there is a negative comparison)

It must be stressed that these effects are only part of the picture. Intention to act on environmental issues depends on the individual's attitudes, culture and beliefs. However, we do see that the framing of the comparison does have an impact on intention, and this is something that we should be aware of if we want our communications to be as effective as possible.

19.4 Vivid messages and attention

In order for our messages to be effective we need to engage the "slow brain" or System 2 thought processes, which consider information more carefully than the "fast brain" decision-making system. Our ability to gain the attention of the audience is key to this.

19.4 **A compelling headline**
Following Hurricane Sandy in 2012, Bloomberg Businessweek made the connection between the storm and climate change. If the intention was to be provocative then the cover was a great success, as it sparked numerous blogs and discussions online. The editor Josh Tyrangiel was unrepentant in a tweet: *"Our cover story this week may generate controversy, but only among the stupid".* To some extent, this cover may have been a message to the mainstream cable news networks which covered the impact of the storm widely but did not discuss the possible connection with climate change.
Source: © Bloomberg Businessweek 2012

In our earlier piece on gaining management commitment, I described the sequence AIDA (attention, interest, desire, action) as a tool to engage people (see page 331). No matter how cleverly we have framed our call to action, it will have no effect unless we can capture the attention of our audience and we can communicate the message in the first place.

One technique for gaining attention is to develop compelling headlines and words. No one can argue that the Businessweek cover, left, caught people's attention because of its boldness. However, we don't need to be provocative to create a powerful message. Take, for example, the descriptions below.

> *"Your attic totally lacks insulation. We call that a 'naked' attic. It's as if your home is facing winter not just without an overcoat, but without any clothing at all."*

> *"If you were to add up all the cracks around and under the doors in your home, you'd have the equivalent of a hole the size of a football in your living room wall. Think for a moment about all the heat that would escape from a hole that size."*

These examples were taken from an experiment[326] where energy auditors in the US were trained to use vivid messages when presenting their results (as well as using techniques such as loss-framing described in the previous section). There were many interrelated factors in this experiment, so the effect of framing and vivid language could not be separated, but the authors concluded that: *"These findings demonstrate the potential for using principles uncovered in the social psychological laboratory to design applied interventions."*

Vivid messaging is not just about creating a compelling story for individuals; these techniques can also be used at an organizational level.

> *"If we consider the resources we wasted last year, that adds up to the profit for two entire weeks of operation. What is a crying shame is that this loss is entirely avoidable. Today we score a C- when we could so easily achieve an A+ if only we put some effort behind this."*

Vivid messaging builds on the notion of urgency, which we explored earlier (see page 321). Vivid messages may startle, shock, surprise, disquiet, amuse and inform. Above all, they should shift us from System 1 into System 2 thinking.

19.5 Giving feedback

Feedback is a key element of any change process. We should note, though,that feedback rarely works alone and even when there is an effect, sustaining the response over time can be challenging.

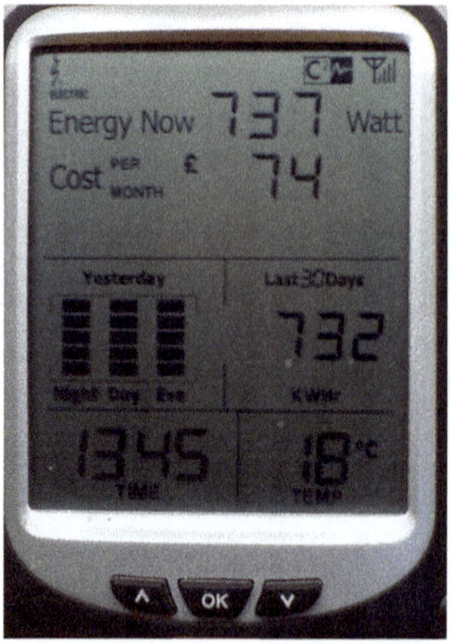

19.5 **A feedback device**
Although now slightly dated, my own home energy display provides a wealth of information on my electricity consumption. I can see the instantaneous electricity use, expressed as watts and, more importantly, the cost in £ per day or per month that this consumption represents. There is also a simple chart of the previous day's consumption pattern. I must admit that when I first received this device, I paid close attention to the information presented, but over time it has been somewhat overlooked. Maintaining interest in the absence of additional prompts or incentives is a real challenge for most feedback systems. If I were to look for a new system, I would want it to upload my data to the internet so that I could carry out an analysis of use over time, applying some of the techniques described earlier.
Source: Niall Enright

Consider the contrast between vehicle fuels and electricity. The car driver has an instantaneous display of the quantity of fuel they have available through their fuel gauge. They get a very strong price signal not just when they have to fill up, which may be fairly often, but every time they drive past a filling station with a large illuminated display of today's fuel costs. When the driver runs low on fuel a light illuminates to remind them that they need to fill up.

Electricity, in comparison, seems never to run out, has very complex pricing structures and a price signal which usually arrives several weeks, or more, after the consumption. Add to this the fact that most electricity users don't understand what a kWh is and are uncertain about the relative power consumption of many of the appliances they own (see page 178), and we can see why it's hard to create urgency to reduce electricity use.

So, one of our first potential points of impact with people is simply to make consumption or waste more visible. There is a lot of evidence[273] that at the level of residential energy users, feedback can have a significant effect on consumption, leading to a decrease of about 7%. Some of the feedback devices are simply the equivalent of a car dashboard (see left), showing instantaneous energy use and the weekly/monthly cost equivalent. Others are quite funky, such as the Energy Orb, which glows red to alert electricity consumers in the US to periods of peak price, and so encourages demand reduction (better referred to as load-shifting) during these high-cost hours.

In many organizations, the issue of invisibility of electricity use often applies to other resources: stream and condensate drain away in hidden corners; compressed air leaks hiss at frequencies above the audible range; water runs invisibly away through sewerage pipes; and few folks look at the contents of waste bins.

We should, however, be very cautious in assuming what works in changing consumer or household behaviours will also be effective in organizations. Making these wastes visible, on its own, does not, in my experience, necessarily lead to spontaneous improvement. Some of the reasons for this have been covered earlier. In many organizations, there are separations of responsibilities which means that those who are close to the point of the waste may not actually pay for it, so they may have no price signal and thus little motivation to act. Increasing visibility in this cohort would have little effect.

> **Real World:** *The Hawthorne Effect*
>
> Back in the 1990s, the consultancy I worked with had the government contract to support an Energy Efficiency Best Practice Programme in the UK dairy sector. So I met with many dairy managers promoting a Monitoring and Targeting software tool to provide feedback.
>
> One particular meeting remains in my memory. The plant manager was very engaged and enthusiastic about the idea of giving his team feedback on their energy use. So you can imagine my disappointment at the end of the meeting when he said to me *"thanks for the information, but we won't be buying your software"*. He went on to tell me that the quality of the numbers is not what matters, but the fact that the team are expected to respond to them. He could obtain sufficient, but lower quality numbers from his existing systems, he said.
>
> What he was describing has been called the Hawthorne effect.[651] This states that people's behaviour changes simply as a result of being observed. While the act of measurement and feedback may lead to increase attention, and therefore engagement of System 2, I don't buy the ideas any observation will sustain improvement.
>
> The purpose of feedback is to inform decisions and so lead to change. So it is not simply enough to say *"you used X kWh this week compared to Y kWh last week"*. The chapter on data analysis (see page 444) makes it clear that we need to isolate the controllable elements of our resource use from the uncontrollable ones for our feedback to have value. In resource efficiency, very careful thought indeed needs to be paid to the feedback metrics used. Observation enhances motivation, but quality data increases capability.

Functional separation can also lead to an inability to respond effectively to feedback. Thus, dealing with compressed air leaks may be seen as a maintenance issue so giving the line operators feedback on compressed air use may be useless as a strategy for improvement. Remember feedback is a device to increase motivation to act, but at the outset, we stated that *both* motivation and capability were required. Motivating the wrong folks is a waste of effort.

So if we are pinning our hopes on feedback as a means of driving improvement, we need to approach the subject with some thought. The keys to successful feedback are as follows (the more of these that you can meet, the more likely you are to succeed):

1. Direct the feedback to the folks who have the capability to act. By all means include others if it will increase the visibility of the process and if it will enable those who act to receive recognition and celebrate success.

2. Target the feedback at the precise change or action you want. A rookie mistake is to give folks lots of irrelevant feedback. Just because you have data does not mean to say you should communicate it. For example, if you want to reduce demand (kW) you will give different feedback to different people at a different frequency than if you want to reduce consumption (kWh).

3. Make the form of the feedback useful and meaningful. Although simplicity is usually the key, just giving a kWh number can be meaningless. Consider a traffic light system based on the variance of actual versus predicted or historical use. Allow time to ensure recipients understand the measure.

4. Give the feedback at an appropriate frequency. Too infrequent and folks won't be able to remember what it is that they did to influence use and periods of abnormal use will remain uncontrolled for longer, but too frequent and people's motivation and ability to respond may be depleted.

5. Monitor the results of the feedback. You may have chosen an inappropriate metric or perhaps, over time, people become desensitized to the feedback. Accept at the outset that you will need to change the content and targets for the feedback as your understanding of your organization's response grows.

6. Avoid using feedback alone. Feedback is just one element of a message or communication. Consider the other aspects covered in this chapter, such as vivid messaging, incentives, norms and commitment. These can all increase the motivation and enhance the capability to act.

A recent study[337] in a US multi-tenanted office reinforced the fact that energy use information has little impact when individuals only receive feedback on their own use alone. It was clear that comparative feedback where individuals receive information about their own use and the consumption and activities of others has an effect. Which brings us to the important topic of norms.

19.6 How to use norms

Norms are powerful tools to drive motivation. Used effectively, they can significantly boost participation in a change process, but used poorly they can lead to the exactly the opposite outcome than that intended.

19.6 **An injunctive norm**
This poster conveys clear disapproval of energy waste, but in a humorous way.
Source: Unknown

Human beings are social creatures. We have evolved to work in groups, and so it is no surprise to learn that our behaviour is strongly influenced by the need to function effectively in a group. A big part of this is an entirely unconscious process of conforming to the group's beliefs and behaviours.

A norm is a word that psychologists and social scientists use to describe the perceived expectations of the group. Norms are the reference points for our actions and judgements, and they are incredibly powerful.

In our resource efficiency programme, we are interested in two types of norms. The first are descriptive norms, which, as the name implies, simply describe what is expected, or normal, in terms of behaviour within the group. The second type is an injunctive norm, which conveys approval or disapproval for a particular behaviour. A descriptive norm might be "*most people pick up litter*", whereas an injunctive norm would say "*littering is simply not acceptable*". As one would imagine, an injunctive norm carries more weight than a descriptive one.

Real World: ***A room with a viewpoint***

This case study takes its name from a 2006 paper[325] by Goldstein and colleagues, where they experimented with three forms of the familiar hotel room message "*please reuse your towel*". The response rates to each version of the message are shown in red text in square brackets:

1. *"HELP SAVE THE ENVIRONMENT. You can show your respect for nature and help save the environment by reusing your towels during your stay."* [37%] See left.

2. *"JOIN YOUR FELLOW GUESTS IN HELPING TO SAVE THE ENVIRONMENT. Almost 75% of guests who are asked to participate in our new resource savings programme do help by using their towels more than once. You can join your fellow guests in this programme to help save the environment by reusing your towels during your stay."* [44%]

They experimented with variations of the second message which contains a descriptive norm, by changing the norm from "*75% of fellow guests*" to "*75% of guests in room XXX*" where XXX is the room number where the sign was posted [49%] or "*75% of fellow citizens*" [44%] or "*76% of women and 74% of men*" [41%].

The first, least effective, message did not contain a norm, while the most effective had a norm that related very specifically to the room that the guest was staying in. The gender, general guest and citizen norm groups had roughly the same result in terms of uptake. The difference between a 37% response and a 49% response is considerable. Norms have a powerful effect, which we should use as appropriate.

Real World: *The power of norms*

A widely quoted study in household energy use is that of Wes Schultz in 2007.[650] It is highly illuminating in terms of the effect of norms on energy efficiency behaviour.

In this study, 290 households in San Marcos, California, received feedback on their energy consumption. The first group received a message detailing their home's energy use, the average use by the other homes and some tips on energy conservation. These were descriptive norms.

In this group, for those who consumed more than the average, the feedback lead to a reduction in energy use of around 1.2 kWh a day. But for those who consumed less than the average, *the feedback led to an increase in energy use* of around 0.9 kWh. This boomerang effect is an unintended consequence of people's desire to conform to the norm. It is unlikely that the folks here consciously thought *"hurrah! let's burn more energy"*, but the information that they were using less than average may well have reduced their motivation to save energy further.

Another key finding of the study comes from the second group of households who received the same messages as the first with the addition of, if they were doing worse than average, a sad face ☹, and if they were below average, a happy face ☺. These faces are injunctive norms as they describe approval.

The result was remarkable. Those higher-than-average energy-consuming households delivered the same level of savings as in the first group, but in the better-than-average homes, the boomerang effect was virtually eliminated.

We have already seen from our piece on the psychology of comparisons (page 660) that norms (i.e. our choice of the normative item in the comparison) unconsciously influence our decision-making. Simply by ordering the comparison in the right way, we can enhance people's motivation to act.

It is as if we are designed to seek confirmation norms and evaluate ourselves against this information. That may well be the reason why the feedback in the office energy efficiency study, quoted on the previous page, had an impact only when people were given data about what they and what *other* individuals in the building where doing. The comparison with others is what is driving the behaviour change not the awareness of the energy use *per se*.

We can think of a norm as a benchmark for behaviour. For it to be most effective, a norm needs to relate to a group with which the target audience identifies - the greater the identification, the more powerful the effect. We can see from the example opposite that people identify more closely with other guests who occupied *their* room than with *fellow citizens*.

Particular care needs to be taken when using descriptive norms, as they may unwittingly sanction undesirable behaviour. An example is a programme by the Petrified Forest National Park, in Arizona, to reduce the amount of petrified wood (fossilized ancient trees) that visitors take away with them. They put up signs that stated:

> *"Your heritage is being vandalized every day by theft losses of petrified wood of 14 tons a year, mostly a small piece at a time."*

In fact, research by Griskevicius[335] and his colleagues demonstrated that this sign actually *encouraged* greater theft of petrified wood as it implied that it was much more commonplace than the 2% incidence per visitor, and therefore acceptable. Thus, descriptive norms should be used with care as they could unwittingly legitimize undesirable behaviour or reduce motivation to act, as explored in the case study on the left.

The examples I have given are of norms being communicated through formal messaging. However, a very powerful method of communicating norms is by example. In experiments, people are much more prone to pick up litter[146] or turn down showers[35] if they have observed someone else do it just before them. These studies are evidence why it is important to get folks to lead by example in any change process we are instigating and - of course - the more closely our target audience can relate to that person, the more powerful the effect.

As well as similarity, or the ability for the individual to identify with the group to whom the norm applies, it has been shown that norms have a greater effect in conditions of uncertainty[145] and where they reinforce a positive self-image. For this reason, norms work best when they are salient, placed in a context or decision that is unfamiliar or non-routine, and where they encourage a positive action rather than discourage a negative one. If possible, injunctive norms should be used with descriptive norms to reduce the boomerang effect.

19.7 Effective league tables

A well-designed league table can be a very useful tool to encourage resource efficiency. There are, however, many potential pitfalls, which if not avoided, can leave us with feedback that demotivates people rather than enthusing them.

We have seen from the preceding discussion of feedback and norms that if we introduce a comparative performance element into our communication, this can increase motivation to act. The comparison can be with a peer group average (in which case it is a norm) or against the performance of other teams.

A very widely used form of comparative feedback is league tables. Nowadays, there has been somewhat of a backlash against performance tables in the UK, especially in the public sector, where they have been used extensively to rate a broad range of services from medical care to policing and education. Many of the objections that can be levelled at these tables should also be borne in mind when constructing league tables to drive resource efficiency.

Common criticisms of these league table are.

1. They can have simply too many individual targets and indicators, some of which may be directly contradictory. This *overload* is a common criticism of the targets for the National Health Service.

2. Indicators may not reflect actual performance. For example, schools may be ranked on final results, without taking into account the initial capabilities of their intake (e.g. in some schools in the UK a significant proportion of students may not have English as a native language). In this case staff in the *"poor"* performing schools may be demotivated by their inability to achieve the highest ranking. In resource efficiency programmes, a typical example of this is when buildings are ranked by kWh/m^2 when in reality some will be constructed under modern efficiency standards, while others may have older and less efficiency equipment and fabric.

3. Measurements may not be accurate. This is a criticism often levelled at crime statistics. Greater focus on and awareness of certain offences leads to increased reporting and an apparent increase in incidence. We have seen how, in resource efficiency, widely used specific ratios (page 458) are a poor measure where there is a significant baseload effect.

4. Behaviours are distorted towards achieving the league table performance rather than the overall mission of the organization. An example would be schools that push students towards *"softer"* qualifications which score the same points but are easier to pass, or who withhold opportunities from marginal students to enter exams fearing they will drag results down.

League tables are universal and easily understood - just ask any sports fan.

In Numbers: *Performance indicators*

Suppose we have a group of buildings where we want a performance indicator for the electricity use. The table below shows the possibilities.

Measure	*Description*
Absolute kWh	This is the absolute use. This value is usually meaningless for comparison as a bigger building will use more electricity.
Normalized Use kWh/m2	This is the specific ratio. There can be problems with this if there is a fixed baseload or if electricity use is driven by other factors like cooling degree days.
Variance from Target (kWh) Actual-Target	This is the target variance. It can take into account m^2 and degree days. But larger users will tend to have larger variances.
Variance from Target (%) (Actual-Target)/Target	This is the percentage variance. It is good because it is dimensionless: that is to say, I could compare percentage variance across a range of resources which are measured in different units. This is the most common performance indicator used in resource efficiency.
Variance from Target (z-score)	The z-score variance takes into account the natural variability (or level of control) of the resource use, so is statistically more valid (see page 492). In practice this is almost never used because it is a difficult concept to explain.

5. League tables can be self-reinforcing. Thus, better-performing schools are over-subscribed and can, therefore, select more able students and so go on to perform better in subsequent tables. In resource efficiency, for example, teams who demonstrate greater success are often in a better position to argue for resources than teams that appear to be struggling, but in the latter case, the resources could have a much greater impact.

If we can design our league tables to avoid the pitfalls above, they can nevertheless be a useful tool to motivate people to act. There are several reasons why we should consider league tables, despite the problems identified.

1. Comparative feedback boost motivation to act. In the earlier office study, feedback alone has little impact on efficiency, but when combined with information about how others are doing it was much more motivational. The impact of comparison is backed up by a recent study that showed that social feedback sustains behaviour change among employees much more effectively than monetary rewards.[345]

2. League tables can provide positive visibility for the resource efficiency programme. With the right emphasis, we can use these as a basis to validate and celebrate success, rather than to *"name and shame"* failure. For example, we can indicate improvement with green *"traffic lights"*, and if everyone makes some improvement, they will all have positive feedback regardless of their relative position in the table.

3. League tables are universal and easily understood. Every sports fan will no doubt be tracking their team's performance on a league table, music and movies are ranked on league tables called charts and so forth, which means that people can instantly relate to these when used in resource efficiency. Indeed, a lot of information can be conveyed on a league table because of people's familiarity with the form.

4. Well-designed league tables can cascade. For example, we can have a league table comparing shifts within a department; comparing departments within a site; sites within a region; and regions within an organization. These overlapping layers of visibility of our programme are a key ingredient to sustain the effort in the long term. By cascading upwards, we can create interest, accountability and motivation to succeed among management tiers remote from our programme.

The choice of the performance indicator for our league table is critical. On the one hand, it needs to be simple to understand and on the other a fair measure of performance. For these reasons, by far the most common indicator, widely used in Monitoring and Targeting (M&T), is variance from target (and the CUSUM from baseline or original). When this performance measure is being used, we should be aware of the benefits of engaging with teams in setting these targets. Only by being closely involved in the target-setting process will individuals feel ownership for, and understand, the goal which is being tracked in the league table.

19.8 Starting small

There is a real tension between making a request to people that is easy for them to achieve and delivering the scale of change is required. In reality we often have no choice but to start small, and in these circumstances we should design our initial interactions so that we prime further action rather than enable participants to "declare victory" and cease participating.

Real World: *Attention deficit*

©MonkeyBusiness. Fotolia.com

In a 2012 survey by Accenture[5] covering 19 countries, it was discovered that more half of electricity consumers had spent *no time at all* interacting with their utility suppliers in the last 12 months. The average interaction time was just 9 minutes with South Africa topping the list at 19 minutes and Japan just 3 minutes.

If people spend such little time engaging with their electricity suppliers, then it is no wonder that attention to this in the workplace is also low.

Unless we can give the issue real prominence with a "burning platform" (page 319), asking for a large up-front commitment of time is unrealistic.

In these circumstances, we have no choice but to develop strategies where we can incrementally grow from an initial small interaction to the point at which we achieve our desired change. This requires some clever planning and psychology.

I started this book (page 9) with a traditional saying: *"How do you eat an elephant? Why, one bite at a time, of course"*. In the section on availability barriers to resource efficiency, on page 155, I argued that we can drive a successful efficiency programme by getting a lot of people to regularly dedicate a little time rather than by getting a few people to commit a lot of time.

Clearly, we need to start with where people are at and it is often unrealistic to ask someone to make a large change in their behaviour from the outset (see box left). Indeed, asking for too much or holding back for *"perfection"* are the root causes of many programme failures I have observed with my own eyes (see page 206). So starting small is a reasonable strategy.

Is that true, though? Some argue that if all that we request in terms of change is a minor action, then this will result in - surprise, surprise - a small result! Folks like Donella Meadows, Bob Doppelt and many others have reasoned eloquently that no less than a fundamental change to our underlying systems will deliver the scale of change needed to address the magnitude of the problems we face. Similarly, Cambridge Professor David MacKay, in his fantastic book *Sustainable Energy — without the hot air,*[491] asserts:

> *"Modern phone chargers, when left plugged in with no phone attached, use about half a watt. In our preferred units, this is a power consumption of about 0.01 kWh per day. For anyone whose consumption stack is over 100 kWh per day, the BBC's advice, always unplug the phone charger, could potentially reduce their energy consumption by one hundredth of one percent (if only they would do it).*
>
> *Every little helps!*
>
> *I don't think so. Obsessively switching off the phone charger is like bailing the Titanic with a teaspoon. Do switch it off, but please be aware how tiny a gesture it is. Let me put it this way: All the energy saved in switching off your charger for one day is used up in one second of car driving. The energy saved in switching off the charger for one year is equal to the energy in a single hot bath."*

Wow! Therein lies the issue with these small actions, that is that they make us think that we have done our bit, that we have made our contribution to solving the problem. In short, they enable us to *rest on our laurels.*

Real World: *Self-perception*

There is evidence that, used appropriately, small initial actions can facilitate a larger change. Some foot in the door techniques do genuinely appear to work.

The classic case study of these techniques is that of the Canadian Cancer Society, which found that people who had agreed to wear a lapel badge with their logo were subsequently twice as likely to donate than those who hadn't been asked to wear the badge.[510]

One explanation for this behaviour dates back to the 1970s where Daryl Bem developed self-perception theory.[67]

As a result of the action of putting on the badge, the theory proposes our own self-view is modified. Now we see ourselves as supporters of the Canadian Cancer Society, and so subsequent actions that we take strive to maintain consistency with this self-image. Thus, when asked to make a donation we are much more likely to do so.

The study above is one where people have been asked to make a commitment or take ownership, something that we shall explore further later. However, this priming effect does not necessarily require a commitment. For example, simply carrying our a survey of people's intentions to vote increases their likelihood of doing so.

Central to the debate over small actions is the notion of spillover. That is whether these small actions will lead to others in a "*virtuous escalator*" (positive spillover) or whether people will declare victory and cease to take an interest in further actions (negative spillover).

Tom Crompton and John Thørgersen have written a strong critique of the notion of spillover in their paper *Simple and painless?*[175] produced for WWF. Their argument, which I have much sympathy with, is that many organizations, individuals and policymakers are in denial of the scale of change needed to fix environmental challenges and that promoting marginal improvements simply will not deliver change at a sufficient scale. A consumer who installs an LED lamp will not go on to install solar PV (although installing solar PV would most likely encourage the adoption of more LEDs).

We have already seen some examples of organizations which are making ambitious commitments to change from the outset, for example, Interface's *Mission Zero* (page 226). This is the ideal.

However, we need to be realistic. Where we don't have a strong mandate, we need to recognize that our request to staff in our organization is in competition with other demands on their time. We may well have simply no choice but to *start small*. It is better to do something than to do nothing, because, if we design our initial engagement carefully we can ensure that it grows over time.

Where we don't have a strong mandate, motivation to act will most likely be low, so we need to focus on capability - we need to make the initial action easy to accomplish. Critically, though, we need to consciously design the initial action so that it will lead to further action.

The initial action might simply be to observe or passively participate in an activity carried out by someone else. We know that when householders were asked to peer into their attics to observe the lack of insulation or to feel for themselves the heat loss from a water tank, they are more motivated to take action[510] themselves. Or we may start with an activity that creates a psychological connection with the change we want to achieve (see right).

We may put together a series of interactions that lead to an increased probability of participation. For example, if we are looking to promote recycling we might prime the process by carrying out a survey of our staff to ask them if they think that providing more facilities for recycling would be a positive step (most would agree with this proposition). We could then follow up with a descriptive norm: "*75% of the team in our department agree that recycling is a good idea.*" At this point, we should be in a position to launch the change, with a much greater probability of commitment by staff.

We must avoid processes where folks can select actions involving minimum effort, which may have small results or, worse, let them conclude that the task is complete. Resource efficiency, we must establish from the outset, is about continual improvement. There is always another next step (or bite!). ⇒ page 672.

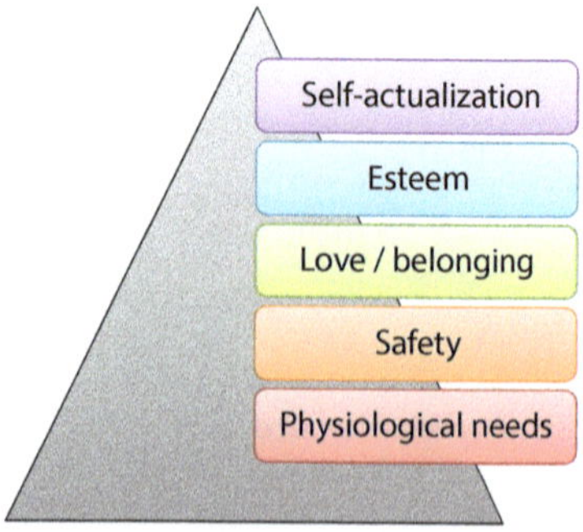

19.7 **Maslow's hierarchy of needs**
Maslow himself never drew this hierarchy as a pyramid, so I have drawn this as a stack of needs with the pyramid in the back indicating the sequence, from the most immediate needs at the bottom to the highest needs at the top.
Source: Niall Enright. This image, and many of the illustrations in this chapter, are available in the companion file pack

		Motivation Factors	
		Low	High
Hygiene Factors	High	Employees well rewarded, but not especially motivated.	Employees highly motivated with few complaints.
	Low	Low motivation and high levels of complaint.	Employees highly motivated, but unhappy with salaries and conditions.

19.8 **Employee types based on Herzberg's two factors**
The bottom right quadrant often describes Public Sector or vocational fields where pay is low but motivation is very high.
Source: Niall Enright

Exploration: ***What motivates people at work?***

Many of the techniques covered in this chapter are derived from studies of individuals. As we move into techniques that are organizational in nature, it is helpful to take a step back and review theories of workplace motivation.

A good starting point is a basic categorization of human motivation, known as Abraham Maslow's hierarchy of needs, shown left. This model suggests that what drives us at a most fundamental level is our physiological needs (oxygen, water, food) and we will satisfy these requirements first and then move on to the next level up the hierarchy.

This is an positive view of mankind, where our ultimate objective is self-actualization, or the realization of our full potential as humans. Actualization is a self-defined goal, for example, one person may wish to become a great athlete, another a good parent, another to make a contribution to their field of science. In later years, Maslow suggested that there might be a higher level still, self-transcendence, which describes aspirations beyond the individual, altruistic or religious in nature.

Maslow believed that until a lower need could be satisfied it would be difficult to engage with higher ones, and the failure to satisfy these needs was a cause of anxiety and unhappiness. While recent work has questioned the precise divisions of the hierarchy (survival, safety, social and psychological) and their cultural consistency, a key notion is that the root of motivation is meeting these needs.

Following on from Maslow's general work on the motivation of humans is work by Herzberg on satisfaction in the workplace, which has led to two-factor theory, sometimes called Frederick Herzberg's motivation-hygiene theory. This theory has had a huge impact in business management circles and Herzberg's paper[366] on the subject in the Harvard Business Review has been the single most reproduced paper, with 1.2m reprints.

This is no surprise, because what Herzberg discovered stunned the business community. He examined what motivated employees and realized that all those factors that management in particular though motivational, such as salary, holidays, fringe benefits, status, security and processes in the company actually have relatively little impact on employee motivation. All that these factors could do was dissatisfy or demotivate. That is to say that an employee with a poor salary or holiday package is likely to be dissatisfied with their work, but once the salary reaches the point where the employee regards it as acceptable, further increases will not lead to significant increases in motivation.

Conversely, there were a series of other factors which seemed to drive motivation: recognition, advancement, the challenging or stimulating nature of the work itself, responsibility, growth and, above all, achievement. The insight is that providing lots of these things increases motivation, but their absence does not necessarily lead to greater dissatisfaction or complaints.

The key lesson then is that the two aspect of employees' attitudes to work - motivation and satisfaction - are two different things, driven by two independent sets of factors. Thus, the two-factor theory. The factors which lead to dissatisfaction are called *"hygiene"* factors in the sense that they are maintenance factors needed to keep the organization functioning. Management need to ensure that these factors are not interfering with the smooth operation of the business and leading to complaints. On the other hand if management want to drive productivity or innovation through greater motivation, they should look to increasing the motivation factors rather than salaries or fringe benefits.

One way of thinking of these factors is as extrinsic or intrinsic to the employee. The hygiene factors are all extrinsic, whereas the motivators are intrinsic. As if to emphasize the external nature of the hygiene factors, Herzberg called them KIKA factors (*"Kick In The Ass"*) to emphasize their limited effectiveness on motivation:

> *"I have a year-old schnauzer. When it was a small puppy and I wanted it to move, I kicked it in the rear and it moved. Now that I have finished its obedience training, I hold up a dog biscuit when I want the schnauzer to move. In this instance, who is motivated–I or the dog? The dog wants the biscuit, but it is I who wants it to move. Again, I am the one who is motivated, and the dog is the one who moves."*

Thus, in Herzberg's analysis, the human equivalent of dog biscuits, salary and perks, are simply a tool to get us to perform our work. The salary does not create an intrinsic desire to work, a love of work or a passion to do better. These things all come from within ourselves. Sure, if we throw enough money at people we can get them to do horrible jobs, but that does not necessarily make them happier or more motivated when doing the work, it does not mean that they will take on any extra activity, or push themselves to deliver more for their employers.

We can see how two-factor theory fits in with Maslow's hierachy of needs. The motivating factors all support the higher needs. Just as we aspire to social and psychological fulfilment as individuals, so too do we aspire to this as employees. Hence achievement, recognition, the stimulating nature of the work are all more motivational once we have met our basic physical needs, which lie at the bottom of the hierarchy.

The implication of Herzberg's analysis is that the key to driving greater performance, once the hygiene factors are met, is to provide job enrichment. This enrichment could involve decreasing micro-management, giving greater authority, making the work more rewarding and complex, enabling them to develop expertise and gain recognition. When these strategies were tested in real workplace situations,[366] they were found to significantly increase performance - although in the short term there was actually a dip as employees came to grips with the new approach.

Building on these ideas is the work by Victor Vroom of Yale who turned his attention to how the desire for reward influences employees' decision-making. This concept was called expectancy theory, and stated that there are three factors that are considered whenever a reward is available (rewards are not just a hygiene factor like money, but motivation factors like recognition).

1. Valence. This is the degree to which the employee wants or values a particular reward. As we have seen from the preceding text, non-monetary rewards may well be valued more highly than monetary ones.
2. Expectancy: This describes the employee's confidence that they can deliver the outcome needed to achieve the reward. In some ways this is related to the capability dimension of any change. Issues here include the employee's assessment of their own skills, the degree of control they have in the process, the inherent difficulty of the goal.
3. Instrumentality: This describes the certainty the individual has that they will be rewarded if the objective is met. It relates to their level of trust in the organization, the strength of the promise of reward, the degree to which the bonus is pegged to a tangible, measurable outcome.

Designing an appropriate reward scheme that effectively motivates change depends on the employer selecting a reward that the employees values (valence), ensuring that the scheme is fair and transparent (instrumentality) and that the employee believes that they can achieve the goal because they have the necessary skill and resources (expectancy). These principles will be revisited when we discuss the design of "honest, fair and achievable targets" which are key to success in M&T.

As we shall see in the following pages, these studies have a very real connection to our techniques to engage folks in resource efficiency. What this emphasizes above all is that participating in a well-designed resource efficiency programme offers huge job enrichment opportunities. This real benefit, in turn, means that motivation is bound to increase which will affect all the work of the employee, not just those activities focused on resource use. However, to access these enrichment opportunities, we need to ensure that the employees believe that they can gain the rewards on offer. In particular, we need to be conscious that motivation is just one aspect that we need to address: the other is capability.

19.9 How to harness knowledge

Our organization is essentially the sum of the knowledge it possesses. Improvement comes from harnessing that knowledge using the most appropriate techniques.

Real World: ***Brainstorming correctly***

There are few better ways to harness knowledge, create ownership, foster teamwork and find solutions than brainstorming. But the rules must be followed for it to be effective.

No criticism. This is the key to a successful brainstorm. Nothing is guaranteed more to switch off the contributions and creativity of the participants than people throwing cold water on ideas.

The wackier, the better. Brainstorming is about connecting with our intuitive tacit knowledge as much as about our explicit knowledge. It is a lateral thinking technique.

Build on others. It is ok to expand on previous ideas, to make them even better. Don't wait to add the idea or you will forget it! Ensure everyone has an equal chance to contribute.

Maintain the flow. Don't ask for clarification. The more ideas we have, the merrier. Explanations come later.

Good facilitation is key to the brainstorming session. The session should start with a clear description of the problem we want to solve. In some ways, a brainstorm runs counter to etiquette and rules of behaviour we usually adopt. One way of overcoming this is to do a "*warm up*": for example, asking people to think of as many uses for a paperclip as they can in 60 seconds. Get folks relaxed and having fun.

The greatest source of improvement in our organization is the knowledge it possesses. This principle applies to both sources of improvement: corrections of design, systems and processes that are understood but which are not operating as specified or from innovation leading to new designs, systems or processes.

It turns out that knowledge exists in several forms. The first categorization of knowledge is divided into tacit or explicit. Tacit knowledge is sometimes referred to as "*know-how*" and is intuitive, learned through practice or repetition, and is hard to write down or explain. When people follow a *hunch*, they may well be harnessing their tacit knowledge, which may be entirely unconscious. Explicit knowledge has been called "*know-what*" and is knowledge that can be readily written down and explained. Taking the example of a bakery, the recipe for a loaf of bread is explicit knowledge, while the precise feel of the dough, which means it has been kneaded just enough, is tacit knowledge. We must not underestimate tacit knowledge as it can form the majority of an organization's knowledge and that part which gives it competitive advantage ("*the way we do things around here*").

The second dimension of knowledge is individual or collective. That is to say that the knowledge might reside in a single person, or in a team or institution. Typically, team knowledge is greater than the sum of the individual knowledge as it includes experience that arises from the interactions of the team members. While all forms of knowledge exist in all organizations, there can be a dominant form used in decision-making, as illustrated in the diagram below. A university, for example, has a large number of individuals with knowledge that can be communicated in written or oral form (publishing papers and teaching is, after all, their primary activity). By contrast, a circus troupe is characterized by individuals with unique skills which cannot be communicated easily i.e. tacit knowledge. On the other hand in most western firms, the dominant knowledge form involves people operating in teams within strictly documented systems (especially so in pharmaceuticals whose

Organization Example	Dominant Knowledge Type	Technical name
University	Individual - Explicit	Embrained
Circus Troupe	Individual - Tacit	Embodied
Pharmaceuticals Manufacturer	Collective - Explicit	Encoded
Japanese Corporation (J-Firm)	Collective - Tacit	Embedded

Real World: *Breadmaking*

©ra3m. Fotolia.com

Ikujiru Nonaka is considered one of the key figures in developing an understand of tacit knowledge in organizations. In his article for the Harvard Business Review,[560] he describes how Matsushita Electric Company struggled to harness tacit knowledge to develop a new home breadmaking appliance.

In the first prototypes of their machines, the outside crust of the loaf was hard and overdone while the inside was undercooked. They could see that there was a problem with the kneading of the dough. No amount of scientific analysis, including taking X-rays of loaves at various stages of production, could solve the problem.

Eventually, a breakthrough came when Ikuko Tanaka, a software developer on the team, decided to volunteer as an apprentice to the head baker at Osaka Hotel, who reputedly made the very best bread in the area. After much observation, one day Tanaka noted that the baker was not only stretching but also twisting the dough. This movement, it seemed, was the key to getting it to rise properly.

Once the solution was understood, the engineers were able to incorporate this twisting motion into the breadmaker (not just in the movement of the hook but also by ridging the inside to keep the dough from moving). The appliance was a huge success.

processes are strictly defined by the Food and Drugs Administration). In Japanese companies, the systems may be as well-documented as in the western businesses, but the team approach and culture means that decisions arise from collective decision-making processes which are more tacit than explicit in nature. Another organization with collective tacit knowledge is an orchestra, whose individuals have the tacit skill of music-making but must employ this collectively to play in harmony.

While these characterizations are obvious stereotypes, they nevertheless help us choose the techniques for knowledge-gathering and problem-solving that best fit our organization.

Knowledge Yype	Knowledge Sharing Technique
Individual - Explicit	Suggestion schemes are good at capturing explicit ideas that can be written down. They are lousy at improving processes which rely on tacit knowledge because the improvement is hard to describe in words.
Individual - Tacit	Techniques to harness individual tacit knowledge include mentoring, storytelling, solving-by-doing.
Collective - Explicit	Team problem-solving processes, such as Six Sigma or CDR workshops (see page 404), work well where we need team input into a problem that can be written down.
Collective - Tacit	Techniques to harness collective tacit knowledge input include brainstorming, shared spaces.

Suggestion schemes, discussed in the next section, provide a common tool to harness the explicit knowledge of individuals to develop improvements. Structured problem-solving workshops of various kinds will do the same to harness the explicit knowledge of a team (see page 404 for an example).

Tacit knowledge is harder to share and focus[471] or to convert to explicit knowledge (see right). Because tacit knowledge generally requires practice and repetition to develop, mentoring is a common way of transferring this from one person to another (that is why craftsmen have apprentices or understudies to pass on their skills). When describing tacit knowledge people tend to use metaphors, analogies and demonstrations to try to explain things so storytelling or open-question interviews are ways of getting input. I also find that asking an individual to improve something by working at it (sort of giving them a consulting brief to go and sort something out), which I call solving-by-doing, engages their tacit knowledge, but I have to make sure I can capture the change that solves the problem. At a collective level, brainstorming (if done correctly, see box opposite) can harness unconscious knowledge, while creating shared spaces[417] where communities of practice can meet and interact in an unstructured way to share knowledge or solve problems is a tool to unlock collective tacit knowledge. Where the knowledge involves a manual or physical skill these shared spaces need to be physical spaces, in other cases they can be online. A combination of these techniques can be used to reach all employees, regardless of organization type, over time.

19.10 Suggestion schemes

Suggestion schemes are often the default employee engagement technique for resource efficiency, but we should be aware that there are many factors to take into consideration if we are to have a successful scheme.

Real World: *A penny for your thoughts*

ideasUK is a not-for-profit organization which promotes employee suggestion schemes.

Their 2005 member survey[388] gives some idea of the extent of these schemes in the UK.

Number of organizations	*52*
Number of employees	*781,000*
Number of ideas received	*113,600*
Participation ratio	*14.5%*
Ideas implemented	*30,500*
Implementation rate	*27%*
Overall savings	*£42 million*
Awards paid	*£1.4 million*
Average clearance time	*49 days*

Although schemes will differ from organization to organization, these data can give us a baseline from which to estimate the volume of responses we might get.

One fun aspect of suggestion schemes is coming up with a name for the scheme. Here are some examples:

Big Ideas	*Marks & Spencer*
AIM (all ideas matter)	*Boots*
First Steps	*Tesco*
One Life	*BUPA*
bRight ideas	*Aviva Trains*
Partner Ideas	*Waitrose*

Almost always the first technique that comes to mind when an improvement process is suggested or a means to engage staff is desired, is a suggestion scheme. Such schemes are good at capturing the explicit knowledge of individuals, but can also help to involve and motivate people more generally.

The illustration opposite sets out the main elements of a suggestion scheme. It is important that we consider all these factors very carefully as there are some pitfalls we need to avoid to make sure our scheme delivers.

One of the recommendations is to define the goals of the scheme very clearly. Suggestion schemes often unwittingly emphasize cost-cutting, when revenue-generation, health and safety, quality and customer satisfaction may be equally value-adding to the organization.

Participants: We need to define who can submit suggestions. In many schemes, those employees who are expected to innovate or problem-solve as part of their jobs, such as managers, designers or engineers, are often excluded from the scheme. Be very careful when making this choice, as excluding some participants, such as contract staff, can be divisive. We should also consider whether or not "team" suggestions are acceptable - as suggestion schemes taking ideas from individuals alone can sometimes work against teamwork.

Drivers: These are the three major factors that encourage people to submit ideas: the desire for a better work experience, the desire for reward and their confidence in their job (self-efficacy). A good understanding of these is critical to success - in particular getting the rewards element wrong can spell disaster (see the next section). Too large a reward can lead to friction and jealousy between employees, too small a reward can result in indifference to the scheme. Remember, rewards do not need to be financial. We should bear in mind that many people are not motivated by money, but by factors such as a better work experience or greater control over their work. Some schemes shortlist ideas and allow the best to be presented in person to the decision-making panel, which may be considered part of the reward of the process - recognition among senior peers is often very highly valued.

Mechanics: Here, we need to design the practical aspects of the scheme which the participants will interact with, such as how ideas are submitted, and who reviews the ideas and how the successful ideas will then be implemented. One of the most critical parts of this process is the cycle time it takes to get ideas

Exploration: *Continual involvement*

One of the problems of suggestions schemes is that they foster discrete involvement in the improvement process. The participant has an idea, and then sits back and waits some time to see if anything follows. *"My job is done"* and it is now up to others to follow up.

The same criticism can be levelled at other problem-solving activities, such as workshops or brainstorming which are episodic in nature. Indeed, the classical, prescriptive planning processes (page 339) with strategy followed by tactics also hinders continual involvement and innovation.

To truly harness our organization's knowledge we need a broader process in which we can place these particular improvement techniques. One such framework, specific to resource efficiency, is M&T, which is discussed in the next chapter.

A notion derived from military thinking that has popped up in many recent management handbooks is VUCA (volatility, uncertainty, complexity and ambiguity). Although used in different ways by different authors the basic concept is one of unpredictabilily and thus the greater importance of continual learning, planning, anticipation and adaptation. Unexpected and undesirable things will happen in a VUCA world. The fact that they happen is not a fault; what is an error is a failure to learn from them.

There is no doubt that we are in a fast-changing world. The more we can ensure that our resource efficiency programme can respond to these constant changes, the more value it will add.

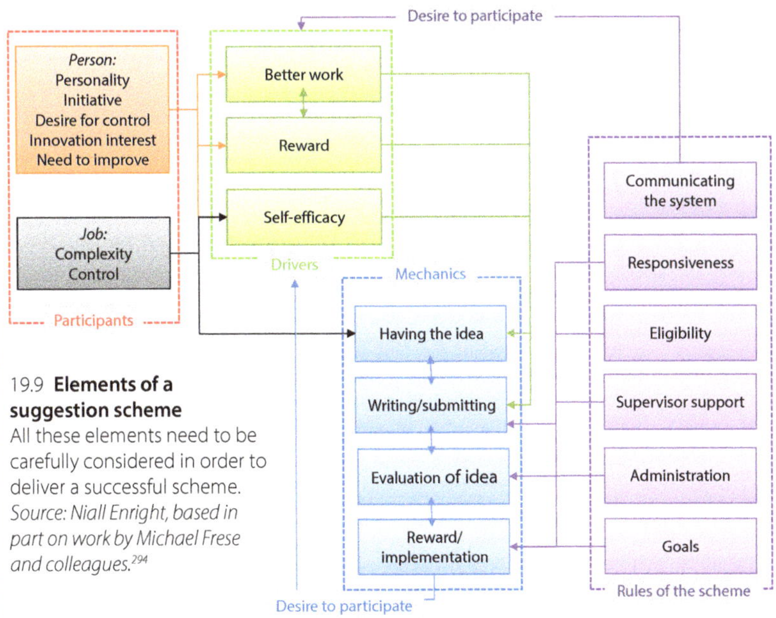

19.9 **Elements of a suggestion scheme**
All these elements need to be carefully considered in order to deliver a successful scheme.
Source: Niall Enright, based in part on work by Michael Frese and colleagues.[294]

through the system (usually the shorter, the better) and the effectiveness with which the participants are given feedback on their ideas (which is essential). Organizations are increasingly turning to software products to make the mechanics of the suggestion scheme easier. There are several of these to choose from.

The rules of the scheme: Again there is a lot of detail to get right here. For example, many schemes prohibit ideas that relate to salaries, terms of employment or management from being raised. Often there will be rules that require ideas to be original and possibly to exclude ideas that would be part of an employee's usual duties and responsibilities. The balance here is between setting out some principles and appearing arbitrary. The nature of the decision-making panel is also important: they need to be objective and also have sufficient expertise to judge the ideas fairly. In the UK, payments of up to £5,000 as part of a suggestion scheme may be tax-free if the scheme meets certain criteria, so it makes sense to consider the tax implications of any scheme. The degree of support provided by supervisors can be a critical success factor, so it is important that they are fully committed to the process, especially if managers are excluded from direct participation by the rules of the scheme, as is often the case.

Although they are very common, suggestion schemes are just one tool to drive employee participation in a resource efficiency process. Suggestion schemes are not good at capturing tacit knowledge, and they are also *"episodic"* in nature (see box left). Often suggestion schemes work best if combined with other techniques as part of a much wider engagement process which can reach those folks who are not comfortable putting forward ideas.

19.11 Using rewards

Rewards can influence behaviour. However, work in the psychology and social sciences is showing that this is far from straightforward.

DJSI Rank	Percent of Award
1	150%
2	125%
3	100%
4-6	75%
7-10	50%
11-15	25%
below 15	0%

19.10 **Schedule of vesting of 50% of the conditional grant of long term incentive shares for the AkzoNobel board**
Note that a top position in the Dow Jones Sustainability Index ranking by RobeccoSAM, leads to an award of 150% of the allocation, so there is a very strong incentive to exceed the target rank of 3rd in the Index.
Source: AkzoNobel Remuneration Report, 2014.[16]

Incentives or rewards are a form of control. They are put in place to influence behaviours and are a widely used tool to drive performance in organizations. We have already seen that one of the important considerations when designing a suggestion scheme is the reward that will be given for either submitting an idea or for an idea being implemented. But rewards for resource efficiency can take all sorts of forms.

Earlier, I described the example of BP (page 239), where the bonus received by business unit leaders was in part dependent on performance on reducing CO_2 emissions. In this example, the incentive had little effect because other, conflicting, performance measures were perceived to be more important. One company that seems to have got this right is AkzoNobel whose long-term incentive plan for the board is related to the company's ranking on the Dow Jones Sustainability Index (DJSI), as measured by RobecoSAM, a respected investment group based in Switzerland. A full 50% of the long term incentive scheme share allocations (valued for the chief executive, for example, at over €1,000,000 in 2014)[16] is contingent on AkzoNobel's position in the Index (see table left). It is not just the board whose bonuses are linked to sustainability performance, but also that of the top 600 managers.[760]

For those jobs where compensation is normally linked to performance, as in the case of these senior managers, it makes great sense to include in the measurement of the performance aspects of sustainability and resource efficiency. As these are known to contribute to shareholder value in the longer term this form of incentive is entirely commensurate with the fiduciary responsibilities of the directors and will act as a counter-balance to financial performance targets which often emphasize short-term returns at the expense of environmental performance. Of course, we need to be sure that the measurement of performance is objective, as human nature being what it is, people will be tempted to game the system to maximize their rewards. Using an external measure such as the DJSI can help to avoid this problem.

Incentives work on the basis that people are rational beings who are always seeking to maximize the economic value of any activity they undertake. This belief underpins many of the strategies that regulators use to change people's behaviours. For example, one way to boost recycling is to give householders one labelled bag a week for general household refuse and then to ask them to pay a fee for further bags. In one town in Canada, this strategy led to a

The big challenge for reward systems is that the **desired behaviour will decrease when the incentive is withdrawn.**

sharp fall in waste (46%) and increases in composting (50%) and recycled materials (26%).[510] There are many examples of incentives around, for example, householders and organizations have been strongly incentivized to take up solar photovoltaic generation through generous subsidies. We can see from the uptake of these technologies that these incentives do influence behaviour. But we can also see the big weakness with incentives; we only get the desired behaviour while the reward is present - withdraw the incentive and the behaviour will decrease.

When discussing incentives, we should also think about disincentives. We have already seen in Chapter 4 on Barriers that there are numerous disincentives to resource efficiency, many of which are due to distorted pricing. Examples include the tiered charge for electricity which gets cheaper the more you use. Companies that consume their own waste products as fuel often see this as *"free"* energy, whereas it actually may have a real cost in handling and emissions terms or an opportunity cost as the waste could be sold to others as a fuel.

Correct price signalling is clearly important. In fact, one key behaviour change strategy is to modify the $, £, ¥ or € component in the value calculation that the individual is making. We have seen earlier the notion of whole life costing (page 302) as a means of working within the same decision framework, but altering the relative value of the resource-efficient option. This information is not providing a direct reward to the individual, of course, but does help them make better decisions, which in turn should be rewarded by their organizations.

Amid all this discussion of motivation and rewards, we should not lose sight of the role of penalties. These, too, are control tools. I have seen plenty of cases of shop floor and executive behaviour which leads to scandalous waste and for which it is perfectly reasonable to impose a sanction if it persists. "*My way, or the highway*" as they say, or, more crudely, "*J*DI*" (think of Nike's "*Just Do It*" and add an encouraging word in the place of the *). Not every behaviour change needs to be approached with kid gloves, and we should be willing to communicate disapproval (injunctive norms) and establish clear consequences if the wasteful behaviour continues. In recent years I have seen an increasing use of sanctions within organizations to drive behavioural safety programme compliance, ensuring that people understand that a failure to adhere to systems will have significant consequences. It may be appropriate to consider this approach for some aspects of resource efficiency if getting change by other means appears intractable. That having been said, we should always seek positive encouragement (rewards) before negative (punishment), as the latter is known to be less effective at changing behaviour.

People recognize these incentives as control devices. This is one reason that incentives can be counterproductive as people resist control (in fact they often crave the precise opposite: greater autonomy and influence over their work). The Exploration piece on the next page sets out some of the contradictions and problems related to rewards and punishment. Although incentives do work, they may not be the most effective behaviour change tools. ⇒ page 680.

19.11 **Five principles for monetary rewards**
These principles are drawn from a wide range of studies on monetary rewards.
Source: Aguinis et al, 2014.[11]

Exploration: ***Key principles of reward***

1. Define and measure performance accurately.
2. Make rewards contingent on performance.
3. Reward employees in a timely manner.
4. Maintain justice in the reward system.
5. Use monetary and non-monetary rewards.

The strongest motivations come from within the person, from **intrinsic factors** such as the love of a job, a desire to do good, personal growth and attainment. Monetary or other external incentives can reduce intrinsic motivation.

Exploration: *The problem with rewards and punishment*

©federherz. Fotolia.com

Rewards and penalties have mixed results. We need to recognize these problems as regulators, for example, are increasingly turning to sanctions to drive environmental compliance (such as obligations to report on emissions or to carry out mandatory audits). These punishments (fines, reputational damage, etc.) are intended to ensure compliance.

We can see that punishment does, as expected, lead to behaviour change desired by the regulators, but for punishments, the primary focus of the change is on avoiding the sanction. This focus on the penalty means that many organizations may simply put their efforts into not being caught breaking the regulations or into exploiting loopholes in the rules, as this may be seen as easier than compliance. The punishment does not induce a greater desire to deliver the underlying improvement. Indeed, compelling organizations to report environmental performance does not make them any better disposed to the environment per se.

Similarly, the use of incentives is also problematic. As we have seen from Herzberg's work on motivation, incentives (such as pay, bonuses or increased holidays) are extrinsic or external influencers of satisfaction. But we have seen that the strongest motivation comes from within the person, from intrinsic factors, such as the love of the job, a desire to do good, personal growth and attainment.

Introducing monetary or other external incentives can reduce intrinsic motivation, as an experiment[198] by Edward Deci in 1971 showed. Two groups of students were invited to participate in a psychology experiment in which they were asked to solve a puzzle. Half the students were promised money for working on the task, the others not. The experimenters then told the students that this phase of the experiment was over and that they could do what they wanted as the next step was prepared, and then left the room (although the students were being observed). Those who were not promised a monetary reward continued to work on the puzzle for longer than those who had been offered money. It seems that the addition of a financial reward reduced the students' innate or intrinsic desire to solve the puzzle.

Another experiment, at around the same time, by Mark Lepper, with very young children who were rewarded for using magic markers showed the same effect. Those who received a reward (not money in this case but a certificate) were seen to use the markers less than the unrewarded children when observed two weeks after the initial experiment.

If we do something for ourselves rather than for reward, we take ownership and responsibility for it. Robert Cialdini draws some interesting example of this in his book *Influence: The Psychology of Persuasion,*[143] ranging from the indoctrination practices of the Chinese Communist Party to initiation rituals of college fraternities. These organizations share a common approach in that they all give minimal or no rewards for the behaviour they seek. By doing so, the individuals concerned were undertaking the action for themselves and so were demonstrating strong personal commitment.

Cialdini goes on to make the broader observation, which calls into question how we treat our employees:

> *"All this has implications for rearing children. It suggests that we should never heavily bribe or threaten our children to do the things we want them to truly believe in".*

So, we can see that rewards are problematic since they are usually extrinsic factors such as money, gifts or prizes.

Ideally, our incentives will all be geared to the intrinsic motivators - thus we can offer recognition, opportunities for advancement, greater control over the participant's work. But these incentives are much harder, as one can imagine, to design.

In his comprehensive book on the subject, *Punished by Rewards: The trouble with gold stars, incentive plans, As, praise and other bribes,*[444] Alfie Kohn suggests some practical steps to counter the negative impacts of rewards on motivation, which I have paraphrased below.

- Get rewards out of people's faces. The more fuss and prominence we give to the rewards the more they reduce motivation.
- Offer rewards after the fact, as a surprise. In this way, the reward is not designed to direct the behaviour, but rather to recognize the achievement. Be aware, though, that if repeated, some people may come to expect the "surprise".
- Never turn the quest for reward into a contest. If rewards are limited, they become even greater inhibitors of motivation as the gap between the rewarded behaviour, and the unrewarded behaviour can be small. Indeed, a prize-giving event instantly turns the majority of those present into losers.
- Make rewards as similar as possible to the task. For example, reward a child for reading by giving them another book. This will reinforce the motivation that existed to complete the task in the first place.
- Give people as much choice about how rewards are used. By involving people in the nature and criteria for rewards and in the evaluation of their own performance, the controlling nature of external rewards can be reduced. However, undue involvement can increase the salience of the rewards, so this is a fine balancing act.
- Try to immunize people against the motivation-killing effects of rewards. You can remind the recipients of the reasons they like the task or the positive environmental consequences of the actions they are taking.

Of course, according to Kohn, the best way to handle the problems of rewards is simply to avoid them altogether. I am not sure that is entirely right - in many cases, we need to do something to influence behaviours, and if we cannot target true inner motivation then we may as well resort to simple satisfaction! But the points above should cause us to consider carefully the pros and cons of the rewards we give and the manner in which we give them.

Another criticism of all reward systems is that they can reduce risk-taking and innovation. People *play it safe* to ensure that they get that reward. But the ability to take measured risks and to fail without blame are central to a culture of innovation. In some organizations I have worked in - naming no names - this risk aversion is the biggest barrier to change. Managers know that they have a five-year cycle period in any given role and the "*name of the game is not to mess up and to move onwards and upwards on the greasy pole*". Here, we will need some very powerful motivators indeed to overcome this lack of willingness to make just about any decisions that could get them in trouble. Decisions to change systems and processes in favour of greater efficiency run counter to an "*if it works don't fix it*" mentality. These are organizations whose primary reward is for inaction.

At the risk of being dismissed for political correctness in this competition-dominated culture, I would urge anyone considering an incentive scheme as a technique to drive resource efficiency to review the scientific evidence on the pros and cons. We have seen from the AkzoNobel example that rewards do work, and I think that such schemes do have a real role to play. But wouldn't it be great if all your employees were motivated instead by an internal appreciation that resource efficiency is a good, noble, stimulating, interesting and hugely satisfying endeavour in its own right. Wouldn't those employees take real ownership for the process and create a culture where resource efficiency is in the collective DNA of your organization? Wouldn't those employees be more willing to go the extra mile, to take risks to innovate, to challenge your organization to go further? Wouldn't those employees be great ambassadors for your organization and connect better with your customers and stakeholders? I believe they would. Ownership and internal motivation are generally superior to external incentives.

19.12 Changing habits

Much of what people do involves little conscious decision-making. As a result, rational arguments for behaviour change may be ineffective because our actions are driven by habit. In order to change habits we need to understand them.

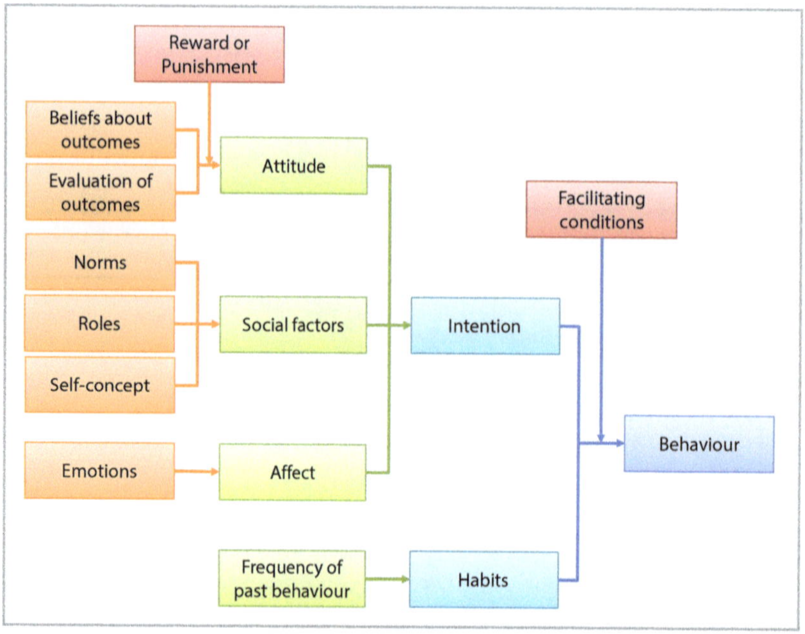

19.12 **Triandis' model of behaviour**
The blue boxes *Intention* and *Habit* equate to motivation, while the red box, *Facilitating Conditions* is an expression of capability.
Source: Adapted by Niall Enright from Tim Jackson[420] and original work by Harry Triandis.[716]

In our discussion of rewards and punishments, we have looked at these as inputs into conscious decision-making or System 2 (slow brain) processes. In Triandis'[716] model of behaviour, left, we can see that there are two ways that a behaviour can arise. System 2, at the top of the chart, leads to a deliberate, conscious intention to act one way or another. Our attitude (rational view) at the top is influenced by rewards and punishments which make us more or less disposed to an action. We have already described the influence of norms and a desire to maintain a consistent self-view as additional drivers of behaviour. Roles describes the role we are adopting at the time of the intention to act - people can have multiple roles such as *employee*, *parent*, *team member* which influence decisions and behaviours. Finally, our mood affects our conscious intentions.

Depending on who you believe, somewhere between 45%[372] and 95%[463] of all thoughts and actions are not considered or conscious at all. These are the System 1 automatic processes in our brain, and are shown by Triandis at the bottom of his diagram. The first few times we have to make decisions about an action these are considered, but as the number of repetitions increases we internalize these into habits which are automatic and do not require conscious thought. The important part of a habitual action is that it is created by repetition and persists as long as the outcomes are the approximately the same.

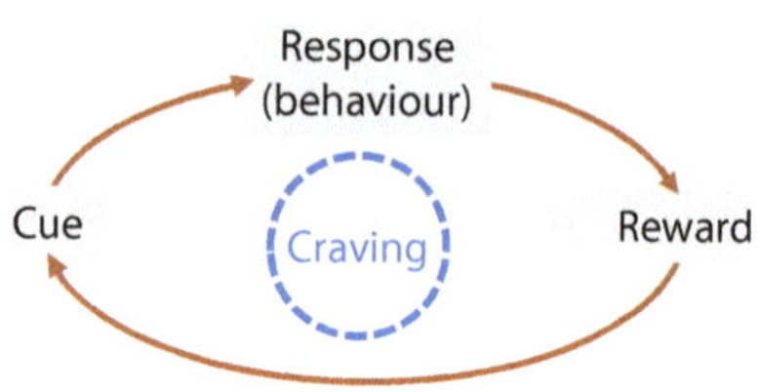

19.13 **The cue-response-reward cycle**
Source: Charles Duhigg, the Power of Habit.[230]

An excellent primer on habits is the book *The Power of Habit*[230] by Charles Duhigg from which the illustration, left, is taken. According to Duhigg, and the researchers he is quoting, habits arise from the *anticipation* of a reward which should follow a behaviour, which in turn has been triggered by a cue. In fact, the reward centres of the brain are activated before the reward arrives. If the reward is removed or deferred, a feeling of dissatisfaction or craving arises,

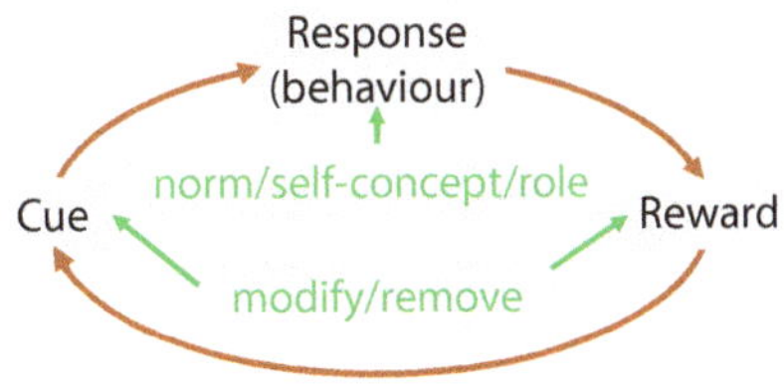

19.14 **Interrupting or redefining the cue-response-reward cycle**
Much of these efforts are designed to disrupt the automaticity of the behaviour by changing the cue or reward, thus forcing us back into System 2 thinking.
Source: Niall Enright

which reinforces the behaviour. At this point, the habit is difficult to break because there is a positive effect of the reward itself and a negative effect if the reward is withheld.

The personal habits of our workforce have a significant impact on efficiency. For example, the Carbon Trust in the UK quotes that of workers it surveyed:

> *"96% are willing to regularly turn off lights in unoccupied rooms or areas, but only 52% actually are."*[115]

What makes the difference between the willingness to switch off a light and the act of switching off the lights is not intention, it is the fact that the behaviour is largely habit. People have become accustomed, or *habituated*, to not turning off the lights, and they need help to break that habit (many of our habits we don't necessarily *like*, such as this one). Given that the value of reducing lighting by, say, 10% would be worth £155 million, this is a habit we would wish to change.

One way to change the behaviour is to modify the cue so that we disengage our System 1 response and engage our System 2. One way we can change cues is to create prompts for the new behaviour - we have all seen the "SWITCH IT OFF" stickers next to light switches, for example (page 687). We can also tackle the behaviour by changing the norm that says "*it is OK to leave lights on*" to *"it not acceptable to leave lights on"*. In this case, the previous reward for the act of leaving the light on, conforming to the dominant behaviour, becomes negative due to the disapproval of the peer group that we have introduced.

The UK's Department for Energy Food and Rural Affairs (Defra) developed a very handy guideline to the type of intervention that would work best, based on the motivation of the individuals and the strength of the habit, shown right. Where the motivation and the strength of the habit are both low, rewards may work best (carrots) but if habit strength increases then some sanction may also be required (sticks). For those who have high motivation to change, and the habit is not strong, then simply emphasizing the environmental or financial impact may be enough (sermon). If the strength of the habit is high and motivation is high we might need to have an intensive engagement with the individual to help them recognize why the habit is hard to break and work on identifying why the cues and rewards are making the habit persist.

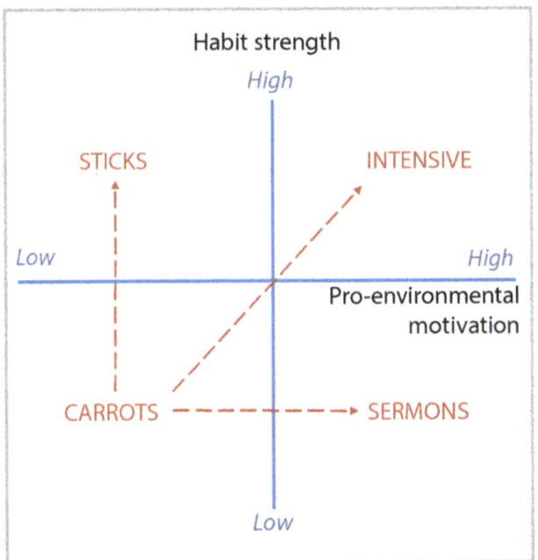

19.15 **Strategies to change habits in relation to habit strength and motivation**
Source: DEFRA.[183]

A good point to modify habits is when there is a major change in the life or situation of the individual. In these circumstances, they are much more open to change as their System 2 thinking is powered up to manage the disruption to their normal activities or patterns.

Of course, we don't just want to change habits, we want to create new positive ones. One way to start is to look at what resource-use cues exist in our workplace: the sound of compressed air leaking, or the red traffic light on an M&T report, or the waste-disposal bill. Then we would design rewards and behaviours around those cues. Changes might involve norms (which modify the rewards) or feedback (e.g. price signals) to enhance the cues.

Change

19.13 Changing practices

Organizations, too, have habits, called practices. As with habits, if we want to change a practice we need to analyse it and understand it.

The cue-response-reward cycle provides a useful description of habits in individuals. What about organizations? Well, we have also seen that habits can have a social context. Norms, in particular, exert their influence on habitual actions. The light-switching injunctive norm is an example of organizational norms intended to influence *individual* behaviour.

However, there are other habits that organizations possess which rely on the combined response of several people. These are more complicated than the simple cue-response-reward habits of individuals.

To distinguish between the habit of an individual and that of an organization, I will use the word practice to describe an organization's habit. There is a multitude of practices within organizations. Like the habits of individuals, practices emerge through repetition and apply to activities that occur with reasonably high frequency.

We can think of the practices of an organization as "*the way we do things around here*". Practices often combine explicit knowledge, the procedures that are documented, and tacit knowledge, the customs, shortcuts and difficult-to-describe skills of the people who carry out the task. In most organizations, practices are informal and not clearly documented.

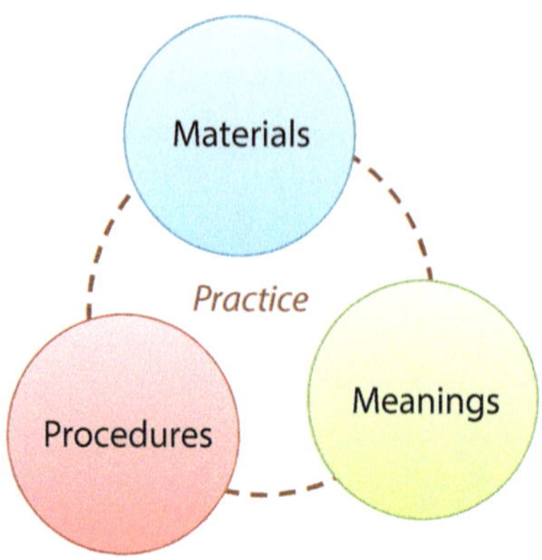

19.16 **3-Elements model**
Elizabeth Shove's model helps us identify the different aspects of a practice, which in turn can provide insights into how it can be changed.
Source: Taken from "Habits Routines and Sustainable Lifestyles", DEFRA.[183]

Elizabeth Shove's 3-Elements model, shown left, can help us understand how to change practices. In this model, three things together form the practice:

- Materials: this is the physical equipment, space and materials which are involved in the practice.
- Meanings: these include perceptions, conventions, ideas and interpretation as well as the norms of the group and values of the individuals.
- Procedures: such as rules, processes and competencies or skills.

So, let's explore these three elements through a practical example of a practice that we may want to change in an organization. Recall the night-shift operators of the Berkshire Brewery (page 182) who were overriding the automatic control system and running the huge chiller plants in manual mode (and wasting a lot of energy). This is an example of a practice we want to change. The starting point would be to create an inventory the three elements of the current practice.

Real World: ***Transforming Alcoa***

What chief executive Paul O'Neil did when he joined a failing Alcoa in 1987 took investors completely by surprise.[230] In his first major presentation to analysts, instead of talking about profits and strategy, he said he was going to focus unremittingly on making Alcoa a zero injury company.

O'Neil saw that safety was a representation of a wider malaise within the company. By focusing on driving a collective effort on safety, he forced management and staff to respond to a change process.

O'Neil even published his own phone number so that employees could ring him at any time with concerns.

Safety became a *keystone habit* - responding to incidents was non-negotiable. When a board executive tried to cover up a safety incident, he was fired within days. People responded to the signal that people *mattered* to O'Neil. The strategy recognized that people can't be forced to change - they have to want it for themselves - and safety was something that everyone agreed needed changing. The result was lost workday incidents fell from 1.89 to 0.2 per 100 workers over the 13 years that O'Neil was in charge.[51]

The key is that other stuff changed too. The lines of communication and insistence on perfection around safety spread, and created *habitual excellence*[632] around everything that Alcoa did, and the business was turned around spectacularly.

Resource efficiency has the same characteristics as safety. It is an activity everyone can get behind. Thus, a focus on driving excellence in efficiency has the power to energize, enthuse and transform the whole organization.

- Materials: Design of chiller control system, switch that allows change from automatic to manual mode, indicators of performance, control room layout, visibility of current mode of operation, feedback on energy use, noise and vibration (especially as the chillers ramp-up).
- Meaning: Fear of unexpected equipment failure, anxiety about multiple concurrent chiller starts (town of Reading plunged into darkness), lack of control when the automatic system is in operation, perception of irrational equipment choice by the automated system, intuitive selection *"lead"* unit from many years' experience of the equipment.
- Procedures: Autonomy given to night shift, lack of knowledge about automated control logic, chief engineer rarely visits the site at night, the cost of electricity and who pays for it, informal rule to minimize the number of plant starts per hour, staff schedules and rotas.

This analysis (which has been abbreviated here - there would be many more possible items) has identified many different aspects that influence the current practice and the changes that we might need to make. In terms of stopping the night shift operators switching from the automatic mode of operation to manual control, we could just disconnect the manual override switch (a material action). However, we should have spotted among the meanings that there are fears and anxieties among the operators about the automatic mode. So, in our practice change, we would do well to recognize this and work on the underlying lack of knowledge or training that is reinforcing these concerns (a procedures issue), even if we have a *magic bullet* solution to deliver the practice change. Indeed, we must not discount the possibility that the basis for the concerns are genuine, and the automated system has missed some important aspect of the operation of the plant.

UK government departments have widely used the 3-elements model in developing sustainable best practices.[183] In many ways, it echoes and expands on my earlier characterization of the "levers" for change as People (meanings), Systems (procedures) and Technology (materials) (page 265).

Changing practices usually works best when the change targets multiple elements at the same time. Thus, we need the physical aspects of the change to be put in place, we need people fully on board and we need to ensure that the rules and processes are supportive and reinforcing.

Many organizations believe that a *best-practices* programme will deliver voluntary change to practices simply by publishing the ideal practice. Alternatively, they may *impose* the best-practice by creating rules, standards or regulations. Where the practice is in the context of a clearly documented process, then this may well work, but we should remind ourselves that most practices in organizations are not documented and are so are not easily modifiable by checklists or rules. We should never underestimate the ability of human beings to bypass rules and standards if they feel that the requirements are at odds with the meanings they have developed.

19.14 Creating ownership

Ownership is shorthand for the degree of motivation that a team has to undertake an activity. There are ways that we engage people that can increase the ownership. But this usually comes at a price; we may have to sacrifice control over how people achieve the desired results.

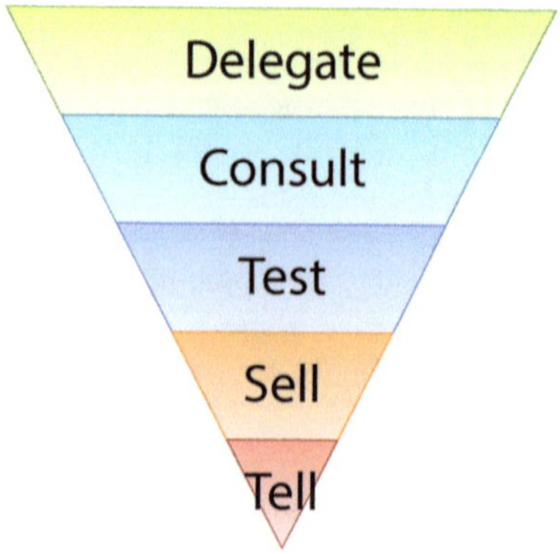

19.17 **Involvement arising from various forms of participation in change**
There is greater involvement as we move up this hierarchy of engagement techniques.
Source: Adapted from "The Fifth Discipline" by Peter Senge.

Ownership describes the situation where an individual or a team *consciously* decide to make a change of some sort. Ownership, by our definition, is thus a System 2 (slow brain) process, not a System 1 (fast brain) habit. That does not mean that unconscious cues cannot influence ownership. We have already seen from self-perception theory (page 669) that a small commitment can prime us to make a bigger commitment subsequently. Even a very subtle commitment (see box below) can influence subsequent behaviour. However, in this section I want to explore how we can get an explicit, conscious decision to change by engaging with people.

We know we want to engage System 2 and that we want people to draw on their intrinsic motivation (personal aspirations, desire for growth, achievement) which is stronger than the extrinsic drivers of reward and punishment. Thus, we want them to decide *for themselves* to act. For this to happen, we need their involvement in the change process. We need to get their time and attention and then use one of the engagement techniques illustrated, left.

First, we can tell people what to do, shown at the bottom, which results in the least involvement ("*You are going to do this because I tell you so*"). This instruction is all extrinsic motivation - people will do what is asked of them only because they want to keep their job. Similarly, we can sell, which also has a low level of involvement ("*This is why you should adopt my good idea for improvement*"). If we

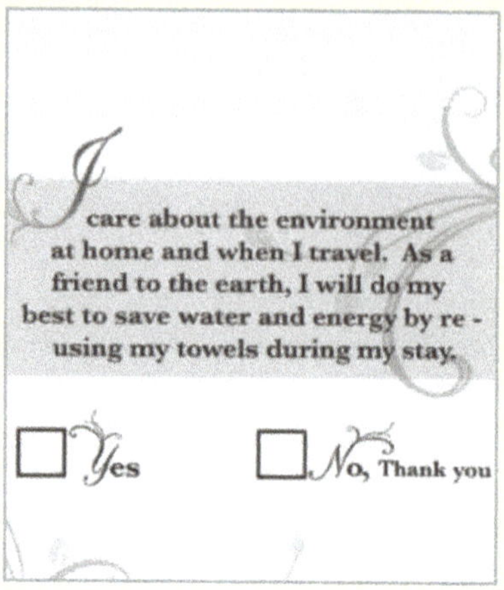

Real World: ***Hotel towels revisited***

Baca-Motes and her colleagues carried out a study[50] where some hotel guests were told at check-in that the hotel makes efforts to reduce water use (message only) and others we offered the chance to make a commitment such as the one shown left. In addition, some guests, regardless of whether they made the pledge, also received a badge stating that they were a "*friend of the earth*". Guest were unaware that their actual towel reuse was being monitored, so there was no external pressure to change behaviour (i.e. no Hawthorne effect, page 663).

Those that made the commitment and received a badge hung up 40% more towels than the control group without messages or badges. A badge alone has no effect, possibly because no commitment to wear the badge was made and there was no information to explain the badge's meaning. It should be noted that the value of savings that would arise from the commitment and badges is US$51,000 a year, so this is not just an interesting academic study, but a real opportunity.

Real World: *Timing*

When I was a tour guide, many years ago, one of my jobs was to encourage the travellers to take up optional activities (e.g. a *"Gondola tour of the canals of Venice with opera singer!"*).

I learnt very quickly that the best time to pitch was after the evening meal when everyone was together, they had eaten a pudding and were feeling replete and content.

It turns out that my own observations are supported by research. For example, a study of decision-making by Israeli parole judges, whom one would expect to be objective, considered and, above all, consistent in their decision-making, showed that timing had a huge influence on their judgements. On average around one in three cases were in favour of the petitioner. However, the variation of the decisions was remarkable. The decisions in favour dropped from around two-thirds in favour immediately after a break or a meal to almost zero at the end of the session.[182] The researchers looked for but could not find any alternative explanation for the bias observed.

This bias has been attributed to a *"depletion effect".* System 2 thinking, it is argued, requires more resources, so as we grow fatigued or short of nutrients, we revert increasingly to System 1 processes that have an inherent bias towards the *status quo*, in this case denying the petitioner's request for parole or modified probation terms.

We can only assume that this effect applies to all those who make sequential decisions, not just judges. Where we are seeking decisions that lead to a change in status quo, then we should do everything possible to ensure that those we are trying to influence are refreshed and fed!

test an idea with them we will get a slightly greater level of involvement (*"Here is what I propose, can we improve on any details?"*). If we consult we create increased participation (*"These are the areas we should focus our improvement on, how do you think we can approach them?"*). Finally, if we delegate we get the highest level of involvement (*"If I were to ask you to come up with a plan for improvement, what would that be?"*). In this last approach, the intrinsic motivators are very high, hence the much greater ownership.

We need to be very careful in the way that we use these techniques. If we intend to consult our staff, then we should be prepared to let them make critical decisions. Far too often there are business-change "*consultations*" whose outcomes are already predetermined and which end up demotivating the staff involved. Better not to consult at all in these circumstances.

Sometimes involvement is not the right approach. In a crisis, we would expect leaders to be telling or selling. Decisiveness is important, as in the case of the Alcoa safety programme. I am a strong advocate of a mixed approach. The Leader should tell the organization what the overall goal is for the improvement process, and sell the reasons why. Achieving the goal is then delegated down to the teams to set their own targets and tactics that work for them and which they take ownership over.

We can think of a commitment as a promise. When folks have set themselves a target, they are making a commitment. Commitments work more effectively when they are public, i.e they are shared. Written commitments are more effective than verbal ones.[589] In fact, there was a vogue a few years ago to get chief executives to publicize their organization's *carbon reduction commitment*.

Making a commitment public can have a dramatic effect. In an energy efficiency study[587] of homeowners in the US, participants were divided into two groups - one where a private commitment to use less energy was sought and one where a public commitment was sought (the participants were told their names would be publicized in the local paper). What is remarkable is that when the public commitment group were told that, in fact, their names would not be published, they still went on to use 20% less electricity compared to the private commitment group - an effect that persisted well over 12 months later. The more visible a promise is, the more likely we are to honour it, possibly because the degree of identification with the improvement process is that much greater.

We should sound a note of caution, though, about making pledges public. In another study[661] of small firms which were asked to make a "*mild commitment*" - to have their names published as participants in an energy efficiency programme - or a "*strong commitment*" which involved publication of the names and actual energy use data. In this case, it turned out that the strong commitment group performed less well because the business owners did not have immediate success to show and were demotivated by the public disclosure of this. The discouraging effect of stretching goals should be considered when using league tables, as discussed earlier.

19.15 Prompts and nudges

A nudge is the name that is given to techniques designed to modify behaviours in a non-coercive way. Prompts encourage people to make resource-efficient choices by catching their attention, providing appropriate cues at the time the decision is being taken.

It would be helpful if we can draw together the strands of the various behaviour techniques discussed so far into an overarching checklist we can use in the design of our own programme. Fortunately, we can borrow work commissioned by the UK government, an institution that has been at the forefront of adopting behavioural techniques to achieve policy objectives. In a 2010 report,[221] Paul Dolan and colleagues for the Cabinet Office and Institute for government came up with the mnemonic MINDSPACE[221] as a checklist of behaviour influencers.

Messenger	We are heavily influenced by who communicates the information.
Incentives	Our responses to incentives are shaped by predictable mental shortcuts such as strongly avoiding losses.
Norms	We are strongly influenced by what others do.
Defaults	We "go with the flow" of preset options.
Salience	Our attention is drawn to what is novel and seems relevant to us.
Priming	Our acts are often influenced by subconscious cues.
Affect	Our emotional associations can powerfully shape our actions.
Commitments	We seek to be consistent with our public promises and reciprocate acts.
Ego	We act in ways that make us feel better about ourselves.

If we bear in mind each of the points above when designing our programme, then we are likely to increase its effectiveness. Following on from this report, the Behavioural Insights Team[62] was established: *"The world's first government institution dedicated to the application of behavioural sciences"*. One question that persists about a government, or any organization, using these techniques is whether it is entirely ethical to manipulate people in this way (see left).

Exploration: *Nudge or shove?*

The proponents of behavioural science reject the notion that these techniques are, at best paternalistic, or at worst coercive. Rather than compelling people to do things against their will, they argue, they are choice architects, simply "*making the right choice easier*".

The term "*nudge*" was popularized in a book[702] of the same title by Richard Thaler, a Chicago economist and Cass Sunstein, a Harvard academic lawyer. They had originally planned to title their book *Libertarian Paternalism*, emphasizing their view that these are benign interventions.

However, David Halpern, chief executive of the Behavioural Insights Team, raises three potential concerns about *nudge* techniques:

- A lack of transparency, in that these methods are used without people's knowledge;
- A lack of efficacy that assumes the problems targeted are due to people's behaviour, allowing organizations to abrogate their responsibilities; and
- A lack of accountability meaning that folks using or sanctioning these techniques are not answerable for their effects.

When using these techniques, we clearly must have safeguards in place to address these legitimate concerns.

Real World: *Prompts*

A prompt is a generic term for cues that remind us to act in a resource-efficient way. Some of these cues may be supporting good habits, and so we may not be conscious of them; others are intended to engage System 2 thinking and force us to override bad habits. Prompts help folks remember to act.

Many prompts rely on increasing visibility. In Copenhagen, for example, they painted all their litter bins in bright colours and painted footsteps leading to them, which resulted in an increase in litter deposits of 45%. Footprints have been used with great success to as prompts for people to use stairs (for example, at MediaCityUK where I have been working on energy efficiency with the team). A fun variant of this makes stairs look like black or white piano keys!

Another study in Australia showed that labelling increased water conservation by 23%, but had no effect on energy saving. The authors speculate that this was due to two factors: the prominence of water use in the media due to recent droughts (which primed the response), and the fact that the resource water can be seen as it comes out of the tap, whereas energy cannot.[297]

A widespread prompt in resource efficiency is a reminder to switch off equipment. These kinds of stickers are widely available from energy efficiency programmes run by governments and utilities. Most tend to be a simple reminder like the illustration, below left. We can make these more effective if we combine normative information (recall the Carbon Trust data earlier that shows that 96% of employees are willing to switch off lights). These are also shown below. The example on the right is intended to personalize the norm to users of the actual room where the sign is displayed. Of course, the design need not be quite so text-oriented as the powerful image right shows. Sometimes a visual prompt is needed to tell people that it is NOT OK to switch off equipment, usually in the form of a red dot, ensuring the cleaner turns of the PC monitor but not the 24/7 server.

In his outstanding book, *Fostering Sustainable Behaviour,*[510] Doug McKenzie-Mohr offers some very practical advice about using prompts.

1. Make the prompts noticeable or prominent.
2. The prompt should explain exactly what the person should do.
3. The prompts should be visible at the time the action is required.
4. Encourage people to do positive things, rather than avoid negative ones.

19.18 **Various switch it off prompts**
A combination of good visuals (below) or norms (bottom) can make switch-off messages more effective. Immediately below is a light switch decal I placed on a corridor light in my home to encourage my teenagers and their friends to switch off the lights. Below that are a selection of switch-off messages.
Source: image below Niall Enright, design ©HU2[380]*; images bottom Niall Enright, available in companion file pack*

19.16 Fostering teamwork

Effective team working will be important in most resource efficiency programmes. Here, we explore some of the challenges and techniques that will influence our ability to get people working together.

Collaboration across functions and teams is important in most resource efficiency programmes because so many people can influence resource use (page 276) and improvements need to be disseminated across the organization.

However, developing teamwork depends a lot on the existing culture in an organization. It can be particularly hard in some organizations where competition exists between business units (see left) or where staff have different status and stakeholding in the firm (e.g. in some facilities, a large proportion of staff may be hourly-paid employees contracted through agencies). At the other end of the spectrum, some organizations may be very open to collaboration and sharing best practice. In these organizations, the desire to be seen to innovate and show leadership can be a powerful motivator for action.

Whatever the starting point in terms of an organization's culture, it is important to take active steps to encourage collaboration, and not just to assume it will come about naturally. Active interventions are particularly important where we need to encourage collaboration across functional divisions.

In my experience, the *best way to get folks to collaborate, is to get folks to collaborate*: you need to get people working as equals on solving a common problem, with goals set by the team, and an overarching mandate and appropriate resources provided by the organization. Critically, the measure of success is the collective output of the team (which is quite different from the sum of individual outputs). In these circumstances, organizational barriers and politics tend to disappear as folks develop mutual accountability and appreciate the inputs that they can each provide to achieve the goal.

Many organizational conflicts are not rooted in the feelings of individuals towards each other, but in the desire for existing business units to demonstrate ascendancy (aka office politics). In the field of resource efficiency, this can be particularly acute: it is not uncommon for procurement, engineering, production, maintenance and finance to each feel that they are in charge and call the shots about resource use. By creating a starting point that says "*none of you will be in charge, so you need to solve the problem together*", we can reset the relationship and harness these diverse skills. Of course, strong leadership is needed in the first place to set these ground rules, and thought needs to be given to the rewards systems that may favour individual over group performance.

Real World: *Competition*

Some global organizations I have worked with, such as Unilever, have a highly competitive culture which inhibits collaboration across sites.

The rationale is that it is up to local management to determine how to run the operations in the most profitable way. When it comes to the allocation of manufacturing (e.g. deciding where the new model car is built, or which plant will produce a particular brand of soap powder) the centre will then favour the most efficient units. In effect plants "bid" against each other to be the location chosen for that particular production.

This drive to be the lowest-cost producer raises resource efficiency up the agenda. However, it also acts as a barrier to teamwork. There have been several occasions when a site I have been working with has identified an excellent energy-saving measure, which could be replicated elsewhere, but I have been asked to withhold the details from other sites! In a competitive situation, one does not want to help the competitor. In these cases, it was usually OK to report the quantity of savings (indeed there may be league tables setting this out) made but not *how* the savings were realized.

Competitiveness is not a universal issue and some industries, such as brewing, are well known for sharing ideas both internally and externally.

Exploration: *Creating powerful teams*

In *The Discipline of Teams,*[437] Jon Katzenbach and Douglas Smith give the following definition:

> *A team is a small number of people with complementary skills who are committed to a common purpose, set of performance goals, and approach for which they hold themselves mutually accountable.*

In their follow-up book, *The Wisdom of Teams,*[438] they provide some very practical advice on developing teams.

1. Teamwork starts with a clear purpose, set by the team, and which is distinct from the purpose of the organization.
2. The purpose will be measured by mutually agreed performance goals.
3. Team success depends on the right mix of skills, not personalities.
4. Teams work best in small numbers (say no more than 25, but probably closer to 10)
5. The team must agree on a common approach and way of working to meet their purpose.
6. The team will be mutually accountable for the results.

The team is seen by Katzenbach and Smith as the primary unit for organizational improvement as challenges become too complex for managers to analyse and solve alone. The good news is that teamwork can be developed by following the principles above; the bad news is that this will require discipline and time. The process is to some extent emergent rather than prescriptive as teams find their own way of achieving peak performance, rather than following a rigid template.

Real World: *The power of checklists*

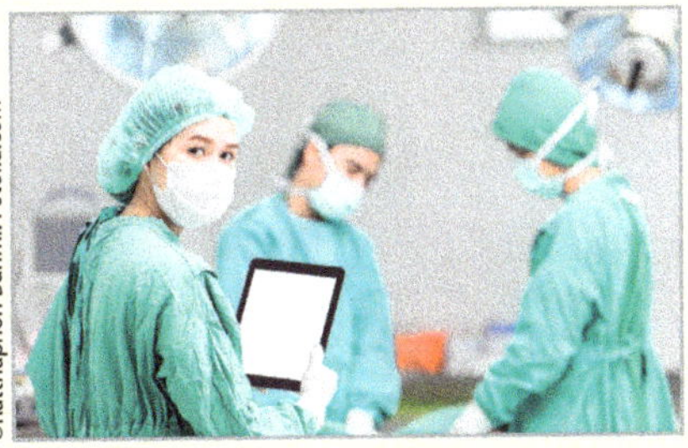

It may seem strange to have an item about checklists when we are discussing teamwork. But the two things are closely related.

In 2009, a ground-breaking paper[360] by Alex Haynes and colleagues was published in the New England Journal of Medicine. It demonstrated that checklists covering the main activities in anaesthesia, infection control and teamwork reduced complications by one-third and mortality by almost half. This improvement was observed across a wide range of hospitals in the US, Canada, UK, New Zealand, India, Tanzania, Jordan and the Philippines.

Checklists work where individuals with high levels of expertise (anaesthetists, surgeons, nurses) need to collaborate on complex processes. They ensure that basic steps are not overlooked and the team have core processes to check against. Checklists have been around a long time in medicine - I remember when I was a medic the rigorous counting in and counting out process in operating theatres to ensure that swabs were not left in patients!

There is a good argument for introducing more checklists as a resource efficiency tool in the operation of critical equipment or processes in buildings and industry. Checklists will counter the, often false, assumption that control systems are controlling and will ensure that the best expertise in the team is encapsulated in instructions that less-expert colleagues can follow. Areas that are obvious candidates for checklists are the operations of boilers, chillers, steam systems, compressed air systems, shut-down and start-up activities. The list is potentially limitless and each organization will know which activities are 1) material, 2) are somewhat complex (in that they require a sequence of tasks), 3) involve several people (acting individually or in teams) and 4) can be written down.

The benefits of checklists in teams is demonstrated in aviation where just about every key activity in flying (and in emergencies) is managed with checklists. Used correctly, these checklists can reduce negotiation, support collaboration and improve communication between team members.

The design of the checklist should be carefully considered. Checklists should only be used for activities where their introduction can lead to significant benefits; they should be developed with the participation of the staff involved; leaders should take the time to explain why they are being introduced, and they should be evaluated and modified if they are not working.[28]

Checklists can fall into two types: a DO-CONFIRM checklist, where you carry out a series of actions and then check that all the steps were completed (much like the swabs process I learnt), and a READ-DO checklist where you read out each item in turn and then do the action. The checklist should not be too lengthy (one task could if necessary be broken down into sub-tasks), typically 5-9 items in length. They are not complete HOW-TO lists - instead, they should focus on the key things that people can get wrong, ideally, fit on one page of paper without distracting colours or clutter, and the wording should be precise and exact.[305]

19.17 Working with consultants

Consultants can provide valuable additional expertise and effort in a resource efficiency programme. Understanding how they work and what motivates them can increase the effectiveness of their contribution to the programme.

Real World: *Stating the obvious?*

© gustavofrazao. Fotolia.com

The old chestnut, attributed to Robert Townsend, former CEO of AVIS, goes something along the lines that "*a consultant is someone who borrows your watch to tell you the time*". In other words, the consultant doesn't bring any new knowledge to an organization, but simply repackages the existing knowledge.

This criticism arises from a misunderstanding of the role of a consultant. As a consultant, I would *not* expect to bring new knowledge about a client's organization - after all, how could I possibly know more than the existing staff?

What I would bring is a set of skills (problem-solving or data analysis), experience (e.g. best practices observed in similar organizations) and objectivity (an ability to formulate a conclusion in an unbiased way) which will help the client gain new insight through a collaborative process. If I have done a good job, the client will very much *feel* as if it is their own insight, and they will take ownership for the outputs. Rather than being a criticism, this is a measure of success.

Consultants are often involved in resource efficiency programmes, bringing external expertise or just providing additional effort to get the programme up and running. The knowledge tends to fall into two areas: engineering expertise on equipment, controls and processes or management expertise on systems, change processes and business cases. Often one wants expertise in both areas. The big management consultants can lack technical know-how, while many engineering consultancies don't really get the people or change aspects. For this reason, it may be necessary to identify *"boutique"* firms specializing in resource efficiency rather than rely on more generic consultancies.

Some consultancy may be available free of charge from government agencies or utilities. Be aware that generally in consultancy, "*you get what you pay for*".

Most consultants are not, despite common assumptions to the contrary, motivated by money. Their biggest reward comes from the variety of the work, their ability to become an expert in their field and the professional satisfaction they get from a job well done. It really can be a terrific job for those prepared to work hard. Speaking of hard work, consultants often put in many more hours than they are contracted to work for, due to several reasons.

1. Consultants are usually managed on utilization. That is the percentage of their working hours that they are billable. Utilizations of 80-100% are common. Any project overruns, time unproductive because clients make last-minute changes to plans, training time or administration duties falls into the remaining 20-0%. To maintain good figures, consultants often absorb these overruns into their personal time.

2. Many firms pitch the work very competitively. This tight pricing means that there are too few hours in the job in the first place.

3. Clients often expect consultants to start work at 9 AM at their premises and put in a whole day's work, so travelling time is not properly accounted for.

One thing that many consultants appreciate is feedback on the end result of the project. Even a brief email will be greatly valued and will ensure that the next time you employ the same consultant, they will be even more enthusiastic and committed. If you address what motivates the experts you hire, you will inevitably get more from them. By following the suggestions opposite, you should have a positive outcome.

Real World: *10 tips for a successful consulting engagement*

These are some of my suggestions for getting the best out of consultants.

1. Make the brief as detailed as possible, so there is no misunderstanding. Spell out the goals of the project and the expected timescales. If the output is a report, set out the scope and the depth of the report, and the review and acceptance processes. Once the project has started, don't expect to *move the goalposts* without a full discussion with the consultants and a possible renegotiation of fees.
2. Make sure that you know precisely who will be delivering the service to you, and meet them in advance. Don't accept the A-team for the pitch and the B-team for the project. Know who will do the actual work.
3. If you want the consultants to address uncomfortable issues then you need to be clear about this, otherwise they may avoid questioning the status quo.
4. Agree on a regular progress reporting process and stick with it.
5. Make sure that all decisions and actions are documented. If you think that the notes are incomplete or inaccurate, raise this as soon as possible.
6. If problems arise or you are unhappy with anything, tackle it early.
7. Meet your obligations to the consultants:
    - Ensure the health and safety of the consultants, and ensure that they know their responsibilities to their own and your people's health and safety;
    - Provide all the information, data and access to people and places that the consultants request;
    - Try not to change schedules (particularly at short notice) as this can lead to the consultant being unable to bill their time and lots of stress;
    - Be honest (most good consultants will pick up on the true situation so misleading only wastes valuable project time and goodwill);
    - Trust the expertise of the consultant, do not micromanage or second-guess them, give them space to work;
    - Try to create a good working environment for the consultants at your premises, as this will make them more productive;
    - Share final outcomes with the consultants even after their input has ended, they really appreciate it.
8. Ensure that the bases for calculations to be included in the service are understood (e.g. estimation accuracy, whether or not you get the detailed workings of calculations or just the headline results).
9. Be clear about the ownership of any intellectual property related to the project (e.g. spreadsheet tools, benchmark data).
10. Don't ask for excessive indemnity insurance and warranties, this can add substantially to the price you pay or deter some consultants from the engagement altogether. Ensure that you understand the scope of the advice (e.g. consultants will often be reluctant to provide third-party guarantees).

The biggest pleasure for most consultants comes from the variety of the work, their ability to become an expert in their chosen field and the professional satisfaction they get from a job well done.

Exploration: *Becoming a consultant*

A former colleague, Adrian Partridge, has written an excellent book, *Consulting Made Easy*,[591] which provides a great introduction to anyone who is interested in becoming a consultant.

This book offers a great insight into consulting work from a very experienced resource efficiency consultant. It is written in the style of a warm, encouraging, honest, straightforward and illuminating conversation with a real pro. I highly recommend it, not only for any budding consultants but for those who regularly employ external experts.

19.18 Using games

In our increasingly digital world, games are seen as a way of influencing behaviour. To date games in the resource efficiency field have been largely educational, although there no inherent reason why more behaviour-focused games could not be created.

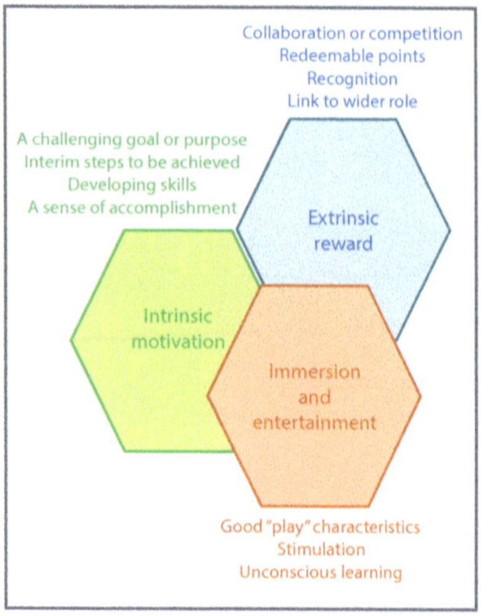

19.19 **Game elements**
Games designed to change behaviour focus primarily on increasing intrinsic motivation, but will also incorporate other elements. Note that intrinsic motivation and extrinsic reward do not overlap to any extent.
Source: Niall Enright

Motivation, we will recall, can come about because of extrinsic factors, such as pay and punishment, or from intrinsic factors such as autonomy, mastery and purpose. Extrinsic motivation is essentially a rational transaction: "*if I do x, you will do y*". Intrinsic rewards, by contrast, work at primarily an emotional level.

A great way of harnessing intrinsic motivation is through games. A well-designed game will have a purpose or goal and, as we develop our skills and advance through the game, we feel a sense of accomplishment. Some games will add a social dimension by allowing us to interact and play or compete with others, gaining feedback and encouragement in the process.

Using gameplay to drive behaviour change aligned to specific social or organizational objectives has been called gamification. As with many fashions, the potential for gamification to drive change may have been somewhat over-hyped. Professor of entertainment technology at Carnegie-Mellon University, Jesse Schell, dismisses the idea of adding game mechanics to activities which aren't games as equivalent to adding chocolate to everything.[645] Just because chocolate enhances ice cream doesn't mean to say it will improve cottage cheese! A requirement of a successful game is to is to put the player's motivation and goals first, according to Brian Burke in his book, *Gamify.*[106] *"One of the key problems in many gamified solutions is that they are focused on getting the player to achieve the organization's goals rather than the player's goals"*.

The illustration left, shows some of the elements of a gamified approach to behaviour change. The primary focus is on increasing the intrinsic reward that the player feels. They should have a goal and purpose which is challenging but achievable (usually through interim steps that become progressively more challenging). If the game is too easy the sense of accomplishment and mastery will be reduced, but if too hard then the player will ultimately become demotivated.

Have fun at every opportunity. Few things are more motivational.

The second element is the immersive and entertaining quality of the game. The game should be stimulating or pleasurable - something that those driven to resource efficiency by moral principles sometimes struggle to accept. The ability to create a convincing and absorbing environment is an important aspect of a game's success. Entertainment is a positive feature, but not the main feature. That is what distinguishes a standard videogame, whose primary aim is entertainment, from a social change-focused game.

Real World: *Idea Street*

DWP Department for Work and Pensions Idea Street

As well as education and behaviour change, games can also support innovation within organizations.

The UK's Department for Work and Pensions' Idea Street is a gamified version of a traditional suggestion scheme.

Participants in the game can make suggestions for improvement, for which they receive points called DWPeas. The important thing is that these points can be earned through participation at all stages of the idea development process. Thus, commenting on an idea will earn points. If the idea goes on to be approved, additional points are awarded, so it makes sense for employees to identify and contribute to successful ideas.

If they think an idea is especially good, individuals can also *"invest"* their points directly in an idea. If the idea is taken forward for implementation, these *"investor"* employees who contributed will receive additional points, but if the idea is rejected then the *"investors"* lose their investment.

Ideas that are most actively supported or commented rise to the top of the *"buzzfeed"* and gain prominence, encouraging further contributions.

In effect, there is a dynamic market for ideas which harness the collective knowledge and innovation capability of the DWP employees. Unlike classic suggestion schemes, the employees do not just originate ideas, but actively develop them and signpost the most feasible through their *investments*. In the first phase of the scheme, 1,400 ideas were proposed, by 4,500 users and 63 ideas worth £21 million were formally taken forward.[107]

Extrinsic rewards do play a part in many games. In some cases, this can be driven by recognition and status from peers; in other cases, game points are convertible to real world currency (such as the linden$ in the massive game Second Life, which can be earned in-game and then exchanged for actual dollars). Note that the hexagon denoting the extrinsic rewards in figure 19.19, opposite, does not overlap the intrinsic motivation one, indicating that these are somewhat mutually exclusive aspects of a game design.

One resource efficiency game we have already encountered is *My2050* from the UK's Department of Energy and Climate Change (see page 40). This simulation is a great education tool exploring alternatives ways of achieving the UK's emissions reduction targets. A similar game is *The Climate Challenge*,[616] where a player takes on the role of President of Europe and must formulate policies to reduce climate change while ensuring that there is enough electricity, food and water for the people.

Unlike escapist online games where people are adopting a new persona, many of the education games focusing on resource efficiency are alternate reality games. Here, the game is all about remaining in your own character solving the resource problems faced by the world today. A celebrated example of this genre was World Without Oil, an idea conceived by Ken Eklund and implemented by Jane McGonical. This game was played out over 30 days and players (there were over 1,000) were asked to participate in a scenario where the world ran out of oil. These players were able to contribute their broadcasts, blogs, videos, reactions and responses to the story that was unfolding. Each day the game designers would add more items of information (the petrol price rises to over US$6.00 a gallon, for example) and the players would then blog about the inability of public transport systems to cope with the demand, or of hoarding and speculation. Ministers would deposit prayers and sermons on how to act compassionately in the face of the suffering created. The important thing about this game is that folks were interacting with each other and responding to each other's posts. There is an excellent description of the game in Jane McGonigal's book *Reality is Broken*,[508] which observes that folks at first were highly pessimistic in their contributions but, as time passed, they grew more positive and focused on solutions. This crowd-sourced scenario development is one of the great legacies of the game, and many of the 100,000 contributions made are accessible in the chronology of the game.[813]

There are plenty of games focusing on particular behaviour, such as StepJockey, which encourages folks to use the stairs rather than lifts, as a part of a healthy lifestyle. One example of a game targeting resource efficiency, called Lost Joules, is given by Jane McGonical. This describes an app for smart-meter equipped folks to place bets with each other about their potential to reduce energy use. However, this game appears not to have gone into production. One can assume that it will only be a matter of time before we see more games directed at specific behaviour change in resource efficiency. For those interested in this area, there are several websites which track feedback, such as *Ecofeedback*[236] and social games such as *Games for Good*.[302]

19.19 Dealing with resistance

There are many forms of resistance that can arise in a resource efficiency programme and to find a solution it is necessary to find a root cause.

> *"People do not resist change, they resist being changed."*
>
> *- Peter M. Senge*

The resistance we will examine here is taken to be more than a passive disinterest in resource efficiency, but active opposition to it. Resistance can take many forms, such as an unwillingness to engage, dissent or deliberate sabotage.

The first thing that we should note is that resistance is rarely malicious. On examination, there is almost always a rational explanation for the individual's negative response to our programme. Without understanding the root cause we won't be able to address the problem.

Let's first of all, think in terms of internal or intrinsic causes of resistance. Here, the resistance arises from emotions, attitudes, beliefs or character traits. The psychological aspects we have explored earlier can cause of resistance when they work against our programme, such as loss-aversion, contradicting norms, habits, lack of ownership, inappropriate incentives or lack of motivation.

More generally we need to understand that it is perfectly natural to resist change. It is only because people inherently favour the status quo that organizations can function at all. The cultures and systems of most organizations, and the mindset of those who work in them, are set up to maintain the present state of affairs, a form of organizational homoeostasis, which ensures order, continuity and effectiveness.

Change is generally seen as *threatening*. Resistance is a reaction to that threat. Resistance can take the form of passivity - an unwillingness to actively support a programme, to share information, to work as a team, slow responses or failures to meet deadlines. This covert resistance can be very difficult to identify as the people concerned may be outwardly expressing support for the changes, whereas they are, in fact, distancing themselves.

In many ways, visible resistance is much easier to deal with. Forms of open resistance include complaints about the programme, expressing doubts, falsifying data, removing resources or an outright refusal to participate.

Sometimes an individual's character is to blame. Some people are just so cynical or pessimistic that they bring down those around them. Here, I make a distinction between a *realist,* who accepts that a problem like man-made climate change poses severe challenges and risks, and a *pessimist,* who ignores any positive action that we can take to reduce the danger. A pessimist who promotes inaction on resource efficiency through a belief that we cannot be

Real World: *Removing the human*

Sometimes people can seem so difficult to understand and change that it is very tempting to just bypass them altogether.

Eliminating people's ability to influence the resource use altogether can seem an easier approach. However, as a general rule, one should not use control systems to fix behaviour problems. This advice is because people are very adept at bypassing control systems.

A classic example is the lengths to which office employees will go to take control over the heating and cooling systems. They might, for example, put an ice-pack on a thermistor to force the heating up or conversely place a hot lamp beside it to force additional cooling.[641]

Because of the ingenuity of folks seeking to override setpoint, manufacturers have resorted to installing dummy or placebo thermostats. These appear to be fully functioning, and may even induce noise to give the impression of an actuator being started, but in fact, they are not connected to the actual control system. There are varying opinions on the effectiveness of this approach.[135] Of course, men and women have different thermal needs[442] and this too can be a source of dissatisfaction even when the thermostat is connected.

effective can cause so much damage that active steps need to be taken to counter their influence. In many cases, the only solution may be to exclude the individuals from the process or sanction the negative behaviour.

Selfish motives for resistance can be seen where individuals feel that the proposed changes will get in the way of promotion or advancement or favour rivals or diminish power. The transparency that the resource efficiency techniques described here can bring is not necessarily welcomed by all, especially if it leads to greater accountability and workload.

Extrinsic sources of resistance arise where people do not have the ability to bring about the desired change. Examples of extrinsic sources of resistance are conflicting objectives, inadequate resources or knowledge and unrealistic goals. One tool that we can use to identify conflicting objectives is pairwise comparison, described on page 167.

Intrinsic resistance is about motivation or willingness to act, whereas extrinsic resistance if often about capability to act.

There are several techniques that we can use to reduce all forms of resistance.

- We can ensure that the call to action is compelling and visibly supported by top management and key influencers.
- We should encourage active discussion about the proposed changes we wish to make. This dialogue can "*smoke out*" covert resistance, and equally important, give space for folks to articulate any genuine practical challenges that we may not have thought about.
- We need to give people some control over the changes where possible. People do not resist change, they resist being changed. If folks feel that they can manage this change, then they are much more likely to accept it.
- We should recognize that change can be hard and can have an adverse effect on some people. We need to ensure that there is adequate time, resources and support available.
- We should act decisively on resistance that arises from selfish motives. Inaction sends a signal that the programme does not really matter and people can bring their own agendas to bear with impunity.

By and large, people do not come to work with the intention to pollute the planet. The vast majority of folks agree that resource efficiency is a good thing and are willing to support reasonable changes. For this reason, if we encounter resistance is it vital that we do not simply come out "*all guns blazing*" and attack those who are not supporting the process. It is entirely possible that the root cause of resistance is our own failure to design a process that people can get behind. The one area where behaviour is not necessarily rational concerns man-made climate change. Climate change denial brings some specific challenges, which are covered in the next section.

19.20 Climate change denial

Supporting energy and resource efficiency should not be presented as an ideological choice, but rather as a means of increasing value for the organization. Unfortunately denial of man-made climate change may create resistance to our programme and so we need to understand it and address it.

There are many folks who, despite overwhelming evidence to the contrary, feel that man-made climate change is controversial. Unfortunately, a number of vested interests have spent a large amount of time and money creating this impression to serve their own financial or political agendas (for a great exposition of this subject see Oreskes, and Conway's *Merchants of Doubt*[576]).

It is simplistic, however, to think that climate change denial is just a consequence of folks being taken in by false information. In many cases, the root cause of an inability to accept man-made climate change is much deeper. For example, someone may not wish to acknowledge the need to address the issue of climate change because the implications are too horrible to contemplate. Individuals may be in a state of avoidance, because of the powerlessness that they feel or the magnitude of the consequences. Accepting climate change, and their role in it, may invoke a host of undesirable emotions: anxiety, feelings of powerlessness, guilt or shame. This state of mind is called cognitive dissonance, where one part of our brain actively suppresses facts that another part of the brain knows to be true. Death is another subject which invokes cognitive dissonance.

In these circumstances, it is key that we engage with people in a non-threatening way, that the messages we send about addressing climate change are hopeful and the call to action is positive. Effort should be made to put the contribution of the individual in the greater context of the organization's collective impact and the long-term, incremental nature of the change needed. We need to counter feelings of helplessness and hopelessness.

As well as denial arising from the suppression of uncomfortable truths, resistance also occurs when the social or cultural basis of people's identities are threatened. This phenomenon has become particularly evident in the US, where issues such as man-made climate change have become intrinsically (dare I suggest it, *deliberately* and *cynically*) aligned with broader social issues on the conservative/liberal spectrum.[431] This cultural alignment is one of the features exploited by those who intentionally propagate doubt about man-made climate change. In these circumstances, folks may not want to acknowledge the reality of climate change because it will bring them into conflict with the (artificially created) norms of their peer groups. This polarization gets in the way of an honest and open dialogue because unrelated core beliefs about liberty and so forth have become attached to the subject of man-made climate change.

Exploration: ***Sceptics and deniers***

In some dictionaries, a sceptic is "*someone who rejects habitually held beliefs*" (Collins) and in others they are "*a seeker after truth, an inquirer who has not yet arrived at a definite conclusion*" (Oxford English Dictionary).

Scepticism is a good thing in the sense that only by challenging conventions can we advance. Galileo was a sceptic for not accepting that the sun rotated around the earth, as was Columbus for not believing that the world was flat. In fact, scepticism lies at the heart of the modern scientific method, where scientists posit and then test hypotheses which deliberately question the existing paradigms.

Those who reject man-made climate change like to refer to themselves as sceptics as this imbues them with the respectability of honest scientific enquiry. Those who believe in climate change prefer the label of deniers, or contrarians, to emphasize the irrational basis for the rejection.

By and large, if we are seeking to persuade we should avoid calling each other names. So I tend to avoid the word denier, with its pejorative overtones. That is not to say that there are not people who should be referred to as such, particularly those who deliberately and maliciously distort information to diminish understanding of the issues, or those who simply mistake opinion for fact.

> **Real World:** *Five deceptions*
>
> Some individuals and organizations deliberately set out to cloud public understanding of man-made climate change.
>
> In their book *Climate Change Denial*[773] Haydn Washington and John Cook describe the five most common denial arguments.
>
> 1. Conspiracy theories - where man-made climate change is presented as a fabrication by greedy scientists looking to line their pockets with more research grants.
> 2. Fake experts - where people with no standing in climatology or related fields are brought in to refute the work of genuine experts.
> 3. Impossible expectations. For example, *"we can't predict the weather next week so how can we predict climate change?"*, which is irrelevant since climate and weather are not the same, but nevertheless creates doubt.
> 4. Misrepresentations and logical fallacies: the classic example is that "*climate has changed in the past*" implying that change in the future is inconsequential or that the current changes we are experiencing are due to the same processes as in the past.
> 5. Cherry-picking: the recent "*flat spot*" of climate temperatures (now shown to be illusory), or "*climate change is good for agriculture*" are just two examples of cherry-picking data or outcomes to suit denial.
>
> By recognizing these tactics, we can better marshal our arguments against the misinformation.

George Marshall's outstanding book on climate change denial, *Don't Even Think About It*,[500] goes into great depth on these and other social and cultural dimensions. The final chapter provides many concrete recommendations for how we might create a dialogue that can overcome some of these barriers. I have summarized, hopefully correctly, some of the suggestions he makes to those seeking to overcome resistance.

- Be positive and inclusive: Create a narrative of positive change where the response to climate change leads to a better world; emphasize cooperation; stress what groups have in common and build on this; avoid demonizing and defining enemies; invoke sacred values (e.g. universal concern for our children) and affirm common moral ground.
- Be honest: emphasize the present problems rather than the distant future impacts; avoid the frames that others put on it (e.g. it is an environmental problem, it is a lifestyle issue); accept uncertainty about the detail and bias in some sources while asserting the overwhelming need to act; acknowledge that there will be losses (e.g. coal mining communities will lose jobs).
- Be personal. Don't be afraid to show your own conviction; articulate the pride you feel acting on climate change; acknowledge your own emissions and their impacts; present climate change action as a matter of conviction; accept others have different motives to act and don't assume that what works for you will work for them; be prepared to tell personal stories; keep an open mind and be aware of your own bias.

Another technique taken from the scientific literature[157] is to create opportunities for people to affirm their self-identity before challenging the validity of their beliefs. Self-affirmation would, for example, be a reflection of personal values or a discussion of a particular skill the individual has. This will, the researchers argue, make folks more receptive and open to ideas that would otherwise prove threatening.

Like most people, I suspect, I often avoid confrontation with people who don't accept climate change. Instead, I usually propose that "*we accept we have different perspectives*". Indeed, if the basis for denial is anxiety about the topic, pressing a debate may well be unkind. A strategy I sometimes use is to have an alternative discussion about why responding to climate change *as if it is real* is a good thing, regardless of whether we think man-made climate change is happening or not (see page 80). In this way, I am not challenging someone's fundamental views, but can still have a conversation about taking action that has a positive effect.

That having been said, we should not hide the value that resource efficiency brings in terms of man-made climate change, and we should do our bit to change the norms around this subject by registering our view. In circumstances where a conversation about the myths and realities of man-made climate change is unavoidable, we can use the strategies in the next section to help us.

19.21 Using facts to deny myths

The basis for understanding is facts. Unfortunately, incorrect facts can influence people's motivation to support resource efficiency. Here, we explore how myths can be corrected.

It takes much more than a **killer fact** to correct misinformation. We need to structure our communication very carefully.

In a rational world, the outcome of a debate on a subject like man-made climate change would depend purely on the quality of the facts that are put forward for or against. The presumption by those who accept that man-made climate change is real, is: "*If people knew the same facts as I do, they would support my position.*" In reality, no amount of data on global temperatures, charts of global CO_2 levels or pronouncements from eminent scientists will change a convinced climate change denier, because for them it comes down to a matter of *opinion*, *faith* or *belief*, not fact.

Research[564] shows that presenting facts that contradict ideologically based views can reinforce the myths. The researchers, Brendan Nyhan and Jason Reifler, called this the backfire effect,[565] which comes in several variants. When we *repeat* a myth in order to dispel it, we may unwittingly strengthen it:[666] the familiarity backfire effect. This occurs because repetition makes the claim itself more memorable while reducing recall of the context (the falsehood) of the claim. We tend to remember the myth and forget the fact that it is false.

Because the myth is something which deniers are familiar with, they find it easier to recall and process than a new fact that is presented to them.[652] This effect is fluency bias, which leads to the overkill backfire effect - if we present too many counter-arguments to a myth, these are more difficult to process and so the myth persists. "*Less is more*" when it comes to debunking myths.[166]

Many of the most challenging myths are articulated in terms that reinforce the beliefs of climate change deniers. We instinctively favour facts that reinforce existing beliefs, a process called confirmation bias.

Facts about climate change are often framed in ways that distort their meaning. In *Don't Even Think About It*,[500] George Marshall advises:

> *"Never accept your opponent's frames. Don't negate them, or repeat them, or structure your arguments to counter them."*

Thus if a charge for emissions is presented as a *"tax"*, those on the conservative political spectrum who disapprove of taxation will respond negatively, whereas if exactly the same charge is portrayed as *"offsetting"* the negative view disappears.[166] Framing not only relates to language but also to who conveys the fact/myth. People favour information received from those who share their beliefs rather than from those who do not.

19.20 **Debunking (opposite)**
Debunking the Petition Project's implication that there is no consensus on man-made climate change requires carefully structured argument and presentation.
Source: The Debunking Handbook[166] by John Cook and Stephan Lewandowsky, modified by Niall Enright. Available in the companion file pack

Real World: *Refuting myths without reinforcing them*

The Debunking Handbook[166] by John Cook and Stephan Lewandowsky is an excellent seven-page explanation of an effective way to counter a myth using facts. The technique is described as follows:

> *Debunking myths is problematic. Unless great care is taken, any effort to debunk misinformation can inadvertently reinforce the very myths one seeks to correct. To avoid these "backfire effects", an effective debunking requires three major elements. First, the refutation must focus on core facts rather than the myth to avoid the misinformation becoming more familiar. Second, any mention of a myth should be preceded by explicit warnings to notify the reader that the upcoming information is false. Finally, the refutation should include an alternative explanation that accounts for important qualities in the original misinformation.*

I have slightly modified an example given by Cook and Lewandowsky, below, while following their structure exactly. Apart from some wording changes, I have expanded the misinformation techniques listed from one to two (adding *the use of figures that appear significant when they are not*), and so the final panel is longer. Cook and Lewandowsky explicitly warn against the "*more is better*" approach when seeking to debunk a myth, stating that if you provide too many counter-arguments your explanation may not get across, but I couldn't resist adding one more rebuttal of the myth! The website www.skepticalscience.com (with a "k" not a "c") provides a lot of responses to specific man-made climate change myths. By the way, the source of the 97% *"core fact"* below is a paper,[165] also by John Cook with colleagues, while the Petition Project can be found at www.petitionproject.org and the population data I have used is from the US census[737] and a study of licensed physicians.

97 out of 100 climate scientists agree that humans are causing global warming

Core fact in headline

An independent survey has found that 97% of climate scientists who are actively publishing peer-reviewed climate research agree that humans are causing global warming.

Reinforcing text providing more details about the core fact

Infographic providing a visual version of the core fact

However, movements that deny a scientific consensus have sought to cast doubt on the fact that a consensus exists using techniques designed to mislead. First is the use of fake experts, citing scientists who have little to no expertise in the field, and second the use of figures that seem significant when they are not.

Warning to the reader that misinformation is coming and the nature of the misinformation

For example, the Oregon Institute of Science and Medicine's 1998 Petition Project claims 31,487 scientists disagree with the scientific consensus on global warming and think that increases in CO_2 will have beneficial effects.

The misinformation or "myth" being debunked

Fewer than 0.1 % of the scientists listed in the Petition Project are climatologists. The petition is open to anyone with a Bachelor of Science degree or higher and includes medical doctors, mechanical engineers, metallurgists and computer scientists. In the US alone, there are over 16 million employees with science, technology, engineering, maths and medicine degrees, so the number of signatories is a miniscule proportion (less than 0.2%) of those who could sign the petition.

Filling the gap with two items of information; most of the signatories are not climate scientists and very few scientists have signed the petition

19.22 Helping people to learn

Change is about learning to do things in a different way. By using the most effective learning techniques, we can get the most from our resource efficiency programme.

The challenge with learning is that individuals are quite varied in the ways in which they assimilate and retain information. I, for example, will always write when I am learning, not to create notes that I can reference later, but simply because the act of writing something down helps me commit it to memory.

A now discredited model, Dale Edgar's cone of learning, stated that the more active the learning process, the greater the retention. Thus, it was said that we retain 10% of what we read, 20% of what we hear, 30% of what we see, 50% of what we see and hear, 70% of what we write and 90% of what they say they do. These percentages, suspiciously rounded as they are, have no scientific basis at all, but the notion that the more actively people engage with the learning, the more they remember does seem logical. Studies[285] in the 1990s have suggested that students prefer one of four modes of learning.

Visual	Favours images and symbols. Use charts, diagrams, illustrations to aid learning.
Aural (or Auditory)	A preference for listening to information. Podcasts or recordings of lectures will help this form of learning.
Reading or Writing	Some people, like me, find that writing down something commits it to memory. A similar process can occur when reading.
Kinaesthetic	This essentially means "doing". Although you use the other modes of learning when you carry out the action, the distinction is that the learned experience is put to use, either for real or in a simulation.

19.21 **The experiential learning loop** According to this model, individuals favour one of four particular learning styles as their starting point, but they will then move around the loop, using different methods to consolidate the learning. These learning styles are given different names, shown in green.[376] *Source: Based on Experiential Learning[446] by Kolb,[166] modified by Niall Enright*

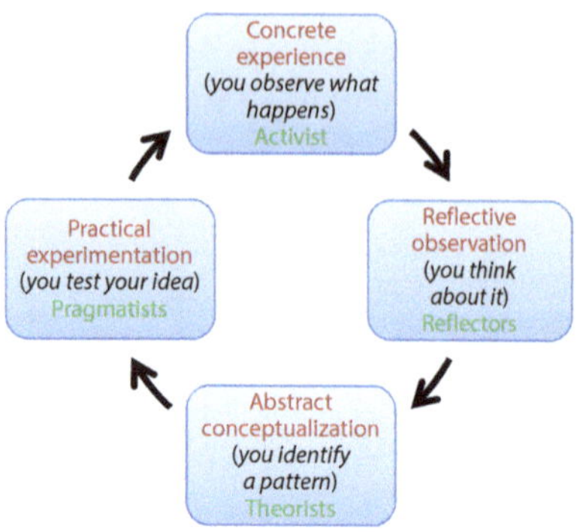

The VARK idea is complemented by another model of adult learning called the learning loop, described by Kolb.[446] In this model, all modes of learning are employed, but people have a preferred *starting point*. Activists are "*hands-on*" learners: they prefer on-the-job learning, trial and error, group learning and hands-on workshops. Reflectors are "*tell me*" learners: they like to watch others, take notes and need time to assimilate what they are learning. They like to come to the class prepared. Theorists are "*convince me*" learners: they will gather relevant information and may well undertake personal research, they like lectures, presentations and textbooks like this (which contain references they can follow up). Theorists like to have a clear framework in which to place the facts that they are learning. Pragmatists are "*show me*" learners: they like on-the-job learning and will want to test out what they are learning with real situations or well-designed simulations. Pragmatists appreciate an expert who can demonstrate the skills or tasks they are about to learn.

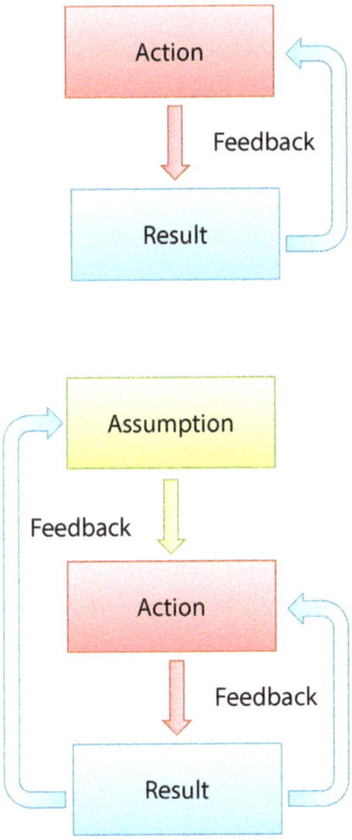

19.22 **Active learning**
In programmed knowledge there is a single feedback loop, top, which modifies the actions leading to results. In questioning insight, bottom, there is a second feedback loop that influences the underlying assumptions that determine the action.
Source: Based on Active Learning[620] *by Reg Revans*

Neuroscientists, such as Susan Greenfield[80] and others[775], have called into question the notion of learning modes, arguing that human beings make sense of the world through senses working in unison, rather than separately. Educators, too, have criticized Kolb's model as being too theoretical, abstract and prescriptive,[216] and ignoring the importance of emotion on learning.

Without going into the debate too deeply, we should reflect on the fact that most conventional teaching uses the *chalk and talk* method, with a presenter at the front of the room talking to/at their audience. We need to recognize that this is not a wholly effective way for people to learn.

Activity-based or experiential learning, where people solve complex problems, through a process of experimentation and feedback - i.e. by *doing* - seems to be much more powerful. Indeed, the energy efficiency courses that I deliver through SustainSuccess are *workshops*, as I have found that people learn much better by doing things for themselves, rather than being told how to do things.

In the context of an organization, the vast majority of learning does not take place in a formal teaching context, such as a classroom. It occurs from the first day that we join the organization and start to learn about the specifics of our role, the systems within which we operate, the tools we have at our disposal, the relationships that will help or hinder our goals. This form of learning is called experiential learning, is largely unstructured and informal, and is particularly important when we are developing tacit knowledge (page 672).

A pioneer in the study of learning, Reg Revans, made the distinction between traditional learning, which he called programmed knowledge, and a more profound form of learning, questioning insight.[620] Learning, he argued, arises from the result of an action, whose results provides feedback. In programmed knowledge, the feedback merely serves to better calibrate the original action to fine tune the results, whereas in questioning insight there is an additional feedback loop which influences the underlying assumptions that prompt and shape the action, as shown left. Questioning insight, according to Revans, is critical where organizations are experiencing rapid change.

The best resource efficiency practitioners are reflective people; they are experimentalists, the results that they observe will not only calibrate their actions but will lead to an examination of the assumptions that underpin the action. In the very best, this questioning extends beyond the technical domain to challenge assumptions about the relationships, system, culture, values, business models and aspirations of the organization. Developing questioning insight in employees is not easy. Experts need to stand aside and give people the opportunity to solve problems for themselves. Instead of being simply taught tasks, people need to develop the skills necessary to *create insights* - skills like the data analysis techniques described earlier. Active learning can be encouraged by creating opportunities for problem-solving such as the CDR workshops at TRW Automotive (page 404). Questioning insight is also important because it supports innovation, discussed next.

19.23 Imagination & innovation

An important part of any change process is the ability to innovate. While organizations can have a big impact by creating the right conditions for innovation, ideas are produced by people, whether working alone or in teams. Here, we look at some techniques to develop imagination and innovation.

"To build a better world, we need to imagine it first."

- Alex Steffen

Innovation is the process of introducing new methods, ideas or products to an organization or market. While considerable gains may be achieved simply by improving existing systems and processes, most organizations will need to innovate to respond to the scale of change that they will experience. Of course, innovation does not mean *invention*, in practice very few companies are the originators of truly new ideas. Everett Rogers, a professor of communication studies, wrote a very influential book in 1962 (since updated) called *Diffusion of Innovations*. It is this book which describes the familiar categories of adopters (and the percentage of the total in each category): innovators (2.5%), early adopters (13.5%), early majority (34%), late majority (34%) and laggards (16%).

The reason that so few organizations are innovators is that innovation is usually risky (there is no previous history to provide guidance); it is disruptive; it usually requires an advanced understanding of the issues involved; as well as people who can *think outside the box* and come up with radical changes. For innovation to work we need to address these barriers, in particular, we need to create a culture where it is ok to take measured risks and to fail occasionally.

Not only is innovation challenging, but even when it is seen to work, its adoption can be held back by a wide range of factors: hostile attitudes such as "*not invented here*"; inflexible standards (such as in the use of cement - page 184); an insufficient appreciation of the value or importance of change; and uncertainty, fear or aversion to risk, among others.

If we want our organizations to be able to innovate or to be early/early majority adopters of innovations to improve our efficiency, then we need to take active steps to make this happen.

Innovation requires us to be able to imagine alternatives to the current way of doing things. There are several tools that can stimulate people to think outside the box. The creativity technique called problem reversal asks people to think about the precise opposite of what they want to achieve - thus to improve the energy efficiency of a building, you could get people to think about steps that they could take to make the efficiency *worse*. Not only may we get a greater insight into the categories of actions that lead to inefficiency, but we will engage the participants in an unexpected way which will enhance their creativity and willingness to suggest new ideas.

Innovation is essentially a people-centred activity. While organizations, famously GE in the US, can have a strong commitment to innovation, the process of innovation depends on the creativity of individuals. To be an innovator, the person must not fear failure, and they must possess a good understanding of the current system together with insight on how it can be improved. However, they must not have *too much* of a commitment to the existing approach as this will stifle originality and result in minor, incremental change rather than breakthrough innovation. The oft-quoted example of this is that the design of the high-speed Japanese *shinkansen* bullet trains was made possible only because of the aviation engineers employed by the Japanese railways had no preconceptions about the maximum speed a train could achieve.

An unwillingness to believe in the existing paradigm is clearly a help for innovators. Often the biggest challenge in making breakthrough change is to convince people that there are other possibilities - especially in mature industries. This tendency to dismiss novel ideas is one reason why many large organizations put their more radical projects into *incubator* businesses where they can be sheltered from the overbearing weight of custom and habit which can so quickly kill new ideas. GE developed a process called reverse innovation where the challenge of meeting the needs of emerging markets can act as a great stimulus for innovation. For example, GE's traditional medical ultrasound equipment was too expensive, bulky and complex for the Chinese market, so the local business developed a much simpler, more portable and cheaper solution using a laptop with some special peripherals and software. These innovations, forced on the business by the tough requirements of the Chinese market, then found their way back into the premium ultrasound products in the core GE business. The important thing to note is that organizations can create these conditions for innovation by giving staff greater autonomy and seeking out challenging or unusual conditions in which to experiment. In larger organizations, the lessons from the most resource-constrained businesses can provide best practice examples of what can really be achieved in terms of resource efficiency.

19.23 **Backcasting and forecasting**
Backcasting can help assess the feasibility of a desired future, whereas forecasting is generally used to predict the most likely future state.
Source: Niall Enright, image in companion file pack

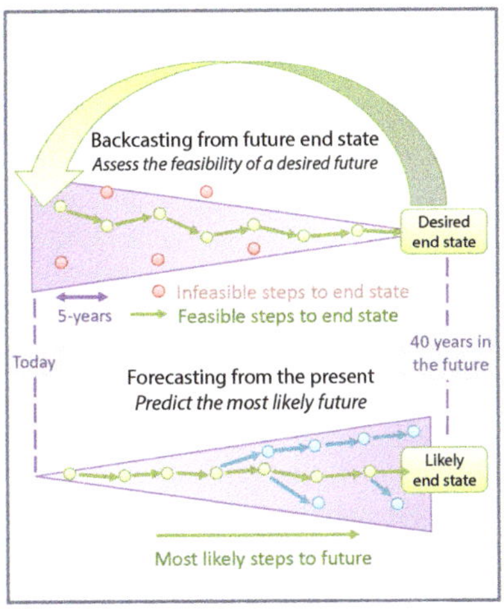

Imagination is another key part of innovation. One great technique for this called backcasting. Here, a group of people are set a desirable future scenario - such as a world where fossil fuels no longer used, or a world where everyone has a personal carbon budget. This scenario might be posed as being 40 years in the future and the exercise is to look forwards, say in five-year steps, to establish what changes will be necessary to achieve the end state. Backcasting is a well-documented process[226,624] which is especially suited to complex issues such as sustainability. The important thing is to start out with a *desirable* end-state. It is a particularly useful tool for energy and resource planning because it focuses attention on the incremental changes that could explain how the desired status comes about, unlike forecasting which seeks to understand the probability of changes that could arise from present conditions. More innovation techniques can be found in Chapter 23, including biomimicry, TRIZ and other techniques specifically related to design.

19.24 Empathy

A characteristic that makes us human and motivates us to act on resource efficiency is empathy. If we can understand and share the feelings of other people then we can truly grasp the impact that our decisions - on resource use, for example - can have on their situation.

19.24 **Empathy is a business skill**
The ability of individuals to empathize with each other is important for effective communications and decision-making. This is particularly important in international organizations.
Source: photo © iofoto, fotolia.com

This chapter started with my recollections of a conference presentation at the Dubai Sustainable Cities Conference on the theme of behaviour change. On the same panel, Dr Abdulla Al Karam spoke powerfully of how real change will only come about when we all develop our ability to empathize with others. It is only when we truly acknowledge that our over-consumption of resources damages the lives of other people and living creatures that we will be motivated to change our lifestyles sufficiently. A lack of empathy lies at the heart of many of the greatest challenges faced by humanity today.

So the solution, then, is to develop greater empathy. Here, Dr Karam spoke of the value of our children being exposed to outsiders: people of different cultures, faiths and race. By experiencing the world together, by recognizing that our many similarities overshadow our superficial differences, we can begin to see the world through the eyes of others. We can start to imagine the effects that the decisions we make have upon others. We can become wiser and more connected to the planet we inhabit and the future generations who will occupy this precious place after us.

Amid all the nudge, motivation, reward and capacity-building techniques set out in this chapter, it is important not to lose sight of this truth. If we want change, we should develop our own empathy and encourage it in others.

Techniques to develop empathy and understanding include putting yourself in someone else's shoes, active listening, willingness to be open, adult-to-adult and non-judgemental discourse, reflection. Empathy is most definitely not to be confused with pity, which is condescending and demeaning. Leaders, in particular, should reinforce the importance of empathy towards others both in the way they behave and the way they describe the programme goals. Far from being *"touchy-feely"*, empathy is a key business skill and there is considerable research to confirm its benefits.[164] Curiosity about strangers, understanding how your customers, competitors and employees feel, being able to see - and solve - the *"pebble in the shoe"* are all characteristics of successful entrepreneurs. Empathy underpins teamwork and great leadership.

That does not mean to say that the world's problems can be changed through self-improvement alone. We also have to engage effectively with our organizations, institutions and systems to change the rules of the games, to overcome the barriers that impede greater efficiency.

The most powerful component of our resource efficiency programme is the people involved.

Just as our programme is likely to have a capital investment plan, we should have a ***people investment plan.***

Summary:

1. Two aspects determine the effectiveness of people's response to change - their motivation to change and their capability to change. There is no point devoting excessive effort on motivation when capability is weak.
2. Many engagement and change programmes fail because of over-optimistic assumptions about people's responses. If you follow a formal design process, the risks of failure will be reduced.
3. Little is new. Examine how similar organizations and projects have fared when designing your own programme.
4. Many unconscious psychological traits influence how we respond to communications. These need to be understood in order to motivate change.
5. Framing, norms, comparison and language all influence how we interpret a call to action. In some cases, these can lead to unintended consequences such as increased energy use by legitimizing undesirable behaviour.
6. Starting small may be the only viable approach, but we need to consciously plan our engagement so that folks don't feel that this one initial action is enough.
7. People are more strongly motivated by internal factors such as growth, responsibility, recognition and achievement than by external factors such as money, holidays, perks. Sometimes, rewards can be counter-productive.
8. Knowledge is key to change, and processes to capture and share ideas, to develop learning and encourage innovation should form part of most resource efficiency programmes.
9. Many actions and decisions are entirely unconscious, driven by our fast brain, System 1, processes. We should not underestimate the difficulty of changing habits - although promising techniques are emerging.
10. Teamwork is an important part of most resource efficiency programmes. Active steps are usually needed to get people to work together collaboratively as most resource efficiency programmes are cross-functional, cross-location.
11. Resistance to change is normal. Denial of climate change is often a product of cultural and social issues. Successfully dealing with either of these is rarely a matter of shouting louder, but of understanding the root causes.
12. The most powerful component of our resource efficiency programme is the people involved. We need to invest in them, support their participation, develop their knowledge, foster their imagination, encourage innovation, develop empathy and promote teamwork.
13. Remember, have fun and celebrate success!

19.25 **(below) An ironic take on motivation**

Source: Reproduced with permission from a series of articles entitled "How to Waste Energy" by Vilnis Vesma. See www.vesma.com

Real World: *"How to waste energy" No. 5: motivation and awareness*

People are your greatest asset in the battle against energy efficiency. Here are my top tips for disengaging your workforce.

- Focus on trivial behaviours like leaving phone chargers plugged in.
- Position climate change as a key consideration in order to maximize time-wasting and unproductive debate. Remember also that a message of fear will paralyse rather than stimulate action.
- Over-promise with slogans like *"together we can save the planet"*.
- Give away branded mugs, coasters and other merchandise to enrage anyone bothered by waste of resources.
- Do not canvass people for their opinions or ideas; remember the best instrument of communication is a megaphone.
- If you do an opinion survey, use online techniques to be certain of reaching only those with computer access.
- Use multiple-choice questions to be sure of missing responses you did not expect (obvious missing options also infuriate and alienate people).
- Mount a high-profile launch event before you are ready with follow-on activities.
- Appoint energy champions and leave them to sink or swim.
- Be slow responding to staff suggestions.
- If a suggestion does win an award, do not implement it.
- Give individual cash awards: they can be wonderfully divisive if they are perceived as having gone to an undeserving winner.
- If payouts are a share of savings, be ready to reduce the share for really successful ideas.
- Don't forget everybody loves to be awarded a T-shirt with an energy-saving slogan on it.
- Have a poster campaign.

Further Reading:

For those interested in the relationship between motivation and capability BJ Fogg has some interesting work in this area, which takes a slightly different but complementary approach. See http://www.behaviormodel.org/index.html.

A great starting point on climate change denial is George Marshal's excellent *Don't Even Think about it: Why Our Brains are Wired to Ignore Climate Change.*[500]

For those who enjoyed reading about how norms, vivid messages, loss-aversion, social diffusion, etc. all affect behaviour, I highly recommend Doug McKenzie-Morh's outstanding book *Fostering Sustainable Behaviour.*[510] It is remarkably easy to read and packed with real-world examples and links to the scientific literature.

In his book *Punished by Rewards,*[444] Alfie Kohn provides a powerful critique of reward systems. If you are interested in checklists, then Atul Gawande's *The Checklist Manifesto*[305] is an excellent introduction.

Susan Mazur-Stommen and Kate Farley have written an excellent overview of utility-run behaviour programmes for the American Council for an Energy Efficient Economy.[684]

Questions:

1. Examine *"How to waste energy"*, left. Describe why these tips will have a negative impact on people's motivation or capacity to act more efficiently. Is there an alternative action in each case which would have a positive effect? Why?
2. How do heuristics influence decision-making?
3. What evidence is there that norms can change behaviour? What do we need to consider when using norms in a change programme?
4. How should comparisons be framed to encourage action? How would this affect the design of a performance league table?
5. Are "nudge" and other psychological techniques appropriate or an unacceptable manipulation of people's behaviour? Discuss.
6. How do you change people's behaviour?

20 Driving Improvement

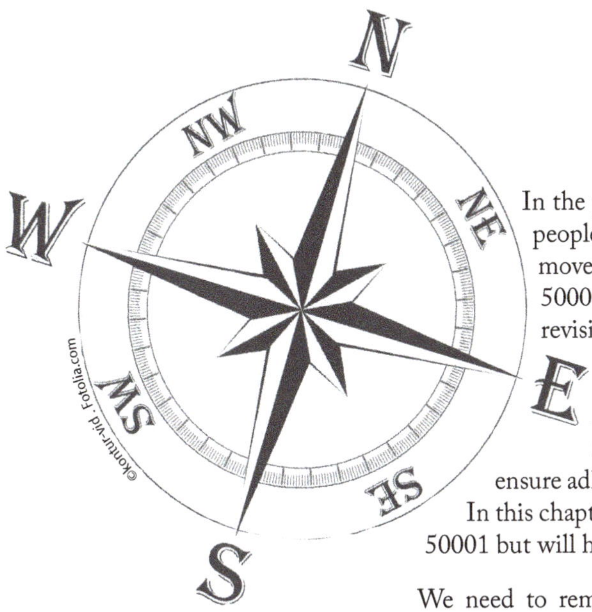

In the previous chapter, we looked at a range of techniques to help people get behind our resource efficiency programme. Before we move on to look at how the energy management system ISO 50001 can help us mandate behaviours and actions, it is useful to revisit some broader concepts about driving improvement in our organizations.

Although systems like ISO 50001 provide a useful structure for a programme, as well as audit and review processes that ensure adherence, the reality is that they can be somewhat mechanistic. In this chapter, we will touch on techniques that are compatible with ISO 50001 but will help us connect our people to our systems.

We need to remind ourselves that most of the improvement methods to deliver greater resource efficiency are not closed-loop systems. They rely on people responding to *something* in order for change to happen.

By asking ourselves *who?*, *why?*, *what?*, *when?*, and *how?*, we can begin to answer the question about just what *exactly* we need people to do. What will they be accountable for and to whom? If these aspects of our programme are unclear, then it is hard to see how we can actually drive the improvement, rather than being passengers in a process we don't know the destination of.

Just as we need a compass to guide our travels, we also need a speedometer to tell us how fast we are moving. Two instruments will provide this functionality: a Monitoring and Targeting (M&T) system and an Opportunities Database. I will touch again on their central role as tools to *motivate* as well as report.

Stretching the travel analogy even further, we need to think of our resource efficiency process as a journey with a bus full of people. The more folks on board our bus, the more likely we are to get long-term change. We want the folks to be looking out of the windows, asking how fast we are going and what the stops are along the journey. A key way we do this is by asking them open questions.

Of course, our journey, especially if it is a long one like resource efficiency, may encounter some bumps in the road, so I will touch on how we manage unexpected developments: the risks that our organization encounters and how these can be turned into opportunities.

20.1 Clear accountability

Our resource efficiency programme needs to ask people to make changes. By answering some very simple questions, we can understand what, exactly, we need people to do.

Upward delegation is an important aspect of our programme. Management need to be given clear instructions on what they need to do, by when and how the outcome will be measured.

The design of our resource efficiency programme should provide the answer to five fundamental questions:

- Who do I need to involve in my programme?
- Why do they need to be or want to be involved?
- What do they need to do?
- When do we want them to do that?
- How will we all know that it has been done?

A good planning technique is to create a table with the five simple columns above and, starting with the folks you feel are most important, answer each question. This will help understand not only the capability and motivation of those involved, but also the organization and structure of our programme.

We must consider leaders and management in the process. A common failing of Champions in energy and resource efficiency programmes is the inability to effectively "*delegate upwards*", that is to give Leaders clear and precise instructions on what they are expected to do. It seems that many folks are understandably uncomfortable telling superiors what to do. Instructing more senior folks is clearly not an easy thing to do, but getting this aspect right could be the difference between success and failure.

If we do not feel we have sufficient status to "*instruct*" our leaders, then perhaps we can involve an intermediary or senior supporter or consultant to present the required actions. While we need to be clear about what we want management to do, we should recognize that they are busy people and structure their input accordingly (see *the 15-minute pitch* on page 195).

In the next section, we will consider ISO 50001:2011, the standard for energy management systems. One reason that this standard works is that there is an unambiguous requirement for *top management* to do certain very specific things (see the box right). Unless top management can demonstrate that they are fulfilling their obligations the system will not be certified.

Of course, we would ideally prefer the support of senior management to be driven by much more than an obligation to achieve certification. We would

Real World: *A lot of people, for a little time, adds up to a lot of effort*

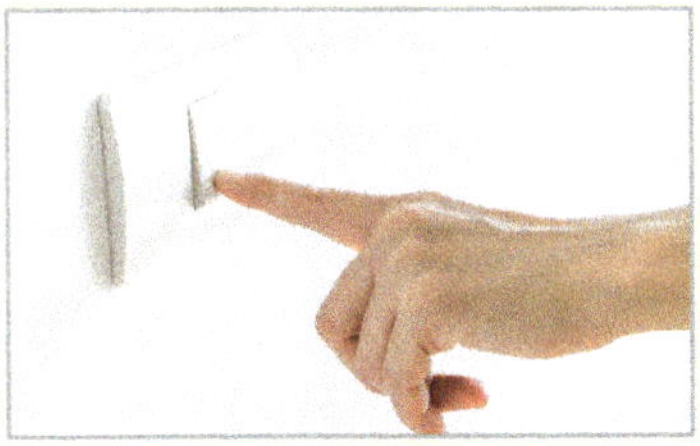

©weedezign, Fotolia.com

One of the principles that can help us design an effective efficiency programme is to consider how we can "*work smarter, not harder*".

This approach is a practical response to a major availability barrier, people's time (see page 193). But it is also a reflection of a philosophy of change management which is about disseminating the change so widely across the organization that it becomes permanent. Where we have a programme driven by just a few dedicated individuals, then it is much more likely to come to an end if some of them move on.

In order to get as many people involved as possible, we need to:

- Give everyone feedback - that is to say that we should let all teams know about their resource use and if it is improving we should celebrate it (our M&T system at work);
- Feedback gives rise to engagement, which in turn gives rise to ideas. We need to offer everyone the opportunity to contribute ideas for improvement - not just "experts" (our Opportunities Database at work).

The expectation is that folks will dedicate a few minutes to the process in their typical working week. If we can spread this widely, the collective effort will add up to a considerable impact.

much rather their participation arises from a personal passion and commitment to the cause of resource efficiency. That is why we have focused so much on articulating the Value, capital V, that resource efficiency can provide, or by establishing if we can eliminate a "*pebble in the shoe*" (page 319). Clearly, intrinsic motivators are better than extrinsic ones, but we do need motivation from the management participants.

It is self-evident, but worth restating, that success will only come about only when everyone involved understands their role. Everyone should be able to answer these five questions: *who?*, *what?*, *why?*, *when?* and *how?*

The obligation to act comes from accountability:

> *Accountable: 1. responsible to someone or for some action.*
> *(Collins Concise Dictionary of the English Language, 1984)*

Accountability connects the participants in our programme. As a Champion I may be responsible for delivering a system, e.g. an Opportunities Database, that is needed by someone to play their role in the process. I am accountable to them, so that they may complete their actions. There is a sense of *mutuality* that speaks of a collective effort or contribution to a bigger goal.

Accountability also brings negative associations. It has overtones of potential punishment for failure or at the very least, embarrassment. To balance these negatives, we need to reinforce the support that is available and to visibly celebrate success as early and often as possible. That is not to say that, like in ISO 50001, there shouldn't be a harder "*just do it*" reinforcing our expectations, for those who may be reluctant recruits to the cause.

Interestingly, Collins provides a second definition of accountable; "*2. able to be explained*" - i.e. something that can be conveyed to others. We shall see that accounting for variation in resource use - that is being able to understand and then explain the variation - is a key part of what we expect people to do.

I should emphasize here that accountability is about actions or outcomes, not about micro-management. By and large, folks will only feel ownership if they have some control over the best way to achieve the objective. That is why "*How it should be done*" is not listed in my five basic questions - the detail of that should generally be left as open as possible.

"How the result is measured", on the other hand, is perfectly acceptable to prescribe, so that there is no ambiguity or uncertainty about what success looks like. The more that we can introduce visibility and transparency around our programme outcomes, the more difficult it will be for the programme to be wound down in some future change of priorities.

We shall look at the measurement process next - in particular, the role of the two key systems: M&T and the Opportunities Database. These, we shall see, are not just tools for measurement but also for motivation.

20.2 You can't manage...

... what you can't measure. There are two essential measurement tools in resource efficiency: an M&T system and an Opportunities Database. Not only will these systems track performance, but used effectively they can greatly motivate individuals.

*A **Monitoring and Targeting system** and an **Opportunities Database** are complementary tools, not alternatives. Both should be used in every energy or resource efficiency programme.*

Don't be thrown by its name - M&T is simply a system for measuring energy and resource use before and after the changes that our programme is driving. The system should offer the analysis capabilities described in our earlier Chapter 14, Analysing Data (page 431). While there are many bespoke M&T systems available, a spreadsheet like Excel could be considered an M&T system. The critical aspect is that it can model the energy and resource performance before improvement measures, so that the actual use after the measures provides the basis for a record of the improvement made.

Not only is the M&T system the scorecard for our resource efficiency programme, but it is also a vital motivation tool. This arises because the system can record the effect of lots of small, seemingly inconsequential, changes working together to reduce resource use. The new-found habit of switching off computers at the end of the day, the fitting of a motion sensor to the lights in the washrooms, raising the interior thermostat setpoint by one half a degree (for cooling, the opposite for heating) - all of these will combine to visibly *move the dial* on electricity use in a building, and provide a means to give those responsible positive feedback.

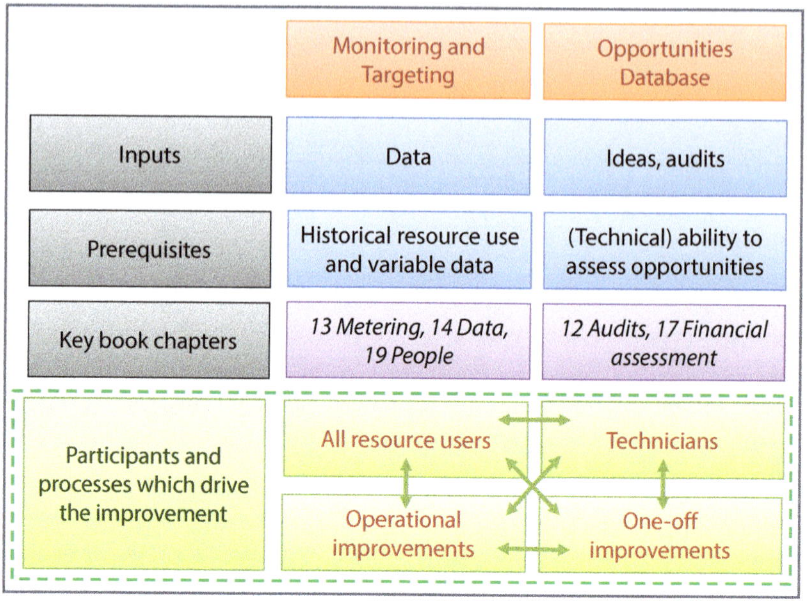

20.1 **Core improvement tools (right)** Although M&T and the Opportunities Database are shown as discrete systems to drive improvement, they should ideally be implemented as a single process. The arrows in the illustration indicate the interaction between all the participants and processes. *Source: Niall Enright, available in companion file pack.*

20.2 **Opportunities GANNT (opposite)** The illustration is from an Opportunities Database in "Carbon Desktop", a product of Verco Ltd, which also provides M&T functionality *Source: Reproduced with kind permission from Peel Land & Property Group.*

The M&T system allows all resource users to become involved in the process, through the way that they interact with the existing systems and processes. By eliminating the effect of uncontrollable factors (e.g. weather or activity) the impact of these individual decisions becomes visible, and positive changes can be celebrated. The M&T system drives the Optimize opportunities, so it is usually deployed very early on in our programme, but is vital for all stages of improvement. The M&T system breaks our performance down to individual items of equipment, teams and departments (usually following our existing management structure) so that the effects of everyone's actions can be seen and appreciated. Outcomes can be "rolled up" at higher levels so that more senior folks at a site or regional level can see how their part of the organization is performing.

Complementing the M&T system is a second tool, which is desirable in every energy and resource efficiency programme. This tool is the Opportunities Database, which is about capturing and driving discrete projects, ideas or initiatives. As with the M&T system, the Opportunities Database does not have to be a database (again, Excel can be used quite effectively). The key is that the status of a project can be recorded and monitored.

The Opportunities Database is sometimes seen as a tool for engineering folks as it captures projects of an engineering nature (e.g. equipment upgrades or process changes). While the Opportunities Database is key to their contribution, it should not be *exclusive* to the technical teams - after all, everyone can originate a good idea even if the feasibility assessment has to be undertaken by an expert, and many opportunities involve behaviour not equipment.

In the Opportunities Database, we capture all ideas for improvement (whether they are operational no-cost ideas or whether they are equipment or process changes, which may require capital). Every idea, however far-fetched, should be recorded with the name of the originator. The opportunity is then *"study required"* or, when it is being assessed, *"study begun"*, and then becomes *"feasible"* or *"infeasible"*, and then *"in progress"* and finally *"achieved"*. Some systems have a category *"archived"*, for ideas which cannot be considered at this time for some reason. As the opportunities are assessed, their costs and savings become known and refined. As with the M&T system, people and teams get credit for ideas and feedback when they are implemented. Because the initial ideas are just as likely to involve Optimize-type interventions, the Opportunities Database should be implemented as early as possible in our programme, to capture these and the early Modify and Transform ideas that we also want to work on right from the start (page 242). The schematic opposite shows how these tools complement each other and how all participants in our programme can gain value from them.

Real World: *Tracking opportunities*

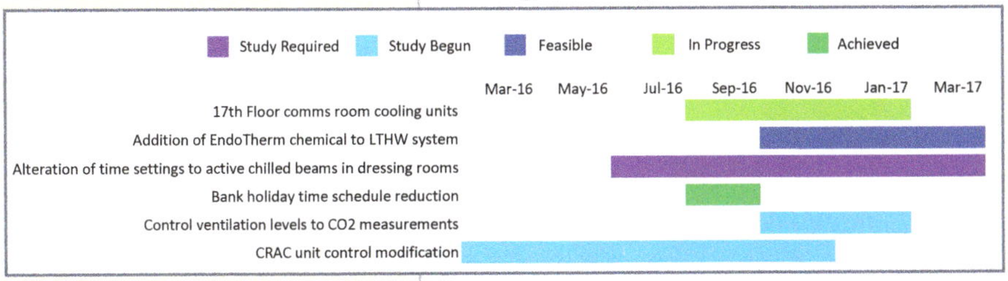

Because the M&T system reports on outcomes, i.e. it is looking back at the actual resource use compared to that predicted, it is a lagging indicator.

The Opportunities Database, however, tells us how many projects we have for future improvement, a leading indicator (see page 378). It is a *flow meter* for our project implementation activities. In particular, it ensures that small projects, which taken collectively have a big impact on savings, are driven to completion.

At Peel Land & Property Group the team have implemented an Opportunities Database, which has, at the time of writing, 605 opportunities of which 330 have been completed, delivering savings of £1.4 million *a year* (~35% of the total energy costs) for an expenditure of £1.5 million. Knowing which of these projects are complete, who came up with the idea and who implemented them enables us to celebrate success.

The Opportunities Database is also forward-looking. We know, too, that we have 65 projects *In Progress* or *Feasible* which should deliver £103,000 of savings and 93 projects *Study Required* or *Study Underway*, which could provide another £190,000 of savings. This insight allows us to intervene. For example, if finance is an issue, the projects will begin to pile up at the *Feasible* Stage and we can work to release extra funding.

20.3 Reinforcing action

A great way to reinforce action on resource efficiency is to engage people in coming up with ideas for improvement. By asking open questions, management at all levels can foster participation in the programme and encourage folks to think for themselves about changes.

Asking **open questions** is at the centre of the resource efficiency process. It is what **reinforces action.**

In the *"a lot of people for a little time"* approach to resource efficiency, we need to make sure that what we are asking people to do can comfortably fit within their regular work programme. Ideally, what we ask them to do should become *part* of their ordinary work activities. Returning to our five questions:

> *Who?: I need all folks who influence energy use to take part, which is why we are asking you and your team to get behind our efficiency programme.*
> *Why?: Because we need to reduce our energy and resource use to lower our costs, preserve jobs and help the environment.*
> *What?: You will get a regular report on your resource use. If it varies from what is expected, then your team need to work out why this is and repeat the good and eliminate the bad. If it is something outside your control, then we ask you to give us suggestions for improvement, which will be investigated.*
> *When?: At least once a week, at your team meetings.*
> *How?: We can measure the effect of quite small changes and will give you a regular progress report, so everyone who contributes will be recognized.*

In this dialogue, it is important that folks feel supported. That is to say that they know that they can get help if they are struggling with understanding a variance in their resource use. However, the bottom line is that they *must account for variances in their use.* In other words, they are *accountable* for *knowing if their use is good or bad* in any given week and, where the variance is greater than that considered normal, they need to put in place the necessary work to *explain it and correct it*. By *correct*, I do, of course, mean *repeat the good and eliminate the bad* - we need to look at positive variance as rigorously as negative variance.

Accountability for understanding the resource use can be as individuals or as teams, or it can sit with a team leader. Ideally, accountability will follow the existing management structure so that employee A is responsible to their boss for accounting for their resource use, who is in turn accountable to their boss for accounting for their resource use, and so forth, up the organization.

This approach avoids telling people how to respond to variance. It might be the case that folks need specific training and methods on how to investigate variance. It might be that a checklist or flowchart can be developed to help diagnose or correct a particular process. Where possible we want the resource users to come up with the tools so that they feel in control and take ownership.

Exploration: *M&T failures*

In the mid-1990s there was a boom in M&T programmes in the UK as a result of active government promotion.

I was with March Consulting Group at the time, responsible for a leading M&T software application, Montage. From where I sat in the organization, I could see that a significant percentage of the M&T programmes were failing (as measured by organizations discontinuing their software use just few years after the launch of M&T).

This is a reminder that tools and systems alone will not sustain change.

As a result of the failure rate in the traditional M&T programmes, colleagues at March and I developed a methodology called enManage™, which anticipated many of the features of the latest management standard, ISO 50001 (for more on this history, see page 293).

The enManage™ process, for example, formally documented the involvement of management in the process and required management's ongoing participation at key times. In many ways the key development was to describe the *accountability* of management within the programme.

But, as well as great tools and great systems, our programme also needs curiosity, inspiration, celebration, fun, tenacity, fearlessness, optimism, creativity, adaptability and joy. These are not provided by systems but by people.

What we should be trying to achieve is a thinking approach to resource use. We need to ask folks open questions that put the ball firmly in their court: "*How are we doing on energy use?*" We know from our previous chapter on engaging people, that this is a way of harnessing the tacit knowledge of individuals and in the absence of any immediate reward other than the intrinsic satisfaction of solving a problem, for the folks to take ownership and personal responsibility. We should use this open questions approach to reinforce anticipatory thinking (see page 292): *"So what do you think this change means for our future resource use?"*

Asking open questions is at the centre of the resource efficiency process. It is what reinforces action and encourages people to access the M&T system and Opportunities Database for answers. ISO 50001 talks about the *"Plan, Do, Check, Act"* cycle, but doesn't really say how you do this. In fact, the process starts with shared information (such as our variance against target given by the M&T system) which tells folks how they are doing - Check. We then ask open questions, which leads them to identify possible actions - Act. These are either done immediately (or marked for further investigation or future action in the Opportunities Database) - Do. And so the cycle begins again - but it is the questions that generate the actions that drive the process forward.

By asking a question, we show we are engaged and interested in a subject. If my boss asks me a question, then I usually want to give a good answer. If my boss regularly asks me about the same things, then I am likely to prepare for the question. If my boss is asked the same question by their boss, then I know that my answer is important to them so that they can respond to their boss.

So one skill that we need to develop, especially among the managers who are supporting our programme, is how to ask open questions and listen effectively. Open questions initiate a discussion and are used to get the subject to think about their response. A closed question is the opposite since it can be answered in a single word or short sentence - they are usually used to gain facts.

- Open questions start with *"why, how, describe....what do you think about... or tell me about.... "*
- Closed questions start with words like *"did, are, when, where, will, won't, didn't, aren't, would, if, is"*.

In any open conversation, we can specify the boundaries of the question, "*What kind of things can we do to stop water losses?*", and we can ask follow-up questions to get more detail. For example, we can probe for clarity: "*What sorts of behaviour changes do you have in mind?*" Or we can ask: "*What else do you thing we can do?*" (Avoid "*Is there anything else we can do?*" as this is a closed question, potentially leading to a one-word answer, "*No*"). Invite creativity: "*If money was no object...*", "*Give me any ideas, no matter how wacky...*". Try to avoid questions which imply a correct answer, keep it neutral e.g. "*How could we change the shift clean-up routine?*", instead of "*What's wrong with our clean-up processes?*", which may prompt a defensive answer.

20.4 Risk and opportunity

Many people see risk as something to be avoided. For most organizations, however, risk is inevitable, so we may as well use it to our advantage. With the right approach, we can treat risk as an opportunity to drive further improvements in our organization.

Risk is something to be embraced and understood. For those who are prepared, it offers an opportunity to drive greater improvements than would otherwise be possible.

Risk is a major force for driving improvement in organizations. For example the risk of accelerated depreciation, aka *stranded assets* (see page 111), is the primary reason why property companies invest in developing ever more sustainable and resource-efficient buildings. We saw in the research by Catherine Cooremans (page 598), evidence that, for financial decision-makers, risk is a greater driver than cost savings in efficiency investment.

When we examined creating a mandate for our programme (page 316), we observed that a crisis presents a great opportunity for change. This opportunity is in part because there is not the same degree justification constraint around issues of survival, licence to operate or reputation, and also because the jolt to an organization that comes from a crisis often forces folks to think outside the box and be open to more radical solutions.

So, an important aspect of our resource efficiency programme is to continuously scan the horizon for risks and the opportunities for change that they might present. The flowchart opposite provides an overview of how risk management could be incorporated into our resource efficiency process.

There are two forms of risk that we need to manage. The first is programme risk, which are risks that our resource efficiency process might fail. These could arise from internal or external events, such as the departure of a senior sponsor or the emergence of a new initiative which competes for attention and resources. Maybe our programme design has failed in some way, so that we have not set the correct expectations, or the return on investments falls short, or staff are resistant, or we lag behind our competitors. There are many reasons that a programme can fail: remember, a rule of thumb is that about a third of programmes achieve their goal, a third fail completely and the remainder deliver some value, but short of what was anticipated.

The second source of risk, which is likely to be much more significant, is that of resource risk. This are risks related to the direct and indirect effects that resources have on our organization's ability to operate and generate Value. Obvious examples of this would be things like water scarcity affecting golf courses, but less obvious are examples like the manufacturer of snow-chains for car tyres which has seen their business decline dramatically as a warming climate means that the demand for their products has dramatically declined. Note that other types of risks can also represent an opportunity for greater

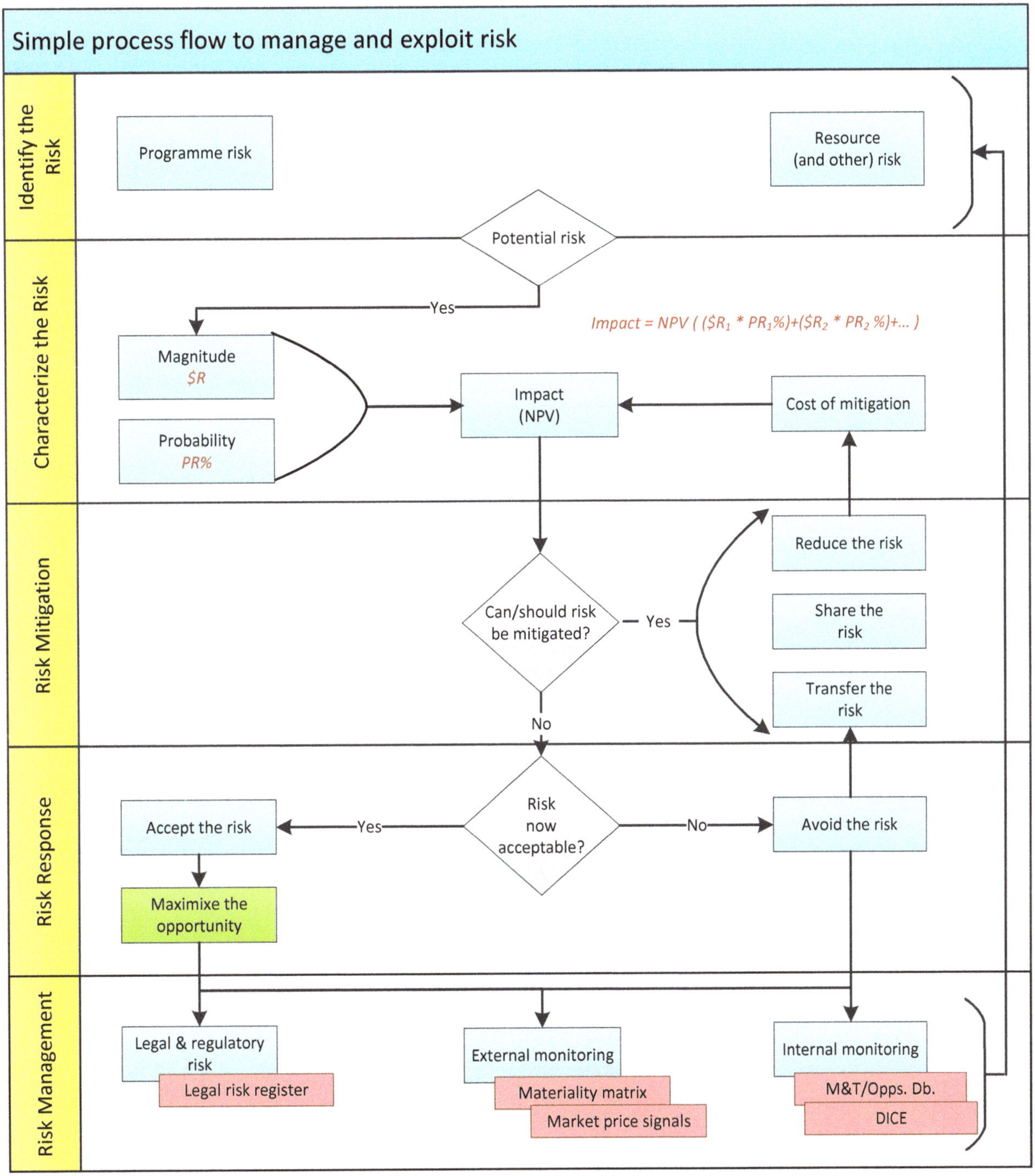

Change

20.3 **Process flow for analysing and managing resource risk**

Two forms of risks that are considered in this process flowchart. First we have programme risk - that is risks associated with our programme failing to achieve its objectives; then we have direct effects of resources on our organization e.g. profits, reputation, licence to operate etc. *Source: Niall Enright, inspired by Bekefi and Epstein, Integrating Social and Political Risk into Management Decision-Making.*[64] *In file pack.*

Summary

1. We must not forget that people lie at the heart of our efficiency process.

2. To make informed decisions, people need to have the correct information.

3. Information alone will not drive change - people also need to have the motivation and capability to act on this information.

4. Being clear about people's roles is critical - in an organizational context, this is called accountability.

5. The backward-looking tool to understand variance in resource use against a target or original performance is called Monitoring and Targeting.

6. The tool that captures and drives discrete ideas for improvement is called the Opportunities Database.

7. We want folks to approach these tools as sources of inspiration and insight. Leaders can encourage this by asking open questions about what the tools are showing.

8. Most systems and standards are designed on the assumption that the organization is in a stable condition. In practice, this is rarely the case, and we must come to see risk and change as opportunities for improvement, not as threats. Risk is not something that *happens*, it is something you *anticipate* and gain advantage from.

resource efficiency, not just risk due to resource use. I am thinking here of risk from increased competition in a market, which may favour differentiating products by their environmental performance (as in the earlier examples of Method, L'Oreal, Body Shop and Interface).

In order to manage risk, we first need to identify it. For the programme risk, we have our core M&T system and Opportunities Database, which should tell us if we are on track. A more strategic assessment of the programme can use techniques like the DICE tool (page 209). For resource risk, our monitoring tools include a materiality matrix (page 362), or a legal risk register, as well as a wide range of signals from markets - such as increasing emissions costs, or declining sales where products fall out of favour.

When a risk has been identified, the most common way of assessing the response is to determine the impact, usually as the net present value of the risk calculated as the product of financial cost times the probability of the risk occurring. Of course, many risks are by their nature uncertain, so some form of sensitivity analysis may be needed in order to arrive at a reasonable estimate.

The next step in the process, having identified that there is a risk, is to assess if the risk can be mitigated or reduced in any way, and what the costs of these mitigation measures might be. Risks can be reduced (for example, in the case of the golf course, by moving to more drought-tolerant grass varieties) or they can be transferred (the golf course could take insurance to cover the cost of water in the event of shortages) or the risk can be shared (the golf course could develop water storage along with other water users). The various mitigation options can be priced up and then a final decision made as to whether the risk is acceptable or not. If the risk is still not palatable, then the organization needs should be eliminated (e.g. by developing their own wells to access underground water) or accepted (e.g. do nothing and accept the consequence of drought).

The mitigation options described above may reduce the risk to the point where it becomes acceptable to our organization. In selecting our specific response, we have an opportunity to choose/investigate/develop those that offer more radical or permanent changes to our resource use which have the benefit of eliminating the risk, compared to measures, such as insurance, which merely mitigate the harm in the event that the risk materializes. For the golf course, permanent measures would be using new varieties of grass which demand less water, or developing systems to reuse *grey* water. In other words, we may have solutions which mask the symptoms and others which cure the disease. This *maximize the opportunity* task is shown in green in the flowchart to emphasize how important this step is.

Our risk management process now involves executing the desired outcome and ensuring that our monitoring systems and processes are updated to track the results. This approach to risk management is appropriate even if we have adopted a rigorous system, such as ISO 50001, discussed next, as these formal systems are just as prone to unexpected shocks which give rise to opportunities.

21 ISO 50001:2018

Photo: © laufer, Fotolia

The International Standards Organization (ISO) develops and publishes international standards. ISO 50001:2018 is properly called "*Energy management systems - Requirements with guidance for use*". The "2018" in the standard reference is the year in which it was approved. This is the second edition of the standard, the first having been released in 2011.

These standards are written by technical committees, in this case, ISO/TC 301, which has members from national standards organizations on the panel as well as industry representatives, academic experts and energy efficiency practitioners, among others. Although the ISO is based in Switzerland, the secretariat for ISO 50001 is provided by the American National Standards Institution (ANSI) and Standardization Administration of China (SAC).

At this point, I must make a confession. In the past, I have been somewhat underwhelmed by the effectiveness of these standards in driving real improvement. The old quality standards joke goes that it is perfectly OK to design a bicycle with square wheels, as long as *every* bike has square wheels! Imagine then my surprise when the predecessor of ISO 50001, BS EN 16001 was first published in the UK and it had at its heart not *consistency* but *improvement*.

Indeed, ISO 50001 has gone even further to reinforce the notion that an effective system is one that is embedded and owned right across the organization. Furthermore, the standard supports my own thinking that energy efficiency is more about management than it is about technology.

This chapter is not a detailed guide to the standard. For that, you will need to turn to one of the publications listed at the end of this chapter. My focus here will be helping you decide if the standard is right for your organization and, if you choose to go ahead and use the standard, what are the techniques that will maximize the value from the system. I will also share my own case study from implementing 50001 in a large and complex organization, as well as some advice from an ISO 50001 assessor. There will also be links between the Framework with the standard.

Although ISO 50001:2018 is an energy management system (EnMS) specification, it is perfectly suitable as a wider *resource* management system. There is nothing to stop you incorporating resources such as water, waste and raw materials into your own system and then certifying it for the energy parts.

Editions: 2011 or 2018?

This chapter covers the second edition of the standard released in 2018. However you can download the text for the 2011 version from the iwik website, In reality, there aren't that many changes between the editions.

Standards: *ISO 50001:2018 Energy Management Systems*

ISO 50001 is part of a suite of four certifiable management standards which include ISO 90001 for quality management, ISO 14001 for environmental management and ISO 22000 for food safety. With the 2018 update, 50001 has been aligned with the other standards through common structures, terminology and requirements. ISO 50001 is based on its predecessor BS EN 16001, which was developed in the UK. Improvements made in the transition to ISO 50001:2018 include an emphasis on *"top management"*, more detail around "*energy reviews*", normalisation of energy performance indicators (EnPIs) and baselines, and a greater emphasis on demonstrating real improvement.

ISO 50001 is a very well-considered standard designed to drive continual improvement and overcome many of the factors that lead to energy efficiency programmes being abandoned over time. A plan, do, check, act cycle, shown in green, is at the heart of the process, illustrated below. There are regular activities shown in purple.

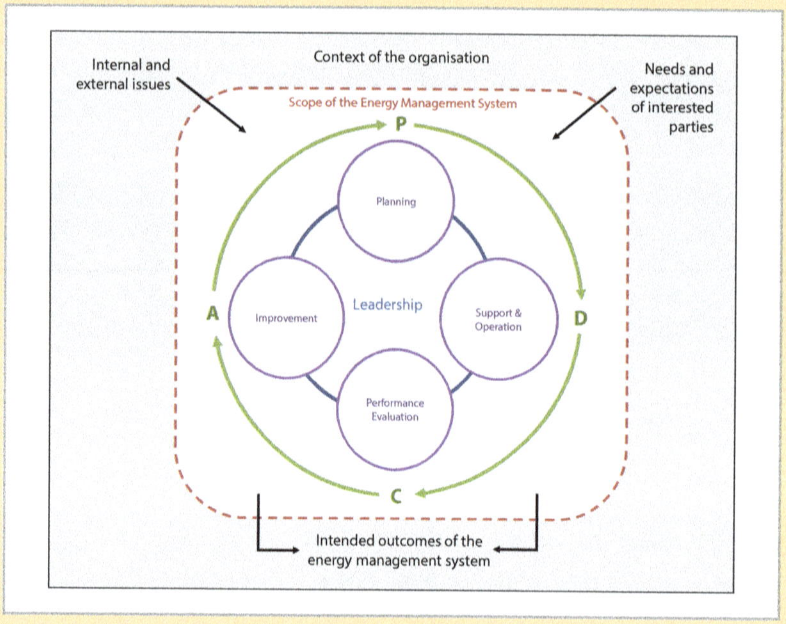

Everything derives from the organization's Leadership (aka *"Top Management"*), shown in blue. Management must establish an energy policy, based on the context of the organisation and needs of interested parties. This policy may well set the intended outcomes of the Energy Management System (EnMS), against which performance should be regularly evaluated.

At the outset of the process, there is an energy planning process. In this activity, there will be an energy review, which assesses the total energy consumption and defines what the significant energy uses are (the organization can set its own criteria as to what constitutes significant). This energy review must identify interested parties and their needs in terms of energy performance improvement. The review needs to be repeated on a regular basis or if there are significant changes to the energy using activities.

The planning will state the baseline performance for each significant energy user and determine the energy performance indicators that will monitor the performance in future. For these EnPIs and other energy activities that impact energy use, we will establish objectives and targets for improvement. We will also create energy management action plans (EMAPS) which include the specifics of what will be done, by whom, when, with what resources and how the improvement and action will be verified. The planning process will also establish any legal and other requirements (e.g. voluntary performance standards) that we need to adhere to and how this will be achieved (this is recorded in the legal register, which needs to be maintained on an ongoing basis). Other activities will include defining people's roles and developing a competencies and training plan and a broader communications and reporting plan to enable all employees to contribute to the EnMS. We also need to set out processes to ensure we achieve operational control of our energy use, and that our procurement and design activities incorporate energy efficiency where appropriate.

The documentation of these data and tasks, along with the supporting evidence, is our EnMS. We need to establish an internal audit process which checks, on a regular basis, that we are following our own documented procedures and that the procedures meet the requirements of the standard. Failures to meet the ISO 50001 standard are labelled as major or minor nonconformities which, together with other actions to address energy use or improve the ISO 50001 system, may lead to corrective or preventive measures. All these need to be documented and addressed. On a regular basis, typically annually, there must be a management review of the operation of the EnMS.

21.1 The fit with this Framework

ISO 50001 doesn't actually tell you how to save energy, which this book and Framework goes into at great depth. However, in respect of the systems and processes to achieve continual improvement, the ISO 5001 standard and the Framework are highly compatible and complementary.

The Framework set out in this book is deliberately not prescriptive, recognizing that different organizations will have different needs, opportunities and starting points. If the processes set out in the Framework were to be adopted in full, then this would deliver the core continual improvement systems needed for ISO 50001 certification and go well beyond the formal requirements of the standard in many areas. Taking just one example, the Opportunities Database system recommended in this Framework is a much more powerful way to drive discrete actions than the ISO 50001 Energy Management Action Plans (EMAP). However, if an Opportunities Database is adopted, it would be an easy task to create ISO 50001 compatible EMAPs from it.

In several areas, ISO 50001 goes beyond this Framework by incorporating requirements for documentation, internal audit and a legal registers, all of which will help to ensure that the continual improvement process is maintained in the long run. Thus, we can say that both systems are overlapping and complementary, not substitutes for one another.

21.1 **ISO 50001 requirements and the relevant Framework chapters**
The ISO 50001 process, summarized opposite, has a number of mandatory elements, most of which will be achieved if the Framework and Methodology described in these pages are implemented. The table below indicates which volumes and chapters address requirements of the ISO 50001 standard.
Source: Niall Enright

ISO 50001:2018 Requirements	Volumes and chapters of this Framework related to the ISO 50001 requirement
5.1 Leadership and commitment	*I-6 Mandate*: The importance and role of Leaders. *II-9 Creating a Mandate:* how to get top management commitment *II-11 Setting Goals:* is relevant to policy
5.2 Energy policy	
6.2 Objectives, targets, action plans	*II-12 Discovery Processes:* how to do an audit goes beyond the requirements of the energy review; intensity use can help define significance *II-14 Analysing Data:* covers all the techniques needed to establish effective EnPIs *II-17 Financial Analysis:* provides the tools to determine which opportunities are feasible and should appear on the action plans
6.3 Energy review	
6.4 Energy performance indicators	
6.5 Energy baselines	
7.2 Competence	*I-6 Mandate*: Some key roles. *II-19 Engaging People:* Training and learning
7.3 Communication and 7.4 Awareness	*II-19 Engaging and Empowering People:* has a wide range of relevant techniques
7.5 Documentation	This is specific to ISO 50001
8.1 Operational control	*II-20 Driving Improvement:* covers some of the systematic approaches to resource efficiency. Specific improvement measures available through operation, CAPEX procurement and design are covered throughout this book.
8.2 Design	
8.3 Procurement	
9.1 Monitoring, etc.	*II-14 Analysing Data:* has techniques to truly understand performance
9.1.2 Legal and other compliance	*I-3 Value:* covers the licence to operate benefits of efficiency
9.2 Internal audit, conformities	This is specific to ISO 50001
9.3 Management review	This is specific to ISO 50001 but *II-9 Creating a Mandate* has some relevance

21.2 Establish if 50001 adds value

ISO 50001 offers some great benefits, but as with all systems there can also be drawbacks. Here, we consider the pros and cons to help determine if the standard is right for your organization.

There are some very compelling reasons for implementing an EnMS that conforms to the ISO 50001 standard:

- It genuinely addresses many of the difficulties resource efficiency practitioners have in sustaining improvement programmes in the long term. In particular, the standard focuses on continual improvement, which is key to driving value from energy and resource efficiency.
- In many countries, there are explicit incentives to implement ISO 50001. For example, most EU countries implementing the Energy Efficiency Directive requirements for mandatory audits have made ISO 50001 an alternative - potentially less costly or less prescriptive - route to compliance. In Germany, tax incentives mean that it leads in terms of the number of ISO 50001 implementations. Incentives also exist in France, Denmark and Switzerland.
- Meeting the standard provides an acknowledgement that your organization takes energy efficiency seriously and has implemented a robust process to drive this forward.

The latter two benefits will almost certainly require that you have your EnMS independently assessed by an accredited body qualified to do so. In this case, your system will be *certified* (assuming you pass) and you will be given a formal document proving this. It is perfectly acceptable to self-declare or *attest* that you are operating a system in conformity with ISO 50001, or to get customers or other external bodies with an interest to review your system and vouch that it conforms to the standard. If you don't make any form of declaration of conformity, you could choose to implement just those parts of the standard that you feel will add the most value.

Although there are many compelling reasons to consider ISO 50001 there are some disadvantages that you should be aware of:

- While ISO 50001 is flexible and adaptable, there is a minimum level of effort that will be needed which means that it is not suitable for small organizations.
- Taking a complex organization to certification can be an expensive process, not just in terms of money but the time needed from internal staff.

Standards: ***Related standards***

There are several related standards produced by ISO/TC 301.

Unlike ISO 50001, these are not certifiable standards.

- ISO 50002:2014 *"Energy audits - Requirements with guidance for use".* This is a very generic document, of little value to anyone with a modicum of experience carrying out energy audits.
- ISO 50003:2014 *"Energy management systems - Requirements for bodies providing audit and certification of energy management systems".* If you want to be sure that your ISO 50001 certifier is following the correct process, this is the document for you. This standard is also useful in that it sets out the method to determine the amount of time the certification and surveillance audits should require based on the number of personnel involved in the EnMS and the complexity of your site.
- ISO 50004:2014 *"Energy management systems - Guidance for the implementation, maintenance and improvement of an energy management system".* This is complementary to ISO 50001, and provides some good tips aimed at those who are inexperienced in implementing an ISO:50001 system. ISO are at pains to point out that the suggestions are not "*intended to represent the only possibilities, nor are they necessarily suitable for every organization*" and furthermore "*it is not intended to provide interpretations of the requirements of ISO 50001.*"

It is entirely possible that implementing ISO 50001 in an organization that is not receptive could make a good energy management process ***worse.***

- There is an overhead involved. That is to say that a proportion of the effort devoted to ISO 50001 is not directly driving improvement *per se* but is devoted to documentation and administration, which could distract from more constructive uses for the resources.

- In an extreme case of overhead, some organizations I have encountered seem to have developed hugely bureaucratic management systems, requiring large amounts of paper and data. These systems can end up meeting the needs of management systems managers rather than the organization. If this is the case, alarm bells should be ringing about whether ISO 50001 is appropriate or whether the process will be hijacked by people who have little direct influence on - or interest in - energy use.

- At the opposite end of the spectrum, for some organizations, the requirements of documentation, verification and internal audit are culturally alien. If people are not accustomed to working within formal systems, attempting to implement ISO 50001 could be demotivational and lead to resistance, if not outright failure. It is entirely possible that implementing ISO 50001 in an organization that is not receptive could make a good energy management process *worse*.

- ISO 50001 cannot be implemented solo or by just one function. No single Energy Champion, however competent or enthusiastic, can make it a success alone, because the continual improvement process requires the active and engaged participation of many people.

- Unless there is the prospect of senior management support for the system on an ongoing basis, not just at the beginning, the probability of success is very low.

- If you are interested just in the badge that 50001 can bring, do consider alternatives such as Energy Star, Carbon Trust Certification or any one of the plethora of other schemes which impose a much smaller implementation burden. Where the standard is sought merely as a tick box or as a means to compliance, there is a strong probability that it will be implemented in a token way, at a high level and in a manner that adds little or nothing to the energy saving process in the organization. These may be valid reasons to proceed, but if the focus is on these benefits (branding or compliance), then the energy savings benefits may become secondary.

Although many energy consultants and government agencies promote ISO 50001 as the panacea for organizations seeking to drive improvement, we should recognize that these are usually not dispassionate or objective recommendations.

Embarking on an ISO 50001 process is not to be taken lightly, and as with many other change systems, there is a not insubstantial failure rate - not so much in terms of an inability to achieve certification, but rather in terms of a failure to realize the value expected. *Caveat emptor* or *buyer beware*.

Standards: 能源管理体系 要 求

The Standardization Administration of China adopted ISO 50001:2011 as the national standard for Energy Management Systems in 2012, replacing the previous 2009 standard.

The designation of the standard is GB/T 23331-2012 , where "GB" indicates that this is a national standard and the "T" means it is a recommendation.

21.3 How to start ISO 50001

They way most organizations plan for ISO 50001 is to compare the fit between their current energy efficiency activities, and the requirements of the standard, taking into account the specific areas where the organization sees the standard adding value.

In an ideal world, an organization will already have an existing energy management process in place before implementing ISO 50001. This will mean that some roles are defined, systems in place, data available and opportunities for improvement identified. The implementation of ISO 50001 becomes a process of establishing where the existing systems meet the requirement or where additional work needs to be done (see Gap Analysis, left).

If you have an existing system, then you will know what specifically you want to improve in the system, you will have some idea what aspects of ISO 50001 will add value (which will be emphasized in your process) and which not (so be implemented in the easiest possible way to meet the requirements).

If you don't have an existing system in place, then I would strongly recommend that you implement ISO 50001 in a staged way. That is to say, you engage with top management first and do some basic analysis that can contribute to your energy review. Only when you have defined the scope of and significant energy uses, should you go about recruiting the energy management team.

In both cases, once the energy management team has been identified, I would involve them extensively in documenting the processes that they will follow to drive improvement. The energy management team, and all involved, should feel that they *own* the process.

Whoever designs the system has a lot of power. Because people's conformance with the documented procedures will be checked regularly, the actions we are setting out become compulsory. If not handled carefully, this could lead to resentment and/or resistance. It could also be very demotivational.

On the other hand, it is the very nature of the threat that those not adhering to the system will be found out, that gives ISO 50001 the ability to overcome some of the shortcomings of conventional resource efficiency programmes. If we fail certification because someone was not cooperating, top management are likely to be unimpressed with that person. This is the "*stick*" that ISO 50001 provides - but it must be used with wisdom and judgement.

An ISO 50001 process is only ultimately as good as the internal audit processes that are put in place around it. These audits not only ensure that we adhere to our process, but also play a critical role in the certification of our system, as described on the page opposite.

Real World: ***Gap analysis***

Photo: © Lee Avison Photography, Fotolia

The most common starting point when contemplating implementing ISO 50001 is a *"gap analysis"*.

On one level this simply involves listing all the requirements and recommendations in the 50001 process and examining the extent to which our existing activities address these or if we will need to modify or add procedures.

However, those undertaking the gap analysis need to appreciate the adaptable nature of ISO 50001. The scope of the system, the objectives set in policy, what is or is not economically feasible, where the greatest value is to be gained, should all have a significant bearing on the proposed effort and focus.

Given this, I would recommend that whoever is carrying out the gap analysis has a good understanding of the ISO standard and your organization. Ideally, the assessor appointed by the organization will be involved so there is certainty that the actions proposed will result in certification.

*Internal audits are a sort of "**get out of jail free card**" of ISO 50001. If you acknowledge a minor nonconformity and have a credible plan to fix it, you shouldn't be failed on that item, as long as it is not stopping the EnMS from working altogether.*

Real World: *The importance of internal audits*

Photo: © kemaltaner, Fotolia

One of the requirements of ISO 50001 that seems to add the least value, because it is not directly related to saving energy, is the internal audit process. Actually, this is one of the most important parts of the EnMS.

As noted opposite, it is the internal audit process that keeps folks on track. The audits are there to examine if people are doing what the documentation of the EnMS says they should be doing. Every part of the EnMS should be audited by someone independent of the process at least once a year. If there are multiple businesses or locations, then efforts should be made to get around all these.

Audits are not surprise inspections. The internal audit documentation of the EnMS should include an audit schedule or calendar saying when different locations or aspects are being audited. In advance of a visit, an audit plan should be produced explaining exactly what parts of the EnMS will be reviewed when and with whom.

So, apart from keeping people on track, why are audits so important? First of all, they are key to the certification process. Without internal audits, assessors would need to establish that a system was working 100% correctly 100% of the time in order to recommend an organization for certification. With internal audits, however, an organization can identify itself that there are minor nonconformities and can put a plan in place to remedy these. The certification body's certification decision process should ensure that *"for any minor nonconformities it has reviewed and accepted the client's plan for correction and corrective action."* (ISO/IEC 17021-1:2015 9.5.2c). Internal audits are a sort of *"get out of jail free card"* of ISO 50001. If you acknowledge a nonconformity and have a *credible* plan to fix it, you shouldn't be failed on that item, as long as it is not stopping the EnMS from working altogether.

This nonconformity process also applies where an organization has a unit or location that has just joined in the ISO 50001 process (for example, many organizations make acquisitions which need to be integrated with their own internal processes). In this case, the organization would raise a nonconformity and develop an action plan for the new business to meet all the 50001 requirements. The fact that at certification some are not complete is acceptable because this shortcoming is acknowledged and is being addressed. This process was followed in the Peel Land & Property Group certification, where the two regional airports were new to 50001, and so there was an item in the internal audit records identifying current compliance against the standard and the remaining tasks.

One of the very first things that an external assessor is likely to look at is the internal audit records of the organization. That way they will quickly establish if there is a robust process in place to correct deficiencies in the EnMS, and if there are some ongoing improvement plans which need to be taken into account in the audit.

Finally, the internal audit records are also critical to the management review, which occurs at least once a year with top management. The audit process is there to give top management the assurance that the EnMS is working.

21.4 Getting the documents right

When people think about a standard like ISO 50001, the first thing that comes to mind is paperwork. Here, we explore the documentation that is needed and how you can approach this so that the system works for you.

The first thing to note about ISO 50001:2018 is that it is a standard for an energy management *system*, not for energy management. Meeting the requirements of this standard tells us nothing about whether our energy use is good or bad relative to our peers, just whether our processes and systems are delivering continual improvement.

The standard is very prescriptive in *what* an organization must do. The key word here is "*shall*", as in "*the organization shall communicate internally with regards to its energy performance and EnMS...*". The word "*shall*" appears 9 times in Section 4 onwards of the standard setting out the requirements.

However, the standard is rarely prescriptive about *how* the requirement is met. Thus the nature and frequency of the internal communications are not spelt out. It is up to us to decide if it is in the form of a newsletter, staff briefing, notice boards or whatever we feel is appropriate. This flexibility makes ISO 50001 very adaptable and enables us to implement a process that is suitable and value-adding for our organization, however large or small.

If our system is administratively burdensome or achieves little improvement, it is almost certainly the result of our own ***poor design,*** *rather than the ISO 50001 standard itself.*

Each and every requirement needs to have two forms of documentation. First, we need to explicitly state in our EnMS the "*how*" methods we will use to meet the requirement. This description is the EnMS process documentation.

For internal communication, for example, we could state that "*Individual communications plans may be created as necessary by the energy Champions to reflect the communications priorities for their business units*". Here, we have said what we will do and who will do it. Note the caveats "*may*" and "*as necessary*". The way this statement has been phrased means that not every energy Champion needs to produce a communications plan. So, when the system is audited (for internal audit or certification), the fact that one Champion has not created a plan will not lead to nonconformity.

Wording like this is important. For example, if I state in my process that "*the energy committee will meet every month to review the action plans*", then if we miss one month we will have a nonconformity. On the other hand, if I state "*the energy committee will meet at least once every quarter to review the action plans*", even though the intention is to meet monthly, then missing one meeting will not be an issue. If our system is administratively burdensome or achieves little improvement, it is almost certainly the result of our own poor design, rather than the ISO 50001 standard itself.

Real World: ***Roles in ISO 50001:2018***

The ISO 50001 standard sets out some specific roles in the system.

- Top management: a person or group of people who directs and controls an organization at the highest level. The standard requires that top management *"shall demonstrate leadership and commitment with respect to continual improvement of its energy performance and the effectiveness of the EnMS".* It then goes on to list 13 specific activities (see page 729) that top management must do such as defining a policy and the formation of an:
- Energy management team: comprising *person(s) with responsibility and authority for effective implementation of an energy management system and for delivering energy performance improvement.*

In the 2011 edition of the standard there was reference to a "Management Representative". This role no longer exists in the 2018 version, but it is pointed out that *"A single person can perform the role of the* [energy management] *team"*, and organisations are still perfectly free to appoint someone as the manager of the EnMS.

More broadly, all individuals who have an impact on significant energy uses, need to be identified, their competency assessed, and training provided if required.

In the Peel System, we created specific roles and expectations for business managers, energy Champions, energy influencers, a data administrator, and project directors (who run the large scale construction projects), as well as the roles specified above.

In our documentation we should pay especially close attention to those parts of the standard which are left open to interpretation, such as our determination of a "*significant energy use*". The standard defines significant energy use as *"energy use accounting for substantial energy consumption and/or offering considerable potential for energy performance improvement"*, so we need to document whether we are considering *consumption* or *potential* as our criteria (or both) and how we define *substantial* or *considerable* in each case (e.g. above x kWh or delivering y£ savings). As there are specific requirements in relation to these significant energy uses, setting the threshold too low may lead to an excessive effort in areas where there is little value to be gained.

The second set of documentation in our EnMS relates to verifying the operation of the EnMS. All the "*shall*" requirements, as documented in our process, can be subject to a check in an audit. One way to make this easy is to spell out in the process document itself how the verification is to be achieved. For example, "*the energy committee will meet at least once every quarter to review the action plans and this will be verified by reference to records of the minutes of the meeting, which will be stored in XXXXX location*". The auditor will then know exactly what has been committed to and how it can be confirmed. The documents showing what has been done or achieved are called records.

Verification has two elements, verifying that an action or process *happened* and verifying that it had the *desired effects*. The standard requires that every action in our energy management action plan has a verification plan. Once again it is up to us how difficult we make this: we could state that verification will be "*by taking a photo of the equipment after the installation*", or "*by observing a decrease in the metered energy use in the affected area*".

As you would expect, the methods for verification of standards are documented in yet another standard ISO/IEC 17021. This generic standard states: "*Methods to collect information shall include, but are not limited to: a) interviews; b) observation of processes and activities and c) review of documentation and records*." Thus, an auditor could confirm that there has been communication about the EnMS to employees in the organization by asking people directly, or by observing that there is information about the EnMS on notice boards, as well as by seeing a copy of an email that was sent to all staff. This is sometimes called the evidence triangle: interviews, observations and records.

This focus on documentation tends to reinforce the belief that 50001 is all about paperwork. Actually, it is about continual improvement and the paperwork should be considered a means to an end. At Peel Land & Property Group, where I implemented ISO 50001, one objective we wanted to achieve was to get the business managers to sit down with the energy Champions at least twice a year to discuss priorities and investments. We did through our procedure for developing action plans: "*progress against action plans will be reviewed by the energy Champion and their business unit managers at least once every 6 months and the action plans updated in the light of any changes.*" The paperwork enabled us to create an internal *requirement*. That is the real value of this process.

21.5 How to achieve certification

Certification to the ISO 50001 standard is usually compulsory where the system is linked to compliance or tax benefits. Even if an organization is not obliged to achieve certification, the process can bring many benefits, not least an objective assessment of your system and advice on how it can be improved.

It is entirely possible to use the ISO 50001 standard without making any public declaration. Alternatively, an organization may choose to self-certify, known as attesting, by stating that they run an "*ISO 50001 compliant energy management system*".

If you want to get formal independent certification for the system, then you need to appoint an accredited assessor. By accredited, I mean that they and their organization are formally able to issue 50001 certificates of conformity, which usually means that they are registered with a national standards body such as UKAS in the UK. Note that the assessor works for you, but their assessment must confirm to the assessment standards (ISO/IEC 17021:2015 and ISO 50003:2014).

Although they are notionally following the same rules, not all assessors are the same. I have come across assessors who are over-exacting in their interpretation of the minutiae of standard and incapable of recognizing that the applicability of these requirements will be different between organizations. This rigidity runs counter to the guidance in ISO 50003:

> "*The implementation of an energy management system specified by this International Standard is intended to result in improved energy performance.*" and "*The organization is given flexibility in how it implements the EnMS, e.g. the rate, extent and timescale of the continual improvement process are determined by the organization. The organization can take into account economic and other considerations when determining the rate, extent and timescale of the continual improvement process. The concept of scope and boundaries allows flexibility to the organization to define what is included within the EnMS.*"

I would strongly urge that you engage your potential assessors early on in the implementation of your ISO 50001 programme. I would recommend that you interview the actual assessor who will be doing the work so that you can get an understanding of their approach to certification and the degree to which they are open and receptive when it comes to interpreting the standard.

Good communication and rapport between assessor and client are important. There may be times during the certification when the assessor may be uncertain about whether a particular requirement is being met. In these circumstances, it is important that an open dialogue can take place to establish if this is an

Real World: *Shopping around*

Although I advise organizations seeking certification to select their auditors carefully, the suggestion is not to shop around for an *"easy"* audit.

Choosing an auditor because they are tractable, acquiescent or pliable would be a grave mistake and fundamentally misses the purpose of certification audits.

The audit is there to make the EnMS better, to objectively evaluate whether energy improvements and EnMS improvements are genuinely being achieved. The audit is the activity that gives the EnMS *"teeth"* to drive behaviours and decision-making. By weakening the audit, these benefits are diminished and the drivers for improvement reduced.

Unfortunately, it does seem possible to gain *"token"* certification. I have worked on energy efficiency in at least one European site in a large multinational group certified to ISO 50001 where there was absolutely no evidence or knowledge of the system at the facility level. This business seems to have only been interested in the tax breaks, rather than energy savings, so they must have contracted a certification body which was not going to examine the operation too closely.

This is a pity because this made my subsequent energy efficiency work much harder that it needed to be.

21.2 **The ISO 50001 certification audit at the Lowry Outlet Mall, Manchester, part of Peel Land & Property Group** Chris Dunham, on the left, the facilities manager and energy Champion for the Lowry Outlet Mall, is showing the assessor Tim Watts a lighting upgrade that is in the process of being installed in the service corridors of this popular retail centre. *Source: Niall Enright*

underlying failure to do something, a difference in the terminology employed, a lack of evidence to support something that is happening, or there is a difference in the interpretation/applicability of the standards. A confrontational working relationship with the assessor will not help resolve any misunderstandings. On the other hand, a good and well-informed assessor will provide an audit report with helpful observations about improving your system.

The duration of the certification visit is determined by a formula set out in ISO 50003, which takes into account the organization's complexity and the number of people involved. Where there are multiple similar sites, e.g. a supermarket chain, auditing via sampling is permitted. There is usually an initial documents review, which establishes the conformity of the documentation of the system, and related records (such as EnPIs). This document review is then followed by on-site audits where the assessor will meet with a range of the personnel involved in the system and observe if there is evidence that the system is functioning. At Peel Land & Property Group the initial document review took one day and the on-site audit was three days.

To recommend certification, the assessor needs to be satisfied that there are no major nonconformities in the EnMS. A major nonconformity is a "*nonconformity that affects the capability of the management system to achieve the intended results*" which could, for example, be:

> "*audit evidence that energy performance improvement was not achieved;*" or "*a significant doubt that effective process control is in place;*" or "*a number of minor nonconformities associated with the same requirements or issue could demonstrate a systemic failure and thus constitute a major nonconformity.*"

21.3 **The ISO 50001 certification audit at MediaCityUK, Manchester** Phil Harris, energy engineer from the facilities management company Engie, left, and Derek Elliott, technical services manager and energy Champion for MediaCityUK, centre, share a joke with Tim Watts, during the ISO 50001 certification audit for Peel Land & Property Group. In a certification audit, the assessor will expect to meet with staff and observe improvement that have been made. The sources of evidence used by the assessor will be a mix of interviews, observations and the inspection of records. *Source: Niall Enright*

Note that the auditor is checking *both* the ability of the system to be continually improved through internal audits and controls, etc. ("*improvement in effectiveness of the management system*"), *and* evidence that energy efficiency itself is improving ("*improvement in measurable results related to energy efficiency, energy use, or energy consumption compared to the energy baseline*").

The assessor does not award the certification. They usually have to make a recommendation to a technical or supervisory committee within their own organization, which will review the certification report and recommendations and may, possibly, seek further details before making a determination. This process will take a few weeks, and if any minor nonconformities were found in the audit, the host company would normally be expected to provide a corrective plan for these prior to a certificate being issued.

In Europe and the US, certification lasts for three years so long as annual third-party surveillance audits are carried out. The final surveillance audit is a recertification visit and, although this goes through the formal committee approval process, it is not usually as onerous as an initial certification. However, we should note that "*confirmation of continual energy performance improvement is required for granting the recertification*". In other words, we have to demonstrate that the EnMS is driving improvement. ⇒ page 731.

I hope that the certification process is seen as a positive contribution to gaining a ***better system*** *rather than just a tick in the box or a piece of paper.*

In My Experience: ***Successful certification***

Tim Watts is the Operations Manager - Assurance and Assessment Team Leader for Lucideon, a leading global expert in verification and certification services.

Here, Tim shares his advice on how to get the most from ISO 50001 and successfully certify your EnMS.

I have been working in certification in one form or another for over 25 years - I initially started out focusing on quality, environmental and health and Safety management systems. Since 2014, when Lucideon was part of the accreditation body UKAS' pilot scheme I have been involved with ISO 50001.

Since then, I have certified 21 organizations which involved issuing 52 ISO 50001 certificates (since some organizations can have more than one legal entity). All in all, my ISO 50001 certification activities have covered more than 450 sites, not just in the UK but worldwide.

I must say that this is a fascinating and satisfying job. I firmly believe in the value that ISO 50001 can bring, and I think an assessor can contribute very positively to the continual improvement process.

I should also mention that we, the certification bodies, are ourselves regularly audited by UKAS to ensure that we are maintaining the very highest standards and consistency in our certification. There is a standard, ISO 50003, that we need to follow when carrying out our certification and we are monitored extremely closely against that standard. So we do know what it feels like to have to follow a process and then be at the receiving end of a certification audit to confirm that you have done so correctly.

For those who are close to their first certification visit, I would offer the following advice: relax! I know it can be stressful to have a stranger "*mark your homework*" in such a visible way, but please bear in mind that it is rare for an organization to fail a certification audit. The key is preparation, and if you follow the suggestions in these pages, you should be ok.

Just about every organization I have audited has had some nonconformities to address following the certification visit. This is no surprise given the extensive requirements of the standard. On average I would say I pick up 4-6 minor nonconformities. These won't prevent certification, but they will be noted in the audit report and evidence that these are being addressed will be required on subsequent visits. So one way to take the pressure off is to set the expectation ahead of time that there will definitely be some areas for improvement that come from the audit.

Please bear in mind that I am not there to dig and dig and dig to find something wrong. I am there to check each part of the standard and when I have obtained sufficient evidence to satisfy myself that the requirement is met, I will move on to the next area. Also, every audit should have an audit plan issued in advance which will set out a schedule of when there will be a discussion of specific aspects of the EnMS - once again the intention is not to spring surprises on you.

Exploration: *13 Top management requirements in ISO 50001:2018*

"Top management shall demonstrate leadership and commitment with respect to continual improvement of its energy performance and the effectiveness of the EnMS, by:

a) ensuring that the EnMS scope and boundaries are established;

b) ensuring that the energy policy, objectives and energy targets are established and are compatible with the strategic direction of the organization;

c) ensuring the integration of the EnMS requirements into the organization's business processes;

d) ensuring that action plans are approved and implemented;

e) ensuring that the resources needed for the EnMS are available;

f) communicating the importance of effective energy management and of conforming to the EnMS requirements;

g) ensuring that the EnMS achieves its intended outcome(s);

h) promoting continual improvement of energy performance and the EnMS;

i) ensuring the formation of an energy management team;

j) directing and supporting persons to contribute to the effectiveness of the EnMS and to energy performance improvement;

k) supporting other relevant management roles to demonstrate their leadership as it applies to their areas of responsibility;

l) ensuring that the EnPI(s) appropriately represent(s) energy performance;

m) ensuring that processes are established and implemented to identify and address changes affecting the EnMS and energy performance within the scope and boundary of the EnMS. "

However, if a site is not meeting the requirements of the standard, then they will fail. I remember turning up at one site where the client did not have a copy of the standard and they thought I was going to bring the EnMS with me! At another site the documentation of the EnMS was the same as one I had already assessed for another client - the consultant had not even bothered to change the document referencing system which still had the initials of the previous company.

These are pretty extreme examples where it is clear that the EnMS is not embedded in the organization. In fact, one of the most common problems that we encounter is when an EnMS has been driven almost exclusively by one person. For ISO 50001 to be successful it needs to be a team effort - many people need to be involved in the necessary steps of planning, checking, doing and reviewing that lie at the heart of the process.

ISO 50001 is a "management system". The clue is in the title. What I need to see evidence of in the audit is that there is a structured approach to energy management which addresses issues as they arise and makes people responsible for achieving improvements. There also needs to be clear evidence of the involvement of senior management. It cannot be a solo effort.

Here, I would share an observation that there are broadly two different approaches that organizations take to implementing ISO 50001, depending on who has been responsible for the system: the energy manager or the management systems manager. In my experience, energy managers tend to be very strong on the technical issues but weaker on the people and management aspects, and vice versa. In reality, you need to involve both people and skill sets as appropriate.

Note that I said "*as appropriate*". ISO 50001, like all the management systems standards, is quite adaptable and none contain limits, durations or frequencies. For a small organization, it may be perfectly reasonable for an energy team to meet every three months to discuss progress; for a large, energy-intense company more frequent attention would be expected. It is the role of the assessor to interpret what the client does in a common sense way and evaluate it against the standard.

This is where I could be critical of some of my fellow assessors. Too often, especially if they are inexperienced, they have a tendency to be single-minded and expect to see what they have seen before. In reality, ISO 50001 is applicable to all organizations, some of which may have already "done" energy management and some of which are starting out. Some organizations will be large and complex and others very straightforward. Assessors must be able to recognize these differences and engage constructively with the client as to why they have taken a particular approach and understand if the intentions of the standard are being reflected in that particular case.

While the most common reason for an audit failure is an inability to demonstrate that the EnMS is embedded in the organization; the two specific aspects that lead to the most minor nonconformities in audits are verification and a failure to respond to positive deviations in energy use.

The verification nonconformity comes up when reviewing an organization's action plans, which are a requirement of the standard. Verification seems to stump many people, possibly because it is relatively new to management systems, having first made its appearance in ISO 50001. In practice, every action listed on an action plan needs to have a what, who, when and "how verified". There seems to be lots of confusion about the verification. If you have replaced some lights in

21.4 **An ISO 50001 certified system logo**
There is no official ISO 50001 logo but most certification bodies will provide you with a logo that you can use to communicate the fact that your organization has been certified. The example above is from Lucideon. The use of a logo is in addition to a formal company certificate. Some certification bodies also offer *"presentation certificates"* which acknowledge that specific locations or brands fall in the scope of the certified system. These presentation certificates can be useful for display in a site's reception area, notice boards or meeting rooms.
Source: © Lucideon, reproduced here with permission

21.5 **Top 10 tips (opposite)**
These top 10 tips formed part of an article on the implementation of ISO 50001 at Peel Land & Property Group.
Source: the Environmentalist, April 2016, edited by Paul Suff, reproduced here with kind permission

a washroom, then verification could simply be *"visual inspection that the work is complete"*. It does not have to be complicated. On the other hand, if a major item of equipment has been installed the verification of the impact might require some energy monitoring or commissioning data. Once again it is a question of common sense - just make sure you have a reasonable means of verification for *every* action.

The second common nonconformity (which was the only one I picked up when I audited Niall Enright and his colleagues at Peel Land & Property Group [*see page 733*]) is a lack of evidence that there is always a response to a significant positive variation in energy use. Almost all folks I audit can demonstrate that they respond to a negative variance when they have used more energy than expected, but the standard requires evidence of a response to positive variance too. Of course, this makes sense - if you have a particularly good period of energy use you would want to know why that was and repeat that again.

Finally, I can't emphasize enough how important it is to have a good internal audit process in place, with evidence that the auditor has appropriate training. This is the other major area where nonconformities can arise. One site I visited had only completed one internal audit - which was more of a tick-box sheet than a proper review. Internal audit records are one of the key outputs that I would need to see at a certification visit so, depending on the organization and its starting point, I would say the absolute minimum time required to run an ISO 50001 system before certification is six months.

During my visit to a site, as well as nonconformities, my audit report may contain some observations. These are not failures of the system, but rather opportunities - often based on best practice I have seen elsewhere - about how the system could be further strengthened and additional energy savings realized. It is entirely up to the client whether or not they respond to these. For my part being able to identify opportunities for improvements is one of the most satisfying aspects of being an ISO 50001 auditor.

A professional assessor should be able to put auditees at ease and develop an open communication process. This will facilitate the dialogue which allows the common sense interpretation of action versus standard to work. It also enables the client to get the most from the auditors' experience concerning how the EnMS can be improved and to realize greater value from the investment made in the system. Don't be too defensive if the assessor queries something - I recently had to explain that it was highly improbable that a company would have used just 3 kWh of energy to make £3 million of products, but the client who had entered the data incorrectly was determined to argue that this was the case!

On a practical note, having the documentation available, ensuring it is easy to locate evidence of conformity and providing opportunities to see projects in situ will all make the certification easier. The site tours are particularly important, in terms of seeing evidence that the EnMS is working. Recently I was given a guided tour to explain work being done on freezers and cool rooms in a laboratory when I realized that it was storing slices of frozen human brains! I must admit I was taken aback a little but did appreciate the opportunity to see the energy savings made!

No doubt ISO 50001 will change and evolve further over time. It seems clear that there are a lot of incentives for organizations to take this system up, and I hope that the certification process is seen as a positive contribution to gaining a better system rather than just a tick in the box or a piece of paper.

For those who are close to their first certification visit, I would offer the following advice: ***relax!***

21.6 10 top tips for success

ISO 50001 should not be implemented in a "one size fits all fashion". Every organization will need to tune the system to their needs and priorities. However, there are some basic principles that will help all programmes succeed.

The first recommendation is to write down in a simple sentence why your organization is choosing to implement ISO 50001 and what it hopes to achieve as a result. Top management and the key folks on the energy management team need to agree with this statement, or help create a consensus statement if they don't. One way to get a great result is not to think about implementing ISO 50001, but think about implementing great energy management, and then identifying how what you are doing fits the various sections of the standard.

The biggest demotivator is a lack of involvement, so engaging people in the design of the system is critical. We should ask people how they think great energy management can be achieved and then how that fits the requirement of 50001. Where there are processes that are required but appear to add little value, these should be designed in a way that minimizes effort on them.

Keep the documentation (and process) as simple as possible. Remember to describe what you are going to do, who will do it, when (or how frequently), and how you will verify that the action has a) been done and/or b) achieved the desired results. Give yourself "*wiggle room*" - for example, commit to meeting every three months rather than every month, just in case a meeting is cancelled for any reason. You may want to adopt some of the ISO 50001 vocabulary in your process so that it is clear to the auditor what the connections are.

Audit and certification are the essential parts of the ISO 50001 system and the one that creates most anxiety. If your organization or the participants in 50001 are unfamiliar with audits, then you should take great pains to reassure them and get them on board. Manage expectations concerning nonconformities - these are positive things to find as they indicate that there is a process to strive for continuous improvement that works. Before certification, you should probably have run the ISO 50001 process for at least a year so that there is a full cycle complete, including the management review, and also so that there are plenty of closed nonconformities, corrective or preventive actions.

Remember, too, that all systems change and adapt. If something is not working well in your ISO 50001 system, then there is no reason why it should not be modified, so long as the change still meets the requirement of the standard and is properly documented. Finally, don't forget to celebrate! Make a big fuss of gaining certification and be sure to thank all those involved in the process.

10 TOP TIPS for 50001 Success

1. Be clear why you are doing 50001
2. Senior management commitment is critical to success
3. Beware of the demotivating potential of imposed systems
4. Ideally, put great energy management in place, then do 50001
5. Describe your process, don't just repeat the standard
6. Make the document as concise as possible
7. Remember to document who, what, when and how verified
8. Meet the prospective certifier before appointing them
9. Run the system for a year (minimum) before certification
10. Make sure you complete your internal audits

	ISO 50001 clauses setting out requirements	Principal focus of the requirement: Define	Identify	Document	Communicate	Record	What will be recorded by the EnMS
Plan	*4 Context of the Organisation*	X		X			*The EnMS process*
			X	X			*The interested Parties and their requirements*
		X		X			*Scope and boundaries*
	5 Leadership				X	X	*Commitment to improvement & effectiveness*
	5.2 Energy policy	X		X	X		*Energy policy*
	5.3 Roles, responsibilities and authorities	X		X	X	X	*Assignment of roles and responsibilities*
	6.1 Energy planning		X	X			*Risks and opportunities consistent with policy*
	6.2 Objectives, energy targets and planning	X		X			*Objectives and target by function/level*
	6.2.3 Establish and maintain action plans	X		X		X	*What, how, who, when, how verified*
	6.3 Energy review		X	X		X	*Past and current energy use*
			X			X	*Areas of significant energy use (SEU)*
			X	X		X	*Opportunities for improving performance*
				X		X	*Forecast of future energy use*
	6.4 Energy performance indicators			X		X	*Normalised EnPIs (and variables/models used)*
	6.5 Energy performance indicators			X		X	*Normalised baselines for SEUs (EnBs)*
	6.6 Planning for collection of data	X		X			*An energy data collection plan*
				X			*Calibration of monitoring equipment*
Do	*7.1 Resources*		X			X	*Determination & provision of EnMS resources*
	7.2 and 7.3 Competence and awareness	X	X	X		X	*Competency requirement and training plans*
	7.4 Communication				X	X	*What, when, whom, how, verification, reliable*
	7.5 Documented Information			X		X	*All EnMS process and operation documents*
	7.5.2 and 7.5.3 Creation & Control of documents			X			*Procedure to approve and review documents*
	8.1 Operational control	X	X		X	X	*Operation and maintenance criteria/activities*
	8.2 Design	X		X		X	*Impact and result of design activity*
	8.3 Procurement of energy services, products, equipment and energy	X		X			*Establish energy equipment purchasing processes, criteria & specifications*
					X	X	*Communications with suppliers and records of criteria being applied*
Check	*9.1.1 Monitoring, measurement and analysis*					X	*Effectiveness of plans; EnPIs; SEU operations; and response to variances. EnB comparison.*
	9.1.2 Legal & other requirements					X	*Detail of compliance and records of actions*
	9.2 Internal audit of the EnMS	X		X			*Audit plan, process and schedule*
						X	*Audit results*
Act	*9.3 Management Review*	X		X		X	*Inputs and outputs of the review*
	10.1 Nonconformities					X	*Corrective and preventive actions*
	10.2 Continual improvement		X			X	*Evidence of continual energy improvement*

21.6 **The environmentalist**
Source: This article written by Niall Enright first appeared in the Institute of Environmental Management and Assessment's house journal, "the environmentalist", April 2016, edited by Paul Suff, reproduced here with kind permission

21.7 **Requirements of ISO 50001 (opposite)**
The table opposite lists the main clauses of the ISO system requirements. Against each clause is a verb, describing the kind of action needed: **Define** - establish your own criteria, conditions or scope; **Identify** - using the criteria, conditions or scope you have set, describe the specific items that need to be addressed; **Document** - write down your organization's approach or process for this clause; **Communicate** - there is a specific communications requirement for this clause; **Record** - maintain a record of an event or action or the result of an action (e.g. *"retain documented information"*). Only the main action type is listed although most clauses have more than one requirement.
Source: Inspired by a table in "ISO 50001: Energy management systems – A practical guide for SMEs"[513] by Liam McLaughlin, adapted by Niall Enright for 2018 edition. Available in the companion file pack.

Real World: ***A complex implementation of ISO 50001 at Peel Land & Property Group***

Until recently, few UK organizations contemplated achieving the international energy management standard. That changed in 2014 when the government included ISO 50001 as one of two main routes to complying with the mandatory energy savings opportunity scheme (ESOS).

Reaching 50001 is a demanding process and, to date, only 217 out of 5,938 organizations, or 3.6%, have taken this route to achieve ESOS compliance.

One organization that has is Peel Land & Property Group (PL&P), which owns and manages 1.2 million m^2 of property and 15,000 hectares of land and water worth £2.3 billion. Properties include the MediaCityUK development in Salford, the nearby EventCity exhibition complex and two outlet malls, Salford's Lowry and Gloucester Quays. PL&P is part of The Peel Group, one of the UK's largest investors in real estate, infrastructure and transport, with assets over £5 billion.

Going for the standard

The Peel Group has had an overarching energy and carbon policy since 2009. It includes an emissions reduction target of 3% year on year, adjusted for activity. It is a goal that Peel has achieved every year.

Achieving 50001 was first considered in 2012 when the company was updating the policy, but it took a further year before it decided to go for the standard. A key attraction was the independent assessment of performance. However, there were misgivings among the management team about implementing 50001 in an organization where people were accustomed to working with a high degree of trust, autonomy and professional freedom. David Glover, the operations director and the board member driving Peel's sustainability agenda, says:

> *"One of my biggest concerns about the process was the impact that this would have on the enthusiasm and motivation of our 'energy Champions', who were doing such an outstanding job."*

When the company decided to go for 50001 there was already a strong record on energy management in place. Between 2009 and 2013 it was certified to the Carbon Trust Standard. Every major asset in its portfolio had an energy champion and was subject to a full audit to the CIBSE TM22 assessment and reporting standard. At the same time, sophisticated performance targets were established to account for measures such as weather and occupancy. Performance was monitored using Carbon Desktop from Verco, while an Opportunities Database was created to drive improvement. Between 2010 and 2013 more than £500,000 savings were achieved each year against costs of around £4 million - a reduction of more than 12.5%. The Carbon Trust certification ranked Peel top of 29 UK property companies benchmarked.

So there was a concern that 50001 would undermine a process that was delivering good results for the company and its tenants. Glover says:

> *"What decided it for us was the realization that the standard route to ESOS compliance through audits offered little value as we had already audited all our major assets and quantified our use as a result of the CRC [carbon reduction commitment].So we decided to proceed with 50001, but in a way that would put the energy champions front and centre."*

Working with the champions

With this direction, getting started was easy. As "*management representative*", I worked with the energy champions to examine their processes and devise a model for best practice that involved improvement opportunities and target-setting. These became the core 50001 activities, with some changes in terminology to make it easier for auditors to relate them to the ISO standard. The champions were clear about which aspects of 50001 they believed added value and those they considered burdensome. We emphasized the former and reduced the latter to a minimum.

Formulating the 50001 process in terms that reflect an organization's activities is critical. Although it may appear easier to copy and paste from the standard, there is a risk of failing to document what the organization does, a shortcoming that would be uncovered during the audit.

As well as tailoring the ISO process to the organization, it is also important to keep it simple. Peel's 50001 process applied to a range of facilities: offices, exhibition centres, studios, car parks, three regional airports, outlet malls, environmental assets, energy generation and retail and industrial parks. The description of the entire Peel ISO 50001 process, including all the forms, schedules of businesses and people, and the audit calendar, fits on just 24 pages of A4. A concise process is easier for staff to follow and is simpler to maintain.

However, brevity should not lead to ambiguity. It is important not only to spell out what the process is, but who will do it, by when and how it will be verified and recorded. If this is clear, it will be easy for the certification body to validate a system.

After three months working on the document, Peel's system went live in April 2014. Since so much of the process was based on the existing approach to energy efficiency, the energy champions took it all in their stride.

Learning points

One early learning point from PL&P was the importance of the internal audits. About six months after the launch of 50001, these took centre stage. First, the audits provided confirmation that everyone understood the processes. Critically the internal audits enabled those working on certification to flag up where work was still needed. For example, one relatively minor aspect of the standard involves checking the calibration records for all meters in the system. However, in some older properties it was difficult to find these. An action request as part of the audit/nonconformity process signalled to the certifiers that PL&P had a programme of work in place to fix it so they could not fail the company on this.

Another tip for anyone looking to have a 50001 system externally certified (a requirement for ESOS) is that some auditors adopt a different approach. I would strongly urge anyone seeking certification to interview auditors they are considering appointing. Although the certification bodies are UKAS-accredited and notionally work to the same standard, individual auditors can fall into several camps. We found some took a "literalist" approach to the standard, nitpicking yet lacking in knowledge about energy management. One certifier even suggested that we "dumb down" the targets to a simpler kWh/ m² because they found correlation with heating and cooling degree days too complicated. Others, usually more experienced, took a "key principles" approach and would test the system

21.8 **The properties**
The complex portfolio of properties in the Peel Land & Property Group ISO 50001 scheme includes EventCity in Manchester, shown below, host to many of the UK's largest public and trade exhibitions as well as media, sporting and leisure events.
Source: © EventCity

21.9 **A good reason to celebrate**
David Glover, operations director, Peel Land & Property Group, left, has just been informed by Lucideon auditor Tim Watts, right, that Peel Land & Property Group will be recommended for certification to the ISO 50001 Standard.
Source: Niall Enright

against the broad intentions of the standard, such as continual improvement, measurement, accountability, verification and so forth.

The chosen certifier, Tim Watts from Lucideon, showed lots of experience in auditing a complex organization such as Peel. He also had first-hand knowledge of energy systems and construction.

The audit process, between April and July 2015, was thorough. An initial one-day document review was followed by an intense three-day audit at the Peel offices, with every document, action plan,the organizational structure, EnPIs and internal audit records scrutinized. Watts met most of the energy champions and several business managers. He also inspected actual energy efficiency measures in situ. Every single aspect of the standard, no matter how small, was checked. At the end of the process, Watts identified one minor nonconformity: the teams did not respond as vigorously to green exceptions (when PL&P was performing better that expected) than to red exceptions (when it was performing worse).

> *"In the course of the three-day on-site certification audit, it became clear to me that the energy management system at Peel was a long-established process, which has widespread support from the boardroom to the facilities teams,"* says Watts. *"This was reflected in the very considerable number of completed energy-efficient projects, which I was able to observe during my visit, and the professionalism and dedication of the staff involved."*

Going forward

So how do staff at PL&P involved in the 50001 process feel now the company has its certification? It was certainly hard work to launch the system, and the certification process itself was stressful. But it has improved the company's energy management in two ways: first, the action plans formalized input from business managers into the energy management process; second, the EnMS has enabled the firm to take action in parts of the business that were yet to adopt best practice. In terms of a statement of quality, the fact that 50001 is difficult to achieve added to the rigour of the certification process and makes this a benchmark the company values. It is one badge that Peel wears with pride and is happy to communicate to its customers, tenants and partners.

> *"This recognizes the commitments we have made over a number of years in terms of our energy use and carbon emissions,"* says Glover.

How have the energy champions reacted? They are going from strength to strength. Today, the savings exceed £1.3 million a year, reducing energy by almost 30%. Peel is not resting on its laurels, however.

> *"We are determined to continue to lead the way in the design of new low-carbon facilities as well as review how we operate our existing properties,"* says Glover.

Update: *In July 2018 Peel Land and Property Group were formally recertified to ISO 50001 for a further three years. Amazingly, at the final annual recertification audit, no major or minor nonconformities were noted, nor any observations. This is a credit to the strong leadership support from David Glover, the new Management Representative, Joanne Holden, who approached her role with great energy and enthusiasm and the continuing dedication and professionalism of the energy Champions and Influencers in the diverse portfolio of properties.*

Standards: ***The Superior Energy Performance standard, going beyond ISO 50001***

The US Department of Energy has developed a programme for industrial companies and commercial buildings to achieve recognition for Superior Energy Performance® (SEP™).[742] This programme is set out in the ANSI/MSE 50021 standard (and its normative references) and follows the ISO 50001 standard, with the same titles, section numbers, etc., but adds some energy management system and energy performance improvement requirements beyond ISO 50001. In order to achieve SEP certification, a site needs to meet the requirement of both ISO 50001 and ANSI/ MSE 50021. SEP applies at the facility level. However, an organization can implement SEP at multiple sites to benefit from economies of scale. A central office would implement ISO 50001 at the enterprise-level, while each participating facility would need to achieve SEP. A facility could, however, be a campus or location with multiple buildings.

Certification to the SEP standard requires that a facility demonstrates a specified level of energy performance improvement. A facility needs to calculate a superior energy performance indicator (SEnPI). Although the ANSI 50021 standard established the requirement for an SEnPI, the methodology that must be followed is set out in *"The Measurement and Verification Protocol"*[751] The SEnPI is a ratio of the reporting period energy consumption compared to the baseline period. The energy performance improvement is simply one minus this ratio as a percentage: (1-SEnPI) * 100.

The SEnPI should be normalized to take account of production, weather, or product characteristics, such as moisture, so that the baseline period data and reporting period data are on an equivalent basis (the adjustments can be made to the baseline, reporting period or both). The models for normalization include simple intensity ratios, single or multiple linear regression, and non-linear polynomial models (see Chapter 14 on page 431 for more about these analysis techniques).

Facilities are recognized on the basis of the percentage of energy performance improvement over number of years, compared to a baseline period. Depending on the improvement and number of years in the achievement period, companies can receive bronze, silver, gold or platinum ratings. Companies with long-standing energy management programmes, which may be unable to demonstrate high levels of short-term improvement can achieve gold or platinum levels using the SEP Scorecard to demonstrate innovation in energy management best practices.[750]

Performance Levels Summer 2017 (proposed)		Bronze	Silver	Gold	Platinum
Energy Improvement	**Certification**	1% over 1-3 years	*Sliding scale:* Between 5% in 1-3 years, and 16% in 10 years.		
	Recertification	1% over last 3 years	3% over last 3 years		
SEP Scorecard				Minimum 40 Scorecard Credits *including* 20 points for Energy Management System	Minimum 60 SEP Scorecard credits, *including* 35 points for Energy Management System *and* 10 points for Advanced Practices and Additional Energy Performance

This scorecard is a fascinating list of good practices in energy management under the following headings: data monitoring and measurement; identification, control and procurement of significant energy users; demand management and involvement of supply personnel; the management of energy projects including life cycle costing and preferential financing; the programme sustainability including involvement of top management, training etc.; additional energy performance improvements; and innovation. These are all best practices recommended elsewhere in this Framework.

At present, the SEP certification can be undertaken at the same time as ISO 50001 certification, or organizations that already have 50001 certification can take steps to *"upgrade"* to SEP. DOE collaborates internationally with other governments to promote ISO 50001 best practices through the Clean Energy Ministerial Energy Management Working Group (EMWG). The following countries are represented on the EMWG:[254] Australia, Canada, Chile, China (observer), Denmark (observer) the European Commission, Finland, Germany, India, Indonesia, Japan, Korea, Mexico, Saudi Arabia, South Africa, Sweden, United Arab Emirates and the US.

Real World: ***The business case for ISO 50001 and Superior Energy Performance***

One of the key questions for an organization seeking to implement ISO 50001 is the costs and benefits that they should expect. Given the variety of organizations and the scope and depth of implementation, there is not a simple answer to this question. In the case of Peel Land & Property Group, the internal staff time, systems, certification and consultant costs were under £80,000 for the two years leading to certification, during which an additional £400,000 of energy savings were achieved.

Company	Facility	SEP Rating	Annual Saving US$	% Saving	SEP Costs	Operational Savings (SEP) US$	SEP/ISO payback
Nissan (1st certification)[749]	Smyrna, TN	Silver	$2.4 million	7%	$116,000	$938,000	6 weeks
Nissan (recertification)[749]	Smyrna, TN	Platinum		17%	$71,000	$748,000	5 weeks
Cummins[744]	Rocky Mount, NC	Gold	$ 716,000	12.6%	$248,000	$281,000	11 months
General Dynamics[747]	Scranton, PA	Gold	$956,000	12%	$255,000	$558,000	6 Months
HARBEC (1st certification)[748]	Ontario, NY	Platinum	$52,000	16.5%	$127,000	$52,000	2.4 years

In the US SEP programme case studies listed above, the direct savings attributable to the SEP process were those which were operational in nature, in other words, those improvements which did not require any capital expenditure. These show returns on investment ranging from five weeks to 2.4 years. The latter example is of a company, HARBEC, which put achieving carbon neutrality as a core objective rather than costs savings, so was willing to contemplate longer paybacks.

Schneider Electric is another company which has published its SEP costs.[792] It estimates that for each plant they needed 1.75 person years equivalent for ISO 50001 and SEP implementation (0.57 person equivalents of effort to meet the additional requirements of SEP). However, this effort has reduced rapidly since much of the documentation and system can be reused at subsequent plants. Interestingly, Schneider Electric takes around six months from start to certification audits, which is very quick.

A study of the cost-benefits of the SEP programme by the DOE[703] in 2015 established that the average cost was US$180,000 per industrial facility, based on data from 13 facilities. As in Schneider Electric, the costs were seen to be diminishing significantly over time, with early projects costing around US$300,000 per facility in 2011 which then falls to around $100,000 by 2014. The majority of the costs were related to the staff time required to implement the EnMS (see right).

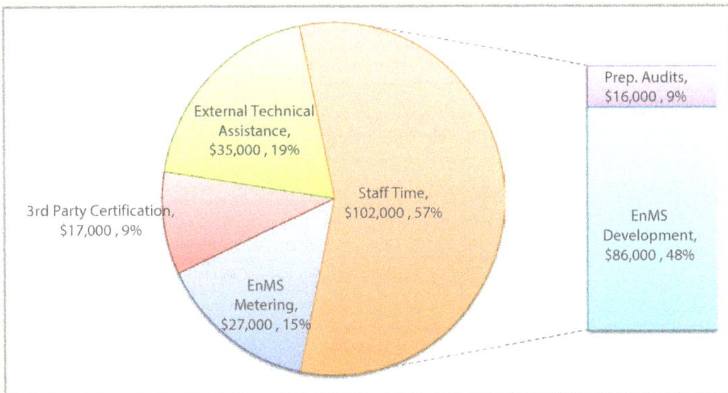

In terms of the benefits arising from this investment, the average additional annual energy cost reduction based on data from 10 facilities was 10% attributable to SEP over and above the Business As Usual rate of savings of 2% (bottom right). It should be noted that there is a distinct ramp up to this level of saving - the first quarter post-initiation of the SEP programme shows virtually no additional savings, and it can be seen to take three quarters for the improvements to begin to be felt.

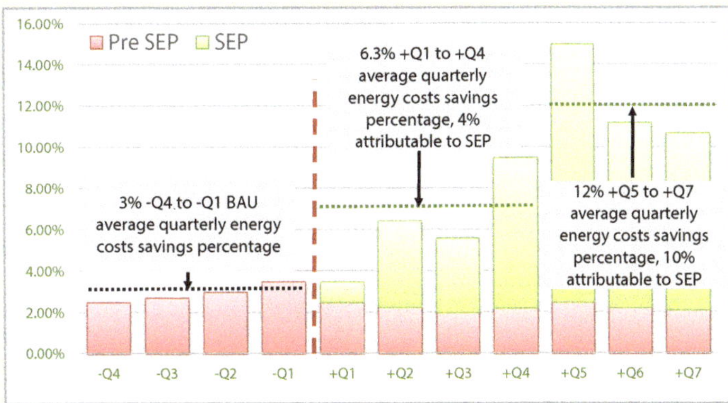

This data is drawn from a wide mix of sites with energy bills ranging from US$0.3 million to US$21.7 million. Clearly, the costs for the smaller sites are likely to be greater in relation to the benefits, and the study reports that under US$2 million energy spend, facilities saw payback ranging from two to eight years, while all the facilities over US$2 million energy spend had a payback of under two years, with the average at 1.5 years.

Real World: *Buying the standard*

Standards can be expensive if your organization is not a member of the local standards institution, so it can pay to shop around the various bodies to see which one offers the cheapest version. The English text for the standard will always be the same. At the time of writing (August 2018), not all stores offered the 2018 edition. The cost of the ISO 50001 standard, in descending order, was:

- BSI charges £160 to non-members, [US$205], and £80 to members [US$102]
- ANSI charges US$162 to non-members and US$129 to members
- ISO charges CHF138 [US$140], for the paper or PDF version
- CSA charges CAD$183 [US$140]
- *[2011 ed.]* The New Zealand Standards store charges NZ$181 [US$120]
- *[2011 ed.]* The NSAI in Ireland charges €64 [US$74]
- *[2011 ed.]* The Bureau of Indian Standards charges 430RS within India [US$6] or 4300RS outside India [US$62]
- *[2011 ed.]* The Malaysian Standards online store offers ISO 50001 (in English) for 40 Malaysian Ringgit [US$9].

SAI Global InfoStore lists the 50001:2011 standards (usually hardcover) for 20 different National Standards institutes. When purchasing the standard in electronic form (as a PDF file), be aware that you are normally purchasing a single licence for the document for individual use. The PDF file often has the name of the purchaser on the margin on every page, in order to discourage copies being circulated.

Further Reading:

Clearly, the ISO 50001 standard itself needs to be read prior to implementing the system. Copies of the standard are available from national quality institutions like ANSI or the British Standards Institution, or from ISO direct. These are expensive documents so shopping around is recommended (see left).

ISO provide a general introduction to the 50001 standard: *Win the Energy Challenge with ISO 50001.*[413] The Hong Kong government has published[377] the *Guidebook for ISO 50001* for small and medium sized companies, which includes a useful self-evaluation checklist in the appendices. A similar version is available from TUV-Nord[722]

ISO has also published an excellent guide for smaller organizations: *Energy management systems - A practical guide for SMEs*[513] by Liam McLaughlin (a member of ISO/TC 242), which is available in both English and Spanish. It takes the form of a checklist of 76 of the requirements with ample explanations. If you download the PDF version from the ISO website (around US$40) there is a handy electronic version of the checklist.

A slightly different angle is taken in another great resource, *Implementing and Improving an Energy Management System*[803] by Graham Wooding and Kit Oung (another ISO /TC 242 member who also reviewed the book above). This takes a much more detailed look at the standard in general, the certification and the technical aspects. It focuses on the nuances and interpretation of the standard.

Finally, *Effective Implementation of an ISO 50001 Energy Management System (EnMS)*[379] by Marvin T. Howell describes a straightforward implementation of ISO 50001 in a fictional business, QVS. Not as detailed as the title above, but for uncomplicated organizations, this is a good *"how to"* guide.

Resources:

There are various free checklists to help assess how ready your organization is to implement ISO 50001 from consultancies such as BSI[94] and LRQA[477] (an online tool). The Advanced Manufacturing Office of the US Department of Energy has a combined online checklist[739] for both ISO 50001 and SEP.

Peel Land & Property Group kindly agreed to make its full ISO 50001 process documentation, which I wrote, available in the companion file pack (this is the ISO 50001 system in the earlier case study). The documents have had details of named individuals removed, but are otherwise complete. This resource is provided as an *example* of a very concise set of documents. If used as a starting point for your own system, please ensure that it is modified to reflect your organization and priorities and that your certifier is happy with these.

Some of the forms in the Peel System have been adapted from those in the fabulous US Department of Energy eGuide to ISO 50001,[741] which has 110 forms and examples of records of ISO 50001 activities. It is a very comprehensive system, although I would recommend only using those documents for processes which are important to your organization - if you were to use every form provided, the paperwork could be significant.

22 Disclosure

Photo: © Polarpx, Fotolia.com

Right at the outset of our exploration of resource efficiency, disclosure was described as one of four principal sources of value for an organization. The three other sources of value - strategic value, financial value and the licence to operate - are all strongly influenced by the statements our organization makes and the perceptions that others have of us, which is why disclosure was placed at the centre of the value pyramid.

This chapter complements the earlier broad exploration of disclosure with some practical techniques. First, we will consider the types of disclosure and then a process to follow to ensure that our declarations are beyond reproach. I will then follow up with a discussion on mandatory disclosure and how this can drive more radical change in organizations.

The subject of disclosure is vast and complex. Given the limitations of space available here, the intention is not to provide an exhaustive description of all the issues, but rather to give the reader an overview from which to access the links and resources provided for further detail.

While disclosure is a source of great value to our organization and can move us to make improvements that would otherwise be difficult to justify, we need to be aware of its potential for damage. With several scandals in this area, we need to ensure that our approach exhibits the highest levels of integrity.

22.1 **A classification of disclosure**
There are many types of disclosure. This classification starts by determining whether the disclosure is made by the organization or by a third party. Then the focus of the disclosure is considered (for product, read Product or Service). Some labels consider multiple factors while others may look at single issues. The International Standards Organization (ISO) categorizes disclosure as Type I - Environmental Labelling, Type II - Self-Declarations and Type III - Certified Environmental Declarations, for which there are different standards. Disclosure can also be mandatory.
Source: Niall Enright

By	Focus	Classification	Example	Information	Verification
First Party	Product - Claim	ISO Type II	Recyclable	Qualitative/Quantitative	Optional
	Product - Cause	ISO Type II	US$x to conservation	Qualitative/Quantitative	Optional
	Organization - Claim	ISO Type II	ISO50001 compliant	Qualitative/Quantitative	Optional
	Organization - Cause	ISO Type II	Supports WWF	Qualitative/Quantitative	Optional
Third Party	Information Disclosure	Mandatory	EPC rating of C	Quantitative	Regulator
	Hazard Disclosure	Mandatory	Do not dispose of in household waste	Qualitative/Quantitative	Regulator
	Product - Label (multiple)	ISO Type I	Good Environmental Choice Australia	Qualitative (pass or fail)	Label Producer
	Product - Label (single)	ISO Type I	Energy Star label	Qualitative (pass or fail)	Label Producer
	Product - Label (multiple)	ISO Type I	BREEAM Outstanding or LEED Platinum	Quantitative	Label Producer
	Organization	ISO Type I	B Corporation	Quantitative	Label Producer
	Product	ISO Type III	Environmental Product Declaration	Quantitative	Certification

22.1 Voluntary disclosure

Many organizations want to publicize and celebrate their good decisions around resource efficiency either by adopting a recognized label or making a declaration. Here, we will examine how this can be approached.

- Be honest and truthful
- Detail the specific part of the product or process it is referring to
- Use language which the average member of the public can understand
- Explain the significance of the benefit
- Be able to be substantiate the claim.

22.2 **Basic principles of disclosure**
Most jurisdictions have guidance to organizations and advertisers on making environmental claims. They all share the same basic principles.
Source: Green Marketing and Australian Consumer Law.[46]

22.3 **Checklist for voluntary declaration (opposite)**
Source: Niall Enright, available in the companion file pack.

The illustration opposite describes a process that any sustainability claim by an organization should go through. The three tasks illustrated in green are taken from the UK government's excellent *Green Claims Guide.*[199] I have expanded these tasks to include definition, test and monitoring stages.

We start the process by first being absolutely clear why we are making a claim, ensuring that this does not contradict other communications by our organization and that it aligns with our brand and reputation strategy. We need to be precise about the item we are making the claim about - it could be an ingredient we are using, a product or service we provide or all or part of our organization. Also, we can only take credit for something we have done or a decision we have taken. Finally, we need to be sure that the purpose of the claim is defensible - i.e. the intention of the claim, however genuine, is not to deflect criticism for poor performance elsewhere.

Having identified a possible declaration, we need to ensure that the claim itself is valid. It is important that we understand the full impact of the item in question - for example, it is not acceptable to claim that a fridge has been manufactured in an energy-efficient way when its energy performance in use is poor. Neither is it acceptable to make spurious claims about irrelevant issues - e.g. stating that a paint is lead-free when all lead use in paint has been banned so it is not an issue. Here, the concept of additionality (see page 374) is important; we cannot, for example, make a claim about our emissions decreasing if this arises purely because of the obligation by suppliers to invest in renewable energy and not as a result of our own actions. Nor should we rely on stating a fact alone - many coffee chains claim that their cups are recyclable when in reality the facilities for recycling are non-existent. It is self-evident that the positive action we are claiming must go beyond that which is legally required; if all we are doing is following the law then we have no basis for a claim.

Where we are making a comparison with another organization or product, that comparison must be valid and fair. For example, we should not compare the energy use in our hotel chain with that in other countries where the climate may require much more cooling, however similar the organizations are. Unfortunately, there are far too many examples of these false comparisons - but few more shocking that VW's cheat devices designed to manipulate the emissions tests of their diesel motors, which allowed the company to make false *"clean diesel"* claims of superior performance compared to competitors.[30]

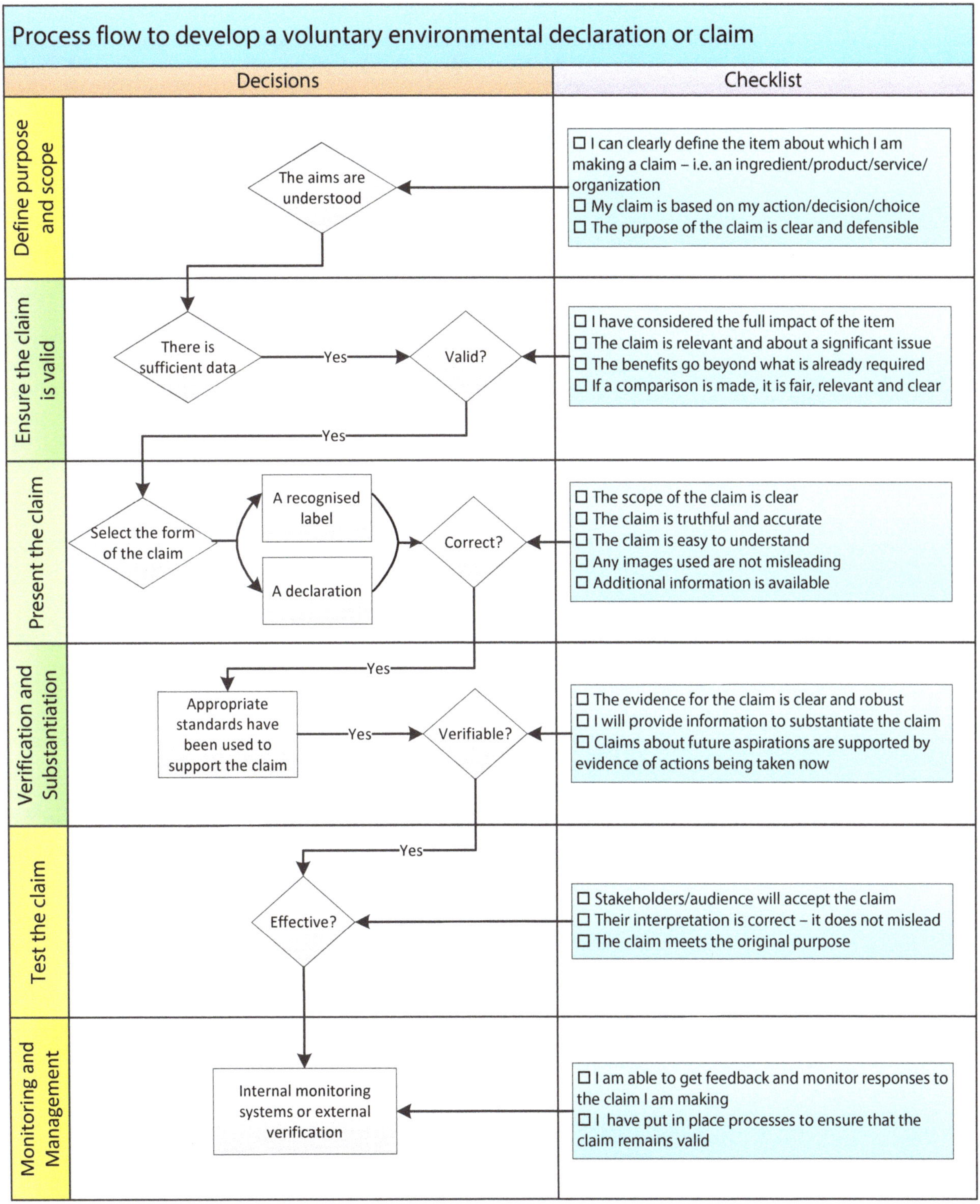
Process flow to develop a voluntary environmental declaration or claim
Decisions
Checklist
Define purpose and scope
The aims are understood
☐ I can clearly define the item about which I am making a claim – i.e. an ingredient/product/service/organization
☐ My claim is based on my action/decision/choice
☐ The purpose of the claim is clear and defensible
Ensure the claim is valid
There is sufficient data
Yes
Valid?
Yes
☐ I have considered the full impact of the item
☐ The claim is relevant and about a significant issue
☐ The benefits go beyond what is already required
☐ If a comparison is made, it is fair, relevant and clear
Present the claim
Select the form of the claim
A recognised label
A declaration
Correct?
☐ The scope of the claim is clear
☐ The claim is truthful and accurate
☐ The claim is easy to understand
☐ Any images used are not misleading
☐ Additional information is available
Verification and Substantiation
Yes
Appropriate standards have been used to support the claim
Yes
Verifiable?
☐ The evidence for the claim is clear and robust
☐ I will provide information to substantiate the claim
☐ Claims about future aspirations are supported by evidence of actions being taken now
Test the claim
Yes
Effective?
☐ Stakeholders/audience will accept the claim
☐ Their interpretation is correct – it does not mislead
☐ The claim meets the original purpose
Monitoring and Management
Internal monitoring systems or external verification
☐ I am able to get feedback and monitor responses to the claim I am making
☐ I have put in place processes to ensure that the claim remains valid

A claim could be **factually correct** but still be **misleading.**

If we have a valid claim, then we need to determine how we are going to communicate it. We could, for example, apply to use a recognized label that denotes the characteristic we want to communicate. The website Ecolabel Index,[237] for example, lists 465 eco-labels in 199 countries and 25 industry sectors. The site provides lots of details of each label, such as the countries in which it is found and the product categories covered. While the proliferation of green labels is a positive indication that organizations want to differentiate their goods or services on the grounds of environmental performance, the selection of a label needs careful consideration. Some of these have very low standards and little verification. There is also a problem of the sheer number of labels and their similarity leading to confusion - there are 42 labels starting with the words *"Green"* and 21 labels starting with *"Eco"* in the Ecolabel Index.

An alternative to using a label is to make a declaration. In both cases, the test to apply is whether the claim is correct in the form and context in which it is presented. We must ensure that the scope of the claim is correctly described; the claim is truthful and accurate and can be easily understood; any images used are pertinent and appropriate (polar bear images are out unless your claim specifically relates to habitat preservation or bear conservation!). Particular care needs to be taken with percentages, where a lot of liberties are taken. For example, a claim of *"a 50% increase in recycled content"* may be factually correct if previously the product had 10% recycled content which had increased to 15%, but misleading since the claim could easily be misunderstood as stating the recycled content is 50%. Remember, when examining the correctness of a claim, it is not sufficient to be factually correct. Words like *"safe"* and *"friendly"* are meaningless and potentially misleading.

As well as being valid (there is a real benefit arising from our actions) and correct (it is truthful, as presented), our claim also needs to be verifiable. This is the next stage of the process. Part of this is ensuring that the appropriate standards are applied when arriving at the claim, or that the labelling organization has properly assessed conformity its requirements. Problems can occur when standards are weak and participants *game* the results. An example of this was the energy labelling standard of halogen lamps, which allowed manufacturers a degree of tolerance in the measured light output, leading to an overstatement of the brightness of the lamps which flattered the energy consumption data.[548]

I have introduced two final further steps in the process. Prior to releasing a claim it makes sense to test the claim, to obtain confirmation that it will achieve the purpose set out in step one and to verify that the recipients of the communication will respond as expected. Finally, as in any communication process, there needs to be a method to monitor the response of our target audience to the messages. We need to regularly revisit the claim to ensure that its purpose is still relevant, that it remains valid, correct, verifiable and effective.

The ISO 14025 standard, and others mentioned opposite, provide much the same guidance as the flowchart, and any significant claim should also be assessed against these standards, as well as the guidance mentioned earlier.

Standards: *Voluntary declarations*

There are a number of standards regarding environmental declarations issued by the International Standards Organization, illustrated below, and the ISO has provided a useful guide[408] to these. The first principle set out in ISO 14020:2001 is:

> *"labels and declarations shall be accurate, verifiable, relevant and not misleading",*

reminding us that factual accuracy alone is not enough when making a claim of any kind. As well as the ISO standards, many jurisdictions publish their own legal requirements in relation to marketing, as well as national guidance on specific claims. As a result, claims could have to comply with many regulations:

1. Item-specific guidance applicable in the locality (e.g. PAS 2060 "*Specification for demonstration of carbon neutrality*" in the UK);[96] and
2. The ISO standards below (or other requirements of label providers); and
3. Market-specific regulations e.g. regarding advertising to consumers.

22.4 **ISO standards**
There are three ISO Standards for environmental labels and declarations, depending on the type of declaration. ISO 14020 sets out general principles to be followed in all cases.
Source: Niall Enright

Standard/scope	Title	Summary
ISO 14020:2001 All	*Environmental labels and declarations — general principles*	*Provides* 9 core principles*: 1 - integrity (see above); 2 - label must not be a barrier to international trade; 3 - it must be based on accurate and reproducible science; 4 - information on the methodology available on request; 5 - must consider all aspects of the life cycle; 6 - declaration must not inhibit environmental innovation; 7- admin demands shall be reasonable (to allow all organizations to use the label); 8 - the label shall be developed through open consultation and participation; 9 - purchasers of products can ask for information from the party using or running the label.*
ISO 14021:2016 Type II	*Environmental labels and declarations — self-declared environmental claims (Type II environmental labelling)*	*Sets out* criteria for 16 standard claims*: 1 - Compostable; 2 - Degradable; 3 - Designed for disassembly; 4 - Extended life product; 5 - Recovered energy; 6 - Recyclable; 7 - Recycled content; 8 - Reduced energy consumption; 9 - Reduced resource use; 10 - Reduced water consumption; 11 - Reusable and refillable; 12 - Waste reduction; 13 - Renewable material; 14 - Renewable energy; 15 - Sustainable; 16 - Greenhouse gas emissions, in some cases with an accompanying logos (e.g. the "Mobius loop" for recyclable products). This standard must be used* in conjunction with the general principles set out in ISO 14020:2001*. Vague and non-specific terms are forbidden, e.g. ""environmentally safe", "environmentally friendly", "earth friendly", "non-polluting", "green", "nature's friend" and "ozone friendly" etc, nor should there be claims of "sustainability", which the standard says is currently too complex to measure. The standard then sets out 18 specific "dos" and "don'ts" which expand on the ISO14020 principles. There is a discussion about the evaluation of comparative claims and assessment methods.*
ISO 14024:1999 Type I	*Environmental labels and declarations — Type I environmental labelling — principles and procedures*	*This standard sets out the criteria for developing a labelling system - i.e. it is the standard that applies to the providers of the label. This standard must be used* in conjunction with the general principles set out in ISO 14020:2001*. It provides information on the selection of product categories and the related environmental criteria, and their publication. Then it considers general principles of awarding a label (e.g. the costs associated shall be kept as low as possible).*
ISO 14025:2006 Type III	*ISO 14025:2006, Environmental labels and declarations – Type III declarations – principles and procedures*	*Covers the criteria for issuing quantified environmental data on products based on life cycle data (called Environmental Product Declarations). This standard must be used* in conjunction with the general principles set out in ISO 14020:2001*. As would be expected, these forms of declarations are more complex since they cover multiple factors, and this is reflected in the standard. Where a product contains several components the declarations for these "modules" can be incorporated into the overall declaration. Since the standard itself cannot possibly hope to define the criteria for evaluating all product types it sets up the notion of* Product Category Rules *(PCRs) which are used for Type III declarations. So this is very much about the "rules about the rules". The standard sets out the ISO life cycle assessment standards that must be used in developing the PCRs and how a consistent approach can be taken across products. Then it states what the Type III declaration should contain. Finally, it sets out a requirement for independent verification of the declaration by competent verifiers.*

22.2 Reporting GHG emissions

No other aspect of corporate sustainability performance is more carefully scrutinized than emissions. There are a number of well-established methods that must be used when making declarations about energy or emissions.

Insofar as voluntary emissions reporting is concerned, there is really only one show in town - the CDP,[126] which has over 5,000 companies reporting. The CDP was previously called the Carbon Disclosure Project, but renamed since it now runs initiatives for organizations to disclose information on other areas, such as water and forestry. Most major corporations listed on leading stock markets regard *"voluntary"* carbon reporting as essential, as there is an expectation by investors and other stakeholders that they will provide this information. In fact, the CDP's data comes from "*Climate Change Information Requests*" that are sent out in the name of the 822 major institutional investors behind the CDP, representing some US$95 trillion in assets. The responses organizations make to the CDP request are publicly accessible, and it is clear from these that many of them invest a large amount of time and effort responding. All emissions data are expected to follow the Greenhouse Gas Reporting Protocol (see opposite).

In my experience, though, there remains quite a lot of room for organizations to game the system (in particular with the use of green electricity tariffs in situations where additionality cannot be demonstrated, a particular bug-bear of mine - see Exploration, opposite). In some ways, this desire to flatter the submissions is positive - it demonstrates that organizations feel that there is an important reputational effect that arises from their performance.

The CDP deliberately uses reputation as a driver for improvement. The submissions are rated on their response to climate change issues: the lowest score is D for organizations that are just beginning to *"Disclose"*; C for those which exhibit some *"Awareness"* of climate issues; B for those which have *"Management"* processes in place to address the issues; and A for *"Leaders"* who can demonstrate some best-practices. To achieve an A score, at least 70% of the emissions need to have been independently verified and the CDP will also carry out additional reputational checks on the organization. Within these categories a "-" rating is possible, so an organization could be C- or A-.

Any organization can create an account and submit a response (for which there is a fee of US$975 in most of the EU and US, unless the submission is part of a supply chain request). The response is electronic and the route through the questions depends on the preceding answers - some industries such as Oil and Gas or Electricity Utilities may need to answer extra questions. The details of the scoring methodology are publicly available and many firms will use external consultants to assist in preparing their response and help maximize their score.

Standards: ***ISO 14064***

The International Standards Organization has a series of standards covering emissions reporting.

- ISO 14064-1:2006 (UK date 2012) *Greenhouse gases -- Part 1: Specification with guidance at the organization level for quantification and reporting of greenhouse gas emissions and removals.*
- ISO 14064-2:2006 (UK date 2012) *Greenhouse gases -- Part 2: Specification with guidance at the project level for quantification, monitoring and reporting of greenhouse gas emission reductions or removal enhancements.*
- ISO 14064-3:2006 (UK date 2012) *Greenhouse gases -- Part 3: Specification with guidance for the validation and verification of greenhouse gas assertions.*

These standards provide a very broad framework, e.g. on how to quantify emissions ISO 14064-1 states: *"the organization shall quantify and document... the selection of quantification methodology".*

As a result, ISO 14064 is compatible with more detailed approaches like the GHG Protocol and the ISO has signed an agreement with the GHG Protocol authors (the WBCSD and WRI), to ensure this remains the case.

Exploration: *Failed additionality*

In 2015 a significant change was introduced in how Scope 2 emissions from electricity could be calculated in the GHG Protocol. Before the change, all calculations used a location method, where the conversion factor from kWh to CO_{2e} was based on the location of the electricity supply. In the change, a new market method was introduced which allows organizations to use different emissions factors based on their specific supplier emissions data.

Assuming the data meet the quality requirement, e.g. by providing certification such as Guarantees of Origin (GOs) in Europe and Renewable Electricity Certificates (RECs) in the US, it is now possible for organizations to use low or zero emissions factors for their Scope 2 calculations.

The well-intentioned, but misplaced, idea was that by creating a demand for GOs and RECs there will be a stimulus for renewable electricity. However, the current prices for GOs (€0.35 per MWh[563]) fall far short of the extra cost (€40+ per MWh) that renewable electricity needs to compete with fossil-fuel generation.

The claim by organizations purchasing a *"renewable tariff"*, that they have zero emissions is not valid because the principle of additionality (page 374) is not met. These organizations are free riders on the electricity users or taxpayers who actually foot the bill to enable the renewable generation.

This matters because organizations are jumping on a *"100% renewable"* bandwagon to declare dramatic reductions in emissions, when in fact their electricity use may actually have risen. Not only is this misleading but it distracts from the hard work of actually reducing consumption.

Standards: *The Greenhouse Gas Protocol*

The GHG Protocol is the global standard for reporting greenhouse gas emissions. The protocol covers seven gases or groups of gases: carbon dioxide (CO_2), methane (CH_4), nitrous oxide (N_2O), hydrofluorocarbons (HFCs), perfluorocarbons (PFCs), sulphur hexafluoride (SF_6) and nitrogen trifluoride (NF_3).

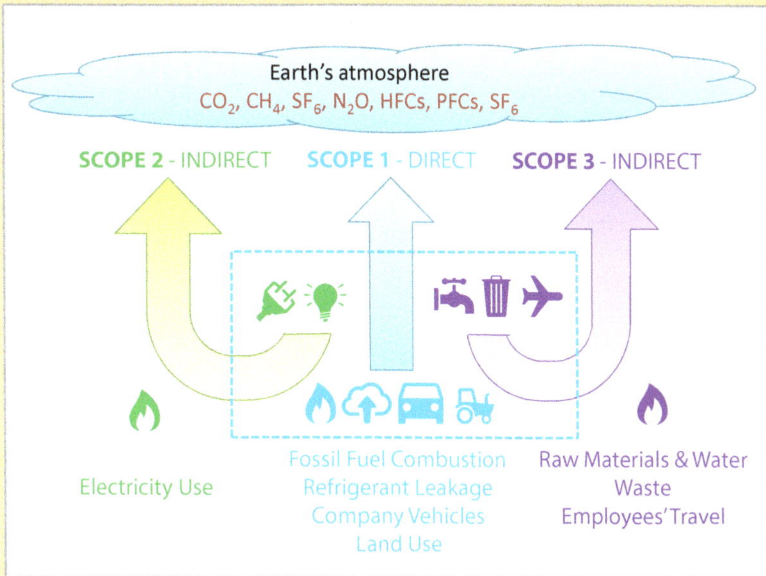

The standard describes three *Scopes* of emissions, illustrated above, where the dashed blue box shows our organization. Scope 1 direct emissions from our facilities, typically due to our combustion of fossil fuels (which releases CO_2), releases of refrigerants or releases due to land use changes. Scope 2 indirect emissions due to electricity consumption on our sites, reflects any remote emissions due to the generation of the electricity. Finally, we have Scope 3 indirect emissions, which are due to the emissions related to other activities such as the raw materials and water we consume, business travel or arising from the waste we create.

The GHG Protocol consists of several standards. *The Corporate Accounting and Reporting Standard*[613] (and the new *Scope 2 Guidance*),[670] which cover Scope 1 and 2 emissions, and the *Corporate Value Chain (Scope 3) Standard*, which as its name implies, covers the Scope 3 emissions. These are the standards used by virtually all organizations reporting their emissions, for example, to the CDP.

In addition, there are standards for product life cycles, determining city-level emissions, for determining progress against specific emissions goals and for determining the emissions impacts of policies or actions.

These are long documents, with the *Corporate Accounting and Reporting Standard* and the new *Scope 2 Guidance* each being over 100 pages. However, they are clearly written and there is plenty of excellent supplementary material on the GHG Protocol website (http://ghgprotocol.org/), so the principles are fairly easy to get to grips with. Unlike the ISO standards, these are also free of charge.

22.3 Mandatory reporting

Recognizing the power of disclosure to drive change, many policymakers have instituted mandatory reporting on a range or energy and resource issues, with a view to enhancing their own knowledge and encouraging action.

*The best responses to mandatory reporting involve **teamwork** across the legal, corporate affairs, and efficiency professionals.*

There is a plethora of mandatory accounting requirements, and it goes without saying that organizations should ensure that they comply with these.

By far the most significant reporting requirements involve emissions. According to the World Resources Institute, at least 40 countries[663] - both developed and developing—currently have mandatory emissions reporting programmes in place, with further programmes at a regional level, such as in California. In the UK, companies quoted on the stock exchange are required to measure and report the current and prior year's Scope 1 and Scope 2 emissions. Many emissions reporting schemes are installation-based, that is to say, that participation is determined by the level of emissions (25,000 tonnes or more of carbon dioxide emissions from a facility annually is the figure that is often used).

In the US, the Securities and Exchange Commission has long had an expectation that corporations would publish information (in their "10-K" filings) about their climate change risks so that investors could make better-informed decisions. There are moves now to greatly strengthen the enforcement of this requirement, as institutions like the G20 Financial Stability Board come to recognize that the direct effects of global warming or rapid decarbonization pose material risks to certain industries and the global economy.

Inadequate disclosures by organizations may also provide the basis for investor or regulator lawsuits. In November 2015 Peabody Energy, a coal company, settled a lawsuit brought by the New York Attorney General that accused the company of misleading investors by not disclosing internal studies that showed substantially material financial impacts from climate change regulations. As part of the settlement, Peabody agreed to make more complete disclosures.

What we need to take from this is that those responding to mandatory reporting and those running resource efficiency programmes need to work closely together. Not only is there an issue of ensuring consistency in the information disclosed by either activity, but there needs to be an acknowledgement by the legal and corporate affairs folks that the only real way to manage the risks related to mandatory reporting is to understand all the impacts of resource use on the organization and to have systems and processes in place that actively address these. This approach requires a much greater degree of sophistication in the organizations' thinking on sustainability and an abandonment of a culture of "*the minimum response compatible with compliance*".

Standards: *Verification and assurance*

Verification is the process of determining that a fact or statement is correct. Assurance, on the other hand, introduces the notion of a *level* of assessment of one or more statements. In the financial world, these processes are governed by the International Standards on Auditing (ISAs) set by the International Federation of Accountants (IFAC) and the International Auditing and Assurance Standards Board (IAASB).

For limited assurance the auditor would typically state, in the negative, that "'*nothing has come to our attention that causes us to believe that the values reported are not correct and have been prepared in accordance with the appropriate legislation and standards*'". For reasonable assurance, on the other hand, the statement would be in the positive "*the statements are correct and have been prepared in accordance with the legislation and standards.*" Limited assurance tends to be a sampling approach, whereas reasonable assurance should examine all the material data.

The AA1000AS (2008) is an internationally accepted, freely available standard that provides the requirements for conducting sustainability assurance developed by AccountAbility.[7] Type 1 Assurance tests the organization against the fundamental Principle of Inclusivity, and two related principles: materiality (that significant issues are considered) and responsiveness (that the needs of stakeholders are responded to). Type 2 Assurance goes on to check actual performance data. High Assurance is similar to reasonable assurance where there is a low risk or error, while Limited Assurance only confirms that no errors were encountered for the data or statements examined using the processes described.

Exploration: *Internalizing externalities*

The central justification for obliging organizations to report on their carbon emissions - and pay carbon taxes - is the notion that this will *disclose and internalize* hitherto *hidden external* costs, with a presumption that, once these impacts and costs are understood, and transparent to stakeholders, the capabilities of business will be mobilized to identify solutions.

In a fascinating examination[475] of this process of *ecological modernization*, Ingmar Lippert examines the main tools used to internalize impacts - environmental management systems (EMS) and carbon accounting - and concludes that the way that these are implemented in organizations is much more to do with managing inconvenient realities than with acting on the real environment.

Lippert observes that the carbon models used in his study organization are abstract and disconnected from the reality of the environment in which the organization operates. This is in part because they have formal constraints imposed by regulators and standards as to what is or is not permitted, but also because carbon is traded in a market which is entirely disconnected from the daily reality of emissions on the ground. The carbon market has *"futures"*; *"offsets"*; *"carbon sequestration by forestry"* whose longevity is uncertain; *"conversion factors"* which may, or may not, reflect reality; emissions *"equivalences"* expressed by the unit of measure CO_{2e}, which may be far from equivalent in the real environment; and the market ignores some GHGs such as hydrofluoroethers, which are not covered by the GHG Protocol or Kyoto mechanisms. As Arthur Mol[527] puts it:

> *...the abstract carbon markets increasingly become subject to and partly dominated by instruments, practices and products of creative investors, banks, traders, brokers, and speculators who see these GHG emission rights and offsets just as financial products, as a means of profit-making.*

These complexities and abstractions are great news for consultants like myself and also for software systems vendors. In fact, a huge industry has been created around managing these data - not surprising given that the carbon market worldwide in 2016 was worth €48 billion.[706]

In this book, too, it has been argued that the internalization of emissions costs and environmental impacts (see page 124 or page 600, for example) are desirable. However, we need to ground our approach on the fundamental understanding that nothing other than substantial absolute emissions reductions by our own organizations will address the risk of climate change.

The problem is that, for some sustainability managers, getting a number on a spreadsheet to what it needs to be has become the goal. Organizations, as Lippert's ethnographic study shows, become focused on showing dedication to the cause, rather than necessarily doing something. This leads to a mentality of quick fixes such as buying *green electricity* or spinning data (see the case study on page 370) that simply *"kicks the can down the road"*. Indeed, it may be the case that a Leader has established a stretching goal for all the right reasons which is undermined by their very own organization's gaming of the system.

Internal sustainability staff and their consulting firms are sometimes, sadly, the instigators of this corporate self-deception. Absent a Hippocratic oath for sustainability practitioners, we rely on the judgements of the individual in the hope that they will do what is right rather than what is easy.

22.4 Gaining trust in disclosures

The value of disclosure depends on whether it will be accepted by our audience. Here are some techniques that can increase trust.

Activist NGOs play a helpful role for committed organizations by "naming and shaming" those who are not addressing environmental issues.

Almost all organizations have fans and critics. How an organization interacts with these stakeholders depends in part on their influence and the culture of the organization itself.

A very important stakeholder group consists of non-governmental organizations (NGOs), which are largely not-for-profit organizations promoting a specific sustainability issue. Some of these NGOs, such as Greenpeace, have a reputation for taking very robust positions on issues and being willing to challenge organizations through direct action. Because of the often imaginative nature of their *stunts,* and also because of their large following, a protest by Greenpeace is regarded as highly undesirable by many organizations - especially if they have consumer-facing brands to protect.

Over the years, though, many NGOs have come to the conclusion that direct action is not necessarily the most effective approach, and so they have become much more actively engaged with corporations. The Environmental Defence Fund (EDF), is just one example of an NGO which regularly partners with companies on specific campaigns. EDF is clear that its relationship remains independent: "*Because we accept no funding from our corporate partners, we're free to set aggressive goals and influence entire industries,*" says Tom Murray, vice-president of EDF's corporate partnerships.[261] This emphasis on independence is important to NGOs as otherwise their individual members or donors may feel that their principles have been compromised.

22.5 **Talking is best**
We can all learn from open dialogue. Issues such as resource efficiency and sustainability are complex and nuanced, so we should endeavour to engage with all stakeholders, whether we agree with their views or not. This dialogue can be public - through the megaphone of the press - or a face-to-face process where all parties are genuinely listening to each other.
Source: Niall Enright, drawn using Pixton. Image available in the companion file pack.

The 2013 KPMG Survey of Corporate Social Responsibility (CSR) Reporting[449] indicated that among the third of companies that include external stakeholder voices in their CSR report, just 6% were from individual NGOs but 26% had a stakeholder panel to gather inputs. Although the numbers remain small, the trend of engaging with NGOs when making disclosures seems to be growing. This is understandable; an independent voice is much more credible when making a statement about an organization's performance than self-declaration.

Care needs to be taken that the NGO is not seen as simply legitimizing an organization. This suspicion will be created where an organization quotes small international NGOs from remote operations which are unlikely to have a big-picture view. Thus, the selection of NGO partners needs to be made with care and the impact of their involvement in our communications properly considered - both for them and for us. Indeed, activist NGOs play a very helpful role for those organizations which are truly committed to addressing environmental impacts, by *"naming and shaming"* those organizations which are gaming the system, or whose efforts are ineffective at best or positively obstructionist at worst.

Any form of disclosure can and should lead to community or stakeholder engagement. When I worked with the multinational consultancy RPS, my colleagues in the Republic of Ireland were involved in a consultation with communities about the development of the Corrib Onshore Gas Pipeline by Shell, which was quite a controversial project. One great lesson I got from this team is that engaging stakeholders works best when the process is treated as a dialogue and not as a PR exercise. The RPS team largely consisted of bright young science graduates who did not have an ounce of "*spin*" in them, but simply understood the facts of the matter and were willing to chat with anyone about the project. Because they were not polished corporate communications experts and clearly were not starting from a perspective of "*persuasion at all costs*", these young consultants were seen as accessible, honest and credible, which ensured an effective dialogue. What the RPS team understood is that good stakeholder engagement is not about "*selling*" at all. I have seen a similar approach being equally successful at Peel Land & Property Group where the Peel Energy team have held consultation meetings with communities about onshore wind developments, so I am concluding that this is a general principle.

Making a disclosure with the involvement of an NGO or other stakeholders is a great way to build trust in what we are communicating. Another way is to have our communications, e.g. our CSR report or emissions data, independently verified. KPMG's 2015 CSR reporting survey[450] showed that nearly two-thirds of the companies analysed used independent third-party assurance of their emissions and CSR data. The vast majority of CSR reports use some elements of the Global Reporting Initiative which will incorporate other standards as required: e.g. emissions should be calculated using the GHG Protocol. The KPMG survey indicates that use of the GRI, while still high, is declining as organizations integrate CSR disclosure into their financial reports - reflecting the fact that the GRI was originally designed for standalone CSR reporting.

Standards: ***CSR reporting***

The Global Reporting Initiative (GRI)[321] is a standard that has been in use to report on corporate sustainability since the late 1990s. The GRI provides comprehensive guidance, starting with some *"universal"* standards to be applied in all cases:

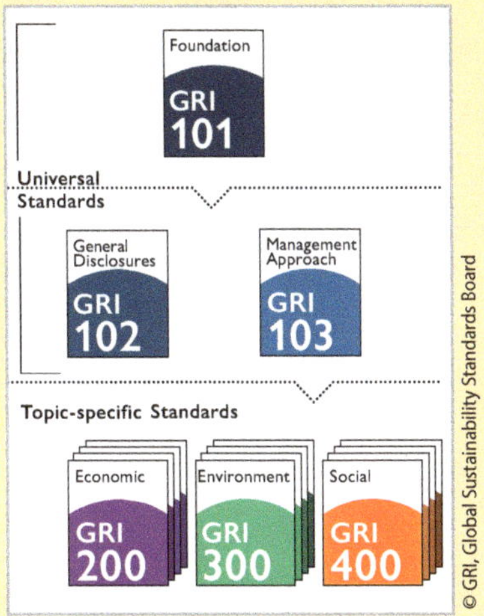

Individual elements from three topic standards can be used as appropriate.

Economic: Economic and market performance; indirect economic impacts; procurement practices; anti-corruption; anti-competitive behaviour

Environmental: Materials; energy; water; biodiversity; emissions; effluent and waste; environmental compliance; supplier compliance

Social: Employment; labour; H&S; training and education; diversity and equal opportunities; non-discrimination; freedom of association and collective bargaining; child labour; forced labour; security practices; indigenous rights; human rights; local communities; suppliers; public policy; customer H&S; marketing and labelling; privacy; socioeconomic compliance.

Further Reading and Resources

There are several organizations which provide information about third-party labels (ISO Type I declarations). The Global Ecolabelling Network is an association of label-providers which conform to the ISO 14024 Type 1 standard. They have a helpful list of members by category of products.[318] The Ecolabel Index[237] provides basic information on 465 labels free of charge, although extended information requires a modest payment.

The International Social and Environmental Accreditation and Labelling Alliance (now just referred to as the ISEAL Alliance)[418] sets out codes of good practice for sustainability standards. Its *Challenge the Label*[419] information provides guidance on making sustainability claims.

For product and location-specific sustainability standards, codes of conduct and audit protocols, the International Trade Centre's Standards Map[414] is an outstanding resource. It focuses mainly on goods traded internationally. For example, if you wanted to see if there is a standard for biomass products originating in Asia for the North America market, you can enter these criteria and find that there are 27 relevant standards, which can be further filtered and compared. Some standards, such as for building (see the next chapter) are not covered, presumably because buildings are not traded!

The ISO has produced a helpful summary of its label standards: *Environmental labels and declarations - How ISO standards help.*[408]

For those interested in the notion of *"greenwash"* - false or unsubstantiated environmental claims and associations - then the Greenwashing Index website (www.greenwashingindex.com) is a reminder of how NOT to make an environmental declaration.

Jacquelyn Ottman's *New Rules of Green Marketing*[579] is a good resource for organizations seeking to position their products or services on the basis of environmental performance. She advises that organizations wishing to avoid greenwash claims should follow five basic principles: 1. Walk your talk; 2. Be transparent; 3. Don't mislead; 4. Enlist the support of third parties; and 5. Support responsible consumption.

Chapter 5 of *The Business Student's Guide to Sustainable Management,*[530] on sustainability reporting by Christian Herzig and Biswaraj Ghosh, has an excellent historical overview of the topic, current approaches and discusses problem areas.

23 Systems and Design

© peshkov @ fotolia.com

This concluding chapter provides some practical techniques and insights to help us change the key systems in our organization that influence energy and resource efficiency. These are huge subjects which we can only begin to scratch the surface of.

Our starting point is design - the choices that take place when we make things. These decisions can be guided by standards, such as building codes, or concepts, such as Design for Environment.

However, our products will ultimately only be as good as our ability to innovate. Radical change requires radical solutions. We shall see how we can draw inspiration from nature and also how a once-secret Russian technique can enable anyone (engineers at least) to become inventive.

Another system that we need to influence is that which govern the flows of information in our organizations. We need to be able to overcome the barrage of messages people receive in order to give prominence to our resource efficiency communications.

Dialogue is an important aspect, too, of how we engage outside our organization, with stakeholders, customers and regulators. When considering how we drive change in our supply chain, we need to decide whether we based our approach on the impact our suppliers have on our resource use or if we should take into consideration our broader ability to influence.

Then we will turn to the finance systems that control our actions, from an international and national level, right down to the discrete investment choices our own organizations make on a day-to-day basis. The challenge here is huge and may seem somewhat abstract to a resource efficiency practitioner, but it is worthwhile taking a step back and contemplating the bigger picture because this also must change in order to make our goals achievable. Donnella Meadows offers some guidance on where system changes have the greatest impact.

23.1 Design standards

Design is an activity that has a great bearing on energy and resource use. Here we consider how we can influence this process within our organization.

There are professions more harmful than ***design,*** *but not many.*

-Victor Papanek

Design describes the decisions about function, materials and form which together determine how something is made and used. All designs involve choices about resources - even *"virtual"* products such as software consume resources in the form of electricity to power the computers on which they run.

Every energy and resource efficiency programme Champion should, at some point, understand the extent of design within their organization or which is influenced by the organization, the range of alternatives that are available in the design, and the scope to influence the choices being made.

As noted earlier (page 567), most investment decisions about energy and resource efficiency involve making a choice between efficient and inefficient options *within investments that are being made for other purposes.* Consequently, the primary challenge for most efficiency Champions is to intercede in decisions driven by considerations other than resources - something that may well be met with resistance by those making the decisions. So we may need a combination of a strong Mandate and an easy process to overcome this resistance.

A simple way to modify our design process is to set clear parameters for the design:

1. We can institute measurements and targets of the desired qualities (e.g. energy use and or water consumption) and leave it to the designers to ensure that these are assessed when alternatives are considered.

2. Virtually all design involves conforming to standards. So what we can do is to introduce new standards which modify the selections of alternatives to favour those with a lower resource use.

If we are taking this route, we must be clear about which aspects of the life cycle we are concerned with when making these changes. Choices of materials will have a big impact on the resource use earlier in the supply chain; decisions around function may greatly influence the in-use phase; and choices around function, materials and form can all affect the resource use in manufacture.

Sometimes the most efficient design is not about what happens in our organization; the formulation of our soap powder to clean better at 30° C may be the most far-reaching decision we can make.

Real World: *Eco-efficient products*

© Resevoir Dots @ fotolia.com

Measurement is important. If an organization can differentiate the environmental performance of the products in its portfolio, it can set objectives to favour these in the long run. This should improve competitiveness, as these products are likely to be favoured by the customers in the future.

That is exactly what AkzoNobel has done with what it calls its eco-premium products, which consider the following criteria: toxicity; energy efficiency; use of natural resources/raw materials; emissions and waste; land use; and risks/accidents. The eco-premium solution must be significantly better in terms of at least one criterion and have no significant adverse effects with respect to any of the other criteria. Since the decision to increase the proportion of these products, AkzoNobel has seen eco-premium products increase from 18% of sales in 2007 to 24% in 2015.[15]

Another company that has set a clear objective for the value of resource-efficient technologies sold is GE. In 2005 it set itself the goal of achieving US$20 billion of revenues from Ecomagination products. In 2010, the sales of these products was US$85 billion and in 2015 it was US$160 billion.

Most methods to determine the environmental impact of products rely on life cycle assessment (LCA), which was discussed on page 440.

Standards: *Green building frameworks and standards*

There are many important building sustainability standards, which are based on design. The leading examples are:

Leadership in Energy and Environmental Design (LEED)[755] is a standard that originated in the US but is now used internationally. The criteria it considers are: Integrative design (see next page); Location & Transportation; Sustainable Sites; Water Efficiency; Energy & Atmosphere; Material & Resources; Indoor environmental quality; Innovation; Regional priority (a placeholder for location-specific credits). Some credits are mandatory. Different levels of certification are awarded based on the score: 40 points - Certified; 50 points - Silver; 60 points - Gold; 80 points - Platinum. At the time of writing, there were 79,000 LEED-certified projects worldwide and it is the dominant standard in North America.

The Building Research Establishment's Environmental Assessment Methodology (BREEAM)[105] family of standards has been developed in the UK, but like LEED is used internationally. Issues rated are: Management; Health and Wellbeing; Hazards; Energy (the most significant); Transport; Water; Materials; Waste; Land use and ecology, and Pollution. Extra credits are available for Innovation. BREEAM is scored on a scale: Pass: Good; Very Good; Excellent; Outstanding (for some scores to be achieved mandatory credits must be earned). There are 552,000 BREEAM-certified developments worldwide, and this is the leading standard in Europe.

Although LEED and BREEAM are the most widely used, many other standards have been developed to respond to the priorities in different locations. For example Estidama[266] (aka the Pearl Rating System) is used in Abu Dhabi and emphasizes water conservation. Japan's Comprehensive Assessment System for Built Environment Efficiency (CASBEE)[121] majors on land use. The Building Environmental Assessment Method (BEAM)[61] is suitable for the high-rise, high-density built environment and sub-tropical climate in Hong Kong.

Other notable standards are Green Tag in Australia, New Zealand and South Africa, Deutsche Gesellschaft fur Nachhaltiges Bauen (DGNB)[210] in Germany, Building Owners and Managers Building Environmental Standards (BOMA BEST)[104] in Canada, Green Globes[332] in North America, Green Mark[103] in Singapore and the Global Sustainability Assessment System (GSAS),[338] formerly the Qatar Sustainability Assessment System (QSAS). Then there are many other sector-specific building standards such as for schools, retail and so forth, too numerous to mention here.

As well as standards based on the design of the building, sometimes called asset ratings, there are also certification schemes that are based on the performance of a building, many of which are mandatory since they are linked to minimum energy performance standards. These are usually focused on just one or a few aspects of a building.

In Australia, we have the government-administered National Australian Built Environment Rating System (NABERS)[539] which allows buildings to be rated in terms of energy, water, indoor environment and waste and achieve a one to six star rating. Europe has the mandatory Display Energy Certificate in the EU (there is an equivalent Energy Performance Certificate which is a mandatory asset ratings for certain buildings). Energy Star[255] is a voluntary US scheme to rate buildings based on performance - to gain certification a building must be in the best 25% of buildings.

23.2 Design concepts

Measurements and standards enable us to assess our design against the objective of greater efficiency. Here, we explore some general design concepts that support more efficient design.

When considering how to influence design, an alternative to imposing a standard is to introduce a design philosophy. This approach can result in a much more profound change, but clearly involves considerably more effort.

There are several techniques which are grouped together under the heading design for environment (DfE). These practices address different parts of the product life cycle such as:

- Materials and manufacture: lightweighting (reducing materials to the minimum), hazardous material minimization, modern methods of construction (buildings);
- Use: design for energy efficiency, design for durability (to extend product life and reduce obsolescence), lightweighting (e.g. reduce car weight for reduced fuel consumption);
- End of life: design for recycling, design for disassembly, design for disposability, design for remanufacture (see page 60).

Since the idea was first introduced in the 1990s, DfE has become especially prevalent in some industries such as defence equipment manufacture (see this example from Airbus),[12] but the concept is applicable across the board. Many good designers may not put their process into one of the formal categories above, but they will still think in this way as they make choices between different alternatives.

As well as techniques to address resource use throughout the life cycle of products, the way that designers work is also highly relevant to the outcome achieved. When designing complex products, like a building, where many systems interact, it is clear that optimizing single components does not necessarily lead to the optimum performance overall. There is now a strong emphasis on integrated design (see opposite), where the optimization of the whole system takes precedence over the optimization of individual elements. This approach is considered so important in building design, that standards such as LEED (see the previous page) explicitly reward it. Our earlier chapter on discovery emphasized "*starting with demand*" (page 396), which is another reflection of the need to consider the performance of a system holistically and question why there is a requirement for a resource in a particular form at a given place, before optimizing individual items of equipment.

Real World: *Obsolescence*

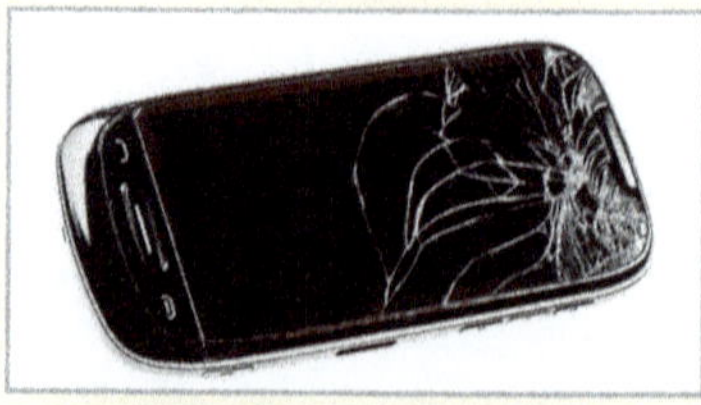

© Grigory Bruev, Fotolia.com

Life extension is a key efficiency concept that is not universally embraced in industrial design. One of the most infamous examples of deliberate obsolescence was the Phoebus Cartel[451] between Osram, GE International and Phillips which limited incandescent light bulbs to 1,000 hours of operation.

A market where obsolescence drives sales is mobile phones. A 2015 Gallup survey[697] among US phone users found that half of upgrades were because the phone stopped working or became obsolescent.

Phone manufacturers know this. Features such as stronger glass or waterproofing, while desirable to consumers, are not helpful to manufacturers because they slow the replacement cycle. The less direct competition a manufacturer has, and the more saturated their market, the greater their reliance on replacement sales and the greater the incentive to build in obsolescence into their products.

The solution, in this case, is either regulation to protect consumers or greater competition.

Real World: ***Breakthroughs by thinking holistically***

We tend to think of investments as offering diminishing returns and our job as designers to establish where the best cost to benefit point lies.

The problem with this approach is that we do this on each design element independently. Take, for example, insulation; we know that there is an inverse power law that describes the heat losses relative to the thickness of insulation - the more insulation we add, the less additional energy we will save. So the designer of the fabric of the building will ensure that we insulate the walls just enough (usually to comply with building regulations), but not too much.

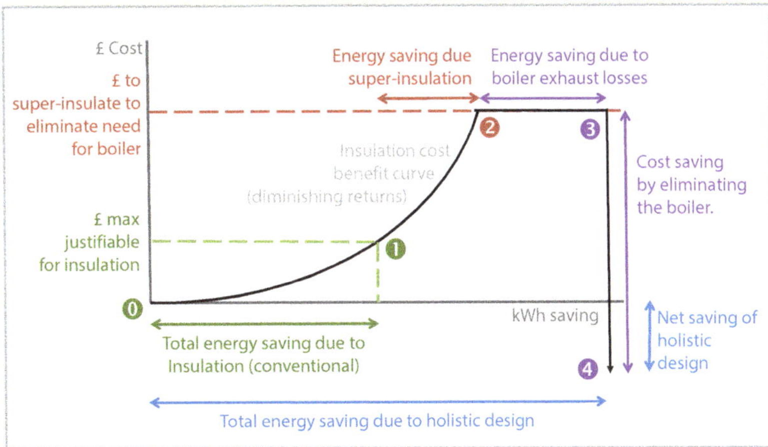

23.6 **Integrated design**
Considering equipment in isolation is not the best way to achieve maximum efficiency.
Source: Niall Enright inspired by Amory Lovins "Energy Efficiency, Taxonomic Overview", Fig 3.[484]
Image available in companion file pack.

Consider the chart above. The vertical axis is the cost and the horizontal axis the energy savings. The curved line between points ⓿ and point ❷ represents the insulation cost-benefit curve. In a conventional design process, illustrated in green, the designer of the building fabric would track along this curve until the point indicated by ❶, where further investment in insulation could not be justified. Our total energy savings would be shown on the horizontal axis.

In a holistic design process, we would think about what other systems are affected by insulation. In this example, we see that by super-insulating our walls we can get rid of our gas boiler altogether. We can calculate the effect of this by first tracking along the insulation cost-benefit curve from point ⓿ to point ❷. At this point, we can add a further energy saving to the total since, by eliminating the boiler, we also eliminate the flue heat losses, so we track horizontally from point ❷ to point ❸. Finally, we need to take into account the cost saving on the boiler (which exceeds the cost of the insulation) by tracking from point ❸ to point ❹. As we can see, we have arrived at a lower net cost and considerably higher net energy saving for our holistic design process compared to our conventional design process.

In a traditional building design process, the fabric and insulation are selected by the architect and/or specialist fabric contractor, while the boiler is chosen by the M&E (mechanical and electrical) designer. Usually, the detailed M&E design happens late in the building design process, by which point the chief elements of the fabric have been fixed. The solution to this is obvious: get all designers to work together at the outset and establish where their elements interact to optimize the whole, not the components. This is called integrated design.

Optimizing every single component does not necessarily lead to the optimum performance overall.

23.3 Design innovation

Given the dramatic reduction in resource use we will need to achieve, small step-by-step changes in design may not be enough. As they say, "a light bulb did not arise from incremental improvements to candles". Here, we explore innovation techniques that can help us come up with more radical solutions to design problems.

The standards and design concepts that we have discussed provide us with a structure and process to consider design. However, our improvements will only be as good as our ability to come up with new ideas to solve old problems.

At the risk of grossly over-simplifying, businesses have two basic strategies to increase profit: sell more for less or sell less for more. The second strategy has implicit improvement in resource use, the problem is with the first, dominant approach. Most organizations are geared up to sell more, more, more.

In order to achieve the necessary substantial reductions in resource use we either need to abandon this business model (unlikely) or we need to radically reduce the resource used in the products - not just to meet the necessary deep cuts per product, but also to offset the desired volume growth. This increase in production is inevitable and *good*, as more people are lifted from poverty and can enjoy the material benefits of progress.

Success in creating more from less will come down to our ability to radically innovate. Sometimes we need to be able to see a problem from a different perspective in order to find a design solution. In the cartoon below, it is the outsider who spots the obvious strategy to increase the volume of toothpaste sold. The ability to innovate is often thought of as something that is intrinsic to people. Nothing could be further from the truth - there are many of techniques that help us to *think outside the box*.

23.7 **Out of the box**
There is a story - probably a myth - that one day the executives of a company were meeting to work out how to sell more toothpaste. The meeting wasn't going especially well, when a member of the cleaning staff happened to overhear them and suggested that they "make the hole bigger". This story illustrates the idea that a solution to a problem often requires the ability to think unconventionally. By the way, the idea is not bad: an increase in the diameter of the orifice of a toothpaste tube from, say, 5mm to 6mm would increase the flow rate by a staggering 44%. This is because *flow = velocity * area* so if the velocity is constant the increase in flow is [36-25]/25 = 44%.
Source: Niall Enright, drawn using Pixton. The image is available in the companion file pack.

Real World: *Shinkansen*

© Scartech, Fotolia.com

The design of the Japanese bullet train, or shinkansen, is an example of biomimicry.

As speeds increased, there was a real problem with the sonic boom the trains created when moving from open air to enclosed tunnels, caused by the shift from low to high drag.

Eiji Nakatsu, the Shinkansen 500 train's chief engineer and a keen bird-watcher, asked: *"Is there something in nature that travels quickly and smoothly between two very different mediums?"* He hit upon the kingfisher which dives from air (low drag) into water (which is 800 times denser and a high drag environment) and yet barely creates a ripple. The key is the shape of the beak, which is now incorporated into the design of the Shinkansen 500 and has reduced tunnel noise and improved electricity use by 15%.

© Rsaが新大阪駅で撮影, CC Share Alike.

Biomimicry was also used to reduce the sound of the pantographs, which connect the train to the overhead power lines. These use shapes that disrupt the airflow, inspired by the way an owl disrupts the airflow over its wings to create near-silent flight.

The first technique for innovation that I would propose is reverse brainstorming. This tool is exactly like traditional brainstorming (see page 672), only you will be searching for ideas that have the opposite to the intended effect and, when you have identified the possible changes, you can think about how these can be reversed. Reverse brainstorming is useful when we are seeking solutions that run counter to conventional thinking (e.g. how to sell less and make more money). There is the drawback that brainstorming relies on the knowledge and experience of the participants.

The Russian engineer and science-fiction writer Genrich Altshuller certainly felt that innovation was not a question of intuition but could be systematized. He developed "теория решения изобретательских задач" (Teoriya Resheniya Izobretatelskikh Zadach), or the Theory of Inventive Problem Solving, known as TRIZ. Altshuller was a patent administrator who was intrigued by the nature of inventions, and by 1969 he had reviewed 40,000 patent abstracts to come up with some profound general principles of innovation.

The core insight was that most innovations involve a trade-off between two contradictory elements of a design. Take an umbrella, you want it to be big when it is raining and small when you need to pack it away. The solution, which has been known for years, is to make the umbrella *collapsible*. This is a generic solution to a physical contradiction. What TRIZ does is to provide the methods to help people turn their particular problem into a general concept, such as "*big yet small*" in the case of an umbrella or a tent. Having identified the specific contradiction(s) or concepts inherent in a design, TRIZ then lists which of 40 Inventive Principles have been used most often to solve that specific contradiction. TRIZ then offers further tools to distil these down into the most effective design for the specific case.

In her terrific book[299] and training, Karen Gadd illustrates the power of the TRIZ toolkits. She asks people to brainstorm ways of removing water from a glass. Some will think of putting stones in the glass, others of putting another glass inside the original (both general concepts of displacement); physicists may come up with heating the liquid; engineers may drill a hole in the glass; chemists could consider changing the water chemically. All in all, there might be 15-25 solutions using 5-8 general concepts. TRIZ, on the other hand, defines the problem very precisely as "*move a liquid*" and provides an Effects Database of 99 different techniques to achieve this function which can be tested systematically to see if they provide the answer to the specific case. With TRIZ, innovation becomes a systematic process that builds on prior invention. It takes time to learn, but it is worthwhile if design is your core job.

Another important approach to innovation in design is called biomimicry, which takes as its premise the idea that the constant improvement that occurs through natural selection in biological systems leads to highly efficient solutions to practical design problems. Thus the skin of a shark tells us how to design ships' hulls or swimwear with dramatically less resistance, or the beak of a kingfisher shows us how to design a better train (see case study left).

23.4 Communication!

Effective communication is what drives change. Here, we will consider communication as a system and what can be done to change the system so that it aligns with our resource efficiency objectives.

*Communication is about the **flow of information** in our organization. This needs to align with our efficiency objectives in order to succeed.*

The ability to innovate successfully depends on talent and hard work. Change, we have seen, is driven by people who are both motivated and capable.

Communication underpins motivation and capability. By communication I mean the wider information flows within our organization, not just the formal messages that are sent to employees, but the data, goals, rewards, values, knowledge and associations that bind and align people to a common purpose.

First, let us consider communication as a process. Every day, folks in our organizations are inundated by messages. A 2013 briefing by consultants Bain and Company called the flood of data that most employees and managers are subjected to, infobesity.[627] This information overload will inhibit any effort to communicate resource efficiency messages within our organization, as the messages are in competition for attention with the hundreds of pieces of information each one of us receive each day.

So the first thing we need to consider in our organizations is how to make the value of efficiency stand out from the constant background flow of information and data. Part of this is about ensuring that the messages are aligned with the core values of our organization. There is also something to do with vivid messaging (see page 661), and also about getting the leaders to communicate the message and create urgency (page 216).

The bottom line is we need to ensure that the information flow related to resource efficiency is effective. By this, I don't just mean the call to action, but all the information that supports better decision-making, such as the performance metrics that will determine how we are doing against our goals. So one of the first tasks is to do a programme information audit:

- Are the communications sufficiently prominent and accessible? That is to say, are they reaching the people they are intended for? Are our messages clear and unambiguous? Are we varying the delivery for different audiences?
- Are the communications focused, in other words, are we sharing the stuff that matters to decision-making and not overloading folks or broadcasting in a scattergun manner?
- Is our information presented in a standard and consistent way which makes it easy to collate and understand?

Real World: *A journey not a destination*

Thinking about a communications plan compels us to recognize the nature of our messages.

- Resource efficiency is a core source of value - not an incidental activity. It should be managed accordingly.
- Resource efficiency is not a project, it is a way of thinking, it is a habit, it must be integral to all major decisions we take. It is a continuous improvement process which will take time to embed in our organization.
- Resource efficiency is not optional. Whether we like it or not, we collectively need to achieve a radical reduction in our resource use. There are no substitutes for natural capital - once it is gone, it is gone.
- Resource efficiency is urgent. The risks and opportunities are increasing rapidly - we cannot delay. The work of lifting millions out of poverty must be accomplished as soon as possible, but without destroying the foundations of our future prosperity or wellbeing.
- Resource efficiency is not a solitary journey. Just about every important improvement will require us to work with suppliers, customers, investors, staff and stakeholders. We should embrace dialogue. We should celebrate success. We should form coalitions with like-minded organizations.

Our communications should avoid simplistic, catchy slogans - *"People, Planet & Profit"*, *"Triple Bottom Line"* etc. and instead focus on the true nature of the challenge. Interface's *"Mission Zero"* comes to mind.

- Are our communications and information timely?
- Are we structuring our communications effectively? For example, when I was a medic, I was taught what is now called SBAR - communicating the situation, background, assessment and recommendation (think of the reality TV shows where an ambulance crew hand over a patient to a medical team). Another variant on this is "*bring me a solution, not a problem*".
- Is our communication flow two-way? Passive messages do not create an emotional commitment - what drives and motivates people is their ability to participate and contribute.

Thinking of communication as a system, or series of systems, there are many processes where we can incorporate our messages about resource efficiency. Each of these systems should be considered in a communications plan.

- First of all, we need to consider internal communications. One key process is staff inductions, which will inform new arrivals in the organization. We need to consider training as a key system to align with our programme. We need to think of the overarching goals and how these translate to local targets and performance indicators, and how performance against targets is celebrated (e.g. through newsletters, noticeboards and other staff communications) and how success is linked to appraisal and reward systems. We need to think about how people can become involved and contribute (e.g. through suggestion schemes). Often overlooked in communications plans are the strategic and financial planning systems in our organization as well as research and development of new products and services.
- Then we need to consider our external communications. Critical decisions include what are we going to say, and how transparent are we going to be with stakeholders. These choices may require us to engage with various systems and processes such as marketing, corporate affairs, legal and regulatory affairs, charity and community engagement and investor relations. Important considerations here are the degree of transparency we are willing to consider. Open acknowledgement of failings or areas for improvement can be difficult for organizations, but in the long run this is what builds trust with stakeholders, facilitates dialogue and breaks down barriers. In this context we also need to think about how we interact with academic institutions, standards setters, trade bodies and sector representatives - are we supporting improvement within our industries and supporting the efforts of our peers? We also need an external communications plan for our customers, which may go beyond marketing and include things such as the guidance that accompanies our products. Finally, as discussed in the next section, we need to understand how we should communicate with our supply chain, which will involve our procurement function in our programme.

23.5 Supply chain systems

In this section, we will consider how our supply chain processes can be harnessed to improve resource efficiency. This involves some deep questions about how we wish to exert influence.

Exploration: ***Sphere of influence***

The expression sphere of influence was introduced into the field of corporate social responsibility by the UN Global Compact of 2000.[317] It

> *"asks companies to embrace, support and enact, within their sphere of influence, a set of core values in the areas of human rights, labour standards, the environment, and anti-corruption".*

This notion caused controversy[478,801] about whether an organization's responsibility should be based on its capacity to influence other parties or only on its actual contribution to social and environmental outcomes.

In normative terms, just because an organization *"can"* do something does this mean it *"ought"* to? The wording has now been dropped from the main UN website.[316]

The alternative approach, that an organization should base its actions on its real impact by exercising due diligence is criticized for being insufficiently precise - leaving the organization plenty of wriggle-room to determine what is appropriate to examine. This approach is related to the notion that executives have a fiduciary duty (page 218) to assess their organization's impact, but within that, they can determine what is, or is not, relevant, and are only dictated by the established practices of the sector in which they operate.

How we decide to spend our money affects our suppliers. If we communicate that resource efficiency matters to us, then we will encourage our suppliers to take this into account in their operations. We have influence, and we should make that influence support our resource efficiency objectives.

Indeed, we have seen from our earlier examples of a circular economy (page 63), Walmart's supply chain initiative (page 138) and life cycle assessment (page 440) that our supply chains and customers are often material to the value and opportunity that resource efficiency brings. The process of aligning our interests with that of our supply chain is called sustainable supply chain management.

Historically, supply chain management has been much more about social aspects, such as human rights, worker exploitation and fair trade. Child labour scandals in suppliers to companies like Nike have shown that brand value is especially vulnerable to consumer sentiment about wrongdoing in the supply chain. In many consumers' eyes, it is not good enough for companies to claim ignorance; there is an expectation that they will ensure that their suppliers behave as expected. Now similar concerns are being expressed about environmental performance both downstream (e.g. palm oil leading to deforestation) and upstream (e.g. plastic microbeads polluting water).

Standards such ISO 26000:2010 Guidance on social responsibility are setting the key principles of supply chain engagement. A key concept is that organizations will acknowledge the influence that they have and use this influence to promote positive social and environmental outcomes. This notion of *sphere of influence*, is used in the standard in the sense of being able to influence other parties' decisions to act or refrain from acting in certain ways. This is not without controversy (see panel left), but as long as it remains central to the ISO standard (it is mentioned no less than 33 times in ISO 26000), organizations need to decide if they will adopt this approach in assessing what parts of their supply chain they wish to engage with.

The reason that this debate is important is that it will determine how we approach our supply chain management plan. Assuming that we want our approach to be standards-based, we will shortly be able to turn to ISO 20400 Sustainable Procurement, still in draft form,[409] but should be published in 2017,[540] which relies heavily on the notion of the sphere of influence.

> **Real World:** *Design vs designer*
>
> At Peel Land & Property Group I was closely involved in the development of a sustainable procurement policy, a process led by a supervisory board director, David Glover and Dale Mullane, the procurement manager, with inputs from Barry Collins of CSR consultancy Collins McHugh as well as ActionSustainability and Envirolink.
>
> This process started with a workshop to which Peel invited suppliers across a range of disciplines and sought their views on how they could help Peel be more sustainable and resource-efficient.
>
> These suppliers were the folks who collectively designed the Peel developments: architects, mechanical and electrical (M&E) consultants, civil engineers and facilities management contractors.
>
> What was remarkable from these presentations is that every single advisor thought that sustainability was about their behaviour rather than their advice. They told us about their travel policies, about their use of green electricity, about their recycling rates. While it is laudable that the architects travel to the Peel offices by bike rather than a Hummer, there is the danger that focusing on the organization's behaviour misses the important point that a supplier's biggest impact usually arises from the quality of their advice, service or product.
>
> Reflecting this duality, the Peel L&P Sustainable Procurement Policy includes both a commitment to *"procure goods and services that help us to embed sustainability in our business activities"*, as well as the broader aim to *"engage with suppliers that can demonstrate their sustainability credentials"*

The traditional first step in formulating our plan is to map our supply chain, which involves creating an inventory of suppliers, identifying the relevance and significance of their environmental and resource impacts to us, and prioritizing these. If we fully embrace the ISO 20400 approach, then our map should consider "exercising influence" (bearing in mind this is *draft* text):

> *An organization and its procurement function should:*
> *a) assess its sphere of influence within the supply chain;*
> *b) promote sustainability in the supply chain;*
> *c) exercise influence within the supply chain.*

Having identified what parts of the supply chain are important to us and the issues that we want to influence, we are in a position to engage with suppliers in a dialogue about expectations. I recommend that this is an open discussion rather than an edict - our suppliers can help us refine our requirements so that they are achievable. Out of this dialogue, we should develop baseline performance measures, against which we can track progress. I would urge organizations to set any emissions requirements using the GHG Protocol location method of reporting (for an explanation see *Failed additionality* on page 745).The product of these discussions is often a supplier charter which sets out goals, our organization's expectations of suppliers and, equally, what we commit to our suppliers to help them achieve the goals. The charter should explain how procurement will take efficiency into consideration, and offer means for suppliers to raise grievances or challenge our processes.

While resource efficiency in the supply chain has traditionally focused on downstream sourcing (making sure that the supplies are causing no harm) and logistics (ensuring that they reach us in the most efficient manner), there is an increasing concern with upstream issues, once we have supplied our customers. One area of great interest is reverse supply chains which enable reuse, remanufacture or recycling of resources.

23.8 **Global Reach** In today's highly connected world it is possible for even small organizations to have a global supply chain, with the complexity that implies. *Source:©zmkstudio, Fotolia.com*

23.6 Finance systems

In this section, we shall see that our finance systems, both in our organizations and more broadly, are flawed. Only by acknowledging these flaws we can we begin to imagine alternatives. The bottom line is that we have an obligation to act.

We have to acknowledge that achieving dramatic improvement in resource efficiency ***calls into question the functioning of many aspects of our finance systems.***

A recurring theme in this book has been financing. It is sometimes assumed that implementing a resource efficiency process is simply a question of allocating funds to this task. The reality, as we have seen, is quite different. Driving resource efficiency is as much about the *function* of our organization's financial systems, as it is the *provision* of funds to our programme.

For example, we have seen that cost allocation (see page 624) is a key step to engage people in conserving resources. We have seen that the details of how we calculate costs, such as whole life costing (page 571) and marginal return (page 567) greatly influence the attractiveness of an investment. Even the prices that we choose to use, affect the outcome (see page 576). All these techniques are practical ways to harness finance systems for resource efficiency.

However, there are some profound problems with finance systems in most organizations which we need to understand and change if we are to drive the scale of efficiency that is required. The default position of most - not all, I hasten to add - finance decision-makers is to say "*no*" to any initiative that is not perceived to be core to the organization.

Without care, finance can destroy innovation. For example, I have emphasized discounted cash flow as a tool to gain investment (see page 584), but we need to understand its inherent flaws. While the notion of discounting itself is impeccable, it suffers some key problems, as pointed out by a paper in the Harvard Business Review.[139]

First, there is the *business as usual trap* in which it is assumed that the most likely future cash flow, absent investment, is the same as the current cash flow. In reality, without change, most future cash flows will inevitably decline as markets saturate and sales decrease, as competition increases and disruptive changes emerge. We all know that products have a cycle of eventual decline, but organizations don't seem to take this into account when assessing the value of investments. This illusion of a state of perpetual constancy has a name: it is called the Parmenides fallacy and it causes us to underestimate the return on innovation. It is another form of status quo bias.

The second flaw in discounting future cash flows is our ability to accurately assess future value, in particular when we are dealing with disruptive and uncertain technology. The problem, according to the Harvard Business Review paper, is especially pronounced by an inability to assess the terminal value

Real World: *Sunk costs*

© MIRACLE MOMENTS, Fotolia.com

Why is it that established organizations find it so difficult to achieve disruptive innovation? One big reason is that many existing enterprises have sunk costs in older, less efficiency equipment.[139]

Imagine that I am a steel mill, *OldCorp*, whose product sells for US$400 a tonne, and costs, on average US$350 to produce, giving me a reasonable profit of US$50.

However, if my mill has unused capacity, then the marginal cost to produce an extra tonne of steel might only be US$50. So I make US$350 on each *additional* tonne I produce.

This profit compares very favourably with opening a new, more efficient, steel mill where the production cost, including the capital costs, is US$320 per tonne, yielding US$80 profit.

For an existing steel producer, sticking with their old, outdated technology is a *no-brainer* since the additional output from the old plant has a much greater profit margin.

For a new entrant, *NewCorp*, to the market, the US$80 profit on a US$320 cost is very attractive - certainly more than the weighted cost of capital. So the new entrant, *NewCorp*, sets up its factories and, over time, captures sales from *OldCorp*. As a result, *OldCorp's* volumes decrease and the average production cost increases, leaving them in a spiral of decline due to their less-efficient process.

of projects correctly. Uncertainty inevitably leads to undervaluation, which results in a deeply ingrained finance system preference for more certain, but small, incremental change over disruptive change.

Another big problem with finance is that many publicly listed companies are driven by excessively short-term financial performance goals, to the detriment of long-term investment. What matters is the next quarter's earnings, not surviving another decade. This focus on earnings per share (EPS), is another innovation killer. It leads to perverse financial decisions such as share buybacks to *return funds to investors* which, by contracting the pool of shares forming the denominator in the EPS calculation, can dramatically boost EPS. But buybacks have the negative effect of starving the enterprise of investment and do nothing to the *real* value of the business.

These problems are made worse by the innate conservatism of the valuation and risk professions which do not adequately price a sustainability premium in assets like green buildings or investments like green bonds. The self-fulfilling argument is that, since the financial industry (i.e. they) don't recognize the value in reducing resource risk, then resource risks can be considered immaterial. At some point enough of us will realize that the *emperor has no clothes*, and the consequences of stranded assets, such as carbon bubbles on the balance sheets of fossil-fuel businesses, will crystallize.

What we have to remember is that our capitalist (small c) financial systems are artificial constructs (albeit ones that put clothes on our backs and food on the table). Corporations as entities with some of the "rights" of individuals - such as the right to own assets, to sue, to make political contributions, to limit liability; the notion that fiduciary duty demands executives pursue profit regardless of other considerations; the categorization of costs into CAPEX and OPEX and the resulting constraints on investment. These are human ideas. Even the notion of a *free market* depends on contrivances such as money, debt, exchange rates etc. created by us. While capitalism has unleashed huge resources for the improvement of mankind's condition, the system is not without flaws. Its saving grace is that it can adapt over time.

In the broader economy, too, there are countless distortions, such as the huge fossil-fuel subsidies, energy supply tariffs which incentivize consumption, free riding by organizations which are depleting *the commons* of natural resources such as fish, the countless trade barriers that prevent poor countries from processing their commodities (with the resulting inefficiencies that brings).

In considering how our financial system works we need to remind ourselves that the only true foundation for our wealth is the natural capital which our planet provides. It is our fiduciary and moral duty to recognize this and to accept that, by consuming resources, our organization is a part of the problem and part of the solution. While the harm may be due to the actions and choices of others, the fact that we are the ultimate consumers of the resources obliges us to act. Addressing resource efficiency is necessary, justifiable and right.

Exploration: *Where to change the system*

In a thought-provoking paper *Leverage Points, Places to Intervene in a System*[516] Donella Meadows, whom we have already encountered as a co-author of *The Limits to Growth* (see page 18), describes a series of increasingly effective points at which change can take place. I have adapted her 12 Leverage Points to put them in the modern context of an organization seeking to transform its building stock. My précis for the leverage point is in bold, and Donella Meadows' original description in italic.

12	**Metrics**: *Constants, parameters, numbers.* These are the numbers that drive the resource performance of our building, such as the building energy intensity kWh/m2. 90%+ of our resource efficiency efforts are spent addressing these numbers, as they are easy to change, but, according to Meadows this is the least effective place to intervene in the system to deliver improvement if they are not mandated (see Rules, below). The reason is because the many factors below have a greater influence on the system ability to respond to numbers alone.
11	**Buffers:** *Size of buffers and stabilizing flows.* One reason why changes to numbers don't have as big an impact as we expect is because big changes in the energy intensity can only be made when we replace or refurbish a building. In fact, the replacement rate of buildings is around 2% per annum and the refurbishment rate well under 10%. Buffers represent barriers to system changes which maintain the *status quo*.
10	**Stocks:** *The structure of material flows and stocks.* Here, Meadows is referring to the physical constraints of the system. For property developers this could be factors such as the availability of land (which provides a physical constraint on development) or the demographics of the country (which affect demand for property types). These are difficult factors to change as they tend to be intrinsic to the system itself or the design of the system.
9	**Responsiveness:** *The length of delays relative to the system change.* Here, the issue is the delay in our ability to make a change - for example if we want to develop a combined heat and power facility to cost-effectively power our buildings this will take several years to implement. The delay in implementing technology change and verifying the effects is a real barrier change. In these circumstances our buildings cannot respond to short-term changes using these types of technologies even where they are proven and cost-effective. This becomes a real issue if the desired rate of system change exceeds the speed of our response. It is not just technology that creates these delays but also skills - for example, engineering skills shortages are predicted to have a real impact on our ability to transition to some low-carbon generation such as nuclear power. If we can reduce delay then we can increase the rate of change.
8	**Negative Feedback:** *The strength of negative feedback loops relative to the impact they are trying to correct against.* For example, carbon taxes have a relative weak effect on investment. This is a negative loop in the sense that it inhibits the undesirable performance (emitting emission). Its effectiveness is judged in relation to the size of the change needed. At present carbon taxes are generally weak because the negative effect does not lead to significant corrective response (i.e. greater investment in order to get lower emissions). However, if these taxes ramp up they could have a significant effect on emissions.
7	**Positive Feedback:** *The gain around driving positive feedback loops.* An example of a positive feedback loop is one which reinforces itself. For example good building ratings, such as BREEAM, are having an effect on the value of building stock, which in turn is encouraging further investment in getting good ratings. The consequence is that, over time, the median rating increases (so that, for example, for buildings a new top rating of *"Outstanding"* had to be added, or for fridges we have gone from A to A+ to A++ as top ratings). Because of the runaway effects, positive feedback tends to be more powerful points at which to influence a system than negative feedbacks.
6	**Information**: *The structure of information flows.* Here, we can use one of Donella Meadow's own examples as it relates to our theme of buildings: *"There was this sub-division of identical houses, the story goes, except that for some reason the electric meter in some of the houses was installed in the basement and in others it was installed in the front hall, where the residents could see it constantly, going round faster or slower as they used more or less electricity. With no other change, with identical prices, electricity consumption was 30% lower in the houses where the meter was in the front hall. It's not a parameter adjustment, not a strengthening or weakening of an existing loop. It's a new loop, delivering information to a place where it wasn't going before and therefore causing people to behave differently."* This is why, for example, I am a strong advocate of allocating resource costs to users (i.e. moving these away from central overheads), because this new information flow (or feedback) can have a potent effect of the desire for improvement.

5	**Rules**: *The rules of the system*: These are a very powerful point of change. For example we have seen how right-hand only turns make for a dramatic improvement of UPS's fuel efficiency (see page 306). In the EU, installations with large CO_2-emitting capacity which previously were allowed to discharge as much as they wanted now have to buy finite allowance for each tonne of emissions they produce. Rules lead to outright bans on certain products (again keeping to the theme of buildings, in the UK, property owners have been banned from selling or leasing buildings with Energy Performance Certificate ratings of F & G.
4	**Structure**: *The power to add, change, evolve, or self-organize system structure*. For this example, Donella Meadows used the analogy of biological organisms as having structures which are changed by a process called evolution, and it is this evolution which is the change-point in the system. Translating this into the property world, a building can be seen as a part of a finance *system* which has a number of components: investors or owners, tenants, contracts (leases), rents, rates, insurances and asset managers, which in turn are influenced by features such as location, regulation and sentiment. The structures of these finance systems are not fixed, and will change over time. For example, introducing new sources of finance in the form of green property bonds could have a big impact on the valuation, design and operation of buildings because a major component of the finance system has changed.
3	**Goals**: *The goal of the system*. In the preceding example, the goal of the finance system in property is to maximize return. Every higher aspect of the system will be tuned to support that goal. If the goal is changed then the system will inevitably change to reflect that new purpose. This can be seen when long-term investors, such as pension funds, purchase buildings; they tend to invest much more in their sustainability features as these investors are concerned with asset depreciation over the long-term. Changing a goal from a focus on short-term returns to long-term returns has a big impact.
2	**Paradigms**: *The mindset or paradigm out of which the system arises*. This is an even more fundamental point of change. For example, we have said that buildings are physical objects, *assets*, achieving a financial objective, *yield*, on behalf of their owners. An obvious change to this paradigm would be the elimination of the concept of private ownership (e.g. in a Socialist political system) - which would dramatically change the way the building was operated and the way future buildings would be financed or designed. These sorts of changes of paradigm are regarded as hard to achieve, but they can come about when enough people see the world in a different way. Once slavery was acceptable; once people believed in witchcraft. If enough people see a purpose for buildings other than making money, then the system will inevitably change over time to reflect that view.
1	**Transcendence**: *The power to transcend paradigms*. Here, Donella Meadows speaks to an ability to see beyond whatever dominant paradigm exists at any given time - this is much more about an ability to act on what is right at any point in time. Turning again to our property analogy an example would be an owner who invests money to improve the building's energy use, even though this makes no sense financially. That is to say, the owner can transcend the paradigm of *return on investment* because there may be another paradigm at work: *be fair to future generations* or *make the world a better place*. Freedom comes from recognizing that no paradigm is ever true and so we are able to choose the appropriate paradigm for a given situation. The most stubborn paradigm that we have today is the notion of *growth* that is embedded in so many of the systems that affect our lives.

The most striking thing about Donella Meadow's analysis is just how often our interventions are at the least effective point such as *changing metrics* (level 12 of our hierarchy). Although we may have an *aspiration* for a building, say, to achieve a particular energy performance metric in terms of kWh/m^2, unless we intervene at lower levels of our hierarchy, the change we aspire to is unlikely to persist. The sorts of things we also need to put in place include negative (level 8) or positive feedback (level 9) loops or changes to the rules (level 5) or the structure of the system (level 4) to ensure the change in metric is a permanent feature.

If all we do is change the metrics, then we may, for a while, get a false impression that our high-level changes are working. The problem is that as soon as there is any kind of competing change, working on more fundamental aspects of the system, or a shock to the system of some sort, we will see our changes abandoned.

That is why systems like ISO 50001, enManage, QUEST and the Framework set out in this book, will embed change by intervening at multiple levels of a system, usually starting at the rules.

Further Reading and Resources

Joseph Fiksel's *Design for Environment - A Guide to Sustainable Product Development*[283] is a great resource for the techniques within DfE. The US Environmental Protection Agency has several useful reports and tools for Design for Environment, available online.[753]

For a clear, succinct description of TRIZ, see *The Six Sigma Performance Handbook*[340] by Praveen Gupta, pp278-285. There are several excellent books on TRIZ, but one of the most approachable titles must be Karen Gadd's outstanding *TRIZ for Engineers*,[299] which beautifully illustrates key concepts with cartoons. Her organization, Oxford Creativity, also offers courses on TRIZ ranging from a one-day introduction to a five-day immersion in the subject.

For those interested in biomimicry, the work of Janine Benyus is highly recommended. See her book *Biomimicry: Innovation Inspired by Nature*[66] and the website of the Biomimicry Institute at https://biomimicry.org. According to Benyus, there are many reasons why we should copy nature:

Nature runs on sunlight.
Nature uses only the energy it needs.
Nature fits form to function.
Nature recycles everything.
Nature rewards cooperation.
Nature banks on diversity.
Nature demands local expertise.
Nature curbs excesses from within.
Nature taps the power of limits.

Another very influential book on design is *Cradle to Cradle* by William McDonough and Michael Braungart[506] If you buy the physical copy of this book, do check out the material it is made from - a synthetic paper which can be reused many times.

The UN Global Compact has a useful *Practical Guide for Continuous Improvement*[378] in supply chain sustainability. While this is much broader in focus than resource efficiency, the process is entirely applicable.

The Global Environmental Management Initiative (GEMI) has an online tool[319] to help organizations address the environmental aspects of their supply chain.

Standards: *Relevant ISO Standards*

The following International Standards Organization standards are relevant:

- ISO 14006:2011 Environmental management systems — Guidelines for incorporating eco-design
- ISO 14045:2012 Environmental management— Eco-efficiency assessment of product systems — Principles, requirements and guidelines
- ISO/DIS 20400 Sustainable Procurement Guidelines (*draft*)
- ISO 26000: 2010 Guidance on social responsibility

24 Reference Tables

24.1 Statistical formulae

Item	Formula	Notes. Excel Function
Sample Mean	$\overline{x} = \dfrac{\sum x_i}{n}$	Often called the "average". Sum (Σ) of all the values in the sample divided by the number of values, n =AVERAGE(Range)
Population Mean	$\mu = \dfrac{\sum x_i}{N}$	The same calculation as the Mean, only with a different notation. A sample is a subset of the population and if the sample is large enough we can draw conclusions about the population from it. The difference in notation is to inform that a whole population is being used. In almost all cases, the data analysis we will do with resource-use data is on samples.
Weighted Average	$\overline{x} = \dfrac{\sum w_i x_i}{\sum w_i}$	The Sum of the product of the values and their weighting is divided by the sum of the weighting. w[i] is the weight for value i.
Sample Variance	$s^2 = \dfrac{\sum (x_i - \overline{x})^2}{n-1}$	This is a measure of how values in a sample are spread around the Mean. It is always positive as the square of a negative number is a positive value. Note that the units of the Variance are squared, so for electricity we would express the Sample Variance in kWh^2. =VAR.S(Range)
Population Variance	$\sigma^2 = \dfrac{\sum (x_i - \overline{x})^2}{N}$	As above, but for the whole population. =VAR.P(Range)
Sample Standard Deviation	$s = \sqrt{s^2} = \sqrt{\dfrac{\sum (x_i - \overline{x})^2}{n-1}}$	The Standard Deviation is the square root of the Sample Variance, which means that we have a measure expressed in the same units as our data. For a sample this is denoted by the letter s. The Standard Deviation is a very useful value to have if we are looking at normally distributed data. =STDEV.S(Range)
Population Standard Deviation or Standard Error	$\sigma = \sqrt{\sigma^2} = \sqrt{\dfrac{\sum (x_i - \overline{x})^2}{N}}$	As above but for a population, where the Standard Deviation is denoted by the lowercase Greek letter sigma, σ. Also called the standard error. =STDEV.P(Range)

Item	Formula	Notes. Excel Function
Coefficient of Variation	$\left(\frac{s}{\overline{x}}x100\right)\%$	This is a simple way to express how big the Standard Deviation is in comparison to the Mean. It is usually shown as a percentage as it is unitless.
z-score	$z_i = \frac{x_i - \overline{x}}{s}$	This is a measure of how far an individual reading, x_i, is from the Mean, expressed in terms of Standard Deviation. It is really useful for detecting outliers in normally distributed data, especially when the sample size, n, is large.
Sample Covariance	$s_{xy} = \frac{\sum(x_i - \overline{x})(y_i - \overline{y})}{n-1}$	This is a measure of how one sample of data, denoted by x, varies compared to another sample, y. If the number is positive, it means that as x increases then so does y. If it is negative then as x increases, y decreases. Because the Covariance is influenced by the units employed (if x is in Fahrenheit rather than Celsius, the numerator is bigger), the Correlation (below) is a preferred measure of the relationship between two series as it is not affected by units of measurement. =COVARIANCE.S(Array1, Array2)
Pearson Product Moment Correlation or Pearson Correlation Coefficient	$r_{xy} = \frac{s_{xy}}{s_x s_y} = \frac{\sum(x_i - \overline{x})(y_i - \overline{y})}{\sqrt{\sum(x_i - \overline{x})^2 \sum(y_i - \overline{y})^2}}$ or $r_{xy} = \frac{n\sum x_i y_i - \sum x_i \sum y_i}{\sqrt{n\sum(x_i - \overline{x})^2 n\sum(y_i - \overline{y})^2}}$ or $r_{xy} = \sqrt{R^2}$	Note that there are many different rearrangements of these equations to allow r_{xy}, or the Correlation, to be calculated.[788] In the examples provided, s_{xy} is the Sample Covariance while s_x and s_y are the Standard Deviations of x and y. Note that the n-1 terms in the individual formulae cancel each other out. R^2 is the Coefficient of Determination described later. Like the Covariance, the sign of the Correlation indicates whether y increases with x (called a Positive Correlation) or whether it decreases (called a Negative Correlation). r_{xy} can be any value between -1 and +1. A 0 value indicates that the change in y is not related to the change in x, whereas -1 or +1 means that all the change in y can be explained by a change in x. r_{xy} is unitless and is always expressed as a decimal, not a percentage. =PEARSON(Range1, Range2) or =CORREL(Range1, Range2)
Simple Linear Regression model	$\hat{y} = y = mx + c$	This is the "equation of the line", which enables us to calculate a predicted value for y, ŷ based on two statistics: m, the Slope of the line and c, the Intercept. Please note that this is the equation that is used where we have a Linear relationship (that is a straight line) between x and y, but that non-Linear equation are not uncommon in resource-use data. The units of ŷ are the same as y.

Item	Formula	Notes. Excel Function
Slope of the Linear Regression model	$m=\frac{\sum(x_i-\bar{x})(y_i-\bar{y})}{\sum(x_i-\bar{x})^2}$ or $m=\frac{\sum x_i y_i - n(\bar{x}\,\bar{y})}{\sum x_i^2 - n(\bar{x})^2}$	The Slope of the Linear Regression model indicates how much y will increase or decrease for each unit change in x. A Positive Correlation is indicated by a positive value for m and a negative Correlation by a negative value. The units of m are yx^{-1} e.g. for a relationship between electricity (kWh) and number of cars produced (cars) the slope is kWh car^{-1} often expressed as kWh/Car or kWh per Car . =SLOPE(known_ys, known_xs)
Intercept of the Linear Regression model	$c=\bar{y}-m\bar{x}$	The Intercept of the line is the value of y when x is zero. This is the constant that is added to every period in our Linear Regression model, which leads to c being referred to as the baseload. Please note that our confidence in the estimation of c depends on the range of x in our data - if our x-values are clumped together a long way from the y-axis (i.e. far from 0 x) then the confidence in our calculation of c is diminished. The units of c are the same as y. =INTERCEPT(known_ys, known_xs)
Coefficient of Determination	$R^2=r_{xy}^2=\frac{SSR}{SST}=\frac{\sum(\hat{y}_i-\bar{y})^2}{\sum(y_i-\bar{y})^2}$ or $R^2=\left(\frac{\sum(x_i-\bar{x})(y_i-\bar{y})}{\sqrt{\sum(x_i-\bar{x})^2\sum(y_i-\bar{y})^2}}\right)^2$ or $R^2=\frac{\text{explained variation}}{\text{total variation}}$	This is a measure of the goodness of fit of the estimated Linear Regression equation to the data. It can be interpreted as the proportion of the variable y value that is determined by x, as shown by the proportion of the Sum of Squares of the Residuals, SSR, (which use our Linear Regression formula) compared to the Total Sum of Squares, SST (which is based on the mean values of y). SSR and SST are defined below. R^2 can range from 0 to 1. R^2 is unitless but sometimes is shown as a percentage value, in which case the decimal is multiplied by 100. =RSQ(Array1,Array2)
Adjusted Coefficient of Determination	$R^2{}_a=1-(1-R^2)\frac{n-1}{n-k-1}$	As one adds independent variables to a regression the value for R^2 will rise as the coefficients of determination for the additional variables contribute to the total. One could therefore add many variables which individually have little significance but which nevertheless increase R^2. Many analysts prefer to adjust R^2 by the number of variables k. Note that it is possible for R^2_a to have a negative value.
Total Sum of Squares	$SST=SS_{total}=\sum(y_i-\bar{y})^2$	This is the Total Sum of Squares which is the difference between our actual values of y and the Mean of y.
Sum of Squares due to Error	$SSE=SS_{resid}=\sum(y_i-\hat{y}_i)^2$	$\hat{y}$ is the predicted value of y (i.e. mx+c in a Simple Linear Regression). SSE is also called the Sum of the Squares for Residuals. In Monitoring and Targeting, the term Variance is used for the difference between actual and predicted so this statistic is simply the sum of variances squared.

Item	Formula	Notes. Excel Function
Sum of Squares due to Regression	$SSR = SS_{regress} = \sum(\hat{y}_i - \bar{y})^2$	This is the Sum of Squares for the Regression, which is the difference between the value of ŷ predicted by our Linear Regression model and the Mean value of y.
Relationship between the Squares	$SST = SSR + SSE$	Since SSR is the difference between the predicted value ŷ and the Mean of y, and SSE is the difference between the actual value of y and the predicted value of ŷ, and SST is the difference between our actual value of y and the Mean value of y, we can conclude that SST=SSR+SSE.
Mean Square Error	$s^2 = MSE = \frac{SSE}{n-2} = \frac{\sum(y_i - \hat{y}_i)^2}{n-2}$	The Sum of Squares of our Errors, SSE, will grow as the number of points in our series increases. We can get to a Mean of this value by dividing by n-2. The reason we take 2 away from the number of values in our series is because of what is known as the "degree of freedom" in the calculation (in this case, since SSE uses our Linear Regression model there is one estimated value for the intercept and one for the slope, so the number of changeable values is decreased by 2 - hence 2 degrees of freedom less than the total number of values in the series). Because of the adjustment for degrees of freedom, this is sometimes called an unbiased estimator of s^2.
Standard Error or Standard Deviation of the Error	$s = \sqrt{MSE} = \sqrt{\frac{SSE}{n-2}} = \sqrt{\frac{\sum(y_i - \hat{y}_i)^2}{n-2}}$	This is sometimes also called the Root Mean Square Error (RMSE). It is an estimate of the Standard Deviation of the Residuals or Variances of the observed y values compared to the predicted ŷ values. =STEYX(known_y's, known_x's)
t-value	$t = \frac{r_{xy}}{\sqrt{\frac{1-r_{xy}^2}{n-2}}}$ or $t = \frac{r_{xy}\sqrt{n-2}}{\sqrt{1-r_{xy}^2}}$	The t-Test is a test of significance. The formula returns a value, which then needs to be looked up in the t critical values Table (page 781) against the desired probability α, on the basis of which one can conclude that the result meets the level of significance required or does not and so must be rejected as due to random influences. For the interpretation of significance in the analysis of resource data we would propose that the r_{xy} and R^2 critical values tables provided later in this chapter are used instead.
Mean Square Regression	$MSR = \frac{SSR}{k} = \frac{\sum(\hat{y}_i - \bar{y})^2}{k}$	Dividing the Sum of Squares due to Regression by the number of variables in the Regression, k, provides a statistic called the Mean Square due to Regression or MSR.
F-value	$F = \frac{MSR}{MSE}$	The F-value can be used in another statistical test of significance. It is particularly useful when testing the significance of multivariable regressions. As with the t-value, the results need to be looked up on a critical values table.

24.2 Exergy value

24.1 **Exergy values for energy sources.** The quality factors and exergy for different energy sources. The reference temperature Tc is in this case assumed to be 298K (20 °C) *Source: Based on Schijndel et al.*[646] *and Dincer and Rosen,*[214] *table by Niall Enright.*

Energy Source	Energy (kWh)	Quality Factor	Exergy (KWh)
Water (80 °C)	100	0.16	16
Steam (1 bar, 120 °C)	100	0.24	24
Steam (1 bar, 600 °C)	100	0.6	60
Sunlight	100	0.9	90
Nuclear Energy	100	0.95	95
Natural Gas, Gasoline, Kerosene, Fuel Oil	100	0.99	99
Electricity	100	1.00	100

24.3 Excel line fitting

The Excel Trendline tool enables non-linear relationships between variables to be described. Despite being a relatively easy tool to use it provides for a number of different relationships which should meet the needs of most single-variable models.

Excel's line-fitting tools work on data that has been charted. The chart type that we will use for modelling resource use will usually be an X-Y scatter. Since we cannot chart multiple regression these Excel line-fitting routines are strictly single-variable only.

First we want to have our data in two columns, for ease with the Variable (X-Value) first and the resource (Y-Value) second, select the data and select Insert, Charts, X-Y Scatter. This will create a simple scatter chart. We can then follow either of the two methods shown left to insert a trendline.

The default trendline inserted will be a linear trend. If we want to delete this line we simply select it and press the delete key. If we want to modify this line we again select this and in Excel (version 2013 onwards) will display the format Trendline panel, shown opposite (for Excel 2010 and earlier, select the line, right-click and choose Format Trendline, which displays the same settings as the panel but in a dialogue box).

The first line type is Exponential, which will calculate an exponential equation, in the form $y = ce^{nx}$. This equation is based on the exponential of e, Euler's number, approximately 2.712, and is called a natural exponential. The line created by this function tends to have a pronounced curve which increases rapidly as the value of x rises. We have already seen an exponential function in our earlier discussion on compound growth, see *The Limits to Growth (page 18)*, where doubling of value y every n periods is indicated by the equation $y = 2^n$. Another example of an exponential curve is radioactive decay, where the proportion of an isotope decreases exponentially over time.

The next option is the linear equation, which we have already seen in our simple linear regression. The line that will be placed is the Best Fit Line.

We then have the option to place a logarithmic equation which will take the form $y = n \ln(x) - c$. This type of line is most useful when the rate of change in the data increases or decreases quickly and then levels out. For example, if a system becomes capacity constrained its energy use may level out as demand increases. The fourth choice is a power curve which takes the form $y = nx^c$. An example of a power curve in practice is the distance travelled by a car that is accelerating at a constant rate. Please note that a zero x value will prevent an exponential trendline from being created, while a zero x or y value will prevent a power trendline from being drawn.

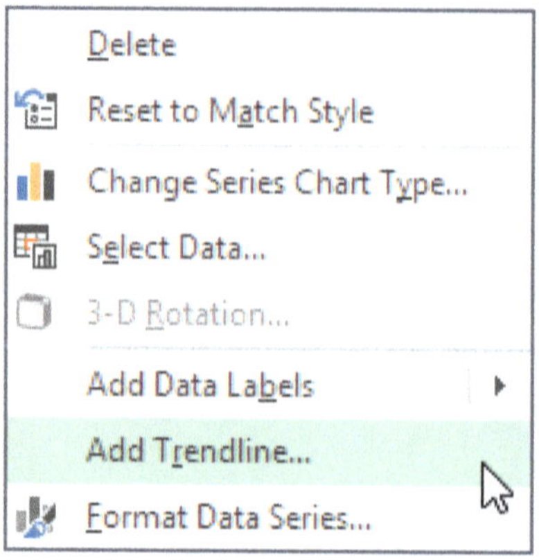

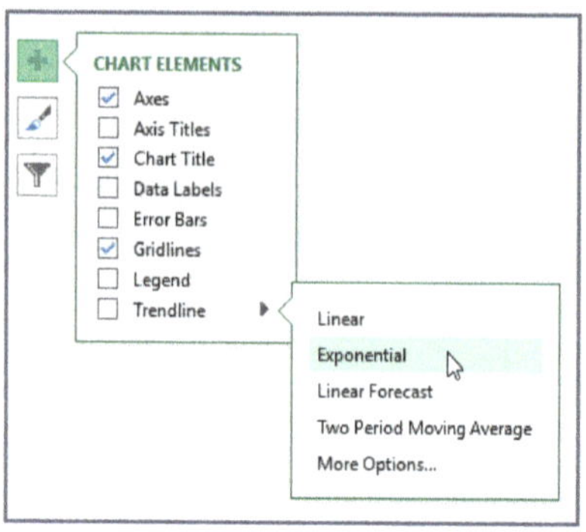

24.2 **Two methods to add a Trendline** The first method, illustrated top, involves selecting the data series in the chart and then clicking the right mouse button and selecting Add Trendline. The second option, shown bottom, (available for Excel 2013 on) is to click on the "+" button that appears to the right of the chart and selecting Trendline and then the type of line. *Source: Niall Enright (image is of Excel 2016)*

Polynomial equations can be used where the data fluctuates. Usually the number of peaks and troughs in the data (hills and valleys) indicates the order that you would choose to get the closest fit, thus Order 2 polynomial trendline generally has only one peak or trough. Order 3 generally has one or two peaks or troughs. Order 4 generally has up to three, and so forth. An example of 2-Order polynomial fit is the relationship between speed and km per litre, which rises until around 70-80 km per hour (when the car is most efficient) and then declines. Here there is basically one peak.

The next option, Moving Average, is not used with X-Y scatter data, although it can be useful with trend data. Other options offered by Excel include setting the name of the Trendline, which will be shown in the legend of the chart. There is an option to forecast forwards or backwards for a number of periods (which means a value of x - the predictor variable). If you select the lowest x-value and put this number in the Forecast Backward box, then the line will be projected back to x = 0 or the Y-axis. This makes it easy to visualize our intercept, although we do need to bear in mind the earlier advice that a relationship only holds good for the range over which we actually have values so if the first data point is far from the Y-axis then forecast is less certain. If we know that our intercept should be a particular value, we can force this with the Set Intercept option.

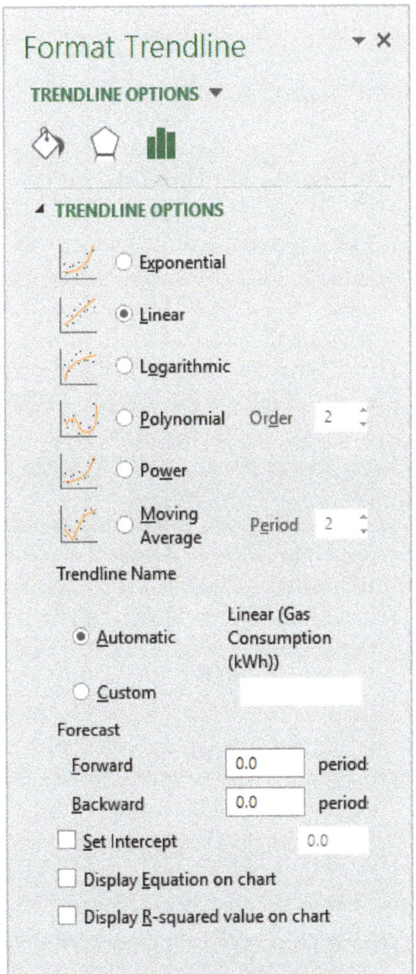

24.3 **Format Trendline**
This panel (or dialogue box) determines the type of line-fitting that Excel will apply to the data and allows a number of other options to be set.
Source: Niall Enright (image is of Excel 2016)

Finally, the two most useful options of the Trendline panel are tucked away right at the bottom. We can choose to display the equation of the line and the Coefficient of Determination, R^2. The R-squared option will allow us to test a number of line types to confirm which works best. If the data fits a generally linear pattern, the curve fitting options, exponential, logarithmic and power, will tend to reduce the Correlation of Determination (although the polynomial one will usually slightly increase it but should only be used where the relationship between data is obviously curved, as it introduces complexity for little benefit).

24.4 Excel data Analysis Toolpak

The Excel Regression statistics tool outputs are described here. The examples given refer to Excel 2016 Desktop, but the functionality is available in earlier version of Excel Desktop.

Excel provides some very powerful regression analysis tools. These are available as part of the Analysis Toolpak, which ships free of charge with Excel, but which may not be installed by default. To establish if the Analysis Toolpak is installed in Excel 2016, you can go to the Data menu on the menu ribbon, illustrated below:

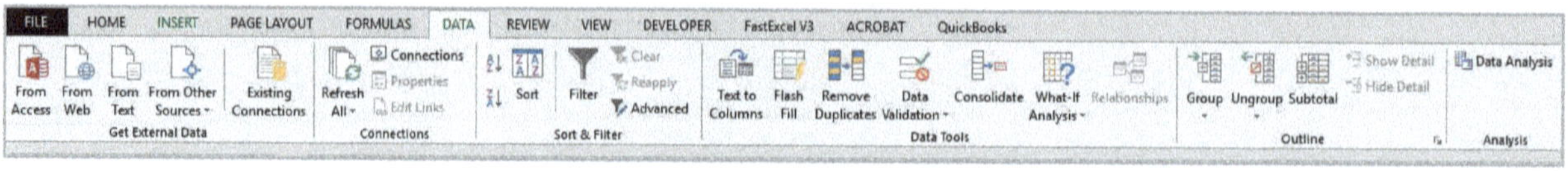

24.4 **The Excel data ribbon**
The statistical functions described in this section can be accessed from the Data Analysis option, usually shown on the far right hand side of the data ribbon.
Source: Niall Enright (image is of Excel 2016)

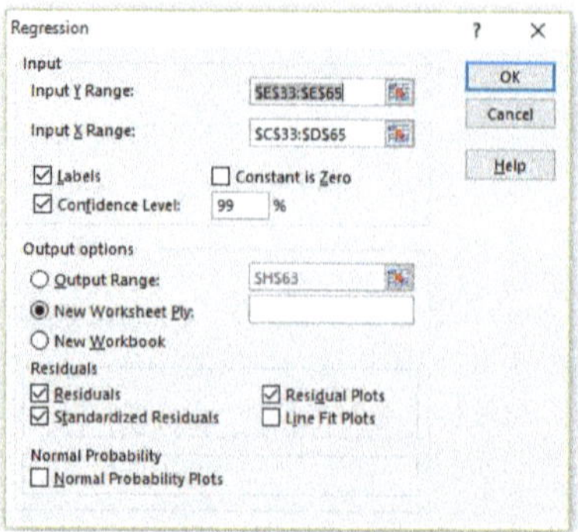

24.5 **(above) Regression dialogue**
Here we set out the inputs and outputs of the Regression Analysis we wish Excel to perform.
Source: Niall Enright

If the Analysis Toolpak is installed, the statistical functions can be accessed by selecting the **Data Analysis** option on the far right of the ribbon as shown above. If the Analysis Toolpak is not installed then you will need to go to **File, Options,** and then select **Add-ins** on the left of the dialogue box which appears. Then at the bottom of the same dialogue, one should select **Excel Add-ins** from the **Manage** drop-down list and then press **Go.** This will display a list of add-ins and we need to ensure the Analysis Toolpak entry is ticked (the Analysis Toolpak - VBA is not necessary) and then press **OK.** It may be possible that the security settings for Microsoft Office prevent you from accessing these choices, in which case you will need to speak to the systems administrators to get the Analysis Toolpak installed.

When the Data Analysis menu choice is selected a dialogue box will be displayed with a number of statistical tools from which we will select **Regression.** This will display an additional dialogue box shown in Figure 24.5. Using this we must, as a minimum, inform Excel of the single Y range for the resource-use data (in a column) and then one or more columns for the X range (variables). The X range does not have to be contiguous as long as the number of data points for each variable matches the number of points in the Y series. Up to 16 variables can be modelled at the same time. The last mandatory choice is to inform Excel where we want the output to go (by default this will be a new worksheet as the output is a range of cells at least 9 columns by 17 rows, but we can set the output location to be a cell in the existing sheet). Please note that I have ticked the Labels box and included the column headers in my X and Y series so that my outputs are labelled and I have additionally requested Confidence Level data for 99% limits, as well as the Residuals, Standard Residuals and Residual plots.

24.6 **(opposite) Excel regression outputs**
These outputs have been colour-coded for clarity with the orange outputs being the standard ones and the residual outputs in blue. The statistical outputs have been labelled and more information about their interpretation can be found in the earlier table of statistics or in Chapter 14. The example provided is for a two-variable analysis whose resulting equation is:
y=(27.9*Prodn.)+(723.7*Cooling DD)+ 523416
The data is available on the website.
Source: Niall Enright

SUMMARY OUTPUT

Regression Statistics		
Multiple R	0.768395536	The **Pearson Correlation** Coefficient, r
R Square	0.590431699	The **Coefficient of Determination**, R^2
Adjusted R Square	0.56218561	The **Adjusted Coefficient of Determination**, R^2_a
Standard Error	100790.1066	The **Standard Error**
Observations	32	The number of points in our series

Sum of Squares of Regression, **SSR**

Sum of Squares of Residuals, **SSE**

ANOVA

	df	SS	MS	F	Significance F
Regression	2	4.24695E+11	2.12347E+11	20.903131	2.39184E-06
Residual	29	2.94601E+11	10158645583		
Total	31	7.19296E+11			

The **F Statistic** and its **significance**

The smaller this is the more significant the correlation. We are usually looking for a value < 0.05 for 95% confidence.

	Coefficients	Standard Error	t Stat	P-value	Lower 95%	Upper 95%	Lower 95.0%	Upper 95.0%
Intercept	523416.4098	92465.43475	5.660671052	4.056E-06	334303.3617	712529.4578	334303.3617	712529.4578
X Variable 1	27.90625267	6.415436996	4.349859985	0.0001536	14.78521076	41.02729458	14.78521076	41.02729458
X Variable 2	723.7050211	169.181706	4.277678942	0.0001874	377.689581	1069.720461	377.689581	1069.720461

Intercept, c

Coefficient, m_1

Coefficient, m_2

The **t-Statistic** and its **significance**.

We are interested in the significance of each variable. We are usually looking for a value < 0.05 for 95% confidence the variable is significant.

Upper and Lower **95%**

This gives the upper and lower boundaries where 95% of the intercept or variable values will lie if the distribution of the data is normal.

RESIDUAL OUTPUT

Observation	Predicted Y	Residuals	Standard Residuals
1	1128999.284	-5546.283987	-0.056893945
2	1144448.418	158195.5822	1.622774948
3	1182322.818	51705.18164	0.530393278
4	1340263.222	-20441.22192	-0.209686657
5	1101470.888	258363.1119	2.650296423
6	1118268.25	90545.74953	0.92882097
7	981425.0169	105157.9831	1.078713692
8	899393.6369	200555.3631	2.05730283
9	995219.9935	-20153.99351	-0.20674026
10	942032.5335	7431.466537	0.076232203
11	996643.2124	-115728.2124	-1.187143416
12	993776.4276	-29176.42764	-0.299292655
13	835049.4636	123497.5364	1.266841371
14	965148.8438	-97289.84381	-0.998002088
15	1014806.89	-100853.8902	-1.03456218
16	1086968.522	-22668.52185	-0.232534366
17	1048652.156	-21319.15609	-0.218692532
18	1011584.375	-44506.37497	-0.456547707
19	938794.7485	-37023.74848	-0.379790704
20	937165.1649	39421.83505	0.404390347
21	894119.275	-28215.27502	-0.289433123
22	945056.9731	-145208.9731	-1.489557929
23	993643.6939	-125950.6939	-1.292005934
24	819847.943	-14362.94305	-0.147335493
25	1023552.203	-140027.2029	-1.436403178
26	1057751.078	-83960.07792	-0.861264956
27	965096.7913	37760.20873	0.387345335
28	1057707.258	-107467.258	-1.102402302
29	1003119.556	-8921.556096	-0.091517586
30	913282.2363	62943.76374	0.645678985
31	845357.9957	-19268.99571	-0.197661926
32	740833.1313	52512.86866	0.538678556

Standard Residuals or **z-score** can help identify outliers. We would expect a t-score of 2.65 once every 125 readings so the data for this observation is worth checking, since we only have 32 values in our series.

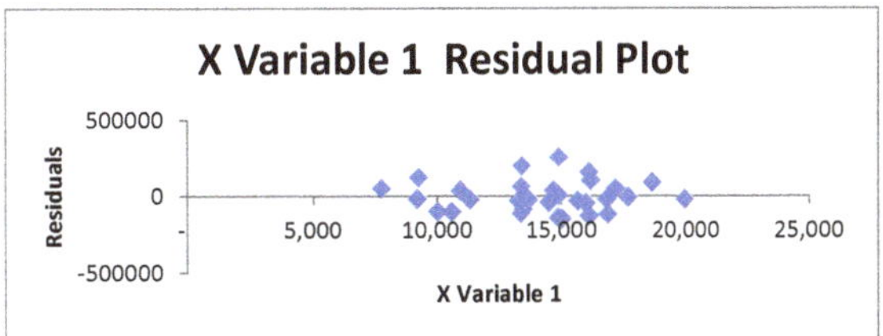

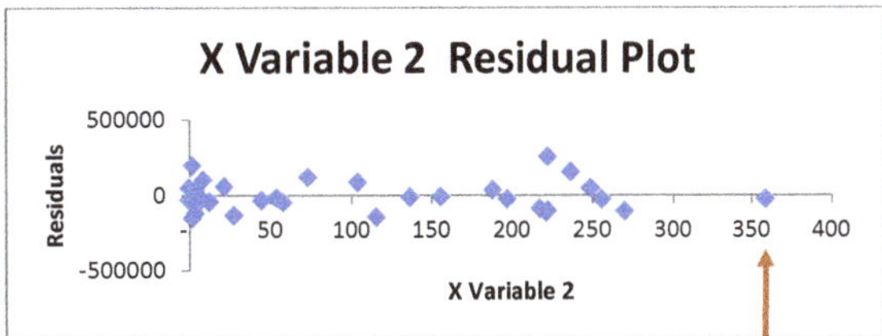

This is what we would call the **variance** in Monitoring and Targeting. I would format this with no decimals and a comma separator to make the differences stand out visually.

Residual Plots for each variable are useful in assessing if there is an outlier. We are looking for points spaced equally around the X-axis. For example, the Cooling DD value at 360 is high but the residual from predicted is minimal so this is likely to be a valid data point.

Reference

24.5 z-distribution tables

The following tables display the probability of a value in a normal distribution falling a given number of standard deviations from the mean (using the z-score of the value).

To save space one looks up the first two digits of the z-score on the left-hand columns of the tables to select the desired row and then the third digit in the column going across the table. The probability is shown at the intersection of the row and column. The colour banding is purely to aid selection of the rows and has no special meaning.

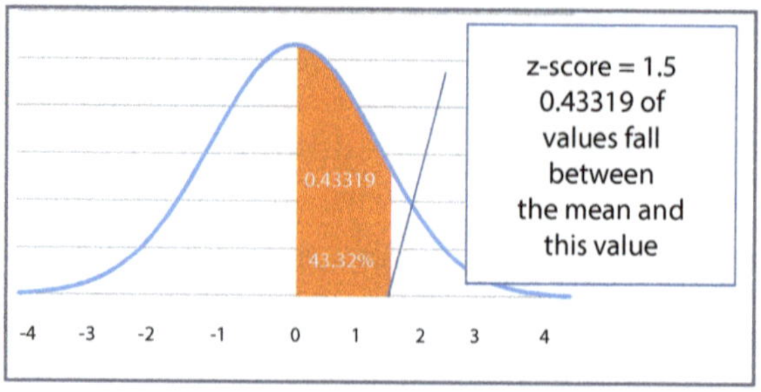

Z-Table 1: shows the probability of a value lying between the mean and a given z-value (whether negative or positive). For example, if we look up z=1.5 we see a probability of 0.43319, which means that 43.319% of the values will fall in the orange-shaded section of the normal distribution.

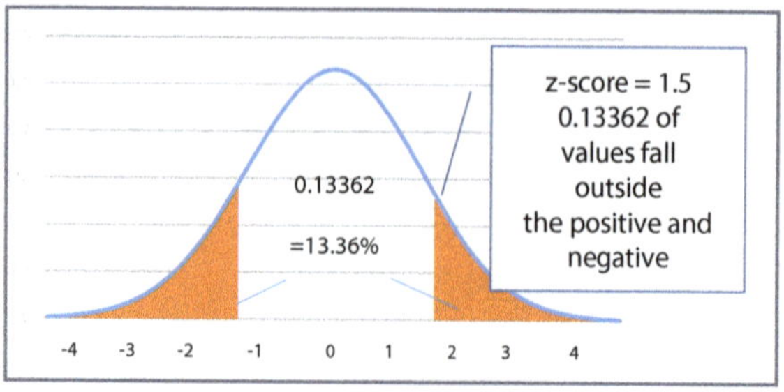

Z-Table 2: shows the probability of a value lying outside both positive and negative of the given z-value. For example, if we look up z=1.5 we see a probability of 0.13362, which means that 13.36% of the values will fall in the orange-shaded section of the normal distribution.

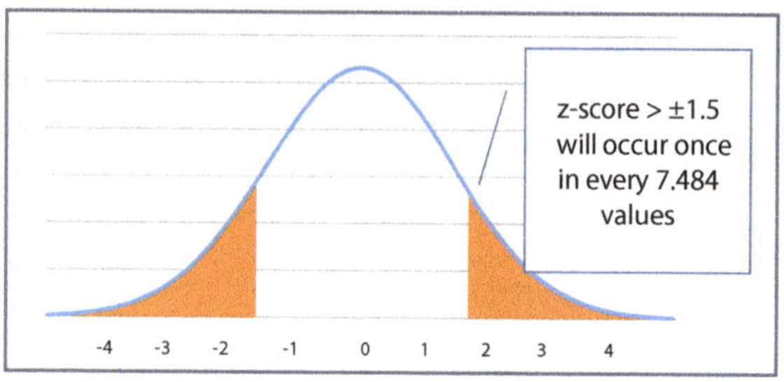

Z-Table 3: shows the same data as Table 2, but in this case as a frequency of a value lying outside both positive and negative of the given z-value. For example, if we look up z=1.5 we see the frequency of 7.484, which means that one out of every 7.5 readings, approximately, will fall in the orange zone of our distribution.

24.7 **Z-Table 1. Standardized normal distribution for one tail**

This shows the probability of values lying between the mean and a positive or negative z-score.
The first two digits of the score are read from the vertical column left, and the third digit from the top row.
Source: Data from NIST/SEMATECH e-Handbook of Statistical Methods[555] modified by Niall Enright

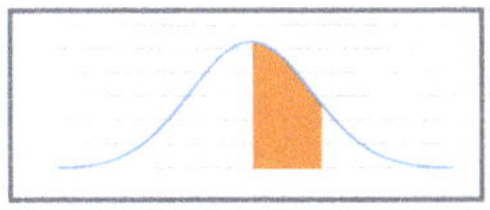

Z-score	0	0.01	0.02	0.03	0.04	0.05	0.06	0.07	0.08	0.09
0	0	0.00399	0.00798	0.01197	0.01595	0.01994	0.02392	0.0279	0.03188	0.03586
0.1	0.03983	0.0438	0.04776	0.05172	0.05567	0.05962	0.06356	0.06749	0.07142	0.07535
0.2	0.07926	0.08317	0.08706	0.09095	0.09483	0.09871	0.10257	0.10642	0.11026	0.11409
0.3	0.11791	0.12172	0.12552	0.1293	0.13307	0.13683	0.14058	0.14431	0.14803	0.15173
0.4	0.15542	0.1591	0.16276	0.1664	0.17003	0.17364	0.17724	0.18082	0.18439	0.18793
0.5	0.19146	0.19497	0.19847	0.20194	0.2054	0.20884	0.21226	0.21566	0.21904	0.2224
0.6	0.22575	0.22907	0.23237	0.23565	0.23891	0.24215	0.24537	0.24857	0.25175	0.2549
0.7	0.25804	0.26115	0.26424	0.2673	0.27035	0.27337	0.27637	0.27935	0.2823	0.28524
0.8	0.28814	0.29103	0.29389	0.29673	0.29955	0.30234	0.30511	0.30785	0.31057	0.31327
0.9	0.31594	0.31859	0.32121	0.32381	0.32639	0.32894	0.33147	0.33398	0.33646	0.33891
1	0.34134	0.34375	0.34614	0.34849	0.35083	0.35314	0.35543	0.35769	0.35993	0.36214
1.1	0.36433	0.3665	0.36864	0.37076	0.37286	0.37493	0.37698	0.379	0.381	0.38298
1.2	0.38493	0.38686	0.38877	0.39065	0.39251	0.39435	0.39617	0.39796	0.39973	0.40147
1.3	0.4032	0.4049	0.40658	0.40824	0.40988	0.41149	0.41308	0.41466	0.41621	0.41774
1.4	0.41924	0.42073	0.4222	0.42364	0.42507	0.42647	0.42785	0.42922	0.43056	0.43189
1.5	0.43319	0.43448	0.43574	0.43699	0.43822	0.43943	0.44062	0.44179	0.44295	0.44408
1.6	0.4452	0.4463	0.44738	0.44845	0.4495	0.45053	0.45154	0.45254	0.45352	0.45449
1.7	0.45543	0.45637	0.45728	0.45818	0.45907	0.45994	0.4608	0.46164	0.46246	0.46327
1.8	0.46407	0.46485	0.46562	0.46638	0.46712	0.46784	0.46856	0.46926	0.46995	0.47062
1.9	0.47128	0.47193	0.47257	0.4732	0.47381	0.47441	0.475	0.47558	0.47615	0.4767
2	0.47725	0.47778	0.47831	0.47882	0.47932	0.47982	0.4803	0.48077	0.48124	0.48169
2.1	0.48214	0.48257	0.483	0.48341	0.48382	0.48422	0.48461	0.485	0.48537	0.48574
2.2	0.4861	0.48645	0.48679	0.48713	0.48745	0.48778	0.48809	0.4884	0.4887	0.48899
2.3	0.48928	0.48956	0.48983	0.4901	0.49036	0.49061	0.49086	0.49111	0.49134	0.49158
2.4	0.4918	0.49202	0.49224	0.49245	0.49266	0.49286	0.49305	0.49324	0.49343	0.49361
2.5	0.49379	0.49396	0.49413	0.4943	0.49446	0.49461	0.49477	0.49492	0.49506	0.4952
2.6	0.49534	0.49547	0.4956	0.49573	0.49585	0.49598	0.49609	0.49621	0.49632	0.49643
2.7	0.49653	0.49664	0.49674	0.49683	0.49693	0.49702	0.49711	0.4972	0.49728	0.49736
2.8	0.49744	0.49752	0.4976	0.49767	0.49774	0.49781	0.49788	0.49795	0.49801	0.49807
2.9	0.49813	0.49819	0.49825	0.49831	0.49836	0.49841	0.49846	0.49851	0.49856	0.49861
3	0.49865	0.49869	0.49874	0.49878	0.49882	0.49886	0.49889	0.49893	0.49896	0.499
3.1	0.49903	0.49906	0.4991	0.49913	0.49916	0.49918	0.49921	0.49924	0.49926	0.49929
3.2	0.49931	0.49934	0.49936	0.49938	0.4994	0.49942	0.49944	0.49946	0.49948	0.4995
3.3	0.49952	0.49953	0.49955	0.49957	0.49958	0.4996	0.49961	0.49962	0.49964	0.49965
3.4	0.49966	0.49968	0.49969	0.4997	0.49971	0.49972	0.49973	0.49974	0.49975	0.49976
3.5	0.49977	0.49978	0.49978	0.49979	0.4998	0.49981	0.49981	0.49982	0.49983	0.49983
3.6	0.49984	0.49985	0.49985	0.49986	0.49986	0.49987	0.49987	0.49988	0.49988	0.49989
3.7	0.49989	0.4999	0.4999	0.4999	0.49991	0.49991	0.49992	0.49992	0.49992	0.49992
3.8	0.49993	0.49993	0.49993	0.49994	0.49994	0.49994	0.49994	0.49995	0.49995	0.49995
3.9	0.49995	0.49995	0.49996	0.49996	0.49996	0.49996	0.49996	0.49996	0.49997	0.49997
4	0.49997	0.49997	0.49997	0.49997	0.49997	0.49997	0.49998	0.49998	0.49998	0.49998

24.8 **Z-Table 2. Standardized normal distribution for two tails (probability)**

This table shows the probability of values lying outside a positive or negative z-score.
The first two digits of the score are read from the vertical column left, and the third digit from the top row.
Source: Data from NIST/SEMATECH e-Handbook of Statistical Methods[555] *modified by Niall Enright*

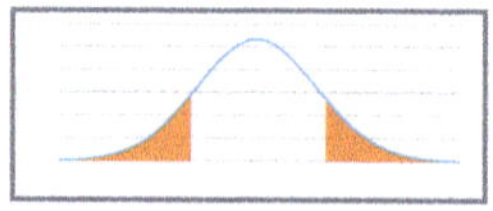

Z-score	0	0.01	0.02	0.03	0.04	0.05	0.06	0.07	0.08	0.09
0	1.00000	0.99202	0.98404	0.97606	0.96810	0.96012	0.95216	0.94420	0.93624	0.92828
0.1	0.92034	0.91240	0.90448	0.89656	0.88866	0.88076	0.87288	0.86502	0.85716	0.84930
0.2	0.84148	0.83366	0.82588	0.81810	0.81034	0.80258	0.79486	0.78716	0.77948	0.77182
0.3	0.76418	0.75656	0.74896	0.74140	0.73386	0.72634	0.71884	0.71138	0.70394	0.69654
0.4	0.68916	0.68180	0.67448	0.66720	0.65994	0.65272	0.64552	0.63836	0.63122	0.62414
0.5	0.61708	0.61006	0.60306	0.59612	0.58920	0.58232	0.57548	0.56868	0.56192	0.55520
0.6	0.54850	0.54186	0.53526	0.52870	0.52218	0.51570	0.50926	0.50286	0.49650	0.49020
0.7	0.48392	0.47770	0.47152	0.46540	0.45930	0.45326	0.44726	0.44130	0.43540	0.42952
0.8	0.42372	0.41794	0.41222	0.40654	0.40090	0.39532	0.38978	0.38430	0.37886	0.37346
0.9	0.36812	0.36282	0.35758	0.35238	0.34722	0.34212	0.33706	0.33204	0.32708	0.32218
1	0.31732	0.31250	0.30772	0.30302	0.29834	0.29372	0.28914	0.28462	0.28014	0.27572
1.1	0.27134	0.26700	0.26272	0.25848	0.25428	0.25014	0.24604	0.24200	0.23800	0.23404
1.2	0.23014	0.22628	0.22246	0.21870	0.21498	0.21130	0.20766	0.20408	0.20054	0.19706
1.3	0.19360	0.19020	0.18684	0.18352	0.18024	0.17702	0.17384	0.17068	0.16758	0.16452
1.4	0.16152	0.15854	0.15560	0.15272	0.14986	0.14706	0.14430	0.14156	0.13888	0.13622
1.5	0.13362	0.13104	0.12852	0.12602	0.12356	0.12114	0.11876	0.11642	0.11410	0.11184
1.6	0.10960	0.10740	0.10524	0.10310	0.10100	0.09894	0.09692	0.09492	0.09296	0.09102
1.7	0.08914	0.08726	0.08544	0.08364	0.08186	0.08012	0.07840	0.07672	0.07508	0.07346
1.8	0.07186	0.07030	0.06876	0.06724	0.06576	0.06432	0.06288	0.06148	0.06010	0.05876
1.9	0.05744	0.05614	0.05486	0.05360	0.05238	0.05118	0.05000	0.04884	0.04770	0.04660
2	0.04550	0.04444	0.04338	0.04236	0.04136	0.04036	0.03940	0.03846	0.03752	0.03662
2.1	0.03572	0.03486	0.03400	0.03318	0.03236	0.03156	0.03078	0.03000	0.02926	0.02852
2.2	0.02780	0.02710	0.02642	0.02574	0.02510	0.02444	0.02382	0.02320	0.02260	0.02202
2.3	0.02144	0.02088	0.02034	0.01980	0.01928	0.01878	0.01828	0.01778	0.01732	0.01684
2.4	0.01640	0.01596	0.01552	0.01510	0.01468	0.01428	0.01390	0.01352	0.01314	0.01278
2.5	0.01242	0.01208	0.01174	0.01140	0.01108	0.01078	0.01046	0.01016	0.00988	0.00960
2.6	0.00932	0.00906	0.00880	0.00854	0.00830	0.00804	0.00782	0.00758	0.00736	0.00714
2.7	0.00694	0.00672	0.00652	0.00634	0.00614	0.00596	0.00578	0.00560	0.00544	0.00528
2.8	0.00512	0.00496	0.00480	0.00466	0.00452	0.00438	0.00424	0.00410	0.00398	0.00386
2.9	0.00374	0.00362	0.00350	0.00338	0.00328	0.00318	0.00308	0.00298	0.00288	0.00278
3	0.00270	0.00262	0.00252	0.00244	0.00236	0.00228	0.00222	0.00214	0.00208	0.00200
3.1	0.00194	0.00188	0.00180	0.00174	0.00168	0.00164	0.00158	0.00152	0.00148	0.00142
3.2	0.00138	0.00132	0.00128	0.00124	0.00120	0.00116	0.00112	0.00108	0.00104	0.00100
3.3	0.00096	0.00094	0.00090	0.00086	0.00084	0.00080	0.00078	0.00076	0.00072	0.00070
3.4	0.00068	0.00064	0.00062	0.00060	0.00058	0.00056	0.00054	0.00052	0.00050	0.00048
3.5	0.00046	0.00044	0.00044	0.00042	0.00040	0.00038	0.00038	0.00036	0.00034	0.00034
3.6	0.00032	0.00030	0.00030	0.00028	0.00028	0.00026	0.00026	0.00024	0.00024	0.00022
3.7	0.00022	0.00020	0.00020	0.00020	0.00018	0.00018	0.00016	0.00016	0.00016	0.00016
3.8	0.00014	0.00014	0.00014	0.00012	0.00012	0.00012	0.00012	0.00010	0.00010	0.00010
3.9	0.00010	0.00010	0.00008	0.00008	0.00008	0.00008	0.00008	0.00008	0.00006	0.00006
4	0.00006	0.00006	0.00006	0.00006	0.00006	0.00006	0.00004	0.00004	0.00004	0.00004

24.9 **Z-Table 3. Standardized normal distribution for two tails (frequency)**
This shows the number of values needed to obtain a value outside a given positive or negative z-score.
The first two digits of the score are read from the vertical column left, and the third digit from the top row.
Source: Data from NIST/SEMATECH e-Handbook of Statistical Methods[555] modified by Niall Enright

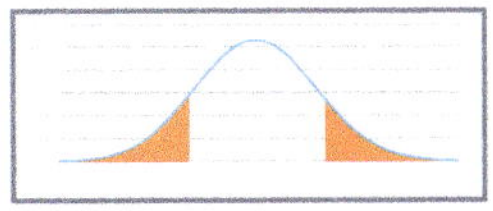

Z-score	0	0.01	0.02	0.03	0.04	0.05	0.06	0.07	0.08	0.09
0	1.000	1.008	1.016	1.025	1.033	1.042	1.050	1.059	1.068	1.077
0.1	1.087	1.096	1.106	1.115	1.125	1.135	1.146	1.156	1.167	1.177
0.2	1.188	1.200	1.211	1.222	1.234	1.246	1.258	1.270	1.283	1.296
0.3	1.309	1.322	1.335	1.349	1.363	1.377	1.391	1.406	1.421	1.436
0.4	1.451	1.467	1.483	1.499	1.515	1.532	1.549	1.567	1.584	1.602
0.5	1.621	1.639	1.658	1.678	1.697	1.717	1.738	1.758	1.780	1.801
0.6	1.823	1.845	1.868	1.891	1.915	1.939	1.964	1.989	2.014	2.040
0.7	2.066	2.093	2.121	2.149	2.177	2.206	2.236	2.266	2.297	2.328
0.8	2.360	2.393	2.426	2.460	2.494	2.530	2.566	2.602	2.639	2.678
0.9	2.717	2.756	2.797	2.838	2.880	2.923	2.967	3.012	3.057	3.104
1	3.151	3.200	3.250	3.300	3.352	3.405	3.459	3.513	3.570	3.627
1.1	3.685	3.745	3.806	3.869	3.933	3.998	4.064	4.132	4.202	4.273
1.2	4.345	4.419	4.495	4.572	4.652	4.733	4.816	4.900	4.987	5.075
1.3	5.165	5.258	5.352	5.449	5.548	5.649	5.752	5.859	5.967	6.078
1.4	6.191	6.308	6.427	6.548	6.673	6.800	6.930	7.064	7.200	7.341
1.5	7.484	7.631	7.781	7.935	8.093	8.255	8.420	8.590	8.764	8.941
1.6	9.124	9.311	9.502	9.699	9.901	10.107	10.318	10.535	10.757	10.987
1.7	11.218	11.460	11.704	11.956	12.216	12.481	12.755	13.034	13.319	13.613
1.8	13.916	14.225	14.543	14.872	15.207	15.547	15.903	16.265	16.639	17.018
1.9	17.409	17.813	18.228	18.657	19.091	19.539	20.000	20.475	20.964	21.459
2	21.978	22.502	23.052	23.607	24.178	24.777	25.381	26.001	26.652	27.307
2.1	27.996	28.686	29.412	30.139	30.902	31.686	32.489	33.333	34.176	35.063
2.2	35.971	36.900	37.850	38.850	39.841	40.917	41.982	43.103	44.248	45.413
2.3	46.642	47.893	49.164	50.505	51.867	53.248	54.705	56.243	57.737	59.382
2.4	60.976	62.657	64.433	66.225	68.120	70.028	71.942	73.964	76.104	78.247
2.5	80.515	82.781	85.179	87.719	90.253	92.764	95.602	98.425	101.2	104.2
2.6	107.3	110.4	113.6	117.1	120.5	124.4	127.9	131.9	135.9	140.1
2.7	144.1	148.8	153.4	157.7	162.9	167.8	173.0	178.6	183.8	189.4
2.8	195.3	201.6	208.3	214.6	221.2	228.3	235.8	243.9	251.3	259.1
2.9	267.4	276.2	285.7	295.9	304.9	314.5	324.7	335.6	347.2	359.7
3	370.4	381.7	396.8	409.8	423.7	438.6	450.5	467.3	480.8	500.0
3.1	515.5	531.9	555.6	574.7	595.2	609.8	632.9	657.9	675.7	704.2
3.2	724.6	757.6	781.3	806.5	833.3	862.1	892.9	925.9	961.5	1,000
3.3	1,042	1,064	1,111	1,163	1,190	1,250	1,282	1,316	1,389	1,429
3.4	1,471	1,563	1,613	1,667	1,724	1,786	1,852	1,923	2,000	2,083
3.5	2,174	2,273	2,273	2,381	2,500	2,632	2,632	2,778	2,941	2,941
3.6	3,125	3,333	3,333	3,571	3,571	3,846	3,846	4,167	4,167	4,545
3.7	4,545	5,000	5,000	5,000	5,556	5,556	6,250	6,250	6,250	6,250
3.8	7,143	7,143	7,143	8,333	8,333	8,333	8,333	10,000	10,000	10,000
3.9	10,000	10,000	12,500	12,500	12,500	12,500	12,500	12,500	16,667	16,667
4	16,667	16,667	16,667	16,667	16,667	16,667	25,000	25,000	25,000	25,000

24.6 t critical values table

Where a t-value is greater than the value indicated in this table, we can conclude that the hypothesis we are testing (that our correlation is due to chance) is false. In other words, a t-value greater than the t-critical lookup value in the table indicates that the result is significant.

The second column in this table shows the degrees of freedom. This should be the number of records in our series less 2. This is the row that we will select for our look-up.

The columns represent the probability that we want to test against (this is the desired significance level, denoted by the letter alpha α). You will note that there are two values for α. This is because we can test if our value falls above or below the expected value (a two tail test) or if it fall on one side only (a one tail test). In the vast majority of cases (e.g. when assessing the data from a regression analysis), we will use a two tail test, and so this is shown highlighted red. For example in a regression analysis some values will fall above the best fit line and some below, so the two tail significance is the one to use. Meter readings can be ± the real value so again a two tail test applies.

To check there is a less than 5% probability that the data we have obtained is due to chance, we would select the column 0.05 (that is the same as saying that we are 95% confident that the result is not due to chance as confidence is 1-α).

For example, if the number of records N in my analysis is 25, I would go along the row labelled 23 (N-2) and see that for 95% confidence the critical t-value that must be exceeded is 2.069 for a two tail test. If my t-value is greater than this I can assert that the hypothesis that the values obtained are due to chance is false, and therefore my correlation is significant (at a 0.05 level).

A general rule of thumb is that for a reasonable data series, (i.e. 60 or greater values) t needs to be greater than 2 to confirm that our results are not due to chance, for a two tail test with 95% confidence.

Please note that colour banding is purely to aid selection of the rows and has no special meaning.

24.10 **t-critical values for one-tailed and two-tailed distributions**

Source: Data from NIST/SEMATECH e-Handbook of Statistical Methods[555] modified by Niall Enright

		Probability α					
Test	One Tail	0.10	0.05	0.025	0.01	0.005	0.001
	Two Tail	0.20	0.10	0.05	0.02	0.01	0.002
Degree of Freedom (N-2)	1	3.078	6.314	12.706	31.821	63.657	318.313
	2	1.886	2.920	4.303	6.965	9.925	22.327
	3	1.638	2.353	3.182	4.541	5.841	10.215
	4	1.533	2.132	2.776	3.747	4.604	7.173
	5	1.476	2.015	2.571	3.365	4.032	5.893
	6	1.440	1.943	2.447	3.143	3.707	5.208
	7	1.415	1.895	2.365	2.998	3.499	4.782
	8	1.397	1.860	2.306	2.896	3.355	4.499
	9	1.383	1.833	2.262	2.821	3.250	4.296
	10	1.372	1.812	2.228	2.764	3.169	4.143
	11	1.363	1.796	2.201	2.718	3.106	4.024
	12	1.356	1.782	2.179	2.681	3.055	3.929
	13	1.350	1.771	2.160	2.650	3.012	3.852
	14	1.345	1.761	2.145	2.624	2.977	3.787
	15	1.341	1.753	2.131	2.602	2.947	3.733
	16	1.337	1.746	2.120	2.583	2.921	3.686
	17	1.333	1.740	2.110	2.567	2.898	3.646
	18	1.330	1.734	2.101	2.552	2.878	3.610
	19	1.328	1.729	2.093	2.539	2.861	3.579
	20	1.325	1.725	2.086	2.528	2.845	3.552
	21	1.323	1.721	2.080	2.518	2.831	3.527
	22	1.321	1.717	2.074	2.508	2.819	3.505
	23	1.319	1.714	2.069	2.500	2.807	3.485
	24	1.318	1.711	2.064	2.492	2.797	3.467
	25	1.316	1.708	2.060	2.485	2.787	3.450
	26	1.315	1.706	2.056	2.479	2.779	3.435
	27	1.314	1.703	2.052	2.473	2.771	3.421
	28	1.313	1.701	2.048	2.467	2.763	3.408
	29	1.311	1.699	2.045	2.462	2.756	3.396
	30	1.310	1.697	2.042	2.457	2.750	3.385
	40	1.303	1.684	2.021	2.423	2.704	3.307
	50	1.299	1.676	2.009	2.403	2.678	3.261
	60	1.296	1.671	2.000	2.390	2.660	3.232
	70	1.294	1.667	1.994	2.381	2.648	3.211
	90	1.291	1.662	1.987	2.368	2.632	3.183
	100	1.290	1.660	1.984	2.364	2.626	3.174
	infinity	1.282	1.645	1.960	2.326	2.576	3.090

24.7 r_{xy} and R^2 critical values tables

These tables give the minimum (or critical) value of the Pearson correlation coefficient r_{xy} and the coefficient of determination R^2, needed to demonstrate significance at a given probability, α, given the number of values in our data.

The first table, headed in green, shows the Pearson correlation coefficient, r_{xy} values, while the second table, headed in orange, shows the coefficient of determination, R^2.

Let's imagine I have two years of monthly readings, 24 values, and I want to know what is the minimum Pearson correlation coefficient needed for significance at a confidence level of 95% (i.e. a probability, α, of 5% or 0.05). I would start by using the column in the table on the left (headed 0.05 in green). Then I would read off the row 22 (since 24 records gives us 22 degrees of freedom, N-2), and see that the minimum r_{xy} value is 0.404.

What this tells us is that a correlation of 0.404 or greater, based on 24 data values, will happen due to chance less than 0.05 (or 5%) of the time. Put another way, a correlation of 0.404 based on 24 data points shows a 95% or better significance. For negative correlations, the sign can be simply changed to a positive value - significance is not affected.

For convenience, the same table has been expressed in terms of the coefficient of determination R^2. In the example above the critical value for R^2 is 0.163 (0.404 squared, found in data column headed 0.05 in orange, using the same row for degrees of freedom = 22).

24.11 **t-critical values for two-tailed distribution**

Highlighted rows, thus, are the degrees of freedom appropriate for monthly data series with 12, 24, 36 points respectively.

Source: Data from Neag School of Education - University of Connecticut[547] and Turner[721] modified by Niall Enright

Degree of Freedom (N-2)	Critical r_{xy}-value at α (level of significance)				Critical R^2 at α (level of significance)			
	0.10	0.05	0.02	0.01	0.10	0.05	0.02	0.01
1	0.988	0.997	1.000	1.000	0.976	0.994	0.999	1.000
2	0.900	0.950	0.980	0.990	0.810	0.903	0.960	0.980
3	0.805	0.878	0.934	0.959	0.648	0.771	0.872	0.920
4	0.729	0.811	0.882	0.917	0.531	0.658	0.778	0.841
5	0.669	0.754	0.833	0.874	0.448	0.569	0.694	0.764
6	0.622	0.707	0.789	0.834	0.387	0.500	0.623	0.696
7	0.582	0.666	0.750	0.798	0.339	0.444	0.563	0.637
8	0.549	0.632	0.716	0.765	0.301	0.399	0.513	0.585
9	0.521	0.602	0.685	0.735	0.271	0.362	0.469	0.540
10	0.497	0.576	0.658	0.708	0.247	0.332	0.433	0.501
11	0.476	0.553	0.634	0.684	0.227	0.306	0.402	0.468
12	0.458	0.532	0.612	0.661	0.210	0.283	0.375	0.437
13	0.441	0.514	0.592	0.641	0.194	0.264	0.350	0.411
14	0.426	0.497	0.574	0.623	0.181	0.247	0.329	0.388
15	0.412	0.482	0.558	0.606	0.170	0.232	0.311	0.367
16	0.400	0.468	0.542	0.590	0.160	0.219	0.294	0.348
17	0.389	0.456	0.528	0.575	0.151	0.208	0.279	0.331
18	0.378	0.444	0.516	0.561	0.143	0.197	0.266	0.315
19	0.369	0.433	0.503	0.549	0.136	0.187	0.253	0.301
20	0.360	0.423	0.492	0.537	0.130	0.179	0.242	0.288
21	0.352	0.413	0.482	0.526	0.124	0.171	0.232	0.277
22	0.344	0.404	0.472	0.515	0.118	0.163	0.223	0.265
23	0.337	0.396	0.462	0.505	0.114	0.157	0.213	0.255
24	0.330	0.388	0.453	0.496	0.109	0.151	0.205	0.246
25	0.323	0.381	0.445	0.487	0.104	0.145	0.198	0.237
26	0.317	0.374	0.437	0.479	0.100	0.140	0.191	0.229
27	0.311	0.367	0.430	0.471	0.097	0.135	0.185	0.222
28	0.306	0.361	0.423	0.463	0.094	0.130	0.179	0.214
29	0.301	0.355	0.416	0.456	0.091	0.126	0.173	0.208
30	0.296	0.349	0.409	0.449	0.088	0.122	0.167	0.202
34	0.279	0.329	0.386	0.424	0.078	0.108	0.149	0.180
35	0.275	0.325	0.381	0.418	0.076	0.106	0.145	0.175
40	0.257	0.304	0.358	0.393	0.066	0.092	0.128	0.154
45	0.243	0.288	0.338	0.372	0.059	0.083	0.114	0.138
50	0.231	0.273	0.322	0.354	0.053	0.075	0.104	0.125
60	0.211	0.250	0.295	0.325	0.045	0.063	0.087	0.106
70	0.195	0.232	0.274	0.303	0.038	0.054	0.075	0.092
80	0.183	0.217	0.256	0.283	0.033	0.047	0.066	0.080
90	0.173	0.205	0.242	0.267	0.030	0.042	0.059	0.071
100	0.164	0.195	0.230	0.254	0.027	0.038	0.053	0.065

24.8 Useful financial formulae

This reference section provides a list of the financial formulae most often used in resource efficiency business cases. Advice on how to use these formulae can be found in Chapter 17 on page 555. Excel functions are shown in green text.

Item	Formula	Notes. Excel Function
Simple payback	$$Payback_{years} = \frac{Initial\ Investment}{Annualized\ Savings}$$	Simple payback is a measure of risk of a project. The annualized savings are net which means that any costs for financing the project, such as interest payments, should be deducted from the savings. If the savings vary from year to year, the annualized savings are the average saving over the study period. If there are annual tax benefits these can be added to the savings, or if the tax effects are one-off then the Initial Investment can be adjusted. Sometimes payback is calculated on a monthly basis, in which case the simple payback would be multiplied by 12.
Accounting rate of return (ARR)	$$ARR_{\%} = \frac{Annualized\ Savings}{Initial\ Investment}$$ $$or\ ARR_{\%} = \frac{1}{Payback} * 100$$	Accounting rate of return is essentially the inverse of payback, expressed as a percentage. As with payback all the figures are net, so we should take into account interest, financing and tax cost, as described above.
Future value	$$FV_t = PV_0\ (1 + r)^t$$	The future value FV_t of a sum of money PV_0 which has been invested at an interest rate r (expressed as a decimal fraction, e.g. 10% = 0.1) for t periods. t is usually years but it does not have to be - the same formula works for daily, weekly or monthly interest rates. **=FV(rate, nperiods, pmt, [pv],[type]). Where rate is r (expressed as a decimal fraction); nperiods is self explanatory; pmt is the equal payments made each period which is entered as negative value, if pmt is omitted then pv is required; pv is an optional present value $\$_0$ that the future payments will produce - if this is omitted it is assumed to be zero; type is 0 if the payment is at the end of the period (the default) and 1 if it is at the beginning.** **To achieve our purposes we will use the Excel formula as follows:** FV(r,t, , -PV_0) remembering to enter PV_0 as a negative value.

Item	Formula	Notes. Excel Function
Present value	$$PV_0 = \frac{1-(1+r)^{-t}}{t}$$ $$or\ PV_0 = FV_t\ (1+r)^{-t}$$ $$or\ PV_0 = \frac{FV_t}{(1+r)^{t}}$$	This is the value today, PV_0 of a sum of money FV_t received in t periods where r is the discount rate. **=PV(rate, nperiods, [pmt], [fv],[type]). Where rate is the discount rate r (expressed as a decimal fraction); nperiods is self explanatory; pmt is the equal payments made each period which is entered as negative value, if pmt is omitted then fv is required; fv is an optional future value $\$_t$ that the payments will produce - if this is omitted it is assumed to be zero; type is 0 if the payment is at the end of the period (the default) and 1 if it is at the beginning.** **To achieve our purposes we will use the Excel formula as follows:** PV(r,t, , -FV_n) remembering to enter FV_t as a negative value.
Capital recovery factor	$$CRF = \frac{1}{PV_0}$$ $$or\ CRF = \frac{t}{1-(1+r)^{-t}}$$ $$or\ CRF = \frac{(1+r)^{t}}{FV_t}$$	Assuming we have a capital cost, with present value PV_0 the capital recovery factor is the amount that PV_0 needs to multiplied by, to get to the t equal payments at a discount rate of r, which will repay the capital in full. Put another way, the discounted value of t payments of PV_0 * CRF using a discount rate of r is PV_0. In Excel, because the capital recovery factor is the inverse of the present value, so you can use the formula 1/PV(r,t,,-1).
Net present value	$$NPV = \sum \frac{Y_t}{(1+r)^t} + Y_0$$	The net present value for a cash flow Y_0 to Y_t is the sum of all the post investment cash flows (Y_1 to Y_t) plus the original investment value Y_0 (which is usually a negative sum). r represents the discount rate and t the total number of periods following the investment period. = NPV (discount rate, cash flow Y_1-Y_t) + Y_0 This formula assumes that there is no discounting of the initial investment. If that is not the case, then the investment will be included with all the subsequent cash flows. = NPV (discount rate, cash flow Y_0 -Y_t)
Inflation and Deflation Factors	$$\text{Inflation Factor Year}_t = (1+p)^t$$ $$\text{Deflation Factor Year}_t = (1+p)^{-t}$$	These factors are used to convert a nominal cash flow to a real cash flow. The variable p is the rate of inflation and t is the number of periods over which inflation has occurred. To convert the nominal cash flow to the real cash flow you *divide* the nominal cash flow values by the inflation factor, or multiply them by the deflation factor. Note that the inflation factor formula is the same as the future value formula, opposite.

Item	Formula	Notes. Excel Function
Fisher equation	$(1+n) = (1+r)*(1+p)$ or $r = \frac{(1+n)}{(1+p)} - 1$	The Fisher equation is used to create a real discount rate, r, using the nominal discount rate n, and the inflation rate, p. Assuming that the inflation rate is positive, the real discount rate will be lower than the nominal discount rate.
Profitability Index	$\text{Profitability Index} = \frac{\text{NPV}(Y_0...Y_t)}{\text{Cost}(-Y_0)}$	The profitability index divides the net present value of the post investment cash flow (Y_0 to Y_t) by the absolute value of the cost (Y_0). NOTE: NPV here is not the Excel Formula as we often use the NPV(Y_1-Y_n)-Y_0 form as the initial investment in not usually discounted. The profitability index is used to compare investments with differing initial investments, where the greater the profitability index, the more attractive the project. Investments with a profitability index of less than one (where the true costs is negative) do not recover the initial investment cost.
Internal rate of return	$\$0 = Y_0 + \frac{Y_1}{(1+IRR)^1} + \frac{Y_2}{(1+IRR)^2} \cdots \frac{Y_t}{(1+IRR)^t}$	The internal rate of return is equivalent to the interest earned by an investment. It is the discount rate where the net present value is zero. It cannot be computed easily as it is the solution of a geometric progression shown left, where Y_0 to Y_n are the investment cash flow in Year zero to Year t (equivalent to FV_0 to FV_t) and IRR is the value being computed. The normal solution is provided by trial and error using a computing package. In Excel the formula is: =IRR(cash flow Y_0-Y_t[, guess])
Cost of Saved Electricity (CSE)	$CSE = \frac{Cost * CRF}{kWh\ Saved}$ or $= \frac{Cost * \frac{r(1+r)^t}{(1+r)^t - 1}}{kWh\ Saved}$ or $= \frac{Cost * \frac{t}{1-(1+r)^t}}{kWh\ Saved}$	This formula is used by the American Council for an Energy Efficient Economy to convert an initial capital cost, *Cost*, to an annualized payment per unit of energy saved, *kWh Saved*. This involves multiplying the cost by the capital recovery factor, *CRF*. There is a table of CRF values on page 790. Note that, while this formula is used as a tool to compare energy savings, the denominator can be any resource, not just electricity. This formula is especially useful when comparing an internal investment option, where the organization bears the capital cost itself, with an external option such as an Energy Performance Contract, where the supplier has quoted an annualized payment.

Item	Formula	Notes. Excel Function
Annualized rate, r, using the goal R.	$r = \sqrt[t]{(1+R)} - 1$	r is the annual rate of improvement needed to achieve an overall improvement of R over t periods.
Annualized rate, r, using future and present value	$r = \sqrt[t]{\frac{FV}{PV}} - 1$	r is the annual rate of improvement needed to achieve a future value, FV, given a present value, PV, over t periods.
Total change, R	$R = (1+r)^t - 1$	R is the total change achieved after t periods at r rate of change.
Marginal Abatement Cost	$\frac{NPV(Costs)}{Emissions\ Reduction^{-y}}$	The marginal abatement cost is the discounted net present value of the project/programme/technology over the analysis period y divided by the average annual emissions reduction achieved over the analysis period.
Marginal Abatement	$Emissions\ Reduction^{-y}$	The marginal abatement is the average annual emissions reduction achieved by the project/programme/technology over the analysis period, y.

Reference

24.9 Discount factors table

	1%	2%	3%	4%	5%	6%	7%	8%	9%	10%	11%	12%	13%	14%	15%
1	0.9901	0.9804	0.9709	0.9615	0.9524	0.9434	0.9346	0.9259	0.9174	0.9091	0.9009	0.8929	0.8850	0.8772	0.8696
2	0.9803	0.9612	0.9426	0.9246	0.9070	0.8900	0.8734	0.8573	0.8417	0.8264	0.8116	0.7972	0.7831	0.7695	0.7561
3	0.9706	0.9423	0.9151	0.8890	0.8638	0.8396	0.8163	0.7938	0.7722	0.7513	0.7312	0.7118	0.6931	0.6750	0.6575
4	0.9610	0.9238	0.8885	0.8548	0.8227	0.7921	0.7629	0.7350	0.7084	0.6830	0.6587	0.6355	0.6133	0.5921	0.5718
5	0.9515	0.9057	0.8626	0.8219	0.7835	0.7473	0.7130	0.6806	0.6499	0.6209	0.5935	0.5674	0.5428	0.5194	0.4972
6	0.9420	0.8880	0.8375	0.7903	0.7462	0.7050	0.6663	0.6302	0.5963	0.5645	0.5346	0.5066	0.4803	0.4556	0.4323
7	0.9327	0.8706	0.8131	0.7599	0.7107	0.6651	0.6227	0.5835	0.5470	0.5132	0.4817	0.4523	0.4251	0.3996	0.3759
8	0.9235	0.8535	0.7894	0.7307	0.6768	0.6274	0.5820	0.5403	0.5019	0.4665	0.4339	0.4039	0.3762	0.3506	0.3269
9	0.9143	0.8368	0.7664	0.7026	0.6446	0.5919	0.5439	0.5002	0.4604	0.4241	0.3909	0.3606	0.3329	0.3075	0.2843
10	0.9053	0.8203	0.7441	0.6756	0.6139	0.5584	0.5083	0.4632	0.4224	0.3855	0.3522	0.3220	0.2946	0.2697	0.2472
11	0.8963	0.8043	0.7224	0.6496	0.5847	0.5268	0.4751	0.4289	0.3875	0.3505	0.3173	0.2875	0.2607	0.2366	0.2149
12	0.8874	0.7885	0.7014	0.6246	0.5568	0.4970	0.4440	0.3971	0.3555	0.3186	0.2858	0.2567	0.2307	0.2076	0.1869
13	0.8787	0.7730	0.6810	0.6006	0.5303	0.4688	0.4150	0.3677	0.3262	0.2897	0.2575	0.2292	0.2042	0.1821	0.1625
14	0.8700	0.7579	0.6611	0.5775	0.5051	0.4423	0.3878	0.3405	0.2992	0.2633	0.2320	0.2046	0.1807	0.1597	0.1413
15	0.8613	0.7430	0.6419	0.5553	0.4810	0.4173	0.3624	0.3152	0.2745	0.2394	0.2090	0.1827	0.1599	0.1401	0.1229
16	0.8528	0.7284	0.6232	0.5339	0.4581	0.3936	0.3387	0.2919	0.2519	0.2176	0.1883	0.1631	0.1415	0.1229	0.1069
17	0.8444	0.7142	0.6050	0.5134	0.4363	0.3714	0.3166	0.2703	0.2311	0.1978	0.1696	0.1456	0.1252	0.1078	0.0929
18	0.8360	0.7002	0.5874	0.4936	0.4155	0.3503	0.2959	0.2502	0.2120	0.1799	0.1528	0.1300	0.1108	0.0946	0.0808
19	0.8277	0.6864	0.5703	0.4746	0.3957	0.3305	0.2765	0.2317	0.1945	0.1635	0.1377	0.1161	0.0981	0.0829	0.0703
20	0.8195	0.6730	0.5537	0.4564	0.3769	0.3118	0.2584	0.2145	0.1784	0.1486	0.1240	0.1037	0.0868	0.0728	0.0611
21	0.8114	0.6598	0.5375	0.4388	0.3589	0.2942	0.2415	0.1987	0.1637	0.1351	0.1117	0.0926	0.0768	0.0638	0.0531
22	0.8034	0.6468	0.5219	0.4220	0.3418	0.2775	0.2257	0.1839	0.1502	0.1228	0.1007	0.0826	0.0680	0.0560	0.0462
23	0.7954	0.6342	0.5067	0.4057	0.3256	0.2618	0.2109	0.1703	0.1378	0.1117	0.0907	0.0738	0.0601	0.0491	0.0402
24	0.7876	0.6217	0.4919	0.3901	0.3101	0.2470	0.1971	0.1577	0.1264	0.1015	0.0817	0.0659	0.0532	0.0431	0.0349
25	0.7798	0.6095	0.4776	0.3751	0.2953	0.2330	0.1842	0.1460	0.1160	0.0923	0.0736	0.0588	0.0471	0.0378	0.0304
26	0.7720	0.5976	0.4637	0.3607	0.2812	0.2198	0.1722	0.1352	0.1064	0.0839	0.0663	0.0525	0.0417	0.0331	0.0264
27	0.7644	0.5859	0.4502	0.3468	0.2678	0.2074	0.1609	0.1252	0.0976	0.0763	0.0597	0.0469	0.0369	0.0291	0.0230
28	0.7568	0.5744	0.4371	0.3335	0.2551	0.1956	0.1504	0.1159	0.0895	0.0693	0.0538	0.0419	0.0326	0.0255	0.0200
29	0.7493	0.5631	0.4243	0.3207	0.2429	0.1846	0.1406	0.1073	0.0822	0.0630	0.0485	0.0374	0.0289	0.0224	0.0174
30	0.7419	0.5521	0.4120	0.3083	0.2314	0.1741	0.1314	0.0994	0.0754	0.0573	0.0437	0.0334	0.0256	0.0196	0.0151
35	0.7059	0.5000	0.3554	0.2534	0.1813	0.1301	0.0937	0.0676	0.0490	0.0356	0.0259	0.0189	0.0139	0.0102	0.0075
40	0.6717	0.4529	0.3066	0.2083	0.1420	0.0972	0.0668	0.0460	0.0318	0.0221	0.0154	0.0107	0.0075	0.0053	0.0037
45	0.6391	0.4102	0.2644	0.1712	0.1113	0.0727	0.0476	0.0313	0.0207	0.0137	0.0091	0.0061	0.0041	0.0027	0.0019
50	0.6080	0.3715	0.2281	0.1407	0.0872	0.0543	0.0339	0.0213	0.0134	0.0085	0.0054	0.0035	0.0022	0.0014	0.0009

24.12 **Discount factors table** (facing and current page)

Discount factors are computed for discount rates shown in columns and number of periods shown in rows.

Shading is to assist in differentiating number of periods and has no special meaning. Values less than 0.0001 are not displayed.

Source: Table by Niall Enright. Excel version in companion file pack.

	16%	17%	18%	19%	20%	21%	22%	23%	24%	25%	30%	35%	40%	45%	50%
1	0.8621	0.8547	0.8475	0.8403	0.8333	0.8264	0.8197	0.8130	0.8065	0.8000	0.7692	0.7407	0.7143	0.6897	0.6667
2	0.7432	0.7305	0.7182	0.7062	0.6944	0.6830	0.6719	0.6610	0.6504	0.6400	0.5917	0.5487	0.5102	0.4756	0.4444
3	0.6407	0.6244	0.6086	0.5934	0.5787	0.5645	0.5507	0.5374	0.5245	0.5120	0.4552	0.4064	0.3644	0.3280	0.2963
4	0.5523	0.5337	0.5158	0.4987	0.4823	0.4665	0.4514	0.4369	0.4230	0.4096	0.3501	0.3011	0.2603	0.2262	0.1975
5	0.4761	0.4561	0.4371	0.4190	0.4019	0.3855	0.3700	0.3552	0.3411	0.3277	0.2693	0.2230	0.1859	0.1560	0.1317
6	0.4104	0.3898	0.3704	0.3521	0.3349	0.3186	0.3033	0.2888	0.2751	0.2621	0.2072	0.1652	0.1328	0.1076	0.0878
7	0.3538	0.3332	0.3139	0.2959	0.2791	0.2633	0.2486	0.2348	0.2218	0.2097	0.1594	0.1224	0.0949	0.0742	0.0585
8	0.3050	0.2848	0.2660	0.2487	0.2326	0.2176	0.2038	0.1909	0.1789	0.1678	0.1226	0.0906	0.0678	0.0512	0.0390
9	0.2630	0.2434	0.2255	0.2090	0.1938	0.1799	0.1670	0.1552	0.1443	0.1342	0.0943	0.0671	0.0484	0.0353	0.0260
10	0.2267	0.2080	0.1911	0.1756	0.1615	0.1486	0.1369	0.1262	0.1164	0.1074	0.0725	0.0497	0.0346	0.0243	0.0173
11	0.1954	0.1778	0.1619	0.1476	0.1346	0.1228	0.1122	0.1026	0.0938	0.0859	0.0558	0.0368	0.0247	0.0168	0.0116
12	0.1685	0.1520	0.1372	0.1240	0.1122	0.1015	0.0920	0.0834	0.0757	0.0687	0.0429	0.0273	0.0176	0.0116	0.0077
13	0.1452	0.1299	0.1163	0.1042	0.0935	0.0839	0.0754	0.0678	0.0610	0.0550	0.0330	0.0202	0.0126	0.0080	0.0051
14	0.1252	0.1110	0.0985	0.0876	0.0779	0.0693	0.0618	0.0551	0.0492	0.0440	0.0254	0.0150	0.0090	0.0055	0.0034
15	0.1079	0.0949	0.0835	0.0736	0.0649	0.0573	0.0507	0.0448	0.0397	0.0352	0.0195	0.0111	0.0064	0.0038	0.0023
16	0.0930	0.0811	0.0708	0.0618	0.0541	0.0474	0.0415	0.0364	0.0320	0.0281	0.0150	0.0082	0.0046	0.0026	0.0015
17	0.0802	0.0693	0.0600	0.0520	0.0451	0.0391	0.0340	0.0296	0.0258	0.0225	0.0116	0.0061	0.0033	0.0018	0.0010
18	0.0691	0.0592	0.0508	0.0437	0.0376	0.0323	0.0279	0.0241	0.0208	0.0180	0.0089	0.0045	0.0023	0.0012	0.0007
19	0.0596	0.0506	0.0431	0.0367	0.0313	0.0267	0.0229	0.0196	0.0168	0.0144	0.0068	0.0033	0.0017	0.0009	0.0005
20	0.0514	0.0433	0.0365	0.0308	0.0261	0.0221	0.0187	0.0159	0.0135	0.0115	0.0053	0.0025	0.0012	0.0006	0.0003
21	0.0443	0.0370	0.0309	0.0259	0.0217	0.0183	0.0154	0.0129	0.0109	0.0092	0.0040	0.0018	0.0009	0.0004	0.0002
22	0.0382	0.0316	0.0262	0.0218	0.0181	0.0151	0.0126	0.0105	0.0088	0.0074	0.0031	0.0014	0.0006	0.0003	0.0001
23	0.0329	0.0270	0.0222	0.0183	0.0151	0.0125	0.0103	0.0086	0.0071	0.0059	0.0024	0.0010	0.0004	0.0002	0.0001
24	0.0284	0.0231	0.0188	0.0154	0.0126	0.0103	0.0085	0.0070	0.0057	0.0047	0.0018	0.0007	0.0003	0.0001	0.0001
25	0.0245	0.0197	0.0160	0.0129	0.0105	0.0085	0.0069	0.0057	0.0046	0.0038	0.0014	0.0006	0.0002	0.0001	0.0000
26	0.0211	0.0169	0.0135	0.0109	0.0087	0.0070	0.0057	0.0046	0.0037	0.0030	0.0011	0.0004	0.0002	0.0001	0.0000
27	0.0182	0.0144	0.0115	0.0091	0.0073	0.0058	0.0047	0.0037	0.0030	0.0024	0.0008	0.0003	0.0001	0.0000	0.0000
28	0.0157	0.0123	0.0097	0.0077	0.0061	0.0048	0.0038	0.0030	0.0024	0.0019	0.0006	0.0002	0.0001	0.0000	0.0000
29	0.0135	0.0105	0.0082	0.0064	0.0051	0.0040	0.0031	0.0025	0.0020	0.0015	0.0005	0.0002	0.0001	0.0000	0.0000
30	0.0116	0.0090	0.0070	0.0054	0.0042	0.0033	0.0026	0.0020	0.0016	0.0012	0.0004	0.0001	0.0000	0.0000	0.0000
35	0.0055	0.0041	0.0030	0.0023	0.0017	0.0013	0.0009	0.0007	0.0005	0.0004	0.0001	0.0000	0.0000	0.0000	0.0000
40	0.0026	0.0019	0.0013	0.0010	0.0007	0.0005	0.0004	0.0003	0.0002	0.0001	0.0000	0.0000	0.0000	0.0000	0.0000
45	0.0013	0.0009	0.0006	0.0004	0.0003	0.0002	0.0001	0.0001	0.0001	0.0000	0.0000	0.0000	0.0000	0.0000	0.0000
50	0.0006	0.0004	0.0003	0.0002	0.0001	0.0001	0.0000	0.0000	0.0000	0.0000	0.0000	0.0000	0.0000	0.0000	0.0000

24.10 Capital recovery factors

	1%	2%	3%	4%	5%	6%	7%	8%	9%	10%	11%	12%	13%	14%	15%
1	1.0100	1.0200	1.0300	1.0400	1.0500	1.0600	1.0700	1.0800	1.0900	1.1000	1.1100	1.1200	1.1300	1.1400	1.1500
2	0.5075	0.5150	0.5226	0.5302	0.5378	0.5454	0.5531	0.5608	0.5685	0.5762	0.5839	0.5917	0.5995	0.6073	0.6151
3	0.3400	0.3468	0.3535	0.3603	0.3672	0.3741	0.3811	0.3880	0.3951	0.4021	0.4092	0.4163	0.4235	0.4307	0.4380
4	0.2563	0.2626	0.2690	0.2755	0.2820	0.2886	0.2952	0.3019	0.3087	0.3155	0.3223	0.3292	0.3362	0.3432	0.3503
5	0.2060	0.2122	0.2184	0.2246	0.2310	0.2374	0.2439	0.2505	0.2571	0.2638	0.2706	0.2774	0.2843	0.2913	0.2983
6	0.1725	0.1785	0.1846	0.1908	0.1970	0.2034	0.2098	0.2163	0.2229	0.2296	0.2364	0.2432	0.2502	0.2572	0.2642
7	0.1486	0.1545	0.1605	0.1666	0.1728	0.1791	0.1856	0.1921	0.1987	0.2054	0.2122	0.2191	0.2261	0.2332	0.2404
8	0.1307	0.1365	0.1425	0.1485	0.1547	0.1610	0.1675	0.1740	0.1807	0.1874	0.1943	0.2013	0.2084	0.2156	0.2229
9	0.1167	0.1225	0.1284	0.1345	0.1407	0.1470	0.1535	0.1601	0.1668	0.1736	0.1806	0.1877	0.1949	0.2022	0.2096
10	0.1056	0.1113	0.1172	0.1233	0.1295	0.1359	0.1424	0.1490	0.1558	0.1627	0.1698	0.1770	0.1843	0.1917	0.1993
11	0.0965	0.1022	0.1081	0.1141	0.1204	0.1268	0.1334	0.1401	0.1469	0.1540	0.1611	0.1684	0.1758	0.1834	0.1911
12	0.0888	0.0946	0.1005	0.1066	0.1128	0.1193	0.1259	0.1327	0.1397	0.1468	0.1540	0.1614	0.1690	0.1767	0.1845
13	0.0824	0.0881	0.0940	0.1001	0.1065	0.1130	0.1197	0.1265	0.1336	0.1408	0.1482	0.1557	0.1634	0.1712	0.1791
14	0.0769	0.0826	0.0885	0.0947	0.1010	0.1076	0.1143	0.1213	0.1284	0.1357	0.1432	0.1509	0.1587	0.1666	0.1747
15	0.0721	0.0778	0.0838	0.0899	0.0963	0.1030	0.1098	0.1168	0.1241	0.1315	0.1391	0.1468	0.1547	0.1628	0.1710
16	0.0679	0.0737	0.0796	0.0858	0.0923	0.0990	0.1059	0.1130	0.1203	0.1278	0.1355	0.1434	0.1514	0.1596	0.1679
17	0.0643	0.0700	0.0760	0.0822	0.0887	0.0954	0.1024	0.1096	0.1170	0.1247	0.1325	0.1405	0.1486	0.1569	0.1654
18	0.0610	0.0667	0.0727	0.0790	0.0855	0.0924	0.0994	0.1067	0.1142	0.1219	0.1298	0.1379	0.1462	0.1546	0.1632
19	0.0581	0.0638	0.0698	0.0761	0.0827	0.0896	0.0968	0.1041	0.1117	0.1195	0.1276	0.1358	0.1441	0.1527	0.1613
20	0.0554	0.0612	0.0672	0.0736	0.0802	0.0872	0.0944	0.1019	0.1095	0.1175	0.1256	0.1339	0.1424	0.1510	0.1598
21	0.0530	0.0588	0.0649	0.0713	0.0780	0.0850	0.0923	0.0998	0.1076	0.1156	0.1238	0.1322	0.1408	0.1495	0.1584
22	0.0509	0.0566	0.0627	0.0692	0.0760	0.0830	0.0904	0.0980	0.1059	0.1140	0.1223	0.1308	0.1395	0.1483	0.1573
23	0.0489	0.0547	0.0608	0.0673	0.0741	0.0813	0.0887	0.0964	0.1044	0.1126	0.1210	0.1296	0.1383	0.1472	0.1563
24	0.0471	0.0529	0.0590	0.0656	0.0725	0.0797	0.0872	0.0950	0.1030	0.1113	0.1198	0.1285	0.1373	0.1463	0.1554
25	0.0454	0.0512	0.0574	0.0640	0.0710	0.0782	0.0858	0.0937	0.1018	0.1102	0.1187	0.1275	0.1364	0.1455	0.1547
26	0.0439	0.0497	0.0559	0.0626	0.0696	0.0769	0.0846	0.0925	0.1007	0.1092	0.1178	0.1267	0.1357	0.1448	0.1541
27	0.0424	0.0483	0.0546	0.0612	0.0683	0.0757	0.0834	0.0914	0.0997	0.1083	0.1170	0.1259	0.1350	0.1442	0.1535
28	0.0411	0.0470	0.0533	0.0600	0.0671	0.0746	0.0824	0.0905	0.0989	0.1075	0.1163	0.1252	0.1344	0.1437	0.1531
29	0.0399	0.0458	0.0521	0.0589	0.0660	0.0736	0.0814	0.0896	0.0981	0.1067	0.1156	0.1247	0.1339	0.1432	0.1527
30	0.0387	0.0446	0.0510	0.0578	0.0651	0.0726	0.0806	0.0888	0.0973	0.1061	0.1150	0.1241	0.1334	0.1428	0.1523
35	0.0340	0.0400	0.0465	0.0536	0.0611	0.0690	0.0772	0.0858	0.0946	0.1037	0.1129	0.1223	0.1318	0.1414	0.1511
40	0.0305	0.0366	0.0433	0.0505	0.0583	0.0665	0.0750	0.0839	0.0930	0.1023	0.1117	0.1213	0.1310	0.1407	0.1506
45	0.0277	0.0339	0.0408	0.0483	0.0563	0.0647	0.0735	0.0826	0.0919	0.1014	0.1110	0.1207	0.1305	0.1404	0.1503
50	0.0255	0.0318	0.0389	0.0466	0.0548	0.0634	0.0725	0.0817	0.0912	0.1009	0.1106	0.1204	0.1303	0.1402	0.1501

24.13 **Capital recovery factors** (facing and current page)

Capital recovery factors are computed for discount rates shown in columns and number of periods shown in rows.

Shading is to assist in differentiating number of periods and has no special meaning.

Source: Table by Niall Enright. Excel version in companion file pack.

	16%	17%	18%	19%	20%	21%	22%	23%	24%	25%	30%	35%	40%	45%	50%
1	1.1600	1.1700	1.1800	1.1900	1.2000	1.2100	1.2200	1.2300	1.2400	1.2500	1.3000	1.3500	1.4000	1.4500	1.5000
2	0.6230	0.6308	0.6387	0.6466	0.6545	0.6625	0.6705	0.6784	0.6864	0.6944	0.7348	0.7755	0.8167	0.8582	0.9000
3	0.4453	0.4526	0.4599	0.4673	0.4747	0.4822	0.4897	0.4972	0.5047	0.5123	0.5506	0.5897	0.6294	0.6697	0.7105
4	0.3574	0.3645	0.3717	0.3790	0.3863	0.3936	0.4010	0.4085	0.4159	0.4234	0.4616	0.5008	0.5408	0.5816	0.6231
5	0.3054	0.3126	0.3198	0.3271	0.3344	0.3418	0.3492	0.3567	0.3642	0.3718	0.4106	0.4505	0.4914	0.5332	0.5758
6	0.2714	0.2786	0.2859	0.2933	0.3007	0.3082	0.3158	0.3234	0.3311	0.3388	0.3784	0.4193	0.4613	0.5043	0.5481
7	0.2476	0.2549	0.2624	0.2699	0.2774	0.2851	0.2928	0.3006	0.3084	0.3163	0.3569	0.3988	0.4419	0.4861	0.5311
8	0.2302	0.2377	0.2452	0.2529	0.2606	0.2684	0.2763	0.2843	0.2923	0.3004	0.3419	0.3849	0.4291	0.4743	0.5203
9	0.2171	0.2247	0.2324	0.2402	0.2481	0.2561	0.2641	0.2722	0.2805	0.2888	0.3312	0.3752	0.4203	0.4665	0.5134
10	0.2069	0.2147	0.2225	0.2305	0.2385	0.2467	0.2549	0.2632	0.2716	0.2801	0.3235	0.3683	0.4143	0.4612	0.5088
11	0.1989	0.2068	0.2148	0.2229	0.2311	0.2394	0.2478	0.2563	0.2649	0.2735	0.3177	0.3634	0.4101	0.4577	0.5058
12	0.1924	0.2005	0.2086	0.2169	0.2253	0.2337	0.2423	0.2509	0.2596	0.2684	0.3135	0.3598	0.4072	0.4553	0.5039
13	0.1872	0.1954	0.2037	0.2121	0.2206	0.2292	0.2379	0.2467	0.2556	0.2645	0.3102	0.3572	0.4051	0.4536	0.5026
14	0.1829	0.1912	0.1997	0.2082	0.2169	0.2256	0.2345	0.2434	0.2524	0.2615	0.3078	0.3553	0.4036	0.4525	0.5017
15	0.1794	0.1878	0.1964	0.2051	0.2139	0.2228	0.2317	0.2408	0.2499	0.2591	0.3060	0.3539	0.4026	0.4517	0.5011
16	0.1764	0.1850	0.1937	0.2025	0.2114	0.2204	0.2295	0.2387	0.2479	0.2572	0.3046	0.3529	0.4018	0.4512	0.5008
17	0.1740	0.1827	0.1915	0.2004	0.2094	0.2186	0.2278	0.2370	0.2464	0.2558	0.3035	0.3521	0.4013	0.4508	0.5005
18	0.1719	0.1807	0.1896	0.1987	0.2078	0.2170	0.2263	0.2357	0.2451	0.2546	0.3027	0.3516	0.4009	0.4506	0.5003
19	0.1701	0.1791	0.1881	0.1972	0.2065	0.2158	0.2251	0.2346	0.2441	0.2537	0.3021	0.3512	0.4007	0.4504	0.5002
20	0.1687	0.1777	0.1868	0.1960	0.2054	0.2147	0.2242	0.2337	0.2433	0.2529	0.3016	0.3509	0.4005	0.4503	0.5002
21	0.1674	0.1765	0.1857	0.1951	0.2044	0.2139	0.2234	0.2330	0.2426	0.2523	0.3012	0.3506	0.4003	0.4502	0.5001
22	0.1664	0.1756	0.1848	0.1942	0.2037	0.2132	0.2228	0.2324	0.2421	0.2519	0.3009	0.3505	0.4002	0.4501	0.5001
23	0.1654	0.1747	0.1841	0.1935	0.2031	0.2127	0.2223	0.2320	0.2417	0.2515	0.3007	0.3504	0.4002	0.4501	0.5000
24	0.1647	0.1740	0.1835	0.1930	0.2025	0.2122	0.2219	0.2316	0.2414	0.2512	0.3006	0.3503	0.4001	0.4501	0.5000
25	0.1640	0.1734	0.1829	0.1925	0.2021	0.2118	0.2215	0.2313	0.2411	0.2509	0.3004	0.3502	0.4001	0.4500	0.5000
26	0.1634	0.1729	0.1825	0.1921	0.2018	0.2115	0.2213	0.2311	0.2409	0.2508	0.3003	0.3501	0.4001	0.4500	0.5000
27	0.1630	0.1725	0.1821	0.1917	0.2015	0.2112	0.2210	0.2309	0.2407	0.2506	0.3003	0.3501	0.4000	0.4500	0.5000
28	0.1625	0.1721	0.1818	0.1915	0.2012	0.2110	0.2208	0.2307	0.2406	0.2505	0.3002	0.3501	0.4000	0.4500	0.5000
29	0.1622	0.1718	0.1815	0.1912	0.2010	0.2108	0.2207	0.2306	0.2405	0.2504	0.3001	0.3501	0.4000	0.4500	0.5000
30	0.1619	0.1715	0.1813	0.1910	0.2008	0.2107	0.2206	0.2305	0.2404	0.2503	0.3001	0.3500	0.4000	0.4500	0.5000
35	0.1609	0.1707	0.1806	0.1904	0.2003	0.2103	0.2202	0.2302	0.2401	0.2501	0.3000	0.3500	0.4000	0.4500	0.5000
40	0.1604	0.1703	0.1802	0.1902	0.2001	0.2101	0.2201	0.2301	0.2400	0.2500	0.3000	0.3500	0.4000	0.4500	0.5000
45	0.1602	0.1701	0.1801	0.1901	0.2001	0.2100	0.2200	0.2300	0.2400	0.2500	0.3000	0.3500	0.4000	0.4500	0.5000
50	0.1601	0.1701	0.1800	0.1900	0.2000	0.2100	0.2200	0.2300	0.2400	0.2500	0.3000	0.3500	0.4000	0.4500	0.5000

24.11 Control chart runs table

24.14 **Runs table**
Runs are used in statistical process control to assess if a series of values is random or not.
Source: Based on Swed and Eisenhart,[695] Table by Niall Enright.

RUNS TABLE					
Total number of data points on the run chart that do not fall on the median	Lower limit for the number of runs (< than this number runs is "too few")	Upper limit for the number of runs (> than this number runs is "too many")	Total number of data points on the run chart that do not fall on the median	Lower limit for the number of runs (< than this number runs is "too few")	Upper limit for the number of runs (> than this number runs is "too many")
10	3	9	34	12	24
11	3	10	35	12	24
12	3	11	36	13	25
13	4	11	37	13	25
14	4	12	38	14	26
15	5	12	39	14	26
16	5	13	40	15	27
17	5	13	41	15	27
18	6	14	42	16	28
19	6	15	43	16	28
20	6	16	44	17	29
21	7	16	45	17	30
22	7	17	46	17	31
23	7	17	47	18	31
24	8	18	48	18	32
25	8	18	49	19	32
26	9	19	50	19	33
27	10	19	60	24	37
28	10	20	70	28	43
29	10	20	80	33	48
30	11	21	90	37	54
31	11	22	100	46	65
32	11	23	110	46	65
33	12	23	120	51	70

25 Bibliography

1. 3M. 3M Corporate Sustainability Reporting (2011) (see http://solutions.3m.com/wps/portal/3M/en_US/3M-Sustainability/Global/Environment/ClimateChange/).
2. 3M. 2012 Sustainability Report (2012) (see http://solutions.3m.com/3MContentRetrievalAPI/BlobServlet?lmd=1338504824000&locale=en_WW&assetType=MMM_Image&assetId=1319229751200&blobAttribute=ImageFile).
3. 3M. 2016 Sustainability Report (2016) (see http://www.3m.com/3M/en_US/sustainability-report/goals-and-results/)
4. ABInBev. Anheuser-Busch InBev. 2015 Global Citizenship Report. (see http://www.ab-inbev.com/content/dam/universaltemplate/ab-inbev/sustainability/document.pdf).
5. Accenture. *Actionable Insights for the New Energy Consumer* (2012) (see https://www.accenture.com/t20151127T210510__w__/us-en/_acnmedia/Accenture/Conversion-Assets/DotCom/Documents/Global/PDF/Industries_9/Accenture-Actionable-Insights-New-Energy-Consumer.pdf).
6. Accenture, Marks & Spencer. *Fortune Favours the Brave* (2013) (see http://www.bitc.org.uk/system/files/fortune_favours_the_brave_3962_rsbr_full_report_screen_only_aw_hi-res.pdf).
7. AccountAbility. The AA1000 Standards. (see http://www.accountability.org/standards/index.html)
8. Ackerman, Frank, and Ramón Bueno. *Use of McKinsey abatement cost curves for climate economics modelling* (2011) (see http://www.sei-international.org/mediamanager/documents/Publications/Climate/sei-workingpaperus-1102.pdf).
9. Adam, David. *ExxonMobil continuing to fund climate sceptic groups, records show*, The Guardian (2009) (see http://www.guardian.co.uk/environment/2009/jul/01/ExxonMobil-climate-change-sceptics-funding).
10. AEPCA. *A Best Practice Guide to Measurement and Verification of Energy Savings*, Australian Energy Performance Contracting Association (2004) (see http://icaen.gencat.cat/web/.content/20_Energia/25_empreses_servei_energetic/arxius/a_best_practice_guide_to_mmeasurement_and_verification_of_energy_savings.pdf).
11. Aguinis, Herman, Harry Joo, and Ryan K Gottfredson. *What monetary rewards can and cannot do: How to show employees the money* (Business Horizons, Elsevier, 2013), pp. 241–249.
12. Airbus Group. ACADEMY - Design for Environment (2008) (see http://ec.europa.eu/environment/life/project/Projects/index.cfm?fuseaction=home.showFile&rep=file&fil=ACADEMY_DesignforEnvironment.pdf).
13. Airbus Group. The nose section of the first A380 at its arrival to Toulouse. (see http://www.airbus.com/galleries/photo-gallery/dg/idp/47540-a380-nose-section-close-up/?share=1).
14. AkzoNobel. Carbon Management through the Value Chain. (see http://www.akzonobel.com/sustainability/managing_sustainability/key_focus_areas/carbon_management_value_chain/).
15. AkzoNobel. Creating Value from Eco-premium Solutions. (see https://www.akzonobel.com/sustainability/managing_sustainability/key_focus_areas/creating_value_eco_premium_solutions/).
16. AkzoNobel. Remuneration Report (2014) (see http://report.akzonobel.com/2014/ar/governance/remuneration-report.html).
17. Al-Shemmeri, Tarik. *Energy Audits - A Workbook for Energy Management in Buildings* (Wiley-Blackwell, 2011).
18. Aldersgate. *Beyond Carbon: Towards a resource efficient future*, Aldersgate Group (2010) (see http://www.aldersgategroup.org.uk/asset/download/114/beyond_carbon.pdf).
19. Aldersgate. Wallace, Steve, and Andrew Raingold. *Resilience in the Round: Seizing the growth opportunities of a circular economy*, Aldersgate Group (2012) (see http://www.aldersgategroup.org.uk/asset/download/458/Report%20-%20Resilience%20in%20the%20Round%20(1).pdf).
20. Allen, David T. et al. *Measurements of methane emissions at natural gas production sites in the United States* (2013)pp. 17768–17773 (available at http://www.pnas.org/content/110/44/17768.full.pdf).
21. All-Party Parliamentary Sustainable Resource Group. *Remanufacturing: towards a resource efficient economy.* (March 2014) (see http://www.policyconnect.org.uk/apsrg/sites/site_apsrg/files/apsrg_-_remanufacturing_report.pdf).
22. Allwood, J. M. et al. *Sustainable Materials — without the hot air.* (2nd Ed. UIT Cambridge Ltd, 2015) ISBN 978-1906860301 (see http://www.withbotheyesopen.com).
23. American Council for an Energy Efficiency Economy, Energy Efficiency Resource Standards (see http://aceee.org/topics/energy-efficiency-resource-standard-eers)

24. American Water Works Association, Water Audit Data Initiative, 2014 Validated Water Audit Data (see http://www.awwa.org/Portals/0/files/resources/water%20knowledge/water%20loss%20control/2014-Validated-Water-Audit-Data.xlsm).

25. Anderson, Ray. *Confessions of a radical Industrialist* (Random House Business Books, 2011).

26. Anderson, Sweeny and Williams, *Essentials of Modern Business Statistics*, pub 2016, Cengage Learning: ISBN 978-1-285-86704-5.

27. Anscombe, Francis J. (1973) *Graphs in statistical analysis*. American Statistician, 27, 17–21.

28. Anthes, Emily. *Hospital checklists are meant to save lives - so why do they often fail?* (2015) (see http://www.nature.com/news/hospital-checklists-are-meant-to-save-lives-so-why-do-they-often-fail-1.18057), p. 516.

29. AP (Associated Press), *IKEA Stops Selling Incandescent Light Bulbs in U.S.*, Associated Press (2010) (see http://www.ikea.com/ms/en_US/about_ikea/press_room/press_release/national/phase_out_incandescent_lighting_press_release.html).

30. AP (Associated Press). *Volkswagen sued by US watchdog over false "Clean Diesel" claims* (29AD) (see https://www.theguardian.com/business/2016/mar/29/volkswagen-vw-sued-federal-trade-commission-clean-diesel-claims).

31. Applin, Mandy, and Catherine Reeves. Xerox's GHG Reduction Initiative (2010) (see http://epa.gov/climateleadership/documents/events/june2010/xerox_2june2010.pdf).

32. Ariely, Dan, George Loewenstein, and Drazen Prelec. *"Coherent arbitrariness": Stable demand curves without stable preferences*, The Quarterly Journal of Economics 118, 73–106 (2003).

33. Arjen Y. Hoekstra, Ashok K. Chapagain Maite M. Aldaya Mesfin M. Mekonnen. *The Water Footprint Assessment Manual* (Earthscan, 2011) (see http://www.waterfootprint.org/?page=files/WaterFootprintAssessmentManual).

34. Arkes, Hal R, and Laura Hutzel. *The role of probability of success estimates in the sunk cost effect.* (Wiley Periodicals Inc., 2000) (see http://www.communicationcache.com/uploads/1/0/8/8/10887248/arkes-hutzel-the-role-of-probability-of-success-estimates-in-the-sunk-cost-effect.pdf), p. 295.

35. Aronson, Elliot, and Michael O'Leary. *The relative effectiveness of models and prompts on energy conservation: A field experiment in a shower room.* (1983) (see https://triggered.clockss.org/ServeContent?url=http://baywood.stanford.clockss.org%2FBWES%2FBAWOOD_BWES_12_3%2FUBD54Y9B61EFWUM6.pdf), pp. 219–224.

36. ASHRAE. *Handbook - Fundamentals*, American Society of Heating, Refrigerating and Air-Conditioning Engineers Inc., Atlanta, GA (2009) (see http://www.techstreet.com/ashrae/products/1859162).

37. ASHRAE. Guideline 14-2014, Measurement of Energy and Demand Savings (2014) (see https://www.researchgate.net/publication/26901656_ASHRAE's_Guideline_14-2002_for_Measurement_of_Energy_and_Demand_Savings_How_to_Determine_What_Was_Really_Saved_by_the_Retrofit).

38. ASHRAE, Procedures for Commercial Buildings Energy Audits. (ASHRAE, 2011) (see https://www.techstreet.com/ashrae/standards/procedures-for-commercial-building-energy-audits-2nd-edition?product_id=1809206).

39. Association for the Advancement of Sustainability in Higher Education (AASHE). Directory of Campus Revolving Funds. (see http://www.aashe.org/resources/campus-sustainability-revolving-loan-funds/all/)

40. Association for the Advancement of Sustainability in Higher Education (AASHE). Directory of Campus Revolving Funds. Harvard University Entry. (see http://www.aashe.org/resources/campus-sustainability-revolving-loan-funds/green-loan-fund/)

41. Association of Energy Engineers. Certified Professionals Directory (see http://www.aeecenter.org/custom/cpdirectory/index.cfm).

42. Association of Energy Engineers. CMVP Body of Knowledge Guide (see http://www.aeecenter.org/files/certification/CMVPStudyGuide).

43. ASTM. E2797-15 *Standard Practice for Building Energy Performance Assessment for a Building Involved in a Real Estate Transaction* (2015) (see http://www.astm.org/Standards/E2797.htm).

44. Attari, S. Z., M. L. DeKay, C. I. Davidson, and W. Bruine de Bruin. *Public perceptions of energy consumption and savings: Supporting Information.* Proceedings of the National Academy of Sciences 107, 6 (2010).

45. Attari, S. Z., M. L. DeKay, C. I. Davidson, and W. Bruine de Bruin. *Public perceptions of energy consumption and savings - Questionnaire.* Proceedings of the National Academy of Sciences 107, 6 (2010).

46. Australian Competition & Consumer Commission. *Green Marketing and Australian Consumer Law.* (2011) (see http://www.accc.gov.au/system/files/Green%20marketing%20and%20the%20ACL.pdf).

47. Autoblog. *Toyota's "green bond" an industry first, quickly rises to $1.75 billion* (see http://www.autoblog.com/2014/03/25/toyotas-green-bond-industry-first-quickly-rises-1-75-billion/).

48. Automotive Industry Action Group, and American Society for Quality Control. Supplier Quality Requirements Task Force. *Fundamental statistical process control: reference manual.* (AIAG, 1991) (see https://books.google.co.uk/books?id=mNMmAQAAMAAJ).

49. Ayres, Robert U. *Sustainability economics: Where do we stand?* (2008) (see http://www.sciencedirect.com/science/article/pii/S0921800907006088), pp. 281–310.

50. Baca-Motes, Katie, Amber Brown, Ayelet Gneezy, Elizabeth A Keenan, and Leif D Nelson. *Commitment and behavior change: Evidence from the field* (The Oxford University Press, 2013) (see http://www.haas.berkeley.edu/groups/online_marketing/facultyCV/papers/nelson_commitment.pdf), pp. 1070–1084.

51. Baer, Drake. *How Changing One Habit Helpled Quintuple Alcoa's Income* (2012) (see http://www.businessinsider.com/how-changing-one-habit-quintupled-alcoas-income-2014-4?IR=T).

52. Bailey, Jakki O. et al. *The Impact of Vivid Messages on Reducing Energy Consumption Related to Hot Water Use* (SAGE Publications, 2015) (see https://vhil.stanford.edu/pubs/2014/bailey-eb-vivid-messages.pdf), pp. 570–592.

53. Ballantyne, A.P., C.B. Alden, J.B. Miller, P.P. Tans, and J.W.C. White. *Increase in observed net carbon dioxide uptake by land and oceans during the past 50 years*, Nature 488, 70–72 (2012).

54. Bank of America. 2014 Sustainability report (see http://about.bankofamerica.com/assets/pdf/Bank-of-America-2014-Corporate-Social-Responsibility-Report.pdf).

55. Bank of International Settlements (see http://www.bis.org/publ/rpfx13.htm).

56. Bardi, U. *The limits to growth revisited* (Springer Verlag, 2011).

57. Barnett, H., and C. Morse. *Scarcity and growth.-The economics of natural resource availability*. (1963).

58. BASF. Annual Report 2011 (2012) (see http://report.basf.com/2011/en/servicepages/welcome.html).

59. Bawden, Tom. *Cost of metals used in hi-tech devices soars as China limits supplies*, The Observer (2011) (see http://www.guardian.co.uk/business/2011/jun/19/rare-earth-minerals-china/).

60. BBC. *China sets up rare earth body to streamline the sector* (see http://www.bbc.co.uk/news/business-17655146).

61. BEAM Society Ltd, Building Environmental Assessment Method (see http://www.beamsociety.org.hk/en_about_us_0.php).

62. Behavioural Insights Team. (see http://www.behaviouralinsights.co.uk/).

63. Beinhocker, E. et al. The carbon productivity challenge, McKinsey Global Institute (2008) (see http://www.mckinsey.com/Insights/MGI/Research/Natural_Resources/The_carbon_productivity_challenge).

64. Bekefi, Tamara, and Marc J. Epstein. *Integrating social and political risk into management decision-making* (2006) (see http://www.cimaglobal.com/Documents/ImportedDocuments/Tech_MAG_Integrating_Social_Political_Risks_Sept06.pdf).

65. Bell, Stephanie. *Measurement Good Practice Guide No. 11* (Issue 2) (2001) (see http://publications.npl.co.uk/npl_web/pdf/mgpg11.pdf).

66. Benyus, Janine. *Biomimicry - Innovation Inspired by Nature* (2009) ISBN: 978-0688160999.

67. Bem, Daryl J. *Self-perception theory* (1973) (see http://www.dbem.ws/SP%20Theory.pdf).

68. Bertoldi, Paolo, Benigna Boza-Kiss Strahil Panev Nicola Labanca. *ESCO Market Report 2013*, European Commission (2014) (see http://iet.jrc.ec.europa.eu/energyefficiency/system/tdf/jrc_89550_the_european_esco_market_report_2013_online.pdf?file=1&type=node&id=8869).

69. Better Building Partnership. *Green Lease Toolkit*. (see http://www.betterbuildingspartnership.co.uk/download/bbp-gltk-2013.pdf).

70. Bicheno, John, and Matthias Holweg. *The Lean Toolbox - the essential guide to Lean transformation* (PICSIE, 2009, ISBN: 978-0-9541244-5-8).

71. Bickman, L. (1972). *Environmental attitudes and actions*. The Journal of Social Psychology, 87(2), 323-324.

72. Biello, David. *Cement from CO2: A Concrete Cure for Global Warming?*, Scientific American (2008) (see http://www.scientificamerican.com/article.cfm?id=cement-from-carbon-dioxide).

73. Blanding, Michael. *Corporate Sustainability Reporting: It's Effective*, Harvard Business School (2011) (see http://hbswk.hbs.edu/cgi-bin/print/6701.html).

74. Bloomberg New Energy Finance and Ernst & Young, *South Korea's Emissions Trading Scheme*, (10 May 2013) (see https://www.bnef.com/InsightDownload/7608/pdf/).

75. Bloomberg New Energy Finance, *Sustainable Energy In America Factbook*. (2016) (see http://www.bcse.org/sustainableenergyfactbook/#).

76. Boyd, Gale, and Mark Curtis. *Evidence of an "Energy-Management Gap" in US Manufacturing: Spillovers from Firm Management Practices to Energy Efficiency*, US Census Bureau Center for Economic Studies Paper No. CES-WP-13-25 (2013).

77. Boyd, Peter. TEDx Vilnius, (25 Feb 2014). *Solving climate change -- society's great opportunity masked as a crisis* (see http://www.youtube.com/watch?v=8O0BPFWFo5Y).

78. BP. 2003 Sustainability Report (2003) (see http://www.socialfunds.com/shared/reports/1210789674_BP_Sustainability_Report_2003.pdf).

79. BP. Direct CO2 Emissions By Segment. (see https://www.bp.com/en/global/corporate/sustainability/performance-data/hse-charting-tool.html).

Reference

80. Brainbox, *Learning Styles, the debate*. (see http://www.brainboxx.co.uk/a2_learnstyles/pages/lstyles_debate.htm)

81. Brand Finance Global 500, Brand Finance (2012) (see http://www.brandfinance.com/images/upload/bf_g500_2012_web_dp.pdf).

82. Brand Finance. Global Top 50 Cosmetics Brands 2013 (2013) (see http://brandfinance.com/images/upload/cosmetics_2013_spc.pdf).

83. Briggs, Helen. *Climate Change: is the Dutch Court's Ruling a Game Changer?* (see http://www.bbc.co.uk/news/science-environment-33261474).

84. British Broadcasting Corporation. *New Green Deal is very generous, says energy secretary*. 9 June 2014 (see http://www.bbc.co.uk/news/business-27759217).

85. British Broadcasting Corporation. *Only four people sign up for flagship Green Deal*. 27 June 2013 (see http://www.bbc.co.uk/news/business-23081896).

86. British Geological Survey, Natural Environment Research Council, *World Mineral Production 2008–2012* (2014) (see https://www.bgs.ac.uk/downloads/start.cfm?id=2897).

87. British Geological Survey. *Risk List 2012*, (2012) (see http://www.bgs.ac.uk/downloads/start.cfm?id=2643).

88. Brat, Ilan. *Caterpillar Gets Bugs Out of Old Equipment*. Wall Street Journal (July 5, 2006) (see http://www.reman.org/Articles/WSJ_Cat_Reman.pdf).

89. Brockway, Paul E., John R. Barrett, Timothy J. Foxon, and Julia K Steinberger. *Divergence of trends in US and UK aggregate exergy efficiencies 1960-2010* (ACS Publications, 2014) (see http://pubs.acs.org/doi/full/10.1021/es501217t), pp. 9874–9881.

90. Brown, Kandi, William Hall, Marjorie Snook, and Kathleen Garvin. *Sustainable Land Development and Restoration - Decision Consequence Analysis - Volume 1* (Elsevier, 2010) (see http://store.elsevier.com/product.jsp?isbn=9781856177979&pagename=search).

91. Browne, John. Energy and the Environment, 10 years on (2007) (see http://www.bp.com/genericarticle.do?categoryId=98&contentId=7032698).

92. Brundtland, G. H. *Our common future* (Oxford University Press, 1987) (see http://conspect.nl/pdf/Our_Common_Future-Brundtland_Report_1987.pdf).

93. Bryan, Gill, Cohen, Robert and Stepan, Paul. Wider Public Sector Emissions Reduction Potential Research, 2011 (see http://www.vercoglobal.com/images/uploads/insights-reports/2261-camco-report-public-sector-emissions.pdf).

94. BSI, Checklist for ISO 50001 (see http://www.bsigroup.com/Documents/iso-50001/resources/BSI-ISO50001-Assessment-Checklist-UK-EN.pdf).

95. BSI. PAS 2050:2011 Specification for the assessment of the life cycle greenhouse gas emissions of goods and services (2011) (see http://shop.bsigroup.com/upload/Shop/Download/PAS/PAS2050.pdf).

96. BSI. PAS 2060:2010 Specification for the demonstration of carbon neutrality, British Standards Institution (2010) (see http://shop.bsigroup.com/en/ProductDetail/?pid=000000000030198309).

97. BT. Delivering Our Purpose Report (2016) (see https://btplc.com/Purposefulbusiness/Deliveringourpurpose/Downloads/2016/2016_BTDeliveringOurPurposeReport.pdf and http://www.btplc.com/Purposefulbusiness/Deliveringourpurpose/Downloads/2016/2016_OurPurposefulTargets.xlsx).

98. BT. Futures (2013) (see https://www.btplc.com/betterfuture/betterfuturereport/pdf/2013/BT_Better_Future_2013_Highlights.pdf).

99. BT. PLT. Carbon Reduction Commitment Performance League Table / Annual Report Publications for 2011-2016. (see https://www.gov.uk/guidance/crc-energy-efficiency-scheme-annual-report-publication#history).

100. BT. Powers on towards global green energy targets. (2016) (see https://www.btplc.com/Purposefulbusiness/Energyandenvironment/Improvingouroperations/RenewablesCaseStudyAWLR.pdf).

101. BT. Purposeful. Net Good Portfolio (2014) (see https://www.btplc.com/Purposefulbusiness/Energyandenvironment/Our31methodology/NetGood2013-14summary.pdf).

102. BT. Purposeful. Net Good - Summary (2014) (see https://www.btplc.com/betterfuture/betterfuturereport/section/summaries/Net%20Good.pdf).

103. Building and Construction Authority, Singapore. Green Mark. (see https://www.bca.gov.sg/GreenMark/green_mark_buildings.html).

104. Building Owners and Managers Building Environmental Standards. (see http://bomacanada.ca/bomabest/aboutbomabest/).

105. Building Research Establishment. BREEAM (see http://www.breeam.com/).

106. Burke, Brian, *Gamify*, 2014, IBN 978-1-937134-85-3.

107. Burke, Brian, and M. Messaglio. *Case Study: Innovation Squared: The Department for Work and Pensions Turns Innovation Into a Game* (Gartner Research, 2010) (see http://innovation360.co/wp-content/uploads/2011/10/Gartner_DWP_Case_Study.pdf).

108. Calrberg, Conrad, *Predictive Analytics*: Microsoft Excel, QUE, ISBN-13: 978-0-7897-4941-3.

109. Calrberg, Conrad, *Statistical Analysis*: Microsoft Excel 2013, QUE, ISBN-13: 978-0-7897-5311-3.

110. Canadian Industry Program for Energy Conservation . Energy Savings Toolbox - An Energy Audit Manual and Tool (CIPEC 2011 est) (see http://www.nrcan.gc.ca/sites/www.nrcan.gc.ca/files/oee/pdf/publications/infosource/pub/cipec/energy-audit-manual-and-tool.pdf).

111. Carbon Desktop. (more information available at https://www.carbondesktop.com/).

112. Carbon Dioxide Information Analysis Centre (CDIAC). (see http://cdiac.ornl.gov/).

113. Carbon Tracker. Grantham Research Institute of Climate Change and the Environment (2013) (see http://carbontracker.live.kiln.it/Unburnable-Carbon-2-Web-Version.pdf).

114. Carbon Trust. *Degree days for energy management — a practical introduction*. 2012. (see https://www.carbontrust.com/media/137002/ctg075-Degree Days-for-energy-management.pdf).

115. Carbon Trust. *Low Carbon Behaviour Change: the £300 million opportunity*, The Carbon Trust (2013) (see https://www.carbontrust.com/media/434481/ctc827-low-carbon-behaviour-change.pdf).

116. Carbon Trust. *Monitoring and Targeting - Techniques to help organizations control and manage their energy use*, The Carbon Trust (2012) (see http://www.carbontrust.com/resources/guides/energy-efficiency/monitoring-and-targeting-download/).

117. Carbon War Room. (see: http://www.carbonwarroom.com/)

118. Cargill's Freshwater Efficiency Goal. (see http://www.cargill.com/corporate-responsibility/environmental-sustainability/environmental-goals-actions/freshwater-efficiency/index.jsp).

119. Carney, Mark. Speech by Mark Carney at the Arthur Burns Memorial Lecture, Berlin, on Thursday 22 September 2016, Bank of England (2016) (see http://www.bankofengland.co.uk/publications/Documents/speeches/2016/speech923.pdf).

120. Carrington, Damian. *Britain's damp, leaky homes among Europe's most costly to heat*, The Guardian 29 Nov 2013 (see http://www.theguardian.com/environment/damian-carrington-blog/2013/nov/29/uk-homes-most-expensive-heat-eu-fuel-poverty).

121. CASBEE, Comprehensive Assessment System for Built Environment Efficiency. (see http://www.ibec.or.jp/CASBEE/english/index.htm).

122. Casey, Alyssa. *Reforming Energy Subsidies Could Curb India's Water Stress* (see http://www.worldwatch.org/reforming-energy-subsidies-could-curb-india%E2%80%99s-water-stress-0).

123. Casten, Thomas, and Robert Ayres. in *Energy and American Society - Thirteen Myths*, Benjamin Sovacool and Marylin Brown, Ed. (Springer, 2007), pp. 201–237.

124. Caterpillar. Sustainability Report (2012) (see http://s7d2.scene7.com/is/content/Caterpillar/C10015259).

125. CBRE. 2010 National Investor Survey (2010) (see http://www.propertymetrics.com/wp-content/uploads/2013/09/CLF_CBRE-Valuation-Advisory-cap-rate-report.pdf).

126. CDP. (see https://www.cdp.net/en).

127. CDP. Carbon Action Report 2012: Carbon reductions generate positive ROI (2012) (previously available at www.cdproject.net/CDPResults/CDP-Carbon-Action-Report-2012.pdf).

128. CDP, and PWC Webinar. CDP Global Forum on forging sustainable profitable business growth (2011) (previously available at www.cdproject.net/en-US/Pages/CDP-Global-Forum-webcast.aspx).

129. CDP. Use of internal carbon price by companies as incentive and strategic planning tool, Carbon Disclosure Project (2013) (see https://www.cdp.net/CDPResults/companies-carbon-pricing-2013.pdf).

130. CDSB. Climate Change Reporting Framework – Edition 1.0, Climate Disclosure Standards Board (2010) (see http://www.cdsb.net/file/8/cdsb_climate_change_reporting_framework_2.pdf).

131. Centre for Remanufacturing and Reuse. Newsletter 34 (2013) (see http://www.remanufacturing.org.uk/pdf/newsletter/ns45.pdf?-session=RemanSession:42F94B7D09f0a03A58JSP15CF285).

132. Centre for Remanufacturing and Reuse. *US - Legal issues of remanufacturing in the automotive sector*. (16 April 2012) (see http://www.remanufacturing.org.uk/centrenews-detail.lasso?report=479&-session=RemanSession:D56978570c48a0F063jtP2811A1A).

133. Cesarotti, Vittorio, Vito Introna, Raffaele Rotunno, and Giulia Scerrato. *An empirical approach to investigate the relationship between Overall Equipment Effectiveness and Energy consumptions.* (2013) (see https://www.academia.edu/7606686/An_empirical_approach_to_investigate_the_relationship_between_Overall_Equipment_Effectiveness_and_Energy_consumptions_an_application_in_the_manufacturing_field).

134. CFA. *ESG at Listed Companies - A Manual for Investors*, CFA Institute (2008) (see http://www.cfapubs.org/toc/ccb/2008/2008/2).

135. Checket-Hanks, Barbara. Placebo Stats (2003) (see http://www.achrnews.com/articles/92414-placebo-stats).

136. Chertow, Marian R. *Industrial symbiosis: literature and taxonomy* (Annual Reviews 4139 El Camino Way, PO Box 10139, Palo Alto, CA 94303-0139, US, 2000), pp. 313–337. (see https://www.researchgate.net/profile/Marian_Chertow/publication/249558396_Industrial_symbiosis_Literature_and_taxonomy/links/0deec53a0ff2f1d91b000000/Industrial-symbiosis-Literature-and-taxonomy.pdf).

137. China Development and Reform Commission. (webpage translated from the Chinese) (14 June 2012) (see http://www.cnstock.com/08chanye/top/201206/2078432.htm).

138. China Council for International Cooperation on Environment and Development. *China's Low Carbon Industrialization Strategy*, (2011) (see http://www.cciced.net/encciced/policyresearch/report/201205/P020120529370737933117.pdf).

139. Christensen, C.M., S.P. Kaufman, and W.C. Shih. *Innovation killers: how financial tools destroy your capacity to do new things.* (2008) (see https://hbr.org/2008/01/innovation-killers-how-financial-tools-destroy-your-capacity-to-do-new-things), pp. 98–105.

140. Christensen, Peter et al. Cost Estimate Classification System 18R-97 Engineering, Procurement, Construction and Process Industries, AACE International Recommended Practices, 10 (2005).

141. Churcher, David. *Whole Life Costing Analysis*, BISRIA (2008).

142. Churcher, David. and Tse Peter, *Life Cycle Life Costing Analysis*, BG 67/2016, BISRIA (2016).

143. Cialdini, Robert B., and Nathalie Garde. *Influence* (A. Michel, 1987).

144. Cialdini, Robert B., et al. *Managing social norms for persuasive impact* (Taylor & Francis, 2006) (see http://www.fs.fed.us/psw/publications/winter/psw_2006_winter001.cialdini.pdf), pp. 3–15.

145. Cialdini, Robert B, Noah J. Goldstein. *Social influence: compliance and conformity.* (2004),(see http://www2.psych.ubc.ca/~schaller/Psyc591Readings/CialdiniGoldstein2004.pdf).

146. Cialdini, Robert B., Raymond R. Reno, and Carl A. Kallgren. *A focus theory of normative conduct: recycling the concept of norms to reduce littering in public places.* (American Psychological Association, 1990) (see http://media.cbsm.com/uploads/1/AFocusTheoryofNormativeConduct.pdf), p. 1015.

147. CIBSE. TM22 Energy Assessment and Reporting Methodology (inc CD-ROM) 2nd Edition (CIBSE, 2012) (see http://www.cibse.org/knowledge/cibse-tm/tm22-energy-assessment-reporting-methodology#1).

148. CIBSE TM41. Day, Tony, and others. *Degree days: theory and application* (2006) (see https://www.cibse.org/Knowledge/knowledge-items/detail?id=a0q20000008I73TAAS).

149. CIBSE TM46: 2008, Butcher, K. Energy Benchmarks, (CIBSE Publications, London, 2008) (see http://www.cibse.org/knowledge/cibse-tm/tm46-energy-benchmarks).

150. Citi Research. Commodities Outlook (2014) (see https://ir.citi.com/gpV66Qk64p1uNrDu8zrJ%2BoggS0GEiCbDfSD5C3XzQb1aJyWXi4psLw%3D%3D).

151. Clarkson, P. M., Y. Li, G. D. Richardson, and F. P. Vasvari. *Does it really pay to be green? Determinants and consequences of proactive environmental strategies*, Journal of Accounting and Public Policy 30, 122–144 (2011).

152. Climate Action Reserve (see http://www.climateactionreserve.org/).

153. Climate Bonds Initiative. *Bonds and Climate Change - the State of the Market in 2016*, Climate Bonds Initiative (2016) (see https://www.climatebonds.net/files/files/CBI%20State%20of%20the%20Market%202016%20A4.pdf).

154. Coalition for Energy Savings (see http://energycoalition.eu/sites/default/files/Poster-CES-final-page-002.jpg).

155. CN. *Coltan, Gorillas and Conflict* (2001) (see http://www.cellular-news.com/coltan/).

156. Cohen, David, and Jack L. Knetsch. *Judicial choice and disparities between measures of economic values*, Pace Law Faculty Publications Paper 419 (1992) (see http://digitalcommons.pace.edu/lawfaculty/419).

157. Cohen, Geoffrey L et al. *Bridging the partisan divide: Self-affirmation reduces ideological closed-mindedness and inflexibility in negotiation.* (American Psychological Association, 2007) (see https://ed.stanford.edu/sites/default/files/bridging_divides1.pdf), p. 415.

158. Cohen, Stanley. *States of denial: Knowing about atrocities and suffering* (John Wiley & Sons, 2013).

159. COI. Communications and Behaviour Change, COI and government Communications Network (2009) (see http://www.behaviourworksaustralia.org/wp-content/uploads/2012/09/commongood-behaviourchange.pdf).

160. CollaborativeConsumption. *Collaborative pioneer: an inside interview with Elina Furman, co-founder of Pley* (2014) (see http://www.collaborativeconsumption.com/2014/03/22/collaborative-pioneer-an-inside-interview-with-elina-furman-founder-of-pley/).

161. Committee on Climate Change, Carbon Budgets and targets (available at https://www.theccc.org.uk/tackling-climate-change/reducing-carbon-emissions/carbon-budgets-and-targets/).

162. Consumer Council for Water. UK Average Water Use. (see http://www.ccwater.org.uk/savewaterandmoney/averagewateruse/).

163. Conniff, Richard. *Putting a price on biodiversity*, The Guardian (2010) (see http://www.guardian.co.uk/environment/2010/sep/28/price-biodiversity-species-worth-cop10).

164. Consortium for Research on Emotional Intelligence in Organizations. (see http://www.eiconsortium.org/).

165. Cook, John et al. *Quantifying the consensus on anthropogenic global warming in the scientific literature* (IOP Publishing, 2013) (see http://iopscience.iop.org/article/10.1088/1748-9326/8/2/024024/pdf).

166. Cook, John, and Stephan Lewandowsky. *The debunking handbook* (Sevloid Art, 2011) (see http://www.skepticalscience.com/docs/Debunking_Handbook.pdf).

167. Cooke, Nancy J., Eduardo Salas, Janis A. Cannon-Bowers, and Renee J. Stout. *Measuring team knowledge* (SAGE Publications, 2000) (see http://www.dtic.mil/cgi-bin/GetTRDoc?AD=ADA527515), pp. 151–173.

168. Cooremans, Catherine. *Investment in energy efficiency: do the characteristics of investments matter?* (Springer, 2012) (see http://fourfact.com/images/uploads/1-476_Cooremans.pdf), pp. 497–518.

169. Corbett, Charles J., and Jeh-Nan Pan. *Evaluating environmental performance using statistical process control techniques* (2002) (see http://www.stat.ncku.edu.tw/faculty_private/jnpan/publication/2002_139_1_EJOR.pdf), pp. 68–83.

170. Costanza, R. et al. *The value of the world's ecosystem services and natural capital*, Nature 387, 253–260 (1997).

171. Costanzo, Mark, Dane Archer, Elliot Aronson, and Thomas Pettigrew. *Energy conservation behavior: The difficult path from information to action.* (American Psychological Association, 1986) (see http://www.researchgate.net/profile/Thomas_Pettigrew/publication/232577216_Energy_conservation_behavior_The_difficult_path_from_information_to_action/links/00b7d5165cba2bdb9e000000.pdf), p. 521.

172. Cowe, Peter. *Subsidising the wrong energy*, The Guardian (2012) (see http://www.guardian.co.uk/sustainable-business/perverse-government-subsidies-renewable-energy-oil).

173. Cox, Prentiss. *Keeping Pace: The Case against Property Assessed Clean Energy Financing Programs* (HeinOnline, 2011) (see http://lawreview.colorado.edu/wp-content/uploads/2013/1 1/9.-Cox-FINAL_s.pdf), p. 83.

174. Creative Commons Attribution 3.0 Unported Licence. (See http://creativecommons.org/licences/by/3.0/).

175. Crompton, Tom, and John Thøgersen. *Simple and painless? The limitations of spillover in environmental campaigning*, WWF-UK Godalming (2009) (see http://assets.wwf.org.uk/downloads/simple_painless_report.pdf).

176. Cropper, Angela. *Decoupling Economic Growth from Environmental Degradation* (2009) (see http://www.unep.fr/scp/rpanel/pdf/opening_angela_cropper.pdf).

177. CT. *Energy Management Strategy* (CTV022), The Carbon Trust (2007) (see www.carbontrust.co.uk/energy).

178. CT. *Canbon Footprints*, The Carbon Trust (2012) (see https://www.carbontrust.com/media/44869/j7912_ctv043_carbon_footprinting_aw_interactive.pdf).

179. CT. *Metering - Introducing the techniques and technology for energy data management*. The Carbon Trust (2012) (see http://www.carbontrust.com/media/31679/ctv027_metering_technology_overview.pdf).

180. Cullen, Jonathan M., and Julian M. Allwood. *Theoretical efficiency limits for energy conversion devices*, Energy 35, 2059–2069 (2010).

181. Daly, H. E. *Beyond growth: the economics of sustainable developmen*t (Beacon Press, 1997).

182. Danziger, Shai, Jonathan Levav, and Liora Avnaim-Pesso. *Extraneous factors in judicial decisions* (National Academy of Sciences, 2011) (see http://www.pnas.org/content/108/17/6889.full.pdf), pp. 6889–6892.

183. Darnton, Andrew, Bas Verplanken, Paul White, and Lorraine Whitmarsh. *Habits, routines and sustainable lifestyles*: A summary report to the Department for Environment, Food and Rural Affairs (2011) (see http://randd.defra.gov.uk/Document.aspx?Document=HabitsRoutinesSustainableLifestylesEVO502FinalSummaryReportNov2011(2).pdf).

184. Damodaran, Aswath. Estimated Cost of Capital (2016 update) (see http://people.stern.nyu.edu/adamodar/New_Home_Page/datacurrent.html).

185. Davis, Joel J. *The effects of message framing on response to environmental communications* (SAGE Publications, 1995) (see http://www.classmatandread.net/daenvf.pdf), pp. 285–299.

186. Davis, Steven J., and Ken Caldeira. *Consumption-based accounting of CO_2 emissions*. (2010), http://www.pnas.org/content/107/12/5687, pp. 5687–92.

187. Dayananda, Don. *Capital Budgeting: financial appraisal of investment projects* (Cambridge University Press, 2002). ISBN 978-0-521-52098-0

188. Dean, Jamie, and Jesse Purewal. Thesis, (2006) (see http://toolbelt.se-alliance.org/system/files/resources/method_sia_462.pdf).

189. DECC. Annual Report 2014-2015 (see https://www.gov.uk/government/uploads/system/uploads/attachment_data/file/447144/6.656_DECC_JG_Annual_Report_2014-15_AW_WEB.pdf).

190. DECC. Energy Consumption in the UK - Domestic Consumption Tables, UK Department for Energy and Climate Change (2015) (see https://www.gov.uk/government/statistics/energy-consumption-in-the-uk).

191. DECC. Plan. *The Carbon Plan: Delivering our low carbon future*, Department of Energy and Climate Change (2011) (see https://assets.publishing.service.gov.uk/government/uploads/system/uploads/attachment_data/file/47613/3702-the-carbon-plan-delivering-our-low-carbon-future.pdf).

192. DECC. Progress towards targets: 2012 provisional figures, UK Department of Energy and Climate Change (2013) (previopusly at https://www.gov.uk/government/uploads/system/uploads/attachment_data/file/211907/Progress_towards_targets_2012_provisional_figures.pdf).

193. DECC. Updated Energy and Emissions Projections, Department for Energy and Climate Change (2011) (see http://webarchive.nationalarchives.gov.uk/20121217150421/http://www.decc.gov.uk/assets/decc/11/about-us/economics-social-research/3134-updated-energy-and-emissions-projections-october.pdf).

194. DECC. Webarchive. Annex C DECC 2011 Energy and Emissions Projections, Department of Energy and Climate Change, UK (2011) (see http://webarchive.nationalarchives.gov.uk/20130106105028/http://www.decc.gov.uk/media/viewfile.ashx?filetype=4&filepath=11/about-us/economics-social-research/3118-annex-c-final-energy-demand.xls).

195. DECC. Webarchive. Annex A DECC Energy and Emissions Projection: 2011, Department of Energy and Climate Change, UK (2011) (see http://webarchive.nationalarchives.gov.uk/20130106105028/http://www.decc.gov.uk/media/viewfile.ashx?filetype=4&filepath=11/about-us/economics-social-research/3115-annex-a-greenhouse-gas-emissions-by-source.xls).

196. DECC. Webarchive. Energy Performance Contract (EPC): contract guidance note and model contract, Department of Energy and Climate Change (2015) (see https://www.gov.uk/government/uploads/system/uploads/attachment_data/file/395240/contract_guidance_note_and_model_contract_energy_performance_contract.doc).

197. DECC. Webarchive. Guide to Energy Performance Contracting Best Practices, Department of Energy and Climate Change (2015) (see https://www.gov.uk/government/uploads/system/uploads/attachment_data/file/395076/guide_to_energy_performance_contracting_best_practices.pdf).

198. Deci, Edward L. *Effects of externally mediated rewards on intrinsic motivation.* (American Psychological Association, 1971) (see http://www.quilageo.com/wp-content/uploads/2013/07/fn103.Deci_.pdf), p. 105.

199. DEFRA. Green Claims Guidance (2011) (see https://www.gov.uk/government/uploads/system/uploads/attachment_data/file/69301/pb13453-green-claims-guidance.pdf).

200. DEFRA, and DECC. GHG Conversion Factors (2011) (see http://webarchive.nationalarchives.gov.uk/20130402151656/http://archive.defra.gov.uk/environment/business/reporting/pdf/110707-guidelines-ghg-conversion-factors.pdf).

201. DegreeDays.Net, explanation of linear regression and obtaining a building's baseline temperature, 2015 (see http://www.degreedays.net/regression-analysis#optimal-base-temperature).

202. De Geus, Arie. *The Living Company* (Harvard Business Press, 2002).

203. De Groot, H. L. F., E. T. Verhoef, and P. Nijkamp. *Energy saving by firms: decision-making, barriers and policies* (Elsevier, 2001) (see https://www.econstor.eu/bitstream/10419/85670/1/99031.pdf), pp. 717–740.

204. Deloitte, and Co. *Life Cycle Assessment: Where is it on your sustainability agenda?* (2009) (see http://odpowiedzialnybiznes.pl/public/files/LifecycleAssessment_deloitte_2009.pdf).

205. Department for Business, Energy & Industrial Strategy, Building Energy Efficiency Survey 2014-2015. Nov 2016. (see https://www.gov.uk/government/publications/building-energy-efficiency-survey-bees).

206. Department for Business, Energy & Industrial Strategy, Greenhouse gas reporting - Conversion factors 2016. (see https://www.gov.uk/government/publications/greenhouse-gas-reporting-conversion-factors-2016).

207. Department for Transport. Making Car-sharing and Car Clubs Work (2005) (see http://www.civilservice.gov.uk/wp-content/uploads/2012/02/makingcarsharingmarch2005.pdf).

208. De Rooij, Ellen. *A brief desk research study into the average life expectancy of companies in a number of countries* (1996).

209. Desrochers, Pierre. *Regulatory Roadblocks to Turning Waste to Wealth* (2003), pp. 20–22. (see http://www.fee.org/files/docLib/desrochers0903.pdf).

210. DGNB, Deutsche Gessellschaft fur Nachhaltiges Bauen (see http://www.dgnb.de/en/).

211. Dicken, Peter. *Global Shift* (Sage, 2010).

212. Diesendort, Mark. *Sustainable Energy Solutions for Climate Change*, 2014, ISBN 978-0-415-70614-8.

213. Dillon, Sir Andrew. Carrying NICE over the threshold, NICE (2015) (see https://www.nice.org.uk/news/blog/carrying-nice-over-the-threshold).

214. Dincer, Ibrahim, and Marc A Rosen. *EXERGY: Energy, Environment and Sustainable Development* (Elsevier, 2013).

215. Dobbs, Richard et al. *Resource Revolution: Tracking Global Commodity Markets* (McKinsey Global Institute, 2013) (see http://www.mckinsey.com/~/media/McKinsey/dotcom/Insights/Energy%20Resources%20Materials/Resource%20revolution%20Tracking%20global%20commodity%20markets/MGI_Resources_survey_Full_report_Sep2013.ashx).

216. Doceo. *The Experiential Learning Cycle.* (see http://doceo.co.uk/l&t/learning/experience.htm).

217. DOE. US Alumina and Aluminum Footprint, Department of Energy (2012) (see https://www.energy.gov/sites/prod/files/2013/11/f4/aluminum_footprint_2012.pdf).

218. DOE. US Department of Energy, Buildings Energy Data Book (see http://buildingsdatabook.eren.doe.gov/TableView.aspx?table=3.1.4).

219. DOE. US Manufacturing Footprint, Department of Energy (2014) (see http://www.energy.gov/eere/amo/manufacturing-energy-and-carbon-footprints-2010-mecs).

220. DOE. US Plastics Footprint, Department of Energy (2010) (see https://www.energy.gov/sites/prod/files/2014/02/f7/2014_plastics_rubber_energy_carbon_footprint.pdf).

221. Dolan, Paul, Michael Hallsworth, David Halpern, Dominic King, and Ivo Vlaev. *Mindspace: influencing behaviour through public policy*. The Cabinet Office and Institute for Government (2014) (see http://www.instituteforgovernment.org.uk/sites/default/files/publications/MINDSPACE.pdf).

222. Doppelt, Bob. *Leading Change toward Sustainability* (Greenleaf Publishing, ed. 2nd, 2010).

223. Dowell, G., S. Hart, and B. Yeung. *Do corporate global environmental standards create or destroy market value?*, Management Science, 1059–1074 (2000).

224. Dow Sustainability Goals for 2015. (see http://www.dow.com/sustainability/goals/).

225. Dow Sustainability Goals for 2025. (see http://www.dow.com/en-us/science-and-sustainability/2025-sustainability-goals/valuing-nature).

226. Dreborg, Karl H. *Essence of backcasting* (Elsevier, 1996) (see http://www.sciencedirect.com/science/article/pii/S0016328796000444), pp. 813–828.

227. DRET. Energy Efficiency Opportunities Assessment Handbook (Department of Resources, Energy and Tourism, Australia, 2011) (see https://www.eex.gov.au/sites/g/files/net1896/f/files/2014/06/Assessment-Handbook.pdf).

228. DRET. Energy Savings Measurement Guide (Department for Resources, Energy and Tourism, Australia, 2008).

229. DRET. Energy Savings Measurement Guide (Department for Resources, Energy and Tourism, Australia, 2013) (see http://eex.gov.au/files/2014/06/ESMG.pdf).

230. Duhigg, Charles. *The power of habit: Why we do what we do in life and business* (Random House, 2012). ISBN: 978-1847946249.

231. DuPont. Improving Energy Efficiency & Profitability with DuPont (2010).

232. DuPont. Letter to Shareholders, 2015 (2015) (see http://investors.dupont.com/investor-relations/investor-news/investor-news-details/2015/DuPont-CEO-Ellen-Kullman-Sends-Letter-to-Shareholders/default.aspx).DuPont, Financial Data Book (2009) (see http://phx.corporate-ir.net/External.File?item=UGFyZW50SUQ9MzkzMDd8Q2hpbGRJRD0tMXxUeXBlPTM=&t=1).

233. DuPont. Sustainability report 2015 (see http://www.dupont.com/content/dam/dupont/corporate/our-approach/sustainability/documents/DuPont-Sustainability-Report-2015_111615.pdf).

234. Eccles, Robert G., Beiting Cheng, and Daniela Saltzman. *The Landscape of Integrated Reporting*, Harvard Business School (2010) (see hbswk.hbs.edu/pdf/The_Landscape_of_Integrated_Reporting.pdf).

235. Eccles, Robert, and George Serafeim. *Leading and Lagging Countries in Contributing to a Sustainable Society*, Harvard Business School (2011) (see http://hbswk.hbs.edu/cgi-bin/print/6716.html).

236. Ecofeedback. (see http://ecofeedback.ca/).

237. Ecolabel Index. (accessible at http://www.ecolabelindex.com/).

238. Economist, The. *Good Policy and Bad: GHG mitigation* (2009).

239. Ecorys. Study on the Competitiveness of European Companies and Resource Efficiency, Ecorys (2011) (see http://ec.europa.eu/enterprise/policies/sustainable-business/files/competitiveness_of_european_companies_150711_en.pdf).

240. Ecosystem Marketplace. Raising Ambition - State of the Voluntary Carbon Market 206 (2016) (see http://www.forest-trends.org/documents/files/doc_5242.pdf).

241. EERE, DOE. 3M's Model Rewards and Recognition Program Engages Employees and Drives Energy Savings Efforts, Industrial Technologies Program (2010) (see http://www1.eere.energy.gov/industry/saveenergynow/pdfs/3m.pdf).

242. EEVS Insight Ltd. Electricity Demand Reduction Pilot Scheme Phase II - Measurement and Verification Manual, Department of Energy and Climate Change (2015) (see https://www.gov.uk/government/uploads/system/uploads/attachment_data/file/435266/M_and_V_manual.pdf).

243. Ehrlich, P. R., J. P. Holdren, and others. *Impact of population growth*, Science 171, 1212–1217 (1971).

244. EIA. Manufacturing Energy Consumption Survey, Energy Information Agency (2010) (see http://www.eia.gov/consumption/manufacturing/data/2010/index.php?view=data).

245. EIA Short Term Energy Outlook, Prices Viewer. (see http://www.eia.gov/forecasts/steo/realprices/).

246. Eichholtz, Piet, Nils Kok, and John M Quigley. Doing well by doing good? Green office buildings (JSTOR, 2010) (see http://escholarship.org/uc/item/4bf4j0gw), pp. 2492–2509.

247. Ekins, P., F. Kesicki, and A. Smith. *Marginal abatement cost curves: A call for caution* (2011) (see https://www.ucl.ac.uk/energy/home-top-cols/image-link-docs/MACCCritGPUKFin.pdf).

248. Ellen MacArthur Foundation. Caterpillar Case Study. (see http://www.ellenmacarthurfoundation.org/case_studies/caterpillar).

249. Ellen MacArthur Foundation, Towards the Circular Economy. (2012) (see http://www.ellenmacarthurfoundation.org/business/reports/ce2012).

250. Enbridge. Unilever's Monitoring and Targeting Program Captures $1 million in Savings (see https://www.enbridgegas.com/businesses/assets/docs/cs_unilever_m%26t_case_study.pdf).

251. Energetics. Energetics One2Five(R) (see http://www.energetics.com.au/Services/One2Five-management-diagnostic-tool/One2Five-Energy#tabs).

252. Energi. Risk Mitigation Insurance in New Energy Finance (2012) (see www.energi.com).

253. Energy and Climate Change Committee: Inquiry into Home Energy Efficiency., The Green Deal Finance Company (2015) (see http://www.tgdfc.org/assets/Energy_and_Climate_Change_Committee_GDFC_written_evidence_final_13Oct15.pdf).

254. Energy Management Working Group. (see http://www.cleanenergyministerial.org/Our-Work/Initiatives/Energy-Management/EMWG-Overview).

255. Energy Star (see https://www.Energy Star.gov/buildings/facility-owners-and-managers/existing-buildings/earn-recognition/energy-star-certification).

256. Energy Technology List, UK Department of Energy and Climate Change (see https://www.gov.uk/guidance/energy-technology-list).

257. Energy Technology Support Unit. *Monitoring and Targeting in Large Companies: Good Practice Guide 122.*, UK Energy Efficiency Best Practice Programme (1998).

258. Energy Transitions Commission, and Ecofys. Pathways from Paris - Assessing the INDC Opportunity (2016) (see http://www.energy-transitions.org/sites/default/files/20160426%20INDC%20analysis%20vF.pdf).

259. Enkvist, P., Tomas Nauclér, and Jerker Rosander. *A cost curve for greenhouse gas reduction* (McKinsey & Co. Inc, 2007) (see http://www.mckinsey.com/business-functions/sustainability-and-resource-productivity/our-insights/a-cost-curve-for-greenhouse-gas-reduction), p. 34.

260. Enkvist, Per-Anders, Jens Dinkel, and Charles Lin. *Impact of the financial crisis on carbon economics: Version 2.1 of the global greenhouse gas abatement cost curve* (2010) (see http://www.mckinsey.com/~/media/McKinsey/Business%20Functions/Sustainability%20and%20Resource%20Productivity/Our%20Insights/Impact%20of%20the%20financial%20crisis%20on%20carbon%20economics%20Version%2021/Impact%20of%20the%20financial%20crisis%20on%20carbon%20economics%20Version%2021.ashx).

261. Environmental Defence Fund. Partnerships: The key to scalable (see https://www.edf.org/approach/partnerships).

262. Enviros. *Review and development of carbon dioxide abatement curves for available technologies as part of the Energy Efficiency Innovation Review*, SKM Enviros Consulting (2005).

263. EPA. Climate Leaders Partnership Directory, US Environmental Protection Agency (2010) (see http://www.epa.gov/climateleadership/documents/directory.pdf).

264. EPRI. U.S. Energy Efficiency Potential Through 2035 (2014) (see http://www.epri.com/abstracts/Pages/ProductAbstract.aspx?ProductId=000000000001025477).

265. Erb, K. H. et al. *Analyzing the global human appropriation of net primary production—processes, trajectories, implications*. An introduction, Ecological Economics 69, 250–259 (2009).

266. Estidama. (see http://estidama.upc.gov.ae/estidama-and-pearl-rating-system.aspx?lang=en-US).

267. ESMAP, MACCTool. (see http://esmap.org/MACTool).

268. Esty, Daniel C, and Andrew S Wilson. *Green to Gold* (Wiley, 2006).

269. Evans, Raymond, Richard Haryott, Norman Haste, and Alan Jones. *The long-term costs of owning and using buildings* (2004) (see http://www.academia.edu/download/32394203/Designing_Better_Buildings.pdf#page=63).

270. European Environment Agency. *How is climate change affecting the energy demand for space heating across Europe?* Trend in heating degree days in the EU-27 (see http://www.eea.europa.eu/data-and-maps/indicators/heating-Degree Days-1/assessment).

271. European Parliament and Council of Ministers. Directive 2012/27/EU on energy efficiency. (see http://eur-lex.europa.eu/legal-content/EN/TXT/?qid=1399375464230&uri=CELEX:32012L0027).

272. Farrell, D., J. Remes, F. Bressand, M. Laabs, and A. Sundaram. *The Case for Investing in Energy Productivity*, McKinsey Global Institute (2008) (see http://www.mckinsey.com/locations/swiss/news_publications/pdf/mgi_carbon_productivity_challenge_report.).

273. Faruqui, Ahmad, Sanem Sergici, and Ahmed Sharif. *The impact of informational feedback on energy consumption—A survey of the experimental evidence* (Elsevier, 2010), pp. 1598–1608.

274. Fawkes, Steven. *Energy Efficiency the definitive guide to the cheapest, cleanest, fastest source of energy*. (Gower, 2013).

275. Fawkes, Steven. *Let's ditch energy efficiency*, 14 December 2015. (see http://www.onlyelevenpercent.com/lets-ditch-energy-efficiency/)

276. Federal Reserve Bank. Mortgage Debt Outstanding. Figures for Q2 2016 used. (see https://www.federalreserve.gov/econresdata/releases/mortoutstand/current.htm).

277. FEMP. M&V Guidelines: Measurement and Verification for Performance-Based Contracts version 4.0, US Department of Energy, Federal Energy Management Programme (2015) (see http://energy.gov/sites/prod/files/2016/01/f28/mv_guide_4_0.pdf).

278. Figge, F., and T. Hahn. *Sustainable Value Added-measuring corporate contributions to sustainability beyond eco-efficiency*, Ecological Economics 48, 173–187 (2004).

279. Figge, Frank, and Tobias Hahn. *The Cost of Sustainability Capital and the Creation of Sustainable Value by Companies* (2005), doi:10.1162/108819805775247936, pp. 47–58. (see http://sustainable-finance.com/downloads/figgehahnthecostofsustainabilitycapital.pdf).

280. Financial Stability Board, Phase I Report of the Task Force on Climate-Related Financial Disclosures., (see https://www.fsb-tcfd.org/wp-content/uploads/2016/03/Phase_I_Report_v15.pdf).

281. Fisher-Vanden, K., and K. S. Thorburn. *Voluntary corporate environmental initiatives and shareholder wealth*, Journal of Environmental Economics and Management (2011) (see http://mba.tuck.dartmouth.edu/initiative/pdf/FVT_WorkingPaperCER-jan_08.pdf).

282. Fisk, William J. *Health and productivity gains from better indoor environments and their relationship with building energy efficiency* (Annual Reviews 4139 El Camino Way, PO Box 10139, Palo Alto, CA 94303-0139, US, 2000) (see http://www2.bren.ucsb.edu/~modular/private/Articles/Fisk%20HealthandProductivity%202000.pdf), pp. 537–566.

283. Fiksel, Joseph. *Design for Environment - A Guide to Sustainable Product Development.* (2012) ISBN 978-0-07-177622-6.

284. Fleming, David, and Shaun Chamberlin. *TEQs (Tradable Energy Quotas): A Policy Framework for Peak Oil and Climate Change*, All Party Parliamentary Group on Peak Oil (2011) (see http://www.teqs.net/APPGOPO_TEQs.pdf).

285. Fleming, Neil D., and Colleen Mills. *Not another inventory, rather a catalyst for reflection* (1992) (see http://digitalcommons.unl.edu/cgi/viewcontent.cgi?article=1245&context=podimproveacad).

286. Flynn, Emily. What corporations can learn from green revolving funds, GreenBiz (2013) (see https://www.greenbiz.com/blog/2013/01/17/what-corporations-can-learn-green-revolving-funds?page=0%2C0).

287. FootprintNetwork. World Footprint (see http://www.footprintnetwork.org/en/index.php/GFN/page/world_footprint/).

288. Ford. Ford's Energy Efficiency Efforts Earn Sixth ENERGY STAR Award (2011) (see https://greenlivingguy.com/2011/03/fords-energy-efficiency-efforts-earn-sixth-energy-star-award/).

289. Ford Sustainability Report 2010-2011. (2011) (see http://corporate.ford.com/doc/sr10.pdf).

290. Ford Sustainability Report 2015-16 (see http://corporate.ford.com/microsites/sustainability-report-2015-16/index.html).

291. Ford Sustainability Report and Goals. (see http://corporate.ford.com/microsites/sustainability-report-2012-13/world.html).

292. Forum For the Future and The Climate Group. *Measuring your way to Net Positive* (2016) (see https://www.forumforthefuture.org/sites/default/files/Measuring%20Net%20Positive.pdf).

293. Freedman Andrew, *Severe U.S. drought sets another record; costs to U.S. economy upward of $1.5 billion.* Washington Post (2 Aug 21011) (see http://www.washingtonpost.com/blogs/capital-weather-gang/post/severe-us-drought-sets-another-record-costs-to-us-economy-upward-of-15-billion/2011/08/01/gIQA7cQbpI_blog.html).

294. Frese, Michael, Eric Teng, and Cees JD Wijnen. *Helping to improve suggestion systems: Predictors of making suggestions in companies* (Journal of Organizational Behaviour, 1999), pp. 1139–1155.

295. Friedman, Milton. *The social responsibility of business is to increase its profits*. Springer Berlin Heidelberg, 2007. Original article in the New York Times, The New York Times Magazine, September 13, 1970. (see http://umich.edu/~thecore/doc/Friedman.pdf).

296. Frosch, R.A., and N.E. Gallopoulos. *Strategies for manufacturing*. (Scientific American, 1989), pp. 144–52. (see http://isfie.onefireplace.com/resources/Documents/Strategies_For_Manufacturing_Sci_American_1989.pdf).

297. Frost & Sullivan, *Car-sharing - Driving the Way to a Greener Future* (2010) (see: http://www.frost.com/prod/servlet/press-release.pag?docid=193331843).

298. Fuller, Sieglinde ,Stephen Petersen. Life Cycle Costing Manual for the Federal Energy Management Programme, US Department of Commerce (1995) (see http://www.eeperformance.org/uploads/8/6/5/0/8650231/lcc_manual_for_femp.pdf).

299. Gadd, Karen. *TRIZ for Engineers - Enabling Inventive Problem Solving* (2011). ISBN: 9780470741887.

300. Galitsky, C., and E. Worrell. Energy Efficiency Improvement and Cost Saving Opportunities for the Vehicle Assembly Industry: An ENERGY STAR Guide for Energy and Plant Managers (2008) (see http://ies.lbl.gov/iespubs/Energy Star/vehicleassembly.pdf).

301. Gallai, N., J. M. Salles, J. Settele, and B. E. Vaissière. *Economic valuation of the vulnerability of world agriculture confronted with pollinator decline*, Ecological Economics 68, 810–821 (2009).

302. Games for Change. (see http://www.gamesforchange.org/play/climate-challenge/).

303. Gap. Sustainability Report 2013-14 (see http://www.gapincsustainability.com/sites/default/files/Gap%20Inc.%202013%20-%2014%20Report.pdf).

Reference

304. Gapminder, A fact-based world view. (accessible at www.gapminder.org, CO2 emissions chart at www.bit.ly/b9p3dA).

305. Gawande, Atul. *The Checklist Manifesto.* (Profile Books, 2010).

306. GE. Annual Report 2011 (2011) (see http://www.ge.com/ar2011/pdf/GE_AR11_EntireReport.pdf).

307. GE Capital. Driving change and continuous process improvement. (2012) (previously at http://www.gecapital.com.au/common/documents/access-ge//GE_capital_HowTo_Driving_Change_and_Continuous_Process_Improvement.pdf now https://pdfs.semanticscholar.org/presentation/d158/a8bd3f5d9e6d84e5868ab4a4e4446ca68f36.pdf).

308. General Motors. Sustainability Report 2016 (see http://www.gmsustainability.com/reporting/generatePDF.php).

309. Generation. Sustainable Capitalism, Generation Investment (2012) (see http://www.generationim.com/media/pdf-generation-sustainable-capitalism-v1.pdf).

310. George, Rowlands, Price and Maxey, *The Lean Six Sigma Pocket Toolbook*, McGraw Hill, ISBN 0-07-144119-0.

311. Georgina Mace, Hillary Masundire, Jonathan Baillie et al. Current States and Trends, *Chapter 4: Biodiversity, Millennium Ecosystems Assessment* (Island Press, Washington, DC, 2005) (see http://www.maweb.org/documents/document.273.aspx.pdf).

312. GHF. *The Anatomy of a Silent Crisis*, Global Humanitarian Forum (2009) (see http://www.ghf-ge.org/human-impact-report.pdf).

313. Gilding, Paul. *The Great Disruption* (Bloomsbury, 2011).

314. Gilligan, Donald. Overview of MUSH Market ESCO Financing, ACEEE Financing Forum (2016) (see http://aceee.org/sites/default/files/pdf/conferences/eeff/2016/Gilligan_Session1D_FF16_5.23.16.pdf).

315. Giutini, Ron, and Kevin Gaudette. *Remanufacturing: The next great opportunity for boosting US productivity* (Elsevier, 2003) (see http://www.cardone.com/docs/environmental-commitment/reman-20next-20great-20opportunity-20for-20productivity.pdf), pp. 41–48.

316. Global Compact. (see http://www.globalcompact.org.uk/issues/).

317. Global Compact Network, UK. (see http://www.globalcompact.org.uk/issues/).

318. Global Ecolabelling Network. Environmental Standards by Product Category. (see http://www.globalecolabelling.net/eco/eco-friendly-products-by-category/).

319. Global Environmental Management Initiative. Forging New Links. (see http://gemi.org/supplychain/wizframe.htm).

320. Global Environmental Product Declarations Network (accessible at http://gednet.org/).

321. Global Reporting Initiative (GRI) (see https://www.globalreporting.org/Pages/default.aspx).

322. Gold Standard (see http://www.goldstandard.org/)

323. Golden, ed Jay. *An Overview of Ecolabels and Sustainability Certification in the Global Marketplace* (2010) (see http://center.sustainability.duke.edu/sites/default/files/documents/ecolabelsreport.pdf).

324. Goldenberg, Suzanne. *Exxon knew of climate change in 1981* (8 July 2015) (see http://www.theguardian.com/environment/2015/jul/08/exxon-climate-change-1981-climate-denier-funding).

325. Goldstein, Noah J., Robert B. Cialdini, and Vladas Griskevicius. *A Room with a Viewpoint: Using Social Norms to Motivate Environmental Conservation in Hotels* (2008) (see http://www.ejcr.org/Curations-PDFs/Curations11/Curations11-5.pdf), pp. 472–482.

326. Gonzales, Marti Hope, Elliot Aronson, and Mark A Costanzo. *Using Social Cognition and Persuasion to Promote Energy Conservation: A Quasi-Experiment* (Wiley Online Library, 1988), pp. 1049–1066.

327. Graham, T. A *comparison of the limits to growth with thirty years of reality*, Socio-Economics and the Environment in Discussion. CSIRO Working Paper Series 9 (2008) (see http://www.ecsim.org/Vista/archivos/TURNER%20G%20-%20TLG%2030%20years%20comparison%20to%20reality.pdf).

328. Granade, Hannah Choi et al. *Unlocking Energy Efficiency in the U.S. Economy*, Report, 165 (2009).

329. Granarolo. Environmental Product Declaration for high quality pasteurised milk. (2014) (see http://environdec.com/en/Detail/epd118#.Vd2RKTZRF_t).

330. Gray, Casper, and Martin Charter. *Remanufacturing and product design: Designing for the 7th generation* (2007) (see http://cfsd.org.uk/Remanufacturing%20and%20Product%20Design.pdf).

331. Green, David I. *Pain-cost and opportunity-cost* (JSTOR, 1894) (see http://www.jstor.org/stable/1883711), pp. 218–229.

332. Green Globes (see http://www.greenglobes.com/home.asp)

333. Green Jobs in the US. (see http://www.ibtimes.co.uk/articles/20121011/green-jobs-growth-usa-manufacturing-leads-way1-million-green-jobs-november.htm).

334. Greenpeace Letter to Fossil Fuel Company Directors. (see http://www.greenpeace.org/international/Global/international/briefings/climate/2014/climate-denialism/Companies-letter-example.pdf).

335. Griskevicius, Vladas, Robert B. Cialdini, and Noah J. Goldstein. *Social norms: An underestimated and underemployed lever for managing climate change* (2008) (see http://www.ecy.wa.gov/programs/wq/stormwater/municipal/MUNIdocs/SocialNormsResearch.pdf).

336. GTAA. Unilever Rexdale's Award-winning Ecoefficiency Strategies (see http://www.partnersinprojectgreen.com/files/CS_Unilever_CSR.pdf).

337. Gulbinas, Rimas, and John E. Taylor. *Effects of real-time eco-feedback and organizational network dynamics on energy efficient behavior in commercial buildings* (Elsevier, 2014), pp. 493–500.

338. Gulf Organizations for Research and Development. Global Sustainability Assessment System. (see http://www.gord.qa/gsas-trust).

339. Gunther, Marc. *Activists attack sellers of bottled water,* CNN Money (2007) (see http://money.cnn.com/2007/04/24/news/economy/pluggedin_gunther_water.fortune/index.htm).

340. Gupta, Praveen. *The Six Sigma performance handbook: a statistical guide to optimizing results,* (McGraw Hill Professional, 2005).

341. Gupta, S., and B. Goldar. *Do stock markets penalize environment-unfriendly behaviour?* Evidence from India, Ecological Economics 52, 81–95 (2005).

342. Haig, M, and M Shapiro. *Financial Institutions: Taking Greenhouse Gases into Account* (2011) (see http://www.cdsb.net/file/56/financial-institutions--taking-greenhouse-gas-emissions-into-account-cdsb-2011.pdf).

343. Haldane, Andrew, and Richard Davies. *The Short Long* (2011) (see http://www.bankofengland.co.uk/archive/Documents/historicpubs/speeches/2011/speech495.pdf).

344. Hamilton, Booz Allen. Process Ownership: The Overlooked Driver of Sustained BPR (2003) (see http://www.boozallen.com/content/dam/boozallen/media/file/process_ownership.pdf).

345. Handgraaf, Michel J. J., Margriet A. Van Lidth de Jeude, and Kirstin C. Appelt. *Public praise vs. private pay: Effects of rewards on energy conservation in the workplace* (2013), pp. 86–92.

346. Hanemann, Michael. Willingness to Pay and Willingness to Accept: How Much Can They Differ? (1991), pp. 635–47.

347. Hanley, Nick, Jason Shogren, and Ben White. *Introduction to environmental economics* (Oxford University Press, 2013). ISBN 978-0-19-956873-4

348. Hansen, J. et al. *Earth's energy imbalance: Confirmation and implications*, Science 308, 1431–1435 (2005).

349. Hansen, J. et al. *Target atmospheric CO2: Where should humanity aim?*, Arxiv preprint arXiv:0804.1126 (2008) (see http://arxiv.org/ftp/arxiv/papers/0804/0804.1126.pdf).

350. Hansen, Shirley J. *ESCOs around the world: lessons learned in 49 countries*. ISBN 978-1-4398-1101-6.

351. Harmann, Anja, Dianna Farrell, Michael Graubner, and Remes Jaana. *Capturing the European energy productivity opportunity,* McKinsey Global Institute (2008) (see http://www.mckinsey.com/insights/energy_resources_materials/capturing_european_energy_productivity).

352. Harrington, Eric, and Cara Carmichael. RetroFit: Project Case Study - Empire State Building, Rocky Mountain Institute (2009) (see www.rmi.org/Content/Files/ESBCaseStudy.pdf).

353. Harrington, L., and M. Damnics. Energy labelling and standards programs throughout the world, The National Appliance and Equipment Energy Efficiency Committee, Australia (2004) (see http://www.energyrating.com.au/library/pubs/200404-internatlabelreview.pdf).

354. Harris, Peter. *Energy monitoring and target setting using CUSUM* (Cheriton Technology Publications, 1989) (see World Cat as a hard copy http://www.worldcat.org/title/energy-monitoring-and-target-setting-using-cusum/oclc/59849181).

355. Harvard Business Review, 2010. *A Market Solution for Achieving "Green"* by Jack D. Hidary (see https://hbr.org/2010/01/the-hbr-list-breakthrough-ideas-for-2010).

356. Harvard's Commitments to Sustainability. (see http://green.harvard.edu/commitments).

357. Harvard University. Green Revolving Fund Overview. (see https://green.harvard.edu/sites/green.harvard.edu/files/GRF%20Overview.pdf).

358. Harvard University. Sustainability Programmes website (see https://green.harvard.edu/programs/green-revolving-fund).

359. Hawken, P. *The ecology of commerce: A declaration of sustainability* (Harper Paperbacks, 1994).

360. Haynes, Alex B. et al. *A surgical safety checklist to reduce morbidity and mortality in a global population* (Mass Medical Soc, 2009) (see http://www.nejm.org/doi/pdf/10.1056/NEJMsa0810119), pp. 491–499.

361. Heck, Stefan, and Matt Rogers. *Resource Revolution - How to capture the biggest business opportunity in a century* (Amazon Publishing, Seattle, 2014).

362. Hedgeland, Tim. *Story Telling with Intent - the sustainability story that makes your point wins*, UC Berkeley Office of Sustainability and Energy (2014) (see http://sustainability.berkeley.edu/sites/default/files/Storytelling%20with%20Intent%20FINAL_0.pdf).

363. Heede, Richard. *Tracing anthropogenic carbon dioxide and methane emissions to fossil fuel and cement producers, 1854-2010* (Springer, 2014) (see http://link.springer.com/article/10.1007/s10584-013-0986-y), pp. 229–241.

364. Heinzle, Stefanie Lena, and Rolf Wüstenhagen. *Dynamic Adjustment of Eco-labeling Schemes and Consumer Choice-the Revision of the EU Energy Label as a Missed Opportunity?* (Wiley Online Library, 2012), pp. 60–70.

Reference

365. Herbert, Simon. *Administrative Behavior* (The Free Press; 4th Revised edition (1 Sep 1997), 1947).

366. Herzberg, Frederick. *One more time: How do you motivate employees?* (1987) (see http://www2.arnes.si/~korpl2/FM/articles/Herzberg_motivate-Damijan%20M.pdf), pp. 109–120.

367. Hewett, M. J. Achieving energy efficiency in a restructured electric utility industry, Minnesotans for an Energy Efficient Economy (1998) (see http://citeseerx.ist.psu.edu/viewdoc/download?doi=10.1.1.195.3605&rep=rep1&type=pdf).

368. Heyn, Gerrit. Companies that Invest in Sustainability Do Better Financially (see http://blogs.hbr.org/cs/2012/09/sustainable_investing_time_to.html).

369. Hilliard, Antony, and Greg A. Jamieson. *Recursive Estimates as an Extension to CUSUM-based Energy Monitoring and Targeting* (2013) (see http://aceee.org/files/proceedings/2013/data/papers/4_094.pdf).

370. Hislop, Hannah and Hill, Julie. *Reinventing the wheel: A circular economy for resource security.* Green Alliance (October 2011) (see http://www.green-alliance.org.uk/uploadedFiles/Publications/reports/Reinventing%20the%20wheel_dbl.pdf).

371. HM Government, Department for Business, Innovation and Skills, "One-in, two-out: statement of new regulation", 2013 (see https://www.gov.uk/government/collections/one-in-two-out-statement-of-new-regulation).

372. Hollingworth, Crawford, and Liz Baker. *Habits, the Holy Grail of Marketing*, The Marketing Society (2014) (see https://www.marketingsociety.com/sites/default/files/thelibrary/Habits%20-%20The%20Behavioural%20Architects_2.pdf).

373. Hollins, Oakdene. *Further Benefits of Business Resource Efficiency*, Defra (2011) (see http://www.oakdenehollins.co.uk/pdf/Further_Benefits_of_RE_Final_report.pdf).

374. Hollmann, John K. Estimate Accuracy: Dealing with Reality (2012) (see http://www.validest.com/index_htm_files/Hollmann_Accuracy.pdf), p. 17.

375. Hollmann, John K. RISK-1027: Estimate Accuracy: Dealing with Reality (2012) (see http://www.nflaace.org/index_files/n_fl_aacei_hollmann_estimating_accuracy.ppsx).

376. Honey, Peter, Alan Mumford, and others. *The manual of learning styles* (Peter Honey Maidenhead, 1992).

377. Hong Kong Electronic Industries Association, Guidebook for ISO 50001 (2013) (see http://www.gmn.hkpc.org/file/file/Guidebook/ISO50001Guide_ENG%2019Aug(Final).pdf).

378. Househam, Anita, Elena Bombis (current ed.) Cody Sisco Blythe Chorn, and Peder Michael Pruzan-Jorgensen (prior ed.). *Supply Chain Sustainability: a Practical Guide for Continuous improvement*, United Nations (2016) (see https://www.unglobalcompact.org/docs/issues_doc/supply_chain/SupplyChainRep_spread.pdf).

379. Howell, Marvin T. *Effective Implementation of an ISO 50001 Energy Management System* (EnMS), (ASQ Quality Press, 2014) ISBN13 978-0873898720.

380. HU2. (see http://www.hu2.com/decals/co2-factory-reminder-sticker/).

381. Huefner, R., and J. Largay. The Effect of the New Goodwill Accounting Rules on Financial Statements (2004) (was previously available at http://www.nysscpa.org/cpajournal/2004/1004/essentials/p30.htm).

382. Hughes, Will P., Debbie Ancell, Stephen Gruneberg, and Luke Hirst. Exposing the myth of the 1: 5: 200 ratio relating initial cost, maintenance and staffing costs of office buildings (2004) (see http://centaur.reading.ac.uk/12142/1/File12142.pdf).

383. IBM. Sustainability Report 2015 (see https://www.ibm.com/ibm/responsibility/2015/assets/downloads/IBM_2015_CR_report.pdf).

384. ICAP. Emissions Trading Worldwide - Status Report 2014, International Carbon Action Partnership (2014) (see https://icapcarbonaction.com/index.php?option=com_attach&task=download&id=152).

385. ICAP. Emissions Trading Worldwide - Status Report 2016, International Carbon Action Partnership (2016) (see https://icapcarbonaction.com/images/StatusReport2016/ICAP_Status_Report_2016_Online.pdf).

386. ICF International. Study on Energy Efficiency and Energy Savings Potential in Industry and Possible Policy Mechanisms (2015) (see https://ec.europa.eu/energy/sites/ener/files/documents/151201%20DG%20ENER%20Industrial%20EE%20study%20-%20final%20report_clean_stc.pdf).

387. ICP. Project Development Specification, Investor Confidence Project (2014) (see http://www.eeperformance.org/uploads/8/6/5/0/8650231/projectdevelopmentspecificationv1.0.pdf).

388. IDS. Implementing a Suggestion Scheme (2005).

389. IEA. World Energy Outlook 2010 - Executive Summary, International Energy Agency (2010) (see http://www.iea.org/Textbase/npsum/weo2010sum.pdf).

390. IEA, Birol, Fatih. *World energy outlook 2012*, IEA, OECD (2012) (see http://www.iea.org/publications/freepublications/publication/WEO2012_free.pdf).

391. IEA. *World Energy Outlook 2013 - Factsheet* (2013) (see http://www.worldenergyoutlook.org/media/weowebsite/2012/factsheets.pdf).

392. IET. *Guide to Metering Systems: Specification, Installation and Use*, ISBN: 978-1-78561-059-2 (see http://www.theiet.org/resources/standards/metering.cfm).
393. IIASA. The GGI Scenario Database (Version 2.0) (see http://www.iiasa.ac.at/web-apps/ggi/GgiDb/dsd?Action=htmlpage&page=about).
394. IIGCC, INCR IGCC. Investor Expectations (2012) (see http://www.iigcc.org/publications/publication/institutional-investors-expectations-of-corporate-climate-risk-management).
395. IIRC. Towards Integrated Reporting Discussions Paper, International Integrated Reporting Council (2011) (see http://theiirc.org/wp-content/uploads/2011/09/IR-Discussion-Paper-2011_spreads.pdf).
396. Incandescents. (2010) (see http://www.tlt.co.jp/tlt/english/info/release/release_20100317.htm and http://www.washingtonpost.com/wp-dyn/content/article/2010/09/07/AR2010090706933.html).
397. Institute for Building Efficiency. 2013 Energy Efficiency Indicators (2013) (see http://navitascap.com/wp-content/uploads/2013/11/061213_IBE_Energy_Efficiency_Indicator_Survey.pdf).
398. Interface's objective to Eliminate Waste. (see http://www.interfaceflor.co.uk/web/sustainability/mission_zero/seven_fronts/eliminate_waste).
399. International Atomic Energy Agency. *Uranium 2011: Resources, Production and Demand (The Red Book)*. (see http://www.iaea.org/OurWork/ST/NE/NEFW/Technical_Areas/NFC/uranium-production-cycle-redbook.html).
400. International Bar Association. *Achieving Justice and Human Rights in an Era of Climate Disruption*. International Bar Association (2014) (see http://www.ibanet.org/PresidentialTaskForceClimateChangeJustice2014Report.aspx).
401. International Capital Markets Association. Green Bond Principles (2016) (see http://www.icmagroup.org/assets/documents/Regulatory/Green-Bonds/GBP-2016-Final-16-June-2016.pdf).
402. International Energy Efficiency Financing Protocol., Efficiency Valuation Organization (2009) (see http://www.evo-world.org/index.php?option=com_content&view=article&id=373&Itemid=473&lang=en).
403. International Environmental Products Declaration System, (accessible at http://www.environdec.com/en/PCR/).
404. International Institute for Applied Systems Analysis, Laxenburg Austria. GEA, 2012: Global Energy Assessment – Toward a Sustainable Future (Cambridge University Press, 2012) (see (Summary) http://webarchive.iiasa.ac.at/Research/ENE/GEA/doc/GEA-Summary-web.pdf and http://www.iiasa.ac.at/web/home/research/researchPrograms/Energy/Home-GEA.en.html).
405. International Performance Measurement and Verification Protocol., Efficiency Valuation Organization (2012), Concepts and Options for Determining Energy and Water Savings. (see http://www.evo-world.org/).
406. International Performance Measurement and Verification Protocol - Core Concepts., Efficiency Valuation Organization (2016) (see http://www.evo-world.org/).
407. IRC. Deception with Purpose: Pepsico's Water Claims in India, India Resource Center and Community Resource Centre (2011) (see http://www.indiaresource.org/news/2011/PepsiPositiveWaterDeception.pdf).
408. International Standards Organization: Environmental labels and declarations - How ISO standards help (2012) (see http://www.iso.org/iso/environmental-labelling.pdf).
409. International Standards Organization: DIS 20400, DRAFT INTERNATIONAL STANDARD, Sustainable procurement — Guidance (see www.jisc.go.jp/international/PC277/E_ISO_DIS_20400.pdf).
410. International Standards Organization: ISO 14046:2014 (see http://www.iso.org/iso/catalogue_detail?csnumber=43263).
411. International Standards Organization:: ISO 5725-1:1994(en) (see https://www.iso.org/obp/ui/#iso:std:11833:en).
412. International Standards Organization: ISO/TS 14067:2013 (see http://www.iso.org/iso/catalogue_detail?csnumber=59521).
413. International Standards Organization. Win the Energy Challenge with ISO 50001 (see http://www.iso.org/iso/iso_50001_energy.pdf).
414. International Trade Centre. Standards Map (see http://www.standardsmap.org/).
415. International vocabulary of metrology – Basic and general concepts and associated terms (VIM)., Bureau International des Poids et Mesures - 3rd Edition (2008) (see http://www.bipm.org/utils/common/documents/jcgm/JCGM_200_2012.pdf).
416. Investor Confidence Protocol (see www.eeperformance.org).
417. Irick, Michael L. *Managing tacit knowledge in organizations* (2007) (see http://www.tlainc.com/articl139.htm), pp. 1–5.
418. ISEAL Alliance. (see http://www.isealalliance.org/).
419. ISEAL Alliance. Challenge the Label (see http://www.isealalliance.org/challenge).
420. Jackson, Tim. *Motivating sustainable consumption* (Citeseer, 2005) (see http://www.sustainablelifestyles.ac.uk/sites/default/files/motivating_sc_final.pdf), p. 30.

421. Jacobs, B. W., V. R. Singhal, and R. Subramanian. *An empirical investigation of environmental performance and the market value of the firm*, Journal of Operations Management 28, 430–441 (2010).

422. James Hansen, Makiko Satoa, and Reto Ruedy. *Perception of climate change*, Proceedings of the National Academy of Sciences (2012) (see http://www.pnas.org/content/early/2012/07/30/1205276109.full.pdf+html?with-ds=yes).

423. Jason Channell, Heath R. Jansen Alastair R. Syme, Sofia Savvantidou, Edward L. Morse, and Anthony Yuen. *Energy Darwinism - The Evolution of the Energy Industry*, Citi (2013) (see https://www.citivelocity.com/citigps/ReportSeries.action?recordId=21).

424. Jenkins, Jacqueline M. *Variation in Energy Audits - A Case Study of Navy Yard Building 101*, Consortium for Building Energy Innovation (2014) (see http://cbei.psu.edu/variation-in-energy-audits/).

425. Johnson and Johnson. Citizenship Report (see http://www.jnj.com/sites/default/files/pdf/cs/2015-JNJ-Citizenship-Sustainability-Report.pdf).

426. Johnson, and Johnson. Johnson and Johnson Healthy Planet 2010 Goals (2012) (previously at http://www.jnj.com/responsibility/Our_Goals/Healthly+Planet+2010+Goals/).

427. Jolly, David. *China Export Restrictions on Metals Violate Global Trade Law, Panel Finds.* New York Times (March 26, 2014) (see http://www.nytimes.com/2014/03/27/business/international/china-export-quotas-on-rare-earths-violate-law-wto-panel-says.html?_r=0).

428. Jones, Phil. Metering energy use in new non-domestic buildings - a guide to Part L2 compliance, Energy Efficiency Best Practice Programme (2002) (see http://www.cibse.org/getmedia/03849a88-7e3e-4ae8-bde5-0ffa6602f6bd/GIL065-Metering-Energy-Use-in-New-Non-domestic-Buildings.pdf.aspx).

429. JP Morgan, *A Brave New World - deep decarbonization of electricity grids.*, JP Morgan (2015) (see https://www.jpmorgan.com/jpmpdf/1320687247153.pdf).

430. Sustainability Report 2015., JX Nippon Mining and Metals Corporation (2015) (see http://www.nmm.jx-group.co.jp/english/sustainability/pdf/report2015_e_full.pdf).

431. Kahan, Dan M. et al. *The polarizing impact of science literacy and numeracy on perceived climate change risks* (Nature Publishing Group, 2012), pp. 732–735.

432. Kahneman, Daniel. *Thinking, fast and slow* (Macmillan, 2011) ISBN: 9781429969352.

433. Kalundborg Symbiosis. (see http://www.symbiosis.dk/en).

434. Kalundborg Symbiosis Diagram. (see http://www.symbiosis.dk/en/diagram).

435. Kang, S. *South Korea Requires 374 Companies to Set Greenhouse Gas-Reduction Targets* (2010) (see http://www.bloomberg.com/news/2010-09-27/south-korea-requires-374-companies-to-set-greenhouse-gas-reduction-targets.html).

436. Karl, T. R., J. M. , and T. C. Petersen. *Global climate change impacts in the United States* (Cambridge Univ. Press, 2009) (see http://downloads.globalchange.gov/usimpacts/pdfs/climate-impacts-report.pdf).

437. Katzenbach, Jon R., and Douglas K. Smith. *The discipline of teams* (Harvard Business Press, 1993).

438. Katzenbach, Jon R., and Douglas K. Smith. *The wisdom of teams: Creating the high-performance organization* (Harvard Business Press, 1993).

439. Kearney, A. T. Texas City Refinery Retrospective Analysis (2002) (see http://www.csb.gov/assets/document/Texas_City_Refinery_Retrospective_Analysis_October_28__2002.pdf).

440. Kesicki, F. in IAEE's 2010 International Conference. Rio de Janeiro, (2010) (see http://www.homepages.ucl.ac.uk/~ucft347/Kesicki_MACC.pdf).

441. Kingfisher plc. Annual Report and Accounts, 2016 (see http://www.kingfisher.com/files/reports/annual_report_2016/files/pdf/annual_report_2016.pdf).

442. Kingma, Boris, and Wouter van Marken Lichtenbelt. *Energy consumption in buildings and female thermal demand* (Nature Publishing Group, 2015) (see http://www.nature.com/articles/nclimate2741.epdf?).

443. Kissock, John Kelly, Jeff S. Haberl, and David E. Claridge. *Inverse modeling toolkit: Numerical algorithms* (RP-1050) (ASHRAE; 1999, 2003) (see http://www.eeperformance.org/uploads/8/6/5/0/8650231/ashrae_-_inverse_modeling_toolkit_-_numerical_algorithms.pdf), pp. 425–434.

444. Kohn, Alfie. *Punished by rewards: The trouble with gold stars, incentive plans, A's, praise, and other bribes* (Houghton Mifflin Harcourt, 1999).

445. Kok, Nils, and Maarten Jennen. *The impact of energy labels and accessibility on office rents* (Elsevier, 2012) (see http://www.corporate-engagement.com/files/publication/KJ_JEPO_2012.pdf), pp. 489–497.

446. Kolb, A., and D.A. Kolb. *Experiential learning theory: A dynamic, holistic approach to management learning*. Education and Development Department of Organizational Behavior. Case, Western Reserve University Working Paper (2008) (see https://www.researchgate.net/profile/David_Kolb/publication/267974468_Experiential_Learning_Theory_A_Dynamic_Holistic_Approach_to_Management_Learning_Education_and_Development/links/5559122408ae6fd2d826eb12.pdf).

447. Konar, S., and M. A. Cohen. *Does the market value environmental performance?* (MIT Press, 2001), pp. 281–289.

448. Kotter, J. P. *Leading change - Why transformation efforts fail*, Harvard Business Review 73 (1995) (see http://89.248.0.102/upload/Topplederprogrammet/Internsider/Kull9/Litteratur/2.1%20Leading%20Change%20-%20Why%20Transformation%20Efforts%20Fail%20by%20JP%20Kotter.pdf).

449. KPMG. The KPMG Survey of Corporate Responsibility Reporting 2013 (KPMG The Netherlands, 2013) (see https://assets.kpmg.com/content/dam/kpmg/pdf/2015/08/kpmg-survey-of-corporate-responsibility-reporting-2013.pdf).

450. KPMG. The KPMG Survey of Corporate Responsibility Reporting 2015 (KPMG The Netherlands, 2016) (see https://assets.kpmg.com/content/dam/kpmg/pdf/2016/02/kpmg-international-survey-of-corporate-responsibility-reporting-2015.pdf).

451. Krajewski, Markus. *The Great Lightbulb Conspiracy* (2014) (see http://spectrum.ieee.org/geek-life/history/the-great-lightbulb-conspiracy).

452. Kramer, Hannah, James Russell Eliot Crowe, and Joan Effinger. Inventory of Commercial Energy Management and Information Systems (EMIS) for M&V Applications, North West Energy Alliance (Jul 2016, 2013) (see http://www.eeperformance.org/uploads/8/6/5/0/8650231/inventory_of_mv_applications.pdf).

453. Krey, Volker et al. IPCC *Fifth Assessment Report, Annex 2-Metrics and methodology* (Cambridge University Press, 2014) (see http://pure.iiasa.ac.at/11109/1/ipcc_wg3_ar5_annex-ii.pdf).

454. Kübler-Ross, Elisabeth, and David Kessler. *On grief and grieving: Finding the meaning of grief through the five stages of loss*. Simon and Schuster, 2014.

455. Kuhn, T. S. *The structure of scientific revolutions* (University of Chicago Press, 1996).

456. Kurz, Tim, Ngaire Donaghue, and Iain Walker. *Utilizing a Social-Ecological Framework to Promote Water and Energy Conservation: A Field Experiment* (Wiley Online Library, 2005) (see https://www.researchgate.net/profile/Iain_Walker/publication/227876463_Utilizing_a_SocialEcological_Framework_to_Promote_Water_and_Energy_Conservation_A_Field_Experiment1/links/0a85e534b692ebe7d0000000.pdf), pp. 1281–1300.

457. L'Oreal. Financial Report (English version) (2012) (see http://www.loreal-finance.com/_docs/pdf/rapport-annuel/2012/LOREAL_Document-de-Reference-2012_GB.pdf).

458. L'Oreal. Financial Report (French version) (2011) (see http://www.loreal-finance.com/_docs/us/2011-annual-report/LOREAL_Document-de-Reference-2011.pdf).

459. L'Oreal. Sustainable Development Summary (2010) (see https://www.unglobalcompact.org/system/attachments/11042/original/2010-interactive-pdf.pdf?1309537237).

460. Lacy, Peter, Tim Cooper, Rob Hayward, and Lisa Neuberger. *A New Era of Sustainability*. UN Global Compact-Accenture CEO Study 2010 (2010) (see https://www.unglobalcompact.org/docs/news_events/8.1/UNGC_Accenture_CEO_Study_2010.pdf).

461. Lafarge. Sustainability Report 2014 (see http://www.lafarge.com/sites/default/files/atoms/files/04302015-publication_sustainable_development-sustainable_report_2014-uk.pdf).

462. Lagi, M., K. Bertrand, and Y. Bar-Yam. *The food crises and political instability in North Africa and the Middle East* (2011) (see http://papers.ssrn.com/sol3/papers.cfm?abstract_id=1910031).

463. Lakoff, George, and Mark Johnson. *Philosophy in the flesh: The embodied mind and its challenge to western thought* (Basic Books, 1999).

464. Laszlo, Chris, and Nadya Zhexembayeva. *Embedded Sustainability: the next big competitive advantage* (Greenleaf Publishing, 2011).

465. Lavery, Dr Greg. Replacing Marginal Abatement Cost Curves (MACCs) with ERIC (2011) (see http://laverypennell.com/replacing-marginal-abatement-cost-curves-maccs-with-erics/).

466. Lee, David. *Apple in conflict mineral 'name and shame' crackdown*. BBC News. (13 Feb 2014) (see http://www.bbc.co.uk/news/technology-26144981).

467. Lee, Jennifer. *City Council Shuns Bottles in Favor of Water From Tap*, New York Times (2008) (see http://www.nytimes.com/2008/06/17/nyregion/17water.html).

468. Lee, P., B. Walsh, and P. Smith. *Quantification of the business benefits of resource efficiency: A research report completed for the Department of Rural Affairs* by Oakdene Hollins and, Defra Report (UK) (2007) (see http://randd.defra.gov.uk/Default.aspx?Menu=Menu&Module=More&Location=None&Completed=0&ProjectID=14609).

469. Legal Principles Relating to Climate Change., International Law Association (2014) (see http://papers.ssrn.com/sol3/papers.cfm?abstract_id=2461556).

470. Lempert, Robert J. *Capital cycles and the timing of climate change policy* (Pew Center on Global Climate Change, 2002).

471. Leonard, Dorothy, and Sylvia Sensiper. The role of tacit knowledge in group innovation (University of California Press Journals, 1998) (see https://www.researchgate.net/profile/Dorothy_Leonard/publication/230557658_The_Role_of_Tacit_Knowledge_in_Group_Innovation/links/5469fed90cf2f5eb180770de.pdf), pp. 112–132.

472. Lepper, Mark R., David Greene, and Richard E. Nisbett. *Undermining children's intrinsic interest with extrinsic reward: A test of the" overjustification" hypothesis.* (American Psychological Association, 1973) (see http://web.mit.edu/curhan/www/docs/Articles/15341_Readings/Motivation/Lepper_et_al_Undermining_Childrens_Intrinsic_Interest.pdf), p. 129.

473. Levine, Stephan and Szabat, *Statistics for Managers using Microsoft Excel*, Pearson, 2014, ISBN 978-0-273-78711-2.

474. Li, Shuo. *China's Wind Power Production Increased More Than Coal Power Did For First Time Ever In 2012*. (20 March 2013) (see http://thinkprogress.org/climate/2013/03/20/1744741/chinas-wind-power-production-increased-more-than-coal-power-did-for-first-time-ever-in-2012/)

475. Lippert, Ingmar. *Enacting environments: an ethnography of the digitalisation and naturalisation of emissions* (2013) (see http://opus.bibliothek.uni-augsburg.de/opus4/files/2199/Lippert2013Enacting_Environments.pdf).

476. LLoyds Bank. Conference on Accelerating the Transition to Greener Buildings, 6 October 2016, Park Lane Sheraton, London.

477. Lloyds Register, PRQA, ISO 50001 Pathway (see http://pathway.lrqa.co.uk/questionnaire-list/environment-and-energy/energy-iso-50001/).

478. Loeve, Boris. *Analysis of the notions of "due diligence" and "sphere of influence" in the context of corporate respect for human rights: the issues for delineating the scope of application of CSR standards*, French Government, Ministry of Foreign and European Affairs, Office of the Ambassador responsible for Bioethics and Corporate Social Responsibility (2010) (see http://www.diplomatie.gouv.fr/en/IMG/pdf/Analysis_of_the_notions_of_due_diligence.pdf).

479. Lomborg, B., and others. *The skeptical environmentalist: measuring the real state of the world.* (2001).

480. Lord, Richard. *Climate change liability: Transnational law and practice* (Cambridge University Press, 2012) (see http://policy-practice.oxfam.org.uk/publications/climate-change-liability-transnational-law-and-practice-191277).

481. Loroz, Peggy Sue. *The interaction of message frames and reference points in prosocial persuasive appeals* (Wiley Online Library, 2007), pp. 1001–1023.

482. Lovell, H., T. Sales de Aguiar, J. Bebbington, and C. Larrinaga-Gonzales. Accounting for Carbon Association of Chartered and Certified Accountants, Ed. (2010) (see http://www.acca.co.uk/pubs/general/activities/research/research_archive/rr-122-001.pdf).

483. Lovins, Amory. *Energy and American Society - Thirteen Myths*, Benjamin Sovacool and Marylin Brown, Ed. (Springer, 2007), pp. 239–263.

484. Lovins, Amory. *Energy efficiency, taxonomic overview* (Rocky Mountain Institute 2004), p. 6 (see http://www.rmi.org/cms/Download.aspx?id=5082&file=E04-02_EnergyEffTax.pdf&title=Energy+Efficiency%2c+Taxonomic+Overview).

485. Lovins, Amory. *Reinventing Fire* (Chelsea Green Publishing Co, 2011).

486. Lovins, Amory (2005), Scientific American, *More Profit with less Carbon*, (September 2005), pp. 74–83. (see http://www.scientificamerican.com/article/more-profit-with-less-car/).

487. Lund, Robert T. The Database of Remanufacturers (2012) (see http://www.bu.edu/reman/The%20Remanufacturing%20Database.pdf).

488. Lüthi, Dieter et al. High-resolution carbon dioxide concentration record 650,000-800,000 years before present (Nature Publishing Group, 2008) (see http://www.nature.com/nature/journal/v453/n7193/abs/nature06949.html), pp. 379–382.

489. Lynch, Richard. *Strategic Management*. Pearson; 7th edition (21 Jan 2015) Pub. ISBN-13 (pbk): 978-1292064666.

490. M&S. How We Do Business Report (2011) (see http://plana.marksandspencer.com/media/pdf/how_we-do_business_report_2011.pdf).

491. MacKay, David. *Sustainable Energy — without the hot air* (UIT Cambridge, 2008). ISBN 978-0954452933.

492. Maddison, Angus. *Historical Statistics of the World Economy: 1-2008 AD* (see http://www.ggdc.net/maddison/oriindex.htm).

493. Mallon, Karl, Greg Bourne, and Richard Mott. *Climate Solutions - WWF's vision for 2050*, WWF International (2007) (see https://www.wwf.org.uk/updates/climate-solutions-wwfs-vision-2050).

494. Management, Osmosis Investment. Resource Efficiency = Shareholder Returns (see http://www.osmosisim.com/resource_efficiency_shareholder/).

495. Managing Change., Government Office for the Southwest & Envirowise (2007) (see http://www.oursouthwest.com/SusBus/mggchange.pdf).

496. Mansfield, Ian. *Apple Starts Verifying Mineral Supplies Don't Come from War Zones* (2014) (see http://www.cellular-news.com/story/Technology/64397.php).

497. Mansfield, Ian. *US Firms Required to disclose if using conflict Coltan* (2012) (see http://www.cellular-news.com/story/Regulatory/56058.php).

498. Markandya, A. et al. *Counting the cost of vulture decline—An appraisal of the human health and other benefits of vultures in India*, Ecological Economics 67, 194–204 (2008).

499. Marks & Spencer. Sustainability report 2016. (see http://planareport.marksandspencer.com/M&S_PlanA_Report_2016_Performance.pdf).

500. Marshall, George. *Don't Even Think about it: Why Our Brains are Wired to Ignore Climate Change* (Bloomsbury Publishing USA, 2014).

501. Marshall, Michael. *Climate change - The great civilization destroyer*, New Scientist, 32–36 (2012).

502. Masaaki, I. *Gemba Kaizen. A Commonsense, Low-cost Approach to Management* (McGraw-Hill, 1997).

503. Matsumoto, Mitsutaka, and Yasushi Umeda. *An analysis of remanufacturing practices in Japan* (Springer, 2011) (see http://www.journalofremanufacturing.com/content/1/1/2/), pp. 1–11.

504. McAfee, R. Preston, Hugo M. Mialon, and Sue H. Mialon. *Do sunk costs matter?* (Wiley Online Library, 2010) (see http://papers.ssrn.com/sol3/Delivery.cfm/SSRN_ID1000988_code389408.pdf?abstractid=1000988&mirid=1), pp. 323–336.

505. McCabe, Joseph. *Salvage Value of Photovoltaic Systems* (2012) (see https://ases.conference-services.net/resources/252/2859/pdf/SOLAR2012_0783_full%20paper.pdf).

506. McDonough, William, and Michael Braungart. *Crade to Cradle: remaking the way we make things* (North Point Press, 2002).

507. McDonough, William. Method Manufacturing Facility. (see http://www.mcdonoughpartners.com/projects/method-home/).

508. McGonical, Jane. *Reality Is Broken: Why Games Make Us Better and How They Can Change the World*, 2011.

509. McGrath, Matt. *Carbon Bubble Threatens Stock Markets, say MPs* (2014) (see http://www.bbc.co.uk/news/science-environment-26455763?print=true).

510. McKenzie-Mohr, Doug. *Fostering sustainable behavior: An introduction to community-based social marketing* (New society publishers, 2013).

511. McKibben, Bill. *Global warming's terrifying new math* (2012) (see http://www.rollingstone.com/politics/news/global-warmings-terrifying-new-math-20120719).

512. McKinsey, and Co. *Impact of the financial crisis on carbon economics - update to MACC curves* (2010) (see http://www.mckinsey.com/~/media/McKinsey/dotcom/client_service/Sustainability/PDFs/ImpactFinancialCrisisCarbonEconomicsGHGcostcurveV21.ashx).

513. McLaughlin, Liam. *ISO 50001 Energy Management Systems - A practical guide for SMEs* (International Standards Organization, 2015) (see http://www.iso.org/iso/home/store/publication_item.htm?pid=PUB100360). ISBN 978-92-67-10634-2.

514. McPherson, Bruce. Bruce McPherson's "Excel Liberation" pages (see http://ramblings.mcpher.com/Home/excelquirks/d3/sankey).

515. MEA. *Ecosystems and human wellbeing*, Millennium Ecosystems Assessment (Island Press Washington, DC, 2005) (see http://www.millenniumassessment.org/documents/document.356.aspx.pdf).

516. Meadows, D. *Leverage Points*, Sustainability Institute (1999) (see http://www.gruposinapsys.com:8080/Plone/website/ue/k12/donella-h.-meadows/Leverage_Points%20Donella%20Meadows.pdf).

517. Meadows, D. H., DL Meadows, J. Randers, and W. W. Behrens III. *The Limits to Growth: A Report to The Club of Rome* (1972) (Universe Books, New York, 1972).

518. Meadows, D. H., J. Randers, and D. L. Meadows. *The limits to growth: the 30-year update* (Chelsea Green, 2004).

519. Medina-Elizalde, M., and E. J. Rohling. *Collapse of Classic Maya Civilization Related to Modest Reduction in Precipitation*, Science 335, 956–959 (2012).

520. Meehl, G. A. et al. *How much more global warming and sea level rise?*, Science 307, 1769–1772 (2005).

521. Meier, Alan Kevin. Supply curves of conserved energy (1982) (see http://escholarship.org/uc/item/20b1j10d#page-1).

522. Meinshausen, M. et al. *Greenhouse-gas emission targets for limiting global warming to 2 C*, Nature 458, 1158–1162 (2009).

523. Mento, A., R. Jones, and W. Dirndorfer. *A change management process: Grounded in both theory and practice*, Journal of Change Management 3, 45–59 (2002).

524. Method. Sustainability One Sheet. (2014) (see http://methodhome.com/wp-content/uploads/method_sustainability_352014.pdf).

525. Michie, Susan, Maartje M van Stralen, and Robert West. *The behaviour change wheel: a new method for characterising and designing behaviour change interventions* (BioMed Central, 2011) (see http://implementationscience.biomedcentral.com/articles/10.1186/1748-5908-6-42).

526. Miles, Jeremy, and Mark Shevlin. *Applying regression and correlation: A guide for students and researchers* (Sage, 2001).

527. Mol, Arthur P.J. *Carbon flows, financial markets and climate change mitigation* (Elsevier, 2012) (see https://www.researchgate.net/profile/Arthur_Mol/publication/257742323_Carbon_flows_financial_markets_and_climate_change_mitigation/links/0deec5279ebaca42fa000000.pdf), pp. 10–24.

528. Molina-Azorin, J. F., E. Claver-Cortés, M. D. López-Gamero, and J. J. Tari. *Green management and financial performance: a literature review*, Management Decision 47, 1080–1100 (2009).

529. Molnar, Mike. Cummins Climate Leaders Progress - a strategic imperative for energy efficiency, EPA Climate Leaders Program (2008) (see http://www.epa.gov/climateleadership/documents/events/oct2008/10_7_partner_progress_report_molnar.pdf).

530. Molthan-Hill, Petra (ed.). *The Business Student's Guide to Sustainable Management - principles and practice* (2014) Greenleaf Publishing, ISBN13: 9781783531202.

531. Monbiot, George. *Selling Indulgences* (2006) (see http://www.monbiot.com/2006/10/19/selling-indulgences/).

532. Monika Dittrich, Stefan Giljum Stephan Lutter, and Christine Polzin. *Green economies around the world?*, Sustainable Europe Research Institute (SERI) (2012) (see http://seri.at/wp-content/uploads/2012/06/green_economies_around_the_world.pdf).

533. Monster. MonsterTRACK Survey Press Release (see http://www.about-monster.com/content/monstertrak-joins-forces-ecoamerica-launch-greencareersmonstertrak).

534. Moodie, Alison. *Can Apple's $1.5bn green bond inspire more environmental investments?* (2016) (see https://www.theguardian.com/sustainable-business/2016/mar/20/apple-green-bond-environment-energy-toyota-climate-change).

535. Moore, Frances C, and Delavane B. Diaz. *Temperature impacts on economic growth warrant stringent mitigation policy* (Nature Publishing Group, 2015), pp. 127–131. (see http://www.naturskyddsforeningen.se/sites/default/files/dokument-media/Nature_Climate_cost_of_CO2_dokKalmar_0.pdf)

536. Mott MacDonald. *Review of large public procurement in the UK* (2002) (see http://www.parliament.vic.gov.au/images/stories/committees/paec/2010-11_Budget_Estimates/Extra_bits/Mott_McDonald_Flyvberg_Blake_Dawson_Waldron_studies.pdf).

537. Mulvey, Kathy, and Shulman, Seth. *The Climate Deception Dossiers*, Union of Concerned Scientists (2015) (see http://www.ucsusa.org/global-warming/fight-misinformation/climate-deception-dossiers-fossil-fuel-industry-memos).

538. Murphy, D. J., and C. A. S. Hall. *Energy return on investment, peak oil, and the end of economic growth*, Annals of the New York Academy of Sciences 1219, 52–72 (2011) (see http://www.researchgate.net/publication/49846798_Energy_return_on_investment_peak_oil_and_the_end_of_economic_growth/file/e0b495267227339b25.pdf).

539. NABERS (see https://nabers.gov.au/AnnualReport/2015-2016/our-purpose.html).

540. Narden, Clare. First International Standard for sustainable procurement nears publication, ISO (2016) (see http://www.iso.org/iso/home/news_index/news_archive/news.htm?refid=Ref2105).

541. NAS, America's Energy Future Energy Efficiency Technologies Subcommittee, 2010. *Real Prospects for Energy Efficiency in the United States*. National Academy of Sciences. (see http://www.nap.edu/catalog/12621.html).

542. National Association of Manufacturers. Efficiency and Innovation In U.S. Manufacturing Energy Use, National Association of Manufacturers and Alliance to Save Energy (2005) (see http://patapsco.nist.gov/mep/documents/pdf/about-mep/reports-studies/energy-nam.pdf).

543. National Audit Office. Green Deal and Energy Company Obligation Review (2016) (see https://www.nao.org.uk/wp-content/uploads/2016/04/Green-Deal-and-Energy-Company-Obligation.pdf).

544. Natural Capital Protocol., The Natural Capital Coalition (2016) (see http://naturalcapitalcoalition.org/protocol/).

545. Nauclér, Tomas, and Per-Anders Enkvist. *Pathways to a low-carbon economy: Version 2 of the global greenhouse gas abatement cost curve* (2009) (see http://www.mckinsey.com/~/media/McKinsey/Business%20Functions/Sustainability%20and%20Resource%20Productivity/Our%20Insights/Pathways%20to%20a%20low%20carbon%20economy/Pathways%20to%20a%20low%20carbon%20economy.ashx).

546. Naughton, Keith, and Craig Trudell. *Vehicle Sales Surge in U.S. as $4 Gas Makes Mileage Vital*, Bloomberg (2012) (see http://www.bloomberg.com/news/2012-04-26/vehicle-sales-surge-in-u-s-as-4-gas-makes-mileage-vital.html).

547. Neag School of Education - University of Connecticut (see http://www.gifted.uconn.edu/siegle/research/Correlation/corrchrt.htm).

548. Neslen, Arthur. *Big brands cheating consumers with false lightbulb efficiency claims* (17AD) (see https://www.theguardian.com/environment/2015/dec/17/leading-lightbulb-brands-making-false-claims-on-energy-efficiency).

549. New Climate Economy. *The Sustainable Infrastructure Imperative* (2016) (see http://newclimateeconomy.report/2016/wp-content/uploads/sites/4/2014/08/NCE_2016Report.pdf).

550. Newell, G., J. MacFarlane, and N. Kok. *Building better returns-A study of the financial performance of green office buildings in Australia* (2011) (see http://www.api.org.au/assets/media_library/000/000/219/original.pdf?1315793106).

551. Newman, D. J., and G. M. Cragg. *Natural Products as Sources of New Drugs over the Last 25 Years*, Journal of Natural Products 70, 461–477 (2007).

552. New South Wales government. Measurement and Verification Operational Guide - Best practice M&V processes, Office of Environment and Heritage NSW (2012) (see http://www.eec.org.au/uploads/images/NEEC/Information%20Tools%20and%20Resources/M&V%20Operational%20Guide.pdf).

553. NHS Carbon Reduction Strategy for England., Sustainable Development Unit, NHS England (2009) (see http://www.sdu.nhs.uk/documents/publications/1237308334_qylG_saving_carbon,_improving_health_nhs_carbon_reducti.pdf).

554. Nidumolu, R., CK Prahalad, and MR Rangaswami. *Why sustainability is now the key driver of innovation* (Harvard Business School Publication Corp., 2009) (see https://hbr.org/2009/09/why-sustainability-is-now-the-key-driver-of-innovation), pp. 57–64.

555. NIST/SEMATECH. eHandbook of Statistics, 2012 (see http://www.itl.nist.gov/div898/handbook/index.htm).

556. NIST/SEMATECH. eHandbook of Statistics, 2012 . Explanation of one-tail probability. (see http://www.itl.nist.gov/div898/handbook/eda/section3/eda3671.htm)

557. NMI. NMI Takes a Global Look at Green Consumers (see http://www.nmisolutions.com/index.php/about-nmi/news-a-publications/published-articles/58-nmi-takes-a-global-look-at-green-consumers).

558. NOOA Earth Systems Research Laboratory, and Scripps Institution of Oceanography. Earth Systems Research Laboratory CO2 data (see http://www.esrl.noaa.gov/gmd/ccgg/trends/weekly.html).

559. Noesis. Energy Projects Survey (2013) (see https://www.noesisenergy.com/site/sites/default/files/noesis_energy_projects_survey_july_2013_0.pdf).

560. Nonaka, Ikujiro. *The knowledge-creating company* (2007) (see https://hbr.org/2007/07/the-knowledge-creating-company#).

561. NRDC. *Better Viewing, Lower Energy Bills and Less Pollution: Improving the efficiency of set-top boxes.* NRDC (2011) (see http://www.nrdc.org/energy/files/settopboxes.pdf).

562. NSW. Telstra Corporation Limited v Hornsby Shire Council [2006] NSWLEC 133 (see http://www.lawlink.nsw.gov.au/lecjudgments/2006nswlec.nsf/c45212a2bef99be4ca256736001f37bd/fdf89ace6e00928bca25713800832056?OpenDocument).

563. NValue - Guarantees of Origin Prices. (see http://www.nvalue.ch/home.php).

564. Nyhan, Brendan, and Jason Reifler. *Misinformation and Fact-checking* (2012) (see http://www.dartmouth.edu/~nyhan/Misinformation_and_Fact-checking.pdf).

565. Nyhan, Brendan, and Jason Reifler. *When corrections fail: The persistence of political misperceptions* (Springer, 2010) (see http://www.uky.edu/AS/PoliSci/Peffley/pdf/nyhan-reifler%202007%20When%20Predictions%20Fail.pdf), pp. 303–330.

566. Obermiller, Carl. *The baby is sick/the baby is well: A test of environmental communication appeals* (Taylor & Francis, 1995), pp. 55–70.

567. O. Davidson, H. C. de Coninck M. Loos, and L. A. Meyer (eds). *Carbon Capture and Storage,* IPCC (Cambridge University Press, 2005) (see http://www.ipcc.ch/pdf/special-reports/srccs/srccs_wholereport.pdf).

568. O'Keefe, Brian. *Exxon's big bet on shale gas*, CNNMoney (2012) (see http://fortune.com/2012/04/16/exxons-big-bet-on-shale-gas/).

569. Ölander, Folke, and John Thøgersen. *Informing versus nudging in environmental policy* (Springer, 2014), pp. 341–356.

570. O'Neil, Jim. *China's 'slowing' economy still holds staggering opportunities.* The Daily Telegraph (24 Jan 2014) (see http://www.telegraph.co.uk/finance/china-business/10595637/Chinas-slowing-economy-still-holds-staggering-opportunities.html).

571. O'Neill, S., and S. Nicholson-Cole. *Fear Won't Do It: Promoting Positive Engagement With Climate Change Through Visual and Iconic Representations*, Science Communication 30, 355–379 (2009).

572. Ocean Tomo. Intangible Asset Market Value Study (see http://www.oceantomo.com/productsandservices/investments/intangible-market-value).

573. Oliver J.G.J., Janssens-Maenhout G., and Peters J.A.H.W. Trends in global CO2 emissions; 2012 Report, PBL Netherlands Environmental Assessment Agency; Ispra: Joint Research Centre. (2012) (see http://edgar.jrc.ec.europa.eu/CO2REPORT2012.pdf).

574. Ölander, Folke, and John Thøgersen. Informing versus nudging in environmental policy (Springer, 2014), pp. 341–356.

575. Orange, Richard. *Waste not want not: Sweden to give tax breaks for repairs* (2016) (see https://www.theguardian.com/world/2016/sep/19/waste-not-want-not-sweden-tax-breaks-repairs).

576. Oreskes, Naomi and Conway, Erik. *Merchants of Doubt: how a handful of scientists obscured the truth on issues from tobacco smoke to global warming*. Bloomsbury, 2011.

577. Osmosis Investment Management. MoRE World. (see http://www.osmosisim.com/investment-strategies/the-model-of-resource-efficiency/more-world/).

578. Osmosis Investment Management. Resource Efficiency = Shareholder Returns. (see http://www.osmosisim.com/resource_efficiency_shareholder/).

579. Ottman, Jacquelyn A. *The New Rules of Green Marketing* (Greenleaf Publishing, 2011).

580. Otto, F.E.L., N. Massey, G.J. van Oldenborgh, R.G. Jones, and M.R. Allen. *Reconciling two approaches to attribution of the 2010 Russian heat wave*, Geophysical Research Letters 39, L04702 (2012).

581. Oung, Kit. *Energy Management in Business: The Manager's Guide to Maximising and Sustaining Energy Reduction* (Gower Publishing, Ltd., 2013).

582. Pacala, S., and R. Socolow. *Stabilization wedges: solving the climate problem for the next 50 years with current technologies,* Science 305, 968–972 (2004).

583. PACENation. C-PACE Q2 Market Update 2016. (2016) (see http://pacenation.us/wp-content/uploads/2016/08/Market-update-Q2-2016.pdf).

584. PACENation. PACE Market Data. (2016) (see http://pacenation.us/pace-market-data/).

585. Pachauri, R. K. et al. Climate Change 2007: Contribution of Working Groups I, II and III to the Fourth Assessment Report of the Intergovernmental Panel on Climate Change, Synthesis Report. (IPCC, 2007) (see http://www.ipcc.ch/pdf/assessment-report/ar4/syr/ar4_syr.pdf).

586. Pachauri, R. K. et al. *Climate Change 2007: Impacts Report. Chapter 4, Ecosystems.* (IPCC, 2007) (see http://www.ipcc.ch/pdf/assessment-report/ar4/wg2/ar4-wg2-chapter4.pdf).

587. Pallak, Michael S, David A. Cook, and John J. Sullivan. *Commitment and energy conservation.* (see https://books.google.co.uk/books?hl=en&lr=&id=TFsnqmk97csC&oi=fnd&pg=PA352&ots=KFvC3NB7AE&sig=7nLveVqO7eG0Q0clPApRvyBwAT8#v=onepage&q&f=false), p. 352.

588. Panev, Strahil, Nicola Labanca Paolo Bertoldi Tiago Serrenho Caiman Cahill Benigna Boza Kiss. *ESCO Market Report for Non-European Countries 2013*, European Commission Joint Research Centre Institute for Energy and Transport. (2014) (see http://iet.jrc.ec.europa.eu/energyefficiency/sites/energyefficiency/files/reqno_jrc91689_ld-na-26886-en-n.pdf).

589. Pardini, Anton U., and Richard D. Katzev. *The effect of strength of commitment on newspaper recycling* (1983) (see http://triggered.clockss.org/ServeContent?url=http://baywood.stanford.clockss.org%2FBWES%2FBAWOOD_BWES_13_3%2F6PN9MXFP3BFFCHHB.pdf), pp. 245–254.

590. Parker, D. *Remanufacturing in the UK: A significant contributor to sustainable development*, Oakdene Hollins and Resource Recovery Forum. (2004) (see http://www.remanufacturing.org.uk/pdf/story/1r17.pdf?-session=RemanSession:42F944CE09d1605421Pjq15A8D06).

591. Partridge, Adrian, *Consulting Made Easy* (2016, iUniverse, ISBN 978-1-4917-9202-5) (see https://www.adrianpartridge.com/pages/consulting-made-easy).

592. Peltier, Greg. Peltier Tech Charts for Excel 3.0. (see http://peltiertech.com/Utility30/).

593. Pennell, NIck, Lavery, Greg, and Steve Evans. *Food and Beverage Sector Non-Labour Resource Efficiency*, Lavery Pennell (2014) (see http://laverypennell.com/wp-content/uploads/2014/03/Food-and-Beverage-Resource-Efficiency.pdf).

594. PepsiCo. PepsiCo Website Water and Sustainability (2012) (see http://www.pepsico.com/Purpose/Environmental-Sustainability/Water.html).

595. PG&E (see http://www.pge.com/en/myhome/myaccount/charges/index.page).

596. Phillips, Tom. *China ratifies Paris climate change agreement ahead of G20.* (2016) (see https://www.theguardian.com/world/2016/sep/03/china-ratifies-paris-climate-change-agreement).

597. PIDC. Building 101 Property Details (2015) (see http://www.pidcphila.com/images/uploads/properties/Building_101_The_Navy_Yard_real_estate_package_web_LL_10.21.15.pdf).

598. Pilkington, Ed. *Saving global fish stocks would cost 20 million jobs*, says UN, The Guardian (2010) (see http://www.guardian.co.uk/environment/2010/may/17/saving-fish-stocks-cost-jobs).

599. Polimeni, John, Kozo Mayumi, Mario Giampetro and Blake Alcott. *The Myth of Resource Efficiency - the Jevons Paradox*, Earthscan (2008) ISBN 978-1-84407-813-4.

600. Porter, Michael E., and Mark R. Kramer. *Creating Shared Value*, Harvard Business Review (2011) (see http://hbr.org/2011/01/the-big-idea-creating-shared-value/ar/1).

601. Potocnik, Janez. Resource efficiency as a driver for growth and jobs (2010).

602. Prescott, D., and D. T. Stibbe. Unleashing the Power of Business: a practical roadmap to systematically engage business as a partner in development. The Partnering Initiative (2015) (see http://www.thepartneringinitiative.org/wp-content/uploads/2015/07/Unleashing-the-Power-of-Business_Roadmap_full_forweb.pdf).

603. Preston, Brian J. Judicial Implementation of the Principles of Ecologically Sustainable Development in Australia and Asia, Law Society of New South Wales (2006) (see http://www.lec.justice.nsw.gov.au/Documents/preston_judicial%20implementation%20of%20the%20principles%20of%20eologically%20sustainable%20development.pdf).

604. Preston, Felix. A Global Redesign? Shaping the Circular Economy (2012) (see https://www.chathamhouse.org/sites/default/files/public/Research/Energy,%20Environment%20and%20Development/bp0312_preston.pdf).

605. Prindle, William, and Andre de Fontaine. A Survey of Corporate Energy Efficiency Strategies, ACEEE Summer Study on Energy Efficiency in Industry 5, 13 (2009).

606. PWC. Barriers to Energy Efficient Lighting (2011) (see http://www.aemc.gov.au/getattachment/499ca4dc-cb38-4e37-b18e-c32b055972e8/PWC-Report-to-Equipment-Energy-Efficiency-Committe.aspx).

607. PWC. CDP Global 500 Report 2011, Carbon Disclosure Project (2011) (see https://www.cdproject.net/CDPResults/CDP-G500-2011-Report.pdf).

608. PWC. CDP Global 500 Report 2013, Carbon Disclosure Project (2013) (see http://www.sustainablebrands.com/digital_learning/white-paper/cdp-global-500-climate-change-report-2013).

609. Quinn, Francis. For a Sustainable Development of our Growth, L'Oreal (2011) (previously at http://www.slideshare.net/KGSglobal/loreal-presentation-at-innocos-europe-paris).

610. Rabin, Emily. *Business/NGO Engagement: All the Rage* (2005) (see http://www.greenbiz.com/news/2005/03/20/businessngo-engagement-all-rage).

611. Rabinovich, Anna, Thomas A. Morton, and Christopher C. Duke. *Collective Self and Individual Choice: The Role of Social Comparisons in Promoting Public Engagement with Climate Change* (Earthscan, 2011), p. 66.

612. Randall, Alan. *Risk and Precaution* (Cambridge University Press, 2011).

613. Ranganathan, J. et al. The Greenhouse Gas Protocol: A Corporate Accounting and Reporting Standard (Revised Edition). World Resources Institute and World Business Council for Sustainable Development (2004) (see http://www.ghgprotocol.org/standards/corporate-standard).

614. Raven, J. et al. *Ocean acidification due to increasing atmospheric carbon dioxide.* (The Royal Society, 2005) (see http://dge.stanford.edu/labs/caldeiralab/Caldeira%20downloads/RoyalSociety_OceanAcidification.pdf).

615. Reddy, T. Agami, Itzhak Maor, and Chanin Panjapornpon. *Calibrating detailed building energy simulation programs with measured data—Part I: General methodology* (RP-1051) (Taylor & Francis, 2007), pp. 221–241.

616. Red Redemption, The Climate Challenge, 2007, (see http://www.bbc.co.uk/sn/hottopics/climatechange/climate_challenge/).

617. REEEP. Renewable Energy and Energy Efficiency Partnership Policy Database (see http://www.reeep.org/policy-database).

618. RE:FIT, Mayor of London and the London Assembly. (see https://www.london.gov.uk/what-we-do/environment/energy/energy-buildings/refit).

619. Rennie, J. *Misleading math about the Earth*, Scientific American 286, 61–61 (2002).

620. Revans, Reg W. *Action learning: Its terms and character* (MCB UP Ltd, 1983) (see http://www.emeraldinsight.com/doi/pdfplus/10.1108/eb001310), pp. 39–50.

621. RICS. *Valuation Information Paper 13 - Sustainability and Commercial Property*, Royal Institution of Chartered Surveyors (ed. 1st, 2009) (see http://www.joinricsineurope.eu/uploads/files/sustainabilityandcommercialvaluation_2.pdf).

622. Rio Tinto 2011 Approved Capital Projects. (see http://www.riotinto.com/annualreport2011/performance/capital_projects.html).

623. Rio Tinto. Diavik Diamond Mine Fact Book (2009) (see http://www.diavik.ca/documents/DiavikFactBook.pdf).

624. Robinson, John B. *Futures under glass: a recipe for people who hate to predict* (Elsevier, 1990) (see https://www.researchgate.net/publication/222076849_Futures_Under_Glass_A_Recipe_for_People_Who_Hate_to_Predict), pp. 820–842.

625. Rockström, J. et al. *A safe operating space for humanity*, Nature 461, 472–475 (2009).

626. Roe v Arch Coal Inc. et al, Case: 4:15-cv-00910-NAB, United States District Court, Eastern District of Missouri, 9 June 2015.

627. Rogers, Paul, Rudy Puryear, and James Roat. *Infobesity: The enemy of good decisions*, Bain and Company (2013) (see http://www.bain.com/publications/articles/infobesity-the-enemy-of-good-decisions.aspx).

628. Rohde, Robert, Richard A. Muller, Robert Jacobsen, Elizabeth Muller, Saul Perlmutter, Arthur Rosenfeld, Jonathan Wurtele, Donald Groom, Charlotte Wickham. *A New Estimate of the Average Earth Surface Land Temperature Spanning 1753 to 2011*, JGR Atmospheres (2012) (see http://berkeleyearth.org/pdf/results-paper-july-8.pdf).

629. Rooney, Brian. *UPS Figure out the "Right Way" to save money, time and gas* (see http://abcnews.go.com/WNT/story?id=3005890&page=1).

630. Ross, M. *Capital budgeting practices of twelve large manufacturers* (JSTOR, 1986) (see http://costkiller.net/tribune/Tribu-PDF/capital-Budgeting-Practices-of-Twelve-Large-Manufacturers.pdf), pp. 15–22.

631. Rossiter Alan, Jones Beth. *Energy Management and Efficiency for the Process Industries.* American Institute of Chemical Engineers, Wiley 2015, ISBN 978-1-118-83825-9.

632. Roth, Mark. *Habitual Excellence: The workplace according to Paul O'Neil* (2012) (see http://www.post-gazette.com/business/businessnews/2012/05/13/Habitual-excellence-The-workplace-according-to-Paul-O-Neill/stories/201205130249).

633. Rowe, Gene, and George Wright. *The Delphi technique as a forecasting tool: issues and analysis* (1999) (see https://www.researchgate.net/profile/Gene_Rowe2/publication/4960099_Rowe_G_Wright_G_The_Delphi_technique_as_a_forecasting_tool_issues_and_analysis/links/00463529eea02558c4000000.pdf), pp. 353–375.

634. Rowe, Gene, and George Wright. *The Delphi technique: Past, present, and future prospects*—Introduction to the special issue (Elsevier, 2011), pp. 1487–1490. (see https://www.researchgate.net/profile/Gene_Rowe2/publication/251496903_The_Delphi_technique_Past_present_and_future_prospects_-_Introduction_to_the_special_issue/links/02e7e53bc45269ada7000000.pdf).

635. Russell, Christopher. Management Books. *Managing energy from the top down* (The Fairmont Press also available on Lulu.com), (2010) (see http://www.lulu.com/product/paperback/the-industrial-energy-harvest/2924044).

636. Russell, Christopher. *Executive Reactions to Energy Efficiency*, Alliance to Save Energy (1850 M Street, N.W., Suite 600, Washington, DC 20036, US, 2006) (see https://www.ase.org/news/leadership-key-controlling-industrial-energy-costs-says-alliance-save-energy).

637. Russell, Christopher. *Monetizing Energy Solutions*, Energy Pathfinder LLC (see http://chollisr.powweb.com/presentations/Russell%20Monetizing%20Energy%20Solutions.pdf).

638. Salinas, Gabriella. *The International Brand Valuation Guide* (Wiley, 2009).

639. Salisbury, Peter. *Behind the Brand: L'Oreal* (2011) (see http://www.theecologist.org/green_green_living/behind_the_label/1008667/behind_the_brand_loral.html).

640. Samuelson, W., and R. Zeckhauser. *Status quo bias in decision-making*, Journal of Risk and Uncertainty 1, 7–59 (1988).

641. Sandberg, Jared. *Employees Only Think They Control Thermostat* (2003) (see https://www.wsj.com/articles/SB1042577628591401304).

642. Sankey Diagrams. (see http://www.sankey-diagrams.com/sankey-diagram-software/).

643. SB-350 Clean Energy and Pollution Reduction Act of 2015. (see http://leginfo.legislature.ca.gov/faces/billNavClient.xhtml?bill_id=201520160SB350).

644. Schandl, Heinz et al. *Resource Efficiency: Economics and Outlook for Asia and the Pacific*, UNEP and CSIRO (2011) (see http://www.unep.org/dewa/Portals/67/pdf/Resource_Efficiency_EOAP_web.pdf).

645. Schell, Jesse. 2013 DICE Summit presentation, (see http://www.dicesummit.org/video_gallery/video_gallery_2013.asp).

646. Schijnel, P.P.A.J.V., J.M.N. Kasteren, and F.J.J.G. Janssen. *Exergy analysis-a tool for sustainable technology in engineering education* (1998) (see http://www.slideshare.net/patrickvanschijndel/exergy-analysis-a-tool-for-sustainable-technology-in-engineering-education).

647. Schneider, Dana, and Paul Rode. Energy Renaissance, High Performing Buildings (2010) (see https://www.ashrae.org/File%20Library/docLib/Public/20100310_empirestatebuilding.pdf).

648. Schmidt, M. *The Sankey diagram in energy and material flow management* (Wiley Online Library, 2008) (see http://onlinelibrary.wiley.com/doi/10.1111/j.1530-9290.2008.00004.x/pdf), pp. 82–94.

649. Schuller, Oliver, Matthias Fischer and Michael Faltenbacher. *Exergy-Efficiency as Enhancement of Energy-Efficiency – an LCA Perspective* (2008) (see http://www.lcacenter.org/LCA8/presentations/323.pdf).

650. Schultz, P. Wesley, Jessica M. Nolan, Robert B. Cialdini, Noah J. Goldstein, and Vladas Griskevicius. *The constructive, destructive, and reconstructive power of social norms.* (SAGE Publications, 2007) (see https://www.jsmf.org/meetings/2008/july/social%20norms%20Cialdini.pdf), pp. 429–434.

651. Schwartz, Daniel, Baruch Fischhoff, Tamar Krishnamurti, and Fallaw Sowell. *The Hawthorne effect and energy awareness.* (2013), (see PNAS http://www.pnas.org/content/110/38/15242.full pp. 15242–6).

652. Schwarz, Norbert, Lawrence J. Sanna, Ian Skurnik, and Carolyn Yoon. *Metacognitive experiences and the intricacies of setting people straight: Implications for debiasing and public information campaigns* (ACADEMIC PRESS INC LTD, 2007)

653. SEAF Unveils First-in-kind Energy Efficiency Financing Platform. (see https://www.seaf-h2020.eu/the-sustainable-energy-asset-framework-unveils-first-in-kind-energy-efficiency-financing-platform/).

654. SEAI. Energy Performance Contract Handbook, Sustainable Energy Authority Ireland (2014) (see http://www.seai.ie/Your_Business/National_Energy_Services_Framework/EPC_Handbook/EPC-Handbook.pdf).

655. SEHN. Wingspread Conference on the Precautionary Principle (see http://www.sehn.org/wing.html).

656. SEI. Greening the Bottom Line - The Trend toward Green Revolving Funds on Campus, Sustainable Endowments Institute (2011) (see http://www.aashe.org/files/documents/resources/greening_bottom_line.pdf).

657. Sells, Bill. in Harvard Business Review, (1994).

658. Sheehan, and Carr. in Air and Waste Management Association, (2010).

659. Shell. Royal Dutch Shell Sustainability Report 2011 - Environment (see http://reports.shell.com/sustainability-report/2011/ourperformance/environmentalperformance.html).

660. Shewhart, Walter Andrew. *Economic control of quality of manufactured product* (ASQ Quality Press, 1931).

661. Shippee, Glenn, and W. Larry Gregory. *Public commitment and energy conservation* (Wiley Online Library, 1982), pp. 81–93.

662. Siegel, R.P. *When Pigs Fly: Halliburton Makes the Dow Jones Sustainability Index* (2010) (see http://www.triplepundit.com/2010/09/when-pigs-fly-halliburton-makes-the-dow-jones-sustainability-index/).

663. Singh, Neelam, and Lindsey Longendyke. World Resources Institute (2015) A Global Look at Mandatory Greenhouse Gas Reporting Programs (see http://www.wri.org/blog/2015/05/global-look-mandatory-greenhouse-gas-reporting-programs).

664. Sioshansi, Fereidoon, Ed. *Energy Efficiency - towards the end of demand growth*, (Academic Press, 2013).

665. Sirkin, Harold L., Perry Keenan, and Alan Jackson. *The hard side of change management*, Harvard Business Review 83, 108 (2005).

666. Skurnik, Ian, Carolyn Yoon, Denise C. Park, and Norbert Schwarz. *How warnings about false claims become recommendations* (The Oxford University Press, 2005) (see http://www.rcgd.isr.umich.edu/news/Skurnik.Yoon.Park.Schwarz.2005.jofconsumer%20resh.pdf), pp. 713–724.

667. Smil, Vaclav. *Making the Modern World: Materials and Dematerialization* (Wiley, 2013).

668. Smith, R., King, D., Sidhu, R. and Skelsey, D. eds. *The Effective Change Manager's Handbook*, Kogan Page, ISBN: 978-0-7494-7307-5.

669. Sorrell, S. et al. 2004. *The economics of energy efficiency: barriers to cost-effective investment*. Edward Elgar Pub. ISBN-13: 978-1840648898.

670. Sotos, Mary. GHG Protocol Scope 2 Guidance: An amendment to the GHG protocol reporting standard, World Resources Institute and World Business Council for Sustainable Development (2015) (see http://ghgprotocol.org/files/ghgp/Scope%202%20Guidance_Final.pdf).

671. Spector, B. *From bogged down to fired up: Inspiring organizational change*, Sloan Management Review 30, 29–34 (1989).

672. Spurgeon, James (lead author). *A Guide to Corporate Ecosystems Valuation*, World Business Council for Sustainable Development (2011) (see http://www.wbcsd.org/Pages/EDocument/EDocumentDetails.aspx?ID=104).

673. Spurgeon, James. *Natural Capital Accounting for Business* (2015) (see http://ec.europa.eu/environment/biodiversity/business/workstreams/workstream1-natural-capital-accounting/index_en.htm).

674. Spurgeon, James. *Natural Capital Accounting for Business - Guide to selecting an approach* (2015) (see http://ec.europa.eu/environment/biodiversity/business/assets/pdf/b-at-b-platform-nca-workstream-final-report.pdf).

675. Stahel, Walter R. in *A New Dynamic. Effective Business in a Circular Economy*, Ellen MacArthur Foundation, Ed. (2013).

676. Stanton, Tracy, Marta Echavarria, Katherine Hamilton, and Caroline Ott. *State of Watershed Payments - An emerging marketplace.* (2010) (see http://www.foresttrends.org/documents/files/doc_2438.pdf).

677. Stuart, Elizabeth, Peter H. Larsen Charles A. Goldman, and Donald Gilligan. *Current Size and Remaining Market Potential of the U.S. Energy Service Company Industry*, U.S. Department of Energy Office of Energy Efficiency and Renewable Energy, Weatherization and Intergovernmental Program (2013) (see https://emp.lbl.gov/sites/all/files/lbnl-6300e_01.pdf).

678. Stein, Norman. *ERISA and the Limits of Equity* (JSTOR, 1993), pp. 71–110. (see http://scholarship.law.duke.edu/cgi/viewcontent.cgi?article=4178&context=lcp).

679. Steinhauser, Gabriele. *Regulators Examine Financial Risks of Climate Change* (2016) (see http://www.wsj.com/articles/regulators-examine-financial-stability-risks-to-climate-change-oil-exposure-1459444688).

680. Stern, N. H. et al. Stern Review: *The economics of climate change* (HM Treasury London, 2006) (see http://webarchive.nationalarchives.gov.uk/+/http:/www.hm-treasury.gov.uk/sternreview_index.htm).

681. Stern, Paul C., and Elliot Aronson. *Energy use: The human dimension* (WH Freeman and Co., New York, NY, 1984) (see http://www.nap.edu/download/9259).

682. Stetz Consulting, LLC. Verification by Equipment or End-Use Metering Protocol, Bonneville Power Administration (2012) (see http://www.eeperformance.org/uploads/8/6/5/0/8650231/bpa_verification_by_equipment_or_end_use_metering_protocol.pdf).

683. Stocker, T. F. D. Qin G. -K. Plattner M. Tignor S. K. Allen J. Boschung A. Nauels Y. Xia V. Bex, and P. M. Midgley (eds.). *Climate change 2013: The physical science basis, Intergovernmental Panel on Climate Change* (Cambridge University Press, 2013) (see http://www.climatechange2013.org/).

684. Stommen, Susan Mazur, and Kate Farley. *ACEEE Field Guide to Utility-Run Behavior Programs*, American Council for an Energy-Efficient Economy (2013) (see http://www.aceee.org/research-report/b132).

685. STRIDE. (see http://www.thestride.co.uk/).

686. Stout, Lynn A. *The shareholder value myth: How putting shareholders first harms investors, corporations, and the public.* Berrett-Koehler Publishers, 2012. ISBN 978-1-60509-813-5.

687. Stroeve, J. C. et al. *The Arctic's rapidly shrinking sea ice cover: a research synthesis*, Climatic Change , 1–23 (2012).

688. Stuart, Elizabeth, Peter H. Larsen, Charles A. Goldman, and Donald Gilligan. *Current Size and Remaining Market Potential of the U.S. Energy Service Company Industry*, U.S. Department of Energy Office of Energy Efficiency and Renewable Energy, Weatherization and Intergovernmental Program (2013) (see https://emp.lbl.gov/sites/all/files/lbnl-6300e_0.pdf).

689. Stuart, Elizabeth, Peter H. Larsen, Juan Pablo Carvallo, Charles A. Goldman, and Donald Gilligan. *U.S. Energy Service Company (ESCO) Industry: Recent Market Trends*, Ernest Orlando Lawrence Berkley National Laboratory (2016) (see http://naesco.org/data/industryreports/esco_recent_market_trends_30sep2016.pdf).

690. Sui-Lee Wee, Reuters (2014) (see: http://in.reuters.com/article/2014/01/16/china-pollution-idINL3N0KQ3CF20140116).

691. Sukhdev, P. et al. *The Economics of Ecosystems & Biodiversity* (TEEB): Mainstreaming the Economics of Nature: A synthesis of the approach, conclusions and recommendations of TEEB (2010) (see http://www.teebweb.org/TEEBSynthesisReport/tabid/29410/Default.aspx).

692. SustainAbility. Rate the Raters - Phase Five, the Investor View (2012) (see http://www.sustainability.com/library/attachment/396).

693. SustainAbility. Signed, Sealed... Delivered? Behind Certifications and Beyond Labels (2011) (see http://www.sustainability.com/library/attachment/363).

694. Sutherland, Ronald J. *Market barriers to energy-efficiency investments*, The Energy Journal (JSTOR, 1991), pp. 15–34.

695. Swed, Frieda S.; Eisenhart, C. *Tables for Testing Randomness of Grouping in a Sequence of Alternatives*. Ann. Math. Statist. 14 (1943), no. 1, 66--87. https://projecteuclid.org/euclid.aoms/1177731494).

696. Sweeter, Richard. Building 101 Instrumentation Project, Consortium for Building Energy Innovation (see http://cbei.psu.edu/instrumenting-navy-yard-building-101/).

697. Swift, Art. *Americans Split on How Often They Upgrade Their Smartphones* (2015) (see http://www.gallup.com/poll/184043/americans-split-often-upgrade-smartphones.aspx?utm_source=Economy&utm_medium=newsfeed&utm_campaign=tiles).

698. Tandberg, and IpsosMORI. *Corporate Environmental Behaviour and the Impact on Brand Values* (2007) (see http://www.tandberg.com/collateral/video_communication/tandberg_environmental_survey_report.pdf).

699. Tesco. Corporate Social Responsibility Report (2011) (see http://www.tescoplc.com/media/60113/tesco_cr_report_2011_final.pdf).

700. Tesco. Tesco Greener Living website (2012) (see http://www.tesco.com/greenerliving/greener_tesco/what_tesco_is_doing/energy_efficient_solutions.page).

701. Tesco. Tesco Society report 2014 (see https://www.tescoplc.com/media/1184/tesco_and_society_review_2014.pdf).

702. Thaler, Richard H., and Cass R. Sunstein. *Nudge: Improving decisions about health, wealth, and happiness* (Penguin Books, 2008).

703. Therkelsen, Peter et al. *Development of an Enhanced Payback Function for the Superior Energy Performance Program* (2015) (see http://energy.gov/sites/prod/files/2015/10/f27/LBNL-190883.pdf).

704. Thiede, S, G Bogdanski, and C Herrmann. *A systematic method for increasing the energy and resource efficiency in manufacturing companies* (Elsevier, 2012) (see http://ac.els-cdn.com/S2212827112001357/1-s2.0-S2212827112001357-main.pdf?_tid=31c64d80-4be9-11e5-b361-00000aab0f02&acdnat=1440590243_d7e8ebdbc575ef8e0fa1110cbc295823), pp. 28–33.

705. Thompson, Fraser. *Resource Revolution*, McKinsey & Company (2012) (see http://www.unido.org/fileadmin/media/documents/pdf/Energy_Environment/Green_Industry/IAB_Meeting/Resource%20Revolution%20-%20Fraser%20Thompson%2C%20McKinsey.pdf).

706. Thompson Reuters. Carbon Market Monitor 2015 (2016) (see http://trmcs-documents.s3.amazonaws.com/3501ec8eae589bfbef9cc1729a7312f0_20160111104949_Carbon%20Market%20Review%202016_1.5.pdf).

707. Thornton, Strum, Kinkel, Water Loss Control, McGraw Hill, 2008, ISBN: 978-0-07-149918-7.

708. Thumann, Albert, and William J. Younger. *Handbook of energy audits* (The Fairmont Press, Inc., 9th Edition., 2012).

709. Times, New York. *China Sets Targets* (2010) (see http://www.nytimes.com/2010/08/10/business/energy-environment/10yuan.html).

710. Toman, M., R. D. Simpson, and R. Ayres. Scarcity and Growth in the New Millennium: Summary, Resources For the Future (2004) (see http://ageconsearch.umn.edu/bitstream/10835/1/dp040001.pdf).

711. Tomo, Ocean. Intangible Asset Market Value (see http://www.oceantomo.com/2013/12/09/Intangible-Asset-Market-Value-Study-Release/).

712. Toyota. Environmental Data 2016 (see http://www.toyota-global.com/sustainability/report/sr/pdf/sdb16_er_f_en.pdf).

713. Transparense. European Code of Conduct for Energy Performance Contracting (2014) (see http://www.transparense.eu/download-library/european-code-of-conduct-for-epc).

714. Treasury, H. M. The green book: appraisal and evaluation in central government (Stationery Office, 2003 with update in 2011). (see https://www.gov.uk/government/uploads/system/uploads/attachment_data/file/220541/green_book_complete.pdf).

715. Tremblay, M., S. Tice, R. Tkaczyk, and M. Bennett. *Evaluation of the one tonne challenge program* (2006) (see http://publications.gc.ca/collections/collection_2013/ec/En4-216-2006-eng.pdf).

716. Triandis, Harry Charalambos. *Interpersonal behavior* (Brooks/Cole Publishing Company Monterey, CA, 1977).

717. TRW. 2010 TRW Annual Report for Sustainability, TRW Automotive (2010) (see http://www.aiag.org/staticcontent/committees/download_files/2010-TRW-Annual-Report-for-Sustainability.pdf).

718. TRW. 2011 TRW Automotive Annual Report for Health, Safety, Environment and Sustainability, TRW Automotive (2011) (see http://www.trw.com/sites/default/files/HSE_DOCS/web-content/pdfs/2011_HSES_Report.pdf).

719. T'serclaes, P., and N. Jollands. *Mind the gap: quantifying principal agent problems in energy efficiency*, International Energy Agency and OECD (2007), p. 224.

720. Turner, Wayne C., and Steve Doty. *Energy management handbook*. The Fairmont Press, Inc., 2007. ISBN-13: 978-1466578289.

721. Turner, Lawrence, Southwestern Adventist University, Table of critical values of r (available at: http://turner.faculty.swau.edu/mathematics/math241/materials/rctable/rc.php).

722. TUV-Nord, ISO 50001 (EnMS) Energy Management Systems (2014) (see http://www.tuv-uk.com/cps/rde/xbcr/SID-A25B9185-102594DE/tng_uk/iso-50001-guide-and-check-list-uk.pdf).

723. Tversky, A., and D. Kahneman. *The framing of decisions and the psychology of choice*, Science 211, 453–458 (1981).

724. UK Government. The National Archives. Open government Licence. (see www.nationalarchives.gov.uk/doc/open-government-licence/version/3/)

725. UK Meteorological Office, UKCP09 gridded observation data sets — frequently asked questions. (2015) (see http://www.metoffice.gov.uk/climatechange/science/monitoring/ukcp09/faq.html#faq1.8).

726. UK Parliament: *Sustainability Targets and Performance* (see http://www.parliament.uk/about/sustainability/targets-and-performance/).

727. UN. *Review of the state of world marine fishery resources*, Marine Resources Service (United Nations) (Food and Agriculture Organization, 2009) (see http://www.fao.org/docrep/015/i2389e/i2389e.pdf).

728. United Nations Security Council. Interim report of the United Nations Expert Panel on the Illegal Exploitation of natural resources and Other Forms of Wealth of the Democratic Republic of the Congo (2001) (see http://www.globalpolicy.org/images/pdfs/0116report.pdf).

729. UNEP. International Resource. Report 2: *Recycling Rates of Metals: A Status Report* (2011) (see http://www.unep.org/resourcepanel/Portals/24102/PDFs/Metals_Recycling_Rates_110412-1.pdf).

730. UNEP. "One Planet Many People" Atlas. Launched to Mark World Environment Day (2005) (see http://www.unep.org/Documents.Multilingual/Default.asp?DocumentID=434&ArticleID=4807&l=en).

731. UNEP. *Towards a Green Economy: Pathways to Sustainable Development and Poverty Eradication*, United Nations (2011) (see https://sustainablede-velopment.un.org/content/documents/126GER_synthesis_en.pdf).

732. UNEPFI. *Fiduciary Responsibility*, United Nations Environment Program Finance Initiative (2009) (see http://www.unepfi.org/fileadmin/documents/fiduciaryII.pdf).

733. Unilever. Sustainable Living Plan 2015 Progress Report (see https://www.unilever.com/Images/uslp-mobilising-collective-action-summary-of-progress-2015_tcm244-424809_en.pdf).

734. UNFCC. List of non-Annex I countries. (2016) (see http://unfccc.int/parties_and_observers/parties/non_annex_i/items/2833.php)

735. *UPS Figure out the "Right Way" to save money, time and gas.* (see http://abcnews.go.com/WNT/story?id=3005890&page=1).

736. US Appeal LHC Dismissed by US Court of Appeals, US Court of Appeals for the 9th Circuit (2010) (see http://www.symmetrymagazine.org/breaking/2010/08/26/lhc-lawsuit-dismissed-by-us-court/).

737. US Data. Census. Data on Industry and Occupation (see http://www.census.gov/people/io/ - see Table 1), and Yound Aaron, et al, A Census of Actively Licenced Physicians in the United States, 2014 (see https://www.fsmb.org/Media/Default/PDF/Census/2014census.pdf). Adding together the STEM professions from the census and the medic numbers from the paper give a total of over 16.1 million individuals, which is a conservative estimate of those with a Bachelor of Science etc. as it only covers employed individuals.

738. US Department of Commerce, Bureau of Economic Analysis (see http://www.bea.gov/index.htm).

739. US DOE. Advanced Manufacturing Office Energy Resources Centre. *Strategic Energy Management Checklist* (see https://ecenter.ee.doe.gov/EM/SPM/Pages/SEM_checklist.aspx).

740. US DOE. *Cost Estimating Guide* (2011) (see http://science.energy.gov/~/media/opa/pdf/processes-and-procedures/Cost_Estimating_G_413_3_21_final_05092011.pdf).

741. US DOE. eGuide resources for ISO 50001 (see http://www1.eere.energy.gov/apps/manufacturing/eguide/eGuide-Resources.zip)

742. US DOE. Superior Energy Performance. (see https://www.energy.gov/eere/amo/superior-energy-performance).

743. US DOE. Superior Energy Performance Case Studies. Cook Composites and Polymers. (see http://energy.gov/sites/prod/files/2014/05/f16/ccp_sep_case_study.pdf).

744. US DOE. Superior Energy Performance Case Studies. Cummins. (see http://energy.gov/sites/prod/files/2015/07/f24/Cummins_case_study_SEP_revised_Final.pdf).

745. US DOE. Superior Energy Performance Case Studies. Daimler Detroit Diesel. (see http://energy.gov/eere/amo/articles/daimlers-detroit-diesel-plant-earns-superior-energy-performance-saves-37m).

746. US DOE. Superior Energy Performance Case Studies. Freescale Semiconductor. (see http://energy.gov/sites/prod/files/2014/05/f16/freescale_case_study.pdf).

747. US DOE. Superior Energy Performance Case Studies. General Dynamics. (see http://energy.gov/sites/prod/files/2014/09/f18/GSEP_EMWG-GD_casestudy.pdf).

748. US DOE. Superior Energy Performance Case Studies. HARBEC. (see http://energy.gov/sites/prod/files/2015/05/f22/SEP_HARBEC_case_study.pdf).

749. US DOE. Superior Energy Performance Case Studies. Nissan. (see https://energy.gov/sites/prod/files/2016/05/f32/Nissan%20case%20study_SEP_recertification_1.pdf).

750. US DOE. Superior Energy Performance, Industrial Facility Best Practice Scorecard. (see http://www.energy.gov/eere/amo/downloads/superior-energy-performance-industrial-facility-best-practice-scorecard).

751. US DOE. Superior Energy Performance, Measurement and Verification Protocol for Industry, November 2012 (see http://www.energy.gov/sites/prod/files/2014/07/f17/sep_mv_protocol.pdf).

752. US Energy Information Administration. Based on work by Richard Bonskowski, William D. Watson Fred Freme. *Coal Production in the United States - A Historical Overview*, (2006) (available at http://www.steelonthenet.com/pdf/EIA_06-Oct-06.pdf).

753. US EPA. Design for the Environment Programs, Initiatives, and Projects (see https://www.epa.gov/saferchoice/design-environment-programs-initiatives-and-projects).

754. United States International Trade Commission. *Remanufactured Goods: An overview of the US and Global Industries, Markets and Trade.*, United States International Trade Commission. (see http://www.usitc.gov/publications/332/pub4356.pdf).

755. USGBC. LEED (see http://www.usgbc.org/leed).

756. USGCRP. *Climate Change Impacts in the United States* (2014) (see http://s3.amazonaws.com/nca2014/high/NCA3_Climate_Change_Impacts_in_the_United%20States_HighRes.pdf?download=1).

757. USGS National Minerals Information Centre. Mineral Commodity Summaries 2014 and 2016 (2016) (see http://minerals.usgs.gov/minerals/pubs/mcs/2016/mcs2016.pdf).

758. Van Beers, Dick, Albena Bossilkov, Glen Corder, and Rene Berkel. *Industrial symbiosis in the Australian minerals industry: the cases of Kwinana and Gladstone.* (Wiley Online Library, 2007) (see http://www.sswm.info/sites/default/files/reference_attachments/VAN%20BEERS%20et%20al%20 2007%20Industrial%20Symbiosis%20in%20the%20Australian%20Mineral%20Industry.pdf), pp. 55–72.

759. Vaughan, Adam. *Tesco drops carbon-label pledge*, The Guardian (2012) (see http://www.guardian.co.uk/environment/2012/jan/30/tesco-drops-carbon-labelling).

760. Veneman, Andre. *The Dow Jones Sustainability Index: Why winning isn't everything*, (2014) (see https://www.theguardian.com/sustainable-business/2014/aug/01/dow-jones-sustainability-index-why-winning-isnt-everything).

761. Verified Carbon Standard. (see http://www.v-c-s.org/).

762. Vigen, Tyler, *Spurious Correlations*, ISBN 978-0-316-33942 (2015).

763. Vitousek, Peter M., Paul R. Ehrlich, Anne H. Ehrlich, and Pamela A. Matson. *Human appropriation of the products of photosynthesis*, BioScience 36, 368–373 (1986).

764. Vogt-Schilb, A., and S. Hallegatte. *When Starting with the Most Expensive Option Makes Sense* (2011) (see https://hal-enpc.archives-ouvertes.fr/hal-00626261v2/document).

765. Vogt-Schilb, Adrien, and Stéphane Hallegatte. *Marginal abatement cost curves and the optimal timing of mitigation measures* (2014) (see https://hal.archives-ouvertes.fr/file/index/docid/916328/filename/MACCs_timing.pdf), pp. 645–653.

766. Vogt-Schilb, Adrien, Stephane Hallegatte, Christophe de Gouvello. *Long-Term Mitigation Strategies and Marginal Abatement Cost Curves - A case study on Brazil.* The World Bank (2014) (see http://documents.worldbank.org/curated/en/2014/03/19258772/long-term-mitigation-strategies-marginal-abatement-cost-curves-case-study-brazil).

767. Walker, Rob. Method, New York Times (2004).

768. Walmart. Sustainability Supplier Assessment FAQs (2011) (see http://walmartstores.com/sustainability/9292.aspx).

769. Walmart Announces New Commitments to Dramatically Increase Energy Efficiency and Renewables. (see http://news.walmart.com/news-archive/2013/04/15/walmart-announces-new-commitments-to-dramatically-increase-energy-efficiency-renewables).

770. Walmart's Sustainability Index. (see http://www.walmartsustainabilityhub.com/app/answers/detail/a_id/242).

771. Walter, K.M., S.A. Zimov, J. P. Chanton, D. Verbyla, and F. S. Chapin. *Methane bubbling from Siberian thaw lakes as a positive feedback to climate warming*, Nature 443, 71–75 (2006).

772. Ward, Karen. The World in 2050, HSBC (2011) (see http://www.research.hsbc.com/midas/Res/RDV?p=pdf&key=ej73gSSJVj&n=282364.PDF).

773. Washington, Haydn and Cook, John, *Climate Change Denial*, Earthscan, 2011.

774. WBCSD. Cement Technology Roadmap, World Council for Sustainable Development and International Energy Agency (2009) (see http://www.wbcsdcement.org/pdf/technology/WBCSD-IEA_Cement%20Roadmap.pdf).

775. Weale, Sally. *Teachers must ditch 'neuromyth' of learning styles, say scientists.* (see https://www.theguardian.com/education/2017/mar/13/teachers-neuromyth-learning-styles-scientists-neuroscience-education).

776. Webster, Ken. *The Decline of the Linear Economy and the Rise of the Circular.* In A New Dynamic. Effective Business in a Circular Economy, Ellen MacArthur Foundation, Ed. (2013).

777. Weizsäcker, von E. U., E. U. Weizsäcker, A. B. Lovins, and L. H. Lovins. *Factor four: doubling wealth-halving resource use: the new report to the Club of Rome* (Earthscan, 1998) (see http://www.ima.kth.se/im/3c1395/Pdf/FactorFour.pdf).

778. Weyzig, Francis, Barbara Kuepper, Jan Willem van Gelder, and Rens van Tilburg. *The Price of Doing too Little too Late - the impact of the carbon bubble on the EU financial system*, Green Europe Foundation (2014) (see http://reinhardbuetikofer.eu/wp-content/uploads/2014/03/GND-Carbon-Bubble-web1.pdf).

779. Wheeler, Donald J. *The six sigma practitioner's guide to data analysis*, (Section 3.3 What Makes the XmR Chart Work?) (SPC Press Knoxville, TN, 2005) (see http://www.spcpress.com/pdf/DJW250.pdf).

780. Wiel, S., and J. E. McMahon. *Energy-Efficiency Labels and Standards: A Guidebook for Appliances, Equipment, and Lighting*, Ernest Orlando Lawrence Berkeley NationalLaboratory, Berkeley, CA (US) (2005) (see http://www.clasponline.org/en/ResourcesTools/Resources/StandardsLabelingResourceLibrary/2005/~/media/Files/SLDocuments/2005_SLGuidebook/English/SLGuidebook_eng_1_FullGuidebook.pdf).

781. Wigley, T. M. L. *The climate change commitment*, Science 307, 1766–1769 (2005).

782. Wikipedia. Calibration (see https://en.wikipedia.org/wiki/Calibration).

783. Wikipedia. Chemical Plant Cost Index (see https://en.wikipedia.org/wiki/Chemical_plant_cost_indexes)

784. Wikipedia. Largest cities of the European Union by population within city limits (see http://en.wikipedia.org/wiki/Largest_cities_of_the_European_Union_by_population_within_city_limits).

785. Wikipedia. List of United States cities by population (see http://en.wikipedia.org/wiki/List_of_United_States_cities_by_population).

786. Wikipedia. Material Flow-cost Accounting (see https://en.wikipedia.org/wiki/Material_Flow_Cost_Accounting).

787. Wikipedia.Minard, Charles, Illustration of Napoleon's Russian Campaign, 1812 (see: https://en.wikipedia.org/wiki/Charles_Joseph_Minard#/media/File:Minard.png).

788. Wikipedia. Pearson product-moment correlation coefficient (see https://en.wikipedia.org/wiki/Pearson_product-moment_correlation_coefficient).

789. Wikipedia. Quality-adjusted Life Year (see https://en.wikipedia.org/wiki/Quality-adjusted_life_year)

790. Wilkipedia. REACH (see http://en.wikipedia.org/wiki/Registration,_Evaluation,_Authorization_and_Restriction_of_Chemicals).

791. Willard, Bob. *Next Sustainability Wave: Building Boardroom Buy-In* (New Society Publishers, 2005).

792. Willhite, Edwin and Tamm, Yannick, Costs and Benefits of Enterprise-wide Implementation of Superior Energy Performance at Schneider Electric (Oct 2015) (see: http://energy.gov/sites/prod/files/2015/10/f27/Schneider%20Electric%20WEEC%202015.pdf).

793. Willis, Peter. Further Benefits of Resource Efficiency Seminar (2011) (see http://www.google.co.uk/url?sa=t&rct=j&q=further%20benefits%20of%20business%20resource%20efficiency&source=web&cd=4&sqi=2&ved=0CDkQFjAD&url=http%3A%2F%2Fwww.eftec.co.uk%2Fkeynotes%2Fseminars-2011%2Fseminar-3-benefits-of-resource-efficiency%2Fseminar-3-2011-willis-presentation%2Fdownload&ei=fWc_T8SeB6qO0AXX1Y2PDw&usg=AFQjCNEISltOnewhbaXHuZk75OnFoPMKYw&sig2=9QOTqTogl55SJiOo5wwy4Q).

794. Willmann, R., K. Kelleher, R. Arnason, and N. Franz. *The sunken billions*, World Bank and FAO (2009) (see http://siteresources.worldbank.org/EXTARD/Resources/336681-1224775570533/SunkenBillionsFinal.pdf).

795. Wingender, John, and Eric Woodroof. in *How to Finance Energy Management Projects, Solving the "Lack of Capital Problem".* (The Fairmont Press, 2013), pp. 157–168.

796. Witkowski, Wallace, Global stock market cap has doubled since QE's start, Market Watch 2015 (see http://www.marketwatch.com/story/global-stock-market-cap-has-doubled-since-qes-start-2015-02-12).

797. Woetzel, Jonathan et al. *Preparing for China's urban billion* (2009) (see http://www.mckinsey.com/~/media/McKinsey/dotcom/Insights%20and%20pubs/MGI/Research/Urbanization/Preparing%20for%20Chinas%20urban%20billion/MGI_Preparing_for_Chinas_Urban_Billion_full_report.ashx).

798. Wolinsky, A. *Competition in markets for credence goods*, Journal of Institutional and Theoretical Economics (JITE)/Zeitschrift für die gesamte Staatswissenschaft , 117–131 (1995).

799. Womack, J. P., and D. T. Jones. *Lean thinking: banish waste and create wealth in your corporation* (Simon and Schuster, 2003).

800. Wong, Shonshan, *China's One-Day Shopping Spree Sets Record in Online Sales*, New York Times (November 11, 2013) (see http://www.nytimes.com/2013/11/12/business/international/online-shopping-marathon-zooms-off-the-blocks-in-china.html).

801. Wood, Stepan. *Four Varieties of Social Responsibility: Making Sense of the "Sphere of Influence" and "Leverage" Debate Via the Case of ISO 26000*, Osgoode Hall Law School of York University, Comparative Research in Law & Political Economy, Research Paper No. 14/2011 (2011) (see http://digitalcommons.osgoode.yorku.ca/cgi/viewcontent.cgi?article=1052&context=clpe).

802. Woodall, William H. *Controversies and contradictions in statistical process control* (ASQC, 2000), pp. 341–350.

803. Wooding, Graham, and Kit Oung. *Implementing and Improving an Energy Management System: How to Meet the Requirements of ISO 50001*, BSI Standards Institution, Ed. (BSI, 2013) (see http://shop.bsigroup.com/ProductDetail/?pid=000000000030252733). ISBN 978-0-580-76613-8.

804. Woodroof, Eric, and Thurman, Albert. *How to Finance Energy Management Projects*. The Fairmont Press, 2013. ISBN 978-1-4665-7153-2.

805. World Bank. World Bank Statistics (2012) (see http://data.worldbank.org/indicator/NY.GDP.MKTP.KD.ZG).

806. World Bank. World Bank Statistics (2013) (see http://data.worldbank.org/indicator/NY.GDP.MKTP.CD).

807. World Economic Forum. *Towards the Circular Economy: Accelerating the scale-up across global supply chains* (2014) (see http://www3.weforum.org/docs/WEF_ENV_TowardsCircularEconomy_Report_2014.pdf).

808. World Energy Council, Enerdata, and Ademe. Energy Efficiency Policies and Measures Database (see http://www.wec-policies.enerdata.eu/). This is an excellent summary of international legislation in relation to energy.

809. World Energy Council, *World Energy Perspective Energy Efficiency Policies – What works and what does not*. (2013) (see http://www.worldenergy.org/wp-content/uploads/2013/09/WEC-Energy-Efficiency-Policies-report.pdf).

810. World Energy Outlook 2013: Energy Subsidies. (see http://www.worldenergyoutlook.org/resources/energysubsidies/).

811. World Energy Perspective Energy Efficiency Policies – What works and what does not. (2013) (see http://www.worldenergy.org/wp-content/uploads/2013/09/WEC-Energy-Efficiency-Policies-report.pdf).

812. World Land Trust - Carbon Balanced Offset Programme. (2016) (see http://www.worldlandtrust.org/eco-services/offsetting/project-design).

813. World Without Oil, Archive (see http://writerguy.com/wwo/archive.htm).

814. Worm, B. et al. *Impacts of biodiversity loss on ocean ecosystem services* (American Association for the Advancement of Science, 2006), pp. 787–790. (see http://grizzlie.utm.utoronto.ca/~w3bio205/Worm_etal_2006Science.pdf).

815. Worrell, Ernst, John A. Laitner, Michael Ruth, and Hodayah Finman. *Productivity benefits of industrial energy efficiency measures* (Elsevier, 2003) (see http://escholarship.org/uc/item/0013f49z.pdf), pp. 1081–1098.

816. WRI. Global Ecolabel Monitor 2010, Big Room and WRI (2010) (see http://pdf.wri.org/2010_global_ecolabel_monitor.pdf).

817. WWF. Living Planet Report, Worldwide Fund for Nature (2012) (see http://awsassets.panda.org/downloads/1_lpr_2012_online_full_size_single_pages_final_120516.pdf).

818. Yang, Ailun. *What Is the Future of King Coal in China?* WRI (October 15, 2012) (see: http://www.wri.org/blog/what-future-king-coal-china).

819. Yamaguchi, K. *Re-examination of stock price reaction to environmental performance*. Ecological Economics 68, 345–352 (2008).

820. York, Emily Bryson. *Nestle, Pepsi and Coke Face Their Waterloo*, Advertising Age (2007) (see http://adage.com/article/news/nestle-pepsi-coke-face-waterloo/120986/).

821. Zaitsau, Viktar, *Energy Audits Undone, How to find 101 simple ways to slash your utility bills*, Ecodo Ltd (2013). ISBN: 978-0-9576410-1-3.

822. ZCH. Defining a Fabric Energy Efficiency Standard, Zero Carbon Hub (2009) (see http://www.zerocarbonhub.org/resourcefiles/ZCH-Defining-A-Fabric-Energy-Efficiency-Standard-Task-Group-Recommendations.pdf).

823. ZCH. Energy Performance of Buildings Directive - Introduction to recast EPBD-2, Zero Carbon Hub, NHCB Foundation (2011) (see http://www.zerocarbonhub.org/resourcefiles/EPBD_Introduction.pdf).

824. Zipcar. *Croydon Council cuts employee car usage in half with Zipcar*. (see http://www.zipcar.co.uk/press/releases/croydon-council-cuts-employee-car-usage-in-half-with-zipcar).

26 Index

*To encourage exploration within themes, the index entries are grouped into a number of top-level topics: **Audit**; **Barriers**; **Communication**; **Data Analysis**; **Disclosure**; **Formulae**; **Funding**; **Meters**; **People; Presenting Data**; **Proposal**; **Psychology** and **Standards**, among others.*

Symbols

A

B

Reference

C

D

E

Reference

F

G

L

M

Reference

P

Q

R

S

T

U

V

W

X

Y

Z